Advanced Techniques in Computing Sciences and Software Engineering

Khaled Elleithy

Advanced Techniques in Computing Sciences and Software Engineering

Editor
Prof. Khaled Elleithy
University of Bridgeport
School of Engineering
221 University Avenue
Bridgeport CT 06604
USA
elleithy@bridgeport.edu

ISBN 978-94-007-9152-7 ISBN 978-90-481-3660-5 (eBook)
DOI 10.1007/978-90-481-3660-5
Springer Dordrecht Heidelberg London New York

Printed on acid-free paper

Springer is part of Springer Science+Business Media (www.springer.com)

Dedication

To my Family

Preface

This book includes Volume II of the proceedings of the 2008 International Conference on Systems, Computing Sciences and Software Engineering (SCSS). SCSS is part of the International Joint Conferences on Computer, Information, and Systems Sciences, and Engineering (CISSE 08). The proceedings are a set of rigorously reviewed world-class manuscripts presenting the state of international practice in Advances and Innovations in Systems, Computing Sciences and Software Engineering.

SCSS 08 was a high-caliber research conference that was conducted online. CISSE 08 received 948 paper submissions and the final program included 390 accepted papers from more than 80 countries, representing the six continents. Each paper received at least two reviews, and authors were required to address review comments prior to presentation and publication.

Conducting SCSS 08 online presented a number of unique advantages, as follows:

- All communications between the authors, reviewers, and conference organizing committee were done on line, which permitted a short six week period from the paper submission deadline to the beginning of the conference.
- PowerPoint presentations, final paper manuscripts were available to registrants for three weeks prior to the start of the conference
- The conference platform allowed live presentations by several presenters from different locations, with the audio and PowerPoint transmitted to attendees throughout the internet, even on dial up connections. Attendees were able to ask both audio and written questions in a chat room format, and presenters could mark up their slides as they deem fit
- The live audio presentations were also recorded and distributed to participants along with the power points presentations and paper manuscripts within the conference DVD.

The conference organizers and I are confident that you will find the papers included in this volume interesting and useful. We believe that technology will continue to infuse education thus enriching the educational experience of both students and teachers.

Khaled Elleithy, Ph.D.
Bridgeport, Connecticut
December 2009

Table of Contents

Acknowledgements

The 2008 International Conference on Systems, Computing Sciences and Software Engineering (SCSS) and the resulting proceedings could not have been organized without the assistance of a large number of individuals. SCSS is part of the International Joint Conferences on Computer, Information, and Systems Sciences, and Engineering (CISSE). CISSE was founded by Professor Tarek Sobh and I in 2005, and we set up mechanisms that put it into action. Andrew Rosca wrote the software that allowed conference management, and interaction between the authors and reviewers online. Mr. Tudor Rosca managed the online conference presentation system and was instrumental in ensuring that the event met the highest professional standards. I also want to acknowledge the roles played by Sarosh Patel and Ms. Susan Kristie, our technical and administrative support team.

The technical co-sponsorship provided by the Institute of Electrical and Electronics Engineers (IEEE) and the University of Bridgeport is gratefully appreciated. I would like to express my thanks to Prof. Toshio Fukuda, Chair of the International Advisory Committee and the members of the SCSS Technical Program Committee including: Abdelaziz AlMulhem, Alex A. Aravind, Anna M. Madueira, Hamid Mcheick, Hani Hagras, Julius Dichter, Low K.S., Marian P. Kazmierkowski, Michael Lemmon, Mohamed Dekhil, Mostafa Aref, Natalia Romalis, Raya Al-Qutaish, Rodney G. Roberts, Sanjiv Rai, Shivakumar Sastry ,Tommaso Mazza, Samir Shah, and Mohammed Younis.

The excellent contributions of the authors made this world-class document possible. Each paper received two to four reviews. The reviewers worked tirelessly under a tight schedule and their important work is gratefully appreciated. In particular, I want to acknowledge the contributions of all the reviewers. A complete list of reviewers is given in page XIX.

Khaled Elleithy, Ph.D.
Bridgeport, Connecticut
December 2009

Reviewers List

Aamir, Wali
Aaron Don, Africa
Abd El-Nasser, Ghareeb
Abdelsalam, Maatuk
Adam, Piorkowski
Adrian, Runceanu
Adriano, Albuquerque
Ahmad Sofian, Shminan
Ahmad, Saifan, 281
Ahmed, Zobaa
Alcides de Jesús, Cañola
Aleksandras Vytautas, Rutkauskas, 417
Alexander, Vaninsky, 1
Alexei, Barbosa de Aguiar
Alice, Arnoldi
Alionte, Cristian Gabriel
Amala V. S., Rajan, 489
Ana María, Moreno
Anna, Derezińska, 543
Antal, Tiberiu Alexandru, 85
Anton, Moiseenko
Anu, Gupta
Asma, Paracha
Atif, Mohammad, 197
Aubrey, Jaffer
Baba Ahmed, Eddine
Biju, Issac
Brana Liliana, Samoila, 287
Buket, Barkana, 355
Cameron, Cooper, 549
Cameron, Hughes, 549
Cecilia, Chan
chetankumar, Patel, 355
Chwen Jen, Chen
Cornelis, Pieters
Craig, Caulfield, 245
Curila, Sorin, 537
Daniel G., Schwartz, 45
Daniela, López De Luise, 51, 339
David, Wyld, 123
Denis, Berthier, 165
Dierk, Langbein
Dil, Hussain, 555
Dmitry, Kuvshinov
D'Nita, Andrews-Graham
Ecilamar, Lima, 561
Edith, Lecourt, 249
Emmanuel Ajayi, Olajubu, 371
Erki, Eessaar, 377
Ernesto, Ocampo
Fernando, Torres, 321
Gennady, Abramov, 435
Ghulam, Rasool
Gururajan, Erode
Hadi, Zahedi
He, xing-hua, 191
Hector, Barbosa Leon
Houming, FAN
Igor, Aguilar Alonso
Ilias, Karasavvidis
Jaakko, Kuusela, 231
James, Feher, 469
Jan, GENCI
Janett, Williams
Jian-Bo, Chen
Jonathan, White
José L., Fuertes
Jozef, Simuth
József, Berke
Juan, Garcia
junqi, liu
Jussi, Koskinen
Jyri, Naarmala
Kenneth, Faller II
Khaled, Elleithy
Krystyna Maria, Noga
Kuderna-Iulian, Benta
Laura, Vallone
Lei, Jiasu
Leszek, Rudak
Leticia, Flores, 399
Liang, Xia, 191
madjid, khalilian, 459
Mandhapati, Raju
Margareth, Stoll, 153
Maria, Pollo Cattaneo, 339
Marina, Müller
Marius, Marcu
Marius Daniel, Marcu, 339
Martina, Hedvicakova
Md. Abdul, Based
Miao, Song, 531
Mircea, Popa, 429
Mohammad Abu, Naser
Morteza, Sargolzaei Javan
Muthu, Ramachandran
Nagm, Mohamed, 441
Nazir, Zafar
Neander, Silva, 525
Nilay, Yajnik
Nita, Sarang, 567
Nova, Ovidiu
Olga, Ormandjieva, 333
Owen, Foley
Paola, Ferrari
Paul, David and Chompu, Nuangjamnong, 249
Peter, Nabende
Petr, Silhavy, 267
PIIA, TINT
Radek, Silhavy, 267
Richard, Barnett, 327
S. R., Kodituwakku, 39
S. Shervin, Ostadzadeh, 495
Sajad, Shirali-Shahreza
Salvador, Bueno
Samir Chandra, Das
Santiago, de Pablo, 321
Šarūnas, Packevičius, 177
Seibu, Mary Jacob
Sergiy, Popov
Serguei, Mokhov, 531
shalini, batra, 311
Sherif, Tawfik, 243
Shini-chi, Sobue, 159
shukor sanim, m. fauzi
Siew Yung, Lau
Soly Mathew, Biju
Somesh, Dewangan
Sridhar, Chandran
Sunil Kumar, Kopparapu
sushil, chandra
Svetlana, Baigozina
Syed Sajjad, Rizvi, 447
Tariq, Abdullah
Thierry, Simonnet, 299
Thomas, Nitsche
Thuan, Nguyen Dinh
Tibor, Csizmadia
Timothy, Ryan
Tobias, Haubold, 393
Tomas, Sochor, 267, 475
Umer, Rashid, 573
Ushasri, anilkumar
Vaddadi, Chandu, 359
Valeriy, Cherkashyn
Veselina, Jecheva
Vikram, Kapila
Xinqi, Zheng, 13
Yaohui, Bai
Yet Chin, Phung
Youming, Li, 383
Young, Lee
Yuval, Cohen, 25
Zeeshan-ul-hassan, Usmani
Zsolt Tibor, Kosztyán, 261

Bridging Calculus and Statistics: Null - Hypotheses Underlain by Functional Equations

Alexander Vaninsky
Hostos Community College of The City University of New York
avaninsky@hostos.cuny.edu

***Abstract*-Statistical interpretation of Cauchy functional equation f(x+y)=f(x)+f(y) and related functional equations is suggested as a tool for generating hypotheses regarding the rate of growth: linear, polynomial, or exponential, respectively. Suggested approach is based on analysis of internal dynamics of the phenomenon, rather than on finding best-fitting regression curve. As a teaching tool, it presents an example of investigation of abstract objects based on their properties and demonstrates opportunities for exploration of the real world based on combining mathematical theory with statistical techniques. Testing Malthusian theory of population growth is considered as an example.**

I. Introduction. History of Cauchy Functional Equation

In 1821, a famous French mathematician Augustin-Louis Cauchy proved that the only continuous function satisfying a condition

$$f(x+y) = f(x) + f(y), \tag{1}$$

is

$$f(x) = Kx, \tag{2}$$

where K is a constant, [1]. The proof provides an interesting example of how the properties of an object, function f(x) in this case, may be investigated without presenting the object explicitly. In what follows, we follow [2]. First, it can be stated that f(0)=0, because for any x, $f(x) = f(x+0)=f(x)+f(0)$, so that

$$f(0)=0. \tag{3}$$

The last was obtained by subtracting f(x) from both sides. From this finding it follows directly that $f(-x) = -f(x)$, because

$$0 = f(0) = f(x + (-x)) = f(x) + f(-x). \tag{4}$$

Next, we can show that $f(nx)= nf(x)$ for any x and natural number n. It can be easily proved by mathematical induction:

$$f(2x) = f(x+x)=f(x)+f(x)= 2f(x), \quad \text{for } n=2, \tag{5}$$

and, provided that it is true for $n-1$, we get

$$f(nx) = f((n-1)x+x)=f((n-1)x)+f(x)= (n-1)f(x) +f(x)= nf(x) \tag{6}$$

for any natural n. Finally, we can show that $f((1/n)x)= (1/n)f(x)$ for any natural number n. This follows directly from

$$f(x) = f(\frac{nx}{n}) = f(n(\frac{1}{n}x)) = nf(\frac{1}{n}x), \tag{7}$$

so that the result follows from division of both sides by n. Summarizing, it can be stated that function $f(x)$ satisfying condition (1) can be represented explicitly as

$$f(r) = Kr \tag{8}$$

for all rational numbers r.

At this point, condition of continuity comes into play to allow for expansion of the result for all real numbers x. Let sequence of rational numbers r_i be converging to x with $i\rightarrow\infty$. Then based on the assumption of continuity of the function $f(x)$, we have

$$f(x) = f(\lim r_i) = \lim f(r_i) = \lim Kr_i = K \lim r_i = Kx, \tag{9}$$

where limit is taken with $i\rightarrow\infty$. Note, that in this process an object, function $f(x)$, was first investigated and only after that described explicitly.

The Cauchy finding caused broad interest in mathematical community and served as a source of fruitful researches. Among their main topics were:

a) Does a discontinuous solution exist?
b) Can the condition of continuity be omitted or relaxed?
c) Can a condition be set locally, at a point or in a neighborhood, rather than on the whole domain of the function?

Weisstein [3] tells the story of the researches regarding these problems. It was shown by Banach [4] and Sierpinski [5] that the Cauchy equation is true for measurable functions. Sierpinski provided direct proof, while Banach showed that measurability implies continuity, so that the result of Cauchy is applicable. Shapiro [6] gave an elegant brief proof for locally integrable functions. Hamel [7] showed that for arbitrary functions, (2) is not true; see also Broggi [8] for detail. Darboux [9] showed that condition of additivity (1) paired with continuity at only one point is sufficient for (2), and later he proved in [10] that it is sufficient to assume that $xf(x)$ is nonnegative or nonpositive in some neighborhood of $x=0$.

Cauchy functional equation (1) was included in the list of principal problems presented by David Hilbert at the Second International Congress in Paris in 1900. The fifth Hilbert's problem that generalizes it reads: "Can the assumption of differentiability for functions defining a continuous transformation group be avoided?" John von Neumann solved this problem in 1930 for bicompact groups, Gleason in 1952, for all locally bicompact groups, and Montgomery, Zipin, and Yamabe in 1953, for the Abelian groups.

For undergraduate research purposes, a simple result obtained by author may be of interest. It states that differentiability at one point is sufficient. Assume that $f'(a)=K$ exists at some point a. Then for any x:

$$f'(x) = \lim_{h\to 0} \frac{f(x+h) - f(x)}{h} =$$

K. Elleithy (ed.), *Advanced Techniques in Computing Sciences and Software Engineering*,
DOI 10.1007/978-90-481-3660-5_1, © Springer Science+Business Media B.V. 2010

$$\lim_{h\to 0}\frac{f(x+a-a+h)-f(x+a-a)}{h}=$$
$$\lim_{h\to 0}\frac{f(x+a-a+h)-f(x+a-a)}{h}=$$
$$\lim_{h\to 0}\frac{f(a+h)+f(x-a)-f(a)-f(x-a)}{h}=$$
$$\lim_{h\to 0}\frac{f(a+h)-f(a)}{h}=f'(a). \quad (10)$$

This means that the derivative function $f'(x)=K$ for any x, so that by the Fundamental Theorem of Calculus we have

$$f(x)=\int_0^x Kdt+f(0)=K(x-0)+0=Kx. \quad (11)$$

Geometric interpretation may be as follows, Fig. 1. Cauchy functional equation without any additional conditions guarantees that every solution passes through the origin and varies proportionally with the argument x. In particular, $f(x) = Kr$ for all rational r. In general, a graph of the function has "dotted-linear" structure that underlies findings of Hamel [7] and Broggi [8]. Additional conditions are aimed at combining all partial dotted lines into one solid line. Cauchy[1] used continuity condition imposed on all values of x. It may be mentioned, however, that linearity of the partial graphs allows imposing conditions locally, because only slopes of the dotted lines should be made equal. Thus, either continuity at one point only, or sign of $xf(x)$ preserving in the neighborhood of the origin, or differentiability at the origin is sufficient. Results of Hamel [7] and Broggi [8] demonstrate that in general the horizontal section of the graph is weird if no conditions are imposed.

II. Functional Equations for Power and Exponential Functions

Functional equation (1) characterizes a linear function (2). For objectives of this paper we need characterizations of power and exponential functions as well. Power function will serve as an indicator of the rate of growth no greater than polynomial one. The last follows from the observation that any power function $g(x)= x^a$, $a>0$, is growing no faster than some polynomial function, say, $h(x)=x^{[a]+1}$,where $[a]$ is the largest integer no greater than a.

Characterization may be obtained based on basic functional equation (1). Following Fichtengoltz [2], we will show that functional equation

$$f(x+y) = f(x)\,f(y), \quad (12)$$

is held for exponential function only

$$f(x) = e^{Kx}, \quad (13)$$

while functional equation

$$f(xy) = f(x)\,f(y), \quad (14)$$

only for power function:

$$f(x) = x^K. \quad (15)$$

Assume for simplicity that functions are continuous and are not equal to zero identically. Then given

$$f(x+y) = f(x)\,f(y), \quad (16)$$

consider some value a, such that $f(a) \neq 0$. From equation (12) and condition $f(a) \neq 0$ we have that for any x

$$f(a) = f((a-x) + x)= f(a-x)f(x) \neq 0, \quad (17)$$

so that both $f(a-x)$ and $f(x)$ are not equal to zero. The last means that function $f(x)$ satisfying equation (16) is not equal to

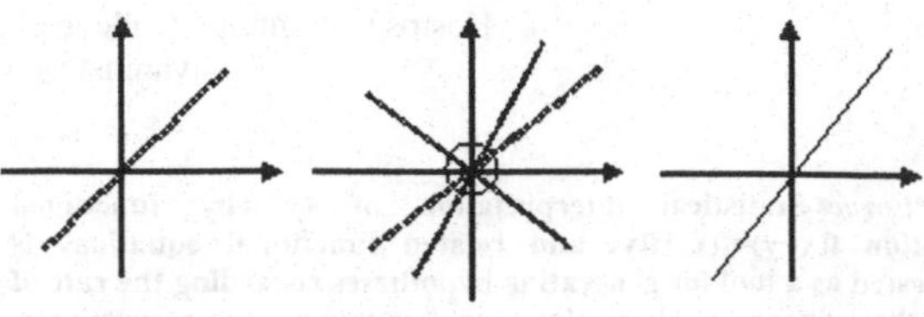

Fig. 1. Geometric Interpretation of Cauchy functional equation.
a) $y=Kr$ for rational r's; b) Possible structure of the f(x) graph; c) Graph with additional conditions imposed

equal to $x/2$, we get

$$f(x) = f(x/2+x/2) = f(x/2)f(x/2) = f(x/2)^2 >0. \quad (18)$$

The last means that $f(x)>0$ for all values of x. This observation allows us taking logarithms from both parts of equation (16):

$$\ln(f(x+y)) = \ln(f(x)\,f(y)) = \ln(f(x)) + \ln(f(y)). \quad (19)$$

From the last equation it follows that a composite function $g(x)= \ln(f(x))$ satisfies equation (1), and thus,

$$g(x) = Kx. \quad (20)$$

Substitution of the function $f(x)$ back gives

$$g(x) = \ln(f(x))=Kx \text{ or } f(x)=e^{Kx} \quad (21)$$

as desired. If we denote $a = e^K$, then function $f(x)$ may be rewritten as

$$f(x)=e^{Kx} =(e^K)^x = a^x,\ a>0. \quad (22)$$

For characterization of power function, consider equation (14) for positive values of x and y only. Given

$$f(xy) = f(x)\,f(y),\ x,y >0, \quad (23)$$

introduce new variables u and v, such that $x=e^u$ and $y=e^v$, so that $u= \ln(x)$, $y = \ln(v)$. Then functions $f(x)$ and $f(y)$ may be rewritten as function of the variables u or v as $f(x)=f(e^u)$ and $f(y)=f(e^v)$, correspondingly. By doing so, we have

$$f(xy)=f(e^u e^v)= f(e^{u+v})=f(x)\,f(y)=f(e^u)f(e^v). \quad (24)$$

Consider composite function $g(x)=f(e^x)$. In terms of $g(x)$, equation (24) may be presented as

$$g(u+v) = g(u)g(v), \quad (25)$$

that is similar to equation (12). The last characterizes the exponential function, so that we have, in accordance with equation (22),

$$g(u) = e^{Ku}. \quad (26)$$

Rewriting equation (26) in terms of $f(x)$, we get

$$g(u)= e^{Ku} = e^{K\ln x} = (e^{\ln x})^K = x^K = g(\ln x)= f(e^{\ln x})= f(x), \quad (27)$$

as desired.

More equations and details may be found in [2], [11] and [12].

III. Statistical Procedures Based on Cauchy Functional Equation

In this paper, our objective is to develop a statistical procedure that allows for determination of a pattern of growth in time: linear, polynomial, or exponential, respectively. Finding patterns of growth is of practical importance. It allows, for instance, early determination of the rates of expansion of unknown illnesses, technological change, or social phenomena. As a result, timely avoidance of undesirable consequences becomes possible. A phenomenon expanding linearly usually does not require any intervention. Periodical observations made from time to time are sufficient to keep it under control. In case of polynomial trend, regulation and control are needed and their implementation is usually feasible. Case of exponential growth is quite different. Nuclear reactions, AIDS, or avalanches may serve as examples. Extraordinary measures should be undertaken timely to keep such phenomena under control or avoid catastrophic consequences.

To determine a pattern, we suggest observing the development of a phenomenon in time and testing statistical hypotheses. A statistical procedure used in this paper is paired t-test, [13]. The procedure assumes that two random variables X_1 and X_2, not necessarily independent, are observed simultaneously at the moments of time $t_1, t_2, ..., t_n$ (paired observations). Let d and $\bar{d}$ be their differences and sample average difference, correspondingly:

$$d_i = X_{1i} - X_{2i}, \quad \bar{d} = \frac{\sum_{i=1}^{n} d_i}{n} \tag{28}$$

Then the test statistic formed as shown below has t-distribution with $(n-1)$ degrees of freedom

$$t = \frac{\bar{d}}{\left(\frac{s_d}{\sqrt{n}}\right)}, \quad s_d = \sqrt{\frac{\sum_{i=1}^{n} d_i^2 - nd}{n-1}}, \tag{29}$$

where s_d is standard deviation of t.

Statistical procedure suggested in this paper is this. A sample of observed values is transformed correspondingly and organized into three sets of sampled pairs corresponding to equations (1), (14), or (12), respectively. The equations are used for statement of the corresponding null hypotheses H_0. The last are the hypotheses of equality of the means calculated for the left and the right hand sides of the equations (1), (14), or (12), respectively. In testing the hypotheses, the objective is to find a unique set for which the corresponding hypothesis cannot be rejected. If such set exists, it provides an estimation of the rate of growth: linear for the set corresponding to (1), polynomial for the set corresponding to (14), or exponential for the set corresponding to (12). It may be noted that suggested approach is preferable over regression for samples of small or medium sizes. It analyzes the internal dynamics of the phenomenon, while exponential regression equation $f(x)=Ae^{Bx}$ may fail because on a range of statistical observations $Ae^{Bx} \approx A(1+(Bx)+(Bx)^2/2!+...+(Bx)^n/n!)$ for some n, that is indistinguishable from a polynomial.

To explain suggested approach in more details, consider an example. Suppose that observations are made in days 2, 3, 5, and 6, and obtained data are as shown in table 1. What may be plausible estimation of the rate of growth? Calculating the values of $f(x) + f(y) = 7 + 10 = 17$ and $f(x) \cdot f(y) = 7 \cdot 10 = 70$, we can see that the only correspondence that is likely held is $f(x \cdot y) = 71 \approx f(x) \cdot f(y) = 70$. Based on this observation, we conclude that the rate of growth is probably polynomial in this case.

TABLE 1
EXAMPLE OF STATISTICAL OBSERVATIONS

$x = 2$	$y = 3$	$x+y = 5$	$x \cdot y = 6$
$f(x) = 7$	$f(y) = 10$	$f(x+y) = 48$	$f(x \cdot y) = 71$

In general, we compare sample means calculated at times x, y, $x+y$, and xy, denoted below as $M(x)$, $M(y)$, $M(x+y)$, and $M(xy)$, respectively, and test the following hypotheses:

$$H_0^{(1)} : M(x) + M(y) = M(x+y), \tag{30}$$

$$H_0^{(2)} : M(x)\, M(y) = M(xy), \tag{31}$$

or

$$H_0^{(2)} : M(x)\, M(y) = M(x+y) \tag{32}$$

It may be noted that these hypotheses allow for testing more general equations than those given by formulas (2), (14), and (12), namely:

$$f(x) = Kx + C, \tag{33}$$

$$f(x) = Cx^K, \tag{34}$$

or

$$f(x) = Ce^{Kx}, \tag{35}$$

where C is an arbitrary constant. To do this, raw data should be adjusted to eliminate constant term C. The appropriate transformations of raw data are as follows. Equation (33) is transformed by subtraction of the first observed value from all other ones. This observation is assigned ordinal number zero, $f(0) = C$:

$$f(x) - f(0) = (Kx + C) - C = Kx. \tag{36}$$

Equation (34), is adjusted by division by $f(1) = C$:

$$\frac{f(x)}{f(1)} = \frac{Cx^K}{C(1)^K} = x^K, \tag{37}$$

and equation (35), by division by the value of the observation $f(0) = C$:

$$\frac{f(x)}{f(0)} = \frac{Ce^{Kx}}{Ce^{K0}} = e^{Kx}. \tag{38}$$

To process comprehensive sets of experimental data, we need to compare many pairs of observations, so that a systematic approach to form the pairs is needed. In this paper, we formed pairs as shown in table 2. The table presents a case of ten observations available, but the process may be continued similarly. We start with the observation number two and pair with each of consequent observations until the last observation is achieved in the process of multiplication. Thus, given ten observations, we pair the observation number two with observations number 3, 4, and 5. Then we continue with observation number 3. In this case, observation number 3 cannot be paired with any of consequent, because $3 \cdot 4 = 12 > 10$. The same is true for the observations 4, 5, etc. For compatibility, additive pairs are chosen the same as multiplicative ones.

Table 3 represents expansion of table 2 using spreadsheets[1]. In the spreadsheet, pairs are ordered by increasing of the product of the ordinal numbers of paired observations, and the set is limited to 25 pairs. As follows from table 3, a critical case is testing the polynomial rate of growth hypothesis that uses observations with ordinal number $i \cdot j$. The last product reaches the set boundary pretty soon. As a consequence, testing the polynomial rate of growth hypothesis in our experiments was performed with smaller number of pairs than those used for testing two other hypotheses. The Excel statistical function used in the spreadsheets is *TTEST*(Range-of-first-elements-in-pair, Range-of-second-elements-in-pair, value-*2*-for-two-tail-test, value-*1*-for-paired-*t*-test).

IV. Example. Testing the Malthusian Theory of Population Growth

In this section, we test a theory of Thomas Malthus, who suggested the Principle of Population, http://en.wikipedia.org/wiki/. The main idea was that population if unchecked increases at a geometric rate (i.e. exponentially, as 2, 4, 8, 16, etc.) whereas the food supply grows at an arithmetic rate (i.e. linearly, as 1, 2, 3, 4, etc.). He wrote: "The power of population is so superior to the power of the earth to produce subsistence for man that premature death must in some shape or other visit the human race." Malthus made a prediction that population would outrun food supply, leading to a decrease in food per person. He even predicted that this must occur by the middle of the 19th century. Fortunately, this prediction failed, in particular, due to his incorrect use of statistical analysis and ignoring development of industrial chemistry, though recently new concerns aroused caused by using of food-generating resources for production of oil-replacing goods, see, for example, *http://blogs.wsj.com/energy/2007/04/16/foreign-affairs-ethanol-will-starve-the-poor/*.

In this paper, we focus on the hypothesis of exponential growth of population using data of the US Census Bureau for 1950 -2050, available on website *http://www.census.gov/ipc/www/idb/worldpop.html*. For calculations, the following groups of observations were formed: 1950 - 2050, 1950 - 2000, 2000 - 2050, 1950 - 1975, 1975 - 2000, 2000 - 2025, and 2025 - 2050. These groups correspond to the whole period, two halves of the period, and its four quarters, respectively. The first observation in each group was assigned an ordinal number 0 and used for normalization.

Obtained results are shown in table 4 and reveal that for the relatively long periods like 1950 - 2050, 1950 -2000, or 2000 - 2050, no one of the hypotheses can be accepted. At the same time, results obtained for shorter periods of 1950 -1975, 1975 - 2000, 2000 - 2025, and 2025 - 2050 lead to different conclusions regarding the rate of population growth. Using the 5% significance level, the following conclusions may be made. For the period of 1950 -1975, polynomial rate of growth cannot be rejected, and for the period of 1975 - 2000, exponential rate. For the periods of 2000 - 2025 and 2025 - 2050 results are inconclusive, though polynomial hypothesis seems more likely than the exponential one. It should be noted that data used for 2002 and later years were just forecasts, and cannot be taken for granted, so that the estimations obtained for the periods of 2000 - 2025 and 2025 - 2050 are just expectations as well.

TABLE 2
FORMING PAIRS OF OBSERVATIONS

Ordinal number of an observation		Observations combined for paired t-test	
x	*y*	*x·y*	*x+y*
2	3	2*3=6	2+3 = 5
2	4	2*4=8	2+4=6
2	5	2*5= 10	2+5 =7

TABLE 4
PROBABILITIES OF LINEAR, POLYNOMIAL, OR EXPONENTIAL GROWTH HYPOTHESES, AS MEASURED BY PAIRED T-TEST, %

Period	Hypothesis of world population growth		
	Linear	Polynomial	Exponential
1950 - 1975	0.000	**8.190**	0.007
1975 - 2000	0.000	4.812	**28.164**
2000 - 2025	0.000	**2.034**	0.000
2025 - 2050	0.000	**0.918**	0.000

Summarizing, we can state that expected situation is not so dangerous as Malthus predicted, because periods of exponential growth of population alternate with the periods of polynomial growth. This alternation together with advances of industrial chemistry allows for the hope that there will be enough food in the world for all.

V. Conclusions

Suggested approach is two-folded. On one hand, it demonstrates close relationships between Calculus and Statistics and allows for inclusion of statistical components into Calculus courses and vice versa. On the other hand, it may serve as a basis for joint undergraduate research that unites students of different specializations, research interests, and levels of preparation. Taken together, the two aspects help the development of students' creativity, raise their interest in studying both subjects, and allow for early enrollment in research work.

References

[1] Cauchy, A. L. Cours d'Analyse de l'Ecole Royale Polytechnique. Chez Debure frères, 1821. Reproduced by Editrice Clueb Bologna, 1992. (In French.)

[2] Fichtengoltz, G. Course on Differential and Integral Calculus. Vol 1. GIFML, Moscow, 1963. (In Russian.)

[3] Weisstein, E. Hilbert's Problems. *http://mathworld.wolfram.com/HilbertsProblems.html.*

[4] Banach, S. Sur l'équation fonctionnelle f(x+y)=f(x)+f(y). Fundamenta Mathematicae, vol.1, 1920, pp.123 -124.

[5] Sierpinski, W. Sur l'équation fonctionnelle f(x+y)=f(x)+f(y).Fundamenta Mathematicae, vol.1, 1920, pp.116-122.

[6] Shapiro, H. A Micronote on a Functional Equation. The American Mathematical Monthly, vol. 80, No. 9, 1973, p.1041.

[7] Hamel, G. Eine Basis aller Zahlen und die unstetigen Lösungen der Funktionalgleichung f(x+y)=f(x)+f(y). Mathematische Annalen, vol. 60, 1905, pp.459-462.

[8] Broggi, U. Sur un théorème de M. Hamel. L'Enseignement Mathématique, vol. 9, 1907, pp.385-387.

[9] Darboux, G. Sur la composition des forces en statique. Bulletin des Sciences Mathématiques, vol. 9, 1875, pp. 281-299.

[10] Darboux, G. Sur le théorème fondamental de la Géométrie projective. Mathematische Annalen, vol. 17, 1880, pp.55-61.

[11] Borwein, J., Bailey, D. and R. Girgensohn. Experimentation in Mathematics: Computational Paths to Discovery. Natick, MA: A. K. Peters, 2004.

[12] Edwards, H. and D. Penney. Calculus. 6th Ed. NJ: Prentice Hall, 2002.

[13] Weiers, R. Introduction to Business Statistics. 4th Ed. CA: Wadsworth Group. Duxbury/Thompson Learning, 2002.

TABLE 3
SPREADSHEET EXAMPLE

Hypothesis testing for the rate of magnitude (linear, polynomial, exponential)

Paired t-test

World population, 1975 - 2000.

Sourse: http://www.census.gov/ipc/www/idb/worldpop.html

				Adjust for actual #Obs!!!						Adjust for actual #Obs!!!				Adjust for actual #Obs!!!			
#Obs =	25			P{T-test	0.0000					P{T-test}	0.0481			P{T-tes	0.2816		
				St. Dev	246.3		252.4			St. Dev	0.0603		0.6262	St. Dev	0.0624		0.0618
		Norm Poly		Avrg	810.6		837.5			Avrg	1.1600		0.8846	Avrg	1.2007		1.2004
		4,155															
		Norm Li	Lin Adj	*Linear*				Poly Adj	Exp Adj	*Poly*				*Exponential*			
	1975	4,084	*Xi=Ri-Norm*					Xi=Ri/No	Xi=Ri/Norm								

#		Raw data	Xi	*i+j*	*Xi+Xj*	*(i+j)*	*Xi+j*	Xi	Xi	*i*j*	*Xi*Xj*	*(i*j)*	*Xi*j*	*(i*j)*	*Xi*Xj*	*(i+j)*	*Xi+j*
1	1976	4,154.7	71.0	**2+3**	356.5	**5**	363.3	1.0000	1.0174	**2*3**	1.0522	**6**	1.0885	**2*3**	1.0891	**5**	1.0890
2	1977	4,226.3	142.5	**2+4**	430.8	**6**	438.8	1.0172	1.0349	**2*4**	1.0704	**8**	1.1271	**2*4**	1.1079	**6**	1.1074
3	1978	4,297.7	214.0	**2+5**	505.9	**7**	517.9	1.0344	1.0524	**2*5**	1.0888	**10**	1.1658	**2*5**	1.1270	**7**	1.1268
4	1979	4,372.0	288.3	**3+4**	502.2	**7**	517.9	1.0523	1.0706	**3*4**	1.0885	**12**	1.2065	**3*4**	1.1267	**7**	1.1268
5	1980	4,447.1	363.3	**2+6**	581.3	**8**	599.0	1.0704	1.0890	**2*6**	1.1073	**12**	1.2065	**2*6**	1.1461	**8**	1.1467
6	1981	4,522.5	438.8	**2+7**	660.4	**9**	678.8	1.0885	1.1074	**2*7**	1.1266	**14**	1.2482	**2*7**	1.1661	**9**	1.1662
7	1982	4,601.6	517.9	**3+5**	577.3	**8**	599.0	1.1076	1.1268	**3*5**	1.1072	**15**	1.2693	**3*5**	1.1460	**8**	1.1467
8	1983	4,682.7	599.0	**2+8**	741.5	**10**	760.0	1.1271	1.1467	**2*8**	1.1465	**16**	1.2894	**2*8**	1.1867	**10**	1.1861
9	1984	4,762.5	678.8	**3+6**	652.7	**9**	678.8	1.1463	1.1662	**3*6**	1.1260	**18**	1.3289	**3*6**	1.1655	**9**	1.1662
10	1985	4,843.7	760.0	**2+9**	821.3	**11**	843.0	1.1658	1.1861	**2*9**	1.1660	**18**	1.3289	**2*9**	1.2069	**11**	1.2064
11	1986	4,926.8	843.0	**4+5**	651.6	**9**	678.8	1.1858	1.2064	**4*5**	1.1263	**20**	1.3675	**4*5**	1.1658	**9**	1.1662
12	1987	5,012.7	929.0	**2+10**	902.5	**12**	929.0	1.2065	1.2275	**2*10**	1.1859	**20**	1.3675	**2*10**	1.2275	**12**	1.2275
13	1988	5,099.3	1015.5	**3+7**	731.9	**10**	760.0	1.2273	1.2487	**3*7**	1.1457	**21**	1.3868	**3*7**	1.1859	**10**	1.1861
14	1989	5,185.7	1102.0	**2+11**	985.5	**13**	1015.5	1.2482	1.2699	**2*11**	1.2062	**22**	1.4058	**2*11**	1.2485	**13**	1.2487
15	1990	5,273.4	1189.7	**3+8**	813.0	**11**	843.0	1.2693	1.2913	**3*8**	1.1659	**24**	1.4431	**3*8**	1.2068	**11**	1.2064
16	1991	5,357.2	1273.5	**4+6**	727.0	**10**	760.0	1.2894	1.3118	**4*6**	1.1454	**24**	1.4431	**4*6**	1.1856	**10**	1.1861
17	1992	5,440.5	1356.8	**2+12**	1071.5	**14**	1102.0	1.3095	1.3322	**2*12**	1.2273	**24**	1.4431	**2*12**	1.2703	**14**	1.2699
18	1993	5,521.3	1437.6	**2+13**	1158.0	**15**	1189.7	1.3289	1.3520	**2*13**	1.2485	**26**	0.0000	**2*13**	1.2923	**15**	1.2913
19	1994	5,601.0	1517.3	**3+9**	892.8	**12**	929.0	1.3481	1.3715	**3*9**	1.1857	**27**	0.0000	**3*9**	1.2273	**12**	1.2275
20	1995	5,681.7	1597.9	**4+7**	806.2	**11**	843.0	1.3675	1.3913	**4*7**	1.1655	**28**	0.0000	**4*7**	1.2064	**11**	1.2064
21	1996	5,761.6	1677.9	**2+14**	1244.5	**16**	1273.5	1.3868	1.4109	**2*14**	1.2697	**28**	0.0000	**2*14**	1.3142	**16**	1.3118
22	1997	5,840.6	1756.8	**5+6**	802.1	**11**	843.0	1.4058	1.4302	**5*6**	1.1651	**30**	0.0000	**5*6**	1.2060	**11**	1.2064
23	1998	5,918.7	1835.0	**3+10**	974.0	**13**	1015.5	1.4246	1.4493	**3*10**	1.2060	**30**	0.0000	**3*10**	1.2482	**13**	1.2487
24	1999	5,995.6	1911.9	**2+15**	1332.2	**17**	1356.8	1.4431	1.4682	**2*15**	1.2911	**30**	0.0000	**2*15**	1.3364	**17**	1.3322
25	2000	6,071.7	1988.0	**4+8**	887.2	**12**	929.0	1.4614	1.4868	**4*8**	1.1860	**32**	0.0000	**4*8**	1.2276	**12**	1.2275

NOTES:

1. Example. The formula in the cell corresponding to the exponential rate of growth hypothesis is

   =TTEST(X14:X38,AA14:AA38,2,1).

   The result is 0.2816.
2. The spreadsheet is available from author upon request.

Application of Indirect Field Oriented Control with Optimum Flux for Induction Machines Drives

S. Grouni [1 2], R. Ibtiouen [2], M. Kidouche [1], O. Touhami [2]
[1] University of Boumerdes, Boumerdes, 35000 ALGERIA
[2] Polytechnics School, El harrach Algeria 16000 ALGERIA
E-mail: Said Grouni (**sgrouni@yahoo.fr**) Rachid Ibtiouen (**ribtiouen@yahoo.fr**)
E-mail: Madjid Kidouche (**kidouche_m@hotmail.com**) Omar Touhami(**omar.touhami@enp.edu.dz**)

Abstract— The rotor flux optimization is crucial parameter in the implementation of the field oriented control. In this paper, we considered the problem of finding optimum flux reference that minimizes the total energy control for induction machine drive under practical constraints: voltage and current. The practical usefulness of this method is evaluated and confirmed through experiments using (1.5kW/380V) induction machine. Simulations and experimental investigation tests are provided to evaluate the consistency and performance of the proposed control model scheme.

Keywords— Indirect Field Oriented Control (IFOC), Induction Machine, Loss Optimization, Optimum Rotor Flux.

I. INTRODUCTION

Induction machines are widely used in various industries as prime workhorses to produce rotational motions and forces. Generally, variable speed drives for induction machines require both wide speed operating range and fast torque response, regardless of load variations. These characteristics make them attractive for use in new generation electrical transportation systems, such as cars and trains. They are also used in ventilation and heating systems and in many other electrical domestic apparatus [8].

By using the advances of power electronics, microprocessors, and digital signal processing (DSP) technologies, the control schemes of induction machines changed from simple scalar or auto-tuning to Field Oriented Control "FOC" and Direct Torque Control "DTC". The FOC is successfully applied in industrials applications on real time control when dealing with high performance induction machines drives [1], [2], [9], [10], [11].

The FOC is the most suitable way in achieving a high performance control for induction machines. Estimating the magnitude and phase of rotor flux is very crucial for the implementation control field oriented method. Direct ways of sensing the rotor flux by implementing suitable hardware around the machine have proved to be inaccurate and impractical at speed and torque. Indirect methods of sensing the rotor flux employ a mathematical model of induction machine by measuring state variables, like currents and voltages. The accuracy of this method will depend mainly on the precise knowledge time constant. This of rotor machine parameter may change during the operation of the drive, which introduce inaccuracies in the flux estimation as both stator and rotor resistance windings change with temperature [2], [7], [9].

With an aim to improve induction machines performance and stability properties, researches have been conducted to design advanced nonlinear control systems [4], [8], [13]. Most of these systems operate with constant flux norms fixed at nominal rate [8], [19].In this situation, maximum efficiency is obtained .However machines do not operate at their nominal rate as the desired torque changes on-line or may depend on system states such as position or velocity. It is then technically and economically interesting to investigate other modes of flux operation seeking to optimize system performance. Aware of these facts, some previous works have already used the reference flux as an additional degree of freedom to increase machine efficiency [6], [9].

This problem has been treated by many researchers. In [13], [18], [20], the heuristics approaches used offer fairly conservative results. They are based on measurement of supplied power in approximating the optimum flux algorithm. The convergences of these algorithms are not guaranteed.
In [17], an applied analytical approach has been used directly in real time to obtain the optimal trajectory equation of the control drives, nevertheless this solution is less robust than the heuristics methods.

In this paper, the objective of the newly developed method is to offer a unified procedure by adding the adaptation parameters and reducing losses. Our work will be structured as follows: In section II, the induction machine model is first presented, then in Section III, we will describe the application of field oriented current and voltage vector control. In section IV, a new optimization approach of optimum flux is described and analyzed in both simulation and practical control. Finally, we have applied several techniques reducing losses with optimum rotor flux in indirect field oriented control for induction machine at variable speed. Simulation and practical results are given to demonstrate the advantages of the proposed scheme. Conclusion and further studies are explained in the last section.

K. Elleithy (ed.), *Advanced Techniques in Computing Sciences and Software Engineering*
DOI 10.1007/978-90-481-3660-5_2, © Springer Science+Business Media B.V. 2010

II. CONTROL PROBLEM FORMULATION

A. Dynamic induction machine model

Mathematical model of induction machine in space vector notation, established in d-q axis coordinates reference rotating system at ω_s speed can be represented in the Park's transformation shown in Fig.1.

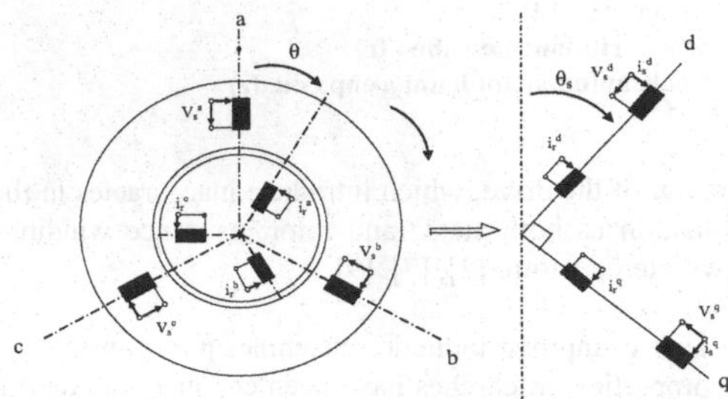

Fig.1. Scheme of Park transformation for induction machines

Standard dynamic model of induction machine are available in the literature. Parasitic effects such as hysteresis, eddy currents, magnetic saturation, and others are generally neglected. The state-space model of the system equations related to the indirect method of vector control is described below. The reference frame with the synchronously rotating speed of ω_s, d-q-axis voltage equations are [15].

$$v_{ds} = R_s i_{ds} + \frac{d\varphi_{ds}}{dt} - \omega_s \varphi_{qs} \tag{1}$$

$$v_{qs} = R_s i_{qs} + \frac{d\varphi_{qs}}{dt} + \omega_s \varphi_{ds} \tag{2}$$

$$0 = R_r i_{dr} + \frac{d\varphi_{dr}}{dt} - (\omega_s - \omega)\varphi_{qr} \tag{3}$$

$$0 = R_r i_{qr} + \frac{d\varphi_{qr}}{dt} + (\omega_s - \omega)\varphi_{dr} \tag{4}$$

Where ω_s, ω are the synchronous and rotor angular speeds.

The stator and rotor fluxes are defined by the following magnetic equations:

$$\varphi_{ds} = L_{ls} i_{ds} + L_m (i_{ds} + i_{dr}) = L_s i_{ds} + L_m i_{dr} \tag{5}$$

$$\varphi_{qs} = L_{ls} i_{qs} + L_m (i_{qs} + i_{qr}) = L_s i_{qs} + L_m i_{qr} \tag{6}$$

$$\varphi_{dr} = L_{lr} i_{dr} + L_m (i_{ds} + i_{dr}) = L_r i_{dr} + L_m i_{ds} \tag{7}$$

$$\varphi_{qr} = L_{lr} i_{qr} + L_m (i_{qs} + i_{qr}) = L_r i_{qr} + L_m i_{qs} \tag{8}$$

L_{ls}, L_{lr} and L_m are the stator leakage inductance, rotor leakage inductance, and mutual inductance, respectively.

The expressions of electromagnetic torque and mechanical speed are stated by:

$$C_{em} = p\frac{L_m}{L_r}(\varphi_{dr} i_{qs} - \varphi_{qr} i_{ds}) \tag{9}$$

$$\frac{d\omega}{dt} = \frac{pL_m}{JL_r}(\varphi_{dr} i_{qs} - \varphi_{qr} i_{ds}) - \frac{p}{J}C_r - \frac{f_r}{J}\omega \tag{10}$$

The difficulty of equation (9), is the strong coupling between flux and current of machine.

B. Indirect Field Oriented Control (IFOC)

For the rotor flux oriented control system, the rotor flux linkage vector has only the real component, which is assumed to be constant in the steady state. From (7) and (8), the rotor currents are given by:

$$i_{dr} = \frac{1}{L_r}(\varphi_{dr} - L_m i_{ds}) \tag{11}$$

$$i_{qr} = \frac{1}{L_r}(\varphi_{qr} - L_m i_{qs}) \tag{12}$$

Substituting (11) and (12) into (3) and (4), we can extract two expressions of dynamic d-q axis rotor flux components are expressed by:

$$\frac{d\varphi_{dr}}{dt} + \frac{R_r}{L_r}\varphi_{dr} - \frac{L_m}{L_r}R_r i_{ds} - \omega_{sl}\varphi_{qr} = 0 \tag{13}$$

$$\frac{d\varphi_{qr}}{dt} + \frac{R_r}{L_r}\varphi_{qr} - \frac{L_m}{L_r}R_r i_{qs} + \omega_{sl}\varphi_{dr} = 0 \tag{14}$$

$\omega_{sl} = \omega_s - \omega$ is the slip angular speed

If the vector control is fulfilled such that q-axis rotor flux can be zero, and d-axis rotor flux can be constant, the electromagnetic torque is controlled only by q-axis stator current, therefore from (9), with $\varphi_{qr} = 0$, $i_{dr} = 0$, yields

$$\frac{d\varphi_{dr}}{dt} = \frac{d\varphi_{qr}}{dt} = 0 \tag{15}$$

$$C_{em} = p\frac{L_m}{L_r}(\varphi_{dr} i_{qs}) \tag{16}$$

Substituting (15) into (3) and (11)-(14) yields

$$i_{qr} = -\frac{L_m}{L_r} i_{qs} \tag{17}$$

$$\varphi_{dr} = L_m i_{ds} \tag{18}$$

$$\omega_{sl} = \frac{L_m}{T_r}\frac{i_{qs}}{\varphi_{dr}} = \frac{1}{T_r}\frac{i_{qs}}{i_{ds}} \tag{19}$$

where $T_r = L_r / R_r$ is the time constant of rotor, $\omega = \dot{\theta}$ with θ is the position of the rotor and i_{ds}, i_{qs} are the direct and quadrant axis components stator currents, where φ_{dr}, φ_{qr} are the two–phase equivalent rotor flux linkages and the rotor speed ω is considered as state variable and the stator voltage v_{ds}, v_{qs} as command variables.

We have shown in equation (9) that the electromagnetic torque expression in the dynamic regime, presents the coupling between stator current and rotor flux. The main objective of the vector control of induction machine is, as in direct current (DC) machines, to control independently the torque and the flux [4]. This can be realized by using d-q axis rotating reference frame synchronously with the rotor flux space vector. The d-axis is aligned with the rotor flux space vector. Under this condition we have; $\varphi_{dr} = \varphi_r$ and $\varphi_{qr} = 0$. In this case the electromagnetic torque of induction machine is given by equation (16). It is understood to adjust the flux while acting on the component $\boldsymbol{i_{ds}}$ of the stator current and adjust the torque while acting on the $\boldsymbol{i_{qs}}$ component. One has two variables of action then as in the case of a DC machine.

Combining equations (13), (15) and (16) we obtain the following d-q-axis stator currents:

$$i_{ds}^* = \frac{1}{L_m}(T_r \frac{d\varphi_r^*}{dt} + \varphi_r^*) \tag{20}$$

$$i_{qs}^* = \frac{L_r}{pL_m} \frac{C_{em}^*}{\varphi_r^*} \tag{21}$$

$$\omega_{sl}^* = \frac{L_m}{T_r} \frac{i_{qs}^*}{\varphi_r^*} \tag{22}$$

The torque C_{em}^* and flux φ_r^* are used as references control and the two stator currents $\boldsymbol{i_{ds}}$, $\boldsymbol{i_{qs}}$ as inputs variables [13], [19].

Combining equations (21)-(22) we obtain the following expression of reference torque as a function of reference slip speed.

$$C_{em}^* = p \frac{\varphi_r^{*2}}{R_r} \omega_{sl}^* \tag{23}$$

with $\omega_s^* = \omega + \omega_{sl}^*$

The references voltages are given in steady state by:

$$v_{ds}^* = R_s i_{ds}^* - \omega_s^* \sigma L_s i_{qs}^* \tag{24}$$

$$v_{qs}^* = R_s i_{qs}^* + \omega_s^* \sigma L_s i_{ds}^* \tag{25}$$

where σ is the total leakage coefficient given by:

$$\sigma = 1 - \frac{L_m^2}{L_s L_r} \tag{26}$$

These equations are functions of some structural electric parameters of the induction machine ($\boldsymbol{R_s}$, $\boldsymbol{R_r}$, $\boldsymbol{L_s}$, $\boldsymbol{L_r}$, $\boldsymbol{L_m}$) which are in reality approximate values.

The rotor flux amplitude is calculated by solving (19), and its spatial position is given by:

$$\theta_s = \int_0^t \left(\omega + \frac{L_m i_{qs}}{T_r \varphi_r} \right) dt \tag{27}$$

C. simulation Study of IFOC voltage and current Control

A simulation study was carried out to investigate the following models controls used on a closed loop IFOC system which depend on the loading conditions. The current and voltage control simulations of IFOC are given by Fig.2 and Fig.3.

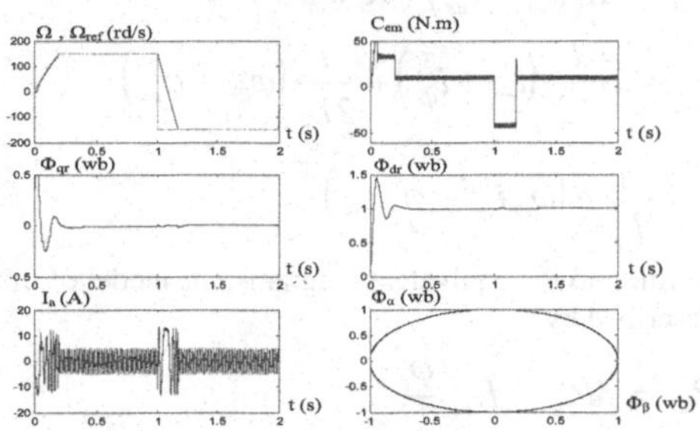

Fig. 2. Simulation of IFOC - IM drives with current control.

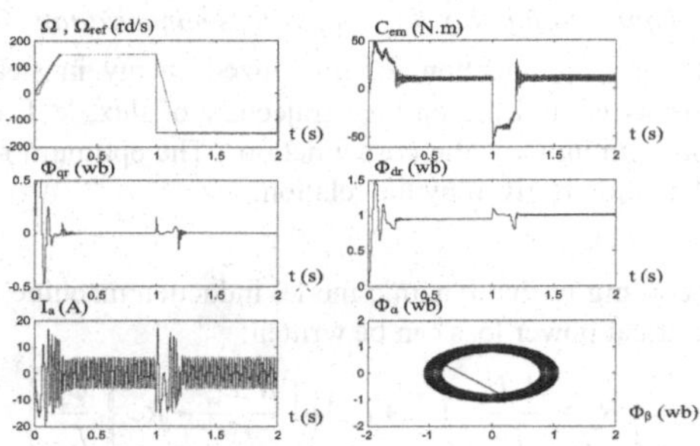

Fig. 3. Simulation of IFOC-IM drives with voltage control.

III. LOSS MINIMISATION IN IFOC INDUCTION MACHINE

A- Power losses of induction machine

In the majority constructions of induction machines, electromagnetic time constant is much smaller than mechanical time constant. For this reason the strategy of torque control which minimizes power losses can be reduced to the steady electromagnetic state. Several loss models are proposed and used in the literature, among this work those of [8], [9] and [10] which take into account the copper losses of the stator , the rotor windings and the iron losses. The total electrical input power of induction machine composed by resistive losses $P_{R,loss}$, power P_{field} stored as magnetic field energy in the windings and mechanical output power P_{mech} expressed in the d,q variables. Applying the transformation to rotating reference frame on d,q variables the total power is given by:

$$\begin{aligned} P_{el,total}(t) &= \sigma L_s\left(i_{ds}\frac{di_{ds}}{dt} + i_{qs}\frac{di_{qs}}{dt}\right) - \frac{1}{L_r}\left(\frac{d\varphi_{dr}}{dt}\varphi_{dr} + \frac{d\varphi_{qr}}{dt}\varphi_{qr}\right) \\ &+ R_s\left(i_{ds}^2 + i_{qs}^2\right) - R_r\left(i_{dr}^2 + i_{qr}^2\right) \\ &+ \frac{L_m}{L_r}\omega\left(\varphi_{dr}i_{qs} - \varphi_{qr}i_{ds}\right) - \frac{2R_rL_m}{L_r}\left(i_{qr}i_{qs} + i_{qs}i_{ds}\right) \end{aligned} \tag{28}$$

This equation can be written in a condensed form as:

$$P_{el,total} = \frac{dP_{field}}{dt} + P_{R,loss} + P_{mech} \tag{29}$$

With:

$$P_{R,loss} = R_s\left(i_{ds}^2 + i_{qs}^2\right) + R_r\left(i_{dr}^2 + i_{qr}^2\right) \quad (30)$$

$$P_{field} = \sigma\frac{L_s}{2}\left(i_{ds}^2 + i_{qs}^2\right) + \frac{1}{2L_r}\left(\varphi_{dr}^2 + \varphi_{qr}^2\right) \quad (31)$$

$$P_m = \frac{L_m}{L_r}\omega\left(\varphi_{dr}i_{qs} - \varphi_{qr}i_{ds}\right) \quad (32)$$

According to the equivalent diagrams, the model of iron losses is described by:

$$\Delta P_{Fe} = \Delta P_{Fe}^s = R_{Fe}\frac{\varphi_r^2}{L_m^2} \quad (33)$$

B. Power loss minimization with level flux control

B.1 Optimization flux for copper loss minimization

The flux optimization for minimized energy in steady state consists to find the optimal trajectory of flux, $\forall$ T $\in$[0,T], which minimizes the cost function. The optimum rotor flux calculation is given by the relation:

$$\varphi_r = f(C_{em}) \quad (34)$$

According to the dynamic model induction machine, the total electrical power loss can be written:

$$\Delta P_t = \left(R_s + \frac{R_r L_m^2}{L_r^2}\right)\left(i_{ds}^2 + i_{qs}^2\right) + \left(\frac{R_r L_m^2}{L_r^2} + R_f\right)\left(\frac{\varphi_r}{L_m}\right)^2 \quad (35)$$

$$-2\frac{R_r L_m}{L_r^2}i_{ds}\varphi_r = \beta_1\varphi_r^2 + \beta_2\frac{C_{em}^2}{\varphi_r^2}$$

Optimum operation point, corresponding to the minimum loss is obtaining by setting:

$$\frac{\partial(\Delta P_t)}{\partial\varphi_r} = 0 \quad (36)$$

The resolution of this equation, gives the optimum rotor flux:

$$\varphi_r^{opt} = \beta\sqrt{\left|C_{em}\right|} \quad (37)$$

Where

$$\beta = \left(\frac{L_r^2 R_s + R_r L_m^2}{\left(R_s + R_f\right)p^2}\right)^{1/4} \quad (38)$$

This parameter depends on machine operating.

From the equations (20) and (21) we deduce that the optimal control in the steady state is given by:

$$u_1^o = \frac{T_r}{L_m}\left(\frac{d\varphi_r^{opt}}{dt} + \frac{1}{T_r}\varphi_r^{opt}\right) \quad (39)$$

The optimization method consists in minimizing the copper losses in steady state while imposing the necessary torque defined by the speed regulator. To check the simulation results and to evaluate the feasibility and the quality of control, we carried out experimental test on the copper loss minimization by the flux variation method. It's noticed that the experimental result, Fig.6 is similar to the result found by simulation Fig.7.

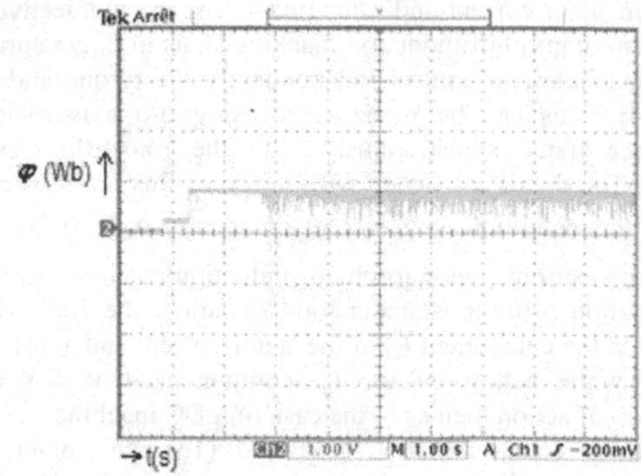

Fig.4. Experimental optimum flux variation

Fig.6 shows a simulation test, the curves of the losses in nominal and optimal mode with the application of a light load (C_r=5N.m), enable us to note that during load torque reduction, this method becomes more effective and it compensates excess loss by a flux reduction.

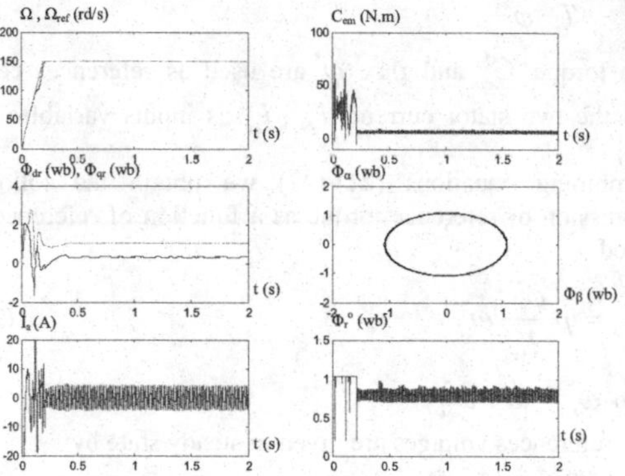

Fig.5. Simulation results of optimal control with flux variation, Cr= 5 N.m

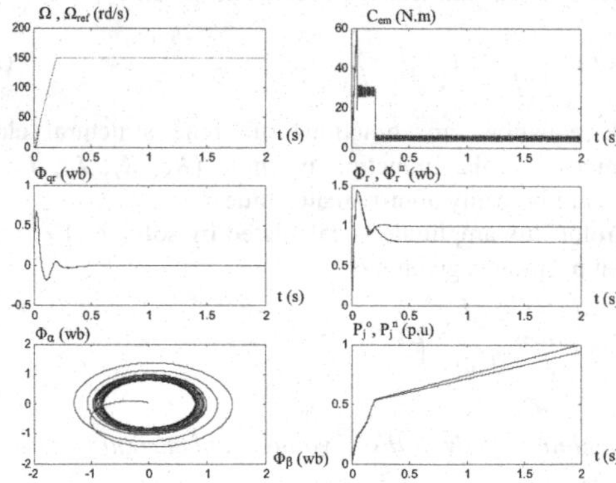

Fig.6. Simulation result of loss minimization, Cr=5N.m Comparison between nominal & optimal

C. Copper loss minimization with $i_{ds} = f(i_{qs})$

The optimization loss is given by using the objective function linking two components of stator current for a copper loss minimization. The expression of power losses is given by:

$$\Delta P_t = \sigma L_s \left(i_{ds} \frac{di_{ds}}{dt} + i_{qs} \frac{di_{qs}}{dt} \right) - \frac{L_m}{T_r L_r} \varphi_{dr} i_{ds} + \left(R_s + \frac{L_m^2}{L_r T_r} \right) \left(i_{ds}^2 + i_{qs}^2 \right) \tag{40}$$

In steady state, the minimum of power losses is reached for:

$$i_{ds} = \left(1 + \frac{L_m^2}{R_s L_r T_r} \right)^{1/2} i_{qs} \tag{41}$$

The behavior of the machine is simulated by using the block diagram of figure 7. The simulation results are presented on figure 8 under light load application. This method shows an important decrease in copper losses under low load torque.

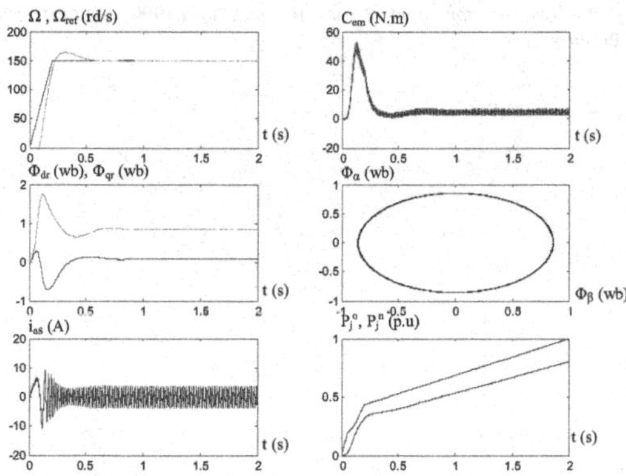

Fig.7. Simulation result of copper loss minimization, Cr=5N.m Comparison between nominal & optimal

D. Copper and iron loss minimization with $i_{ds} = f(i_{qs}, \omega)$

Loss Minimization is presented by introducing the mechanical phenomena. Copper and iron losses are given by:

$$\Delta P_t = R_s \left(i_{ds}^2 + i_{qs}^2 \right) + \frac{R_r R_{Fe}}{R_r + R_{fe}} i_{qs}^2 - \frac{L_m^2 \omega^2}{R_r + R_{fe}} i_{ds}^2 \tag{42}$$

$$i_{ds} = \sqrt{\frac{R_s R_r + R_s R_{fe} + R_r R_{fe}}{L_m^2 \omega^2 - R_s \left(R_r + R_{fe} \right)}} i_{qs} \tag{43}$$

The simulation results of Fig.8 shows a faster response time speed.

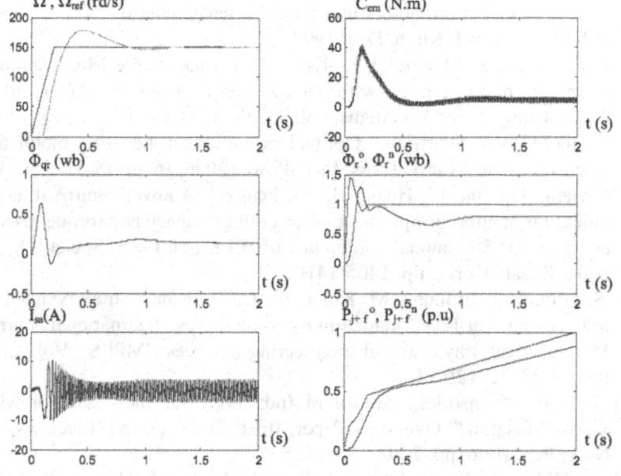

Fig.8. Simulation result of copper and iron loss minimization, Cr=5N.m Comparison between nominal & optimal

IV. CONCLUSION

In this work, we have presented a new method in reducing losses while considering and keeping under control machine parameters. This method could be important for the implementation in real time field oriented control. It has successfully demonstrated the design of vector field oriented control technique with optimum rotor flux using only the stator currents and position measurements. The main advantages of this method is that it is cheap and can be applied in both open and closed loop control.

APPENDICE

induction motor parameters: P_n= 1.5 kw,
U_n= 220 v, Ω_n= 1420 tr/mn, I_n= 3.64 A(Y) 6.31A(Δ),
R_s= 4.85Ω, R_r= 3.805Ω, L_s= 0.274 H, L_r= 0.274 H, p= 2,
L_m= 0.258 H, J= 0.031 kg.m ^{2}, f_r= 0.008 Nm.s/rd.

REFERENCES

[1] F. Abrahamsen, F. Blåbjer, J.K. Pederson, "Analysis of stability in low cost energy optimal controlled PWM VSI fed induction motor drive" in Proc. EPE'97 Trondheim Norway 1997, pp. 3.717-3.723.

[2] F. Abrahamsen, "Energy optimal control of induction motors" Ph.D. dissertation, Aalborg University, institute of energy technology, 1999.

[3] A.S. Bezanella, R. Reginetto, "Robust tuning of the speed loop in indirect field oriented control induction motors" Journal Automatica, Vol. 03, 2001, pp. 1811-1818.

[4] F. Blaschke, "The principle of field orientation as applied to the new transvector closed loop control system for rotating field machines" Siemens review, Vol. 34, May 1972, pp. 217-220.

[5] B.K. Bose, *Modern Power Electronics and AC Drives*. 2002, Prentice Hall.

[6] J. H. Chang, Byung Kook Kim," Minimum- time minimum – loss speed control of induction motors under field oriented control " IEEE Trans. on Ind. Elec. Vol. 44, No. 6, Dec. 1997

[7] A. El-Refaei, S. Mahmoud, R. Kennel, "Torque Ripple Minimization for Induction motor Drives with Direct Torque Control (DTC)" Electric Power Components & systems, vol. 33, No.8, Aug. 2005

[8] J. Faiz, M.B.B. Sharifian," Optimal design of an induction motor for an electric vehicle " Euro. Trans. Elec. Power 2006, 16: pp.15-33

[9] Y. Geng, G. Hua, W. Huangang, G. Pengyi, "A novel control strategy of induction motors for optimization of both efficiency and torque response" in Proc. The 30th annual conference of IEEE Ind. Elect. Society, Nov2-6, 2004, Busan, Korea, pp. 1405-1410.

[10] S. Grouni, R. Ibtiouen, M. Kidouche, O. Touhami, "Improvement rotor time constant in IFOC induction machine drives" International Journal of Mathematical, Physical and Engineering Sciences IJMPES, Vol. 2, No.1, Feb. 2008, pp. 126-131.

[11] J. Holtz, "Sensorless control of Induction Machines- with or without Signal Injection?" Overview Paper, IEEE Trans. on Ind. Elect. , Vol. 53, No.1, Feb. 2006, pp. 7-30.

[12] J. Holtz, J. Quan, "Sensorless control of induction motor drives" Proceedings of IEEE, Vol.(90), No.8, Aug. 2002, pp. 1359-1394.

[13] G.S. Kim, I.J. Ha, M.S. Ko, "Control of induction motors for both high dynamic performance and high power efficiency " IEEE Trans. on Ind. Elect., Vol.39, No.4, 1988, pp. 192-199.

[14] I. Kioskeridis, N. Margaris, "Loss minimization in motor adjustable speed drives" IEEE Trans. Ind. Elec. Vol. 43, No.1, Feb. 1996, pp. 226-231.

[14] R. Krishnan "Electric Motor Drives, Modeling, Analysis and control" 2001, Prentice Hall.

[15] P. Krause, *Analysis of electric Machinery*, Series in Electrical Engineering, 1986, McGraw Hill, USA.

[16] E. Levi, "Impact of iron loss on behavior of vector controlled induction machines" IEEE Trans. on Ind. App., Vol.31, No.6, Nov./Dec. 1995, pp. 1287-1296.

[17] R.D. Lorenz, S.M. Yang, "AC induction servo sizing for motion control application via loss minimizing real time flux control theory" IEEE Trans. on Ind. App., Vol.28, No.3, 1992, pp. 589-593.

[18] J.C. Moreira, T.A. Lipo, V. Blasko, "Simple efficiency maximizer for an adjustable frequency induction motor drive" IEEE Trans. on Ind. App., Vol.27, No.5, 1991, pp. 940-945.

[19] J. J. Martinez Molina, Jose Miguel Ramirez scarpetta," Optimal U/f control for induction motors" 15th Triennial world congress, Barcelona, Spain, IFAC 2002.

[20] M. Rodic, K. Jezernik, "Speed sensorless sliding mode torque control of induction motor" IEEE Trans. on Industry electronics, Feb. 25, 2002, pp.1-15.

[21] T. Stefański, S. Karyś, "Loss minimization control of induction motor drive for electrical vehicle" ISIE'96, Warsaw, Poland, pp. 952-957

[22] S.K. Sul, M.H. Park, "A novel technique for optimal efficiency control of a current-source inverter-fed induction motor" IEEE Trans. on Pow. Elec., Vol.3, No.2, 1988, pp. 940-945.

[23] P. Vas, *Vector control of AC machines*.1990, Oxford Science Publications.

Urban Cluster Layout Based on Voronoi Diagram

—— *A Case of Shandong Province*

ZHENG Xinqi WANG Shuqing FU Meichen ZHAO Lu

School of Land science and Technology,China University of Geosciences (Beijing)

Beijing 100083, China,telephone.+86-10-82322138;email:zxqsd@126.com

YANG Shujia

Shandong Normal University, 250014, Jinan, China

***Abstract:* The optimum layout of urban system is one of important contents in urban planning. To make regional urban system planning more suitable to the requirements of urbanization development, this paper tests a quantitative method based on GIS & Voronoi diagram. Its workflows include calculating city competitiveness data, spreading the city competitiveness analysis by aid of spatial analysis and data mining functions in GIS, getting the structural characteristics of urban system, and proposing the corresponding optimum scheme of the allocation. This method is tested using the data collected from Shandong Province, China.**

Keywords: *City competitiveness; Spatial analysis; Voronoi Diagram; GIS; Optimum layout*

1 INTRODUCTION

The urbanization level embodies comprehensively the city economic and social development level, and is the important mark that city--region modernization level and international competitiveness. Now facing the economical globalization, the cities increasingly becoming the core, carrier and platform of competition in the world.

No matter in a nation or in a region, constructing the urban platform with all strength, optimizing the urban system structure, developing the urban economy, advancing the urbanization level, all of these should be done well, which can strengthen the developing predominance and international competitiveness, further win the initiative in future. Completely flourishing rural economy and accelerating the urbanization course, which are the important contents and striving aims of constructing the well-off society completely that put forward in 16th Party Congress [1]. All of these should be considered from the region angle, and by planning the more reasonable regional urban system to be achieve.

The optimum layout of urban system is always an important content of urbanization construction and urban planning study. Many scholars have conducted studies on this aspect [2,3,4,5,6]. These researches considered more about the urban space structure and evolution than its economic attraction, effectiveness and competitive superiority in the urban system, which did not consider the combined factors when conducting the urban system planning. To solve the problem, we tested a method and used the calculation data of urban competitiveness, by aid of the spatial analysis and data mining functions in GIS, Voronoi Diagram[7,8,9,10],then discussed the analysis and technical method about the optimizing scheme suggestion of urban system layout. Taking Shandong Province as an example and comparing with the past achievements, we found that the method in this paper is more scientifically reasonable.

2 DEFINING CITY COMPETITIVENESS

Competitiveness assessment is necessary in this competitive era. In urban system planning, we also need to think about the urban competitiveness assessment. Michael Porter suggested that the core of competitiveness be the concept of "Clusters" after he analyzed the comparative superiority theory in Nation Competitive Advantage [11]. In fact, the district which only possesses the superior natural resources will not become the prosperous district in the 21st century. The striding development will realize in some districts, which depend on the development of cluster (zone) to some degree.

The most influencing competitiveness assessment now in the world is WPY, founded by the World Economy Forum (WEF) and the Switzerland Lausanne international management and development college (IMD) in 1980. This assessment designed the integrated quantitative assessment index system according to the solid theoretical research with theoretical research, and began to take effect in 1989. The number of the assessment indices which are adopted at present reaches 314, and the number of nations and districts which are evaluated are up to 49. And 36 academic organizations and 3532 experts in the world are invited to participate in cooperating. Many organizations in China have had the research about competitiveness, such as the Renmin

K. Elleithy (ed.), *Advanced Techniques in Computing Sciences and Software Engineering*,
DOI 10.1007/978-90-481-3660-5_3,

university of China, Institute of Finance and Trade Economics (IFTE) of the Chinese Academy of Social Sciences (CASS), Beijing LiYan Institute of Economics, city comprehensive competitiveness compare research center of Shanghai Academy of Social Sciences (SASS), School of Economics & Management of Southeast University, Guangdong Academy of Social Sciences (GDASS) and Hong Kong city competitiveness seminar etc. The most famous competitiveness evaluation in China at present is city competitiveness evaluation, Typical references include "Blue book of city competitiveness: report of the China city competitiveness" [12], "the concept and index system of city competitiveness" [13], and "the index system of China county-level economy basic competitiveness, China county-level economy basic competitiveness evaluation center, etc. Although there are a lot of evaluation methods about the city competitiveness at present, authoritative assessment index system and method has not been formed yet. In view of this, according to the needs of source and evaluation of data, we adopt a comparatively simple competitiveness index to weigh competitiveness.

The urban land is the carrier of all activities in a city. The city development is reflected by the change of the land to a great extent, and it is particularly true under the market economy condition. In order to study city competitiveness, we can take urban land as core and analyze the influencing factors, such as the maturity degree of urban land market and urban land use benefit, etc., and then indirectly evaluate the city competitiveness. If the urban land use condition is the result of the joint interaction of the nature, society, economy and market materials in a city, and if the urban land use benefit in a city is high, its city competitiveness generally is also comparatively high. So we analyzed and compared urban land assessment index system and city competitiveness evaluation system. We found that they have many similar contents, thus decided to use urban land evaluation results to reflect city competitiveness evaluation results, and called them the city competitiveness.

3 METHODS

The city competitiveness factors have great influence on the city development, and embody the city's advantages of economic, social and natural conditions among cities. They are generally divided into two levels: factor and indictor.

3.1 Selection principle

(1) The changes of index have notable influence to city;

(2) The index values have clearer variable range;

(3) The index can reflect the differences among different quality cities;

(4) The indexes reflect the city development trend at present time; and have influence on city development in future.

(5) Making full use of the city statistical data which were published by national bureau of statistics of China;

(6) Adopting the non-statistics investigation data as much as possible that are collected by the authoritative organizations, like construction department and planning bureau etc.

3.2 Method of evaluation selection factors

There are a lot of factors and indicators that influence the synthetic quality of urban land, which include macroscopic and microcosmic, static and dynamic, direct and indirect, and mutually contact and interaction. In order to select representative indices, adopting the following basic procedure: use the existing documents as reference, and investigate and analyze general factors and indicators that probably influence the urban land quality, finally select 82 indexes. Because it is not easy to procure effective data directly by part of the indexes, a total of 68 factors and indicators participate in the beginning of the analysis, then using SPSS10.0 to process principal components analysis. We obtain the principal components used in evaluation. The number of grading indexes is 58.

After removing the similar and repeated indexes, and extensively consulting experts across and outside the province, the final result is that the Shandong provincial city grading index system includes eight factors, 31 sub-factors and 58 indicators. The eight factors include: ① City location, including transportation location, external radiant capacity, urban built-up area; ② City concentration size, including population size, population density and the proportion of the secondary and tertiary industrial production value in GDP; ③ urban infrastructure conditions, including road traffic, water supply, gas supply, sewerage and heat supply; ④ Urban public service conditions, including school condition, medical and hygienic treatment condition; ⑤ Urban land use efficiency, including municipal service condition, capital construction investment intensity, the secondary and tertiary industrial increment intensity, trade intensity and labor input intensity; ⑥ City environmental conditions, including afforestation condition, waste treatment; ⑦ City economy development level, including GDP, financial condition, fixed assets investment condition, commerce activity condition, external trade activity condition and finance condition; ⑧ City development potential, including science and technology level, agriculture population per capita cultivated land and water resources condition etc.

3.3 Identifying the weights of grading factors

The weights of the grading factors and indicators were identified by five criteria: ① Weights are direct ratio to the factors' influence on land quality, which value between 0 and 1 and their sum is 1; ② Weight values of indicators that related with every evaluation factor vary from 0 to 1, and their sum is 1; ③ the invited experts should be related domain technologist, high level administrative decision-maker, who are familiar with urban land and society economy development condition as well as have higher authority. The total is 10 - 40 persons; ④ According to corresponding work background and marking demonstration, experts give mark, and they should independently mark without consulting.

According to the above principles and adopting Delphi method, we identified the weights of grading factors and indicators in Shandong provincial city competitiveness evaluation.

3.4 DATA PROCESSING

In the process of evaluating city competitiveness, datum involved and dimension among each index are different, so we firstly standardized the datum. There are many methods of data standardization. According to data characteristic and research purpose, we selected order standardization law. The formula is as follows:

$$Y_{ij} = 100 \times X_{ij} / n$$

where Y_{ij} is the score of indicator j of city i, X_{ij} is the sequence of city i, which is gotten according to the value of indicator j of every city sequencing, and when the index is positive correlation with city competitiveness, the sequencing is for ascending order, otherwise for descending order. n is the number of evaluated cities.

The concrete process is as follows. Firstly, sequencing every index and counting its value, then counting sub-factors' score through adopting the adding weights to sum method, taking the same step to count score of sub-factors. Finally we take adding weights to sum method to count synthetic score.

We calculated the index scores using the formula:

$$F_{ik} = \sum_{j=1}^{n} (W_{kj} \times Y_{ij})$$

where F_{ik} is the score of index k of city i; W_{kj} is the weight of factor k corresponding with indicator j; Y_{ij} is the score of index i of city j; n is the number of indicators included in factor k.

We calculated the synthetic index scores by the formula:

$$S_i = \sum_{k=1}^{n} (W_k \times F_{ik})$$

where S_i is the synthetic scores of grading object i; W_k is the weight of factor k; F_{ik} is the score of factor k of grading object i; n is the number of factors.

All the provincial cities get their total score respectively based on the calculation results of eight synthetic factors. What needs to be explained is that when taking the place order method to standard, for making the cities that participated in the competitiveness evaluation processes comparable, we carry on the unified standardization according to Shandong regionalism in 2001, which includes 17 cities, 31 county level cities, 60 counties and 12 independent and not entirely linking the piece areas under the jurisdiction of municipality, and amounting to 120 towns.

4 SPATIAL ANALYSIS METHOD

The spatial analysis originated from the "Computation Revolution" in geography and regional science in the 60's. At the beginning stage, it chiefly applied quantitative (chiefly being statistic) methods to analyze the spatial distribution model of point, line and polygon (Hi Prof. Zheng, do you have any reference for this part? Thank you!). Afterwards, it even more emphasized the character of geographical space itself, spatial decision course and the temporal and spatial evolution course analysis of the complicated space system. In fact, from the map aooearance (what is aooearance?), people are carrying on all kinds of spatial analysis all the time consciously or unconsciously.

The spatial analysis is the general designation for analysis technology of spatial data. According to different data quality, it can be divided into: ① the analysis operation based on spatial graph data; ② the data operation based on non-spatial attribute; ③ combined operation of spatial and non-spatial data. The foundation that spatial analysis relies on is the geographical space database, and the mathematic means of its application including the logic operation, quantitative and statistical analysis, algebraic operation, etc. Its final purpose is to solve geographical space factual question that people encountered, draw and transmit geographical spatial information, especially implied information, and assist decision-making.

GIS is a science that integrates the newest technology of many disciplines, such as the relational database management, effective graphic algorithm, inserting value analysis, zoning and network analysis, which provides powerful spatial analysis tools and makes the complicated and difficult assignment become simple and easy to do. At present there are the spatial analysis functions in most GIS software. This article mainly depends on MapInfo Professional 7.0 [14] terrace, and uses the development languages such as Mapbasic [15] to carry on the function module developing, and realize the spatial analysis of competitive assessment in is the town. The process generalized in the following:

(1) digitize 1:500000 "Shandong district map" which is published by Shandong publishing company in the year 2001;

(2) establish a database for town space position and competitiveness power assessment index, and the connection by way of administration code establishment;

(3) build Voronoi Diagram[16], as shown in Fig.1;

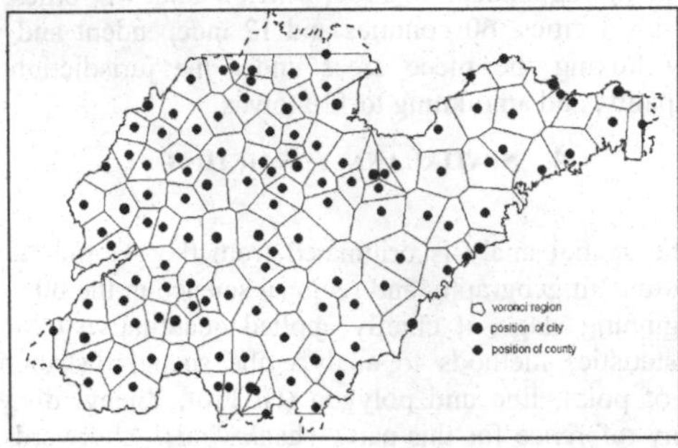

Fig.1 Voronoi diagram of town in Shandong province

(4) Optimize the layout of the town cluster based on Voronoi diagram: calculating the amount of town cluster (zone) (such as 4、5、6、7、8 and so on), calculating the central position of the town cluster (zone) (which is called the center of mass in town cluster) using the total point value of town competition power as the weight of grading. It has the smallest sum of distance from the location to the surrounding towns. Through the trail, the amount of the center of mass chosen this time is 8, as shown in Fig. 2.

(5) use data mining technology to carry on spatial analysis and the knowledge discovery, serve as the new layout object with the group center of mass in town, and establish Voronoi diagram again. So we can obtain the scope of town cluster;

(6) establish competitiveness power evaluation picture (dividing the value with total points value and factor serves as the radius, and the seat is the centre of a circle with the town), which includes the total points value and factor branch value picture, as Fig. 2 and Fig. 3 show;

(7) optimize the layout based on optimizing principle, which means adjusting suitably on the contrary to the facts, like incorporation and displacement, etc., analyzing competing power space analysis picture comparatively and making the final decision for the project of optimizing layout.

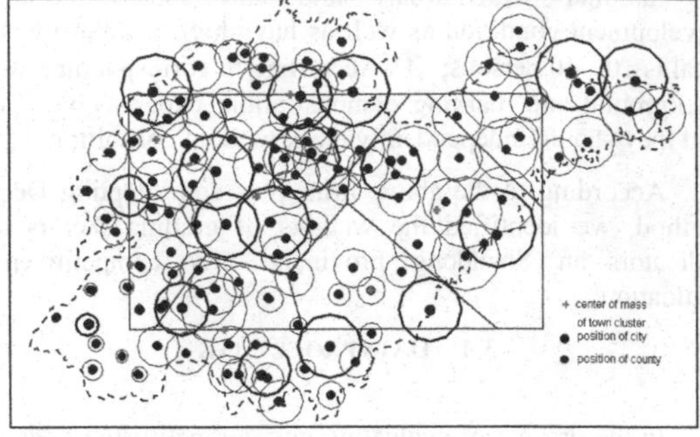

Fig.2 8 centers Voronoi diagram of town in Shandong province

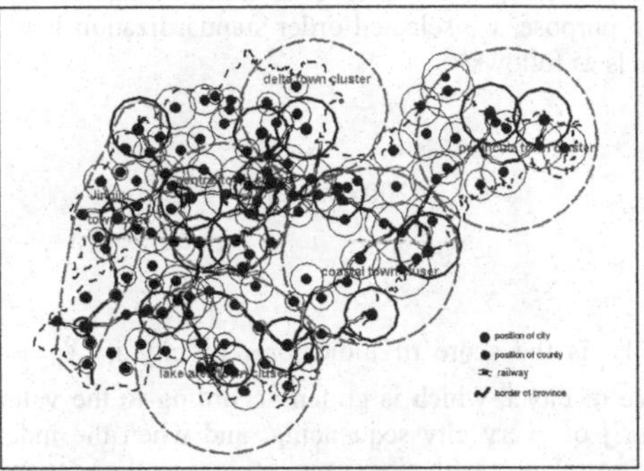

Fig.3 spatial analysis of competition evaluation

5 DISTRIBUTION CHARACTERISTICS AND THE LAYOUT OPTIMIZATION

According to the grand blue print that strides across the century of the planning "Development Report of National Economy and the Society of Distant View Objective Outline of 9th Five-Year Plan and in 2010 in Shandong" [17]: build the necessary mutually modernized town systems of four administrative levels in big or middle or small city and the villages and towns, and form four industry gathering zones Jiaoji, Xinshi, Dedong, Jingjiu based on heavy chemical industry bases and the high, new technology industry and three modernized agriculture areas each having its characteristic in the construction Jiaodong area coast, one of the warring states into which China was divided during the Eastern Zhou period, located in the southern portion of modern Shandong Province Central South mountain area and one of the warring states into which China was divided during the Eastern Zhou period, located in the southern portion of modern Shandong Province northwest China plain are quickened to put into effect Huanghe River delta to develop and "at sea Shandong" to build two striding across the century the engineering greatly, and promote to economize entirely economical to realize division of labor reasonably, harmonious development in the high administrative levels.

According to the planning mentioned above, based on the Shandong province town Voronoi diagram, there are 5 town clusters (zone) of Shandong province by optimizing: the central Shandong Province town cluster (Jinan, Taian, Zibo, Laiwu), the coastal town cluster (Qingdao, Wei Fang, Rizhao), peninsula town cluster (Yantai, Weihai), delta town cluster (Dongying, Binzhou), lake area town cluster (Jining, Linyi, Zaozhuang) and Jingjiu town cluster (Dezhou, He ze, Liaocheng).

The five clusters (zone) cover 85.83% of 120 towns in Shandong province which is known from the competing power percentage of the five cluster (zone) (Fig. 4). The central Shandong town cluster has the strongest competing power, whose number of towns is placed in the middle, but the infrastructure, the efficiency of land use, the economy development level, the integrated strength are all at the first place in Shandong province. This region which aims to be the leader modern province capital in China pays more attention to develop high, new technological industry such as micro-electronics, precise engine instrument, new materials, fine chemical combination, biological project and so on, the third industry which is high level such as finance trade, business service, tourism and entertainment, science and technology consultation, higher education, science research, medical treatment, sanitation and so on by the aid of the persons with ability, developed traffic, the powerful foundation of economy technology in province capital. Industry diffusing and grads transfer should be done better especially with the chance of the region financial centre which Jinan has been made by the country. Solidify and improve the position of region centre, try best to build Jinan to be a modern integrated central city with multi-function.

Both the coastal town cluster and the lake area town cluster have their own characteristics. Qingdao is the central city in the coastal town cluster where the position condition of the town and the ecological environment are at the fist place all over the province. It is superior to other clusters for the lake area town on the centralizing scale of town, the infrastructure of town and the region develop potential. Qingdao is the central city in coastal town cluster which has the superior region position and perfect infrastructure, with the large opening, large economy and trade and extraversion. The coastal town cluster has the biggest port in Shandong and the town system with fine function which should make full use of being open to the outside in order to speed up the develop of the whole province, develop the economy and technology cooperation with other countries energetically, take part in international division of labor, catch the trends of the newest technology home and abroad, putting forth effort development high and new technology industry and the export-oriented economy and makes hard this area become the whole nation to be open one of the highest and economic the strongest export-oriented economy region of Vigour of level. The lake area town cluster is with Jining for the center, and each town distributes in minute hill lake vicinity, and takes resource exploitation and the commodity trade as the major characteristic here is that Shandong develops the largest district of latent capacity in the 21st century.

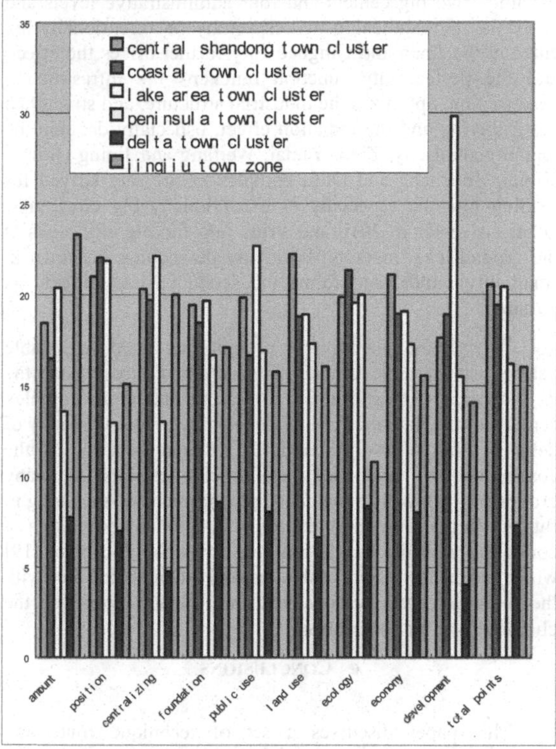

Fig.4 the comparison of the optimum layout factors competing power in Shandong province

The peninsula town cluster and the delta town cluster are two characteristic town clusters of Shandong. The focal point that crowd in delta town is a Shandong is built one of engineering, the very big result had been gained in the town infrastructure construction here, town economy development standard and town public installation constructions etc, the rich superiority of petroleum resource be fully given play to, and the industry made a good job of that the petroleum is processed and the petroleum is formed a complete set or system. At the same time, adopt the method that protection nature has developed, and positively develop using the natural resources such as wasteland etc, and develop petroleum replacement such as foodstuff and agriculture by-product process, salt chemical industry, machinery, builds the material, gently spins and electronics etc the industry energetically, and can go on the development in order to realize. With ground efficiency, infrastructure and publicly installation constructions and economic development standard etc are fairly better to the peninsula town cluster at the town ecology environment and town, and will fully give play to his external radiant potential superiority, and expands the international market, and promote urbanization intensity and standard hard.

At present, the layout planning of town in Shandong is: forming "two big centers and four administrative levels and five axis". Strengthening two especially or exceptionally big cities in the Jinan and Qingdao City further drives the effect, and the perfect city function quickens the infrastructure construction, optimizes the industrial structure, and strengthen the gathering and the radiation effect; especially develop the region central city, Zibo, Yantai, Weifang and Jining (Jining, Yantai, Zoucheng and Qufu complex centre) are strived for striding into the especially or exceptionally big city ranks; Other strive for in 2010 and stride into the big city ranks at the ground level markets; Positively developing the town in small city builds, and forms the scope [18] as quickly as possible.

Though the planning at present has fixed reasonable nature, for example among them implied city cluster the development (Jining, Yantai, Zoucheng and Qufu complex center), meanwhile the coming development latent capacity of delta of the yellow River is huge, and because of the confinement of land use space in city, establishing many groups of town will fit the development trends of shandong in future . At the same time, the result mentioned above is basic consistent with the conclusion consulting the document [19] which means the results of this method are in accordance with the ones of other methods, but this method possesses the characteristic of automation.

6 CONCLUSIONS

This paper discusses a set of technique route and realizing method for town system optimizing plan using the evaluation results of competing power based on GIS tools and space analysis technique of data mining—Voronoi diagram. It is proved to be available through cases research for optimum layout planning of town cluster (zone) in Shandong province.

Certainly, energetically appraising the question still possesses general agreeable justice to the competition in town, and its assessment index system and assessment method still need to be perfected in the process of the research.

Competing the town energetically to appraise the organic combination that with Voronoi figure space was analyzed, this paper puts forward the feasible technology route, and the function achieved the research purpose, but owing to not solve the boundary control question of Voronoi diagram properly when programming, it is not very visually beautiful.

ACKNOWLEDGMENT

Foundation: National Natural Science foundation of China,No.40571119.; the National Social Science Foundation of China under Grant No. 07BZZ015; the National Key Technology R&D Program of China under Grant No. 2006BAB15B03; the Talent Foundation of China University of Geosciences (Beijing) under Grant No. 51900912300

REFERENCES

[1]. Jiang Zemin, 2002, Build a Well-off Society in an All-Round Way and Create a New Situation in Building Socialism with Chinese Characteristics. People Press

[2].Wang Xingping, Li Zhigang, 2001, Study on the Centre-Town Strategy and Choice in County-Planning. Economic Geography, 21(1):61-65

[3].Wang Xinyuan,Fan Xiangtao,Guo Huadong,2001,Pattern Analysis of Spatial Structure of Urban System Dommated by Physical Geographic Factors. Progress In Geography, 20(1):67-72

[4].Wu Guobing, Tan Shengyuan. 2001, Research on Agglomeration Economic Benefit and Evolution and Optimization of the Urban Area Structure. Urban Research, (4):8-11

[5].Huang, B., Yao, L. and Raguman, K., 2006. Bi-level GA and GIS for mult i-objective TSP route planning. Transportation Planning and Technology, 29(2):105-124.

[6].Tetsuo A. 2007,Aspect-ratio Voronoi diagram and its complexity bounds. Information Processing Letters, 105(1): 26-31.

[7].ZHANG Zhe, XUE Huifeng. 2008,D iv ision of Urban Influenced Reg ions of Baoji C ity Ba sed on We ighed Vorono iD iagram. Computer Simulation,25(6):261-264.

[8].WANG Guiyuan,CHEN Meiwu. 2004,Measurement of Urban Hinterland Area Based on GIS, Geography and Geo-Information Science,20(3):69-73.

[9].Ma Juan,Qin Zhiqin. 2007,APPLING THE Voronoi MAP TO THE RESEARCH ON THE URBAN SPACIAL RESOURCE OF SHANDONG PROVINCE, Journal of Shandong Normal University(Natural Science),22(2):96-98.

[10] . Chen Jun, 2002, Spatial Dynamic Data Models of Voronoi,Surveying And Drawing Press

[11].Michael E.Porter., 2002, The Competitive Advantage of Nations. Free Press

[12].Ni Pengfei, 2003, Report of Urban Competitive Capacity in Chian.Social Scientific Document Press

[13].Ning Yuemin,Tang Lizhi, 2001, The Concept and Indicator System of Urban Competitive Capacity. Urban Research, (3):19-22

[14].MapInfo Corporation.User Guide. 2002

[15].MapInfo Corporation.2001, Mapbasic Development Environment Reference Guide Version 6.5

[16].Haowen Y., Robert W.. An algorithm for point cluster generalization based on the Voronoi diagram. Computers & Geosciences,34(8):939-954.

[17]. Li Chunting, 1996-2-15, Development Report of National Economy and the Society of Distant View Objective Outline of 9th Five-Year Plan and in 2010 in Shandong

[18].He Yong, 2002-12-11, The Overall Pattern of Shandong city & Town.People Daily(East China news)

[19]. Zhou Yixing. Wei Xinzhen Feng Changchun.in et al., 2001, Discussions on the Stratecic Development Planning for the Jinng-Qufu Metropolitan Area in Shandong Province, City Plaamng Review , (12):7-12

A Reconfigurable Design and Architecture of the Ethernet and HomePNA3.0 MAC

M. Khalily Dermany[1],M. Hossein Ghadiry[2]
[1]Islamic Azad University, Khomein Branch, Iran
[2] Islamic Azad University, Arak Branch, Iran,
[1]mkhalili@iaukhomein.ac.ir

***Abstract*-In this paper a reconfigurable architecture for Ethernet and HomePNA MAC is presented. By using this new architecture, Ethernet and HomePNA reconfigurable network card can be produced. This architecture has been implemented using VHDL language and after that synthesized on a chip. The differences between HomePNA (synchronized and unsynchronized mode) and Ethernet in collision detection mechanism and priority access to media have caused the need to separate architectures for Ethernet and HomePNA, but by using similarities of them, both the Ethernet and the HomePNA can be implemented in a single chip with a little extra hardware. The number of logical elements of the proposed architecture is increased by 19% in compare to when only an Ethernet MAC is implemented**

I. Introduction

Nowadays with decrease cost and increase use of computer in homes, there is often more than one computer in homes, and users demand to network them. Ethernet with simplicity, reliability and simple management is working well and almost has not a rival between other networking methods. Ethernet benefits from various environments like coaxial, twisted pair and optical fiber. But none of these environments -that Ethernet supports them- exists in home and it is required wiring in home to use Ethernet, that is costly. Wiring cost of Ethernet and also the cost of wireless technology -like HomeRF, WiFi- is a motivation to use exist media in home (ex. Phone lines, mains) for home networking. HomePNA is one of the best methods of home networking [6].

Use of phone line for home networking was presented by TUT System with 1Mbps for the first time. There are several defacto standards for HomePNA that are HomePNA1, HomePNA2, HomePNA3 and HomePNA3.1. In this paper we use the standard that ITU_T has presented. HomePNA standards are available in [1], [2], [3], [4] and the Ethernet in [5].

A reconfigurable system has two parts. A constant (some parts of its algorithm that is never changed) and a hardware reconfigurable part. The structure of the reconfigurable part can be changed via a control unit respect to application and location that system is used. The principal benefits of reconfigurable system are the ability to execute larger hardware designs with fewer gates and to realize the flexibility of a software-based solution while retaining the execution speed of a more traditional, hardware-based approach. This makes doing more with less a reality. Reconfigurable computing is intended to fill the gap between hardware and software, achieving potentially much higher performance than software, while maintaining a higher level of flexibility than hardware.

In this paper two Ethernet MAC and HomePNA methods have been implemented together on a single chip as a reconfigurable architecture. The cost of implementation has been increased in compare to when only one method is used, but the cost is much less than when both are used in separate.

This paper is organized as follows. Section 2 studies protocol stack of HomePNA3 and Ethernet, Section 3 describes proposed architecture, section 4 is results and implementation information and section 5 is conclusion.

II. Protocol stack and frame format of HomePNA3

Protocol stack of HomePNA3 is defined in physical and data link layer which has been illustrated in figure1. In continue each layer is explained.

A. Physical layer

The PHY layer of HomePNA3 provides transmission and reception of physical layer frames using QAM and FDQAM modulation techniques over phone lines. Since phone line is also used by other technologies like DSL and POTS too, HomePNA3 must be adapted with these technologies. Figure2 depicts the spectrum each technology uses. POTS, UADSL[1], Internet connectivity and home phone line networking use the same media, but they operate at different frequencies.

B. Data Link Layer

The data link layer is composed of three sub layers that they are MAC, LLC and convergence. In continue each sublayer is described.

[1] universal asynchronous DSL

K. Elleithy (ed.), *Advanced Techniques in Computing Sciences and Software Engineering*,
DOI 10.1007/978-90-481-3660-5_4, © Springer Science+Business Media B.V. 2010

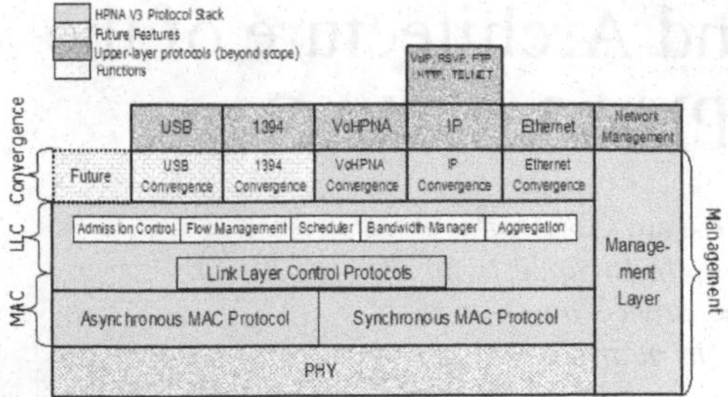

Figure 1: protocol stack of HomePNA3

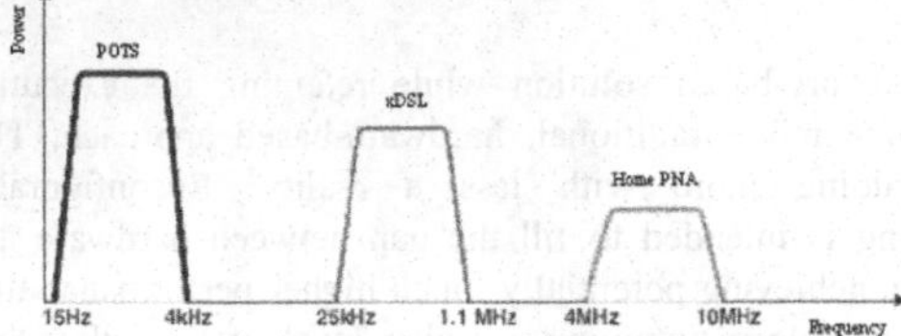

Figure 2: HomePNA Spectral Usage

Convergence sub layer

The convergence sub-layer is a protocol-specific set of sub-layers that maps various transport layer protocols into the native primitives of the LLC sub-layer. It is the responsibility of the convergence sub-layer to translate the native protocol into this underlying framework. The convergence sub-layer may use protocol or configuration specific information to perform the translation. Ethernet and HomePNA3 use NDIS in this sub-layer [4], [17].

LLC Sub-Layer

The LLC sub-layer is responsible for performing link functions control. In particular, it is responsible for managing information concerning network connections, for enforcing Quality of Service (QoS) constraints defined for the various system data flows and for ensuring robust data transmission using Rate Negotiation, Reed-Solomon coding techniques and ARQ (Automatic Repeat Request) techniques. HomePNA3 requires the support of additional link control protocols that manage network admission and flow setup and teardown procedures. These protocols are used to manage the information about connected devices and their associated service flows. The link layer protocols interact with upper convergence protocol layers in order to signal such events as device registration, timing events and flow control operations. In addition to the link layer control protocols required by the G.9954 synchronous MAC, the following link layer functions are required: scheduling, bandwidth management, flow management, network admission and packet aggregation [19], [18].

MAC sub layer

The MAC sub-layer is responsible for managing access to the physical media using a media access protocol. It uses the PHY layer to schedule the transmission of MAC Protocol Data Units (MPDU's) over the physical media within PHY layer transport frames [20].

Ethernet use CSMA/CD techniques to control media, access therefore it doesn't have any priority for nodes to access the media. Ethernet use Binary Exponential backoff for resolving collision [20].

The HomePNA3 MAC sub-layer supports media access according to two different protocol modes: asynchronous mode and synchronous mode. The asynchronous MAC protocol provides priority-based media access that uses CSMA/CD techniques to control media access and a signalling protocol to resolve media collisions. In contrast, the HomePNA3 synchronous MAC protocol uses CSMA/CA techniques, under master control, to avoid collisions by pre-planning the timing of all media-access [4].

Asynchronous mode of HomePNA use CSMA/CD for control media, however this method has been adapted with the challenges of phone equipments. Since HomePNA is used to carry multiple types of data, including media streams such as voice, streaming audio, and video, to reduce the latency variation in these streams, a priority mechanism is defined to allow higher layers to label outgoing frames with priority value, and guarantee that those frames will have preferential access to the channel over lower priority frame [4].

As showed in figure 3.a, slots are numbered in decreasing priority, with the highest priority starting at level 7. Higher priority transmissions commence transmission in earlier slots and acquire the channel without contending with the lower priority traffic. Figure 3.b shows a transition with priority level 5. Thus frames with highest priority are transmitted without contention with lower priority traffic.

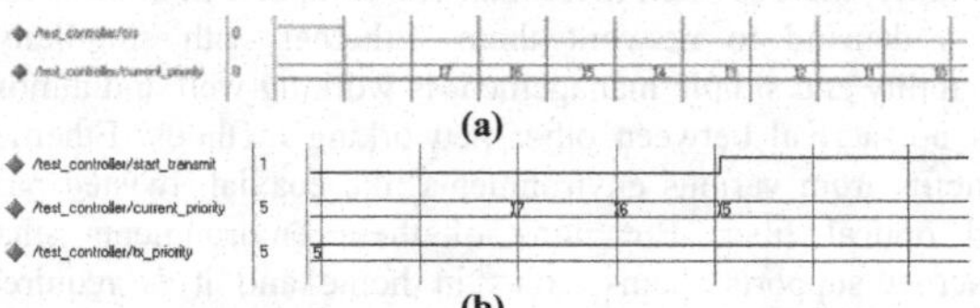

Figure 3: Priority slots of asynchronous mode

Figure 4 is an example of video traffic at priority level 7 gaining access ahead of best effort traffic scheduled at level 2 but if priority method is not implemented, the best effort traffic gaining access ahead of video traffic.

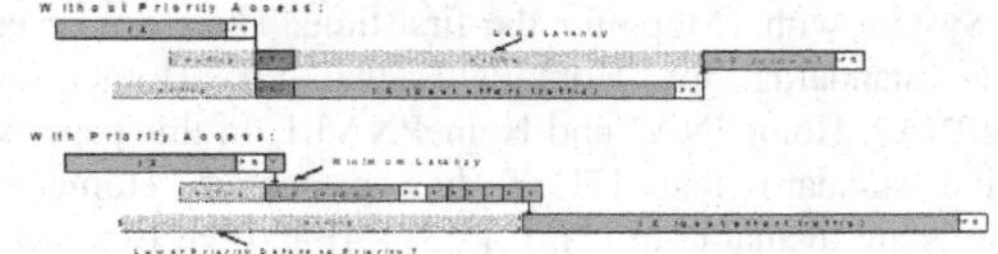

Figure 4: Example of priority access [4]

Two or more stations may begin transmitting in the same priority slot that in this situation collision occurs. All stations monitor the channel to detect the colliding transmissions of other stations. Generally in asynchronous mode of HomePNA, collisions are between frames at the same priority level.

In asynchronous mode of HomePNA, a distributed collision resolution algorithm is run which results in stations becoming ordered into backoff Levels where only one station is at backoff Level 0 and can therefore acquire the channel. All stations, even those with no frame to transmit, monitor the activity on the medium. This results in access latency being tightly bounded.

This mechanism differs from Binary Exponential backoff used in Ethernet which in that, the backoff Level does not determine the contention slot chosen by a station. All stations at a given priority always contend in the slot corresponding to the access priority. The method used is called distributed fair priority queuing (DFPQ). Instead stations at non-zero backoff Levels defer contending until stations that are at zero backoff Level transmit.

After a collision and an IFG, three special backoff signal slots (S0…S2) are present before the normal sequence of priority contention slots occurs. Figure 5 shows signal slots only occur after collisions, they do not follow successful transmissions.

Each station maintains eight backoff level (BL) counters, one for each priority. The backoff level counters are initialized to 0.

Each active station pseudo-randomly chooses one of the slots, and transmits a *backoff20* signal, defined below. More than one station can transmit a backoff20 signal in the same slot. The active stations transmit backoff20 signals to indicate ordering information that determines the new backoff levels to be used.

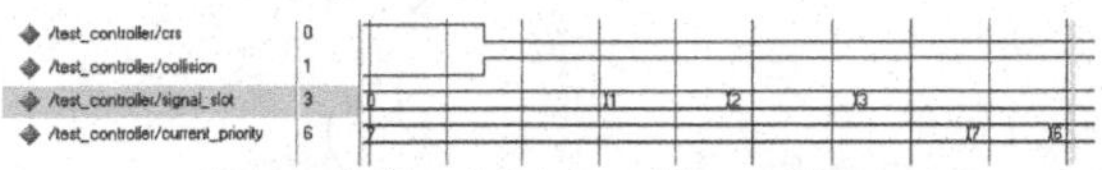

Figure 5: Signal slots providing collision

Synchronous mode of MAC does not use CSMA/CD technique and prevent collision using scheduling of media access. In other word, SMAC mode employs a master. Each station on a HomePNA network segment, in the presence of a master device, shall execute the synchronous MAC function to coordinate access to the shared media.

Media access in a synchronous network is controlled by the master using a media access plan (MAP). The MAP specifies media access timing on the network. Media access time, in the MAP, is broken up into transmission opportunities (TXOPs) of a specified length and start time that are allocated to specific network devices in accordance with their resource demands. Media access timing is planned by the master in such a manner so as to normally avoid collisions. Collisions may, however, still occur within transmission opportunities that are designated as contention periods. Collisions are similarly resolved using the AMAC collision resolution method.

III. Differences and Similarities Between Ethernet and HomePNA

A reason of success of Ethernet is using CSMA/CD mechanism. AMAC mode of HomePNA is used CSMA/CD mechanism too. Hence Ethernet and AMAC mode of HomePNA have a similar mechanism in MAC.

As showed in figure 6, HomePNA and Ethernet have a similar frame format.

There are some differences that are studied in continue:

The differences of HomePNA and Ethernet are:

- AMAC mode of HomePNA has a priority access to media is different from Ethernet.
- Collision resolution method is different from AMAC mode of HomePNA, SMAC mode of HomePNA and Ethernet.

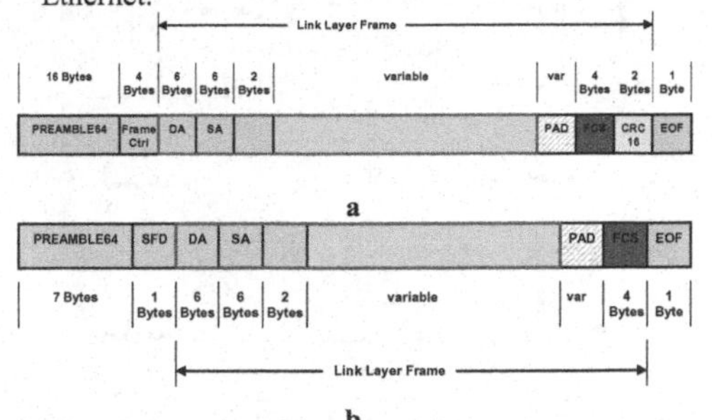

Figure 6: frame format of a-HomePNA b-Ethernet] 4] , [5]

- Parameter like minimum frame length is different from HomePNA to Ethernet.
- Preamble format is different from HomePNA to Ethernet.
- Ethernet use CRC32 for error detection but HomePNA use CRC32 and CRC16.
- Media access control in SMAC mode of HomePNA dose not use CSMA/CD but Ethernet does.

These similarities have been used to reach to a reconfigurable architecture for HomePNA and Ethernet. In following section this architecture is demonstrated.

IV. Proposed Model for Reconfigurable Ethernet and HomePNA Architecture

As figure 7 shows, presented architecture is formed from three basic modules and interface signals. Signal that used in this architecture are demonstrated in below.

- Start_Tx: Permission of transmitting in transmission queue.
- Tx_finished: the transmit queue is finished its frame.
- OP_mod: indicates the operating mode of the controller (HomePNA or Ethernet), that is reported to transmit or receive queue.
- Send_ jam: controller order to transmitter queue to send BACKOFF20 or Jam frame dependent on operating mode.
- TXOP_sig: this signal shows node's TXOP has been started and node can transmit in SMAC mode of HomePNA. LLC layer generated this signal.
- CRC_check: if a CRC error occurred, receiver queue report it to controller via this signal.
- Tx_en, TX_clk, TXD, RXD, CRS, Rx_clk and Rx_dv are used in MII interface to transmit and receive to/from MII.

In continue each module is illustrated.

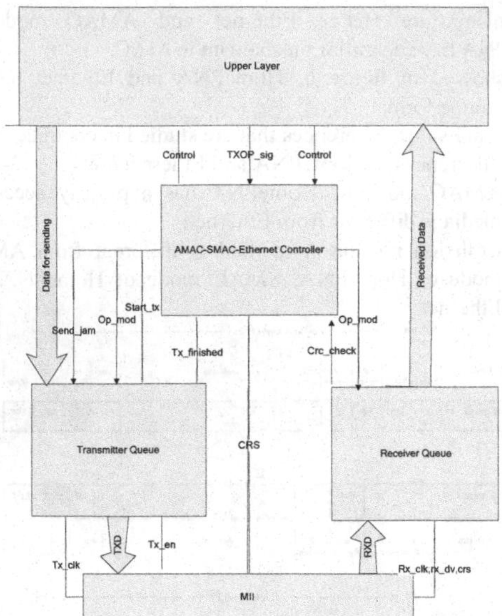

Figure 7: presented reconfigurable architecture of HomePNA and Ethernet

C. *Receiver and transmitter queue*

Receiver queue receives data frame from MII and send it to upper sub layer and transmitter queue does inverse of this function. These modules do below functions too in addition to their main functions. However these functions should be did respect to the modes of operation (Ethernet or HomePNA).

- Detach and attach various field of frame dependent to frame format of Ethernet and HomePNA.
- Receive and transfer of frame based on frame length field dependent to frame size of Ethernet and HomePNA.
- Specification the PAD and detach and attach PAD dependent to frame format of Ethernet and HomePNA.
- Error detection with produce and check CRC (in HomePNA CRC32 and CRC16 but in Ethernet CRC32)
- Insert and study of receiver and transmitter address that is similar in Ethernet and HomePNA.
- Sending and receiving jam or BACKOFF20 signal after collision.
- Receive and transfer frame to/from MII medium. And Adapt with MII

In above different there isn't any difference between AMAC and SMAC modes in transmitter and receiver queues.

The MAC sublayer connects to physical layer via MII standard interface. Since MII transfers 4 bit data to the MAC in each transmission, therefore MAC sublayer should works with 1/4 transmission bit rate. For example when the transmission bit rate is 100Mbps, MAC works with 25 MHz.

D. *Controller*

The controller must controls transmit and receive path, and detect and resolve collision in HomePNA (SMAC and AMAC) and Ethernet. This module permits to transmitter queue based on if media is empty or not and various modes of operation parameters.

There are some differences between SMAC and AMAC modes in controller because in SMAC mode transmitting must occur on allocated TXOP but it is not necessary in AMAC. Figure 8 shows the FSM chart of this module.

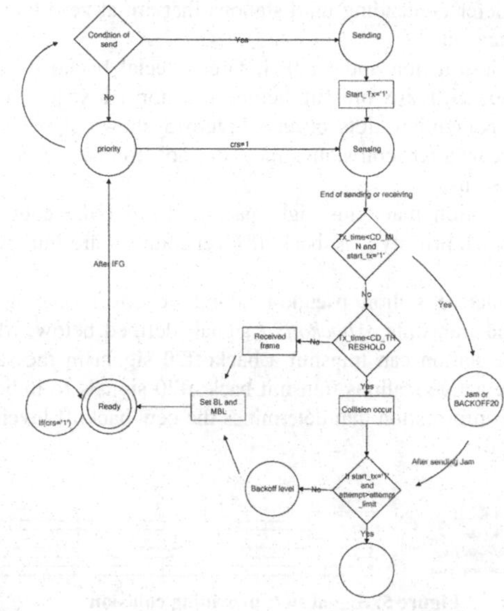

Fig. 4: FSM of controller

The following parameters are used in the FSM. The values of those parameters exist in standards and depend to operation modes.

- Minimum time length of frame (CD_MIN)
- Priority slot time (PRI_SLOT)
- Collision Detection Threshold (CD_THRESHOLD)
- Number of attempt to transmit a frame if collision occur (Attempt_Limit)

The difference between modes of operation is in collision resolution method and condition of transfer. Node should operates depend on the in a mode. In continue each mode and condition of transfer and collision resolution methods are illustrated.

E. *SMAC mode*

In this mode each node can send data if it set on its TXOP. Studying the MAP frame and setting the TXOP_sig is the function of link control sublayer. Node changes its state to sending after TXOP_SIG signal set to logic 1. In fact setting TXOP_SIG signal to 1 indicates the start of a node TXOP. The length of TXOP_SIG signal must be equal to length of a frame. Those frames which needs more time to transmit must not be sent to MAC sublayer.

F. AMAC mode

In this mode, frames transmit inside specific priority slot. If collision is occurred BL counter must be changed with specific algorithm that is described in HomePNA standard.

G. Ethernet mode

In Ethernet mode, node transmits each time the media is free. Exponential back off algorithm is used to resolve the collision.

V. VERIFICATION

There are two approaches to verfy the MAC sublayer functions:

- Loop back test: in this approach output of MAC is connected to its input. Then MAC sends frames and receives them on its input and does as well as when it receives new frames.[21]
- Connection of two MAC: In this approach there is two MAC, that one sends frames and other receives them as figure 9 shows. However collision, transmission error and link errors (ex. disconnection) can be simulated.

In this work both methods are used to verify the MAC functionality.

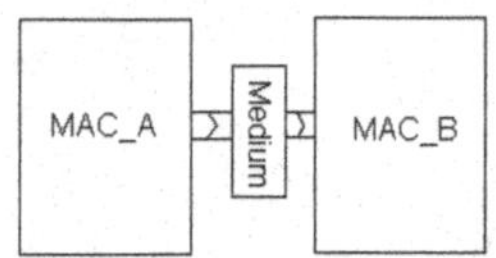

Figure9. Connection of two MAC to test the functionality of the reconfigurable MAC

VI. RESULTS

Proposed architecture has been implemented in VHDL using Quartus compiler and then synthesized on a chip from Altera company with part number EP1S10B672C6 from Startix family. The number of logical element which Quartus reports after synthesis of this architecture is 2561 unit. Ethernet MAC also has been implemented and the required logical element is 2156 unit. Comparing reconfigurable MAC and Ethernet MAC shows the number of logical element increased by 19%.

$$\frac{2561-2154}{2154}\times 100\% = 19\%$$

It is noteworthy that Ethernet MAC should works with 100Mbps rate therefore reconfigurable architecture should work at maximum rate of 25 MHz.

VII. CONCLUSION

In this paper a reconfigurable architecture for HomePNA and Ethernet in MAC sub layer was presented. By using this new architecture reconfigurable Ethernet and Home PNA network card can be manufactured. The cost of manufacturing the new architecture in a single chip is much less than manufacturing each one in separate in two chips. In other hand the new architecture is easier to use for user. The number of logical elements of the new architecture is increased only 19% in compare to an Ethernet MAC.

REFERENCES

[1] ITU-T Recommendation,"Phone Line Networking Transceivers
[2] Foundation," G.989.1, 2001
[3] ITU-T Recommendation,"Phone line Networking Transceivers – Payload Format and Link Layer Requirements," G.989.2, 2001.
[4] ITU-T Recommendation,"Phone line Networking Transceivers – Isolation Function," G.989.3, 2003.
[5] ITU-T Recommendation, "Phone line Networking Transceivers – Enhanced Physical, Media Access, and Link Layer Specifications," G.9954, 2005
[6] IEEE STD 802.3-2002®, "Carrier Sense Multiple Access with Collision Detection (CSMA/CD) Access Method and Physical Layer Specifications," 2002.
[7] T. Zahariadis, K. Pramataris, N. Zervos, "A Comparison of Competing Broadband in Home Technologies," Electronics & Communication Engineering Journal, vol.14, pp: 133- 142, 2002.
[8] S. Kangude, J. Copeland, M. Sherman,"An Analysis of the HomePNA Collision Resolution Mechanism," 28th IEEE Conference on Local Computer Networks LCN, vol.4, pp: 1618-1626, 2003.
[9] S. Kangude, M. Sherman, J. Copeland,"Optimality Analysis of the HomePNA Collision Resolution Mechanism," Georgia Tech CSC Technical Report GIT-CSC-04, vol.10, pp: 11100-1110, 2004.
[10] B. Amodei, L. Costa, O. Duarte,"Increasing the Throughput of the HomePNA MAC Protocol," Local Computer Networks, vol.5, pp: 120-130, 2004
[11] W. Wada, M. Tode, H. Murakami, "Home MAC: QoS-Based MAC Protocol for the Home Network," The 7th International Symposium on Computers and Communications, ISCC, pp: 407- 414, 2002
[12] M. Chung, H. Kim, T. Lee,"HomePNA 2.0-Saturation Throughput Analysis," IEEE Communications Letters, vol.5, pp: 1324-1332, 2003
[13] H. Kim, M. Chung, T. Lee, J. Park, "Saturation Throughput Analysis of Collision Management Protocol in the HomePNA 3.0 Asynchronous MAC Mode," IEEE Communications Letters, vol.8, no.7, pp: 476–478, 2004
[14] P. Bisaglia, R. Castle, "Receiver Architectures for HomePNA 2.0," Global Telecommunications Conference, GLOBECOM, vol.2, pp: 861-866, 2001
[15] P. Bisaglia, R. Castle, S. Baynham, "Channel Modeling and System Performance for HomePNA 2.0," IEEE Journal on Selected Areas in Communications, vol.20, no.5, pp: 913–922, 2002.
[16] Shin, J. Choi, J. Lim, S. Noh, N. Baek, J. Lee, "A 3.3-V analog front-end chip for HomePNA applications," The 2001 IEEE International Symposium on Circuits and Systems ISCAS, vol.4, pp: 698-701, 2001
[17] http://www.homepna.org, 2007.
[18] IEEE STD 802.1H, "IEEE Standards for Local and Metropolitan Area Networks: Recommended Practice for Media Access Control (MAC) Bridging of Ethernet V2.0 in IEEE 802 Local Area Networks," 1995.
[19] IEEE STD 802.2c, "Information technology - telecommunications and information exchange between systems - local and metropolitan area networks - specific requirements. Part 2: logical link control," 1994.
[20] IEEE STD 802.2, "Information technology - telecommunications and information exchange between systems - local and metropolitan area networks - specific requirements. Part 2: logical link control," 1998.
[21] Tanenbaum, "Computer Network," pub: Prentice Hall, Fourth Edition, 2002.
[22] Sh. Mazloman, A. Motamedi. M. Sedighi "design of a Ethernet controller and implementation on FPGA" AmirKabir Uni

Automatic Translation of a Process Level Petri-Net to a Ladder Diagram

Yuval Cohen*

Department of Management and Economics
The Open University of Israel
Raanana, Israel, 43107

Ming-En Wang, Bopaya Bidanda

Department of Industrial Engineering
The University of Pittsburgh
Pittsburgh, USA, 15261

Abstract *– Major part of discrete industrial automation hardware is controlled by Programmable Logic Controllers (PLCs). While Petri-nets (PNs) have been proposed by many researchers for modelling shop floor discrete control, the majority of the world PLCs are programmed using ladder diagrams (LD) and significant portion of them cannot be programmed using another language. This paper proposes hierarchical approach to translating PN to LD: for describing the high level process, it introduces automatic translation technique of process level PN, while utilizing a recently developed method (for translating PN to Boolean logic) for translating the embedded tasks to a LD. Interestingly, the generated LD code enables the reconstruction of the original Petri-net.*

I. Introduction

Petri net (PN) is a discrete event modelling technique thoroughly described in [1] and [2]. Due to its advantages many researchers advocate modelling discrete control using a Petri net [1,2,3,4,5,6].

This paper presents a new technique for systematic translation of a process level Petri-net to ladder diagram (LD). Using this technique, the translation could be easily automated as well as recovery of the model from low level code. The technique presented in this paper not only relieves the automation engineer from dealing with Programmable Logic Controller (PLC) code when constructing the logic, but also in later stages, when changes are required.

PN uses two types of nodes: places (denoted by circles); and transitions (denoted by heavy lines). Directed arrows connect places to transitions and vice versa. Places may denote processing stages (termed tasks) or idle resources (resource/s places). A first serious trial to implement PN on PLCs generated the Grafcet PLC programming language [7]. However, Grafcet has many limitations, and among them inability to modelling part of the non-sequential processes such as state machines. This is also the reason that Grafcet was renamed Sequential Function Chart (SFC) by the international PLC standard: IEC61131-3 [8]. While Grafcet is only prevalent around France, LD is the dominant method in discrete event control of industrial automated systems in most parts of the world [3,6,9]. Thus, the need for translating PN to LD remains. Recently, Lee et. al. [9] proposed a translation method from PN to Boolean logic that could be translated to LD. Their method is suited for small tasks, managing mainly the Boolean relationships between binary inputs and outputs. While this translation technique dealing efficiently with Boolean logic for actuating equipment it is nor geared towards high level process planning. However, most of PN modelling is applied to higher level description: of the process stages, the material moves and the corresponding resource engagement and release. Moreover, the method in [9] also ignores many important issues. For example, it ignores issues related to resource allocation and deadlock avoidance; they ignored the need for model hierarchy; and they ignore the desirability of being able to reconstruct the PN from the generated ladder logic.

To better model the whole process, we propose a hierarchical approach: a new process oriented translation technique is introduced that is geared toward high level process description and the process stages or steps are embedded so that each step could be further translated using the method in [9].

Thus, the main focus of this paper is on the high-level process description. It is assumed the high-level PN is structured as in [10] to model the general stages of the production process and the required resources at each stage. Among the advantages this approach offers, are: enhanced verification and validation, parallel representation of simultaneous activities. We also adopt the convention that there can be only one task node following each transition, and that all the required resources for each task

* Corresponding author: Tel.: (972) 9-778-1883; Fax: (972) 9-778-0668; E-mail: yuvalco@openu.ac.il

K. Elleithy (ed.), *Advanced Techniques in Computing Sciences and Software Engineering*,
DOI 10.1007/978-90-481-3660-5_5,

are inputs to each of the task's preceding transitions.

II. THE EXAMPLE CASE STUDY

Throughout the paper the example depicted in figure 1 is used. Figure 1 describes the layout of a robotic cell. The robot moves the product according to its manufacturing process plan between the machines and buffers until the product is placed at the departure dock.

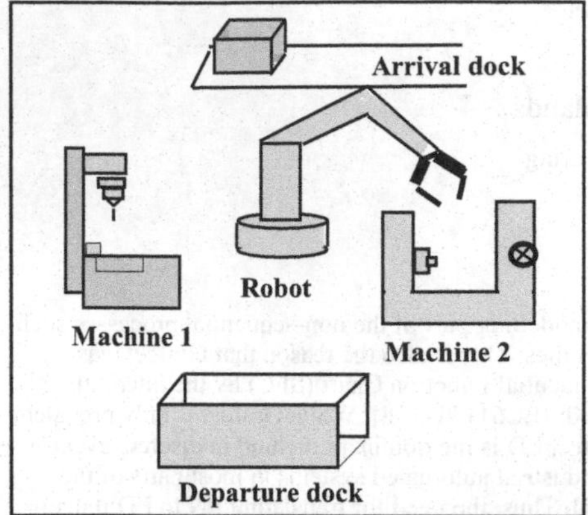

Fig. 1. The example shop floor layout

An example of the high level process description by PN is illustrated in figure 2. The high-level PN describes the stages of the production process and the material moves while the lower level describes the Boolean logic associated with executing each stage. The activation of a production stage of the high-level activates the code of the corresponding embedded task. The PN of figure 2 describes the production process of a product type that is first machined by machine 1, before being processed by machine 2, and then is placed at the departure dock. In figure 2 tasks are shaded to signal embedded levels of detail.

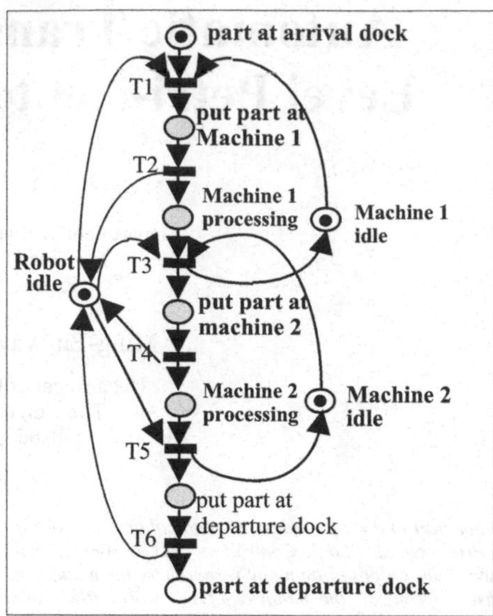

Fig. 2. High-level PN description of the production process

III. ATTACHING VARIABLES

The high level PN scheme should be translated into ladder diagram to enable PLC execution and recovery of the high level status of the system from the PLC code. The first phase in the translation is attaching binary (0/1) variables to the various PN elements. There are only five variable *types* as follows:

1. Process variables: $\boldsymbol{P_i}$ for the ith variable.
2. Process finish variables: $\boldsymbol{F_{ij}}$ for the ith process leading to the jth transition.
3. Resource variables: $\boldsymbol{R_i}$ for the ith variable
4. Transition variables: $\boldsymbol{T_i}$ for the ith variable
5. Sensor variables: $\boldsymbol{S_i}$ for the ith sensor

This translation is facilitated by the one to one correspondence between PN constructs and LD variables as shown in Table 1.

Table 1. The correspondence between the Petri Net constructs and the Ladder Diagram variables

Petri Net Construct	Corresponding Ladder Diagram Variable
Task place	Process variable P_i for the ith task (LD internal variable).
Idle Resource place	Resource variable R_i for the ith resource. (LD internal variable).
Sensor place	Sensor variable S_i for the ith sensor
Transition	Transition variable T_i for the ith transition. (LD internal variable).
Token in a certain place	The variable corresponding to the PN place is *ON*.
Enabled transition	The corresponding transition variable is *ON*.
Arc emanating from a place ***i*** and terminating in transition ***j***	Finish variable F_{ij} leading from the end of task *i* to transition *j*. (LD internal variable).
Arc emanating from a transition ***i*** and terminating in place ***j***	A ladder rung with an input contact T_i, and output coil that corresponds to place *j* (either P_j or R_j).

Legend

S_1= Sensor detects Part at arrival dock
R_0= Robot idle
R_1= Machine 1 idle
T_1= Transition 1
T_2= Transition 2
Task-1 (put part at machine 1):
P_1=Process-1,
F_{12}= Finish P_1 and activate T_2

S_1 R_0 R_1 T_1 P_1 F_{12} T_2

Fig. 3. PN segment extracted from Fig. 2.

Structured steps for attaching variables:

1. List the tasks (i.e., shaded places from a PN similar to the one in Fig. 2) and assign a process variable P_i for each task *i*. For example, in figure 3, task 1 is assigned P_1 variable (corresponding to *process-1*)
2. List the sensors and assign a sensor variable S_i for each sensor *i*. For example, in figure 3, the sensor that detects the part arrival is assigned S_1 variable (corresponding to *sensor-1*)
3. List the resources (e.g., machines) and assign a resource variable R_i for each sensor *i*. For example, in figure 3, the Robot is assigned a resource variable R_0.
4. Each PN arc emanating from a task *i* and terminating at a PN transition *j* is assigned a F_{ij} variable. For example, in figure 3, F_{12} is assigned to the arc leading from task1 to transition 2.
5. Each PN transition is assigned a variable termed T_i for the ith transition. For example, T_1 and T_2 in figure 3 correspond to the upper two transitions of figure 2.

VI. TRNSLATING THE PN TO LD

All the variables defined in section III are included in the LD program. In particular, the transition variables T_i are activated when all their input variables are *ON*. Also, variables associated with places are activated when they are *ON*.

Translating to LD is done by implementing the following logic and conditions:

1. The sensor variables (S_i) are input variables activated by external events.
2. All other variables (P_i, F_{ij}, T_i, R_i) are internal variables activated by internal logic and input variables.
3. External outputs are activated and deactivated only within the embedded tasks (do not appear in the high level PN).
4. An activation of each transition is done at the instant that all its inputs are *ON*.
5. An activation of each transition turns *OFF* all its input resources variables (R_i) signalling that the resources are occupied.
6. An activation of each transition turns *ON* the variables of its PN outbound arcs.
7. An activation of a transition that precedes a task *i*, activates its P_i variable which activates the corresponding embedded task logic.
8. Completion of an embedded task turns *ON* all its F_{ij} variables. A turned ON F_{ij} signals the jth transition that task *i* has been finished.
9. Transition *j* having an F_{ij} input, turns *OFF* (or frees) the F_{ij} the resources of task *i*, and turns *ON* all its resource variables (R_i): signalling the resources availability.

Following the above logic, the LD code of each task *i*, is activated when its corresponding process variable *Pi* turns *ON*. When task *i* is completed, its process variable *Pi* turns *OFF*, and all its finish variables F_{ij} turn *ON*. Activation of a transition *j* turns *OFF* its corresponding F_{ij} . Each F_{ij} variable is turned *OFF* either by transition *j*, or when P_i is turned *ON* again.

The above behaviour can be tracked in figure 4: transition T_1 is activated only when S_1, R_0, R_1 are all *ON*. Corresponding to Sensor-1 detects part at the arrival dock, and both the robot and machine 1 are idle. T_1 dedicate the robot and machine 1 by turning R_0 and R_1 *OFF*. T_1 also *turns* P_1 *ON* to activate the embedded code of task 1. When task 1 is completed the embedded code of task 1 turns *OFF* P_1 and turns *ON* F_{12} to signal the completion of the task to transition 2. Transition 2 is activated (T2 turns *ON*) and when T_2 is finally activated, it turns F_{12} *OFF*.

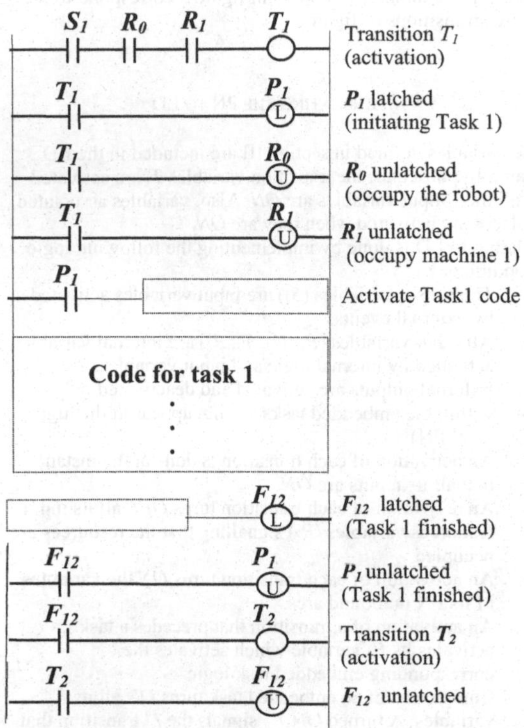

Fig. 4 Ladder logic segment for the PN segment in Fig. 3.

VI. LD REARANGEMENT FOR BACKWARD TRANSLATION

A rearrangement of the LD rungs is necessary for automatic backward translation of this LD code into high level PN. The order of the LD rungs play crucial role in enabling backward translation. The rungs of the actual LD appear in a different order than shown in figure 3. The rungs are rearranged into three groups as follows: 1) transitions triggers, 2) transitions effects, and 3) tasks' code. Figure 5 illustrates the rearrangement scheme, and the mapping of LD segments from figure 4 into the three groups.

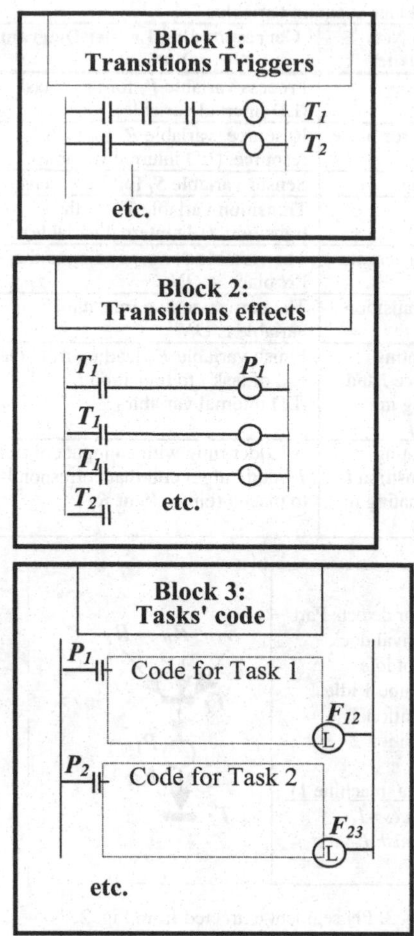

Fig. 5. Rearrangement of LD rungs into the three blocks schema

The three blocks structure as illustrated in figure 5 enable a backward translation from a ladder diagram into the high level Petri Net as discussed below.

When an LD is arranged in three blocks as in figure 5, the blocks can be identified without any additional knowledge as follows:

<u>Rules for the three blocks identification.</u>

1. Each rung in the first block corresponds to a transition and uniquely identifies a transition variable. The transition variable is the rung output. Transition enabling conditions appear as

contacts on the corresponding transition rung.
2. The output variable of the first rung appears again (as an input contact) only in the first rung of the second block. Therefore, all the rungs before this reappearance belong to the first group of transitions triggers. The number of transitions is the number of rungs in the first block.
3. In the second block each rung has an input contact corresponding to a transition. Thus, the second block ends when this condition does not hold.
4. The third block is divided into tasks. Each task *i* begins with the activation of the P_i variable and finishes with the activation of at least one F_{ij} variable. In this way the LD boundaries of each task are identified.

Note that only the first two blocks (transitions triggers and transitions effects) are necessary for constructing the high level PN as follows:

Rules for PN reconstruction from the identified LD blocks
1. Each rung in the first block (transitions triggers) corresponds to a unique PN transition. Each input contact on this rung signify a PN input place to this PN transition (i.e., tokens move from the place to the transition).
2. The second block identifies the effects of the PN transitions. If the rung output is LATCH, the corresponding variable is a PN output place of the transition. If the rung output is UNLATCH, the corresponding variable is a PN input place of the transition.
Although the PN can be recovered from its LD in this way, the association of PN places with their corresponding machines and other resources requires additional information. (i.e., we need some documentation to associate the places of idle machines with the actual machines.) For example, it is sufficient to have a legend similar to the one used in figure 3. Alternatively, remarks available in most LD editors can be used to document the input contacts of the first block rungs, and the outputs of the second block rungs. Any alternative to such documentation involves tedious tinkering with the outputs of each task.

V. Conclusion

This paper presents a new technique for translating process level Petri network to a corresponding ladder diagram code. The process stages or steps are embedded so that each step could be further translated by using the method in [9] (which is based on Boolean logic). The structured steps allow for automatic implementation of this translation. The proposed technique could be easily employed for implementing higher modeling levels of discrete industrial automation such as proposed in [11] and [12]. Moreover, the proposed rearrangement of LD code allows structured reconstruction of the original PN. A future research on a practical case-study may examine the efficiency of the proposed technique.

References

[1] Murata, T., "Petri nets: properties, analysis, and applications", Proceedings of the IEEE, Vol. 77, No. 4, pp. 541-580, 1989.

[2] Zhou, M., and R. Zurawski, " Petri Nets and Industrial Applications: A Tutorial", IEEE Transactions on Industrial Electronics, 41 (6), 567-583, 1994.

[3] Venkatesh, K., Zhou M. C., and Caudill, R. J., " Comparing Ladder Logic Diagrams and Petri Nets for Sequence Controller Design Through a Discrete Manufacturing System", IEEE Transactions on Industrial Electronics. 41 (6), 611-619, 1994.

[4] Jafari, M. A., Meredith G. A. and Boucher T. O., (1995) A Transformation from Boolean equation control specification to a Petri Net, IIE Transactions, 27 (9), 9-22.

[5] Moore, K. E., and Gupta, S.M., "Petri Net Models of Flexible and Automated Manufacturing Systems: A Survey, International Journal of Production Research, 34 (11), 3001-3035, 1996.

[6] Peng S. S., Zhou M. C., "Ladder diagram and Petri-net-based discrete-event control design methods", IEEE Transactions on Systems, Man, and Cybernetics, Part C: Applications and Reviews, 34 (4), 523 – 531, 2004.

[7] David R.," GRAFCET: A powerful tool for specification of logic controllers", IEEE transaction on control systems technology, 3(3), 253-368, 1995.

[8] Le Parc P., L'Her D., Scarbarg J.L., Marcel L., "GRAFCET revisited with a synchronous data-flow language" IEEE Transactions on Systems, Man and Cybernetics, Part A: Systems and Human, 29 (3), 1999.

[9] Lee G.B., Zandong H., Lee J.S., "Automatic generation of ladder diagram with control Petri Net ", Journal of Intelligent Manufacturing, 15(2), 245-252, 2004.

[10] Jeng, M. D. and F. DiCesare, "Synthesis using resource control nets for modeling shared resource systems", IEEE Transactions on Robotics and Automation, 11(3), 317-327, 1995.

[11] Cohen Y., Wang M., and Bidanda B., "Modeling and implementation of agent-based discrete industrial automation", in: Sobh T., Elleithy K., Mahmood A., Karim M. (editors), Innovative algorithms and techniques in automation, industrial electronics and telecommunications, Springer, 535-541, 2007.

[12] Cohen Y., "A modeling technique for execution and simulation of discrete automation" , in: Sobh T., Elleithy K., Mahmood A., Karim M. (editors), Novel algorithms and techniques in telecommunications, automation and industrial electronics, Springer, 273-278, 2008.

Software Quality Perception

Radosław Hofman
Department of Information Systems @ Poznań University of Economics, Polish Standardization Committee
radekh@teycom.pl

***Abstract-* This article analysis non-technical aspects of software quality perception and proposes further research activities for this subject naming this branch *Software Quality Psychology*.**

Cognitive science, psychology, micro economics and other human-oriented sciences do analyze human behavior, cognition and decision processes. On the other hand engineering disciplines, including software engineering, propose many formal and technical approaches for product quality description. Linkage between technical and subjective quality has been subject of research in areas related to food and agricultural applications and in this article we propose analysis of professional product perception which beyond doubt is a software product.

I. Introduction and Motivation

Software Engineering emerged in 1960's as an answer to software quality problems occurring at that time. Software product differed from other human industry products mainly because it was intangible and because its static attributes (attributes that can be marked for product without using it) were irrelevant while dynamic attributes (attributes describing behavior of product when used under certain conditions) were of highest importance. Software products usage grows constantly and currently it is being used in almost every area of human activity. Software quality, as quality definition for common product applicable in various contexts is then important issue not only for software developers but customers, users and people community as a whole (for example: software is in control of traffic lights, airplane steering systems, TV receivers etc.) From 1970's until these days software quality models are being developed – the latest ISO/IEC SQuaRE model is still under development.

Quality models are way of expressing quality, but despite these definitions one has to know what does quality of software product mean. We do know [32] that quality means different set of characteristics for different perspectives, and for different contexts of use (the same product may be perfect for one and useless for another context of use).

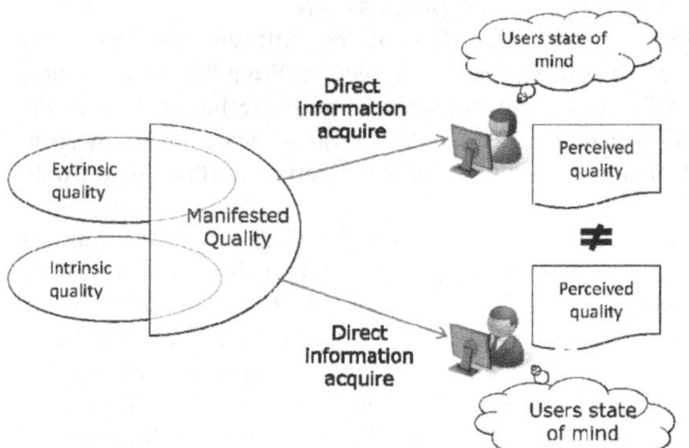

Fig. 1. Different quality concepts in perception processes

De gustibus et coloribus non est disputandum what formally may be understood as experiencing qualia, as defined in philosophy of the mind [16] theory. Qualia are defined as basic properties of sensory perception and therefore cannot be fully explained by person experiencing these sensations. If we assume that quality relates to sensory perception then it may be surprising that authors of software quality models give set of objective measures to express quality.

What is "Software quality"? Does it rely only on internal (static and dynamic) properties of software? Does it rely on way software is being used and maintained? Considering customer or user as ultimate source of software quality measure (compare CMM [46], ISO9001:2000, TQM, [56]) there is another important question: does user quality perception follow objective information about software product (as for professional products), or users follow set of observer biases having their perception unpredictable using psychologically contextless model. If we consider that user is unable to verify quality attributes to the bottom it shall be obvious that user has to rely also on beliefs in his mind.

The same research questions may be asked about what makes customer and user satisfied with the product they have. Certainly there are attributes manifested by product, but if satisfaction is dependant from anticipated attributes or some information not related to product itself (eg. related to users general opinion about job he/she is performing) then it may be possible to influence customer and user perception process significantly changing satisfaction level and quality attributed to software.

Above questions, if answer is yes, seem to be important area for every software project. Efficient methods affecting customer and user perception will increase probability of software delivery success considered as delivery of product having satisfactionary level of quality.

II. Introduction to Cognitive Processes

A. *Cognitive psychology*

Cognitive sciences, as interdisciplinary study, concern on human mind, intelligence, analyzing processes determining human behavior etc. Such studies were present in philosophic works of Plato and Aristotle in ancient Greece, becoming important stream of research in 17th century owing to Descartes. Despite of rapid development of these sciences for

K. Elleithy (ed.), *Advanced Techniques in Computing Sciences and Software Engineering*,
DOI 10.1007/978-90-481-3660-5_6, © Springer Science+Business Media B.V. 2010

past 40 years there are still more questions than answers about explanations of human mind processes [54].

The branch of psychology explaining internal mental processes become named cognitive psychology [41], but unlike many of previous approaches cognitive psychology adopts scientific method rejecting introspection (used in symbol-driven method by S. Freud). Cognitive psychology also gives explicit attention to named mental states such as desire, motivation, belief etc. in contradiction to behaviorist approach.

Methodologically, cognitivism in psychology adopts positivist approach and asserts that human psychology can be fully explained using scientific methods, experiments and measurements assuming that cognitive processes are related to discrete, internal symbols (mental states) and manipulation on these symbols. This approach is criticized by phenomenological approach with arguments that meaning of symbols is strictly related to some context which cannot be simply ignored.

In this article we concentrate on cognitive perception of software, adopting cognitive psychology, but also psychological concepts presented in 18th, 19th and 20th century by Gossen, von Wieser, Kant, Hume, Freud, Maslow and other thinkers.

In our approach it is important not only to understand perception of software quality, but also we are discussing possibility of influencing this perception. In other words we propose set of experiments which could explicitly discover relation between possible attributes of product and influence on environment what in consequence would change quality perception.

B. *History of valuation*

Value of things is one of basic ideas in human perception related processes. Sociological, psychological, economical, ethical value models has been developed from ancient times with attempts to explain why human beings prefer or choose some things over others and how personal behavior may be guided (or fail to be guided) in valuation. In this section we will present short summary of ideas resulting from these works which seem to be applicable in analysis of software quality perception and subjective distinction to software products made by evaluators.

In neoclassical economics or in microeconomics value of object is often described as the price dependant from supply and demand on competitive or non-competitive market.

In classical economics the value is often seen as labor or equivalent for some other goods or discomfort used to produce object (labor theory of price [11] or value). In this approach value of object is not dependant from situation on the market so is different from the price ("natural price" in [55] or "prices of production" in Karl Marx works [51]).

Another classical approach proposed by Karol Marx is perception of value as usability value, meaning that object is valuated according to benefits of his owner. Similar idea may be found in works of Ludwig von Mises [45] whose understanding of value was associated with some measure of utility following consumption of good or service. In [52] another approach was proposed – value was derived from moral issues associated with the product.

In 19th century Herman Gossen proposed law of diminishing marginal utility [19] arguing that if a good satisfies some need, then value of next unit of this good is diminishing. This thought was continued in Austrian school (named "philosophic") and Friderch von Wieser [63] explicitly expressed observation that satisfied needs are of less importance. In 20th century Abraham Maslow proposed pyramid of needs [38]. Basic needs (D–needs), which are not recognized unless not satisfied, and upper needs (B-needs) appear only when lower level needs are satisfied. In addition we may refer to Sigmund Freud thought that world is perceived by humans mainly on sub-conscious level [17].

Scheller [8] and Dewey [23] had proposed their catalogues of value dimensions distinguishing valuation method in dependence on valuation dimension. On contradiction to subjective valuation ideas in 20th century objectivist theory of value was formulated by Ayn Rand [49]. This theory asserts that there exists some reality, and it is independent of human perception arguing that even if humans did not exist then this reality will preserve.

Above milestones in philosophical economy are important clues to understand cognitive processes associated with valuation of goods. Aristotle considered quality as non quantitative factor allowing to distinguish thing among other from the same category [31]. Thinking about valuation in terms of cognitive science it is required to identify mental state of valuator, his needs and level of their satisfaction remembering that satisfaction level is not linearly correlated with saturation of need.

Further on we will discuss "units" but referring to software they should not be considered as quantity of this software. We assume that for every characteristic we may increase or decrease its strength by some "unit" influencing user satisfaction. In this assumption we follow economists defining utility as quantifiable in some units.

III. Software Quality

A. *Software Quality and Quality Models*

From 1960's development of software products was perceived as discipline of engineering. From that time attempts to define goals and measures for software began. One of the most difficult measures to define was software quality measure although it seemed to be highly important attribute of software product.

Software products stated new set of definition requirements in aspect of product measures and quality measures. Any measures known for ages (weight, size, durability, water resistance etc.) could not be applied to get significant information about software. First attempts to state quality measures were done in 1970s by McCall's [39] and Boehm's [9] quality models. Thirteen years later, in 1991, International Standardization Organization published first international standard for software product quality in ISO9126 [28]. After ten years standard was reviewed and new approach was presented. In this model quality issues were divided into quality in use, external quality and internal quality. In the same time new international initiative Software product QUality Requirements and Evaluation (SQuaRE) was set up aiming to develop set of norms ISO/IEC25000 [26]. This new approach is perceived as new generation of software quality models [61] and is being used for decomposition of end users perspective to software components [1].

Quality related issues are often consisting of mixture of different quality meanings. This problem was noticed since publication of first models, containing extremely different perspectives. Important summary of five quality views was made in [32]:

- Quality as an abstract, meta-physical term – unreachable ideal, which shows the direction where to products are heading but will never get there,
- Quality as a perspective of user considering attributes of software in special context of use,
- Quality as a perspective of manufacturer, seen as compliance with stated requirements and following ISO 9001:1994 view,
- Quality as a product perspective understood as internal characteristic, resulting from measures of product attributes,
- Quality as a value based perspective, differing in dependence on stakeholder for whom is defined.

Quality models are important way to express quality in commonly understandable terms. In this article we use ISO/IEC SQuaRE vocabulary and consider software quality in use as representation of user and customer view on software quality.

B. *Quality perception modeling*

The need of measuring quality of the products is a natural consequence of stating quality as thing that matters. First software quality models and need to measure users opinion appeared in McCall [39] and Boehm [9] publications. At that time it was only concepts.

In 1984 Grönroos described quality as function of expectations, dividing perception into three dimensions: functional, technical and image (perception of vendor) [20].

Year later in [44] SERVQUAL model had been proposed. This model, and its successors are the most widely used quality perception models [30] not only for IT products but also for airline services, fast-food, telephony, banking, physiotherapy, web sites, healthcare and many others [5], [40]. SERVQUAL defines 10 (possibly overlapping) dimensions: tangibles, reliability, responsiveness, communication, credibility, assurance, competence, courtesy, understanding/knowing the customer, and access based on focus groups sessions. Model was revised in 1988 into five dimensions: tangibles, reliability, responsiveness, assurance, empathy and in 1991 into two-part model [43].

The main idea of SERVQUAL and successor models is measurement of expectations and perception difference. $Q_i=P_i-E_i$ where Q stands for quality indicator, P is perception mark and E expectations level. Although it was mentioned that this model is one of most popular ones there is still the discussion if such approach is sufficient. In [14] authors argue that perception is a function of its performance defining quality as $Q_i=P_i$, but still follow to use five dimensions defined in revised SERVQUAL model.

Another approach to define software quality perception is based on belief revision theory [47]. This method adopts AGM paradigm [4] or Grove's system of spheres approach [21] and proposes epistemology approach to define beliefs and their revision process following assumption that observer is rational, deductive agent using principle of minimal change. Model based on this approach also states that user perception should be consider in analogy to users qualifications in respect to computers in general [64] following dimensions proposed in [48]. Users have their initial opinion then continuously discover software product gaining new information and reviewing their beliefs reaching their final opinion. In 1997 authors had presented this model as able to use any software quality model McCall, Boehm, FCM, ISO9126 and in 1998 [57] present summary of their research on how user perception changes over time to some level of consensus, meaning that all users final opinion is very similar, and therefore relates to real software quality [56].

Above models uses assumption that users are rational agents following deductive reasoning and that beliefs may be represented in formal system. Authors does not analyze context of user (context of purpose) nor users personal aspects (tiredness, attitude treat evaluation seriously etc). It should be mentioned that authors continue to measure technical quality factors, as defined in ISO9126 although usage of them is commonly regarded as to abstract for users to express user point of view [61]. The most important problem of those results is problem of repetitive observations on the same group of users. In this case we may expect that evaluation experiment was also influencing users opinion and tendency of changing beliefs to similar level could be group thinking effect or could be influenced by large set of external information. In this article we propose much broader view on quality perception not narrowed to intrinsic software attributes.

On the other hand if one rejects anticipated assumption that there exists some real quality (considering real quality as having different value for different user) then supposed correlation should not exist. Rejection of this assumption would be consequence of several theories of truth constructivist theory (arguing that truth is kind of convention related to social experience – compare works of Hegel and Marx) or pragmatic theory (where truth is something that could be verified in practice by some community – compare works of Peirce, James and Dewey). Considering these theories we may conclude that if two similar measurements were conducted in socially deferring groups their final opinions about quality would differ unless there occurred "regression to mean" bias.

It seems useful to analyze non software oriented attempts to define software quality perception. One of attempts to define quality perception in terms of cognitive processes was made by Steenkamp in his dissertation in 1986 [59], [58], revised by Oprel in 1989 [42]. Since his work there were several publications using his model for analysis of food, plants etc. quality perception modeling, including research in area of social background influence on food quality perception [53].

This model conceptualizes relation between intrinsic and extrinsic quality attributes, perception of these attributes and hand preference establishment.

Although some of ideas used in this model are undoubtedly common for perception of food products and software products, there are some important differences. First of all we have to distinguish role of person making valuation of software quality [32], while Steenkamp considers all people as consumers (we can distinguish between consumers and people responsible for food purchasing for some organizations). Second important difference is change over time, what is not addressed in Steenkamp's neither in successor models. In last place we may add that unlike food, software is used mainly by organizations (groups of individuals) and group behavior may influence individual perception of quality.

Concluding this literature review we stress out that there is no commonly accepted method for measuring user perception of software quality nor common understanding what "user perception" is. Perception model presented in this article considers very wide view on users cognitive processes proposing research in areas not related directly to software product but probably influencing perceived quality.

C. *Quality requirements and limitations*

Modern software engineering approaches explicitly state, that software project begins with analysis resulting in establishment of functional and quality requirements [24]. Publications concerning software quality lifecycle accept this approach adding another type of requirements to gather – evaluation process requirements [27], [25]. There are several guidelines how to identify requirements and build comprehensive and correct set of stated requirements [15].

On the other hand it is commonly accepted to define quality as ability to satisfy stated and implied needs [26]. Implied needs remain however unaddressed in software engineering state of art. Main stream approach assumes that users explicitly express they requirements, give them priorities, analysts identify and analyze user requirements and at the end of this phase users accept complete set of requirements [27], [7], [35], [62].

Looking back on great thoughts associated with cognitive psychology we may identify significant limitation associated with such approach. Although users are certainly best source for defining goal and purpose of software usage, they are unable to identify their own basic needs or are unable to assign priority expressing real importance of the need which are being satisfied.

We assert that users defining their requirements for future system also are unable to completely recognize their needs. Analysts gather user requirements, add requirements based on their personal experience and identified formal needs [6]. At that moment specification for system is being closed, and after this moment development begins [24].

Strong alternative for specification closure can be found in agile development methodologies. Most of them stress need for being close to user and allow user to evaluate software product as often as possible [2]. Another example is evolutionary approach assuming deliveries to live environment several times in the project lifecycle [36]. These methods should be much more effective and allow higher success ratio for IT projects.

Although agile methodologies are intended to answer problems with incompleteness of user perspective they seem to be vulnerable from the same cognitive issue. Users evaluating software prototype do not need to employ they complete set of needs having only abstract reason to work with evaluated prototype (this bias is partially evened out in evolutionary approach [18]). We may think about it as having brain storm meeting in subject "are these tools enough for us to live on desolated island?" Unless people were really put in real situation they do evaluate appropriateness of product against their assumptions about imaginary reality. If software is to support new internal organization (planned organizational change is often associated with new software to be implemented) then users have limited ability to evaluate it.

Another problem associated with agile methodologies is that this approaches forces user to exert what leads to discouragement. One has to remember that on early phases user receives non-working or having poor quality software prototypes for evaluation but it may be observed that user prepossess his mind rejecting software on final acceptance.

Concluding this section we shall underline gap between stated requirements and user needs, especially implied needs. We also have to remember that software engineering defines mainly technical (intrinsic) quality attributes, while user valuates only these attributes which are associated with his observations of software and its surrounding from perspective of his everyday use of software and his state of mind.

IV. Software Quality Perception Model

A. *Expectations*

In this section we shall recall important works of two thinkers. Immanuel Kant in his theory of perception argued that human beings adopt their a'priori concepts and knowledge perceiving their thoughts not real attributes of observed reality [22]. David Hume analyzing where human concepts come from observed that people tend to reject observations outstanding from other observations or their beliefs [60]. Both of above works are widely accepted in modern psychology where thesis that human perception is only interpretation of human mind is rather common [37].

Referring to buyer decision process, defined as psychological and cognition model such as motivation and need reduction [12] one may see that before decision of buying new software was made, customer recognized his need, analyzed alternatives and made decision. This is first time when some expectations have appeared – that new product will satisfy customer recognized need (with implied needs), and will still satisfy needs satisfied with old system (this requirement is not often explicitly formulated).

Customer and users perspective may differ at this point – customer has some vision of new system satisfying new needs while users may be reluctant to unknown, new system expecting things to be more difficult.

This attitudes play role of priming processes [33] preparing a'priori concepts as in Kant's theory.

Before software product is seen by customer and users for the first time it has already been evaluated and its quality was assessed. This observation is obvious for brand managers promoting physical goods but seems to be not applied in software industry.

B. *Software quality perception model*

Models are intended to be reliable prediction of future observations, repository of rules collated and integrated from past research [50]. Software quality perception model proposed in this article is designed according to above idea taking into account cognitive psychology, quality perception models for other types of products and commonly accepted software quality model SQuaRE.

Proposed model is omitting elements associated with preference and choice making which may be seen in other quality perception models, as an result on focusing on software produce evaluation purposes. Software in most cases, especially tailored software is fully evaluated after purchase and implementation, so there are no decisions about purchase to be made. Proposed model is presented on fig. 2.

Product attributes (intrinsic attributes) and information associated with product (extrinsic attributes) are at first filtered by observer attention filter. From the mathematical point of view filtering is application of certain function ($f_x(a_x)$) for original attribute. This function is a consequence of observer mental state and knowledge (including past experience with product or similar kinds of products) etc. Evaluation product quality employs establishment of perceived attributes relevant

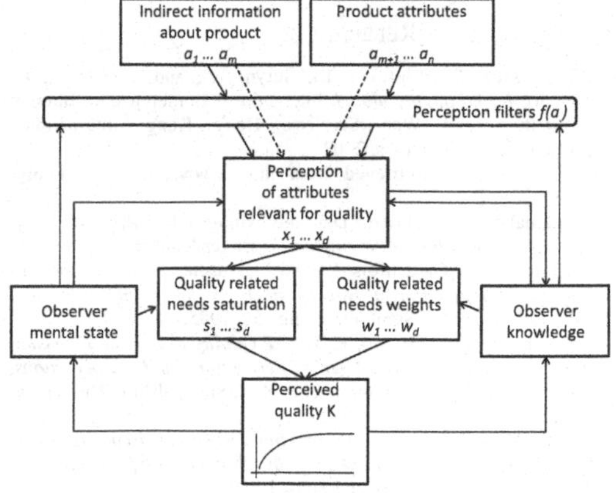

Fig. 2. Software quality perception model

for quality – observer chooses set of attributes expressing quality seen through perception filters. This process is also based on observer mental state and his knowledge. We assume that observation (information about perceived attribute value) augments user knowledge.

In next stage observer relates perceived quality related attributes values through perspective of needs taking general importance of need and subjective saturation of need as the base. Overall quality gives feedback to observer knowledge and his mental state.

x_i variables (perceived value of attributes relevant for quality) may be interpreted as direct measures assigned in observation process (where observation is evaluation of product or processing information associated with product) or derived measures based on simple measures (for example number of software errors and number of lines of code are simple measures; number of software errors per number of lines of codes is derived measure relevant for quality). This concept is similar to Quality Measure Elements idea in ISO/IEC 25021.

Another important issue is evaluation for somebody's else perspective. In this case observer relates needs saturation to imaginary mental stated attributed to future users of the product, but probably using only observer state of knowledge (compare theory of the mind [54]). Such observation method tends to produce quality measures differing from real users evaluation, because evaluators are prone to mistakes in setting up imaginary state of mind of future users (e.g. cognitive bias called conjunction fallacy – tendency to assume that specific conditions are more probable then general ones), and using different knowledge base (user having wider/narrower knowledge about product usage will assign different set of weights).

Calculation of subjective quality value, or in other words conversion to uni-dimensional quality measure, in most of quality models is an addition operation of single measures multiplied by attribute weight: (compare [59]). This approach adopts assumption that each "unit" of attribute value influences overall quality index with same strength.

Quality perceived by human should be considered as latent variable and in consequence one could employ Rash polytomus model [3] designed to measure personal attitudes. Irrespective to mathematical model involved we assume that quality perception of single attribute value follows diminishing marginal value concept, attribute valuation are additive so overall quality value is calculated as:

Where $F_i(s_i,w_i,x_i)$ is monotonic function.

Similar to [59], perceived quality in above model differs from the other approaches in that it regards quality neither as absolute nor as objective.

Model may further extended with quality perception change patterns (quality lifecycle patterns) – at this moment we only assume that due to observer mental state and knowledge change quality attributes perception changes over time.

C. Quality perception model and lifecycle

Software development lifecycles are widely known and accepted in software engineering discipline. User quality perception exceeds time boundaries of software development project as it begins before project is even started and ends in some time after product had been withdrawn (it lasts in user remembering for some time).

We will not give detailed information about perception of quality in certain phases referring to complete version of this article and emphasize difference in perception using vendor and user perspective during project (problem with implied needs), users expectation (quality cues in [59]), disappointment and several biases during evaluation.

D. Affecting quality perception

If user quality is dependant not only from intrinsic product attributes, then there are several methods to influence overall quality perception. Such method are being used mainly for repeatable products (COTS), although there is theoretical possibility to use them in most of software projects.

We may affect quality value during evaluation and long term operation but it is likely that different methods will have different impact on those two values.

Example of positive impact on quality value during evaluation is use of primacy effect presenting high quality product first.

Perception affecting methods are known mainly in marketing and psychology research. In this article we argue the need for further research in area of affecting software quality perception, since according to authors best knowledge, there are no publicly available consideration on this subject.

V. Summary and Further Research

A. Conclusion

In this article we have proposed software perception quality model based on modern quality perception models for other types of products but also extending them with cognitive processes elements. Our models give comprehensive overlook of how software quality is perceived and may be used as starting point for further research, for setting up new branch: **Software Quality Psychology**.

There exists several software development lifecycles, methodologies, approaches and so on. Although most of them concentrates on software production processes resulting in delivery of deliberate set of intrinsic product attributes. Following Kant's thought we may think that observer's ability to fully evaluate those attributes is limited and if we add von Weisser's thought about importance unsaturated needs we may conclude that intrinsic attributes and their evaluation is not enough to understand how evaluators evaluate software quality.

If there it is possible to prepare coordinated actions to prepare designed extrinsic information about the product then software product will be evaluated with higher quality mark even if it still was exactly the same product. Consequence of better quality overall mark is significant – probability of acceptance increases, customer satisfaction increases and in consequence business cooperation between vendor and customer grows.

B. Quality perception scaling - experiments

We propose further research to evaluate importance of below listed set of questions. In general we want to estimate impact of isolated variable on quality perception using [13] as a method.

How post-evaluation quality perception is influenced by primacy effect, contrast effect and maximum intensity [29]. This experiment should give information how important it is to assure quality of first version delivered and how important it is to manage escalation of perception.

How post-evaluation quality perception is influenced by repetitions, authority and group thinking/herd instinct effects. This experiment should advice us how evaluation process may be influenced by information exchange.

How post-evaluation quality perception is influenced by conjunction fallacy, exploitive and impact bias. This experiment will give information how important is education of evaluators before evaluation process to avoid falling into these biases.

How hidden (hidden as in [34], named "second order type attributes" in [10] and "credence attributes" in [59]) attributes evaluation based on information provision impacts the overall quality value in opposition to non-hidden attributes evaluation.

For above area there could be some important quality factor picked up and investigated for ability to influence perception of this attribute using above scenarios. For example for reliability we could prepare two versions "hanging" with the same probability with one having "blue screen" with technical information shown and second with message that this is only prototype, but final version will be reliable.

Further on we suggest that there should be research on other types of influence factors on evaluation process with distinctions to post-evaluation and long-term operation of software using user and evaluator distinction. Another dimension is to investigate correlation between software application and perception impacts or influence of personal and situational characteristics.

References

[1] Abramowicz, W., Hofman, R., Suryn, W., and Zyskowski, D. *SQuaRE based quality model*. International Conference on Internet Computing and Web Services. Hong Kong: International Association of Engineers, 2008.

[2] *Agile Manifesto*. Retrieved from http://www.agilemanifesto.org, 2008.

[3] Alagumalai, S., Curtis, D., and Hungi, N. *Applied Rash Measurement: A book of exemplars*. Springer-Kluwer, 2005.

[4] Alchourron, C., Gardenfors, P., and Makinson, D. *On the Logic of Theory Change: Partial Meet Functions for Contraction and Revision*. Journal of Symbolic Logic , 50, 1985.

[5] Babulak, E., and Carrasco, R. *The IT Quality of Service Provision Analysis in Light of User's Perception and Expectations*. International Symposium on CSNDSP. Staffordshire University, 2002.

[6] Bahill, A., and Dean, F. *Discovering system requirements*. In A. Sage, and R. W., *Handbook of Systems Engineering and Management*. John Wiley and Sons, 1999.

[7] Beck, K. *Extreme programming eXplained: embrace change.* Addison-Wesley, 2000.
[8] Bershady, H. *On Feeling, Knowing, and Valuing. Selected Writings.* Chicago: University of Chicago Press, 1992.
[9] Boehm, B., Brown, J., Lipow, M., and MacCleod, G. *Characteristics of software quality.* New York: American Elsevier, 1978.
[10] Braddon-Mitchel, D., and Jackson, F. *Philosophy of Mind and Cognition.* Oxford: Blackwell, 1996.
[11] Case, K., and Fair, R. Principles of Economics. Prentice-Hall, 1999.
[12] Cheng, M., Luckett, P., and Schulz, A. *The Effects of Cognitive Style Diversity on Decision-Making Dyads: An Empirical Analysis in the Context of a Complex Task.* Behavioral Research in Accounting , 15, 2003.
[13] Coolican, H. *Research Methods and Statistics in Psychology.* Oxford: Hodder and Stoughton, 1999.
[14] Cronin, J., and Taylor, S. *SERVPERF versus SERVQUAL: reconciling performance based and perceptions minus expectations measurement of service quality.* Journal of Marketing , 58 (1), 1994.
[15] Davis, A. Software Requirements: Analysis and Specification. Prentice Hall, 1993.
[16] Dennet, D. *Quining Qualia.* In A. Marcel, and E. Bisiach, *Consciousness in Modern Science.* Oxford University Press, 1988.
[17] Freud, S. Das Ich und das Es (The Ego and the Id), 1923.
[18] Glib, T. *Principles of Software Engineering.* Wokingham, England: Addison-Wesley, 1988.
[19] Gossen, H. *Die Entwicklung der Gesetze des menschlichen Verkehrs und der daraus fließenden Regeln für menschliches Handel* (The Development of the Laws of Human Intercourse and the Consequent Rules of Human Action), 1854.
[20] Grönroos, C. *A service quality model and its marketing implications.* European Journal of Marketing , 18 (4), 1984.
[21] Grove, A. *Two modellings for theory change.* Journal of Philosophical Logic , 17, 1988.
[22] Haden, J., and Körner, S. *Kants Leben und Lehre* (Kant's Life and Thought). (E. Cassirer, Trans.) Yale University Press, 1981.
[23] Hickman, L., and Alexander, T. *The Essential Dewey* (Vol. I and II). Indiana University Press, 1989.
[24] Institute of Electrical and Electronics Engineers, *Software Engineering Body of Knowledge* from http://www.swebok.org/, Retrieved 2008
[25] ISO/IEC14598-1. *Information technology - Software product evaluation.* Geneve: International Standardization Organization, 1999.
[26] ISO/IEC25000 *Software Engineering - Software product Quality Requirements and Evaluation (SQuaRE).* Geneve: International Standardization Organization, 2005.
[27] ISO/IEC25010. *Software Engineering - Software product Quality Requirements and Evaluation (SQuaRE).* Internal ISO/IEC JTC1/SC7 document - Commision Draft, 2007.
[28] ISO9126. *Information Technology - Software Product Quality.* Geneve: International Standardization Organization, 1991.
[29] Kahneman, D. *A perspective on judgment and choice: Mapping bounded rationality.* American Psychologist , 58, 2003.
[30] Kang, G., James, J., and Alexandris, K. Measurement of internal service quality: application of the Servqual battery to internal service quality. Managing Service Quality , 12 (5), 2002.
[31] Kiliński, A. *Jakość.* Warszawa: Wydawnictwo Naukowo Techniczne, 1979.
[32] Kitchenham, B., and Pfleeger, S. *Software Quality: The Elisive Target.* IEEE Software 13, 1996.
[33] Kolb, B., and Whishaw, I. *Fundamentals of Human Neuropsychology.* Freeman, Worth Publishers, 2003.
[34] Kramer, A., and Twigg, B. *Quality control in the food industry* (3rd edition ed.). Avi, 1983.
[35] Kruchten, P. *The Rational Unified Process: An Introduction* (3rd edition ed.). Addison-Wesley Professional, 2003.
[36] Krzanik, L. *Enactable models for quantitative evolutionary software processes.* In C. Tully, Proceedings of the Forth International Software Process Workshop (ISPW '88). Moretonhampstead, Devon, UK: IEEE Computer Society, 1988.
[37] Libet, B., Sutherland, K., and Freeman, A. *The Volitional Brain: Towards a Neuroscience of Free Will.* Thorverton, UK: Imprint Academic, 2000.
[38] Maslow, A. *A Theory of Human Motivation.* Psychological Review (50), 1943.
[39] McCall, J., Richards, P., and Walters, G. Factors In software quality. Griffiths Air Force Base, NY, Rome Air Development Center Air Force Systems Command, 1977.
[40] Miguel, P., daSilva, M., Chiosini, E., and Schützer, K. *Assessment of service quality dimensions: a study in a vehicle repair service chain.* POMS College of Service Operations and EurOMA Conference New Challenges in Service Operations. London, 2007.
[41] Neisser, U. *Cognitive psychology.* New York: Appleton-Century-Crofts, 1967.
[42] Oprel, L. *Kwaliteit in breder perspectief, Proefstation voor de Bloemisterij Aalsmeer.* 1989.
[43] Parasuraman, A., Berry, L., and Zeithaml, V. *Refinement and Reassessment of the SERVQUAL Scale.* Journal of Retailing , 67, 1991.
[44] Parasuraman, A., Zeithaml, V., and Berry, L. *A conceptual model of services quality and its implication for future research.* Journal of Marketing , 49 (4), 1985.
[45] Paul, R. *Mises and Austrian economics: A personal view.* The Ludwig von Mises Institute of Auburn University, 1984.
[46] Paulk, M., Weber, C., Curtis, B., and Chrissis, M. *The Capability Maturity Model: Guidelines for Improvement of the Software Process.* Addison-Wesley, 1995.
[47] Peppas, P., and Williams, M. *Constructive modellings for theory change.* Notre Dame Journal of Formal Logic , 36 (1), 1995.
[48] Pressman, R. *Software Engineering. A Practitioner's Approach.* McGraw Hill, 1992.
[49] Rasmussen, D. *Ayn Rand on Obligation and Value.* Libertarian Alliance, 1990.
[50] Rickert, K. *Beef cattle production.* In L. Tijskens, M. Hertog, and B. Nicolai, *Food process modelling.* Cambridge: Woodhead Publishing, 2001.
[51] Rubel, M. *Marx Without Myth: A Chronological Study of his Life and Work.* Blackwell, 1975.
[52] Ruskin, J. *Unto this last.* 1860.
[53] Sijtsema, S. *Your health!? Transforming health perception into food product characteristics in consumer-oriented product design* (WAU no. 3359). Wageningen University Dissertation, 2003.
[54] Smart, J. *The Identity Theory of Mind.* In Stanford Encyclopedia of Philosophy, 2007.
[55] Smith, A. *An Inquiry into the Nature and Causes of the Wealth of Nations.* 1776.
[56] Stavrinoudis, D., Xenos, M., Peppas, P., and Christodoulakis, D. *Early Estimation of Users' Perception of Software Quality.* Software Quality Journal, 13, 2005.
[57] Stavrinoudis, D., Xenos, M., Peppas, P., and Christodoulakis, D. *Measuring user's perception and opinion of software quality.* Proceedings of the 6th European Conference on Software Quality. Vienna: EOQ-SC, 1998.
[58] Steenkamp, J. *Product quality: an investigation into the concept and how it is perceived by consumers* (WAU no. 1253). Wageningen University Dissertation, 1989.
[59] Steenkamp, J., Wierenga, B., and Meulenberg, M. *Kwali-teits-perceptie van voedingsmiddelen deel 1.* Swoka. Den Haag, 1986.
[60] Stroud, B. *Hume.* London and New York: Routledge, 1977.
[61] Suryn, W., and Abran, A. *ISO/IEC SQuaRE. The seconod generation of standards for software product quality.* IASTED2003, 2003.
[62] Wiegers, K. *Software Requirements 2: Practical techniques for gathering and managing requirements throughout the product development cycle* (second edition ed.). Redmont: Microsoft Press, 2003.
[63] Wieser, F. *Der natürliche Werth* (Natural Value). 1889.
[64] Xenos, M., and Christodoulakis, D. *Software quality: The user's point of view.* In M. Lee, B. Barta, and P. Juliff, *Software Quality and Productivity: Theory, practice, education and training.* Chapman and Hall Publications, 1995.

An Offline Fuzzy Based Approach for Iris Recognition with Enhanced Feature Detection

S. R. Kodituwakku and M. I. M. Fazeen
University of Peradeniya
Sri Lanka
salukak@pdn.ac.lk, mfazeen.pdn@gmail.com

Abstract **- Among many biometric identification methods iris recognition is more attractive due to the unique features of the human eye [1]. There are many proposed algorithms for iris recognition. Although all these methods are based on the properties of the iris, they are subject to some limitations. In this research we attempt to develop an algorithm for iris recognition based on Fuzzy logic incorporated with not only the visible properties of the human iris but also considering the iris function. Visible features of the human iris such as pigment related features, features controlling the size of the pupil, visible rare anomalies, pigment frill and Collarette are considered [2]. This paper presents the algorithm we developed to recognize iris. A prototype system developed is also discussed.**

I. Introduction

Human identification (ID) plays a very important role in the fast moving world. Human ID is useful for authentication, person recognition, distinguishing personals and so forth. Fraud ID is one of the major problems in any part of the world. These frauds can lead to many disasters like unauthorized entry for terrorist act and so on. These factors have led researchers to find out more accurate, reliable and most importantly non-transferable or hard to fake human identification methods. As a result one of the amazing ideas has emerged by means of biometrics human identification.

When using biometrics for identifying humans, it offers some unique advantages. Biometrics can be used to identify you as you. Tokens, such as smart cards, magnetic stripe cards, photo ID cards, physical keys and so forth, can be lost, stolen, duplicated, or left at home. Passwords can be forgotten, shared, or observed. Moreover, in today's fast-paced electronic world, people more often asked to remember multitude of passwords and personal identification numbers (PINs) for computer accounts, bank automated teller machines, e-mail accounts, wireless phones, web sites and so forth. As a solution, Biometrics holds the promise of fast, easy-to-use, accurate, reliable, and less expensive authentication for a variety of applications.

There is no any perfect biometric that fits all needs. All biometric systems have their own advantages and disadvantages. The primary advantage of biometric authentication is that it provides more instances of authentication in such a quick and easy manner without additional requirements. As biometric technologies mature and come into wide-scale commercial use, dealing with multiple levels of authentication or multiple instances of authentication will become less of a burden for users [3].

By definition "Biometrics is automated methods of identifying a person or verifying the identity of a person based on a physiological or behavioral characteristic" [3]. In other words, biometrics is to identify a person by using "Something that they have physically which cannot be changed" rather than "Something they know". For example fingerprint, iris, retina, face and voice are used rather than passwords, names or any other. In this research we use iris patterns of humans to ID each individual uniquely. What made our interest to do this research on iris recognition is the reliability and accuracy of iris patterns is overwhelmingly high due to the randomness of the iris pattern. Furthermore when compared to other biometrics the iris is less vulnerable to aging and external injuries because it is well protected by the anatomy of the eye.

Human eye is basically divided into two parts namely anterior chamber and interior chamber. It is separated by the iris and the lens which is vital and most important section of this research. The iris is a thin circular diaphragm which lies between the cornea and the lens in the direction of light entering in to the eye. When looked from the front of the eye, the iris is the pigmented area between pupil and the sclera. The iris have visible features like pigment related features, features controlling the size of the pupil, visible rare anomalies, pigment frills, crypts, radial furrows and Collarettes. These features of an individual are unique for that person. Hence our proposed system works based on above features to recognize an individual uniquely as these are the features that are mainly responsible for the randomness of the human iris pattern [2].

In this research work we attempted to develop an algorithm for iris recognition based on not only the main function of the human iris but also the outer appearance or visible features. A prototype system was developed based on the main function of the human iris and Artificial Intelligence (AI) techniques. In developing the system first we found a proper algorithm to extract important and essential feature of a human iris image. Secondly, as an AI technique, Fuzzy logic is applied for iris recognition and person identification. Finally a software

K. Elleithy (ed.), *Advanced Techniques in Computing Sciences and Software Engineering*,
DOI 10.1007/978-90-481-3660-5_7,

system is implemented to perform the above two tasks in order to demonstrate the system.

II. METHODOLOGY OVERVIEW AND USED MATERIALS

The system consists of two major phases. First phase is the enrollment (feature extraction) phase and the second phase is the verification & identification phase. In the first phase we detected the features of human iris[1] by using image processing methods. Then we extracted these features in a numeric (or digitized) format called iris code. In the second Phase we designed a fuzzy system which capable of accepting above iris codes as the crisp set input. The fuzzy system compares iris code of an individual with the iris codes from the enrolled iris code database to determine a match or mismatch. The architecture of the system is shown in Fig. 1.

The system is developed in Java programming language. In order to test the system an image database of 756 grayscale eye images are used. The grayscale iris images are collected from the Laboratory of Pattern Recognition (NLPR), Chinese Academy of Sciences, Institute of Automation (CASIA). CASIA Iris Image Database (version 1.0) includes 756 iris images from 108 eyes (hence 108 classes). According to CASIA for each eye, 7 images are captured in two sessions, where three samples are collected in the first session and four in the second session.

PERSON IDENTIFICATION

Identification of persons involves several steps: contrast stretching [4], pupil extraction, smoothing the image, detecting the pupil boundary, restoring the original image, detecting the iris boundary, unrolling the iris [5], Gabor filtering [6], detecting features of the unrolled iris, extracting detected features and identification of person. Additionally an extra step is used for eyelashes and eyelids removal. The algorithm used to perform these tasks is described below.

A. *Apply contrast stretching to enhance the image* [4].

Contrast stretching ensures that the pupil fall in to very dark region of the eye image. This is a way of normalizing the image before processing. See Fig. 6.

B. *Extract the pupil boundary.*

Since the early versions of pupil boundary detection algorithm failed in cases like the eye images with lot of eye lashes covering the pupil region. In order to solve this problem we applied the second form of contrast stretching as shown in Fig. 2 successively several times with different values of r_1 and r_2 where r_1 & r_2 are some arbitrary values each range from 0 to 255. The resulting image is shown in Fig. 7. Equation (1) is used to calculate the new value of the pixel (nv) if old value of the pixel (old), minimum pixel value of the pixels in the image (min) and maximum pixel value of the pixels in the image (max) are known.

$$nv = 255 * (old - (min + r_1)) / ((max - r_2) - (min + r_1)) \quad (1)$$

Steps

1) First contrast stretch the resulting image of above step with r_1=0 and r_2=10
2) Repeat the contrast stretching with r_1=200 and r_2=0, then with r_1=100 and r_2=0, finally with r_1=20 and r_2=0
3) Now apply the contrast stretching three times in backward direction with r_1=0 and r_2=200

1. It is considered that all the input images of the system are grayscale images and they should be in JPEG format.
2. The size of the input image is 320x280 pixels.

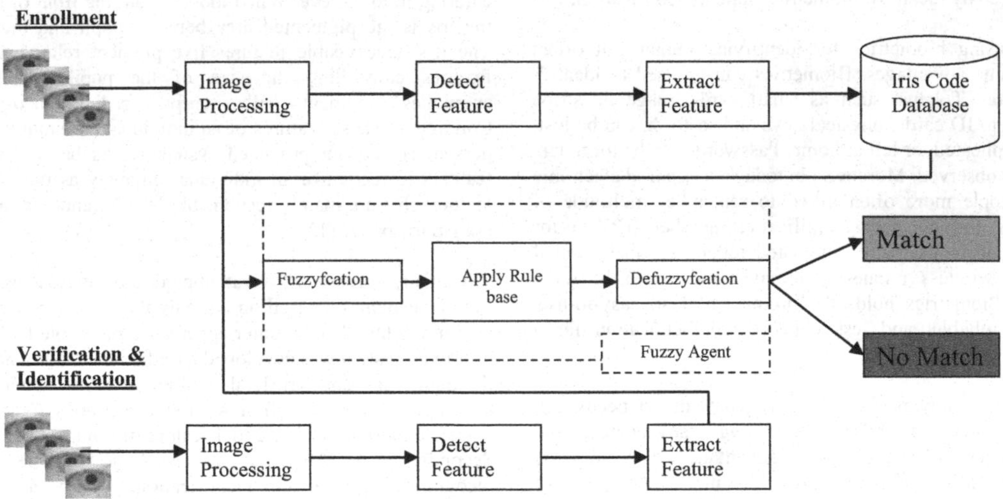

Fig. 1. The architecture of the system

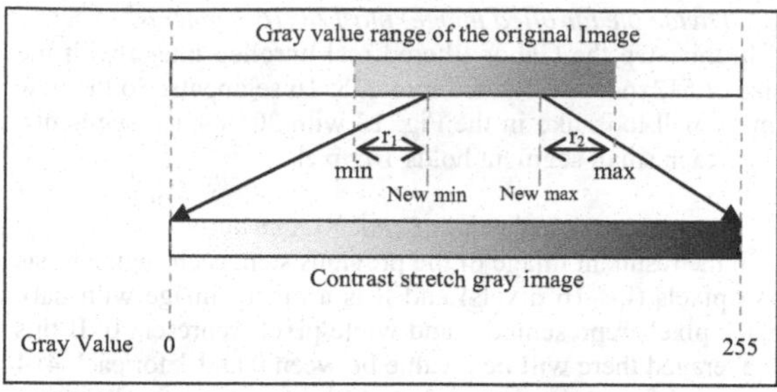

Fig. 2 Second form of contrast stretching.

4) The dominant part of the resulting image would be the pupil region.

C. Smooth the image by applying a smooth filter with a 3 by 3 kernel.

Although the pupil boundary could be detected from the resultant image of the above step, the edge of the pupil is not smooth. This problem can be solved by applying a smooth filter as shown in Fig. 8.

D. Detect the pupil boundary with the following algorithm.

Since the success of the person identification process heavily depends on the accuracy of detecting the pupil boundary, we put more emphasis on this step. As a result we outlined and used a new algorithm for this purpose. The algorithm is specified below. See the result in Fig. 9.

Algorithm

```
Begin
  Find the maximum and minimum pixel grey values in the eye image.

  Set Threshold = (maximum grey value - minimum grey value) * 22%

  Iterate pixel by pixel from left-top corner to the right-bottom corner of
  the image.
      While traversing find what pixels falls under the threshold level.
      If (a pixel falls below the threshold) Then
          Check how many more pixels falls below the threshold just
          after that pixel contiguously in horizontal direction from left
          to right.
          Record that number of pixels and the coordinate of the
          starting pixel.
      End-If
  End-Iterate

  Using the above recorded values, find the longest horizontal stream
  of pixels which falls under the threshold and coordinate of the
  starting pixel.

  Again do the above iteration in the same direction and find the
  contiguous maximum number of pixels and its starting pixel
  coordinate that falls under the threshold in the vertical direction.

  Then repeat the above two approaches from bottom-right to top-left
  corner of the image and find another pair of longest horizontal and
  vertical contiguous stream of pixels which falls under the threshold
  value.

  Finally average the middle points of the horizontal lines to get the x
  value of the pupil center and average the middle points of the
  vertical lines to get the y value of the pupil center.

  Then average the half of the distance of all the 4 lines to get the
  radius of the pupil.
End
```

E. Restore the original image by loading the original image with the detected pupil boundary.

In this step the contrast stretched iris image with the pupil boundary was restored as shown in Fig. 10.

F. Detect the iris boundary using Dynamic Iris Boundary Detection algorithm.

For extracting the iris of an eye image, the pupil boundary alone is not sufficient. The outer boundary of the iris is also very important. Since most of the patterns of the iris rely near the pupillary area like radial furrows and pigment frills, it is not necessary to extract iris boundary as the actual iris boundary all the way up to Sclera from the pupil boundary. The use of a constant radius circular region from the pupil center, which includes only the vital features of the iris, is adequate in settling on the iris boundary. Sometimes the length of the radial furrows may get smaller or larger due to pupil dilation. It is important, therefore, to detect iris boundary such that length of the radial furrows in the detected iris region (i.e. the region between pupil boundary and iris boundary) for a same person should be same even the pupil has dilated. The radius of the iris boundary should change according to the size of the pupil. This part is tricky that, even though the iris boundary should be larger than the size of the pupil boundary, if the pupil boundary increased, the iris boundary should be reduced so that all the vital information of the iris will contained in between those boundaries (See Fig. 12). On the other hand iris boundary should expand if the pupil got smaller (See Fig. 11). To fulfill above phenomena we used equation (2) which calculated the iris boundary according to the size of the pupil.

$$NIrD = ((PD / 2) + 55 - ((PD / 2) * 2 / 100)) * 2 \quad (2)$$

Where, NIrD is the new iris boundary diameter. PD is the diameter of the pupil.

G. Unrolled the iris into a 512x64 rectangular image [5].

The resultant image of the previous step is a radial image. In order to make the processing easier the rounded iris is unrolled to a rectangular 512x64 size image called “unrolled iris” image. The following existing algorithm is used to unroll the image. The resulting image is shown in Fig. 13.

Algorithm

```
Begin

Get the pupil center of the image.
Create a new unwrapped image with the default output image size.
```

```
Iterate over the Y values in the output image.
  Iterate over the X values in the output image.
    Determine the polar angle (ANG) to the current coordinate using
    the following formula:
        ANG = 2 * π * (X output image / width output image)        (3)
    Find the point that is to be 'mapped to' in the output image.

    Find the distance between the radius of the iris and the pupil.

    Compute the relative distance from the pupil radius to the 'map
    from' point.

    The point to map from is the point located along theta at the pupil
    radius plus the relative radius addition.

    Set the pixel in the output image at 'map to' with intensity in
    original image at 'map from'.
  Next X
Next Y
End
```

H. Apply the Gabor filtering [6].

To do the Gabor filter we applied the Gabor wavelet as explained in [6]. The interesting part is each iris has unique texture that is generated through a random process before birth. So this Gabor filter is based on Gabor wavelets turn out to be very good in detecting patterns in images. We have used a fixed frequency one dimension (1D) Gabor filter to look for patterns in the unrolled image. The algorithm is given below.

Algorithm

```
Begin
    Consider one pixel wide column from the unrolled image and
    convolve it with a 1D Gabor wavelet by using a 5x5 kernel.

    Since the Gabor filter is complex, the result will have two parts
    namely real and imaginary which are treated separately.

    // Then the real and Imaginary parts are each quantized.
    If (a given value in the result vector > 0) Then
        Store 1
    Else
        Store 0
    Enf-If

    Once all the columns of the image have been filtered and
    quantized, a new black and white image will be formed by
    arranging all the resulting columns side by side.
End
```

For the parameters of the Gabor wavelet equation explained in [6], we have used following values:

$\mathbf{K} = 0.01$
$(\mathbf{a}, \mathbf{b}) = (1/50, 1/40)$
$\theta = -\pi/2$
$(\mathbf{x_0}, \mathbf{y_0}) = (0, 0)$
$(\mathbf{u_0}, \mathbf{v_0}) = (0.1, 0.1)$
$\mathbf{P} = \pi/2$

These values are entirely arbitrary. These values are tuned such that most of the image details will be fall in the real part of the image rather than in the imaginary part so that it is enough to consider only the real part of the image. See Fig. 14.

I. Divide the unrolled image into 128x16 segments.

In this step the Gabor filtered real unrolled image with the size of 512x64 was divided into 128x16 segments. So the new image will look like in the Fig. 15 with 2048 small segments. Here each small segment holds 16 pixels.

J. Average the pixel values in each segment.

In the resultant image of the previous step, each segment has 4x4 pixels (i.e. 16 pixels) and it is a binary image with dark black pixels represented 1 and white pixels represent 0. If this is averaged there will be a value between 0 and 1 for each 4x4 segment. Since the image has 128x16 segments altogether there are 2048 segments in the image. Therefore we get altogether 2048 averaged segment values for an unrolled, Gabor filtered, real part of the iris image. Now this 2048 decimal valued string is unique code for this iris. So for each and every person this code is saved in the database.

K. Apply AI technique on these average pixel values to detect individuals.

We used Fuzzy Logic [7] to identify persons based on their iris code. For each matching, the iris is unrolled in different angels and applied the steps *J* through *K* to overcome the eye tilting in the image. That is, if the head of the person is tilted when the eye image is acquiring, there will be some amount of rotation with respect to iris images corresponding to the iris codes stored in the database. So the idea is to check the current iris code acquired with different rotation angles with the database iris codes. In the system it uses -7^0 to $+7^0$ angles. That is altogether 15 different rotations.

Agent and System

The system has one fuzzy input, one fuzzy output and the rule base.

Fuzzy Input

The crisp input of the fuzzy system is the magnitude of the difference between the iris code values corresponding to a particular segment of the matching iris code and the iris code from the database. For example, if the first segment value of the matching code is 0.75 and the first segment value of the iris code from the database is 0.65 then the crisp input of the fuzzy system is $(0.75 - 0.65) = 0.1$. This crisp input is amplified by multiplying 100. Hence the input can range from -100 to +100 and the fuzzy input can be shown as follows;

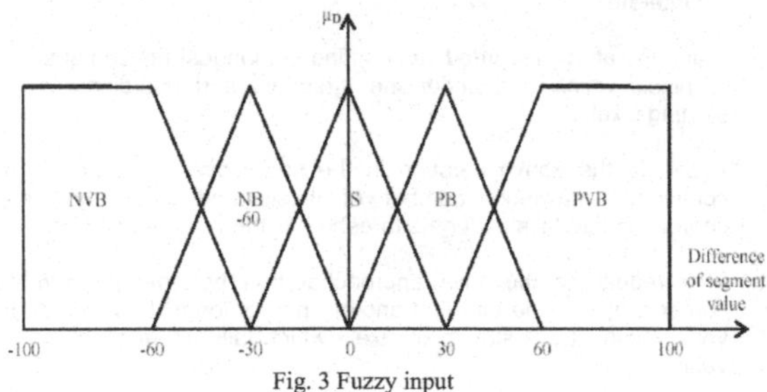

Fig. 3 Fuzzy input

NVB - Negative Very Big
NB - Negative Big
S - Small
PB - Positive Big
PVB - Positive Very Big

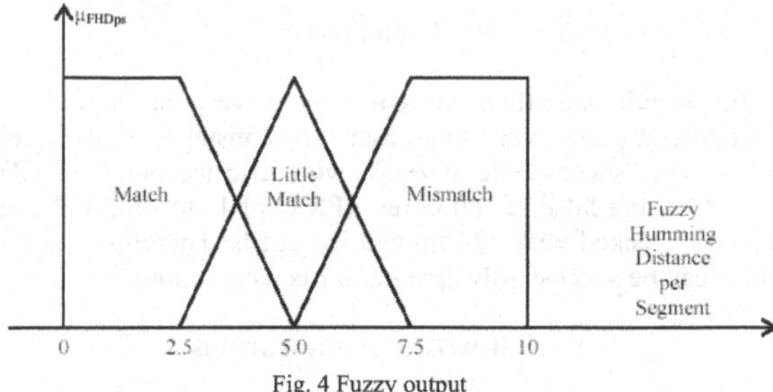

Fig. 4 Fuzzy output

Fuzzy Output

The fuzzy output of the system is called "Fuzzy Humming Distance per Segment" (FHDps). Fuzzy Humming Distance per Segment is the value which represents how close two segments of the two iris codes match each other. If the matching is high this value tend move towards zero. If the matching is very poor the FHDps will be large value which is close to 10. The fuzzy output is shown below.

Rule Base

The rule base is the knowledge of the system. The following rules base was created according to the simple domain knowledge.

IF Difference of segment values is 'Negative Very Big'
THEN the Fuzzy Humming Distance per Segment is 'Mismatch'
IF Difference of segment values is 'Negative Big'
THEN the Fuzzy Humming Distance per Segment is 'Little Match'
IF Difference of segment values is 'Small'
THEN the Fuzzy Humming Distance per Segment is 'Match'
IF Difference of segment values is 'Positive Big'
THEN the Fuzzy Humming Distance per Segment is 'Little Match'
IF Difference of segment values is 'Positive Very Big'
THEN the Fuzzy Humming Distance per Segment is 'Mismatch'

This way the fuzzy system will give a crisp value of ranging from 0 to 10. This value is only for one segment. So to match two Iris Codes we checked pairs of all 2048 segment values. Then the output was obtained. That obtained all the crisp outputs which are greater than 5.0 were added together. In this way outputs less than 5 will not contribute to this addition. Finally this added value is divided by the number of segments checked. It's kind of taking the average of the added output. This added and divided value is I called "Fuzzy Humming Distance" (FHD). Fuzzy Humming Distance is the factor that used to match persons. This is somewhat similar to Hamming Distance (HD) in Dr. Daugman's approach [8]. But the difference between HD and FHD is that FHD is all based on the values of defuzzyfication.

If the FHD is less than 1.5 then the checked iris images are 'Perfect Match'

If the FHD is less than 2.5 then the checked iris images are 'Normal Match'

If the FHD is less than 2.75 then the checked iris images are 'Low Match'

If the FHD is greater than 2.75 then the checked iris images are 'No Match'

Now there is a tread off in 2.5 to 2.75 regions. If a person recognized in the confidence of a FHD value between 2.5 and 2.75 then there is a probability of misidentified.

The following sequence of images depicts the resultant images in applying the above described steps.

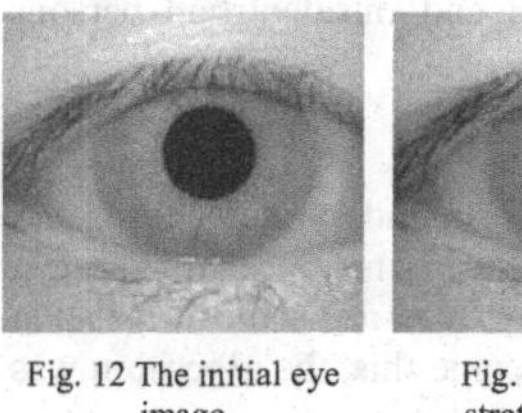

Fig. 12 The initial eye image

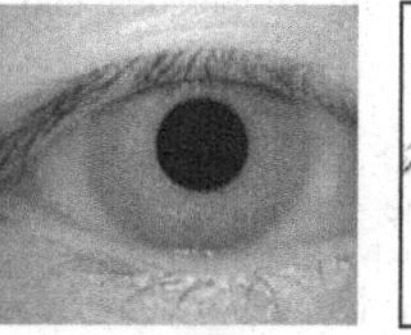

Fig. 10 Contrast stretched image

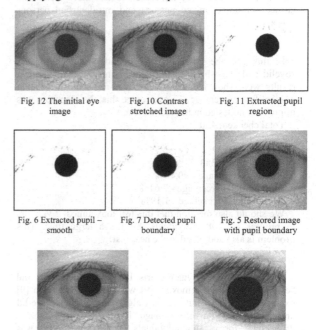

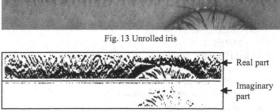

Fig. 11 Extracted pupil region

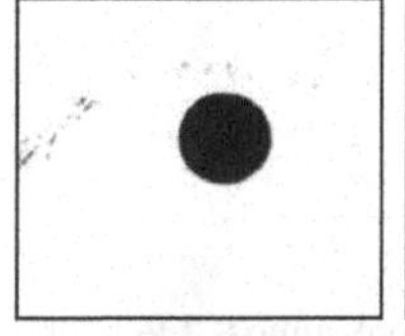

Fig. 6 Extracted pupil – smooth

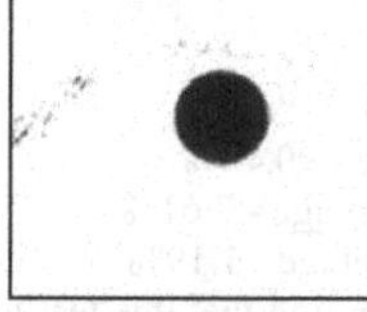

Fig. 7 Detected pupil boundary

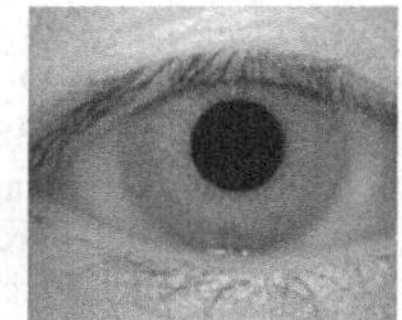

Fig. 5 Restored image with pupil boundary

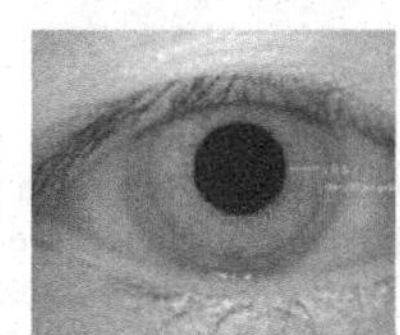

Fig. 8 Dynamic iris boundary detection - sample of contracted pupil

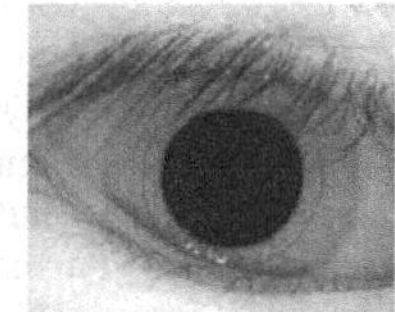

Fig. 9 Dynamic iris boundary detection - sample of expanded pupil

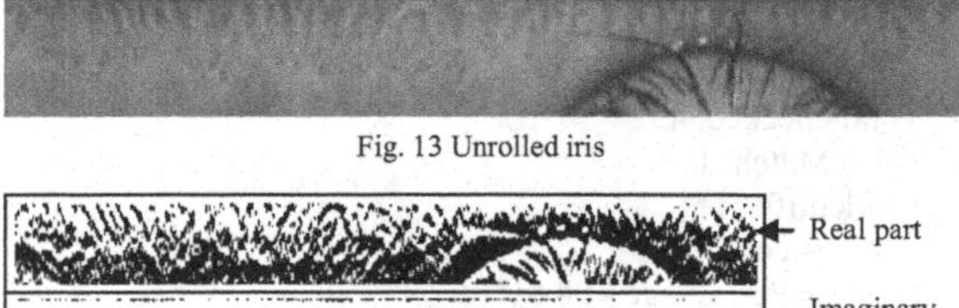

Fig. 13 Unrolled iris

Fig. 14 Gabor filtered unrolled iris

Fig. 15 Gabor filtered, feature detected, real part of unrolled iris image

Fig. 16 Extracted features shown as a binary image

IV. Results

Two tests were carried out with the CASIA image database. In that database, each person has 7 iris images which are divided in to two folders. One folder has 3 images and the other one has 4. In our test, we created a database of an iris code of all 108 individuals where it contained iris code for 3 eye images per person (using 3-image folder). The rest of the images (i.e. 4*108=432) were used to test the system. The number of unrecognized persons and misidentified persons was also computed.

A. *Test 1*

In this test the "dynamic iris boundary detection" and "eyelid and eyelash remover" were [9] not present. The test results were shown below. There was a bit of increase in unidentified percentage. To over come this, the algorithm was improved and tested in the second test.

Total checked: 420
False Match: 5
Unidentified: 32
Success count: 383
Success Percentage - 90.48%
Unidentified Percentage - 7.61%
False Match Percentage - 1.19%

We should also mention that this test took almost 7 ½ hours to process all 108 individuals. So it's a large time. This problem is also addressed in the next test.

B. *Test 2*

In this test the "dynamic iris boundary detection" and "eyelid and eyelash remover" [9] were applied where in [9] explains an accurate but complex algorithm to remove eye lid and eye lashes from an eye image. The test took 169 minutes 25 seconds for all 108 individuals. In this test the iris was rotated for angles -7^0 to $+7^0$ and checked for match. If not found then only the "eyelid and eye lashes remover" was applied and retested due to its time complexity. The test results were shown below.

Total checked: 432
False Match: 1
Unidentified: 5
Success count: 426
Success Percentage - 98.61%
Unidentified Percentage - 1.16%
False Match Percentage - 0.23%

The success percentage rose to 98.6%. The reason for misidentification and 1 false match may be due to the poor quality of the eye image such as very little visibility of the iris in the eye image.

V. Conclusion

In overall the final system was a very successful in recognition except very small number of mistakes. In the final test showed success rate of 98.6% with false accept rate (FAR) of 0.23% and false rejection rate (FRR) of 1.16% for 432 eye images checked with 324 trained iris coeds. Therefore, Fuzzy logic can be successfully applied to iris recognition.

Future Works & Suggestions

Our whole intension was to find a more accurate algorithm for iris recognition using Fuzzy logic and enhanced feature detection. The results showed us our success. But we did not put our intention on the time complexities of the algorithms and more to the point it is not the scope of this research. There is plenty of room to enhance the algorithm on this section. Finally, we did our research based on offline images. So this work can be extended to recognize individuals using iris images in real time by making the system be able to cope with video streams instead of still images.

References

[1] Ryan, M. D., Authentication, 2005, Available from: http://www.cs.bham.ac.uk/~mdr/teaching/modules/security/lectures/biometric.html (Accessed 20 February 2007).

[2] Muroň, A., Jaroslav, P., Human Iris Structure and Its Usage, acta univ. palacki. olomuc., fac. rer. nat. 2000, physica 39, 87-95.

[3] Podio, F. L., and Dunn, J. S., Biometric Authentication Technology: from the Movies to Your Desktop, ITL Bulletin, May 2001. Accessed from: http://www.itl.nist.gov/div893/biometrics/Biometricsfromthemovies.pdf (Accessed 20 February 2007).

[4] Rafael, C. G., and Richard, E. W., Digital Image Processing. New Jersey: Prentice Hall press, 2002.

[5] Atapattu, C., Iris Recognition System, B.Sc. Project Report, University of Peradeniya, 2006, unpublished.

[6] Movellan, J. R., Tutorial on Gabor Filter, 1996, Accessed from: http://mplab.ucsd.edu/wordpress/tutorials/gabor.pdf (Accessed 16 March 2007).

[7] Jamshidi, M., and Zilouchian, A., (Ed.) Intelligent Control Systems Using Soft Computing Methodologies. Boca Raton: CRC Press LLC, 2001. Chapter 8-10.

[8] Daugman, J., How Iris Recognition Works, IEEE Trans. CSVT 14(1). 2004, pp. 21-30.
Available from: http://www.cl.cam.ac.uk/users/jgd1000/csvt.pdf

[9] Kodituwakku, S. R., and Fazeen, M. I. M., Eye Lid and Eye Lash Remover for Iris Recognition Systems, unpublished.

A Logic for Qualified Syllogisms

Daniel G. Schwartz
Department of Computer Science
Florida State University
Tallahassee, FL, 32306-4019
schwartz@cs.fsu.edu

Abstract— **In various works, L.A. Zadeh has introduced fuzzy quantifiers, fuzzy usuality modifiers, and fuzzy likelihood modifiers. This paper provides these notions with a unified semantics and uses this to define a formal logic capable of expressing and validating arguments such as '*Most* birds can fly; Tweety is a bird; therefore, it is *likely* that Tweety can fly'. In effect, these are classical Aristotelean syllogisms that have been "qualified" through the use of fuzzy quantifiers. It is briefly outlined how these, together with some likelihood combination rules, can be used to address some well-known problems in the theory of nonmonotonic reasoning. The work is aimed at future applications in expert systems and robotics, including both hardware and software agents.**

I. Introduction

The notion of *fuzzy quantifier* as a generalization of the classical "for all" and "there exists" was introduced by L.A. Zadeh in 1975 [8]. This provided a semantics for fuzzy modifiers such as *most, many, few, almost all*, etc. and introduced the idea of reasoning with syllogistic arguments along the lines of "*Most* men are vain; Socrates is a man; therefore, it is *likely* that Socrates is vain", where vanity is given as a fuzzy predicate. This and numerous succeeding publications [9], [10], [11], [12], [13], [14] developed well-defined semantics also for *fuzzy probabilities* (e.g., *likely, very likely, uncertain, unlikely*, etc.) and fuzzy *usuality modifiers* (e.g., *usually, often, seldom*, etc.). In addition, Zadeh has argued at numerous conferences over the years that these modifiers offer an appropriate and intuitively correct approach to nonmonotonic reasoning.

The matter of exactly how these various modifiers are interrelated, however, and therefore of a concise semantics for such syllogisms, was not fully explored. Thus while a new methodology for nonmonotonic reasoning was suggested, it was never developed. The present work grew initially out of an effort to realize this goal. What follows here is a thumbnail sketch of a comprehensive reasoning system that has previously been published as [7].

II. Intuitive Motivation

We will define a system **Q** for reasoning with *qualified syllogisms*. In effect, these are classical Aristotelean syllogisms that have been "qualified" through the use of fuzzy quantification, usuality, and likelihood. (The term "fuzzy likelihood" is here preferred over "fuzzy probability", taking the latter to mean a probability that is evaluated as a fuzzy number.) In contrast with the syllogisms originally considered by Zadeh, we here deal only with the case of fuzzy modifiers in application to crisp (nonfuzzy) predicates. Some examples are

Most birds can fly.
Tweety is a bird.
It is *likely* that Tweety can fly.

Usually, if something is a bird, it can fly.
Tweety is a bird.
It is *likely* that Tweety can fly.

Very few cats have no tail.
Felix is a cat.
It is *very unlikely* that Felix has no tail.

From a common-sense perspective, such arguments are certainly intuitively correct. A more detailed analysis is as follows.

First, note that there is a natural connection between fuzzy quantification and fuzzy likelihood. To illustrate, the statement

Most birds can fly.

may be regarded as equivalent with

If x is a bird, then it is *likely* that x can fly.

The implicit connection is provided by the notion of a statistical sampling. In each case one is asserting

> Given a bird randomly selected from the population of all birds, there is a *high probability* that it will be able to fly.

Suppose we express this equivalence as

$$(\textit{Most}\ x)(\mathrm{Bird}(x) \to \mathrm{CanFly}(x)) \leftrightarrow (\mathrm{Bird}(x) \to \textit{Likely}\mathrm{CanFly}(x))$$

Then the first of the two syllogisms involving Tweety can be reduced to an application of this formula, together with the syllogism

$\mathrm{Bird}(x) \to \textit{Likely}\mathrm{CanFly}(x)$
$\mathrm{Bird}(\mathrm{Tweety})$
$\textit{Likely}\mathrm{CanFly}(\mathrm{Tweety})$

This follows because the left side of the equivalence is the first premise of the original syllogism, and the right side of the equivalence is the first premise of the above syllogism. A

K. Elleithy (ed.), *Advanced Techniques in Computing Sciences and Software Engineering*,
DOI 10.1007/978-90-481-3660-5_8,

Quantification	*Usuality*	*Likelihood*
all	always	certainly
almost all	almost always	almost certainly
most	usually	likely
many/about half	frequently/often	uncertain/about 50-50
few/some	occasionally/seldom	unlikely
almost no	almost never/rarely	almost certainly not
no	never	certainly not

TABLE I

INTERRELATIONS ACROSS THE THREE KINDS OF MODIFIERS.

key observation to be made here is that the latter syllogism follows by instantiating x with Tweety and applying ordinary (classical) Modus Ponens. This suggests that the desired formulation of fuzzy quantification and fuzzy likelihood may be obtained by adjoining classical logic with an appropriate set of modifiers. It also suggests that the modifiers of interest may be introduced in the manner of either quantifiers or modal operators, and that the semantics for such a system could be based on some version of probability theory.

A second observation is that there is a similar connection between the foregoing two concepts and the concept of usuality. Based on the same idea of a statistical sampling, one has that

Usually, if something is a bird, then it can fly.

is equivalent with the former two assertions. Thus one should be able to include usuality modifiers along with quantifiers and likelihood modifiers in a similar extension of classical logic.

The system **Q** is an outgrowth of these various insights and reflections. In addition to the syllogisms given above, it allows for expression of all similar syllogisms as represented by the lines of Table 1 (where the two 'Tweety' examples are given by the third line, and the 'Felix' example is given by first and last entry of the sixth line).

III. FORMAL SYNTAX

We shall begin by defining the kind of languages to be employed. Let the modifiers in Table 1, in top-down then left-right order, be represented by $\mathcal{Q}_3, \ldots, \mathcal{Q}_{-3}$, $\mathcal{U}_3, \ldots, \mathcal{U}_{-3}$, $\mathcal{L}_3, \ldots, \mathcal{L}_{-3}$. As *symbols* select: an (*individual*) *variable*, denoted by x; countably infinitely many (*individual*) *constants*, denoted generically by $a, b, \ldots$; countably infinitely many unary *predicate symbols*, denoted generically by $\alpha, \beta, \ldots$; seven *logical connectives*, denoted by $\neg$, $\vee$, $\wedge$, $\rightarrow$, $\dot{\rightarrow}$, $\ddot{\neg}$, and $\ddot{\vee}$; the abovementioned modifiers $\mathcal{Q}_i$, $\mathcal{U}_i$, and $\mathcal{L}_i$; and *parentheses* and *comma*, denoted as usual. Let the *formulas* be the members of the sets

$$\mathrm{F}_1 = \{\alpha(x) | \alpha \text{ is a predicate symbol}\}$$

$$\mathrm{F}_2 = \mathrm{F}_1 \cup \{\neg P, (P \vee Q), (P \wedge Q) | P, Q \in \mathrm{F}_1 \cup \mathrm{F}_2\}^1$$

$$\mathrm{F}_3 = \{(P \rightarrow Q) | P, Q \in \mathrm{F}_2\}$$

$$\begin{aligned}\mathrm{F}_4 = \{&\mathcal{L}_3(P \dot{\rightarrow} \mathcal{L}_i Q), \mathcal{L}_3(P \dot{\rightarrow} \mathcal{Q}_i Q), \mathcal{L}_3(P \dot{\rightarrow} \mathcal{U}_i Q),\\ &\mathcal{Q}_3(P \dot{\rightarrow} \mathcal{L}_i Q), \mathcal{Q}_3(P \dot{\rightarrow} \mathcal{Q}_i Q), \mathcal{Q}_3(P \dot{\rightarrow} \mathcal{U}_i Q),\\ &\mathcal{U}_3(P \dot{\rightarrow} \mathcal{L}_i Q), \mathcal{U}_3(P \dot{\rightarrow} \mathcal{Q}_i Q), \mathcal{U}_3(P \dot{\rightarrow} \mathcal{U}_i Q) |\\ &P, Q \in \mathrm{F}_2 \cup \mathrm{F}_3, i = -3, \ldots, 3\}\end{aligned}$$

$$\mathrm{F}_5 = \{\mathcal{L}_i P, \mathcal{Q}_i P, \mathcal{U}_i P, | P, Q \in \mathrm{F}_2 \cup \mathrm{F}_3, i = -3, \ldots, 3\}$$

$$\mathrm{F}_6 = \mathrm{F}_4 \cup \mathrm{F}_5 \cup \{\ddot{\neg} P, (P \ddot{\vee} Q) | P, Q \in \mathrm{F}_4 \cup \mathrm{F}_5 \cup \mathrm{F}_6\}$$

$$\mathrm{F}_1' = \{P(a/x) | P \in \mathrm{F}_1 \text{ and } a \text{ is an individual constant}\}$$

$$\mathrm{F}_2' = \{P(a/x) | P \in \mathrm{F}_2 \text{ and } a \text{ is an individual constant}\}$$

$$\mathrm{F}_3' = \{P(a/x) | P \in \mathrm{F}_3 \text{ and } a \text{ is an individual constant}\}$$

$$\begin{aligned}\mathrm{F}_4' = \{&\mathcal{L}_3(P \dot{\rightarrow} \mathcal{L}_i Q)(a/x) | \mathcal{L}_3(P \dot{\rightarrow} \mathcal{L}_i Q) \in \mathrm{F}_4, a \text{ is an}\\ &\text{individual constant, and } i = -3, \ldots, 3\}\end{aligned}$$

$$\begin{aligned}\mathrm{F}_5' = \{&\mathcal{L}_i P(a/x) | P \in \mathrm{F}_5, a \text{ is an individual constant,}\\ &\text{and } i = -3, \ldots, 3\}\end{aligned}$$

$$\mathrm{F}_6' = \mathrm{F}_5' \cup \{\ddot{\neg} P, (P \ddot{\vee} Q) | P, Q \in \mathrm{F}_5' \cup \mathrm{F}_6'\}$$

where $P(a/x)$ denotes the formula obtained from P by replacing every occurrence of the variable x with an occurrence of the constant a. As abbreviations take

$$(P \ddot{\wedge} Q) \quad \text{for} \quad \ddot{\neg}(\ddot{\neg} P \ddot{\vee} \ddot{\neg} Q)$$

$$(P \ddot{\rightarrow} Q) \quad \text{for} \quad (\ddot{\neg} P \ddot{\vee} Q)$$

$$(P \ddot{\leftrightarrow} Q) \quad \text{for} \quad ((P \ddot{\rightarrow} Q) \ddot{\wedge} (Q \ddot{\rightarrow} P))$$

Formulas without modifiers are *first*- or *lower-level* formulas, and those with modifiers are *second*- or *upper-level*. The members of the set $\mathrm{F}_1 \cup \mathrm{F}_1'$ are *elementary first*- or *lower-level* formulas, and the members of $\mathrm{F}_4 \cup \mathrm{F}_4' \cup \mathrm{F}_5 \cup \mathrm{F}_5'$ are *elementary second*- or *upper-level* formulas. A formula is *open* if it contains the variable x, and *closed* if not.

By a *language* L is meant any collection of symbols and formulas as described above. Languages differ from one another essentially only in their choice of individual constants and predicate symbols. As an example, the first of the foregoing syllogisms can be written in a language employing the individual constant a for Tweety and the predicate symbols α and β for Bird and CanFly—and for clarity writing these names instead of the symbols—as

$$\begin{array}{l}\mathcal{Q}_1(\mathrm{Bird}(x) \rightarrow \mathrm{CanFly}(x))\\ \mathcal{L}_3\mathrm{Bird}(\mathrm{Tweety})\\ \hline \mathcal{L}_1\mathrm{CanFly}(\mathrm{Tweety})\end{array}$$

In words: For *most* x, if x is a Bird then x CanFly; it is *certain* that Tweety is a Bird; therefore it is *likely* that Tweety CanFly.

[1] This notation abbreviates the usual inductive definition, in this case the smallest class of formulas containing F_1 together with all formulas that can be built up from formulas in F_1 in the three prescribed ways.

IV. The Bayesian Semantics

This section and the next define two alternative semantics for **Q**, one Bayesian and one non-Bayesian. The first will be the more general, but the second will be more useful for certain kinds of applications. In both semantics, an *interpretation* I for a language L will consist of a *likelihood mapping* l_I which associates each lower-level formula with a number in $[0,1]$, and a *truth valuation* v_I which associates each upper-level formula with a *truth value*, T or F. The subscript I will be dropped when the intended meaning is clear.

Here the definition of l is based on the Bayesian subjectivist theory of probability as described in [4], pp. 29–34. A key feature of Bayesian theory is that it takes the notion of conditional probability as primitive. A *likelihood mapping* l_I *for* an interpretation I *of* a language L, will be any function defined on the lower-level formulas P of L, and the ordered pairs $(Q|P)$ of lower-level formulas of L, satisfying: for elementary P,

$$l(P) \in [0,1]$$

for ordered pairs $(Q|P)$ of formulas (elementary or not),

$$l(Q|P) \in [0,1]$$

and, for any P and Q (elementary of not),

$$l(\neg P) = 1 - l(P)$$

$$l(P \wedge Q) = l(Q|P)l(P)$$

$$l(P \vee Q) = l(P) + l(Q) - l(P \wedge Q)$$

$$l(P \rightarrow Q) = l(Q|P)$$

$$\text{if } l(P) = r, \text{ then for any } a,\ l(P(a/x)) = r$$

$$l(Q|P)l(P) = l(P|Q)l(Q)$$

The value $l(P)$ is here taken to be the Bayesian *degree of belief* (in the truth) of P. The value $l(Q|P)$ is taken to be the Bayesian *conditional probability*, which by definition is the degree of belief (in the truth) of P under the assumption that Q is known (to be true) with absolute certainty. Under this interpretation common sense would dictate that, if $l(P) = 0$, then $l(Q|P)$ should be undefined. The last of the above equations is a reconstrual of the familiar "inversion formula" (see [4], p. 32) and ensures that $\wedge$ and $\vee$ are commutative. The second from the last line asserts that, if a formula P involving the variable x is held with a certain degree of belief, then in the absence of any special information about an individual a, the formula $P(a/x)$ will be held to the same degree. The only thing left to make any such l a Bayesian probability function is to agree that "absolute certainty" will be represented by the value 1.

To define the valuation mapping v, one must first select, for each $i = -3, \ldots, 3$, a *likelihood interval* $\iota_i \subseteq [0,1]$ in the manner of

$$\iota_3 = [1,1] \quad \text{(singleton 1)}$$

$$\iota_2 = [\tfrac{4}{5}, 1)$$

$$\iota_1 = [\tfrac{3}{5}, \tfrac{4}{5})$$

$$\iota_0 = (\tfrac{2}{5}, \tfrac{3}{5})$$

$$\iota_{-1} = (\tfrac{1}{5}, \tfrac{2}{5}]$$

$$\iota_{-2} = (0, \tfrac{1}{5}]$$

$$\iota_{-3} = [0,0] \quad \text{(singleton 0)}$$

These intervals then become associated with the corresponding modifiers. Their choice is largely arbitrary, but should in principle be guided either by intuition or experimental results based on psychological studies (see [7] for a discussion and references. The only formal requirement is that they be nonoverlapping and cover the interval $[0,1]$. Given such a set of intervals, the mapping v is defined by, for all $i = -3, \ldots, 3$: for open lower-level P, Q, and with $\mathcal{M}$ being any of $\mathcal{L}$, $\mathcal{Q}$, or $\mathcal{U}$,

$$v(\mathcal{M}_3(P \dot{\rightarrow} \mathcal{M}_i Q) = T \text{ iff } l(P \rightarrow Q) \in \iota_i$$

for closed lower-level P and Q,

$$v(\mathcal{L}_3(P \dot{\rightarrow} \mathcal{L}_i Q)) = T \text{ iff } l(P \rightarrow Q) \in \iota_i$$

for open lower-level P and $\mathcal{M}$ being any of $\mathcal{L}$, $\mathcal{Q}$, or $\mathcal{U}$,

$$v(\mathcal{M}_i P) = T \text{ iff } l(P) \in \iota_i$$

for closed lower-level P,

$$v(\mathcal{L}_i P) = T \text{ iff } l(P) \in \iota_i$$

and for open or closed upper-level P and Q,

$$v(\ddot{\neg} P) = T \text{ iff } v(P) = F$$

$$v(P \ddot{\vee} Q) = T \text{ iff either } v(P) = T \text{ or } v(Q) = T$$

It is straightforward to verify that this provides a well-defined semantics for the languages in concern. Note that a second-level formula is either T or F, so that this part of the system is classical. This justifies introducing $\ddot{\wedge}$, $\ddot{\rightarrow}$, and $\ddot{\leftrightarrow}$ in the manner that is customary for classical logic, i.e., via the abbreviations given in Section III. By contrast, at the lower level there is no similarly convenient syntactical way to express the definition of $l(P \vee Q)$ in terms of $l(P \wedge Q)$, so the two connectives must be defined separately.

To illustrate this semantics, let us verify in detail that the foregoing syllogism regarding Tweety is *valid* in any such interpretation I, i.e. that if the premises of the syllogism are both T in I, then so also will be the conclusion. It will be seen that validity in this example is a direct result of associating $\mathcal{Q}_1$ (*most*) and $\mathcal{L}_1$ (*likely*) with the same likelihood interval. Suppose I is such that

$$v(\mathcal{Q}_1(\mathrm{Bird}(x) \rightarrow \mathrm{CanFly}(x)) = T$$

$$v(\mathcal{L}_3\mathrm{Bird}(\mathrm{Tweety})) = T$$

From the latter we obtain by definition of v that

$$l(\mathrm{Bird}(\mathrm{Tweety})) = 1$$

which means that Bird(Tweety) is absolutely certain. From the former we obtain by definition of v that

$$l(\mathrm{Bird}(x) \rightarrow \mathrm{CanFly}(x)) \in \iota_1$$

By definition of l, this gives

$$l(\text{Bird}(\text{Tweety}) \to \text{CanFly}(\text{Tweety})) \in \iota_1$$

whence

$$l(\text{CanFly}(\text{Tweety})|\text{Bird}(\text{Tweety})) \in \iota_1$$

in accordance with Bayesian theory, the latter means that the degree of belief in CanFly(Tweety), given that Bird(Tweety) is absolutely certain, is in ι_1. This, together with the above certainty about Tweety being a bird, yields that the degree of belief in CanFly(Tweety) must also be in ι_i. Then, by definition of l,

$$l(\text{CanFly}(\text{Tweety})) \in \iota_1$$

giving, by definition of V, that

$$v(\mathcal{L}_1\text{CanFly}(\text{Tweety})) = T$$

This is what we were required to show.

In general, it can be shown that the upper-level validates all the axioms and inference rules of classical propositional calculus, in particular, *Modus Ponens*: From P and $P\ddot{Q}$ infer Q. In addition, it validates the *Substitution Rule*: From P infer $P(a/x)$, for any individual constant a, as well as equivalences of the form discussed in Section II, formally expressed here as

$$\mathcal{Q}_i(\alpha(x) \to \beta(x)) \dddot{\to} (\mathcal{L}_3\alpha(x) \ddot{\to} \mathcal{L}_i\beta(x)) \qquad (*)$$

for all $i = -3, \ldots, 3$. Verification of these items, together with additional formulas of interest validated by this semantics, can be found in [7].

V. The Counting Semantics

Whenever one uses a quantifier in everyday conversation, there is an implicit reference to an underlying domain of discourse. This observation evidently served as the basis for Zadeh's original formulation of fuzzy quantification. For example, "*Most* birds can fly" refers to a domain of individuals which is presumed to include a collection of birds, and an assertion to the effect that there is a "high probability" that a randomly chosen bird will be able to fly (Section II) is represented mathematically by the condition that a "large proportion" of birds are able to fly.

Unfortunately, the semantics developed in the preceding section does not reflect this type of meaning. At the same time, however, while Bayesian theory insists on a purely subjectivist interpretation of probabilities as degrees of belief, there is nothing that rules out the statistical intuitions discussed earlier. Indeed the theory does not say anything about how one's degrees of belief are to be determined; it says only that they must be chosen in such a way that they conform to certain laws.

The present section develops an alternative semantics which explicitly portrays the role of the underlying domain. This *counting semantics* arises by restricting Zadeh's notion of "σ-count" to crisp predicates (see the aforementioned references).

An *interpretation* I for a language L will now consist of: a *universe* U_I of *individuals* (here assume U_I is finite); assignment of a unique individual $a_I \in U_I$ to each individual constant a of L; assignment of a unique unary predicate α_I on U_I to each predicate symbol α of L; a *likelihood mapping* l_I which associates each lower-level formula with a number in $[0, 1]$; and a *truth valuation* v_I which associates each upper-level formula with a *truth value*, T or F. As before, the subscript I will be dropped when the intended meaning is clear.

Given assignments for the individual constants and predicate symbols, the mappings l and v are defined in the following way. Observe that the assignments α_I induce the assignment of a unique subset P_I of U_I to each (open) formula in F_2 according to

$$(\neg P)_I = (P_I)^c$$

$$(P \vee Q)_I = P_I \cup Q_I$$

$$(P \wedge Q)_I = P_I \cap Q_I$$

For subsets $X \subseteq U$, define a *proportional size* σ by

$$\sigma(X) = |X|/|U|$$

where $|\cdot|$ denotes cardinality. Then l is defined by: for $P \in \text{F}_2$,

$$l(P) = \sigma(P_I)$$

for $(P \to Q) \in \text{F}_3$,

$$l(P \to Q) = \sigma(P_I \cap Q_I)/\sigma(P_I)$$

with l undefined if $\sigma(P_I) = 0$; and for $P \in \text{F}_2 \cup \text{F}_3$,

$$\text{if } l(P) = r, \text{ then } l(P(a/x)) = r$$

It is easy to see that σ is a probability function. These definitions merely replicate the standard way of defining probability where events are represented as subsets of a universe of alternative possibilities. The value $\sigma(P_I)$ is defined to be the probability that a randomly selected a_I in U_I will be in P. This means that, for each a and each open $P \in \text{F}_2$, and given no additional information about a, $l(P(a/x))$ is the probability that $a_I \in P_I$. The definition of $l(P \to Q)$ is the traditional (non-Bayesian) way of defining conditional probability in terms of joint events (see [4], p. 31). Thus the value of this ratio is, by definition, the probability that an individual a_I will be in Q_I, given that a_I is known to be in P_I.

Assuming this version of l, the corresponding v is defined exactly as in Section IV. It is a routine matter to verify that this semantics validates all the same syllogisms and formulas as were considered for the Bayesian semantics in [7]. (This is not to say, however, that the two semantics are necessarily equivalent with respect to the given class of languages L, an issue which as yet remains unresolved.) To illustrate, the "Tweety" syllogism can be established as follows. As before, assume that both premises have value T. Letting Pr denote probability, we have

$$v(\mathcal{Q}_1(\text{Bird}(x) \to \text{CanFly}(x)) = T$$

$$\begin{aligned}
&\text{iff } l(\text{Bird}(x) \to \text{CanFly}(x)) \in \iota_1 && \text{(def. } v) \\
&\text{iff } \sigma(\text{Bird}_I \cap \text{CanFly}_I) \in \iota_1 && \text{(def. } l) \\
&\text{iff } \forall a_I, \Pr(a_I \in \text{Bird}_I) = 1 \text{ implies} \\
&\qquad \Pr(a_I \in \text{CanFly}) \in \iota_i && \text{(non-Bayes cond.)} \\
&\text{iff } \forall a, l(\text{Bird}(a)) = 1 \text{ implies} \\
&\qquad l(\text{CanFly}(a)) \in \iota_i && \text{(discussion above)} \\
&\text{iff } \forall a, v(\mathcal{L}_3\text{Bird}(a)) = T \text{ implies} \\
&\qquad v(\mathcal{L}_1\text{CanFly}(a)) = T && \text{(def. } v)
\end{aligned}$$

Then taking the last line with Tweety as an instance of a and combining this with the second premise of the syllogism gives the desired result. As with the Bayesian semantics, the counting semantics also validates classical propositional calculus at the upper level, as well as all of the same additional formulas discussed in [7].

It would be easy to implement such a mapping σ in any database; one need only scan records and perform counts wherever appropriate. In other types of applications, however (e.g., many expert systems), the underlying universe will be such that it is not possible to count the numbers of objects that satisfy certain relations. For example, it is not known exactly how many birds there are in the world, nor how many of them can fly. Hence instead of basing the likelihood valuation l on actual counts, it would be more reasonable to define it in terms of estimates of sizes of populations. Such estimates might be arrived at by means of statistical samplings; alternatively, they might be subjective estimates of relative sizes, essentially educated guesses, not necessarily based on any deeper methodology. In the latter case one is nearing a return to the type of reasoning portrayed by the Bayesian semantics. The counting semantics would nonetheless be useful in this context, inasmuch as the principles of set theory can be used to help ensure that these estimates are selected in intuitively plausible ways. For example, if A's are known to always be B's, then in any valid interpretation the set of A's should be a subset of the set of B's. Such an approach might be characterized as subjective, but non-Bayesian.

The restriction to finite universes was made in order to define the counting semantics in terms of relative cardinalities. It seems reasonable, however, that one could extend to infinite domains via an abstract measure-theoretic formulation of probability as in Kolmogorov [3].

VI. Application to Nonmonotonic Reasoning

In order to apply the logic Q to the tasks of nonmonotonic reasoning, several additional components are required. First is needed a logic for likelihood combination. For example, if by one line of reasoning one derives *LikelyP*, and by another derives *UnlikelyP*, then one would like to combine these to obtain *UncertainP*. In effect, one needs a set of inferences rules covering all possible likelihood combinations. Such a set of rules is described in Table II, where the numbers are subscripts for the likelihood modifiers. To illustrate, the foregoing example is represented by the 0 in the cell at row 3, column -3. (The $*$ in the upper right and lower left corners represent contradictory conclusions, which can be handled by means of a special "reason maintenance" process. This is a form of nonmonotonic reasoning first identified by Doyle [1], [2]. A version of this can be formulated via the notion of "path logic" discussed below. Please see [7] for details.)

	3	2	1	0	-1	-2	-3
3	3	3	3	3	3	3	*
2	3	2	2	2	2	0	-3
1	3	2	1	1	0	-2	-3
0	3	2	1	0	-1	-2	-3
-1	3	2	0	-1	-1	-2	-3
-2	3	0	-2	-2	-2	-2	-3
-3	*	-3	-3	-3	-3	-3	-3

TABLE II

Rules for likelihood combination.

Second is needed a means for providing such inference rules with a well-defined semantics. A problem arises in that simultaneously asserting *LikelyP* and *UnlikelyP* requires that P have two distinct likelihood values. This cannot be accommodated in a conventional formalism. To remedy this is introduced the notion of a *path* logic, which explicitly portrays reasoning as an activity that takes place in *time*. In effect, one distinguishes between different occurrences of P in the derivation path (i.e., the sequence of derivation steps normally regarded as a *proof*) by labeling each of them with a *time stamp* indicating its position in the path. In this manner the likelihood mapping can be defined on labeled formulas, in which case each differently labeled occurrence of P can each have its own well-defined likelihood value.

A third needed component is a means of distinguishing between predicates that represent *kinds* of things and those that represent *properties* of things. To illustrate, in the "Tweety" syllogism, "Bird" represents a kind, whereas "CanFly" represents a property. For this purpose [7] introduces the notion of a *typed predicate*, indicated formally by superscripts as in $\text{Bird}^{(k)}$ and $\text{CanFly}^{(p)}$.

Last is needed a way of expressing a *specificity relation* between kinds of things, together with an associated *specificity rule*. For example, if "$All(\text{Penguin}^{(k)}(x) \to \text{Bird}^{(k)}(x)$" is asserted in the derivation path, asserting in effect that the set of penguins is a subset of the set of birds, then one needs to make an extralogical record that $\text{Penguin}^{(k)}$ is more specific than $\text{Bird}^{(k)}$. Given this, one can define a rule which says that more specific information takes priority over less specific.

Collectively, these various components comprise a system for a style of nonmonotonic reasoning known as as default reasoning with exceptions. The problems associated with formulating this kind of reasoning have been illustrated by a variety of conundrums, the most well-known being the situation of Opus as illustrated in Figure 1. This is from Touretzky [5], which in turn has been cited in numerous other AI publications. As the figure indicates, the situation with Tweety is clear, namely, Tweety can fly; but the situation with Opus is contradictory. By one line of reasoning, Opus is a penguin, penguins are birds, and birds can fly, so Opus can fly, whereas by another line, Opus is a penguin, and penguins cannot fly, so Opus cannot fly.

A way to resolve the conundrum can be shown in terms of Figure 2. The diagram portrays the results of adding the following formulas into the derivation path:

1) $\mathcal{L}_3\text{Bird}(\text{Tweety})$
2) $\mathcal{Q}_1(\text{Bird}(x) \to \text{CanFly}_1(x))$
3) $\mathcal{L}_3\text{Opus}(\text{Penguin})$
4) $\mathcal{Q}_1(\text{Penguin}(x) \to \text{Bird}(x))$
5) $\mathcal{Q}_1(\text{Penguin}(x) \to \neg\text{CanFly}_2(x))$

where the subscripts on CanFly indicate the first and second occurrences of CanFly(x) in the path. Note first that one can obtain the Tweety syllogism described earlier from (1) and (2) as follows. By (*) in Section III, one can add

6) $\mathcal{Q}_1(\text{Bird}(x) \to \text{CanFly}(x)) \ddot{\leftrightarrow}$
$(\mathcal{L}_3\text{Bird}(x) \ddot{\to} \mathcal{L}_1\text{CanFly}(x))$

Then, by classical propositional calculus, (2) and (7) yield

7) $(\mathcal{L}_3\text{Bird}(x) \ddot{\to} \mathcal{L}_1\text{CanFly}(x))$

From this, instantiation of Tweety for x gives

8) $(\mathcal{L}_3\text{Bird}(\text{Tweety}) \ddot{\to} \mathcal{L}_1\text{CanFly}(\text{Tweety}))$

Then, by Modus Ponens, (1) and (8) give

9) $\mathcal{L}_1\text{CanFly}(\text{Tweety}))$

For the case of Opus, one can similarly apply classical propositional calculus to (3), (4), and (2) to derive

10) $\mathcal{L}_1\text{CanFly}(\text{Opus}))$

and to (3) and (5) to derive

11) $\mathcal{L}_3\ddot{\neg}\text{CanFly}(\text{Opus}))$

Then, by the specificity rule, since Penguin is more specific than Bird, its properties take priority, and one concludes

12) $\mathcal{L}_3\ddot{\neg}\text{CanFly}(\text{Opus}))$

This makes use of the derivation path for allowing different occurrences of CanFly to have different likelihood values, but it does not require likelihood combination. A conundrum that employs likelihood combination rules and which also is handled effectively by this reasoning system is the well-known Nixon Diamond [6]. Again refer to [7] for details.

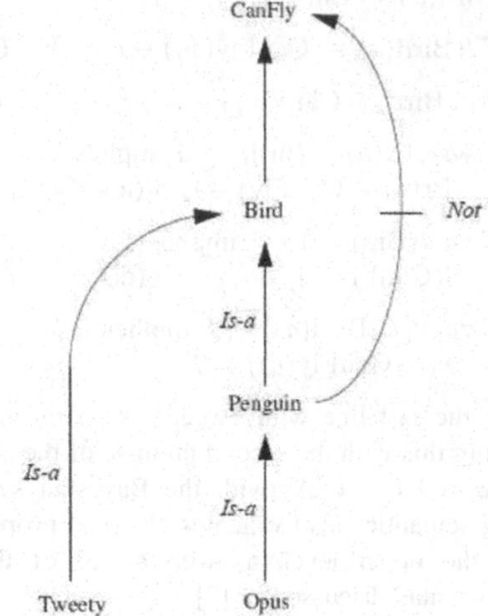

Fig. 1. Tweety can fly, but can Opus?

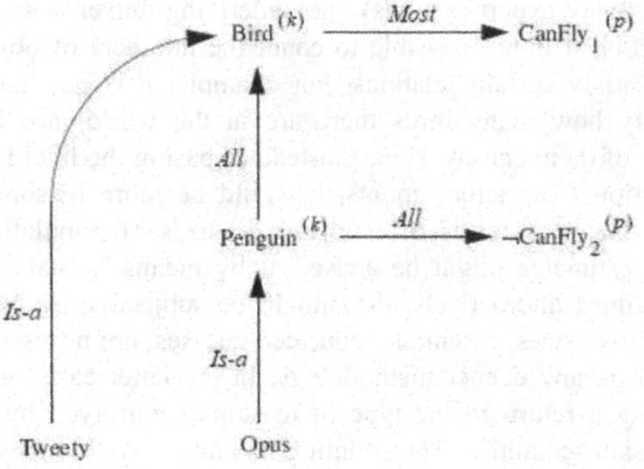

Fig. 2. Tweety *likely* can fly, and Opus *certainly* cannot.

REFERENCES

[1] J. Doyle, A truth maintenance system, *Artificial Intelligence*, **12** (1979) 231–272.

[2] B. Smith and G. Kelleher, eds., *Reason Maintenance Systems and their Applications*, Ellis Horwood, Chichester, England, 1988.

[3] A.N. Kolmogorov, *Foundations of the Theory of Probability, 2nd English Edition*, Chelsea, New York, 1956.

[4] J. Pearl, *Probabilistic Reasoning in Intelligent Systems: Networks of Plausible Inference*, Morgan Kaufmann, San Mateo, CA, 1988.

[5] D. Touretzky, Implicit ordering of defaults in inheritance systems, in: *Proceedings of the Fifth National Conference on Artificial Intelligence, AAAI'84*, Austin, TX (Morgan Kaufmann, Los Altos, CA (1984) 322–325.

[6] D. Touretzky, J.E. Horty, and R.H. Thomason, A clash of intuitions: the current state of nonmonotonic multiple inheritance systems, in *Proceedings of IJCAI-87*, Milan, Italy, 1987, pp. 476–482.

[7] D.G. Schwartz, Dynamic reasoning with qualified syllogisms, *Artificial Intelligence*, **93** (1997) 103–167.

[8] L.A. Zadeh, Fuzzy logic and approximate reasoning (in memory of Grigore Moisil), *Synthese* **30** (1975) 407–428.

[9] L.A. Zadeh, PRUF—a meaning representation language for natural languages, *International Journal of Man-Machine Studies* **10** (1978) 395–460.

[10] L.A. Zadeh, A computational approach to fuzzy quantifiers in natural languages, *Computers and Mathematics* **9** (1983) 149–184.

[11] L.A. Zadeh, Fuzzy probabilities, *Information Processing and Management* **20** (1984) 363–372.

[12] L.A. Zadeh, Syllogistic reasoning in fuzzy logic and its application to usuality and reasoning with dispositions, *IEEE Transactions on Systems, Man, and Cybernetics* **15** (1985) 754–763.

[13] L.A. Zadeh, Outline of a theory of usuality based on fuzzy logic, in: A. Jones, A. Kaufmann, and H.-J. Zimmerman, eds., *Fuzzy Sets Theory and Applications* (Reidel, Dordrecht, 1986) 79–97.

[14] Zadeh and R.E. Bellman, Local and fuzzy logics, in: J.M. Dunn and G. Epstein, eds., *Modern Uses of Multiple-Valued Logic* (Reidel, Dordrecht, 1977) 103–165.

Improved Induction Tree Training for Automatic Lexical Categorization

M. D. López De Luise, M. Soffer, F. Spelanzon - aigroup@palermo.edu
Department of Informatics Engineering, Universidad de Palermo
Av. Córdoba 3501, Capital Federal, C1188AAB, Argentina

Abstract—**This paper studies a tuned version of an induction tree which is used for automatic detection of lexical word category. The database used to train the tree has several fields to describe Spanish words morpho-syntactically. All the processing is performed using only the information of the word and its actual sentence. It will be shown here that this kind of induction is good enough to perform the linguistic categorization.**

Index Term — Machine Learning, lexical categorization, morphology, syntactic analysis.

I Introduction

The lexical category of a word indicates if it is a noun, an article, a verb, etc. This kind of categorization plays an important role in several applications such as automatic error correction in documents[1], classification of documents[2], written text analysis, inflectional language[1] analysis [3], statistical machine translation[4], text summarization[5], automatic grammar and style checking [6] automatic translation[7], statistical modeling of speech[8], etc.

However, it is hard to process natural language due to several phenomena such as synonymy (different words with similar meaning), polysemy (a word with two or more meanings), anaphora (implicit mention by means of demonstrative pronouns), metaphors (use of a word with a meaning or in a context different from the habitual one), metonymy (rhetorical figure that consists of transferring the meaning of a word or phrase to another word or phrase with different meaning, with semantic or logical proximity)[5], misspellings, punctuation, neologisms, foreign words and differences between linguistic competence (based in grammar rules) and actuation (the way grammar is used by a native speaker)[6].

Many approaches have been used to achieve word categorization. Some of these are:
-Lexical knowledge databases such as WordNet[5].
-Normatives[6].
-Contextual information processing[9].
- Morphological rules for prefixes and suffixes[10].

This paper uses an induction tree trained with a set of fields that are morpho-syntactic descriptors of words. This is an extension of the firsts results presented in[11]. The main goal is to find the actual lexical category of certain words with reasonable precision, using only local information.

Therefore, there is a first step to derive the morpho-syntactic characteristics of the actual word. This is an automatic process that uses information from the same sentence or from the word itself. It includes stemming [12] the word. The second step consists in getting the category from a J48 induction tree. Although Induction Trees[2] can be used for learning in many areas[12], they are applied here to word classification. An induction tree is a model of some basic characteristics of a dataset extracted by an induction process on instances. It is used due to its flexibility and its power to apply the acquired knowledge to new concrete instances.

The rest of the paper is divided into the following sections: database structure (section II), sample characteristics (section III) quality metrics used in the context of this study (section IV), main results (section V), conclusions (section VI) and future work (section VII).

II Database Structure

The database was populated with words extracted from Spanish web pages. From a total of 340 pages, 361217 words were extracted with a Java application. The output was saved as 15 plain text files. The files were converted into Excel format to be able to use a form and fill in the field *tipoPalabra* (lexical category of the word) manually. The resulting files were processed with other Java program to introduce the Porter's stemming column and convert it into *csv* format. This conversion made it possible to work with WEKA[3] software. Afterwards, some preliminary statistics were performed with InfoStat[4] to detect the main dataset features. Finally, the *csv* files were processed with WEKA Explorer to build an induction tree model.

III Sample Characteristics

Each row of the database is composed of a Spanish word. For each word, a set of 25 morpho-syntactic description fields were derived. Therefore, each database record represents a word and its characteristics. Fields are detailed below:

-Continue fields: they do not exist.

-Numerable fields: These 10 non-negative integer fields have a big boundary (see Table 1). All of them were discretized into fixed-size intervals with the purpose of

---

[1] Language whose words can take a considerable amount of morphological variations. This is usually done by changing a suffix.

[2] From Mitchell: "Decision Tree Learning is a method to approximate discrete-value target functions, in which the learned function is represented by a decision tree. Learning trees can also be re-represented as sets of *if-then* rules to improve human readability. These learning methods are among the most popular of inductive inference algorithms and have been successfully applied to a broad range of tasks such as learning how to diagnose medical diseases or how to assess credit risk loan applicants.

[3] WEKA: Open source workbench for Data Mining and Machine Learning [14].

[4] InfoStat: Statistical software of a group called InfoStat in Universidad Nacional de Córdoba.

K. Elleithy (ed.), *Advanced Techniques in Computing Sciences and Software Engineering*,
DOI 10.1007/978-90-481-3660-5_9, © Springer Science+Business Media B.V. 2010

categorizing and processing them together with nominal fields. They were divided into 3 or 5 categories. (See TABLE 2).

-Discrete fields: they do not exist.

-Non-numeric fields: There are 15 fields that have a domain composed of a specific set of literals (syllables, punctuation signs, a set of predefined words or the classical binomial Yes/No). See TABLE 3 for details.

-Missing data: They were considered to be a distinct data value that would be processed with the rest of the data.

The dependences of the data fields were studied with correspondence analysis. This task was performed with InfoStat software. All 25 fields were considered but only a random sample of 47820 instances was processed. The independency test was performed with parameter $\alpha = 0.05$, statistic χ^2 y H_0= "independent".

The results obtained show the following:

-tipoPalabra (lexical category of the word): It is independent from *tipoPag* (type of page) and *siguePuntuación* (punctuation that follows after the actual word).

TABLE 1
NUMERABLE FIELDS

Field	Description
Id-caso	We page identifier
Cant-ocurrencias	Times the word is repeated in the page
Cant-pal-pagina	Number of words in the page
Long-palabra	Number of characters in the word
Cant-vocales-fuertes	Number of strong vowels in the word (a, e, o)
Cant-vocales-debiles	Number of weak vowels in the word (i, u)
Long-oracion	Number of words in the sentence
Cant-numeros	Qantity of numbers in the sentence
Cant-signos-especiales	Number of special characters in the sentence

TABLE 2
FIELD CATEGORIES

Field	Categories	Max value
Web-profundidad-pagina	5	7000
Cant-ocurrencias	3	1168000
Long-palabra	5	6792600
Cant-vocales-fuertes	3	11000
Cant-vocales-debiles	3	6000
Long-oracion	5	259000
Cant-numeros	5	842000
Cant-signos-especiales	5	149000

TABLE 3
NON-NUMERIC FIELDS

Field	Description
Id-palabra	Word extracted
Tema	Main topic of the web page
Tipo-pal	One of these: sustantivo, verbo, otro (noun, verb, other respectively)
Pal-anterior	Identifier of the previous word in the sentence
pal-ant-tipo	One of these: ninguna, otro (none, other). This indicates if it is the first word in the sentence
Tipo-pag	One of these: indice, contenido (index, plain text)
Pais-radicacion	Country code of the URL: us, ar, es, cu, mx, etc.
Terminacion	Suffix of the word: ar, er, ir, or, ur, ra, re, ri, ro, ru, s, m, sa, se, si, so, su, an, en, in, on, un, cion, ciones, null
Sigue-puntuacion	One of these: org, com, net, otro. Indicates the kind of domain from the URL.
Empieza-mayuscula	One of these: si, no (Yes, NO). This value is "si" when one of the following is after the word: ",", ".", ";", ":"
Clase-pag	One of these: org, com, net, otro. Indicates the kind of domain from the URL
Empieza-mayuscula	One of these: si, no (yes, no). This value is "si" when the word starts capitalized.
Resaltada	One of these: si, no (yes, no). This value is "si" when the word is enclosed with quotation marks or is capitalized.
Es-titulo	One of these: si, no (yes, no). This value is "si" when the word belongs to the first sentence in the text
Frase-especial	One of these: si, no (yes, no). This value is "si" when the sentence is enclosed with: ¡!, ¿?, <>, (), [], {} or «»
Stem	radix of the word (according to Porter algorithm)

-palAntTipo (type of previous word): It is independent from *cantVocalesFuertes* (number of strong vowels).

-resaltada (highlighted word in the text): It is independent from *cantVocalesFuerte* (number of strong vowels).

IV QUALITY METRICS

In this section, the construction of an induction tree (using WEKA[13] software) is analyzed. The metric used to evaluate error handling, result predictability and confidence validation will be described.

A. *Metrics Used for Error Handling Evaluation*

1) Precision: Metric used in Information Retrieval (IR). It is the rate of relevant instances returned by the total of instances returned.

2) Recall: Metric used in IR. It is the rate of relevant returned by the number of relevant instances.

3) Recall-precision: Plot used in IR with recall (x-axis) and precision (y-axis).

B. *Metric Used for Predictability*

Kappa (κ)*:* Metric used to compare predictability against a random predictor. It can take from 0 to 1, being 0 the random predictor value and 1 the best predictor.

C. *Metric for confidence validation*

Margin curve: A bigger margin denotes a better predictor. It is the difference between the estimated probability of the true class and that of the most likely predicted class other than the true class.

V RESULTS

The J48 algorithm, which is the WEKA implementation of the C4.5 induction algorithm, is used to build an induction tree. The lexical category of Spanish words [11] may be inferred from the resulting model. Next, data fields will be studied in order to determine the best subset of database fields, the most convenient number of categories for numeric fields, the windowing and the sensibility to the number of lexical categories.

A. *Categorizations*

The induction tree works well with nominal attributes. Therefore, a number of categories must be defined for numeric data in order to handle it in this context. The number of categories plays an important role in the predictability and precision of the inductive model. In [11], the descriptor variable cantOcurrencias (number of times the word is detected within the html page) is selected for this study because it is always near the tree-model root; hence, it is important to determine the kind of word. It is evaluated with 3 and 7 bins. The result is that both precision and total error change due to categorization as well as to Margin-curves, showing that it is better to have 7 than 3 categories.

B. *Descriptor subset*

According to [11], alternate descriptor selection criteria can be studied to find out the influence of field selection in the classification power. Some cases are:

1) Low computational-cost field selection: The high-cost fields are taken out whenever the removal does not affect the tree performance.

2) Categorized fields: In this case only categorized fields are considered.

3) Nominal: All the fields processed here are nominal.

4) Independent: Only fields independent from the rest are considered here.

As it can be seen from TABLE 4, the best kappa and classifications correspond to low cost criteria.

C. Lexical categories

The previous analysis was performed using three lexical categories: noun, verb and other. In this section the 12 best descriptors combined with stem are used to build a model based on 6 categories. The dataset has 33824 rows, processed first with three categories (noun, verb, other) and afterwards with six categories (noun, verb, preposition, pronoun, article, other). Results are reflected in TABLE 5 and TABLE 6 respectively: more lexical categories improve classification power and predictability through Kappa statistics.

TABLE 4
DESCRIPTOR SUBSETS

Criteria	Fields	Total	Classif. ok	Kappa
Low cost	tema, palAntTipo, tipoPag, terminacion, empiezaMayuscula, resaltada,esTitulo, CATlongPalabra	12	71.1%	0.4761
Categorized	CATwebProfundidadPag, CATcantOcurrencias, CATcantPalPagina, CATlongPalabra, CATcantVocalesFuertes, CATcantVocalesDebiles, CATlongOracion, CATcantNumeros, CATcantSignosEspeciales	9	63.82%	0.3043
Nominal	tema, tipoPag, palAntTipo, paisRadicacion, terminacion, siguePuntuacion, clasePag, empiezaMayuscula, resaltada, esTitulo, fraseEspecial	12	65.63%	0.34
independent	palAntTipo, tipoPag, siguePuntuacion, resaltada, CATlongPalabra, CATcantVocalesFuertes, CATcantVocalesDebiles, CATlongOracion	8	63.51%	0.3

TABLE 5
CLASSIFICATION WITH 3 CATEGORIES

Split %	Classif. OK %	MSE	Kappa statistic
66	88.17	0.1776	0.7868
70	88.34	0.1759	0.7905
100	97.14	0.0924	0.9506

TABLE 6
CLASSIFICATION WITH 6 CATEGORIES

Split %	Classif. OK %	MSE	Kappa statistic
66	91.09	0.1397	0.8857
70	91.45	0.1370	0.8902
100	97.62	0.0724	0.9696

VI CONCLUSIONS

Induction trees can be used to infer lexical categories for Spanish words with reasonable precision. There is a set of morpho-syntactic word descriptors that could be used to train the induction tree. Several facts influenced the degree of precision of the tree model, among which there is the categorization procedure of numeric fields and the number of lexical categories. Word lexical classifications can be improved by increasing the number of categories in the categorization procedure or by considering more lexical categories in the training step.

It should be noted that the amount and types of lexical categories should be selected according to the objectives of the project for which they may be intended.

VII FUTURE WORK

Some interesting future work will be to evaluate alternate algorithms and compare results with J48 induction tree. The performance of this model with a higher database will also be analyzed. This will be applied to novel web context such as photo-logs, where the Spanish slang is quite different and uses "HOYGAN", a phonetic transcription of the sentences.

Finally, the model obtained will be inserted and tested in the context of a conversational system with automatic Spanish text processing.

REFERENCES

[1] Platzer C., Dustdar S., "A Vector Space Search Engine for Web Services", in Proc. of the Third European Conference on Web Services (ECOWS' 05), Vaxjo, Sweden. 2005.

[2] López De Luise D. "Mejoras en la Usabilidad de la Web a través de una Estructura Complementaria". PhD thesis. Universidad Nacional de La Plata. La Plata. Argentina. 2008.

[3] Trabalka M., Bieliková M., "Using XML and Regular Expressions in the Syntactic Analysis of Inflectional Language", In Proc. of Symposium on Advances in Databases and Information Systems (ADBIS-ASFAA'2000), Praha. pp. 185-194. 2000.

[4] Nießen S., Ney H., "Improving SMT quality with morpho-syntactic analysis", in Proc. of the 18th conference on Computational linguistics – Vol. 2, pp. 1081 – 1085, Saarbrücken, Germany. 2000.

[5] Mateo P.L., González J.C., Villena J., Martínez J.L., "Un sistema para resumen automático de textos en castellano", DAEDALUS S.A., Madrid, España. 2003.

[6] Aldezábal I., "Del analizador morfológico al etiquetador: unidades léxicas complejas y desambiguación". Procesamiento del lenguaje natural. N. 19, pp. 90-100. España.1996.

[7] Fernández Lanza S., "Una contribución al procesamiento automático de sinonimia utilizando Prolog" PhD dissertation, Santiago de Compostela, España. 2003.

[8] Levinson S., "Statistical Modeling and Classification", AT&T Bell Laboratories, Murray Hill, New Jersey, USA. Also available at http://cslu.cse.ogi.edu/HLTSurvey/ch11node4.html. 2006.

[9] Oliveira O.N., Nunes M.G. V., Oliveira M.C. F. "¿Por qué no podemos hablar con una computadora?" Magazine of Sociedad Mexicana de Física., México, v. 12, pp. 1 - 9. 1998.

[10] Mikheev A., "Automatic rule induction for unknown-word guessing". Computational Linguistics, Volume 23 , Issue 3, pp.405 - 423. 1997.

[11] López De Luise D, Ale J, "Induction Trees for Automatic Word Classification". CACIC. XIII Congreso Argentino de Ciencias de la Computación (CACIC07). Corrientes. Argentina. pp. 1702. 2007.

[12] Porter, M. F. "An Algorithm for suffix Stripping", Program, vol. 14 (3), pp. 130-137. 1980.

[13] Mitchell T. Machine Learning, New York: WCB/Mc Graw Hill, pp. 51-80.1997.

[14] Witten I. H., Frank E., "Data Mining – Practical Machine Learning Tools and Techniques", 2nd ed., San Francisco: Morgan Kaufmann Publishers. 2005.

Comparisons and Analysis of DCT-based Image Watermarking Algorithms

Ihab Amer and Pierre Hishmat
The German University in Cairo (GUC)
Main Entrance, Fifth Settlement, Cairo, Egypt
ihab.amer@guc.edu.eg
pierre.hishmat@student.guc.edu.eg

Wael Badawy and Graham Jullien
Advanced Technology Information Processing Systems (ATIPS)
2500 University Drive, NW
Calgary, AB, Canada, T2N 1N4
{badawy, jullien}@atips.ca

***Abstract*—This paper provides analysis and comparisons between three main Discrete Cosine Transform-based watermarking algorithms. The comparisons are made based on the degree of degradation the algorithms introduce after embedding the watermark, as well as the Percentage Error Bits (PEB) of the reconstructed watermark with respect to the original watermark. The algorithms' computational requirements (measured in the total elapsed time on a unified platform) are also considered. Although the three techniques use the same concept, the results show that they behave differently. This variation in the behavior makes each of them suitable for different spectrum of applications.**

***Index Terms*—Benham algorithm, DCT-based techniques, Digital watermarking, Hsu algorithm, Koch algorithm.**

I. Introduction

Watermarking is the concept of embedding hidden data into any media content. The embedding must be done in an effective and efficient way to avoid damaging the media, meanwhile maintaining the robustness. The embedded data must have some characteristics in common as they: should be imperceptible (in-audible and in-visible), should carry retrievable information, should survive under degradations of the content, and they also should be difficult to remove and change by unauthorized users.

A watermark can either be embedded into the *spatial domain* by altering some pixels values or in the *frequency domain* by altering some frequency coefficients of the media being watermarked [1].

In this paper, comparisons and analysis between DCT–based watermarking techniques are introduced. The remainder of the paper is organized as follows: Section II provides an overview of the selected DCT-based techniques, while Section III contains the obtained results and comparisons between the examined algorithms. Finally Section IV concludes the paper.

II. Description of the Chosen DCT-based Techniques

This paper provides comparisons and analysis between three well-known DCT-based watermarking algorithms. The algorithms are compared according to the degree of degradation they introduce after embedding the watermark, as well as the Percentage Error Bits (PEB) of the reconstructed watermark with respect to the original watermark. The algorithms' computational requirements (measured in the total elapsed time on a specific unified platform) are also considered. The three algorithms are mainly based on the concept of embedding the watermark by modifying the transform coefficients (typically middle-region coefficients). Middle-region coefficients are typically chosen for embedding the watermark since modifying the lowest frequency coefficients results in high quality degradation of the media being watermarked, while modifying the high frequency coefficients makes the watermarking less robust as these coefficients have a higher probability of being lost if the media is compressed afterwards (specifically during quantization).

In all the algorithms, for the watermark to be embedded, it is first converted into a sequence of zeros and ones, each denoted by S_i. Assuming that the watermark is a black and white image, a white sample in the watermark corresponds to a zero in the bit sequence, while a black sample corresponds to a one.

A. Koch Algorithm

Koch algorithm has been developed by E. Koch and J. Zhao at the Frauenhofer Institute for Computer Graphics in Darmstadt, Germany. The algorithm is divided into two parts: a part for watermark embedding, and a part for watermark detection [3].

i. Watermark embedding

The image is initially divided into 8×8 blocks and is converted into the frequency domain using 2-D DCT. Two pseudo-randomly selected positions in each of the 8×8 blocks are initially determined to carry the watermark data. For each, block, according to the value of the bit S_i to be inserted, a constraint is forced into the coefficients that are located in the two selected positions, according to the rules in Table 1.

K. Elleithy (ed.), *Advanced Techniques in Computing Sciences and Software Engineering*,
DOI 10.1007/978-90-481-3660-5_10, © Springer Science+Business Media B.V. 2010

Table 1. Rules used to embed and detect the watermark using Koch algorithm

If $S_i = 0$	$\lvert c_b(x_{i,1},y_{i,1})\rvert - \lvert c_b(x_{i,2},y_{i,2})\rvert > k$
If $S_i = 1$	$\lvert c_b(x_{i,1},y_{i,1})\rvert - \lvert c_b(x_{i,2},y_{i,2})\rvert < -k$

where b is the index of the block, and (x,y) is the position of the coefficient in the block b.

In order to obey the constraints, a robustness value (k) is added to the selected coefficients. The (k) can be any number which when added to the coefficients, helps maintaining a certain relation between them. The higher the value of "k" the more robust the watermark detection is expected to be However, a decrease in the quality of the watermarked image should also be expected. In the final stage, the image is converted again into the spatial domain using the 2D IDCT operation.

i) Watermark recovery

The watermarked image is divided again into 8×8 blocks and is converted into the frequency domain using 2-D DCT. The same two pseudo-randomly selected positions in the embedding process are used. Each block is checked for the constraints to be able to get all the embedded bits S_i, according to the rules in Table 1.
After getting all the embedded bits, the watermark image is reconstructed, and becomes ready for evaluation of the algorithm.

B. Benham Algorithm

Benham algorithm has been developed by Dave Benham and Nasir Memon at Northern Illinois University and Boon-Lock Yeo and Miverva Yeung from IBM T.J Watson Research Center. The algorithm is divided into two parts: a part for watermark embedding, and a part for watermark detection [4].

i) Watermark embedding

The image is initially divided into 8x8 blocks and is converted into the frequency domain using 2-D DCT. Three pseudo-randomly selected positions in each of the 8×8 blocks are initially determined to carry the watermark data. Each block is checked where smooth and sharp edged blocks are discarded according to the rules in Table 2.

Table 2. Rules used to discard blocks in Benham algorithm

Block discarded	$c_b(x_{i,1},y_{i,1})$	$c_b(x_{i,2},y_{i,2})$	$c_b(x_{i,3},y_{i,3})$
If	H	L	M
If	L	H	M
If	M	M	M

In the above table H means that this coefficient has the highest value of the three coefficients, M stands for the middle value, L for the lowest value.

For each, block, according to the value of the bit S_i to be inserted, a constraint is forced into the coefficients that are located in the three selected positions, according to the rules in Table 3.

Table 3. Rules used to embed and detect the watermark using Benham algorithm

S_i	$c_b(x_{i,1},y_{i,1})$	$c_b(x_{i,2},y_{i,2})$	$c_b(x_{i,3},y_{i,3})$
1	H	M	L
1	M	H	L
1	H	H	L
0	M	L	H
0	L	M	H
0	L	L	H

The constrains are applied by adding the robustness variable (k) or any of its fractions to the 3 coefficients to maintain ratios according to the constrains. In the final stage the image is converted again into the spatial domain using the 2D IDCT operation.

ii) Watermark recovery

The watermarked image is divided again into 8×8 blocks and is converted into the frequency domain using 2-D DCT. The same three pseudo-randomly selected positions in the embedding process are used. Each block is checked for the constraints to be able to get all the embedded bits S_i, according to the rules in Tables 2 and 3. After getting all the embedded bits, the watermark image is reconstructed, and becomes ready for evaluation of the algorithm.

C. Hsu Algorithm

Hsu algorithm was developed by Chiou-Ting Hsu and Jaling Wu at Taiwan National University. The algorithm is divided into two parts: a part for watermark embedding, and a part for watermark detection [5].

i) Watermark embedding

The image is initially divided into 8x8 blocks and is converted into the frequency domain using 2-D DCT. A pseudo-randomly selected position in each of the 8×8 blocks is initially determined to carry the watermark data. 2 blocks are checked each time, and according to the value of the bit S_i to be inserted, a constraint is applied to one of the two blocks coefficients to make its coefficient greater than the coefficient of the other block, and that by adding the (k) variable. Table 4 shows the rules used.

Table 4. Rules used to embed and detect the watermark using Hsu algorithm

If $S_i = 0$	$\lvert c_b(x_{i,1},y_{i,1})\rvert - \lvert c_{b+1}(x_{i,1},y_{i,1})\rvert > k$
If $S_i = 1$	$\lvert c_b(x_{i,1},y_{i,1})\rvert - \lvert c_{b+1}(x_{i,1},y_{i,1})\rvert < -k$

where b is the index of the block, and (x,y) is the position of the coefficient in the block.

In the final stage, the image is converted again into the spatial domain using the 2D IDCT operation.

ii) Watermark recovery

The watermarked image is again into 8×8 blocks and is converted into the frequency domain using 2-D DCT. The same pseudo-randomly selected position for each block in the embedding process is used. Each block is checked for the constraints to be able to get all the embedded bits S_i, according to the rules in Table 4.
After getting all the embedded bits, the watermark image is reconstructed, and becomes ready for evaluation of the algorithm.

III. COMPARISONS AND RESULTS.

The three considered algorithms are compared according to their influence on the quality of the media, their robustness, and their computational requirements. The quality of the reconstructed image from the three algorithms is measured objectively and subjectively. Figure 1 provides a subjective comparison between the three algorithms by showing the effect on Lena image after embedding the watermark each time.

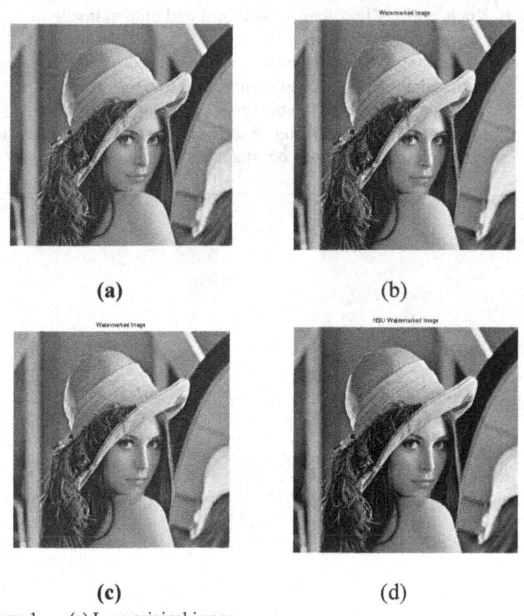

Figure 1. (a) Lena original image
(b) Reconstructed from Koch algorithm (k = 10)
(c) Reconstructed from Benham algorithm (k = 10)
(d) Reconstructed from Hsu algorithm (k = 10)

For the objective comparison, Figure 2 plots the relation between the Peak Signal to Noise Ratio (PSNR) extracted after applying each of the three algorithms at a time on Lena versus the robustness variable (k).

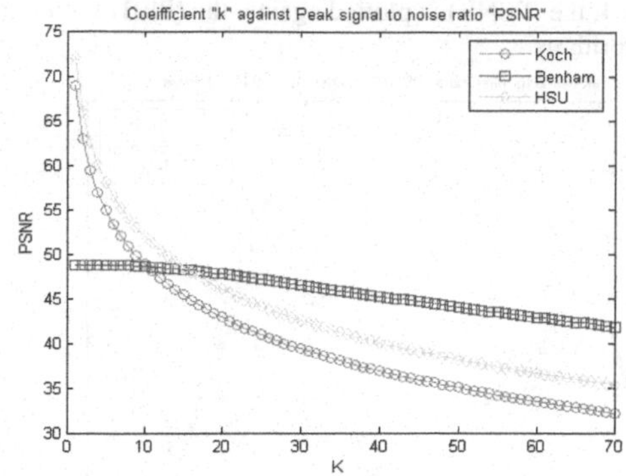

Figure 2. Robustness variable (k) vs. the (PSNR)

It is clear from the figure that the relation between the robustness variable (k) and the (PSNR) is inversely proportional. The results make sense as the more the (k) increases the more the reconstructed image blocks are different from the original image ones. It is also shown from the graph that for small (k) values both Koch, and Hsu algorithms have higher (PSNR) values than Benham which in turn has a higher (PSNR) values if high (k) values are used.

Figure 3 plots the relation between the Percentage of Error Bit (PEB) versus the robustness variable (k) when applying the three algorithms.

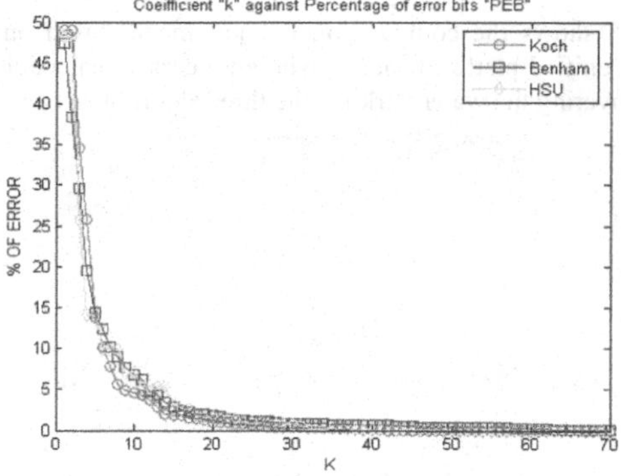

Figure 3 Robustness variable (k) vs. the Percentage of error bit (PEB)

The (PEB) is inversely proportional to the robustness variable (k). The higher the value of (k) the more probability that it will successively achieve the constraint (when its value is higher than the difference between the coefficients), and so decreasing the error between the bits of the original and the reconstructed watermark. The three algorithms showed almost the same results while varying (k) from 0 till 70.

In figure 4, the (PSNR) is plotted against the (PEB) for all the three algorithms.

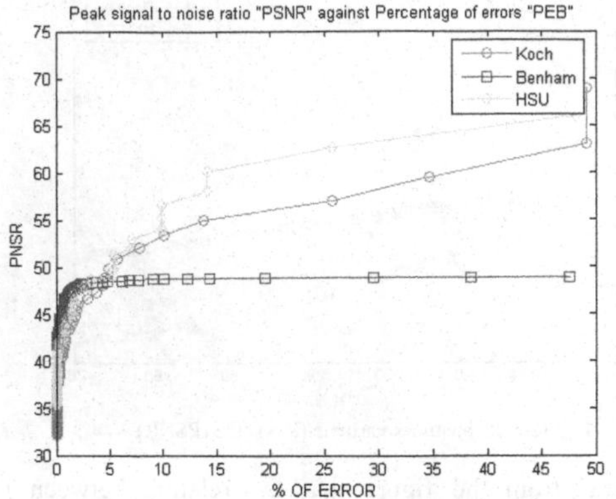

Figure 4. The Percentage of Error (PEB) vs. the Peak Signal to Noise Ratio (PSNR)

The results obtained where expected as a low (PEB) is achieved when using a high (k) value that in turn when added to the coefficients selected to embed the watermark, it will increase the difference between the original and the reconstructed image blocks thus decreasing the (PSNR) of the image. For small (PEB), Benham showed a higher (PSNR) for the reconstructed image than Koch and Hsu, while for high (PEB) Hsu and Koch have higher (PSNR) than Benham.

Figure 5 shows the computational requirements based on a specific unified platform for the whole process of embedding and recovering the watermark for the three algorithms. .

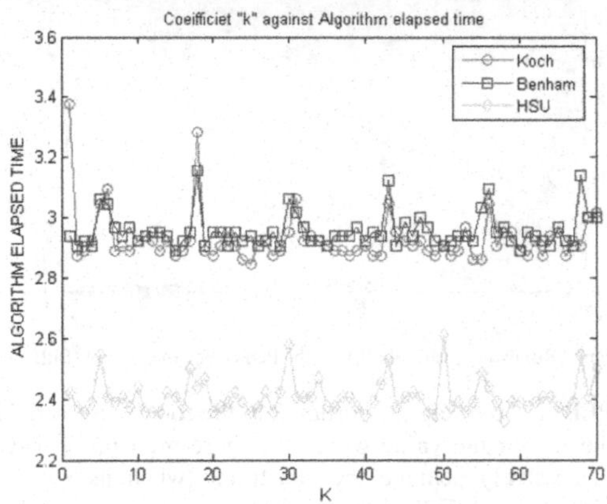

Figure 5 Robustness variable (k) vs. the algorithms elapsed time.

The results show that Hsu algorithm has a lower computational requirement than both Benham and Koch algorithms. This is expected since in Hsu algorithm a bit S_i is embedded in every two blocks while in Koch and Benham algorithms a bit S_i is embedded in every block of the image

IV. Conclusion

In this paper, comparisons and analysis between three main DCT–based watermarking techniques are introduced. Although the three algorithms use the same concept, but the watermark embedding and recovery is different in the three techniques. The results show that the three algorithms exhibit different behaviors. This makes each them suitable for different range of applications. If the target application favors robustness over other parameters Benham algorithm would be the best choice. If the (PSNR) of the reconstructed image, or the computational time are the main issues, Hsu algorithm would be the most suitable to use. Koch algorithm shows a trade off between robustness of the watermark, and quality of the reconstructed image.

Acknowledgment

The authors would like to thank the German University in Cairo (GUC) and the Advanced Technology Information Processing Systems (ATIPS) Laboratory and for supporting this research.

References

[1] Mauro. Barni and Franco. Bartolini. Book: Watermarking Systems Engineering. ISBN:0-8247-4806-9 2004.

[2] S.A. Kyem, "The Discrete Cosine Transform (DCT) : Theory and Application". Michigan State University,2003.

[3] E. Koch, and J. Zhao. Towards Robust and hidden image copyright labeling 1995.

[4] Dave. Benham , Nasir. Memon, and Boon-lock. Yeo. Fast watermarking of DCT-based compressed images Information hiding: First international workshop , 1996.

[5] Chiou-Ting. Hsu and Ja-Ling. Wu. Hidden digital watermarks in images. IEEE Transactions on image processing, 1999.

A Tool for Robustness Evaluation of Image Watermarking Algorithms

Ihab Amer #1, Tarek Sheha #2, Wael Badawy *3, Graham Jullien *4

German University in Cairo (GUC)
Main Entrance, Fifth Settlement, Cairo, Egypt
[1] ihab.amer@guc.edu.eg
[2] tarek.sheha@student.guc.edu.eg
* *ATIPS Laboratories*
ECE Dept. University of Calgary
2500 University Dr. NW, Calgary, AB
[3] badawy@atips.ca
[4] jullien@atips.ca

***Abstract*—Multimedia watermarking started to take place as a hot topic in the hi-tech arena during the last two decades. Proper evaluation of the robustness of different watermarking techniques is highly demanded as it represents a trusted feedback to the algorithm designers, which helps them in enhancing their algorithms. This paper proposes a new tool that can effectively test the robustness of the developed algorithms against a list of well known attacks. A set of case studies has been used to demonstrate the capabilities of the tool to categorize various algorithms in terms of their robustness to different attacks.**

I. Introduction

Nowadays, the World Wide Web and the availability of inexpensive high quality recording, playing, and storing devices have made a huge difference in the Digital Media field. It is easy to copy, store or even transmit digital contents over networks without losing quality. However, this lead to huge financial losses to multimedia content publishing companies as a result from unauthorized copy and transmission of contents, and therefore researches and investigations are carried to resolve this problem. The need for protection of media contents strongly contributed in the emergence of different digital watermarking techniques.

The basic idea of digital watermarking is to generate data that contains information about the digital content to be protected such as owner's contact information, name of content, and year of publishing. This data is called a *watermark* and the digital content to be protected is called a *cover work.* The watermarking procedure occurs in two steps, watermark embedding and watermark detection. A secret key is to be used during the embedding and the detection process to assure privacy. After the recovery procedure, the original embedded watermark w and the extracted one w^* are compared in order to verify if any unauthorized user had modified the content.

The following equation defines a similarity measure between the original message w and the extracted one w^*.

$$sim(w, w^*) = \frac{w^*.w}{\sqrt{w^*.w^*}} \tag{1}$$

A watermarking system consists of three main phases, watermark embedding, watermark extraction, and quality evaluation.

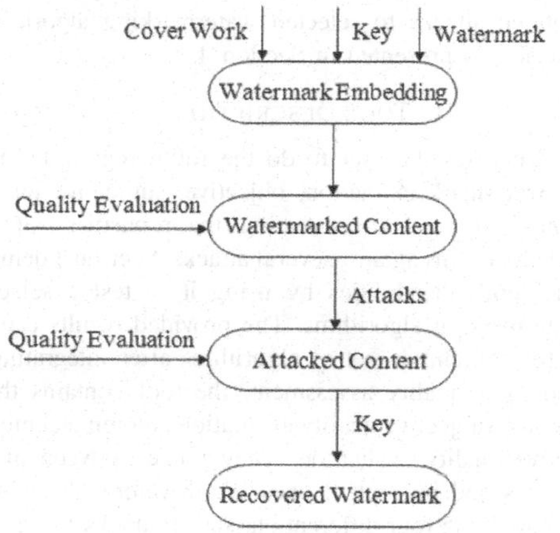

Fig. 1. General Watermarking Scheme

Measuring the quality of the processed visual data can estimate how much quality degradation it faces. The quality evaluation phase in a watermarking system intends to compare the behavior of the developed systems. It consists of three sub phases which are subjective quality evaluation, objective quality evaluation, and robustness against attacks.

An attack is any processing that may intentionally or unintentionally destroy the embedded information [1]. One of the main factors a watermark should have is to be robust against attacks. The watermarked content should survive any processing that may destroy the watermark. If some dispute occurs regarding the ownership of the digital content, and the watermark is lost, then the publisher cannot prove his ownership to the content. During our development, we were keen on making the watermark robust to attacks. This is done

K. Elleithy (ed.), *Advanced Techniques in Computing Sciences and Software Engineering*,
DOI 10.1007/978-90-481-3660-5_11,

by applying different attacks on a watermarked image and then calculate the response between the original and extracted watermark to the attack applied using the similarity measure.

A variety of attacks can be found in [2], [3], [4]. They are classified into two main categories: Intentional and Unintentional attacks. For the intentional ones, the major goal of an attacker is to succeed in altering the media content in a way that removes the watermark with a minimum effect on the quality of the media itself. Unintentional attacks are the ones that are being applied to a media content for the purpose of processing, transmission, and storage. In this case, a watermark is said to be lost or destroyed, unintentionally.

In this paper, a tool is presented that can be used as a stand-alone application that enables developers of watermarking algorithms to evaluate the robustness of their proposed algorithms and compare it to others in the literature. Using such a tool provides an early judgment of the performance of the proposed algorithms, which results in a faster time-to market. The remaining of this paper is organized as follows: Section 2 describes the tool, its uses, and the main goal behind its development. Section 3 introduces the results and analysis of the applied attacks to selected watermarking algorithms. The conclusion is presented in Section 4.

II. Tool Description

The tool enables the user to do the following: watermark insertion, watermark extraction, objective, subjective quality assessment, and as well as testing the robustness of the developed algorithms against several attacks. Section 3 demonstrates the tool's capabilities by using it to test a selected set of watermarking algorithms. The provided results can be obtained for any chosen set of algorithms after integrating it to the tool. For quality assessment, the tool contains three techniques for subjective quality evaluation and ten techniques for objective quality evaluation. They range between pixel-based metrics and perceptual ones. Furthermore, the tool is capable of applying four different classes of attacks for testing the robustness of the developed algorithms. This includes:

A. Applied Attacks

1) Image Enhancement Techniques: Those techniques tend to improve the perception of information in image for human viewers, or to provide better input for other steps in a multi-phase vision system. This includes stretching, shrinking, or sliding the range of intensities in an image to cover another desired range. It can be clearly noticed when analyzing the histogram of the image.

2) JPEG Compression: It is a lossy compression technique where the reconstructed image will never be identical but somehow similar to the original image. This can be controlled by the Quality Level (QL). In our test, we used five different quality levels of compression. They are characterized by a number of coefficients used in the reconstruction process in an image of size 512×512, with JPEG 1 meaning the lowest quality level by a number of 50 thousands coefficients, JPEG 2 by a number of 75 thousands coefficients, JPEG 3 by a number of 120 thousands coefficients, JPEG 4 by a number of 175 thousands coefficients, while JPEG 5 is the highest one by a number of 200 thousands coefficients.

3) Noise Addition: It is adding a certain amount of noise to an image in order to reduce its quality. Different types of noise differ on whether they are dependent on the image itself as the Gaussian Noise or independent as the Salt and Pepper Noise where the corrupted pixels are changed into zeros and ones.

4) Low Pass Filtering: The aim is to reduce the amount of noise in an image by convolving it with a filter function. It can be done in the spatial or frequency domain. It is performed by applying a kernel of size ($M \times N$) to an image and then calculating the mean or median between the pixels of the image and the elements of the applied kernel. The configurable parameter is the kernel size which is preferably 3x3 or 5x5.

A constant threshold T may be deduced based on the obtained results for differentiating between a preserved watermark and a destroyed one (in our case, T is 4.1 and the maximum similarity measure reached is 4.2533). Strength, amount, location of the embedded information are the same throughout all the experiments. It is strictly put in consideration, that any modification of those three parameters may give different results.

Fifteen implemented attacks (**Table** 1) are applied as shown below in Figure 2:

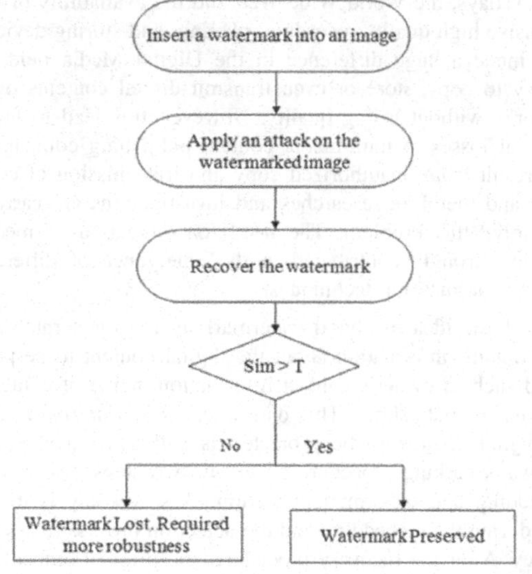

Fig. 2. General Robustness Testing Scheme

III. Results and Analysis

This section demonstrates the tools capabilities by showing its performance with a set of case studies involved in the comparison which are (but not limited to) Cox, Hsu, Benham, and

Koch [5], [6], [7], [8]. They are all DCT Based Techniques. Moreover, this section shows results of the similarity measures for all applied attacks and several chosen snapshots of the recovered watermarks after being attacked. The test images used in our study are *Peppers* and *Lena*. Both images are gray scale with 512×512 pixels. The following watermark (http://www.vu.union.edu/ shoemakc/watermarking/) is used as the information to be embedded and extracted for analysis.

Copyright

Fig. 3. The Original Watermark

A. Robustness to Enhancement Techniques

As shown in Table 1, all algorithms are robust to this kind of attack. The algorithms that showed a high degree of similarity are Koch and Benham, followed by Hsu. Some trials were carried out during the tests that include decreasing the embedding strength, thus improving the visual quality. As a result, the robustness decreased along with a decrease in the similarity measures. Figures $8(a)$ and $8(b)$ show the recovered watermarks when shrinking the histogram of an Hsu and Koch watermarked Lena image. The similarity measure for Koch algorithms is higher than that of the Hsu.

B. Robustness to JPEG Compression

Although Visual quality of Koch's and Benham's watermarked images decreased, yet they show great performance in all levels for both images. Cox failed to survive most of the compression levels. Figures $8(c)$ and $8(d)$ show recovered watermarks after applying JPEG Level 4 compression on Hsu and Koch watermarked Lena image. The similarity measure for Koch is higher than that of the Hsu and hence better quality for the recovered watermark.

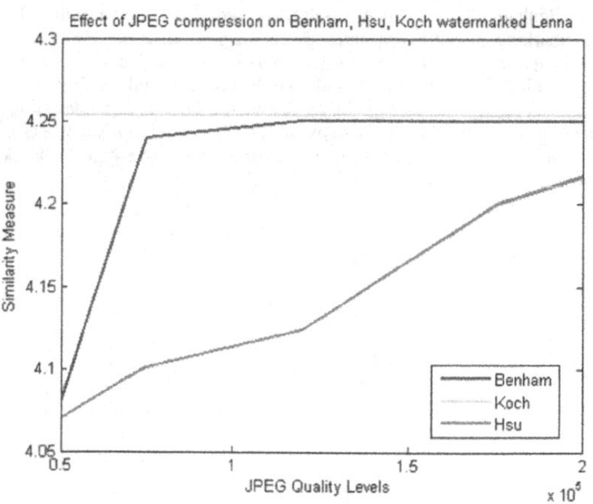

Fig. 4. JPEG Quality Levels vs. Similarity Measure for Lena

Figures 4 and 5 show a relation between different JPEG Quality Levels and Similarity Measures for Lena and Peppers. Both graphs were generated from the tool which later can be used for comparisons with different algorithms and images. Koch showed nearly constant high performance with all quality levels, however, Benham failed to survive the first quality level and as approaching higher quality levels, the extracted watermark was successfully preserved. Finally, Hsu was a moderate case, perfect visual quality that survived three different levels.

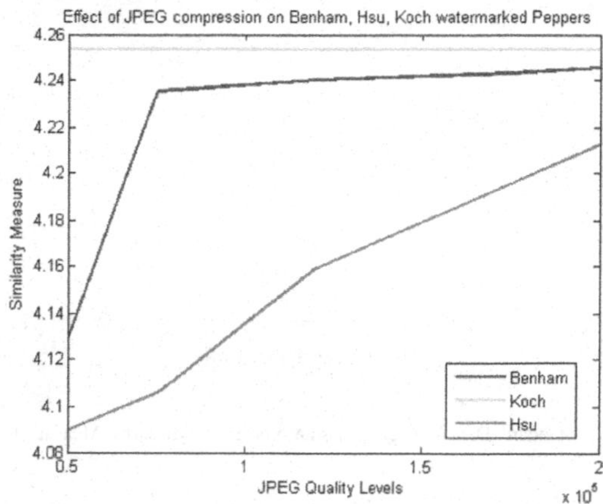

Fig. 5. JPEG Quality Levels vs. Similarity Measure for Peppers

C. Robustness to Noise Addition

Cox algorithm showed high degree of similarity when extracting the watermark and comparing it with the original one. The noisy nature of a Cox watermarked image is behind that performance. However, decreasing the embedding strength, lead to the total destruction of the watermark. Koch performed well in some noise addition attacks; however, Benham did not pass except for "Poisson Noise"since it contains the least amount of noise compared to other types. Moreover, Hsu failed to pass all noise addition tests. Figures $8(e)$ and $8(f)$ show the recovered watermarks when applying the Poisson Noise Addition attack for Lena image. The two algorithms used in watermarking are Benham and Koch. This agrees with the difference in similarity measures in Table 1.

Gaussian Noise Addition attack was chosen for comparison between different algorithms where the configurable parameter is the standard deviation of the noise added. Figures 6 and 7 show a relation between different values of Standard Deviation and Similarity Measures for Lena and Peppers. Both graphs were generated from the tool which later can be used for comparisons with different algorithms and images. Cox showed nearly constant high performance with different Standard Deviations for Lena and Peppers. Since watermarked Koch and Benham images contain acceptable amount of noise, therefore, they show moderate performance as approaching higher values of Standard Deviations in both images. Hsu with the least amount of noise after watermarking, failed to survive any of the Gaussian Noise Addition attacks. To sum up, this proves a direct relation between noise added during watermarking and the one that is added during an attack.

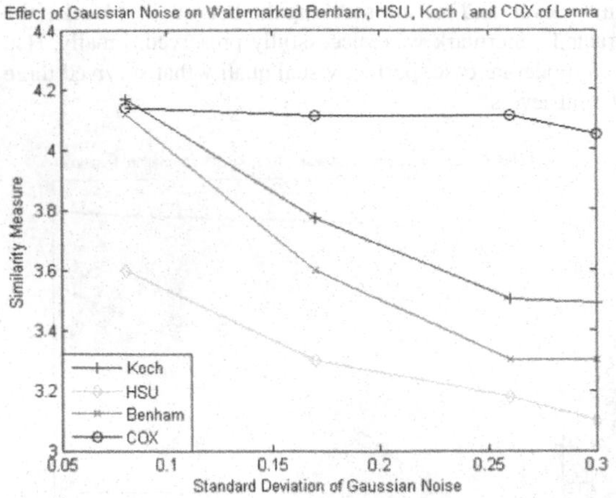

Fig. 6. Standard Deviation of Gaussian Noise vs. Similarity Measure for Lena

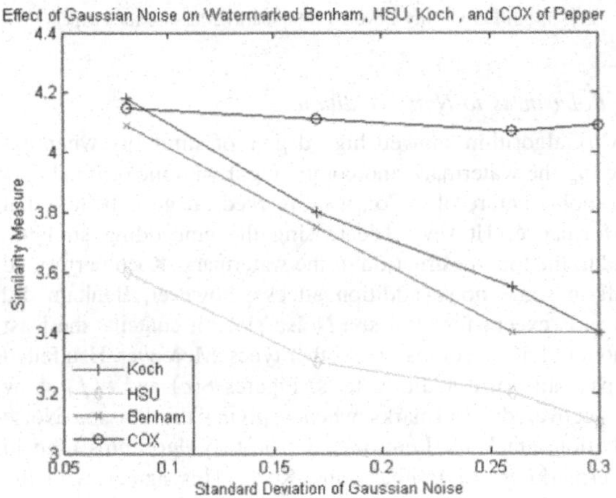

Fig. 7. Standard Deviation of Gaussian Noise vs. Similarity Measure for Peppers

D. Robustness to Low Pass Filtering

Benham and Koch acted in a robust manner towards low pass filtering attacks. Koch showed a high performance for all filter types and closely reached the maximum value a similarity measure can take. This is due to the noisy nature a watermarked Benham and Koch has. Moreover, Cox passed the "Median Filter"attack, while, Hsu performed poorly against it. As shown in Figures $8(g)$ and $8(h)$, the visual quality of the recovered watermarks are nearly the same. This is proven by the nearly equal values of the similarity measures in Table 1.

IV. Conclusion

A tool that is capable of testing the robustness of several selected watermarking algorithms is developed. The tool applies several intentional and unintentional attacks on the watermarked image, extract the distorted watermark, and then calculate the similarity measure between the original watermark and the extracted one. This tool can be used as a stand-alone application for watermark embedding, extraction, and quality evaluation. This includes subjective and objective quality assessment. Any algorithms that need to be inspected for robustness can be embedded into the tool and then tested using the same testing scheme described above.

Acknowledgment

The authors would like to thank the German University in Cairo (GUC) and the Advanced Technology Information Processing Systems (ATIPS) Laboratory for supporting this research.

References

[1] Ton Kalker. Basics of watermarking. Philips Research and University of Eindhoven, February 2004.

[2] Stefan Katzenbeisser and Fabien A. P. Petitcolas. *Information hiding techniques for steganography and digital watermarking*. Artech House, USA, 2000.

[3] Ross J. Anderson, Fabien A. P. Petitcolas, Markus G. Kuhn. Attacks on copyright marking systems. In *Proceedings of the 2nd International Workshop on Information Hiding*, pages 218–238, Springer-Verlag, Portland, Oregon, USA, April 1998.

[4] Martin Kutter and Fabien A. P. Petitcolas. A fair benchmark for image watermarking systems. In *Proc. SPIE Security and Watermarking of Multimedia Contents*, volume 3657, pages 226–239, San Jose, CA, USA, January 1999.

[5] Joe Kilian, Ingemar J. Cox, Tom Leighton, and Talal Shamoon. A secure, robust watermark for multimedia. In *Proceedings of the International Conference on Image, Science, Systems, and Technology (CISST 97)*.

[6] Ja-Ling Wu and Chiou-Ting Hsu. Hidden digital watermarks in images. In *IEEE Transactions on Image Processing*, volume 8, January 1999.

[7] Dave Benham, Nasir Memon, Boon Lock Yeo, and Minerva Young. Fast watermarking of dct-based compressed images. In *Proceedings of the International Conference on Imaging Science, Systems, and Technology*, pages 243–252, Las Vegas, Nevada, USA, June 30 - July 3, 1997.

[8] Jian Zhao and Eckhard Koch. Towards robust and hidden image copyright labeling. In *Proceedings of the IEEE International Workshop on Nonlinear Signal and Image Processing*, pages 452–455, Halkidiki, Marmaras, Greece, June 1995.

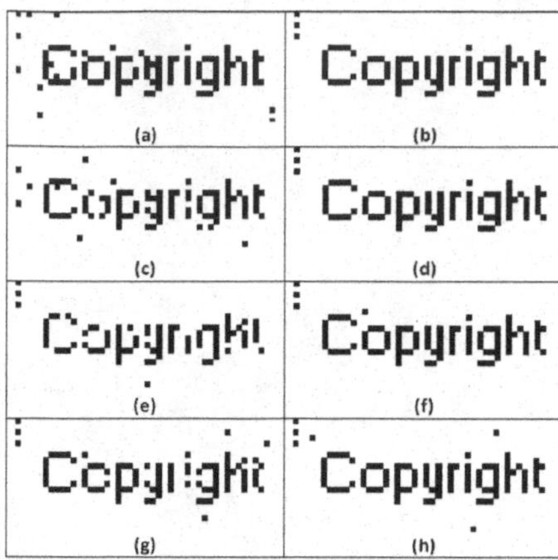

Fig. 8. Recovered watermark after, (a) Shrink attack Hsu (b) Shrink attack Koch (c) JPEG 4 attack Hsu (d) JPEG 4 attack Koch (e) Poisson Noise attack Benham (f) Poisson Noise attack Koch (g) Median attack Benham (h) Median attack Koch, watermarked Lena image

TABLE I
SIMILARITY MEASURES BETWEEN ORIGINAL AND EXTRACTED WATERMARKS OF LENA AND PEPPERS IMAGES

Attacks	Benham	Cox	Hsu	Koch
Stretch	4.2533	4.1584	4.2217	4.2533
Shrink	4.2533	4.138	4.2099	4.2533
Slide	4.25	4.227	4.2397	4.2486
Stretch	4.2432	4.132	4.2133	4.2533
Shrink	4.2456	4.142	4.1744	4.2533
Slide	4.1573	4.08	4.1896	4.2444
JPEG 1	4.08	N/A	4.07	4.2533
JPEG 2	4.24	N/A	4.101	4.2533
JPEG 3	4.25	N/A	4.1236	4.2533
JPEG 4	4.25	4.0255	4.2	4.2533
JPEG 5	4.25	4.0619	4.217	4.2533
JPEG 1	4.13	N/A	4.09	4.2533
JPEG 2	4.2355	N/A	4.1058	4.2533
JPEG 3	4.2405	N/A	4.1596	4.2533
JPEG 4	4.2432	4.0385	4.1956	4.2533
JPEG 5	4.2456	4.1068	4.2124	4.2533
Gaussian	3.9	4.1	3.6	4.09
Salt & Pepper	3.93	4.12	3.5	4.0
Speckle	3.91	4.14	3.4	3.97
Poisson	4.2	4.11	3.7	4.25
Gaussian	3.99	4.12	3.4	4.09
Salt & Pepper	3.94	4.12	3.5	3.95
Speckle	3.95	4.13	3.5	4.03
Poisson	4.19	4.15	3.6	4.24
Gaussian	4.235	4.0	3.5	4.2533
Average	4.209	4.0	2.9	4.2533
Median	4.222	4.118	2.8	4.2465
Gaussian	4.2181	4.08	3.5	4.2533
Average	4.1829	4.05	2.8	4.2506
Median	4.1881	4.1315	2.8	4.2506

Implementation of Efficient seamless non-broadcast Routing algorithm for Wireless Mesh Network

Ghassan Kbar, *Senior Member, IEEE* Wathiq Mansoor, *Member, IEEE*

***Abstract*—Wireless Mesh Networks become popular and are used everywhere as an alternative to broadband connections. The ease of configuration of wireless mesh LAN, the mobility of clients, and the large coverage make it attractive choice for supporting wireless technology in LAN and MAN. However, there are some concerns in assigning the multiple channels for different node and having efficient routing algorithm to route packet seamlessly without affecting the network performance. Multiple channel usage has been addressed in previous research paper, but efficient routing algorithm still to be researched. In this paper an efficient seamless non-broadcast routing algorithm has been developed and implemented in C++ to support the wireless mesh network. This algorithm is based on mapping the mesh wireless routing nodes geographically according to 2 coordinates. Each node will apply this algorithm to find the closet neighboring node that leads to destination based on the mapped network without the need for broadcast such as the one used in traditional routing protocol in RIP, and OSPF.**

***Index Terms*—Routing Algorithm, Wireless Mesh Network, Non-Broadcasting Algorithm, Wireless Routing Protocol.**

I. Introduction

WLAN technology is rapidly becoming a crucial component of computer networks that widely used in the past few years. Wireless LAN technology evolved gradually during the 1990s, and the IEEE 802.11 standard was adopted in 1997 [1]. Companies and organizations are investing in wireless networks at a higher rate to take advantage of mobile, real-time access to information.

Enterprise managers want to deploy wireless networks with several important qualities. These include; high security, highly reliable and available WLANs with very little downtime, and high performance. The ideal wireless network is to have reliability, availability, security, and performance criteria to be similar of wired enterprise networks. In addition, it should be possible to deploy wireless networks very quickly and without the need for extensive and time-consuming site surveys. Furthermore, the networks should have the flexibility needed to support load balance and changes in the radio environment. Radio Resource Management (RRM) forms the basis of Quality of Service (QoS) provisioning for wireless networks [2]. It is an intense research area due to the wireless medium's inherent limitations and the increasing demand for better and cheaper services. Improving the mobility management has been addressed in [3] based on dividing the location management into two levels, intra and inter mobility. This will reduce the amount of signaling traffic, but still didn't address the problem of reliability and availability. Supporting security, reliability and QoS in dynamic environment has been discussed in [4] using modified routing protocol OSPF-MCDS over WLANs. In [5] securing the distributed wireless management has been addressed while supporting reliability and high availability. The above criteria of high security, reliability and availability in evaluating WLAN are important for high performance wireless network. However, easy of configuration, setup and efficient routing algorithm still to be taken care of in wireless network. Multi-hop Wireless Mesh Networks (WMN) address this concern in easily setting up wireless network. This become a popular alternative in extending the typical

Wireless Local Area Networks (WLANs) we use today. Mesh networks consist of heterogeneous wireless clients and stationary wireless mesh routers. Mesh routers are wirelessly linked to one another to form a large scale wireless network to allow mobile clients communicating to each other within large coverage area. Clients in the mesh network can connect to mesh router Access Points in the traditional way. They can also become part of the mesh network by having their traffic forwarded through other wireless access point or client with routing feature. Mesh networks can be used in a variety of applications, including enterprise networking, university campus, Malls, building automation, extending provider service coverage, and wireless community networks [6]. These networks have the advantage of adding new router at any time in proportion to the number of users. In addition, this wireless mesh network can be connected to Internet Service Providers (ISP) to increase the coverage area. Additionally Mesh networks also provide robustness through redundant links between clients and routers.

G. Kbar is with the American University in Dubai, Dubai, UAE (corresponding author to provide phone: 971-4318-3437; fax: 971-4318-3437; e-mail: gkbar@aud.edu).

W. Mansoor was with the American University in Dubai, Dubai, UAE (corresponding author to provide phone: 971-4318-3436; fax: 971-4318-3437; e-mail:wmansoor@aud.edu)..

K. Elleithy (ed.), *Advanced Techniques in Computing Sciences and Software Engineering*, DOI 10.1007/978-90-481-3660-5_12, © Springer Science+Business Media B.V. 2010

The WMN is still to be designed properly to allow easy efficient configuration. Several aspects, such as the number of hops, number of channels, node/antenna placement, efficient routing algorithm, throughput and delay requirements, and other application demands will have an impact on the performance and cost of these networks.

In the past few years, multi-hop approach has been advent to extend network coverage. Many research papers were dedicated to Mobile Ad Hoc Networks (MANETs) with the assumption of using multiple hops to reach far off destinations. Gambiroza et al. [7] used a simulation-based linear topology model. They placed APs in a straight line configuration with multiple clients off of each AP. Thus the farthest AP's client to the sink gets the least amount of throughput. Raman and Chebrolu [8] designed a new MAC protocol for 802.11 networks. Their design uses directional antennas but a single channel for all APs. The new MAC protocol was designed for a mesh backbone. The goal of their work was to increase throughput and lower latency by limiting the collision affects of the IEEE 802.11 MAC protocol. Aguayo et al. [9] deployed and studied an actual 802.11b mesh network. Multiple Channels, Single Hop networks with a slight twist to the regular single hop model have been also researched where there are now commercial products with two or more wireless radios. These radios can be tuned to different frequencies and in-effect gives us a multiple channel network. Proxim Wireless Networks have such an access point [10]. The main advantage of having multiple radios is diversity in channel selection. A user with interference on one channel can tune itself to another for association to the WLAN. Another useful technique is a logical separation of users. With the different MAC addresses on the two radios, an AP can do more sophisticated QoS in admission control, load balancing, etc. D. Multiple Channel, Multiple Hop networks Recently, multiple channel, multiple hop WLANs have become very popular in both academia and industry. Bahl et al. [11] gave a list of standard problems in using multiple channels over a wireless network and experimental results on using multiple channels. Baiocchi et al. [12] simulated a multiple channel wireless network with one control channel and multiple data channels. One very important usage of multiple channels and multiple hops is for Wireless Mesh Networks [6]. There has been much previous work in the simulation of mesh networks [13], [14]. Raniwala and Chiueh [15] designed an architecture for mesh networks and gave some simulation and experimental data on various issues. Navda et al. [16] focused on handoff issues in a mesh network. They chose to use Wireless Distribution System (WDS) links over six hops and compared the handoffs of using Layer 2 (MAC) and Layer 3 (Network) mechanisms. In [17] Leise had designed an efficient WMN with measurement to number of hops, channel allocation and interference, throughput and delay requirement has been addressed. However, they didn't address the issue of efficient routing algorithm which affects the throughput and delay.

An efficient seamless non-broadcast routing algorithm has been developed to support the wireless mesh network [17].In this paper, implementation of this algorithm has been done in C++, where its result shown excellent performance in term of identifying the shortest path without the need for broadcast packets. This algorithm is based on mapping the mesh wireless routing nodes that are geographically spread according to 2 coordinates as described in section 2. In section 3, a flow state diagram illustrates the transition of packets between adjacent cells or node. Each node will apply the algorithm as explained in section 4 to find the closet neighboring node that leads to destination based on the mapped network without the need for broadcast such as the one used in traditional routing protocol in RIP, and OSPF. In section 5, implementation results are illustrated.

II. WMN BASED ON MULTIPLE WIRELESS CELL ROUTING NODES

In order to make an efficient wireless mesh network, network availability, fault tolerance, low interference, high coverage, easy configuration, efficient routing, and high performance would be desirable. To meet this criteria Wireless router APs have been distributed geographically in different cells with minimum coverage overlap as shown in Figure 1. Different frequency channel would be assigned to different APs to avoid interference. These mesh wireless nodes would communicate among each other wirelessly through wireless router at the center of four different APs to form a large mesh network. The protocol used for this communication between the APs, wireless routers and mobile clients is WIFI. Where, mobile clients (Laptops or Mobile phone WIFI-support) would first communicate to the AP that is fallen within its range.

An example mobile A belong to cell $S_{1,1}$ would communicate to other mobile terminal within the same cell $S_{1,1}$ through its AP. However, when A wants to communicate to other mobile terminals within other cells such B (in cell $S_{5,1}$) it will forward the packets to the AP of $S_{1,1}$, which run the routing algorithm described in section 3 to find the next wireless node that lead it to destination (which is $S_{2,1}$ in this scenario). Then the node received this packet ($S_{2,1}$) would identify the next wireless node again by applying the same routing algorithm at $S_{2,1,}$ as explained in section 3, and forward the packet to the next identified node, until packet reaches the last node that contains the destination mobile terminal (which is $S_{5,1}$ in this scenario). In this algorithm the wireless node would only be responsible of finding the next routing node that has the shortest path using its current cell Id and destination cell Id without the need for broadcast such as the one used in traditional routing algorithm in RIP and OSPF. Each cell would be assigned different IP subnet that is mapped to its cell Id. Routing packet based on cell Id would prevent the broadcast which causes low network performance in utilizing the network bandwidth. Note that saving bandwidth in wireless network is critical in improving its performance. In addition it will prevent the looping that might occur in traditional routing such RIP. There is no need to broadcasting packets that might flow through different node and cause looping that is the main cause for low network performance.

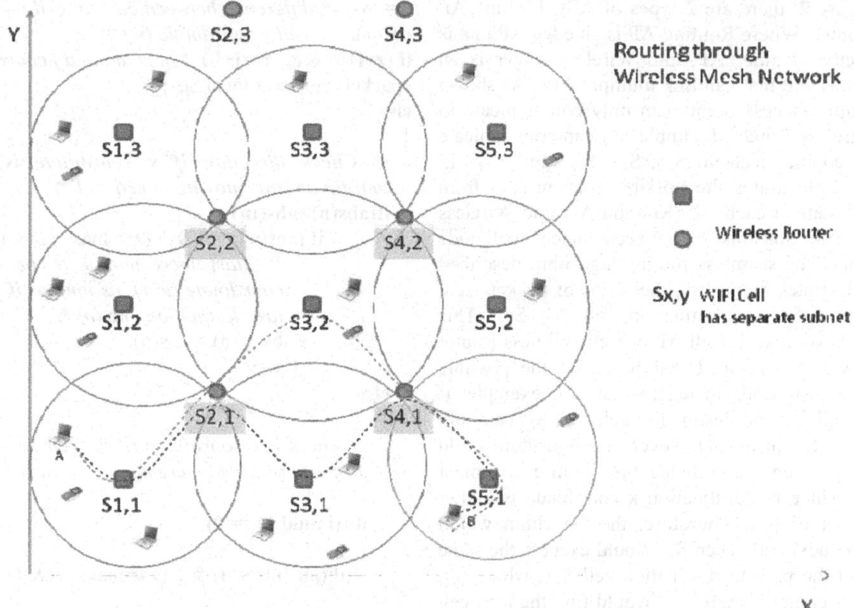

Figure 1, Cells Wireless Mesh Network

III. EFFICIENT WIRELESS ROUTING

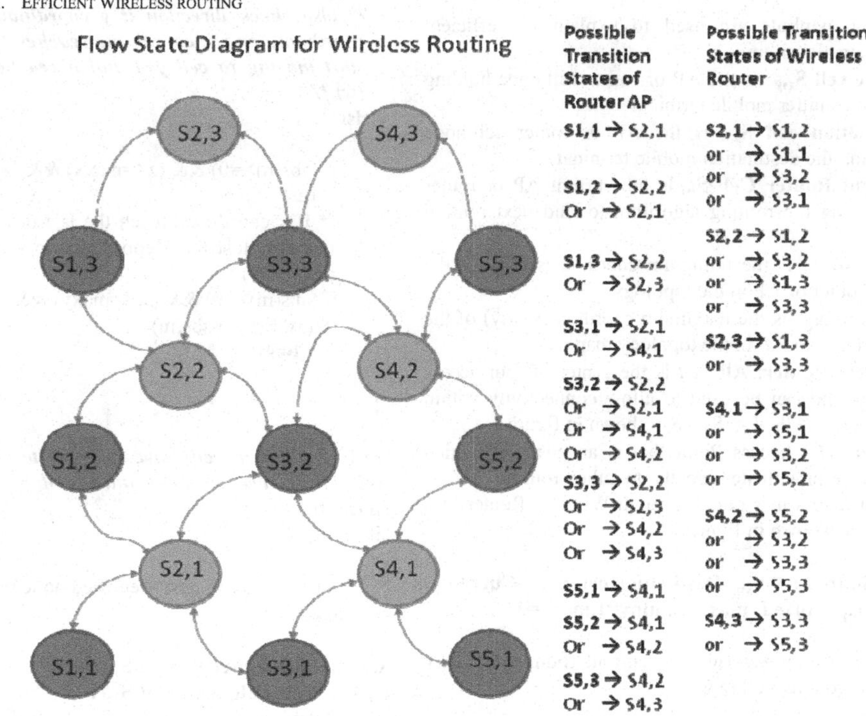

Figure2. Flow State Diagram for Wireless Routing

As shown in figure 2, there are 2 types of APs, Routing AP and Wireless Router. Where Routing AP is an edge AP that is not linked directly to other AP, and Wireless router is an intermediate Router AP that can link multiple APs. As shown in Figure 2, routing access points can only communicate to intermediate Wireless Router. Example $S_{1,1}$ can communicate to $S_{2,1}$. $S_{3,2}$ can communicate to $S_{2,2}$, $S_{2,1}$, $S_{4,1}$, and $S_{4,2}$. In addition, figure 2 illustrates the possible transmission from one state to next state at each AP (Routing AP, and Wireless router). There is only transmission between adjacent cell/node in both directions. The seamless routing algorithm described in section IV illustrates the routing behavior of packets sent from one cell AP $S_{1,1}$ to destination cell AP $S_{5,1}$. This algorithm is executed at each Cell AP or Cell Wireless Router to decide who can be the next cell based on the possible selected cell/node that leads to destination. For example, if packet arrive at cell $S_{2,1}$ and destined to cell $S_{5,1}$, $S_{2,1}$ can only use the path $S_{1,2}$ or $S_{3,2}$ or $S_{3,1}$. However, the algorithm would choose to increment the x coordinate first, before increment the y coordinate since the destination x coordinate is higher than the current x of $S_{2,1}$. Therefore, the algorithm would choose $S_{3,2}$ as the next cell. Then $S_{3,2}$ would execute the same algorithm to find the path to destination cell $S_{5,1}$, where $S_{4,1}$ would be the best choice. Finally $S_{4,1}$ would find the next cell as to be $S_{5,1}$ since x coordinate =5 is the next cell and eventually packet would reach its destination $S_{5,1}$.

IV. Efficient Wireless Routing Algorithm for WMN

The following symbols are used to explain the efficient seamless routing algorithm.

- **Source cell $S_{x,y}$** is the AP or router cell/node holding the transmitter mobile terminal
- **Destination cell $S_{z,w}$** is the AP or router cell/node holding the destination mobile terminal
- **Current Router Cell $S_{i,j}$** is the current AP or router executing the routing algorithm to find next routing cell
- **minx, miny:** is the minimum coordinates (x,y) of the first router or AP in the topology map.
- **maxx, maxy:** is the maximum coordinates (x,y) of the last router or AP in the topology map.
- **Wireless Router**: AP that is the center of four access points and can be used to allow connectivity within each other APs. Eg. $S_{2,1}$, $S_{2,2}$ shown in figure 1.
- **Router AP**: Access Point that is an edge AP which can't communicate directly to other routing Access Point unless it is going through Wireless Router. Eg. $S_{1,1}$, $S_{3,1}$ shown in figure 1.

Algorithm: Source cell $S_{x,y}$ Destination cell $S_{z,w}$ Current Router Cell $S_{i,j}$ minx=1, maxx=5,miny=1,maxy=3.

Analyzing the routing algorithm at current routing cell (i,j), and destination routing cell (z,w):

```
a=0, b=0; /* Temp variables */
n= z-i /* difference between current cell x coordinate (i) and destination cell x coordinate (z) */
m= w-j /* difference between current cell y coordinate (j) and destination cell y coordinate (w) */
if (z=i+a) && (w=j+b) /*start at i and j coordinate*/
{packet remain at local S_i,j }
else
{
  /* Check direction if x coordinate is higher than y coordinate start moving to cell x+1 */
  if(abs(n)>abs(m))
        if (abs(n)>=0) && (z<=maxx) && (z >= minx)
            /* start increment a to be used for next x coordinate (i+1), as long as it doesn't reach the last destination one (n) */
        if abs(++a) > abs(n)
            { a=n}
  else
  {
    /* check if x coordinate (i) is even and still didn't reach last one (n), also increment a for next x coordinate (i+1) */
    if ((i mod 2) ==0)
    {
      if (abs(m)>=0) && (z<=maxx) && (z >= minx)
        ++b;
      if (abs(n)>=0) && (z<=maxx) && (z >= minx)
       if abs(++a) > abs(n)
          { a=n}
    }
    /* else check direction if y coordinate is high than x coordinate and x coordinate reaches the destination (z) start moving to cell y+1until it reaches last coordinate (w) */
    else
    {
      if (abs(n)>=0) && (z<=maxx) && (z >= minx)
       ++a;
      /* if x coordinate reach the last destination coordinate (n), now test the y coordinate to start increment b for next (y+1) */
      if (abs(m)>=0) && (w<=maxy) && (w >= miny)
      if abs(++b) > abs(m)
         { b=m}
    }
  }

  /* if destination cell move in opposite direction to source cell (i,j), then reset a to -a and b to -b */
  if (n<0) { a=-a}
  if (m<0) {b=-b}
}
/* Send the packet to next identified node/cell
Send packet to S_i+a,j+b
```

Example1 send packet from $S_{1,1}$ **to** $S_{5,1}$

Step1: Execute algorithm at $S_{1,1}$:

```
a=0, b=0, x=1, y=1, i=1, j=1, z=5, w=1, n=5-1=4, m=1-1=0,
if (5=1+0) && (1=1+0) {False} //statement is false
else
{ // Only code executed is shown
```

if(4>1)

if (4>=0) && (5<=5) && (5 >= 1)
{ if abs(1) > abs(4)
{ a=n} //set a to 1 since if statement is false

//a and b have been updated so packet moves to next cell until reach final destination
Send packet to $S_{i+a,j+b} = S_{1+1,1+0} = S_{2,1}$

Step2: Execute algorithm at $S_{2,1}$:
a=0, b=0, x=1, y=1, i=2, j=1, z=5, w=1, n=5-2=3, m=1-1=0,
if (5=2+0) && (1=1+0) {False}
else
{ // Only code executed is shown
if(3>1)
{ if (3>=0) && (5<=5) && (5 >= 1)
{ if abs(1) > abs(3)
{ a=n}
Send packet to $S_{i+a,j+b} = S_{2+1,1+0} = S_{3,1}$

Step3: Execute algorithm at $S_{3,1}$:
a=0, b=0, x=1, y=1, i=3, j=1, z=5, w=1, n=5-3=2, m=1-1=0,
if (5=3+0) && (1=1+0) {False}
else
{ // Only code executed is shown
if(2>1)

{ if (2>=0) && (5<=5) && (5 >= 1)
{ if abs(1) > abs(2)
{ a=n}
Send packet to $S_{i+a,j+b} = S_{3+1,1+0} = S_{4,1}$

Step4: Execute algorithm at $S_{4,1}$:
a=0, b=0, x=1, y=1, i=4, j=1, z=5, w=1, n=5-4=1, m=1-1=0,
if (5=4+0) && (1=1+0) {False}
else
{ // Only code executed is shown
if(2>1)

{ if (4>=0) && (5<=5) && (5 >= 1)
{ if abs(1) > abs(4)
{ a=n}

Send packet to $S_{i+a,j+b} = S_{4+1,1+0} = S_{5,1}$
Step5: Execute algorithm at $S_{5,1}$:
a=0, b=0, x=1, y=1, i=5, j=1, z=5, w=1, n=5-5=0, m=1-1=0,
If (5=5+0) && (1=1+0)
{Packet reaches destination network}

Example2 send packet from $S_{5,2}$ to $S_{1,1}$
Step1: Execute algorithm at $S_{5,2}$:
a=0, b=0, x=5, y=2, i=5, j=2, z=1, w=1, n=z-i=1-5=-4, m=w-j=1-2=-1,
if (1=5+0) && (1=2+0) {False}
else
{ // Only code executed is shown
elseif(4>1)

{ if (4>=0) && (1<=5) && (1 >= 1)
{ if abs(1) > abs(4)
{ }
if (n<0) { a=-a} becomes If (-3<0) { a=-a=-1}
if (m<0) {b=-b} becomes If (-1<0) { b=-b=0}
Send packet to $S_{i+a,j+b} = S_{5-1,2-1} = S_{4,2}$

Step2: Execute algorithm at $S_{4,2}$:
a=0, b=0, x=5, y=2, i=4, j=2, z=1, w=1, n=z-i=1-4=-3, m=w-j=1-2=-1,
if (1=4+0) && (1=1+0) {False}
else
{ // Only code executed is shown
elseif(3>1)

{ if (3>=0) && (1<=5) && (1 >= 1)
{ if abs(1) > abs(3)
{ }
}
}
if (n<0) { a=-a} becomes If (-3<0) { a=-a=-1}
if (m<0) {b=-b} becomes If (0<0) { }
Send packet to $S_{i+a,j+b} = S_{4-1,1+0} = S_{3,2}$

Step3: Execute algorithm at $S_{3,2}$:
a=0, b=0, x=5, y=2, i=3, j=2, z=1, w=1, n=z-i=1-3=-2, m=w-j=1-2=-1,
if (1=3+0) && (1=1+0) {False}
else
{ // Only code executed is shown
elseif(2>1)

{ if (2>=0) && (1<=5) && (1 >= 1)
{ if abs(1) > abs(2)
{ }
if (n<0) { a=-a} becomes If (-2<0) { a=-a=-1}
if (m<0) {b=-b} becomes If (0<0) { }
Send packet to $S_{i+a,j+b} = S_{3-1,1+0} = S_{2,2}$

Step4: Execute algorithm at $S_{2,2}$:
a=0, b=0, x=5, y=2, i=2, j=2, z=1, w=1, n=z-i=1-2=-1, m=w-j=1-2=-1,
if (1=2+0) && (1=1+0) {False}
else
{
if (i mod 2) ==0 True
if (abs(-1)>=0) && (z<=maxx) && (z >= minx)
if abs(++a) > abs(n)
{ a=n}

}
if (n<0) { a=-a} becomes If (-1<0) { a=-a=-1}
if (m<0) {b=-b} becomes If (0<0) { }
Send packet to $S_{i+a,j+b} = S_{2-1,1+0} = S_{1,2}$

Step5: Execute algorithm at $S_{1,2}$:
a=0, b=0, x=5, y=2, i=1, j=2, z=1, w=1, n=z-i=1-1=0, m=w-j=1-2=-1,
if (1=2+0) && (1=1+0) {False}

```
else
  If (i mod 2) ==0  False
      if (abs(n)>=0) && (z<=maxx) && (z >= minx)
      if abs(++a) > abs(n)
            { a=n}
  else
      {
      if (abs(n)>=0) && (z<=maxx) && (z >= minx)
      ++a;

      if (abs(-1)>=0) && (w<=maxy) && (w >= miny)
      if abs(++b) > abs(m)
          { b=m}
      }
If (n<0) { a=-a} becomes If (-1<0) { a=-a=-1}
 If (m<0) {b=-b} becomes If (0<0) { }
```

Send packet to $S_{i+a,j+b} = S_{2+1,2-1} = S$

Step6: Execute algorithm at $S_{2,1}$:

a=0, b=0, x=5, y=2, i=2, j=1, z=1, w=1, n=z-i=1-2=-1, m=w-j=1-1=0,

```
if (1=2+0) && (1=1+0) {False}
else
    {
    If (i mod 2) ==0  True
      if (abs(1)>=0) && (z<=maxx) && (z >= minx)
       if abs(++a) > abs(n)
             { a=n}

    }
 If (n<0) { a=-a} becomes If (-1<0) { a=-a=-1}
 If (m<0) {b=-b} becomes If (0<0) { }
```

Send packet to $S_{i+a,j+b} = S_{2-1,1+0} = S_{1,1}$

Step7: Execute algorithm at $S_{1,1}$:

a=0, b=0, x=5, y=2, i=1, j=1, z=1, w=1, n=z-i=1-1=0, m=w-j=1-1=0,

```
if (1=1+0) && (1=1+0)
   {Packet reaches destination network}.
```

V. Implementation

A simulation program written in C++ has been developed to test the proposed algorithm. The test result shows an excellent performance with regards to the shortest path been chosen, and without the need for high overhead. The small overhead is due to the fact that the proposed algorithm is not using broadcasting to advertise its table like in RIP and OSPF. In fact each cell or node would execute the algorithm to find the next cell that leads it to destination.

Scenario1 *(medium mesh network and traffic from lower cell to upper cell)*- The program developed has been run for a mesh network that has multiple cells start from $S_{1,1}$ up to cell $S_{10,10}$. In this example a mobile terminal belong to Cell $S_{2,3}$ wants to communicate with other mobile terminal belong to cell $S_{8,9}$. By running the developed program based on the above algorithm explained in section3, the following result illustrate the path found to send packet to the right destination:

D:\test\routing\Debug>routing.exe
type the Max of x
10
type the Max of y
10
type the xth of current node
2
type the yth of current node
3
type the zth of destination node
8
type the wth of destination node
9
Send S2,3 to S8,9
send packet to S3,4
send packet to S4,5
send packet to S5,6
send packet to S6,7
send packet to S7,8
send packet to S8,9packet reaches its destination S8,9

It is clear from the above result that the path taken the packet from $S_{2,3}$ to destination $S_{8,9}$ is: $S_{3,4}$→ $S_{4,5}$ → $S_{5,6}$ → $S_{6,7}$ → $S_{7,8}$ → $S_{8,9}$.

Scenario2 *(large mesh network and traffic from upper cell to lower cell)*- The program developed has been run for a mesh network that has multiple cells start from $S_{1,1}$ up to cell $S_{15,15}$. In this example a mobile terminal belong to Cell $S_{13,14}$ wants to communicate with other mobile terminal belong to cell $S_{1,2}$. By running the developed program based on the above algorithm explained in section3, the following result illustrate the path found to send packet to the right destination:

D:\test\routing\Debug>routing.exe
type the Max of x
15
type the Max of y
15
type the xth of current node
13
type the yth of current node
14
type the zth of destination node
1
type the wth of destination node
2
Send S13,14 to S1,2

send packet to S12,13
send packet to S11,12
send packet to S10,11
send packet to S9,10
send packet to S8,9
send packet to S7,8

send packet to S6,7
send packet to S5,6
send packet to S4,5
send packet to S3,4
send packet to S2,3
send packet to S1,2packet reaches its destinationS1,2

It is clear from the above result that the path taken for the packet from $S_{13,14}$ to destination $S_{1,2}$ is: $S_{12,13} \rightarrow S_{11,12} \rightarrow S_{10,11} \rightarrow S_{9,10} \rightarrow S_{8,9} \rightarrow S_{7,8} \rightarrow S_{6,7} \rightarrow S_{5,6} \rightarrow S_{4,5} \rightarrow S_{3,4} \rightarrow S_{2,3} \rightarrow S_{1,2}$

As shown in scenario 1 and 2, the implementation of seamless routing algorithm prove working to find the optimum path with low overhead. This low overhead is achieved through mapping the AP to cell ID, and applying the developed algorithm. This algorithm work independently at each node without the need for broadcast which has a downside effect on the wireless network performance.

VI. Conclusion

An excellent algorithm has been proposed for non-broadcast Routing algorithms for wireless mesh networks. This algorithm is based on distributed routing protocol that could find the best path with mesh wireless networks without the need for broadcasting as in other methods. A simulation program has been designed and tested using C++. The testing proved the excellent performance of the proposed algorithm. The efficient and better performance is due to the fact that the algorithm avoids broadcasting and using novel routing algorithm for forwarding the packets to the shortest available path. For future work, a more sophisticated simulation program needed to compare the performance of the proposed algorithm with the existing algorithms to further investigate for an improved algorithm if needed. In addition, further work would be done to dynamically map the new APs intended to join the existing network and to address fault tolerance issue in the case of a failure in one AP.

References

[1] IEEE 802.11, "Wireless LAN Medium Access Control (MAC) and Physical Layer (PHY) Specifications," 1997.

[2] Kyriazakos, S., Karestos, G.,"Practcial Resource Management In Wireless Systems,". *Book Reviews/Edited by Andrzej Jajszczyk IEEE Communications Magazine,* vol. 42, no. 11, pp. 12 – 14, , Nov. 2004.

[3] Ush-Shamszaman Z., Samiul Bashar Md., Abdur Razzaque M., Showkat Ara S., Sumi J. K. "A Mobility Management Scheme in All-IP Integrated Network," A Mobility Management Scheme in All-IP Integrated Network. *in Proceedings of the 23rd IASTED International Multi-Conference Parallel and Distributed Computing And Networks,* Innsbruck, Austria, pp. 32-37, Feb. 2005.

[4] DaSilva, L.A., Midkiff, S.F., Park, J.S., Hadjichristofi, G.C., Davis, N.J.; Phanse, K.S., Tao Lin "Network Mobility and Protocol Interoperability in Ad Hoc Networks", IEEE *Communications Magazine,* vol. 42, no. 11, pp. 88 – 96, , Nov. 2004.

[5] G. *Kbar, W. Mansoor presented paper "Securing the Wireless LANs against internal attacks". The 3rd International Conference on Mobile Ad-hoc and Sensor Networks (MSN 2007), 12-14 December 2007, Beijing, China*

[6] I. F. Akyildiz, X. Wang, and W. Wang, "Wireless mesh networks: A survey," 2005. [Online]. Available: http://www. ece.gatech.edu/research/labs/bwn/mesh.pdf

[7] V. Gambiroza, B. Sadeghi, and E. W. Knightly, "End-to-end performance and fairness in multihop wireless backhaul networks," in MobiCom '04: Proceedings of the 10th annual international conference on Mobile computing and networking. New York, NY, USA: ACM Press, 2004, pp. 287–301.

[8] B. Raman and K. Chebrolu, "Revisiting mac design for an 802.11-based mesh network," in Third Workshop on Hot Topics in Networks, 2004. [Online]. Available: http: //www.cse.iitk.ac.in/users/braman/papers/2p.pdf

[9] D. Aguayo, J. Bicket, S. Biswas, G. Judd, and R. Morris, "Link-level measurements from an 802.11b mesh network," in SIGCOMM '04: Proceedings of the 2004 conference on Applications, technologies, architectures, and protocols for computer communications. New York, NY, USA: ACM Press, 2004, pp. 121–132.

[10] P. W. Networks, "Orinoco ap-2000 access point," 2005. [Online]. Available: http://www.proxim.com/products/wifi/ap/ ap2000/11

[11] P. Bahl, A. Adya, J. Padhye, and A. Walman, "Reconsidering wireless systems with multiple radios," SIGCOMM Comput. Commun. Rev., vol. 34, no. 5, pp. 39–46, 2004.

[12] A. Baiocchi, A. Todini, and A. Valletta, "Why a multichannel protocol can boost ieee 802.11 performance," in MSWiM '04: Proceedings of the 7th ACM international symposium on Modeling, analysis and simulation of wireless and mobile systems. New York, NY, USA: ACM Press, 2004, pp. 143–148.

[13] J. So and N. H. Vaidya, "Multi-channel mac for ad hoc networks: Handling multi-channel hidden terminals using a single transceiver," in MobiHoc '04: Proceedings of the 5th ACM international symposium on Mobile ad hoc networking and computing. New York, NY, USA: ACM Press, 2004, pp. 222–233.

[14] J. Zhu and S. Roy, "802.11 mesh networks with two-radio access points," in IEEE ICC '05: IEEE International Conference on Communications, 2005.

[15] A. Raniwala and T.-c. Chiueh, "Evaluation of a wireless enterprise backbone network architecture," in High Performance Interconnects, 2004. [Online]. Available: http://www.ecsl.cs. sunysb.edu/tr/hyacinth-hoti.pdf

[16] V. Navda, A. Kashyap, and S. Das, "Design and evaluation of imesh: an infrastructure-mode wireless mesh network," in WOWMOM, 2005'

[17] G. Kbar, W. Mansoor, "Efficient Seamless packet routing in Wireless LAN", The second IFIP International Conference on new Technologies, Mobility and Security NTMS 2008, Tanjier Morocco 5-7 November 2008

Investigations into Implementation of an Iterative Feedback Tuning Algorithm into Microcontroller

Grayson Himunzowa

University of Cape Town, Faculty of Engineering & the Built Environment, Rondebosch 7701, South Africa

Email: ghimunzowa@fastmail.fm

Keywords: Implementation of an IFT, Microcontroller Hardware

Abstract
In this paper, implementation of an Iterative Feedback Tuning (IFT) and Myopic Unfalsified Control (MUC) algorithms into a microcontroller is investigated. First step taken was to search for a suitable hardware to accommodate these complex algorithms. The Motorola DSP56F807C and ARM7024 microcontrollers were selected for use in the research. The algorithms were coded in the C language of the respective microcontrollers and were tested by simulation of the DC Motor models obtained from step response of the motor.

1. Introduction

A vast number of IFT and MUC applications currently existing and described in literature are a motivating factor for investigations into the feasibility of implementing IFT and MUC algorithms on a microcontroller as this can lead to the development of a product for industrial use. Some of these applications are outlined in [1-9]. All these applications indicate the wide usage of IFT and MUC techniques suggesting that there is great demand for this technique. But no mention of IFT or MUC hardware development was made so far and this was the main motivating factor of carrying out the research.

A particular case of industrial interest is tuning PI or PID controller since classical approaches contain a number of fundamental problems, such as: The amount of offline tuning required; assumption on the plant structure; the issue of system stability, and the difficulties in dealing with nonlinear, large time delayed and time variant plant. To overcome these problems, it has been proposed that an IFT microcontroller based hardware be developed since IFT has the capacity [1, 2] to mitigate problems highlighted above.

2. Iterative Feedback Tuning and Myopic Unfalsified Control Theory

IFT is a method for tuning parameterized controllers in a feedback loop when a mathematical description of a plant is not available and the controller must be tuned on the basis of input-output measurements. In IFT, tuning of the controller parameters is performed through an iterative procedure where a sequence of parameter updates is calculated and implemented to check the performance until there is convergence in parameter update. While in MUC, input and output data are measured and compared with performance specification to falsify or unfalsify that controller.

2.1 The Idea of Iterative feedback Theory

The concept of the iterative feedback tuning algorithm is illustrated in the block diagram in figure 2.1 [1]. The controller k is type 1 (and represents a one degree of freedom transfer function), g is an unknown model representing the plant and m is the desired model.

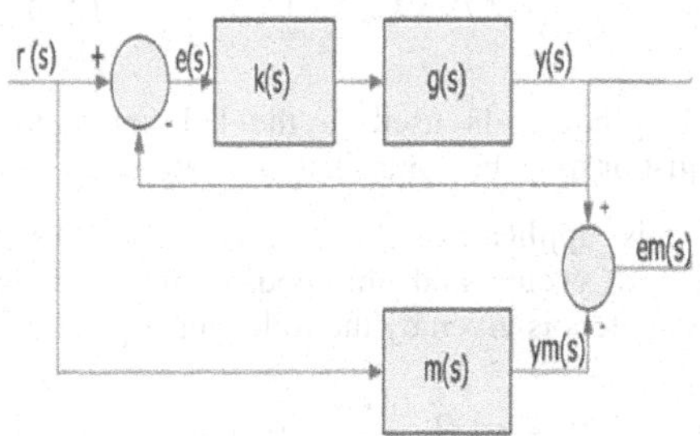

Figure 2.1 Block diagram of a one degree of freedom controller in closed loop system

K. Elleithy (ed.), *Advanced Techniques in Computing Sciences and Software Engineering*,
DOI 10.1007/978-90-481-3660-5_13,

1) The Dynamic Models

The equations governing the closed loop dynamics (assuming the noise and disturbances are zero) are

$$y=Tr \tag{2.1}$$

where T is the transfer function of the closed loop system.

Equations for desired model for open loop system yields:

$$y_m = mr \tag{2.2}$$

$$e_m = y - y_m \tag{2.3}$$

2) The Cost Function

The cost function is a scalar based and is given by

$$J = \frac{1}{2}\sum_{1}^{N} e_m^2(t) \tag{2.4}$$

The model for a controller, k in figure 2.1 is given as

$$k(z) = \frac{\rho_1 z + \rho_0}{(z-1)} \tag{2.5}$$

This was chosen for the sake of simplicity, and to save memory as this is one of the key constraints assumed for the present research project. Secondly, the proportional-plus-Integral (PI) controller is commonly used in industry.

The gradient of cost function criterion is represented by:

$$\frac{\partial J}{\partial \rho} = -\sum_{1}^{N} e_m * \frac{\partial e_m}{\partial \rho} \tag{2.6}$$

$$\frac{\partial y}{\partial \rho} = \frac{1}{k} * \left[T * r - T^2 * r\right] \tag{2.7}$$

This gradient is used in the IFT method to adjust or tune the controller parameters.

In this application the controller parameters form a vector and the update of controller parameters is given by the following equation:

$$\rho_{i+1} = \rho_j - \gamma_j R_j^{-1} \frac{\partial J}{\partial \rho}(\rho_j) \tag{2.8}$$

3) Input signals for the one degree of freedom controller

From the IFT control loop shown in figure 2.1 the process input is obtained from the controller as shown in the transfer function equation:

$$u = k*(r-y) \tag{2.9}$$

The gradient of the input with respect to the controller parameter vector is given by:

$$\frac{\partial u}{\partial \rho} = \frac{1}{k} * \frac{\partial k}{\partial \rho} * u^{(2)} \tag{2.10}$$

The input gradient is used together with output gradient to achieve controller parameter optimization. The superscript (2) indicates the control signal, u existing in procedure 'experiment2'.

2.2 Myopic Unfalsified Control Theory (MUC)

As stated above, MUC and IFT are related in some sense. Both employ a 'myopic' gradient-based steepest descent approach to parameter optimization. MUC is simpler and easier to implement as compared to IFT [6] and for this reason it was also investigated as to whether it can run on to the DSP56F807C or ARM7024 microcontroller.

The concept of MUC algorithm is illustrated in the block diagram in figure 2.2. As before the controller k is type 1 (and a set pre-defined controllers) controller, and g is unknown model representing the plant. Note that, unlike the IFT of figure 2.1, there is no desired model.

A MUC controller is said to be falsified by measurement information if this information is sufficient to deduce that performance specification (r, y, u) ∈ Tspec ∀ r∈ R would be violated if that controller were in feedback loop. Otherwise, the controller is said to be unfalsified [9].

Unfalsified Control method has only three elements: Data is required for observation. It is data that assess whether performance objective is not met and then the controller is tested for falsification; Candidate controller hypotheses are sets of controllers that are tested for falsification or unfalsification. If the controller is falsified it is discarded. In case of Myopic Unfalsified Control, the controller is not discarded but adjusted in the steepest-descent direction $-\nabla \tilde{J}(\theta,t)$ so that the performance specification, $\tilde{J}(\theta,t)$ tends to decrease whenever the currently active controller parameter vector θ is falsified; Goals may be described as control laws or likened to desired models in IFT. Data are observed to be consistency with goals and if there is any discrepancy then particular controller becomes falsified.

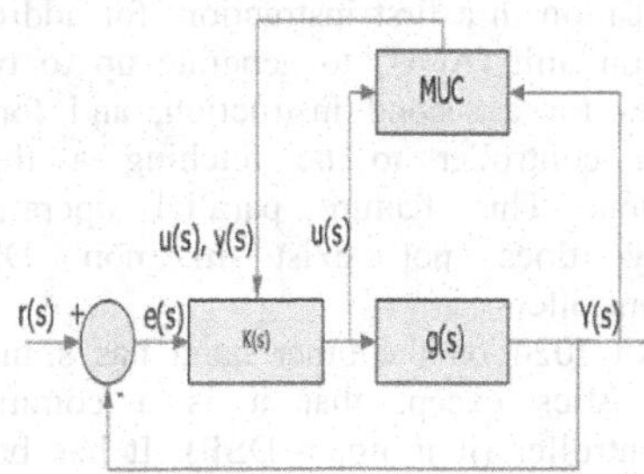

Figure 2.2 Block diagram of a MUC controller in closed loop system

These elements are sieved by the computer so that controllers that falsify data are removed and only those that do not falsify data are kept.

2.3 Myopic Unfalsified Control Algorithm

(1) Measure input, u(t) and output, y(t) from an unknown plant

(2) Use u(t) and y(t) to calculate the cost function, $J(\theta,t)$ thus

$$\tilde{J}(\theta,t) = -\rho + \int T_{spec}(r(\tau),u(\tau),y(\tau))d\tau \qquad (2.11)$$

$\rho \geq 0$ and $T_{spec}(.,.,.)$ are chosen by the designer.

(3) A candidate controller ki becomes unfalsified at time τ by plant data u(t), y(t) if, and only if $\tilde{J}(i,t) \leq 0$.

(4) If the controller ki is falsified then equation 3.11 becomes

$$\nabla\tilde{J}(\theta,t) = \int_0^t T_{spec}(\tilde{r}(\theta,\tau),u(\theta,\tau),y(\theta,\tau)\nabla\tilde{r}(\theta,\tau)d\tau \quad (2.12)$$

$$T_{spec}(\tilde{r}(\theta,\tau),u(\theta,\tau),y(\theta,\tau) =$$

$$|w_1*(r(\tau)-y(\tau))|^2 + |w_2*u(\tau)|^2 - \sigma^2 - |r(t)|^2$$

$$\nabla\tilde{r}(\theta,\tau) = -K(\theta)^{-1}\nabla K(\theta)K(\theta)^{-1}u(\tau)$$

$$\nabla K(\theta) = \left[\frac{\partial K(\theta)}{\partial\theta_1}\frac{\partial K(\theta)}{\partial\theta_2}\cdots\frac{\partial K(\theta)}{\partial\theta_n}\right]^T$$

Operator '*' denotes convolution operation and w_1 and w_2 are filters. In the MUC algorithm implemented onto microcontroller, filters, w_1 and w_2 were ignored.

Therefore controller parameter adaptation equation can be expressed as:

$$\dot{\theta} = \gamma\int_0^t \frac{\partial T_{spec}(r,\tau)}{\partial\tilde{r}}K(\theta)^{-1}\nabla K(\theta)K(\theta)^{-1}u(\tau)d\tau$$

Where $\gamma > 0$ is a design constant that determines the rate of adaptation

(5) If and only if $\tilde{J}(i,t) \leq 0$, then parameter updates otherwise does not update.

2.4 Microcontroller Hardware

In discussing this topic, IFT will be referred to more than MUC since motivation in carrying out this project was as a result of successful results achieved from an IFT work in [1]. Since IFT and MUC have relatively complex algorithms the main task was to identify suitable hardware that could accommodate the algorithms. As given by the title of the project, digital hardware was preferred to analog hardware in that it is impractical to implement such a complex algorithm into an analog hardware. In choosing the hardware the following factors were considered to be critical in ensuring the requirement of IFT algorithm would be met:

(1) Large memory since the IFT visual basic program in [1] had fifteen arrays for data storage and the length of each array was one thousand (1000) elements. These arrays were declared as floats taking four bytes of memory space per element in the array hence a microcontroller with small memory would fail to run the IFT algorithm,

(2) floating point was necessary since a lot of IFT parameters such as error and modeling error. require floating point feature,
(3) high speed of processing since IFT algorithm executes in stages that occupy N-length time periods to collect and process data. Data is first collected and stored in an array for N-length time periods before being passed to the next stage for optimization for N-length time periods. In the case of the IFT visual basic program N was made one thousand (1000) taking considerable amount of time,
(4) a suitable compiler such as C that is capable of handling floating points and high speed of program execution and for ease of coding was preferred to other programming languages,
(5) on board Analog to Digital Converter and Digital to Analog Converter were required to create a hardware feedback loop of an IFT. The output of the motor is fed back to the input of the ADC and the input of the motor is fed by the DAC as will be indicated later.

Such functionality is scarce in common microcontroller hardware. Hence a carefully search for the right hardware with features as shown above was necessary. Criterion used to decide on these features given above for selecting the microcontroller hardware for use in implementation of the IFT was derived from Visual Basic Program used in a previous IFT project [1]. This also included the speed at which IFT ran without causing any problems. This estimate was arrived at by the number of arrays used in the original IFT algorithm coded in Visual basic in [1]. There were about fifteen arrays declared as float variables that take four bytes of memory space each. Considering one thousand iterations required in the original algorithm, means the total memory required was six thousand (6000), four-byte words for all fifteen (15) arrays.

In line with the characteristic requirement of a microcontroller for use in this research derived from above calculation a survey was conducted and the Motorola DSP56F807 and ARM7024 microcontrollers were chosen for use in this research mainly due to its good characteristics as outlined above. The DSP56F807C architecture is a 16-bit multiple-bus processor designed for efficient real-time digital signal processing and general purpose computing hence making it easy to code advanced algorithm such as IFT or MUC. It is composed of functional units that operate in parallel to increase the throughput of the machine hence decreasing the execution time of each instruction. For example it is possible for the data arithmetic logic unit (ALU) to perform a multiplication in a first instruction, for address generation unit (AGU) to generate up to two addresses for a second instruction, and for a program controller to be fetching a third instruction. This feature parallel operation normally does not exist in non DSP microcontrollers.

The ARM7024 on the other hand has similar characteristics except that it is a common microcontroller (it is not a DSP). It has both ADC and DAC imbedded into it.

Research has revealed that microcontrollers are already in use in implementing proportional-integral-differential (PID) controllers. This observation is based on the number of hits (19500 hits) obtained from the Google scholar search engine for the term 'PID' or 'PID and implement'. There were 87 hits for microcontroller based Model Reference Adaptive Control and none for IFT. It is still not known whether a microcontroller implementation of the IFT algorithm does exist indicating that IFT is a new technique still in active research and has not been commercialized yet. Hence, implementation of an IFT algorithm into microcontroller if achieved could lead to the development of an economical and optimal controller. This is an important objective for industrial control and will form a base-case against which other similar techniques could be compared.

3 Simulation of IFT and MUC Algorithms

The simulation of IFT and MUC algorithms were conducted using the ARM7024 microcontroller. The algorithms were coded in C language of the microcontroller.

The IFT control of DC Motor model obtained from the step response of the DC Motor is given in equation (3.1). The graphs of the output response of the said model with its desired model are depicted below to demonstrate the optimization of controller parameters.

The details of various models are being reproduced here for ease of reference.

The process model is

$$g(s) = \frac{17.75}{1+6.7s} \text{ [v/v]} \quad (3.1)$$

The desired response model is

$$m(s) = \frac{1}{1+4s} \text{ [v/v]}$$

The controller was initialized to

$$k(s) = \frac{0.02+0.04s}{s} \text{ [v/v]}$$

and subsequently converged to

$$k(s) = \frac{3.0+3.0s}{s} \text{ [v/v]}$$

The two responses of desired model, ymta in purple and the output response, yta in blue are indicated in figure 3.1. The responses were obtained from the two runs (i.e. a run is a complete cycle of the IFT algorithm) of the IFT algorithm.

3.1 Simulation Results of IFT control of the DC Motor

In the first run Ymta is a fast and damped with a settling time of 10s while yta is slow and damped having a settling time of 54s. the closed loop is approximately 1v. This is true since the heavy disc model has its pole closer to the unit circle in the z-plane compared to the pole of the desired model.

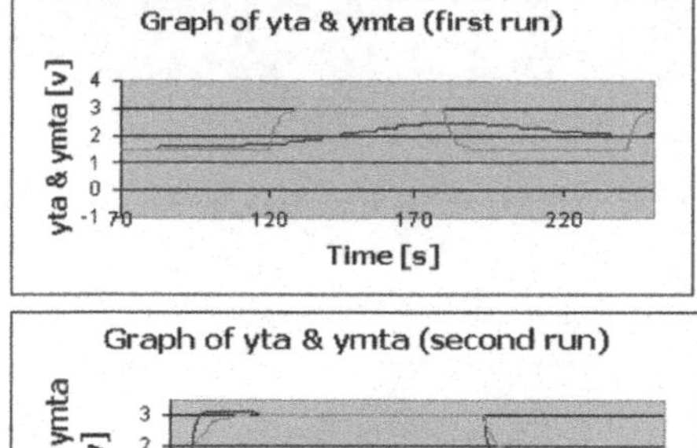

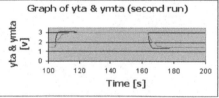

Figure 3.1 Output response, yta & ymta of the DC Motor simulation.

The most important point here is the action of the plant adapting to that of desired model a key concept of the IFT algorithm. This phenomenon begins to manifest itself in the second run of the IFT algorithm because of the action of the IFT. In the second run of the IFT algorithm, yta becomes faster than ymta but settles 4s latter than the desired model (at approximately 14s) an improvement from the response of yta in the first run which takes 54s to settle. This is a demonstration that controller parameters are being optimized.

But it is observed that the model did not converge to the desired model. The reason could be the problem of non-minimum phase pole-zero cancellation that occurs especially when the desired model is fixed and not let to adjust as the controller parameters are being adapted. This phenomenon is easily explained using a PI controller given below:

$$k(s) = \frac{\rho_0+\rho_1 s}{s} \text{ [v/v]} \quad (3.2)$$

Hence

$$k(s)*g(s) = \frac{\rho_0+\rho_1 s}{s} * \frac{1}{1+Ts} \text{ [v/v]} \quad (3.3)$$

Equation (6.2) can result in pole-zero cancellation if $\frac{\rho_1}{\rho_0} = T$ since

$$k(s)*g(s) = \frac{\rho_0(1+\frac{\rho_1}{\rho_0}s)}{s} * \frac{1}{1+Ts} = \frac{\rho_0}{s}$$

[v/v] (3.4)

pushing the pole to the origin therefore making it an integrator and this was the case since the parameters were equal after convergence. This explains the outcome of the response being very fast and not materializes as expected. It entails then that even with many IFT runs the plant model would not have converged to the desired model.

3.2 Simulation Results of MUC control of the DC Motor

Similarly, MUC control of the DC Motor model obtained from the step response of the DC Motor is simulated. The graphs of the output responses of the said model are depicted below. MUC has no desired model in its algorithm. Its convergence is triggered by performance specification, $\tilde{J}(\theta,\tau) \leq 0 \forall \tau \in [0,t)$ when its condition is satisfied. In view of the results that are presented below indicates only the output response, y.

The initial conditions of the simulation for MUC are reproduced here for ease of reference. The process model is

$$g(s)=\frac{17.75}{1+6.7s} \text{ [v/v]}$$

The controller was initialized to

$$k(s)=\frac{10.0+10.0s}{s} \text{ [v/v]}$$

and subsequently converged to

$$k(s)=\frac{4.2+0.3s}{s} \text{ [v/v]}$$

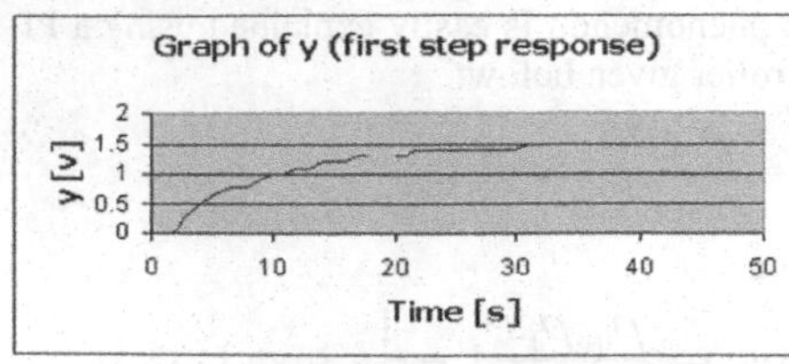

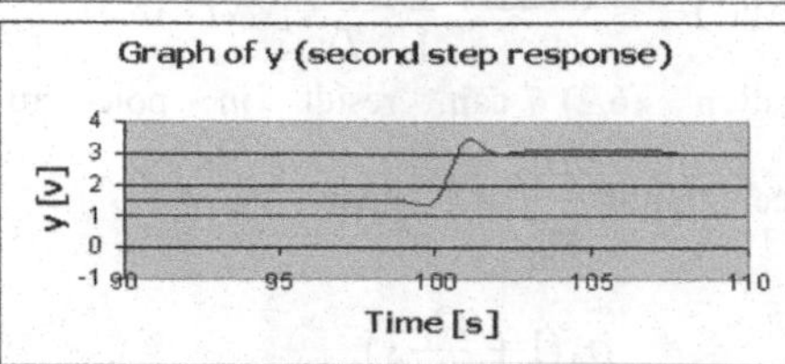

Figure 3.3 MUC Output response, y of the DC Motor simulation

4. Conclusion

This paper has investigated the feasibility of implementing the IFT and MUC algorithms on a microcontroller. The analysis and simulations carried out indicate that both are feasible to be implemented on a microcontroller. MUC converges faster than the IFT.

IFT can make a good automatic controller tuning system as opposed to being applied as a speed control of a DC Motor which is required to run at constant speed.

References

[1] Martin I.Machaba, "Investigations into Iterative Feedback Tuning", Dept. Electrical Eng., Univ. Cape Town, 2004.

[2] Hakan Hjalmarsson, Michel Gevers, Svante Gunnarsson, Olivier Lequin "iterative feedback tuning: Theory and Applications", Control Systems Magazine, v18, pp (26-41), 1998.

[3] K.Hamamoto, T. Fukuda and T. Sugic "Iterative Feedback Tuning of Controllers for two Mass-spring System with friction", Control Engineering Practice, v11, PP (1061-1068), 2002.

[4] A. Karimi, L. MisKovic, D. Bonvin, "Iterative Feedback Tuning correction based controller tuning with application to magnetic suspension system", Control Engineering Practice, v11, PP(1069-1078), 2002.

[5] J. Sjoberg, F.Debruyne, M. Agarwal, B.D.O Anderson, M. Gevers, F.J Kraus and N. Linard, "Iterative Controller Optimization for non linearn systems", Control Engineering

[6] Michael G. Safonov, "The Comparison of Unfalsified Control and Iterative Feedback Tuning", Dept. of Electrical Engineering, University of Southern California, 2002.

[7] M. Jun and M. G. Safonov, "Automatic PID Tuning: An Application of Unfalsified Control", In: proc. of the IEEE int. on CCA/CACSD. PP. 328-333. Honolulu, HI, 1999.

[8] Razavi and Kurfess, "Detection of Wheel and Work piece Contact/Release in Reciprocating Engine", Int. Journal of Machine tools and Manufacture, v43 pp 185-191, 2003.

[9] O. Lequin, M. Gevers, M. Mossberg, E. Bosmans and L.Triest, "Iterative Feedback Tuning of PID parameters: comparison with classical tuning rules", Control Engineering Practice, v11, PP (1023-1033), 2003.

Comparative Study of Distance Functions for Nearest Neighbors

Janett Walters-Williams
School of Computing and Information Technology
University of Technology, Jamaica
Kingston 6, Jamaica W.I.
jwalters@utech.edu.jm

Yan Li
Department of Mathematics and Computing
Centre for Systems Biology
University of Southern Queensland
Toowoomba, Australia
liyan@usq.edu.au

Abstract **- Many learning algorithms rely on distance metrics to receive their input data. Research has shown that these metrics can improve the performance of these algorithms. Over the years an often popular function is the Euclidean function. In this paper, we investigate a number of different metrics proposed by different communities, including Mahalanobis, Euclidean, Kullback-Leibler and Hamming distance. Overall, the best-performing method is the Mahalanobis distance metric.**

Keywords
Kullback-Leibler distance, Euclidean distance, Mahalanobis distance, Manhattan distance, Hamming distance, Minkowski distance, Nearest Neighbor.

I. INTRODUCTION

Nearest Neighbor algorithms are examples of instance-based learning which simply retain the entire training set during learning. Unlike other common classifiers, these algorithms do not build a classifier in advance. When a new sample arrives, the algorithm finds the neighbors nearest to the new sample from the training space based on a distance metric.

Distance functions, or distance metric learning functions are to learn distance metrics for input data from a given collection of pair or similar/dissimilar points that preserves the distance relation among the training data. This paper focuses on local, supervised distance metric learning useful for K nearest neighbor (KNN) classifiers. We are interested in answering the following question: Which distance function should be selected to produce a more accurate output when applied to KNNs? We seek the answer from theoretical analysis and experimental results. Research has shown that Euclidean distance is the mostly widely used function in practice [14, 17, 18], although Cover and Hart [5] state that any function can be used. Choosing the correct function however, ultimately dictates the success or failure of any learning algorithm.

In this paper we focus on distance metrics from two classes: (1) metrics which do not involve any normalization of the components - Euclidean, Mahalanobis, Manhattan (city block), Hamming and Minkowski, and (2) entropy based measures namely Kullback-Leibler, the most widely used theoretical metric [10]. We propose to compare the performance of these six distance metrics when applied to Nearest Neighbor Algorithms. We compute the confusion matrix from each function which is analyzed. We found that the expected performance of each is not the final result. From theoretical analysis and experimental results, we found that there are more similarities among most of the six functions than differences.

The paper is organized as follows. In Section 2, the six distance functions are described. Section 3 highlights work already done in the area. In Section 4 the theoretical analysis and experimental results for the six distances are presented. Finally section 5 discusses the conclusion.

II. DISTANCE

To define a distance is equivalent to defining rules to assign positive numbers between *pairs* of objects. Let, therefore, *a*, *b*, and *c* be three vectors with *j* elements each. A distance is a function which associates to any pair of vectors a real positive number, denoted *d*(**a**,**b**), which has the following properties [1]:-

$$d(a,a) = 0 \quad (1)$$

K. Elleithy (ed.), *Advanced Techniques in Computing Sciences and Software Engineering*,
DOI 10.1007/978-90-481-3660-5_14,

$$d(a,b) = d(b,a)$$
$$d(a,b) \le d(a,c)+d(c,b) \quad (1)$$

There are many learning systems that depend upon a good distance function to be successful. The following defines the six distance metrics used in this paper.

A. Kullback-Leibler Distance

The Kullback-Leibler distance is a natural distance function from a true probability distribution p to a target probability q. It is also known as relative or mutual entropy and is defined as

$$KL(p,q) = \sum_{i=1}^{n} p_i \times \log_2\left(\frac{p_i}{q_i}\right) \quad (2)$$

where n is the number of levels of the variables.

B. Euclidean Distance

The Euclidean distance computes the real straight line distance between two points, i.e. it measures the 'as-the-crow-flies' distance. If $p = \{p_1, \ldots, p_n\}$ and $q=\{q_1, \ldots, q_n\}$ the Euclidean distance is defined as:

$$EUD(p,q) = \sqrt{\sum_{i=1}^{n}(p_i - q_i)^2} \quad (3)$$

C. Manhattan Distance

The Manhattan distance is also known as the "absolute value" or city block distance. It computes the distance that would be traveled to get from one data point to the other if a grid-like path is followed. It is the sum of the differences of their corresponding components. The Manhattan distance is defined as:

$$MN(p,q) = \sum_{i=1}^{n} | p_i - q_i | \quad (4)$$

D. Hamming Distance

Hamming distance or the symmetric difference distance is a set of operations which associates to two sets a new set made of the elements of these sets that belong to only one of them. Elements that belong to both sets are excluded. It gives the number of the elements of the symmetric difference set. It is defined as:

$$HAM(p,q) = \sum_{i=1}^{n} | p_i - q_i | \quad (5)$$

if p and q are vectors consisting of zeroes and ones. Hamming distance is equal to the number of positions where the bit patterns are different.

E. Mahalanobis Distance

The Mahalanobis distance is based on the correlations between variables. It is defined as:

$$MD(p,q) = \sqrt{(p_i - q_i)^T V^{-1} (p_i - q_i)} \quad (6)$$

where V is the covariance matrix of $A_1..A_m$ and A_j is the vector of values for attribute j occurring in the training set instances *1..n*.

F. Minkowski Distance

The Minkowski distance or p-distance between two strings is the geometric distance between two inputs and uses a variable scaling factor, r. It is widely used for measuring similarity between objects (e.g., images) and is defined as:

$$L_m = \left(\sum_{i=1}^{n} | p_i - q_i |^r\right)^{\frac{1}{r}} \quad (7)$$

III. RELATED WORK

Since 1981 researchers have tried to compare different similarity measures. Noreault et al. [11] looked at evaluating the performance of measures, empirically. Further Jones and Furnas [9] studied several similarity measures in the field of information retrieval. In particular, they performed a geometric analysis on continuous measures in order to reveal important differences which would affect retrieval performance. Further comparative studies were done by Zwick et al. [19] focusing on Fuzzy sets.

A detailed study of heterogeneous distance functions (for data with categorical and continuous attributes) was carried out by Wilson and Martinez[16]. They did this for instance based learning. Their study was based on a supervised approach where each data instance had class information in addition to a set of categorical/continuous attributes.

The latest set of research has been done by Qian et al.[12] who compared the Euclidean and Cosine Angle distances for nearest neighbor queries in high dimensional data spaces and Boriah et al. [4] who looked at the performance of a variety of similarity measures in the context of a specific data mining task: outlier detection.

IV. COMPARISON ANALYSIS

A. Theoretical Analysis

Euclidean and Mahalanobis

The Euclidean norm of p yields the equation of a spheroid. This means that all components of an observation p contribute equally to the Euclidean distance of x from the center. Taking variability of that variable into account we get the distance between p and q in Euclidean as:

$$ED(p,q)=\sqrt{\left(\frac{p_i-q_i}{s_i}\right)^2+..+\left(\frac{p_n-q_n}{s_n}\right)^2}=\sqrt{(p-q)^T D^{-1}(p-q)} \quad (8)$$

where $D = diag(s_i^2..s_n^2)$.

We then take the correlation between variables into account. To do this the axes of ellipsoid are used to reflect this correlation. This is obtained by allowing the axes of the ellipsoid at constant distance to rotate. This yields the Mahalanobis distance (fig. 1). Thus if V in (6) becomes a

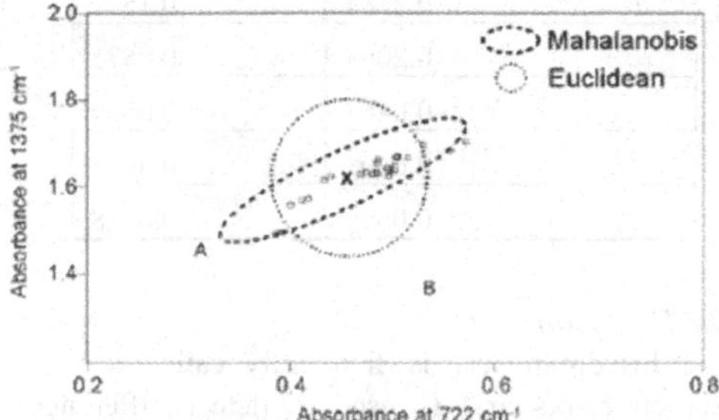

Fig. 1. Conversion of Euclidean to Mahalanobis

d x d identity matrix the Mahalanobis distance is defined as the Euclidean distance. If V is diagonal, then the resulting distance measure is called the normalized Euclidean distance and is defined as:

$$MD(p,q)=\sqrt{\sum_{i=1}^{n}\frac{(p_i-q_i)^2}{\sigma_i^2}} \quad (9)$$

where σ_i is the standard deviation of the x_i over the sample set. Mahalanobis is different from Euclidean because it takes into account the correlations of the data sets and is not dependent on the scale of measurements. It therefore generalizes the Euclidean function [6].

The Minkowski Relation

The degree r in the Minkowski distance (7) can take any number. When $r = 1$ the distance function is called the Manhattan distance. If the vectors, when $r = 1$, are binary numbers, the distance becomes the Hamming distance. When r is equal to 2, we obtain the usual Euclidean distance. Euclidean and Manhattan are therefore apart of the Minkowski family of distance metrics.

In this family the higher the value of r, the greater the importance given to large differences. Thus, when $r = 1$ or L_1 there is equal importance to all differences while when $r = 2$ or L_2 the distance metric takes into account only that component for which the difference is maximum. These are Manhattan and Euclidean respectively. When calculating Manhattan also deals with the sum of distance along each

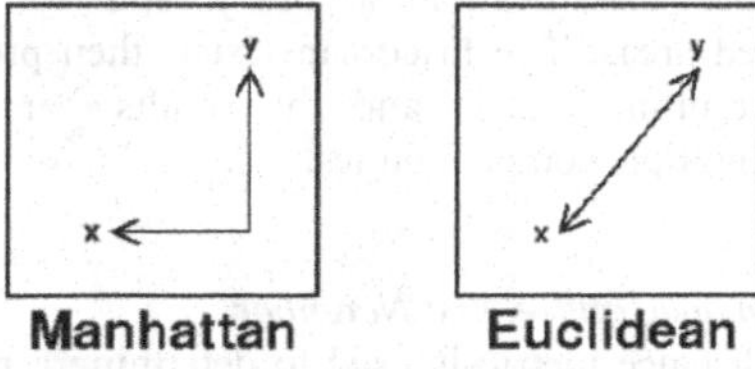

Fig. 2. Difference between Manhattan and Euclidean distances

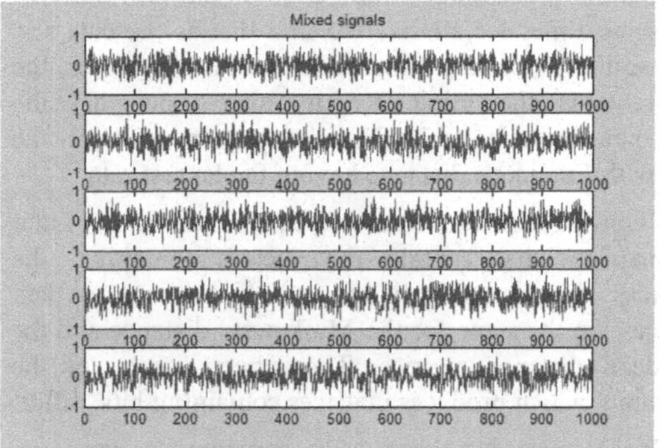

Fig. 3. Sample Signal Set

dimension while Euclidean corresponds to the length of the shortest path between two points as shown in fig. 2.

B. *Experimental Analysis*

In order to do the study effectively data was collected for analysis. This data, taken from different sites are of two types - real and artificial. All data is comprised of EEG data signals. The artificial data is made of six different data sets, each containing at least 1,000 points per vector (fig 3). The data sets are of two types – mixed with noise and independent from noise. These were taken from the RADICAL ICA algorithm site http://www.cs.umass.edu/~elm/ICA/.

Real data sets comprised EEG signals from both human and animals. These data have been acquired using the Neuroscan or Neurofax software. The human data set is a collection of 32-channel data from 14 subjects (7 males, 7 females) who performed a go-nogo categorization task and a go-no recognition task on natural photographs presented very briefly (20 ms). Each subject responded to a total of 2500 trials. The data is CZ referenced and is sampled at 1000 Hz. This data set can be found at http://sccn.ucsd.edu/~arno/fam2data/publicly_available_EEG_data.html.

Each data set passes through the k-nearest neighbor code. This produces a confusion matrix of the set and a classification rate (%). The confusion matrix is used to compute the classification accuracy and to identify misclassified areas. The Friedman test is then preformed on each resulting matrix and the results are passed through a multiple comparison test.

Performance on Nearest Neighbor

Each distance metric is used to determine a resulting nearest neighbor matrix. Each produces such a matrix but at different processing times. Fig. 4 shows results of processing times for all six. We find that as the data sets increase the processing times also increase, however, the rate remains the same. Results also show that the performance rate for each of L_m distances are relatively the same with Euclidean distance having the largest rate.

When the metrics are calculated it is the findings that the matrices that contain the vectors containing the distances between each pair of observations in data matrices are the same for the Minkowski distance and the Euclidean distance metrics. The others except for the Hamming, which produces matrices containing bits, differ.

The Nearest Neighbor code generates a confusion matrix for each distance metric calculation. If the data set is $M \times N$ this matrix is $N \times N$ in size showing in Table 1 with the 'true' class in rows and the 'predicted' class in the columns. The diagonal elements represent correctly classified compounds while the cross-diagonal elements represent misclassified compounds. The table also shows the accuracy of the classifier as the percentage of correctly classified compounds in a given class divided by the total number of compounds in that class. Table 1 shows the first ten rows in the first column of a confusion matrix based on the Hamming Distance.

The classification rate is calculated as

$$C_{rate} = \sum diag\left(c_{mat}\right) x100 / M \qquad (10)$$

where M is the number of rows and c_{mat} is the confusion matrix. It was found that the rate varied and ranged from -3.7899 in Mahalanobis to 47.9836 in Euclidean, using the 2D mixed data set. It was also found that Kullback-Leibler did not produce the lowest rate; it was13.2525, one of the highest rates.

TABLE 1
FIRST 10 ROWS-CONFUSION MATRIX (2 TYPES OF DATA SET)

2 row independent	2 row mixed
0.0938	-0.0758
0.5272	0.6028
0.1873	0.2171
0.6460	0.6357
-0.6883	-0.6491
0.2188	0.1521
0.2063	0.0671
-0.0467	-0.0509
0.6408	0.5912
0.0006	0.0218

The Friedman Test

The Friedman test is frequently called a two-way analysis on ranks and is used to detect differences in treatments across multiple test attempts. It is at the same time a generalization of the Sign-Test and the Spearman Rank Correlation Test and test models the ratings of n (rows) judges on k (columns) "treatments". The test is used to test if the means of the distance functions are totally matched when the distribution of the underlying population is not specified.

The hypothesis being tested is that all the methods have equal mean total matches, and the alternative hypothesis is that all methods do not have equal mean total matches. It is our findings that of the six functions Manhattan and Kullback-Leibler produced slightly higher means in each data set (fig. 5). It also shows that as the data sets increase in the number of vectors the error in each increased, however it showed that it increased more in Euclidean, Manhattan and Hamming as the dimensions increased (fig. 4).

It was also seen that the results changed when the data set types changed. Figure 7 shows the Friedman Test table of a independent data set while figure 6 shows a simpler set that was mixed with noise. It can be seen that the probability of having a Chi-square and the error calculated on the mean increase when an independent dataset is used

Multiple Comparison Test

Once the Friedman test is completed the resulting statistics are used in the Multiple Comparison Test. This test is done using the Tukey-Kramer Method. This method is chosen over Scheffé, Bonferoni and Sidák because it produces smaller critical values and it controls the experiment wise error rate at approximation ε very well [7].

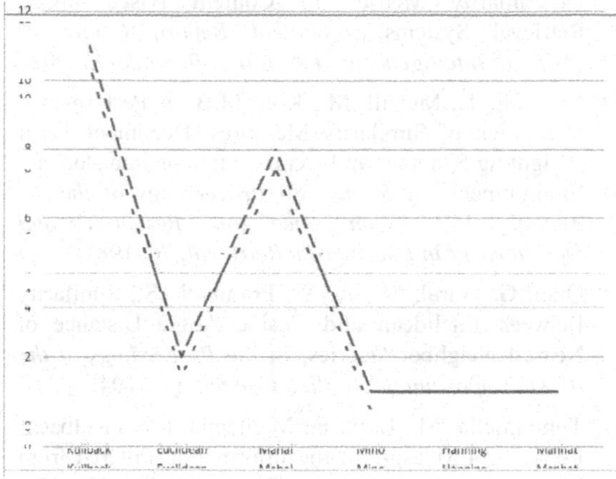
Fig. 4. Performance Rate of The Six Functions

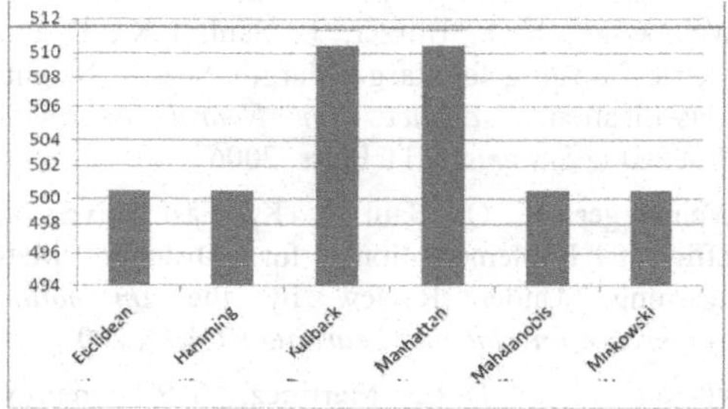

Fig. 5. Mean Values for 2D mixed data set

The hypothesis being tested is based on the results from the Friedman Test. We wish to prove that since there are differences in the mean for the Friedman Test it will be the same behavior in the Comparison Test.

In this test the minimum significant difference (MSD) is calculated for each pair of means. If the observed difference between a pair of means is greater than the MSD, the pair of means is significantly different. It was found that for all distance function, except Euclidean, all the mean column ranks were significantly different. For Euclidean there were only significant difference in a few columns.

When the error rates are examined the Kullback-Leibler is considered to have the worst. It is found that the following is the order:- *Mahalanobis* < *Minkowski family* < *Kullback-Leibler.* Mahalanobis has the best since the error rate is controlled.

Discussion

Most nearest neighbor algorithms are based on the Euclidean distance function. In this paper, we examine six known distance metrics. Is it really necessary to use the Euclidean function? Our results indicate that Euclidean does not have the fastest performance rate or the best error rate. Overall Mahalanobis metric has the best performance when applied to nearest neighbor. It has (1) a low performance rate; (2) performs well when data is controlled; (3) has a low classification rate; and, (4) its mean value is one which has a low increase rate as the number of

Friedman's ANOVA Table

Source	SS	df	MS	Chi-sq	Prob>Chi-sq
Columns	99900	999	100	399.6	1
Error	249650100.01	998001	250.15		
Total	249750000.01	999999			

Test for column effects after row effects are removed

Fig. 6. Friedman Test on a Kullback-Leibler mixed data set

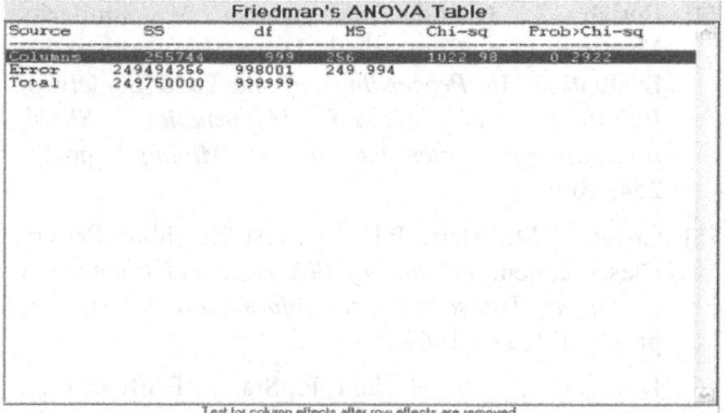
Friedman's ANOVA Table

Source	SS	df	MS	Chi-sq	Prob>Chi-sq
Columns	255744	999	256	1022.98	0.2922
Error	249494256	998001	249.994		
Total	249750000	999999			

Test for column effects after row effects are removed

Fig. 7. Friedman Test on a Kullback-Leibler independent data set

vectors increases. Research has also shown that based on Maximum Likelihood criteria Euclidean and Manhattan are proven to be optimal distances for Gaussian and Exponential data, respectively [9]. Mahalanobis on the other hand, is useful for both Gaussian and non-Gaussian data. It is also scale-invariant, i.e. not dependent on the scale of measurements which makes it approximate for applications with different types of measurements.

V. CONCLUSION

Over the years Euclidean has been the distance metric of choice by most researchers, however we have observed from our experiments that Mahalanobis has the best performance of the six metrics studied. The Minkowski family of distances is not suitable for all applications. So why is Euclidean the distance of choice? This maybe because of (1) the ease of implementing the Euclidean distance and (2) researchers tend to assume data to be

Gaussian in distribution. Although Mahalanobis is the best of the six researchers may choose their distance metric based on their personal choice and the size and type of the datasets been used. For example if one does not have any prior knowledge the Euclidean function is usually recommended. If there is the need to capitalize on statistical regularities in data that maybe estimated from a large training set then Mahalanobis is best.

REFERENCES

[1] Abdi, H., *Encyclopedia of Measurement and Statistics*, 2007

[2] Bar-Hillel, A., *Learning from Weak Representations using Distance Functions and Generative Models*, Ph.D. Thesis, Hebrew University of Jerusalem, 2006.

[3] Beitao L., Chang, E., Wu, C., DPF – A Perceptual Distance Function for Image Retrieval. In *Proceedings of the IEEE conference on Image Processing,* Sept 2002.

[4] Boriah, S., Chandola, V. Kumar, V. Similarity Measures for Categorical Data: A Comparative Evaluation, In *Proceedings of the 2008 Society of Industrial and Applied Mathematics (SIAM) International Conference on Data Mining.,* pp.23-254, 2008.

[5] Cover, T.M., Hart, P.E., Nearest Neighbor Pattern Classification. *Institute of Electrical and Electronics Engineers Transactions on Information Theory, 13,* pp. 21-271, Jan. 1967.

[6] Davis, J.V., Kulis, B., Jain, P., Sra, S., Dhillon, I. S., Information-Theoretic Metric Learning, In the *Proceedings of the 24th International Conference on Machine Learning*, 2007.

[7] Griffiths, R. Multiple Comparison Methods for Data Review of Census for Agriculture Press Releases, In the *Proceedings of the Survey Research Methods Section of the American Statistical Association,* 1992.

[8] Jensen, D.D., Cohen, P.R., Multiple Comparisons in Induction Algorithms, *Klumer Academic Publishers*, pp. 1-33, 2002.

[9] Jones, W.P., Furnas, G.W., Pictures of Relevance: A Geometric Analysis of similarity Measures, *Journal of American Society of Information Science* vol. 38, issue 6, pp. 420-442, 1987.

[10] Kamichety, H.M., Natarajan, P., Rakshit S., An Empirical Framework to Evaluate Performance of Dissimilarity Metrics in Content Based Image Retrieval Systems, *Technical Report, Center of Artificial Intelligence and Robotics, Bangalore,* 2002.

[11] Noreault, T., McGill, M., Koll, M.B., A Performance Evaluation of Similarity Measures, Document Term Weighting Schemes and Representations in a Boolean Environment, In *SIGIR '80 Proceedings of the 3rd Annual ACM Conference on Research and Development in Information Retrieval*, 76, 1981.

[12] Qian, G., Sural, S., Gu, Y., Pramanik, S., Similarity Between Euclidean and Cosine Angle Distance of Nearest Neighbor Queries, In *the Proceedings of the ACM Symposium on Applied Computing,* 2004.

[13] Tumminello, M., Lillo, F., Mantegna, R.N., Kulback-Leiber as a Measure of the Information Filtered from Multivariate Data, Physical Review E. 76, 031123 , 2007.

[14] Weinberger, K.Q., Blitzer, J., Saul, L.K., Distance Metric Learning for Large Margin Nearest Neighbor Classification, *Advances in Neural Information Processing Systems*, MIT Press, 2006.

[15] Weinberger, K. Q., Saul, L. K., Fast Solvers and Efficient Implementations for Distance Metric Learning, Under Review by the *International Conference on Machine Learning (ICML)*, 2007.

[16] Wilson, D.R., Martinez, T.R., Improved Heterogeneous Distance Functions, *Journal of Artificial Intelligence Research (JAIR),* vol. 6, issue 1, pp. 1-34, 1997.

[17] Wilson, D.R., *Advances in Instance-Based Learning Algorithms*, Ph.D. Thesis, Brigham Young University, 1997.

[18] Wölfel, M., Ekenel,H. K., Feature Weighted Mahalanobis Distance: Improved Robustness for Gaussian Classifiers, In the Proceedings of the 13th European Signal Processing Conference (EUSIPCO 2005), Sept 2005.

[19] Zwick, R., Carlstein, E., Budescu, D.V., *Measures of Similarity among Fuzzy Concepts: A Comparative Analysis, International Journal of Approximate Reasoning* 1, 2, pp. 221-242, 1987.

Determination of the geometrical dimensions of the helical gears with addendum modifications based on the specific sliding equalization model

Antal Tiberiu Alexandru[1], Antal Adalbert
[1]The Technical University from Cluj-Napoca, str. C. Daicoviciu, nr. 15, 400020, Cluj-Napoca, Romania
Tiberiu.Alexandru.Antal@mep.utcluj.ro

***Abstract* - The paper gives a new computational method for the determination of the geometrical dimensions of the helical gears with addendum modification based on the model of the sliding equalization at the beginning and at the ending of the meshing. In this model the sliding between the teeth's flanks during the meshing is used to increase the lifetime of the gears by uniformization at the points where the differences are highest. The variations of the addendum modification values, at different axis angles and axis distances, are determined using the MATLAB computing environment.**

I. Introduction

The technical literature from the filed of gears gives different models or methods to determine the specific addendum modification x_1 and x_2 in the case of planar gears [2], [3], [4] and [5]. The role of these addendum modifications is to rise the lifetime of the gear. In order to improve of the operating conditions, the models used at planar gears may be extended to spatial gears or helical gears. For the helical gears, the authors, have created two new computational models, that of the specific sliding equalization and that of the relative velocities equalization [6] at the points where de meshing starts and ends. These, are used to determine the addendum modifications and the rest of the geometrical parameters of the wheels of the helical gear.

II. Sliding coefficients at helical gears

Considering the drawing from Figure 1, the gearing line is obtained by the intersection of the P_1 and P_2 tangent plans to the base cylinders. On this line, we have the A point, where the meshing begins, and the E point where the meshing ends. During the meshing process between the teeth flanks, on the gearing line, the relative velocity of the 1 tooth is varying in regard to the 2 tooth. Because of this the sliding comes between the teeth flanks. The sliding between the teeth flanks can be evaluated with the help of the ζ_{12} and the ζ_{21} sliding coefficients [4]. ζ_{12} measures the sliding of the 1 flank in regard to the 2 flank, while ζ_{21} measures the sliding of the 2 flank in regard to the 1 flank. The sliding coefficients are determined at A and E points which are the most far meshing points from the C point (pitch point) as here the sliding coefficients have the highest values. The determination of the sliding coefficients for the helical gear from Figure 1 at the point A, where the meshing begins, and at point E, where the meshing ends, was proved in [1].

Considering the following expressions:

$$\left.\begin{aligned}
A_A &= y_A[1-u_{21}\cos(\Sigma)]+r_{w1}+r_{w2}u_{21}\cos(\Sigma)\\
B_A &= x_A[1-u_{21}\cos(\Sigma)]+z_A u_{21}\sin(\Sigma)\\
C_A &= (y_A-r_{w2})u_{21}\sin(\Sigma)\\
A_E &= y_E[1-u_{21}\cos(\Sigma)]+r_{w1}+r_{w2}\cos(\Sigma)\\
B_E &= x_E[1-u_{21}\cos(\Sigma)]+z_E u_{21}\sin(\Sigma)\\
C_E &= (y_E-r_{w2})u_{21}\sin(\Sigma)\\
u_{21} &= \frac{\omega_2}{\omega_1}=\frac{z_1}{z_2}
\end{aligned}\right\} \tag{1}$$

The sliding coefficients are determined in [1] with the (2) and the (3) expressions:

$$\varsigma_{12A}=\frac{A_A^2+B_A^2+C_A^2}{(y_A+r_{w1})A_A+x_A B_A} \tag{2}$$

and

$$\varsigma_{21E}=\frac{A_E^2+B_E^2+C_E^2}{C_E^2-(y_E+r_{w2})u_{21}A_E\cos\Sigma-u_{21}[x_E\cos(\Sigma)-z_E\sin(\Sigma)]B_E} \tag{3}$$

The x_A, y_A, z_A and the x_E, y_E, z_E used in (1), (2) and (3) are the coordinates of the A and E points from Figure 1 having the following expressions:

$$\left.\begin{aligned}
x_A &= r_{b2}\cos(\alpha_{tw2})[\tan(\alpha_{ta2})-\tan(\alpha_{tw2})]\frac{\tan(\alpha_{tw2})}{\tan(\alpha_{tw1})}\\
y_A &= -r_{b2}\sin(\alpha_{tw2})[\tan(\alpha_{ta2})-\tan(\alpha_{tw2})]\\
z_A &= -r_{b2}\cos(\alpha_{tw2})[\tan(\alpha_{ta2})-\tan(\alpha_{tw2})]\times\\
&\quad\times\frac{\tan(\alpha_{tw1})+\cos(\Sigma)\tan(\alpha_{tw2})}{\sin(\Sigma)\tan(\alpha_{tw1})}
\end{aligned}\right\} \tag{4}$$

and

K. Elleithy (ed.), *Advanced Techniques in Computing Sciences and Software Engineering*,
DOI 10.1007/978-90-481-3660-5_15,

$$\left.\begin{aligned} x_E &= r_{b1}\cos(\alpha_{tw1})[\tan(\alpha_{ta1}) - \tan(\alpha_{tw1})] \\ y_E &= r_{b1}\sin(\alpha_{tw1})[\tan(\alpha_{ta1}) - \tan(\alpha_{tw1})] \\ z_E &= r_{b1}\cos(\alpha_{tw1})[\tan(\alpha_{ta1}) - \tan(\alpha_{tw1})]\times \\ &\times\frac{\tan(\alpha_{tw1}) + \cos(\Sigma)\tan(\alpha_{tw2})}{\sin(\Sigma)\tan(\alpha_{tw2})} \end{aligned}\right\} \quad (5)$$

Here, r_{b1} and r_{b2} are the base circles radii;

α_{tw1} and α_{tw2} are the pressure angles in the frontal plans of the gear wheels;

α_{ta1} and α_{ta2} are the pressure angles in the frontal plans on the addendum circles determined by the expressions $\alpha_{ta1}=\arccos(r_{b1}/r_{a1})$ and $\alpha_{ta2}=\arccos(r_{b2}/r_{a2})$;

r_{a1} and r_{a2} are the radii of the teeth head circles;

Σ the angle between the helical gears axes.

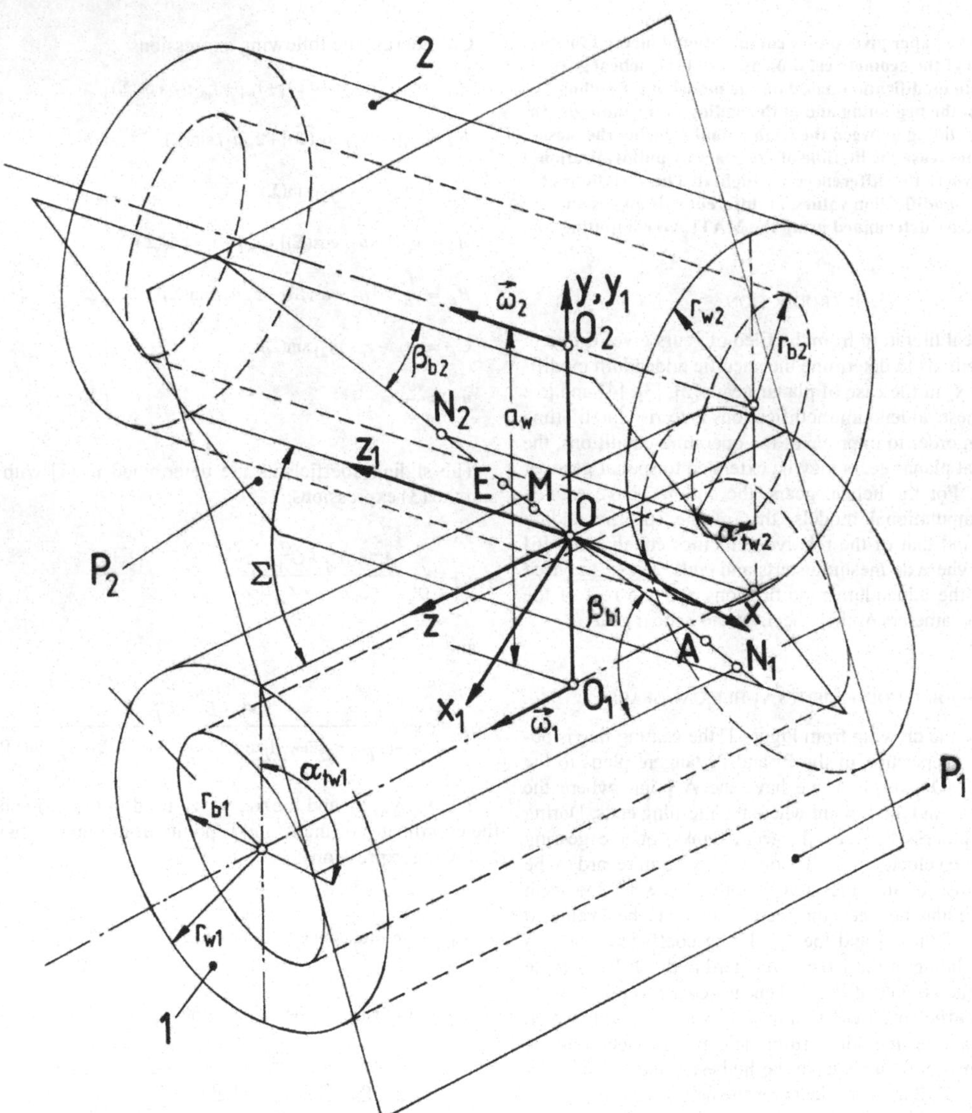

Figure 1. The helical gear schema.

III. EQUALIZATION OF THE ζ_{12} AND ζ_{21} SPECIFIC SLIDING

In order to equalize the specific sliding, in the A and the E points of the helical gear, we have the make the expressions from (2) and (3) equal.

As the equalization criterion must be obtained for different Σ angles, we obtain the following system of nonlinear equations:

$$\left.\begin{aligned}\varsigma_{12A}(x_1,x_2,\beta_1,\beta_2)&=\varsigma_{21E}(x_1,x_2,\beta_1,\beta_2)\\ \beta_{w1}(x_1,x_2,\beta_1,\beta_2)+\beta_{w2}(x_1,x_2,\beta_1,\beta_2)&=\Sigma\end{aligned}\right\} \tag{6}$$

where β_{w1} and β_{w2} the helix angles on the rolling cylinders.

The system obtained at (6) allows computing the variations of the helix angles on the pitch cylinders β_1 and β_2 with respect of the specific addendum modifications x_1 and x_2 at given values of the Σ angle between the axes of the helical gears.

III. DETEREMINATION OF THE GEOMETRICAL DIMENSIONS OF THE HELICAL GEAR

The geometrical relations used to compute the dimensions of the 1 and the 2 helical wheels are:

- the diameters of the pitch circles:

$$\begin{aligned}d_1&=m_n\frac{z_1}{\cos(\beta_1)}\\ d_2&=m_n\frac{z_2}{\cos(\beta_2)}\end{aligned} \tag{7}$$

- the teeth declination angles on the base cylinders:

$$\begin{aligned}\sin(\beta_{b1})&=\sin(\beta_1)\cos(\alpha_n)\\ \sin(\beta_{b2})&=\sin(\beta_2)\cos(\alpha_n)\end{aligned} \tag{8}$$

- the number of the teeth of the equivalent wheels:

$$\begin{aligned}z_{n1}&=\frac{z_1}{\cos^2(\beta_{b1})\cos(\beta_1)}\\ z_{n2}&=\frac{z_2}{\cos^2(\beta_{b2})\cos(\beta_2)}\end{aligned} \tag{9}$$

- the meshing angle at normal plan:

$$inv(\alpha_{wn})\approx 2\frac{x_1+x_2}{z_{n1}+z_{n2}}\tan(\alpha_n)+inv(\alpha_n) \tag{10}$$

- the teeth declination angles on the rolling cylinders:

$$\begin{aligned}\sin(\beta_{w1})&=\frac{\sin(\beta_{b1})}{\cos(\alpha_{wn})}\\ \sin(\beta_{w2})&=\frac{\sin(\beta_{b2})}{\cos(\alpha_{wn})}\end{aligned} \tag{11}$$

- the meshing angles at frontal plans:

$$\begin{aligned}\cos(\alpha_{tw1})&=\cos(\alpha_{wn})\frac{\cos(\beta_{w1})}{\cos(\alpha_{b1})}\\ \cos(\alpha_{tw2})&=\cos(\alpha_{wn})\frac{\cos(\beta_{w2})}{\cos(\alpha_{b2})}\end{aligned} \tag{12}$$

- the specific cutback of the tooth head:

$$k=x_1+x_2+\frac{\dfrac{d_1}{\cos^2(\beta_{b1})}\left(1-\dfrac{\cos(\alpha_n)}{\cos(\alpha_{wn})}\right)+\dfrac{d_2}{\cos^2(\beta_{b2})}\left(1-\dfrac{\cos(\alpha_n)}{\cos(\alpha_{wn})}\right)}{2m_n} \tag{13}$$

- the profile angles of the basic rack at frontal plans of the 1 and 2 helical wheels:

$$\begin{aligned}\cos(\alpha_{t1})&=\cos(\alpha_n)\frac{\cos(\beta_1)}{\cos(\alpha_{b1})}\\ \cos(\alpha_{t2})&=\cos(\alpha_n)\frac{\cos(\beta_2)}{\cos(\alpha_{b2})}\end{aligned} \tag{14}$$

- the diameters of the base circles for the 1 and the 2 wheel:

$$\begin{aligned}d_{b1}&=d_1\cos(\alpha_{t1})\\ d_{b2}&=d_2\cos(\alpha_{t2})\end{aligned} \tag{15}$$

- the diameters of the rolling circles for the 1 and the 2 wheel:

$$\begin{aligned}d_{w1}&=d_1\frac{\cos(\alpha_{t1})}{\cos(\alpha_{wt1})}\\ d_{w2}&=d_2\frac{\cos(\alpha_{t2})}{\cos(\alpha_{wt2})}\end{aligned} \tag{16}$$

- the diameters of the head circles for the 1 and the 2 wheel:

$$\begin{aligned}d_{a1}&=m_n\left(\frac{z_1}{\cos(\beta_1)}+2h_a^*+2x_1-2k\right)\\ d_{a2}&=m_n\left(\frac{z_2}{\cos(\beta_2)}+2h_a^*+2x_2-2k\right)\end{aligned} \tag{17}$$

- the diameters of the foot circles for the 1 and the 2 wheel:

$$d_{f1} = m_n \left(\frac{z_1}{\cos(\beta_1)} - 2h_a^* - 2c^* + 2x_1 \right)$$
$$d_{f2} = m_n \left(\frac{z_2}{\cos(\beta_2)} - 2h_a^* - 2c^* + 2x_2 \right) \quad (18)$$

In order to solve the nonlinear system from (6) the expressions from (7) - (25) must be computed. If the x_1, x_2, z_1, z_2, α_n, h_a^*, c^*, m_n, and the Σ values are given, a set of the β_1 and β_2 values are obtained while the ζ_{12A} and ζ_{21E} values are the same, and the sum of β_{w1} and β_{w2} is equal to Σ. When obtaining the numerical solutions the distance between the wheels ($a = r_{w1} + r_{w2}$) axes however is not an integer number. When manufacturing the wheels this would be one of the desired conditions. For this the (6) system can be rewritten in two ways:

$$\left.\begin{array}{l} \varsigma_{12A}(x_1,x_2,\beta_1,\beta_2) = \varsigma_{21E}(x_1,x_2,\beta_1,\beta_2) \\ \beta_{w1}(x_1,x_2,\beta_1,\beta_2) + \beta_{w2}(x_1,x_2,\beta_1,\beta_2) = \Sigma \\ r_{w1}(x_1,x_2,\beta_1,\beta_2) + r_{w2}(x_1,x_2,\beta_1,\beta_2) = a \end{array}\right\} \quad (19)$$

$$\left.\begin{array}{l} \varsigma_{12A}(x_1,x_2,\beta_1,\beta_2) = \varsigma_{21E}(x_1,x_2,\beta_1,\beta_2) \\ \beta_{w1}(x_1,x_2,\beta_1,\beta_2) + \beta_{w2}(x_1,x_2,\beta_1,\beta_2) = \Sigma \\ r_{w1}(x_1,x_2,\beta_1,\beta_2) + r_{w2}(x_1,x_2,\beta_1,\beta_2) = \text{int} \end{array}\right\} \quad (20)$$

where *a* from (19) is a given distance between the axes, or an integer value at (20).

IV. Numerical Rezults with and without the Equalizations of the Specific Sliding

The data from Table I - Table IV are computed, based on a MATLAB application, for the following input data: z_1=20, z_2=40 (the number of the teeth), α_n=20^0, h_a^*=1 (the height coefficient from head of the tooth), c^*=0.25 (the clearance coefficient from the head of the tooth), m_n=4mm (the normal module of the gear), Σ=90^0. The values for the specific x_1 and x_2 addendum modifications were chosen so that no interferences would appear at the head and at the based of the teeth.

The values from Table I are showing the results obtained for the specific sliding without equalization. The (19) system is solved without the first equation. The x_1 and x_2 values are given, and the β_1 and β_2 values are obtained, while Σ, the angle between the axes, is 90^0 and a, the distance between the axes, is 170 mm. Table I shows clearly that the ζ_{12A} and ζ_{21E} values are not equal and because of this difference the wearing of the wheels will not be uniform. Always, the wearing of one wheel will be higher and this way that wheel will crash faster.

The values from Table II are computed from the (6) system. The x_1 and x_2 values are given, and the β_1 and β_2 values are obtained, while Σ, the angle between the axes, is 90^0 and the equalization of the specific sliding is made. Although the equalization succeeds, the distance between the axes of the wheels is not an integer values (and as sometimes we prefer integer values instead of real ones).

The values from Table III are computed from the (19) system. The x_1 value is given, and the x_2, β_1 and β_2 values are obtained, while Σ, the angle between the axes, is 90^0 and a, the distance between the axes, is 170 mm.

The values from Table IV are computed from the (20) system. The x_1 value is given, and the x_2, β_1 and β_2 values are obtained, while Σ, the angle between the axes, is 90^0 and a, the distance between the axes, is an integer value. As shown in Table IV the distance between the axes will have some variation (170 mm - 175 mm) and this way we can cover an integer domain of distances.

The fsolve() function from MATLAB's Optimization Toolbox is used to find the roots of each of the nonlinear systems. MATLAB allows the solving of the nonlinear equations using more algorithms. This application is using the 'Trust-region dogleg algorithm'. The solutions from the tables were found for the following start values: x_2 = 0.0, β_1 = 45^0, β_2 = 45^0.

TABLE I

VARIATION OF THE SPECIFIC SLINDING AT THE A AND THE E POINTS FOR Σ=90^0 AND a = 170 mm

x_1	x_2	$\beta_1(^0)$	$\beta_2(^0)$	ζ_{12A}	ζ_{21E}	$\beta_{w1}(^0)$	$\beta_{w2}(^0)$	$\Sigma(^0)$	$a = r_{w1}+r_{w2}$ (mm)
-0.20	-0.400	43.954	46.949	2.353	2.099	43.526	46.474	90	170
-0.15	-0.300	44.127	46.549	2.324	2.122	43.803	46.197	90	170
-0.10	-0.200	44.310	46.141	2.294	2.146	44.091	45.909	90	170
-0.05	-0.100	44.503	45.722	2.265	2.170	44.393	45.607	90	170
0	0	44.709	45.291	2.234	2.195	44.709	45.291	90	170
0.05	0.100	44.928	44.847	2.204	2.221	45.041	44.959	90	170
0.10	0.200	45.164	44.386	2.173	2.249	45.392	44.608	90	170
0.15	0.300	45.419	43.906	2.141	2.278	45.765	44.235	90	170
0.20	0.400	45.697	43.402	2.108	2.308	46.165	43.835	90	170
0.25	0.500	46.004	42.870	2.074	2.341	46.598	43.402	90	170
0.30	0.600	46.349	42.299	2.039	2.378	47.073	42.927	90	170
0.35	0.700	46.745	41.677	2.001	2.419	47.604	42.396	90	170
0.40	0.800	47.215	40.980	1.960	2.468	48.217	41.783	90	170
0.45	0.900	47.810	40.157	1.913	2.530	48.965	41.035	90	170

0.50	1.000	48.693	39.044	1.851	2.622	50.018	39.982	90	170
0.55	1.100	50.101	37.307	1.764	2.780	51.632	38.273	90	170
0.60	1.200	49.862	37.087	1.766	2.760	51.527	38.137	90	170
0.65	0.325	46.539	41.996	1.962	2.486	47.331	42.669	90	170
0.70	0.350	46.745	41.677	1.940	2.515	47.604	42.396	90	170
0.75	0.375	46.968	41.340	1.917	2.547	47.898	42.102	90	170
0.80	0.400	47.215	40.980	1.893	2.582	48.217	41.783	90	170
0.85	0.425	47.492	40.590	1.868	2.620	48.568	41.432	90	170
0.90	0.450	47.810	40.157	1.842	2.664	48.965	41.035	90	170
0.95	0.475	48.193	39.659	1.813	2.715	49.429	40.571	90	170
1	0.500	48.693	39.044	1.778	2.781	50.018	39.982	90	170

TABLE II

EQUALIZATION OF THE SPECIFIC SLINDING FOR $\Sigma=90^0$ AND a IS REAL (mm)

x_1	x_2	$\beta_1(^0)$	$\beta_2(^0)$	ζ_{12A}	ζ_{21E}	$\beta_{w1}(^0)$	$\beta_{w2}(^0)$	$\Sigma(^0)$	$a = r_{w1}+r_{w2}$ (mm)
-0.20	-0.400	45.416	45.509	2.218	2.218	44.954	45.046	90	168.3974
-0.15	-0.300	45.294	45.396	2.218	2.218	44.949	45.051	90	168.7433
-0.10	-0.200	45.173	45.284	2.217	2.217	44.945	45.055	90	169.0867
-0.05	-0.100	45.055	45.172	2.216	2.216	44.942	45.058	90	169.4278
0	0	44.938	45.062	2.215	2.215	44.938	45.062	90	169.7668
0.05	0.100	44.823	44.952	2.213	2.213	44.935	45.065	90	170.1038
0.10	0.200	44.710	44.843	2.211	2.211	44.933	45.067	90	170.4389
0.15	0.300	44.598	44.734	2.208	2.208	44.931	45.069	90	170.7724
0.20	0.400	44.487	44.627	2.205	2.205	44.929	45.071	90	171.1042
0.25	0.500	44.378	44.520	2.202	2.202	44.928	45.072	90	171.4346
0.30	0.600	44.270	44.413	2.199	2.199	44.927	45.073	90	171.7635
0.35	0.700	44.163	44.308	2.195	2.195	44.926	45.074	90	172.0911
0.40	0.800	44.058	44.202	2.191	2.191	44.925	45.075	90	172.4175
0.45	0.900	43.953	44.098	2.187	2.187	44.925	45.075	90	172.7427
0.50	1.000	43.850	43.993	2.183	2.183	44.926	45.074	90	173.0668
0.55	1.100	43.748	43.890	2.178	2.178	44.926	45.074	90	173.3898
0.60	1.200	43.647	43.786	2.174	2.174	44.927	45.073	90	173.7119
0.65	0.325	43.323	45.273	2.198	2.198	44.001	45.999	90	172.9000
0.70	0.350	43.196	45.297	2.196	2.196	43.921	46.079	90	173.1515
0.75	0.375	43.069	45.323	2.194	2.194	43.841	46.159	90	173.4047
0.80	0.400	42.941	45.349	2.192	2.192	43.759	46.241	90	173.6597
0.85	0.425	42.813	45.376	2.189	2.189	43.677	46.323	90	173.9166
0.90	0.450	42.684	45.405	2.187	2.187	43.593	46.407	90	174.1754
0.95	0.475	42.555	45.435	2.184	2.184	43.508	46.492	90	174.4362
1	0.500	42.425	45.466	2.182	2.182	43.423	46.577	90	174.6991

TABLE III

EQUALIZATION OF THE SPECIFIC SLINDING FOR $\Sigma=90^0$ AND a = 170mm

x_1	x_2	$\beta_1(^0)$	$\beta_2(^0)$	ζ_{12A}	ζ_{21E}	$\beta_{w1}(^0)$	$\beta_{w2}(^0)$	$\Sigma(^0)$	$a = r_{w1}+r_{w2}$ (mm)
-0.20	0.707	45.521	43.718	2.201	2.201	45.913	44.087	90	170
-0.15	0.577	45.377	43.982	2.204	2.204	45.705	44.295	90	170
-0.10	0.443	45.235	44.250	2.207	2.207	45.497	44.503	90	170
-0.05	0.307	45.094	44.521	2.209	2.209	45.289	44.711	90	170
0	0.167	44.954	44.795	2.212	2.212	45.080	44.920	90	170
0.05	0.025	44.816	45.071	2.214	2.214	44.872	45.128	90	170
0.10	-0.120	44.680	45.351	2.216	2.216	44.665	45.335	90	170
0.15	-0.269	44.545	45.633	2.218	2.218	44.458	45.542	90	170
0.20	-0.419	44.412	45.917	2.219	2.219	44.252	45.748	90	170
0.25	-0.573	44.281	46.204	2.221	2.221	44.047	45.953	90	170
0.30	-0.729	44.152	46.492	2.222	2.222	43.843	46.157	90	170
0.35	-0.887	44.025	46.782	2.222	2.222	43.641	46.359	90	170
0.40	-1.048	43.900	47.074	2.223	2.223	43.440	46.560	90	170
0.45	-1.210	43.778	47.367	2.222	2.222	43.241	46.759	90	170
0.50	-1.375	43.657	47.661	2.222	2.222	43.043	46.957	90	170
0.55	-1.543	43.538	47.957	2.221	2.221	42.847	47.153	90	170
0.60	-1.712	43.421	48.254	2.219	2.219	42.653	47.347	90	170
0.65	-1.883	43.306	48.553	2.217	2.217	42.460	47.540	90	170
0.70	-2.057	43.193	48.854	2.215	2.215	42.269	47.731	90	170
0.75	-2.233	43.081	49.157	2.212	2.212	42.078	47.922	90	170
0.80	-2.411	42.970	49.463	2.208	2.208	41.889	48.111	90	170
0.85	-2.592	42.861	49.773	2.203	2.203	41.699	48.301	90	170

0.90	-2.777	42.752	50.087	2.198	2.198	41.509	48.491	90	170
0.95	-2.965	42.643	50.409	2.191	2.191	41.318	48.682	90	170
1	-3.159	42.534	50.739	2.183	2.183	41.125	48.875	90	170

TABLE IV

EQUALIZATION OF THE SPECIFIC SLINDING FOR $\Sigma=90^0$ AND a IS INTEGER (mm)

x_1	x_2	$\beta_1(^0)$	$\beta_2(^0)$	ζ_{12A}	ζ_{21E}	$\beta_{w1}(^0)$	$\beta_{w2}(^0)$	$\Sigma(^0)$	$a = r_{w1}+r_{w2}$ (mm)
-0.20	0.707	45.521	43.718	2.201	2.201	45.913	44.087	90	170
-0.15	0.577	45.377	43.982	2.204	2.204	45.705	44.295	90	170
-0.1	0.443	45.235	44.250	2.207	2.207	45.497	44.503	90	170
-0.05	0.307	45.094	44.521	2.209	2.209	45.289	44.711	90	170
0	0.901	45.076	43.592	2.195	2.195	45.759	44.241	90	171
0.05	0.761	44.924	43.876	2.198	2.198	45.535	44.465	90	171
0.10	0.618	44.773	44.165	2.201	2.201	45.310	44.690	90	171
0.15	0.471	44.623	44.458	2.204	2.204	45.084	44.916	90	171
0.20	0.321	44.475	44.754	2.207	2.207	44.859	45.141	90	171
0.25	0.941	44.467	43.796	2.190	2.190	45.346	44.654	90	172
0.30	0.786	44.307	44.108	2.194	2.194	45.102	44.898	90	172
0.35	0.627	44.149	44.426	2.197	2.197	44.858	45.142	90	172
0.40	0.464	43.992	44.748	2.200	2.200	44.614	45.386	90	172
0.45	1.112	44.006	43.746	2.181	2.181	45.136	44.864	90	173
0.50	0.944	43.836	44.086	2.185	2.185	44.871	45.129	90	173
0.55	0.771	43.668	44.431	2.189	2.189	44.605	45.395	90	173
0.60	0.593	43.502	44.783	2.192	2.192	44.340	45.660	90	173
0.65	0.409	43.337	45.139	2.196	2.196	44.075	45.925	90	173
0.70	1.088	43.358	44.103	2.174	2.174	44.610	45.390	90	174
0.75	0.898	43.178	44.482	2.179	2.179	44.321	45.679	90	174
0.80	0.702	43.001	44.867	2.183	2.183	44.032	45.968	90	174
0.85	0.500	42.827	45.258	2.187	2.187	43.743	46.257	90	174
0.90	0.292	42.655	45.654	2.191	2.191	43.456	46.544	90	174
0.95	1.002	42.676	44.588	2.169	2.169	43.997	46.003	90	175
1	0.784	42.487	45.012	2.173	2.173	43.681	46.319	90	175

VI. CONCLUSIONS

Analyzing the data sequences obtained in the four tables we must observe that not all of the solutions from Table III can be used. Some of the x_2 values have high negative values that could lead to interferences at the base on the tooth.

The design of the helical gears based on the model of specific sliding equalization at the points where the meshing begins and ends can be made with success based on the data from Table II and Table IV. However, the results from Table IV should be preferred, as here, the distance between the axes, is obtained as an integer value. The equalization of the sliding coefficient conduces to uniform lubrication conditions and uniform wearing at the teeth flanks.

The application created in MATLAB allows the study of different helical gears where the specific addendum modifications are obtained based on the model of the specific sliding equalizations at the beginning and at the end of the meshing. As the more solutions are obtained these are given in a tabular form to help the designer to choose the most appropriate one for his project. If the designer has no special needs, the solution with the lowest equalized sliding coefficient values is recommended to be chosen in order to rise the lifetime of the gears.

REFERENCES

[1] A. Antal, T. A. Antal, T., "A computer program for the calculus of the sliding at the helical gears." *PRASIC '98 – NATIONAL SIMPOSIUM with international participation, Braşov, Romania*, Vol. II – Machine Elements. Mechanical Transmissions, 5-7 November 1998, p.235-238, ISBN 973-98796-0-8.

[2] I. A. Bolotovskii, "Spravocinik po gheometriceskomu rasciotu evolventnih zubciatih I cerviacinih peredaci.", Moskva, *Masinostroenie*, 1986.

[3] D. Maros, "Cinematica rotilor dintate", Bucuresti, *Editura Tehnica*, 1958, p. 140-146,187-194.

[4] G. Nimann, H. Winter, "Maschinenelemente". Band II, Berlin, *Springer Verlag*, 1983, p.273-275.

[5] MAAG – Taschenbuch. MAAG – Zahnrader. Aktiengesellschaft CH – 8023, Zurich, Schweizm 1985, p.72-76.

[6] Antal, T. A., Antal, A., Arghir, M., Determination of the addendum modification at helical gears at the point where the meshing stars and ends, based on the relative velocity equalization criterion, GAMM 2008 - 79th Annual Meeting of the International Association of Applied Mathematics and Mechanics at the University of Bremen, 31 March – 4 April 2008, Germany, http://www.zarm.uni-bremen.de/gamm2008/?nav=body_view_all.php.

Is Manual Data Collection Hampered by the Presence of Inner Classes or Class Size?

Steve Counsell
School of Information Systems, Computing and Mathematics
Brunel University, Uxbridge, Middlesex, UB8 3PH, UK

George Loizou and Rajaa Najjar
School of Computer Science and Information Systems
Birkbeck, University of London, Malet Street, London WC1E 7HX, UK

Abstract. In this paper, we present an empirical study in which we hypothesize that the existence of Java 'inner classes' and class size are strong impediments to the data collector during manual data collection. We collected inner class and class size data from the classes of four Java open-source systems - first manually and then automatically (after the manual collection had finished) using a bespoke software tool. The data collected by the tool provided the benchmark against which errors and oversights in the manual data collection of these two features could be recorded. Results showed our initial hypotheses to be refuted – manual errors in data collection from the four Java systems arose *not* from the presence of inner classes *or* from class size but from variations in coding style, lack of standards, class layout and disparateness of class feature declarations.

I. INTRODUCTION

Knowledge of source code traits is often assisted by the collection of data. For example, collection of metrics whether automatically (using a software tool) or manually (using human subjects) and then analyzing the collected data can provide insights into trends in the construction and quality of software [1, 2, 3, 11]. For example, elimination of 'bad smells' in code [9] often requires a visual search amongst multiple declarations for groups of, or individual, types and their associated dependencies; scanning code for security vulnerabilities was the subject of a study by Viega et al. [22] with inner classes the key feature under scrutiny. In theory therefore, the quality of OO systems is aided by ensuring that their features conform to sound and accepted practice [13, 14]. Anecdotal evidence would thus suggest that the existence of constructs such as Java *inner classes* and large size classes (herein, size of a class is given by its number of attributes and methods) may adversely influence the effectiveness and accuracy of manual data collection from code by a developer. Inner classes allow a nested class access to the attributes of the enclosing class and have been the subject of certain criticism because they add a level of complexity to the system [15, 20, 22]. In this paper, we explore whether the existence of inner classes and class size are impediments to the data collector during manual data collection of class features. Data from four systems was collected both manually and automatically (using a tool) and results compared using statistical analysis.

II. RELATED WORK

The work described in this paper stems from three key motivating factors. First, evaluation of Java class features is an important and potentially valuable activity from a software engineering perspective. Yet, to our knowledge, no studies have attempted to evaluate the influence that inner classes and class size may have on the effectiveness of the manual data collection process. Second, the research follows two previous studies - a pilot study which found that the work of a data collector was severely hampered, and more errors were made in the data collection, from systems which consistently violated sound OO practice; for example, the frequent declaration of protected features in classes with no explicit inheritance links [17]. Also, a further study by the same authors in which a direct comparison between automatic and manual data collection found that the latter was significantly less error-prone than hypothesized [4]. Third, OSSs are now playing an increasing role in industrial practice and academic research [6, 21, 12]. While research into OSS is gaining momentum, we are still only starting to address important issues associated with software development [18]. In Counsell et al. [5], the question of inheritance and encapsulation in five C++ systems was addressed. Metrics were automatically collected for three of the systems analyzed therein. The remaining two systems were not available electronically and, as a result, manual collection from paper sources was necessary. Common trends were found across all five systems, suggesting that manual data collection does have credible uses as well as that described hereafter. Fenton and Pfleeger [7] describe manual recording as subject to bias,

K. Elleithy (ed.), *Advanced Techniques in Computing Sciences and Software Engineering*,
DOI 10.1007/978-90-481-3660-5_16, © Springer Science+Business Media B.V. 2010

whether deliberate or unconscious, but they do admit that sometimes there is no alternative to manual data collection.

III. PRELIMINARIES

A. Inner Classes

Inner classes potentially add a level of complexity to a class by requiring the developer to follow the explicit nesting. Fig. 1 shows the basic structure of an inner class and how it fits inside an outer class. Class 'Foo' does not know that 'This_is_the_InnerClass' exists. Methods in an inner class can access private attributes and methods of the outer class.

```
class This_is_the_OuterClass{
...............

// The enclosing class
        class This_is_the_InnerClass{

        ..........
        // The enclosed class
        }
}

class Foo{
................
 // Class Foo has no knowledge of class
 // This_is_the_InnerClass
}
```

Fig. 1. Structure inner class, an outer class and scope

B. Application Domains Studied

The choice of the Java systems was influenced strongly by previous research into four corresponding C++ systems. In Counsell et al. [5], a C++ framework, a C++ compiler and two C++ libraries were investigated for trends in their encapsulation and inheritance patterns. For consistency across empirical studies and comparison purposes, we selected the four Java systems on the same basis. Finally, we note that the same Java systems were used in previous studies [5, 22, 23]. The systems investigated were:

1. System One: the Bean Scripting Framework (BSF). an architecture for incorporating scripting into Java applications/applets; contains 65 classes.
2. System Two: Barat. A compiler front-end for Java. A framework that supports static analysis of Java programs. It parses Java source code files and class files and builds a complete abstract syntax tree from Java source code files, enriched with name and type analysis information; contains 407 classes.
3. System Three: Libjava. The language sub-library set of 89 Java classes available from the public domain at the Gnu gcc website (`www.gnu.org`).
4. System Four: the Swing Java Package Library. This system provides a set of Java components that, to the maximum degree possible, execute consistently on all platforms; contains 1248 classes.

C. Data Collected

In total, fifteen class-level metrics were collected for each of the classes, in each of the four systems as a basis for our analysis and comparison. The metrics represent three Java categories: metrics related to *private* features, metrics related to *protected* features and metrics related to *public* features. We consider every attribute and return type of a method as a primitive attribute or primitive method, respectively, if their types are taken from one of the following: `int, long, float, double, short, char, byte, boolean, void`. A non-primitive attribute is one that is defined as a class; again this is normally another class, but can equally be the class in which the attribute is defined. We define a non-primitive method as one whose return type is that of a class (which is normally a different class, but can equally be the same class in which the method is defined). The class-level metrics collected were:

1. Number of private primitive attributes.
2. Number of private non-primitive attributes.
3. Number of private constructors.
4. Number of private primitive methods.
5. Number of private non-primitive methods.
6. Number of protected primitive attributes.
7. Number of protected non-primitive attributes.
8. Number of protected constructors.
9. Number of protected primitive methods.
10. Number of protected non-primitive methods.
11. Number of public primitive attributes.
12. Number of public non-primitive attributes.
13. Number of public constructors.
14. Number of public primitive methods.
15. Number of public non-primitive methods.

An assumption in this paper is that, for conciseness, *no* distinction is made between *under-counting* and *over-counting* of a class feature by the data collector. The former represents the case where, for example, there are 5 primitive private attributes in a class and the data collector manually records only 3; the latter represents the case where, for example, for the 5 primitive private attributes, incorrectly, the collector manually records 6. A further assumption that we make in our study of the four systems is that the automatic data collection represents the *correct* values of the metrics. To support this assumption, results from the automatic collection were meticulously verified by visual inspection; both the automatic and manual collections were carried out by the same investigator consistently over the course of six months. As well as the data upon which our error analysis could take place, the following data was also collected from each class (after the automatic and manual data collections had taken place):

a) The number of inner classes within each class. We hypothesize that the larger the number of inner classes belonging to an outer class is, the larger the number of manual errors made will be; b) The total number of attributes and methods for each class. We hypothesize that the greater the size of a class is, the larger the number of manual errors made by the data collector will be.

D. Summary Data

Table 1 summarizes the number of classes in each of the four systems for which at least one manual error was made by the data collector in the collection of the fifteen metrics (and the percentage that this number represents out of the total number of classes in the system). It also shows the number of classes, from those containing at least one error, which contained at least one inner class (and the percentage of total classes in the system that this value represents).

Table 1. Errors found in each of the four Java systems

System	No.≥ 1 Error	% of Total	No.≥ 1 Inner	% of Total	Max Error	Max Inner
BSF	8	12.31	2	3.08	2	1
Barat	64	15.72	13	3.19	3	6
Libjava	15	16.85	4	4.49	4	2
Swing	104	8.33	81	6.49	10	16

Finally, Table 1 shows the maximum value for, first, errors made by the data collector and, second, the number of inner classes in each system. For example, the BSF system contains 8 classes, where at least one manual error was made, on the collection of the fifteen metrics previously described (i.e., 12.31% of the total number of classes in BSF). Of those 8 classes, 2 contained at least one inner class (i.e., 3.08% of the total number of classes in BSF). The maximum number of errors made for any single class was 2 and the maximum number of inner classes was 1. The most striking feature of Table 1 is the high proportion of classes belonging to BSF, Barat and Libjava, where errors were made (12.31%, 15.72% and 16.85%, respectively). For the Libjava system, this means that approximately one in every 6 classes caused an error to be made by the data collector. This contrasts with the corresponding value for the Swing system, where one in 12 classes (8.83%) caused a manual collection error. This result indicates that size of a system does not have any significant bearing on the propensity of the data collector to make manual data collection errors. Swing clearly has the highest number of classes with *at least* one inner class and also has the largest error made across all four systems (10).

IV. DATA ANALYSIS

In the sequel, we describe the two hypotheses upon which our analysis rests. For each hypothesis, we state the null (H_{01} and H_{02}) and alternative (H_{A1} and H_{A2}) hypotheses.

E. Hypothesis One

Hypothesis H_{01}: the number of inner classes belonging to an outer class has no effect on the number of errors made by the data collector. Hypothesis H_{A1}: the larger the number of inner classes belonging to an outer class, the larger the number of errors made by the data collector.

Table 2 shows the median, mean and standard deviation (SD) values for errors in classes containing at least one inner class (I) and for classes without any inner classes (NI). For example, for classes in the Barat system, the median number of errors for classes with at least one inner class was 2 with mean 1.85 and standard deviation 0.69. For classes not containing any inner classes, the median was 1, with mean 1.43 and standard deviation 0.57; 'NC' stands for 'Not Computable'. (From Table 2, there appears to be a trend of fewer errors made by the data collector in classes without inner classes, deduced by comparing the mean values from Table 2.) Comparing the median values from Table 2 shows no particular trend towards classes with inner classes inducing any more errors than classes without inner classes. We note that the standard deviation values for NI in column 7 show a remarkable similarity unshared with the corresponding values for I (column 4).

Table 2. Statistical summary values for errors

System	Median I	Mean I	SD I	Median NI	Mean NI	SD NI
BSF	NC	1.50	0.71	1.5	1.50	0.55
Barat	2	1.85	0.69	1	1.43	0.57
Libjava	1.5	2.00	1.41	2	1.55	0.52
Swing	1	1.90	1.73	1	1.18	0.39

Table 3 shows the correlation values (non-parametric) for Barat and Swing of: number of errors made by the data collector versus number of inner classes per outer class.

(We exclude correlation analysis of the other two systems because of their small sample size.) Both Kendall's and Spearman's correlation values are non-parametric (make no assumptions about the data distribution).

Table 3. Correlation of errors versus inner classes

System	Kendall's	Spearman's
Barat	-0.63**	-0.70**
Swing	0.07	0.09

For Barat, the correlation values are negative and significant at the 1% level (denoted by '**' beside the correlation values). For Swing, the values are only marginally positive. The results suggest that hypothesis H_{01} is not supported and the opposite of H_{A1} was found; inner classes were not an impediment to the data collector despite the high numbers of inner classes in each of the two systems. Less data collection errors were made by the manual data collector in the presence of inner classes. To strengthen the case for rejecting both H_{01} and H_{A1}, we computed the Wilcoxon Signed-Rank test for Barat. We compared the errors for classes with at least one inner class with the errors for classes without any inner classes (we make no assumption about the distribution of the two samples). The Z value of -2.60 was only found to be significant at the 1% level. (We observe that Z is bigger than 1.96 (ignoring the minus), so the test is significant at $p < 0.05$, [8]) The same computation for Swing revealed a Z value of -1.91, significant only at the 10% level. For the Swing system, and in harmony with the result for Barat, the presence of a high number of inner classes implies a very low number of errors were made by the data collector, in complete contrast to H_{A1}. On balance, on the evidence presented, we reject H_{A1} and yet we cannot find support for H_{01} either - the number of inner classes belonging to an outer class actually has a negative correlation (significant for one system) on the number of errors made by the data collector.

F. Hypothesis Two

Hypothesis H_{02}: the size of a class has no bearing on the number of errors made by the data collector. Hypothesis H_{A2}: the greater the size of a class, the larger the number of errors made by the data collector.

Table 4 shows summary data of class size for all four systems. Looking at each system in turn, we see that the BSF system has minimum number of class features 2 and maximum 130. The mean number of features per class is 30.23, the median 14 and the standard deviation 32.40. The values for the Barat system are comparable to those of BSF. From Table 4, we note that Libjava has, by far, the highest median value of 33, yet its mean class size is comparable to that of the other three systems. This suggests that classes in this system had relatively few large classes (and this is supported by the relatively low StD value for this system (24.64)). Finally, the wide variety of class size in the Swing system is illustrated by its low median, yet high mean and standard deviation values.

Table 4. Summary data for class size in all four systems

System	Min	Max	Mean	Median	SD
BSF	2	130	30.23	14	35.85
Barat	2	130	28.12	13	32.40
Libjava	2	90	35.00	33	24.64
Swing	1	215	35.16	18	39.74

Table 5 shows, for Barat and Swing, the correlation of the number of manual errors made by the data collector versus class size (again, we omit correlation analysis for the BSF and Libjava systems because of the small sample size).

Table 5. Correlation of errors versus class size

System	Kendall's	Spearman's
Barat	-0.02	-0.03
Swing	0.06	0.07

Clearly, evidence from Table 5 suggests that class size seems to have little or no impact on the number of manual errors made by the data collector. The correlation coefficients for both Swing and Barat are very low. Based on Table 5, we cannot find any support for Hypothesis H_{A2}. We conclude in favour of H_{02} - the size of a class has no bearing on the number of errors made by the data collector.

G. Inner Classes and Class Size

One aspect of the analysis that we have not yet considered is the relationship between the number of inner classes and the size of a class. One question that this might address is whether the presence of inner classes influences, or is influenced by, class size. Table 6 shows the correlation between inner classes and class size for Barat and Swing (the analysis includes classes with 0 inner classes).

Table 6. Correlation of inner classes versus class size

System	Kendall's	Spearman's
Barat	0.01	0.00
Swing	0.06	0.08

As suggested by the analysis in the previous sections, there is no apparent relationship between the number of inner classes and class size for either system according to the correlation values. In theory, we could argue that a class with many inner classes is likely to be a relatively cohesive class, since we would expect the functionality of the class as a whole to be focused on a specific task. We would also expect the same class to have low coupling, since much of the coupling that the class needed would realistically be satisfied by its inner class or classes.

V. DATA COLLECTION IMPEDIMENTS

Both hypotheses H_{A1} and H_{A2} were unsupported by our analysis. This was a surprising, yet interesting result to emerge from our analysis. If two seemingly obvious possibilities for the magnitude of manual errors have been discounted, then the natural question that arises is: '*what class features did cause errors in the manual data collection to be made by the data collector?*' Subsequent visual inspection and recourse to notes made by the data collector (of the classes where most of the errors were made) identified a number of key trends which *may* account for the underlying reasons pertaining to data collection errors; we next discuss each of the four systems in turn, pointing to specific features that made the manual data collection process problematic in each case.

H. The BSF System

The BSF system has consistently shown itself to be a well-written system that conforms to OO principles [5, 16]; the errors in BSF were, in fact, the lowest in terms of frequency and number of errors from all four systems (as a percentage of total classes, Swing was the least erroneous). Visually, classes in BSF tended to be well and consistently commented, well indented and well laid out. Collection of both private and public class features showed few errors; collection of private attributes was virtually error-free; only two errors were made in total during the collection of private features throughout the whole system, suggesting an easily comprehensible and easily digested layout at the head of each class. Moreover, the errors made in BSF did not discriminate against size; two errors were made in a class with just 2 features (i.e., sum of attributes and methods), namely, class *IOUtils*), as well as a class with 74 features (class *CodeBuffer*). In contrast to the other three systems, the data for BSF suggests no specific or obvious reason for the errors made by the data collector for this system. To reinforce the relatively low numbers of errors for this system, only six of the fifteen metrics collected provided the source of manual errors. For any manual process there will always be some natural element of error. Of the four systems analyzed, we believe that the majority of errors made for BSF fall into this category.

I. The Barat System

First, comment lines seem to exist in a haphazard style, with different commenting syntax and the tendency of developers to embed comments at inconsistent places within the code. One class: *InnerClassVisitor* contained three types of comment line style within the space of four lines of code/declarations. While we could, in theory, suggest that the more comments there are in code, the better the aid is to code comprehension by a developer, an alternative viewpoint is that comments can clutter up the code; Fowler [9] suggests that comment lines can also be a bad smell if their purpose for existing is not obvious. In the Barat system, the inconsistency of commenting seems to have had a strong influence on the tendency of the data collector to make data collection errors. Second, there was a tendency for the developers to interleave definitions of private, protected and public features. In other words, in contrast to the BSF system, there was a tendency in this system for the developers to add declarations in a haphazard and illogical manner. For example, a series of public declarations would be followed by a series of private declarations, followed by more public declarations and so on. Third, many errors could be attributed to an inconsistent style of class definition. For example, contrast the *ClassImpl* class of Barat (1 attributable manual error), containing zero code comments, with the *InnerClassVisitor* class of the same system (3 attributable manual errors) containing copious numbers of comment lines; we note that each of these classes was written by a separate developer).

J. The Libjava System

Interestingly, many of the errors made for Libjava were mainly on public features. Only five of the fifteen metrics listed in Section 3.3 were the cause of any single error by the data collector (three of these were for public features). One explanation for the relatively large number of errors made in the public features of Libjava classes is that being a library-based system, as much scope as possible is given to the developer to tailor the classes used from the Library for his/her own purposes. Making declarations public in the first instance facilitates this tailoring process more easily. Consequently, we suggest that there are likely to be more public features in a library-based system than in a non-library one and, *mutatis mutandis*, relatively more errors. Apart from this feature, Libjava is close to BSF in terms of accuracy of collection of all other metrics. While the research in this paper condones grouping of identical type declarations, e.g., bunching of private, protected and public declarations separately, one drawback of large numbers of grouped declarations is that (as we have previously noted for BSF) human errors will inevitably creep in as a result. This seems to be a feature of Libjava also.

K. The Swing System

The Swing system is the largest of the four systems and the most error-prone in terms of manual errors. Thirteen of the fifteen metrics caused at least one error to be made by the data collector (private and protected constructors were the only two metrics that did not cause an error to be made). In common with the Libjava system, many of the errors for this system were for the public features of classes. In fact, over half of all errors for this system were made on public constructors and methods. Overall, the Swing system exhibited the largest number of manual data collection errors. It is interesting that the results for Swing are in complete contrast to Libjava, the other library-based system. One suggestion as to why Swing was so error-prone (and hence why it was so different to Libjava) is that the system may well have evolved to a greater extent and has thus been subjected to more maintenance effort. We note that the two occurrences of the aforesaid ten errors (Section 3.4) were for the classes *MutableCaretEvent* and *JTextComponent.* These two classes contained 16 and 106 class members, respectively, again reinforcing the claim (of Hypothesis H_{02}) that size of a class did not seem to unduly influence the propensity for errors to be made by the data collector.

VI CONCLUSIONS

In this paper, we investigated the propensity of a data collector to make manual data collection errors whilst collecting fifteen class metrics from four Java open-source systems. We hypothesized that inner classes and class size would be the main causes of such data collection errors – this was not found to be the case. Inconsistency in coding styles lay at the heart of the problem. The study thus highlights the problem that adoption of OSS presents, that is, different Java developers working on different classes and all adopting their own style of coding. We suggest that an important aspect of managing and coping with OSS software is to enforce a consistent style and arrangement of declarations in classes. It also highlights the importance of manual data collection as a vehicle for demonstrating software engineering 'good' and 'bad' practice, and the importance of studies focussing on data collection as an important software engineering discipline that can inform our understanding of software and its characteristics.

REFERENCES

[1] Basili, V.R., Briand, L.C., and Melo, W.L. A validation of object-oriented design metrics as quality indicators. IEEE Trans. on Software Engineering, 22(10):751-761, 1996.

[2] Briand, L., Bunse, J., Daly, J., and Differding, C. An experimental comparison of the maintainability of object-oriented and structured design documents. Empirical Soft. Eng.: An International Journal, 2(3):291-312, 1997.

[3] Briand, L., Devanbu, P., and Melo., W. An investigation into coupling measures for C++. Proceedings of the 19th IEEE International Conference on Software Engineering (ICSE 97), Boston, USA, pages 412-421, 1997.

[4] Counsell, S., Loizou, G., and Najjar, R. Quality of manual data collection in Java software: an empirical investigation. Empirical Software Engineering: An International Journal, 12(3):275-293, 2007.

[5] Counsell, S., Loizou, G., Najjar, R., and Mannock, K. On the relationship between encapsulation, inheritance and friends in C++ software. Proc. Intl. Conf. Soft. and Systems Eng. and Applications (ICSSEA'02), Paris, France, 2002.

[6] Dinh-Trong, T., and Bieman, J. Open source software development: A case study of FreeBSD. Proc. IEEE Intl. Symp. Software Metrics, Chicago, USA, pp. 96-105, 2004.

[7] Fenton, N., and Pfleeger, S. L. *Software Metrics. A Rigorous and Practical Approach*. Thomson International Computer Press, 1996.

[8] Field, A. *Discovering Statistics Using SPSS*. Sage Publications, 2005.

[9] Fowler, M. *Refactoring: Improving the Design of Existing Code*. Addison Wesley, 1999.

[10] Gnu at: http://www.gnu.org/. Accessed 4/12/07.

[11] Harrison, R., Counsell, S., and Nithi, R. An investigation into the applicability and validity of object-oriented design metrics. Empirical Software Engineering: An International Journal, 3:255-273, 1998.

[12] Izurieta, C., and Bieman, J. The evolution of FreeBSD and Linux. Proceedings of the ACM/IEEE International Symposium on Empirical Software Engineering (ISESE 2006), Rio de Janeiro, Brazil, pages 204-211, 2006.

[13] Kitchenham, B., Pfleeger, S., Pickard, L., Jones, P., Hoaglin, D., El Emam, K., and Rosenberg, J. Preliminary guidelines for empirical research in software engineering. IEEE Trans. on Soft. Engineering, 28(8):721-734, 2002.

[14] Kitchenham, B., and Pfleeger, S. Software quality: The elusive target. IEEE Software, 13(1):12-21, 1996.

[15] McGraw, G., and Felten, E. Twelve rules for developing more secure Java. Java World, 12/01/1998. http://www.javaworld.com/javaworld/jw-12-1998/jw-12-securityrules.html?page=1; Accessed 30/12/06.

[16] Najjar, R., Counsell, S., Loizou, G., and Mannock, K. The role of constructors in the context of refactoring object-oriented systems. Proc. European Conf. on Soft. Maint. and Reengineering, Benevento, Italy, pp. 111-120, 2003.

[17] Najjar, R., Counsell, S., Loizou, G., and Hassoun, Y. The quality of automated and manual data collection processes in Java software: an empirical comparison. Intl. Conference Advanced Inf. Systems Engineering (CAiSE Workshops (2)), Riga, Latvia, pages 101-112, 2004.

[18] Rosenberg, S. *Dreaming in Code*. Crown Pub., 2007.

[19] Siegel, S. and Castellan, N.J. (1988) *Nonparametric Stat. for the Behavioural Sciences*. McGraw-Hill, NY.
[20] Sintes, T. So what are inner classes good for anyway? http://www.javaworld.com/javaworld/javaqa/2000-03/02-qa-innerclass.html?page=1; Accessed 30/12/06.
[21] Spinellis, D., and Szyperski, C. How is open source affecting soft. development? IEEE Soft., 21(1):28-33, 2004.
[22] Viega, J., McGraw, G., Mutdosch, T., and Felten, E.W. Statically scanning Java code: finding security vulnerabilities. IEEE Software, 17(5): 68-74 2000.

An Empirical Study of "Removed" Classes in Java Open-Source Systems

Asma Mubarak, Steve Counsell and Robert M. Hierons
Department of Information Systems and Computing
Brunel University, Uxbridge, Middlesex, UB8 3PH, UK
{asma.mubarak, steve.counsell, rob.hierons}@brunel.ac.uk

***Abstract*-Coupling is an omni-present and necessary feature of OO systems; ideally, classes with excessive coupling should be either refactored and/or removed from the system. However, a problem that immediately arises is the practical difficulty of effecting the removal of such classes due to the many coupling dependencies they have; it is often easier to leave classes where they are and 'work around' the problem. In this paper, we describe empirical coupling and size data of classes removed from multiple versions of four open-source systems. We investigated three related, research questions. First, does the amount of coupling influence the choice of removed class? Second, does class size play a role in that choice? Finally, is there a relationship between the frequency with which a class is changed and its point of removal from a system? Results showed a strong tendency for classes with low 'fan-in' and 'fan-out' to be candidates for removal. Evidence was also found of class types with high imported package and external call functionality being removed; finally, size, in terms of methods and lines of code did not seem to be a contributing factor to class removal. The research addresses an area that is often overlooked in the study of evolving systems, notably the characteristics and features of classes that disappear from a system.**

I. Introduction

Excessive class coupling has often been related to the tendency for faults in software [1]. It is widely believed in the OO software engineering community that excessive coupling between classes creates a level of complexity that can complicate subsequent maintenance and potentially lead to the seeding of (further) faults. In practice, a class that is highly coupled to many other classes is an ideal candidate for re-engineering or removal from the system to mitigate current and potential future problems. Moreover, a highly coupled class is, other things remaining equal, likely to have grown to be a relatively large class (making it *even more* suitable theoretically for removal from the system). The paradox that immediately arises, however, is that it is often easier to leave a highly coupled class undisturbed, than to attempt to remove it. In other words, the disadvantages associated with its removal (i.e., the side-effects, re-work and re-test) outweigh the disadvantages of simply leaving the class where it is. In this paper, we investigate versions of four Open Source Systems (OSSs) with particular reference to classes 'removed' during their evolution. We answer three related research questions. First, are classes removed from the system lowly or highly coupled relative to other classes in the same package? Second, are the same classes excessively large compared with the remaining classes in the package? Third, are removed classes changed frequently before they are removed? The paper is organized as follows: Section 2 describes the motivation for the research and related previous work. In Section 3, the systems under study are introduced together with an overview of the metrics collected. Section 4 presents an analysis of the data collected; Section 5 provides a discussion of the points raised by the study and finally, conclusions and future research are presented (Section 6).

II. Motivation and Related Work

The research in this paper is motivated by a number of factors. First, we would always expect potentially problematic classes to be re-engineered by developers through techniques such as refactoring [2]; however, practical realities (limited time and resources) mean that only when classes exhibit particularly bad 'smells' [2] are they dealt with. This paper explores the characteristics of classes removed from a system, research that has not been touched on in any previous work that we know of. Throughout, we interpret the term 'removed' to mean that either a class has been a) decomposed to form one or more newly named classes, b) moved to a different package and renamed or c) simply removed from the system because it is moribund. Second, there is no prior study that we know of which suggests large classes with high coupling are removed any more or less frequently than small, low-coupled classes. Large classes may be a maintenance problem and hence candidates to be decomposed. On the other hand however, small classes are more portable (and hence moved more easily). Finally, while there has been some work on finding the optimal size of class [3], very little empirical research has investigated whether through analysis of removed classes, there is a coupling level beyond which action by the developer is usually triggered. The research described in this paper relates to areas of software evolution, coupling metrics, and the use of open-source software [4], [5]. In terms of software evolution, the laws proposed by Lehman [6] have provided the basis for many evolutionary studies in the past. Evolution has also been the subject of simulation studies [7] and this has allowed OSS evolution to be studied in a contrasting way to that empirically. In terms of coupling, a framework for its measurement was introduced in

K. Elleithy (ed.), *Advanced Techniques in Computing Sciences and Software Engineering*,
DOI 10.1007/978-90-481-3660-5_17,

[8]; variations for different programming styles have also been proposed [9]. Li and Henry [10] support the view that excessive coupling makes maintenance and tracing more difficult. Chidamber and Kemerer proposed six OO metrics amongst which were the Response for a Class and Coupling between Objects coupling metrics [11]. Finally, this study contributes to an empirical body of knowledge on coupling and longitudinal analysis of which more studies have been urged [12], [13].

III. Systems and Metrics

Four systems were used as a basis of our study:

1) Jasmin. A Java assembler which takes ASCII descriptions of Java classes and converts them into binary Java .class files suitable for loading into a Java Virtual Machine. The system comprised 5 versions. It started with 5 packages and 110 classes in the first version and had 5 packages and 130 classes by the latest version.

2) DjVu. Provides an applet and desktop viewer Java virtual machine. The system comprised 8 versions. It started with 12 packages and 77 classes in the first version with 14 packages and 79 classes in the latest version.

3) pBeans. Provides automatic object/relational mapping (ORM) of Java objects to database tables. The system comprised 10 versions, with 4 packages and 36 classes in the first version and 10 packages and 69 classes in the latest version.

4) Asterisk. The Asterisk Java system consists of a set of Java classes that allow you to easily build Java applications that interact with an Asterisk PBX Server. It supports the FastAGI protocol and the Manager API. This system includes 6 versions. It started with 12 packages and 222 classes in the first version and ended with 14 packages and 277 classes in the final version.

OO metrics usually capture properties of OO systems such as cohesion, inheritance, encapsulation, polymorphism, size or coupling [14]. For this study, the JHawk [15] tool was used to collect five coupling metrics. For each of the four systems, we collected five coupling metrics 1) Message Passing Coupling (MPC): The number of messages passed among objects of a class; 2) PACK: Number of imported packages; 3) Number of EXTernal methods called (EXT): The more external methods that a class calls the more tightly bound that class is to other classes; 4) Fan In (FIN) and Fan Out (FOUT): FIN of a function is the number of unique functions that call the function. FOUT counts the number of distinct non-inheritance related class hierarchies on which a class depends. We also collected for each removed class the total number of methods (private, protected and public) and the lines of code (LOC) in each class as size measures.

IV. Data Analysis

The study comprises three research questions (RQ1, RQ2 and RQ3), stated as follows:

1) RQ1: Do removed classes contain significantly more or less coupling than other classes in the same package? This question is based on the belief that removed classes will tend to contain relatively small amounts of coupling when compared with other classes in the same package. We take the median coupling values of each metric within each package as basis for our comparison. The median represents the midpoint of all values for that metric. All values below the median would be 'relatively' low values and values above, relatively 'high' values by comparison.

2) RQ2: Are removed classes significantly 'larger' than other classes in the same package? This question is based on the belief that removed classes will tend to be small (in terms of their number of methods and LOC) when compared with other classes in the same package. Again, we take the median value for methods and LOC as a basis of our comparison.

3) RQ3: Do removed classes tend to be modified significantly before they are removed? This question is based on the belief that classes which are modified significantly through versions of the systems studied are more likely to be removed because they are causing frequent maintenance problems in the system.

A. Research Question 1 (RQ1)

Table 1 shows the values for the five coupling metrics expressed as values plus or minus the median for that package and in the version where the class was removed (the median metric value for the package and for that version of the system is shown in brackets after each value in each case; if classes are removed in different versions, the median values for that particular version are shown). For example, the MPC value for class *StackMapAttr* was 25 greater than the median value of 6 for that package (i.e., it had value 31). Equally, the MPC for class **Signed_num_token** was 4.5 less than the median MPC of 4.5 in that package (i.e., zero). Since both *StackMapAttr* and *StackMapFrame* classes were removed in the same version they share the same set of median values given in the first row (this is not always the case). We have also highlighted the fact that classes are taken from different packages by alternating *italicized* class values with bold un-italicized values. So, the first two classes in Table 1 are from one package and the third class from a different package. We note also that the values in brackets represent the median for the *whole* package and therefore apply to *all* similar rows below it in the same table. For the Jasmin system, the three removed classes were all found in the fourth version (out of five). The first two removed classes are higher than the median for the coupling metrics. For the third class, all but one the same metrics are below the median. Clearly, for this system, coupling exceeds the median in the majority of cases.

TABLE 1
REMOVED CLASSES COMPARED TO MEDIAN (JASMIN)

Removed Class	MPC	EXT	PACK	FOUT	FIN
StackMapAttr	*25 (6)*	*13(5)*	*2 (1)*	*6 (0)*	*0 (0)*
StackMapFrame	*15*	*7*	*2*	*2*	*3*
Signed_num_token	**-4.5 (4.5)**	**-3.5 (3.5)**	**-0.5 (1.5)**	**-1.5 (1.5)**	**0 (0)**

For DjVu (Table 2), a number of different patterns emerge. First, it seems that when a class was removed, it tended to have relatively low (i.e. minus) MPC, EXT and PACK values compared with other classes (given by the median). For example, seven of the twelve removed classes contained values of MPC significantly less than the median; five of the twelve removed classes contained 10 or less EXT values than the median. The same trend applies to the PACK metric. (It is relatively easy to remove a class that is lowly-coupled in terms of message passing and external calls.) Equally, with the exception of one class, the values of FOUT for this system are either 0 or negative. This is not always the case for FIN, suggesting a difference in emphasis between these two metrics when removing classes. A class with a higher FOUT than FIN is, in theory, easier to remove because it has fewer incoming dependencies than outgoing. Interestingly, only in four of the twelve cases does this occur in Table 2. Nonetheless, the values of FIN and FOUT are generally low; for two of the packages every FOUT value of removed classes is less than or identical to the median value. Also of note are the exceptionally low values of MPC and EXT for the third package (each of the three classes in this package was removed in different versions because they each have their own set of median values). Overall, of the sixty values for all metrics in the table, 33 were negative and 7 equivalent to the median (value of zero in the table).

TABLE 2
REMOVED CLASSES COMPARED TO MEDIAN (DJVU)

Removed Class	MPC	EXT	PACK	FOUT	FIN
GMapRect	*12 (20)*	*3 (15)*	*0 (1)*	*-1 (5)*	*4 (10)*
GRectMapper	*7*	*1*	*-1*	*-5*	*-10*
LibRect	*-15*	*-10*	*1*	*-5*	*4*
Annotation	*-20*	*-15*	*-1*	*-5*	*0*
ByteVector	**5.5 (20)**	**2(15)**	**0 (2)**	**0 (5)**	**-2.5 (10.5)**
DataPool$ CachedInput-Stream	**1.5**	**1**	**-2**	**0**	**-10.5**
IFFContext	**-9.5**	**-8**	**-1**	**-3**	**-7.5**
GMapOval	*-17 (27)*	*-7 (17)*	*1 (1)*	*0 (4)*	*3 (0)*
GMapPoly	*51*	*20*	*1*	*1*	*3*
BoundImage	**-41(42)**	**-27.5 (28.5)**	**0 (5)**	**-7.5 (7.5)**	**1 (3)**
DjVuBean$ HyperlinkListener	*-46 (59)*	*-30 (43)*	*-5 (5)*	*-7 (11)*	*-5 (5)*
SimpleArea	**-50 (75)**	**-34 (48)**	**-4 (6)**	**-10.5 (12)**	**0.5 (4.5)**

Table 3 shows the coupling metrics for the pBeans system expressed as values plus or minus the median. Eleven classes were removed from three different packages. In common with the DjVu system, the FIN values seem to be low compared with the median value and in this case, so too the FOUT. The most notable feature of Table 3 is the fact that all six classes removed from the second package (bolded) relate explicitly to databases. Moreover, the MPC and EXT values are exceptionally high for these classes. There is a plausible explanation for this feature.

Database classes are more likely to be used extensively by other classes and that could explain the high MPC and EXT values (the PACK values for the same classes are relatively low). It might be the case that these six classes may not have been removed necessarily, but simply 'moved' *en masse* as part of an 'Extract Package' refactoring to re-locate database classes where they are most needed [2]. It is interesting that not all of the same six classes had low FIN and FOUT values, suggesting that only some forms of coupling may be relevant or considered by a developer when deciding on class removal. Of the 55 values in Table 3, only 21 values were negative. The majority of positive values were accounted for by the database classes.

For the Asterisk system (Table 4), eight classes were removed from four packages. The Asterisk system exhibits a similar pattern to the DjVu system in terms of the FIN and FOUT values, the majority of which were either zero or negative when compared with the median. In keeping with the pBeans system, the MPC and EXT values for removed classes are quite large in many cases.

Consider for example the classes **ReplyBuilderImpl**, **ReplyBuilderImplTest**, **RequestBuilderImpl** and **RequestBuilderImplTest** – all of which have high MPC and EXT values. Finally, the two classes **ServerSocketFacadeImpl** and **SocketConnectionFacadeImpl** are related to patterns and, in particular, the façade pattern (evidence in [16] suggests that pattern classes are susceptible to change). The same phenomenon of moving related classes such as that for the database classes of pBeans may apply here.

TABLE 3
REMOVED CLASSES COMPARED TO MEDIAN (pBeans)

Removed Class	MPC	EXT	PACK	FOUT	FIN
ObjectClass	*-2 (2)*	*-2 (2)*	*0 (0)*	*-1.5 (1.5)*	*1 (5)*
ObjectClass_StoreInfo	*2*	*2*	*2*	*0.5*	*-5*
PersistentMap Entry_StoreInfo	*-1*	*-1*	*0*	*-0.5*	*-5*
PersistentMap_StoreInfo	*-1*	*-1*	*0*	*-1.5*	*-5*
HsqlDatabase	**33.5 (0.5)**	**27.5 (0.5)**	**4.5 (0.5)**	**15.5 (0.5)**	**-1.5 (1.5)**
HsqlDatabase$ UpperCaseMap	**3.5**	**2.5**	**-0.5**	**0.5**	**-1.5**
PostgreSQLDatabase	**36.5**	**32.5**	**3.5**	**17.5**	**-1.5**
PostgreSQLDatabase$ LowerCaseMap	**3.5**	**2.5**	**-0.5**	**0.5**	**-1.5**
MySQLDatabase	**9.5**	**6.5**	**2.5**	**3.5**	**-1.5**
SQLServerDatabase	**18.5**	**14.5**	**3.5**	**10.5**	**-1.5**
InitFilter	*-1(10)*	*0 (8)*	*3 (3)*	*1(5)*	*0 (2)*

TABLE 4
REMOVED CLASSES COMPARED TO MEDIAN (ASTERISK)

Removed Class	MPC	EXT	PACK	FOUT	FIN
ReplyBuilder	*0 (0)*	*0 (0)*	*1.5 (0.5)*	*0 (0)*	*-1 (1)*
RequestBuilder	*0*	*0*	*0.5*	*0*	*-1*
ReplyBuilderImpl	**17 (12)**	**15 (10)**	**-2 (5)**	**9 (7)**	**0 (0)**
ReplyBuilderImplTest	**36**	**3**	**-3**	**-1**	**0**
Request BuilderImpl	**55**	**41**	**5**	**7**	**0**
RequestBuilder ImplTest	**101**	**19**	**-2**	**0**	**0**
CommonsLoggingLog	*5 (2)*	*3 (2)*	*0.5 (0.5)*	*-1(2)*	*0 (0)*
NullLog	*-2*	*-2*	*-0.5*	*-2*	*0*
ServerSocket FacadeImpl	**4 (0)**	**4 (0)**	**2.5 (1.5)**	**2 (0)**	**-7 (7)**
SocketConnection FacadeImpl	**18**	**12**	**4.5**	**1**	**-7**
Util	*5.5 (2.5)*	*4 (2)*	*-2 (2)*	*0 (1)*	*7 (0)*

In response to RQ1, we suggest that FIN and FOUT coupling may be strong determinant of whether a class is removed – low values of each may help the removal of a class; equally, high amounts of MPC and EXT may actually be one stimulus for moving a class. However, the key driver for removing classes as noted for classes in pBeans and Asterisk may be the need to remove related classes to a more convenient location.

B. Research Question 2 (RQ2)

Research question 2 attempts to answer the question whether removed classes were significantly 'larger' than other classes in the same package? Figure 1 shows the values of LOC for classes for each system and Figure 2 shows the NOM for the same four systems. The 'zero' vertical axis represents the median value of NOM and LOC in the four systems. Hence, plotted values represent NOM and LOC values above (plus) or below (minus) the median. Figure 1 seems to show that a similar number of the 37 removed classes had LOC values below the median as above it; in fact 21 of the 37 were either zero or above and therefore 16 were below the median.

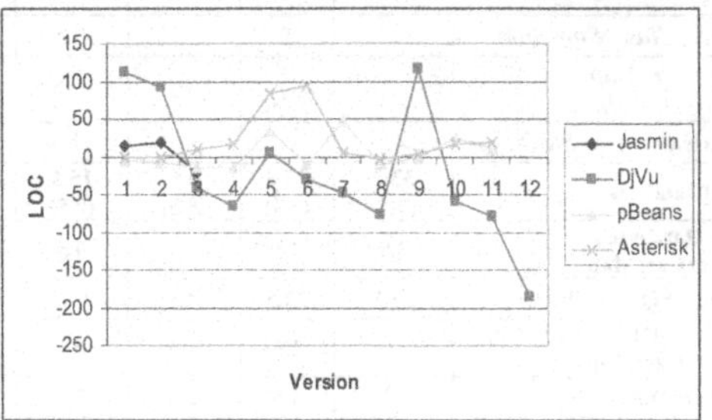

Fig. 1. LOC in removed classes

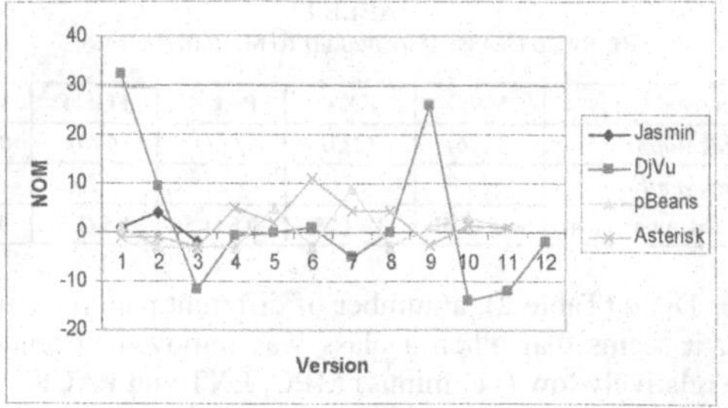

Fig. 2. NOM in removed classes

Figure 2 shows a similar pattern to Figure 1. Of the 37 values, 19 were zero or above (18 value were therefore below). In both figures, the DjVu system seems to be the system where both large and small classes were removed from the system (given by the erratic peaks). These results suggest that size, both in terms of NOM or LOC, seemed to have little bearing on the choice of removal of a class. A similar effect appears to take place for the pBeans and Asterisk systems, but to a lesser extent. For the Asterisk system, the peak in NOM and LOC coincides with the high values for the second package in Table 4. This implies that for this system, removed classes were both highly coupled and relatively large.

Based on the evidence presented, and in response to RQ2, size does not seem to be a key determinant in the removal of a class. Both small and large classes were removed (coupling may be a far greater determinant). This result is supported by previous work [17] where size was found to be a poor predictor of OO cohesion; coupling was a far better determinant.

C. Research Question 3 (RQ3)

Research question three aims to answer the question whether removed classes were also the subject of significant changes over the course of the versions studied (prior to being removed)? Table 5 shows the number of classes of the set of removed classes that were the subject of changes during the five versions of the system studied. (As before, the values in the table are relative to the median.) For example, two of the three removed classes in the Jasmin system had had changes applied to them. Class *StackMapAttr* was removed 'In' version 5 and had changes applied to it between version 2 and 3.

For the DjVu system (Table 6), six out of the twelve classes were changed and these changes occurred between the second and the third versions in every case. Only one of these six classes was removed in the following version; the remaining classes were removed much later in the sixth, seventh and eighth versions.

For the pBeans system (Table 7), there were changes in just three classes out of eleven. However, most of these changes were in the first three versions and they were all removed in the ninth version. The class MySQLDatabase was changed twice during the period studied and the class SQLServerDatabase modified three times over the course of the versions studied (they thus have two and three entries in Table 7, respectively).

TABLE 5
CHANGES FOR REMOVED CLASSES (JASMIN)

Removed Classes	In	Changes		NOM	NOS	FOUT	FIN
StackMapAttr	*V5*	*V2-V3*	*0*	*0*	*-1*	0	0
StackMapFrame	*V5*	*V2-V3*	*2*	*6*	*-1*		

TABLE 6
CHANGES FOR REMOVED CLASSES (DJVU)

Removed classes	In	Changes	NOM	NOS	FOUT	FIN
ByteVector	*V8*	*V2-V3*	*0*	*-4*	*-3*	*-1*
DataPool$ CachedInputStream	*V8*	*V2-V3*	*0*	*0*	*0*	*0*
IFFContext	*V8*	*V2-V3*	*0*	*0*	*0*	*-1*
BoundImage	**V4**	**V2-V3**	**0**	**0**	**0**	**-6**
DjVuBean$ HyperlinkListener	*V6*	*V2-V3*	*0*	*2*	*0*	*0*
SimpleArea	**V7**	**V2-V3**	**0**	**0**	**0**	**-6**

TABLE 7
CHANGES FOR REMOVED CLASSES (PBEANS)

Removed classes	In	Changes	NOM	NOS	FOUT	FIN
PostgreSQLDatabase	**V9**	**V3-V4**	**2**	**8**	**1**	**0**
MySQLDatabase	**V9**	**V1-V2**	**1**	**6**	**1**	**0**
MySQLDatabase	**V9**	**V6-V7**	**0**	**3**	**1**	**0**
SQLServerDatabase	**V9**	**V1-V2**	**1**	**13**	**4**	**0**
SQLServerDatabase	**V9**	**V2-V3**	**0**	**6**	**1**	**0**
SQLServerDatabase	**V9**	**V3-V4**	**2**	**8**	**1**	**0**

TABLE 8
CHANGES FOR REMOVED CLASSES (ASTERISK)

Removed classes	In	Changes	NOM	NOS	FOUT	FIN
SocketConnection FacadeImpl	**V5**	**V2-V3**	**1**	**2**	**0**	**0**
Util	*V5*	*V1-V2*	*1*	*6*	*0*	*1*
Util	*V5*	*V2-V3*	*0*	*0*	*0*	*2*

Finally for the Asterisk system (Table 8), there were changes in just three classes out of eleven. However, these changes were in the first two versions and it was not until the fifth version that they were removed.

The conclusion we can draw in response to RQ3 is that first, removed classes are not necessarily changed significantly prior to their removal for the systems analyzed. Second, that removal of the classes took place at a later date to that of change in all cases investigated. This was a surprising result to emerge from the analysis. Finally, we note that for all the changed systems in Tables 5-8, the FIN and FOUT values are small even when compared with the other FIN and FOUT values in Tables 1-4.

D. Study Validity

One threat to the validity of the study is that we have only used four OSSs as part of our study. While that provides a cross-sectional view of systems, we accept that this limited number threatens the generalizability of the results. Second, we have identified removed classes but could not say whether these classes were simply moved to a different package and renamed (we would expect most removed classes to be decomposed and for the subsequent classes to be renamed). To counter this threat to validity, we did search for classes with different names but which had identical compositions to those removed classes, but found very little evidence to suggest that classes are actually moved and renamed (i.e., they tend to be decomposed or simply removed from the system). Finally, we have focused on coupling and size as the basis of our analysis. We could have used many other features of classes as a basis; for example, their cohesion or their position in the inheritance hierarchy [18]. We leave such analyses for future work, however.

V. CONCLUSIONS AND FUTURE WORK

In this paper, we have investigated the removed classes in four Java systems. Five coupling metrics were collected from four Java open-source systems using the JHawk tool. The study investigated three research questions. First, we investigated whether the extent of coupling influenced the removal of classes from a system. We found that the fan-in and fan-out metrics tended to be relatively small for removed classes. Moreover, that imported functionality (packages) and external calls play a role in certain cases (we found evidence of movement of database classes with high levels of message passing and external references). Second, we explored whether size was an influence on removed classes. We found little evidence that size did influence that choice. Finally, the expectation that removed classes were changed significantly before being removed was ill-founded; changes for most of the classes were made in early versions and removed relatively later on. In terms of future work, we want to extend our analysis to more systems and more versions in the spirit of [19]. We also want to investigate the link that removal of classes may have with refactoring, and specifically with respect to package refactoring in follow-up work to [20].

REFERENCES

[1] L. Briand, P. Devanbu, and W. Melo, "An investigation into coupling measures for C++," in *19th International Conference on Software Engineering (ICSE 97),* Boston, USA, 1997, pp. 412-421.

[2] M. Fowler, *Refactoring: Improving the Design of Existing Code.* New York: Pearson Education, 1999.

[3] K. El Emam, S. Benlarbi, N. Goel, and S. Rai, "The confounding effect of class size on the validity of object-oriented metrics," *IEEE Transactions on Software Engineering,* vol. 27 pp. 630-650, 2001.

[4] T. Dinh-Trong, and J. Bieman, "Open source software development: a case study of FreeBSD," in *10th IEEE International Symposium on Software Metrics,* Chicago, USA, 2004, pp. 96-105.

[5] R. Ferenc, I. Siket, T. Gyimothy, "Extracting facts from open source software," in *20th International Conference on Software Maintenance (ICSM 2004),* Chicago, USA, 11-17 September 2004, pp. 60-69.

[6] L. Belady, and M.A. Lehman, Model of Large Program Development. *IBM Sys. Jour.*, vol. 15, pp. 225-252, 1976.

[7] N. Smith, A. Capiluppi, and J. Fernandez-Ramil, "Agent-based simulation of open source evolution," *Journal of Software Process - Improvement and Practice,* vol. 11, pp. 423-434, 2006.

[8] L. Briand, J. Daly, and J. Wust, "A unified framework for coupling measurement in object-oriented systems," *IEEE Transactions on Software Engineering,* vol. 25, pp. 91-121, 1999.

[9] M. Bartsch, and R. Harrison, "A coupling framework for AspectJ," in *10th International Conference on Evaluation and Assessment in Software Engineering (EASE)*, Keele, UK, 10-11 April 2006.

[10] W. Li, and S. Henry, "Object oriented metrics that predict maintainability," *Journal of Systems and Software,* vol. 23, pp. 112-122, 1993.

[11] S.R. Chidamber, C.F. Kemerer, "A metrics suite for object oriented design," *IEEE Transactions on Software Engineering,* vol. 20, pp.476-493, 1994.

[12] C.F. Kemerer, and S. Slaughter, "Need for more longitudinal studies of software maintenance," *Empirical Software Engineering: An International Journal,* vol. 2, pp. 109-118, 1999.

[13] C.F. Kemerer, and S. Slaughter, "An empirical approach to studying software evolution," *IEEE Transactions on Software Engineering,* vol. 25, pp. 493-509, 1999.

[14] N. Fenton, and S. Pfleeger, *Software Metrics: A Rigorous and Practical Approach, 2nd ed.* Boston: Mass, 1997.

[15] JHawk tool: (http://www.virtualmachinery.com/jhawkprod.htm).

[16] J. Bieman, G. Straw, H. Wang, P. Munger, and R. Alexander, "Design patterns and change proneness: an examination of five evolving systems," in *Ninth International Software Metrics Symposium,* 2003, pp. 40-49.

[17] S. Counsell, "Do student developers differ from industrial developers?" in *Proceedings Information Technology Interfaces (ITI) Conference*, Dubrovnik, Croatia, June 2008.

[18] M. Cartwright, and M. Shepperd, "An empirical investigation of an object-oriented (OO) system," *IEEE Transactions on Software Engineering,* vol. 26, pp. 786-796, 2000.

[19] T. Girba, M. Lanza, and S. Ducasse, "Characterizing the evolution of class hierarchies," in *Ninth European Conference on Software Maintenance and Reengineering,* Manchester, UK, 2005, pp. 2-11.

[20] A. Mubarak, S. Counsell, R. Hierons, and Y. Hassoun, "Package evolvability and its relationship with refactoring," in *Third International ERCIM Symposium on Software Evolution,* Paris, France, 2007.

Aspect Modification of an EAR Application [a]

Ilona Bluemke, Konrad Billewicz
Institute of Computer Science, Warsaw University of Technology
Nowowiejska 15/19, 00-665 Warsaw, Poland
I.Bluemke@ii.pw.edu.pl

***Abstract*-We propose aspect modification of compiled Java programs for which source code and documentation are not available. Aspect oriented programming is used to trace the program execution and identify points, in which aspects implementing new functionalities should be applied. A special tool for aspect oriented program tracing was designed and implemented. A modification in an real Enterprise Application Archive (EAR), compiled, without source code and documentation is presented in this paper. Advantages and disadvantages of described concepts are pointed out.**

I. INTRODUCTION

In the maintenance of software one of the severe problems is the lack of documentation. Often a new functionality has to be added to a program or some bugs should be corrected and the technical documentation of this program is not available. If sources of this program are available, this problem can be solved by code reviewing, debugging, or by adding some logs to the source code and later analyzing them. These activities are time consuming but may result in the requested new functionalities or corrected errors. More severe problem in the maintenance occurs if the program sources are unavailable. In such case the program can be decompiled, some logs are added to the decompiled version, then the program execution should be observed and analyzed. For such programs modification or error correction is very difficult and risky.

We propose a new approach to the modification of programs with unavailable source code and documentation i.e. aspects oriented programming [1,2,3]. Aspect oriented programming (AOP) is a new paradigm in programming, extending traditional programming techniques, first introduced in [2]. Its basic idea is to encapsulate concerns which influence many modules of a given software system, so called *crosscutting concerns*, in a new module called *aspect*. This encapsulation improves separation of concerns and can avoid invasive changes of a program if crosscutting concerns are affected by system evolution. The functionality defined in the aspect is *woven* into the base system with the so called *aspect weaver* in the place called *joinpoint*. Aspects may be used to trace the program execution and identify points, in which aspects implementing new functionalities may be applied.

Some general information concerning tracing, tracking and monitoring of software can be found in [4]. Aspects oriented approach were used in program tracing [4,5,6,7] but the goal of tracing was slightly different than in our approach. Aspects can introduce new functions into a program. An example of adding new functionalities by aspects can be found in [8].

In this paper new domains for aspect oriented programming are proposed i.e. aspect oriented program tracing and aspect oriented program modification. Both are operating mainly on compiled programs without source code. A special tool – Crusader [9] for aspect oriented program tracing was designed and implemented at the Institute of Computer Science Warsaw University of Technology. Aspect approach to program tracing is new and we were not able to find a tool supporting this process. Crusader is weaving aspects into compiled Java programs. Tool consists two components. One of them is a desktop Java program used for tracing. The second is a traced program with aspects woven. Every program is woven with the same aspects. Aspects are sending messages to the Crusader console informing about actions executed in the traced program. To limit the number of send messages several filters can be applied e.g. on classes, packages, methods calls, arguments. The main ideas of aspect tracing and modification are described accordingly in sections II.A and II.B.

The properties of aspect modifications of compiled programs were examined on different programs types e.g. with console and graphical interfaces, in programs logic and in the interfaces. We used some programs available on LGPL (GNU Lesser General Public License) license from SourceForge.net [10]. For these programs the documentation were not accessible, we didn't use the code sources either. In [9] we shown, that modification in program logic e.g. adding new functionalities, can be successfully achieved with aspect modification. As an example we used program *JMathLib* [11] to which a new mathematical function was added. In other experiment, described in [12], we shown that modification in business logic of complex WAR (Web Application Archive) J2EE application are also possible. We also examined aspect modification in graphical interface. Aspect modification in GUI of *ClassEditor* [13] program we presented in [14].

---

[a] This work was supported by Dean of the Department of Electronics and Information Technology Warsaw University of Technology under grant no 503/G/1032/4300/008

K. Elleithy (ed.), *Advanced Techniques in Computing Sciences and Software Engineering*,
DOI 10.1007/978-90-481-3660-5_18, © Springer Science+Business Media B.V. 2010

In this paper we concentrate on aspect modification in business logic of EAR [15] application. Section II contains description of aspect tracing and modifications. In section III modification of EAR application – *timetowork* [16] from SorceForge.net site [10] is presented. Section IV contains some conclusions.

II. Aspect Tracing and Modification

To modify compiled Java programs without source code and documentation by aspect approach we need to:

1) *Trace the execution of program* to find appropriate places in which modification should be incorporated.

2) *Prepare aspect* introducing required changes and insert it into the program.

In our approach aspects are used to trace the program execution. They are automatically woven into the traced, compiled program by the Crusader tool. To every traced program are applied the same aspects, dynamically configured during run time. The tracing aspects are producing messages after each operation executed by the program under trace. The number of generated such messages can be huge, so the next step in the tracing process is to find a filter, limiting the number of messages. Following types of filters are provided in Crusader:

1. type of a *joinpoint* and calls from that *joinpoints*,
2. package and calls from it (a name or part of the package name can be given),
3. class and calls from this class,
4. method and method's arguments,
5. calls from a specified method .

The filters can be selected in several consecutive steps, in each step limiting the number of considered messages. Our goal is to use aspect tracing to find *joinpoints* in which aspect modifying program should be applied. In aspect tracing neither the source code, nor the documentation, of a program is required, so this method can be successfully used in maintaining and tracing compiled programs. Other advantage of this approach is the automation of this process. In aspect tracing every program operation can be traced. Such thorough tracing can be a tedious work, if the logs would be applied manually. The disadvantage of this approach is the effort needed to choose the message filters. The process of selecting filters can not be automated, human inventiveness is needed. Several iterations may be necessary to configure the filters.

Aspect tracing as an approach to trace program is known. Some examples can be found in [5, 6, 7]. In [5] aspects are used to built monitoring system for legacy applications. In this paper monitoring is treated as a separate concern and aspects are used to provide application state information.

Stamey and Saunders [6] use aspects to for tracing the execution of loops, methods. They show how aspects can be used to implement debugging of Java programs.

In [7] aspects are used for program instrumentation to get the traces. These traces are used in the analyze of aspect application. The traces for the base and woven system with identical inputs are compared. The comparison process is able to distinguish effects which are due to aspect application and due to behavioral changes. This approach can also be used to validate, that refactoring does not change the system behavior.

Aspects can be used to add new functionality to an existing system. Aspects, woven into a program, can change its functionality or add new ones. This approach is known but not widely used. An idea of aspect modification of a program can be found in [8]. The authors of this paper are using aspects to incorporate new, collaborative features into existing applications.

Aspects can be woven into a program treated as a "black box". Such approach can be used if aspects have some generic function e.g. logging all operations executed by program. We used this approach in the aspect tracing and in the implementation of Crusader [9]

Normally aspects can not be "blindly" applied. The *joinpoint* should be carefully detected e.g. by aspect tracing.

The main advantages of aspect modification are similar to aspect tracing i.e. no need for source code and documentation of the program and automatic application of aspects. Other advantage of this approach is that once prepared an aspect, may be used in new versions of the same program, or even in other programs. The disadvantage of aspect modification is the effort needed to write the code of aspect and especially to find the appropriate *joinpoint* for it. The logic of an aspect is usually more sophisticated than the logic of a modification inserted into the source code. During some experiments with aspect modification we noticed some problems with the effort needed to accomplish them. Currently, we are not aware of any research comparing the effort and efficiency in aspect and traditional modifications. The other problem with aspect modifications is that aspect can unintentionally change the behaviour of other parts of the program. This problem is studied in several research papers e.g. [7, 17].

III. Aspect Modification of an EAR Application

Enterprise Application Archive (EAR) [15] are advanced components in J2EE [18] standards. Currently, many complex applications are produced in this technology. It can be expected that in the near future these applications should be modified, to meet the changed user needs. We wanted to check how these applications could be modified by aspect approach. The structure of EAR application is not as simple as java program or even WAR application. An EAR application contains several components of different types e.g. archive, application or library. WAR applications could also be one of EAR components. We decided to modify an EAR application with only two components types i.e. WAR application and EJB component [19]. Aspect modification of an WAR applications we presented in [12]. In this paper we concentrate on modification of an EJB component. As an example of EAR application ***timetowork*** [16] was chosen. This application is available on BSD license in SourceForge.net [10]. The source code of this application is not provided. In its WAR part this application uses Struts [20] technology. *timetowork* application can be used in the time management process. The user fills the table with different activities. In figure 1, the table with activities and theirs durations from *timetowork* screen is presented.

In the timetable, shown in fig.1, the user writes the name of an activity, its durations and comments. One form is created for one day. In *timetowork* application the data inserted by user are not semantically checked. The comment field may contain any character but the hours field not. On the "hour" field several constraints may be put. We decided to check, if the number of hours, given as activities duration, does not exceed twenty four hours. This type of a constraint will prevent the user to write incorrect data. The modification is in the logic of application. The EJB component which was modified has no GUI. The GUI is implemented in WAR component of this EAR application. To fully introduce functional modification changes in both components i.e. EJB and WAR are necessary.

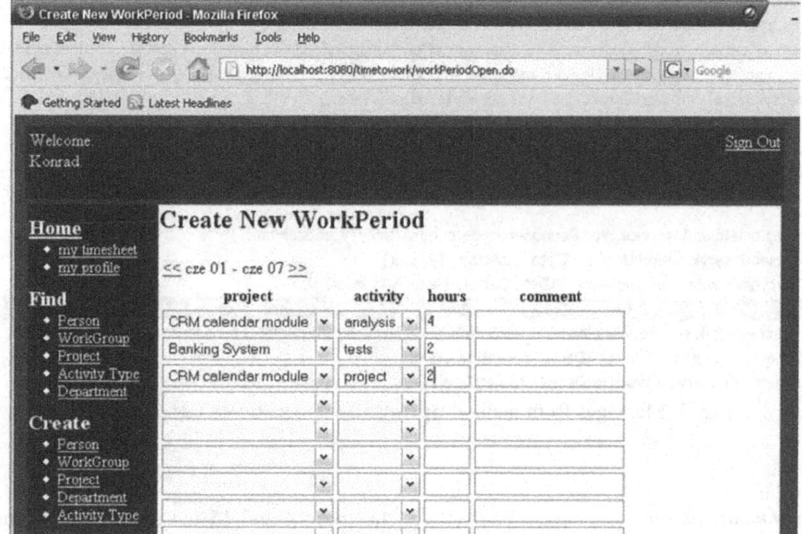

Fig. 1. Screen from *timetowork* application

As was mentioned at the beginning of this section, in this paper only modification in EJB component will be described. The EJB component after recognizing incorrect data will notify other application components by throwing an exception. In next sections the identification of modification point and modifying aspect will be presented.

A. Finding modification point

The modification points are located in the tracing process described in section II. Aspects woven into the application code by Crusader will produce tracing messages. The scenario to find the modification point is following:

1. Crusader adds tracing aspect to the EJB component of the EAR application.
2. EAR application with EJB is installed on application server.
3. Crusader is started with tracing option off.
4. Application server and *timetowork* application are started.
5. In *timetowork* user goes to the edition of timetable form.
6. In Crusader the tracing option is switched to on.
7. The time form is introduced into application
8. Tracing messages are observed on Crusader console.

As a result of the above described scenario about 5000 tracing messages were generated. It is not possible to analyze as many messages so filters on packages `java.lang`, `java.util` and `java.naming` were added. Two types of joinpoints i.e. field-read and field–write were also filtered. The number of messages was reduced to about 1000. It was observed, that classes from package `org.timetowork.domain` were found in methods type `get` and `set`. Objects from this package were `ValueObject` [21], they were also filtered. Filter set on package `org.timetowork.domain` reduced the number of messages to about 300. The analysis of the messages was difficult. It occurred, that *timetowork* application was using the CMP (Container Manager Persistence) [22] strategy. The person performing the aspect tracing was not familiar with the CMP strategy, so it was very difficult to guess the program logic. The idea to find in messages word "update" was successful. Often this word is used in the name of method responsible for some changes. Two messages shown in figure 2 and figure 3 have "update" as a part of the name. The analysis of methods containing in the name word "update" identified method `updateWorkPeriod()` from class `WorkPeriodServiceBeanImpl` as the method responsible for updating data in users form. In the analysis JAD [23] decompiler was also used to show the contents of objects present in the tracing messages. The analysis of methods containing in the name "update"word was time consuming, but succeeded in finding the modification point.

```
method-execution :: org.timetowork.service.WorkPeriodServiceUtil.getHome()
 method-call :: javax.rmi.PortableRemoteObject.narrow(org.timetowork.service/WorkPeriodService/HomeHome, interface org.timetowork.service.WorkPe
method-execution :: org.timetowork.service.WorkPeriodServiceBeanImpl.updateWorkPeriod(192c8d9:112a39fdcde:-7ff9, org.timetowork.service.WorkPer
 method-call :: org.timetowork.service.WorkPeriodServiceBean.getWorkPeriodLocalHome()
  method-execution :: org.timetowork.service.WorkPeriodServiceBean.getWorkPeriodLocalHome()
 method-call :: org.timetowork.service.WorkPeriodServiceBeanImpl.createNewWorkUnits(org.timetowork.service.WorkPeriodDetailData@e1e567)
  method-execution :: org.timetowork.service.WorkPeriodServiceBeanImpl.createNewWorkUnits(org.timetowork.service.WorkPeriodDetailData@e1e567)
   method-call :: org.timetowork.service.WorkPeriodDetailData.getWorkUnitDetailDatas()
    method-execution :: org.timetowork.service.WorkPeriodDetailData.getWorkUnitDetailDatas()
```

Fig. 2. Tracing modification point – messages from `updateWorkPeriod()`

```
    method-execution :: org.timetowork.service.WorkPeriodServiceBean.getActivityTypeLocalHome()
   method-call :: org.timetowork.service.WorkUnitDetailData.getActivityTypeId()
    method-execution :: org.timetowork.service.WorkUnitDetailData.getActivityTypeId()
 method-call :: org.timetowork.service.WorkPeriodServiceBeanImpl.updateCompositeRelation(org.jboss.ejb.plugins.cmp.jdbc.bridge.RelationSet@16c14e7
  method-execution :: org.timetowork.service.WorkPeriodServiceBeanImpl.updateCompositeRelation(org.jboss.ejb.plugins.cmp.jdbc.bridge.RelationSet@:
method-execution :: org.timetowork.service.WorkUnitDetailData.setProjectId()
method-execution :: org.timetowork.service.WorkUnitDetailData.getData()
```

Fig. 3. Messages from method `updateCompositeRelation()`

B. *Implementation of modification*

The process described in section III.A located the method `updateWorkPeriod()` in class `WorkPeriodServiceBeanImpl` as appropriate to introduce modification. The results produced by JAD decompiler comprised also the contents of simple classes `WorkPeriodDetailData` and `WorkUnitDetailData`, which were used in the modification. Aspect implementing the modification is presented in figure 4. The modification points are located in the tracing process described in section III.A.

In lines 14 and 15 the declaration of the joinpoint – the execution of method `updateWorkPeriod()` from class `WorkPeriodServiceBeanImpl`, is given. In lines 18 to 26 the data written by user are extracted. The hours input by user are summed and it is checked (lines 28, 29), if the sum is greater than 24. If it is, an exception, notifying the object calling this method, is thrown. The logs from *timetowork* program in which the sum of hours written is greater than 24, is presented in figure 5, the thrown exception can be seen.

```
package eu.billewicz.aspects;

import java.util.Collection;
import java.util.Iterator;

import org.timetowork.domain.WorkUnitValue;
import org.timetowork.service.WorkPeriodDetailData;
import org.timetowork.service.WorkUnitDetailData;



public aspect LimitHoursInTheDayAspect {

    before(WorkPeriodDetailData workPeriodDetailData)
        : execution(* org.timetowork.service.WorkPeriodServiceBeanImpl.updateWorkPeriod(..))
        && args(String, workPeriodDetailData) {

        Collection workUnitDetailDatas = workPeriodDetailData.getWorkUnitDetailDatas();

        int hoursSum = 0;
        for (Iterator i = workUnitDetailDatas.iterator(); i.hasNext(); ) {
            WorkUnitDetailData data = (WorkUnitDetailData)i.next();
            WorkUnitValue value = data.getData();

            hoursSum += value.getHours();
        }

        if (hoursSum > 24)
            throw new IllegalArgumentException("Hours sum > 24");
    }
}
```

Fig. 4. Aspect restricting the number of written hours

```
8:13:27,718 INFO  [STDOUT] Update... 192c8d9:112a39fdcde:-7ff9
18:13:27,718 ERROR [LogInterceptor] RuntimeException:
java.lang.IllegalArgumentException: Hours sum > 24 at
eu.billewicz.aspects.LimitHoursInTheDayAspect.ajc$before$eu_bilwicz_aspects_LimitHoursInTh
eDayAspect$1$fdc41f41 (LimitHoursInTheDayAspect.aj:29) at
org.timetowork.service.WorkPeriodServiceBeanImpl.updateWorkPeriod(Unknown Source)
18:13:27,734 INFO  [STDOUT] java.rmi.ServerException: RuntimeException; nested exception
is:
     java.lang.IllegalArgumentException: Hours sum > 24
```

Fig. 5. Part of log from modified timetowork application

C. *Remarks from the modification process*

In sections III.A and III.B we described how to modify complex EAR application without the source code and documentation by aspect approach. The modification process was difficult, time consuming, other tools i.e. JAD decompiler, were also used. However we have already some experiences in aspect modification (described in [9, 12, 14]), this modification was the most difficult one. The difficulty of this process was caused by many reasons. One of these reasons was the complexity of application logic, next one was the incompatibility of Java versions. Problems were also caused by obsolete technology used in *timetowork*. It can be expected that the same problems will appear in the near future in the maintenance of applications which are implemented nowadays.

The modification process of an EAR application has shown also the importance of proper naming in methods. The guess, that a method responsible for filling the form contains the word "update" in its name, was proved by tracing. Without this guess, finding the modification point would be much more difficult. With even basic knowledge of the technology, used in application, the tracing process would not be so long.

IV. CONCLUSIONS

We shown that modification in program logic of compiled simple program with console interface [9] and in more complicated WAR (Web Application Archive) [12] application can be made by aspect modification, even if the program source code and documentation is not available. In this paper we show that using aspects it is also possible to modify business logic of complex EAR application without the source code and documentation. We managed to successfully modify EJB components in *timetowork* application without any knowledge of the program logic. We introduced a change in the functionality of this application. Intuitive naming of methods and arguments helped in the process of aspect tracing. Familiarity with technology used in application would be also very helpful, unfortunately we were not familiar with CMP (Container Managed Persistance) [22] strategy used in *timetowork.* During software evolution it can be expected, that we will have to change application made in an obsolete technology, not known to us, and with unavailable source code and documentation. We measured the installation time for the original and modified applications. In the measurements process we used logs from JBoss application server. We repeated the measurements ten times. The times are almost the same i.e. 0.895 and 0.897s. We also measured the execution time (time to introduce new schedule) in *timetowork*. The execution times in original and modified by aspect application are almost the same i.e. 0.220s and 0.221 accordingly. The difference is so small that may be caused by measurements inaccuracy. The difference in the execution time is small, because modifications made to the application are small – only summarization and check, if the sum is greater than 24, were added.

Tracing aspects significantly slow the application. In application with tracing aspects and without filtering the time mentioned above was more than 54 s. When filters were applied the time was reduced to 8 s. Tracing aspects are used only to find the modifications point so the loss of effectiveness will not influence the regular application operation.

REFERENCES

[1] C. Lopes, *AOP: A Historical Perspective? (What's in a Name?). Aspect-Oriented Software Development,* Addison Wesley, 2004.

[2] G. Kiczales, J. Lamping, A. Mendhekar, Ch. Maeda, C. Lopes, J. Loingtier, J. Irwin, "Aspect-Oriented Programming", *Proceedings European Conference on Object-Oriented Programming*, vol. 1241, Springer-Verlag, 1997, pp. 220–242.

[3] page of Aspect Oriented Software Association http://www.aosd.net , 2008.

[4] J. Gao, E. Y. Zhu, S. Shim, "Monitoring software components and component-based software", http://www.engr.sjsu.edu/gaojerry/report/compsac2000.pdf.

[5] Hoi Chan, Trieu C. Chieu, "An approach to monitor application states for self-managing (autonomic) systems", *OOPSLA '03*, Anaheim, USA, 2003, pp.312-313.

[6] J. Stamey, B. Saunders, "Unit testing and debugging with aspects", *Journal of Computing Sciences in College*, vol. 20, issue 5, 2005, pp. 47-55.

[7] M. Storzer, J. Krinke, S. Breu, "Trace Analysis for Aspect Application", 2003, http://citeseerx.ist.psu.edu/viewdoc/summary? doi=10.1.1.15.270 .

[8] L. Cheng, J. Patterson, S. Rohall, S. Hupfer, S. Ross, "Weaving a Social Fabric into Existing Software", *Proceedings of the 5th International conference on Aspect-oriented software development AOSD'05*, March, Chicago, USA, 2005,pp. 147-159.

[9] K. Billewicz, I. Bluemke, "Aspect oriented programming in program tracing and modification", in "*Agility and discipline in software engineering*", Nakom, Poznań, 2007, pp. 23-34 (in polish).

[10] page of SourceForge service http://sourceforge.net, 2008.

[11] page of JMathLib http://sourceforge.net/projects/mathlib/, 2008

[12] I. Bluemke, K. Billewicz, "Aspects modification in business logic of compiled Java programs", *IEEE First International Conference on Information Technologies*, Gdansk, Poland, May 2008, pp. 409-412.

[13] page of ClassEditor http://sourceforge.net/projects/classeditor, 2008.

[14] I. Bluemke, K. Billewicz, "Aspects in the maintenance of compiled programs", *IEEE 3rd International Conference on Dependability of Computer Systems DepCoS 2008*, pp. 253-260.

[15] EAR http://docs.sun.com/app/docs/doc/8193875/6n62klump?a=view#jesgl-aoh, 2008

[16] page of timetowork http://sourceforge.net/projects/timetowork, 2008

[17] Dufour et al., "Measuring the dynamic behaviour of AspectJ programs", *OOPSLA '04*, Vancouver, Canada, 2004.

[18] page of J2EE http://java.sun.com/javaee/, 2008

[19] EJB http://docs.sun.com/app/docs/doc/8193875/6n62klump?a=view#jesgl-bxx, 2008

[20] page of Struts - http://struts.apache.org, 2008.

[21] M. Fowler, *Patterns of Enterprice Application Architectures,* Addison Wesley Signature series, pp. 486–487.

[22] CMP http://docs.jboss.org/jbossas/getting_started/v4/html/index.html , 2008

[23] page of JAD - http://www.kdpus.com/jad.html , 2008.

A Model of Organizational Politics Impact on Information Systems Success

Ismail M. Romi
Faculty of Informatics
Palestine Polytechnic University, Palestine
E-Mail: ismailr@ppu.edu

Prof Dr. Ibrahim A. Awad
Faculty of Engineering
Alexandria University, Egypt
E-Mail: i_abdelsalam@yahoo.com

Dr. Manal Elkordy
Faculty of Commerce
Alexandria University, Egypt
E-Mail: melkordy@yahoo.com

Abstract

The extent of the information systems success is based on various determinants including the organizational politics. This area wasn't discussed deeply by researchers. Thus a comprehensive understanding of information systems remains fuzzy and elusive in this area.

In an attempt to address this situation a comprehensive model was developed. Although this model is not empirically tested, it tries to explain the impact of organizational politics forms (avoiding, competing, accommodating, compromising, and collaborative) on the information systems success dimensions (system quality, information quality, service quality, use, user satisfaction, and net benefits).

Keywords: Information systems success, organizational politics, conflict.

1. Introduction

Nowadays Information Systems (IS) has a critical role to organizations success, where globalization, digital economics, and digital organizations took place. So the weakness of IS considered as a dangerous phenomenon on organizational success in general [6]. Furthermore, the question is not whether the organizations should have IS or not, but it should have an effective IS [8].

Therefore IS success and it's determinants considered to be critical in the field of Information Systems [12, 5, 33]. However, empirical results in this area are inconsistent [32], and an overall synthesis across the numerous empirical studies seems lacking [32, 29]. In addition to excluding many determinants that affects IS success [7, 29, 32]. Organizational politics is one of the important determinants of IS success that not deeply discusses by researchers [25], Thus a comprehensive understanding of Information Systems success remains fuzzy and elusive in this field.

The main objective of this study is to provide a further insight in to IS success and the organizational politics as a determinant of IS success, and integrating the results with the prior researches in this area. During this study a comprehensive model will be developed in order to enable the evaluation of information systems success along with the organizational politics, therefore this study addresses the following questions:

1. What are the dimensions of IS success?
2. How does the dimensions of IS success depend on organizational politics?

To pursue these questions, a comprehensive model was developed in this study, including the dimensions of IS success as dependent variables, and the organizational politics as an independent variable, after that an empirical study to be held to examine the developed model.

2. Theoretical Framework

2.1 The General Model

The general model is presented (Fig. 1) to identify the relationship between information systems success as a dependent variable and the organizational politics as an independent variable. The detailed research model is developed by using the prior literature on IS success and the organizational politics as a determinant to this success.

2.2 Information Systems Success Factors

Information system is an integrated computer-based system that utilizes computer hardware, software, users, procedures, models, and database which interacts to produce the suitable information at the appropriate time, to support organizational activities [1, 4, 9, 21, 22, 37].

Where information systems success factors are a set of dimensions and aspects that produces the net benefits of information system, which includes system quality, information quality, service quality, system usefulness, and user satisfaction [10, 11, 12, 34, 35].

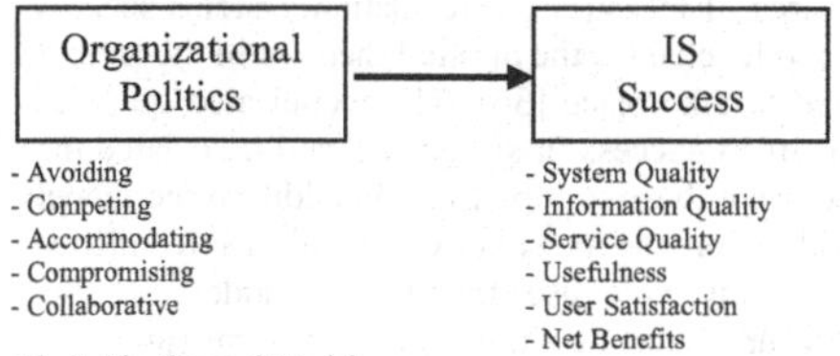

Fig.1: The General Model

K. Elleithy (ed.), *Advanced Techniques in Computing Sciences and Software Engineering*
DOI 10.1007/978-90-481-3660-5_19,

System Quality

System Quality is defined as a set of constructs related to information system that determines the quality of the systems [10, 35]. Those constructs are system reliability, ease of use, response time, relevance, timeliness, accuracy of information, and system productivity [32].

Information quality

Information quality is the degree to which information presents the required benefits [2, 10, 35]. Where Kahn et al [20] indicates a set of characteristics for information which are accessibility, appropriateness, believability, completeness, concise representation, ease of manipulation, free from errors, interpretability, objectivity, relevancy, reputation and security.

Service quality

Service quality was discussed by Parasurman et al, Delone and Mclean, Seddon, Hochstein et al, and Kim et al, [10, 16, 20, 27, 35] as a set of characteristics related to the services submitted by IS to customers which includes system reliability, assurance, empathy, timeliness, and security.

System usability

System usability is the consumption of IS output by the recipient of IS [10, 35]. This usability indicated with learnability, flexibility, and robustness [13]. Where learnability includes predictability, synthesizability, familiarity, generalizability, consistency. Flexibility includes dialogue initiative, multi-threading, substitutivity, customizability. Robustness includes observability, recoverability, responsiveness, task conformance.

User satisfaction

User satisfaction refers to the recipient response to the use of IS output [10, 35]. User satisfaction is associated with attitudes towards IS which depends on system availability, accuracy, completeness, consistency, robustness, flexibility, unobtrusiveness.

Delone and Mclean [12] model (D&M) presented in 2003 (Fig. 2) can be classified as one of the most interested models in the area of measuring information systems success [34, 15]. This model enables the applied theories in the area of IS measurement to take place [36]. The model also specifies the dimensions of IS success in six groups and associated the relationship between those groups [31]. In addition the model takes in consideration the perspectives of all IS recipients. Therefore this study will use the D&M model and the developed measures for its dimensions to measure the information systems success.

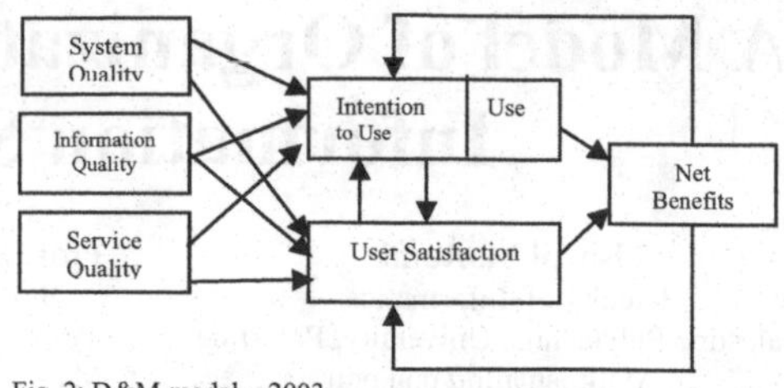

Fig. 2: D&M model – 2003

Source: (Delone and Mclean, 2003, p.24) [12]

2.3 Organizational Politics

Organizational politics are those activities and actions that are not required as part of the formal roles in organizations, and influence decision making and the distribution of benefits and resources in the organization [18, 23, 24, 28, 30].

The main driver of organizational politics is conflict of interest [28]. Miles [23] found that political activities occur in the presence of ambiguous goals, scarce resources, changes in technology or environment, non-programmed decisions, or organizational change. Political activities aim to developing and using power and other resources to obtain the preferred outcomes at the interpersonal level or intergroup level [18, 26, 39, 38].

Kacmar and Ferris [18] conceptualized organizational politics as a three-dimensional construct. This construct has become popular for measuring the level of perceived organizational politics in organizations [26]. The three dimensions of their model are general political behavior, getting along to get ahead, and pay and promotion.

General Political Behavior

General political behavior associated with the organizational politics in such that it involves the development of coalitions within a system that compete for scarce resources. The competition underscores the use of power to gain and maintain control of the political system. Power is expressed by who is allowed to participate in decision-making [28].

Getting Along to Get Ahead

This dimension relates to the existence of group loyalties. Compliance with group norms is valued, while dissenting opinions are not only discouraged, but met with sanctions [39]. Group cohesiveness, conformity, can lead policy-making bodies toward normative behaviors, critical thinking and objections to the majority view [17].

Pay and Promotion

This dimension is related to the concept that people who are a good "fit" for the organization are likely to be promoted. While the need to find individuals who will be a good fit for the organization should not be overlooked. The ability for individuals to use political means to gain promotion creates an environment where politically active individuals are promoted at the expense of others in the organization [39].

Political behavior within an organization can be either positive or negative based on how the coalitions and individuals manage conflict [24]. While managing conflict takes two dimensions which are assertiveness, or cooperativeness [3]. The interaction between these two dimensions (Fig.3) is reflected in a set of political forms, this includes avoiding, competing, accommodating, compromising, or collaborative.

Avoiding

An unassertive, uncooperative approach in which both groups neglect the concerns involved by sidestepping the issue or postponing the conflict by choosing not deal with it.

Competing

An assertive, uncooperative form in which each group attempts to achieve its own goals at the expense of the other through argument, authority, threat, or even physical force.

Accommodating

An unassertive, cooperative position where one group attempts to satisfy the concerns of the other by neglecting its own concerns or goals.

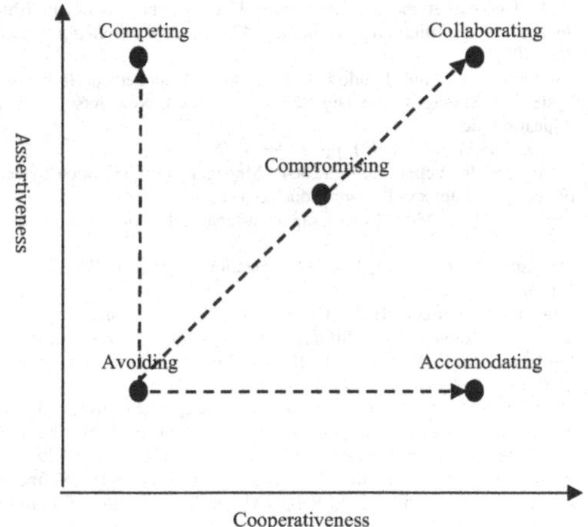

Fig. 3: Matrix of Political Forms

Source: (Bowditch and Buono, 1997, p.183) [3].

Compromising

An intermediate approach in which partial satisfaction is sought for both groups, where each group makes some concessions but also receives some concessions from the other.

Collaborative

An assertive, cooperative form that attempts to satisfy the concerns of both groups. This involves an agreement to confront the conflict, identification of the concerns of the different groups, and problem solving to find alternatives that would satisfy both groups.

2.4 Organizational politics and IS success

Organizational politics is one of the important determinants of information systems success [25]. In his study Warne [38] finds that there is a strong impact of organizational politics on IS success, where the greater the extent of conflict the greater the negative impact on IS success. Where the conflict involving users has the greatest impact on IS success.

In addition to those findings Chang et al [6] finds that organizational politics plays a very important role on IS success, and the greater the resources sharing the greater the conflict and the greater the negative impact on IS success.

3. The research model

The detailed research model (fig.4) incorporates IS success dimensions as dependent variables, and organizational politics forms as independent variables.

Thus the model includes eleven constructs, six constructs of the IS success factors which are system quality, information quality, service quality, system usefulness, user satisfaction, and net benefits. In addition to other five constructs of the political forms which are avoiding, competing, accommodating, compromising, and collaborative.

Therefore the model reflects a comprehensive insight for the impact of organizational politics forms on the IS success dimensions.

4. Conclusion and Recommendations

Prior studies indicate that in general there is an impact of organizational politics on information systems success. Where the greater the conflict, the greater the impact on information systems success.

Organizational politics plays a major role on the information systems success. This paper presents a comprehensive model that enables examining the impact of organizational politics with its various forms (avoiding, competing, accommodating, compromising, and collaborative) on information systems success dimension (system quality, information quality, service quality, system usability, user satisfaction, and net benefits), in order to classify the various constructs of

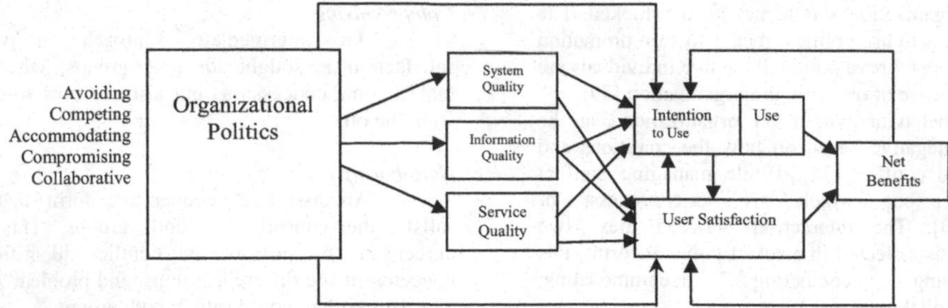

Fig.4: The Detailed Research Model

organizational politics in terms of their impact on information systems success.

A future research will include empirical testing of the model in the Palestinian General Universities in Palestine.

References

1. Alter, S. (1996), Information Systems - A Management Perspective (2nd ed.), New York: The Benjamin/Commings Publishing Company, Inc.
2. Barners, S.J., Vidgen, R.T, (2002), "An Integrative Approach to the Assessment of E-Commerce Quality", *Journal of Electronic Commerce Research,* 3: 11-127.
3. Bowditch, J. L., Buono, A. F. (1997), "A primer on organizational behavior", (4th ed.), *New York: John Wiley & Sons.*
4. Cashman, S., Vermaat,R. (1999), "Discovering Computers 2000- Concepts for A Connected World Retrieved: Feb, 2, 2007, From: http://www.belkcollege.uncc.edu/jwgeurin/istypes.doc.
5. Chang, J. C., King, W. R., (2005), "Measuring the Performance of Information Systems: A Functional Scorecard", *Journal of Management Information Systems,* 22(1): pp. 85-115.
6. Chang, L., Lin, T., Wu, S, (2002), "The Study of Information System Development (ISD) Process from the Perspective of Power Development Stage and Organizational Politics", *IEEE, Proceedings of the 35th Annual Hawaii International Conference on System Sciences*: pp. 265-275.
7. Chrusciel, D., Field, D. W., (2003), "From Critical Success Factors into Criterie for Performance Excellence- An Organizational Change Strategy", *Journal of Industrial Technology,* 19(4): pp.1-11.
8. Cragg, P., Tagliavini, M., (2006) "Evaluating Information Systems Alignment in Small Firms", *13th European Conference on Information Technology Evaluation, Italy.*
9. Davis, G.B., and Olson, M.H. (1985), Management Information Systems (2nd ed.), New York: McGraw-Hill Book Company.
10. DeLone, W.H., McLean, E.R.,(1992), "Information systems success: The quest for the dependen variable", *Information Systems Research,* 3(1): pp. 60–95.
11. DeLone, W.H., McLean, E.R.,(2002), "Information systems success Revisited", *IEEE,* 0-7695-1435-9/20.
12. Delone, W.H, Mclean, E.R, (2003), "The Delone and Mclean Model of Information Systems Success: a Ten- Year Update", *Journal of Information Systems,* 19: pp. 9-30.
13. Dix, A. J., Finlay, J. E., Abowd, G.D., Beale, D., (2004), Human Computer Interaction, (2th ed.), *New Jersey: Pearson Education Inc.*
14. Ferris, G. R., Kacmar, K. M, (1992), "Perceptions of Organizational Politics", *Journal of Management,* 18(1): pp. 93-116.
15. Gable, G.G, Sedera, D., Chan, T., (2003), "Enterprise Systems Success: Ameasurement Model", *Twenty-Fourth International Conference on Information System,* pp. 576-591.
16. Hochstein, A., Zarnekow, R., Brenner, W., (2005), "Managing Service Quality as Perceived by the Customer: The Service Oriented ServQual", *Wirtschafts Informatik,* 46: 382-389.
17. Janis, I.L. (1983) "Groupthink: Psychological Studies of Policy Decisions and Fiascoes", (2nd Ed., Revised), *Boston: Houghton Mifflin.*
18. Kacmar, K.M., Ferris, G.R. (1991). "Perceptions of Organizational Politics Scale (POPS): Development and Construct Validation" , *Educational and Psychological Measurement,* 51(1): pp. 193-205.
19. Kahn,B.K, Strong, D.M.,Wang, R.y., (2002), "Information Quality Benchmarks: Products and Service Performance", *ACM,* 45(4): pp. 184-192.
20. Kim, J.K, Eom, M.T, Ahn, J.H, (2005), "Measuring IS Service Quality in the Context of the Service Quality-User Satisfaction Relationship", *Journal of Information Technology Theory and Application(JITTA)* 7:53-70.
21. Laudon, K. C. and Laudon J. P. (2004), Management Information Systems – Managing the Digital Firm, (8th ed.), New Jersey: Pearson Education Inc.
Management Review 34(2): pp. 29-50.
22. Mcleod, R. Jr., Schell, G. P., (2004), Management Information Systems, (9th ed.), New Jersey: Pearson Education Inc.
23. Miles, R. H., (1980) "Macro organizational behavior", *Santa Monica, CA.*
24. Morgan,G.(1997), "Images of Organization" (2nd ed.). *Newbury Park, CA: Sage*
25. Mukama, F., Kimaro, H. C., Gregory, J., (2005), "Organization Culture and its Impact in Information Systems Development and Implementation: A case study from the health information system in Tanzania", *Proceedings of IRIS 28.*
26. Nye, L.G. &Witt, L.A.,(1993) "Dimensionality and Construct Validity of the Perceptions of Organizational Politics Scale (POPS)", *Educational and Psychological Measurement,* 53(3): pp. 821-829.
27. Parasuraman, A., Zeithaml, V. A., & Berry, L. L., (1991), "Refinement and reassessment of the SERVQUAL scale", *Journal of Retailing,* 67(4): pp. 420-450.
28. Pfeffer, J.,(1992), "Understanding Power in Organizations", *California Publications.*
29. Rai, A., Lang, S. S., Welker, R. B., (2002), "Assessing the Validity of IS Success Models: An Empirical Test and Theoretical Analysis", Information Systems Research, 13(1): pp. 50-69.

30. Robbins, S. P., Judge, T. A., (2007). "Organizational Behavior" (12th ed.), New Jersey: Pearson Education Inc.
31. Roldan, J.L, Leal, A., (2003), "A Validation Test of an Adaptation of the Delone and Mclean's Model in the Spanish EIS Field", Spain: Idea Group Publishing.
32. Sabherwal, R., Jeyarajm A., Chowa, C., (2006), "Information Systems Success: Individual and Organizational Determinants", *Management Scienc, 52(12: pp. 1849-1864.*
33. Seddon, P. B., Staples, S., Patnayakuni, R., Bowtell, M., (1999), "Dimensions of Information Systems Success", Communication of the Association for Information Systems, 2(20).
34. Seddon, P.B, (1997), "A Respecification and Extension of the Delone and Mclean's Model of IS Success", *Information Systems Research,* 18(3): pp. 240-253.
35. Seddon, P.B., (1994), "A partial Test Development of Delone and Mclean's Model of IS Success", *The international Conference of Information Systems, Vanconver.*
36. Seen, M., Rouse, A., Beaumont, N. Mingins, C, (2006), "The Information Systems Acceptance Model: A Meta Model to Explain and Predict Information Systems Acceptance and Success", *Manosh University, ISSN 1327-5216.*
37. Stair, R. M. (1996), Principles of Information Systems - A Managerial Approach (2nd ed.), Cambridge: An International Thomson Publishing Company, *Upper Saddle River, NJ: Prentice Hall.*
38. Warne, L., (1998) "Organizational Politics and Information Systems Development- a Model of Conflict", *IEEE*, 1060-3425/98.
39. Witt, L. A. (1995)., "Influences of supervisor behaviors on the levels and effects of workplace politics. In R. Cropanzano and M. Kacmar (Eds.), Organizational politics, justice, and support: Managing social climate at work", *Quorum Press: Westport, CT*: pp. 37-53.

An Efficient Randomized Algorithm for Real-Time Process Scheduling in PicOS Operating System

Tarek Helmy* **Anifowose Fatai**
College of Computer Science and Engineering,
King Fahd University of Petroleum and Minerals,
Dhahran 31261, Mail Box 413, Saudi Arabia,
Tel: 966-3-860-1967 & Fax: 966-3-860-2174,
{helmy, anifowo}@kfupm.edu.sa

El-Sayed Sallam
College of Engineering,
Tanta University,
Tanta, Egypt,
Tel & Fax: 20-40-3453861,
sallam@eng.tanta.edu.eg

Abstract - **PicOS is an event-driven operating environment designed for use with embedded networked sensors. More specifically, it is designed to support the concurrency in intensive operations required by networked sensors with minimal hardware requirements. Existing process scheduling algorithms of PicOS; a commercial tiny, low-footprint, real-time operating system; have their associated drawbacks. An efficient, alternative algorithm, based on a randomized selection policy, has been proposed, demonstrated, confirmed for efficiency and fairness, on the average, and has been recommended for implementation in PicOS. Simulations were carried out and performance measures such as Average Waiting Time (AWT) and Average Turn-around Time (ATT) were used to assess the efficiency of the proposed randomized version over the existing ones. The results prove that Randomized algorithm is the best and most attractive for implementation in PicOS, since it is most fair and has the least AWT and ATT on average over the other non-preemptive scheduling algorithms implemented in this paper.**

I. Introduction

Each of the existing process scheduling algorithms; First-Come, First-Served (FCFS), Shortest Job First (SJF), Shortest Remaining Time First (SRTF), Round Robin (RR), and Priority-based, has its associated drawbacks such as starvation (e.g. SJF and Priority-based scheduling algorithm), complexity of implementation, long waiting time and high average turn-around time. Fairness, efficiency, high throughput, low turnaround time, low waiting time, low response time, low frequency of context switches, and simplicity of the scheduling algorithm are the goals of any real-time process scheduling algorithm. Randomness has been known for its fairness, on the average [9] and faster speed [15]. A process in the ready queue is selected and dispatched to the CPU by random sampling while assuming the default quantum time is optimal. When a process is switched to waiting state due to I/O requests, another process is dispatched from the ready queue in a similar manner. Previous studies [9, 10, 11, 12, 13, 15] have suggested that randomized scheduling algorithms give fair selection of processes without any of them having the tendency to be starved, improved average throughput and minimum average turn-around and response times. PicOS [1] was based on event-driven process scheduling, only when there are events and processes awaiting such events. At a point, when there is no event being awaited or when there is no process awaiting any event, scheduling becomes FCFS. FCFS is known to be associated with a long ATT and AWT. This is not good for PicOS, being a Real-Time OS. SJF is attractive but with the starvation problem and difficulty in real-life implementation. This paper presents another confirmation of earlier proofs of the efficiency of randomized algorithms and the results presented a good premise to recommend this technique to *Olsonet Corporations* (Trademark Owner of PicOS) for a possible implementation in PicOS in addition to present event-driven technique being used.

The rest of this paper is organized as follows: Section 2 presents an overview of PicOS, Section 3 discusses the drawbacks of existing scheduling algorithms. An overview of randomized algorithm is presented in Section 4, Section 5 describes the approach used in the experimentation of this work. Results are presented and discussed in Section 6 and finally conclusions are drawn from the work in Section 7.

II. Overview Of Picos

A. Description of PicOS

PicOS, a tiny operating system for extremely small embedded platforms [1], was inspired by a project to develop a simple credit-card-size device with a low-bandwidth and a short-range wireless transceiver to exchange information with other devices in the same area. It was initially planned to be programmed in SMURPH (System for Modeling Unslotted Real-time PHenomena) [3], a

* On leave from College of Engineering, Department of Computer and Automatic Control, Tanta University, Egypt.

K. Elleithy (ed.), *Advanced Techniques in Computing Sciences and Software Engineering*,
DOI 10.1007/978-90-481-3660-5_20, © Springer Science+Business Media B.V. 2010

modeling system for low-level communication protocols that was later modified to a new version called LANSF (Local Area Network Simulation Facility). The implementation plan was to organize the functions as simple threads presented as finite state machines. After undergoing some further refinements, it was perfected using a "standard" operating system as a multithreaded execution platform in order to increase the application footprint beyond the capabilities of the original design. Using a programming environment similar to SMURPH, was believed to allow direct porting to the actual hardware while preserving its structure and verified properties.

PicOS is a collection of functions for organizing multiple activities of "reactive" applications executed on a small CPU with limited resources. It provides multitasking (implementable within very small RAM) and has tools for inter-process communication. On the commercial scale, it was first implemented on Cyan Technology eCOG1 microcontroller with 4KB on-chip extensible RAM. The goal was to achieve non-trivial multithreading with no additional memory resources. Different tasks share the same global stack and act as co-routines with multiple entry points like an event-driven Finite States Machine (FSM) [2].

B. Operations of PicOS and the Problem Statement

The FSM states can be viewed as checkpoints at which PicOS tasks can be preempted. A process is described by its code and data. Processes are identified by Process Ids, assigned when they are created. Whenever a process is assigned to the CPU, its code function is called in the current state at one specific entry point. When a process is run for the first time after its creation, its code function is entered at state 0, viewed as an initial state. A running process may be blocked indicating events to resume it in the future by issuing a wait request before releasing the CPU. If a process releases the CPU without issuing a single wait request, it will be unconditionally resumed in its last state. The non-preemptive nature of PicOS is demonstrated by the fact that from the moment a process is resumed until it releases the CPU, it cannot lose the CPU to another process, although, it may lose the CPU to the kernel through an interrupt. When a running process goes to waiting state, the scheduler is free to allocate the CPU to another ready process that is either not waiting for any event or, at least, has one pending awaited event [1].

The problem with the process of scheduling the next process when there is no event being awaited by any process is that, by default and intuition, it is based on FCFS policy. FCFS is known to be associated with a long ATT and AWT and is not recommended for Real-Time OSs [15]. SJF is attractive but with the starvation problem and difficulty in real-life implementation due to process length prediction. This paper presents a confirmation of earlier proofs of the efficiency of randomized algorithms and proposed a randomized algorithm to be used in PicOS in order to reduce the ATT and AWT.

III. Scheduling Algorithms Used in Experimentation

The drawbacks that are associated with each of the existing scheduling algorithms have discouraged their use in real-time operating systems environments, since the objective of existing scheduling algorithms for real-time applications is to execute real-time tasks in a timely fashion [12]. Since the scheduling of processes in PicOS is non-preemptive, the non-preemptive algorithms that will be discussed in this section include: First-Come First-Served (FCFS), Shortest Job First (SJF), and Priority-based algorithms.

In FCFS, the process that arrives first is allocated the CPU first while the ready queue is a FIFO queue. Though, it is simple and easy to implement, since scheduling is done at constant time, independent of the number of processes in the ready queue, it is often associated with a long average waiting, response times and turnaround times. This will not be appropriate for PicOS, being a real-time operating system that needs to be a little mindful of time [15].

In SJF, the next process to be scheduled is selected based on the shortest burst time, which is linear with respect to the number of processes in the ready queue. Though, it gives minimum average waiting time for a given set of processes but selection is more complex than FCFS and starvation is possible since if new, short processes keep on arriving, old, long processes may never be served [15].

In priority-based, a priority number is associated with each process. The CPU is allocated to the process with the highest priority. In the non-preemptive version of Priority-based scheduling policy, the CPU is allocated to the process with the highest priority and equal priority processes are scheduled in FCFS order. Also, SJF is like a priority-based scheduling where priority is the next predicted CPU burst time. This policy is associated with the problem of starvation as low priority processes may never be executed or will be executed last in the non-preemptive version. It is interesting to note that priority-based policy boils down eventually to FCFS [15].

A randomized or probabilistic algorithm employs a degree of randomness as part of its logic. This means that the machine implementing the algorithm has access to a pseudo-random number generator. The algorithm typically uses the random bits as an auxiliary input to guide its behavior, in the hope of achieving good performance in the "average case". If there are 10 processes, a number will be generated between 1 and 10 inclusive. If that number has not been selected before, the process at the location corresponding to that number is selected. Otherwise, it will be ignored and another one will be generated. Formally, the algorithm's performance will be a random variable determined by the random bits, with good expected value. The "worst case" is typically so unlikely to occur that it can safely be ignored.

IV. Overview Of Randomized Algorithms

A randomized algorithm can be defined as one that receives, in addition to its input, a stream of random bits that it can use in the course of its action for the purpose of making random choices. A randomized algorithm may give different results when applied to the same input in different runs. It is recognized now that, in a wide range of applications, randomization is an extremely important tool for the construction of algorithms [14]. Randomized algorithms are usually characterized with smaller execution time or space requirement than that of the best deterministic algorithm for the same problem and are often simple and easy to implement [11].

In literature, [11] proved lower bounds on the competitive ratio of randomized algorithms for several on-line scheduling problems. Their results showed that a randomized algorithm is the best possible and most applicable for the problem. Authors of [9] proposed a suite of randomized algorithms for input-queued (IQ) switches for finding a good matching between inputs and outputs to transfer packets at high line rates or in large switches, which is usually a complicated task. They reported that the performance of the randomized algorithms is comparable to that of well-known, effective matching algorithms, yet are simple to implement. Authors of [10] provided a methodology for analyzing the impact of perturbations on performance for a generic Lebesgue-measurable computation. They concluded that the associated robustness problem, whose solution is computationally intractable, could be nicely addressed with a poly-time procedure based on Randomized algorithms. In order to satisfy given real-time constraints, maximize reliability and minimize inter-communication costs, authors of [12] proposed and developed an objective function approach to schedule real-time task graphs that need to satisfy multiple criteria simultaneously. They used two different searching techniques, a heuristic search technique and a random search technique, that use the objective function. They concluded that the random search technique achieves better performance.

V. Experimental Approach

We have simulated the operations of PicOS that were based on First Come First Served to demonstrate the efficiency and fairness of using randomization policy in real-time process scheduling. The simulator was developed in Microsoft Visual Basic and used to implement three non-preemptive scheduling algorithms (FCFS, SJF and Randomized) in order to establish a valid premise for effective comparison. The simulator takes process IDs as integer input, estimated burst times and their position in terms of their order of execution. Each of the datasets was used for the three non-preemptive scheduling algorithms in order to observe the behavior of each algorithm under the same data environment. For simplicity, the simulator was built on two major assumptions: the scheduling policies are non-preemptive and the processes arrive at the same time. The simulator was run on with six different datasets containing processes with different burst times that have been positioned in different ways in order to create different scenarios. This is in order to determine, as part of the experiment, whether the location of a process in a queue will affect the results of the simulation, especially, the behavior of the randomized algorithm. The simulation was run several times to ensure fairness to all datasets and the final results were recorded and presented for each algorithm using waiting and turn-around times (AWT and ATT) as performance evaluation indices. Tables I – X show the datasets with the burst times, in seconds, as input to the simulator and their different locations.

TABLE I
DATA 1 (3 PROCESSES)

Processes	P1	P2	P3
Burst Time (s)	**44**	9	13

TABLE II
DATA 2 (3 PROCESSES)

Processes	P1	P2	P3
Burst Time (s)	**24**	3	3

TABLE III
DATA 3 (3 PROCESSES)

Processes	P1	P2	P3
Burst Time (s)	**3**	3	24

TABLE IV
DATA 4 (3 PROCESSES)

Processes	P1	P2	P3
Burst Time (s)	3	24	3

TABLE V
DATA 5 (5 PROCESSES)

Processes	P1	P2	P3	P4	P5
Burst Time (s)	6	15	3	4	2

TABLE VI
DATA 6 (5 PROCESSES)

Processes	P1	P2	P3	P4	P5
Burst Time (s)	11	20	5	9	7

TABLE VII
DATA 7 (7 PROCESSES)

Processes	P1	P2	P3	P4	P5	P6	P7
Burst Time (s)	11	20	5	9	7	2	4

TABLE VIII
DATA 8 (10 PROCESSES)

Processes	P1	P2	P3	P4	P5	P6	P7	P8	P9	P10
Burst Time (s)	11	20	5	9	7	2	4	15	1	8

TABLE IX
DATA 9 (15 PROCESSES)

Processes	P1	P2	P3	P4	P5	P6	P7	P8	P9	P10
Burst Time (s)	11	20	5	9	7	2	4	15	1	8
Processes	P11	P12	P13	P14	P15					
Burst Time (s)	18	3	12	6	19					

TABLE X
DATA 10 (20 PROCESSES)

Processes	P1	P2	P3	P4	P5	P6	P7	P8	P9	P10
Burst Time (s)	11	10	20	5	16	9	7	17	2	14
Processes	P11	P12	P13	P14	P15	P16	P17	P18	P19	P20
Burst Time (s)	4	15	1	13	8	18	3	12	6	19

VI. Results And Discussion

The results of the simulation run on the ten datasets, with their corresponding waiting times, turnaround times, as well as their respective averages are plotted in Fig. 1 - 10. Out of the ten cases, randomized scheduling algorithm performed better than FCFS in nine cases. The only case where FCFS performed better than randomized algorithm is a very rare case. That is why the good performance of randomized algorithms has always been described as average. It can be inferred from these plots that the randomized algorithm, on the average, is better than FCFS (except with data 4 in Figure IV) while SJF remains the best. Even though SJF gave the optimal average waiting and turnaround times, that does not make it attractive and appealing for implementation in PicOS due to the problems and difficulties associated with it. The real difficulty with the SJF algorithm is, estimating the length of the next CPU request. Consequently, despite the optimality of the SJF algorithm, it cannot be implemented at the level of short-term CPU scheduling. Also, there is the problem of starvation, since in real life, long processes will be starved and may never be executed, if short jobs keep coming. In some cases, long jobs could be more important than shorter ones.

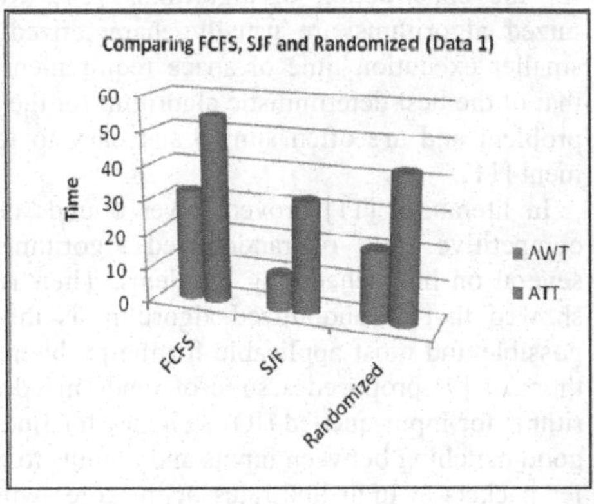

Fig. 1. Average waiting and turnaround times for Data 1.

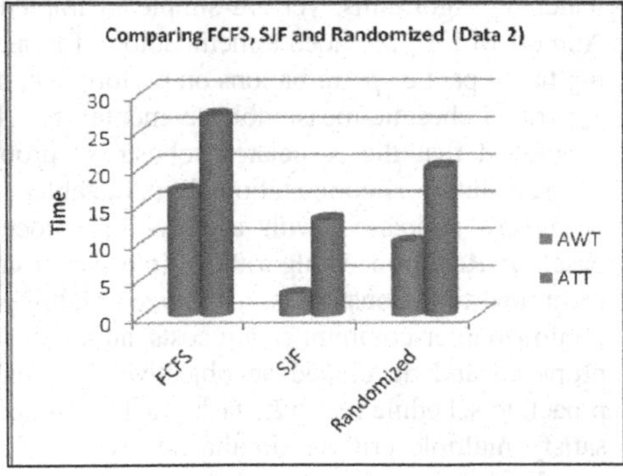

Fig. 2. Average waiting and turnaround times for Data 2.

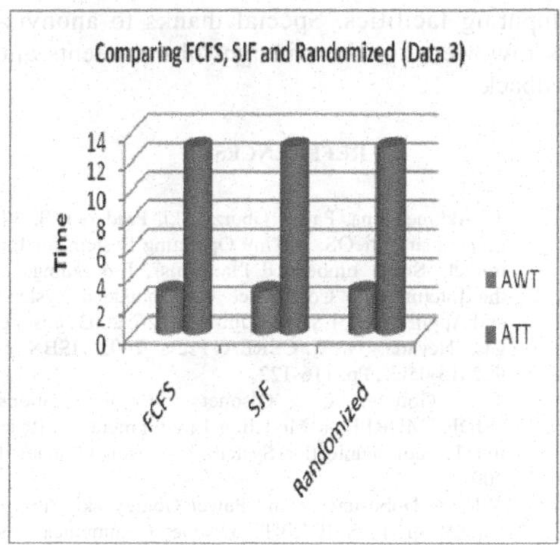

Fig. 3. Average waiting and turnaround times for Data 3.

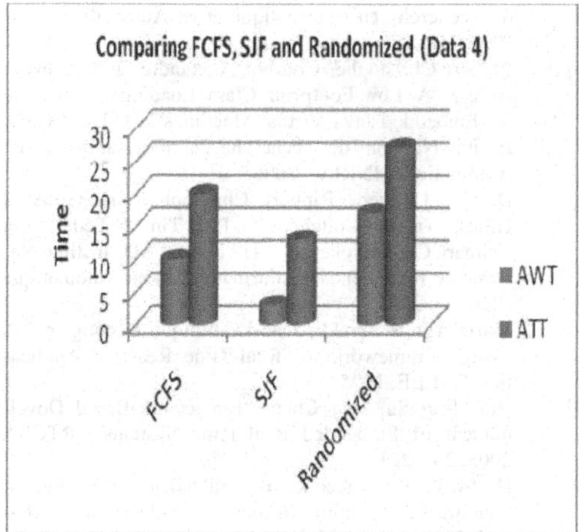

Fig. 4. Average waiting and turnaround times for Data 4.

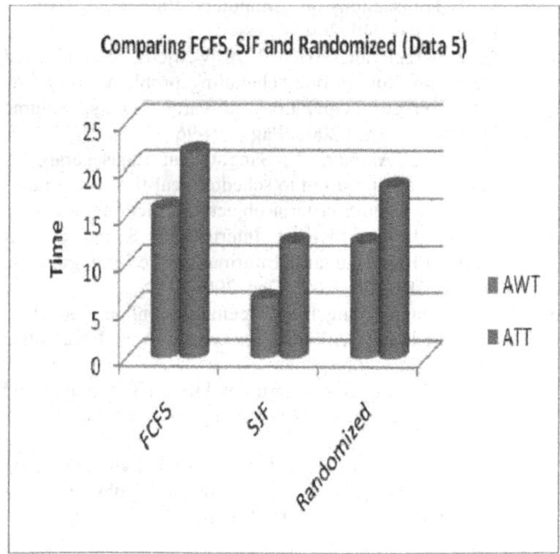

Fig. 5. Average waiting and turnaround times for Data 5.

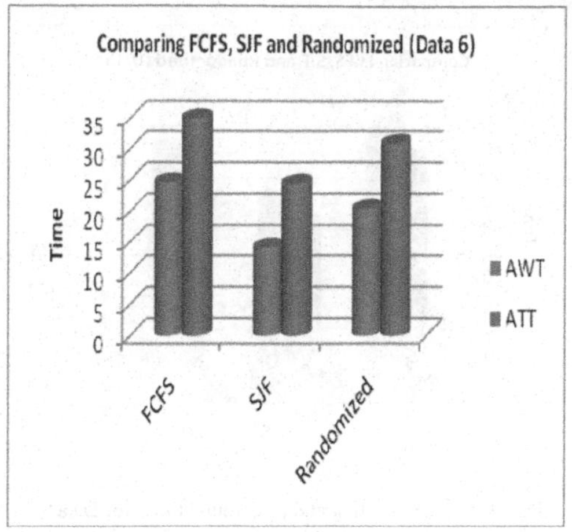

Fig. 6. Average waiting and turnaround times for Data 6.

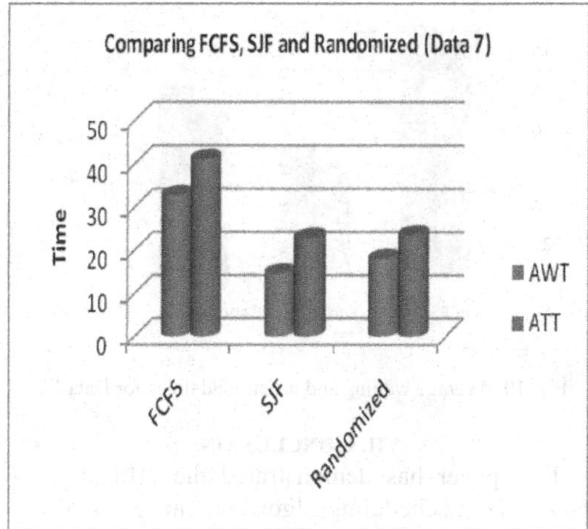

Fig. 7. Average waiting and turnaround times for Data 7.

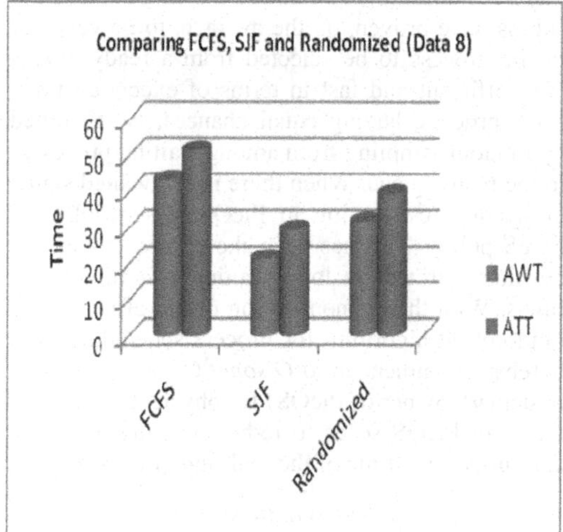

Fig. 8. Average waiting and turnaround times for Data 8.

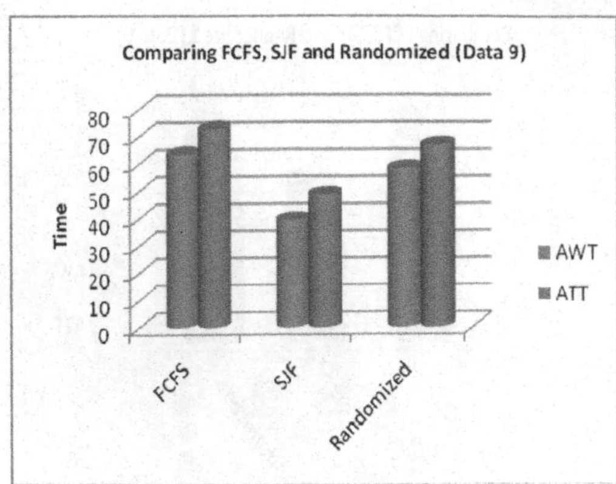

Fig. 9. Average waiting and turnaround times for Data 9.

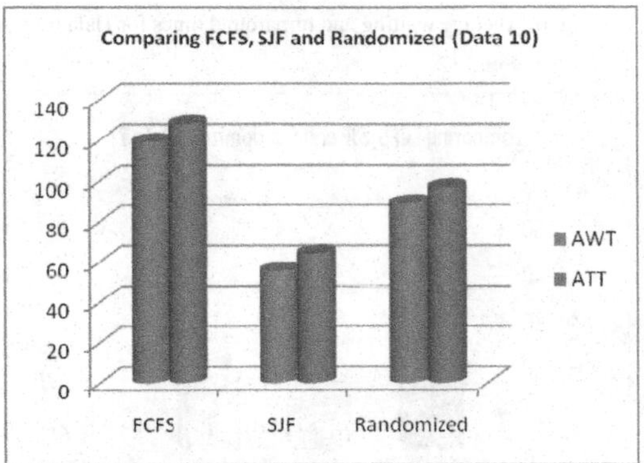

Fig. 10. Average waiting and turnaround times for Data 10.

VII. CONCLUSION

This paper has demonstrated the efficiency of randomized scheduling algorithms in terms of average waiting and turnaround times, compared with FCFS and SJF scheduling algorithms by using a tiny custom-built simulator. Randomized algorithms have proven, on the average, to be very fair to the process to be selected from a ready queue, very efficient and fast in terms of execution time. Each process, having equal chances, is scheduled by random sampling from among waiting processes in the ready queue. When there is no awaited signal or event, the next job in PicOS is scheduled by FCFS policy. This has been shown in this paper to be characterized by long waiting and turnaround times. With this demonstration of the efficiency of randomized algorithm for process scheduling, it is hereby recommended to *Olsonet Corporations* (the trademark owner of PicOS) for physical implementation in PicOS so as to reduce waiting, response and turnaround time of the real-time processes.

ACKNOWLEDGMENTS

We would like to thank King Fahd University of Petroleum and Minerals for providing the computing facilities. Special thanks to anonymous reviewers for their insightful comments and feedback

REFERENCES

[1] E. Akhmetshina, Pawel Gburzynski, Frederick S. Vizeacoumar, "PicOS: A Tiny Operating System for Extremely Small Embedded Platforms", Proceedings of the International Conference on Embedded Systems and Applications, ESA '03, June 23 - 26, 2003, Las Vegas, Nevada, USA. CSREA Press 2003, ISBN 1-932415-05-X, Pp. 116-122.

[2] P. Gburzynski, Olsonet Communications, "SIDE/SMURPH: a Modeling Environment for Reactive Telecommunication Systems", Version 3.0 manual, 2007.

[3] Wlodek Dobosiewicz and Pawel Gburzynski, "Protocol Design in SMURPH", Olsonet Communications, 2007.

[4] Rippert Christophe, Deville Damien and Grimaud Gilles, "Alternative schemes for low-footprint operating systems building", HAL - CCSD, Institut National de Recherche en Informatique et en Automatique (INRIA), 2007.

[5] Rippert Christophe, Courbot Alexandre, and Grimaud Gilles, "A Low-Footprint Class Loading Mechanism for Embedded Java Virtual Machines", HAL - CCSD, Institut National de Recherche en Informatique et en Automatique (INRIA), 2006.

[6] Deville Damien, Rippert Christophe and Grimaud Gilles, "Trusted Collaborative Real Time Scheduling in a Smart Card Exokernel", HAL - CCSD, Institut National de Recherche en Informatique et en Automatique (INRIA), 2007.

[7] Marie-Agnits Pkraldi, Jean-Dominique Decotignie, "A Design Framework for Real-Time Reactive Applications", IEEE, 1995.

[8] Hui-Ming Su, Jing Chen, "Framework-Based Development of Embedded Real-Time Systems", RTCSA 2003: 244-253

[9] D. Shah, P. Giaccone, B. Prabhakar, "An efficient randomized algorithm for input-queued switch scheduling", IEEE Micro, 22(1):19-25, January-February 2002.

[10] Cesare Alippi, "Randomized Algorithms: A System-Level, Poly-Time Analysis of Robust Computation", IEEE Transactions on Computers, Volume 51, Issue 7, Pages: 740 - 749, 2002.

[11] Leen Stougie and Arjen P. A. Vestjens, "Randomized algorithms for on-line scheduling problems: how low can't you go?" Operations Research Letters, Volume 30, Issue 2, April 2002, Pages 89-96.

[12] Amin, A.; Ammar, R.; Sanguthevar Rajasekaran, "A randomized algorithm to schedule real-time task graphs to satisfy a multi-criteria objective function", Proceedings of the Fourth IEEE International Symposium on Signal Processing and Information Technology, vol., no., pp. 381-386, 18-21 Dec. 2004.

[13] Yiwei Jiang, Yong He, "Preemptive online algorithms for scheduling with machine cost", Acta Informatica 41, 315–340 (2005).

[14] M.H. Alsuwaiyel, "Algorithms Design Techniques and Analysis", Lecture Notes Series on Computing, Vol. 7, Pp. 371-392, 1999.

[15] Abraham Silberschatz, Peter Baer Galvin and Greg Gagne, "Operating System Concepts", John Wiley & Sons, Inc., 3rd Edition, Pp. 158-164, 2005.

Managing in the Virtual World: How Second Life is Rewriting the Rules of "Real Life" Business

David C. Wyld
Southeastern Louisiana University
Department of Management – Box 10350
Hammond, LA 70402-0350

***Abstract*-In this paper, we will explore the growth of virtual worlds – one of the most exciting and fast-growing concepts in the Web 2.0 era. We will see that while there has been significant growth across all demographic groups, online gaming in MMOGs (Massively Multiplayer Online Games) are finding particular appeal in today's youth – the so-called "digital native" generation. We then overview the today's virtual world marketplace, both in the youth and adult-oriented markets. Second Life is emerging as the most important virtual world today, due to the intense interest amongst both large organizations and individual entrepreneurs to conduct real business in the virtual environment. Due to its prominence today and its forecasted growth over the next decade, we take a look at the unscripted world of Second Life, examining the corporate presence in-world, as well as the economic, technical, legal, ethical and security issues involved for companies doing business in the virtual world. In conclusion, we present an analysis of where we stand in terms of virtual world development today and a projection of where we will be heading in the near future. Finally, we present advice to management practitioners and academicians on how to learn about virtual worlds and explore the world of opportunities in them.**

I. INTRODUCTION

> "Every human being is interested in two kinds of worlds: the Primary, everyday world which he knows through his senses, and a Secondary world or worlds which he not only can create in his imagination, but which he cannot stop himself creating [1]."

A. *Virtual Worlds 101*

Analysts have predicted that by 2020, "virtual worlds will be as widespread as the World Wide Web is now" [2], and there is a growing belief that virtual worlds may well "replace the web browser as the way we interface with the Internet" [3]. Indeed, some predict that virtual worlds will be as significant a technological disruptor as the invention of the personal computer or the advent of the Internet [4]. In late 2007, Gartner predicted that by the end of 2011, fully 80% of all active Internet users "will have a 'second life'" in the developing sphere of virtual worlds, "but not necessarily in Second Life" [5]. While some have criticized Gartner's 80% projection for being overly optimistic [6], there can be no doubt that the way we interact via the Internet is undergoing a profound change with the advent of virtual worlds.

Virtual worlds have been hailed as "the next great information frontiers" [7]. They are known synonymously as:

- MMOGs (massively multiplayer online games)
- MMORPGs (massively multi-player online role playing games)
- MUVEs (multi-user online virtual environments)
- NVEs (networked virtual environments).

Massively Multiplayer Online Games (MMOGs) – the umbrella term that will be used in this report - can be defined as being: "graphical two-dimensional (2-D) or three-dimensional (3D) videogames played online, allowing individuals, through their self-created digital characters or 'avatars,' to interact not only with the gaming software but with other players" [8]. These are the fastest growing category of online gaming, the total number of MMOG players estimated to be in excess of 150 million worldwide [9]. In these virtual worlds, players from around the world take-on fictional characters to play roles in the games. These are labeled "massively" for good reason. Unlike a traditional computer game (which has a single player) or a home console video game (which may have at most typically four players), MMOGs are likely to have many players from around the globe, perhaps even speaking different languages, engaged in the game simultaneously. They commonly involve a large number of players simultaneously – hundreds and quite commonly, even thousands - playing simultaneously. And, unlike our individual video and computer games, the virtual worlds of MMOGs are persistent environments, operating on a 24/7/365, "never off" basis.

Today's youth are coming to see virtual worlds as one of their primary ways to interact over the Internet. According to research from eMarketer, the percentage of all Internet users between the ages of 3 and 17 that access virtual worlds at least once a month has risen by over 50% in year over year figures (from 5.3 million in 2006 to 8.2 million in 2007). And, if this trend continues, by 2011 – the same year targeted by Gartner when virtual worlds will reach a true tipping point – well over half of all young Internet users – approximately 20 million – will visit virtual worlds [10]. As they interact with virtual worlds, this changes their perspective on what their Internet experience should be, as: "They are growing up not only with social networking but also with the ability to interact with people, shop, learn and play in a graphic environment" [10]. In comparison, "flat" websites may seem boring, paling in comparison to what they see as truly a Web 2.0 - a "new and improved" version of the Internet.

B. *The Virtual Worlds Marketplace*

The growth of virtual worlds is not just an American phenomenon. In fact, it has been said that "for adults,

K. Elleithy (ed.), *Advanced Techniques in Computing Sciences and Software Engineering*,
DOI 10.1007/978-90-481-3660-5_21, © Springer Science+Business Media B.V. 2010

virtuality is an acquired taste" – as fewer than 10% of American adults having a presence in a virtual world environment (compared with almost half of the population of South Korea) [11]. According to the most recent available data, there are an estimated 73 million online game players around the world today, with the worldwide market for MMOGs being estimated to be in excess of $350 million [12].

Virtual worlds are fast becoming an environment of choice for millions of individuals – and a very big business. As can be seen in Tables 1 (*Major Virtual Worlds for Kids and Teens*) and 2 (*Major Virtual Worlds for Adults*), there are a panoply of virtual worlds online today, with hundreds more reportedly in various stages of development, looking to come online in the next few years. Steve Prentice, Gartner Group's Chief of Research, characterized the teen virtual world marketplace as exploding, as indeed, "there just seems to be another coming along every week" [13].

Virtual World	URL
BarbieGirls World	http://www.barbiegirls.com/
Be-Bratz!	http://www.be-bratz.com/
Club Penguin	http://www.clubpenguin.com/
Disney's Fairies	http://www.disney.co.uk/DisneyOnline/fairies/
Disney's Toontown	http://play.toontown.com/
Disney's Virtual Magic Kingdom	http://vmk.disney.go.com/
Dofus	http://www.dofus.com/
Flowplay	http://www.flowplay.com/
Lego Universe	http://universe.lego.com/en-US/default.aspx
MapleStory	http://www.maplestory.com/
Mokitown	http://www.mobile-kids.net/
Nicktropolis	http://www.nick.com/nicktropolis/
Planet Cazmo	http://www.planetcazmo.com/
Puzzle Pirates	http://www.puzzlepirates.com/
Spine World	http://www.spineworld.com/s/start
Teen Second Life	http://teen.secondlife.com/
WebbliWorld Home	http://www.webbliworld.com/
Webkinz	http://www.webkinz.com/
Whirled	http://www.threerings.net/whirled/
Whyville	http://www.whyville.net/
Xivio	http://www.xivio.com/bftq5/index.cfm
Zoodaloo	http://www.zoodaloo.com/

Tab. 1. Major Virtual Worlds for Kids and Teens

Second Life is, in truth, but one slice – albeit a tremendously important one – of the overall virtual worlds' marketplace. In fact, both in terms of population and revenue, Second Life is dwarfed in size by what have been aptly termed "men in tights" games, medieval-styled fantasy games such as -- World of Warcraft, Runescape, Lineage, Ragnarok, and Everquest [14]. In fact, in January 2008, World of Warcraft – the largest MMOG – surpassed the astonishing mark of having 10 million active subscribers – at least a quarter of which are based in the U.S. and Canada and almost half of whom are based in China [15]. Some commentators have likened the explosion of participation to make virtual worlds "the new TV" (as a form of entertainment) [16] or "the new golf," due to the social nature of the online gaming environment [17].

Virtual World	URL
AlphaWorld	http://www.activeworlds.com/
BOTS	http://bots.acclaim.com/
City of Heroes/City of Villains	http://uk.cityofheroes.com/
City Pixel	http://www.citypixel.com/
Coke Studios	http://www.mycoke.com/
Cybertown	http://www.cybertown.com/
Dark Age of Camelot	http://www.darkageofcamelot.com/
Dubit	http://www.dubitchat.com/
Entropia Universe	http://www.entropiauniverse.com/
Eve Online	http://www.eve-online.com/
EverQuest / EverQuest II	http://everquest.station.sony.com/
Faketown	http://www.faketown.com/
Final Fantasy XI: Online	http://www.playonline.com/ff11eu/
Gaia Online	http://www.gaiaonline.com/
Guild Wars	http://www.guildwars.com/
Habbo Hotel	http://www.habbo.com/
HiPiHi	http://www.hipihi.com/index_english.html
Knight Online	http://www.knightonlineworld.com
Lineage/Lineage II	http://www.lineage.com/
Runescape	http://www.runescape.com/
Star Wars Galaxies	http://starwarsgalaxies.station.sony.com/
The Lord of the Rings Online: Shadows of Angmar	http://www.lotro.com/
The Sims/The Sims2	http://www.ea.com/official/thesims/
There	http://www.there.com/
Ultima Online	http://www.uo.com/
Virtual MTV	http://www.vmtv.com/
Virtual World of Kaneva	http://www.kaneva.com/
vSide	http://www.vside.com/
World of Warcraft	http://www.worldofwarcraft.com/

Tab. 2. Major Virtual Worlds for Adults

II. SECOND LIFE

"Once we have enough computing power, we can remake the world using simulation."
----- Philip Rosedale, Linden Lab founder [18]

A. *Overview of Second Life*

Philip Rosedale, the founder and CEO (until May 2008) of Linden Labs, remarked that: "The intention behind Second Life was to create a world not just where a bunch of people were connected together in some way, but a world in which everything was built by us, by the people who were there in a kind of Lego block sort of way to rebuild the laws of physics" [19]. He had been the Chief Technical Officer for RealNetworks in its formative days, and he was known for being an extremely creative, forward thinking type – thinking far beyond the business of streaming audio [20].

The birth of Second Life can be traced to a now legendary night in 1999, when Rosedale and some of his friends from RealNetworks went to see the movie, *The Matrix,* in a local theater. According to the tale, after the movie, the friends sat in a bar discussing the film and its rather dark vision of an alternate, virtual reality. Rosedale announced then and there: "I'm going to build that! And it's not going to turn out that way" [18]. A short time later, Rosedale took his payout from the Yahoo buyout of RealNetworks to form Linden Labs, based in San Francisco [20].

Second Life was first launched in June 2003. It was unique in that is was intended to be "a platform, not a game" – and approximately a thousand users created content in the virtual environment before the site went dark due to Linden Labs funding problems. Then, in January 2004, the site was relaunched, allowing for users to own their own land [21]. Since that time, the number of residents in Second Life has grown rapidly – to over 13 million in 2008 [22].

B. *"Life" in Second Life*

Second Life has been differentiated from MMOGs, in that: "unlike online games, virtual social worlds lack structured, mission-oriented narratives; defined character roles; and explicit goals" [23]. In the virtual social world of Second Life, there are no quests, no scripted play and no top down game plan [24]. There is no embedded objective or narrative to follow. There are no levels, no targets, and no dragons to slay. It has been hailed as nothing less than the "evolution of the computer game," as rather than having a ready-made character with a fixed purpose, one creates his or her own avatar with an open-ended existence [25]. Thus, rather than being a Star Wars-like character or an armed, rogue warrior whose mission it is to shoot as many other characters as possible or to collect enough points or tokens to advance to the next level, the Second Life avatar wonders a virtual world, with the ability to "teleport" – in a "Star Trek"-like manner – from virtual place to virtual place.

Edward Castronova, author of *Synthetic Worlds: The Business and Culture of Online Games*, believes that the growth of virtual social worlds is a sign that "We have an emerging industry of people making fun things" – an "economics of fun." He added that Second Life is itself not "fun," just as the Internet is not inherently "fun." However, there are aspects of Second Life that are very fun, and as such, attract a sizeable number of participants to the site [26]. Justin Bovington, the Chief Executive Officer of Rivers Run Red, a London-based Second Life consultancy, observed that: "Second Life is proof that we are entering a much more immersive era for entertainment. It offers a whole new level of experience with something that you're interested in" [27].

Second Life has been described in myriad ways. In a *Business Week* cover story on virtual worlds, Second Life was portrayed as "something a lot stranger than fiction.....some unholy offspring of the movie *The Matrix*, the social networking site MySpace.com, and the online marketplace eBay" [28]. In *Fortune*, Second Life was painted as "a brightly colored, three-dimensional world that resembles *Grand Theft Auto* crossed with *Lord of the Rings*" [29]. Second Life has also been been described as resembling an idealized version of the world, filled with what has been described as "futuristic Blade Runner-style cities and places that resemble ancient Japan" [27]. In this idealized environment, there are shopping malls, colleges and universities, museums, beaches, cinemas, and nightclubs – lots and lots of nightclubs. There are business conferences and consumer marketing events taking place "in world," as well as real university classes and other types of training sessions. There are also recruitment fairs being held in Second Life, as well as "real" job interviews between employers and prospective workers. As one can constantly remake one's appearance by changing your on-screen character – your avatar – avatar's tend to be idealized versions of ourselves, described as being "all nipped, tucked and primped to perfection" [2].

C. *Participation in Second Life*

John Gage, the chief researcher for Sun Microsystems, described the virtual world of Second Life as "a community built entirely on participation" [30]. The primary means of communications within the Second Life environment is text-based instant messaging chat, leading experienced users to often develop typing speeds rivaling the finest executive secretary. Voice chat was introduced in Second Life in late 2007, and while some believe this to be a development that will attract more regular users to the virtual world, others feel that having virtual voice can have opposite the intended effect – making users "clam up" [31].

The demographics of Second Life are indeed quite interesting. Second Life is not at all representative of the population as a whole. In fact, due to the need for broadband Internet connection and a fairly sophisticated computer to run the application, Second Life is populated by an overly middle class audience with the time – and money - to spend in world [32]. While the virtual world began as a majority female-user site, since 2005, the site has trended to have more male than female residents. In regards to age, since those under 18 are prohibited from entering the main Second Life grid, the user base is relatively normally distributed, with the largest age group being in the 25-34 age range. However, while younger users (under the age of 35) make-up over sixty percent of Second Life's resident population, older users tend to put in a disproportionate amount of time in-world. In just the past year, the number of hours residents spent in-world has *doubled* to over 30 million hours each month, and residents aged 35 and older put-in approximately half of all user hours in the virtual environment [22]. The in-world environment of Second Life has also been rather accurately described as being overly white in demographic (at least in the representation of the avatars)[32]. While Second Life draws users from around the world, it is – by and large – today a North American and European-driven phenomenon [33].

D. *The Virtual Economy of Second Life*

What has propelled the rapid growth of Second Life? Analysts point to the fact that it is an entrepreneurial, free market-based virtual economy [24]. As such, virtual worlds can be viewed as "essentially unregulated playgrounds for economic organization" [28]. Futurist Thomas Frey, a former IBM executive who presently serves as the Executive Director of the DaVinci Institute, recently observed: "I see Second Life as the next generation of social networking, but so much more. It has its own currency, land to buy and sell, and free enterprise systems that allow entrepreneurial-minded people the ability to build new countries and new kinds of business" [34]. The nascent Second Life economy has been described as a "highly entrepreneurial environment," with the attendant opportunities and issues that were encountered in the early days of the Web [24].

Second Life has been described as nothing less than a virtual laboratory for helping to better understand how both the virtual and real world economies work: "This virtual world has become a petri dish of loosely regulated markets and has begun to pose some profound questions about the meaning of assets, currencies, and intellectual property" [35]. In its short existence, Second Life has proven that the "invisible hand" of the economy works reasonably well in the virtual world as well. Second Life has been described as a world built on both a laissez-faire economy, where the invisible hand of capitalism has proven to work in the virtual environment, as well as a laissez-faire attitude toward many activities that could not be tolerated in the real world, which many point to as "unsustainable" [36]. This has proven both advantageous and troublesome, as gambling bans [37] and banking crises [38] have both disrupted the Second Life economy.

A whole host of major companies have established Second Life presences, including retailers, car companies, consumer products firms, and even service providers (including H.R. Block for tax help). For companies, the prospect of ROI in their Second Life ventures is very long term in nature at best. As Greg Verdino, who is Vice President of Emerging Channels for Digitas, concluded: "Can you, as a business, look at the Second Life of today and say it's a viable marketing channel? Can you draw direct lines between what people do in Second Life and what they do in real life? No, you can't. Certainly, Second Life is innovative, but it's far too early to start calculating ROI, or expect any real-world deliverables to come of it" [39]. What has been driving companies to invest in Second Life and other virtual worlds? One analyst recently remarked: "I see FUD ('Fear, Uncertainty and Doubt') hard at work, with advertisers fearing that they'll miss out on a new way to communicate if they don't jump on the Second Life bandwagon soon. It reminds me of the early dot com land grab" [40]. Still, one has to be concerned that several high profile companies have closed-up their virtual world sites in Second Life, including American Apparel (which was the first "real-world" major company to establish a storefront in Second Life) and Starwood Hotels (which shuttered its "Aloft Hotel" and turned over its island for use by a real-world nonprofit group) [41].

However, there is indeed potential for companies to produce real world results from their Second Life operations beyond any in-world sales. Companies can look at a variety of indirect return measures, including media mentions of their Second Life operations and potential cross-promotions aimed at in-world residents to shop in the real world [42]. Virtual worlds can also be seen as a test-bed for new ideas, new products, and new designs, as well as a forum for user-generated ideas to rise to the attention of corporate designers and marketing executives for use in real-life products and services. An example of this can be found in Toyota's Scion City, Second Life, where not only can residents configure and "drive" their own Scion, but also make customizations that Toyota will evaluate for use on the real-life version of the car – with rewards going to those in Second Life who make worthy suggestions [19]. There have also been instances where virtual "products" have become real-world items as well. One Second Life entrepreneur developed a game he called Tringo, which was a cross between Tetris and Bingo. Nintendo licensed the popular in-world game for development as a real-life video game [29].

E. *Law and Disorder in Second Life*

Henry Kelly, who is the President of The Federation of American Scientists, observed: "The ancient tension between liberty and order plays out as vividly in Virtual Worlds as in the real" [43]. One of the significant downsides to participating in Second Life is the sense of lawlessness that has largely pervaded the online world. The laissez-faire attitude that has historically prevailed is encapsulated in the observation of an anonymous user, whose avatar, Bobo Decosta, proclaimed last year: "I thought Second Life was another world where real life laws didn't apply" [37].

Observers have cautioned organizations that while Second Life may be a virtual world, "the trouble it could cause you is real" [44]. Indeed, Second Life has not been immune to the intervention of real world laws and real world legal issues. Indeed, one of the more interesting discussions is who, exactly, should "police" the virtual worlds [45]. While Second Life has drawn attention from the FBI and other agencies (on matters such as gambling and money laundering), there are legal, jurisdictional, and ethical issues over applying real world laws to conduct in virtual worlds [46]. As Professor Beth Simone Noveck of the New York Law School, who is also the founder of the State of Play: Law and Virtual Worlds Annual Conference, recently observed: "People are appearing with new visual identities and likenesses, trademarked brands are appearing in virtual shop windows, and ownership of avatars (is) being settled in divorce and probate proceedings. The law is having to grapple with questions that are similar to, but not all the same as, what we've confronted in two-dimensional cyberspace" [47]. In Second Life for instance, there was a threat to virtual property when, in late 2006, a program known as CopyBot allowed residents to copy any object in Second Life, including clothing, buildings, and other items that had been "bought" legitimately in-world [48].

Analysts have predicted that as virtual worlds and their populations, economies, and complexities grew, "terrestrial governments" would begin to be both compelled and enticed to step-in and begin monitoring the online domains [49]. And, as we have seen with the crackdown on gambling and banking in Second Life, as Professor Joshua Fairfield of Indiana University put it, "governance in the virtual world is already in place and it's the real world governance" [31].

IV. ANALYSIS

Overall, Second Life and the entirety of virtual worlds are still very much in their infancies, and analysts have predicted that we are not even at "the DOS era of virtual worlds" [50]. However, we are shifting from today's 2D Web to the future – what has been aptly described by analysts as the "3D Internet." The Former Speaker of the U.S. House of Representatives Newt Gingrich recently observed that the: "3D Internet represent(s) a brand-new approach to collaboration. While it is true that it's still fairly primitive by the standards of having holograms or being face-to-face in more complex ways. . . . I think that 3D Internet in all of its various forms is going to become one of the great breakthroughs of the next ten years" [51].

What does all this mean…and where does it lead? Observers have declared that: "the virtual world is rapidly evolving into a close representation of the real world with all the opportunities and consequences of the real world" [17]. Harvard Business School Professor Stephen Kaufman, an expert on disruptive technologies, cautions that: "You can't tell where these things will go, and they will probably go places you'd never expect them to" [52]. Still, *Popular Science* and others have predicted that virtual worlds may well be "a window onto the Web's future" [53]. Indeed, the *Metaverse Roadmap Overview* [54] predicted that many of our 2D Internet activities will migrate to the 3D spaces of the Metaverse, envisaging that: "Although the 'Web' technically refers to a particular set of protocols and online applications, the term has become shorthand for online life. It's possible that 'Metaverse' will come to have this same duality: referring to both a particular set of virtualizing and 3D web technologies, and the standard way in which we think of life online. Like the Web, the Metaverse wouldn't be the entirety of the Internet—but like the Web, it would be seen by many as the most important part." As such, virtual environments won't replace today's Internet in a wholesale manner. However, in the future, we will find the best uses of both, as tasks will migrate to and from the Web and virtual worlds – wherever they are best suited for.

V. Conclusion

As virtual worlds increase in both utility and usage, it will become more and more important for executives and academicians alike to know about what is going on in this "second place." Thus, here is a 5-step action plan to help you learn more about Second Life and the rest of the 3D Internet:

1. Join Second Life and create an avatar (be careful – pick a name that your constituents and your Mom - would be happy with). Teleport around to islands of various types (the corporate, university, museum, and governmental sites will make for a good start).
2. Pick at least two other virtual worlds to join and explore – whether you are more of a World of Warcraft or Virtual Magic Kingdom type.
3. Ask your kids (whether they are in first grade or in college) what they are doing in virtual worlds. Don't do this in an accusatory way – they might be able to teach you a great deal.
4. Ask your staff what they are doing in virtual worlds (again, not acting as the IT police). Find out who are "the experts" in your office and which staffers might be helpful in working on your organization's virtual world project.
5. Bookmark or RSS several virtual worlds news sites and/or blogs (CNN's I-Reports on Second Life and Reuter's News Bureau in Second Life are great places to start). Set-up a "Google Alert" for virtual worlds topics (you may want to focus it more narrowly on your market and on Second Life as well if that is your targeted venue).

References

[1] W.H. Auden, *Secondary Worlds*. New York: *Random House*, 1968.

[2] L. Rawlinson, "Virtual worlds: The next Facebook?" *CNN.com*, August 8, 2007. Retrieved October 1, 2007, from http://edition.cnn.com/2007/TECH/08/07/virtual.living/index.html.

[3] J.V. Last, "Get a (Second) Life!: The avatars are coming," *The Weekly Standard*, October 1, 2007. Retrieved October 16, 2007, from http://weeklystandard.com/Content/Public/Articles/000/000/014/145mliuh.asp.

[4] C. Mims, "Second Life chairman's stump speech takes us down the rabbit hole," *Scientific American Observations*, June 29, 2007. Retrieved July 15, 2007, from http://blog.sciam.com/index.php?title=second_life_chairman_s_stump_speech_take&more=1&c=1&tb=1&pb=1.

[5] Gartner, "Gartner says 80 percent of active Internet users will have a 'Second Life' in the virtual world by the end of 2011," April 24, 2007. Retrieved September 13, 2007, from http://www.gartner.com/it/page.jsp?id=503861.

[6] N. Wilson, "Simplicity and the virtual tipping point," *Metaversed*, November 6, 2007. Retrieved January 19, 2008, from http://metaversed.com/tags/news/virtual-worlds.

[7] R. Bush and K. Kenneth Kisiel, *Information & behavior exploitation in virtual worlds: An overview*, July 2007. Retrieved April 13, 2008, from http://blog.wired.com/27bstroke6/files/info_exploitation_in_virtual_worldsiarpanov071.pdf.

[8] C. Steinkuehler and D. Williams, "Where everybody knows your (screen) name: Online games as 'third places,'" *Journal of Computer-Mediated Communication*, vol. 11, no. 4, article 1. Retrieved January 6, 2008, from http://jcmc.indiana.edu/vol11/issue4/steinkuehler.html.

[9] M. Varkey, "Gamers freak out on cyberspace for adventure," *The Economic Times*, February 1, 2008. Retrieved February 6, 2008, from http://economictimes.indiatimes.com/articleshow/msid-2747511,prtpage-1.cms.

[10] D.A. Williamson, "Kids, teens and virtual worlds," *eMarketer*, September 25, 2007. Retrieved May 6, 2008, from http://www.emarketer.com/Article.aspx?id=1005410&src=article1_newsltr.

[11] T.T. Ahonen, "Ten related trends for 2008," January 4, 2008. Retrieved January 6, 2008, from http://communities-dominate.blogs.com/brands/2008/01/ten-related-tre.html.

[12] C. Gaylord, "Can Web-based worlds teach us about the real one?," *The Christian Science Monitor*, January 23, 2008. Retrieved April 2, 2008, from http://www.csmonitor.com/2008/0123/p13s01-stct.htm.

[13] L. Lorek, "Kids flock to sites especially designed for them," *San AntonioExpress-News*, July 2, 2007. Retrieved August 19, 2007, from http://www.mysanantonio.com/business/stories/MYSA070107.1R.VW4Kidsmain.2a76afe.html.

[14] M. Sellers, "A virtual world winter?" *Terra Nova*, December 20, 2007. Retrieved June 17, 2008, from http://terranova.blogs.com/terra_nova/2007/12/a-virtual-world.html.

[15] W.J. Au, "China plays duplicitous game with online gamers," *GigaOM*, January 17, 2008. Retrieved April 2, 2008, from http://gigaom.com/2008/01/17/china-plays-duplicitous-game-with-online-gamers/.

[16] M. Sarvary, "Breakthough ideas for 2008: The metaverse - TV of the future?" *Harvard Business Review*, February 2008. Retrieved July 30, 2008, from http://thelist.hbr.org/.

[17] J. Pinckard, "Is World of Warcraft the new golf? *1up*, February 8, 2006.

Retrieved September 15, 2007, from http://www.1up.com/do/newsStory?cId=3147826.

[18] D. Kushner, "Inside Second Life: Is the hottest spot on the Net paradise on Earth or something a little more sordid?" *Rolling Stone*, April 26, 2007. Retrieved May 13, 2007, from http://www.rollingstone.com/politics/story/14306294/second_life_the_nets_virtual_paradise_heats_up.

[19] K.L. Stout, "A virtual me?: I'll second that," *CNN.com*, January 18, 2007. Retrieved October 1, 2007, from http://www.cnn.com/2007/TECH/01/18/global.office.secondlife/index.html?iref=werecommend.

[20] K. Joly, "A Second Life for higher education?" *University Business*, June 2007. Retrieved July 6, 2007 from http://www.universitybusiness.com/ViewArticle.aspx?articleid=797.

[21] D. Kirkpatrick, "Web 2.0 gets down to business," *Fortune*, March 25, 2008. Retrieved August 18, 2008, from http://money.cnn.com/2008/03/19/technology/web2.0_goofing.fortune/.

[22] Linden Labs, "Second Life: Economic statistics - April 2008." Retrieved April 16, 2008, from http://secondlife.com/whatis/economy_stats.php.

[23] B. Reeves, T.W. Malone, and T. O'Driscoll, "Leadership's online labs," *Harvard Business Review*, vol. 86, no. 5, pp. 58-66.

[24] D. Sharp and M. Salomon, *White paper: User-led innovation - A new framework for co-creating business and social value.* Swinburne University of Technology, January 2008. Retrieved September 18, 2008, from http://smartinternet.com.au/ArticleDocuments/121/User_Led_Innovation_A_New_Framework_for_Co-creating_Business_and_Social_Value.pdf.aspx.

[25] R. Hutchinson, "Reality bytes - Review of *Second Lives: A Journey through Virtual Worlds*," *The Scotsman*, June 30, 2007. Retrieved August 29, 2007, from http://living.scotsman.com/books.cfm?id=1021002007&format=print.

[26] C. Booker, "The economics of fun," *Metaversed*, November 19, 2007. Retrieved November 21, 2007, from http://metaversed.com/21-nov-2007/economics-fun.

[27] S. Kurs, "Welcome to the unreal world: The hugely successful online game Second Life has developed an intriguing ability to blur the lines of reality as we know it," *The Sunday Times*, July 1, 2007. Retrieved August 29, 2007, from http://technology.timesonline.co.uk/tol/news/tech_and_web/personal_tech/article2004489.ece.

[28] R.D. Hof, "My virtual life: A journey into a place in cyberspace where thousands of people have imaginary lives," *Business Week*, May 1, 2006. Retrieved February 13, 2007, from http://www.businessweek.com/magazine/content/06_18/b3982001.htm.

[29] D. Kirkpatrick, "It's not a game: The 3-D online experience Second Life is a hit with users. IBM'S Sam Palmisano and other tech leaders think it could be a gold mine," *Fortune*, February 5, 2007, 155(2): 56-62.

[30] L. Zimmer, "Galveston, (oh Galveston) CVB comes to Second Life," *Business Communicators of Second Life*, January 15, 2007. Retrieved November 29, 2007, from http://freshtakes.typepad.com/sl_communicators/2007/01/galveston_oh_ga.html.

[31] L. Dignan, "Who will govern virtual worlds?" *ZDNet.com*, December 1, 2006. Retrieved May 3, 2007, from http://blogs.zdnet.com/BTL/?p=4041.

[32] D. Welles, "My big, fat, lily-white Second Life," *The Register*, January 30, 2007. Retrieved May 19, 2007, from http://www.theregister.co.uk/2007/01/30/lily_white_and_not_loving_it/.

[33] D. Riley, "Second Life: Europeans outnumber Americans 3 to 1," *Tech Crunch*, May 5, 2007. Retrieved June 16, 2007, from http://www.techcrunch.com/2007/05/05/second-life-europeans-outnumber-americans-3-to-1/.

[34] S. Bhartiya, "Thomas Frey: People are giving birth to empire of one," *EFY Times*, June 30, 2007. Retrieved August 26, 2008, from http://www.efytimes.com/efytimes/fullnews.asp?edid=20045&magid=27.

[35] A. Rappeport, "Second Life's John Zdanowski: The CFO of Second Life explains what the online virtual world is all about, and how his real-world company measures and rewards employee performance with the Love Machine," *CFO Magazine*, October 1, 2007. Retrieved October 16, 2007, from http://www.cfo.com/printable/article.cfm/9858165/c_9891771?f=options.

[36] J. Bennett and M. Beith, "Why millions are living virtual lives online: Second Life is emerging as a powerful new medium for social interactions of all sorts, from romance to making money. It may be the Internet's next big thing," *Newsweek*, July 30, 2007. Retrieved August 5, 2007, from http://www.msnbc.msn.com/id/19876812/site/newsweek/.

[37] A. Greenberg, "Digital media: Second Life goes legit," *Forbes*, July 26, 2007. Retrieved September 6, 2007, from http://www.forbes.com/technology/2007/07/26/second-life-gambling-tech-security-cx_ag_0726secondlife.html.

[38] R., Sidel, "Cheer up, Ben: Your economy isn't as bad as this one. In the make-believe world of 'Second Life,' banks are really collapsing," *The Wall Street Journal*, January 23, 2008. Retrieved January 24, 2008, from http://online.wsj.com/public/article_print/SB120104351064608025.htm.

[39] C. Metz, "The Emperor's new Web," *PC Magazine*, April 24, 2007, vol. 26, no. 9, pp. 70-77.

[40] T. Hespos, "Is Second Life right for your brand? *iMedia Connection*, March 1, 2007. Retrieved September 11, 2007, from http://www.imediaconnection.com/content/13872.asp.

[41] P. Hemp, "The demise of Second Life? *Harvard Business School Press Conversations*, July 26, 2007. Retrieved August 6, 2007, from http://conversationstarter.hbsp.com/2007/07/the_demise_of_second_life.html.

[42] R. Jana and A. McConnon, "Going virtual - How to get a Second Life," *Business Week*, October 30, 2006. Retrieved January 16, 2007, from http://www.businessweek.com/playbook/06/1030_1.htm.

[43] H. Kelly, *Building virtual worlds for education and training: A white paper from The Federation of American Scientists*, November 2007. Retrieved November 19, 2007, from http://vworld.fas.org/bin/view/Main/WhitePaper.

[44] F. Hayes, "Virtual trouble," *Computerworld*, August 13, 2007, vol. 41, no. 33, p. 52.

[45] A. Wade, "Blurred boundaries. In Second Life, players can shop, gamble, buy land or even have sex. But this freedom to do virtually anything raises some increasingly difficult legal questions," *The Guardian Unlimited*, December 10, 2007. Retrieved December 30, 2007, from http://media.guardian.co.uk/digitallaw/story/0,,2225125,00.html.

[46] R. McMillan, "Does Al-Qaeda need a Second Life?" *Network World*, January 7, 2008. Retrieved July 10, 2008, from http://www.networkworld.com/news/2008/010708-does-al-qaeda-need-a-second.html.

[47] C. Svetvilas, "Real law in the virtual world: The future of the Internet may be in 3-D," *California Lawyer Magazine*, January 2008. Retrieved January 28, 2008, from http://californialawyermagazine.com/story.cfm?eid=890855&evid=1.

[48] D.F. Carr, "Second Life: Is business ready for virtual worlds?" *Baseline*, March 1, 2007. Retrieved June 2, 2007, from http://www.baselinemag.com/print_article2/0,1217,a=202017,00.asp.

[49] E. Felten, "Virtual worlds: Only a game?" *Freedom-to-tinker.com*, October 12, 2005. Retrieved May 3, 2007, from http://www.freedom-to-tinker.com/?p=909.

[50] I. Lamont, "Second Life: What's there is potential," *Computerworld*, November 14, 2007. Retrieved November 20, 2007, from http://www.computerworld.com/action/article.do?command=viewArticleBasic&articleId=9045079.

[51] J.V. Last, "Newt's Second Life: On the cyber-stump," *The Weekly Standard*, October 3, 2007. Retrieved October 16, 2007, from http://www.theweeklystandard.com/Content/Public/Articles/000/000/014/182satlx.asp.

[52] H. Clark, "Who's afraid of Second Life?" *Forbes*, January 27, 2007. Retrieved July 1, 2007, from http://www.forbes.com/corporatecitizenship/2007/01/27/second-life-marketing-lead-citizen-cx_hc_davos07_0127disrupt.html.

[53] A. Newitz, "Your Second Life is ready," *Popular Science*, September 2006. Retrieved February 4, 2007, from http://www.popsci.com/popsci/technology/7ba1af8f3812d010vgnvcm1000004eecbccdrcrd.html.

[54] J.M. Smart, J. Cascio, and J. Paffendorf, *Metaverse Roadmap Overview: Pathways to the 3D Web – November 2007*. Retrieved February 14, 2008, from http://www.metaverseroadmap.org/overview/.

Unified Multimodal Search Framework for Multimedia Information Retrieval

Umer Rashid
Faculty of Computing
Riphah International University,
Islamabad, Pakistan
UmerR@riphah.edu.pk

Iftikhar Azim Niaz
Faculty of Computing
Riphah International University,
Islamabad, Pakistan
ianiaz@riphah.edu.pk

Muhammad Afzal Bhatti
Department of Computer Science
Quaid-i-Azam University,
Islamabad, Pakistan
mabhatti@.qau.edu.pk

***Abstract*-There is a trend towards construction of multimedia digital information resources which may hold diverse data types in the form of image, graphics, audio, video, and text based retrieval artifacts or objects. WWW is a huge multimedia information resource. Existing search mechanisms available on WWW are mostly mono-modal. Multimedia information needs are partially satisfied by using mono-modal search mechanisms. Multiple modalities of information are associated with multimedia retrieval object so multimodal search mechanisms are required for searching multimedia information resources. Existing search mechanisms available on WWW whether they are general purpose, domain specific or media specific partially fulfil multimedia information needs because they are mostly mono-modal and able to perform search with in one media type at a time. We explore with the help of multimedia object analysis that search with in all the subsets of multimedia object types is possible by using only four index types, two models of interactions and fifteen possible modes at most. A framework based on an architectural approach is proposed for multimedia information retrieval. Proposed retrieval framework is implemented. Present implementation gives query formulation and information presentation for all modes and search functionality with in few modes of multimedia search framework.**

Keywords: Multimedia Systems, framework, architecture, multimodal, Word Wide Web, mono-modal, digital information resources, search mechanism, multimedia object types.

I. INTRODUCTION

Digital resources are constructed for educational, social, business and organizational purposes. A digital resource mostly consists of a search mechanism for the exploration of information along with data collection. Digital resources show convergence towards knowledge management and e-Learning facilities and acts like knowledge source or knowledge centres [1]. Educational digital libraries and museum digital collections are best examples of digital resources.

World Wide Web (WWW) is also a huge multimedia digital information resource or multimedia database [2] accessible via search engines. In early years digital resources hold only text documents. Single source of information or modality is associated with text documents. A single indexing and retrieval mechanism is sufficient for the retrieval of documents from text based digital information resources. Such type of systems that operate on a single modality for indexing and searching are called mono-modal information retrieval systems [3]. In text retrieval systems user information needs are satisfied by mono-modal search mechanisms. Now information is available in multiple media types. These multiple media types include text, audio, video, image and graphic objects or documents. Mono-modal search mechanisms are satisfactory for the retrieval of text but not for image, audio, video and graphics. Multiple modalities of information are associated with each media type except text. Information retrieval systems that operate on multiple media type objects require multiple modalities of information for indexing and searching. In such systems ordinary user information needs are mostly satisfied by multimodal search mechanisms [3]. A multimodal search mechanism interacts with more than one modalities of information for multimedia information retrieval. In multimodal search mechanisms indexing and searching of more than one modalities of information is required. Unification of multiple indexing techniques and user interaction in a framework is essential for multimodal search.

In this article we discuss web based and non web based search mechanisms and investigate their limitations for multimedia information needs. We propose and implement a Unified Multimodal Search Framework (UMSF).

II. RELATED WORK

Multimedia information retrieval system research shows great advancements in recent years [4] [5] [6]. Researchers deviate from mono-modal search to multimodal search [3] [7]. Interaction and modelling of a single modality is not sufficient to fulfil multimedia information needs. Recent research in multimedia information retrieval systems can be broadly classified in web based and non web based search mechanisms.

A. Web Based Search Mechanisms

Web based search mechanisms are accessible via web. They can be broadly classified into research in general purpose, media specific and domain specific search mechanisms for digital information resources available on WWW. Information seekers are aware of this research

K. Elleithy (ed.), *Advanced Techniques in Computing Sciences and Software Engineering*,
DOI 10.1007/978-90-481-3660-5_22, © Springer Science+Business Media B.V. 2010

because they mostly interact with information retrieval systems accessible via WWW.

General Purpose Web Based Search Mechanisms: Goggle, altheweb, AltaVista and lycose are examples of general purpose information retrieval systems. All these information retrieval systems operate on web information resources. They provide search mechanism for more than one multimedia object types.

Evaluation: By using general purpose search mechanisms like Google, altheweb, AltaVista and lycose user is able to formulate query and visualize results for one media type at a time. Our investigation reveals that indicated general purpose search mechanisms adopt mono-modal search mechanisms. They can only perform search with in text modality. Information retrieval functionality is mostly provided by using text present on page having multimedia object and file attributes associated with multimedia objects like file type, size, format and colour. Existing general purpose search engines give illusion that they can perform search within modalities of multimedia objects but their working is totally dependent on text retrieval techniques and pre-recorded attributes of multimedia objects. Due to this reason information needs are partially satisfied by using these search mechanisms [8].

Media and Domain Specific Web Based Search Mechanisms: Media specific and domain specific search mechanisms are also available on the WWW. They mostly provide mono-modal search mechanisms for specific domains [3].

ACM[1] and IEEE[2]: are educational digital libraries. They provide specialized search mechanism for text retrieval with in particular educational domains.

Terra Galleria[3]: provides content base image retrieval along with textual search.

Digital South Asia Library[4]: provides retrieval facility for cultural heritage data of South Asia. Data is mostly stored in the form of images and search facility is provided by annotated text modality.

Open Video Digital Library (OVDL)[5]: provides search and browse mechanisms for video documents for pre-stored specified video repository. Video documents are organized in the form of cluster of genres like documentaries, educational, lectures, ephemerals, and historical. Search mechanisms are provided by using text modality of speech transcripts and bibliographic records. OVDL is more focused towards human computer interaction issues in the retrieval of video documents [9].

Hermitage Museum[6]: is accessible via web and provides searching and browsing of image based museum objects. User is able to browse collections and search using texture and color of image modality.

WOMDA[7]: operates on multimedia data like manuscripts of poems, diaries, letters and information available in the form of audio, video and images revolves in the age of First World War poet Wilfred Own. WOMDA holds its own archives of data [10]. Data is managed in the form of browse-able collections. WOMDA provides search facilities of audio, video, images and manuscripts using annotated text modality.

Evaluation: Domain specific and media specific search mechanisms provide advance search mechanisms. They provide search with in different multimedia modalities and information domains. Media specific search mechanism provides search facility with in a particular type of multimedia. User is mostly able to formulate query and visualize results for a specific multimedia type. They cannot discuss unification of search approaches for all multimedia types. They are usually not expandable for multiple multimedia object type's retrieval. Their integration in general purpose search mechanism that facilitates unified retrieval of all multimedia object types is not approachable. Domain specific search mechanisms rarely provide search with in multiple multimedia object types. Mostly they perform search with in specified domains and particular media types. It is explored that their retrieval model is totally based on accompanying text modality of multimedia object types. Due to their operability for a specific type of media or with in specified domains user information needs are partially satisfied.

B. Non Web Based Search Mechanisms

Research in search mechanisms not associable via WWW includes information retrieval systems designed, developed, investigated and evaluated in research laboratories. These systems are mostly not available on the WWW, some products like greenstone are download-able via web; however their detailed information is available from the related research papers [4, 5, 6, 11, 12, 13, 14, 15]. Example of these systems includes informedia [4], Combinformation [12], greenstone [13], VULDA [4], M-Space Browser [6, 15] and EVIADA [10].

Informedia: is a multimedia information retrieval system for modeling, searching and visualization of video documents [5]. It is designed for searching video news articles. Informedia provides multimodal search mechanism for the retrieval of video documents. User is able to perform search by using text and image modalities. Informedia project is more focused towards human computer interaction aspects in searching and browsing of only video documents.

Combinformation: multimedia document collections are represented by image and text surrogates so this mechanism facilitates browsing of multimedia collections [12].

Greenstone: is open source software designed and developed for construction of personal digital libraries [14] [13] Greenstone provides services for the construction of digital library collections and searching, browsing facilities. Greenstone provides collection construction facility using documents in different formats like HTML, DOC, PDF,

1 http://portal.acm.org
2 http://www.ieeexplorer.org
3 http://www.terragalleria.com
4 http://dsal.uchicago.edu
5 http://www.open-video.org
6 http://www.hermitagemuseum.org

7 http://www.hcu.ox.ac.uk/jtap/

multi format images, MP3 and MIDI audio. This software also provides searching and browsing facilities using full text search and metadata browsing mechanisms.

Video University Digital Library for All (VULDA): provides all functionalities of video streaming, segmentation, scene detection and automatic speech recognition [4]. VULDA provides content based indexing mechanism using speech transcripts and image modalities. Retrieval is also facilitated by metadata of video documents. VULDA provides multimodal query formulation mechanism using text and image modalities.

M-Space browser: discusses human computer interaction issues in video searching and browsing. It does not provide query processing and retrieval models for video documents [6, 15].

The Ethnomusicological Video for Instruction and Analysis Digital Archive (EVIADA): provides video document management facilities like video segmentation, manual video annotation and metadata creation for documents in video archives [11].

Evaluation: These advance university projects provide valuable multimodal search for multimedia information retrieval. They mostly address search with in one multimedia type. Complex document modeling and retrieval mechanisms are investigated in these projects. These research projects mostly investigate some specified retrieval problems for predefined domain. Due to specified domains and operability with in particular multimedia type these research projects are mostly not applicable for unified multimedia retrieval.

III. Unified Framework

Existing multimedia information retrieval systems whether they are web based or non web based have usability issues because user is not able to perform search with in multiple multimedia types [8]. Advance multimedia information retrieval techniques [16, 17] have been investigated in recent years. It is required that researchers should enhance these techniques by investigating and resolving the limitations. These advance retrieval techniques discuss basic indexing and retrieval mechanism for multimedia data types. Their unification with in information retrieval frameworks is mostly not addressed, so satisfactory multimedia retrieval is not possible. Unification of these techniques in proper information retrieval framework should provide enhancement in multimedia information retrieval. To overcome problems [8] in existing search mechanisms we investigate and propose a **U**nified **M**ultimodal **S**earch **F**ramework (UMSF). The framework is explained with the help of following multimedia object analysis [8].

A. Multimedia Object Analysis

Multimedia object or document in context of multimodal search mechanism is an interactive artifact that can be modeled or indexed, searched, browsed, visualized and retrieved. Audio, video, image, graphics and text are five basic multimedia object types. Image and graphics are interchangeable. We place graphics in broad category of image types. In this article hereafter refer multimedia object types as image, audio, video and text. They can be represented interchangeable. Multimedia objects can be expressed as supersets and subsets of each other. Video can be decomposed into audio objects and visual streams. Audio object is composition of speech and music. Speech can be represented as text objects and their context. Text consists of keywords and their context. Music portion of audio object is decomposable into musical instruments and their context.

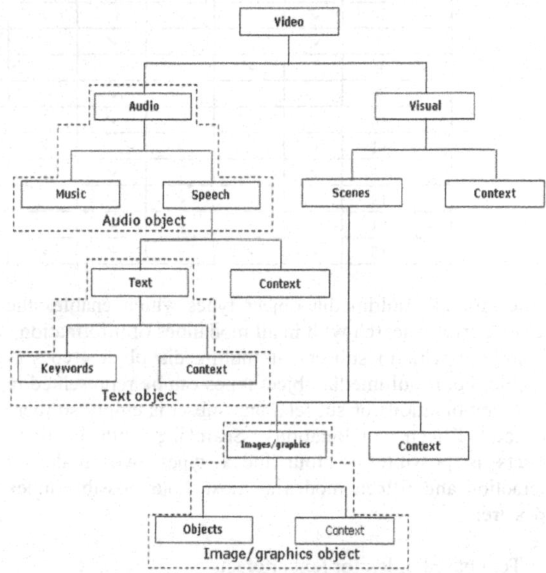

Fig. 1. Multimedia Object Decomposition Hierarchy, dotted region represents occurrence of text, audio, image/graphics object types with in video object type

Visual stream of video can be decomposed into scenes and their context. Scene is a video clip of continuous images and graphics having some specific context. Image and graphic objects can be decomposed into objects and their context. "Fig. 1." depicts that video artifact is composition of text; image/graphics and audio object types. Video object owns all features of other multimedia object types. Top down analysis of above hierarchy shows that video object type is decomposed into audio, image and text multimedia object types at intermediate levels and finally represented as

1. Text keywords and their context
2. Image/graphics objects and their context

Bottom up analysis of Figure 1 show initially text keywords, their context and image/graphics objects, their context composes image/graphics, audio and text object types and finally interpreted as video object type [8].

Indexing text, accompanying text, speech transcript and image modalities are sufficient for searching with in all subsets of multimedia object types. By using image and text modalities ordinary multimedia information user formulate

TABLE 1
POSSIBLE MODES, INDEXES AND MODELS OF INTERACTION USED IN MULTIMODAL SEARCH FRAMEWORK

Modes	Multimedia object types				Indexes Involved in search				Modals of Interaction	
	Text	Image	Audio	Video	Text Based	Accompanying Text	Speech transcript	Image feature	Keywords	Visual Image
1	✓				✓				✓	
2		✓				✓		✓	✓	✓
3			✓			✓	✓		✓	
4				✓		✓	✓	✓	✓	✓
5	✓	✓				✓	✓	✓	✓	✓
6	✓		✓		✓	✓	✓		✓	
7	✓			✓	✓	✓	✓	✓	✓	✓
8		✓	✓			✓	✓	✓	✓	✓
9		✓		✓		✓	✓	✓	✓	✓
10			✓	✓		✓	✓	✓	✓	✓
11	✓	✓	✓		✓	✓	✓	✓	✓	✓
12	✓	✓		✓	✓	✓	✓	✓	✓	✓
13		✓	✓	✓		✓	✓	✓	✓	✓
14	✓		✓	✓	✓	✓	✓	✓	✓	✓
15	✓	✓	✓	✓	✓	✓	✓	✓	✓	✓

queries for all multimedia object types which enables the user to perform search with in all modalities of information.

Searching with in subsets of multimedia object types is possible. Four multimedia object types can be represented in 2^4=16 combinations or subsets, one subset is empty so there are total fifteen combinations. Searching with in these subsets is possible by four index types two modal of interaction and fifteen modes at most. Four possible index types are:

1. Text based index for text artefacts
2. Speech transcript text index for audio and video object types
3. Image feature based index for image and video object types
4. Accompanying text based index for audio, video and image objects having accompanying textual information.

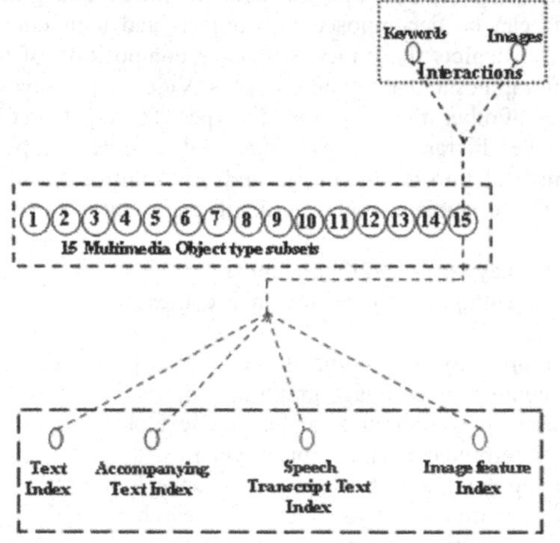

Fig. 2. Framework for Multimodal Search Mechanism

We explain fifteen combinations in Table 1. User is able to formulate query for any subset of multimedia object types by using only text based and image based interactions or query formulation mechanisms. Our proposed multimodal search mechanism provides search mechanism with in all the subsets of multimedia objects types by using fifteen possible modes or layers. Table 1 demonstrates modes, possible index types and interactions against each mode of multimodal search mechanism.

A search mechanism that incorporates specified four possible index types and two models of interaction provides search facility with in all possible subsets of multimedia object types. Figure 2 demonstrates a multimodal search mechanism that has capability of searching with in any subset of multimedia objects types. One combination of multimedia object types, mode or a subset is activated at a time using four possible index types and two models of interaction at most. "Fig. 2." demonstrates this multimodal search framework.

First dotted rectangle represents two interaction mechanisms via keywords and images; second dashed rectangle represents possible modes or subsets. One mode is activated at a time. User is able to perform search with in any subset of multimedia object types by using these fifteen modes. Third rectangle represents possible four index types. Search is always performed with in theses four index types at most [8].

IV. DESIGN AND IMPLEMENTATION OF MULTIMODAL SEARCH FRAMEWORK

Unified Multimodal Search Framework is designed and implemented. Design and implementation contains a framework based on an architectural approach and interactive interface for multimodal query formulation and information presentation.

A. M^3L Architecture for UMSF

Decomposition of system into subsystems or components establishes a framework. Communications among components discuss architectural design [22]. Unified Multimodal Search Framework distributes components among three main categories, modes, models and modalities.

Mode represents possible ways of interaction with multimedia object types. There are fifteen modes that provide search facility with in all subsets of multimedia object types. Fifteen modes of UMSF are given in table 1.

Model is basic retrieval function that provides search facility with in multimedia based documents. There are four retrieval functions for text, image, audio, and video based documents. Text retrieval function search in text modality, audio retrieval in accompanying text and speech transcript text modality, image retrieval function in accompanying text and image modality, and video retrieval function in accompanying text modality, speech text transcript modality and image modality.

Modality contains representations or indexes of multimedia based documents.

In recent years object based architectures appeared in the form of frameworks [23]. Unified Multimodal Search Framework requires an architectural design that identifies necessary objects and specifies their communication. Our framework is based on proposed Modes, Models, and Modalities Layered architecture (M^3L).

M^3L exhibits layered based architectural approach and distribute framework objects in three adjacent layers which are modes, models, and modalities. Possible objects in M^3L are activation, interaction, indexing, presentation, retrieval, and processing objects. We briefly discuss their basic functionalities.

Activation Objects: Activate passive objects. Passive objects must be activated and updated as active objects before use. There are two types of activation objects, mode and index activation objects. Fifteen mode activation objects make functioning of fifteen modes of UMSF. Each mode activation object activates necessary interaction objects for a particular mode. A mode activation object also invokes required retrieval functions regarding each mode of framework. An index activation object activates representations of multimedia based documents before use by loading them in temporary memory. Query processing objects only search information contents in index objects when they are active.

Interaction Objects: An interaction object provides user interaction for multimodal query formulation using image and text based interactions. They describe interaction interfaces for each specified mode. Fifteen specified modes of UMSF have been discussed. Each mode interaction can be collectively represented by a set of interaction objects. Interaction in each mode can be represented by composition of one or more than one interaction objects. Activation objects activate specified interaction objects that collectively provide interaction interfaces regarding each mode of UMSF. Each interaction object appears in one or more than one interaction interfaces concerning each mode.

Index Objects: Each multimedia document can be represented by index object. Index objects are document representations. Representations are initially passive. In our search framework there are four types' index objects for representation of text, audio, image, and video based documents. They are permanently stored on secondary storage. When they are stored on secondary storage they are in passive state. Index objects must be activated before searching by loading them in temporary memory. A retrieval object only interacts with index objects when they are in active state.

Presentation Objects: Provide presentation interfaces for results returned from retrieval objects. There are four presentation objects each one is responsible for presentation of results regarding a specific multimedia type. Each retrieval object returns results to a specific presentation object for result presentation. In each mode it might be possible that more than one presentation objects collectively provides results interface for more than one multimedia document types.

Retrieval Object: Holds information retrieval functions for audio, video, image and text retrieval. Text, audio, image, and video retrieval functions have been discussed in previous chapters. Retrieval functions are encapsulated in distinct retrieval objects. They calculate probability of similarity of a document with respect to a query formulated by interaction objects. Retrieval objects take input query from interaction objects and return result to presentation objects. Retrieval objects are activated by mode activation objects. One mode activation object may activate more than one retrieval objects. One retrieval object may be activated by more than one mode activation objects.

Processing Objects: there are three processing objects used for information search in textual modality, image modality and video modality. Three processing objects encapsulate processing algorithms. Three types of processing objects are capable of searching information contents with in all index objects. Retrieval objects always invoke one or more than one retrieval objects for document similarity calculation by using relevancies or approximations returned by retrieval objects.

M^3L organizes Unified Multimodal Search Framework objects in three layers. Each layer collectively performs a specific type of task and interacts only with adjacent layers. Now we discuss three layers of M^3L architecture for proposed search framework.

M^3L organizes Unified Multimodal Search Framework objects in three layers. Each layer collectively performs a specific type of task and interacts only with adjacent layers. Now we discuss three layers of M^3L architecture for proposed search framework.

Layer 1 represents possible modes of interactions present in search framework. Layer 1 is top most layer of M^3L architecture and provides interactions and presentations. It is composed of interaction, presentation and mode activation objects. Interaction object provides interface elements for multimedia query formulation. Presentation object provides information presentation mechanism. There are four interaction objects each one represent specified interface element for more than one mode. A mode activation object activates required interaction objects for each mode in same layer and information retrieval objects in layer 2.

Layer 2 mostly contains implementation of basic query processing algorithms and retrieval functions. It is composed of retrieval objects, processing objects, and index activation objects. Retrieval object invokes index activation objects with in same layer. Index activation objects activate index objects in layer 3 by loading them in temporary memory. Retrieval objects in layer 2 also interact with processing objects in same layer for basic relevancy and approximation calculations. Fusion of relevancies and approximations of the same document from multiple modalities are performed by retrieval objects. Retrieval objects also interact with interaction objects. They return document's overall probabilistic similarities to presentation objects for result presentations.

Layer 3 provides index storage for framework. It contains index objects, which exits in passive state when stored on permanent storage. Their state will be changed from passive to active when they are activated by index activation objects. Activation objects load them in temporary memory before processing. After loading of required index objects in temporary memory, processing objects interacts with them for relevancy as well as approximation calculations. Active index objects always exit on temporary memory while passive index objects always exist on permanent storage. "Fig. 3." demonstrates layered M^3L architecture.

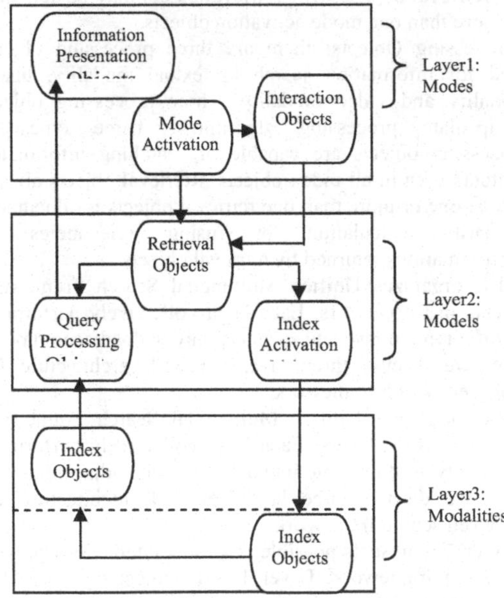

Fig. 3. M3L architecture for proposed search framework

B. Interactive Interface for UMSF

We explore resource modeling and layered based approach for interface design [24, 25, 26].query formulation and information presentation components are system resources. They are distributed in fifteen possible layers. Each layer provides query formulation and information presentation mechanism for one mode of UMSF. Interface is divided into mode specification, query formulation and information presentation regions as shown in "fig. 4.".

Mode specification Region: This region of interface is used for mode specification. Users explicitly specify multimedia types for which searches are required. Mode specification region allows the user to select any subset of multimedia object types. Mode specification region allows any combinations of four possible choices for text, audio, image, and video based documents. Any combination of multimedia object types from fifteen possible combinations is selected by using these four possible choices. Each combination represents some specific mode of UMSF and activates some specific layer of search for interactions.

Query Formulation Region: query formulation region is constituted by fifteen distinct layers. Each layer contains query formulation interactions for some specific mode of UMSF. One layer is active at a time for interactions. Users at any instant of time have exclusive control of only one layer for interaction.

Information Presentation Region: results are returned from one or more than one multimedia object types. Results from each multimedia type are represented in its specified layer. Text, image, audio, and video based retrieved documents are presented in their specified layers for presentation. Four distinct colors are used for each layer. So user can easily distinguish multiple information presentation layers.

Now we discuss necessary steps of user interactions and system behaviour regarding mode specification, query formulation, and information presentation.

- Mode specification region contains four labels for specification of text, audio, image, and video based multimedia document types in the search process. When multimedia document type is active it is represented by binary "1" below its label surrounded by some specific color used for representation of that multimedia type. When multimedia type is not active it is represented by binary "0" below its label. When curser is moved above a label its presentation is changed. That tells user that some event is associated with label. A mouse click activates or deactivates multimedia document types in the search process. Binary 0's and 1's below four labels are arranged such that their binary representation represents some mode of UMSF; however decimal representation is also presented.
- When ever labels for multimedia document type speciation are clicked search mode is changed. Changed mode number is represented as binary and decimal format in mode specification region. When searched mode is changed corresponding query formulation region is also changed and provides necessary query formulation interactions for user specified search mode of UMSF.

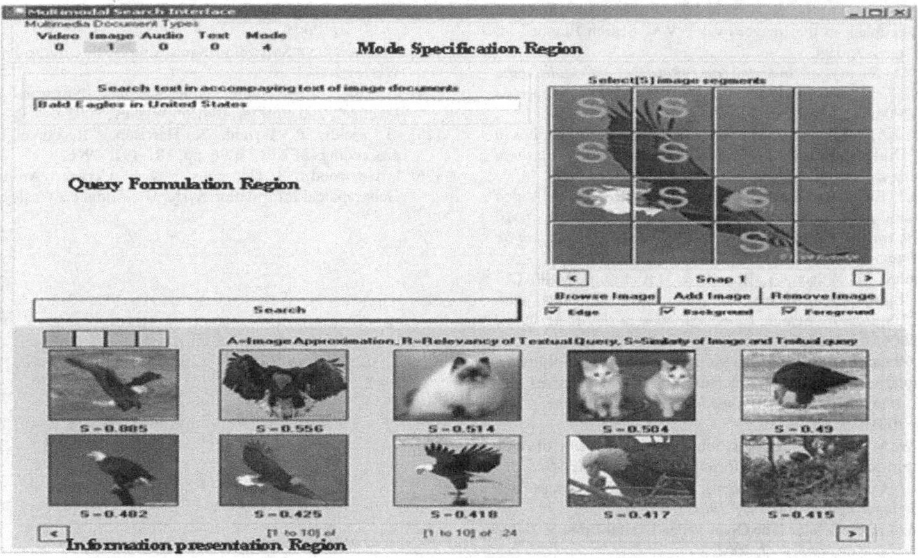

Fig. 4. Interface for UMSF

- Active multimedia document types that are involved in the search process are represented by distinct colors below labels in mode specification region. Information presentation regions also contain a clickable strip of four colors. Colors strip in information presentation region have same colors, and sequence used in mode specification region. Colors with in strip are clickable. Information presentation region contains four layers for information presentation for possible multimedia document types. Only information presentation layers corresponding to each user specified mode are active at a time. Use is able to visualize results for one multimedia type at time by clicking its corresponding color code with in color strip.

Distribution of system resources among different layers using resource modeling have following advantages.

1. User is able to select a mode in which his information needs are fully satisfied. E.g. if user wants to search information with in image and text based document. Multimodal search interface will provides search facility with in text and image based documents.
2. One query formulation layer is activated against each user specified mode of UMSF. Each layer activated by mode specification contains specified interaction objects required for that particular mode. User has only exclusive control of interaction objects necessary for specified mode. By organizing query formulation interface in layers provides a sensible way for multimedia information specification. User is able to search information with in a particular set of multimedia object types.
3. More than one information presentation objects are active at a time. Each mode contains one or more than one multimedia types on which search is performed. Number of active information presentation layers at a time is equal to the number of multimedia types on which search is performed; However user is able to visualize results for one multimedia type at a time. By activating multiple and providing visualization of one layer of information presentation for searched multimedia document types user better concentrates with in search results against each multimedia type given in specified mode of UMSF.

4. CONCLUSION

Existing search systems whether they are general purpose, media specific, domain specific or non web based partially fulfill multimedia information needs. User is able to perform search with in one media type at a time and they are mostly mono-modal. We investigate from multimedia object analysis that searching with in multiple multimedia types is possible. We have proposed a unified multimodal search framework by using two modalities of interaction, fifteen modes and four possible index types at most. By integrating our proposed search framework in multimedia information retrieval systems information users are able to perform multimodal search in all possible subsets of multimedia object types. Proposed search framework is designed. Interactive interface for multimodal search is investigated and implemented. Investigated interface provides query formulation and information presentation for all modes of UMSF. Full search is available with in three modes which are all subsets of text and image based objects. By exploring solution for other two multimedia object types which are audio and video object type's full search can be explored with in all modes.

REFERENCES

[1] B. Marshall, Y. Zhang, R. Shen, E. Fox & L.N. Cassel, "Convergence of Knowledge Management and E-Learning: the GetSmart Experience", *In IEEE'03, 3rd ACM/IEEE-CS Joint Conference on Digital Libraries Digital Libraries,* pp. 135-146, 2003.

[2] L. Taycher, M. L. Cascia, & S. Sclaroff " Image Digestion and Relevance Feedback in the ImageRover WWW Search Engine", In *VISUAL'06*, pp.1-7, 1997.

[3] N. Chen, "*A Survey of Indexing and Retrieval of Multimodal Documents: Text and Images", (2006-505)*. Canada, Queen's University. 2006.

[4] J.A. Arias & J.A. Sánchez, Content-Based Search and Annotations in Multimedia Digital Libraries. *In ENC'03, 4th Mexican International Conference on Computer Science*, pp. 109-116, 2003.

[5] M.G. Christel. Evaluation and User Studies with Respect to Video Summarization and Browsing. In *Multimedia Content Analysis, Management, and Retrieval", part of the IS&T/SPIE Symposium on Electronic Imaging*, Vol. 607, 2006.

[6] M.C. Schraefel, M. Wilson, A. Russel, & D.A. Smith. MSPACE: Improving Information Access to Multimedia Domains with Multimodal Exploratory Search. *Communication of the ACM, 49(4)*, pp. 47-49, 2006.

[7] J. Yang, Y. Zhuang, & Q. Li. Multi-Modal Retrieval for Multimedia Digital Libraries: Issues, Architecture, and Mechanisms. *In ICME'01, Workshop in Multimedia Communications and Computing*, 2001.

[8] U. Rashid, & M.A Bhatti. "Exploration and Management of Web Based Multimedia Information Resources". In *SCS2 07, 3rd International Conference on Systems, Computing Sciences and Software Engineering*, pp. 500-507, *2007.*

[9] G. Marchionini & G. Geisler. The Open Video Digital Library. *D-Lib Magazine, Volume 8 Number 12*, 2002.

[10] Wilfred Owen digital archive to expand. *BLUEPRINT the newsletter of the University of Oxford, Oxford Press*, pp. 4, 1993.

[11] W.Dunn . EVIADA: Ethnomusicological Video for Instruction and Analysis Digital Archive. In *JCDL'05, Joint Conference on Digital Libraries*, pp. 40, 2005.

[12] A. Kerne, E Koh, B. Dworaczyk, J.M. Mistrot, H. Choi, S.M. Smith, R. Graeber, D. Caruso, A. Webb, R. Hill & J. Albea. CombinFormation: A Mixed-Initiative System for Representing Collections as Compositions of Image and Text Surrogates. In *JCDL'06, 6th ACM/IEEE-CS Joint Conference on Digital Libraries*, pp. 11-20, 2006.

[13] I.H. Witten & D. Bainbridge. Building Digital Library Collections with Greenstone, In *JCDL'05, 5th ACM/IEEE-CS Joint Conference on Digital Libraries, pp. 425-425*, 2005.

[14] D.Bainbridge, J.Thompson & I.H. Witten. Assembling and Enriching Digital Library Collections. In IEEE *'03, Joint Conference on Digital Libraries*, pp. 323- 334, 2003.

[15] Wilson,. Advanced Search Interfaces considering Information Retrieval and Human Computer Interaction. *Agents and Multimedia Group School of Electronics and Computer Science Faculty of Engineering and Applied Science University of Southampton.* Retrieved from web site: www.iam.ecs.soton.ac.uk, 2007.

[16] E. Bruno & N.M. Loccoz. Interactive video retrieval based on multimodal dissimilarity representation. *In MLMM'05, workshop on Machine Learning Techniques for Processing Multimedia Conten*, 2005.

[17] T. Westerveld, A.P.D. Vries, A.R.V. Ballegooij, F.M.G.D. Jong & D. Hiemstra. A Probabilistic Multimedia Retrieval Model and its Evaluation. In *EURASIP'03, Journal on Applied Signal Processing, special issue on Unstructured Information Management from Multimedia Data Sources*, 2003(3), pp. 186-19, 2003.

[18] W Agheal, & A.S Kellerman. *Distributed Multimedia Technologies, Applications, and Opportunities in the Digital Information Industry*, New York, ADDISON-WESLEY PUPLISHING COMPANY, 1996.

[21] J.C. French, A.C. Chapin & W.N. Martin. An Application of Multiple Viewpoints to Content-Based Image Retrieval. In IEEE *'03, Joint Conference on Digital Libraries*, pp. 128-130, 2003.

[22] S.Y. Neo, J. Zhao, M.Y. Kan, & T.S. Chua. Video Retrieval using High Level Features: Exploiting Query Matching and Confidence-based Weighting. In *CIVR'06, Proceedings of the Conference on Image and Video Retrieval*, pp. 143-15, 2006.

[19] U. Rashid, & M.A. Bhatti. 2007. Enhanced Term-specific Smoothing using Document World Clusters. In *CITA'07,5th International Conference on Information Technology in Asia*, pp. 140-144, 2007.

[20] D.Hiemstra. Term-Specific Smoothing for the Language Modelling Approach to Information Retrieval: The Importance of a Query Term. In *SIGIR'02, Special Interest Group on Information Retrieval, pp. 35-41, 2006.*

[23] I. Somerville. "Software Engineering", 7th edition, Pearson Addison Wesley.2004.

[24] L. Bass, P. Clements, , & R. Kazman. "Software Architecture in Practice", 2nd edition, Addison Wesley, 2002.

[25] B. Fields, P. Wright, & Harrison. "Resource Modeling". In proceedings of APCHI'96, pp. 181-191, 1996.

[26] I. Heywood, S. Cornelius, , & S. Cerver. "An Introduction to Geographical Information Systems", Addision Wesley, 1998.

Informational analysis involving application of complex information system.

Clébia Ciupak (Unemat - Unic) - clebia.ciupak@terra.com.br
Adolfo Alberto Vanti (Unisinos) - avanti@unisinos.br
Antonio José Balloni (CTI) - antonio.balloni@cti.gov.br
Rafael Espin (Cujae) - espin@tesla.cujae.edu.cu

SUMMARY – The aim of the present research is performing an informal analysis for *internal audit* involving the application of complex information system based on *fuzzy logic*. The same has been applied in internal audit involving the integration of the accounting field into the information systems field. The technological advancements can provide improvements to the work performed by the internal audit. Thus we aim to find, in the complex information systems, priorities for the work of internal audit of a high importance Private Institution of Higher Education. The applied method is quali-quantitative, as from the definition of strategic linguistic variables it was possible to transform them into quantitative with the matrix intersection. By means of a case study, where data were collected via interview with the Administrative Pro-Rector, who takes part at the elaboration of the strategic planning of the institution, it was possible to infer analysis concerning points which must be prioritized at the internal audit work. We emphasize that the priorities were identified when processed in a system (of academic use). From the study we can conclude that, starting from these information systems, audit can identify priorities on its work program. Along with plans and strategic objectives of the enterprise, the internal auditor can define operational procedures to work in favor of the attainment of the objectives of the organization.

I. INTRODUCTION

The current economic scenario in which the organizations are inserted is dynamic and demands from the managers the knowledge of the environment under a systemic approach. Such environment characterizes itself by the identification of opportunities and threats, both internal and external. Overseeing the organization as a whole may provide the managers more accurate decision-making, attending the organizational objectives.

The growing technological advancements have made viable the utilization of computing tools or applications which propitiate to the user the access to information with higher dynamicity. Among the most recent ones, the complex information systems such as extensible Business Related Language (XBRL) which discloses financial information (Mauss, et al, 2007), *Business Intelligence* which helps decision making (Balloni, 2006) and *fuzzy logic* stand out. Fuzzy logic and BI are used in this work. Both approached at the present work, being the first so as to contextualize theoretically the complex systems and the latter with practical application at internal audit.

Cerchiari and Erdmann (2005, p. 4) state that "auditing is nowadays seen as an instrument of administrative control which stands out as a converging point of all feats, facts and information so as they are reliable, suitable, absolute and secure". Thus, one may state that internal audit must be provided with useful, reliable and opportune information. To attend to such necessity one may make use of the internal controls system and of the information technology resources so as to elaborate an audit plan which, by means of operational procedures, may collaborate with the goals of the institution, defined at the strategic planning.

The problem question implicated on the present work is: How to define strategic priorities for the internal audit work via complex information systems? For such, firstly issues which contextualize the accounting audit are broached, right afterwards the focus turns to the complex information systems and their application at the strategic level of an IHE (Institution of Higher Education). Thus a case study was performed at a private institution of higher education, where by means of interviews with the administration pro-rector, the data for posterior analysis were obtained.

II. AUDITING

Crepaldi (2000, p. 27) defines auditing as being "the gathering, the study, and the systematic analysis of the transactions, procedures, operations, routines, and financial statements of an entity". It is thus understood that auditing is a set of detailed studies of all operations which are part of an organization, preventing situations which generate frauds or may incur in error through the application of periodic tests on the internal controls from the companies.

Auditing may also be seen as a systematic observation of internal controls as well as of the documentation from the enterprise. However, at modernity the concept of auditing is more dynamic, attributing to it functions which comprise the entire administration of the enterprise, initiating a policy towards orientation, interpretation, and forecasting of facts. From the exposed, one notices that auditing has moved from being seen as a simple verification of accounting documents to assuming a more significant attitude at the management of the enterprise, thus aiming at aiding the aforementioned at the acquirement of positive results. It is emphasized still, that auditing may be external or internal, in accordance to the presented as follows.

K. Elleithy (ed.), *Advanced Techniques in Computing Sciences and Software Engineering*,
DOI 10.1007/978-90-481-3660-5_23, © Springer Science+Business Media B.V. 2010

Regarding the external audit, Crepaldi (2000) highlights its main purpose is to verify the financial and patrimonial situation of the enterprise. The operation results in the net wealth, and the origins and applications of resources in accordance to the Brazilian accounting norms. It is performed by an external auditor, with no employment relationship to the audited enterprise. Franco and Marra (1995), present the taxation of income as one of the causes which lead the businessmen to have a higher control of the accounting records.

Sirikulvadhana (2002) mentions that technology impacts in the format at which auditors perform their task. They believe, still, that it is important for the auditors to convey and combine their knowledge and technical skills with the specialists in the field of information technology, following the evolution of such field, so as to accomplish their task in an efficient manner. For such the present work directs informational analysis to internal audit with the application of complex information system based in *fuzzy logic*.

Oliveira *et al.* (2006) affirm the internal audit task comprises all existing services, programs, operations, and controls at the institution so as to advise management in the examination and evaluation of the adequacy, efficiency, and efficacy of the activities, being able to evaluate, inclusive, the quality of performance of the areas in relation to the attributions and to the plans, goals, objectives, and policies adopted by the enterprise. For Paula (1999), the huge progress in communications made the world more competitive, which causes competence, technology, and much creativity to prevail in the elaboration of the audit.

Attie (2006) emphasizes, still, the importance of internal auditing, mentioning that it is responsible for ensuring, to the enterprise's management, if the plans, policies, accounting systems, and internal controls are being followed in accordance to what determined by the organization. On the other hand, Mautz (1985) believes that in order to make the internal audit work more efficient, qualified professionals which possess accounting and auditing techniques knowledge are in need, and most important, that they know how to utilize such knowledge.

The task of internal audit has its freedom somewhat restricted to the management of the enterprise, but such does not difficult its performance if it possesses free will so as to investigate and develop its activities. Planning may be outlined, in accordance to a work plan which may involve the identification of work, the execution time, the personnel involved, the verification of controls, the determination of weak and strong spots, the conclusion of work, and the final report. The significant advancement of technologies enables accounting audit the utilization of technological resources which favor the execution of their works, among which the information technology, specialist systems, and intelligent systems stand out or even complex of information from which the work emphasizes the analysis of application of the latter.

III. COMPLEX INFORMATION SYSTEMS – *BUSINESS INTELLIGENCE* AND FUZZY LOGIC

For Cameira (2003) (apud RAUTER and VANTI, 2005), the *Business Intelligence* BI systems correspond to a conceptual evolution of the systems in support of decision. Barbieri (2001) understands that BI is the utilization of varied sources of information to define strategies of competitiveness at the enterprise's businesses, with the definition of regulations and techniques for adequate formatting of high data volume, aiming at transforming them into structured depositories of information. These data may be originated from various techniques of "prospecting" of information via Competitive Intelligence, or of conceptual sources as Knowledge Management. Dresner (apud Baum, 2001) classifies *Business Intelligence* as a 'conceptual umbrella', under which a variety of technologies which assist the end user in accessing and analyzing the sources of quantitative are sheltered.

According to Birman (2003), under the concept of BI there is a spectrum of technology solutions that attend some of the vital necessities of the organizations. They are constituted of a combination of already known concepts with technological evolution capable of quickly absorbing gigantic masses of data, with a presentation of outcomes through graphics, instant reports, flexible simulations, strategic information, dynamic tables and others.

Geiger (2001) postulates *Business Intelligence* is the whole set of processes and data structure utilized for understanding the enterprise's business environment with purpose of supporting the strategic analysis and decision making. The principal components of BI are *Data Warehouse*, *Data Mart*, interfaces for decision support, and processes for data collection, which must be incorporated into a huge database and integrated into the business community. The major benefits of the BI environment, besides technical reasoning represented by simplification, global vision of consolidated data in a *Data Warehouse*, relate to the possibility of reasoning about the strategic questions from the organization and forecasting for its future through scenarios established in precise data from the present and past.

The executive seeks through *Business Intelligence*, to access and integrate development and tendency metric indicators, with different degrees of synthesis, capable of aiding them at the conveyance of the business. The present work analyses the theoretical and practical applications of the information system through matrix structure prioritizing the strategic variables from the studied case for posterior BI utilization.

The *fuzzy* logic is seen as a concepts, principles, and methods system for dealing with reasoning methods which are more approximate than exact. It allows for the representation of intermediate pertinence values between the true and false values from the classic logic, enabling application at the construction of systems for describing of imprecise elements as solution of problems in the management area of large corporations, in Economy, Social Sciences, Psychology, etc.

According to Oliveira Jr. (1999), Fuzzy Logic is a set of methods based on the fuzzy set and diffuse operations concept, which enable the realistic and flexible modeling of systems. The diffuse set which may be named *fuzzy set*, was adapted with the intention of generalizing the ideas, which are represented by the ordinary sets, called *abrupts* or *crisp sets* in international literature. Such sets, which are as a variety of logic predicate and run through the interval [0,1]. Espin and Vanti (2005) applied a *fuzzy logic* methodology and developed a system applied to strategic planning and to the *Balanced Scorecard* (BSC).The present system obtains truth values for calculating predicates which must possess sensitivity to changes of truth values with "verbal significance". With such system, the fulfillment of the classical conjunction and disjunction properties are renounced to, opposing to such the idea that the increase or decrease of the conjunction or disjunction truth value is compensated by the other one's corresponding decrease or increase. In compensatory fuzzy logic supported by the authors, the operations of "y" are defined as follows:

$$v(p_1 \wedge p_2 \wedge ... \wedge p_n) = (v(p_1).v(p_2)...v(p_n))^{1/n}$$

$$v(p_1 \vee p_2 \vee ... \vee p_n) = 1 - ((1 - v(p_1)).(1 - v(p_2))...(1 - v(p_n)))^{1/n}$$

Such system references the practical application of the present work for internal audit at the identification of strategic priorities. The benefits of utilizing the *Fuzzy* methodology for the decision making is that it "can be structured in a developed system, thus considering possible answers of uncertain reasoning with the representation of possibilities in degrees of uncertainty" (KLIR *et al.*, 1997, p. 218). The benefits of the *fuzzy* or fuzzy logic, according to Mukaidono (2001), are that in many cases, the mathematic process model may not exist or may be exceedingly "expensive" in terms of computing processing power and the memory and system based on empyrean rules may be more effective.

IV. METHODOLOGY

The quantitative-qualitative method of research was utilized as it provides a more detailed study of the issue in evidence. Richardson (1999, p. 80) mentions that "the studies which employ a qualitative methodology can describe the complexity of certain issue, analyze the interaction of certain variables, understand, and classify dynamic processes experienced by social groups". It is quantitative as it makes use, for data collection, of the establishment of values [0,1 to 1] for the definition of the informational priorities, based on the *fuzzy logic* concept. Richardson (1999, p. 70) emphasizes still that the quantitative research characterizes itself "for the employment of quantifications in the conditions of information collection and their treatment by technical statistics to the same extent". It may still be investigative and descriptive, the first, for believing that there is little systemized knowledge in regards to the issue studied, and the latter for providing the observation, record, classification, and interpretation of the facts, with no interference from the researcher.

As for the technical procedures, the research classifies as documental and also as case study. It is documental for utilizing internal documentation from the institution, which may be related to the objectives and issue researched. The case study, for being rich in details of a specific case analyzed, which can assist at the configuration of information for the field object of the present study. For the data collection, a few methodological stages were defined and may be seen on Figure 1.

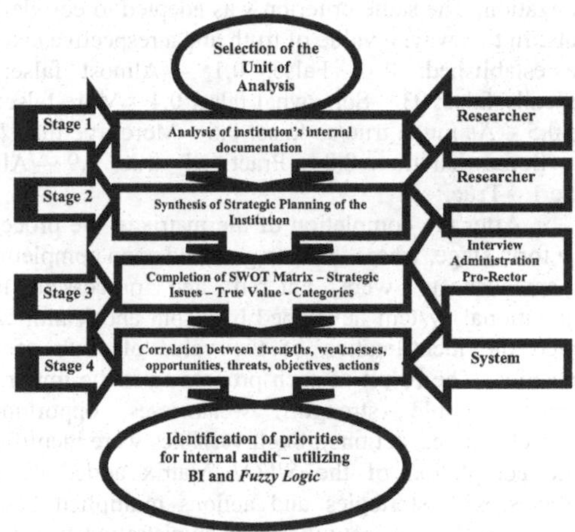

Figure 1 – Methodological Steps for Data Collection.

The unit of analysis was the internal audit of a higher education institution located at the metropolitan region of Porto Alegre. This is considered an assistance agency to the end-activities and to management. It is a new sector inside the institution and it is in phase of structuring, lacking planning which may assist the auditors on the performance of their works with success.

V. ANALYSIS OF RESULTS

The internal auditors need to direct their work to what is most strategic and key to success in the organization, otherwise they run the risk of auditing accounts which do not aggregate value to the undertaking, such as telephone calls control or other operational applications, where the expenditure may easily be limited via technological obstruction. They are result of efforts directed principally to the identification of frauds and error. The first stage performed was the analysis of the internal documents from the institution, and the second was the synthesis of its strategic planning which enabled the identification of strengths, weaknesses, opportunities, threats (*SWOT*) and Strategic Objectives and Actions (*Balanced Scorecard*).

In possession of such information, an interview was carried out with the Administration Pro-Rector, where it was verified, in face of the organization's strategy, the priorities for the internal audit resulting from the complex information system processing. The interviewed attributed "weights" in scale of [0 to 1], so as to establish degrees of truth for the questions, for example, how much is true to say that the strengths (tradition, ethical standards and credibility) relate to the opportunities, threats, and, at what level this is present at the institution. Likewise, one can identify if the proposed objectives and actions which were defined at the strategic planning, are in accordance so as to attend the purposes of the organization. The same criterion was adopted to correlate the threats. In this way, a value of truth and a respective category were established: 0 – False: 0,1 – Almost false: 0,2 Practically false: 03 – Somewhat false; 0,4 – More false than true; 0,5 – As much true as false; 0,6 – More true than false; 0,7 – Somewhat true: 0,8 – Practically true; 0,9 – Almost true; e 1 – True.

After the completion of the matrixes, we proceeded to the forth stage, where the data gathered upon completion of the spreadsheets were entered and processed in a computational system developed by Espín and Vanti, which allowed the identification of the value of truth of each intersection. The result of such processing is the importance of each variable (strengths, weaknesses, opportunities, threats, objectives, actions). Such variables were identified as of the completion of the *SWOT* Matrix added of more objectives with strategies and actions multiplied between them and fed by the institution's Administration Pro-Rector. So, the variables of which value of truth are between [0,8 e 1] were considered, remaining thus identified the points of immediate concern so that the internal audit may define operational procedures so as to attend such priorities.

DESCRIPTION	IMPORTANCE
CHARACTERISTICS OF THE ORGANIZATION	
Potential for attraction of qualified HR	1
Capacity of incrementing partnerships	1
Opening for changes	1
CHARACTERISTICS OF THE ENVIRONMENT	
Inter-institutional relations	1
Distance learning	0,8323
STRATEGIC OBJECTIVES	
Achieving excellence at education for all fields, levels, and modalities	1
Promoting growth and expansion in and out of the headquarters	1
ACTIONS	
Product innovation	1
Qualification of management	0,8992
Sustainability	0,8944

Table 1 – Quantitative Identification of Proprietary Spots for Internal Audit.

With such results, the informational priorities were identified so as the internal audit of the institution object of the present study may define its work plan so as to attend and work in accordance to the enterprise's strategies.

VI. CONCLUSION

The present study configured itself as an informational analysis to define priorities for internal audit through complex information system processing. Such system was based on *fuzzy logic* and BI. The first to priorize the most important strategic variables within the studied institution and the latter so as to then apply in databases, normally denominated *Data Warehouse* (DW) where it is possible to perform intersection of necessary information so as to detect frauds, errors and other non expected and possible behaviors in organizations of considerable size.

The information having priority for the internal audit work allows the auditors to define, on their work plan, operational procedures which may be developed in face of the strategy of the institution. It is believed that with the results from the data processing the IES may then better direct the task of the auditors, thus contemplating more strategic priorities.

Also as consequence, one may utilize the database systems and the BI systems aimed at the applications resulting from the *fuzzy* system processing, therefore eliminating unnecessary costs in activities that may possess low or null aggregate value. It is known that the internal audit aims at the verification of the internal controls and becomes involved with the operational process; however, it is believed that such may act in an operational manner in face of what was determined at the institution's strategic plan. The present work exhibited, by means of complex information systems, how it is possible to identify, among the priorities, those which deserve more notability at the internal audit work.

References

ATTIE, William. Auditoria: conceitos e aplicações. 3. ed. São Paulo: Atlas, 2006.

BALLONI, A.J. Por que GESITI? (Por que GEstão dos Sistemas e Tecnologias de Informação?). 1. ed. CTI/Campinas / Edt. Komedi, 2006.

BARBIERI, Carlos. *BI – Business intelligence*: modelagem e tecnologia. Rio de Janeiro: Axcel Books, 2001.

BAUM, D. Gartner Group's Howard Dresner. *Information Builders Magazine*, p. 26-28, Winter 2001.

BIRMAN, F. Simplesmente BI. *IDG Computerworld do Brasil Serviços e Publicações Ltda.*, São Paulo, ed. 383, mar. 2003.

CAMEIRA, R. F. *Hiper-Integração*: Engenharia de Processos, Arquitetura Integrada de Sistemas Componentizados com Agentes e Modelos de Negócios Tecnologicamente Habilitados. Rio de Janeiro: UFRJ,2003. Tese (Doutorado em Engenharia de Produção), Universidade Federal do Rio de Janeiro, 2003).

CERCHIARI, Giovanna S.F.; ERDMANN, Rolf Hermann. Sistemas de informações para acompanhamento, controle e auditoria em saúde pública. In: Congresso Internacional de Custos, 9., Florianópolis, 2005. p. 1-15. Disponível em: < http://www. abcustos.org.br.>. Acesso em: 14 out. 2006.

CREPALDI, Silvio Aparecido. Auditoria contábil: teoria e prática. São Paulo: Atlas, 2000.

ESPIN, R.; VANTI, A.A. 2005. *Administración Lógica: Um estudio de caso en empresa de comercio exterior*. **Revista BASE**. São Leopoldo, RS- Brasil, ago, 1(3), pp.4-22.

FRANCO, Hilário; MARRA, Ernesto. Auditoria contábil. 2. ed. São Paulo: Atlas, 1992-1995.

GEIGER, J. G., Data Warehousing: Supporting Business Intelligence. *Executive Report.* 2001. Disponível em: <http://www.cutter.com/freestuff/biareport.html>Acesso em: 30 abr. 2003.

KLIR, George J; Clair, U.H.S; Yuan, B. 1997. Fuzzy Set Theory: Foundations and Applications. 11. London: Prentice-Hall.

MAUSS, C.; Biehl, C.; Vanti, A.; Balloni, A. XBRL in public administration as a way to evince and scale the use of information. Innovations and Advanced Techniques in Systems, Computing Sciences and Software Engineering / International Joint Conferences on Computer, Information, and Systems Sciences, and Engineering (CIS^{2}E 07), 2007

MAUTZ, Robert K. Princípios de auditoria. 4. ed. São Paulo: Atlas, 1985.

MUKAIDONO, M. *Fuzzy Logic for Beginners*. Singapore: World Scientific. 2001.

OLIVEIRA Jr, H.A. 1999. Lógica Difusa: Aspectos Práticos e Aplicações. Rio de Janeiro: Interciência.

OLIVEIRA, Terezinha B. A. *et al.* A importância da Auditoria Interna em Uma Instituição Federal de Ensino Superior como Melhoria na Gestão dos Custos dos Serviços Públicos. In: Congresso Brasileiro de Custos, 13., Belo Horizonte, 2006. Disponível em: <http://www.abcustos.org.br>. Acesso: 01 dez. 2006.

PAULA, Maria Goreth Miranda Almeida. Auditoria interna: embasamento conceitual e suporte tecnológico. São Paulo: Atlas, 1999.

RAUTER, A. ;VANTI, A.A. *Configuração Informacional para a Gestão Administrativa do Negócio Educacional com a Análise da Tecnologia da Informação "Business Intelligence (BI) – Um estudo de caso"* CATI/FGV, SP, 2005.

RICHARDSON, Roberto J. Pesquisa social: métodos e técnicas. 3 ed. São Paulo: Atlas, 1999.

SIRIKULVADHANA, Supatcharee. *Data mining as a financial auditing tool.* 2002. M.Sc. (Thesis in Accounting). The Swedish School of Economics and Business Administration. Department: Accounting. Disponível em: <http://www.pafis.shh. fi/graduates/supir01.pdf>. Acesso em: 15 set. 2006.

Creation of a 3D Robot Model and its Integration to a Microsoft Robotics Studio Simulation

M. Alejandra Menéndez O.[1], D. Sosa G.[1], M. Arias E.[2], A. Espinosa M.[2], J. E. Lara R.[1]

[1] Universidad del Mar. Campus Puerto Escondido.
[2] Instituto Nacional de Astrofísica Óptica y Electrónica (INAOE).

***Abstract* - The objective of this paper is to publish the results obtained in a research project developed at INAOE. This project entails the construction of a manipulator-robot three-dimensional model and its incorporation into a simulation environment. The purpose is to analyze the behavior of the virtual prototype in order to adjust details of its operation before using an actual robot.**

I. INTRODUCTION

In general, experimenting on physical robotic mechanisms is complicated or impossible because most of these devices are too expensive. Furthermore, during the experimental stage models might be damaged or it is not possible to keep them completely available at anytime; hence, the need to find new strategies to perform tests without putting the hardware at risk. A good option for these strategies is simulation using 3D modeling.

A 3D model is the virtual representation of an object and the relationship between the values of the variables associated with its articulations [1] [5].

In this experiment the 3D model is created first, then it is tested in a simulation environment allowing the execution of diverse tests on the mechanisms without the need of performing these on the actual device. As a result, we avoided complex work of reprogramming the robot every time it fails or when bugs are detected, or software adaptations are needed.

Simulation is the process of designing a real system to perform experiments over it, with the purpose of understanding the behavior of the system or to evaluate new strategies – within the limits imposed by a certain criteria or a set of these – to know the functioning of the system [6].

The chosen software for this task was Microsoft Robotics Studio (MRS), a Windows®-based environment for hobbyists, academic and commercial developers to create robotics applications for a variety of hardware platforms [2]. MRS brings a graphical data-flow-based programming model, which makes the programming process easier as much for the simulation as for the robot. It allows the creation of behavior routines that the model will use in the virtual simulation environment. Also, it allows the possibility of uploading the application into a physical device.

The project was divided into two main stages: the construction of the 3D robot model, with the integration of each one of its parts; and the incorporation of the model into a virtual simulation environment developed in MRS.

II. CONSTRUCTION OF THE 3D MODEL AND THE INTEGRATION OF ITS PARTS

This stage consisted on creating the 3D robot model in order to incorporate it into MRS. So it became necessary to

K. Elleithy (ed.), *Advanced Techniques in Computing Sciences and Software Engineering*,
DOI 10.1007/978-90-481-3660-5_24, © Springer Science+Business Media B.V. 2010

use a 3D modeling software that would help in the construction of the robot model, which was later inserted into the MRS simulator.

To create the 3D model, the modeling software Autodesk 3Ds Max was used, because it has the capacity to generate files with the characteristics required for MRS, that is, static objects description files (*.OBJ), which can be later transformed into dynamic objects description files (*.BOS). These last ones contain the characteristics that a model must have.

In this phase, the modeling of five robot components was performed: base, pedestal, robotic arm, tray arm and camera.

Base

It is the element in charge of the robot's navigation. Taking as a reference the iCreate robot of iRobot company (Fig. 1), this robot has: an omnidirectional receptor for infrared signals, four sensors located in the bumper and two in the main wheels. Figure 2 shows the 3D model of the base.

Fig. 1: iCreate robot from IRobot Company.

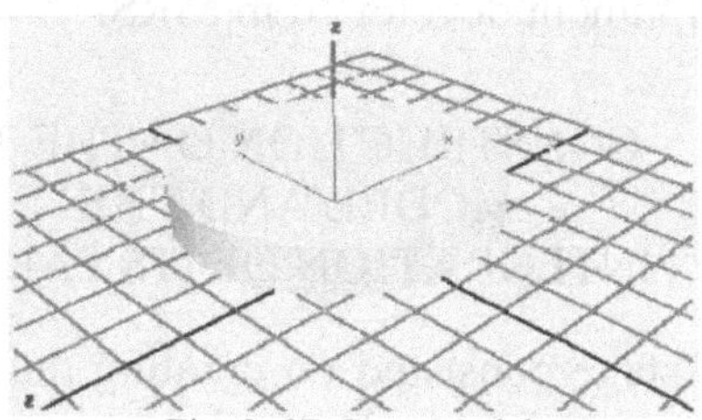

Fig. 2: 3D Base Model.

Pedestal

It is located on the central part of the base; it acts as the base for the robotic arm, the tray arm and the camera. Figure 3 shows the pedestal model, located in the center of the base.

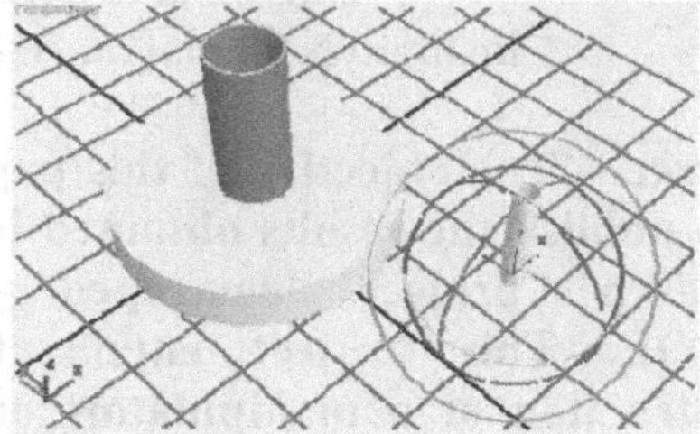

Fig. 3: 3D model pedestal, located over the base.

Robotic Arm

It is the element responsible of objects manipulation. To created it, the robot arm SES of the Lynxmotion company was used as reference (Figure 4). It contains: a clamp, used to hold objects, and six servomotors, which make possible the mobility of the arm, providing it with a freedom of five degrees. All these were considered for the 3D modeling. To design the robotic arm, it was necessary to assemble several objects of the Autodesk 3Ds Max software to model each one of the vital parts. Figure 5 shows the model of the robotic arm.

Fig. 4: The robot arm SES from Lynxmotion Company.

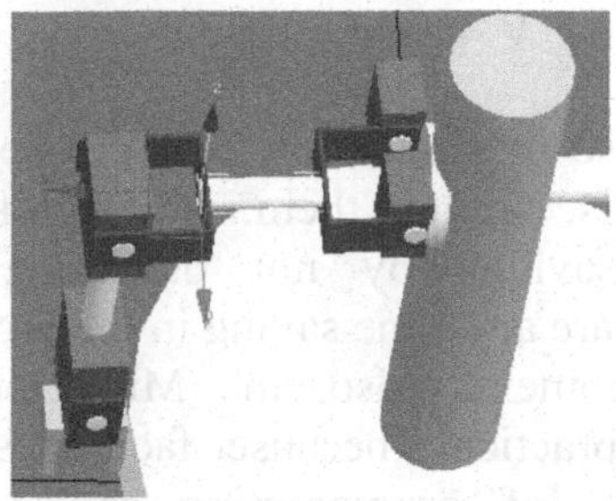

Fig. 5: 3D robot arm model.

Tray Arm

It is the element responsible for carrying objects taken by the robotic arm. For the construction of the virtual tray arm, it was used a group of objects from Autodesk 3Ds Max, with the objective of providing the looks of a carrying tray. Figure 6 shows the three-dimensional model of the tray arm.

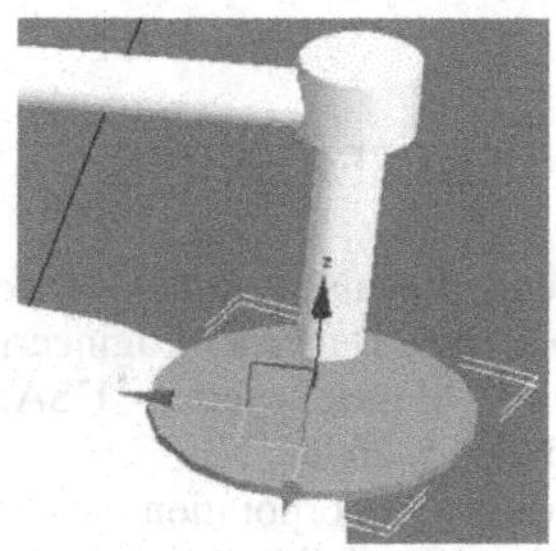

Fig. 6: 3D tray arm model.

Camera

Its role is to provide the vision of the robot and it is located at the top of the pedestal. In the 3D model is represented by a sphere. Figure 7 shows the 3D model of the camera.

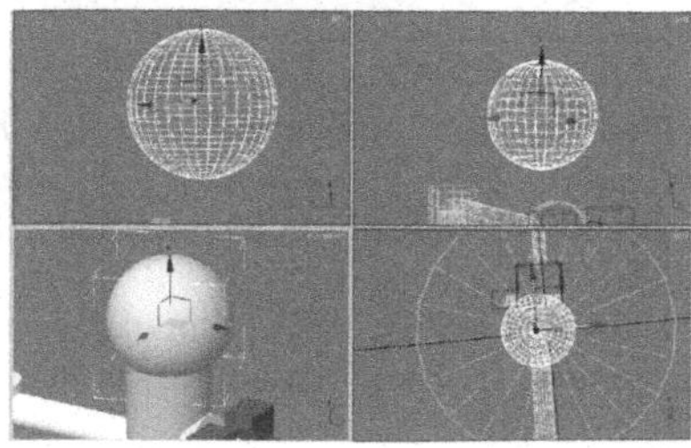

Fig. 7: 3D camera model.

When the models of each one of the robot elements were created, the final model was obtained by the union of each of the elements described above. Figure 8 shows the complete model of the robot.

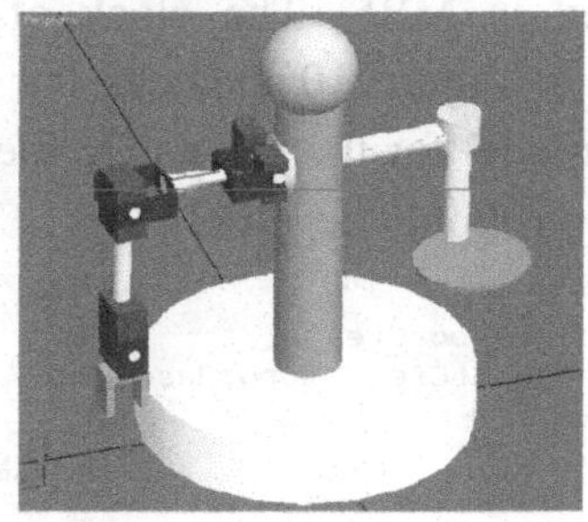

Fig. 8: Final representation of the 3D robot model.

III. INCORPORATION OF THE MODEL IN A VIRTUAL SIMULATION ENVIRONMENT DEVELOPED IN MRS®

Once the 3D model was created as a static model by Autodesk 3Ds Max, it was converted to a dynamic model using the obj2bos tool [3], an aplication that is part of MRS.

The objective of incorporating the dynamic model of the robot into the simulation environment in MRS, was to optimize its behavior in the virtual environment. To do so, a software component (entity) was created, which interacts with the physics engine and the rendering engine. This provides the appropriate high-level interfaces to emulate hardware, and hides the specific use of physical APIs [4]. This proved to be a complicated task, because of the over-encapsulated structure that MRS has in their entities and because the robot support is only defined for those robots already included in their simulations.

IV. RESULTS

To incorporate the 3D model of the robot into the simulation environment it was necessary to use C# code, which is part of the entity for the iRobot robot included in MRS. The block of code shown in Figure 10 includes the name of the model for the robot (the filename extension should be .BOS).

```
private IRobotCreate
CreateIRobotCreateMotorBase(ref Vector3
position)

{

        IRobotCreate robotBaseEntity = new
        IRobotCreate(position);
        robotBaseEntity.State.Assets.Mesh =
        "iRobot-Create-LOD11.bos";

        robotBaseEntity.State.Name =
        "IRobotCreateMotorBase";

        CreateService(drive.Contract.Identif
        ier,Microsoft.Robotics.Simulation.Pa
        rtners.CreateEntityPartner("http://l
        ocalhost/" +
        robotBaseEntity.State.Name));

        return robotBaseEntity;

}
```

Fig. 10: Adding the 3D model of the robot in the entity code of iRobot.

After the name of the robot was included, the next task was to compile it. Because of this, it was possible to view the 3D model of the robot interacting with the simulation environment of MRS.

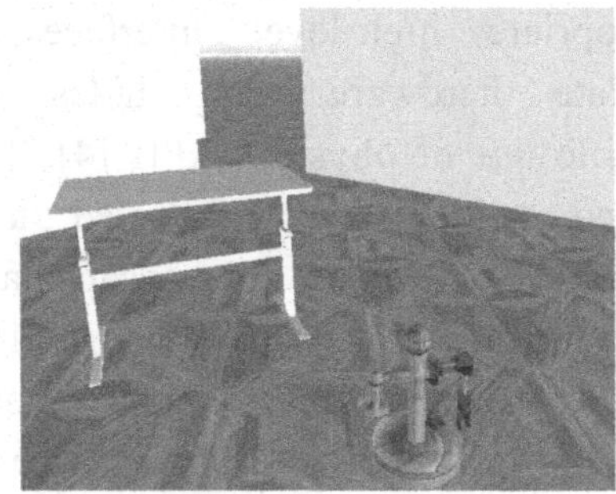

Fig. 11: The 3D model of the robot navigating in the simulation environment of MRS.

V. CONCLUSIONS

The use of simulation provides several benefits, among them: the significant cost savings by not damaging the hardware and time saving in the projects development. Also, the MRS tool is very practical, because facilitates the process of programming robots and motivates beginners or those who do not have the necessary equipment for development to be part of research projects related to robotics and robotics vision.

VI. ACKNOWLEDGEMENTS

We thank the INAOE for the resources and support provided, especially to Miguel Arias Estrada, PhD for the support and advice in conducting this project.

REFERENCES

[1] Manseur, Rachid; *"Robot Modeling and Kinematics"*; Da Vinci Engineering Press; Boston, Massachusetts, USA; ISBN: 1584508515, 2006.

[2] Microsoft Corporation; "Microsoft Robotics Studio", "Microsoft Robotics Studio Developer center"; URL: http://msdn2.microsoft.com/en-us/robotics/default.aspx, 2008

[3] Microsoft Corporation; "Obj-to-Bos File Converter Tool", "Microsoft Robotics Studio Developer center"; URL: http://msdn2.microsoft.com/en-us/library/bb483033.aspx, 2008

[4] Microsoft Corporation, "Simulation Overview", "Microsoft Robotics *Studio Developer center"*, URL: http://msdn2.microsoft.com/en-us/library/bb483076.aspx, 2008.

[5] Ollero B., Aníbal; *"Robótica Manipuladores y robots móviles";* Marcombo; Barcelona, Spain; ISBN: 84-267-1313-0, 2001.

[6] Shannon, Robert E.; "Simulación de Sistemas"; Trillas; Mexico, ISBN: 968-24-2673-1,1988.

A Proposed Framework for Collaborative Design in a Virtual Environment

Jason S. Breland & Mohd Fairuz Shiratuddin
School of Construction, The University of Southern Mississippi, Hattiesburg, MS 39406, USA

***Abstract*-This paper describes a proposed framework for a collaborative design in a virtual environment. The framework consists of components that support a true collaborative design in a real-time 3D virtual environment. In support of the proposed framework, a prototype application is being developed. The authors envision the framework will have, but not limited to the following features: (1) real-time manipulation of 3D objects across the network, (2) support for multi-designer activities and information access, (3) co-existence within same virtual space, etc. This paper also discusses a proposed testing to determine the possible benefits of a collaborative design in a virtual environment over other forms of collaboration, and results from a pilot test.**

Keywords: architecture, collaborative, design, virtual environment

I. INTRODUCTION

A typical construction project requires multi-disciplinary collaboration and expertise of the design team to materialize the owner's intent and vision into constructible designs. The construction industry is constituted of "high level of complexity, uncertainty, discontinuity, with many restrictions" [1]. Typically, the level of collaboration among design team members varies depending on the complexity of a project, and the organizational arrangement the design team members are bound to.

A construction project is a complex endeavor and its success is most likely only when different professionals collaborate especially from the early stages [3]. During a study of computer mediated architectural design, three forms of collaboration were observed [2]; "mutual collaboration" which involves designers equally working together on the same aspect of the task; "exclusive collaboration" is when designers working on separate aspects of the same problem with occasional time for consultation; and "dictatorial collaboration" when by appointment or naturally, emerge a "designer in charge" who makes all the design decisions.

Collaboration among the design team members is limited due to the nature of current design execution which is linear in nature [4], [5], [6] & [7] and restrictions of 2D design tools used to create and communicate the designs [8], [9] & [10]. Each designer performs his/her own discipline-specific element of the design, then passes the design to the next discipline-specific team member (such as a structural designer's starts designing after the architectural designer has completed the architectural design of the facility). The final design of a construction project is complete when each team member collaborates in terms of completing their specific task in furnishing every part of the facility's design.

The computer has been found to play an active role in collaborative design because it provides designers with the visualization support for 3D models, assistance in generating alternative designs, and a platform for design critiques [2]. It can be further argued that early collaboration with the support of computer technology benefits designers and also the consequent processes in construction project designs that are often complex [2],[3]. Computer technology such as virtual environment (VE) provides virtual spaces that can be crafted to support collaborative work and social interplay.

As such, we propose a Collaborative Virtual Environment (CVEs) that enhances collaborative ability for designers, to a more desired level of interaction whereby allowing them to work on the same design within the same space, without interfering any one's assigned task. We define CVE as a VE that supports synchronized multi-participants

K. Elleithy (ed.), *Advanced Techniques in Computing Sciences and Software Engineering*,
DOI 10.1007/978-90-481-3660-5_25,

activities; participants are able to co-exist within the same virtual time and space, and able to see what others are doing in the VE. A CVE that is specifically for architectural design would be a feasible solution to enhance collaboration among designers such as architects and engineers. Not only a CVE allows for real-time 3D viewing and displays of designs, but also permits multiple designers to work on the same design that can be part of a larger project.

We have developed a working prototype i.e. the Collaborative Design Tool in a Virtual Environment (CDT-ve) utilizing the Torque 3D Game Engine from GarageGames. The prototype is capable to stand on its own as a working model. However, there are still a few features which have minor problems, and several new features we would like to add. Our end goal is to provide a robust framework to support architectural design activities. In this paper we describe some of the key components of the framework, a proposed testing procedure and brief results on a pilot study we undertook.

II. MAIN COMPONENTS OF THE FRAMEWORK

Unaltered Torque 3D Game Engine (TGE) uses a fairly straight forward code structure. The computer hosting the VE holds the current environment's data as well as performing collision calculations, and items and characters' tracking. Each visible object within the host's VE is then transmitted to every client's computer as an 'invisible' or 'ghost' object. We have made modifications to the original TGE to allow for real-time collaborative design in a VE.

The TGE supports multiplayer online gaming, thus, the networks speed is always a top priority. The data sent out from the server to clients is optimized to not only reduce the number of packets that need to be sent but also to avoid cheating on the client side. Information could be gleaned from the update messages to give an unfair advantage such as player locations, etc. With a collaborative design environment, the client computers are programmed into having the same information available to them as the host. Object names, properties, and other useful information such as attached data should be sent out to the clients. In addition, the data structures that hold the objects on client computers is different and had to be altered to hold additional information and allow access and manipulation. Fig. 1 shows the changes made to the message system (shown in red). These changes leave the server ultimately in control so that it can ignore changes if necessary as well as simplifying the synchronization process.

The Editor's code was modified as well. Every time an object is manipulated in the Editor, a different command will be executed depending on if the CDT-ve is running as a host or a client. If it is running as a host, the original code usually executes. For every possible manipulation of the VE, new commands had to be written to access the expanded message system. Thus we have almost doubled the Editor's code which handles object manipulation in the VE. For every movement, rotation, scaling, creation, deletion, copying, pasting etc., new code had to be written. Different object types are often handled differently as well as different code for a single or multiple objects. In summary, there exist a myriad of different scenarios when manipulating objects in the Editor. Each scenario had to be independently considered and expanded to include alternate code when the application is running as a client.

The Host and Client Design Update Model

Fig. 1 also shows the additions and alterations made to the prototype CDT-ve (shown in red). Clients can now enter the real-time design environment. As clients create, manipulate, or delete objects, commands are sent to the hosting computer where the objects are altered per the client's request messages. The altered objects are then updated to all clients. If a ghost object (client side object) is being moved, rotated, or scaled, it is immediately updated by the client's real-time design environment. While a ghost object is being moved, rotated, or scaled, the client who is currently moving the object will ignore updates from the host for that object until manipulation is complete. If the host and a client are simultaneously moving an object, the client has precedence.

The Server Structure

The server structure of the CDT-ve remains largely unchanged, (we leave the server in control) only the message system however has doubled in size to allow more detailed communication between servers and clients. A dedicated server is not necessary as the application itself is all that is needed for hosting and to start collaborative work with other users. Whichever computer chooses to host, the design will act as the central hub for all information including the design file. Based on the client-server architecture shown in Fig. 2, there can never be an occasion when clients will need to communicate directly with another client. The messages will always pass through the host. The host also acts as a moderator when it comes to resource control.

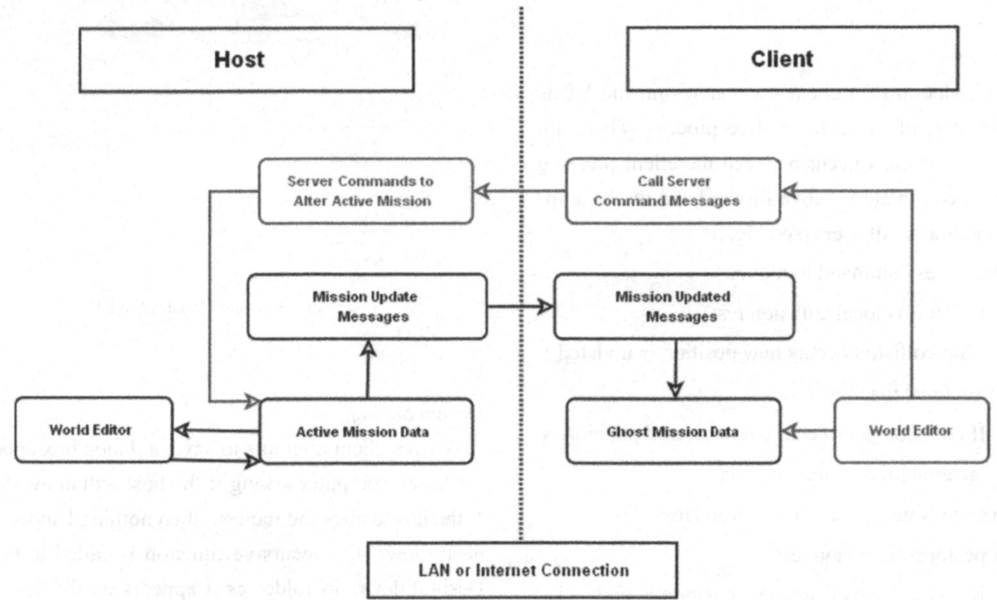

Fig. 1. The messaging update system (additions show in red)

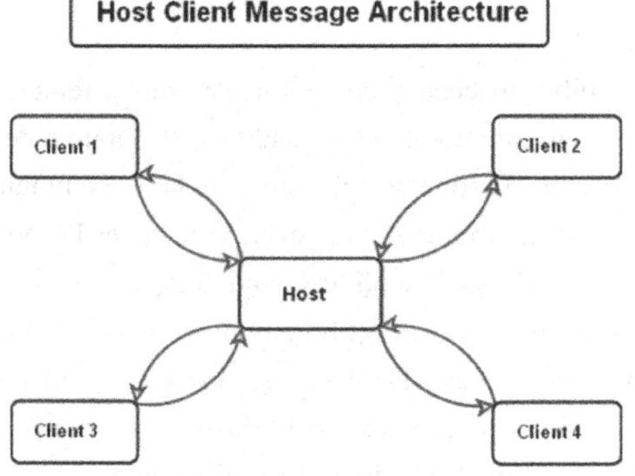

Fig. 2. The CVE client-server architecture

The host user is required to assign a password when the design file is first hosted. This ensures that design resources are secured and once authenticated; the resources are automatically uploaded to the client if the client is missing them. As a form of version control and resource management, if a client wishes to save the current design that everyone is working on collaboratively, the host will receive a message asking for confirmation and permission to save for that particular client. This action occurs every time a client wishes to save, and a negative response from the host denies the client saving privileges for that moment in time.

III. THE FRAMEWORK SUB-COMPONENTS

In addition to the primary collaborative functions of the CDT-ve, there are many sub-components to the prototype that are for the most part handled automatically. Though they are passive, they are still the key features to the collaborative design process of the CDT-ve. Some of the key sub-components are discussed below.

Resource Request

As a client PC joins a host PC's VE, objects such as the construction site (or terrain), buildings, audio cues such as sounds, and even different weather conditions are constructed on the client's computer as ghost objects. If a resource is called and is not found in the client PCs' resources, the host will then send that resource to the client and it will exist as a permanent object to the client's computer permanent pool of resources.

Avatar Updates

The manipulation of a client's avatar within the VE is an essential part of the collaborative process. There are several steps that must occur between the client pressing the keyboard command to move forward and the host updating that action to all users (see Fig. 3).

- Client issues command to move.
- Client performs local collision test.
  - If no collision occurs new position is updated on local machine
  - If collision does occur a new altered position is generated to avoid collision
- Client sends new position vector to Host
- Host performs collision test
  - If no collision occurs new position is updated to all users
  - If collision does occur altered position is updated to all users
- If Client's position differs from that sent by the Host, it is overwritten

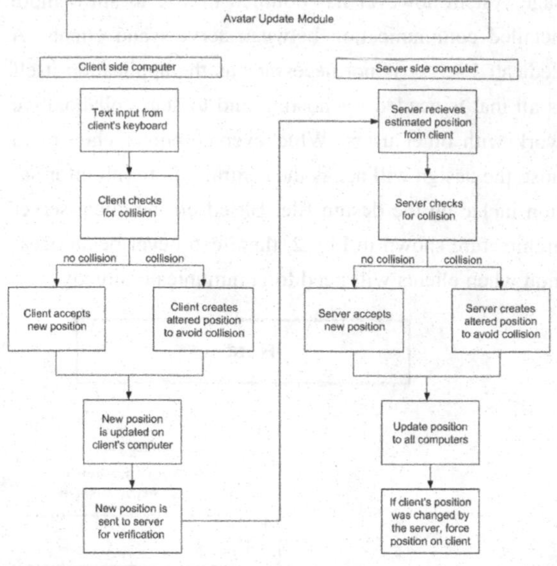

Fig. 3. Avatar Updates Model

Design Saving

When a client attempts to save, a dialog box appears on the host's computer asking if the host will allow the save. If the host denies the request, then nothing happens. If the host allows it, a recursive function is called to build the DesignFileGroup folder as it appears on the host's computer identically on the client's side including subfolders and mission info. The save file dialog box then appears for the client PC to name the design and save the file.

Texting

The ability to clearly communicate among remote designers is an important aspect within collaborative design environment. Texting is the only available communication tool at the moment. However, voice-over-IP (VOIP) programs such as TeamSpeak and Skype can be used along with the tool. The texting model is shown in Fig. 4. When a client sends a text message, the server will ensure that all clients, except the one who sent it receives the message along with that users identification profile.

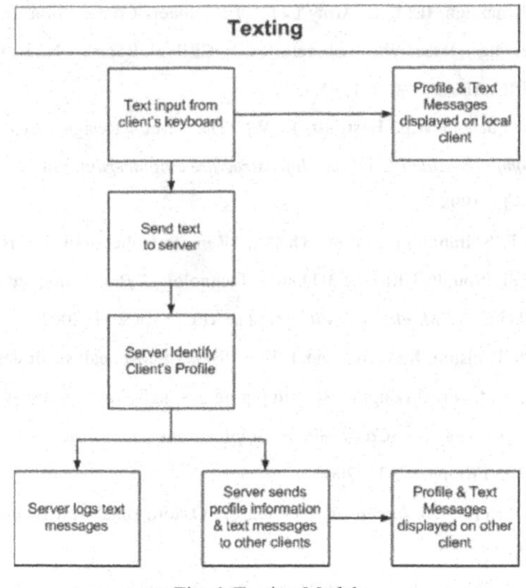

Fig. 4. Texting Model

IV. PROPOSED TESTING & PILOT TEST

Subjects will be broken down into groups of one, two, or three. The subjects who are by themselves will serve as a control group. Subjects who are in groups of two or three will be randomly divided into two sets of testing conditions. We will call these "conditions A" and "conditions B". Subjects under conditions A will be working face to face with each other and use a single non-collaborative computer with our prototype installed. Subjects under conditions B will be working in separate rooms each with a collaborative computer equipped with the prototype and a third party voice-over-IP program (see Fig. 5).

All groups will be presented with specifications and 2d plans, and shown briefly how to use and navigate the prototype. They will then be asked to construct the commercial building that appears in their plans using the prototype collaborative virtual environment. Note that only groups working under conditions B will actually use the collaborative capabilities. All virtual resources necessary to complete tasks will be provided. The prototype program will automatically record the actions of each subject, as well as log technical data (such as number of moves, rotations, etc). The conversations of subjects under conditions B will be recorded, and video of those under conditions A will be recorded. The data will then be analyzed to determine the effects of working collaboratively in a virtual environment as opposed to working face to face with a single interface into the environment.

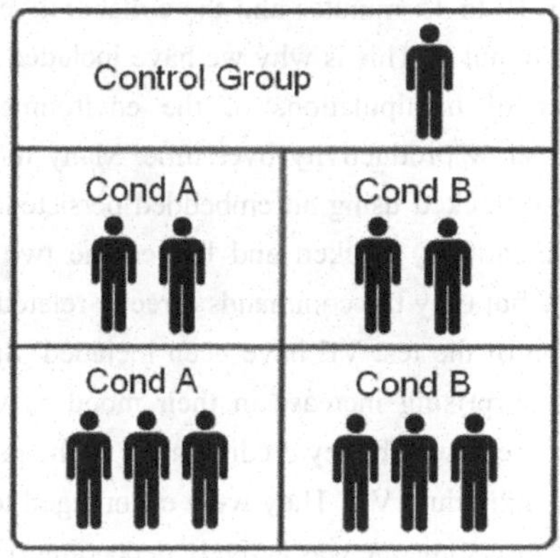

Fig. 5. Proposed testing condition

Pilot Test

We have conducted a brief pilot test using two subjects. First, one user constructed the VE by himself. Then he worked together collaboratively with the second subject to construct the VE. Subjects were in the same room and communicated verbally as well as with visual cues in the VE. Lastly the second subject constructed the VE by himself. The order the subjects constructed the VE was staggered like this to reduce the bias created from users more quickly finishing a VE they have constructed before. The results can be seen below in Table I.

TABLE I

RESULT FROM THE PILOT TEST

	Total Time	Move Count	Rotate Count*	Delete Count	Total Edits Per Minute
Sub. 1 solo	1:50	218	14	13	2.35
Sub. 2 solo	1:28	225	12	8	2.92
Sub. 1 collab	0:50	189	13	24	4.78
Sub. 2 collab	0:50	217	8	3	4.72

The results indicated in Table I was very promising. Total time was greatly reduced while manipulations per minute increased. It should be taken into account though that total time spent in the VE is not an appropriate measure of productivity as a couple of minor crashes set back subject 1 (solo) by 10 to 15 minutes and the collaborative group by 5 to 10 minutes. This is why we have included the total number of manipulations of the environment per minute: to show productivity over time. Many more variables were tracked using an embedded persistent script which automatically tracked and logged the two user's movements but only the commands directly related to the construction of the test VE have been included. Subjects reported a surprising increase in their mood as well as their work speed which they credit wholly to the presence of another individual VE. They were encouraged to work harder because someone was actively depending on them. Subjects also reported that they had an idea of the other subject's location and work area through the experiment even though each of them was in free camera mode (invisible and flying). We plan to research the psychological implications of group work and space perception in greater detail in the future.

ACKNOWLEDGEMENT

The authors would like to thank all the team members at GarageGames for their work on the Torque 3D Game Engine.

REFERENCES

[1] H.R. Hiremath, and M.J. Skibniewskib, "Object-oriented modeling of construction processes by Unified Modeling Language," *Automation in Construction*, vol. 13, pp. 447-468, 2004.

[2] M.L. Maher, A. Cicognani, and S. Simoff, (1996). "An Experimental Study of Computer Mediated Collaborative Design," *Proceedings of the 5th International Workshops on Enabling Technologies: Infrastructure for Collaborative Enterprises (WET ICE'96)*, p 268, 1996.

[3] H.C. Howard, R.E. Levitt, B.C. Paulson, J.G. Pohl, and C.B. Tatum, (1989). "Computer integration: reducing fragmentation in AEC industry," *Journal of Computing in Civil Engineering*, vol. 3(1), pp. 18–21, 1989.

[4] R.S. Spillinger, *Adding value to the facility acquisition process: best practices for reviewing facility designs*. National Academies Press, 2000, http://www.nap.edu/catalog.php?record_id=9769#toc

[5] W.E. East, T.L. Roessler, M.D. Lustig, and M. Fu, "The Reviewer's Assistant system: system design analysis and description," Champaign, IL: U.S. Army Corp of Engineers Construction Engineering Research Laboratories. (CERL Report No. FF-95/09/ADA294605), 1995.

[6] M. Fu, and W.E. EastEast, E. W, "The Virtual Design Review," *Computer-Aided Civil and Infrastructure Engineering*, vol.14, pp. 25-35, 1998.

[7] M.F. Shiratuddin, and W. Thabet, "Remote Collaborative Virtual Walkthroughs Utilizing 3D Game Technology," *Proceedings of the 2002 ECPPM: eWork & eBusiness in AEC*, Sept 9-11, 2002.

[8] P.S. Dunston, X. Wang, and T. Lee, "Benefit/cost analysis of desktop virtual reality application in piping design," *Proceedings of the Conference on Construction Applications of Virtual Reality (CONVR)*, pp. 22-31, 2003.

[9] S. Emmitt, *Architectural Technology*. Oxford:Blackwell Science, 2002.

[10] M.F. Shiratuddin, and W. Thabet, "A Framework for a Collaborative Design Review System Utilizing the Unreal Tournament (UT) Game Development Tool," *Proceedings of the 20th CIB W78 Conference on Information Technology in Construction.* Pp. 23-25, April 2003.

System Development by Process Integrated Knowledge Management

Dr. PhD Margareth Stoll Margareth.stoll@dnet.it
Dr. Dietmar Laner dlaner@unibz.it
Free University of Bozen, Sernesistraße, 1
39100 Bozen, South Tyrol, Italy

***Abstract*-Due to globalization and ever shorter change cycle's organizations improve increasingly faster their products, services, technologies, IT and organization according to customer requirements, optimize their efficiency, effectiveness and reduce costs. Thus the largest potential is the continually improvement and the management of information, data and knowledge. Long time organizations had developed lot separate and frequently independent IT applications. In the last years they were integrated by interfaces and always more by common databases. In large sized enterprises or in the public administration IT must operate various different applications, which requires a lot of personal and cost. Many organizations improve their IT starting from the lived processes using new technologies, but ask not, how they can use technology to support new processes.**

Many organizations of different sizes are implementing already for several years process oriented standard based management systems, such as quality ISO9001, environmental ISO14001, information security ISO/IEC27001, IT service ISO/IEC 20000-1, hygiene management systems ISO 22000 or others, which are based on common principles: objectives and strategies, business processes, resource management and continuously optimization.

Due to this situation we used in different case studies as basis for system development a the organization adapted, holistic, interdisciplinary, integrated, standard based management system to analyze customer requirements and integrate, optimize and harmonize processes and services, documents and concepts. This promotes efficiency, effectiveness and organizational development to guarantee sustainable organization success.

Keywords: process management, organization development, knowledge management, service management, standard based management system, quality management

I. INTRODUCTION

A Starting Situation

Due to globalization and ever shorter change cycle's organizations must improve increasingly faster their products, services, technologies, IT and organization according to customer requirements and optimize their efficiency, effectiveness and reduce costs. Thus the largest potential is the continually improvement by organizational learning based on individual learning, the management of information, data and knowledge and a holistic interdisciplinary approach.

Long time organizations had developed various separate and frequently independent IT applications. In the last years they were integrated by interfaces and always more by common databases. In large sized enterprises or in the public administration IT must operate many different applications, which requires a lot of personal and cost.

Many organizations of different size and scopes are implementing already for several years process oriented management system as quality ISO9001, environmental ISO14001, information security ISO/IEC27001, IT service ISO/IEC 20000-1, hygiene management systems ISO 22000 or others or others. These systems are implemented more frequently holistic, whereby are integrated according with the organizational purpose and objectives different aspects, like quality, environment, hygiene, data security, occupational health and safety, as well as human resource development, resource management, IT - management, communication management, controlling and also knowledge management. The established management system must be documented, communicated, implemented and continuously improved. Thereby the system documentation contains the entire actual explicit knowledge and supports individual and thus organizational learning, whereby management systems push constantly the knowledge and learning spiral, change organizational knowledge and promote sustainable the organizational development.

B Purpose of the article

In large organizations or the public administration different projects and actions are taken sometimes independently of each other and frequently for defined sub organizations to promote efficiency, effectiveness and reduce cost. Thereby sub organizations are optimized, but the holistic system and organization is neglected. Consequently different departments use for the same business process and service different procedures, regulations, forms, concepts, methods and IT solutions.

Organizations commit a radical error and analyze their organization by the lens of their lived processes. They ask themselves "how can we use new technology to optimize our

K. Elleithy (ed.), *Advanced Techniques in Computing Sciences and Software Engineering*,
DOI 10.1007/978-90-481-3660-5_26, © Springer Science+Business Media B.V. 2010

processes" instead of, "how we can do something new with the technology automation instead innovation" [1]. IT objective must be to promote in the best way possible the organization objectives.

Thus we need a holistic, systematic approach, which is oriented to the organization objectives and promote a constantly development to fulfill customer requirements and ambient changes, to guarantee a sustainable organizational development by increased efficiently and effectiveness and reduced costs.

C *Research Approach*

The always stronger requests for stakeholder orientation, efficiency, effectiveness, innovation, shorter change cycles, process orientation, process and service improvement, cost reduction and the increasing implementation of holistic, process oriented management systems, the great importance of the continually organization optimization, the individual and organizational learning and the IT involvement leaded us to introduce as basis for system development a the organization adapted, holistic, interdisciplinary, integrated, standard based management system to analyze customer requirements, integrate, optimize and harmonize processes, services, documents and concepts to promote efficiency, effectiveness and organizational development to guarantee sustainable organization success.

D *Structure of the article*

Firstly we describe based on the starting situation the project objectives [II] and illustrate the common requirements of international standards for management systems [III]. Afterwards we explain our approach [IV]. Finally we document the project experiences and results of the implementation in different organizations with distinct management systems [V] including the achievement of the project principles [V A] and a reflection about success factors [V B] and at the end we express an outlook [VI] and our conclusion [VII].

II. PROJECT OBJECTIVES

Due to this starting situation we have to resolve following questions in order to contribute to organization success by using IT:

How we can establish organization objectives and strategies regarding the requirements of all stakeholders (shareholder, customer, collaborators, supplier, environment and society)? How can we structure and integrate the processes and services to a few main business processes or main services? How can we harmonize documents and regulations?

How can we promote sustainable continually organization improvement and optimization in accordance to stakeholder requirements and established organization objectives?

By implementing a holistic, interdisciplinary, integrated, standard based management system using workflow oriented IT support we expect to foster:

Effectiveness: the whole organization will be managed by figures to fulfil effectively the established objectives.

Efficiently and cost reduction: all processes and services will be constantly monitored by the established figures and optimized in accordance to the organization objectives.

Organizational development by employee involvement and knowledge generation.

Continually organization development promotes sustainable stakeholder orientation, quality and cost effectiveness.

III. MAIN REQUIREMENTS OF STANDARD BASED MANAGEMENT SYSTEMS

The standards for management systems have different specialized focuses, but are all based on common principles [2], [3]:

Organizations objectives and strategies must be established regarding stakeholder requirements.

All business processes including management process, support processes, resource processes and optimization processes must be defined and promote the optimized fulfilment of the organization objectives under the focus of the respective standard.

Process oriented resource management must be promoted including human resource development, IT – management and other infrastructures, tools and instruments.

The organization, their objectives and strategies, services/ products and processes must be continually optimized according to established processes in sense of a PDCA cycle (plan, do, check, act).

Fig. 1. Main requirements of standard based management systems

The established management system must be structured and systematically documented and communicated within the organization and the collaborators must be continually motivated for implementing the system and for recommending improvements.

These standard based management systems are implemented more frequently in a holistic way. In accordance with the organizational purposes and objectives are integrated in a management system different aspects like quality, environment, hygiene, data security, occupational health and safety, as well as human resource development, resource management, IT - management, communication management, controlling and also knowledge management.

IV. APPROACH

Due to large relevance and range of a project for implementing a holistic, interdisciplinary, integrated, process oriented, standard based management system as basis for system development should be used project management methods for planning, adapting optimally to customer requirements, implementing and the objective fulfillment should be controlled constantly by concrete figures deduced from the stakeholder requirements.

A Stakeholder oriented organization objectives

For establishing the vision, organization objectives and strategies we use hoshin kanri [4]. With the concerned collaborators including management we define and prioritize the stakeholders. Due to experiences from intensive contact between stakeholder and collaborators and / or due to interviews, literature research, market research are identified stakeholders requirements and expectations. Subsequently they are prioritized by means of Quality Function Deployment [5] and thus vision, policy and organization objectives are deduced from them. Establishing the policy, objectives and strategies we consider also the knowledge objectives including innovation and knowledge generation and determine therefore knowledge strategies [6]. Due to standard requirements we deduce from defined vision, policy and organization objectives longer-term strategic objectives and concrete full year objectives with appropriate programs and projects, figures, responsible, deadlines and necessary resources. Thus the entire organization is focused on stakeholder requirements.

B Process analysis and process improvement

For process steps are analyzed with associated responsible persons "bottom up" the necessary information, data and documents [1]. Thereby we consider all organizational processes beginning from the management process, all business processes including the development and design process, as well all supporting processes, resources processes and optimization processes. In accordance with the organizational purposes and objectives are also considered different aspects like quality, information security, data protection, ambient, environment, hygiene, occupational health and safety, as well as human resource development, resource management, IT - management, communication management, controlling and also knowledge management and integrated into a holistic model. All processes are analyzed, optimized regarding the established organization objectives and this aspects, as well efficiency, effectiveness, harmonized regulations, legal and regulatory interpretations, information flow, collection and passing necessary knowledge by checklist, regulations, forms and workflow based databases and optimized, if possible [7], [8]. Thereby implicit knowledge is externalized, knowledge identified and possible optimizations (knowledge generation) are discussed.

To harmonize the processes we define in the first step the main services with their processes, analyze these in the single departments, harmonize common process segments (apart from specialist data and calculations), form structures, organizational glossary and others and thereby we establish common, harmonized, integrated main processes and services. Afterwards parameterization resolves specialist differences.

Analyzing the process steps we recognize these, in which experiences and special legal or regulation interpretations (lessons learned) are developed and these experiences will be documented for later use [9], [10], [11]. Thus for all necessary documents and forms e.g. grant application forms, templates are developed, whereby these must be examined only once by specialized experts and afterwards every collaborator is able to fill them with the personal data of applicant. Appropriate applies to checklists, into which the specialized knowledge of the experts is integrated and thereby implicit knowledge is externalized. Further these checklists are continually extended or changed regarding the defined optimization processes according to the experiences during the application in the work everyday life or if special cases appears, which are not clearly regulated until now. Also experiences and necessary interpretations of laws and regulations are integrated and the knowledge base and processes changed flexible. In the same way knowledge, which is no more needed, is removed in time. Thereby we integrate optimal process modeling, process standardization and transparency with need and objective oriented flexible process implementation.

Accordingly also the necessary information about customer/citizen should be analyzed and their collection and processing should be planed and implemented systematically and structured. In the public administration can be integrated directly into the workflow a user-oriented, context sensitive interface to the legal requirements and regulations, which are relevant for the single process step. Thereby a need oriented, context sensitive knowledge source is integrated directly into the working process. It is also integrated by input of personal notes into auxiliary systems.

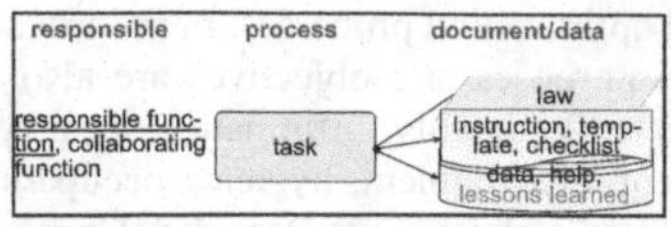

Fig. 2. Applied process documentation method

For the process documentation we use a very simple form of flow charts, which is modeled by graphic support tools [Fig.3]. By process modeling for all process steps the responsible functions, as well as documents, information, tools, IT - applications including the observing laws and regulations are defined. For all documents we deduce from the process model the responsible persons for establishing, generating, changing, handling, as well as the read entitled functions, also possible external information sources or receiver (document logistic), the data protection class in accordance to the data protection law and further the required archiving methods (for current, intermediate and historical archiving). Thereby we define necessary and licit access rights, archiving procedures, information security requirements regarding confidentiality, availability and integrity, signature rights and signature procedures as basis for the implementation of the signature regulations and required data encryption with necessary encryption procedures. Thereby all information security requirements and necessary treatments in accordance to a holistic information security management system regarding ISO/IEC 27001 information security management can be established and afterwards implemented systematically and optimized constantly. Due to established external sources and receivers and the necessary data exchange between the different IT - applications are defined also all required interfaces. All treated data are examined for it necessity and lawfulness.

To promote efficiency already existing customer data should be accessed regarding data protection law and not collected another time. If there are changes, the data must be changed immediately in all applications and all necessary subsequent procedures must be taken. To guarantee the necessary data integrity we need clear harmonized regulations.

For standardizing forms and terms a central glossary for the entire organization including specialist terms of single departments is elaborated. Further also clear guidelines for the linguistic design of masks/forms, the corporate design and a harmonized structure for all masks/forms are established (e.g. personal data). To promote efficiency, service quality and customer orientation we pay specially attention for an accurate and easily understandable language. Check routines and check criteria must be defined and applied uniformly. This requires sometimes also the revision of existing internal regulations.

Function profiles and competences are deduced from the definition of the responsible persons for the single process steps. Thus knowledge carriers are specified and knowledge maps - yellow pages- are constructed. For all function profiles the necessary requirement profiles can be intended due to the posed process steps and the necessary knowledge [10], [12], [13].

C *Resource management*

The organization must determine and provide due to standard requirements necessary resources, tools and instruments to obtain the established objectives and to continually improve the organization. Therefore also optimal IT – systems are promoted.

In the resource management also training and competence objectives must be planed, realized according to defined processes and their effectiveness has to be evaluated. Thus the continual improvement of the organizational knowledge and the organizational knowledge base is promoted systematically and structured, their effectiveness evaluated and possibly necessary corrective actions are taken. The strengthened IT – applications effect task modifications, redistribution of responsibilities, job enlargement, stronger customer orientation, increased service quality and service orientation and this requires self and social competencies, communication skills, media competences, IT – competences, interest in new knowledge, change and learning willingness, team ability, openness and tolerance, empathy, as well as autonomous, entrepreneurial spirit, self-driven, objective oriented acting and strategic thinking in accordance to established regulations and also system and process oriented thinking. Therefore suitable human resource development projects must be developed, implemented and their effectiveness evaluated. Further usually the personnel order (like career profiles, requirement profiles, selective procedures) must be changed.

D *Continually improvement*

In management systems the ability to achieve planned objectives are evaluated periodically in base of established measurements (process and organization measurements including customers satisfaction) and consequently determined, implemented and evaluated necessary corrective, optimization or preventive actions. Integrating knowledge objectives and strategies into the objectives and strategies of the whole organization also the knowledge is thereby constantly evaluated and if necessary changed and optimized. For problem solving, the implementation of recognized optimizations and for preventive actions must be determined and implemented appropriate systematically and structured problem solving, optimization, change and documentation processes. Other optimizations and / or preventive actions are introduced by means of collaborators ideas and suggestions and systematically handled. Also periodically internal and external audits, feedbacks from stakeholders, the collaboration with supplier and praxis experiences promote possible optimizations, which are handled systematically. These promotes a systematically organization development, service, product and process innovations and a continually organization improvement in accordance to established objectives and stakeholder requirements. Therefore the organization and the organizational knowledge base are continuously coordinated optimized according to organization objectives and experiences. Theoretical considerations are tested in practice.

Thereby we integrate also optimal process modeling, process standardization and transparency with need and objective oriented flexible process implementation. Changing organizational knowledge new individual learning becomes possible. Thus new knowledge will be generated, represented, discussed, communicated and implemented. Thereby the knowledge and learning spiral is constantly pushed again and the organizational knowledge base is extended continually [14].

E IT Implementation

A established, implemented and optimized, standard based, the organization optimally adapted management system offers by its strong stakeholder and objective orientation, the systematically, structured and holistic approach, the collaborator involvement, the harmonized, simplified, in the practice "tested" and optimized processes and forms an optimal basis for a stakeholder and objective oriented, efficiently, effectively, holistic and lower-cost IT system using workflow based database. By strong and early user involvement and a systematically, holistic, optimized requirement analysis the IT will best adapted to the organization and stakeholder requirements and promotes a sustainable organization optimization.

V. PROJECT EXPERIENCES AND RESULTS

This concept for establishing and implementing a holistic, integrated, standard based management system for system development is implemented successfully in several organizations with different management systems. Thereby implementing process thinking, harmonizing and simplifying processes and process controlling were great challenges.

A Achieving the project objectives

The described concept leads the following case study results collected by measuring the established process and system measurements and interviewing the managers and collaborators:

- Effectiveness: the fulfilment of the established objectives is periodically evaluated by defined measurements. The customer satisfaction and the fulfilment of the objectives were increased constantly over more years. By an optimal IT support the measurements can be continually evaluated with no additional expenditure. Many collaborators and specially managers felt the clear definition of objectives and aligned measurements as a great help and assistance.
- Efficiently and cost reduction: By constantly optimizing processes and services in accordance to established stakeholder oriented objectives using the management system could be reduced or kept the numbers of collaborators increasing services.
- Organizational development by employee involvement and knowledge generation: Prior to introducing the management system the collaborators ideas and suggestions were occasionally missing on the way through the different decision levels. Now all ideas are documented, handle in accordance to established processes and all collaborators receive a reliable answer. The collaborators appreciate it and bring in a lot of ideas and suggestions.

Establishing and optimizing the processes in accordance with the defined organization objectives involving early the collaborators and the following IT implementation using workflow based systems of the optimized processes promotes the user system acceptance and they feel it as helpful tool and instrument to mange their job. Thereby are realized optimal starting conditions for system development.

Standard based, holistic management systems are by there clear structured, holistic, systemic approach, there strong orientation to stakeholders, there systematic process approach and the continually optimization through measurement and facts an excellent reference models for a holistic development and implementation of IT systems.

By optimizing information and communication flows, the organizational relationship promoted by process thinking, the implemented human resource development, knowledge exchange and knowledge generation and the improved collaborators involvement increased the collaborators satisfaction.

The new IT system is felt from the collaborators not as the introduction of something new one, but seen as a helpful, useful, best need adapted, powerful tool to accomplish their job efficiently and effective.

Developing an IT system by this approach the IT system is adapted optimal to the requirements of the user and the organization and can contribute thereby in the best way possible also to the continually optimization of the organization.

Changes and planned actions must be communicated and implemented systematically and constantly by all collaborators. Therefore the knowledge application is very strongly promoted.

In the periodically internal and external audits the compliance between lived practice and the system documentation is examined. This supports also the constantly updating and change of the system documentation and thereby we integrate optimal process modeling, process standardization and transparency with need and objective oriented flexible process implementation.

B Success factors

Corporate culture processes and IT technologies must be integrated optimally according to the organization objectives and to collaborators needs and requirements in order a to successfully system development based on holistic, integrated standard based management system. The system is thereby only a tool, which supports the optimization of the organization so far as this is also admitted by the culture. Therefore we need an open, confident based, fault-tolerant, objective oriented, innovative, flexible, cooperative, participative, esteeming, stakeholder, customer, service oriented corporate and learning culture with criticism and change readiness, which promotes self organizing units, autonomous, objectives oriented acting, as well as personality development of all collaborators regarding self, learn and social competences. The collaborators should be interested in new knowledge, have personal employment, team ability and change and learning willingness apart from necessary IT-competences. All managers must use constantly and actively the system and motivate their collaborators in following these principles. A strengthening point for achieving this is maintaining an optimal communication internally as well as with external partners.

Introducing holistic standard based management systems as basis for system development requires the organization best adapted systems, which is impossible buying external sample manuals, which do not correspond with the lived processes and objectives. The management systems must be continually best implemented and lived and can not be only an alibi for the auditor for maintaining the certificate.

By introducing this concept for system development the system manager and/or IT analyst extend their own job, needing apart from the technical skills also the necessary skills about standard requirements, organizational development, change management, business process management, business reengineering, organizational learning, knowledge management, controlling , as well as experience in information technology (system manager), information management, IT – service management, information security management, data protection and knowledge about all other relevant laws, standards and regulations.

Sufficient IT-infrastructure and IT-support are also very important for the project success. Only by promoting a need and objective oriented workflow based integrated, holistic database supported system a continuously optimization of the organization, in accordance with its objectives, and a sustainable organization success are secured.

VI. OUTLOOK

Due to these excellent project experiences in several organizations with different management systems there should be used enhanced a holistic, integrated, standard based management systems regarding all success factors [V B] as basis for system development.

IT analyst trainings should inform about the principles of standard based management systems and management system training should inform also about the workflow based database systems.

The management systems standards should emphasize the importance of IT support and an open, confident based, fault tolerant corporate and learning culture with criticism and change readiness.

VII. CONCLUSION

Establishing and implementing a holistic, integrated, standard based, individual, the organization best adapted management system as basis for system development to analyze customer requirements, integrate, optimize and harmonize processes, services, documents and concepts promotes efficiency, effectiveness, stakeholder orientation, high service quality and the continually organizational development by means of collaborators ideas and suggestions to guarantee sustainable stakeholder orientation, service quality, profitability and sustainable organization success.

REFERENCES

[1] M. Hammer, *Beyond reengineering*, HarperCollins Business, London, 1996.

[2] ISO, *EN/ISO 9001:2000 Quality Managemnt Systems* – requirements. ISO 17.12.2000.

[3] P. Osanna, M. Durakbasa and A. Afjehi-Sada, *Quality in Industry*, Vienna University of Technology, 2004.

[4] Y. Akao, *Hoshin Kanri*, policy deployment for successful TQM. Productivity Press, Portland, 1991.

[5] Y. Akao, Hoshin Kanri, *Quality Function Deployment*, integrating customer requirements into product design. Productivity Press, Portland, 1990.

[6] M. Stoll *Wissensmanagement in KMU durch die Erweiterung normbasierter Managementsysteme* in N. Gronau Eds. Proc. 4th Conference on Professional Knowledge Management – Experiences and Visions, Gito, Berlin (2007), volume 1 pp. 105-113.

[7] A. Abecker, K. Hinkelmann, H. Maus, H. Müller, et al. *Geschäftsprozessorientiertes Wissensmanagement,* Springer, Berlin, 2002.

[8] T. Davenport and L. Prusak, *Working Knowledge*, Harvard Business School Press, Boston, 1998.

[9] R. Maier, *Knowledge management systems* - information and communication technologies for knowledge management. Springer, Berlin, 2002.

[10] G. Riempp, *Integrierte Wissensmanagementsysteme*: Architektur und praktische Anwendung. Springer, Berlin, 2004.

[11] G. Probst, S. Raub, K. Romhardt *Wissen managen*. Gabler, Wiesbaden, 1999.

[12] F. Lehner, *Wissensmanagement*: Grundlagen, Methoden und technische Unterstützung. Hanser, Munich, 2006.

[13] S. Güldenberg, *Wissensmanagement und Wissenscontrolling in lernenden Organisationen,* Deutscher Universitäts-Verlag, Wiesbaden, 1997.

[14] P. Pawlowsky, *Wissensmanagement*, Erfahrungen und Perspektiven. Gabler, Wiesbaden, 1998.

An Application of Lunar GIS with Visualized and Auditory Japan's Lunar Explorer "KAGUYA" Data

Shin-ichi Sobue*1, Hiroshi Araki*2, Seiichi Tazawa*2, Hirotomo Noda*2, Izumi Kamiya*3,
Aya Yamamoto *4, Takeo Fujita*4, Ichiro Higashiizumi*5 and Hayato Okumura*1

*1 Japan Aerospace Exploration Agency
2-1-1 Sengen, Tsukuba, Ibaraki 305-8505, Japan
*2 National Astronomical Observation of Japan
*3 Geographical Survey Institute
*4 Remote Sensing Technology Center of Japan
*5 Moonbell project

Abstract – This paper describes an application of a geographical information system with visualized and sonification lunar remote sensing data provided by Japan's lunar explorer (SELENE "KAGUYA"). Web based GIS is a very powerful tool which lunar scientists can use to visualize and access remote sensing data with other geospatial information. We discuss enhancement of the pseudo-colored visual map presentation of lunar topographical altimetry data derived from LALT and the map of the data to several sound parameters (Interval, harmony, and tempo). This paper describes an overview of this GIS with a sonification system, called "Moonbell".

I. INTRODUCTION

In this work, we present our experience of utilizing audio-visual data mappings for GIS-based information visualization. The application we choose is a GIS-based system for visualizing lunar topography maps derived from the laser altimeter (LALT) onboard the Japanese Aerospace Exploration Agency (JAXA)'s lunar explorer (KAGUYA). In this application, we enhance the pseudo-colored visual map presentation of contour line information of lunar topographical data derived from LALT and this altimeter data by LALT is mapped to several sound parameters (Interval, harmony, and tempo). In this application development, we invited five planetary scientists from Kaguya science team and three engineers to review and evaluate the usefulness. Our motivation for choosing sound in addition to vision is guided by our belief that data quantities mapped to various colors in a coloring scheme do not always clearly describe the information being presented for many different tasks that visualization is expected to support. Additional data characteristics can be conveyed to the user through sound to enhance the performance of the user on those tasks. We have conducted experiments with human users to compare the performance of users on visual data mapping alone vs. visual and sound data mappings together and, in more than 70 percent cases, we found that the use of bi-modal visual and sound data mappings together provided a more accurate understanding of data displays. This paper describes the overview of this GIS with a sonification system, called "Moonbell".

II. OVERVIEW OF KAGUYA

KAGYA (SELENE) is the most sophisticated lunar exploration mission in the post-Apollo era and consists of the

K. Elleithy (ed.), *Advanced Techniques in Computing Sciences and Software Engineering*,
DOI 10.1007/978-90-481-3660-5_27, © Springer Science+Business Media B.V. 2010

main orbiter and two small satellites – the Relay satellite (OKINA) and the VRAD satellite (OUNA). SELENE was successful launched on September 14, 2007 from Tanegashima Space Center of JAXA and it entered polar orbit with about 100km altitude on December 21, 2007.

The major objectives of the SELENE mission are the global observation of the Moon to research lunar origin and evolution. KAGUYA observes the distribution of the elements and minerals on the surface, the topography, geological structure, the gravity field, the remnants of the magnetic field and the environment of energetic particles and plasma of the Moon. The scientific data will also be used for exploring the possibilities for the future utilization of the Moon. JAXA will also establish the basic technologies for future Moon exploration, such as, lunar polar orbit insertion, 3-axis attitude control and thermal control in the lunar orbit. In addition, KAGUYA takes pictures and movies of the beautiful Earth-rise from the Moon horizon. Table 1 shows the mission instruments of KAGUYA. KAGUYA Level-2 processed data (standard product) are archived in "Level-2 Data Archive and Distribution System" (L2DB) at SOAC (SELENE Operation and Analysis Center) at Sagamihara-campus of JAXA. Users can search and download Level-2 processed data via Web browser. KAGUYA data will be available for users through the internet from 1 year after the nominal mission phase (about 2 years after the KAGUYA launch). The KAGUYA L2 DB HP Web site (under construction) will be linked to the KAGUYA HP (http://selene. jaxa.jp/index_e.html). Since KAGUYA L2 data format is PDS-like format, prototype PDS interoperable catalogue system is also developing with PDAP (Planetary Data Access Protocol) [1][2].

Before opening KAGUYA L2 products to public, we developed a KAGUYA image gallery to provide visualized images derived from KAGUYA L2 products with Open GIS Web Map Service (OGC/WMS) technology to promote KAGUYA observation data to public (URL is http://wms.kaguya.jaxa.jp).

We already posted more than 30 images and movies on our image gallery site by August, 2008. Visualized images are produced not only from optical sensors but also from geophysical parameter information derived from KAGUYA scientific instruments such as Spectral Profiler. Fig. 1 shows the snap shot of image gallery.

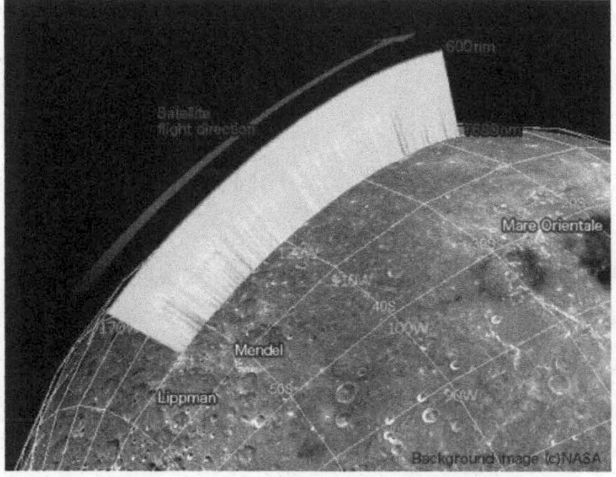

Fig.1 Spectral Profiler visualized image on KAGUYA image gallery

KAGUYA image gallery is only capable of providing visualized image with a 2 dimensional display but we also developed and operate 3 dimensional GIS system by using OGC/WMS to provide visualized images to our KAGUYA science teams for their research. We already uploaded entire lunar topography map by LALT, entire lunar gravity anomaly map by RSAT/VRAD, HDTV coverage map, MI image, etc. We plan to post more geographical parameter data on a KAGUYA WMS server. Our science team members can access those 3D visualized images by using GIS browsers including NASA world wind (NWW) etc. On a KAGUYA WMS server, we can provide a link capability to access multimedia files including movies, sound files, etc. easily when users select visualized images on their GIS browsers. Fig.2 shows the example to show HDTV movie coverage with video icon. So, user can access HDTV movies without search and order operation to show those movies.

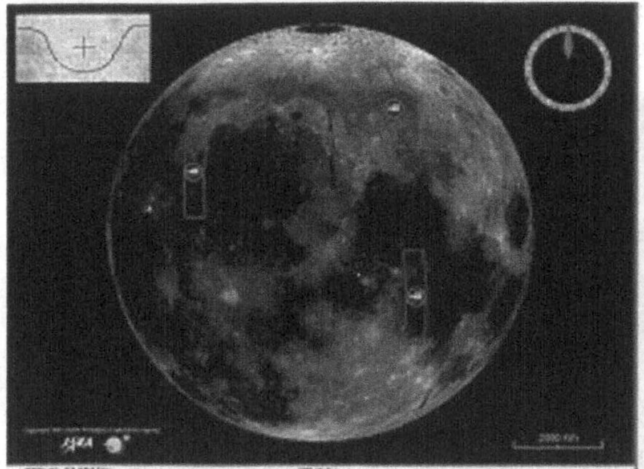

Fig.2 3D visualized image (HDTV coverage with movie icon) on WMS by NWW as client

III. LALT VISUALIZATION and SONIFICATION

LALT is one of the instruments on board the KAGUYA main orbiter and is for laser ranging from the satellite to the surface of the Moon in the nadir direction Fig 3 shows lunar topographical map of Mare Orientale derived from LALT.

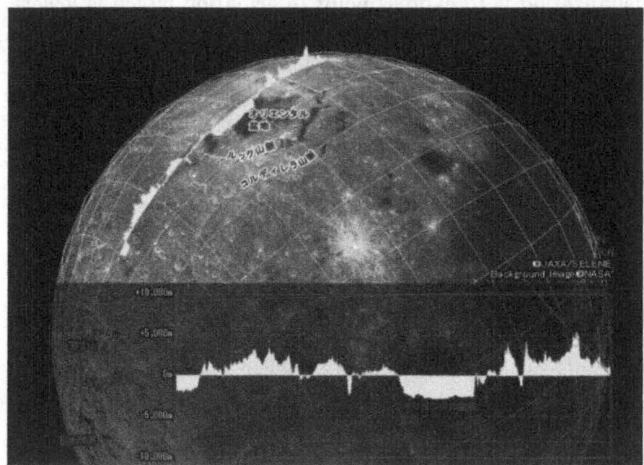

Fig 3 Topography of Mare Orientale derived from LALT

By using LALT, the lunar topographical map was produced by the Geographical Survey Institute (GSI) from the LALT product of the National Astronomical Observatory (NAOJ). The LALT is able to obtain a range of data on a global scale along the satellite's trajectory including the high latitude region above 75 degrees that has never been measured by an altimeter. The number of measurement points as of this March is about 6 million and it is more than 10 times larger than the number for the previous topographical map, named "ULCN 2005" by USGS.

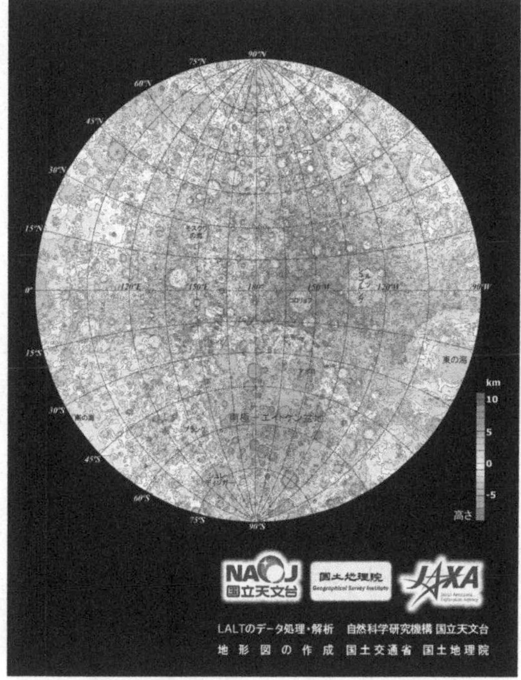

Fig.4 Lunar far side topographical map by LALT (by NAOJ/GSI/JAXA)

Fig. 4 shows the global topographical map of the far side of the Moon created by LALT and it is the most significant topographical lunar map available to date. We can see detailed topographical patterns of the far side of the moon with less than 100 meter diameter class craters including South Pole Aitkin Basin area, etc.

In parallel with the WMS system, we developed a sonification system, named "Moonbell" (voice of moon system) because data quantities mapped to various colors in a coloring scheme do not always clearly describe the information being presented for many different tasks that visualization is expected to support. In many cases additional data characteristics can be conveyed to the user through sound to enhance the performance of the user on those tasks. We have conducted experiments with human users to compare the performance of users on visual data mapping alone vs. visual and sound data mappings and we found that the use of bi-modal visual and sound data mappings together provided more accurate understanding of data displays [3].

This Moonbell system is capable of converting altitude information into sound (interval) with a selected musical instrument and a given tempo by using a MIDI sound device on PC.

As a default setting in "Moonbell", the marimba was selected to play sounds for local features and the piano was selected to play sounds for regional features as a base code. In this case, regional interval was allocated to regional average information with 30 altimeter observation points' data as base sound. On the other hand, the difference of highly developed relativity from the average altitude of 30 points was calculated, and the interval was allocated in the difference additionally about a local interval. However, it was set that only the interval that played harmony (diminish sounds) was allocated between the previous sounds when all intervals of all sounds were not used in default but the sound changed so that the change in the interval should not allow it to feel to jarring about the allocated interval. Therefore, the music that the marimba plays is always composed only because of the sound of constant harmony. "Moonbell" users can freely select interval, musical instruments and tempo by using Moonbell user interface. Fig. 5 shows the Moonbell user interface screen. In this application development, we invited five planetary scientists from Kaguya science team and three engineers to review and evaluate the usefulness. Our motivation for choosing sound (sonification) in addition to vision (visualization) is guided by our belief that data quantities mapped to various colors in a coloring scheme do not always clearly describe the information being presented for many different tasks that visualization is expected to support.

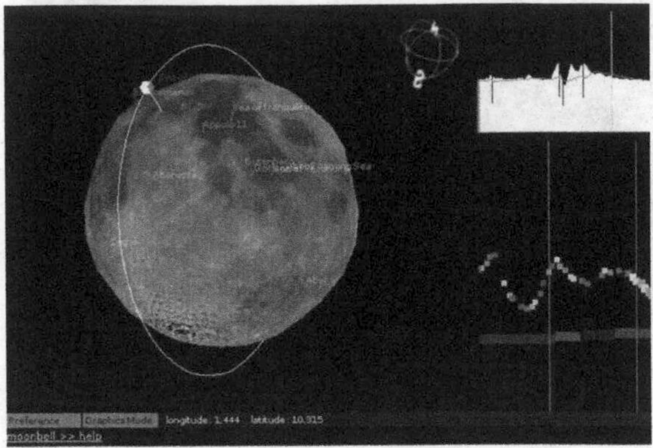

Fig 5 Moonbell user interface screen

In many cases, additional data characteristics can be conveyed to the user through sound to enhance the performance of the user on those tasks [4][5].

One of the good examples of use of sound instead of 3D visualization is to express point information which is lower than the average height. In the expression that uses a spheroid body in 3D lunar GIS, it is difficult to come in succession to visually identify it lower than the average altitude (for instance, Mare Orientale area shown in Fig 4). In contrast, it is easy to show lower altitude with regional and local sound by using sonification since we can show average sound with lower and upper sound. In addition, it is also difficult to show local features and regional features simultaneously by using 3D GIS although 3D GIS can show local features with zooming / panning original observation data as well as regional features by using level slice and rainbow color visualization (aka contour line map) independently. By using sonification with Moonbell, we can show local and regional feature since Moonbell can play two sounds (base line as regional feature and main melody as local feature) simultaneously. In more than 70 percent cases, we found that the use of bi-modal visual and sound data mappings together provided a more accurate understanding of data displays. However, there are disadvantages in the use of sonification. Since sound mapping process is a linear mapping process, it is difficult for sonification to express two and three dimensional information. This means that contour line of topography maps cannot be mapped to two dimensional planes in sonification mapping directly by using "Moonbell". Moonbell can only map altitude information under the orbit of KAGUYA to sounds.

To solve both 3D GIS and sonification issues, we propose 3D GIS WMS with sonification system (Moonbell) in this study. Since WMS is easy to link to a3D Spheroid moon body with multimedia files including movies, sounds, documents etc, we plan to prepare wav files derived from the Moonbell system and link those wav sound files to a predefined target area with that area images on 3D moon provided by WMS server. This idea is an expansion of the linkage between HDTV coverage with HDTV movies shown in Fig. 2. By using this proposed system, we believe that scientists can improve the performance of the user on those research tasks.

IV. CONCLUSION

In this study, we propose an integration system of 3D Web-based lunar GIS with a lunar observation data sonification system, called "Moonbell" and show the possible enhancement of the performance of the user on their research tasks.

Our next step is to expand our sonification system to manipulate other KAGUYA observation data such as Spectral Profiler, Multi-band imager, etc and integrate with 3D Web GIS system. We also believe that it is also very useful for planetary scientists to use visualized images with HDTV camera movies on KAGUYA and sonification data on our integrated 3D Web-based GIS system through our experiment using this application. In addition, we also think that this integration system is useful for 3D GIS system for Earth and other planets.

ACKNOWLEDGEMENT

We thank all the contributors to the SELENE (KAGUYA) project, LALT science team and Moonbell development team for this work.

REFERENCE

(1) S. Sobue, H. Araki, S. Tazawa, H. Noda, H. Okumura, I. Higashiizumi, T. Shimada, S. Higa, and E. Kobayashi, The oralization system of KAGUYA LALT, vol 28 No. 3, Japan remote sensing society, 2008

(2) S. Sobue, M. Kato, H. Hoshino, H. Okumura, and Y. Takizawa, KAGUYA(SELENE) data delivery and EPO plan, ISTS2008, 2008-K-25, 2008

(3) S. Sobue, Y. Takizawa, M. Kato, and S. Sasaki, The overview of KAGUYA, vol 28 No.1, pp44-46, Japan remote sensing society, 2008

(4) Ryan MacVeigh, and R. Daniel Jacobson, Increasing the dimensionality of a Geographic Information System (GIS) using auditory display, pp530-534, proceeding of the 13th international conference on auditory display (ICAD), 2007

(5) Suresh K. Lodha, Abigail J. Joseph, and Jose C. Renterial, Audio-visual data mapping for GIS-based data: an experimental evaluation, pp41-48, proceeding of the 8th ACM international conference on Information and knowledge management, 1999

From Constraints to Resolution Rules
Part I : conceptual framework

Denis Berthier
Institut Telecom ; Telecom & Management SudParis
9 rue Charles Fourier, 91011 Evry Cedex, France

Abstract: Many real world problems appear naturally as constraints satisfaction problems (CSP), for which very efficient algorithms are known. Most of these involve the combination of two techniques: some direct propagation of constraints between variables (with the goal of reducing their sets of possible values) and some kind of structured search (depth-first, breadth-first,...). But when such blind search is not possible or not allowed or when one wants a "constructive" or a "pattern-based" solution, one must devise more complex propagation rules instead. In this case, one can introduce the notion of a candidate (a "still possible" value for a variable). Here, we give this intuitive notion a well defined logical status, from which we can define the concepts of a resolution rule and a resolution theory. In order to keep our analysis as concrete as possible, we illustrate each definition with the well known Sudoku example. Part I proposes a general conceptual framework based on first order logic; with the introduction of chains and braids, Part II will give much deeper results.

Keywords: constraint satisfaction problem, knowledge engineering, production system, resolution rule, strategy, Sudoku solving.

I. Introduction

Many real world problems, such as resource allocation, temporal reasoning or scheduling, naturally appear as constraint satisfaction problems (CSP) [1, 2]. Such problems constitute a main sub-area of Artificial Intelligence (AI). A CSP is defined by a finite number of variables with values in some fixed domains and a finite set of constraints (i.e. of relations they must satisfy); it consists of finding a value for each of these variables, such that they globally satisfy all the constraints.

A CSP states the constraints a solution must satisfy, i.e. it says *what* is desired. It does not say anything about *how* a solution can be obtained. But very efficient general purpose algorithms are known [1], which guarantee that they will find a solution if any. Most of these algorithms involve the combination of two very different techniques: some direct propagation of constraints between variables (in order to reduce their sets of possible values) and some kind of structured search with "backtracking" (depth-first, breadth-first,...), consisting of trying (recursively if necessary) a value for a variable, propagating the consequences of this tentative choice and eventually reaching a solution or a contradiction allowing to conclude that this value is impossible.

But, in some cases, such blind search is not possible (for practical reasons, e.g. one wants to simulate human behaviour or one is not in a simulator but in real life) or not allowed (for theoretical or æsthetic reasons, or because one wants to understand what happens, as is the case with most Sudoku players) or one wants a "constructive" solution.

In such situations, it is convenient to introduce the notion of a candidate, i.e. of a "still possible" value for a variable. But a clear definition and a logical status must first be given to this intuitive notion. When this is done, one can define the concepts of a *resolution rule* (a logical formula in the "condition => action" form, which says what to do in some observable situation described by the condition pattern), a *resolution theory*, a *resolution strategy*. One can then study the relation between the original CSP problem and various of its resolution theories. One can also introduce several properties a resolution theory can have, such as confluence (in Part II) and completeness (contrary to general purpose algorithms, a resolution theory cannot in general solve all the instances of a given CSP; evaluating its scope is thus a new topic in its own). This "pattern-based" approach was first introduced in [3], in the limited context of Sudoku solving.

Notice that resolution rules are typical of the kind of rules that can be implemented in an inference engine and resolution theories can be seen as "production systems" [4]. See Part II.

In this paper, we deal only with the case of a finite number of variables with ranges in finite domains and with first order constraints (i.e. constraints between the variables, not between subsets of variables).

This paper is self-contained, both for the general concepts and for their illustrations with the Sudoku example, although deeper results specific to the introduction of chains or to this example will appear in Part II. Section II introduces the first order logical formulation of a general CSP. Section III defines the notion of a candidate and analyses its logical status. Section IV can then define resolution rules, resolution paths, resolution theories and the notion of a pure logic constructive solution. Section V explains in what sense a resolution theory can be incomplete even if its rules seem to express all the constraints in the CSP.

K. Elleithy (ed.), *Advanced Techniques in Computing Sciences and Software Engineering*, DOI 10.1007/978-90-481-3660-5_28,

II. The Logical Theory Associated with a CSP

Consider a fixed CSP for n variables $x_1, x_2, \ldots, x_n$ in finite domains $X_1, X_2, \ldots, X_n$, with first order constraints. The CSP can obviously be written as a First Order Logic (FOL) theory (i.e. as a set of FOL axioms expressing the constraints) [1]. Thanks to the equivalence between FOL and Multi-Sorted First Order Logic (MS-FOL) [5], it can also be written as an MS-FOL theory. CSP solutions are in one-to-one correspondence with MS-FOL models of this theory.

In MS-FOL, for each domain X_k, one introduces a sort (i.e. a type) X_k (there can be no confusion in using the same letter for the sort and the domain) and a predicate $\text{value}_k(x_k)$, with intended meaning "the value of the k-th variable is x_k". All the basic functions and predicates necessary to express the given constraints are defined formally as being sorted, so that one doesn't have to write explicit conditions about the sorts of the variables mentioned in a formulæ. This has many advantages in practice (such as keeping formulæ short). The formulæ of our MS-FOL theory are defined as usual, by induction (combining atomic formulæ built on the above basic predicates and functions with logical connectives: and, or, not, typed quantifiers). We can always suppose that, for each value a variable can have, there is a constant symbol of the appropriate sort to name it. We can also adopt a *unique names assumption* for constant symbols: two different constant symbols of the same sort do not designate the same entity. However, no unique names assumption is made for variables. With the following Sudoku example, details missing in the above two paragraphs will hopefully become clearer than through additional logical formalism.

A. The Sudoku CSP

Sudoku is generally presented as follows (Fig. 1): given a 9x9 *grid*, partially filled with *numbers* from 1 to 9 (the "entries" or "clues" or "givens" of the problem), complete it with numbers from 1 to 9 in such a way that in each of the nine *rows*, in each of the nine *columns* and in each of the nine disjoint *blocks* of 3x3 contiguous *cells*, the following property holds: there is at most one occurrence of each of these numbers. Notice that this is a special case of the Latin Squares problem (which has no constraints on blocks).

It is natural to consider the three dimensional space with coordinates (n, r, c) and any of the 2D spaces: rc, rn and cn. Moreover, in rc-space, due to the constraint on blocks, it is convenient to introduce an alternative block-square coordinate system [b, s] and variables X_{bs} such that $X_{rc} = X_{bs}$ whenever (r, c) and [b, s] are the coordinates of the same cell. For symmetry reasons, in addition to these variables with values in Numbers, we define additional X_{rn}, X_{cn} and X_{bn} variables, with values, respectively in Rows, Columns and Squares, and such that:

$X_{rc} = n \Leftrightarrow X_{rn} = c \Leftrightarrow X_{cn} = r \Leftrightarrow X_{bn} = s$.

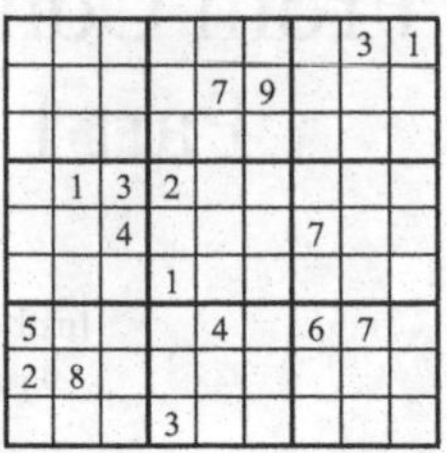

Figure 1: A typical Sudoku puzzle

Since rows, columns and blocks play similar roles in the defining constraints, they will naturally appear to do so in many other places and it is convenient to introduce a word that makes no difference between them: a *unit* is either a row or a column or a block. And we say that two rc-cells *share a unit* if they are either in the same row or in the same column or in the same block (where "or" is non exclusive). We also say that these two cells are *linked*. It should be noticed that this (symmetric) relation between two cells does not depend in any way on the content of these cells but only on their place in the grid; it is therefore a straightforward and quasi physical notion.

Formulating the Sudoku CSP as an MS-FOL theory is done in three stages: Grid Theory, General Sudoku Theory, Specific Sudoku Puzzle Theory. For definiteness, we consider standard Sudoku only, on a 9x9 grid.

Most CSP problems can similarly be decomposed into three components: axioms for a general and static context (here, the grid) valid for all the instances of the problem, axioms for the general CSP constraints expressed in this context (here, the Sudoku constraints) and axioms for specific instances of the problem (here, the entries of a puzzle).

B. Grid Theory

B.1 The sorts in Grid Theory

The characteristic of MS-FOL is that it assumes the world of interest is composed of different types of objects, called sorts. In the very limited world of Grid Theory (GT) and of Sudoku Theory (ST), we need only five sorts: Number, Row, Column, Block, Square. Row, Column and Block correspond in the obvious way to rows, columns and blocks, whereas Square corresponds to the relative position of a cell in a block.

Attached to each sort, there are two sets of symbols, one for naming constant objects of this sort, and one for naming variables of this sort. In the GT case, the variables for Numbers are n, n', n'', n_0, n_1, n_2, …; the constants for Numbers are 1_n, 2_n, 3_n, 4_n, 5_n, 6_n, 7_n, 8_n, 9_n. The variables for Rows are , r', r'', r_0, r_1, r_2, ……; the constants for Rows are 1_r, 2_r, 3_r, 4_r, 5_r, 6_r, 7_r, 8_r, 9_r. And similarly for the other sorts. For each of the first five sorts, obvious axioms can express the range of the variable of this sort and the unique names assumption.

In conformance with the MS-FOL conventions, a quantifier such as "∀r" (resp. "∀c", "∀n", ...) will always mean "for any row r" (resp. "for any column c", "for any number n", ...)

B.2 Function and predicate symbols of Grid Theory

Grid Theory has no function symbol. In addition to the five equality predicate symbols ($=_n$, $=_r$,... one for each sort), it has only one predicate symbol: *correspondence*, with arity 4 and signature (Row, Column, Block, Square), with intended meaning for atomic formulæ "correspondence(r, c, b, s)" the natural one. Given these basic predicates, one can define auxiliary predicates, considered as shorthands for longer logical formulæ:

– *same-row*, with signature (Row, Column, Row, Column); "same-row(r_1, c_1, r_2, c_2)" is defined as a shorthand for: $r_1 =_r r_2$;

– and similarly for *same-column* and *same-block*;

– *same-cell*, with signature (Row, Column, Row, Column); "same-cell(r_1, c_1, r_2, c_2)" is defined as a shorthand for: $r_1 =_r r_2$ & $c_1 =_c c_2$;

As they have been defined, the auxiliary predicates same-row, same-column, same-block and same-cell all have the same arity and signature: informally, they all apply to couples of cells with row-column coordinates. This is very important because it allows to define an auxiliary predicate with the same arity and signature, applying to couples of cells with row-column coordinates, independent of the type of unit they share (we shall see that, most of the time, this type is irrelevant):

– *share-a-unit*, with signature (Row, Column, Row, Column); "share-a-unit(r_1, c_1, r_2, c_2)" is defined as a shorthand for: ¬same-cell(r_1, c_1, r_2, c_2) & [same-row(r_1, c_1, r_2, c_2) or same-column(r_1, c_1, r_2, c_2) or same-block(r_1, c_1, r_2, c_2)].

Of course, the intended meaning of this predicate is that suggested by its name: the two cells share either a row or a column or a block; notice that a cell is not considered as sharing a unit with itself.

B.3 Axioms of Grid Theory

In addition to the 5x36 sort axioms, Grid Theory has 81 axioms expressing the (r, c) to [b, s] correspondence of coordinate systems, such as "correspondence(1_r, 1_c, 1_b, 1_s)".

As an exercise, one can check that this is enough to define the grid (modulo renamings of rows, columns, ...). One can also check that the following formula expresses that row r intersects block b: ∃c∃s correspondence(r, c, b, s).

C. General Sudoku Theory

General Sudoku Theory (ST) is defined as an extension of Grid Theory. It has the same sorts as GT. In addition to the predicates of GT, it has the following one: *value*, with signature (Number, Row, Column); the intended meaning of atomic formula "value(n, r, c)" is that number n is the value of cell (r, c), i.e. indifferently: $X_{rc} = n$, $X_{rn} = c$, $X_{cn} = r$ or $X_{bn} = s$. It is convenient to introduce an auxiliary predicate value[], written with square braces, with signature (Number, Block, Square), with the same meaning as value, but in [b, s] instead of (r, c) coordinates; "value[n, b, s]" is defined as a shorthand for:

∃r∃c [correspondence(r, c, b, s) & value(n, r, c)]

C.1 Axioms of Sudoku Theory

ST contains the axioms of GT, plus the following, written in a symmetrical form that will be useful in the sequel.

ST_{rc}: ∀r∀c∀n_1∀n_2{value(n_1, r, c) & value(n_2, r, c) ⇒ $n_1 = n_2$}

ST_{rn}: ∀r∀n∀c_1∀c_2{value(n, r, c_1) & value(n, r, c_2) ⇒ $c_1 = c_2$}

ST_{cn}: ∀c∀n∀r_1∀r_2{value(n, r_1, c) & value(n, r_2, c) ⇒ $r_1 = r_2$}

ST_{bn}: ∀b∀n∀s_1∀s_2{value[n, b, s_1] & value[n, b, s_2] ⇒ $s_1 = s_2$}

EV_{rc}: ∀r∀c∃n value(n, r, c)

EV_{rn}: ∀r∀n∃c value(n, r, c)

EV_{cn}: ∀c∀n∃r value(n, r, c)

EV_{bn}: ∀b∀n∃s value[n, b, s].

The formal symmetries inside each of these two groups of four axioms must be noticed. ST_{rc} expresses that an rc-cell can have only one value (this is never stated explicitly, but this should not be forgotten). ST_{rn} (resp. ST_{cn}, ST_{bn}) expresses that a value can appear only once in a row (resp. a column, a block); these are the standard constraints. Axiom EV_{rc} (resp. EV_{rn}, EV_{cn} and EV_{bn}) expresses that, in a solution, every rc- (resp. rn-, cn and bn-) cell must have a value; these conditions generally remain implicit in the usual formulation of Sudoku.

D. Specific Sudoku Puzzle Theory

In order to be consistent with various sets of entries, ST includes no axioms on specific values. With any specific puzzle P we can associate the axiom E_P defined as the finite conjunction of the set of all the ground atomic formulæ "value(n_k, r_i, c_j)" such that there is an entry of P asserting that number n_k must occupy cell (r_i, c_j). Then, when added to the axioms of ST, axiom E_P defines the MS-FOL theory of the specific puzzle P.

From the point of view of first order logic, everything is said. A solution of puzzle P (if any) is a model (if any) of theory ST + E_P. The only problem is that nothing yet is said about how a solution can be found. This is the reason for introducing candidates and resolution rules.

III. Candidates and Their Logical Status

A. Candidates

If one considers the way Sudoku players solve puzzles, it appears that most of them introduce candidates in the form of "pencil marks" in the cells of the grid. Intuitively, a candidate is a "still possible" value for a cell; candidates in each cell are progressively eliminated during the resolution process.

This very general notion can be introduced for any CSP problem. Unfortunately, it has no *a priori* meaning from the MS-FOL point of view. The reason is not the non-monotonicity of candidates, i.e. that they are progressively withdrawn whereas one can only add information by applying the axioms of a FOL theory: this could easily be dealt with by introducing non-candidates (or impossible values) instead. The real reason is that the intuitive notion of a candidate (as a "still possible" value) and the way it is used in practice suppose a logic in which this idea of "still possible" is formalised. Different "states of knowledge" must then be considered – and this is typically the domain of epistemic logic. We shall therefore adopt *a priori* the following framework, supporting a natural epistemic interpretation of a candidate.

For each variable x_k of the CSP, let us introduce a predicate *$cand_k(x_k)$ with intended meaning "the value x_k from domain X_k is not yet known to be impossible for the k-th variable".*

The interesting point is that we shall be able to come back to ordinary (though constructivist or intuitionistic) logic for candidates (and thus forget the complexities of epistemic logic).

B. Knowledge states and knowledge space

Given a fixed CSP, define a *knowledge state* as any set of values and candidates (formally written as $value_k$ and $cand_k$ predicates). A knowledge state is intended to represent the totality of the ground atomic facts (in terms of values and candidates) that are present in *some* possible state of reasoning for some instance of the CSP. (Invariant background knowledge, such as grid facts in Sudoku, is not explicitly included).

It should be underlined that this notion of a knowledge state has a very concrete and intuitive meaning: for instance, in Sudoku, it represents the situation on a grid with candidates at some point in some resolution process for some puzzle; this is usually named the PM, the "Pencil Marks". (Notice that some knowledge states may be contradictory – so that inconsistent sets of entries can be dealt with).

Let **KS** be the (possibly large, but always finite) set of all possible knowledge states. On **KS**, we define the following order relation: $KS_1 \leq KS_2$ if and only if, for any constant x° (of sort X) one has:

– if $value_X(x°)$ is in KS_1, then $value_X(x°)$ is in KS_2,

– if $cand_X(x°)$ is in KS_2, then $cand_X(x°)$ is in KS_1.

If "$KS_1 \leq KS_2$" is intuitively interpreted as "KS_2 may appear after KS_1 in some resolution process", these conditions express the very intuitive idea that values can only be added and candidates can only be deleted during a resolution process.

For any instance P of the CSP (e.g. for any puzzle P), one can also define the initial knowledge state KS_P corresponding to the starting point of any resolution process for P. Its values are all the entries of P; its candidates are all the possible values of the remaining variables. The set $\mathbf{KS_P}$ = {KS / $KS_P \leq$ KS} is thus the set of knowledge states one can reach when starting from P; we call it the *epistemic model* of P.

C. Knowledge states and epistemic logic

The above notion of a knowledge state appears to be a particular case of the general concept of a possible world in modal logic; the order relation on the set of knowledge states corresponds to the accessibility relation between possible worlds and our notion of an epistemic model coincides with that of a Kripke model [6]. Let K be the "epistemic operator", i.e. the formal logical operator corresponding to knowing (for any proposition A, KA denotes the proposition "it is known that A" or "the agent under consideration knows that A"). Then, for any proposition A, we have Hintikka's interpretation of KA [7]: in any possible world compatible with what is known (i.e. accessible from the current one), it is the case that A.

Several axiom systems have appeared for epistemic logic (in increasing order of strength: S4 < S4.2 < S4.3 < S4.4 < S5). Moreover, it is known that there is a correspondence between the axioms on the epistemic operator K and the properties of the accessibility relation between possible worlds (this is a form of the classical relationship between syntax and semantics). As the weakest S4 logic is enough for our purposes, we won't get involved in the debates about the best axiomatisation. S4 formalises the following three axioms:

– KA ⇒ A: "if a proposition is known then it is true" or "only true propositions can be known"; it means that we are speaking of knowledge and not of belief and this supposes the agent (our CSP solver) does not make false inferences; this axiom corresponds to the accessibility relation being reflexive (for all KS in **KS**, one has: KS ≤ KS);

– KA ⇒ KKA: (reflection) if a proposition is known then it is known to be known (one is aware of what one knows); this axiom corresponds to the accessibility relation being transitive (for all KS_1, KS_2 and KS_3 in **KS**, one has: if $KS_1 \leq KS_2$ and $KS_2 \leq KS_3$, then $KS_1 \leq KS_3$);

– K(A ⇒ B) ⇒ (KA ⇒ KB): (limited deductive closure of knowledge) if it is known that A ⇒ B, then if it is known that A, then it is known that B. In the case of CSP, this will be applied as follows: when a resolution rule [A ⇒ B] is known [K(A ⇒ B)], if its conditions [A] are known to be satisfied [KA] then its conclusions [B] are known to be satisfied [KB].

D. Values and candidates

As we want our resolution rules to deal with candidates, all our initial MS-FOL concepts must be re-interpreted in the context of epistemic logic.

The entries of the problem P are not only true in the initial knowledge state KS_P, they are known to be true in this state:

they must be written as $Kvalue_k$; similarly, the initial candidates for a variable are not only the *a priori* possible values for it; they must be interpreted as not yet known to be impossible: $\neg K \neg cand$.

Moreover, as a resolution rule must be effective, it must satisfy the following: a condition on the absence of a candidate must mean that it is effectively known to be impossible: $K \neg cand$; a condition on the presence of a candidate must mean that it is not effectively known to be impossible: $\neg K \neg cand$; a conclusion on the assertion of a value must mean that this value becomes effectively known to be true: Kvalue; a conclusion on the negation of a candidate must mean that this candidate becomes effectively known to be impossible: $K \neg cand$.

As a result, in a resolution rule, a predicate "$value_k$" will never appear alone but only in the construct "$Kvalue_k(x_k)$"; a predicate "$cand_k$" will never appear alone but only in the construct $\neg K \neg cand_k$ (given that $K \neg cand_k$ is equivalent, in any modal theory, to $\neg\neg K \neg cand_k$).

All this entails that we can use well known correspondences of modal logic S4 with intuitionistic logic [9] and constructive logic [10] to "forget" the K operator (thus merely replacing everywhere "$Kvalue_k$" with "$value_k$" and "$\neg K \neg cand_k$" with "$cand_k$"), provided that we consider that we are now using intuitionistic or constructive logic. We have thus eliminated the epistemic operator that first appeared necessary to give the notion of a candidate a well defined logical status. Said otherwise: *at the very moderate price of using intuitionnistic or constructive logic, in spite of the fact that candidates can be given an epistemic status, no explicit epistemic operator will ever be needed in the logical formulation of resolution rules.*

One thing remains to be clarified: the relation between values and candidates. In the epistemic interpretation, a value a_k for a variable x_k is known to be true if and only if all the other possible values for this variable are known to be false:

$\forall x_k\ [Kvalue_k(x_k) \Leftrightarrow \forall x'_k \neq x_k\ K \neg cand_k(x'_k)]$.

Using the equivalence between $K\neg$ and $\neg\neg K \neg$ and forgetting the K operator as explained above, we get the value-to-candidate-relation intuitionistic axiom, for each variable x_k:

VCR_k: $\forall x_k\ [value_k(x_k) \Leftrightarrow \forall x'_k \neq x_k \neg cand_k(x'_k)]$.

IV. Resolution Rules and Resolution Theories

A. General definitions

Definiton: a formula in the MS-FOL language of a CSP is in *the condition-action form* if it is written as $A \Rightarrow B$, possibly surrounded with quantifiers, where A does not contain explicitly the "$\Rightarrow$" sign and B is a conjunction of value predicates and of negated cand predicates (no disjunction is allowed in B); all the variables appearing in B must already appear in A and be universally quantified.

Definitons: a formula in the condition-action form is a *resolution rule* for a CSP if it is an intuitionistically (or constructively) valid consequence of the CSP axioms and of VCR. A *resolution theory* for a CSP is a set of resolution rules. Given a resolution theory T, a *resolution path* in T for an instance P of the CSP is a sequence of knowledge states starting with P and such that each step is justified by a rule in T. A *resolution theory T solves an instance P of the CSP* if one can exhibit a resolution path in T leading to a solution. Notice that, contrary to the general notion of a solution of a CSP as any model of the associated FOL or MS-FOL theory, this is a restrictive definition of a solution; it can be called a *constructive definition* of a solution: a resolution theory solves an instance of the CSP only if it does so in the constructive way defined above.

B. The Basic Resolution Theory of a CSP

For any CSP, there is a Universal Resolution Theory, URT, defined as the union of the following three types of rules, which are the mere re-writing of each of the VCR_k axioms, for each sort:

- Elementary Constraints Propagation rule for sort X_k:

ECP_k: $\forall x_k \forall x_{k1} \neq x_k\ \{value_k(x_k) \Rightarrow \neg cand_k(x_{k1})\}$;

- "Singles" rule for sort X_k:

S_k: $\forall x_k \{[cand_k(x_{k1})\ \&\ \forall x_{k1} \neq x_k \neg cand_k(x_{k1})] \Rightarrow value_k(x_k)\}$.

- Contradiction Detection for sort X_k:

CD_k: $\forall x_k(\neg value(x_k)\ \&\ \neg cand_k(x_k)) \Rightarrow \bot$, where "$\bot$" is any false formula.

But for any CSP, there is also a Basic Resolution Theory, BRT, which is the union of URT with all the rules specific to this CSP expressing the direct contradictions (if any) between its different variables (see Part II). All the resolution theories we shall consider will be extensions of this BRT.

C. Example from Sudoku

The BRT of the Sudoku CSP (say BSRT) consists of the following rules. The elementary constraints propagation (ECP) rules are the re-writing of the left-to-right part of axioms ST_{rc}, ST_{bn}, ST_{cn} and ST_{bn} in the condition-action form:

ECP1: $\forall r \forall c \forall n \forall n_1 \neq n\ \{value(n, r, c) \Rightarrow \neg cand(n_1, r, c)\}$

ECP2: $\forall r \forall n \forall c \forall c_1 \neq c\ \{value(n, r, c) \Rightarrow \neg cand(n, r, c_1)\}$

ECP3: $\forall c \forall n \forall r \forall r_1 \neq r\ \{value(n, r, c) \Rightarrow \neg cand(n, r_1, c)\}$

ECP4: $\forall b \forall n \forall s \forall s_1 \neq s\ \{value[n, b, s] \Rightarrow \neg cand[n, b, s_1]\}$

Here cand[] is related to cand() in the same way as value[] was related to value().

The well-known rules for Singles ("Naked-Singles" and "Hidden-Singles") are the re-writing of the right-to-left part of the same axioms:

NS: $\forall r \forall c \forall n \{\ [cand(n, r, c)\ \&\ \forall n_1 \neq n\ \neg cand(n_1, r, c)] \Rightarrow value(n, r, c)\}$

HS_{rn}: $\forall r \forall n \forall c\{ [cand(n, r, c) \ \& \ \forall c_1 \neq c \ \neg cand(n, r, c_1)] \Rightarrow value(n, r, c)\}$
HS_{cn}: $\forall c \forall n \forall r\{ [cand(n, r, c) \ \& \ \forall r_1 \neq r \ \neg cand(n, r_1, c)] \Rightarrow value(n, r, c)\}$
HS_{bn}: $\forall b \forall n \forall s\{ [cand'[n, b, s] \ \& \ \forall s_1 \neq s \ \neg \ cand'[n, b, s_1] \ ] \Rightarrow value'[n, b, s]\}$

Axioms EV have a special translation, with meaning: if there is a (rc-, rn- cn- or bn-) cell for which no value remains possible, then the problem has no solution:
CD_{rc}: $\exists r \exists c \forall n[\neg value(n, r, c) \ \& \ \neg cand(n, r, c)] \Rightarrow \bot$.
CD_{rn}: $\exists r \exists n \forall c[\neg value(n, r, c) \ \& \ \neg cand(n, r, c)] \Rightarrow \bot$.
CD_{cn}: $\exists c \exists n \forall r[\neg value(n, r, c) \ \& \ \neg cand(n, r, c)] \Rightarrow \bot$.
CD_{bn}: $\exists b \exists n \forall s[\neg value'(n, b, s) \ \& \ \neg cand'(n, b, s)] \Rightarrow \bot$.

V. Completeness

A. Completeness

Now that the main concepts are defined, we can ask: what does it mean for a Resolution Theory T for a given CSP to be "complete"? Notice that all the results that can be produced (i.e. all the values that can be asserted and all the candidates that can be eliminated) when a resolution theory T is applied to a given instance P of the CSP are logical consequences of theory $T \cup E_P$ (where E_P is the conjunction of the entries for P); these results must be valid for any solution for P (i.e. for any model of $T \cup E_P$). Therefore a resolution theory can only solve instances of the CSP that have a unique solution and one can give three sensible definitions of the completeness of T: 1) it solves all the instances that have a unique solution; 2) for any instance, it finds all the values common to all its solutions; 3) for any instance, it finds all the values common to all its solutions and it eliminates all the candidates that are excluded by any solution.

Obviously, the third definition implies the second, which implies the first, but whether the converse of any of these two implications is true in general remains an open question.

B. Why a Basic Resolution Theory may not be enough

In the case of Sudoku, one may think that the obvious resolution rules of BSRT are enough to solve any puzzle. After all, don't they express all that there is in the axioms? It is important to understand in what sense they are not enough and why.

These rules are not enough because our notion of a solution within a resolution theory T a priori restricts them to being used constructively; said otherwise, we look only for models of T obtained constructively from the rules in T; but a solution of a CSP is any model of the original axioms, whether it is obtained in a constructive way or not (it may need some "guessing").

To evaluate how far these rules are from being enough, they have been implemented in an inference engine (CLIPS) and a statistical analysis has been made on tens of thousands of randomly generated puzzles. It shows that they can solve 42% of the minimal puzzles ("minimal" means "has a unique solution and has several solutions if any entry is deleted"; statistics would be meaningless without this condition).

VI. Conclusion

We have proposed a general conceptual framework for approximating a constraint satisfaction problem with constructive resolution theories in which the intuitive notion of a candidate is given a simple and well defined logical status with an underlying epistemic meaning. We have given a detailed illustration of these concepts with the Sudoku CSP. We have explained why a resolution theory, even though it seems to express all the constraints of the CSP, may not be complete.

One may ask: is using a resolution theory more efficient (from a computational point of view) than combining elementary constraints propagation with blind search? In the Sudoku example, our simulations (see Part II) show that the answer is clearly negative; moreover, as there exist very efficient general purpose search algorithms, we think this answer is general. But, instead of setting the focus on computational efficiency, as is generally the case, our approach sets the focus on constructiveness of the solution. Another general question remains open: how "close" can one approximate a CSP with a well chosen resolution theory? We have no general answer. But, in Part II of this paper, we shall define elaborated resolution theories, give a meaning to the word "close" and provide detailed results for the Sudoku CSP.

References

[1] E.P.K. Tsang, *Foundations of Constraint Satisfaction*, Academic Press, 1993.

[2] H.W. Guesgen & J. Herztberg, *A Perspective of Constraint-Based Reasoning*, Lecture Notes in Artificial Intelligence, Springer, 1992.

[3] D. Berthier: *The Hidden Logic of Sudoku*, Lulu Publishers, May 2007.

[4] L. Brownston, R. Farrell & E. Kant, *Programming Expert Systems in OPS5*, Addison-Wesley, 1985.

[5] K. Meinke & J. Tucker: *Many-Sorted Logic and its Applications*, Wiley, 1993.

[6] S. Kripke: Semantical Analysis of Modal Logic, *Zeitchrift für Mathematische Logic und Grundlagen der Matematik*, Vol. 9, pp. 67-96, 1973.

[7] J. Hintikka: *Knowledge and Belief: an Introduction to the Logic of the Two Notions*, Cornell University Press, 1962.

[8] J. Moschovakis: Intuitionistic Logic, *Stanford Encyclopedia of Phylosophy*, 2006.

[9] M.C. Fititng, *Intuitionistic Logic, Model Theory and Forcing*, North Holland, 1969.

[10] D. Bridges & L. Vita: *Techniques of Constructive Analysis*, Springer, 2006.

From Constraints to Resolution Rules
Part II : chains, braids, confluence and T&E

Denis Berthier
Institut Telecom ; Telecom & Management SudParis
9 rue Charles Fourier, 91011 Evry Cedex, France

Abstract: In this Part II, we apply the general theory developed in Part I to a detailed analysis of the Constraint Satisfaction Problem (CSP). We show how specific types of resolution rules can be defined. In particular, we introduce the general notions of a chain and a braid. As in Part I, these notions are illustrated in detail with the Sudoku example - a problem known to be NP-complete and which is therefore typical of a broad class of hard problems. For Sudoku, we also show how far one can go in "approximating" a CSP with a resolution theory and we give an empirical statistical analysis of how the various puzzles, corresponding to different sets of entries, can be classified along a natural scale of complexity. For any CSP, we also prove the confluence property of some Resolution Theories based on braids and we show how it can be used to define different resolution strategies. Finally, we prove that, in any CSP, braids have the same solving capacity as Trial-and-Error (T&E) with no guessing and we comment this result in the Sudoku case.

Keywords: constraint satisfaction problem, knowledge engineering, modelling and simulation, production system, resolution rule, chains, braids, confluence, Trial-and Error, Sudoku solving, Sudoku rating.

I. INTRODUCTION

In Part I of this paper, which is an inescapable pre-requisite to the present Part II, the Constraint Satisfaction Problem (CSP) [1, 2] was analysed in a new general framework based on the idea of a constructive, pattern-based solution and on the concepts of a candidate and a resolution rule. Here we introduce several additional notions valid for any CSP, such as those of a chain, a whip and a braid. We show how these patterns can be the basis for new general and powerful kinds of resolution rules. All of the concepts defined here are straightforward generalisations (and formalisations) of those we introduced in the Sudoku case [3, 4]. Because of space constraints, we formulate our concepts only in plain English but they can easily be formalised with logical formulæ using the basic concepts introduced in Part I.

We give a detailed account of how these general notions can be applied to Sudoku solving. Sudoku is a very interesting problem for several reasons: 1) it is known to be NP-complete [5] (more precisely, the CSP family Sudoku(n) on square grids of size n for all n is NP-complete); 2) nevertheless, it is much easier to study than Chess or Go; 3) a Sudoku grid is a particular case of Latin Squares; Latin Squares are more elegant, from a mathematical point of view, because there is a complete symmetry between all the variables: rows, columns, numbers; in Sudoku, the constraint on blocks introduces some apparently mild complexity which makes it more exciting for players; 4) there are millions of Sudoku players all around the world and many forums, with a lot of cumulated experience available – including generators of random puzzles. For all these reasons, we chose the Sudoku example instead of the more "mathematically correct" Latin Squares CSP.

Whereas sections II and III define the general chains and the elementary bivalue chains, sections IV and V introduce three powerful generalisations of bivalue chains: zt-chains, zt-whips and zt-braids. Section VI defines the very important property of confluence and the notion of a resolution strategy; it proves the confluence property of natural braid resolution theories. Finally, section VII proves that braids have the same solving potential as Trial-and-Error with no guessing.

II. CHAINS IN A GENERAL CSP

Definition: two different candidates of a CSP are *linked* by a direct contradiction (or simply linked) if some of the constraints of the CSP directly prevents them from being true at the same time *in any knowledge state* in which they are present (the fact that this notion does not depend on the knowledge state is fundamental for the sequel). For any CSP, two different candidates for the same variable are always linked; but there are generally additional direct contradictions; as expliciting them is part of modelling the CSP, we consider them as givens of the CSP and we introduce a basic predicate "$\text{linked}_{ij}(x_i, x_j)$" to express them, for each couple of CSP variables X_i and X_j. In Sudoku, two different candidates $n_1r_1c_1$ and $n_2r_2c_2$ are linked and we write linked($n_1r_1c_1$, $n_2r_2c_2$), if:

$(n_1 \neq n_2$ & $r_1c_1 = r_2c_2)$ or $(n_1 = n_2$ & share-a-unit$(r_1c_1, r_2c_2))$.

Definition: an *Elementary Constraint Propagation* rule is a resolution rule expressing such a direct contradiction. For any CSP, we note ECP the set of all its elementary constraints propagation rules. An ECP rule has the general form:

$\forall x_i \forall x_j$ $\text{value}_i(x_i)$ & $\text{linked}_{ij}(x_i, x_j) \Rightarrow \neg\text{cand}_j (x_j)$.

Chains (together with whips and braids) appear to be the main tool for dealing with hard instances of a CSP.

K. Elleithy (ed.), *Advanced Techniques in Computing Sciences and Software Engineering*,
DOI 10.1007/978-90-481-3660-5_29,

Definitions: *a chain of length n* is a sequence L_1, R_1, L_2, R_2, … L_n, R_n, of 2n *different* candidates for possibly different variables such that: for any $1 \le k \le n$, R_k is linked to L_k and for any $1 \le k \le n$, L_k is linked to R_{k-1}. A *target of a chain* is any candidate that is linked to both its first and its last candidates.

Of course, these conditions are not enough to ensure the existence of an associated resolution rule concluding that the target can be eliminated. Our goal is now to define more specific types of chains allowing such a conclusion.

III. Bivalue-chains in a general CSP

A. Bivalue-chains in a general CSP

Definition: a variable is called *bivalue* in a knowledge state KS if it has exactly two candidates in KS.

Definition and notation: in any CSP, *a bivalue-chain of length n* is a chain of length n: L_1, R_1, L_2, R_2, …. L_n, R_n, such that, additionally: for any $1 \le k \le n$, L_k and R_k are candidates for the same variable, and this variable is bivalue. A bivalue-chain is written symbolicaly as: $\{L_1\ R_1\}$ - $\{L_2\ R_2\}$ - …. - $\{L_n\ R_n\}$, where the curly braces recall that the two candidates are relative to the same variable.

bivalue-chain rule for a general CSP: in any knowledge state of any CSP, if Z is a target of a bivalue-chain, then it can be eliminated (formally, this rule concludes ¬Z).

Proof: the proof is short and obvious but it will be the basis for all our forthcoming chain and braid rules.

If Z was true, then L_1 would be false; therefore R_1 would have to be the true value of the first variable; but then L_2 would be an impossible value for the second variable and R_2 would be its true value….; finally R_n would be true in the last cell; which contradicts Z being true. Therefore Z can only be false. qed.

B. xy-chains in Sudoku

We shall adopt the following definitions [1]. Two different rc-cells are linked if they share a unit (i.e. they are in the same row, column or block). A *bivalue cell* is an rc-cell in which there are exactly two candidates (here considered as numbers in these cells). An *xy-chain of length n* is a sequence of n different bivalue rc-cells (each represented by a set notation: {… }) such that each (but the first) is linked to the previous one (represented by a "-") , with contents: $\{a_1\ a_2\}$ - $\{a_2\ a_3\}$ - …… $\{a_n\ a_1\}$. A *target* of the above xy-chain is a number a_1 in a cell that is linked to the first and last ones. xy-chains are the most classical and basic type of chains in Sudoku. Our presentation is non standard, but equivalent to the usual ones [6, 7].

Classical xy-chain rule in Sudoku: if Z is a target of an xy-chain, then it can be eliminated.

C. nrc-chains in Sudoku

The above definition of an xy-chain in Sudoku is the traditional one and it corresponds to the general notion of a bivalue-chain in any CSP, when we consider only the natural variables X_{rc} and X_{bs} of the Sudoku CSP. But it is not as general as it could be. To get the most general definition, we must consider not only the "natural" X_{rc} variables but also the corresponding X_{rn}, X_{cn} and X_{bn} variables, as introduced in Part I, with $X_{rc} = n \Leftrightarrow X_{rn} = c \Leftrightarrow X_{cn} = r \Leftrightarrow X_{bn} = s$, whenever correspondence(r, c, b, s) is true. The notion of bivalue is meaningful for each of these variables. And, when we use all these variables instead of only the X_{rc}, we get a more general concept of bivalue-chains, which we called nrc-chains in [4] and which are a different view of some classical Nice Loops [6, 7]. The notion of "bivalue" for these non-standard variables corresponds to the classical notion of conjugacy in Sudoku – but, from the point of view of the general theory, there is no reason to make any difference between "bivalue" and "conjugate". In the sequel, we suppose that we use all the above variables.

Classical nrc-chain rule in Sudoku: any target of an nrc-chain can be eliminated.

IV. The z- and t- extensions of bivalue-chains in a CSP

We first introduced the following generalisations of bivalue-chains in [3], in the Sudoku context. But everything works similarly for any CSP. It is convenient to say that a candidate C is *compatible* with a set S of candidates if it is not linked to any element of S.

A. t-chains, z-whips and zt-whips in a general CSP

The definition of a bivalue-chain can be extended in different ways, as follows.

Definition: a *t-chain* of length n is a chain L_1, R_1, L_2, R_2, …. L_n, R_n, such that, additionally, for each $1 \le k \le n$:

- L_k and R_k are candidates for the same variable,
- R_k is the only candidate for this variable compatible with the previous right-linking candidates.

t-chain rule for a general CSP: in any knowledge state of any CSP, any target of a t-chain can be eliminated (formally, this rule concludes ¬Z).

For the z- extension, it is natural to introduce *whips* instead of chains. Whips are also more general, because they are able to catch more contradictions than chains. A *target of a whip* is required to be linked to its first candidate, not necessarily to its last.

Definition: given a candidate Z (which will be the target), a *z-whip* of length n built on Z is a chain L_1, R_1, L_2, R_2, …, L_n (notice that there is no R_n), such that, additionally:

– for each $1 \leq k < n$, L_k and R_k are candidates for the same variable,

– R_k is the only candidate for this variable compatible with Z (apart possibly for L_k),

– for the same variable as L_n, there is no candidate compatible with the target.

Definition: given a candidate Z (which will be the target), a *zt-whip* of length n built on Z is a chain L_1, R_1, L_2, R_2, ... L_n (notice that there is no R_n), such that, additionally:

– for each $1 \leq k < n$, L_k and R_k are candidates for the same variable,

– R_k is the only candidate for this variable compatible with Z and the previous right-linking candidates,

– for the same variable as L_n, there is no candidate compatible with the target and the previous right-linking candidates.

z- and zt-whip rules for a general CSP: in any knowledge state of any CSP, if Z is a target of a z- or a zt- whip, then it can be eliminated (formally, this rule concludes ¬Z).

Proof: the proof can be copied from that for the bivalue-chains. Only the end is slightly different. When variable L_n is reached, it has negative valence. With the last condition on the whip, it entails that, if the target was true, there would be no possible value for the last variable.

Remark: although these new chains or whips seem to be straightforward generalisations of bivalue-chains, their solving potential is much higher. Soon, we'll illustrate this with the Sudoku example.

Definition: in any of the above chains or whips, a value of the variable corresponding to candidate L_k is called a t- (resp. z-) candidate if it is incompatible with the previous right-linking (i.e. the R_i) candidates (resp. with the target).

B. zt-whip resolution theories in a general CSP

We are now in a position to define an increasing sequence of resolution theories based on zt-whips: BRT is the Basic Resolution Theory defined in Part I. L_1 is the union of BRT and the rule for zt-whips of length 1. For any n, L_{n+1} is the union of L_n with the rule for zt-whips of length n+1. L_∞ is also defined, as the union of all the L_n. In practice, as we have a finite number of variables in finite domains, L_∞ will be equal to some L_n.

C. t-whips, z-whips and zt-whips in Sudoku

In Sudoku, depending on whether we consider only the "natural" X_{rc} and X_{bs} variables or also the corresponding X_{rn}, X_{cn} and X_{bn} variables, we get xyt-, xyz- and xyzt- whips or nrct-, nrcz- and nrczt- whips. In the Sudoku case, we have programmed all the above defined rules for whips in our SudoRules solver, a knowledge based system, running indifferently on the CLIPS [8] or the JESS [9] inference engine.

This allowed us to obtain the following statiscal results.

D. Statistical results for the Sudoku nrczt-whips

Definition: a puzzle is *minimal* if it has one and only one solution and it would have several solutions if any of its entries was deleted. In statistical analyses, only samples of minimal puzzles are meaningful because adding extra entries would multiply the number of easy puzzles. In general, puzzles proposed to players are minimal.

One advantage of taking Sudoku as our standard example (instead of e.g. Latin Squares) is that there are generators of random minimal puzzles. Before giving our results, it is necessary to mention that there are puzzles of extremely different complexities. With respect to several natural measures of complexity one can use (number of partial chains met in the solution, computation time, ...), provided that they are based on resolution rules (instead of e.g. blind search with backtracking), different puzzles will be rated in a range of several orders of magnitude (beyond 13 orders in Sudoku).

The following statistics are relative to a sample of 10,000 puzzles obtained with the suexg [10] random generator. Row 3 of Table 1 gives the total number of puzzles solved when whips of length ≤ n (corresponding to resolution theory L_n) are allowed; row 2 gives the difference between L_n and L_{n-1}. (Of course, in any L_n, the rules of BSRT, consisting of ECP, NS, HS and CD are allowed in addition to whips).

BSRT	L1	L2	L3	L4	L5	L6	L7
4247	1135	1408	1659	1241	239	56	10
4247	5382	6790	8449	9690	9929	9985	9995

Table 1: Number of puzzles solved with nrczt-whips of length ≤ n. The 5 remaining puzzles can also be solved with whips, although longer ones.

As these results are obtained from a very large random sample, they show that almost all the minimal puzzles can be solved with nrczt-whips. But they don't allow to conclude for all the puzzles. Indeed, extremely rare cases are known which are not solvable with nrczt-whips only. They are currently the puzzles of interest for researchers in Sudoku solving. But, for the Sudoku player, they are very likely to be beyond his reach, unless radically new types of rules are devised.

V. ZT-BRAIDS IN A GENERAL CSP

We now introduce a further generalisation of whips: braids. Whereas whips have a linear structure (a chain structure), braids have a (restricted) net structure. In any CSP, braids are interesting for three reasons: 1) they have a greater solving potential than whips (at the cost of a more complex structure); 2) resolution theories based on them can be proven to have the very important confluence property, allowing to introduce various resolution strategies based on them; and 3) their scope can be defined very precisely; they can eliminate any candidate that can be eliminated by pure Trial-and-Error (T&E); they can therefore solve any puzzle that can be solved by T&E.

A. Definition of zt-braids

Definition: given a target Z, a *zt-braid* of length n built on Z is a sequence of different candidates L_1, R_1, L_2, R_2, …. L_n (notice that there is no R_n), such that:

– for each $1 \leq k \leq n$, L_k is linked either to a previous right-linking candidate (some R_l, $l < k$) or to the target (this is the main structural difference with whips),

– for each $1 \leq k < n$, L_k and R_k are candidates for the same variable (they are therefore linked),

– R_k is the only candidate for this variable compatible with the target and the previous right-linking candidates,

– for the variable corresponding to candidate L_n, there is no candidate compatible with the target and the previous right-linking candidates.

In order to show the kind of restriction this definition entails, the first of the following two structures can be part of a braid starting with $\{L_1\ R_1\}$ - $\{L_2\ R_2\}$ -… , whereas the second can't:

$\{L_1\ R_1\}$ - $\{L_2\ R_2\ A_2\}$ - … where A_2 is linked to R_1;

$\{L_1\ R_1\ A_1\}$ - $\{L_2\ R_2\ A_2\}$ - … where A_1 is linked to R_2 and A_2 is linked to R_1 but none of them is linked to Z. The only thing that could be concluded from this pattern if Z was true is (R_1 & R_2) or (A_1 &A_2), whereas a braid should allow to conclude R_1 & R_2.

The proof of the following theorem is exactly the same as for whips, thanks to the linear order of the candidates.

zt-braid rule for a general CSP: in any knowledge state of any CSP, if Z is a target of a zt-braid, then it can be eliminated (formally, this rule concludes ¬Z).

Braids are a true generalisation of whips. Even in the Sudoku case (for which whips solve almost any puzzle), examples can be given of puzzles that can be solved with braids but not with whips. This will be a consequence of our T&E vs braid theorem.

VI. Confluence Property, Braids, Resolution Strategies

A. The confluence property

Given a resolution theory T, consider all the strategies that can be built on it, e.g. by defining various priorities on the rules in T. Given an instance P of the CSP and starting from the corresponding knowledge state KS_P, the resolution process associated with a stategy S built on T consists of repeatedly applying resolution rules from T according to the additional conditions (e.g. the priorities) introduced by S. Considering that, at any point in the resolution process, different rules from T may be applicable (and different rules will be applied) depending on the chosen stategy S, we may obtain different resolution paths starting from KS_P when we vary S.

Let us define the *confluence property* as follows: a Resolution Theory T for a CSP has the confluence property if, for any instance P of the CSP, any two resolution paths can be extended to meet in a common knowledge sate. In this case, all the resolution paths starting from KS_P and associated with all the stategies built on T will lead to the same final state in $\boldsymbol{KS_P}$ (all explicitly inconsistent states are considered as identical; they mean contradictory constraints). If a resolution theory T doesn't have the confluence property, one must be careful about the order in which he applies the rules. But if T has this property, one may choose any resolution strategy, which makes finding a solution much easier.

B. The confluence property of zt-braid resolution theories

As for whips, one can define an increasing sequence of resolution theories based on zt-braids: M_1 is the union of BRT and the rule for zt-braids of length 1. (Notice that $M_1 = L_1$). For any n, M_{n+1} is the union of M_n with the rule for zt-braids of length n+1. M_∞ is defined as the union of all the M_n.

Theorem: any of the above zt-braid theories has the confluence property.

Before proving this theorem, we must give a precison about candidates. When one is asserted, its status changes: it becomes a value and it is deleted as a candidate. (The theorem doesn't depend on this but the proof should have to be slightly modified with other conventions).

Let n be fixed. What our proof will show is the following much stronger stability property: for any knowledge state KS, any elimination of a candidate Z that might have been done in KS by a zt-braid B of length n and target Z will always be possible in any further knowledge state (in which Z is still a candidate) using rules from M_n (i.e. for zt-braids of length n or less, together with BRT). For this, we must consider all that can happen to B. Let B be:

$\{L_1\ R_1\}$ - $\{L_2\ R_2\}$ - …. - $\{L_p\ R_p\}$ - $\{L_{p+1}\ R_{p+1}\}$ - … - L_n.

If the target Z is eliminated, then our job is done. If Z is asserted, then the instance of the CSP is contradictory. This contradiction will be detected by CD after a series of ECP and S following the braid structure.

If a right-linking candidate, say R_p, is eliminated, the corresponding variable has no possible value and we get the shorter braid with target Z: $\{L_1\ R_1\}$ - $\{L_2\ R_2\}$ - …. - L_p. If a left-linking candidate, say L_{p+1}, is asserted, then R_p can be eliminated by ECP, and we are in the previous case.

If a right-linking candidate, say R_p, is asserted, it can no longer be used as an element of a braid. Notice that L_{p+1} and all the t-candidates in cells of B after p that were incompatible with R_p, i.e. linked to it, can be eliminated by ECP. Let q be the smallest number greater than p such that, after all these eliminations, cell number q still has a t- or a z- candidate C_q; notice that the right-linking candidates in all the cells between p and q-1 can be asserted by S, all the t-candidates in cells after q that were incompatible with either of them can be eliminated by ECP and all the left-linking candidates in all the cells between p and q can be eliminated by ECP. Let k be the largest number k ≤ p such that C_q is incompatible with R_k (or q = 0 if C is incompatible only with Z). Then the shorter braid obtained from B by excising cells p+1 to q and by replacing L_q by C_q still has Z has its target and can be used to eliminate it.

Suppose now a left-linking candidate, say L_p, is eliminated. Either {L_p R_p} was bivalue, in which case R_p can be asserted by S and we are in the previous case. Or there remains some t- or z-candidate C for this variable and we can consider the braid, with target Z, obtained by replacing L_p by C. Notice that, even if L_p was linked to R_{p-1}, this may not be the case for C; therefore trying to prove a similar theorem for whips would fail here.

If any t- or z- candidate is eliminated, then the basic structure of B is unchanged. If any t- or z- candidate is asserted as a value, then the right-linking candidate of its cell can be eliminated by ECP and we are in one of the previous cases.

As all the cases have been considered, the proof can be iterated in case several of these events have happened to B. Notice that this proof works only because the notion of being linked doesn't depend on the knowledge state.

C. Resolution strategies

There are the Resolution Theories defined above and there are the many ways one can use them in practice to solve real instances of a CSP. From a strict logical standpoint, all the rules in a Resolution Theory are on an equal footing, which leaves no possibility of ordering them. But, when it comes to the practical exploitation of resolution theories and in particular to their implementation, e.g. in an inference engine as in our SudoRules solver, one question remains unanswered: can superimposing some ordering on the set of rules (using priorities or "saliences") prevent us from reaching a solution that the choice of another ordering might have made accessible? With resolution theories that have the confluence property such problems cannot appear and one can take advantage of this to define different resolution strategies.

Resolution strategies based on a resolution theory T can be defined in different ways and may correspond to different goals:

- implementation efficiency;
- giving a preference to some patterns over other ones: preference for chains over zt-whips and/or for whips over braids;
- allowing the use of heuristics, such as focusing the search on the elimination of some candidates (e.g. because they correspond to a bivalue variable or because they seem to be the key for further eliminations); but good heursitics are hard to define.

VII. Braids vs Trial-and-Error in a general CSP

A. Definition of the Trial and Error procedure (T&E)

Definition: given a resolution theory T, a knowledge state KS and a candidate Z, *Trial and Error based on T for Z, T&E(T, Z)*, is the following procedure (notice: a procedure, not a resolution rule): make a copy KS' of KS; in KS', delete Z as a candidate and assert it as a value; in KS', apply repeatedly all the rules in T until quiescence; if a contradiction is obtained in KS', then delete Z from KS; otherwise, do nothing.

Given a fixed resolution theory T and any instance P of a CSP, one can try to solve it using only T&E(T). We say that P can be solved by T&E(T) if, using the rules in T any time they can be applied plus the procedure T&E(T, Z) for some remaining candidate Z every time no rule from T can be applied, a solution of P can be obtained. When T is the BRT of our CSP, we simply write T&E instead of T&E(T).

As using T&E leads to examining arbitrary hypotheses, it is often considered as blind search. But notice nevertheless that it includes no "guessing": if a solution is obtained in an auxiliary state KS', then it is not taken into account, as it would in standard structured search algorithms.

B. zt-braids versus T&E theorem

It is obvious that any elimination that can be made by a zt-braid can be made by T&E. The converse is more interesting.

Theorem: for any instance of any CSP, any elimination that can be made by T&E can be made by a zt-braid. Any instance of a CSP that can be solved by T&E can be solved by zt-braids.

Proof: Let Z be a candidate eliminated by T&E using some auxiliary knowledge state KS'. Following the steps of T&E in KS', we progressively build a zt-braid in KS with target Z. First, remember that BRT contains three types of rules: ECP (which eliminates candidates), S_k (which asserts a value for the k-th variable of the CSP) and CD_k (which detects a contradiction on variable X_k). Consider the first step of T&E which is the application of some S_k in KS', thus asserting some R_1. As R_1 was not in KS, there must have been some elimination of a candidate, say L_1, made possible in KS' by the assertion of Z, which in turn made the assertion of R_1 possible in KS'. But if L_1 has been eliminated in KS', it can only be by ECP and because it is linked to Z. Then {L_1 R_1} is the first cell of our zt-braid in KS. (Notice that there may be other z-candidates in cell {L_1 R_1}, but this is pointless, we can choose any of them as L_1 and consider the remaining ones as z-candidates). The sequel is done by recursion. Suppose we have built a zt-braid in KS corresponding to the part of the T&E procedure in KS' until its n-th assertion step. Let R_{n+1} be the next candidate asserted in KS'. As R_{n+1} was not asserted in KS, there must have been some elimination in KS' of a candidate, say L_{n+1}, made possible by the assertion in KS' of Z or of some of the previous R_k, which in turn made the assertion of R_{n+1} possible in KS'. But if L_{n+1} has been eliminated in KS', it can only be by ECP and because it is linked to Z or to some of the previous R_k, say C. Then our partial braid in KS can be extended with cell $\{L_{n+1}\ R_{n+1}\}$, with L_{n+1} linked to C.

End of the procedure: either no contradiction is obtained by T&E and we don't have to care about any braid in KS, or a contradiction is obtained. As only ECP can eliminate a candidate, a contradiction is obtained when the last asserted value, say R_{n-1}, eliminates (via ECP) a candidate, say L_n, which was the last one for the corresponding variable. L_n is thus the last candidate of the braid in KS we were looking for.

Here again, notice that this proof works only because the existence of a link between two candidates doesn't depend on the knowledge state.

C. Comments on the braids vs T&E theorem

T&E is a form of blind search that is generally not accepted by advocates of pattern-based solutions (even when it allows no guessing, as in our definition of this procedure). But this theorem shows that T&E can always be replaced with a pattern based solution, more precisely with braids. The question naturally arises: can one reject T&E and nevertheless accept solutions based on braids?

As shown in section VI, resolution theories based on braids have the confluence property and many different resolution strategies can be super-imposed on them. One can decide to prefer a solution with the shorter braids available. T&E doesn't provide this (unless it is drastically modified, in ways that would make it computationally very inefficient).

Moreover, in each of these resolution theories based on braids, one can add rules corresponding to special cases, such as whips of the same lengths, and one can decide to give a natural preference to such special cases. In Sudoku, this would entail that braids which are not whips would appear in the solution of almost no random puzzle.

D. The resolution potential of zt-braids in Sudoku

For any CSP, the T&E vs braids theorem gives a clear theoretical answer to the question about the potential of resolution theories based on zt-braids. As the T&E procedure is very easy to implement, it also allows practical computations. We have done this for Sudoku.

We have generated 1,000,000 minimal puzzles: all of them can be solved by T&E and therefore by nrczt-braids. We already knew that nrczt-whips were enough to solve the first 10,000; checking the same thing for 1,000,000 puzzles would be too long; but it becomes easy if we consider braids instead of whips.

One should not conclude that zt-braids are a useless extension of zt-whips. We have shown that there are puzzles that cannot be solved with whips only but can be solved with braids. Said otherwise, whips are not equivalent to T&E.

VIII. Conclusion

Most of the general CSP solving methods [7, 8] combine a blind search algorithm with some kind of pattern-based pruning of the search graph. Here, instead of trying to solve all the instances of a CSP, as is generally the case in these methods, we have tried to push the purely pattern-based approach to its limits. In Part I, we have defined a general framework for this purpose and in Part II, we have introduced three powerful patterns, bivalue-chains, zt-whips and zt-braids. We have shown that, for any CSP, zt-braids are able to replace one level of Trial-and-Error.

We have applied this framework to the Sudoku CSP and shown that whips (resp. braids) can solve all the puzzles taken from a random sample of 10,000 (resp. 1,000,000). Nevertheless a few puzzles are known to defy both of these patterns.

References

[1] E.P.K. Tsang, *Foundations of Constraint Satisfaction*, Academic Press, 1993.

[2] H.W. Guesgen & J. Hertzberg, *A Perspective of Constraint-Based Reasoning*, Lecture Notes in Artificial Intelligence, Springer, 1992.

[3] D. Berthier: *The Hidden Logic of Sudoku*, Lulu.com Publishers, May 2007.

[4] D. Berthier: *The Hidden Logic of Sudoku (Second Edition)*, Lulu.com Publishers, December 2007.

[5] M. Gary & D. Johnson, *Computers and Intractability: A Guide to the Theory of NP-Completeness*, Freeman, 1979.

[6] Sadman software, http://www.sadmansoftware.com/sudoku /techniques. htm, 2000-2008.

[7] R. van der Werf, *Sudocue Solving Guide*, http://www.sudocue.net/ guide.php, 2005-2007

[8] G. Riley, *CLIPS online documentation*, http://clipsrules.sourceforge.net/ OnlineDocs.html, 2008.

[9] E.J. Friedmann-Hill, *JESS Manual*, http://www.jessrules.com/jess/docs/ 71, 2008.

[10] Sterten/dukuso, *suexg*, http://www.setbb.com/phpbb/viewtopic.php?t= 206 &mforum=sudoku, 2005.

Platform Independent Unit Tests Generator

Šarūnas Packevičius*, Andrej Ušaniov*, Eduardas Bareiša*
*Department of Software Engineering, Kaunas University of Technology
{packsaru, andrej.usaniov, edas}@soften.ktu.lt

Unit tests are viewed as a coding result of software developers. These unit tests are usually created by developers and implemented directly using specific language and unit testing framework. The existing unit test generation tools usually do the same thing – generate tests for specific language using a specific unit testing framework. Thus such a generator is suitable for only one programming language and unit testing framework. Another drawback of these generators – they use the software code as a source for generation mainly.

In this paper we present a tests generator model which could be able to generate unit tests for any language using any unit testing framework. It will be able to use not only software under test code, but the other artifacts, too: models, specifications.

1. INTRODUCTION

Software testing automation is seen as a mean for reducing software construction costs by eliminating or reducing manual software testing phase. Software tests generators are used for software testing automation. Software tests generators have to fulfill such goals:

1. Create repeatable tests.
2. Create tests for bug detection.
3. Create self-checking tests.

Current unit tests generators (for example Parasoft JTest) fulfill only some of these goals. They usually generate repeatable tests, but a tester has to specify the tests oracle (a generator generates test inputs, but provides no way to verify if test execution outcomes are correct) – tests are not self-checking. Tests generators usually target only one programming language and are able to generate tests using only one specific unit testing framework. These generators usually use a source code of software under test as an input for tests generation. For example, Parasoft JTest tests generator generates tests for software which are implemented using Java and unit tests use JUnit testing framework only.

2. RELATED WORKS

A. Krass et al. [4] proposed a way to generate tests from UML models. They were using UML models as an input for the tests generator and have generated tests which are stored as XML documents. These XML documents could be transformed later into a unit test code using a specific unit testing framework (for example, JUnit [5] or TTCN-3). The drawback of this approach is that the tests generator abstracts the generated tests from test frameworks only, and is able to generate tests using only UML models as an input.

Other authors have proposed generators which take UML models directly [2, 3] or the program code in specific programming language [1, 9] and generate the testing code using a specific unit testing framework directly. Thus their generator can produce tests for one specific unit testing framework only.

In this paper, we propose a modified unit tests generator model which will allow generating a test in any programming language using any unit testing framework and will be able to use software under test implemented or modeled in any language.

The remaining part of this paper is organized as follows: The tests generator architecture is presented in Chapter 3. A generator example is presented in Chapter 4. Finally conclusions and the future work are given in Section 5.

3. PLATFORM INDEPENDENT UNIT TESTS GENERATOR

We are proposing a model of the unit tests generator which would be able to generate tests using any unit testing framework and will take not only software's under test code as an input, but also can use other artifacts as an input for tests generation. Artifacts could be: UML models, OCL constraints, Business Rules. The generators model is based on Model Driven Engineering (MDE) ideas [8].

3.1. *Tests Generation Using MDE Ideas*

The MDE idea is based on the fact that a developer does not write the software code. He or she only models it. Models are transformed to software implementation later using any chosen programming language, platform and/or framework. MDE targets implementation code development. But its principles can be used for tests generation also. Tests could be created not only as a code using a selected unit testing framework; they can be created as tests models and transformed later into any selected programming language and unit testing framework. For example, we could model tests as UML diagrams and OCL constraints and transform the modeled tests later into the test code which uses JUnit testing framework; the code is generated in Java programming language.

Such tests generator can be roughly depicted in Figure 1.

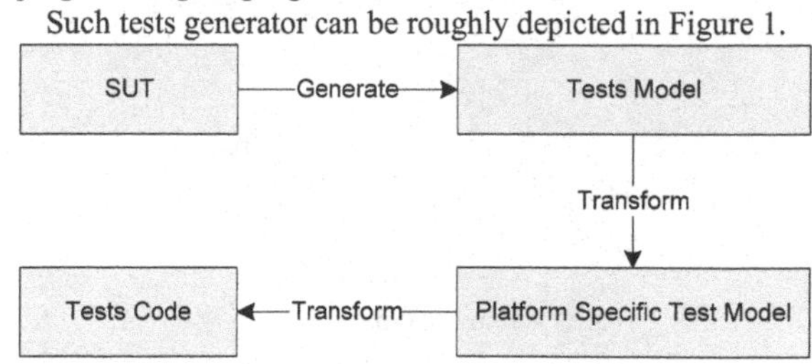

Figure 1 : Tests generator model

The tests generator takes software under test (SUT - software under test in Figure 1) as an input and generates tests which are represented as a model. For example, the UML diagram using Testing profile stereotype could be such a model. After that the tests model is transformed into the

K. Elleithy (ed.), *Advanced Techniques in Computing Sciences and Software Engineering*,
DOI 10.1007/978-90-481-3660-5_30,

platform specific test model. The platform specific test model represents tests as a model also, but this model uses platform specific elements. For example, test classes extend the required unit testing framework base classes, implement the required methods, sequence diagrams show calls to specific unit testing framework methods. The final phase is to transform the platform specific model into tests implementation (test code). This transformation could be no more different than the ones used today for code generation from a model (usually such transformations are used in many UML diagramming tools).

The benefit of this generator is that tests can be transformed to any selected unit testing framework and implementation language. We just have to select a different transformation if we want to generate tests for another programming language or/and unit testing framework.

3.2. *Software Under Test Meta-Model*

The tests generator does not take software under test code or model as tests generation input directly. The tests generator transforms SUT into a software model firstly (Figure 2). That model represents software independently from its implementation language or modeling language. For example, if SUT is implemented using Java programming language, the generator does reverse code engineering. If the program is modeled using OMT notation its model is converted into our model (Similar to UML).

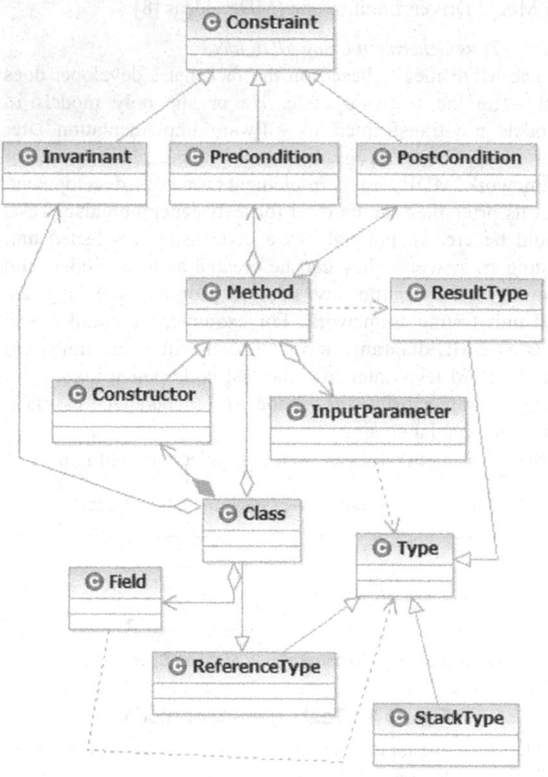

Figure 2 : SUT Meta-model

Figure 2 presents a meta-model of software under test. This model is similar to UML static structure meta-model. We have extended UML meta-model by adding the elements "Constraints" into meta-model. This addition allows us to store software under test model, code, OCL constrains or any combination of them into one generic model.

SUT meta-model has all information about software under test; in order to generate tests it contains information about classes, methods, fields in this software under test. If software was modeled using OCL language, the meta-model links OCL constraints to associated classes, methods, fields, method parameters.

3.3. *Tests Meta-Model*

The tests meta-model represents the generated tests in an abstracted from implementation form. This meta-model is presented in Figure 3.

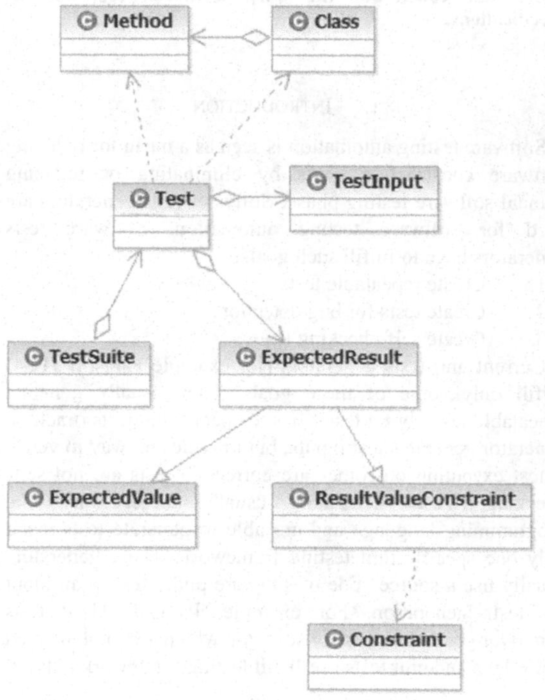

Figure 3 : Tests Meta-model

This model stores unit tests. The meta-model has classes to store the generated input values; it references classes, methods, fields from software under test meta-model. It carries the expected test execution values and/or references to OCL constraints which are used as a test oracle [7]. When implementing our unit tests generator we can store the generated tests in a database. And then the required tests from database can be transformed into a test code using the selected programming language and unit testing framework. Transformations into the test code can be performed using usual code generation techniques.

4. GENERATOR EXAMPLE

We have an implementation of the unit tests generator based on our idea. Its source code is available at http://atf.sourceforge.net/ web page. The generator's structure is represented in Figure 4.

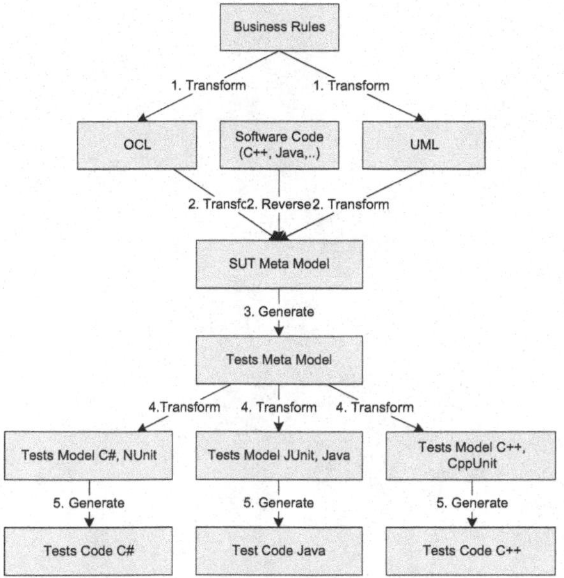

Figure 4 : Tests generator structure

Our generator is able to generate tests for software which is modeled in UML and/or has OCL constraints, requirements expressed as Business Rules (We have presented how to transform Business Rules into UML model and OCL constraints for this generator [6]).

The tests generation procedure if we have a software model is such:

1. Transform the software model into its SUT model. (Step 2)
2. Generate tests using the software SUT model and store tests as a tests model. (Step 3)
3. Generate the test code for the selected unit testing framework using the tests model. (Steps 4, 5)

If software under test is represented as a code, the test generating procedure is such:

1. Reverse the code into the SUT model. (Step 2)
2. Generate tests using the software SUT model and store the tests as a tests model. (Step 3)
3. Generate the test code for the selected unit testing framework using the tests model. (Steps 4, 5)

If we have software specified as Business Rules, the tests generation procedure is:

1. Transform Business Rules into UML, OCL models. (Step 1)
2. Transform the software model into its SUT model. (Step 2)
3. Generate tests using the software SUT model and store the tests as a tests model. (Step 3)
4. Generate the test code for the selected testing framework using the tests model. (Steps 4, 5)

The generator has to generate tests using always the same type of data and has to produce the same type of results (Step 3). All we have to do is just to implement different transformations, if we want to target another unit testing framework or we want to generate tests for programs which are implemented in other programming languages or modeled using other modeling languages.

5. CONCLUSION

The suggested abstract unit tests generator is able to generate tests for software implemented in any programming language and/or modeled using any modeling language. It can generate the test code in any programming language using any unit testing framework.

The tests generator mechanism is independent from the unit testing framework and software under test implementation or modeling language.

This generator model can be used for benchmarking tests generators when we have benchmarks implemented in various programming languages.

REFERENCES

[1] Chandrasekhar, B., K. Sarfraz, and M. Darko, Korat: automated testing based on Java predicates, in Proceedings of the 2002 ACM SIGSOFT international symposium on Software testing and analysis. 2002, ACM Press: Roma, Italy.

[2] Kim, S.K., L. Wildman, and R. Duke. A UML approach to the generation of test sequences for Java-based concurrent systems. In 2005 Australian Software Engineering Conference (ASWEC'05) 2005.

[3] Kim, Y.G., H.S. Hong, D.H. Bae, and S.D. Cha, Test cases generation from UML state diagrams. IEE Proceedings on Software Engineering, 1999. **146**(4): p. 187-192.

[4] Kraas, A., M. Kruger, P. Aschenbrenner, M. Hudler, and Y. Lei. A Generic toolchain for model-based test generation and selection. In TESTCOM / FATES 2007
2007. Tallinn, Estonia.

[5] Louridas, P., JUnit: unit testing and coding in tandem. Software, IEEE, 2005. **22**(4): p. 12-15.

[6] Packevičius Š., A. Ušaniov, and E. Bareiša. Creating unit tests using business rules. In Information Technologies' 2008: 14th International Conference on Information and Software Technologies. 2008. Kaunas, Lithuania.

[7] Packevičius Š., A. Ušaniov, and E. Bareiša, Using Models Constraints as Imprecise Software Test Oracles. Information Technology and Control, 2007..

[8] Schmidt, D.C., Guest Editor's Introduction: Model-Driven Engineering. Computer, 2006. **39**(2): p. 25-31.

[9] Willem, V., S.P. Corina, S. Reanu, and K. Sarfraz, Test input generation with java PathFinder, in Proceedings of the 2004 ACM SIGSOFT international symposium on Software testing and analysis. 2004, ACM Press: Boston, Massachusetts, USA.

Fuzzy Document Clustering Approach using WordNet Lexical Categories

Tarek F. Gharib
Faculty of Computer and Information Sciences,
Ain Shams University
Cairo, Egypt

Mohammed M. Fouad
Akhbar El-Yom Academy
Cairo, Egypt

Mostafa M. Aref
Faculty of Computer and Information Sciences,
Ain Shams University
Cairo, Egypt

Abstract- **Text mining refers generally to the process of extracting interesting information and knowledge from unstructured text. This area is growing rapidly mainly because of the strong need for analysing the huge and large amount of textual data that reside on internal file systems and the Web. Text document clustering provides an effective navigation mechanism to organize this large amount of data by grouping their documents into a small number of meaningful classes. In this paper we proposed a fuzzy text document clustering approach using WordNet lexical categories and Fuzzy *c*-Means algorithm. Some experiments are performed to compare efficiency of the proposed approach with the recently reported approaches. Experimental results show that Fuzzy clustering leads to great performance results. Fuzzy c-means algorithm overcomes other classical clustering algorithms like k-means and bisecting k-means in both clustering quality and running time efficiency.**

I. Introduction

With the growth of World Wide Web and information society, more information is available and accessible. The main problem is how to find the truly relevant data among these huge and large data sources. Most of the applications that querying a search engine obtained a large number of irrelevant results and a small number of relevant pages that meet the keyword typed by the user. Text document clustering can be used here to solve this problem by organizing this large amount of retrieval results.

Text document clustering provides an effective navigation mechanism to organize this large amount of data by grouping their documents into a small number of meaningful classes. Text document clustering can be defined as the process of grouping of text documents into semantically related groups [16]. Most of the current methods for text clustering are based on the similarity between the text sources. The similarity measures work on the syntactically relationships between these sources and neglect the semantic information in them. By using the vector-space model in which each document is represented as a vector or 'bag of words', i.e., by the words (terms) it contains and their weights regardless of their order [3].

Many well-known methods of text clustering have two problems: first, they don't consider semantically related words/terms (e.g., synonyms or hyper/hyponyms) in the document. For instance, they treat {Vehicle, Car, and Automobile} as different terms even though all these words have very similar meaning. This problem may lead to a very low relevance score for relevant documents because the documents do not always contain the same forms of words/terms.

Second, on vector representations of documents based on the bag-of-words model, text clustering methods tend to use all the words/terms in the documents after removing the stop-words. This leads to thousands of dimensions in the vector representation of documents; this is called the "Curse of Dimensionality". However, it is well known that only a very small number of words/terms in documents have distinguishable power on clustering documents [19] and become the key elements of text summaries. Those words/terms are normally the concepts in the domain related to the documents.

Recent studies for fuzzy clustering algorithms [4, 21, 22, 23] have proposed a new approach for using fuzzy clustering algorithms in document clustering process. But these studies neglect the lexical information that can be extracted from text as we used WordNet lexical categories in our proposed approach.

In this paper, we propose fuzzy text document clustering approach using WordNet lexical categories and Fuzzy c-Means algorithm. The proposed approach uses WordNet words lexical categories information to reduce the size of vector space and present semantic relationships between words. The generated document vectors will be input for fuzzy c-means algorithm in the clustering process to increase the clustering accuracy.

The rest of this paper is organized as following; section II show the proposed fuzzy text clustering approach. In section III a set of experiments is presented to compare the performance of the proposed approach with current text clustering methods. Related work is discussed and presented in section IV. Finally, conclusion and future work is given in section V.

II. Fuzzy Text Documents Clustering

In this section we describe in details the components of the proposed fuzzy text clustering approach. There are two main processes: first is Feature Extraction that generated output document vectors from input text documents using WordNet [12] lexical information. The second process is Document

K. Elleithy (ed.), *Advanced Techniques in Computing Sciences and Software Engineering*,
DOI 10.1007/978-90-481-3660-5_31,

Clustering that applies fuzzy c-means algorithm on document vectors to obtain output clusters as illustrated in fig. 1.

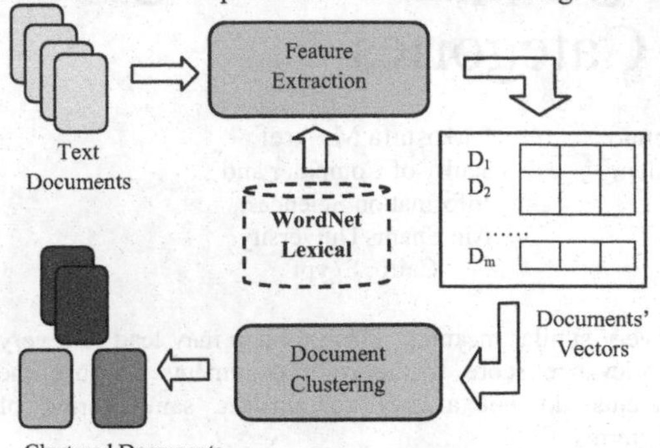

Fig. 1. Fuzzy Text Documents Clustering Approach

A. *Feature Extraction*

The first step in the proposed approach is feature extraction or documents preprocessing which aims to represent the corpus (input documents collection) into vector space model. In this model a set of words (terms) is extracted from the corpus called "bag-of-words" and represent each document as a vector by the words (terms) it contains and their weights regardless of their order. Documents preprocessing step contains four sub-steps: PoS Tagging, Stopword Removal, Stemming, and WordNet Lexical Category Mapping.

1. *PoS Tagging*

The first preprocessing step is to PoS tag the corpus. The PoS tagger relies on the text structure and morphological differences to determine the appropriate part-of-speech. This requires the words to be in their original order. This process is to be done before any other modifications on the corpora. For this reason, PoS tagging is the first step to be carried out on the corpus documents as proposed in [16].

2. *Stopwords Removal*

Stopwords, i.e. words thought not to convey any meaning, are removed from the text. In this work, the proposed approach uses a static list of stopwords with PoS information about all tokens. This process removes all words that are not nouns, verbs or adjectives. For example, stopwords removal process will remove all the words like: he, all, his, from, is, an, of, your, and so on.

3. *Stemming*

The stem is the common root-form of the words with the same meaning appear in various morphological forms (e.g. player, played, plays from stem play). In the proposed approach, we use the morphology function provided with WordNet is used for stemming process. Stemming will find the stems of the output terms to enhance term frequency counting process because terms like "learners" and "learning" come down from the same stem "learn". This process will output all the stems of extracted terms.

The frequency of each stemmed word across the corpus can be counted and every word occurring less often than the pre-specified threshold (called Minimum Support) is pruned, i.e. removed from the words vector, to reduce the document vector dimension. In our implementation we use minimum support value set to 10%, which means that the words found in less than 10% of the input documents is removed from the output vector.

4. *WordNet Lexical Category Mapping*

As proposed in [15], we use WordNet lexical categories to map all the stemmed words in all documents into their lexical categories. We use WordNet 2.1 that has 41 lexical categories for nouns and verbs. For example, the word "dog" and "cat" both belong to the same category "noun.animal". Some words also has multiple categories like word "Washington" has 3 categories (noun.location, noun.group, noun.person) because it can be the name of the American president, the city place, or a group in the concept of capital.

Some word disambiguation techniques are used to remove the resulting noise added by multiple categories mapping which are: disambiguation by context and concept map which are discussed in details in [15].

B. *Document Clustering*

After generating the documents' vectors for all the input documents using feature extraction process, we continue with the clustering process as shown in fig. 1.

The problem of document clustering is defined as follows. Given a set of n documents called DS, DS is clustered into a user-defined number of k document clusters D1, D2,…Dk, (i.e. {D1, D2,…Dk} = DS) so that the documents in a document cluster are similar to one another while documents from different clusters are dissimilar.

There are two main approaches to document clustering, hierarchical clustering (agglomerative and divisive) and partitioning clustering algorithms [17]. In this process we apply three different clustering algorithms which are k-means (partitioning clustering), bisecting k-means (hierarchical clustering) and fuzzy c-means (fuzzy clustering).

1. *K-means and Bisecting k-means*

We have implemented the k-means and bisecting k-means algorithms as introduced in [17]. We will state some details on bisecting k-means algorithm that begins with all data as one cluster then perform the following steps:

Step1: Choose the largest cluster to split.

Step2: Use k-means to split this cluster into two sub-clusters. (Bisecting step)

Step3: Repeat step 2 for some iterations (in our case 10 times) and choose the split with the highest clustering overall similarity.

Step4: Go to step 1 again until the desired k clusters are obtained.

2. *Fuzzy c-means*

Fuzzy c-means is a data clustering technique wherein each data point belongs to a cluster to some degree that is specified by a membership grade while other classical clustering algorithms assign each data point to exactly one cluster. This technique was originally introduced by Bezdec [2] as an improvement on earlier clustering methods. It provides a method that shows how to group data points that populate some multidimensional space into a specific number of different clusters.

Most fuzzy clustering algorithms are objective function based: they determine an optimal (fuzzy) partition of a given data set c clusters by minimizing an objective function with some constraints [4].

In our proposed approach, we use the implementation of fuzzy c-means algorithm in MATLAB (**fcm** function). This function takes the document vectors as a matrix and the desired number of clusters and outputs the clusters centers and the optimal objective function values.

3. *Silhouette Coefficient (SC) for Clustering Evaluation*

For clustering, two measures of cluster "goodness" or quality are used. One type of measure allows us to compare different sets of clusters without reference to external knowledge and is called an internal quality measure. The other type of measures lets us evaluate how well the clustering is working by comparing the groups produced by the clustering techniques to known classes which called an external quality measure [17].

In our application of document clustering, we don't have the knowledge of document classes in order to use external quality measures. We will investigate silhouette coefficient (SC Measure) as one of the main internal quality measures.

To measure the similarity between two documents d_1 and d_2 we use the cosine of the angle between the two document vectors. This measure tries to approach the semantic closeness of documents through the size of the angle between vectors associated to them as in (1).

$$dist(d_1, d_2) = \frac{d_1 \bullet d_2}{|d_1| \cdot |d_2|} \tag{1}$$

Where $(\bullet)$ denotes vector dot product and $(|\ |)$ is the dimension of the vector. A cosine measure of 0 means the two documents are unrelated whereas value closed to 1 means that the documents are closely related [15].

Let $D_M = \{D_1, ..., D_k\}$ describe a clustering result, i.e. it is an exhaustive partitioning of the set of documents DS. The distance of a document $d \in DS$ to a cluster $D_i \in D_M$ is given as in (2).

$$dist(d, D_i) = \frac{\sum_{p \in D_i} dist(d, p)}{|D_i|} \tag{2}$$

Let further consider $a(d, D_M) = dist(d, D_i)$ being the distance of document d to its cluster D_i where $(d \in D_i)$. $b(d, D_M) = \min_{d \notin D_i} dist(d, D_i)\ \forall D_i \in D_M$ is the distance of document d to the nearest neighbor cluster. The silhouette S (d, D_M) of a document d is then defined as in (3).

$$S(d, D_M) = \frac{b(d, D_M) - a(d, D_M)}{\max(b(d, D_M), a(d, D_M))} \tag{3}$$

The silhouette coefficient (SC Measure) is defined as shown in (4).

$$SC(D_M) = \frac{\sum_{p \in DS} S(p, D_M)}{|DS|} \tag{4}$$

The silhouette coefficient is a measure for the clustering quality that is rather independent from the number of clusters. Experiences, such as documented in [11], show that values between 0.7 and 1.0 indicate clustering results with excellent separation between clusters, viz. data points are very close to the center of their cluster and remote from the next nearest cluster. For the range from 0.5 to 0.7 one finds that data points are clearly assigned to cluster centers. Values from 0.25 to 0.5 indicate that cluster centers can be found, though there is considerable "noise". Below a value of 0.25 it becomes practically impossible to find significant cluster centers and to definitely assign the majority of data points.

III. EPERIMENTS AND DISCUSSION

Some experiments are performed on some real text documents to compare the performance of three text clustering algorithms which are k-means, bisecting k-means and fuzzy c-means. There are two main parameters to evaluate the performance of the proposed approach which are clustering quality and running time.

Fuzzy c-means algorithm is implemented in MATLAB and k-means, bisecting k-means algorithms are implemented in Java. Feature extraction process is also implemented in Java using Java NetBeans 5.5.1 and Java API for WordNet Searching (JAWS Library) to access WordNet 2.1.

All experiments were done on Processor P4 (3GHz) machine with 1GB main memory, running the Windows XP Professional® operating system and all times are reported in seconds.

A. *Text Document Datasets*

We evaluate the proposed semantic text document clustering approach on three text document datasets: EMail1200, SCOTS and Reuters text corpuses.

EMail1200 corpus contains test email documents for spam email detection with about 1,245 documents with about 550 words per document. **SCOTS** corpus (Scottish Corpus Of Text and Speech) contains over 1100 written and spoken texts, with about 4 million words of running text. 80% of this total is made up of written texts and 20% is made up of spoken texts. SCOTS dataset contains about 3,425 words per document. **Reuters** corpus contains about 21,578 documents that appeared on the Reuters newswire in 1987. The documents were assembled and indexed with categories by personnel from Reuters Ltd. and Carnegie Group, Inc. in 1987. All the three datasets are used in the text mining testing studies and they are available for download at [24, 25, 26] respectively.

B. Results

First, we pass the three document datasets into the feature extraction process to generate the corresponding document vectors. The vectors are then used as an input for each clustering algorithm. We used the following list (2, 5, 10, 20, 50, 70, and 100) as the desired number of clusters for each algorithm. The output clusters for each one are measured using silhouette coefficient (SC Measure) and report the total running time of the whole process.

1. Clustering Quality

Fig. 2, 3, and 4 show the silhouette coefficient values for the three datasets respectively. In all experiments fuzzy c-means algorithm outperforms bisecting k-means and k-means algorithms in the overall clustering quality using silhouette measure. This experiment shows that the fuzzy clustering is more suitable for the unstructured nature of the text document clustering process itself.

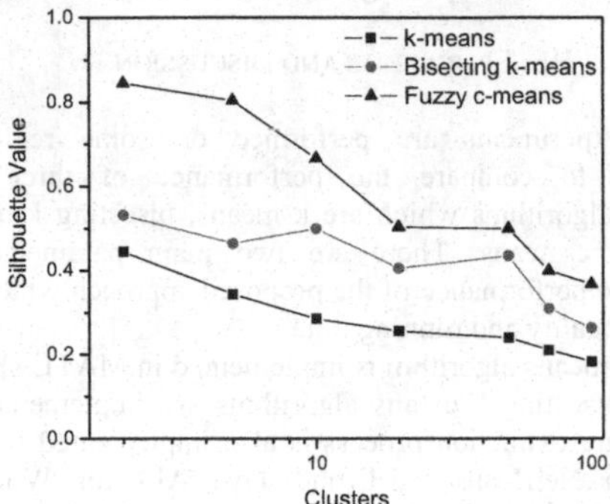

Fig. 2. Silhouette values comparing all clustering algorithms – EMail1200

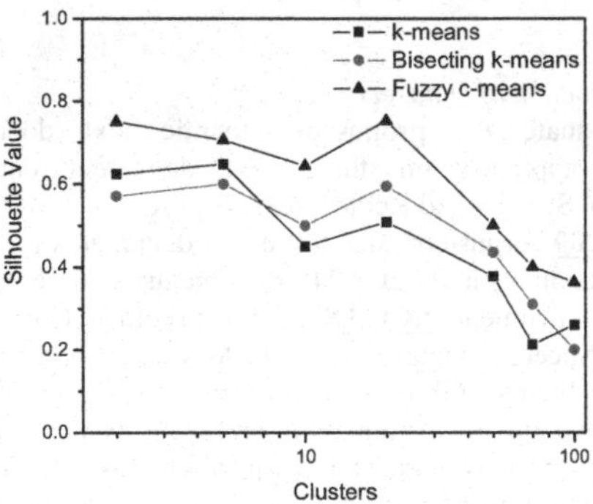

Fig. 3. Silhouette values comparing all clustering algorithms – SCOTS

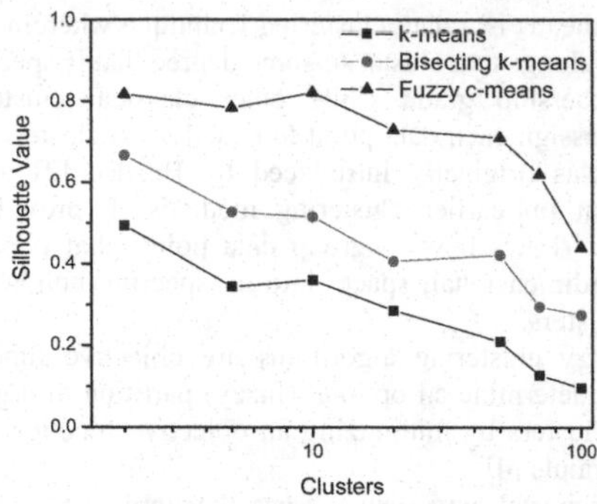

Fig. 4. Silhouette values comparing all clustering algorithms – Reuters

Fig. 5 and 6 show the comparison of using fuzzy c-means clustering algorithm on both SCOTS and Reuters datasets in case of using WordNet lexical categories and not using them. This experiment shows that using WordNet lexical categories in the feature extraction process improves the overall clustering quality of the input dataset document.

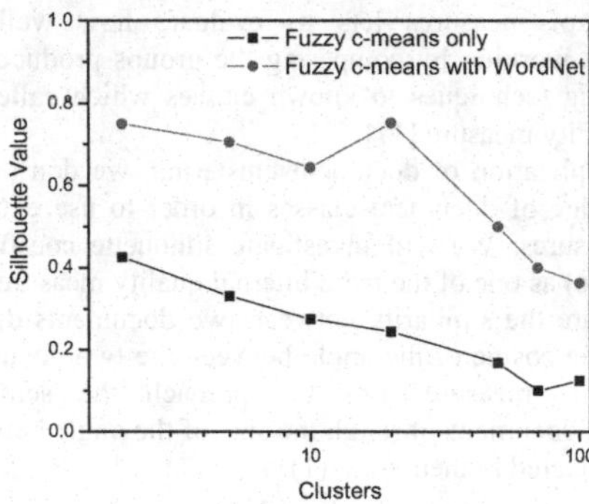

Fig. 5. WordNet improves fuzzy clustering results using SCOTS dataset

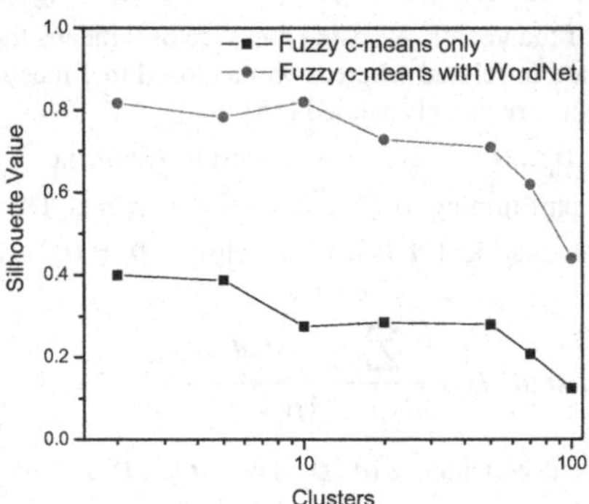

Fig. 6. WordNet improves fuzzy clustering results using Reuters dataset

2. Running Time

Reuters dataset, as mentioned early in this section, contains about 21,578 documents. This is considered a real challenge task that faces any clustering approach because of "Scalability". Some clustering techniques that are helpful for small data sets can be overwhelmed by large data sets to the point that they are no longer helpful.

For that reason we test the scalability of our proposed approach with the different algorithms using Reuters dataset. This experiment shows that the fuzzy c-means performs a great running time optimization with comparison to other two algorithms. Also, according to the huge size of Reuters dataset, the proposed approach shows very good scalability against document size.

Fig. 7 depicts the running time of the different clustering algorithms using Reuters dataset with respect to different values of desired clusters.

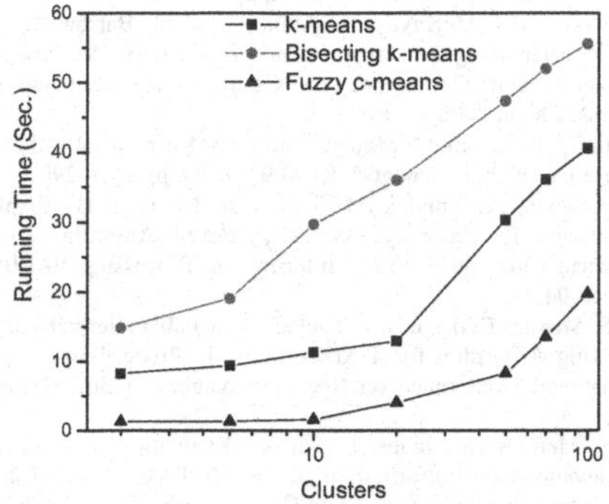

Fig. 7. Scalability of all clustering algorithms on Reuters dataset

IV. Related Work

In the recent years, text document clustering has been introduced as an efficient method for navigating and browsing large document collections and organizing the results returned by search engines in response to user queries [20]. Many clustering techniques are proposed like k-secting k-means [13] and bisecting k-means [17], FTC and HFTC [1] and many others. From the performed experiments in [17] bisecting k-means overcomes all these algorithms in the performance although FTC and HTFC allows to reduce the dimensionality if the data when working with large datasets.

WordNet is used by Green [8, 9] to construct lexical chains from the occurrences of terms in a document: WordNet senses that are related receive high higher weights than senses that appear in isolation from others in the same document. The senses with the best weights are selected and the corresponding weighted term frequencies constitute a base vector representation of a document.

Other works [14, 18] have explored the possibility to use WordNet for retrieving documents by carefully choosing a search keyword. Dave and Lawrence [7] use WordNet synsets as features for document representation and subsequent clustering. But the word sense disambiguation has not been performed showing that WordNet synsets decreases clustering performance in all the experiments. Hotho et al. [10] use WordNet in a unsupervised scenario taking into account the WordNet ontology and lexicon and some strategy for word sense disambiguation achieving improvements of the clustering results.

A technique for feature selection by using WordNet to discover synonymous terms based on cross-referencing is introduced in [6]. First, terms with overlapping word senses co-occurring in a category are selected. A signature for a sense is a synset containing synonyms. Then, the list of noun synsets is checked for all senses for signatures similarity. The semantic context of a category is aggregated by overlapping synsets of different terms senses. The original terms from the category that belongs to the similar synsets will be finally added as features for category representation.

In [16] the authors explore the benefits of partial disambiguation of words by their PoS and the inclusion of WordNet concepts; they show how taking into account synonyms and hypernyms, disambiguated only by PoS tags, is not successful in improving clustering effectiveness because the noise produced by all the incorrect senses extracted from WordNet. Adding all synonyms and all hypernyms into the document vectors seems to increase the noise.

Reforgiato [15] presented a new unsupervised method for document clustering by using WordNet lexical and conceptual relations. In this work, Reforgiato uses WordNet lexical categories and WordNet ontology in order to create a well structured document vector space whose low dimensionality allows common clustering algorithms to perform well. For the clustering step he has chosen the bisecting k-means and the Multipole tree algorithms for their accuracy and speed.

Friedman et al. [21] introduced FDCM algorithm for clustering documents that are represented by vectors of variable size. The algorithm utilizes fuzzy logic to construct the cluster center and introduces a fuzzy based similarity measure which provided reasonably good results in the area of web document monitoring.

Rodrigues and Sacks [23] have modified the Fuzzy c-means algorithm for clustering text documents based on the cosine similarity coefficient rather than on the Euclidean distance. The modified algorithm works with normalized k-dimensional data vectors that lie in hyper-sphere of unit radius and hence has been named Hyper-spherical Fuzzy c-means (H-FCM). Also they proposed a hierarchical fuzzy clustering algorithm (H^2-FCM) to discover relationships between information resources based on their textual content, as well as to represent knowledge through the association of topics covered by those resources [22].

The main problem about these studies is that they neglect any lexical information about the textual data in the documents. This information helps in improving clustering quality as shown in fig. 5 and 6.

V. Conclusion and Future Work

In this paper we proposed a fuzzy document clustering approach based on the WordNet lexical categories and fuzzy c-means clustering algorithm. The proposed approach generates documents vectors using the lexical category mapping of WordNet after preprocessing the input documents. We apply three different clustering algorithms, k-means, bisecting k-means and fuzzy c-means, to the generated documents vectors for clustering process to test the performance of fuzzy c-means.

There are some points that appeared in this work which are:

- Using word sense disambiguation technique reduces the noise of the generated documents vectors and achieves higher clustering quality.
- Fuzzy c-means clustering algorithm achieves higher clustering quality than classical clustering algorithms like k-means (partitioning clustering), and bisecting k-means (hierarchical clustering).
- Using WordNet lexical categories in the feature extraction process for text documents improves the overall clustering quality.

From the shown experimental results, we found that the proposed approach shows good scalability against the huge number of documents in the Reuters dataset along with different values of desired clusters.

For our future work there are two points to investigate which are:

- Using WordNet ontology for generating features vectors [15] for text document along with fuzzy clustering may improve the overall clustering quality.
- Apply the proposed approach on the web documents to solve the problem of web content mining as addressed in [5].

References

[1] F. Beil, M. Ester, and X. Xu, "Frequent term-based text clustering." KDD 02, 2002, pp. 436–442.

[2] J.C. Bezdec, "Pattern Recognition with Fuzzy Objective Function Algorithms", Plenum Press, New York, 1981.

[3] M. Lan, C.L. Tan, H.B. Low, and S.Y. Sung, "A Comprehensive Comparative Study on Term Weighting Schemes", 14th International World Wide Web (WWW2005) Conference, Japan 2005

[4] C. Borgelt, and A. Nurnberger, "Fast Fuzzy Clustering of Web Page Collections". PKDD Workshop on Statistical Approaches for Web Mining, 2004.

[5] S. Chakrabarti, "Mining the Web: Discovering Knowledge from Hypertext Data". Morgan Kaufmann Publishers, 2002.

[6] S. Chua, and N. Kulathuramaiyer, "Semantic feature selection using wordnet". In Proceedings of the 2004 IEEE/WIC/ACM International Conference on Web Intelligence, 2004, pp. 166–172.

[7] D.M.P.K. Dave, & S. Lawrence, "Mining the peanut gallery: Opinion extraction and semantic classification of product reviews" WWW 03 ACM, 2003, pp. 519–528.

[8] S.J. Green, "Building hypertext links in newspaper articles using semantic similarity". NLDB 97, 1997, pp. 178–190.

[9] S.J. Green, "Building hypertext links by computing semantic similarity". TKDE 1999, 11(5), pp.50–57.

[10] A. Hotho, S. Staab, and G. Stumme, "Wordnet improves text document clustering". ACM SIGIR Workshop on Semantic Web 2003.

[11] L. Kaufman, and P.J. Rousseeuw, "Finding Groups in Data: an Introduction to Cluster Analysis", John Wiley & Sons, 1999.

[12] WordNet project available at: http://wordnet.princeton.edu/

[13] B. Larsen, and C. Aone, "Fast and effective text mining using linear-time document clustering". The 5th ACM SIGKDD international conference on knowledge discovery and data mining, 1999, pp. 16–22.

[14] D.I. Moldovan, and R. Mihalcea, "Using wordnet and lexical operators to improve internet searches". IEEE Internet Computing 2000, 4(1), pp. 34–43.

[15] D. Reforgiato, "A new unsupervised method for document clustering by using WordNet lexical and conceptual relations". Journal of Information Retrieval, Vol (10), 2007, pp.563–579.

[16] J. Sedding, and D. Kazakov, "WordNet-based Text Document Clustering", COLING 3rd Workshop on Robust Methods in Analysis of Natural Language Data, 2004.

[17] M. Steinbach, G. Karypis, and V. Kumar, "A Comparison of Document Clustering Techniques", Department of Computer Science and Engineering, University of Minnesota, Technical Report #00-034, 2000

[18] E.M. Voorhees, "Query expansion using lexical-semantic relations". In Proceedings of ACM-SIGIR, 1994, pp. 61–69.

[19] B.B. Wang, R.I. McKay, H.A. Abbass, and M. Barlow, "Learning text classifier using the domain concept hierarchy." In Proceedings of International Conference on Communications, Circuits and Systems, China, 2002.

[20] O. Zamir, O. Etzioni, O. Madani, and R.M. Karp, "Fast and intuitive clustering of web documents". KDD 97, 1997, pp. 287–290.

[21] M. Friedman, A. Kandel, M. Schneider, M. Last, B. Shapka, Y. Elovici and O. Zaafrany, "A Fuzzy-Based Algorithm for Web Document Clustering". Fuzzy Information, Processing NAFIPS '04, IEEE, 2004.

[22] M.E.S. Mendes Rodrigues, L. Sacks, "A Scalable Hierarchical Fuzzy Clustering Algorithm for Text Mining". In Proceedings of the 5th International Conference on Recent Advances in Soft Computing, 2004.

[23] M.E.S. Mendes Rodrigues, L. Sacks, "Evaluating fuzzy clustering for relevance-based information access". In Proceedings of the 12th IEEE International Conference on Fuzzy Systems, FUZZ-IEEE 2003

[24] EMail1200 dataset available at: http://boole.cs.iastate.edu/book/acad/bag/data/lingspam

[25] SCOTS dataset available at: http://www.scottishcorpus.ac.uk/

[26] Reuters dataset available at: http://www.daviddlewis.com/resources/testcollections/reuters21578/

The Study on the Penalty Function of the Insurance Company When the Stock Price Follows Exponential Lévy Process

ZHAO Wu[1] WANG Ding-cheng[2, 3,1] ZENG Yong[1]

(1.School of Management, UESTC, Chengdu, 610054 ,China)

(2.Center of Financial Mathematics, Australian National University, ACT 0200, Australia)

(3.School of Applied Mathematics, UESTC, Chengdu, 610054 ,China)

Abstract- This paper investigates the penalty function under the condition that the insurance company is allowed to invest certain amount of money in some stock market and the remaining reserve in the bond with constant interest force. Through the properties of exponential Lévy process and discrete embedded method, the integral equations for penalty function is derived under the assumption that the stock price follows exponential Lévy process. The method for explicitly computing the ruin quantities is obtained.

I INTRODUCTION

Since 1903, When F. Lundberg introduced a collective risk model based on a homogeneous Poisson claims process, the estimation of ruin probabilities has been a central topic in risk theory. It is known that, if the claim sizes have exponential moment, the ruin probability decreases with the initial surplus; see for instance the book by and Asmussen [1].

It has only been recently that a more general question has been asked: If an insurer additionally has the opportunity to invest in a risky asset modeled, what is the minimal ruin probability she can obtain? In particular, can she do better than keeping the funds in the cash amount? And if yes, how much can she do better? Paulsen et al.[2] have investigated this question, but under the additional assumption that all the surplus is invested in the risky asset. Frovola[3] looked at the case where a constant fraction of wealth is invested in the stock described by geometric Brownian motion. In both cases it was shown that, even if the claim size has exponential moments, the ruin probability decreases only with some negative power of the initial reserve. Browne [4] investigate the case that the insurance business is modeled by a Brownian motion with drift, and the risky asset is modeled as a geometric Brownian motion. Without a budget constraint, he arrives at the following surprising result: the optimal strategy is the investment of a constant amount of money in the risky asset, irrespectively of size of the surplus. In [5], Hipp and Plum consider the general case and analyze the trading strategy which is optimal with respect to the criterion of minimizing the ruin probability. They derive the Bellman equation corresponding to the problem, prove the existence of a solution and a verification theorem. Gaier,Grandits and Schachermayer[6] show that in the case of exponential claims the minimal ruin probability for an insurer with a risky investment possibility can be bounded from above and from below by an exponential function with a greater exponent than the classical Lundberg exponent without investment strategy, which is holding a fixed amount of the wealth invested in the risky asset, such that the corresponding ruin probability is between the derived bounds.

With respect to the investment model, an important issue is whether the geometric Brownian motion appropriately describes the development of the prices of the risky assets. In fact, the prices of many stocks have sudden downward (or upward) jumps, which cannot be explained by the continuous geometric Brownian motion. One way to handle this problem is to model the price of the risky asset by a more general exponential L é vy process with jumps. Paulsen and Gjessing [7] investigate the asymptotic behavior for large initial capital of the infinite time ruin probability when the investment process is a general exponential Lévy process. The results indicate that the ruin probability behaves like a Pareto function of the initial capital. Cai [8] studies ruin probabilities and penalty functions with the assumption that the reserve is invested in a stochastic interest process which is a Lévy process and constructs an integro-differential equation for this model.

Except for the ruin probabilities, other important ruin quantities in ruin theory include the time of ruin, the surplus immediately before ruin, the deficit at ruin, and the amount of claim causing ruin, and so on. A unified method

K. Elleithy (ed.), *Advanced Techniques in Computing Sciences and Software Engineering*,
DOI 10.1007/978-90-481-3660-5_32, © Springer Science+Business Media B.V. 2010

to study these ruin quantities is to consider the (expected discounted) penalty function associated with the time of ruin. See Gerber and Shiu [9] for details. The construction of the penalty function is complex, so it is difficult to calculate it directly. We want to derive a recursive integral equation for the penalty function in this paper. Different from Cai[8], we consider that the insurance company investment a fixed amount of money to risky assets.

THE MODEL

We model the risk process of an insurance company in the classical way: the surplus process R is given by a compound Poisson process:

$$R(t,u) = u + ct - \sum_{i=1}^{N(t)} X_i \,. \quad (1)$$

Where $x \geq 0$ is the initial reserve of the insurance company, $c \in \mathbb{R}$ is the premium rate over time and X_i is an i.i.d.sequence of copies of X with distribution function F , modeling the size of i th claim incurred by the insurer. $N = (N(t))_{t\geq 0}$ is a homogeneous Poisson process with intensity λ . The claims arrive at random time points $0 < T_1 < T_2 < ... < T_n < ...$ and T_1 has exponential distribution with parameter λ .

Now we deviate from the classical setting and assume that the company may also invest in a stock or market index. The price of stock price $S(t)$ described by exponential Lévy process $S(t) = e^{L(t)}$. $L(t)$ is a Lévy process with $L(0) = 0$. Let us assume that the process $S(t)$ is independent of $R(t,u)$. If at the moment t the insurer has wealth $U(t)$, and invests an amount $U(t)$ of money in the stock and the remaining reserve $U(t) - K(t)$ in the bond with zero interest rate, when the interest force on the bond is equal to the inflation force. The wealth process $U(t)$ can be written as

$$U(t) \equiv U(t,u,K) = R(t,u) + \int_0^t \frac{K(t)}{S(t)} dS(t) \,. \quad (2)$$

In this paper, we adopt a special investment strategy that the insurer only invests in the risky market at the time when each claim arrives and the amount of investment is constant denoted by k .

Then, $K(t) = \sum_{n=1}^{\infty} kI_{\{t=T_{n-1}\}}$ .That the constant investment strategy is asymptotic optimal is proved by Gaier,Grandits and Schachermayer[6].

The infinite time ruin probability of the insurance company, defined by

$$\Psi(u,K) = P(U(t,u,K) < 0, \textit{for some } t \geq 0) \,,(3)$$

depending on the initial wealth u and the investment strategy K of the insurer. We further define the time of the ruin $T \equiv T(u,K) \triangleq \inf\{t : U(t,u,K) < 0\}$. Next, we give the definition of the penalty function associated with the time of ruin by defining

$$\Phi_\alpha(u) = E[g(U(T-), |U(T)|) e^{-\alpha T} I(T < \infty)] \,, (4)$$

where $g(x,y)$, $x \geq 0$, $y \geq 0$,is a nonnegative function such that $\Phi_\alpha(u)$ exists; $\alpha \geq 0$;and $I(C)$ is the indicator function of a set C .

A simple and sufficient condition on g for $\Phi_\alpha(u) < \infty$ is that g is a bounded function. With suitable choices of g , $\Phi_\alpha(u)$ will yield different ruin quantities. For example, if $g = 1$ and $\alpha = 0$, then $\Phi_\alpha(u) = \Psi(u,K)$ is the ruin probability; if $g = 1$ and $\alpha > 0$, then $\Phi_\alpha(u) = E(e^{-\alpha T})$ gives the Laplace transform of the time of ruin; if $\alpha = 0$ and $g(x_1, x_2) = I(x_2 \leq y)$ then

$\Phi_\alpha(u) = P(|U(T)| \leq y, T < \infty)$ denotes the distribution function of the deficit at ruin; if $g(x_1, x_2) \equiv I(x_1 + x_2 \leq y)$ and $\alpha = 0$, then $\Phi_\alpha(u) = P(U(T-) + |U(T)| \leq y, T < \infty)$ represents the distribution function of the amount of claim causing ruin. One of the commom research methods used in ruin theory is first to derive integral equations for ruin quantities. In this paper, we derive a recursive integral equation for $\Phi_\alpha(u)$, then we can use numerical methods to solve the equations and obtain numerical solutions.

III MAIN RESULTS

At the beginning of this section, we introduce the discrete embedded method. Let

$$U_n = U(T_n, u, K), n = 1,2,..., U_0 = u, \quad (5)$$

U_n denotes the surplus at the time of the n th claim arrives, the process given by (5) is discrete embedded process. Further, U_n satisfied the following recursive equation

$$U_n = U_{n-1} + c(T_n - T_{n-1}) - X_n + k(e^{L(T_n)-L(T_{n-1})} - 1),$$
$$n = 1,2,..., \quad (6)$$

$(T_n - T_{n-1}, e^{L(T_n)-L(T_{n-1})})$ is a sequence of i.i.d. two dimensional random variables with since $L(t)$ is a Levy process, and has the same two dimension joint distribution as $(T_1, e^{L(T_1)})$, the density distribution function is $p(t,y), t \geq 0, y \geq 0$. With the surplus process $U(t)$, ruin can occur only at claims times, so

$$\Psi(x,K) = P(\bigcup_{n=1}^{\infty}(U_n < 0)).$$

Further, let $\Psi_n(x,K) = P(\bigcup_{i=1}^{n}(U_i < 0))$ be the probability that ruin occurs before or on the n th claim with an initial surplus u. Clearly,

$$0 \leq \Psi_1(u,K) \leq \Psi_2(u,K) \leq ... \leq \Psi_n(u,K) \leq ..., \quad (7)$$

thus, $\lim_{n\to\infty} \Psi_n(u,K) = \Psi(u,K)$ .The following result gives a recursive integral equation for $\Phi_\alpha(u)$.

Main Theorem. $\Phi_\alpha(u)$ satisfies the following integral equation

$$\Phi_\alpha(u) = \int_0^\infty \lambda e^{-(\alpha+\lambda)t} \int_0^\infty \int_0^{u+ct+k(y-1)} \Phi_\alpha(u+ct-x+k(y-1)) \times dF(x)p(t,y)dydt + \int_0^\infty \lambda e^{-(\alpha+\lambda)t} \int_0^\infty A(u+ct-k(y-1))p(t,y)dydt \quad (8)$$

where

$$A(u) = \int_u^\infty g(u, x-u)dF(x)$$

Proof. Conditioning on $T_1 = t$, $X_1 = x$, $e^{L(T_1)} = y$,if $x \leq u+ct+k(y-1)$, then, ruin does not occur, $U_1 = u+ct-x+k(y-1)$ and $\Phi_\alpha(u+ct-x+k(y-1))$ is the expected discounted value at time t, hence, $e^{-\alpha t}\Phi_\alpha(u+ct-x+k(y-1))$ gives the discounted value at time 0. If $x > u+ct+k(y-1)$, then ruin occurs with $T = T_1 = t$, $U(T-) = u+ct+k(y-1)$, and $|U(T)| = x-[u+ct+k(y-1)]$. Thus, noting that $(T_1, e^{L(T_1)})$ and X_1 are independent, we have

$$\begin{aligned}
&\Phi_\alpha(u) \\
&= \int_0^\infty \lambda e^{-\lambda t} E[g(U(T-), |U(T)|)e^{-\alpha T} \\
&I(T<\infty) \mid X_1 = x, T_1 = t, e^{L(T_1)} = y] \times dF(x)g(t,y)dydt \\
&= \int_0^\infty \lambda e^{-(\alpha+\lambda)t} \int_0^\infty \int_0^{u+ct+k(y-1)} \Phi_\alpha(u+ct-x+k(y-1)) \\
&\times dF(x)p(t,y)dydt \\
&+ \int_0^\infty \lambda e^{-(\alpha+\lambda)t} \int_0^\infty \int_{u+ct+k(y-1)}^\infty g(u+ct+k(y-1) \\
&, x-(u+ct+k(y-1))) \times dF(x)p(t,y)dydt
\end{aligned}$$

which implies that (8) holds. The proof is end.

If we set $g = 1$ and $\alpha = 0$, we can get the integral

equation of ruin probability given by the corollary.

Corollary $\Psi(u,K)$ satisfies the following integral equation

$$\Psi(u,K)=\int_0^\infty\int_0^\infty \overline{F}(u+ct+k(y-1))p(t,y)dtdy$$

$$+\int_0^\infty\int_0^\infty\int_0^{u+ct+k(y-1)}\Psi(u+ct+k(y-1)-x,k)p(t,y)$$

$$dF(x)dtdy.$$

References

[1] S.Asmussen, Ruin probabilities, World Scientific Press,2000

[2] J. Paulsen, Ruin theory with compounding assets-a survey. Insurance: Mathematics and Economics 1998.(22),3-16

[3] A.G.Frovola, Yu.M.Kabanov, and S.M. Pergamenshikov, In the insurance business risky investments are dangerous, Preprint,to appear in Finance and Stochastics.

[4] Browne,S.Optimal investment policies for a firm with a random risk process:exponential utility and minimizing the probability of ruin. Mathematics of Operations Research 1995(20),937-958.

[5] C.Hipp and H.K.Gjessing,Ruin theory with stochastic return on investments, Advances in Applied Probability 1997,(29) 965-985.

[6] Gaier J, Grandits P, Schachermayer W. Asymptotic ruin probabilities and optimal investment. Ann.Appl.Probab, 2003,(13):1054-1076.

[7] Paulsen J, Gjessing H H. Ruin theory with stochastic return on investments, Advances in Applied Probability 1997,(29):965-985.

[8] Cai J, Dickson D C M. Upper bounds for ultimate ruin probabilities in the Sparre Andersen model with interest. Insurance: Mathematics and Economics, 2003. (32): 61-71

[9] Gerber,H.U. and Shiu, E.S.W. The joint distribution of the time of ruin, the surplus immediately before ruin, and the deficit at ruin. Insurance: Mathematics and Economics, 1997,(21):129-137.

Studies on SEE Characteristic and Hardening Techniques of CMOS SRAM with Sub-micro Feature Sizes

HE Xing-hua Zhang Cong ZHANG Yong-liang LU Huan-zhang
ATR Key Lab, National University of Defense Technology
Changsha China 410073
huaxinghe@nudt.edu.cn

Abstract—**The single event effects (SEE) characteristic and hardening techniques of CMOS SRAM with sub-micron feature size are studied in the paper. After introducing the relationship SEE with the structure of memory cell, the rate of read-write, the feature sizes and the power supply, the SEE hardening techniques for the COMS SRAM are given from tow aspect: device-level hardening techniques and system-level hardening techniques. Finally, an error detection and correction (EDAC) design based on high reliability anti-fused FPGA is presented, this design has special real-time performance and high reliability, and has been adopted in a space-bone integrated processor platform, which works well in all kinds of environmental experiments.**

Keywords— **COMS SRAM, Space Electronic System, Sub-micron, SEE, Cross-section, EDAC, TMR.**

I. INTRODUCTION

COMS SRAM is a fundamental component of space electronic systems. Comparing to earlier generations with larger features, the new family of COMS SRAM with sub-micron feature sizes have produced enormous gains in speed, array sizes, cell density, and reduction in power consumption. The space application and the ground-based experiment have demonstrated that the SEE of COMS SRAM with sub-micron feature sizes is more sensitive, the cross section increased and the threshold decreased [1]-[2]. Therefore, the reliability of COMS SRAM with sub-micron feature sizes used in space electronic systems has reduced dramatically. Nowadays, studies on the SEE Characteristic and the hardening techniques of COMS SRAM with sub-micron feature sizes are concentrated.

The contributions of this paper are:

•Introducing SEE mechanism.

•Analyzing the relationship SEE with the structure of memory cell, the rate of read-write, the feature size and the power supply.

•Discussing the system-level and chip-level techniques for anti-SEE of COMS SRAM, analyzing the performance of these techniques.

•Presenting an actual implementation, and demonstrating its effectiveness by an actual experiment..

II. SEE MECHANISM

SEE is the main concern of COMS SRAM in space application. Generally, the SEE include the single event upset (SEU) and the single event latch-up (SEL). SEU is a soft error caused by an energetic heavy ion or proton as it travels through the transistor substrate. SEU typically appear as bit-flips in SRAM. SEUs are typically non-destructive in SRAM and are cleared by rewriting of the device. SEL results in a high operating current, usually far above device specifications. Latch-ups are potentially destructive and may cause permanent damage within a device.

A. SEU Mechanism

Double stable state trigger is the basic memory cell of COMS SRAM. There are many structures of double stable state trigger, such as 6T structures, 4T-2R structures, 4T-2TFT structures etc. *Fig.1* shows the structures of 6T and 4T-2R. memory cell.

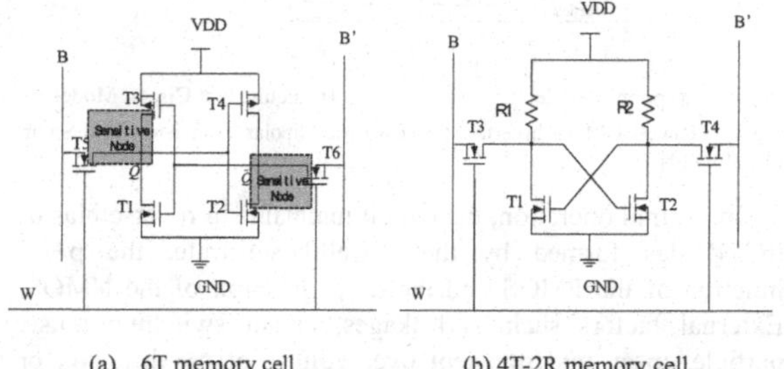

(a) 6T memory cell (b) 4T-2R memory cell

Fig. 1 Structure of CMOS SRAM memory cells

When a charged particle strikes one of the sensitive nodes of a memory cell, such as a drain in an off state transistor, it generates a transient current pulse that can turn on the gate of the opposite transistor. The effect can produce an inversion in the stored value, in other words, a bit flip in the memory cell. Memory cells have two stable states, one that represents a stored '0' and one that represents a stored '1'. For instance, the 6T structure, in each state, two transistors are turned on and two are turned off (SEU target drains). A bit-flip in the memory element occurs when an energetic particle causes the state of the transistors in the circuit to reverse, as illustrated in *Fig.2*. This effect is SEU, and it is one of the major concerns in COMS SRAM memory cell.

B. SEL Mechanism

The basic switching circuit of COMS SRAM is the inverter, which is made from a complementary pair of MOS transistors, one NMOS and one PMOS. Electrical isolation is achieved by using both dielectric and pn junction diodes. P-type doped

K. Elleithy (ed.), *Advanced Techniques in Computing Sciences and Software Engineering*,
DOI 10.1007/978-90-481-3660-5_33, © Springer Science+Business Media B.V. 2010

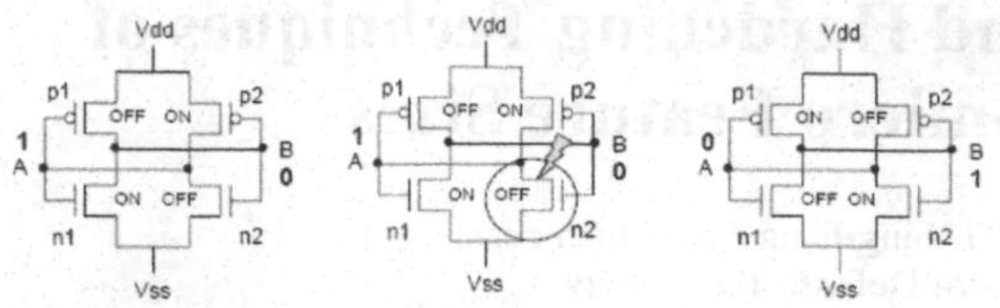

Fig.2 SEU Effect In a SRAM Memory Cell

regions (p-wells) isolate NMOS transistors, while N-type doped regions (n-wells) isolate PMOS transistors. Unfortunately, the twin well CMOS structure always contain a pair of parasitic bipolar transistors (*Fig.3(a)*).This network forms the *pnpn* Silicon Controlled Rectifier (SCR) power device which can be unintentionally biased into a high current, low impedance state. *Fig.3(b)* shows the equivalent circuit model *pnpn* structure.

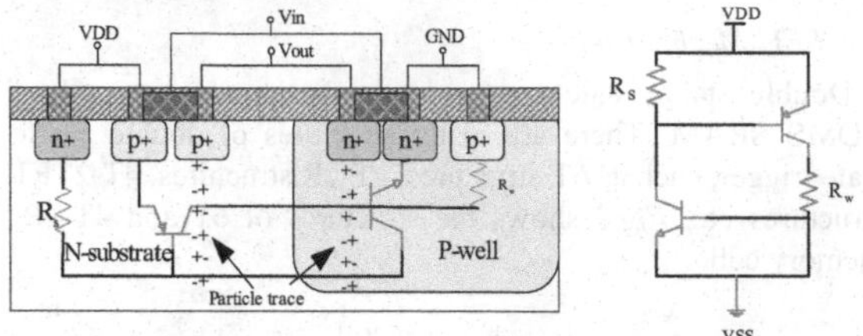

(a) pnpn Model (b) Equivalent Circuit Model

Fig.3 P-well CMOS Inverter with Parasitic Bipolar Transistors and Shunt Resistors [3]

In normal operation, the circuit maintains in reverse bias of the diodes formed by the P-well/N-substrate, the p+/n- junction of the PMOS and the n+/p- junction of the NMOS. External factors such as leakages, circuit switching noise, particle upset, and transient over voltage at p+ diffusion or under voltage at the n+ diffusion can trigger one or both of the parasitic bipolar devices into the active state. If either of the emitter/base junction is forward biased, emitter currents will be sourced to base/collector regions, immediately shifting the local potential. If the current is of sufficient magnitude the blocking diodes will be forward biased, thus shifting the *pnpn* SCR from the blocking state (high impedance, low current) to the on state (low impedance, high current).This is known as latch-up. This results from the single event is known as SEE.

III. SEE Characteristic

In this section we will introduce the relationship of SEE characteristic with the structure of memory cell, the rate of read-write, the density of memory cell and the power supply.

A. SEU Characteristic

1) SEU vs. Feature Size

As described in last section, an energetic particle striking the silicon, it loses its energy via the production of free electron-hole pairs, resulting in a dense ionized track in the local region. Protons and neutrons can cause nuclear reaction when passing through the material, the recoil also produces ionization. The ionization generates a charge deposition that can be modeled by transient current pulse that can be interpreted as a signal in the circuit causing an upset. When the feature size of the cells decreased, the energetic particles can easily pass through the sensitive region, the rate of SEU is increased; on the other hand, with the feature size of the cells decreased, the trace passing through the sensitive region become short, the energy via the production of free electron-hole pairs is decreased, thus the charge deposition is decreased, and the transient current pulse that can be interpreted as a signal in the circuit is decreased, the rate of SEU is decreased. But can be affirmed that the relationship between SEU and feature size is not liner, [2] shows former is the prime suspect.

2) SEU vs. Structure of Memory Cell

There are many structures of double stable state trigger, such as 6T structures, 4T-2R structures, 4T-2TFT structures etc. [3] shows 6T structures is less sensitive than the other structures; the sensitivities of 4T-2R and 4T-2TFT structures are very similar, and this conclusion is independent of the power supply.

3) SEU vs. Read-write Rate

Memory tests include static test and dynamic test. In the static test, the data in the memory is not changed, after irradiation, read back and compare the data and accumulate upsets; In the dynamic test, the data in the memory is changed, during the course of irradiation, read back and compare the data and accumulate upsets. Many tests show the SEU rate of the dynamic test is higher than the static test [4].And with the rate of read-write increased, the SEU rate is also increased, but the mechanism is not clear, the extension of the work can be studying in the future.

4) SEU vs. Power Supply

As the power supply increases, Irradiation testing shows the SEU rate decreased, the testing data are given in *table 1.*

Table 1 Ratio of SEU rate @5V and 3.3V

	SAMSUN4M	MHS 1M	MITSUBISH 4M	HITACHI 4M
SEU(3.3V) / SEU(5V)	2.6	2.2	2.0	1.7

This result can be understood using the Q_c-critical charge concept. We define the deposited ionizing charge Q_{in}. If this charge is greater than the critical charge Q_c defined as the charge necessary to flip the memory bit, then the memory will show errors. For CMOS SRAM, the Q_c can be expressed as the product of a capacitor C by a voltage V-threshold. If the power supply increases, the V-threshold and Q_c increases and then the SEU rate decreases.

B. SEL Characteristic

1) SEL vs. Feature Size

Tests on CMOS SRAM showed the latest generation of CMOS SRAM are high vulnerability to SEL, the smaller feature sizes seem to play a role in some cases, *Fig.4* illustrates a coarse trend observed among the commercial 1M and 4M CMOS SRAM devices that tested: with the feature size decreased the SEL of COMS SRAM with sub-micron feature sizes is more sensitive, the threshold decreased, however, the IDT CMOS SRAM data illustrate the fact that feature size is not the only factor influencing latch-up sensitivity. This trend, based only upon the parts we tested, suggests that CMOS SRAM parts with features sizes as small as 0.09micron (already being produced) and 0.06micron (already being planned) – from certain manufacturers and without deliberate latch-up prevention designed in – might suffer latch-up at even lower LETs than what we observed in the other testing.

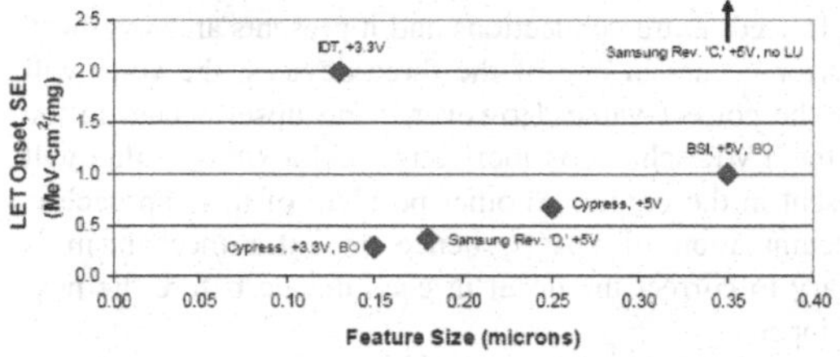

Fig.4 SEL Susceptibility vs. Feature Size

2) SEL vs. Temperatures

The cross section for single event latch-up is expected to increase with temperature [8]. *Fig.5* shows the SEL cross section curves for the IDT 71V67603 as a function of temperature.

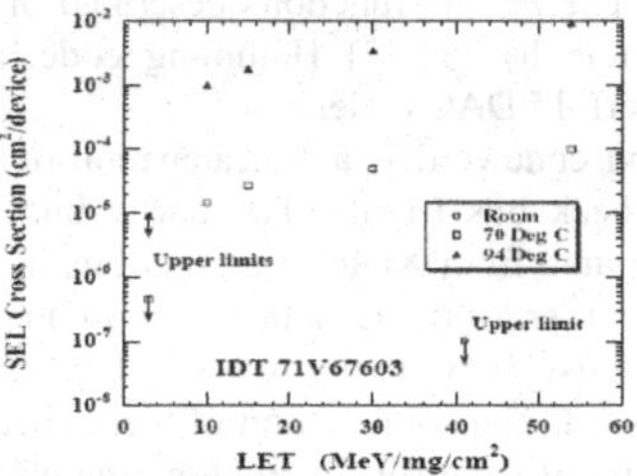

Fig. 5 SEL Cross Section At Various Temperatures for the IDT71V67603

At low LETs, the device did not latch-up at room temperature or at higher temperature, and only upper limits were obtained. Above the threshold, it is clear that the cross section of SEL is increasing with temperature.

3) SEL vs. Power Supply

One expects that the sensitivity to latch-up will decrease with lower bias voltage. Many present high density CMOS SRAM devices studied here use a low bias voltage, compared to a higher bias voltage typically used in previous generations. The parts tested [5] showed no latch-up at an LET of 40 MeV/(mg/cm^2) and fluence greater than 10^7 particles/cm^2. But it is clear that the latch-up is still a risk even with a lower bias voltage.

IV. SEE Hardening Techniques

There has been a great deal of work on radiation hardened circuit design approaches. The SEE hardening techniques for SRAM can be divided into two types: hardening techniques based on device level and techniques based on device level.

A. Device Level Hardening Techniques

1) SOI, SOS

The silicon on insulator (SOI) and silicon son sapphire (SOS) process technology have the good Characteristic about anti-SEE. However, this approach limits the ability to reduce the size of the storage element with decreasing process technology and consumes more power than a commercial memory cell, at the same time, the time-to-market and cost are increased.

2) Hardened Memory Cells

Recently, a novel approach for fabricating radiation-hardened components at commercial CMOS foundries has been developed. In this approach, radiation hardness is designed into the component using nonstandard transistor topologies, which can be resistors or transistors, able to recover the stored value if an upset strikes one of the drains of a transistor in "off" state. These cells are called hardened memory cells and can avoid the occurrence of a SEU by design, according to the flux and to the charge of the particle.

The SEU tolerant memory cell protected by resistors [12] was the first proposed solution in this matter, Fig. 6. The decoupling resistor slows the regenerative feedback response of the cell, so the cell can discriminate between an upset caused by a voltage transient pulse and a real write signal. It provides a high silicon density, for example, the gate resistor can be built using two levels of poly silicon. The main draw backs are temperature sensitivity, performance vulnerability in low temperatures, and an extra mask in the fabrication process for the gate resistor. However, a transistor controlled by the bulk can also implement the resistor avoiding the extra mask in the fabrication process. In this case, the gate resistor layout has a small impact in the circuit density.

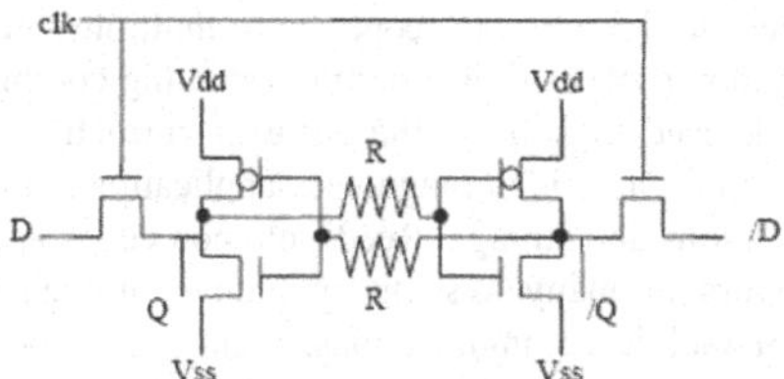

Fig. 6 Resistor Hardened Memory Cell

Memory cells can also be protected by an appropriate feedback devoted to restore the data when it is corrupted by an ion hit. The main problems are the placement of the extra transistors in the feedback in order to restore the upset and the influence of the new sensitive nodes. Such as IBM hardened memory cells [13], HIT cells [14].The main advantages of this method are temperature, voltage supply and technology process independence, and good SEU immunity. The main drawback is silicon area overhead that is due to the extra transistors and their extra size.

B. System Level Hardening Techniques

With the COTS used in space electronic system, studies on hardening techniques based on system level are concentrated. The SEL hardening techniques are easily do by limiting the current. We introduce the SEU hardening techniques: error detection and correction coding (EDAC) and Triple Modular Redundancy (TMR).

1) EDAC

Bit-flips caused by SEU are a well-known problem in SRAM; EDAC code is an effective system level solution to this problem. EDAC codes module include encoding/decoding circuitry. The encoding and decoding functions can be done in hardware, software, or a combination of both.

There are many codes are used to protect the systems against single and multiple SEUs. An example of EDAC is the hamming code [16], in its simplest version. It is an error-detecting and error correcting binary code that can detect all single- and double-bit errors and correct all single-bit errors (SEC-DED). This coding method is recommended for systems with low probabilities of multiple errors in a single data structure (e.g., only a single bit error in a byte of data). Hamming code increases area by requiring additional storage cells (check bits), plus the encoder and the decoder blocks. For an n bit word, there are approximately $\log_2 n$ more storage cells. However, the encoder and decoder blocks may add a more significant area increase, thanks for the extra XOR gates. Regarding performance, the delay of the encoder and decoder block is added in the critical path. The delay gets more critical when the number of bits in the coded word increases. The number of XOR gates in serial can directly proportional to the number of bits in the coded word.

The problem of hamming code is that it can not correct double bit upsets, which can be very important for very deep sub-micron technologies, especially in memories because of the high density of the cells [15]. Other codes must be investigated to be able to cope with multiple bit upsets. Reed-Solomon (RS) [16]. is an error-correcting coding system that was devised to address the issue of correcting multiple errors. It has a wide range of applications in digital communications and storage. Reed-Solomon codes are used to correct errors in many systems including: storage devices, wireless or mobile communications, high-speed modems and others. RS encoding and decoding is commonly carried out in software, and for this reason the RS implementations normally found in the literature do not take into account area and performance effects for hardware implementation. However, the RS code hardware implementation as presented in [17] is an efficient solution to protect memories against multiple SEU.

2) TMR

In the case of the full hardware redundancy, the well-known TMR approach, the logic or device is triplicated and voters are placed at the output to identify the correct value. The first possibility that was largely used in space applications is the triplication of the entire device, *Fig.7* .

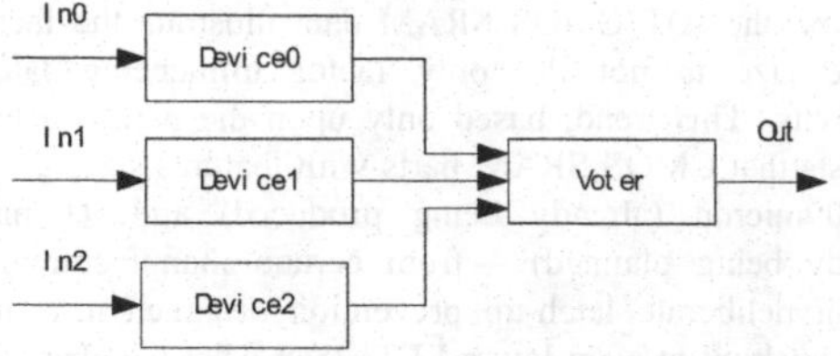

Fig. 7 TMR Implemented In The Entire Device

This approach uses a voter as a fourth component in the board. It needs extra connections and it presents area overhead. If an error occurs in one of the three devices, the voter will choose the correct value. However, if an upset occurs in the voter, the TMR scheme is ineffective and a wrong value will be present in the output. Another problem of this approach is the accumulation of upsets, hence an extra mechanism is necessary to correct the upset in each device before the next SEU happens.

V. A Hardening Design Based FPGA

The structures of DSP+FPGA+SRAM are used in many real-time space electronic system, SRAMs are used as the data or code memories. In these systems, the read-write rate of data and code are very high, the data and code byte can not have bit error. The EDAC functions described in this design are made possible by (39,32) Hamming code, a relatively simple yet powerful EDAC code.

The Hamming codeword is a concatenation of the original data and the check bits (parity).The check bits are parallel parity bits generated from XORing certain bits in the original data word. If bit error(s) are introduced in the codeword, several check bits show parity errors after decoding the retrieved codeword. The combination of these check-bit errors display the nature of the error. In addition, the position of any single bit error is identified from the check bits. *Fig.8* illustrate the hardware relationship of SRAM EDAC based FPGA.

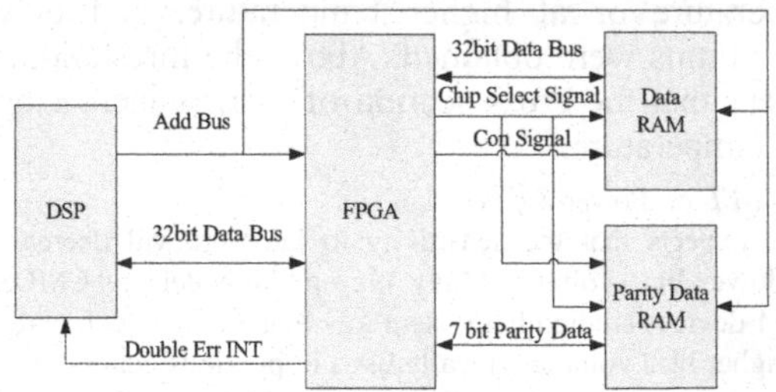

Fig.8 Hardware Structure of SRAM EDAC Based FPGA.

In this design, DSP and FPGA are SEE hardened devices. The Data SRAM and Parity SRAM are connected with the same address bus and control bus. *Fig.9* shows a block diagram using a memory controller with EDAC functions.

This design has a parity encoder and parity decoder unit. The encoder implements the function of the generator matrix, while the decoder is responsible for error detection and correction. In addition to displaying the error type, the reference design also supports diagnostic mode. Single, multiple, and triple bit errors can be introduced to the output codeword. When the ERROR port is "00", no single, two, or greater bit error is detected. When the ERROR port is "01", it indicates single bit error occurred within the 39-bit code word.

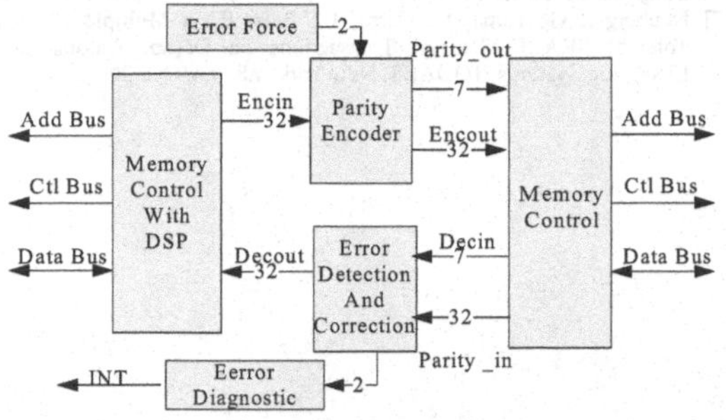

Fig.9 Block Diagram of EDAC

In addition, the error is corrected, and the data is error free. When the ERROR port is "10", a two bit error has occurred within the codeword. In this case, no error correction is possible in this case. When the ERROR port is "11", errors beyond the detection capability can occur within the codeword and no error correction is possible. This is an invalid error type.

A deliberate bit error can be injected in the codeword at the output of the encoder as a way to test the system. Force_error provides several types of error modes.

Force_error = 00

This is the normal operation mode. No bit error has been imposed on the output of the encoder.

Force_error = 01

Single bit error mode. One bit is reversed ("0" to "1" or "1" to "0") in the codeword at every rising edge of the clock. The single bit error follows the sequence moving from bit "0" of the code word to bit 39. The sequence is repeated as long as this error mode is active.

Force_error = 10

Termed double bit error mode. Two consecutive bits are reversed ("0" becomes "1" or "1" becomes "0") in the codeword at every rising edge of the clock. The double bit error follows the sequence moving from bit (0,1) of the codeword to bit (38, 39). The sequence repeats as long as this error mode is active.

Force_error = 11

Termed triple-error mode. Three-bits are reversed ("0" becomes "1" or "1" becomes "0") in a codeword generated at every rising edge of the clock. The double bit error follows the sequence moving from bit (0,1,2) of the codeword together to bit (37,38,39) sequentially. The sequence repeats as long as this error mode is active.

The design utilizes a minimum amount of resources and has high performance. Compare to the read-write rate of not using EDAC, there are two CLK cycles latencies, as illustrated in *Fig.10*.

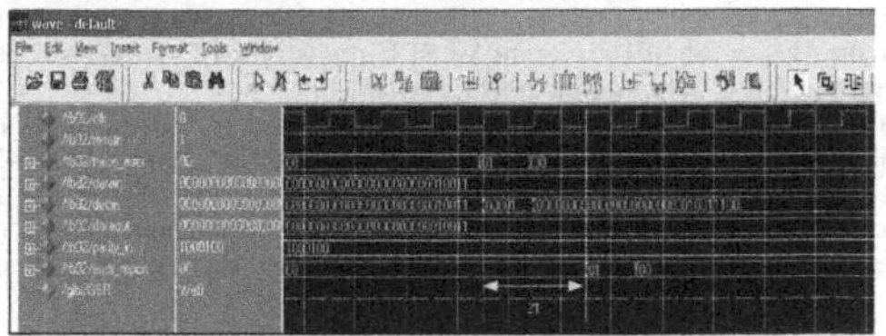

Fig.10 Timing Diagram of SRAM EDAC Based FPGA

Here, the CLK is 133Mhz, CMOS SRAM used as the external memory of DSP, when there is no EDAC, the write-read rate is 33Mhz(4 CLK cycles); if add the EDAC, there are two CLK cycles latencies, the write-read rate is about 22Mhz(6 CLK cycles). If add the pipe-line, the latencies of EDAC can be done in one CLK cycle, the write-read rate is about 26Mhz(5 CLK cycles), it is very high and can be used in real-time system. This design has high reliability, and has been adopted in a space-bone integrated processor platform, which works well in all kinds of environmental experiments.

VI. Conclusion

We have studied the SEE characteristic and the SEE hardening techniques of COMS SRAM with sub-micro feature sizes.

The dependences of the SEU rate with memory cell, feature size, rate of read-write, power supply are found: 6T memory cell is less sensitive than 4T-2R or 4T-2TFT structures; the SEU rate increased with the feature sizes decreased, increased with the rate of data rear-write increased and increased with the power supply decreased. The dependences of the SEL rate with feature size, temperature, power supply are also found: the SEL rate increased with the feature sizes decreased, increased with the temperature increased and increased with the power supply increased.

Techniques based on SOI, SOS, hardened memory cells, EDAC, TMR can anti-SEE effectively. For real-time space electronic system, the EDAC based FPGA is a better choice.

References

[1] Scheick,L.Z, Swift,G.M, Guertin, S.M. SEU Evaluation of SRAM memories for Space Applications[C]. The IEEE Nuclear and Space Radiation Effects Conference, 24-28 July 2000 .p61-63.

[2] Johnston, A.H. Radiation Effects in Advance Microelectronics Technologies [J] IEEE Trans. Nuclear Science, Vol. 45, No. 3, June 1998:1339-1354.

[3] Lai Zu-wu.. Radiation hardening electronics-Radiation effects and hardening techniques [M] (In Chinese), National Defence Industry Press. 1998 Bei-Jing p152-158.

[4] Thouvenot,D, Trochet,P, Gaillard,R. Neutron Single Event Effect Test Results For Various SRAM Memories [C]. The IEEE Nuclear and Space Radiation Effects Conference, 24 July 1997 .p61-66.

[5] George,J, Koga,R, Crawford, K. SEE Sensitivity Trends in Non-hardened High Density SRAM With Sub-micron Feature Sizes[C]. The IEEE Nuclear and Space Radiation Effects Conference, 21-25 July 2003 .p83--88

[6] Cheng hong,Yi fang Research of CMOS SRAM Radiation Harden Circuit Design Technology [J] (In Chinese). Microprocessors ,No.5 Oct, 2005.

[7] Zhang Xiao ping, The Analysis and Design of Radiation-Hardened CMOS Integrated Circuits [D](In Chinese). Shan Xi , Xi'an Xi ' an University of Technology, 2003.

[8] Velazco,R, Bessot,D, Duzellier,S. Two CMOS Memory Cells Suitable for the Design of SEU-Tolerant VLSI Circuits[J] IEEE Trans. Nuclear Science, Vol. 41, No. 6, Dec.1994:2229-2234.

[9] Calin,T, Nicolaidis,M, Velazco, R. Upset Hardened Memory Design for Submicron CMOS technology[J] IEEE Trans. Nuclear Science, Vol. 43, No. 6, Dec. 1996:2874-2878.

[10] Shaneyfelt, M. R, Fleetwood, D. M. ,T. L. Meisenheimer. Effects of Device Scaling and Geometry on MOS Radiation Hardness Assurance [J] IEEE Trans. Nuclear Science, Vol. 40, No. 6, Dec. 1993:1678-1685.

[11] Simon Tam. XAPP645 (v2.2), Single Error Correction and Double Error Detection, August 9, 2006.

[12] Weaver, H.; et al. An SEU Tolerant Memory Cell Derived from Fundamental Studies of SEU Mechanisms in SRAM. IEEE Transactions on Nuclear Science, New York, v.34, n.6, Dec. 1987.

[13] Rocket, L. R. An SEU-hardened CMOS data latch design. IEEE Transactions on Nuclear Science, New York, v.35, n.6, p. 1682-1687, Dec. 1988.

[14] Calin, T, Nicolaidis, M, Velazco, R. Upset hardened memory design for submicron CMOS technology. IEEE Transactions on Nuclear Science, New York, v.43, n.6, p. 2874 -2878, Dec. 1996.

[15] Reed, R. A. et al. Heavy ion and proton-induced single event multiple upset. IEEE Transactions on Nuclear Science, New York, v.44, n.6, p. 2224-2229, Dec. 1997.

[16] Houghton, A. D. The Engineer's Error Coding Handbook. London: Chapman & Hall,1997.

[17] Neuberger, G.; Lima, F.; Carro, L.; Reis, R. A Multiple Bit Upset Tolerant SRAM Memory. Transactions on Design Automation of Electronic Systems, TODAES, New York, v.8, n.4, Oct. 2003.

Automating The Work at The Skin and Allergy Private Clinic : A Case Study on Using an Imaging Database to Manage Patients Records

Mohammad AbdulRahman ALGhalayini [1]

[1]Director, Computer and Information Unit, Vice Rectorate for Knowledge Exchange and Technology Transfer
King Saud University, Riyadh, Saudi Arabia

Abstract

Today, many institutions and organizations are facing serious problem due to the tremendously increasing size of documents, and this problem is further triggering the storage and retrieval problems due to the continuously growing space and efficiency requirements. This problem is becoming more complex with time and the increase in the size and number of documents in an organization; therefore, there is a world wide growing demand to address this problem. This demand and challenge can be met by converting the tremendous amount of paper documents to images using a process to enable specialized document imaging people to select the most suitable image type and scanning resolution to use when there is a need for storing documents images. This documents management process, if applied, attempts to solve the problem of the image storage type and size to some extent. In this paper, we present a case study resembling an applied process to manage the registration of new patients in a private clinic and to optimize following up the registered patients after having their information records stored in an imaging database system; therefore, through this automation approach, we optimize the work process and maximize the efficiency of the Skin and Allergy Clinic tasks.

Problem Definition

Skin and Allergy Clinic is a private medical clinic which operates all the medical and administrative work manually at it's starting phase. Practically, it was facing difficulty in organizing all the patients files which were piling up quickly as the patients come to this clinic for medical consultations.

The need for a database system was urgent to make it easy to store the patient's information when they open new files and when they come for a visit periodically in addition to all administrative related information.

The need for searching for a specific patient information quickly and the need to take a look at his/her previous medical reports raised the need to have a customized imaging database system.

Solution Steps

1 - A form of two pages was properly designed for each new patient who intends to open a new patient file.
2 - A document scanner and a high specs PC were bought in preparation to be used when the patient form was filled and submitted.
3 - A database was designed, implemented, and tested to be used to store the forms information as well as the scanned documents belonging to each patient.
4 - By the time the database system was ready, a dedicated clinic employee was trained on using the database system.
5 - Each time the data entry employee used to come and scan the patient form then store the form in a specific folder.
6 - This hard disk arrangement made it easy to quickly reference any stored document images and retrieve them when necessary.
7 - When the patient data was ready to be entered, all what the data entry person worried about is storing the information of the form as text in the database system corresponding fields, and storing the previously saved scanned image of the patient information form in the OLE field (which was scanned and stored previously as .max file or as a PDF file on the local hard disk) in the proper OLE[1] field.

[1] OLE is Object Linking and Embedding.

K. Elleithy (ed.), *Advanced Techniques in Computing Sciences and Software Engineering*,
DOI 10.1007/978-90-481-3660-5_34, © Springer Science+Business Media B.V. 2010

8 - After storing one full form information and storing scanned documents belonging to that record, the form is signed and dated by the data entry person, then stored back in the paper folder it belongs to. This step was necessary to easily figure out which forms where completely entered and which were not.
9 - Each week end, a complete copy of this database system was backed up on the central file server to be accessed by clinic authorized persons. In addition to saving a backup copy on a writable CD.

Special Notes About The Skin & Allergy Clinic Imaging Database System

1 - This system was designed using MS-ACCESS RDBMS which is one of the MS-OFFICE applications.
2 - It was taken into consideration making this system a very user friendly and easy to use as much as possible.
3 - All clinic PC's were updated to have 1024 MB of RAM at least and 128 MB display memory to access the system conveniently.
4 - All clinic PC's were connected through network to being able to connect to the dedicated central file server (CLINIC_SERVER) with a special user name and password for all authorized clinic employees.
5 - The system is accessible through a password for security reasons.
6 - It is possible to print any page or pages of the documents stored in database system on the central color laser printer, which is directly connected to the server.

Contents and Data Security

Since the Skin and Allergy Clinic system may contain some private information which could be confidential, it was necessary to the clinic member including the data entry employee, to keep a special system password to use it to access the system whether from his own pc in the office or from connecting to the server from outside the clinic.

The server was not accessible to any unauthorized person; therefore, the system was also not accessible to any unauthorized person.

One more thing regarding data security is the issue of weekly backup of the whole database system including the stored documents images on the server hard disk and on a removable writable CD.

Hardware Used

1 - PC with the following specifications :
2.8 Mhz Intel processor
1024 MB DDR RAM
80 GB WD Hard Disk
128 MB ATI display adaptor
10/100 3com network adaptor
52 X CD writer
17 inch display monitor
2 - Image scanner with the following specs :
40-60 pages per minute
Simplex / duplex
Black and white and color scanning
3 - Central file server with the following specs :
2.8 Mhz Intel processor
1024 MB DDR RAM
80 GB WD Hard Disk
128 MB ATI display adaptor
10/100 3com network adaptor
52 X CD writer
17 inch display monitor

Software Used

Windows XP on data entry PC
Windows 2003 server on the file server
PaperPort[2] 11.0 professional version for scanning images on the data entry PC.

Time Consumed

As an automation project, there was a dedicated and limited time to finalize the project. The central database containing the images of the filled information forms and the stored documents images were ready to service The Skin and Allergy Clinic users after the first week of data entry phase. The task of filling up the patients information each on a separate form was assigned to a couple of clinic employees. This process was performed simultaneously with the process of designing, implementing, and testing the required imaging database system. The following table shows the time consumed to finalize such an automation project.

No.	Details	Time consumed
1	Designing the form	1 week
2	Buying and setting up the hardware.	2 week
3	Database system design and implementation and testing.	4 weeks (simultaneous with phase 2)
4	Data entry	10 – 15 complete records per day

Table (1) shows the time consumed for each process of the specified project

[2] PaperPort is a trade name for ScanSoft Inc.

Screen Photos of The Skin & Allergy Clinic Imaging Database System

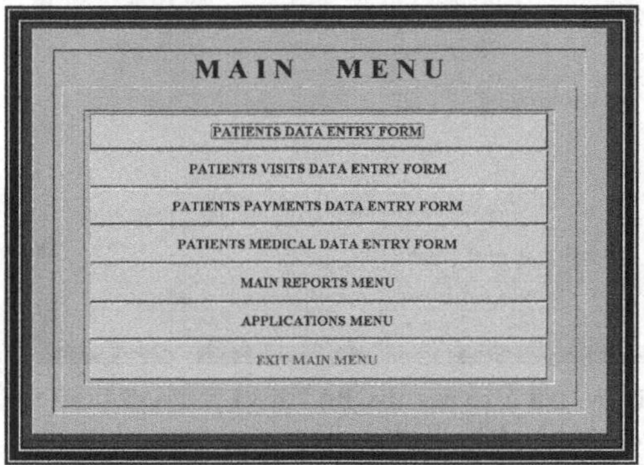

Figure (2) shows the main menu screen. The user should press or click on the any required button using the mouse left click.

Figure (3) shows the PATIENT DATA ENTRY FORM screen which is accessed when the user chooses the first option in the main menu displayed in **Figure (2)**. The user may choose any one of the optional buttons at the bottom of this form and when finished with entering new patient records or manipulating existing records, the user may press or click on the CLOSE button. The user may select any of the pre-saved values to be entered in the fields which have combo boxes provided, (The combo box is the field with a side arrow and when it is pressed a list of pre-saved values is displayed).

The patient ID (or file number) field is represented with a different color to remind the user that this value has to be unique. It is the primary key[3] of the table containing the patients information. If the user mistakenly enters an existing number, this new record would not be saved.

Any modification to any pre-saved data is automatically saved without pressing the SAVE RECORD button, however, it is provided to confirm saving the record or any data manipulation.

OLE means Object Linking or Embedding. It is a type of data provided to allow the user to store (or link to) an image or any type of file in the OLE field. If there is any OLE item to be entered in the OLE field, in this case the patient photo, the user should right-click on the field to be able to continue this process. And the related menu will appear as shown in the next figure.

Figure (4) shows the menu displayed when the user right-click on the OLE field. At this point the user may select the Insert Object selection.

Figure (5) shows the screen prompting the user select the OLE application to be used. For document image storage, the user should select PaperPort Document application then select one of the two options: *Create new* is used to activate the scanner and scan a new image of the document to be stored. *Create from file* is used to specify the location of the image (or any other OLE object) to be linked or embedded in the OLE field. Then the user should press or click on the CLOSE button.

[3] Primary keys are used in databases tables to avoid duplication of records and to ensure data integrity and consistency.

Figure (6) shows the search screen prompting the user for the value to be searched for. The use should make the needed field current then press on the FIND RECORD button. When the search screen appears, the user should type in the value to be search for and select whether this value is to be search for in any part of the field or in the whole field or at the start of the field. After selecting the search criteria, the user should press or click on the Find Next button. If the value is matched in any of the records, the record is displayed instantly. If the record displayed is not the exact record, the user should press or click on Find Next button.

Figure (7) shows the PATIENT'S VISITS DATA ENTRY FORM. The user may select the patient file number from the drop-down list of patients file numbers appearing along with their names. In addition, the user may enter an existing file number, then enter the remaining fields values appearing in this form.

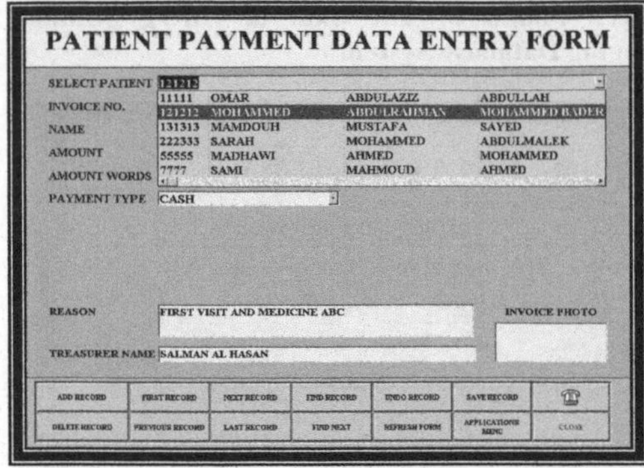

Figure (8) shows the PATIENT'S PAYMENTS DATA ENTRY FORM. The user may select the patient file number from the drop-down list of patients file numbers appearing along with their names. In addition, the user may enter an existing file number, then enter the remaining fields values appearing in this form.

Figure (9) shows the PATIENT'S PAYMENTS DATA ENTRY FORM. The user may select the patient file number from the drop-down list of patients file numbers appearing along with their names. In addition, the user may enter an existing file number, then enter the remaining fields values appearing in this form. If the user selects "CHECK" as the payment type, extra check- related fields appear and the user should enter their values.

PATIENT PAYMENT DATA ENTRY FORM

SELECT PATIENT 121212
INVOICE NO. 12345678 INVOICE DATE 01/01/1423
NAME Dr. OMAR BIN ABDUL AZIZ AAL ALSHAIKH
AMOUNT 195
AMOUNT WORDS ONE HUNDRED AND NINETY FIVE SAUDI RIYALS
PAYMENT TYPE CREDIT CARD
CREDIT CARD VISA EXPIRATION DATE
NAME ON CARD CREDIT CARD NO.
REASON FIRST VISIT AND MEDICINE ABC INVOICE PHOTO
TREASURER NAME SALMAN AL HASAN

ADD RECORD | FIRST RECORD | NEXT RECORD | FIND RECORD | UNDO RECORD | SAVE RECORD
DELETE RECORD | PREVIOUS RECORD | LAST RECORD | FIND NEXT | REFRESH FORM | APPLICATIONS MENU | CLOSE

Figure (10) shows the PATIENT'S PAYMENTS DATA ENTRY FORM. The user may select the patient file number from the drop-down list of patients file numbers appearing along with their names. In addition, the user may enter an existing file number, then enter the remaining fields values appearing in this form. If the user selects "CREDIT CARD" as the payment type, extra credit card-related fields appear and the user should enter their values. When finished, the user may press or click on the CLOSE button.

PATIENT MEDICAL HISTORY DATA ENTRY FORM

FILE ID 11111
PREVIOUS OPERATIONS
PREVIOUS DISEASES
CURRENT DISEASES
CURRENT SYMPTOMS SKIN COLOR IS TURNING BLACK
ALLERGIC DISEASES PENCILLIN
DRUG ALLERGY PENCILLIN
NOTES REPORT PHOTO

11111	OMAR	ABDUL AZIZ	ABDULLAH
121212	MOHAMMED	ABDULRAHMAN	MOHAMMED BADER
131313	MAMDOUH	MUSTAFA	SAYED
222333	SARAH	MOHAMMED	ABDULMALEK
55555	MADHAWI	AHMED	MOHAMMED
7777	SAMI	MAHMOUD	AHMED

ADD RECORD | FIRST RECORD | NEXT RECORD | FIND RECORD | UNDO RECORD | SAVE RECORD
DELETE RECORD | PREVIOUS RECORD | LAST RECORD | FIND NEXT | APPLICATIONS MENU | CLOSE

Figure (11) shows the PATIENT'S MEDICAL HISTORY DATA ENTRY FORM. The user may select the patient file number from the drop-down list of patients file numbers appearing along with their names. In addition, the user may enter an existing file number, and then enter the remaining fields values appearing in this form.

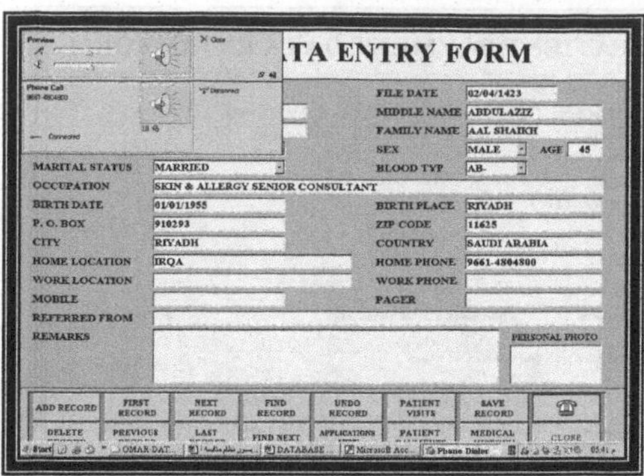

TA ENTRY FORM
FILE DATE 02/04/1423
MIDDLE NAME ABDULAZIZ
FAMILY NAME AAL SHAIKH
SEX MALE AGE 45
MARITAL STATUS MARRIED BLOOD TYP AB-
OCCUPATION SKIN & ALLERGY SENIOR CONSULTANT
BIRTH DATE 01/01/1955 BIRTH PLACE RIYADH
P. O. BOX 910293 ZIP CODE 11635
CITY RIYADH COUNTRY SAUDI ARABIA
HOME LOCATION IRQA HOME PHONE 9661-4804800
WORK LOCATION WORK PHONE
MOBILE PAGER
REFERRED FROM
REMARKS PERSONAL PHOTO

ADD RECORD | FIRST RECORD | NEXT RECORD | FIND RECORD | UNDO RECORD | PATIENT VISITS | SAVE RECORD
DELETE | PREVIOUS | LAST | FIND NEXT | APPLICATIONS | PATIENT | MEDICAL | CLOSE

Figure (12) shows the automatic dialer screen which would appear when the use clicks on the dialer button at the bottom of the form. If a modem is connected, the selected phone number is dialed automatically.

PATIENT VISITS DATA ENTRY FORM

SELECT PATIENT 11111
VISIT DATE 01/01/1423 VISIT TIME
PHYSICIAN Dr. ABDULLAH BIN MOHAMMED ALSEMAARI
DIAGNOSIS
TREATMENT
REMARKS REPORT PHOTO

ADD RECORD | FIRST RECORD | NEXT RECORD | FIND RECORD | UNDO RECORD | SAVE RECORD
DELETE RECORD | PREVIOUS RECORD | LAST RECORD | FIND NEXT | REFRESH FORM | APPLICATIONS MENU | CLOSE

Figure (13) shows the PATIENTS VISITS DATA ENTRY FORM which is displayed always when the user clicks it's button from anywhere in the application's forms. The user may add a new record (which means a new visit of an existing patient) or manipulate an existing record by searching for it based on certain search criteria, or by browsing through the registered visits records. The user may store an image of any report or x-rays used in this session in the OLE field provided for this purpose.

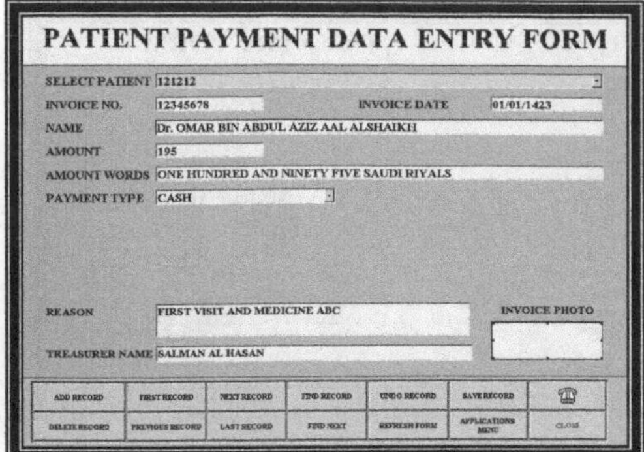

Figure (14) shows the PATIENTS PAYMENTS DATA ENTRY FORM which is displayed always when the user clicks it's button from anywhere in the application's forms. The user may add a new record (which means a new visit of an existing patient) or manipulate an existing record by searching for it based on certain search criteria, or by browsing through the registered payment records. The user may also store an image of the payment invoice for later review in the OLE field provided for this purpose.

PATIENT MEDICAL HISTORY DATA ENTRY FORM
FILE ID 11111
PREVIOUS OPERATIONS HEART TRANSPLANT
PREVIOUS DISEASES CARDIAC RELATED
CURRENT DISEASES SKIN PEALING
CURRENT SYMPTOMS SKIN COLOR IS TURNING BLACK
ALLERGIC DISEASES PENCILLIN
DRUG ALLERGY PENCILLIN
NOTES
REPORT PHOTO
ADD RECORD | FIRST RECORD | NEXT RECORD | FIND RECORD | UNDO RECORD | SAVE RECORD
DELETE RECORD | PREVIOUS RECORD | LAST RECORD | FIND NEXT | APPLICATIONS MENU | CLOSE

Figure (15) shows the PATIENTS MEDICAL HISTORY DATA ENTRY FORM which is displayed always when the user clicks it's button from anywhere in the application's forms. The user may add a new record (which means a new visit of an existing patient) or manipulate an existing record by searching for it based on certain search criteria, or by browsing through the registered patients medical history records.

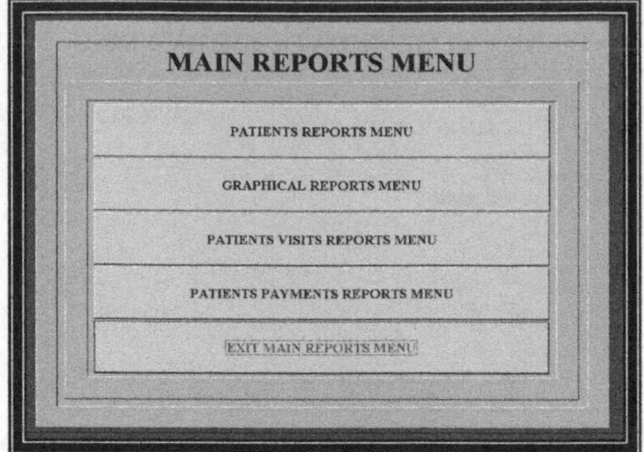

Figure (16) shows the MAIN REPORTS MENU screen. The user may select any of the available menu options or when finished, the user should press or click on the EXIT MAIN REPORTS MENU button.

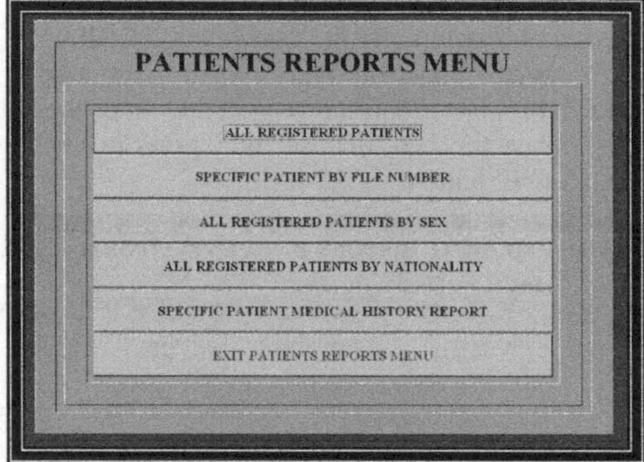

Figure (17) shows the PATIENTS REPORTS MENU. If the user selects the first option, a report of ALL REGISTERED PATIENTS main information will be displayed as shown below in **Figure (18)**.

If the user selects the second option, a report of SPECIFIC PATIENT BY FILE NUMBER, the user will get a screen prompting for the patient file number as shown below in **Figure (20)**, then after displaying the report, the information of the selected patient will be displayed (**Figure (21)**).

If the user selects the third option, a report of SPECIFIC PATIENT BY SEX, the user will get a screen prompting for the patient sex as shown below in **Figure (22)** then after displaying the report, the information of all patients with the selected sex will be displayed (**Figure (23)**).

If the user selects the fourth option, a report of SPECIFIC PATIENT BY NATIONALITY, the user will get a screen prompting for the patient nationality as shown below in **Figure (24)**, then after displaying the report, the information of the patients with the selected nationality will be displayed (**Figure (25)**) the user should press or click on the OK button.

If the user selects the fifth option, a report of SPECIFIC PATIENT MEDICAL HISTORY REPORT, the user will get a screen prompting for the patient file number as shown below in **Figure (26)**, then after displaying the report, the information of the selected patient will be displayed (**Figure (27)**).

When finished, the user should press or click on EXIT MAIN REPORTS MENU button.

Figure (18) shows the report displayed for ALL REGISTERED PATIENTS. This report is read only which means that user may not manipulate the data being displayed. To display the detailed record of any patient, the user may click on the DETAIL button appearing at the left of each displayed record as shown below in **Figure (19)**. The user may also click on the PRINT THIS REPORT button to get a hard copy of this report on the attached printer. When finished the user should press or click on the CLOSE button.

Figure (19) shows the detailed record information for the selected patient from **Figure (18)**.

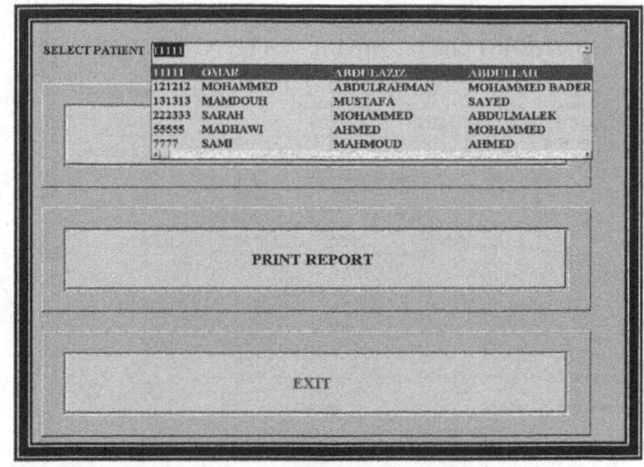

Figure (20) shows the screen prompting the user to select the patient file number to be displayed as a report (in this example the user selected the patient with file number 11111).

Figure (21) shows the screen displaying all data for the selected patient with file number 11111.

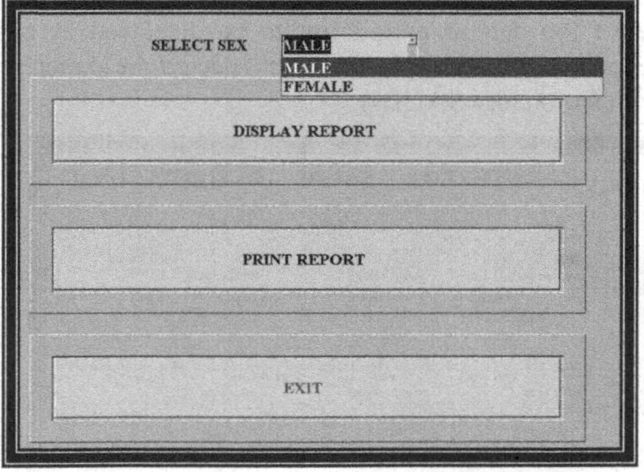

Figure (22) shows the screen prompting the user to select the patient sex to be displayed as a report (in this example the user selected the patient with sex MALE).

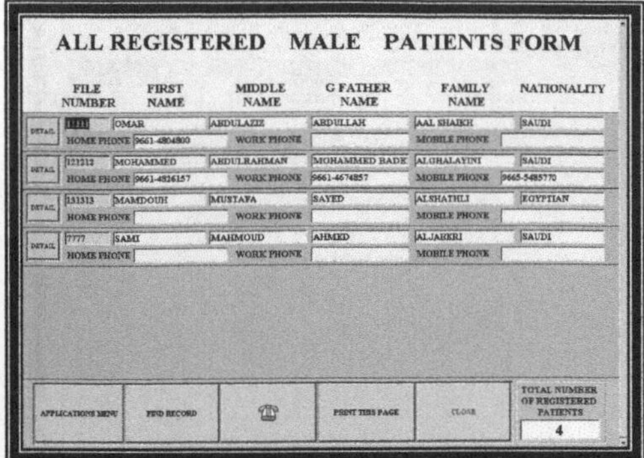
ALL REGISTERED MALE PATIENTS FORM

FILE NUMBER	FIRST NAME	MIDDLE NAME	G FATHER NAME	FAMILY NAME	NATIONALITY
11111	OMAR	ABDULAZIZ	ABDULLAH	AAL SHAIKH	SAUDI
121212	MOHAMMED	ABDULRAHMAN	MOHAMMED BADER	ALGHALAYINI	SAUDI
131313	MAMDOUH	MUSTAFA	SAYED	ALSHATHLI	EGYPTIAN
7777	SAMI	MAHMOUD	AHMED	ALJABERI	SAUDI

APPLICATIONS MENU | FIND RECORD | PRINT THIS PAGE | CLOSE | TOTAL NUMBER OF REGISTERED PATIENTS 4

Figure (23) shows the screen displaying the main data of all registered MALE patients as selected previously. The user may also click on the DETAIL button appearing on the left of each record to display a specific patient detailed information screen.

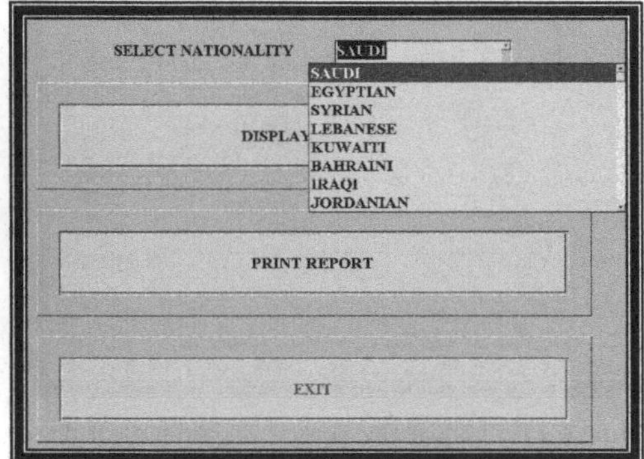
SELECT NATIONALITY SAUDI

SAUDI
EGYPTIAN
SYRIAN
LEBANESE
KUWAITI
BAHRAINI
IRAQI
JORDANIAN

DISPLAY

PRINT REPORT

EXIT

Figure (24) shows the screen prompting the user to select the patient nationality to be displayed as a report (in this example the user selected the patient with nationality SAUDI).

ALL REGISTERED SAUDI PATIENTS FORM

FILE NUMBER	FIRST NAME	MIDDLE NAME	G FATHER NAME	FAMILY NAME
11111	OMAR	ABDULAZIZ	ABDULLAH	AAL SHAIKH
121212	MOHAMMED	ABDULRAHMAN	MOHAMMED BADER	ALGHALAYINI
222333	SARAH	MOHAMMED	ABDULMALEK	AAL ALSHAIKH
55555	MADHAWI	AHMED	MOHAMMED	AAL SAUD
7777	SAMI	MAHMOUD	AHMED	ALJABERI

APPLICATIONS MENU | FIND RECORD | PRINT THIS PAGE | CLOSE | TOTAL NUMBER OF REGISTERED PATIENTS 5

Figure (25) shows the screen displaying the main data of all registered SAUDI patients as selected previously. The user may also click on the DETAIL button appearing on the left of each record to display a specific patient detailed information screen.

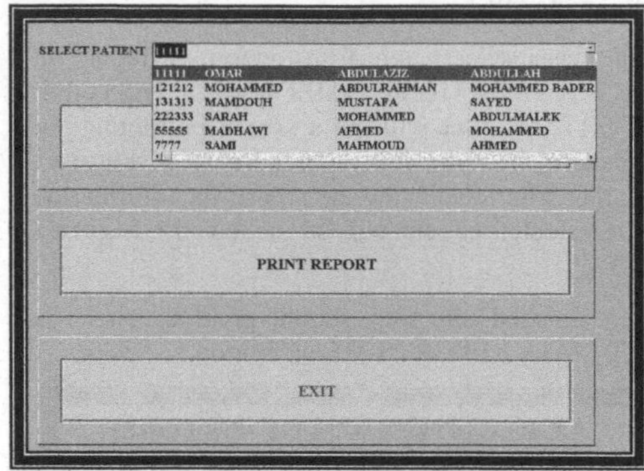
SELECT PATIENT 11111

11111	OMAR	ABDULAZIZ	ABDULLAH
121212	MOHAMMED	ABDULRAHMAN	MOHAMMED BADER
131313	MAMDOUH	MUSTAFA	SAYED
222333	SARAH	MOHAMMED	ABDULMALEK
55555	MADHAWI	AHMED	MOHAMMED
7777	SAMI	MAHMOUD	AHMED

PRINT REPORT

EXIT

Figure (26) shows screen prompting the user to select the patient file number to be displayed as a medical history report (in this example the user selected the patient with file number 11111).

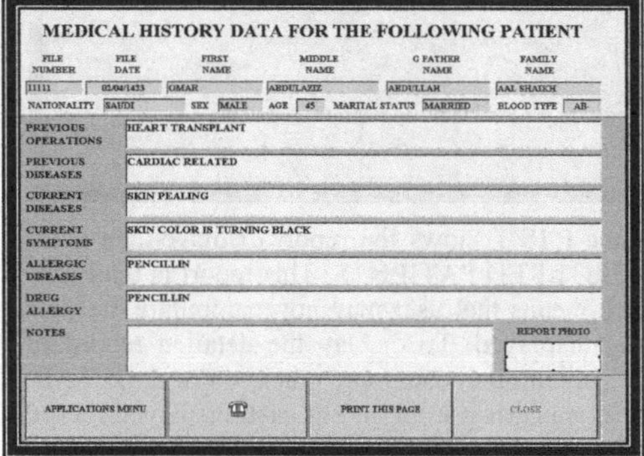
MEDICAL HISTORY DATA FOR THE FOLLOWING PATIENT

FILE NUMBER	FILE DATE	FIRST NAME	MIDDLE NAME	G FATHER NAME	FAMILY NAME
11111	02/04/1423	OMAR	ABDULAZIZ	ABDULLAH	AAL SHAIKH

NATIONALITY SAUDI SEX MALE AGE 45 MARITAL STATUS MARRIED BLOOD TYPE AB

PREVIOUS OPERATIONS: HEART TRANSPLANT
PREVIOUS DISEASES: CARDIAC RELATED
CURRENT DISEASES: SKIN PEALING
CURRENT SYMPTOMS: SKIN COLOR IS TURNING BLACK
ALLERGIC DISEASES: PENCILLIN
DRUG ALLERGY: PENCILLIN
NOTES
REPORT PHOTO

APPLICATIONS MENU | PRINT THIS PAGE | CLOSE

Figure (27) shows screen displaying the medical history data of the patients with file number 11111 as selected previously. The user may preview the stored medical report(s) if there is any stored in the OLE field by clicking twice on the OLE field to activate the application used when the report was saved.

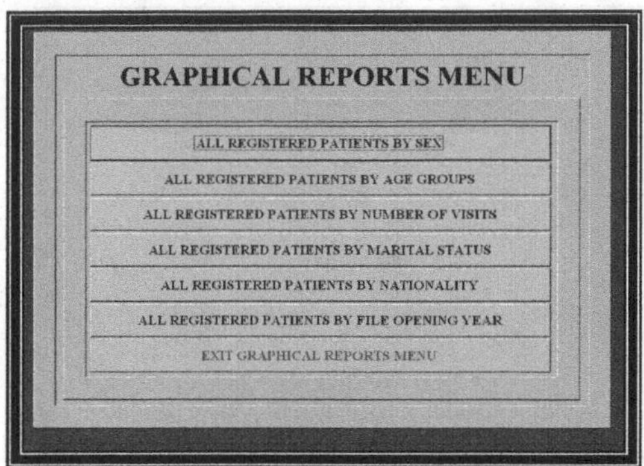

Figure (28) shows the GRAPHICAL PORTS MENU. If the user selects the first option, a report of ALL REGISTERED PATIENTS BY SEX, a graph will be displayed showing the sex distribution for all registered patients as shown below in **Figure (29)**.

If the user selects the second option, a report of ALL REGISTERED PATIENTS BY AGE GROUPS, a graph will be displayed showing the age groups distribution for all registered patients as shown below in **Figure (30)**.

If the user selects the third option, a report of ALL REGISTERED PATIENTS BY NUMBER OF VISITS, a graph will be displayed showing the number of visits distribution for all registered patients. This graph is essential for business statistical reasons at the end of each fiscal year.

If the user selects the fourth option, a report of ALL REGISTERED PATIENTS BY MARITAL STATUS, a graph will be displayed showing the marital status distribution for all registered patients as shown below in **Figure (31)**.

If the user selects the fifth option, a report of ALL REGISTERED PATIENTS BY NATIONALITY, a graph will be displayed showing the nationality distribution for all registered patients **Figure (32)**.

If the user selects the first option, a report of ALL REGISTERED PATIENTS BY FILE OPENING YEAR, a graph will be displayed showing the file opening year distribution for all registered patients as shown below in **Figure (33)**.

When finished, the user should press or click on EXIT GRAPHICAL REPORTS MENU button.

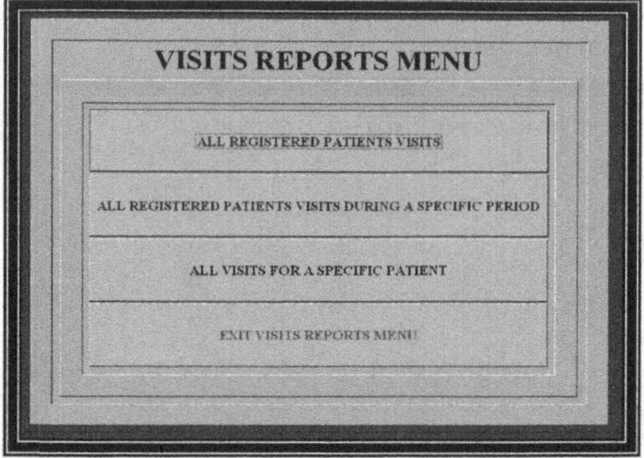

Figure (34) shows the VISITS REPORTS MENU screen. If the user selects the first option, a report of ALL REGISTERED PATIENTS VISITS report will be displayed as shown below in **Figure (35).**

If the user selects the second option, a report of ALL REGISTERED PATIENTS VISITS DURING A SPECIFIC PERIOD will be displayed, the user will get a screen prompting for the starting and ending dates of this time period as shown below in **Figure (36)**, then after displaying the report, the information of the selected time period visits will be displayed (**Figure (37)**).

If the user selects the third option, a report of ALL VISITS FOR A SPECIFIC PATIENT will be displayed. The user will get a screen prompting for the patient file number as shown below in **Figure (38)**, then after displaying the report, the information of all visits of the selected patient will be displayed (**Figure (39)**).

When finished, the user should press or click on EXIT VISITS REPORTS MENU button.

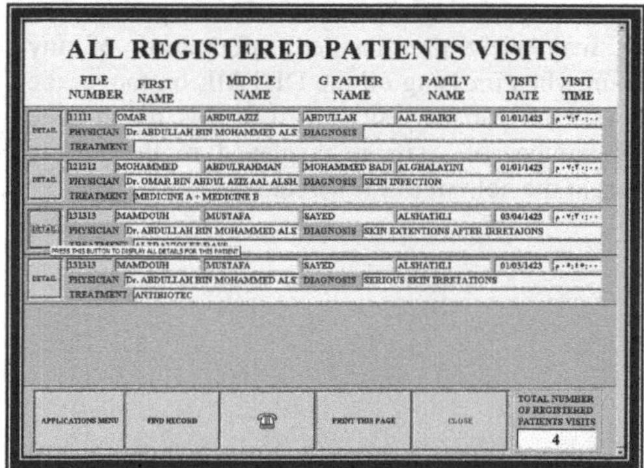

Figure (35) shows the report of all registered patients visits. The user may view the details of any specific patient if the button of DETAIL on the left of each patient record is pressed. The total number of visits is displayed in the yellow boxes at the bottom of the screen.

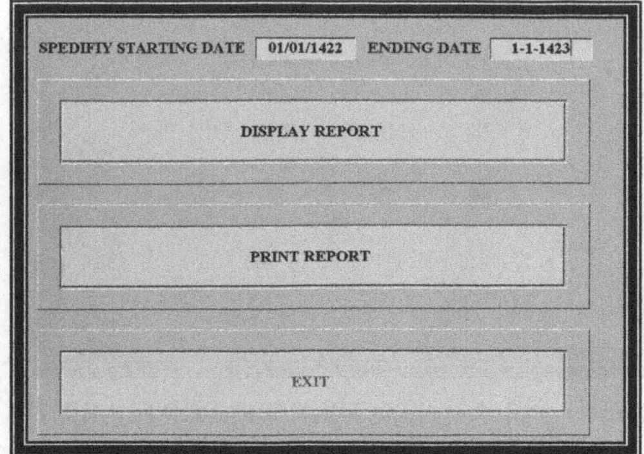

Figure (36) shows the screen prompting the user to enter the starting and the ending dates to display the report of all visits during this specified period. When finished, the user should press or click on the EXIT button.

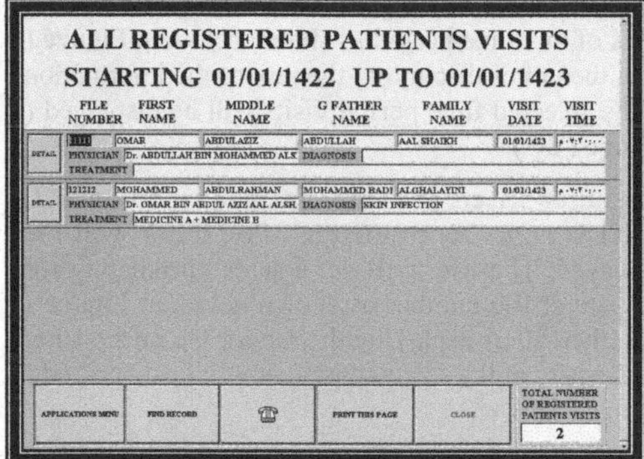

Figure (37) shows the report of all registered patient visits during the specified time period. The user may view the detailed information of any patient when pressing on the DETAIL button on the left of each patient visit record in this report. The total number of visits is displayed in the yellow boxes at the bottom of the screen.

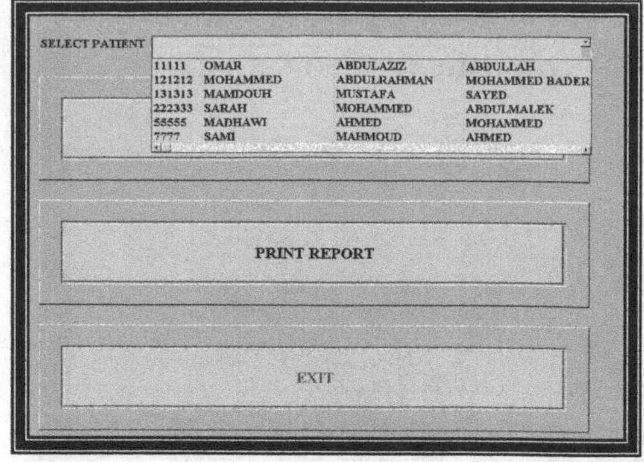

Figure (38) shows the screen prompting the user to select a specific patient file number to display all his/her registered visits report.

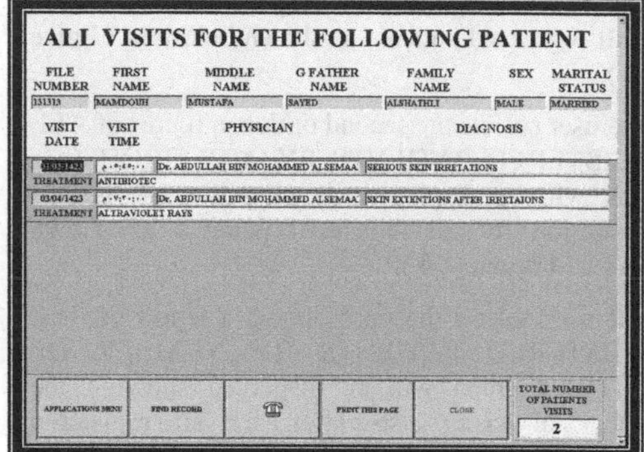

Figure (39) shows the report of all visits of a selected patient (in this example the patient with file number 131313). The total number of visits is displayed in the yellow boxes at the bottom of the screen.

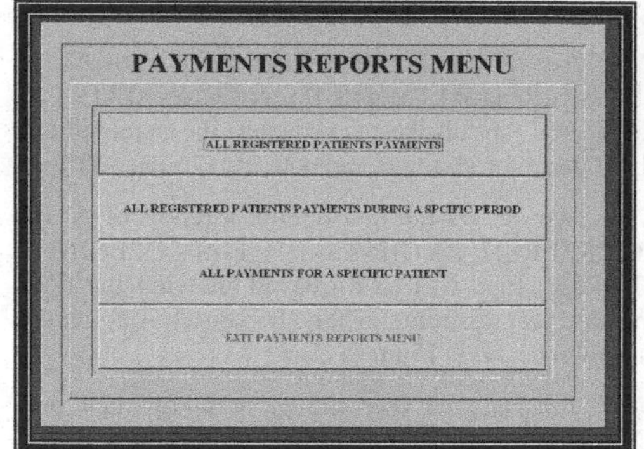

Figure (40) shows the PAYMENTS REPORTS MENU screen. If the user selects the first option, a report of ALL REGISTERED PATIENTS PAYMENTS report will be displayed as shown below in **Figure (41)**.

If the user selects the second option, a report of ALL REGISTERED PATIENTS PAYMENTS DURING A SPECIFIC PERIOD will be displayed, the user will get a screen prompting for the starting and ending dates of this time period as shown below in **Figure (42)**, then after displaying the report, the information of the selected time period patients visits will be displayed (**Figure (43)**).

If the user selects the third option, a report of ALL PAYMENTS FOR A SPECIFIC PATIENT will be displayed. The user will get a screen prompting for the patient file number as shown below in **Figure (44)**, then after displaying the report, the information of all visits of the selected patient will be displayed (**Figure (45)**).

When finished, the user should press or click on EXIT PAYMENTS REPORTS MENU button.

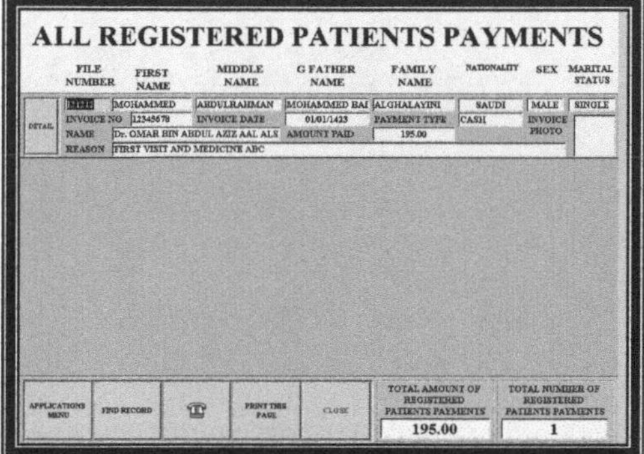

Figure (41) shows the report of all registered patients payments. The user may also view the details of any specific patient when the DETAIL button on the left of each record is pressed. The total number of payments and the total amount of payments are displayed in the yellow boxes at the bottom of the screen.

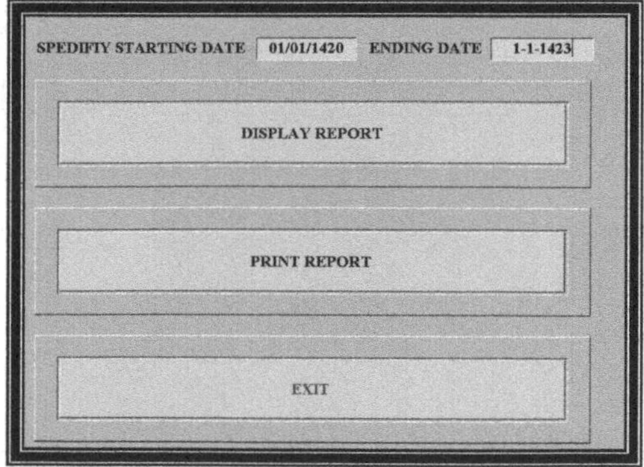

Figure (42) shows the screen prompting the user to enter the starting and the ending dates to display the report of all patients payments during this specified period..

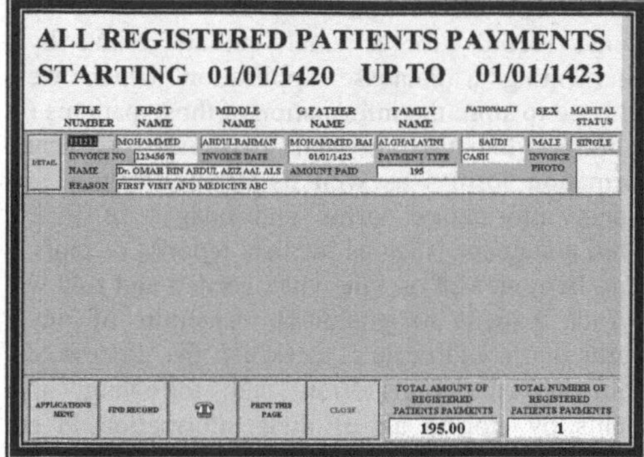

Figure (43) shows the screen report for all the registered patients payments during a specified time period (in this example the starting date is 01/01/1420 and the ending date is 01/01/1423). The total number of payments and the total amount of payments are displayed in the yellow boxes at the bottom of the screen.

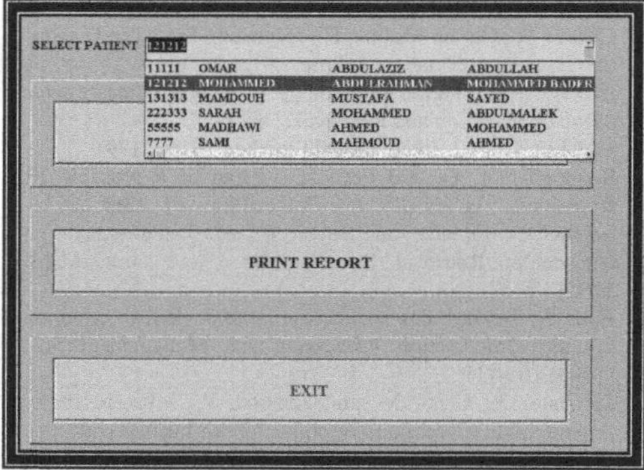

Figure (44) shows the screen prompting the user to select the patient file number to display the report of his/her payments.

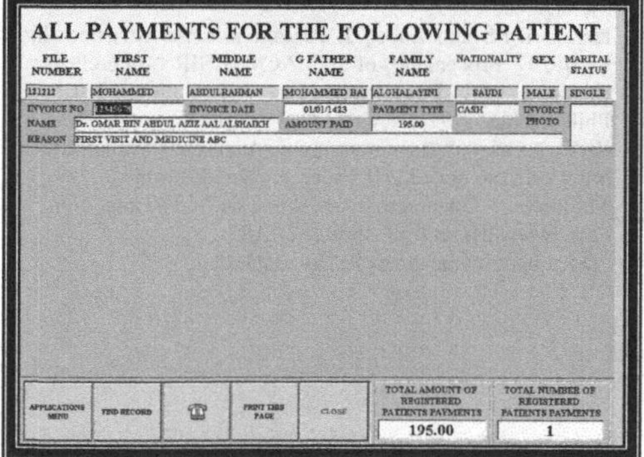

Figure (45) shows the report of all payments made by a specific patient (in this example the patient with file number 121212). The total number of payments and the total amount of payments are displayed in the yellow boxes at the bottom of the screen.

Conclusion and Summary

Our goal in this paper was to show how we could solve a problem of managing and following up patients records for a small private clinic by implementing a database application which we could use to store the information of those patients (according to the official designed clinical information form) as well as storing images of patients information forms and images of their related documents such as medical reports, receipts, etc. to be reviewed on line when needed and follow up each patient to guarantee detection of any failures that could occur. Practically, the process of automating the registration and following up patients by storing all of their information in a database system, tremendously reduced the effort and time previously consumed which added a major efficiency credit to the role of The Skin and Allergy Clinic and it's "automation of all tasks" trend.

References

1 - Frei, H.P. Information retrieval - from academic research to practical applications. In: Proceedings of the 5th Annual Symposium on Document Analysis and Information Retrieval, Las Vegas, April 1996.

2 - Clifton, H., Garcia-Molina, H., Hagmann, R.: ''The Design of a Document Database'', Proc. of the ACM

3 - ISO/IEC, Information Technology - Text and office systems - Document Filing and Retrieval (DFR) -

4 - Part 1 and Part 2, International Standard 10166, 1991

5 - Sonnenberger, G. and Frei, H. Design of a reusable IR framework. In: SIGIR'95: Proceedings of 18th ACM-SIGIR Conference on Research and Development in Information Retrieval, Seattle, 1995. (New York: ACM, 1995). 49-57.

6 - Fuhr, N. Toward data abstraction in networked information systems. Information Processing and Management 5(2) (1999) 101-119.

7 - Jacobson, I., Griss, M. and Jonsson, P. Software reuse: Architecture, process and organization for business success. (New York: ACM Press, 1997).

8 - Davenport, T.H. Information ecology: Mastering the information and knowledge environment. (Oxford: Oxford University Press, 1997).

9 - Bocij, P. et al., Business information systems: Technology, development and management. (London: Financial Times Management, 1999).

10 - Kaszkiel, M. and Zobel, J. Passage retrieval revisited. In: SIGIR'97: Proceedings of 20th ACM-SIGIR Conference on Research and Development in Information Retrieval Philadelphia, 1997. (New York: ACM, 1997).

11 - Richard Casey, " Document Image Analysis" , available at: http://cslu.cse.ogi.edu/HLTsurvey/ch2node4.html

12 - M Simone, " Document Image Analysis And Recognition" http://www.dsi.unifi.it/~simone/DIAR/

13 - "Document Image Analysis" available at : http://elib.cs.berkeley.edu/dia.html

14 - "Read and Display an Image" available at : http://www.mathworks.com/access/helpdesk/help/toolbox/images/getting8.html

15 - ALGhalayini M., Shah A., "Introducing The (POSSDI) Process : The Process of Optimizing the Selection of The Scanned Document Images". International Joint Conferences on Computer, Information, Systems Sciences, and Engineering (CIS^{2}E 06) December 4 - 14, 2006

Automating The Work at KSU Scientific Council : A Case Study on Using an Imaging Database to Manage the Weekly Scientific Council Meetings

Mohammad AbdulRahman ALGhalayini [1]

[1]Director, Computer and Information Unit, Vice Rectorate for Knowledge Exchange and Technology Transfer
King Saud University, Riyadh, Saudi Arabia

Abstract

Today, many institutions and organizations are facing serious problem due to the tremendously increasing size of documents, and this problem is further triggering the storage and retrieval problems due to the continuously growing space and efficiency requirements. This problem is becoming more complex with time and the increase in the size and number of documents in an organization. Therefore, there is a growing and continuous demand to address this problem. This demand and challenge can be met by converting the tremendous amount of paper documents to images using a process to enable specialized document imaging people to select the most suitable image type and scanning resolution to use when there is a need for storing documents images. This documents management process, if applied, attempts to solve the problem of the image storage type and size to some extent. In this paper, we present a case study resembling an applied process to manage the documents in the scientific council in King Saud University and store them in an imaging database to make the retrieval of information and various scanned images easier and faster; therefore, we optimize the work process and maximize the efficiency of the scientific council tasks.

Introduction

Problem Definition

The King Saud University Scientific Council is a vital unit in the university which is responsible for the academic grading and evaluation of all university academic members. This council meets weekly and the committee is formed of approximately 28 members[1], and the vice rector for research and graduate studies who chairs the council. For years, this council operates all the administrative work manually which is collecting all the documents submitted by the college deans and makes 28 copies in folders to be distributed on all council members. Practically, it was facing difficulty in organizing all the documents and making 28 copies (each of them has approximately 500 pages) weekly then dispose those folders just after the meeting. Files and papers were piling up quickly and consuming a lot of work and space for storing them in addition to the man power needed for copying and organizing all those folders.

The need for a shared database system was urgent to make it easy to store the submitted documents information and images as they arrive to be reviewed in the meeting.

The need for searching for specific document information quickly and the need to take a look at the attached documents from outside the meeting room and in different times in addition to the need for reviewing and referencing older documents quickly raised the need to have a customized imaging database system to manage all this.

Solution Steps

1 - A document scanner, 28 high specs table PC (as shown in Scientific Council meeting room photo (**Figure 1**), and a central server were bought in preparation to be used when the documents are submitted.

2 - A database was designed, implemented, and tested to be used to store the documents information and images.

[1] One representative member from each academic college

K. Elleithy (ed.), *Advanced Techniques in Computing Sciences and Software Engineering*,
DOI 10.1007/978-90-481-3660-5_35, © Springer Science+Business Media B.V. 2010

3 - By the time the database system was ready, a dedicated employee was trained on using the database system.
4 - Each time the data entry employee used to come and scan the documents images then store the scanned documents in a specific folder dated per academic year[2].
5 - This hard disk arrangement made it easy to quickly reference any stored document images and retrieve them when necessary.
6 - When the document data was ready to be entered, all what the data entry person worried about is storing the information of the main document as text in the database system corresponding fields, and storing the previously saved scanned image of the form in the OLE[3] field (which was scanned and stored previously as .max file or as a PDF file on the local hard disk) in the proper OLE field.
7 - After storing one full document information and storing scanned documents belonging to that record, the document is signed and dated by the data entry person, then stored back in the paper folder it belongs to. This step was necessary to easily figure out which documents where completely entered and which were not.
8 - Each week end, a complete copy of this database system was backed up on the central file server to be accessed by scientific council authorized persons. In addition to saving a backup copy on a writable CD.

Special Notes About KSU Scientific Council Document Management Database System

1 - This system was designed using MS-ACCESS RDBMS which is one of the MS-OFFICE 2003 applications.
2 - It was taken into consideration making this system a very user friendly and fully in Arabic to make easy to use as much as possible.
3 - All Scientific Council meeting room PC's were updated to have 2 GB of RAM at least and 128 MB display memory to access the system conveniently.
4 - All scientific Council meeting room PC's were connected through a local area network to being able to connect to the dedicated central file server (SCOUNCIL_SERVER) with a special user name and password for all authorized council committee members.
5 - The system is accessible from outside the meeting room (Scientific Council Members Offices) through a password for security reasons.
6 - It is possible to print any page or pages of the documents stored in database system on the central color laser printer, which is directly connected to the server.

Contents And Data Security

Since the Scientific Council document management database system may contain very sensitive information which could be very confidential, it was necessary to the scientific council member including the data entry employee, to keep a special system password to use it to access the system whether from his own pc in the office or from connecting to the server from outside the room.

Even though the Scientific Council server was appearing on the university network, the server was not accessible to any unauthorized person; therefore, the system was also not accessible to any unauthorized person.

One more thing regarding data security is the issue of weekly backup of the whole database system including the stored documents images on the server hard disk and on a removable writable CD.

Figure (1) Shows the scientific council meeting room setup with the piles of document folders and the central server used next to the image scanner.

Hardware Used

1 – PC's with the following specifications :
3 Mhz Intel processor
1 GB DDR RAM
80 GB WD Hard Disk
128 MB ATI display adaptor
10/100 3com network adaptor

[2] Each Academic year the council meets approximately 60 times.
[3] OLE is Object Linking and Embedding

40 X CD player
17 inch display monitor

2 - Image scanner with the following specifications :
40-60 pages per minute
Simplex / duplex
Black and white and color scanning

3 – Central file server with the following specifications :
3 Mhz Intel processor
1024 MB DDR RAM
2 X 70 GB SCSI Hard Disk
64 MB ATI display adaptor
10/100 3com network adaptor
52 X CD writer
17 inch display monitor

Software Used

Windows XP on data entry PC
Windows 2003 server on the file server
PaperPort 11.0 Professional[4] version for scanning images on the data entry PC.

TIME CONSUMED

No.	Details	Time consumed
1	Setting up the meeting room network and connections.	2 week
2	Buying and setting up the hardware.	2 weeks
3	Database system design and implementation and testing.	4 weeks
4	Data entry (as the documents arrive to the scientific council)	Approximately 10 complete records per day

SCREEN PHOTOS OF THE KSU SCIENTIFIC COUNCIL DOCUMENT MANAGEMENT DATABASE SYSTEM

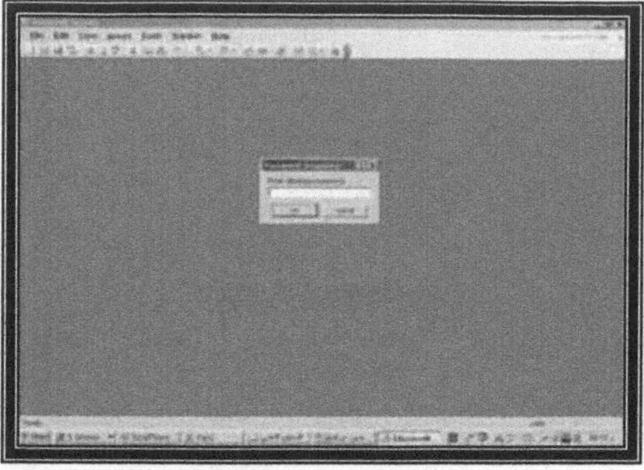

Figure (2) Shows the starting screen prompting the user for the pre-programmed password. After the user enters the password provided by the programmer, the user should press or click on the OK button.

Figure (3) Shows the starting screen of the Scientific Council Document Management Database System.

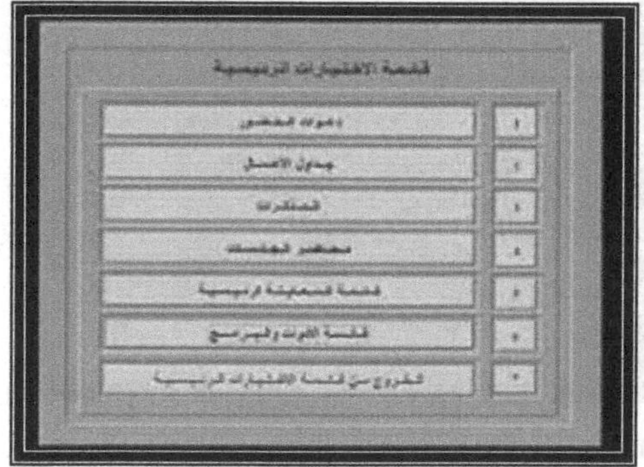

Figure (4) Shows the main selections menu to be chosen from by the user.

[4] PaperPort is a trade name of ScanSoft Inc.

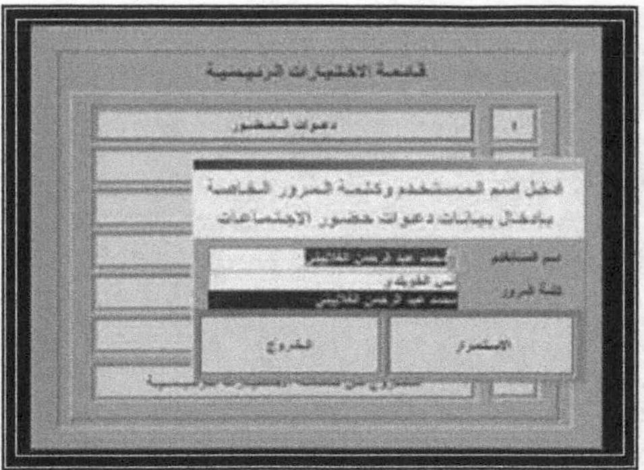

Figure (5) Shows the authorization screen for data entry. Only two persons are authorized to access the data entry forms.

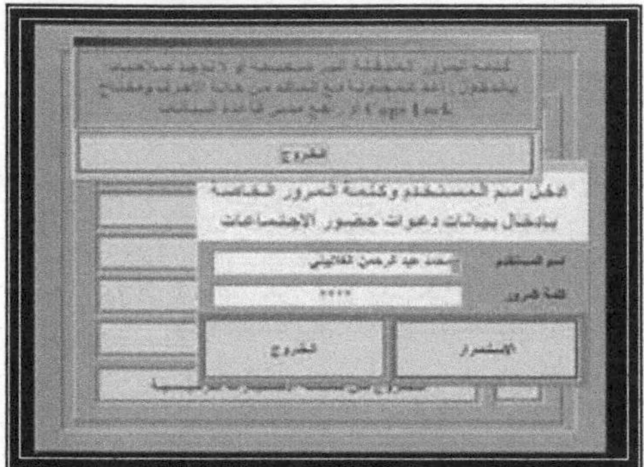

Figure (6) Shows the rejection of the database when a wrong password is entered to access the data entry forms.

Figure (7) Shows the call for a meeting information form which includes an image of the letter to be sent via email to all members.

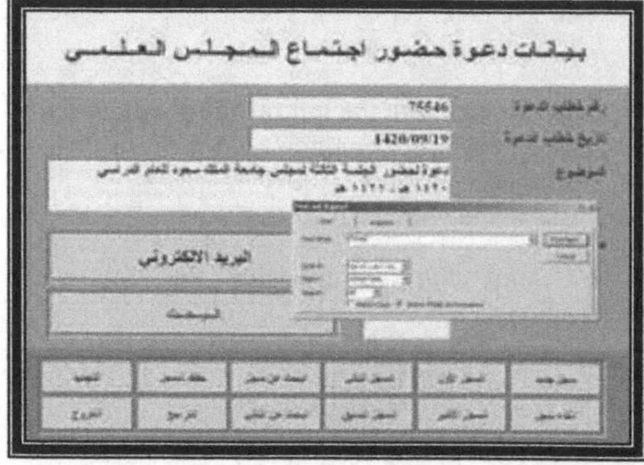

Figure (8) Shows the ability to search for a specific meeting information by accessing the built-in search command button and entering a meeting letter numbers.

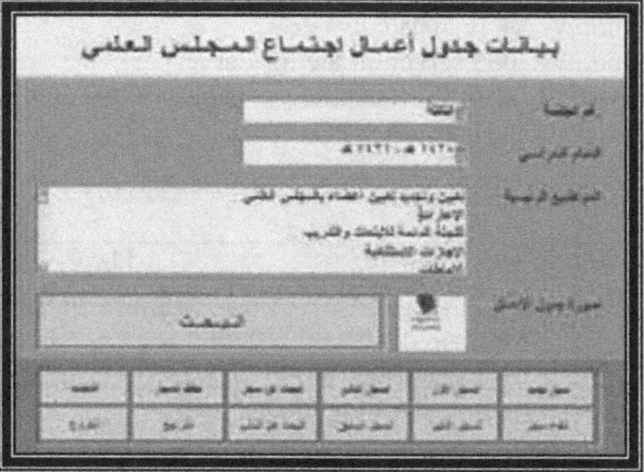

Figure (9) Shows the meeting agenda information form which includes an image of the agenda letter(s) to be reviewed by all members prior to the meeting. In addition the figure shows the categorization field of topics to be discussed during the meeting.

Figure (10) Shows the documents to be reviewed and discussed information form in addition to the image of all the related documents for this displayed record.

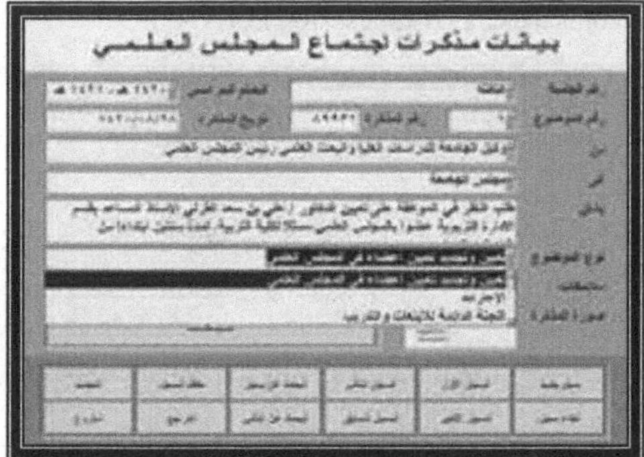

Figure (11) Shows the ability to select the topic of the documents to filter out the documents to be reviewed and discussed. This is some times necessary for priority reasons during the meeting.

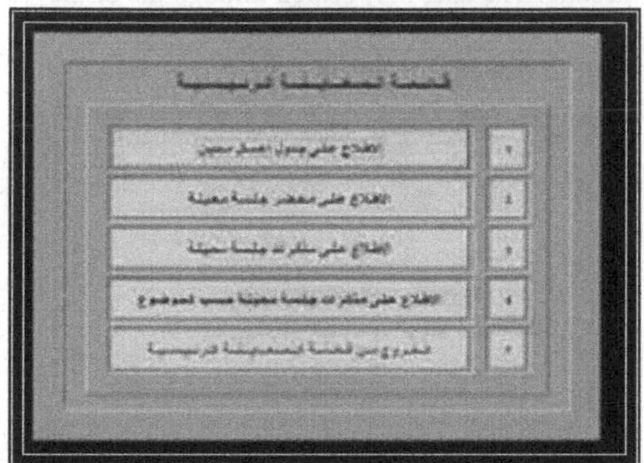

Figure (12) Shows the review main selections menu which is accessible to all council members.

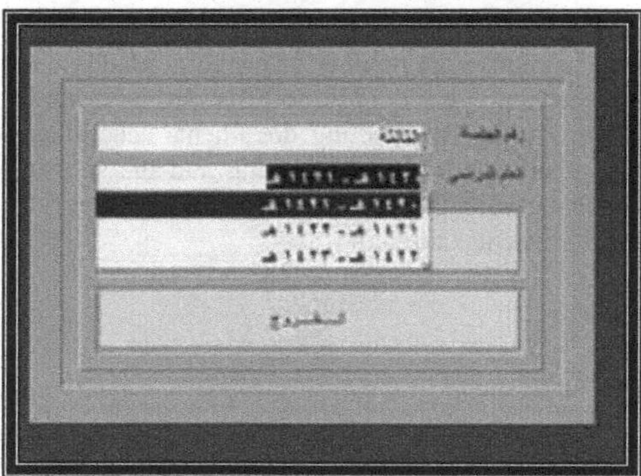

Figure (13) Shows the screen which allows the user to choose the number of meeting in addition to the academic year in order to filter out the result list.

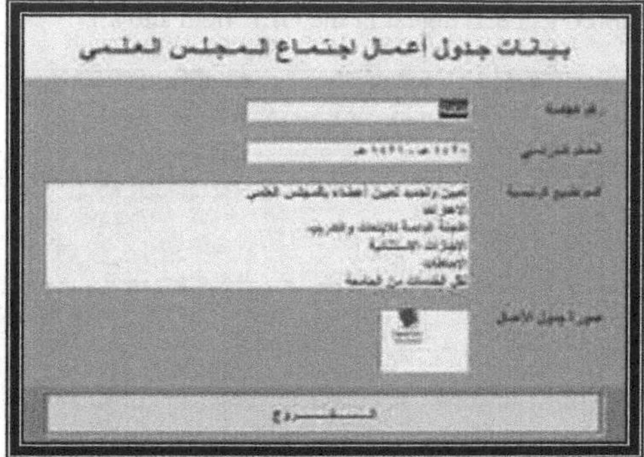

Figure (14) Shows the meeting agenda screen information form which includes an image of the letter to be sent via email to all members.

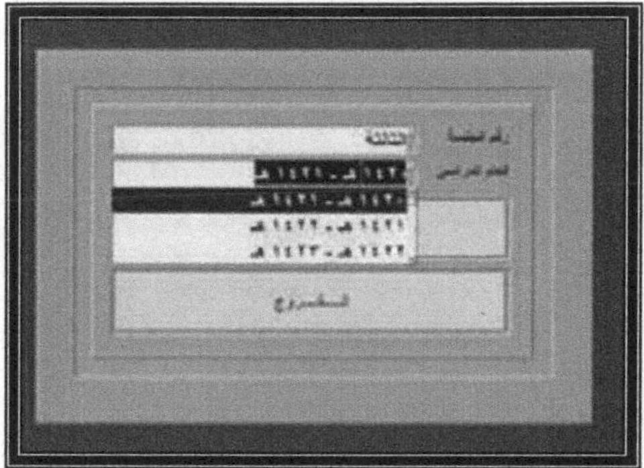

Figure (15) Shows the screen which allows the user to choose the number of meeting in addition to the academic year in order to filter out the result list for reviewing the meeting documents according to the specified number of meeting and the specified academic year.

Figure (16) Shows the screen which displays the result of **Figure (15)** query. The user is able to review the specified meeting documents images stored as OLE object in the OLE field shown.

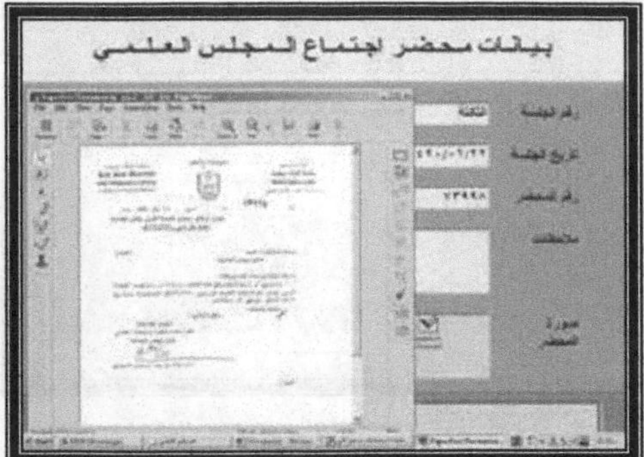

Figure (17) Shows the display of the meeting documents when activated by the application used for storing, here it was PaperPort image .max

Figure (18) Shows the screen which allows the user to choose the number of meeting in addition to the academic year in order to filter out the result list for reviewing the meeting documents according to the specified number of meeting and the specified academic year.

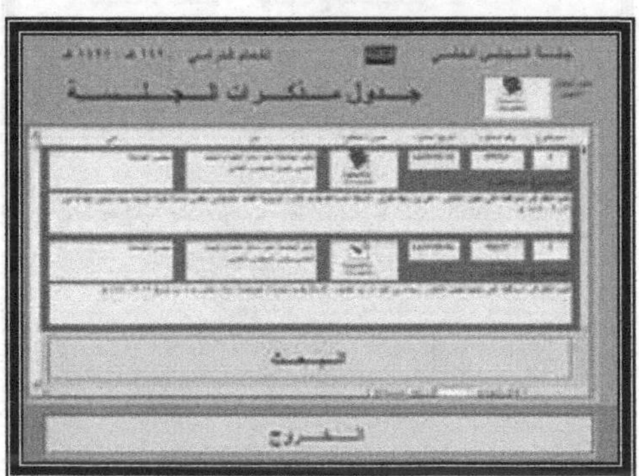

Figure (19) Shows the filtered result displayed for all the submitted documents for the specified meeting. The user may search through the displayed records using a custom search key.

Figure (20) Shows the filtered result displayed for all the submitted documents for the specified meeting. The user may search through the displayed records using the built-in search key, here searching for a specific text in the topic field.

Figure (21) Shows the screen which allows the user to choose the number of meeting in addition to the academic year in order to filter out the result list for reviewing the meeting documents according to the specified number of meeting and the specified academic year in addition to the topic category of the documents stored.

Figure (22) Shows the filtered result displayed for all the submitted documents for the specified meeting and according to the topic category specified in **Figure (21).** The user may search through the displayed records using a custom search key.

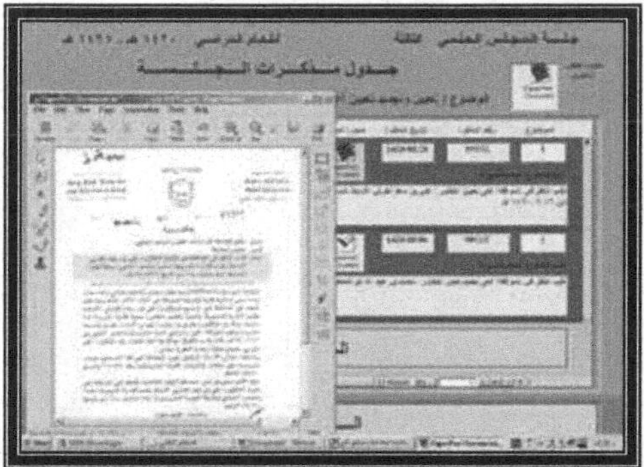

Figure (23) Shows the display of the meeting documents when activated by the application used for storing, here it was PaperPort image .max . The user may browse through the document pages using the "Next Page" or "Previous Page" buttons appearing at the top of the PaperPort Application document viewer.

Conclusion and Summary

Our goal in this paper was to show how we could solve a problem of managing piles of documents by implementing a relatively small imaging database application which we could use to store the information of thousands of documents and their scanned images to be reviewed during the weekly meetings and accessed at different times as well by a group of faculty members. Practically, this imaging database management was able to save the KSU Scientific Council members a lot of effort in referring to any specific document and reviewing it in addition to the amount of tremendous reduction in time of document retrieval which added a major efficiency credit to the role of KSU Scientific Council and it's "automation of all tasks" trend.

References

1 - Frei, H.P. Information retrieval - from academic research to practical applications. In: Proceedings of the 5th Annual Symposium on Document Analysis and Information Retrieval, Las Vegas, April 1996.

2 - Clifton, H., Garcia-Molina, H., Hagmann, R.: ''The Design of a Document Database'', Proc. of the ACM

3 - ISO/IEC, Information Technology - Text and office systems - Document Filing and Retrieval (DFR) -

4 - Fuhr, N. Toward data abstraction in networked information systems. Information Processing and Management 5(2) (1999) 101-119.

5 - Jacobson, I., Griss, M. and Jonsson, P. Software reuse: Architecture, process and organization for business success. (New York: ACM Press, 1997).

6 - Davenport, T.H. Information ecology: Mastering the information and knowledge environment. (Oxford: Oxford University Press, 1997).

7 - Bocij, P. et al., Business information systems: Technology, development and management. (London: Financial Times Management, 1999).

8 - M Simone, " Document Image Analysis And Recognition" , available at : http://www.dsi.unifi.it/~simone/DIAR/

9 - Richard Casey, " Document Image Analysis" , available at: http://cslu.cse.ogi.edu/HLTsurvey/ch2node4.html

10 - M Simone, " Document Image Analysis And Recognition" http://www.dsi.unifi.it/~simone/DIAR/

11 - "Read and Display an Image" available at : http://www.mathworks.com/access/helpdesk/help/toolbox/images/getting8.html

12 - ALGhalayini M., Shah A., "Introducing The (POSSDI) Process : The Process of Optimizing the Selection of The Scanned Document Images". International Joint Conferences on Computer, Information, Systems Sciences, and Engineering (CIS^{2}E 06) December 4 - 14, 2006

Automating The Work at KSU Rector's Office : A Case Study on Using an Imaging Database System to Manage and Follow-up Incoming / Outgoing Documents

Mohammad A. ALGhalayini [1]

[1]Director, Computer and Information Unit, Vice Rectorate for Knowledge Exchange and Technology Transfer
King Saud University, Riyadh, Saudi Arabia

Abstract

Today, many institutions and organizations are facing serious problem due to the tremendously increasing size of flowing documents, and this problem is further triggering the storage and retrieval problems due to the continuously growing space and efficiency requirements. This problem is becoming more complex with time and the increase in the size and number of documents in an organization; therefore, there is a growing demand to address this problem. This demand and challenge can be met by converting the tremendous amount of paper documents to images using a defined mechanism to enable specialized document imaging people to select the most suitable image type and scanning resolution to use when there is a need for storing documents images. This documents management mechanism, if applied, attempts to solve the problem of the image storage type and size to some extent. In this paper, we present a case study resembling an applied process to manage the documents in King Saud University rector's office and store them in an imaging database system to make the retrieval of information and various scanned images easier and faster; therefore, we optimize the work process and maximize the efficiency of the KSU rector's office tasks.

Problem Definition

Introduction

KSU is King Saud University which is one of the biggest universities in the Middle East region and is the oldest and largest government university in the Kingdom of Saudi Arabia[1]. In KSU there are 28 Academic colleges (each has several departments) in addition to 5 assisting deanships and about 70 administrative departments. KSU has about 70,000 students in addition to around 14,000 employees.

The rector office is the office of the university top official and top decision making personnel. Each day a big amount of documents flow in the rector office to be signed or reviewed by him, some of those letters are kept in the office, others are redirected to other university officials[2] for further actions. In addition, there are also a number of daily letters that get initiated from the rector office and start their flow path out side to other departments in KSU.

The average amount of the daily incoming letters is around 250 letters, some of them come with many attachments such as reports or other official documents. And the average amount of outgoing letters is about 150 letters as well.

Each incoming or outgoing letter is photocopied and stored in paper folder on shelves in a dedicated archive room. With the accumulating number of papers stored, there was a serious problem in retrieving a photo copy of any previously saved letter especially if it was a rather old one. The archive room employee used to spend an hour in allocating the needed letter to be re-presented if the subject is to be re-discussed. Hence, the need for a faster retrieval mechanism was rising quickly especially in this sensitive office which was totally un-automated when this problem was taking place.

In 1997, there came a suggestion to design and implement a dedicated database system to store the information and the images of all incoming and outgoing letters and attached documents in addition to assist the office employees to follow-up the letters as they flow in the various KSU offices, and

[1] More information about King Saud University is available at www.ksu.edu.sa

[2] Other KSU officials could be Vice Rectors, College Deans, or General Administrative Directors.

K. Elleithy (ed.), *Advanced Techniques in Computing Sciences and Software Engineering*,
DOI 10.1007/978-90-481-3660-5_36, © Springer Science+Business Media B.V. 2010

to accomplish this there was a need to automate the office and getting it ready for such a shift.

Solution Steps

1 - A local area network was designed and implemented which connected 20 PC's together to a central file server.
2 - A document scanner and a high specs PC were bought in preparation to be used when those letters and attached documents start to arrive in the rector's office.
3 - A database was designed, implemented, and tested to be used to store the incoming and outgoing letters information as well as the scanned documents belonging to each KSU department or college in addition to the personal individuals letters submitted to the rector's office.
4 - By the time the database system was ready, a dedicated rector's office employee started to collect the letters and the attached documents in the office according to a predefined time schedule and to a special office document flow.
5 - The data entry person who was previously trained on using the database system was ready to receive the letters with any accompanying documents and arranging them in folders categorized by colleges and departments for easier referencing.
6 - Each time the data entry employee used to come and scan the letters and the accompanying documents then store them in a folder on the local hard disk.
7 - This hard disk arrangement made it easy to quickly reference any stored document images and retrieve them when necessary.
8 - When the letter data was ready to be entered, all what the data entry person worried about is storing the information of the letter as text in the database system corresponding fields and storing the previously saved scanned image of the letter in the OLE field in addition to storing the attached documents file (which was scanned and stored previously as .max file on local hard disk) in the proper OLE field.
9 - After storing one full letter information and storing scanned documents belonging to that record, the letter was stored back in the paper folder it belongs to. This step was necessary to easily figure out which documents where completely entered and which were not.
10 - Each week end, a complete copy of this database system was backed up on the central file server to be accessed by the authorized persons. In addition to saving a backup copy on a writable CD.

Special Notes About The KSU Incoming/Outgoing Documents Management & Follow-up Imaging Database System

1 - This system was first designed using MS-ACCESS 97 RDBMS which is one of the MS-OFFICE 1997 applications. Then each time a new MS-OFFICE edition came out, an upgrading modification was done to the database system till now which is being used using MS-ACCESS 2003.
2 - It was taken into consideration making this system a very user friendly and totally in Arabic language and easy to use as much as possible.
3 - All rector's office PC's were updated to have 2048 MB of RAM at least and 256 MB display memory to access the system conveniently.
4 - All rector's PC's were connected through network to being able to connect to the dedicated central file server (RECTOR_SERVER) with a special user name and password for all authorized rector's office employees.
5 - The system is accessible from all KSU campus offices through a password for security reasons.
6 - It is possible to print any page or pages of the documents stored in database system on the central color laser printer, which is directly connected to the server.

Contents and Data Security

Since the nature of the rector's office flowing letters and documents may contain some private information which could be very confidential, it was necessary to the rector's office member including the data entry employee, to keep a special system password to use it to access the system whether from his own pc in the office or from connecting to the server from outside KSU campus. And even though the server could be seen to others on KSU network but it was not accessible to any unauthorized person; therefore, the system was also not accessible to any unauthorized person.

One more thing regarding data security is the issue of weekly backup of the whole database system including the stored documents images on the server hard disk and on a removable writable CD.

Hardware Used

1 - PC with the following specifications :
   3.2 Mhz Intel processor
   2048 MB DDR RAM
   80 GB WD Hard Disk
   256 MB ATI display adaptor
   10/100 3Com network adaptor
   52 X CD writer
   17 inch display monitor

2 - Image scanner with the following specifications :
 40-60 pages per minute
 Simplex / duplex
 Black and white and color scanning

3 – Central file server with the following specifications :
 2.8 Mhz Intel processor
 2048 MB DDR RAM
 80 GB WD Hard Disk
 128 MB ATI display adaptor
 10/100 3com network adaptor
 52 X CD writer
 17 inch display monitor

Software Used

1 - Windows 1995 through 2002 (XP) on data entry PC
2 - Windows NT, Windows 2000 server, and Windows 2003 server on the file server
3 - PaperPort 2.0 through 11.0 Professional version for scanning images on the data entry PC

Time Consumed

No.	Details	Time consumed
1	Designing and implementing the LAN	3 weeks
2	Upgrading the office PC's	1 weeks
3	Buying and setting up the hardware.	2 weeks
4	Database system design , implementation, and testing.	4 weeks
5	Data entry	30 – 50 complete records / day

Screen Photos Of The KSU Rector's Office Incoming / Outgoing Documents Management & Follow-up Database System

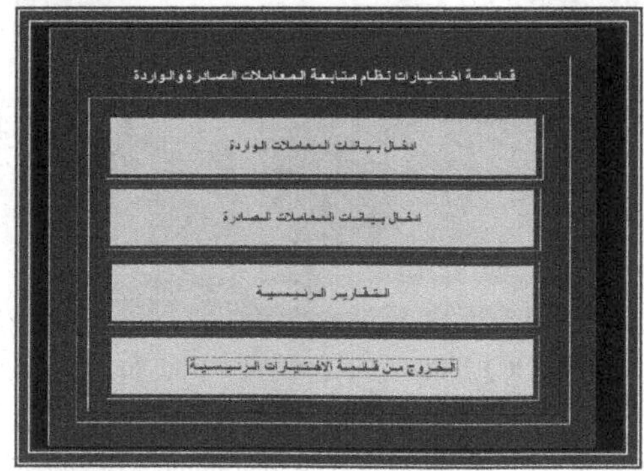

Figure (1) Shows the main selections menu screen

Figure (2) Shows the main data entry form for the incoming documents after passing through the authorization screen. The green field indicates the primary field. The user may use the command keys on the bottom of the form to manipulate the data or/and add new records, In addition to the search keys to search through the previously stored faculty records.

Figure (3) Shows the form displaying the information of the stored record when entering a duplicate document ID number which belongs to a previously stored document record since the document ID number field is the primary key.

Figure (4) Shows the lower part of the main data entry form which includes entering the actions been taken on this specific document for following up the flow of it in addition to the information of the data entry person.

Figure (5) Shows the display of the stored sample of the PDF file as an OLE object. The user may browse through multiple pages here as well.

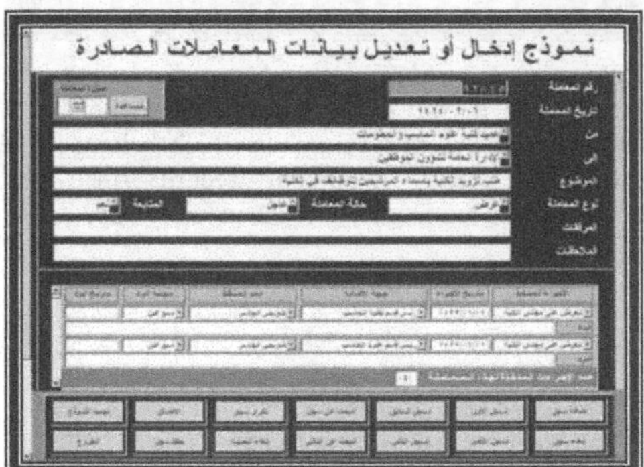

Figure (6) Shows the main data entry form for the outgoing. The blue field indicates the primary field. The user may use the command keys on the bottom of the form to manipulate the data or/and add new records, In addition to the search keys to search through the previously stored faculty records. It also shows the sub-form to enter all the actions taken for the outgoing document for follow up reasons which is very necessary in such systems.

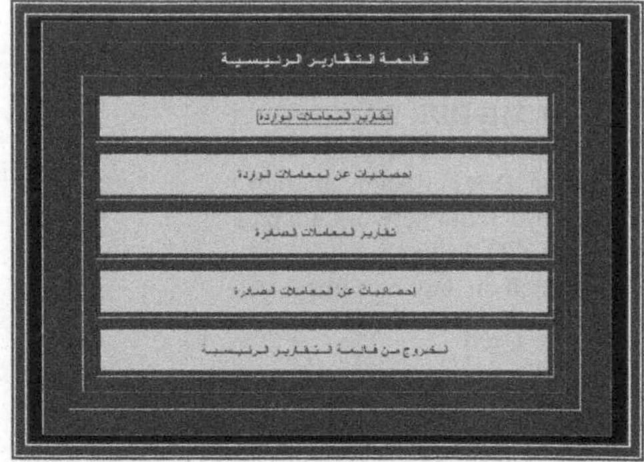

Figure (7) Shows the main reports selection menu.

Figure (8) Shows the main reports selection menu for the incoming documents.

Figure (9) Shows the general report for all the incoming documents. Each documents record shows the image field on it's left side and it details screen button on it's right side. The total number of stored incoming documents is displayed at the bottom of the screen.

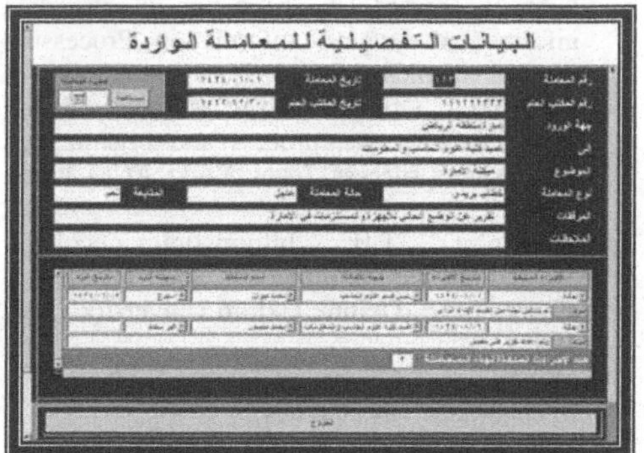

Figure (10) Shows the details screen form for the document record selected after pressing on the button on the right of the record in **Figure (9).**

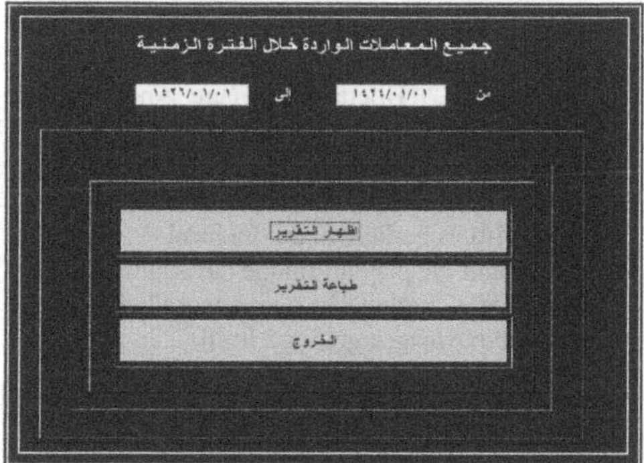

Figure (11) Shows the time period selection form which is displayed prior to displaying the report for all the incoming documents during the specified period.

Figure (12) Shows the general report display form for all the incoming documents during the specified period in **Figure (11).**

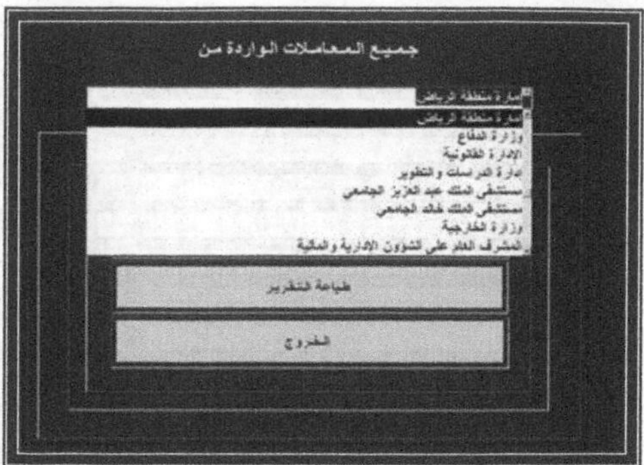

Figure (13) Shows the combo box selections screen to enable the user to choose the institution which the incoming documents came from. This screen is displayed prior to displaying the filtered report for all those yield documents.

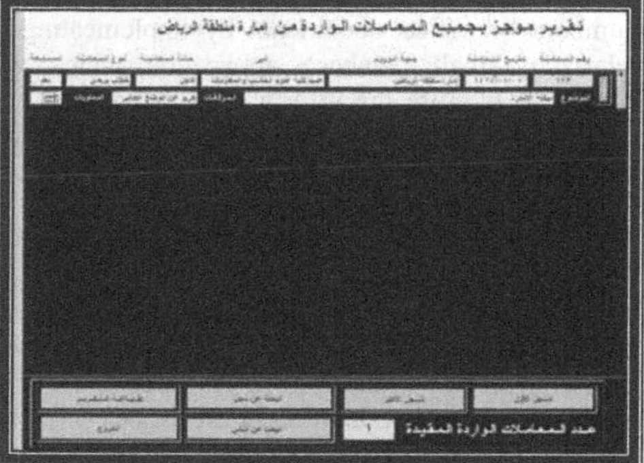

Figure (14) Shows the filtered report form screen for all the incoming documents from the institution selected **Figure (13).**

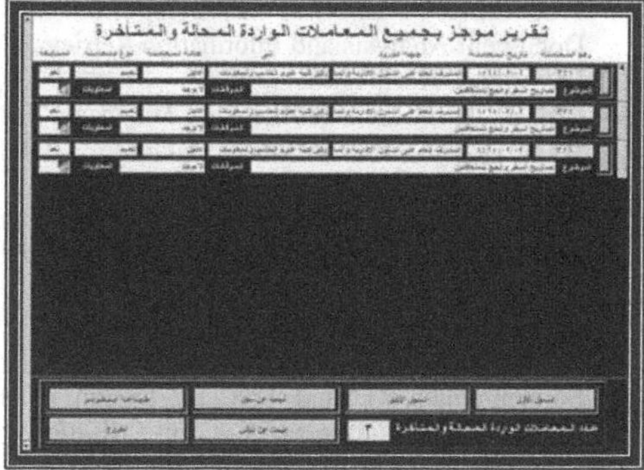

Figure (15) Shows the report screen for all the incoming documents that where redirected for certain actions and are over due (late). This report is necessary for following up reason in such an office.

Figure (16) Shows the main reports selection menu for the outgoing documents[3].

Conclusion and Summary

Our goal in this paper was to show how we could solve a problem of managing and archiving large numbers of office documents by implementing a relatively small database application which we could use to store the information of thousands of circulating letters and documents as well as storing their images to be reviewed on line when needed. Practically, the time and effort consumed of retrieving the document information and reviewing it's corresponding image were reduced tremendously from 1 hour to 1 minute per request; hence, we added a major optimization and efficiency credit to the role of KSU rector's office and it's "automation of all tasks" trend.

References

1 - Frei, H.P. Information retrieval - from academic research to practical applications. In: Proceedings of the 5th Annual Symposium on Document Analysis and Information Retrieval, Las Vegas, April 1996.

2 - Clifton, H., Garcia-Molina, H., Hagmann, R.: ''The Design of a Document Database'', Proc. of the ACM

3 - ISO/IEC, Information Technology - Text and office systems - Document Filing and Retrieval (DFR) -

4 - Part 1 and Part 2, International Standard 10166, 1991

5 - Sonnenberger, G. and Frei, H. Design of a reusable IR framework. In: SIGIR'95: Proceedings of 18th ACM-SIGIR Conference on Research and Development in Information Retrieval, Seattle, 1995. (New York: ACM, 1995). 49-57.

6 - Fuhr, N. Toward data abstraction in networked information systems. Information Processing and Management 5(2) (1999) 101-119.

7 - Jacobson, I., Griss, M. and Jonsson, P. Software reuse: Architecture, process and organization for business success. (New York: ACM Press, 1997).

8 - Davenport, T.H. Information ecology: Mastering the information and knowledge environment. (Oxford: Oxford University Press, 1997).

9 - M Simone, " Document Image Analysis And Recognition" , available at : http://www.dsi.unifi.it/~simone/DIAR/

10 - "Document Image Analysis", avalable at : http://elib.cs.berkeley.edu/dia.html

11 - Richard Casey, " Document Image Analysis" , available at: http://cslu.cse.ogi.edu/HLTsurvey/ch2node4.html

12 - M Simone, " Document Image Analysis And Recognition" http://www.dsi.unifi.it/~simone/DIAR/

13 - "Document Image Analysis" available at : http://elib.cs.berkeley.edu/dia.html

14 - "Read and Display an Image" available at : http://www.mathworks.com/access/helpdesk/help/toolbox/images/getting8.html

15 - ALGhalayini M., Shah A., "Introducing The (POSSDI) Process : The Process of Optimizing the Selection of The Scanned Document Images". International Joint Conferences on Computer, Information, Systems Sciences, and Engineering (CIS^2E 06) December 4 - 14, 2006

[3] Screen Figures of all Outgoing documents are very similar to those presented for the Incoming documents and were omitted for briefing reasons.

Using Clinical Decision Support Software in Health Insurance Company

R. Konovalov
Aroma Software
Dubai, United Arab Emirates,
e-mail: roman.konovalov@gmail.com

Deniss Kumlander
Department of Informatics, TTU
Tallinn, Estonia
e-mail: kumlander@gmail.ee

***Abstract* — This paper proposes the idea to use Clinical Decision Support software in Health Insurance Company as a tool to reduce the expenses related to Medication Errors. As a prove that this class of software will help insurance companies reducing the expenses, the research was conducted in eight hospitals in United Arab Emirates to analyze the amount of preventable common Medication Errors in drug prescription.**

Keywords-decision support software; health industry

I. Introduction

Healthcare System in any country is created to help an individual person obtaining better health services. A person can be a patient who has an encounter with a provider. Provider is any healthcare organization like hospital, clinic, or pharmacy. The Provider then claims some or all the charges from Health Insurance Company that is called Payer [1].

The Payer in its turn collects insurance premiums from individual persons. The relationships between Patient, Provider and Payer are shown on Figure 1.

There is a certain number of Medical Errors occurs while patient is treated. As estimated by [2], up to 98,000 Americans die every year due to medical errors – it is more than number of people dying from AIDS, breast cancer or car accidents.

While Medical Error is any mistake that might happen during patient treatment (e.g. during surgery, drug prescription, wrong actions of patient, etc), Medication Errors are mistakes related to prescribing and consuming drugs. This article discusses the usage of software to degrees the number of Medication Errors in drug prescription.

When Providers sends claim to the Payer, the data set must contain information about the drugs prescribed to a patient, if any. It is very common scenario when Provider has no Healthcare Information System implemented, or they do not support Clinical Decision Support. As results, all drugs are prescribed manually and doctor relies only on his memory. In this case the following issues are possible in the drug prescriptions that might cause Medication Errors:

1. Duplications in Drug Therapy – the situation when two or more drugs with the same therapeutic effect are prescribed to the same patient, which might lead to over dosing. Payer may lose money on paying for duplicated drugs that patient might not need. Also, payer will pay for eventual worsened health condition of the patient due to that duplication. As an example, patient might get two drugs DISPRIL and JUSPRIN, but both of them contain Acetylsalicylic Acid as active chemical ingredient. If both drugs are prescribed to the same patient, it might cause overdose.

2. Precautions – when drugs are prescribed to a patient that may not take these drugs due to his or her condition. The examples are pregnancy, breast-feeding women, elderly patients and etc.

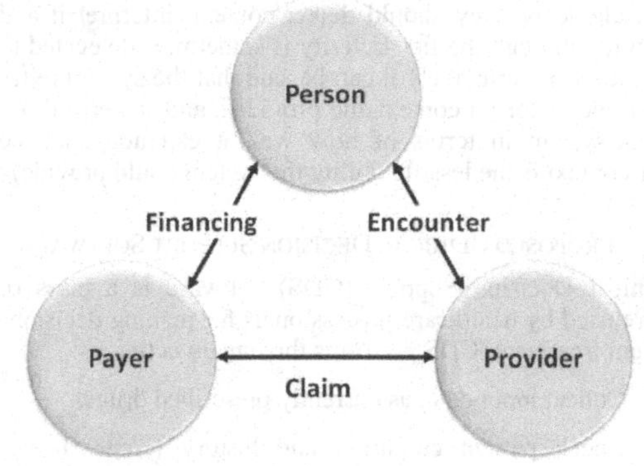

Figure 1. Relationships between Patient, Provider and Payer [1].

Consumption of drugs in these conditions might lead to the certain complications for the patient, hence causing additional expenses for the Payer.

3. Drug-Drug Interactions – when prescribed drugs may interact to each other affecting patient condition. There are five commonly known significance numbers for drug-drug interactions. The most dangerous are drug-drug interactions with significations number one and two – they both might lead to patient death or disability respectively.

Health Insurance Company would be able to analyze the potential Medication Errors in each drug prescription. Knowing the potential Medication Errors, Health Insurance Company may adjust its insurance terms and conditions to force doctors avoiding certain Medication Errors that might lead to patient injury, disability or even death. Since Medication Errors is one of the reasons of repeated patient visits to a doctor or even

K. Elleithy (ed.), *Advanced Techniques in Computing Sciences and Software Engineering*
DOI 10.1007/978-90-481-3660-5_37, © Springer Science+Business Media B.V. 2010

patient hospitalization, Health Insurance Company will reduce its expenses associated with such repeated patient visits.

In addition, Health Insurance Company will be able to find drugs in each prescription that have the same therapeutic effect, hence duplicated drugs. Normally, Health Insurance Company would pay for all drugs in the prescriptions, but using Clinical Decision Support software the usage of duplicated drugs will be reduced. This will lead to the reducing cost for each claim coming to the insurance company.

At the same time Health Insurance Company will help the community by forcing doctors and pharmacists to avoid certain Medication Errors, which will lead to healthier and longer life of the nation.

II. Introduction to Context-Aware Systems

As stated in [3], a context aware software development principal defines the computer reaction can be organized using more than just direct data input. It could and probably should consider circumstances under which the input is defined, sense the environment in which the request is formulated and so forth. In the result the system response will be different for the same request depending on those extra (request context) parameters.

Generally context-aware application contain 3 types of sub-activity elements: they should detect context, interpret it and react on it, although the first activity is sometimes delegated to other systems (alternatively it can be said that the system relies on other one to collect context and provide it and so depends on the front-system in terms of how well it can adapt as the smaller context is the less flexibility the system could provide).

III. Proposed Clinical Decision Support Software

Clinical Decision Support (CDS) software is a class of software used by healthcare professionals for making decisions on patient treatment. CDS performs three main activities:

a. Collect input data, as currently prescribed drugs;

b. Check patient condition and history (context), e.g. previously prescribed drugs, known allergies, pregnancy or breastfeeding, age, etc.

c. Interpret the context to see if there are any conflicts might be detected, and generate the respective alerts.

The system proposed in this paper is based on Greenrain Drug Database developed for the market of United Arab Emirates. It contains information about all drugs existed in United Arab Emirates and approved by Ministry of Health. Article [4] describes the proposed system in more details.

The system contains information about each drug, for example:

a. Active chemical ingredients of each drug;

b. Property of such active ingredient like if it is allowed to be used during pregnancy or breastfeeding, if it causes habits, how safe it is for elder patients, etc.

c. Any other known interactions with other chemical elements.

The proposed system generates alerts if prescribed drugs have any precautions or drug-drug interactions. These alerts might be then used by Health Insurance Companies to prevent possible Medication Errors that might cause additional expenses for the company.

IV. Medication Errors Research

A. *Study Methodology*

To prove that the proposed system will help Health Insurance Companies to save the expenses associated with Medication Errors, the Medication Errors Research was conducted.

The study had a goal to analyze 2,800 prescriptions randomly selected from eight hospitals in United Arab Emirates in the period between September 2007 and February 2008.

The study was made using the proposed Greenrain Drug Database. After the prescriptions were received by researchers, the prescribed drugs were entered into the drug database for further analysis for Medication Errors.

B. *Study Results*

The study showed the following results:

The average cost of a prescription is $40.75.

The average lost on duplicated drugs in a prescription is $15.51.

The serious Medication Errors that might lead to patient disability or even death exist in 6.68% of all prescriptions.

Totally the Medication Errors due to drug-drug interactions were found in 17.79% of all prescriptions.

Total results of the study are presented in Table I.

C. *Data Extrapolation for Larger Amount of Prescriptions*

Normally, Health Insurance Company might have thousands of prescriptions coming to the Company within a month. To be able to calculate the possible amount of losses in this case, we have to apply the results of our research to a large amount of data.

For the purpose of this paper we will consider one of the largest Health Insurance Company in United Arab Emirates. This company has 5,000 prescriptions coming every day, 25 days a month. Table II presents extrapolated data from the research for this Health Insurance Company.

TABLE I. STUDY RESULTS.

Pre-Study Summary	Amount	%
Total Number of Prescriptions	2,800	
Total Price for All Prescriptions	$135,340.68	
Average Price for a Prescription	$40.75	
Total Price for Prescriptions with Duplication in Drug Therapy	$77,003.84	56.90%
Lost Expenses from the Duplication in Drug Therapy	$43,427.40	32.08%
Lost Expenses for each Prescription	$15.51	
Analysis Details	**Amount**	**%**
Number of Prescriptions with Drugs of the same Therapeutic Class	1,286	45.93%
Number of Prescriptions with Drugs of the same Generic Drugs	94	3.36%
Number of Prescriptions with Drug-Drug Interactions (Significance No. 1) – might lead to patient death	40	1.43%
Number of Prescriptions with Drug-Drug Interactions (Significance No. 2) – might lead to patient disability	147	5.25%
Number of Prescriptions with Drug-Drug Interactions (Significance No. 3)	97	3.46%
Number of Prescriptions with Drug-Drug Interactions (Significance No. 4)	86	3.07%
Number of Prescriptions with Drug-Drug Interactions (Significance No. 5)	128	4.57%
Number of Prescriptions with Pregnancy Precautions of Category X, D or C	1,872	76.70%
Number of Prescriptions with Breastfeeding Precautions	2,370	93.59%
Number of Prescriptions with Liver Disease Precautions	2,418	86.36%
Number of Prescriptions with Renal Impairment Precautions	2,388	85.29%

TABLE II. EXTRAPOLATED DATA FOR REAL HEALTH INSURANCE COMPANY.

1	**Lost due to Duplication of Therapy**		
	Lost due to Duplication per Prescription	=$43,427.4 / 2,800 prescriptions	$15.51
	No. of Prescription per month	= 5,000 x 25	125,000
	Total Loss per month (25 days)	=125,000 x $15.51	$1,938,723.09
2	**Lost due to drug-drug interaction combination (Significant No. 1), assuming one day in Intensive Care Unit (ICT), and 3 days hospitalization**		
	No. of affected patients per month	125,000 x 0.0143	1,786 Patients
	Cost of ICU	1,786 x $270 / day	$482,142.86
	Cost of Hospitalization	1,786 x 3 days x $80 /day	$428,571.43
	Total Loss per month		$910,714.29
3	**Lost due to drug-drug interaction combination (Significant No. 2), assuming 3 days hospitalization**		
	No. of affected patients	125,000 x 0.0525	6,563 Patients
	Cost of Hospitalization	6,563 x 3days x $80 / day	$1,525,000
	Total Loss per month		$1,525,000
4	**Lost due to drug-drug interaction combination (Significant No. 3-5), assuming extra doctor visit and medications**		
	No. of affected patients	125,000 x 0.1111	13,884
	Cost of Extra Medications	13,884 x $40.75	$565,760.49
	Cost of Extra Visit	13,884 x $14	$194,375
	Total Loss per month		$760,135.49
Total All Losses per month			$5,184,572.86

V. Conclusion

The research presented in this papers shows that number of Medication Errors are very large: 6.68% of all prescriptions contain errors that might lead to patient disability or even death. This research confirms the extrapolations of possible Medication Errors made in work [5]. Unfortunately, for insurance companies it is not enough to tell that certain solution will save lives and health of the people – as commercial organizations they understand financial calculations that show the consequences of Medication Errors considered in terms of losses and profit.

The proposed solution will allow Health Insurance Companies to analyze all prescriptions coming to them with the payment claims to be able to recognize and react on any possible Medication Errors. It will lead to the saved expenses for Insurance Company associated with Medication Errors, and hence it will improve the quality of healthcare services for the country as it helps preventing mistakes made by healthcare professionals.

The research will have two major steps to continue. First, the larger amount of prescription will be analyzed, e.g. more than 100,000 to compare the analysis result with the presented research. And second, is to compare the number of Medication Errors found in a particular hospital before and after implementation of Clinical Decision Support system.

References

[1] Health Authority Abu Dhabi, "Standards & Procedures Guidance," 24 January 2008, http://www.haad.ae/datadictionary.

[2] L. T. Kohn, J. M. Corrigan, and M. S. Donaldson, *To Err Is Human: Building a Safer Health System*, Washington, DC: National Academies Press, 2000.

[3] W.Y. Lum and F.C.M. Lau, "A context-aware decision engine for content adaptation," *IEEE Pervasive Computing*, vol. 1, no.3, 2002, pp. 41–49.

[4] R. Konovalov and R R. A. Abou-Shaaban, "Assure Safer Drug Therapy in the Middle East," *Middle East Journal of Family Medicine*, vol 4 (6), pp 34-38, 2006.

[5] R. R. A. Abou-Shaaban, "Mortality and Cost of Medication Errors in UAE: A Vision For Making Health Care Safer," *Proceedings of the DUPHAT Conference*, Dubai, pp 23-28, 2002.

A new Artificial Vision Method for Bad Atmospheric Conditions

M. Curilă
Faculty of Environmental Protection
University of Oradea
26, Gen. Magheru Street, Oradea, România
mcurila@uoradea.ro

S. Curilă O. Novac Mihaela Novac
Faculty of Electrical Engineering and Information Technology
University of Oradea
str. Universitatii nr. 1, zip code 410087 Oradea, România
scurila@uoradea.ro, onovac@uoradea.ro, mnovac@uoradea.ro

Abstract **We propose in this paper a method for enhancing vision through fog, based on Blind Source Separation (BSS). BSS method recovers independent source signals from a set of their linear mixtures, where the mixing matrix is unknown. The mixtures are represented in our work by the natural logarithm of the degraded image at different wavelength. These provide an additive mixture of transmittivity coefficient (related to fog) and reflectivity coefficient (related to each point of the scene).**

I. INTRODUCTION

It is well known that traffic accidents are responsible for a large number of deaths and injuries all over the world. The main causes of accidents can be addressed to the driver's objective and subjective reduced driving capabilities. The first case includes difficult environmental conditions (reduced visibility at night, adverse weather conditions). Statistics have shown that from the total number of injury accidents on the roads, around 37% occur in conditions of limited visibility, like darkness and fog. The second case includes, as main causes of accidents, the driver's reduced level of attention due to fatigue or to the habitude to the task.

Really heavy snow and rain can cause a decrease in the image quality, big droplets scatter the light beam, thus decreasing image quality. Fog is determined by a suspension of very small water droplets in the air, reducing visibility at ground level to less than a kilometer [10]. This appears when the humidity of an air region reaches a saturation limit and thus some of the nuclei turn into water droplets.

Intense research and technological efforts are developed to enable applications for detecting objects in bad weather conditions: "Dumont [5]", "Nayar [6]", "Oakley [7]", "MODTRAN report [8]", "National Oceanic and Atmospheric Administration (NOAA) Science Center [9]", etc.

II. SECOND-ORDER NONSTATIONARY SOURCE WITH ROBUST ORTHOGONALIZATION

In many practical problems the processed data consists of multidimensional observations, that has the form:

$$\boldsymbol{x}(k) = \boldsymbol{A}\,\boldsymbol{s}(k) + \boldsymbol{n}(k) \tag{1}$$

where the N-dimensional vector $\boldsymbol{x}(k) = [x_1(k), x_2(k), \ldots, x_N(k)]^T$ is an instantaneous linear mixture of source signals, the M-dimensional vector $\boldsymbol{s}(k) = [s_1(k), s_2(k), \ldots, s_M(k)]^T$ contains the source signals sampled at $1 \le k \le K$, $\boldsymbol{n}(k)=[n_1(k), n_2(k), \ldots, n_N(k)]^T$ is the additive noise vector that is assumed to be statistically independent of source signals and the matrix $\boldsymbol{A}$ called mixing matrix is the transfer function between sources and sensors. The source signals $s_i(k)$, $1 \le i \le M$ $(M < N)$, are assumed to be independent, and additive noises $n_i(k)$, $1 \le i \le N$, can be spatially correlated but temporarily white.

To obtain source signals from observations one utilizes *Blind Sources Separation* (*BSS*) algorithm entitled *Second-Order Nonstationary Source* (*SONS*), developed by A. Cichocki [3] on the foundation of A. Belouchrani s algorithm entitled *Second Order Blind Identification* (*SOBI*)[1]. This algorithm consists of an orthogonalization stage followed by a unitary transform.

Orthogonalization stage is performed by ***Robust-orthogonalization*** algorithm, described by A. Cichocki [3], which ensures that positive definite covariance matrix is not sensitive to the additive white noise. In order to perform ***Robust- orthogonalization*** for nonstationary sources, one divides the sensor data $\boldsymbol{x}(k)$ into B non-overlapping blocks (time windows T_b). At each block for preselected delays $(p_1, p_2, \ldots, p_J)$ one estimates a set of symmetric delayed covariance matrices of sensor signals:

$$\tilde{\boldsymbol{R}}_x(T_b, p_j) = \frac{\boldsymbol{R}_x(T_b, p_J) + \boldsymbol{R}_x^T(T_b, p_J)}{2}, \quad \text{with } j = 1,\ldots,J; b = 1,\ldots,B \tag{2}$$

where $\boldsymbol{R}_x(T_b, p)$ is the delayed covariance matrix of the observation vector from the block T_b computed as:

$$\boldsymbol{R}_x(T_b, p) = E\left[\boldsymbol{x}(k)\boldsymbol{x}^T(k-p)\right] = \frac{1}{K_B}\sum_{k=1}^{K_B} \boldsymbol{x}(k)\ \boldsymbol{x}^T(k-p) \tag{3}$$

and one constructs an $NxNJB$ matrix:

$$R = [\,\tilde{\boldsymbol{R}}_x(T_1, p_1), \ldots, \tilde{\boldsymbol{R}}_x(T_1, p_J), \ldots, \tilde{\boldsymbol{R}}_x(T_B, p_1), \ldots, \tilde{\boldsymbol{R}}_x(T_B, p_J)\,] \tag{4}$$

K. Elleithy (ed.), *Advanced Techniques in Computing Sciences and Software Engineering*,
DOI 10.1007/978-90-481-3660-5_38, © Springer Science+Business Media B.V. 2010

Then it is performed a singular values decomposition of matrix $\boldsymbol{R}$:

$$\boldsymbol{R} = \boldsymbol{Q}\,\boldsymbol{\Sigma}\,\boldsymbol{W}^T \tag{5}$$

where NxN matrix $\boldsymbol{Q} = [\boldsymbol{Q}_s\ \boldsymbol{Q}_n]$ (with NxM matrix $\boldsymbol{Q}_s=[\boldsymbol{q}_1 \ldots \boldsymbol{q}_M]$) and NJBxNJB matrix $\boldsymbol{W}$ are orthogonal, and $\boldsymbol{\Sigma}$ is an NxNJB matrix whose left M columns contain $diag[\sigma_1, \sigma_2, \ldots, \sigma_M]$ (with non increasing singular values) and whose right $NJB\text{-}M$ columns are zero. For a non-zero initial vector of parameters $\boldsymbol{\alpha}=[\alpha_{11},\ldots,\alpha_{1J},\ldots,\alpha_{B1},\ldots,\alpha_{BJ}]^T$ one computes the linear combination:

$$\overline{\boldsymbol{R}} = \sum_{b=1}^{B}\sum_{j=1}^{J} \alpha_{bj}\, \boldsymbol{Q}_s^T \tilde{\boldsymbol{R}}_x(T_b, p_j)\, \boldsymbol{Q}_s = \sum_{b=1}^{B}\sum_{j=1}^{J} \alpha_{bj} \boldsymbol{R}_{bj} \tag{6}$$

One checks if $\overline{\boldsymbol{R}}$ is positive definite ($\overline{\boldsymbol{R}}>0$) and one performs the eigenvalues decomposition of $\overline{\boldsymbol{R}}$. If $\overline{\boldsymbol{R}}$ isn't positive definite one chooses an eigenvector $\boldsymbol{v}$ corresponding to the smallest eigenvalue of $\overline{\boldsymbol{R}}$ and one updates $\boldsymbol{\alpha}$ by $\boldsymbol{\alpha}+\boldsymbol{\delta}$, where

$$\boldsymbol{\delta} = \frac{\left[\boldsymbol{v}^T \boldsymbol{R}_{11} \boldsymbol{v} \ \ldots\ \boldsymbol{v}^T \boldsymbol{R}_{BJ} \boldsymbol{v}\right]^T}{\left\|\left[\boldsymbol{v}^T \boldsymbol{R}_{11} \boldsymbol{v} \ \ldots\ \boldsymbol{v}^T \boldsymbol{R}_{BJ} \boldsymbol{v}\right]\right\|} \tag{7}$$

and with new vector $\boldsymbol{\alpha}$ one returns to compute the linear combination $\overline{\boldsymbol{R}}$. Otherwise, one performs the eigenvalues decomposition of symmetric positive definite matrix:

$$\overline{\boldsymbol{R}}_x(\boldsymbol{\alpha}) = \sum_{b=1}^{B}\sum_{j=1}^{J} \alpha_{bj}\, \tilde{\boldsymbol{R}}(T_b, p_j) \tag{8}$$

as follows:

$$\overline{\boldsymbol{R}}_x(\boldsymbol{\alpha}) \dagger \boldsymbol{V}^T \boldsymbol{A}\, \boldsymbol{V} \tag{9}$$

where $\boldsymbol{\alpha}$ is the set of parameters α_i after the algorithm achieves convergence positive definiteness of the matrix $\overline{\boldsymbol{R}}$, NxM matrix $\boldsymbol{V} = [\boldsymbol{v}_1, \boldsymbol{v}_2, \ldots, \boldsymbol{v}_M]$ contains the eigenvectors corresponding to the largest M eigenvalues of $\overline{\boldsymbol{R}}$, and $\boldsymbol{A} = diag[\lambda_1 \geq \lambda_2 \geq \ldots \geq \lambda_M]$ contains the eigenvalues arranged in decreasing order. The *Robust orthogonalization* transformation is realized by a linear transformation with matrix $\boldsymbol{W}$:

$$\boldsymbol{y}(k) = \boldsymbol{W}\boldsymbol{x}(k) \tag{10}$$

where the matrix $\boldsymbol{W}$ has the form:

$$\boldsymbol{W} = \boldsymbol{A}^{-0,5}\, \boldsymbol{V}^T \tag{11}$$

The covariance matrices of the observed vector can be rewritten as:

$$\boldsymbol{R}_x(p) = \boldsymbol{A}\,\boldsymbol{R}_s(p)\,\boldsymbol{A}^T \tag{12}$$

Because the source signals have unit variance and are assumed to be uncorrelated, the covariance matrix of the sources vector equals the unit matrix:

$$\boldsymbol{R}_s(0) = E[\boldsymbol{s}(k)\,\boldsymbol{s}^T(k)] = \boldsymbol{I} \tag{13}$$

Consequently, $\boldsymbol{R}_s(p)=E[\boldsymbol{s}(k)\ \boldsymbol{s}^T(k - p)]$ are non-zero distinct diagonal matrices, and it follows that:

$$\boldsymbol{R}_x(0) = \boldsymbol{A}\,\boldsymbol{A}^T \tag{14}$$

The components of the orthogonalized vector $\boldsymbol{y}(k)$ are mutually uncorrelated and they have unit variance. The orthogonalized covariance matrices are given by:

$$\boldsymbol{R}_y(0) = \frac{1}{K}\sum_{k=1}^{K} \boldsymbol{y}(k)\,\boldsymbol{y}^T(k) = \boldsymbol{W}\boldsymbol{R}_x(0)\,\boldsymbol{W}^T = \boldsymbol{I} \tag{15}$$

$$\boldsymbol{R}_y(p) = \frac{1}{K}\sum_{k=1}^{K} \boldsymbol{y}(k)\,\boldsymbol{y}^T(k-p) = \boldsymbol{W}\,\boldsymbol{R}_x(p)\,\boldsymbol{W}^T, p \neq 0 \tag{16}$$

From equations (15) and (16) it results:

$$\boldsymbol{R}_y(0) = \boldsymbol{W}\boldsymbol{A}\,\boldsymbol{A}^T\boldsymbol{W}^T = \boldsymbol{W}\boldsymbol{A}\,(\boldsymbol{W}\boldsymbol{A})^T = \boldsymbol{I} \tag{17}$$

Thus, it follows that $\boldsymbol{U} = \boldsymbol{W}\boldsymbol{A}$ is an N x N unitary matrix. Consequently, the determination of M x N mixing matrix $\boldsymbol{A}$ is reduced to that of a unitary N x N matrix $\boldsymbol{U}$. From equations (13) and (17) it results:

$$\boldsymbol{R}_y(p) = \boldsymbol{W}\boldsymbol{A}\,\boldsymbol{R}_s(p)\,\boldsymbol{A}^T\boldsymbol{W}^T = \boldsymbol{W}\boldsymbol{A}\,\boldsymbol{R}_s(p)\,(\boldsymbol{W}\boldsymbol{A})^T, p \neq 0 \tag{18}$$

Since $\boldsymbol{R}_s(p)$ is diagonal, any orthogonalized covariance matrix $\boldsymbol{R}_y(p)$ with $p \neq 0$ is diagonalized by the unitary transform $\boldsymbol{U}$. This transform is founded through a Joint Approximative Diagonalization method.

SONS algorithm exploits the nonstationarity of signals by partitioning the orthogonalized sensor data into non-overlapping blocks (time windows T_b), for which one estimates time-delayed covariance matrices. The method jointly exploits the nonstationarity and the temporal structure of the sources under assumption that additive noise is white or the undesirable interference and noise are stationary signals. Consequently, the orthogonalized observations $\boldsymbol{y}(k)$ are divided into B non-overlapping blocks (time windows T_b) and at each block one computes a set of covariance matrices $\boldsymbol{R}_y(T_b, p_l)$ for b=1,…,B and l=1,…,L. Then, the algorithm retrieves the unitary matrix $\boldsymbol{U}$ by jointly diagonalizing a set of delayed covariance matrices. This matrix jointly diagonalizes the set $M_{\boldsymbol{R}}=\{\boldsymbol{R}_y(T_b, p_l), b=1,\ldots,B\ ;\ l=1,\ldots, L\}$ when the next criterion is minimized:

$$C(M_R, U) = \sum_{b=1}^{B} \sum_{l=1}^{L} off\left(U^T \boldsymbol{R}_y(T_b, p_l)\, U\right) \tag{19}$$

where *off* operator is defined as:

$$off\,(M) = \sum_{1 \le i \ne j \le N} |Mij|^2 \tag{20}$$

The unitary matrix $\boldsymbol{U}$ is computed as product of Givens rotations [1]. When the unitary matrix $\boldsymbol{U}$ is obtained, the mixing matrix is estimated by $\boldsymbol{A} = \boldsymbol{W}^{+} \cdot \boldsymbol{U}$ and the unmixing matrix is then given by $\boldsymbol{U}^T \boldsymbol{W}$, where + denotes the pseudo-inverse.

III. TRANSMISSION OF RADIATION

A general formula for radiation transmission, implemented on a real satellite (the parameters are presented in [8]), contains three terms:

$$L_\nu = \underbrace{\varepsilon_e B_\nu \left(T_e\right) e^{-\left(\tau_1/\mu\right)}}_{\textit{surface radiation}} + \underbrace{\int_{\tau}^{\tau_1} \varepsilon_a B_\nu \left(T_a\right) e^{-(t/\mu)} \frac{dt}{\mu}}_{\textit{atmospheric radiation}} + \underbrace{\int_{\tau}^{\tau_1} \left[k \int_0^{2\pi} \int_{-1}^{1} P_\nu(\mu', \phi') L_\nu(\tau, \mu', \phi')\, d\mu' d\phi'\right] e^{-(t/\mu)} \frac{dt}{\mu}}_{\textit{scattered light by the earth surface}} \tag{21}$$

where L_ν represents the radiance observed by the satellite.

We chose to use a very simple description, due to a small distance between the scene and camera (10 - 20 m). The radiance of the observed image is described for each wavelength λ and at each pixel k by the equation:

$$L(\lambda,k) = L_S(\lambda)\; r_0(\lambda,k)\; \tau + L_S(\lambda)\; (1-\tau(\lambda,k)) \tag{22}$$

where β is the scattering coefficient, L_S is the sky radiance, $\tau(\lambda,k)=exp(-\beta(\lambda)\ d(k))$, d is the distance between the scene and camera, L_0 is the object radiance and $r_0=L_0/L_S$. The first term of the right hand side corresponds to the surface radiation and the second term corresponds to the atmospheric radiation.

IV. DISCRETE WAVELET TRANSFORM AND BLIND SOURCE SEPARATION

A wavelet is, as the name might suggest, a little piece of a wave. A wavelet transform involves convolving the signal against particular instances of the wavelet at various time scales and positions. Since we can model changes in frequency by changing the time scale, and model time changes by shifting the position of the wavelet, we can model both frequency and location of frequency.

When we actually perform the Discrete Wavelet Transform, the common approach involves using the dyadic scheme. This is where we increase the scale and step spacing between wavelets by a factor of two at each step. e.g., in the first pass, we would use wavelets of scale 1 and space them 1 sample apart. Then in the second pass, we'd scale them to two samples, and space them apart by two samplepoints. What we do is take a lowpass and a highpass version of the curve and separate the highpass and lowpass information. In short, we treat the wavelet transform as if it is a filter bank. And we iterate downward along the lowpass subband. The lowpass data is treated as a signal in its own right and it is subdivided into its own low and high subbands.

These properties are essential for noise reduction in real images, these encompass the scene (low frequency components bound to the environment distribution) and of an additive noise in high frequency. These features are exploited to proceed to a blind separation in low frequencies and to de-noise images in high frequencies. The proposed method requires the following steps:

1. calculate a wavelet transform and order the coefficients by increasing frequency;
2. apply a BSS method on the average coefficients of low frequencies in order to separate the scattering environment and the scene;
3. calculate the *median absolute deviation* on the largest coefficient spectrum. The median is calculated from the absolute value of the coefficients. Apply a thresholding algorithm to the wavelet coefficients of high frequencies. Wavelet coefficients below a threshold are left out and only the largest ones correspondent to features of the scene are kept;
4. perform the inverse DWT from the low resolution scene obtained in 2 and thresholded coefficients obtained in 3 in order to restore the de-noised image. Using the inverse wavelet transform the original data can be perfectly reconstructed.

V. EXPERIMENTAL RESULTS

In this work, we proposed a technique, based on Blind Source Separation, for enhancing visibility when fog reduces it.

The basic model that we consider here express a statistic image $L(\lambda,k)$ degraded by fog, using the model described in section III. Equation (22) can be rewritten as:

$$L(\lambda,k)=L_S[1 + \tau(\lambda, k)(r_0(\lambda, k)-1)] \tag{23}$$

Since the sources are determined within a scale factor, an added constant does not change their shape, we arrive at the basic model of source separation, with the mixtures $ln(L(\lambda, k))$, and sources $ln(\tau)$ and $ln(r_0 + constant)$. The exponential of the second restored source gives a gray level enhanced image (reflectivity coefficient r_0).

First we use a synthetic image that contains 200x200 pixels and 16 color squares. Each square has its known depth and also its own color. The synthetic degradation follows the model described in section III *(β=0.7, L_s=110)*. The following objective measure estimates the quality of the enhancement:

$$E = \frac{e_1 - e_2}{e_1} * 100 \quad [\%] \tag{24}$$

where e_1 and e_2 are normalized mean squared errors (nmse) of the degraded and the enhanced images respectively, reported to the original image. In "Fig. 1. d, e", BSS method displays the following extracted sources: two irradiance profiles - one very close to the original image "Fig. 1. d" and the other related to the depth distribution "Fig. 1. e". The enhanced image in "Fig. 1. d" is obtained with a high ratio E = 93. 71%.

To get more data we use a pre-processing technique exploiting information of Hue, Saturation, Luminance. Scattering is characterized by a strong irradiance J (with in 80%) and an extremely low saturation level S (<10%). *Scattering quantity* β is thus estimated in every pixel (k) by:

$$\beta(k) = \min\left(f_1(k)/\max(f_1), f_2(k)/\max(f_2)\right) \tag{25}$$

where $f_1(k)$=L(k)^2, $f_2(k)$=(max(S)-S(k))^2.

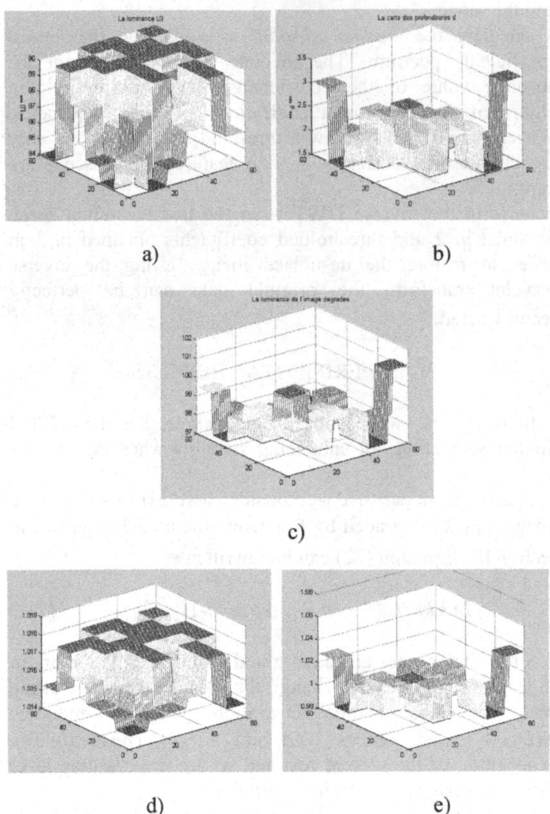

Fig. 1: Mesh representation of the method results (a-original image; b-depth distribution; c-degraded image; and BSS restored sources (d-S_1; e-S_2)

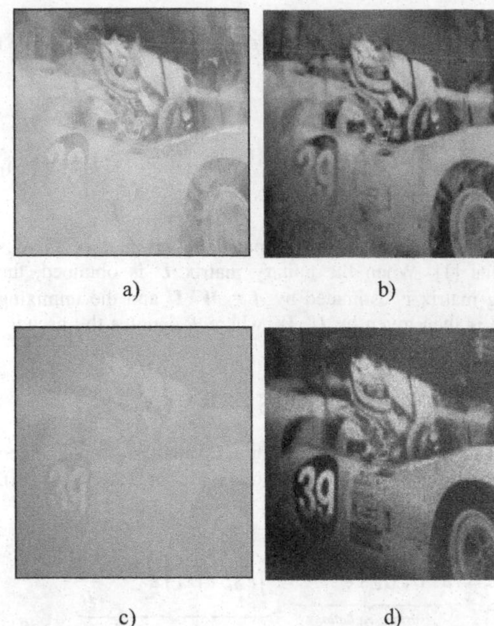

Fig. 2. Non-uniform fog: -a) degraded image, -b) enhanced image; Uniform fog: -c) degraded image. -d) enhanced image.

A minimum of these functions permits to exclude saturated luminous pixels (for example the high yellow) and unsaturated pixels whose luminance is very weak (ex.: black), these colors being signs of no fog. Using *scattering coefficient* β and the *scattering environment color* $C=\{C_R, C_G, C_B\}$ (determined using the mean values of RGB histograms or homogeneous zone), it is possible to recover approximately the original image:

$$J_0 = X(k) - \beta(k) \cdot C / (1 - \beta(k)) \tag{26}$$

These three RGB components complete the original data set to provide a set of six mixtures to be processed.

Next, we experimented two types of artificial fog simulated in the laboratory, the first one is uniform (Fig. 2a) and the second one is non-uniform (Fig. 2c). The data was processed by SONS algorithm and visual quality enhancement can be perceived in Fig. 2 b, d.

Next figure represents an image degraded by real fog. The result confirms the feasibility of the Blind separation process.

Fig. 3. Real conditions: -a) degraded image, -b) enhanced image.

VI. CONCLUSIONS

A new numerical method for artificial vision using Blind Source Separations technique is constructed. This method combines source separation and Discrete Wavelet Transform due to presence of the noise in high frequency. Blind Source Separation recovers the transmittivity coefficient and the reflectivity coefficient of the scene – the enhanced image.

REFERENCES

[1] A. Belouchrani, K. Abdel-Meraim, J.-F. Cardoso, E. Moulines, "A blind source separation technique using second-order statistics", *IEEE Trans. Signal Processing*, vol. 45, pp. 434-444, 1997.

[2] K. Tan and J.P. Oakley, "Enhancement of Color Images in Poor Visibility Conditions," *Proc. Int'l Conf. Image Processing*, vol. 2, Sept. 2000.

[3] A. Cichocki , S. Amari, 2002, "Adaptive Blind Signal and Image Processing. Learning Algorithms and Applications", *John Wiley and Sons*, Ltd Baffins Lane, Chichester West Sussex, PO19 1UD.

[4] D. L. Donoho, "De-noising by soft-thresholding", *IEEE Trans. on Inf. Theory*, 41, 3, pp. 613-627, 1995.

[5] E. Dumont, E. Follin, G. Gallee, 2001, "Improved fog rendering for driving simulation", *Rapport technique LCPC (2001)*, http://www.lcpc.fr/LCPC, dumont@lcpc.fr.

[6] S. G. Narasimhan, S. K. Nayar, "Chromatic Framework for Vision in Bad Weather", *IEEE Conference on Computer Vision and Pattern Recognition (CVPR)*, Vol.1, pp.598-605, Jun, 2000

[7] J. P. Oakley, B. L. Satherley, Improving Image Quality in Poor Visibility Conditions Using a Physical Model for Contrast Degradation, *IEEE Trans. Image Process.*, vol. 7, pp. 167-179, 1998.

[8] http://imk-isys.fzk.de/isys-public/Software-tools/Modtran/science/modrept.ps.

[9] http://orbit-net.nesdis.noaa.gov/arad/fpdt/fog.html.

[10] http://www.bom.gov.au/info/wwords

[11] Website. http://www.cs.columbia.edu/cave/research /publications/vision weather.html. 2000.

[12] Y.Y. Schechner, S.G. Narasimhan and S.K. Nayar, "Instant Dehazing of Images using Polarization," *IEEE Conference on Computer Vision and Pattern Recognition (CVPR)*, Vol.I, pp.325-332, Dec, 2001

[13] Y.Y. Schechner, S.G. Narasimhan and S.K. Nayar, "Polarization-Based Vision through Haze", *Applied Optics*, Special issue, Vol.42, No.3, pp.511-525, Jan, 2003.

[14] Alex Rav-Acha, Shmuel Peleg, "Two motion-blurred images are better than one", *Elsevier* , Pattern Recognition Letters 26 , pp. 311–317, 2005

[15] J.G. Walker, P.C.Y. Chang, and K.I. Hopcraft, "Visibility Depth Improvement in Active Polarization Imaging in Scattering Media", *Applied Optics*, vol. 39, 1995.

[16] L.L. Grewe and R.R. Brooks, "Atmospheric Attenuation Reduction through Multisensor Fusion Sensor Fusion: Architectures, Algorithms, and Applications II", *Proc. SPIE,* vol. 3376, Apr. 1998.

[17] H. Farid and E. H. Adelson, "Separating reflections and lighting using independent components analysis", *Proc. CVPR*, pp. 262-267, 1999.

[18] L.B. Wolff, T.A. Mancini, P. Pouliquen, and A.G. Andreou. "Liquid crystal polarization camera", *IEEE Transactions on Robotics and Automation*, 13(2):195–203, 1997.

[19] P. Premaratne, I. Burnett, and C.D. Liyanage, "Blur retrieval via separation of zeros sheets from noisy blurred images", Proceedings of 2004 International Symposium on Intelligent Multimedia, Video and Speech, pp. 559-562, 20-22 October, 2004.

Aspect-Oriented Approach to Operating System Development Empirical Study

M. Sc. Jaakko Kuusela
Helsinki University of Technology
Poutapolku 3 C 43
02110 Espoo, Finland

Dr. Tech. Harri Tuominen
Sampo Bank plc
Vanha Sveinsintie 27 B
02620 Espoo, Finland

This paper presents a case-study where a new programming technique is applied to an established branch of software development. The purpose of the study was to test whether or not aspect-oriented programming (AOP) could be used in operating systems development. Instead of any real world operating system an educational OS with the name Nachos was used. This was because Nachos is written in Java which makes it easy to introduce aspect-oriented techniques. In this paper a new file system for the Nachos OS is developed and then it is analyzed by profiling and metrics. The results show that it is possible to use AOP in OS development and that it is also beneficial to do so.

I. Introduction

This article is based on a master's thesis with the same title by the author. The main points of the thesis are introduced and the most important results are summarized. The original thesis can be found in [1].

The operating system development has mainly been done in plain old C. This causes many problems, because C is a procedural language and the structural programming paradigm it is based on has its own problems. Better languages for development have emerged but in the area of OS programming they have been neglected. Still, some experimental and research operating systems have been coded in C++ and other object oriented languages. These operating systems however still have problematic issues with their source code. This is because even with the object oriented paradigm there are certain parts of program logic that are deeply intertwined with the other modules. These parts of the program are included in the code of almost every module and changing them is very error prone. Also understanding the program flow becomes very difficult with lots of tangled code that affects all parts of the program. Examples of this kind of program logic include logging, security and virtual memory management. For example, to do logging, every function that has to be logged must call the logging functions when it starts executing and when it is done and finding all these function calls can quickly become a maintenance nightmare.

There is however a solution for this problem of so called crosscutting concerns. An extension to object oriented paradigm called Aspect Oriented Programming (AOP) makes it possible to define these crosscutting concerns in their own modules. This separates their implementation from the rest of the code, the so called core functionality and makes it easy to change their implementation without affecting other code.

In this document AOP is applied to an OS kernel to develop a new better file system to replace the original simple one. This is achieved by using the AspectJ compiler, which is an addition to standard Java compiler that makes it possible to use aspects, the modules defining crosscutting concerns, with Java programming language. The performance of the implementation with aspects is compared with the standard Java implementation by profiling. This with code metrics is used to determine how the aspect oriented technology affects the performance of the program and size of the source code. The original hypothesis was that the aspect version would be somewhat slower but also smaller in size. Also to assess the goodness of the object model and maintainability of the software, metrics related to coupling between classes are calculated. The hypothesis here was that the use of aspects would greatly reduce coupling.

The operating system used for this paper is called Nachos and it is designed for educational uses. It is a rather small and simple operating system but still has all the important features of a real operating system. Further information on Nachos can be found in [2].

II. Other Related Research

Other papers related to using AOP with operating systems have been published in different OOP conferences. Four such papers will be shortly discussed here.

The first paper by Netinant et al. [8] was written in 2000. The paper's goal is to represent how operating systems could be made so from the beginning that they would better support aspect-oriented programming. This is an interesting and useful view point, because during the programming work done by the author for his thesis, it was many times noticed how difficult it is to use aspects when the original code is not designed to support them.

The two other papers were published under the a-kernel project. The a-kernel project is a project that aims to study whether the aspect-oriented programming paradigm can be used to modularize operating systems in an efficient way.

The first of these papers by Coady et al. [9] describes an experiment with a real operating system. In this experiment aspect-oriented techniques are applied to a well-known crosscutting concern, namely the prefetching feature of the virtual memory system. This feature is responsible for loading parts of mapped files to memory before they are actually used. As

K. Elleithy (ed.), *Advanced Techniques in Computing Sciences and Software Engineering*,
DOI 10.1007/978-90-481-3660-5_39, © Springer Science+Business Media B.V. 2010

the programming language this experiment uses AspectC which is a subset of AspectJ that can be used with the C programming language. This experiment concludes that using aspects makes the formerly very unclear prefetching implementation easy to understand and modify.

The second paper [10] describes another experiment. This experiment was done with the well-known network file system called NFS. The goal of the study was to implement server replication functionality in to a Java version of the NFS file system. This was done both by using AspectJ and by using traditional object-oriented method of layering. The results show that aspect-oriented techniques seem to be better than the traditional approach. However one has to take into account the possible performance degradation with aspects.

The fourth paper [11] differs from the aforementioned papers in the way that it includes quantitative experiments. The actual performance degradation is measured. This paper introduces a dynamic aspect weaving system that can be used with the Linux kernel. Its main use is to make kernel profiling and logging applications. The dynamic weaver is accomplished by using a modified version of the GCC-compiler and a kernel modification tool called Kerninst. The results show that the performance differences between a normal kernel and an aspect enabled kernel are acceptable.

III. Building the Filesystem

The Nachos originally includes a simple basic file system. In this paper that file system is extended with new useful features using aspect oriented techniques. This is done according to the Theme/UML software process [3]. Theme/UML makes the point that aspects should only be used when appropriate and that aspects should be included in the software process from the beginning, i.e. when doing requirements analysis. The process followed in this representation is not exactly the Theme/UML as it is introduced in the book. Rather it is a version where several steps were omitted to make it simpler.

TABLE I
FILESYSTEM REQUIREMENTS

R1	User can create, open, close, delete, read and write files through the interface.
R2	There is a logical file system on top of the physical file system.
R3	User communicates with the logical file system.
R4	User can create and delete directories.
R5	User can list the contents of a directory.
R6	Directories and files are represented logically in a composite hierarchy.
R7	User can create symbolic links.
R8	The synchronic disk uses caching to increase performance.
R9	Logical file system controls access based on the ACLs in the physical file system.
R10	Physical file system is implemented with the synchronized disk.
R11	Symbolic link is a file representing a file or directory located in another place.
R12	Access control system gets the ACL from the physical file system.
R13	Use of logical file system triggers the access control system.
R14	The logical file system must conform to the Nachos file system interfaces.

First the requirements for the new file system were captured and collected in table I.

Next the themes were extracted from the requirements. The themes are the main concerns of the program. They can be either cross-cutting concerns or core concerns. The cross-cutting ones are good candidates for using aspects. The themes of the file system are presented in table II.

In the following paragraphs the aspect related themes of the file system are introduced and their implementations are described.

A. Symbolic Links

The symbolic-links theme is implemented by using an aspect oriented Proxy design pattern. The proxy pattern is originally described in [4] and its aspect version in [5].

The aspect Proxy pattern consists of a concrete proxy aspect that inherits from an abstract proxy aspect. It also includes an interface named Subject that is used to mark those file system nodes that are actually proxies, i.e. symbolic links in the file system. The abstract proxy aspect includes point cuts that catch method calls that should be delegated from the proxy object to the real subject and the concrete proxy aspect defines what method calls should actually be intercepted. When a method is called on a proxy object the aspect uses reflection to redirect the call to the real target, i.e. the target of the symbolic link. The pattern is different from the traditional Proxy pattern because in this version of the pattern there is not really a separate proxy object. Instead the proxy object is just a normal object that has been marked with the Subject interface.

B. Access Control

This theme uses the same aspect version of Proxy pattern as the previously introduced theme. This time however different methods of the Proxy aspect are utilized. The aspect Proxy pattern also includes rejection point cuts that catch those method calls that should possibly not be allowed to execute. These actual methods are defined in the concrete proxy aspect. When such a method is called the proxy aspect checks whether the call should be allowed or disallowed. If

TABLE II
FILESYSTEM THEMES

logical-file system	The file system that the user interacts with. Supports the full set of operations.
physical-file system	The underlying file system that supports only limited set of operations.
file-operations	The operations used with file-objects.
directory-operations	The operations used with directory-objects.
file system-hierarchy	The hierarchical structure of file system implemented with composite pattern.
symbolic-links	An aspect for implementing the symbolic links.
caching	An aspect for implementing caching.
access-control	An aspect for implementing access control.
synchronized-disk	An aspect for synchronizing the usage of physical disk.
concurrency-control	An aspect for controlling the concurrent access to file system.

the access control list of the file does not permit the operation to be executed the method call is rejected.

This access control solution was not meant to be a real working solution in the sense that it would be secure. The model is chosen mainly based on ease of implementation and many security flaws can be found on this implementation. For example the principle of complete mediation [7] is not followed. This means that access control list is not loaded from the disk during each file operation but a cached version is used instead. So this solution is not correct but it is good enough for demonstrating the use of aspects for what it was meant.

C. Disk Synchronization

To add synchronization to the physical disk a refactoring was made to the SynchDisk class. It is a class that builds a synchronous interface on top of the Disk class that emulates a physical disk. The Nachos operating system uses only the synchronous disk implementation so the original Disk class is never referenced. This made it possible to remove the entire SynchDisk class and implement the synchronization with aspects that were bound to the Disk class. The Nachos implementation was changed to use the original Disk class and it did not need to know anything about synchronization. The synchronization could be turned on or off by selecting whether to compile the associated aspect into the build.

D. Concurrency Issues

The concurrency theme implements the readers and writers locking paradigm to the logical file system. According to this paradigm only one writer can concurrently access a file while multiple readers can read the file simultaneously. This paradigm and examples of it are described in [6].

The FileSystemConcurrency aspect catches the read and write operations of FileSystemNode class. Then it uses the LockHandler singleton class to acquire the right kind of lock. There are two kinds of locks: read lock and write lock. After the operation the appropriate lock is released.

The LockHandler is implemented as a monitor that is simulated by a mutex and two condition variables. The implementation is based on an example in [6]. The synchronization mechanisms included in the Nachos implementation are used instead of core Java synchronization because this makes the implementation much more portable for example to languages that don't have monitors or locks.

E. Disk Caching

The caching is a very simple feature to implement. The abstract base aspect AbstractCache contains the HashMap used for storing the cached values. The base aspect also contains the method that fetches an object from cache if one is stored there. The concrete aspect DiskCache defines the associated point cuts for read and write operations and around advices for handling these operations.

The result from a disk operation is not returned as a return value but is instead stored to an array given by a reference parameter. This somewhat complicates the caching functionality because data has to be transferred between the result array and the array stored in cache.

It is easier to place the caching operation to the lowest level possible instead of placing it to for example file and directory level. Caching the physical disk is easy because reads and writes are always done in constant size blocks (one disk sector) and there are only two operations to consider (readRequest and writeRequest).

Caching greatly reduces the running time of disk operations because the physical disk has been made very slow in Nachos by simulating the latencies with sleep operation. With many successive read operations the cache works optimally but because of write-through policy write operations are not faster than without cache.

IV. RESULTS

The main object of the study was to make measurements of the aspect oriented code as opposed to a version that was developed using traditional object oriented methods. The metrics had to be chosen so that they were simple enough to implement or do by hand, because of the lack of tools for aspect oriented code measurement. Three different metrics were chosen: lines of code, coupling and performance. The performance was measured by means of a profiler.

When looking at the results it should be noted that these results include also the two other aspect-oriented features that were developed in addition to the new file system. These features are virtual memory and synchronization, which were left out of this paper. The interested should refer to [1].

A. Lines of Code Metrics

The lines of code were measured for both the aspect version of Nachos and the non-aspect version. The results are presented in table III.

As can be seen from the table the general trend seems to be that when aspects are removed from the code and the same functionality is developed with traditional methods the lines of code count actually decreases. This is in contradiction to the original hypothesis that aspects would lead to more compact code. In the original hypothesis it was assumed that the amount of repeated code would be very large without aspects, but that is not the case, because usually the only repeated piece of code is a single method call. Thus the abstraction mechanisms of the base language already eliminate the duplication of code efficiently.

TABLE III
LOC COUNTS FOR THE SYSTEM

full aspect version	9995
LOC of aspect files only	1284
singleton w/o aspects	9882
basic synchronization w/o aspects	9759
proxy pattern w/o aspects	9794
disk synchronization w/o aspects	9767
disk cache w/o aspects	9706
disk concurrency w/o aspects	9641
virtual memory w/o aspects	9498
system with no aspects	9467

The only case where the aspect version is more compact than the traditional version is the aspect Proxy pattern. So using aspects for implementing the Proxy pattern seems like a good idea. There is however the drawback that the aspect pattern is bit harder to understand and this has to be taken into account when deciding whether or not to use the aspect version of the pattern.

B. Coupling Metrics

The original hypothesis was that the use of aspects should dramatically decrease the coupling between core classes. This is because the coupling is moved from the classes itself into the aspect code. The aspect code works like glue that binds together the classes that collaborate to implement some feature, functionality or a pattern.

Three different coupling metrics were calculated. The first is forward coupling, i.e. the number of references going from class to other classes. The second is backward coupling, i.e. the number of references coming into a class from other classes. The third is package coupling, i.e. number of references that go from one package into another. The forward coupling mostly affects reuse, while the backward and package couplings affect ease of maintenance. The names of the metrics were given by the author for the purposes of this paper. Similar metrics with generally accepted names can be found in the Internet and literature.

Table IV shows the number of classes that one class has to know about to handle its responsibilities.

As can be seen from table IV the forward coupling metric is smaller for all the relevant classes when aspects are used.

Table V shows the number of classes that know about a certain class. This is the backward coupling metric.

As can be seen from table V also the backward coupling metric with aspects is smaller for all the relevant classes.

Table VI shows the number of couplings between different packages. The columns are non-aspect, aspect and debugging with aspects, which will be explained below.

From table VI it can be seen that with this implementation the use of aspects seems to have very little effect on package coupling. Indeed between most of the packages the coupling does not change at all. Furthermore, even if only the packages that are affected by aspects are taken into account the average decrease is 17 %. This is still not very significant.

TABLE IV
FORWARD COUPLING METRIC

Class	Coupling w/o aspects	Coupling with aspects
LogicalDirectory	7	5
Disk	6	4
Condition	5	4
LogicalFile	5	3
LogicalFileSystem	5	4
FileSystemNode	4	2
Semaphore	3	2
FileSystemNodeProxy	3	-
Lock	2	1

TABLE V
BACKWARD COUPLING METRIC

Class	Coupling w/o aspects	Coupling with aspects
Interrupt	10	8
FileSystemNode	4	3
Lock	4	2
AccessController	3	1
LockHandler	2	0
LogicalFile	2	1
OpenFileReal	2	1
Condition	2	1
Semaphore	2	0
AddrSpace	2	1
FileSystemNodeProxy	2	0

When studying the original data in the table and thinking about the causes for the large coupling numbers it was realized that the Debug class in the threads package was the cause of most inter-package coupling. Indeed most of the coupling is from the other packages to the thread package.

Implementing the debugging information with aspects would of course be a textbook example of using aspects but because it was not part of the original plan it was not done. However the results in table VI were calculated by ignoring the references to the Debug in the original data. It can be seen from the table that this modification would noticeably reduce the coupling between certain packages.

Summary of the statistics related to coupling is given in table VII. The first column is the name of the metric, the second is metric without aspects, the third is metric with aspects, the fourth is the decrease in percents and the fifth is the standard deviation of the decrease.

TABLE VI
PACKAGE COUPLING METRIC

Package 1	Package 2	Non-Aspect	Aspect	Debug with Aspect
filesys	machine	4	4	4
machine	filesys	0	0	0
filesys	threads	11	11	6
threads	filesys	5	5	5
filesys	userprog	2	2	2
userprog	filesys	1	1	1
machine	threads	21	19	12
threads	machine	13	9	9
machine	userprog	0	0	0
userprog	machine	5	5	5
threads	userprog	4	3	3
userprog	threads	9	8	3

TABLE VII
STATISTICS FOR COUPLING METRICS

Metric	Non-Aspect	Aspect	Decrease	StDev
Forward	4.44	2.78	37 %	0.71
Backward	3.18	1.63	49 %	0.52
Package	6.25	5.58	11 %	1.23
Package with Debug Aspects	6.25	4.17	33 %	3.12

The table confirms the hypothesis that the backward coupling would be the one that is most affected by aspecting the software. With backward coupling the standard deviation of the decrease is also the smallest. This means that the decrease in coupling is more uniformly distributed over the classes than with the forward coupling. So it seems that using aspects clearly makes maintenance easier for all the classes that are part of the aspected feature.

With forward coupling the decrease is not as dramatic as with backward coupling but it still makes a difference. The standard deviation of the decrease is slightly larger than with the backward coupling, which means that the decrease in coupling is more concentrated on certain classes, but not noticeably so. From this the conclusion can be made that using aspects promotes reuse for all the classes that are part of the aspected feature.

The results for package coupling are not as impressive as for the other forms of coupling. This is mostly implementation dependent. The features that were selected for aspecting were mostly localized to their own packages, so there was not crosscutting over package boundaries. Thus for most of the package combinations the aspect implementation had no effect at all to the coupling. This also shows up in larger standard deviation.

To see the effect of implementing a package crosscutting feature with aspects, the effect of using aspects to log debug information was inspected. As table VII shows now the effect of using aspects is almost as significant as with other forms of coupling. Of course now the standard deviation is still larger, because the coupling reduction is concentrated to couplings with the threads package.

C. Performance Evaluation

Two different tests were used to analyze the performance of the aspect and non-aspect implementations. The first test was the file system performance test that writes and reads a 3k file in ten byte chunks and thus stresses the file system. Three kilobytes is the maximum file size in the current implementation. It is too small for any practical purposes but still a performance difference can be seen in the test between the aspect and non-aspect versions. The second test was a virtual memory stress test. The test multiplies two matrices that are too big to fit in the memory, so page faults occur constantly. Table VIII summarizes the results.

Table VIII shows that the aspect version runs slower in both tests. With file system the aspect version is 77 % slower and in the virtual memory test only 6 % slower.

This difference can be explained by the fact that in the file system implementation there are lots of features that are implemented as aspects and thus the aspect overhead is quite large. On the other hand the virtual memory implementation has only a couple of aspects and so the performance difference is smaller.

Table IX shows the results of running the file system performance test in a profiler.

The conclusion is that not only is the aspect version slower but that the aspect features use a larger relative portion of the total running time than their non-aspect counterparts. So

TABLE VIII
RUNNING TIMES FOR PERFORMANCE TESTS

	Aspect version	Non-aspect version
File system performance	140 ms	79 ms
Virtual memory performance	1470 ms	1390 ms

aspects definitely have the drawback of slowing the program down, which could be expected because of mechanisms that the working of aspects is based on. It was known at the start of this project that there is a performance penalty. Thus the relevant question is whether the good sides of aspects balance the bad sides. The authors think that this is the case.

V. CONCLUSIONS

This paper has studied applying the aspect-oriented programming techniques to a traditional software development branch of operating systems. The measurements made to the code show that the use of aspects both slows the program down and increases the size of the code. On the other hand the use of aspects greatly decreases coupling between classes and thus makes the software more maintainable. Nowadays the performance and size of code are not as important as they were before, so the authors think that aspects are a viable solution even to operating systems development. Most operating systems are still written in procedural languages and thus are not very good candidates for aspect-oriented solutions, but hopefully in the future also the operating systems development will move to the direction of object-oriented languages and make using the full power of aspects possible.

VI. FUTURE RESEARCH

As mentioned earlier in this study, the main object of the study was to make measurements of the aspect oriented code as opposed to a version that was developed using traditional object oriented methods.

In spite of some results, which do not alone support decisions to start application/software development using aspect oriented approach it will be anyhow interesting and scientific opportunity to add and diversify research work's focus on software engineering.

TABLE IX
PROFILER RESULTS FOR FILE SYSTEM TEST

	Aspect	**Non-Aspect**
Time spent in write test	62 %	44 %
Time spent in read test	29 %	23 %
Time spent in creating a file	8 %	2 %
Time spent in acquiring and releasing file system locks	41 %	30 %
Time spent with synchronized disk	33 %	31 %

Requirements of software engineering for end user or enterprises are not only running time efficiency or quick response rate. In business world it is important to calculate total costs and productivity – add value - as well. This means finally how to appraise application's applicability for considered purposes from customer or end user point of view.

To solve our further research problem, it will be interesting to assess empirically how well the aspect oriented paradigm fits in the development of software, for which the usage counts are large and algorithms are computationally heavy. Such problems are often encountered in the field of information sciences, so we will be developing software that uses learning methods such as neural networks or Bayesian classifiers.

In our further research our approach will be to create, using aspect oriented code, some specific applications, which means new contribution partly for software engineering as well as for software technical aspects.

Our purpose is to address, could it be profitable or not to create new innovations and decision support systems using aspect oriented approach

It will be challenging to us to create a formula, which shows economically, when the aspect oriented approach is gainful.

The first draft for the formula could be as follows:

$$G_{devel} = C_{devel} \, (T_{develasp} - T_{develtrad}) \quad (1)$$

G_{devel} is the gain for development
C_{devel} is the cost for development in eur/h
$T_{develasp}$ is the time used for development with aspects
$T_{develtrad}$ is the time used for traditional development

$$G_{maint} = C_{maint} \, (T_{maintasp} - T_{mainttrad}) \quad (2)$$

G_{maint} is the gain for maintenance
C_{maint} is the cost for maintenance in eur/h
$T_{maintasp}$ is the time used for maintenance with aspects
$T_{mainttrad}$ is the time used for maintenance with traditional methods

$$G_{resp} = C_{resp} \, (\Sigma T_{iasp} - \Sigma T_{itrad}) \quad (3)$$

G_{resp} is the gain for response time
C_{resp} is the cost for response time in eur/ms
T_{iasp} is the response time for ith operation with aspects in ms
T_{itrad} is the response time for ith operation with traditional methods in ms

$$G_{total} = G_{devel} + G_{maint} + G_{resp} \quad (4)$$

G_{total} is the total gain

Our hypothesis is that the total gain G_{total} would be negative, i.e. using aspects would lead to cost-savings. This is what we are going to test in our further work.

References

[1] Jaakko Kuusela. 2005. Master's Thesis: Aspect-Oriented Approach to OS Development. Empirical Study. http://users.tkk.fi/~jkuusela/dippa.pdf

[2] T. Narten. 1997. A Road Map Through Nachos. URL: http://www.cs.duke.edu/\\$\sim$narten/110/nachos/main/main.html

[3] S. Clarke, E. Baniassad. 2005. Aspect-Oriented Analysis and Design: The Theme Approach. Addison Wesley Professional. 400 pages. ISBN: 0321246748.

[4] E. Gamma, R. Helm, R. Johnson, J. Vlissides. 1995. Design Patterns: Elements of Reusable Object-Oriented Software. Addison-Wesley Professional. 395 pages. ISBN: 0201633612.

[5] Russ Miles. 2004. AspectJ Cookbook. O'Reilly Media, Inc. 354 pages. ISBN: 0596006543.

[6] Gregory R. Andrews. 1999. Foundations of Multithreaded, Parallel and Distributed Programming. Addison Wesley. 496 pages. ISBN: 0201357526.

[7] Matt Bishop. 2002. Computer Security: Art and Science. Addison-Wesley Professional. 1136 pages. ISBN: 0201440997.

[8] P. Netinant, C. A. Constantinides, T. Elrad, M. E. Fayad. Supporting Aspectual Decomposition in the Design of Operating Systems. 3rd ECOOP Workshop on Object-Orientation and Operating Systems. Sophia Antipolis, June 12, 2000.

[9] Y. Coady, G. Kiczales, M. Feeley, N. Hutchinson, J. S. Ong, et al. Exploring an Aspect-Oriented Approach to OS Code. ECOOP 2001.

[10] A. Brodsky, D. Brodsky, I. Chan, Y. Coady, S. Gudmundson, et al. Coping with Evolution: Aspects vs Aspirin?. ASOC Workshop at OOPSLA 2001.

[11] Y. Yanagisawa, K. Kourai, S. Chiba, R. Ishikawa. 2006. A dynamic aspect-oriented system for OS kernels. Proceedings of the 5th international conference on Generative programming and component engineering. Pages: 69-78. ISBN: 1-59593-237-2.

STUDY AND ANALYSIS OF THE INTERNET PROTOCOL SECURITY AND ITS IMPACT ON INTERACTIVE COMMUNICATIONS

Arshi Khan and Seema Ansari
College of Engineering
Pakistan Air Force-Karachi Institute of Economics and Technology, Korangi Creek
Karachi -75190, (Pakistan)
arshibabble@yahoo.com ; sansari@pafkiet.edu.pk

Abstract
Internet Protocol Security (IPSec) is the defacto standard, which offers secured Internet communications, providing traffic integrity, confidentiality and authentication. Besides this, it is assumed that IPSec is not suitable for the protection of real-time audio transmissions as the IPSec related enlargement of packets and the usage of the Cipher Block Chaining (CBC) mode contradict stringent requirements. IPSec overhead of at least 44 bytes for each Internet Protocol (IP)-packet cannot guarantee Quality of Service (QOS) due to a bad wireless link by which the Ethernet flow control intercepts and makes a real time transmission impossible.

This paper presents a survey and analysis study of IPSec and its impact on interactive communication for securing voice and video packets with emphasis on three main domains: Implementation issues of IPSec, Computational and Protocol overhead of IPSec and finally discusses about the scalability issues of IPSec.

Keywords: *IPSec, QOS, WLAN, IKE, MTU, RSA, NAT*

I. Introduction

IPSec is a combination of protocols, which work together to assist in protecting communications over IP networks. IPSec can be used to secure all communications, for example, by using authentication and encryption without requiring any further modifications to applications or protocols. If we need to ensure that a message is not modified in transit or that they are unreadable to network intruders, IPSec provides a solution, which can be used to achieve these ends. Although IPSec can be apply to a variety of situations for example real-time interactive communication. The major disadvantage of IPSec is the lack of protection in real-time multimedia and the protocol overhead.

IPSec is a security mechanism implemented in the IP layer, which offers per-packet basis authentication, privacy and integrity. To enable IPSec, both ends must exchange IPSec Security Associations (SA), which includes IP addresses of ends, algorithms, shared secret and the lifetime of SA. as shown in Fig. 1, IPSec architecture is composed of layered components: two security protocols, Authentication Header (AH) and Encapsulating Security Payload (ESP). Both Protocols perform main role for access control, based on the security associations (SA) and key management.

AH and ESP both are used to protect the IP header and encrypt the data. However, this level of protection is hardly ever used because of the increased overhead that AH would acquire for packets that are already protected by ESP. ESP protects everything but the IP header, and modifying the IP header does not provide a valuable target for attackers. Generally, the only valuable information in the header is the address, and this cannot be spoofed effectively because ESP guarantees data origin authentication for the packets.

For protecting IP packets, IPSec describes "Security Association" (SA) which is defined by the packet's destination IP address and a 32-bit Security Parameter Index (SPI). Furthermore, SA can be encapsulated within SA, to form SA bundles, allowing layered IPSEC protection. For example, one SA might protect all traffic through a gateway, while another SA would protect all traffic to a particular host. The packets finally routed across the network would be encapsulated in an SA bundle consisting of both SA. The other side of the connection could be identical in design, consisting of a gateway implementing a tunnel SA, followed by a host implementing a transport SA, or the entire bundle could be terminated in a single host, which would then implement both SAs. The IP protocol is stateless and connectionless, hence cannot guarantee a secure delivery of the information. To overcome this limitation and offer a stateful security, IPSec introduces logical connections, indicated as Security Associations (SA), able to provide security services to the

K. Elleithy (ed.), *Advanced Techniques in Computing Sciences and Software Engineering*,
DOI 10.1007/978-90-481-3660-5_40, © Springer Science+Business Media B.V. 2010

traffic, which flows through them. Each IPSec node keeps a SA Database (SAD) and a Security Policy Database (SPD). SPD contains a set of rules with selectors used to match the traffic being processed by a policy; each policy is identified by a Security Policy Index (SPI). The SPI is used to identify which SAD entry contains the encryption key, IP source(src)/destination(dst) address (or address range) to be handled by IPSec, encryption algorithm, encryption mode (transport or tunnel), and so forth. The management of SAD and SPD, which includes creation, activation, destruction of entries, can be performed manually by an operator, but due to the overwhelming complexity of such approach, which raises obvious scalability problems, it is often delegated to an automated process. The Internet Security Association and Key Management Protocol (ISAKMP) has been designed to be a generic framework for authentication and key exchange; amongst the others, the Internet Key Exchange (IKE) the most widely adopted protocol for SA negotiation and keying material provisioning [8]. Shortly, IKE consists of two Phases; During Phase 1, two ISAKMP peers establish a secured channel to communicate. During Phase 2, the secured channel is used to negotiate IPSec SA, exchange keying codes, and so on [10].

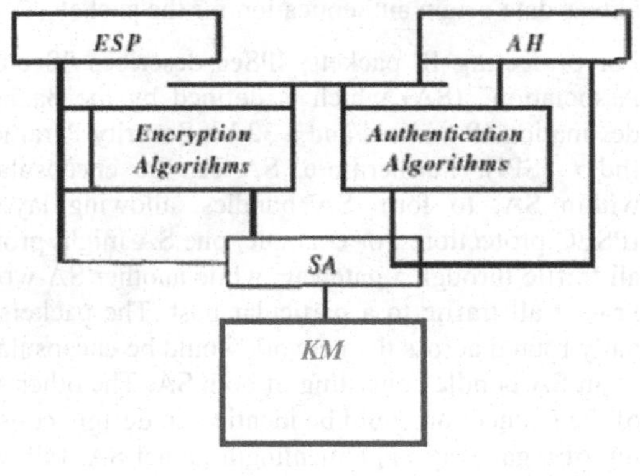

Fig. 1. IPSec Security architecture in layers [6]

II. ISSUES AND PROBLEMS

Although IPSec can be applied to a variety of applications, this paper focuses on issues and problems that are encountered mostly in real-time applications, but not in all environments. Followings are some common issues related to IPSec:

Computational and protocol overhead.

Data interruption threats.

High cost of computation load.

Scalability.

A. *Computational and protocol overhead*

IPSec is not suitable for the protection of real-time audio transmissions, because the IPSec related enlargement of packets and the usage of the Cipher Block Chaining (CBC) mode contradict the stringent requirements. IPSec overhead of at least 44 bytes for each IP-packet noticeably decreases the transmission quality. Moreover, the fact that the CBC mode combines the previous block of cipher text with the current block of plaintext before encrypting it, results in a higher error rate [1].

B. *Data interruption threats*

IPSec allows some traffic to pass unprotected, such as Broadcast, Multicast, Internet Key Exchange (IKE), and Kerberos. Most studies indicate that it is impossible to utilize strong cryptographic functions for implementing security protocols on handheld devices [5].

Attackers could potentially use this knowledge to their advantage to send unauthorized malicious traffic through the IPSec filters. Carefully monitor the traffic that is passing through the IPSec tunnel, as well as that which is bypassing it. For example, network-based intrusion detection system (IDS) or intrusion prevention system (IPS) devices can be typically configured to alert on non-tunneled traffic.

C. *High cost of computation load*

IPSec provides the security services of data confidentiality, message entity authentication, data integrity and protection against replay attacks. On the other hand produces overheads, both computational and protocol related. Consequently, could cause additional IP fragmentation, since the protocol overhead could overflow the Maximum Transmission Unit (MTU) [1].

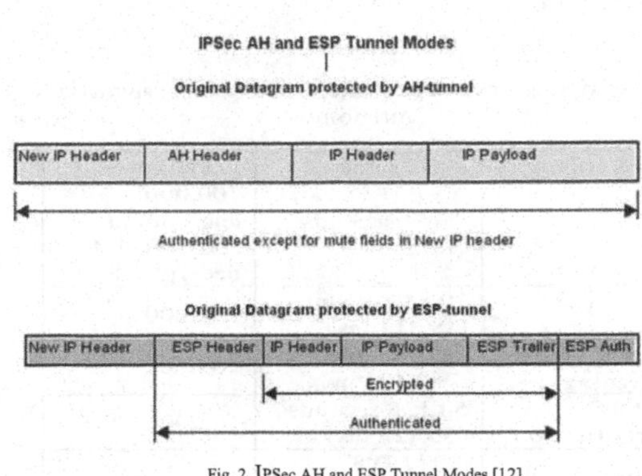

Fig. 2. IPSec AH and ESP Tunnel Modes [12]

D. Scalability

IPSec is being increasingly deployed in the context of networked embedded systems. The resource-constrained nature of embedded systems makes it challenging to achieve satisfactory performance while executing security protocols [2].

III. Approaches And Methodologies

A. Computational and protocol overhead

To measure the overheads of IPSec over WLAN and its impact on a video-conferencing system and on Voice Over Internet Protocol (VoIP) telephony, two video sources with different dynamics have been used on test-bed, one video with highly dynamic contents (actually a cut-out from the action movie "Mission Impossible and one with low dynamics (a student reading-out a newspaper). For monitoring loss-rate, jitter and delay, the videos transmitted Harmonic distortion Inequalities (HMI), Linear Matrix Inequalities (LMI), Harmonic Distortion2 (HM2) and linear matrix2 (LM2) over the test-bed. All frames were treated equally throughout the evaluation of the overall frame loss-rate (no differentiation between Intra (I), Predicted (P) and Bi-Directional (B) frames). The results showed a higher loss-rate for I-frames for both (IPSec on and off) setups. The relative differences showed the same picture as the overall loss-rates. For packet sizes below the threshold value of 256 bytes, the influence of IPSec increases significantly [1].

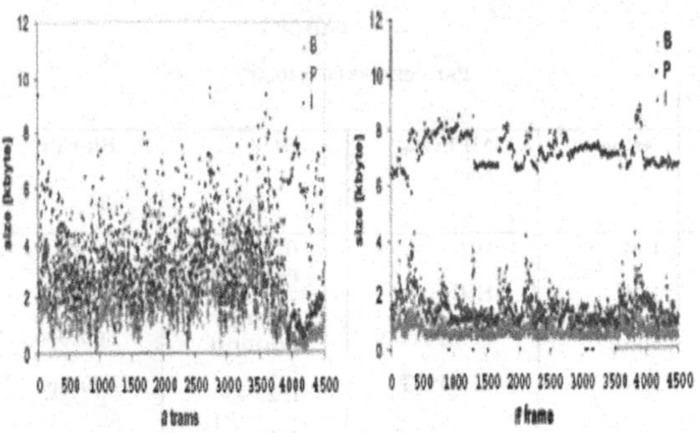

Fig. 3. High motion video (left) and low motion (right) using B-Frames [1]

B. Data interruption threats

When a host wishes to communicate with another host with whom it does not share a Security Association it has to negotiate one using IKE. Each peer needs to perform one Republic of South Africa (RSA) signing and one RSA verification operation, as well as one SHA message hashing operation. The second phase handles the establishment of security associations, between two hosts that have completed the first phase, for a specific type of traffic. The proposed calculations based on HP (Compaq) iPAQ H3630, determined that an IPSec handshake should take approximately 0.16 seconds for a 1,024 bits key and just over a second for a 2,048 bits key on mobile constrained devices [5] as shown in Table II.

C. High cost of computation load

The system administrator [4, 8] bases the protection offered by IPSec to certain traffic on requirements defined by security policy rules defined and maintained. In general, packets are selected for a packet protection mode based on network and transport layer header information matched against rules in the policy, i.e., transport protocol, source address port number, and destination address and port number. To define traffic protection rules, the IPSec standard specifies the policy operational guidelines that should be implemented by vendors rather than a specific policy model. In this work, use a generic policy format that resembles the format used in a wide range of IPSec implementations [7]. This policy model is composed of two lists of packet-filtering rules:

TABLE I

PARAMETERS OF VIDEO SOURCES [1]

Name	MPEG-4 Enc	GOP	Bit-rate
HM1	HHI	IBBPBB PBBPBB	476kbps
LM1	HHI	IBBPBB	276kbps
HM2	FFMPEG	PBBPBB	500kbps
LM2	FFMPEG	IP25	500kbps
		IP25	

Crypto-access list: consists of ordered filtering rules that specify required actions for packets that match the rule conditions. All traffic is matched against the access rules sequentially until a matching rule is found. The matching rule action is either "protect" for secure transmission, "bypass" for insecure transmission, or "discard" to drop the traffic.

Crypto-map list: consists of prioritized filtering rules that determine the cryptographic transformations required to protect the traffic selected for protection by the access list. Traffic may match multiple rules resulting in applying more than one transformation on the same traffic such that higher priority transformations are applied first.

The access list is used to define IPSec protection rules, while the map list is use to define IPSec transformation rules. A transform is any cryptographic service that can be used to protect network traffic. These security services are IPSec, AH and ESP protocols operating either in transport or tunnel mode along with the cryptographic algorithm and the necessary cryptographic parameters. Fig. 4. shows an example of a typical outbound IPSec policy. The policy at each device is defined in terms of the access-list (upper section) and the map-list (lower section). In this work, we consider that inbound traffic arriving at a device interface is matched against a mirror image of the outbound IPSec policy of this interface, i.e., the inbound policy is similar to the outbound policy after swapping the packet filters for source and destination addresses [7].

TABLE II.

TIMING MEASUREMENTS OF LOW-LEVEL CRYPTOGRAPHIC PRIMITIVES ON AN IPAQ H3630

Operation	Time	Iterations
DES	7.354 seconds (7,354 ms)	100,000 encryptions and 100,000 decryptions
SHA	19.111 seconds (19,111 ms)	100,000
1,024 bits RSA signing	782.593 seconds (782,593 ms)	10,000
1,024 bits RSA verification	50.125 seconds (50,125 ms)	10,000
2,048 bits RSA signing	4,972.798 seconds (4,972,798 ms)	10,000
2,048 bits RSA verification	156.006 seconds (156,006 ms)	10,000

D. *Scalability*

The design space of a configurable and extensible embedded processor platform named (Xtensa) for improving the performance of IPSec-like security protocols. The analysis identifies points in the design space that simultaneously improve the performance of both the compute-intensive cryptographic components, and the memory behavior of the protocol processing part, in a cost effective manner on an embedded processor. To the best of our knowledge, this is the first work to present a systematic analysis of IPSec execution on configurable and extensible processors which provides a maximum speed-up of 6X for IPSec processing.

In order to improve the performance of IPSec processing, compute-intensive operations in IPSec protocol and cryptographic processing are converted into special-purpose instructions. These compute-intensive operations (also called hotspots) are identified by studying the performance profiles of IPSec protocol and cryptographic processing [2].

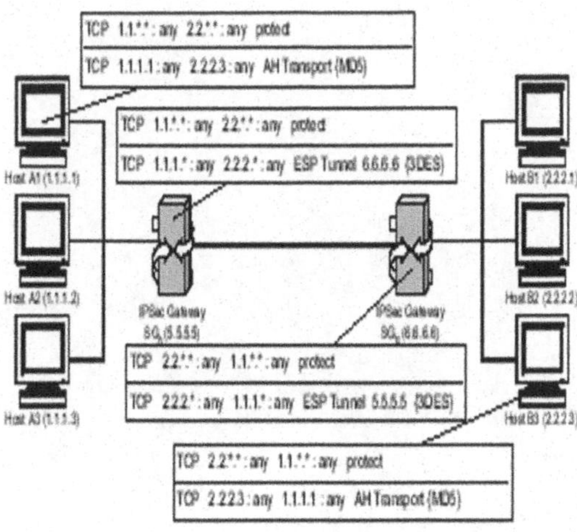

Fig. 4. Example of an IPSec configuration [7]

IV. Conclusions And Future Work

In this paper we present analysis and study of IPSec and demonstrate the feasibility of using IPSec on Interactive communication for the protection of real-time data; like voice and video, even when using a wireless link. Additionally, it has to be taken into account that CBC can be used to remove the packet enlargement issue related with IPSec. It has been shown that the IPSec protocol can be used to secure communication between two ends. The IPSec drawbacks, with respect to interactive communications, namely the enlargement of the IP packets and the consecutive faults (caused by Cipher Block Chaining) have an impact on the perceptual quality while much progress is required to identify the flaws (loss of packets, jitter, and delays) of IPSec.

The paper discusses the general issues related with IPSec and describes mechanisms for secured communication. Further research is required in the use of performance enhancing techniques with an IPSec protected network layer.

Future research plan includes online discovery and recovery of conflicts between IPSec devices and other security devices like firewalls and Network Address Translation (NATs).

References

[1] Jirka Hess,Technical, On the Impact of IPSec on Interactive Communications, *In Proceedings of the 19th IEEE International Parallel and Distributed Processing Symposium*, vol. 18, pp. 4-8, 2005.

[2] Nachiketh R. Potlapally, Srivaths Ravi, Anand Raghunathan, Ruby B. Lee, and Niraj K. Jha, Impact of Configurability and extensibility on IPSec Protocol Execution on Embedded Processors, *In Proceedings of the 19th International Conference on VLSI Design*, vol. 1, pp. 299-204, 2006.

[3] Nobuo Okabe Shoichi Sakane Kazunori Miyazawa Kenichi Kamada, Security Architecture for Control Networks using IPsec and KINK, *In Proceedings of the 2005 Symposium on Applications and the Internet*, vol. 1, pp. 414-420, 2005.

[4] Félix J. García Clemente, Gabriel López Millán, Jesús D. Jiménez Re, Deployment of a Policy-Based Management System for the Dynamic Provision of IPsec-based VPNs in IPv6 Networks, *In Proceedings of the The 2005 Symposium on Applications and the Internet Workshops*, vol. 1, pp.10-13, 2005.

[5] Patroklos G. Argyroudis, Raja Verma, Hitesh Tewari, Donal O'Mahony, Performance Analysis of Cryptographic Protocols on Handheld Devices, *In Proceedings of the Third IEEE International Symposium on Network Computing and Applications*, vol. 2, pp.169-174, 2005.

[6] Hung-Ching Chang, Chun-Chin Chen, and Chih-Feng Lin, *Xscale* Hardware Acceleration on Cryptographic Algorithms for IPSec Applications, *In Proceedings of the International Conference on Information Technology: Coding and Computing*, vol. 2, pp. 592-597,2005.

[7] Hazem Hamed, Ehab Al-Shaer and Will Marrero, Modeling and Veri.cation of IPSec and VPN Security Policies, *In Proceedings of the 13th IEEE International Conference on Network Protocols*, vol. 1, pp. 259-278, 2005.

[8] Shingo FUJIMOTO, Masahiko TAKENAKA, Adoption of the IPsec-VPN for the ubiquitous network, *In Proceedings of the 2005 Symposium on Applications and the Internet*, vol. 2, pp. 78-81, 2006.

[9] J. Arturo Pérez, Víctor Zárate, Ángel Montes, Carlos García, Quality of Service Analysis of IPSec VPNs for Voice and Video Traffic, *In Proceedings of the Advanced International Conference on Telecommunications and International Conference on Internet and Web Applications and Services*, vol. 1, pp. 43, 2006.

[10] C. Floridia, S. Giordano, S. Lucetti, A. Tomasi, An Experience in IPv6 Networking supporting *Ecumene* Web Information System for Cultural Heritage, *In Proceedings of the First International Conference on Testbeds and Research Infrastructures for the Development of Networks and Communities*, vol. 1, pp. 32-41, 2005.

[11] Li Zhitang, Cui Xue, Chen Lin, Analysis And Classification of IPSec Security Policy Conflicts, In Proceedings of the Japan-China Joint Workshop on Frontier of Computer Science and Technology, vol.2, pp. 83-88, 2006.

[12] www.ml-ip.com/assets/

Investigating Software Requirements Through Developed Questionnaires To Satisfy The Desired Quality Systems (Security Attribute Example)

Sherif M. Tawfik, Marwa M. Abd-Elghany
Arab Academy for Science and Technology and Maritime Transport, P.O. Box: 1029 Miami, Alexandria, Egypt
Sherif226@hotmail.com, marwam@aast.edu

Abstract- **It is well recognized in software industry that requirements engineering is critical to the success of any major development project. Quality attributes could not be achieved without certain requirements specified by project managers that should be exhibited within the system. Thus, further research is needed to calculate the weighting factors of software quality attributes in an attempt to quantify, or in other words, to measure a software quality attribute from software project specified document. The aim of this paper is to propose a questionnaire that is designed to model one of the software quality attribute which is the security attribute and to illustrate the method used in determining the weighting factor for each question in the questionnaire. The proposed questionnaire is to elicit security requirements and its relative importance in the project under consideration for example: security could not be fulfilled without the presence of appropriate security mechanisms such as authentication, access control, and encryption, and to give a measurement for development efforts on security related feature.**

I. INTRODUCTION

In a previous work, the authors utilized the non-functional requirements i.e. quality attributes defined by the project managers, as the basis for software cost estimation (SCE) in order to provide enhanced and more realistic results when undertaking the cost estimation process [1]. The authors introduced a new cost estimation (CE) model as seen in "Fig 1"; using case-based reasoning, with "just enough" of each attribute to satisfy the requirements of the software system.

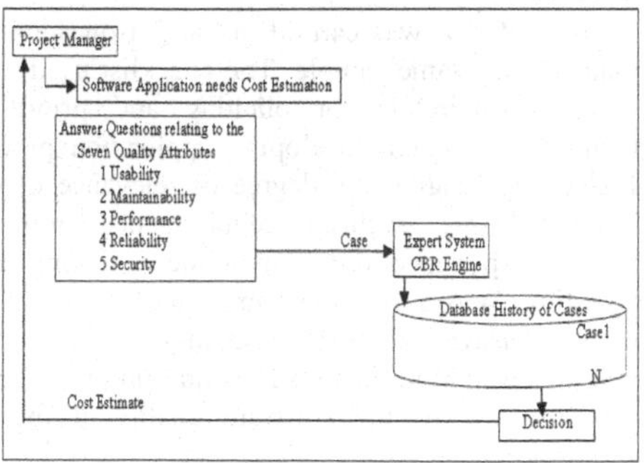

Fig. 1. The Proposed Model

As seen from "Fig 1" that the first step in the cost estimation process is to try to collect the relevant data through answering a group of questionnaires that reflect the quality characteristics that are required in the new software system, and try to quantify these attributes [1].

So that, The main objective of this paper is to give a clear illustration about one of the designed questionnaire that will be utilized to develop the previous work taking the security quality attribute as an example. As an example for the importance of the security in the desired software system, developing web based catalogue services which have on line payment options should be accompanied with secure electronic transaction payment requirements that would acquire policies like cryptographic controls for the protection of the information transmission. It is hoped that this work will act as a useful example to practitioners in the near future for other software quality attributes (like performance, reliability, usability and so on ...), but with different questioning criteria.

In this paper, the second section is dedicated to discuss the software requirement specifications. Subsequently in the third section, focusing on security requirement and then in the fourth section the security questionnaire is presented with the explanation of the questionnaire contents and the justification for choosing these questions, also with the demonstration of the data gathering technique

II. SOFTWARE REQUIREMENTS SPECIFICATIONS

Software Requirement Specification (SRS) is the initial product development phase in which information is gathered about the requirements. This information-gathering stage can include onsite visits, questionnaires, surveys, interviews, and perhaps a Return-on-Investment analysis or Needs Analysis of the customer or client's current business environment. SRS is basically an organization's understanding of a customer's or client's system requirements and dependencies at a certain time prior to any actual design or development work. It's a two-way insurance policy that assures that both the client and the organization understand the other's requirements [2].

The SRS document itself states in precisely and explicitly those functions and capabilities that the software system (i.e. a

978-1-4244-1736-0/08/$25.00 ©2008 IEEE

K. Elleithy (ed.), *Advanced Techniques in Computing Sciences and Software Engineering*,
DOI 10.1007/978-90-481-3660-5_41, © Springer Science+Business Media B.V. 2010

software application, an e-commerce web site, and so on) must provide, as well as any required constraints by which the software system must abide [3]. The SRS is often referred to as the "parent" document because all subsequent project management documents, such as design specifications, work statements, testing and validation plans, and documentation plans are related to it [2]. It is important to note that SRS contains only functional and non-functional requirements; it does not offer design suggestions, possible solutions to technology or business issues, or any other information other than the understanding of the development team of what the customer's system requirements meant to be [4].

III. Security Requirements

When discussing security requirements, they often tend to be general mechanisms such as password protection, firewalls, virus detection tools, and the like. Studies show that attention to security can save the economy billions of dollars, yet security concerns are often treated as an afterthought to functional requirements.

A recent study found that the Return on Investment when security analysis and secure engineering practices are introduced early in the development cycle ranges from 12 to 21 percent [6]. As reported by [7], software with security faults and poor reliability costs the economy $59.5 billion annually in breakdowns and repairs. [8] defined security requirements as "restrictions or constraints" on system services. [9] described a security requirement as "a manifestation of a high-level organizational policy into the detailed requirements of a specific system" and they remarked that security requirements are a kind of non-functional requirement. [10] stated that a security policy is a document that expresses what protection mechanisms are to achieve and that the process of developing a security policy is the process of requirements engineering. [11] appears to take a similar view, stating that "security requirements mostly concern what must not happen". [12] affirmed that "security constraints define the system's security requirements". [13] expressed a security requirement as "a quality requirement that specifies a required amount of security in terms of a system-specific criterion and a minimum level that is necessary to meet one or more security policies". These multiple definitions of security requirements are difficult to understand satisfaction criteria, and lack a clear track for deriving security requirements from business goals.

Generally, requirements should specify what data privileges should be granted to the various roles at various times in the life of the resource, and what mechanisms should be in place to enforce the policy [14]. The integrity of the data is determined if the data origin is validated i.e. to ensure that the data arrived unaltered (whether accidental or malicious) therefore, integrity is handled as data origin authentication where a failure in authentication can lead to a violation of access control policy [14]. Confidentiality mechanisms are used to enforce authorization i.e. when a resource is exposed to a user, what exactly is exposed, the actual resource, or some transformation or proxy? This involves encryption, algorithms and parameters for initialization [14]. Resources can be any piece of data or functionality that can be used by a program, including not only application data such as personal information of users, but also many kinds of resources that are often implicit or overlooked in specifying a software system such as: databases, cryptographic key stores, registry keys, web pages (static and dynamic), audit logs, network sockets, any other files and directories [14].

From all of the above it can be said that security requirements represent constraints on functional requirements that are needed to satisfy applicable system's security goals and they express these goals in operational terms, precise enough to be given to a designer.

IV. Security Questionnaire

The authors report here a survey to examine software security requirements at the early stage of the software development life cycle. Interviews handled with a sample size of 40 participants (including requirements' engineers, system designers, software developers, system operational support, software maintenance specialists, testers, etc.) whom currently and previously were engaged in managing software development projects. The participants are staff members from the Information and Documentation Center and also from the Faculty of Computing & Information Technology, within the collaborating institution (The Arab Academy for Science and Technology and Maritime Transport (AASTMT). The participants were selected based on their expertise in programming and analysis.

The questionnaire was passed through two steps. The aim of the first step was to get the potential responses to each interview question to ensure that each question sought sufficient and appropriate to the security field, taking into consideration that these security requirements does not involve hardware nor networking requirements, it does only involve the software security modules. The output from this step was resultant list of questions that was reviewed and modified to respond to comments.

Then, the second step was carried out and a questionnaire was distributed to the same sample. The checklist used in the distributed questionnaire is for eliciting and prioritizing security requirements when developing application projects, This is obtained by judging the degree of relevance of each statement to software security requirements from the participants' real experience to determine the weighting factor of each question $Wf(q_i)$ = (answer value to question i /summation of all answer values to all questions).

Then using this weighting factor of the question number i to calculate the measure of the software quality attribute in percentage as follows:

$$\sum_{j=1}^{n}[wf(q_i) * wf(ans)_{q_i}] \qquad (1)$$

$Wf(ans)_{qi}$ is the weighting factor of the answer to question number i that would vary according to the responding answers of the project managers depending on the project application type the manager desires to develop; its value would be 0 if he does not want this feature to be included in his required software so it won't cost him and would be equal to (nominal) = 1 if he seeks to apply this feature hence it would cost him.

The questionnaire (see table 1) was divided into four parts as illustrated consecutively below. The participant is being presented to a number of statements to indicate his appropriate response to the importance degree of each statement to security requirements specification by ticking ✓ the suitable number:

5 =Very High, 4 =High, 3 =Fair, 2 =Low, 1 =Very Low. In case of inapplicability of the statement he/she can tick ✓ the Not Applicable (NA) column. For briefing the required system is the new software application under development which:

At first, it should be noted that the Information Systems (IS) that are used to capture, create, store, process or distribute classified information must be properly managed to protect against unauthorized disclosure of information, loss of data integrity, and to ensure the availability of the data and system. Protection requires a balanced approach including IS security features to include administrative, operational, computer, communications, and personnel controls.

Technical protection measures depend on the required system whether it is a single user system or a multi user system. Systems that have one user at a time, but have a total of more than one user with no sanitization between users, are multi-user systems, in which it is allowed that each user of the system to have private files that the other users cannot tamper with or read. Extensive measures are usually inappropriate and inordinately expensive for single-user, stand-alone systems.

As the complexity of a specific IS and the associated residual risk for this system increase, the need for identification and authentication of users and process becomes more significant. Identification and authentication controls are required to ensure that users have the appropriate clearances and need-to-know for the information on a particular system and those controls are divided into two kinds: (a) Unique Identification: Each user shall be uniquely identified and that identity shall be associated with all auditable actions taken by that individual. (b)Authentication at Logon: Users shall be required to authenticate their identities at "logon" time by supplying their authenticator, such as a password, or smart card prior to the execution of any application or utility on the system.

Policies and procedures to detect and deter incidents caused by malicious code, such as viruses or unauthorized modification to software, shall be implemented or not. It is supposed that all files should be checked for viruses before being introduced on an information system and checked for other malicious code as feasible.

Session controls shall be required over identification and authentication or not, for controlling the establishment of a user's session. As for example if the software system provides the capability of tracking successive logon attempts, the following should be done: (a) access denial after multiple repeated unsuccessful attempts on the same user ID, (b) and limitation of the number of access attempts in a specified time period. Furthermore, the required system shall detect an interval of user inactivity then disable any future user activity until the user re-establishes the correct identity with a valid authenticator. All of the previously mentioned requirements are relating to the systems users.

Then coming to the second part, which is dealing with the integrity of the system information sensitivity that would be preserved, the following must be asked. The privileged users required to have access to IS controlling, monitoring or administrative functions. For example: users having "super user", "root", or equivalent access to a system like system administrator, i.e. with complete control of an IS, set up and administer users' accounts and authenticators, users who are given the authority to control and change other users' access to data or program files like database managers, and users who are given special access for troubleshooting or monitoring an IS' security functions like analysts.

After the determination of the system information sensitivity, next requirements should be inquired.

Control of changes to data may range from simply detecting a change attempt to the ability to ensure that only authorized changes are allowed in order to preserve data integrity. Procedures and features are to be implemented to ensure that changes to data are executed only by technically qualified authorized personnel then a transaction log shall be available to allow the immediate correction of all unauthorized data changes at all times. In other words, system recovery functions shall be addressed to respond to failures or interruptions in operation in order to ensure that the system is returned to a condition where all security-relevant functions are operational. If disaster recovery planning is contractually mandated, as in the facility's mission essential applications and information, procedures for the backup of all essential information and software should be identified and the testing procedures as well.

Following program related requirements, the data security requirements should be involved. Security auditing involves recognizing, recording, storing, and analyzing information related to security-relevant activities.

The audit records can be used to determine what activities had occurred and which user or process was responsible for them. Audit records shall be created for the required system to record the following: (a) Enough information to determine the date and time of action, the resources and the action involved; (b) Successful and unsuccessful logons and logoffs; (c) Successful and unsuccessful accesses to directories, including creation, opening, closing, modification, and deletion; (d) Changes in user authenticators; (e) The blocking or blacklisting of a user ID, terminal or access port and the reason for the action; (f) Denial of access resulting from an excessive number of unsuccessful logon attempts.

These contents of audit trails shall be protected against unauthorized access, modification, or detection. Audit analysis and reporting shall also be scheduled, and performed.

TABLE I
Produced Questionnaire from Conducted Interviews

Question Element		Importance Degree				
First: Questions relating to System Users	NA	1	2	3	4	5
1. Supports single user or multi users.						
2. Supports that all the authorized users are uniquely identified before granting access to the system or that all the authorized users are globally identified.						
3. Is capable of stating the number of invalid access attempts that may occur for a given user identifier or access location (terminal or port) and describing the actions taken when that limit has exceeded.						
4. Blocks an account if the password has not been changed within the time limit or the account has remained unused.						
5. Generates logs that contain information about security relevant events such as detection of malicious code, viruses, and intruders (hackers) for example.						
6. Generates logs that contain information about Users relevant events (i.e. identification and documentation of allowed access).						
7. Requires audit logs to be protected from unauthorized access or destruction by means of access controls based on the user.						
Second: Questions relating to Security Administrator by whom the integrity of the sensitivity of all information internal to the system would be preserved	NA	1	2	3	4	5
8. Supports that the security administrator has a choice of enabling or disabling of Users' Identifications.						
9. Supports that the security administrator has the authority of giving grant for accessing specific modules of the systems.						
10. Supports that the security administrator has the authority of giving grant for accessing specific tasks of the systems.						
11. Supports that the security administrator has the authority of giving grant for accessing specific functions of the systems.						
Third: Questions relating to Programs	NA	1	2	3	4	5
12. Requires access control to be established over the system modules.						
13. Requires access control to be established over the system tasks.						
14. Requires access control to be established over the system functions.						
15. Requires any routine program in order to ensure the consistency of the data and its synchronization with the audit logs data.						
16. Requires a policy in case of cryptographic controls for protection of information (i.e. algorithms to transform, validate, authenticate, encrypt or decrypt data).						
17. Requires digital signatures to protect the authenticity and integrity of electronic documents.						
Fourth: Questions relating to Data	NA	1	2	3	4	5
18. Requires its data to be stand alone or shared with other system.						
19. Requires an audit.						
20. Requires an audit for a few of the system tasks or some of the system tasks or a lot of the system tasks or most of the system tasks or all of the system tasks.						
21. Provides audit logs for the capability to investigate unauthorized activities after their occurrences so that proper corrective actions can be taken.						

V. STUDY RESULT

The pilot study yielded the following results after the data analysis of the distributed questionnaires as shown in table 2. Also, the table contains a scenario for assuming that the project manager conduct a meeting with the user and collect the answers for the security requirements desired in the new software system.

The data in column 2 in table 2 shows the weight for each question which, is supposed to be constants and obtained from the distributed questionnaires as described earlier in the proposed formula in the previous section. Also, the score in the last column illustrates how a software quality attribute such as security would be represented in percentage (i.e. to be put into a quantity); taking into consideration that the given values of each question was calculated based on the answer of each question as shown in the answer column.

The last row in the table represents the final result (60%) after conducting the scenario of answering all the questions.

Notes that, this percentage will be varying from system to other as a result of the answers of the questions.

The use of % unit is important since it shows the relative 'saturation' in the attributes with respect to current user requirements thus reflecting the 'maturing' user needs.

VI. Conclusion

This work portrays an unambiguous recognition of the importance of security; the software world within which the argument exist for validating whether or not the system can satisfy the security requirements. Yet, the main contribution of this paper remains in establishing a list of 21 questions that helps the software project manager to cover and acquire all the information needed for the security requirement matter of the required new software system. Moreover, a method for giving a weighing-style quantification of security requirements was presented with a clear example. In the future work, similar questionnaires would be developed to convey the related software features to other software quality attributes in order to quantify them. So that, the main goal for establishing a CBR system that use the quality attribute as a case features for the process of software cost estimation will be achieved.

References

[1] S. M. Tawfiq, M. M. Abd Elghany, S. Green, A Software Cost Estimation Model Based on Quality Characteristics. In MeReP: Workshop on Measuring Requirements for Project and Product Success, IWSM-Mensura, IWSM (International Workshop in Software Measurement) and MENSURA (International Conference on Software Process and Product Measurement), Palma de Mallorca, Spain. November 2007, 13-31. Available at: http://www-swe.informatik.uni-heidelberg.de/home/events/MeRePDocs/paper2.pdf

[2] Jr. Donn Le Vie, Writing Software Requirements Specifications. *Technical Communication Community* ,2007, Available online at: http://www.techwr-l.com/articles/writing/softwarerequirementspecs

[3] S. R. Faulk, Software Requirements: A Tutorial. In Software Requirements Engineering (2nd ed.), *R. Thayer, M. Dorfman (eds.), IEEE Computer Society Press* , Los Alamitos, CA, 1997, 7-22. Available online at: http://www.cs.umd.edu/class/spring2004/cmsc838p/Requirements/Faulk_Req_Tut.pdf

[4] D. E. Jenz, Requirements Packages. *Jenz & Partner* July 2000. Available online at :http://www.bpiresearch.com/Resources/Techniques/requirements_packages.htm

[5] B. Li, Y. Wei, B. Huang, M. Li, M. Rodríguez, C. Smidts, Integrating Software into Probabilistic Risk Assessment. *Final NASA Report*, Center for Risk and Reliability Engineering, University of Maryland, 2006.

[6] K. S. Hoo, J. W. Sudbury, J. R. Jaquith, Tangible ROI through Secure Software Engineering, *Secure Business Quarterly, 1*(2),2001.

[7] National Institute of Standards and Technology: Software Errors Cost U.S. Economy $59.5 Billion Annually. (NIST 2002-10). Available online at: http://www.nist.gov/public_affairs/releases/n02-10.htm

[8] G. Kotonya, I. Sommerville, *Requirements engineering: processes and techniques*. (United Kingdom, John Wiley and Sons, 1998).

[9] P. Devanbu, S. Stubblebine, Software Engineering for Security: A Roadmap. Proceedings of the Conference on The Future of Software Engineering, Limerick, Ireland, 2000, 227-239.

[10] R. Anderson, Security Engineering: A guide to building dependable distributed systems, (Wiley, 2001).

[11] J. Rushby, Security Requirements Specifications: How and What?. Invited paper, In Proceedings of the Symposium on Requirements Engineering for Information Security (SREIS), Indianapolis, USA, 2001.

[12] H. Mouratidis, P. Giorgini, G. Manson, Integrating Security and Systems Engineering: Towards the Modelling of Secure Information Systems. In Proceedings of the 15th Conference on Advanced Information Systems Engineering (CAISE'03). 2003, 63-78.

[13] D. Firesmith, Specifying Reusable Security Requirements. Journal of Object Technology, 3(1), 2004, 61-75.

[14] J. Viega, Building Security Requirements with CLASP, *Proceedings of the 2005 workshop on Software engineering for secure systems—building trustworthy applications* (SESS'05) , St. Louis, MO, USA, 2005, 1-7.

TABLE II
Pilot Study Results

Question Number Q_i	Question Weight W_if	Description / Selected Feature	Answer	Score
Q_1	W_1f= 15%	*a)* single user= (0.2) *b)* multi users= (0.8)	b)✓	(15*0.8) = 12%
Q_2	W_2f= 10%	*a)* uniquely identified= (0.8) *b)* globally identified= (0.2)	a)✓	(10*0.8) = 8%
Q_3	W_3f= 5%	a) capable of stating the number of invalid access attempts= (1) b) incapable of stating the number of invalid access attempts= (0)	a)✓	(5*1) = 5%
Q_4	W_4f= 2%	a) blocks an account if it has remained unused= (1) b) does not block= (0)	a)✓	(2*1) = 2%
Q_5	W_5f= 6%	a) generates logs that contain information about security relevant events= (1) b) does not generate= (0)	a)✓	(6*1) = 6%
Q_6	W_6f= 4%	a) generates logs that contain information about Users relevant events= (1) b) does not generate= (0)	a)✓	(4*1) = 4%
Q_7	W_{7f}= 5%	a) requires audit logs to be protected from unauthorized access= (1) b) does not require= (0)	a)✓	(5*1) = 5%
Q_8	W_8f= 2%	a) permits the security administrator to enable or disable Users' Identifications= (1) b) does not permit= (0)	a)✓	(2*1) = 2%
Q_9	W_9f= 3%	a) allows the security administrator to have the authority of giving grant for accessing specific modules of the systems= (1) b) does not allow= (0)	a)✓	(3*1) = 3%
Q_{10}	$W_{10}f$= 3%	a) allows the security administrator to have the authority of giving grant for accessing specific tasks of the systems= (1) b) does not allow= (0)	a)✓	(3*1) = 3%
Q_{11}	$W_{11}f$= 3%	a) supports that the security administrator has the authority of giving grant for accessing specific functions of the systems= (1) b) does not support=(0)	a)✓	(3*1) = 3%
Q_{12}	$W_{12}f$= 4%	a) requires access control to be established over the system modules= (1) b) does not require= (0)	b)✓	(4*0) = 0%
Q_{13}	$W_{13}f$= 5%	a) requires access control to be established over the system tasks= (1) b) does not require= (0)	b)✓	(5*0) = 0%
Q_{14}	$W_{14}f$= 6%	a) requires access control to be established over the system functions =(1) b) does not require =(0)	b)✓	(6*0) = 0%
Q_{15}	$W_{15}f$= 2%	a) requires any routine program in order to ensure the consistency of the data and its synchronization with the audit logs data= (1) b) does not require= (0)	b)✓	(2*0) = 0%
Q_{16}	$W_{16}f$= 7%	a) requires a policy in use of cryptographic controls for protection of information =(1) b) does not require =(0)	b)✓	(7*0) = 0%
Q_{17}	$W_{17}f$= 4%	a) requires digital signatures to protect the authenticity and integrity of electronic documents =(1) b) does not require= (0)	b)✓	(4*0) = 0%
Q_{18}	$W_{18}f$= 4%	a) requires its data to be stand alone or shared with other system= (1) b) does not require= (0)	b)✓	(4*0) = 0%
Q_{19}	$W_{19}f$= 3%	a) requires an audit= (1) b) does not require= (0)	a)✓	(3*1) = 3%
Q_{20}	$W_{20}f$= 5%	Requires an audit: a) for a few of the system tasks or some of the system tasks= (0.1) b) or a lot of the system tasks= (0.2) c) or most of the system tasks= (0.3) d) or all of the system tasks= (0.4)	d)✓	(5*0.4) = 2%
Q_{21}	$W_{21}f$= 2%	a) provides audit logs= (1) b) does not provide= (0)	a)✓	(2*1) = 2%
Total	100%	Software Security Quality Attribute Percentage		60%

Learning Java with Sun SPOTs

Craig Caulfield
S. Paul Maj
David Veal

School of Computer and Information Science
Edith Cowan University
Perth, Western Australia

***Abstract*- Small Programmable Object Technology devices from Sun Microsystems (Sun SPOTs) are small wireless devices that can run Java programs. In the on-going research project described in this paper, Sun SPOTs have been used in conjunction with model-driven software development techniques to develop a tool that can be used to introduce new software developers to object-oriented programming in general and Java in particular in a new and interesting way.**

The tool, a graphical user interface application, allows users to quickly design, build, and deploy Java applications to the Sun SPOTs without using a conventional integrated development environment or low level commands. In this way new developers aren't so overwhelmed by the wealth of technologies, commands, and possibilities offered by the current sophisticated Java development tools. However, the generated Java source code will be available as either an Eclipse or NetBeans project for those who would like to study it.

I. INTRODUCTION

In their seminal books on the C programming language, Kernighan and Ritchie [1] used a simple program to print the text "Hello World" as a way of introducing some basic features of a C program. Hello World programs have now become something of a standard way, if tired and uninspiring way, of introducing a programming language.

This paper describes an on-going research project at Edith Cowan University aimed at introducing people to the Java programming language and object-oriented concepts while at the same time giving them something useful and sufficiently challenging to work with. Rather than a Hello World program, the project is based around Small Programmable Technology devices from Sun Microsystems (Sun SPOTs) (http://www.sunspotworld.com/).

For someone just starting out in software development, and Java in particular, Sun SPOTs have some unique advantages:

- They use a stripped-down version of Java so there's less to get in the way. But there's still enough left to show good design principles and how an application consisting of a number of classes and other files is bundled into a final, executable product.
- Developing for SPOTs can be done through an integrated development environment (IDE), like NetBeans or Eclipse, but there's not dumb reliance on the IDE. The same tasks can be done through a command-line and the developer learns about fundamental tools like Ant (http://ant.apache.org/).
- One of the hardest decisions to make when creating a new Java application, is where to start. This includes managing the resources used by the program (images, configuration files, databases) and lifecycle events (startup, shut down, pausing and the like). That's why application frameworks have become so popular: they do all the mundane tasks and let developers get down to work more quickly. Likewise, Sun SPOT programs are written according to a specific interface, with specific lifecycle events. So, it's like writing a program using a simple template.
- Most importantly, the SPOTs show output, things happening, meaningful things. Starting with a HelloWorld program that prints a line to screen is of questionable value. Even the most basic Java web application or desktop GUI contains more technologies and code than a new developer can reasonably absorb; and then the resulting program is probably trivial and does nothing. Meanwhile a SPOT can be made to do interesting and useful things with just a single screen of code.

Sun SPOTs are an example of devices known more broadly as cyber physical systems (CPS): "such systems use computations and communication deeply embedded in an interacting with physical processes to add new capabilities to physical systems" (http://www.cra.org/ccc/cps.php). Because of the pervasiveness and economic opportunities presented by CPS, a recent President's Council of Advisors on Science and Technology (PCAST) said that there were few other current areas of research more critical (http://www.cra.org/govaffairs/blog/archives/000620.html).

The approach that has been taken for this research project is to create a graphical user interface application that allows the user to quickly design, build, and deploy SPOT applications. Using model-driven software development techniques [2-4], the Java source for the application is generated from the design decisions made by the user. With a working application and generated code, users are better able to connect theory with practice.

II. WHAT ARE SPOTS?

SPOTs are small wireless devices that run Java programs (see Fig. 1). They can sample their environment through a

K. Elleithy (ed.), *Advanced Techniques in Computing Sciences and Software Engineering*,
DOI 10.1007/978-90-481-3660-5_42,

range of sensors and each has LEDs and push buttons to provide basic feedback.

Some SPOTs can be tethered to workstations via a USB cable and these act as base stations through which other SPOTs can access resources such as databases or web applications.

In terms of hardware, SPOTs are made up of the following elements:

- Main processor is an Atmel AT91RM9200 system on chip.
- Each has 4MB flash RAM and 512K pseudo-static RAM.
- Power is supplied by an internal rechargeable battery (lithium-ion prismatic cell), external voltage, or through the USB host.
- The battery has a charge life of about 3 hours with continuous use, but it will hibernate when nothing is happening which can extend the life.
- The demonstration daughterboard contains temperature and light sensors, a three-axis accelerometer, eight tri-colour LEDs, and two push-button switches. Additional daughter boards can be added if necessary.
- Wireless communication is through an IEEE 802.15.4 compliant transceiver that operates in the 2.4 GHz to 2.4835 GHz unlicenced bands.

SPOTs run a small-footprint Java Virtual Machine (JVM), called Squawk [5], which is written almost entirely in Java. Squawk is compliant with the Connected Limited Device Configuration (CLDC) 1.1 Java Micro Edition (Java ME) configuration (http://java.sun.com/products/cldc/) and runs without the need for an underlying operation system, something called "running on the bare metal".

Applications for SPOTs are written according to Java ME MIDlet programming model [6]. This means that the JVM on each SPOT manages the life cycle the MIDlet (creation, start, pause, and exit). As long as the developer conforms to the simple requirements of the MIDlet programming model, they can concentrate on the logic of their program and leave some of the mundane duties to the JVM.

Java is now a mature programming language with type safety, exception handling, garbage collection, pointer safety, and a mature thread library. But, given the constrained environment in which SPOT programs run, the full

Fig1. A Sun SPOT device

breadth of the typical Java development kit (JDK) is not necessary.For example, SPOT programs don't have access to file streams; there are no object/filter/string readers and writers; no reflection; no serialisation; no classloading; no native methods; no regular expressions; and the only available collection data structures are vectors and hash tables [6, 7].

In standard Java ME, only one application can be run in the JVM although the application may consists of many threads. Squawk allows for multiple applications to be run together in a single SPOT and uses a special Isolate class to prevent applications from interfering with each other. Each MIDlet-based application is run in a separate isolate. While one isolate cannot directly access the instances on another, they all share the same underlying resources.

For any remote device, security is a major concern: applications running on the devices need to be protected from tampering and the data stream to and from the devices needs to be secure. To address these issues, each SPOT SDK installation creates its own public-private key pair. In effect, the first user to deploy an application to a SPOT becomes its owner. Only the owner is allowed to install new applications or make configuration changes wirelessly [8].

III. MODEL-DRIVEN SOFTWARE DEVELOPMENT

For all but the most trivial programs, most software development projects start with a model. A model is typically "a miniature representation of a complex reality. A model reflects certain selected characteristics of the system it stands for. A model is useful to the extent that it portrays accurately those characteristics that happen to be of interest at the moment" [9]. In software development, the Unified Modeling Language (UML) has become the standard modelling language [10], but a model may also be given verbally or mathematically or by using some other symbols.

Eventually, the model of a software application must be turned into source code in some programming language that can be compiled and executed on a computer. It's at this point that the model and the source code start to diverge: if changes are made to the source code without updating the model, the model can quickly become stale and its value as a reflection of some complex reality is diminished.

The disconnect between the model of a software application its source code and the problems this causes has been one of the driving forces behind model-driven software development [2-4]. Under this paradigm:

- Developers create a high-level platform independent model (PIM) of their application using a modelling language such as UML, but in reality any suitable modelling tool could be used. At this stage in the modelling process, the developers concentrate on the logic of the application without regard to programming language or operating systems.

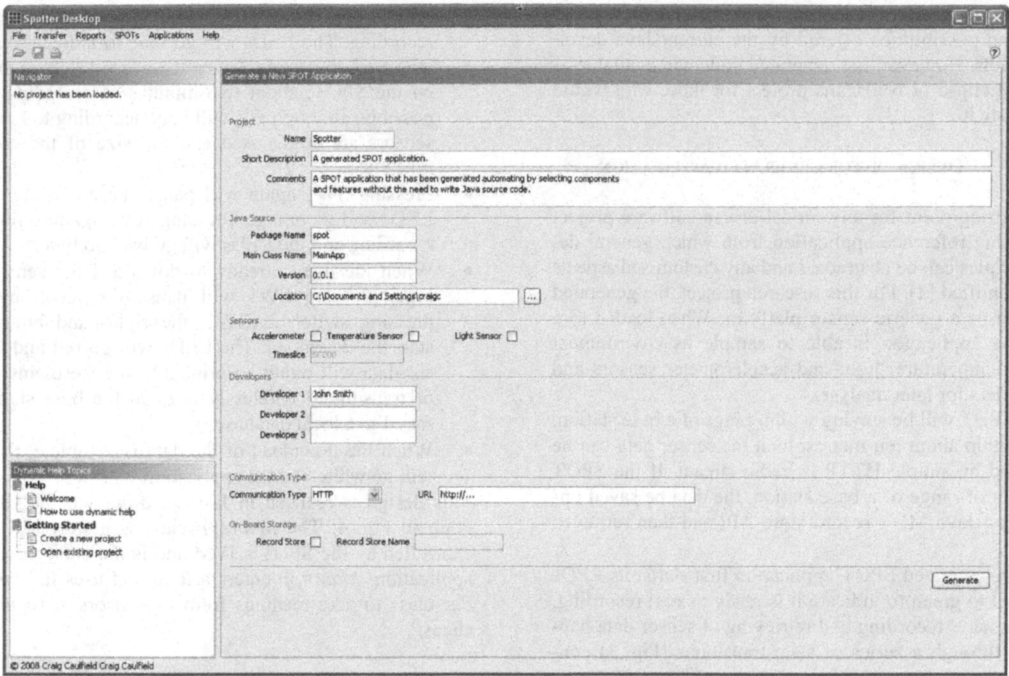

Fig. 2. The Spotter Desktop application through the features of the SPOT application can be specified.

- The PIM is transformed into a platform specific model (PSM) which more closely relates to the target platform. For example, if the target platform is a Java Enterprise Edition application server, the PSM would represent the PIM as, say, Enterprise Java Beans.
- Finally, from the PIM, the specific source code for a particular platform is generated.

Depending on the modelling tool, it may be possible for changes made at any one of the layers (PIM, PSM, or source code) to be propagated automatically to the other layers.

Model-driven software development has a number of advantages [2]:

- Productivity: developers can concentrate on the logic of their programs and leave all, or most, of the mundane coding to the transformation tools.
- Portability and interoperability: decisions about programming languages and operating systems can be delayed until the last moment.
- Uniform implementation according to design patterns. The transformation from the PSM to source code can capture good design principles and patterns [11] and ensure that these are applied consistently for all applications.
- Maintenance and documentation: the PIM and PSM models are high-level models of the application which are easier to read and understand than low-level source code.

IV. THE SPOTTER DESKTOP

For our research project, users work with their SPOTs through a Java application called Spotter Desktop (Fig. 2). Ultimately, Spotter Desktop will allow users to:

- Design a Java ME SPOT application by selecting and configuring the on-board sensors they want to use.
- Configure other behaviour such as how the SPOT will communicate with the outside world (HTTP, radiostream, or radiogram), and whether the SPOT uses an on-board record store.
- Automatically generate the complete Java source code for the application.
- Build and deploy the application to one or remote SPOTs without having to use Ant commands or an IDE such as Eclipse or NetBeans.
- Retrieve, save, display, and analyse the data collected by the SPOT.

All this will be done without writing a single line of Java source code or executing a low-level command.

For now, Spotter Desktop allows users to design, generate, and build a simple Java ME application. Users fill in the details of their application and press the Generate button. The source code is generated using a template engine called FreeMarker (http://freemarker.org/) and compiled into the selected directory.

Using Spotter Desktop rather than a traditional IDE allows users to quickly develop working applications without being overwhelmed by the wealth of technologies, commands, and possibilities offered by the current Java development tools. However, the generated code will available as either an Eclipse or NetBeans project for those who would like to study it.

V. DESIGN OF THE GENERATED APPLICATION

The starting point for any model-driven software project is a working reference application from which general design principles can be abstracted and any customised aspects can be identified [2]. For this research project, the generated application is a generic sensor platform. When loaded to a SPOT, the application is able to sample its environment using its temperature, light and accelerometer sensors and save this data for later analysis.

If the SPOT will be staying within range of a base station, that is, within about ten metres, then the sensor data can be transmitted by simple HTTP or radio stream. If the SPOT will be out of range of a base station, the data be saved onboard using Java ME's record store API and then retrieved later.

- When generated SPOT application first starts, its LEDs are set to green to indicate it is ready to start recording. From here, recording and retrieving of sensor data happens through a series of state transitions (Fig. 3) controlled by the two switches on the face of the SPOT:
- Pressing switch 1 (SW1, the left-hand button) starts recording. The LEDs will go blue to indicate that readings are being taken. Given the limited storage available on the SPOT, about two minutes of recording time is possible, although this will vary according to how many sensors are being used and the size of the recording time slice.
- Pressing SW1 again will pause the recording and the LEDs will go orange. Pressing SW1 again to restart the recording and the LEDs will go back to blue.
- When the user is ready to download the sensor readings, pressing SW1 will pause the recording. Then pressing switch 2 (SW2, the right-hand button) will start the download. The LEDs will go red and the sensor data will either be printed to the NetBeans console or transmitted wirelessly through the base station and saved to a local database.
- When the download of the data is complete, the LEDs will go white and the application will stop.

This design is realised in Java according to the class diagram in Fig. 4. The *MainApp* class is a Java ME MIDlet controlled by the SPOT's JVM and is the entry point of the application. *MainApp* enters a loop and uses the *SpotUtilities* class to take readings from its sensors at regular time slices.

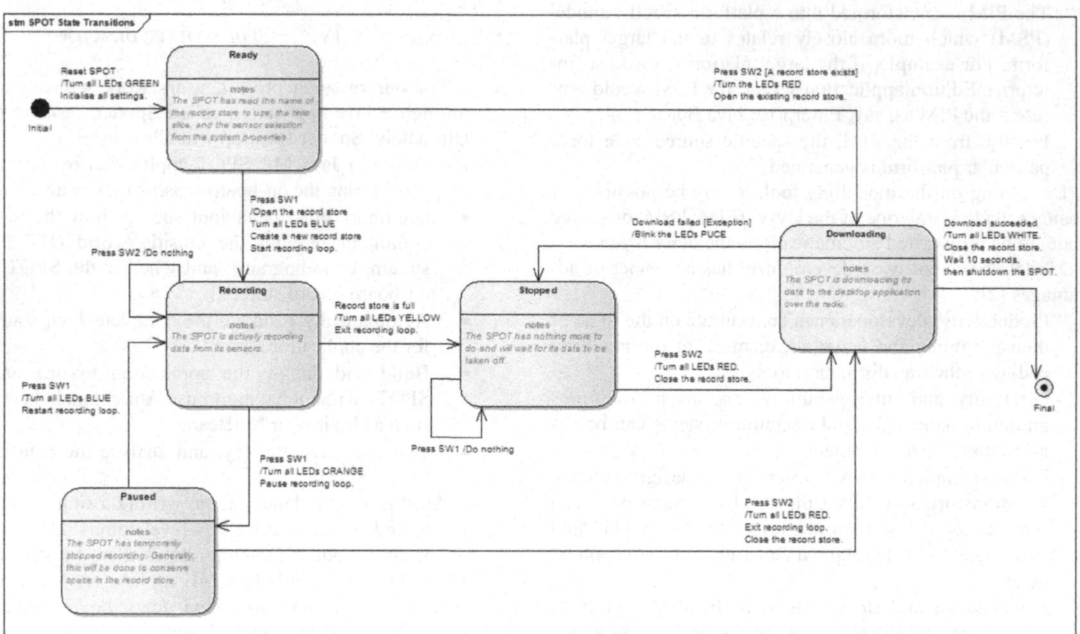

Fig. 3. The state transitions of the generated SPOT application as it responds to the two buttons (SW1 and SW2 on the face of the SPOT.

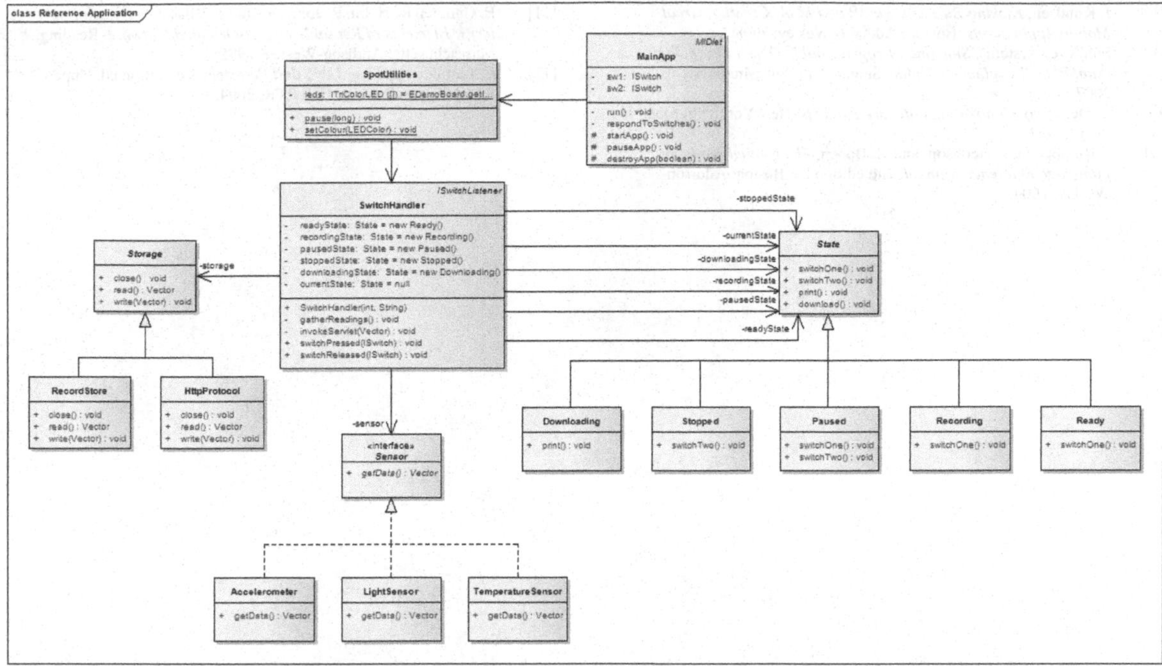

Fig. 4. The UML class diagram of the generated SPOT application.

The *Storage* and *Sensor* classes are high-level interfaces. Depending on the options chosen by the user through the Spotter Desktop interface, there may be one or more implementations of these classes. For example, if the application is to take readings from its accelerometer and light sensor, there will *Accelerometer* and *LightSensor* implementations of *Sensor*. Similarly, if the user elects to have the sensor data stored on-board the SPOT there will be a *RecordStore* implementation of *Storage*.

A key aspect of the application is how it reacts to the switches which control the recording of data. The state diagram in Figure 4 is implemented by the State class and its subclasses according to the State design pattern [11, 12].

VI. FURTHER RESEARCH

The reference implementation described in Section V has been trialled as part of a secondary school physics course called "Physics in Motion". Students attached the SPOTs to rockets, small aircraft, and other devices and used the accelerometer to study Newton's Law of Gravity. Based on these field trials, changes have been made to the reference implementation to make it more robust and to accommodate new sensors.

The next stage of the research project involves a closer examination of the generated Java source code and how it relates to the design choices made by the users through Spotter Desktop.

VII. CONCLUSION

Sun SPOTs are a unique technology that present some interesting opportunities. But, in order to take advantage of these opportunities, we need to know how to program these devices. By providing users with practical devices and generated applications that follow sound design principles, such as those created by Spotter Desktop, new developers can more readily connect the theory and practice of Java software development.

REFERENCES

[1] B. W. Kernighan and D. M. Ritchie, *The C Programming Language*, 2nd edition ed. Englewood Cliffs: Prentice Hall, 1988.

[2] T. Stahl, M. Volter, J. Bettin, A. Haase, and S. Helsen, *Model-Driven Software Development*. Hoboken: John Wiley & Sons, 2006.

[3] D. S. Frankel, *Model Driven Architecture: Applying MDA to Enterprise Computing*. Indianapolis: Wiley Publishing, 2003.

[4] A. Kleppe, J. Warmer, and W. Bast, *MDA Explained: The Model Driven Architecture: Practice and Promise*. Boston: Addison-Wesley, 2003.

[5] D. Simon, C. Cifuentes, D. Cleal, J. Daniels, and D. White, "Java on the Bare Metal of Wireless Sensor Devices: The Squawk Java Virtual Machine," presented at ACM/Usenix International Conference On Virtual Execution Environments, Ottawa, Ontario, Canada, 2006.

[6] S. Li and J. Knudsen, *Beginning J2ME: From Novice to Professional*, 3rd edition ed. Berkeley: Apress, 2005.

[7] J. Knudsen, *Kicking Butt with MIDP and MSA: Creating Great Mobile Applications*. Boston: Addison-Wesley, 2008.

[8] Sun Microsystems, *Sun Small Programmable Object Technology (SunSPOT) Developer's Guide*. Santa Clara: Sun Microsystems, 2007.

[9] T. DeMarco, *Controlling Software Projects*. New York: Yourdon Press, 1982.

[10] J. Rumbaugh, I. Jacobson, and G. Booch, *The Unified Modeling Language Reference Manual*, 2nd edition ed. Boston: Addison-Wesley, 2004.

[11] E. Gamma, R. Helm, R. Johnson, and J. Vlissides, *Design Patterns: Elements of Reusable Object-Oriented Software*. Reading, Massachusetts: Addison-Wesley, 1995.

[12] C. Larman, *Applying UML and Patterns*, 3rd edition ed. Upper Saddle River: Prentice Hall PTR, 2004.

Semi- and Fully Self-Organised Teams

Deniss Kumlander
Department of Informatics, TTU
Tallinn, Estonia
e-mail: kumlander@gmail.ee

***Abstract*—Most modern companies realise that the best way to improve stability and earning in the global, rapidly changing world is to be innovating and produce software that will be fully used and appreciated by customers. The key aspect on this road is personnel and processes. In the paper we review self-organised teams proposing several new approaches and constraints ensuring such teams' stability and efficiency. The paper also introduce a semi-self organised teams, which are in the short-term time perspective as the same reliable as fully self-organised teams and much simpler to organise and support.**

Keywords-self-organised teams; semi-self-organised teams; software engineering methods; software engineering practises

I. Introduction

The number of projects failures is very high in nowadays software development despite all modern approaches. Those failed projects' costs are carried by customers and add a lot of extra cost to successful projects. Unfortunately the situation with "successful" projects is not very different as well: just one fifth of the developed functionality is used "often" or "always" and 16% more "sometimes". The remaining functionality represents improperly spent development resources as it is either never used or used extremely occasionally [1]. The global market increasing competition between software vendors and much more demanding markets force companies to stabilize their productivity and improve their development process in all possible ways [2]. The key factor on this road is personnel, which is recognised as a very important aspect of any company success [3]. Unfortunately the software industry is a highly technological sector [4] with a shortness of personnel resources in many countries. Therefore it is important to motivate company employees. Unfortunately this task is not as easy as it looks like at the first glance – skilled professionals are still migrating from one company to another despite all modern motivating approaches [5, 6, 7], like good salaries, friendly working environment etc. It happens mainly because workers are looking for something more challenging or bored to do the same work using the same tools. Therefore, it is important to address such technological needs in addition to common motivating approaches or build the team in a novel way to increase employees' attachment to the organisation.

The software development process' quality depends on many aspects and one of those is the team performance. An improved communication between team members and different organisation hierarchy level members is one of the key issues [8]. Another one is the level of technology knowledge and the level of motivation. Many modern software development methodologies rely on the advanced teamwork, which is stimulated by a high level of freedom in several types of decisions. The freedom in decisions is usually described by using an "autonomous" team term, which is defined as a team performing its' tasks independently and therefore are given a significant respect within the organisation [9].

II. Self Organised Teams

Some authors see either the autonomous team or self-management team to be a direct synonym for a self-organised team [9, 10], which is generally defined as team able to act autonomously without the normal supervision.

The author experience dealing with self-organising teams started during personnel motivating projects [11, 12] that were executed in several places. A side effect during such motivational software development was nearly always an autonomous team capable to act without any sufficient supervision moving toward the predefined (project) goal. In other words such team will not require everyday management efforts and will need just occasional checks, plan and specifications of tasks included into their plans.

Autonomous teams stimulate participation and involvement, and an effect of this is increased emotional attachment to the organization, resulting in greater commitment, motivation to perform and desire for responsibility. As a result, employees care more about their work as so more motivated and therefore capable to act with greater creativity, productivity and work quality [13].

There are different types of autonomy and those include external autonomy, internal or individual. Alternatively the autonomy could be applied to different subjects (or set of subjects) like people, planning, goals, products decisions and so forth. External autonomy is defined [14] as the level influence of management and other individuals (outside the team) on the team's activities and the smaller the influence is the higher external autonomy is owned by the team. The external autonomy can be:

1. Obtained by the team as a compromise between organizational (i.e. hierarchical) management style and a need to run projects effectively;
2. Granted to the team by the management deliberately in order to force team independency and stimulate innovating and creative thinking;
3. Occur due any gaps in the management hierarchy.

K. Elleithy (ed.), *Advanced Techniques in Computing Sciences and Software Engineering*,
DOI 10.1007/978-90-481-3660-5_43,

The internal autonomy defines in what degree all members of the team are involved into making decisions, i.e. where decisions are made jointly or by a very restricted set of chosen members of the team. The individual autonomy refers to the level of independency or each individual member of the group defining what control s/he has over his/her duties/tasks and freedom to re-organise those.

III. Buidling Self Organised Teams

Classically the only requirements for converting a team into a self-organising team are to share a common goal, believe that the communication is the best way to achieve it and understand that each member should do his best in order to complete the project in the fastest and the most effective way. Although this requirement is generally sufficient to make a team successful, there are still a lot of faulty attempts on this road and therefore the self-organised teams is mostly used in software development based on the agile methodology [10].

The paper proposes that the sometimes failures can be avoided if two more requirements are defined. In the previous statement "sometimes" refers to the fact that it is not always possible to set restrictions described below without excluding from the selection the major part of organisation members and so cannot be followed in any blind manner.

The first requirement is defined for persons included into the self-organised teams – they should correspond to the following criteria:

- Has the ability to think independently;
- Has the high level of knowledge or education;
- Has an ability to learn;

The second promotes the need for an effective collaboration, i.e. a quick and efficient exchange of the information within the team, and is formulated as "The team should be relatively small".

Moreover it is advisable to include into the team individuals having the same level of respect by others and consequently to each other. Basically it is already formulated by the requirement of having the high level of knowledge, but it is not always a matter of just knowledge and ability to learn. If any team person will not correspond to this, then other will tend to skip him in internal discussions breaking the rule of efficient internal collaboration from the classical requirement. Moreover other team members will keep such person on second roles managing his/her work and this could potentially lead to conflicts as all developers hold exactly the same official position. In result this sufficient difference of knowledge, respect, and involvement levels could finally unbalance the team and blow it up. This situation is not purely theoretical and is reported by several authors [15]. For example there is a natural conflict between required autonomy levels to move effectively toward the team goal. On one hand the team requires group autonomy, on another hand individual autonomy and those could conflict sufficiently decreasing the team effectiveness if some individuals dominate in their individual autonomy over others.

Fortunately it is not always that much guaranteed to happen as greatly depends on the low skills' members' ego and approach to the work. If they are not egoistic and are ready to accept the second level position then the team could become externally self-organised and hierarchical internally. Notice that all other papers of the self-organised papers always assume that the self-organised teams internal structure is always flat and so monotonous, although it is not always so. In this article we would like to bring a software development community attention to the fact that internally the self-managing team can cluster members by access to discussions into standard members and low level members.

IV. Semi Self Organised Teams

In this chapter we would like to propose a new type of teams, which is a "semi-self-organised" team – it is a team that is able to act as semi-organised during a short period of time and so is able to survive a temporary lack of management without decreasing its performance. Moreover in the long term such teams could required quite a minimal management efforts and the strength of "semi" part could vary basing on independency level given to the team versus the amount of management efforts required to keep the team moving toward milestones defined by the organisation.

There are a lot of articles where this concept was nearly acquired. Typically a self-organised team article will start defining such team as an independent one and later will fall into a discussion whether the manager role is completely eliminated or the range of autonomy if such that an organisational force is still presented and conclude that managers are still required at least to control the team and prevent any unauthorised evolution processes. This clearly shows that quite many teams titled as self-organising are not such and therefore a new type of team located in-between ordinal and truly self-organised is required.

The self-organised team requires less in order to be produced and is the same efficient and powerful in the short term and is inexpensive to keep as still requires quite minimal managers attentions. Considering that many crisis and pilot projects are not long lasting we could propose that the semi-self-organised teams can be used successfully within any organisations and be a real alternative to the self-organised teams. Notice also that most advantages described below will be shared by this type of the team with the self-organised teams.

V. Activities Ensuring Self-organised or Semi Self-organised Teams Stability

In this chapter we would like to propose some novel approaches ensuring stability of self- or semi-self-organised teams.

The following extra activities are advisable to be applied in the semi self organised teams to increase the level of capability to act autonomously and in the self-organised teams to ensure externally that those will stay stable for very long time.

A. *Rotation of the Team manager role*

Each person acting by him/her-self without communicating much to others will behave and build understanding of the general environment purely basing on own experience and information and so will easily miss all activities, information flows, processes that are not attached to him/her. Therefore it is important to involve each member of such advance team into others work and keep him/her at least informed. As the major information pipeline goes via the team leader it can be achieved by assigning one by one all members of the team into that supervisor position for a short period of time. The best timing for the semi-self-organised teams would be yearly vacations of the team leader that could be divided into several frames or absence of the team leader due travelling to other company branches (like the parent / head office).

Expanding knowledge base for each person is not the only reason while this approach is applied. The following extra reasons will form a complete list of advantages to be gained.

1. Involve the person into the process and highlighted processes that are spanning over the team and how each step of those processes is organised, linked to others, what conditions choosing one or another path are;

2. Increase knowledge base of technologies used by the team, all aspects including benefits and potential troubles of those;

3. Involve the person into making team lead decisions and so highlight to him what kind decisions should be made before one or another thing will happen, what information should be collected to making the decision (showing the reason why one or another bit of information is asked from him) etc.;

4. Demonstrate teams we are working with including the way, reasons and actions triggering it. This will show dependencies between teams, position the team inside of different work-flows and provide extended information on responsibilities areas of different teams. As a side effect, the fresh person could some new bright ideas on improving one or another process been ineffectively done so far. Obviously all those ideas should be communicated to and only to the team leader in order to avoid faulty ideas brought due the lack of knowledge of processes, work load or resources of different teams.

5. Manage work of other team members and so highlight what they are doing. Mostly it means not just what they are doing, but how much they are doing (which is normally stays invisible been involved only in own tasks and processes).

6. The earlier mentioned expansion of knowledge.

B. *Co-work*

The vital components of self organised work are respect and involvement. Therefore the co-work should be also motivated. There are several levels of the co-work that can be used.

The first and the minimal level of co-work will include only collaborative work of the final stage of the process (development, specification writing etc) when the final result is reviewed together with somebody else. The first advantage of such review process will be an improved quality of the work as the fresh person could have a lot of new suggestions, notice mistakes that you are not able to identify after working with the document/code etc so long or during discussions on the best way to achieve the desired goal. The second advantage is in increased collaboration between team members and better transparency of tasks and work progress for each member and the whole team. Finally the collaborative review will serve as a small knowledge transfer, which will secure future of the team/organisation as more than one person could continue developing the project in the future, allow to handle cross-functionality in the best possible way and generalise functionality used in many places of the product.

The maximum level of the co-work is an everyday work together as is promoted by some modern development techniques like, for example, extreme programming. The constant collaboration will improve quality of the product and serves as permanent knowledge transfer process. There are several possible ways to implement such co-work varying by the length of it or by the involved partners. It is possible to implement rotating rules in order to maximise effect either in a certain group of selective (compatible) partners or in the wide scope of the entire team.

Notice also that the co-work lately has crossed bounds of a single organisation and is effectively used and therefore quite popular among independent contractors physically locating (been) in the same room, but involved in completely different projects and so working for completely different organisations. Nevertheless the main idea stays the same – increase base of knowledge for each co-working team member during informal conversation by providing more information for each one obtained from very different areas of responsibilities/ technologies/business etc, which the person will never be involved into in normal circumstances.

The final mark on the co-work should be done by mentioning attempts to simulate the same effect by constant rotation of work-places within the single company. Unfortunately this is not always advisable as greatly depends on abilities of persons to move around, communicate to others and adapt to new working environment. A lot of people will hate such constant changes and therefore will be stressed. The traditional co-work routine should ensure both: stability of the environment (read the set of people you communicate to) and constant incoming flow of new information by involving parties from areas different enough to have a lot of new, specific problems and similar enough to be interesting for each co-working person.

VI. Advantages

In this chapter we would like to extend the list of classical advantages [10, 15] by adding some new, which are related to the motivational projects, new requirements to build the team the paper defined earlier and so forth.

A. *Avoiding conflicts*

Nowadays a lot of companies are trying to build innovating organisations composed of high level professionals and dealing with high end technologies. Notice that a team of professionals is required to keep up with constantly changing technologies, programming languages etc, i.e. a set of persons able to learn constantly and quickly. Unfortunately it means that they also could have a lot of independent opinions, which in case of central management, are posted to him and finally evolves into conflicts within the team and bring the team to the complex to manage level. If the manager is involved much less in doing decisions (of different type) then persons will discuss opinions rather than present to the high level judgement and so differences in opinions will not arrive to conflicts, or it will not happen too often.

Besides, the self-organising team is a great approach in case the team is composed of highly skilled professionals of the same age and environment. The common problem here - nobody would like to report to other similar persons as there is no obvious difference between us except the position. Having a manager similar to the team members will mean that each member could be stressed by inequality and manager will be stressed by the lack of respect to his decisions. Finally some reports will tend to be sent over him to the higher management, who are normally less visible and so more respectful.

B. Motivating

- New technologies

Highly professional and motivated personnel are a key factor of the company success and unmotivated personnel can be seen as a major risk factor [5, 7] since workers either decrease their productivity or are going to leave the company. Unfortunately professionals are still migrating from one company to another mainly because workers are looking for something more challenging, bored to do the same work using the same tools or stacked in their self development at the current workplace. It many cases they would like to use other, modern methodologies and techniques to move themselves forward, ensure their future and try something else. Software engineering using novel approaches can be seen as an important motivation tool, which can be applied only in semi-fully self-organised teams. Alternatively, self-organised teams are rare within any organisation as are composed of highly skilled professionals and therefore such teams are mainly used in different pilot project with new techniques and approaches and this already motivates individuals included into such teams as they see that the organisation highly appreciate their work and skills.

- Involvement

Semi or fully self-organised team typically have a high level of group autonomy, which means that team arrive to decisions by a consensus. This highlights another motivating factor as each member of the team is involved into the discussions and the process of making decisions and therefore each member see that s/he has a real opportunity to affect the product evolution.

- Visibility of work results.

The self-organised teams are generally very effective and quick to produce the result. Considering the fact that this is accomplished by the high level of autonomy meaning enough information about deadlines, results etc. produce a good visibility for the team of their work results. This removes a common de-motivation factor existing in many organisations when the person doesn't see any results of this work producing a believe that he is dealing with problems so unimportant that even if he will go somewhere else, nobody will mark or care about that.

- Acting in the team of professionals

A huge inequality of team members' skills sometimes produces a de-motivating factor of having to explain too much to less experienced colleges. It will be stressing for many employees if this happens rapidly and therefore self-organised teams composed of highly and equivalently skilled members is more likely to be stable during very long time than unequal teams. If more stressing will be to be stacked of other inability to fulfil a simple task.

Besides, a lot of persons would like to transfer their knowledge to others and much more will like to acquire it from others. The degradation of skills will be very much remarkable if the person have nobody among colleges to discuss novel approaches and so forth and therefore the earlier discussed co-work if required to keep the team stable and motivate to stay working for the current organisation.

C. Efficiency

- Cooperation

The types of teams described in the work actually promote a new way of working, which is highly collaborative. This constant collaboration sufficiently increase efficiency, eliminates duplications (of work, research, extra tasks etc) and therefore moves operations to completely new level of quality.

- Improved speed of writing code

The earlier discussed improve visibility of results can be also mentioned as improved speed of writing code, although the last one have larger effect. First of all it improves the time to market making it possible to start more projects that the company did in the past. Besides the quicker delivery cycle allows rapidly response to new market challenges. Moreover this can also serve as a base for a constant collaboration with customers that are interesting in the product evolution and constant development of it.

- Discussions

Unlike the standard hierarchical organisation of teams, where most discussions are started by managers, all types of self-organising teams have an ability to start those by themselves. This makes meeting and discussions start in the right time avoiding delays and consequently avoiding late decisions. Moreover a lot of problems can be solved just by informal meetings which are hardly possible in standard teams due lack of information or management efforts organising meetings and proper delivery of required information. Generally saying self-organised teams' meetings are much less formal as there is no manager ego leading the meeting and this promotes a free and open discussion. Employees are not afraid to make statements and come up with ideas and so the overall speed of moving toward the right decision is much higher than in formal meetings lead by a manager who is normally not an expert in the discussed topic.

D. *Organisational*

Generally saying all earlier mentioned advantages can be attached to both individuals and the organisation, but the organisation ahs also specific advantages. First of all the collaborative highly autonomous work means that there will be a group of persons able to complete a task rather than the experience will be stacked in each individual (doing one or another specific task). It means that loosing whatever person will be a huge problem in the standard hierarchical team and will not be a problem at all in the semi or fully self-organised teams. It includes loosing a manager as the wrong person assigned to the team in hurry could destroy the entire team. Semi autonomous teams allow delaying the manager appointment decision to make the right enough choice without decreasing the team performance.

VII. Cases when the approach cannot be applied

Unfortunately building such sophisticated teams requires certain company vision on the hierarchy and processes to be applied. Many organisations do like the desired goal that can be achieved by this type of teams' organisations – mobility of the team, efficiency and productivity and try to derive benefits without understanding efforts to be applied. It is impossible to build such teams in companies having strict hierarchy policies and a traditional command line to be followed. It is impossible to organise such teams using ordinal persons liking to do their work with minimal efforts and go home as soon as possible, i.e. of those who do not see the work they are doing to be attractive or challenging.

VIII. Conclusion

In the paper self-organised teams were revised. First of all the paper proposed to change the basis of organising such teams in order to increase their stability and attach to the reasons they were produced. Besides some additional novel approaches ensuring performance of this type of teams were proposed together with possible advantages that could be gained in the result.

The paper also introduced a new type of teams locating between standard and self-organised. The essential parts of the new type were described in many previous articles without extracting it to a standalone type and so authors were straggling a lot between self-organised autonomy and managers' efforts that still were required. The semi-self-organised teams are team still requiring mangers attentions, but on sufficiently lower level than standard hierarchical teams. At the same time, the new type of teams is the same effective as the fully self-organised in the short–term perspective and much less demanding from resources point of view. Therefore those teams and approaches to stabilise those can be advised for classical organisation which are not yet read to adopt agile methodologies or not having sophisticated enough professionals to build exclusively self-organised teams.

References

[1] A.A. Khan, "Tale of two methodologies for web development: heavyweight vs agile," *Postgraduate Minor Research Project*, 2004, pp. 619-690.

[2] D. Kumlander, "Software design by uncertain requirements," *Proceedings of the IASTED International Conference on Software Engineering*, 2006, pp. 224-2296.

[3] M. Armstrong, A *Handbook of Personnel Management Practice*, London, UK: Kogan Page 1991.

[4] M. Rauterberg and O. Strohm, "Work organisation and software Development," *Annual Review of Automatic Programming*, vol. 16, pp. 121-128, 1992.

[5] D. Daly and B.H. Kleiner, "How to motivate problem employees," *Work Study*, vol. 44(2), pp. 5-7, 1995.

[6] B. Gerhart, "How important are dispositional factors as determinants of job satisfaction? Implications for job design and other personnel programs," *Journal of Applied Psychology*, vol. 72(3), pp. 366-373, 1987.

[7] F. Herzberg, "One more time: How do you motivate employees?" *Harvard Bus. Rev.*, vol. 65(5), pp. 109-120, 1987

[8] R. E. Miles, C. C. Snow, and G. Miles, "TheFuture.org," *Long range planning*, vol. 33 (3), pp. 300-321, 2000.

[9] R.A. Guzzo and M.W. Dickson, "Teams in organizations: Recent research on performance and effectiveness," *Annual Review of Psychology*, vol. 47, pp. 307-338, 1997.

[10] N.B. Moe, T. Dingsoyr, and T. Dyba, "Understanding Self-Organizing Teams in Agile Software Development," *Australian Software Engineering Conference*, 2008, pp. 76-85.

[11] D. Kumlander, "On using software engineering projects as an additional personnel motivating factor," *WSEAS Transactions on Business and Economics*, vol. 3(4), pp. 261-267, 2006.

[12] D. Kumlander, "Key success factors in personnel motivating projects," *Proceedings of the 7th Conference on 7th WSEAS International Conference on Applied Informatics and Communications*, pp 200-205, 2007

[13] M. Fenton-O'Creevy, "Employee involvement and the middle manager: evidence from a survey of organizations," *J. of Organizational Behavior*, vol. 19(1), pp. 67-84, 1998.

[14] M. Hoegl and K.P. Parboteeah, "Autonomy and teamwork in innovative projects," *Human Resource Management*, vol. 45(1), pp. 67-79, 2006.

[15] C.W. Langfred, "The paradox of self-management: Individual and group autonomy in work groups," *J. of Organizational Behavior*, vol. 21(5), pp. 563-585, 2000.

Stochastic Network Planning Method

Zs. T. Kosztyán*, J. Kiss*
*Department of Management, University of Pannonia, Veszprém, Hungary

Abstract – **The success of the realisation of a project depends greatly on the efficiency of the planning phase. This study presents a new technology supporting the planning phase.**

While projects can differ greatly from one to the other and thus require separate models and considerations, there are some questions that are always applicable. Is this the most efficient realizing sequence of tasks? Have all the possible solutions been taken into consideration before the final schedule was identified? In the course of our work, we searched for answers to these questions. The method under review (SNPM: Stochastic Network Planning Method) is a general technique which is adaptable to solve scheduling tasks. The advantages of the SNPM over already known methods (e.g. PERT, GERT, etc.) are that it identifies possible solutions with the help of stochastic variables and that it takes into consideration all of the possible successor relations. With this method, the parameters can be changed if the impacts on the project change (e.g. due to tendencies of the market, changes of technological conditions). Thus the SNPM could be useful as a module of an expert system.
The steps of the SNPM are introduced through a few examples to show how it works.

Keywords: stochastic network planning method, stochastic relations, stochastic logical planning

I. Introduction

The planning phase has accentuated importance for the project realisation because any changes in this phase affect subsequent processes [1],[2]. Studies show that most unsuccessful projects fail as a result of problems during the planning phase (see e.g. [3][4][5],[6][7]) with the two biggest risk factors being uncertainty and the use of inaccurate estimates [8],[9][10]. The well known tools of project planning are the methods and models for scheduling and the so-called network planning (see e.g. [9],[6][12][13][14][15][16][17][22]19]). But it is not enough in itself to the success of the project if the model is applicable. It is important for the model to be adequate. Thus the results of the finished project will fit not only the model but also reality. For the model to yield results useful in practice, it is very important that the model be adequate to the modelled project.

Though scheduling problems are in theory well wrought, a profound difference can often be noticed between plan and fact when using scheduling methods in practice. The reason for this is uncertainty, which is one of the most unmanageable factors [6]. This uncertainty can be divided into two types: uncertainty in estimating and uncertainty in planning.

To handle uncertainty, the changing parameters should be described by stochastic variables. At present one of the most popular method is the PERT method, which can handle task duration time changes (see e.g. [5],[6][17][18]) as a probability value.

The planning phase contains specific steps. The realising sequence is identified in most cases by the technological sequence. In developing the SNPM, we approached this problem from a logical planning. The sequence of tasks depends mainly on technology [1][22]. This is correct. But we suspect that there are several important variables that have not yet been accounted for as in the definition of the realisation sequence. Examples of such parameters include the rate of return, efficiency, tendencies of the market and good will of the company, which can affect the combination of possible solutions as a target function. Feasible solution means all of scheduled variations of realizable tasks. To select the optimal solution from all the possible solutions is possible with decisions made by the management or by defining another target function (e.g. minimal TPT /Total Project Time/, minimal total cost etc.).

For a given project there are several different methods and techniques for the scheduling of project tasks and processes. One could use a deterministic method like CPM or MPM or a stochastic technique such as the PERT or the GERT. There is a general technique called GERT method as well, which though is similar to the PERT method, is in addition able to handle the possible solution variants [6][14] .

The GERT is a widely used network planning technique method, which can handle the successive probability of possible project outputs. Its main advantage is that the expected duration time associated with a given probability level can be identified relatively early, during the planning phase. Additionally, uncertain successors can be calculated and their rate is known, allowing for preparation for the possible volatile contingency factors. Its disadvantage is that the GERT method manages only task durations and the possible project output is regarded as a variable and it ignores the relation between tasks [23]. In contrast, the SNPM is able to identify the feasible solutions while taking into consideration the intensity of the relation between tasks. The probability variables of the intensity of the relation between the tasks show the preferences of the decision makers. But this model with some restrictions can also use the management preferences.

PERT and GERT methods are useful if we take into consideration the uncertainty of the estimated duration time, which can be defined in turn by analytical and simulation (e.g. Monte Carlo) techniques.

These methods have two difficulties: one is related to the exact estimation of durations and the other problem is the correct estimation of processes [18]. A new method was

K. Elleithy (ed.), *Advanced Techniques in Computing Sciences and Software Engineering*,
DOI 10.1007/978-90-481-3660-5_44,

intended to be developed, which would be able to manage the uncertainties related to estimation and planning that affect the structure of a project.

In some studies [10], the tools of knowledge management and simulating methods based on past records were used successfully to manage the problems related to uncertainty, but these methods were adapted for managing only a small group of uncertainties because they are only able to use in their calculation prior events which had had sufficient applicable data prepared.

There is another technique for scheduling processes called DSM (Dependency Structure Matrix), which is useful for identifying the implementation sequence of tasks or rather coordinating the information flows. The DSM method is similar to the SNPM method. The DSM method gave us the idea to use a matrix to describe the relationship between tasks. Both methods are used for defining the realizing sequence of tasks. Although the SNPM uses a stochastic variable to identify the sequence of tasks, it also takes into consideration the intensity of the relation between tasks and applies the so-called speed of information stretch; the DSM only makes use of the latter [12,25,26]. There are some methods, which use stochastic variables to identify sequence of tasks. Tasks, which have to be realized, are the results of decisions. However usually in case of project scheduling all tasks have to be completed, but the sequences of tasks can be varied. [25,26] The SNPM parameters can be varied and adjusted and they also allow a much wider range of applicability.

Henceforward, the probability variables of the relation between the tasks will provide all the possible solutions. If there is a target function (e.g. minimal TPT, minimal total cost, etc.) a project with minimal total cost or minimal project time can be changed. That could be used in an expert system to generate feedback, in which – if we estimate the project – the intensity of possible sequence relations can be weighted again. A project is a unique range of tasks. In regard to all the steps of the tasks they can not be repeated. But the decomposition of processes generates sub-processes which can be repeated in other cases. So those data could be applicable later on. E.g. during any building projects concreting and masonry are repetitive tasks which are independent of the type of building under construction. Optimal solutions can be found from all of the possible solutions by target functions defined by the management.

II. INTRODUCING SNPM-METHOD

While presenting our method, we will use the AoN technique because most project management software also uses AoN and because in AoN-nets arrows represent the relationships between tasks. In our model we deal with the intensity of the relation in details.

Regular project management software allows using networks that have more than one start and finish points. However, our method allows only those networks in which graphs are directed and acyclic with only one start and finish point.

In this method the intensity of the relation is introduced as a new term. It is a scalar between -1 and 1. (Values between -1 and 0 will be discussed later)

Definition: The intensity of the (directed) relation between task A and task B is represented by the notation $\rho(A,B)$. The value of $\rho(A,B)$ can be between -1 and 1, ($\rho(A,B)\in[-1,1]\subset R$). The probability of the task A is in relation with task B (task A is a predecessor of task B/task B is a successor of task A) and is calculated as follows: $p(\rho(A,B))=|\rho(A,B)|\in[0,1]$. The probability of task A is not in relation with task B and is calculated as follows: $p(\neg\rho(A,B))=1-|\rho(A,B)|\in[0,1]\subset P$.

Note: The intensity of the (directed) relation between task A and task B is usually different from the intensity of the (directed) relation between task B and task A. Therefore, this relation is not reflexive.

Note: If the intensity of the relation between task A and task B is equal to 1, then it is certain that task A is a predecessor of task B (task B is a successor of task A; for instance, task B has to start after the realisation of task A). In this case the probability that task A is not a predecessor of task B is equal to 0. $p(\neg\rho(A,B))=1-|\rho(A,B)|=1-1=0$. If the intensity of relation between task A and task B is 0 then task B is independent from task A and the relation between task A and task B is not notated.

Definition: Let A be the task list, where $A_1, A_2,...A_n$ are tasks. At this time $\boldsymbol{\rho}\in[-1,1]^{n\times n}$ is defined as a matrix of (directed) relations, where the intensity of the (directed) relation between task A_i and task A_j is notated $\boldsymbol{\rho}(A_i,A_j)\in[-1,1]$ $(1\le i,j\le n,\ i\ne j)$.

Note: In this phase, the type of successors / predecessors (e.g. Finish-to-Start, Start-to-Start etc.) is not investigated, because the type of successors/predecessors does not influence the set of feasible networks. For simplicity, it is assumed that between task A_i and task A_j can be defined only one (directed) relation at the same time.

Definition: A graph is a topologically feasible solution if and only if the graph is net (directed acyclic graph with only one source (start point) and only one sink (finish point)). Hereafter, these feasible solutions will be called feasible nets.

Note: Ensue from the definition that only one feasible solution can be permitted only if it is topologically ordered reducible (it is directed and contains no circles). The graph we got has only one start and finish point.

***Note*:** Not every feasible net (topologically feasible solution) is suitable for the management. For instance, feasible nets with too many redundant relations between tasks (too many predecessors/successors) are not suitable for the realisation of the project. These feasible solutions can be ignored.

Example 1 Take a simple range of tasks that contains four tasks. If we pay attention to the technological requirements, we can realize tasks in sequence, but parallel as well. In this case the following relation matrix can be solved as:

$$\rho = \begin{bmatrix} 0 & 1 & 0.5 & 0 \\ 0 & 0 & 0.5 & 0.5 \\ 0 & 0 & 0 & 1 \\ 0 & 0 & 0 & 0 \end{bmatrix} \quad (1)$$

Note: $\rho = 0.5$ means that 0.5 is the probability if there is (or there is not) relation between two tasks, so it does not represent the probability in each solution alternatives.

In this example the solution variants applicable for the management are defined as restrictive conditions:

1. There is only one start and one finish point in the graph,
2. There is not redundant relation.

From the relation matrix these possible graph variants can be drawn in Table 1.

TABLE 1
POSSIBLE GRAPH VARIANTS

Figure 1/a: topologically feasible solution and applicable for the management. Every task is performed in sequence. The probability is: $=p_1= \rho(A_1,A_2) * \neg\, \rho(A_1,A_3) * \rho(A_2,A_3) * \neg\, \rho(A_2,A_4) * \rho(A_3,A_4) =$ $=1*(1-0.5)*0.5*(1-0.5)*1=0.125$	Figure 1/b: topologically feasible solution and applicable for the management (parallel realisation).
Figure 1/c: topologically feasible, not applicable for the management (unfavourable).	Figure 1/d: topologically infeasible solution.
Figure 1/e: topologically feasible, but not applicable solution for the management (unfavourable).	Figure 1/f: topologically infeasible solution.

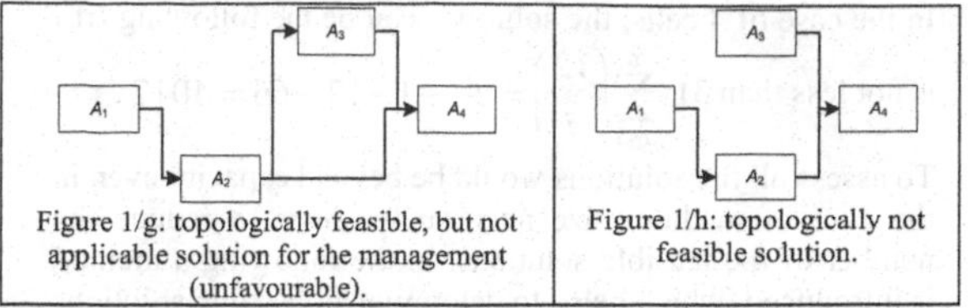
Figure 1/g: topologically feasible, but not applicable solution for the management (unfavourable).
Figure 1/h: topologically not feasible solution.

Solutions *d*), *f*), *h*) can be automatically ignored, since topologically infeasible. Solutions *c*), *e*), *g*) can be ignored in further investigation upon management decision.

To be able to choose from permitted solutions, we must define a target function, e.g.: the shortest total project time. In this case, if we know the duration time of tasks and the parameters of relations, the best solution can be chosen of the permitted ones.

Definition: The reduced relation matrix $\mathbf{r} \in \{0,1\}^{n \times n}$ of a (directed) relation matrix $\boldsymbol{\rho} \in [-1,1]^{n \times n}$ if satisfies every $1 \le \forall i, j \le n$, that $\mathbf{r} \ni r(A_i, A_j) = \lfloor \rho(A_i, A_j) \rfloor$.

***Note*:** Reduced relation matrix of Example 1 according to definition

$$\mathbf{r} = \begin{bmatrix} 0 & 1 & 0 & 0 \\ 0 & 0 & 0 & 0 \\ 0 & 0 & 0 & 1 \\ 0 & 0 & 0 & 0 \end{bmatrix} \quad (2)$$

.

***Note*:** Equation (3) shows the number of all possible solutions in case of *n* node:

$$\sum_{i=0}^{n(n-1)} \binom{n(n-1)}{i} = 2^{n(n-1)} \quad (3)$$

Among *n* nodes maximum *n*(*n*-1) edges are possible (feasible solutions can be edges, those indicated from *i* to *j* or from *j* to *i*, if loops are not permitted). We have to choose a number of *i* edges, where 0≤*i*≤*n*(*n*-1). We can do it as follows: $\binom{n(n-1)}{i}$. If we add possible solutions we will get the scheme above.

Among *n* nodes maximum *n*(*n*-1) edges are possible (feasible solutions can be edges, those indicate from *i* to *j* or from *j* to *i*, if loops are not permitted). We have to choose a number of *i* edges, where $0 \le i \le n(n-1)$. We can do it.

***Note*:** It is not suggested to determine the value of *i* less than *n*-1 (see Figure1/*d*), while graph is technologically not permitted.

In the case of 4 cells, the solutions can be the following (if i is not less than 3): $\sum_{i=3}^{6}\binom{12}{i}=2^{12}-1-12-66=4017$.

To assess all the solutions would be beyond capacity even in the case of medium size nets, so we have to reduce the number of the feasible solutions. In the following a method is introduced, which helps to determine all feasible solutions and reduce the necessary steps.

In a graph that contains n elements, all the feasible solutions can be counted with the above mentioned formula. Only if there are possible relations among the nodes such large number can exist. If the intensity of the relation between two tasks in the relationship matrix cannot be determined (there is no successor relation between the tasks), or the intensity of the relation between the two tasks is 1 (there must be a successor relation between them), there is no feasible variation. Let $k \leq n(n-1)$ be the number of relations where the intensity of relation is between 0 and 1. In this case the number of feasible graphs are $2^k \leq 2^{n(n-1)}$ (In Example 1 k=3, so the number of feasible solutions is 2^3=8). A graph where the number of edges is less than n-1 cannot be among the solutions (so the feasible solutions in the previously mentioned example are 8-1=7).

If there are a lot of feasible relations between the tasks, even if the number of elements k=100 the counting of possible solutions seems to be hopeless. In case of a feasible solution, we need to use topological ordering to decide whether the graph-variant (net) is a feasible solution or not. The run time of topological ordering is $O(n\log(n))$. So if $k \leq n(n-1)$, then it takes $O(2^k n\log(n))$ to evaluate the feasible nets. So it is necessary to reduce the aggregation of feasible solutions.

In our method we benefit from all the attributes of the net. A net is not topologically ordered if it contains a directed circle, so we can not involve those cases where there is a circle in the network. We start out from the premise, that in the case where there is no circle in a simple graph, the adjacent matrix of the graph can be rearranged to a so-called upper triangle matrix. Our method is a back-tracking procedure, discussing all the feasible solutions and leaving out of consideration all the incorrect ones.

III. Steps of the Method

Step No.0.: Definition of the reduced relation matrix: the rearrangement of the matrix to an upper triangle matrix. If it is not possible to rearrange it, the task has no solution -> STOP, otherwise move to Step No.1.

Step No.1.: Should the number of edges be (m) larger than that of nodes n-1, topological ordering has to be done. If not, then move to Step No.2/b. If the solution is allowed (possible to reduce topologically, graph contains one start and finish point) than move to Step No.2/a.

Step No.2/a: Print the solution: choose the next edge. Only an edge can be chosen, that is located in the upper triangle and had not been chosen before. In case of (A_i, A_j) $i>j$ if there is no edge to select, than move to Step No.3, otherwise move to Step No.1.

Step No.2/b: Choosing the possible task. It can be chosen if the adjacent matrix can be reduced to an upper triangle matrix and this edge had not been chosen before. If nothing can be chosen, then move to Step No.3, if there is a selected edge then move to Step No.1.

Step No.3.: Reduction of a chosen relation: if the graph is already among the solutions, or after the reduction the graph is not allowed, then quit the branch. Otherwise move to Step No.1.

Example 1 *(continuation)* Let us see the above mentioned example. In this case the relation matrix is the following:

$$\rho=\begin{bmatrix}0 & 1 & 0.5 & 0\\ 0 & 0 & 0.5 & 0.5\\ 0 & 0 & 0 & 1\\ 0 & 0 & 0 & 0\end{bmatrix} \tag{4}$$

The **r** reduced relation matrix is:

$$\mathbf{r}=\begin{bmatrix}0 & 1 & 0 & 0\\ 0 & 0 & 0 & 0\\ 0 & 0 & 0 & 1\\ 0 & 0 & 0 & 0\end{bmatrix} \tag{5}$$

The number of the possible solutions should be 2^3=8. As in the reduced relation matrix the number of edges is less than 3, it is not a possible solution (see figure 1/d.). According to the method, a new edge is necessary. A possible edge is the following: $\rho(A_1,A_3)=0.5$ (see figure 1/f.). Here, the number of the chosen edges is 3. However, after the topological ordering, the graph does not comply with the definition of the net. So a new edge has to be chosen. That can be the following: $\rho(A_2,A_3)=0.5$. The number of the chosen edges is 4 (see figure 1/e.). Graph is topologically ordered and complies with the definition of the net. After a new choice: $\rho(A_2,A_4)=0.5$, we get figure 1/c. It is possible to order this graph topologically and it complies with the definition of the net as well. No other edge can be chosen, so we have to take an edge of the graph. If we take edge $\rho(A_2,A_3)=0.5$, we get the graph shown in Figure 1/b. It is a permitted solution. We can not take another edge. If we do and take $\rho(A_2,A_4)=0.5$ we get *1/f* again. It is not a feasible solution so there is no point making other reductions; should we take $\rho(A_1,A_3)=0.5$ we would get *1/h* which is not an allowed solution, so we quit the branch. So we get back to solution *1/c*, if we choose $\rho(A_1,A_3)=0.5$ we get to Figure *1/g* which is a feasible solution. If we leave $\rho(A_2,A_4)=0.5$, we get Figure *1/a* which is also a feasible solution. In the example above many nets were not permitted so in the future we only have to work with results that are technologically feasible and also applicable for the management. If we pay attention to these criterions, *a* and *b* variants have equal feasibility to be realized.

Note: ***P*** stands for the aggregation of all the solutions. ***Q*** (***Q***⊆***P***) shows the technologically feasible solutions, which are applicable for the management as well. The probabilities of all solutions create an event algebra: $p=1$. The sum of the probability of the solutions, which are technologically feasible and applicable for the management (project solutions), is noted q, $q \leq p=1$. (In the Example 1: $p=1$, $q=0.125+0.125=0.25$)

Definition: The q_i probability of a „No.i." feasible project (applicable for management and technologically too): the probability of all project solutions divided by the probability total feasible projects, $q_i=p_i/q$.

Example 2 *Determining the optimal solution in a target function.*
The intensity of the relation should designate the preferences of the management.
Take a look at the example above. Let be given the following durations of tasks: d_{A_1} = 5 days, d_{A_2} = 10 days, d_{A_3} = 7 days, d_{A_4} = 6 days. Let target function be the minimal TPT. Restriction: let the successor chance of the project be at least 0.4.

In case of solution *a*) and *b*) the probability of project solution is 0.5. In the other cases, it is 0. In case of solution *a*) duration time = $d_{A_1} + d_{A_2} + d_{A_3} + d_{A_4}$ = = 5 days + 10 days + 7 days + 6 days = 28 days; in case of solution *b*) duration time = $d_{A_1} + \max(d_{A_2}, d_{A_3}) + d_{A_4}$ = = 5 days + max (10 days, 7 days) + 6 days = 21 days. According to the above *b*) is chosen. Tasks proceed without any disturbances so the tasks list is considered to be successful and the value of doubtful relations that participate in the solution is increased by 10 percent. This time $\rho(A_1, A_3) = 0.5 + 0.1 = 0.6$ and $\rho(A_2, A_4) = 0.6$.

In the next project we get two feasible solutions again. The probability of solution *a*) is

$$p_1 = \rho(A_1, A_2) * \neg\rho(A_1, A_3) * \rho(A_2, A_3) * \neg\rho(A_2, A_4) * \rho(A_3, A_4) = \\ =1*(1-0.6)*0.5*(1-0.6)*1=0.08;$$

while the probability of solution *b*) is

$$p_2 = \rho(A_1, A_2) * \rho(A_1, A_3) * \neg\rho(A_2, A_3) * \rho(A_2, A_4) * \rho(A_3, A_4) = \\ =1*0.6*(1-0.5)*0.6*1=0.18.$$

In this case $q=0.08+0.18=0{,}26$; $q_1=0.08/0.26=0.3077$; $q_2=0.18/0.26=0.6923$. This time, due to restrictions, solution a) is eliminated, because it is less than the prescribed value: 0.4.

Example 3 *Representation of the SNPM*
Hereafter the possibility to create the conversion between the SNPM and GERT net is presented.

In this case the intensity of the relation designates the probability of occurrence and not the preferred decisions.
Let us look at the following simple example, on condition that after tasks A, the successors of only two alternatives are possible. According to this, there are two tasks to be done, B or C. The GERT net of this is represented by Figure *3.a.* Figure *3.b* represents how to show it in an AoN. From this, the relation matrix and the reduced relation matrix can be written down as follows:

$$\rho = \begin{bmatrix} 0 & -0.5 & -0.5 \\ 0 & 0 & 0 \\ 0 & 0 & 0 \end{bmatrix} \qquad (4)$$

$$\mathrm{r} = \begin{bmatrix} 0 & 0 & 0 \\ 0 & 0 & 0 \\ 0 & 0 & 0 \end{bmatrix} \qquad (5)$$

Negative values here stand for the following: the given node and the next task in the topologically ordered graph can be left out from the graph. The intensity of relation can be between -1 and 1. If $\rho(A_i, A_j)$ is less than 0, task A_j and the next tasks can be left out from the graph. Henceforth it is right, that $p(\rho(A_i, A_j)) = |\rho(A_i, A_j)|$.

TABLE 2
REPRESENTATION OF SNPM

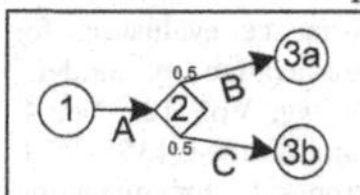 1, A, 2, 0.5, B, 3a, 0.5, C, 3b Figure 3.a: GERT- net	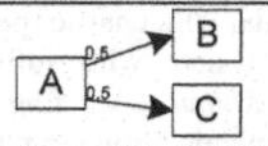 A, 0.5, B, 0.5, C Figure 3.b: GERT converted to AoN	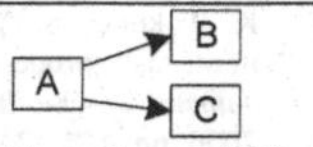 A, B, C Figure 3.c: p_1=0.25 Technologically not feasible
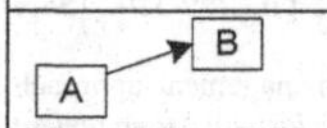 A, B Figure 3.d: p_2=0.25 Permitted solution, q_2=0.25/0.5=0.5	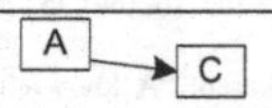 A, C Figure 3.e: p_3=0.25 Permitted solution, q_3=0.25/0.5=0.5	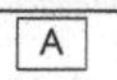 A Figure 3.f: p_4=0.25 Not permitted solution for the management

IV. SUMMARY

The presented method can be used in logical planning. With this method, GERT nets are also able to be modelled. The experiences about the realisation of the project can modify the intensity of the relation, and in this way can modify the number of feasible nets. Logical net can be used or reused not only as a project template, but also indicating the intensity of the relation and this way all the alternative solutions can also be determined. Besides finding the relevant logical plan and determining the duration of tasks, the cost- and resource demands (e.g. minimal total cost and minimal total project time, etc.) can be useful to find the best project plan. This method can even help the project manager to rank the feasible solutions sorted by TPT, total cost etc. This method can be used for a decision support module for project management software.

ACKNOWLEDGMENTS

The authors would like to thank Réka Polák-Weldon for her support.

REFERENCES

[1] PMBOK guide: *A guide to the project management body of knowledge*, Project Management Institute, Newtown Square, Penn., 2000, ISBN 1-880410-23-0, pp. 120-156

[2] M. Casinelli, "Project Schedule Delay vs. Strategic Project Planning". *AACE International Transactions*, 2005, PS.19

[3] D. Cherubini, A. Fanni, A. Montisci, P. Testoni, "A fast algorithm for inversion of MLP networks in design problems". *The International Journal for Computation and Mathematics in Electrical and Electronic Engineering*, Vol. 24 No. 3., 2005, pp. 906-920, Emerald Group Publishing Limited 0332-1649

[4] D. Dvir, T. Raz, AJ. Shenhar, "An empirical analysis of the relationship between project planning and project success" *International Journal of Project Management*, Volume 21, 2003, pp. 89–95, ISSN 0263-7863

[5] G.Mummolo, "Measuring uncertainty and criticality in network planning by PERT-path technique." *International Journal of Project Management,* Volume 15, Issue 6, 1997, pp. 377-387, ISSN 0263-7863

[6] PPM. Stoop, VCS. Wiers, "The complexity of scheduling in practice," *International Journal of Operations & Production Management*, Vol. 16. No. 10., 1996, pp. 37-53. © MCB University Press, 0144-3577

[7] L. Szabó, Z. Gaál, "Project Success and Project Excellence, Sharing Knowledge and Success for the Future," *Congress reports of 18th EUROMAINTENANCE* 2006, 3rd WORLDCONGRESS of MAINTENANCE 20.-22. June 2006 Basle / Switzerland, pp. 194-198,

[8] K. Tokuno, S. Yamada, "Stochastic performance evaluation for multi-task processing system with software availability model," *Journal of Quality in Maintenance Engineering*, Vol. 12., No. 4, 2006, pp. 412-424 © Emerald Group Publishing Limited, 1355-2511

[9] RJ. Dawson, CW. Dawson, "Practical proposals for managing uncertainty and risk in project planning," *International Journal of Project Management*, Vol. 16, No. 5, 1998, pp. 299-310, ISSN 0263-7863

[10] SCL. Koh, A. Gunasekaran, "A knowledge management approach for managing uncertainty in manufacturing". *Industrial Management & Data Systems*, Vol. 106 No. 4., 2006, pp. 439-459, Emerald Group Publishing Limited, 0263-5577

[11] J. Kamburowski, "New Validations of PERT Times," *Omega* Volume 25, Issue 3, Elsevier Science Ltd., 1997, pp. 323-328

[12] JU. Maheswari, K. Varghese, "Project Scheduling using Dependency Structure Matrix," *International Journal of Project Management*, Vol. 23., 2005, pp. 223–230

[13] JWM. Bertrand, JC. Fransoo, "Operations management research methodologies using quantitative modeling," *International Journal of Operations & Production Management*, Vol. 22. No. 2., 2002, pp. 241-264 © MCB University Press Limited, 0144-3577

[14] ZsT. Kosztyán, "Optimal Resource Allocation," PhD thesis, *Doctoral School of Management and Business Studies*, Veszprém, Hungary, 2005

[15] M. Schneider, T. Behr, "Topological Relationships Between Complex Spatial Objects," *ACM Transactions on Database Systems*, Vol. 31., No. 1, March 2006, pp. 39–81.

[16] SA. Oke, OE. Charles-Owaba, "Application of fuzzy logic control model to Gantt charting preventive maintenance scheduling," *International Journal of Quality & Reliability Management*, Vol. 23., No. 4, 2006, pp. 441-459

[17] SMT. Fatemi Ghomi, M. Rabbani, "A new structural mechanism for reducibility of stochastic PERT networks," *European Journal of Operational Research*, Vol. 145., 2003, pp. 394-402,

[18] W. Herroelen, R. Leus, "On the merits and pitfalls of critical chain scheduling." *Journal of Operations Management*, Volume 19, Issue 5, October 2001, pp. 559-577

[19] X. Lin, Q. Liu, Y. Yuan, X. Zhou, H. Lu, "Summarizing Level-Two Topological Relations in Large Spatial Datasets," *ACM Transactions on Database Systems,* Vol. 31., No. 2, June 2006, pp. 584–630.

[20] P. Pontrandolfo, "Project duration in stochastic networks by the PERT-path technique," *International Journal of Project Management*, Volume 18, 2000, pp. 215-222, ISSN 0263-7863

[21] CH. Wang, SL. Hwang, "A stochastic maintenance management model with recovery factor," *Journal of Quality in Maintenance Engineering*, Vol. 10., No. 2, 2004, pp. 154-164, © Emerald Group Publishing Limited, ISSN 1355-2511

[22] E. Menipaz, A. Ben-Yair, "Harmonization simulation model for managing several stochastic projects," *Mathematics and Computers in Simulation*, Vol. 61., 2002, pp. 61–66

[23] A.A.B. Pritsker, W.W. Happ, "GERT: Graphical Evaluation and Review Technique," *The Journal of Industrial Engineering,* 1966

[24] WJ. Gutjahr, C. Strauss, M. Toth, "Crashing of stochastic processes by sampling and optimization," *Business Process Management Journal*, Vol. 6. No. 1., 2000, pp. 65-83, © MCB University Press, 1463-7154

[25] SD. Eppinger, DE. Whitney, RP. Smith, DA. Gebala, "A modelbased method for organizing tasks in product development," *Res Eng Des* 1994;6(1):13.

[26] DV. Steward, "The design structure system: a method for managing the design of complex systems," *IEEE Trans Eng Manage* 1981; EM-28(3):71–4.

Web-based Service Portal in Healthcare

Petr Silhavy, Radek Silhavy, Zdenka Prokopova
Faculty of Applied Informatics, Tomas Bata University in Zlin
Nad Stranemi 4511, 76005 Zlin, Czech Republic
psilhavy@fai.utb.cz; rsilhavy@fai.utb.cz; prokopova@fai.utb.cz

***Abstract*-Information delivery is one the most important task in healthcare. The growing sector of electronic healthcare has an important impact on the information delivery. There are two basic approaches towards information delivering. The first is web portal and second is touch-screen terminal. The aim of this paper is to investigate the web-based service portal. The most important advantage of web-based portal in the field of healthcare is an independent access for patients. This paper deals with the conditions and frameworks for healthcare portals**

I. INTRODUCTION

The web-based portals take an important role in the medical application development. Their importance and usability is rapidly increasing. The web is becoming appropriate eco-system for the medical information portals. These systems use client-server approach, because of their support height level of productivity and scalability.

The web is only one of many possible ways for delivering information. For medical information delivering is very popular the touch-screen kiosk. Discussion about the advantages and disadvantages is held by medical professionals and by academics researchers. This discussion is closely connected to the development of information technology and patients claim to receive more and more information. Touch screen kiosk are usually installed in health centres or hospitals.

Web-based portal is based on web technologies. For communication between patient and application is used Internet network. The most important advantage of web-based portal is the possibility to connect using personal computers at homes or in offices. In web portals is the possibility to integrate several information sources.

Designing of such systems is usually understood as the design of HTML pages and databases. But issue of designing web-based healthcare portal should be more complex.

The aim of this paper is to investigate the principles of the healthcare medical portal. By using results of this investigation it is possible to improve web-based frameworks for portal in healthcare sectors.

Bashshur [1] describes telemedicine in healthcare. His results should be used for internet service in general. Kohane et al. [2] describes methods of building electronic health record system. These findings are useful for the understanding of web technologies in healthcare, but they describe only one part of web-based portal. Therefore, in this paper, their research is develop.

The organization of this contribution is as follows. Chapter 2 describes the need for developing portals. Chapter 3 describes internet penetration and importance for healthcare information portals. Chapter 4 describes web portal framework architecture Finally chapter 5 is the discussion.

II. NEED FOR PORTAL, PATIENT SURVEY

Web portals are well-known and used by companies and institutions for information sharing and communications. This contribution deals with two basics processes which are realized in the healthcare sectors.

Common process of ordering extends of using modern ICT in healthcare sector. Therefore electronic ordering is one of the most important possibilities. Nowadays process of order is realized by standard telephone call, as can be seen in the Figure 1

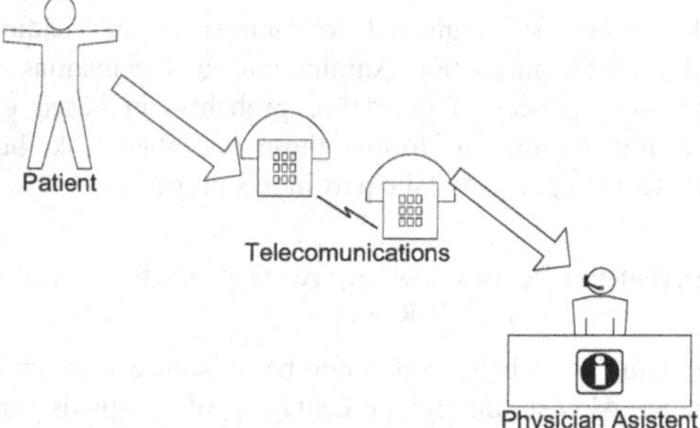

Fig. 1. Common Ordering Process

This process is ineffective and is not used very often. In the healthcare sector is more popular FIFO based system, chronology attendance of waiting patients. These cause long queues in the waiting-rooms. There is no chance to control time of patients and physicians. Although, there is possibility of telephone orders, very often patient missed their examination.

The electronic prescription is specific form of physician-patient communication. This form is based on electronic request for medicament prescription. Nowadays, only physical visit of physician office leads for medicament prescription. This is time consuming. The second aspect is control of interference among medicaments which are prescribed.

Possibility of the electronic consultation is the third carrying pillar in of using web portals and technology. Web portals as usual web-based application allows transferring of multimedia information – photographs or text information only. The telephone-based consultation in ineffective on the score of stress. In written form patient and physician should rethink better their questions and answers.

The main advantage of solving mentioned topics by web portals is possibility of asynchronous communication. The

K. Elleithy (ed.), *Advanced Techniques in Computing Sciences and Software Engineering*,
DOI 10.1007/978-90-481-3660-5_45, © Springer Science+Business Media B.V. 2010

asynchronous communication brings time to the conversations. The main purpose of the contribution is to introduce research goals and this application is asked by patient. This resulting from the study, which were prepared by authors. Patient's claims should be concluded in these areas:

1. Shortening waiting time for the examination
2. Simplification of prescription.
3. Reminder services and invitation for prevention examination.

The first requirement should be achieved by built-in scheduler, which allow choose time of examination which is the best for patient and physician. If is reserved accurate time for examination number of patient in queue is rapidly decreased. Patient should no come in advance.

The second, simplification of prescription, is based on electronic request of prescription. The prescription should be delivered to the selected chemist or it should be collected in the physician office.

The last claim is connected to automation of sending invitation for the prevention examination, for vaccinations or as reminder services. This claim, probably, in based on patients impossibility to follow these important task, but patients feel obligation to take part in this processes.

III. Internet Penetration and Impact on Web-Based Portals

In the Figure 1 can be seen that number of subjects, which is appropriate. Almost ninety percentages of hospitals are equipped by personal computer and at least two thirds of them are equipped by internet connections. Very important is that fifty and more percents of other healthcare subjects are equipped by personal computer and internet connections too.

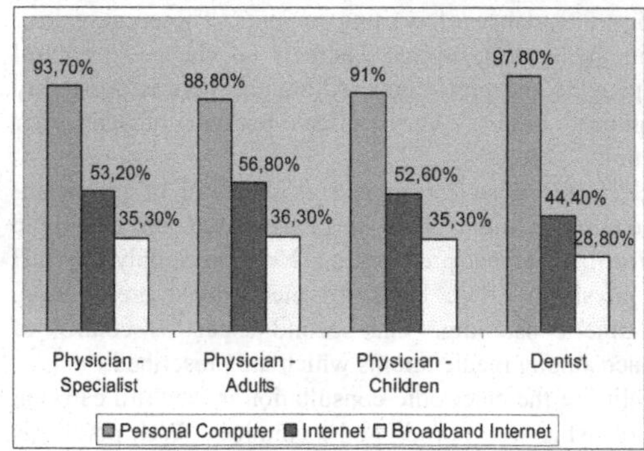

Fig. 2. Information and Communication Technology in Healthcare sector

The Second significant condition for researching of Web-based Services in the Healthcare Sector is number of computers and internet connection in households. In the Figure 2 can be seen that number of these to options is rapidly increased. In the second quarter of the 2007 thirty-two percents of personal computer and internet connections was achieved. More significant than number itself is the trend.

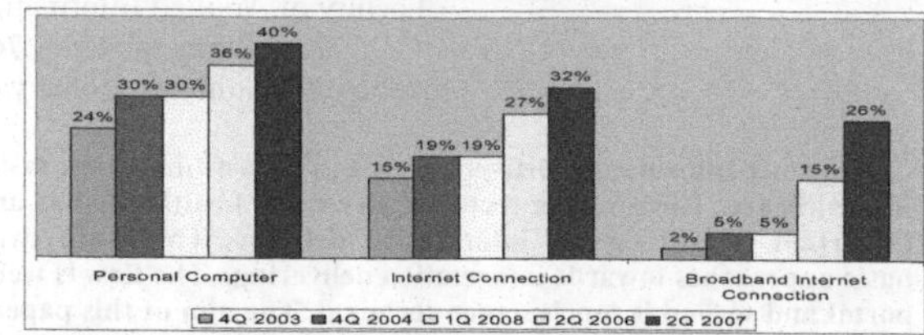

Fig. 3. Information Communication Technology in Households.

The third condition is relatively low penetration of web-based portals in the healthcare sector. In the Figure 4 can be seen eighty-two percents of hospitals have their own web pages. But only nearly eight percents of the independent physicians are used web pages.

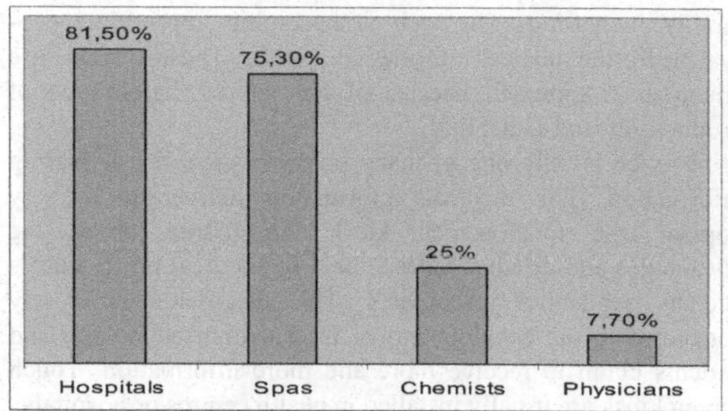

Fig. 4. Web-sites in Healthcare

IV. Web Portal Framework Architecture

The proposed system is based as web-based portal. The basic structure of the portal is shown in the Figure 5. This structure brings accessibility from any computer, which is connected to the internet. Classical three-layer architecture was used for the proposed system. Data layer is used for storing data and information in the system. The data layer is based on relation database and group of stored procedures.

Business logic is in the form of web-based application, which is based on .Net framework. Use interface is created by using HTML and webforms technology. Webforms are expanded by using asynchronous javascript and xml. The web portal is unpretentious of client technology. It should be expanded during implementation.

Communication in web portal will be based on the internet – by using webforms. Other possibility is e-mail which will be process automatically or short text message. Short test messages are planed for reminder services.

The proposed system is based on webform filling. Patient should choose type of request and this request is stored into database. Patient chooses term of meeting too. Patient chooses from free term and checked type of request. If there is no corresponding request, term is assigned by physician's office.

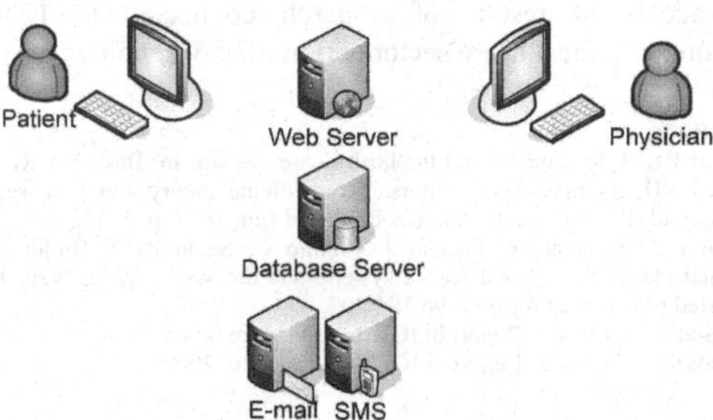

Fig. 4. System Model, all parts

Patient should also cancel the meeting, but only in defined time in advance. Text case is reminder services, which informs patients that his meeting will be. In time, when patient comes to the waiting room, electronic attendance will be used by him.

Figure 5 brings illustration of the system model, which contain all of necessary parts of the proposed system.

Both patients and physicians use only internet explorer for accessing system. The application is based on Model-View-Presenter (MVP) design pattern.

This design pattern is based on commonly known Model-View-Controller (MVC). Basics are the same as in MVC. In MVP data are stored in Model, representation of the model can be found in View and the Presenter control communications between the data and business logic layers.

In the Figure 5 can be seen the structure of MVP applications and its location in each components of the proposed system.

Services, which are offered by the portal should be illustrated on examination ordering request, which can be seen in Figure 6.

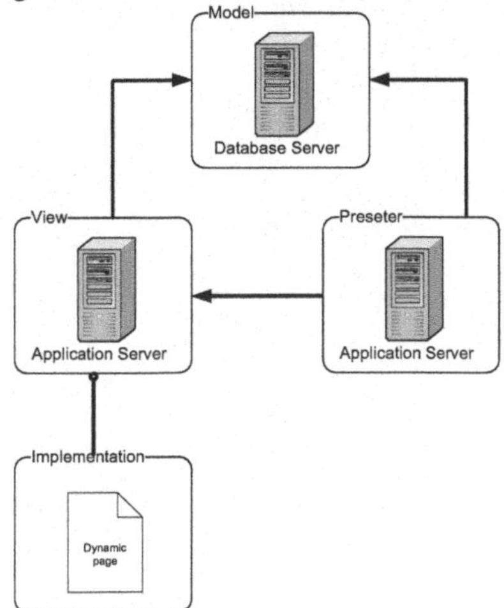

Fig. 5. Model-View-Presenter and its decompositions

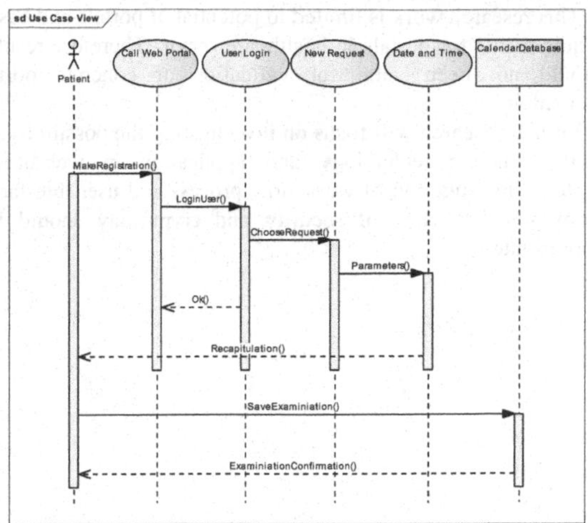

Fig. 6. Sequence diagram of the ordering process

The ordering process, which is described is based on using internet browser. User of the web portal – patient is logged into the portal application and choose the activity. In the shown case patient order an examination.

If his login name and password are valid, he should choose term of the examination and leave brief description of the reasons of examination.

The patient works with scheduler, which contains all physician activity. For individual patients are shown only free frames, which are assigned as the examination time. Ordering process is based on visual environment, which is user friendly and allows, in natural form, filling all necessary information. In the Figure 6 can be shown that two stage confirmations are necessary for storing and booking term.

In there is no confirmation, physician should not be protected before testing users.

V. Discussion

In this contribution, web-based portal and its Model-View-Presenter framework were investigated. The web-based applications are usually used in different locations and are focused on different users. Web-based portals is usually used by well-situated and has connection to internet at home. The healthcare portal brings by electronic healthcare era. E-healthcare is closely connected to telemedicine. Therefore portal takes key role in modern healthcare. Motivation for this development is in penetration of internet connection. Other important reasons are in case, that by using portals should be solved other important issues in the medical informatics.

This research shows possibility of using both portals for information delivering and for patients' management. These two views are not concurrent. Decrease of the cost of medical information is import too.

This research work is limited to potential of portals and basic requirements for portals in healthcare sector. Therefore results should not been valid for validation of concrete portal technology.

Further research will focus on investigating the possibility of using concrete technology and application in web-based system. Investigation of work-flow process and user interface, form point of view of security and cryptology should be appropriate

ACKNOWLEDGMENT

Authors would like to thank Czech Statistical Office, which granted access to results of research connected to ICT penetration in the healthcare sector and in the households.

References

[1] Bashshur RL. Telemedicine and the health care system. In: Bashshur RL, Sanders JH, Shannon GW, editors. Telemedicine theory and practice. Springfield (IL): Charles C. Thomas Publisher Ltd; 1997. p. 5–35.

[2] Kohane IS, Greenspun P, Fackler J, Cimino C, Szolovits P. Building national electronic medical record systems via the World Wide Web. J Am Med Informatics Assoc 1996;191:207.

[1] Czech Statistical Office.: Report of ICT in Healthcare. 2008

[2] Czech Statistical Office.: Report of ICT in Household. 2008

Decomposition of Head-Related Transfer Functions into Multiple Damped and Delayed Sinusoidals

Kenneth John Faller II, Armando Barreto and Malek Adjouadi

Electrical and Computer Engineering Department
Florida International University
Miami, FL 33174 USA

***Abstract*-There are currently two options to achieve binaural sound spatialization using Head-Related Impulse Responses (HRIRs): measure every intended listener's HRIR or use generic HRIRs. However, measuring HRIRs requires expensive and specialized equipment, which removes its availability to the general public. In contrast, use of generic HRIRs results in higher localization errors. Another possibility that researchers, including our group, are pursuing is the customization of HRIRs. Our group is pursuing this by developing a structural model in which the physical measurements of a new intended listener could be used to synthesize his/her custom-fitted HRIRs, to achieve spatialization equivalent to measured HRIRs. However, this approach requires that HRIRs from multiple subjects be initially broken down in order to reveal the parameters of the corresponding structural models. This paper presents a new method for decomposing HRIRs and tests its performance on simulated examples and actual HRIRs.**

Keywords-Head-Related Transfer Functions (HRTFs), Damped and Delayed Sinusoidals (DDS), Hankel Total Least Squares (HTLS), Signal Decomposition

I. INTRODUCTION

Spatial audio, which is often times referred to as surround sound, can be created in two ways: a multi-channel approach or a two-channel approach. One method of creating spatial audio using the multi-channel approach is to record a sound source(s) using surround sound microphone techniques [1]. The recording is then mixed for playback on multiple speakers which encircle the listener (e.g., Dolby 5.1). This method is effective but its availability is restricted by the requirement of expensive and specialized equipment for the recording and playback setups. Another multi-channel approach is Wave Field Synthesis (WFS) which is based on the Huygens-Fresnel principle [2]. This method essentially produces "artificial" wave fronts that are synthesized by a large number of individually driven speakers. These wave fronts will be perceived by the listener to originate from a virtual source. One benefit of this method is the localization of the virtual sources does not depend on or change with the listener's position. However, this method also suffers from the same problems the surround sound microphone technique does.

The two-channel approach uses digital signal processing (DSP) techniques to create binaural (left and right channel) virtual spatial audio from a monaural source that can be delivered to a prospective listener through headphones or speakers. DSP-based azimuthal spatialization can be achieved, on the basis of Lord Raleigh's Duplex Theory [3], by simply manipulating two binaural cues called interaural time difference (ITD) and interaural intensity difference (IID). However, these cues do not account for the elevation information of the sound source.

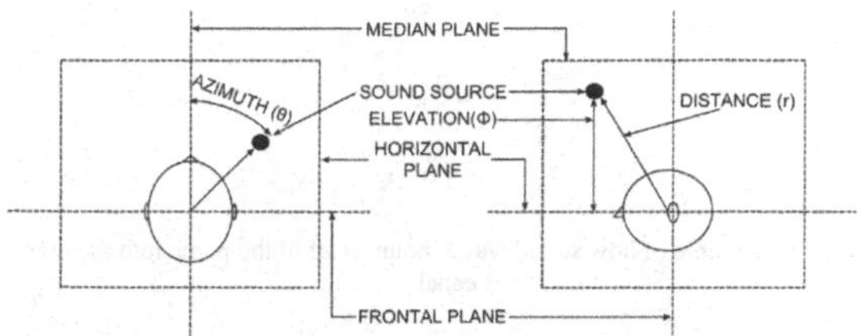

Fig. 1: Azimuth (θ), elevation (Φ) and distance (r) of a sound source with median, horizontal and frontal planes noted.

It is well established that the primary cues for elevation come from the pinna or outer ear [4, 5]. The elevation cues are derived from the modifications a sound experiences as it bounces-off of the various structures of the pinna before entering into the ear canal. A way in which both the azimuth and elevation cues (i.e., modifications from the pinna reflections, IIDs and ITDs) are modeled is by means of a pair (Left and Right) of Head-Related Impulse Responses (HRIRs) for each desired position around the listener's head. The frequency domain representation, referred to as the Head-Related Transfer Functions (HRTFs), is obtained by taking the Fourier transform of the HRIRs. Once these HRIRs are obtained, a sound source can be convolved with the HRIRs that correspond to a desired virtual position. When playing the left and right signals resulting from this convolution to a listener, he/she will perceive the sound as if it emanated from the desired location in 3D space, specified by azimuth (θ), elevation (Φ) and distance (r) (see Fig. 1). However, the dependence of the cues on anatomical features implies that the HRIRs will vary from listener to listener.

Originally, the only method to measure HRIRs for each intended listener was in an anechoic chamber or equivalent facility using expensive and cumbersome equipment. An alternative solution to this is the utilization of "generic" HRIRs (e.g., MIT's measurements of a KEMAR Head Microphone [6] or the CIPIC Database [7]). Due to its requirements, the use of individually measured HRIRs is confined to specialized groups researching 3D audio such as military personnel or academic researchers. On the other hand,

K. Elleithy (ed.), *Advanced Techniques in Computing Sciences and Software Engineering*,
DOI 10.1007/978-90-481-3660-5_46,

it is known that the use of generic HRIRs suffers from accuracy issues such as increased front/back reversals and elevation errors in the perception of the sound location [8]. Therefore, none of the two approaches currently available is simultaneously accurate and simply to implement.

A potential solution that our group is pursuing is to find a method to "customize" HRIRs for each listener utilizing relevant geometrical measurements of his/her head and outer ears (pinnae). Unfortunately, the way in which current HRIR measurement systems report the HRIR samples is in terms of long (128, 256, 512, etc.) sequences, which are not easily associated with the physical characteristics of the listener. As a result, the overall purpose of our research is to develop customizable HRIRs from a generic dynamic model that involves a smaller number of anatomically-related parameters.

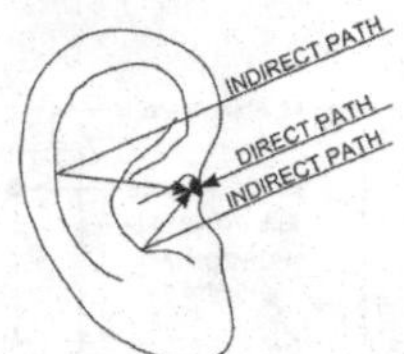

Fig. 2: Example of how sound waves bounce-off of the pinna into the ear canal.

II. HRIR CUSTOMIZATION

Several approaches have been tried to generate HRIRs without requiring inconvenient measurements while still achieving comparable performance to measured HRIRs, including numerical computational methods, physical models and structural models (see [9] for a complete list of customization methods/models). Amongst these categories, our group found the simplicity of a structural pinna model approach appealing, and developed our work from it.

One example of a structural pinna model is the one proposed by Batteau in [4]. He believed that the cues for spatialization result from the way in which the sound waves bounce-off of the convex protrusions of the pinna, which causes delay and intensity differences between the direct and indirect sound paths (see Fig. 2). He developed a customization model that assigned parameters to the protruding structures of each pinna. However, Batteau's model was limited to two reflections which does not represent all the delays that occur [10]. Furthermore, he did not offer a method to estimate the parameters for the model.

In [11], we proposed a model that represents the HRIRs as a summation of multiple Damped and Delayed Sinusoidals (DDSs) with different magnitudes and delays. As in Batteau's model, the changes in magnitude and delay between the DDSs are believed to be the result of the waves bouncing-off of the geometrical structures of the pinnae before entering each of the ear canals (see Fig. 2). However, the method proposed to obtain the parameters for the model was formulated on a manual basis and required a large amount of human intervention. This prompted our group to develop an automated method of estimating the DDSs and their parameters.

III. AUTOMATED METHOD OF PARAMETER ESTIMATION

Initially, the parameters for our model were obtained manually. The HRIR was sequentially segmented in windows such that the single window under analysis at any given time contained a single DDS and then an approximation method, such as Prony [12], was applied to the window. The result of the approximation was then extrapolated and removed from the overall HRIR. The process was repeated to obtain all the parameters needed for the model. The results of this manual windowing process yielded reconstructed HRIRs that closely resembled the original HRIRs. But this parameter estimation method required a large amount of human interaction, and an automated method of estimating the parameters was pursued.

In [13], our group proposed an iterative method of decomposing HRIRs. Essentially, this method automated the same steps used in the manual decomposition method. Although this method resulted in good approximations of the original HRIR, it was shown in [14] that the iterative method could fail to accurately approximate the HRIRs if the latencies between the DDSs are small. As a result, an alternate HRIR decomposition method to handle cases with small latencies between the DDSs was devised.

A method exist, called the Hankel Total Least Squares (HTLS) decomposition method, which can decompose a summation of real or complex exponentially damped sinusoidals [15, 16]. This algorithm uses Total Least Square (TLS) and Singular Value Decomposition (SVD) to estimate the parameters (frequency, phase, amplitude and damping factor) of K individual sinusoidals. A higher order is usually selected to capture what is referred to in [16] as "correct events" (i.e., the actual sinusoidals a signal is comprised of) along with "false detections" (i.e., spurious sinusoidals). After all the sinusoidals are obtained, a classification of the correct events is used to eliminate the false detections. Unfortunately, prior knowledge of how many sinusoidals a signal is comprised of would be needed in order to estimate the model order K. Additionally, one of the assumptions of the HTLS method is that the sinusoidals that make up the signal to be decomposed are not delayed. The following subsections outline how these issues are addressed such that the HTLS method can be used to decompose HRIRs into DDSs.

A. Criteria for Spurious Signal Elimination

The HTLS method is parametric and requires that an order of K be specified such that the results contain both the actual and spurious sinusoidals. The model order K can be set high such that all the actual and spurious sinusoidals are captured. Then criteria are used to eliminate the spurious sinusoidals leaving only the actual sinusoidals. However, it was shown in [15] that if K is set too high then the HTLS method breaks down. To remedy this, we attempted to define an empirical value for K that would be appropriate for HRIR analysis by decomposing synthetic mixtures of damped sinusoidals.

In order to define the model order K, a synthetic signal composed of four non-delayed exponentially damped sinusoidals was created and decomposed using the HTLS method. Equation (1) was used to generate the ith synthetic exponentially damped sinusoidal where F_S is the sampling frequency, d_i is the damping factor, f_i is the frequency in Hertz, φ_i is the phase in radians and a_i is the amplitude. Each sinusoidal was N=256 points in length, $n=0,\ldots,N-1$ and the simulation used a sampling frequency of F_S=96 kHz.

$$x_i(n) = a_i \cdot e^{-d_i \cdot n} \cdot \sin\left(2 \cdot \pi \cdot \left(\frac{f_i}{F_S}\right) \cdot n + \varphi_i\right) \quad (1)$$

After decomposing several synthetic sinusoidal mixtures with various values of K, the value of K=20 resulted in the highest number of correct decompositions. Another issue is the definition of criteria to eliminate the spurious sinusoidals from the HTLS decomposition results. The following criteria were developed to eliminate the spurious sinusoidals:

1. Eliminate sinusoidals that have $f_i > 20$ kHz.
2. Eliminate sinusoidals that have $d_i > 2$.
3. Eliminate sinusoidals that do not have complex-conjugate poles within the unit circle.

The justifications for the criterion one is based on what is accepted by spatial audio researchers as the "relevant" frequency range (from about 1 to 14 kHz) for localization in the median plane (see Fig. 1) [9]. The precise frequency range is, however, unknown. Hence, 20 kHz was selected to ensure that all the important frequencies are captured. The limit for criterion 2 was empirically defined to be $d_i > 2$ because sinusoidals with damping factors greater than this limit were heavily damped and resembled an impulse which contradicts our model premise that an HRIR is comprised of resonances. Criterion 3 is based on the knowledge that impulse response components exhibit damped oscillatory behavior if their complex-conjugate poles lie within the unit circle. Therefore, sinusoidals that did not have complex-conjugate poles within the unit circle were eliminated. The only parameter that is not statically defined in these criteria is the amplitude, a_i.

In previous studies, the sinusoidal had to have $|a_i| > 0.01$ to be considered an actual sinusoidal. This worked well for the synthetic examples. However, when HTLS was used to decompose HRIRs and this criterion was included, no sinusoidals would be retained. It was reasoned that this could be too high of an amplitude value, especially for contralateral (the ear farthest from the sound source) HRIRs were the amplitude is expected to be low and the Signal-to-Noise (SNR) ratio is lower. To remedy this, the a_i parameter is set to a reasonable start value ($a_i = 0.2$) and decremented by 0.01 until the HTLS results retain at least one sinusoidal.

It should be noted here that when HTLS was used to decompose HRIRs, it would occasionally fail to return any sinusoidals when decomposing contralateral HRIRs with K=20. After several trials using contralateral HRIRs, the model order of K=10 would result in at least one sinusoidal. Hence, for the synthetic examples and ipsilateral (the ear closest to the sound source) HRIRs a model order of K=20 is used and a model order of K=10 is used for contralateral HRIRs.

B. Iterative Zeroing Algorithm

As mentioned before, originally HTLS can only decompose signals that are comprised of non-delayed exponentially damped sinusoidals. In the previous section we proposed appropriate model orders (K) and criteria to eliminate spurious sinusoidals from the HTLS decomposition results. To fully adapt HTLS for decomposition of signals that are comprised of DDSs, as we believe is the case for HRIRs, an additional time delay estimation method must be developed. Methods do exist that estimate time delay (see [17] for a list of common methods) but none of them are well suited for the scenario proposed in this paper.

HTLS, however, can fit a non-delayed exponentially damped sinusoidal to a DDS with a time delay that is less than or equal to 5 samples (i.e., 52.0833 μs). To illustrate this, a synthetic DDS (x_{t0}) was created, using Equation (2), and then decomposed using the HTLS decomposition. Equation (2) is essentially Equation (1) multiplied by the unit step sequence (Equation (3)) [18]. The parameter values for the synthetic signal used, x_{t0}, are f_0=8 kHz, $\varphi_0=\pi/9$, d_0=0.1, a_0=2 and t_0=3 samples. The result of this process is shown in Fig. 3.

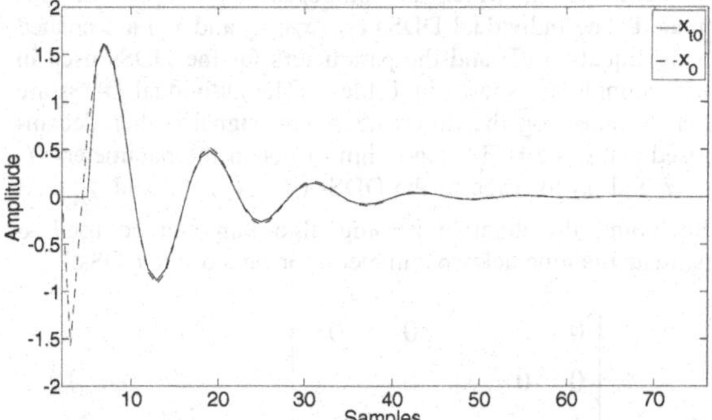

Fig. 3: A DDS (x_{t0}) versus the results from the HTLS method (x_0).

$$x_{ti}(n) = x_i(n - t_i) \cdot u(n - t_i) \quad (2)$$

$$u(n) = \begin{cases} 0 & n < 0 \\ 1 & n \geq 0 \end{cases} \quad (3)$$

In Fig. 3, it can be seen that the resulting sinusoidal from the HTLS decomposition (x_o) captures the overall shape of the synthetic DDS (x_{t0}). Only the time delay, t_i, must be estimated. An algorithm is needed that zeros-out the initial samples (for this example that would be the first 3 samples) of the resulting sinusoidal from the HTLS decomposition (x_o). The resulting algorithm, referred to as the iterative zeroing algorithm, is a general method that performs this task.

The iterative zeroing algorithm is described by Equations (5)-(10) where p is the maximum samples that will be zeroed at the beginning of each of the i damped sinusoidals obtained from the HTLS decomposition. The original incarnation of the iterative zeroing algorithm used the matrix $T(r,i)$ in Equation (4) which is a $(p+1)^i$ x i matrix storing in each of its rows the amount of zeros that should be overwritten at the beginning of the damped sinusoidals. However, the memory requirements for this grows rapidly as the number of damped sinusoidals i is increased. Therefore, Equation (5) is used to extract a single row at a time of the original $T(r,i)$ and is stored in a vector $TR(r,i)$ where $\lfloor \cdot \rfloor$ is the floor operator and mod denotes the modulus operator. In Equation (6), $\hat{x}(r,n)$ is the composite sequence which is a candidate to be the approximation of the original sequence $x(n)$. In all cases, $\hat{x}(r,n)$ is built from the i damped sinusoidals obtained from HTLS but the pattern of zeros imposed at the beginning of each of them is determined by the row r in $TR(r,i)$. For each value of $1 \le r \le (p+1)^i$, the error $(e(r,n))$ sequence and corresponding "fit" $(f(r))$ are calculated (Equations (7) and (8)) and the r^* that maximizes $f(r)$ is chosen. The corresponding composite $\hat{x}(r,n)$ is kept as the best reconstruction of $x(n)$ and the corresponding zeroing pattern is kept as t_i^*.

As mentioned in section II, it is believed that HRIRs are comprised of multiple DDSs [4, 9, 11, 13, 14]. Hence, to evaluate the iterative zeroing algorithm's performance, a synthetic signal, which is composed of four DDSs, was created. The individual DDSs (x_{t1}, x_{t2}, x_{t3} and x_{t4}) are created using Equation (2) and the parameters for the DDSs used in this example are shown in Table 1. The individual DDSs are then summed together to create x. This signal is then decomposed using the HTLS algorithm to obtain the parameters (f_i, φ_i, d_i and a_i) for each of the DDSs ($\hat{x}_{t1}$, $\hat{x}_{t2}$, $\hat{x}_{t3}$ and $\hat{x}_{t4}$). At this point, the iterative zeroing algorithm can be used to estimate the time delay parameter t_i for each of the DDSs.

$$T(r,i)=\begin{bmatrix} 0 & 0 & \cdots & 0 & 0 \\ 0 & 0 & \cdots & 0 & 1 \\ \vdots & \vdots & \cdots & \vdots & \vdots \\ p & p & \cdots & p & p-1 \\ p & p & \cdots & p & p \end{bmatrix} \tag{4}$$

$$TR(r,i)=\left\lfloor \frac{(r-1)}{(p+1)^{(i-1)}} \right\rfloor \bmod (p+1) \tag{5}$$

$$\hat{x}(r,n)=\sum_{i=1}^{I} x_i(n)\cdot u(n-TR(r,i)) \tag{6}$$

$$e(r,n)=x(n)-\hat{x}(r,n) \tag{7}$$

$$f(r)=\left(1-\frac{\frac{1}{N}\sum_{n=1}^{N}\left(e(r,n)^2\right)}{\frac{1}{N}\sum_{n=1}^{N}\left(x(n)^2\right)}\right)\times 100\% \tag{8}$$

$$r^*=\underset{1\le r\le (p+1)^I}{\arg\max}(f(r)) \tag{9}$$

$$t_i^*=TR(r^*,i) \tag{10}$$

After all the parameters are obtained, then the estimated DDSs ($\hat{x}_{t1}$, $\hat{x}_{t2}$, $\hat{x}_{t3}$ and $\hat{x}_{t4}$) are summed together to obtain $\hat{x}$. Equations (11)-(12) are used to measure the similarity of two signals in the time domain. Equation (13) measures the similarity of two magnitude responses which is often referred to as the Spectral Distortion (SD) score. For the SD score, $|H_i|$ is the magnitude response at the ith frequency of x, $|\hat{H}_i|$ is the magnitude response at the ith frequency of $\hat{x}$ and I is the number of frequencies. The lower the SD score, the more similar the magnitude responses are to each other.

In Fig. 4, a plot displaying the time and frequency plots of x versus $\hat{x}$ indicates that the reconstruction was almost perfect. This is confirmed by the high fit (99.86%) and low SD score (0.4229 dB). Additionally, the original DDSs (x_{ti}) were compared to the estimated DDSs ($\hat{x}_{ti}$). This resulted in an average fit of 99.66% and an average SD score of 0.3140 dB. The next section applies this process to real HRIRs.

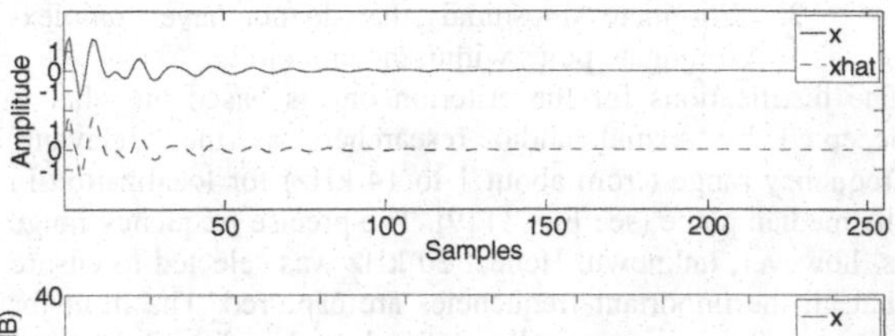

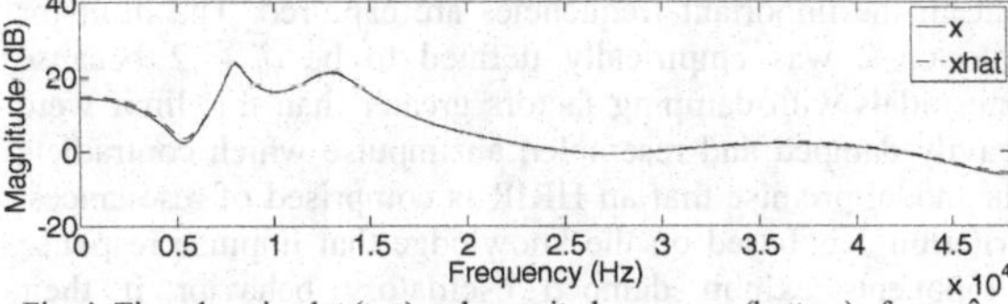

Fig. 4: Time domain plot (top) and magnitude response (bottom) of x vs. $\hat{x}$.

Table 1: The parameters used in example.

x_{ti}	f_i (kHz)	φ_i (radians)	d_i	a_i	t_i (samples)
x_{t1}	13	0	0.09	2.05	0
x_{t2}	9	π/3	0.3	1.2	3
x_{t3}	8	π/33	0.03	-0.8	2
x_{t4}	4	π/11	0.26	1.5	5

$$e(n)=x(n)-\hat{x}(n) \tag{11}$$

$$f=\left(1-\frac{\frac{1}{N}\sum_{n=1}^{N}\left(e(n)^2\right)}{\frac{1}{N}\sum_{n=1}^{N}\left(x(n)^2\right)}\right)\times 100\% \tag{12}$$

$$SD=\sqrt{\frac{1}{I}\sum_{i=1}^{I}\left(20\log_{10}\frac{|H_i|}{|\hat{H}_i|}\right)^2}\ [dB] \tag{13}$$

IV. HRIR DECOMPOSITION

In this section, actual measured HRIRs will be decomposed by the HTLS method and the iterative zeroing method. The results will then be evaluated in both time (using the fit measure in Equations (11) and (12)) and frequency (using the SD score in Equation (13)). The HTLS method, augmented by the iterative zeroing algorithm, is proposed for decomposition of DDS mixtures with short intervening latencies. Therefore, the method was tested only with HRIRs from locations where short latencies are expected [4, 19]. The HRIRs used were measured on a KEMAR mannequin using the AuSIM HeadZap system at a sampling frequency of 96 kHz. Azimuths of 90° and -90° and elevations from -20° to 20° at increments of 10° (see Fig. 5) were decomposed, for a total of 20 HRIRs.

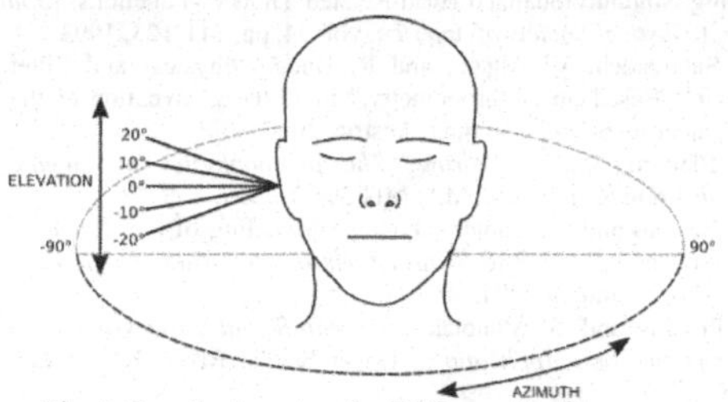

Fig. 5: Coordinate system for HRIRs used in this paper.

Table 2: Mean fits and SD scores (for elevations -20°, -10°, 0°, 10° and 20°)

Ear	Azimuth (°)	Mean Fit (%)	Mean SD Score (dB)
Left	90	95.75	2.4410
Left	-90	79.80	5.1708
Right	90	76.38	6.3480
Right	-90	96.38	3.3698

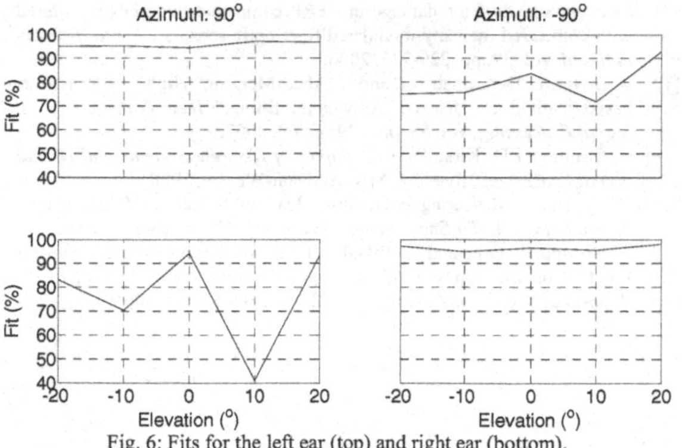

Fig. 6: Fits for the left ear (top) and right ear (bottom).

As mentioned in section III-A, it is known that the most relevant frequencies for localization in the median plane (see Fig. 1) lies between 1 kHz to 14 kHz. Hence, the SD score analysis was restricted to a frequency range of 0 kHz to 20 kHz. The results for the time domain comparisons (fits) are shown in Fig. 6 and the frequency domain comparisons (SD scores) are shown in Fig. 7. Table 2 shows the means for the fits and SD scores.

The reconstructed HRIRs that achieved high and low fits and SD scores are plotted below to examine the results in more detail. Fig. 8 shows the case where a low fit in the time domain (83.17%) and a high SD score (7.7344 dB) were achieved. Fig. 9 shows the case where a high fit in the time domain (97.14%) and a low SD score (2.0533 dB) were achieved. The analysis of these plots confirms the relationship between the fit in the time domain and the SD score comparison in the frequency domain.

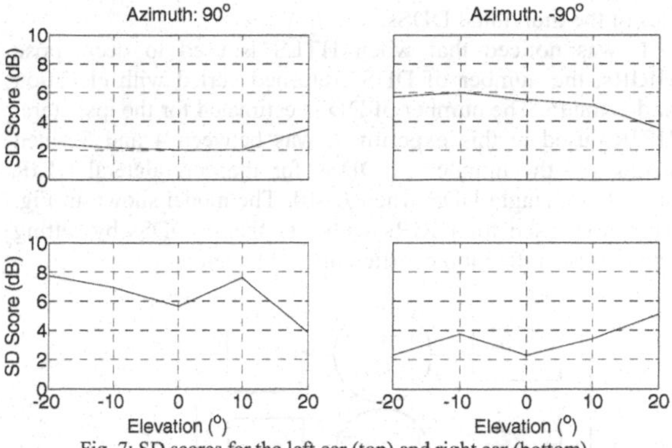

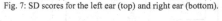

Fig. 7: SD scores for the left ear (top) and right ear (bottom).

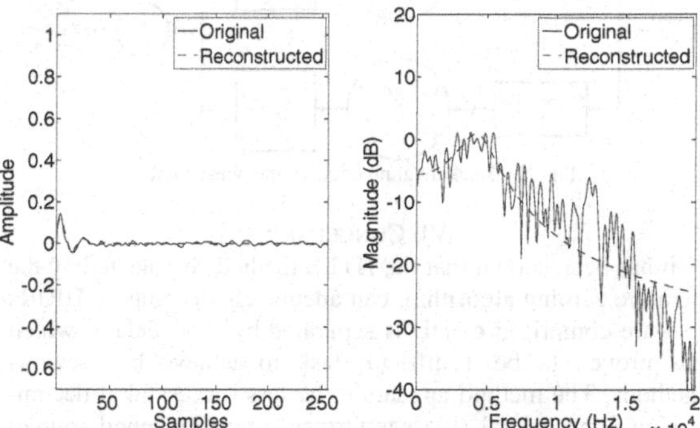

Fig. 8: Time domain (left) and magnitude response (right) plots of original (solid) and reconstructed (dashed) HRIRs for the right ear at azimuth = 90°, elevation = -20°.

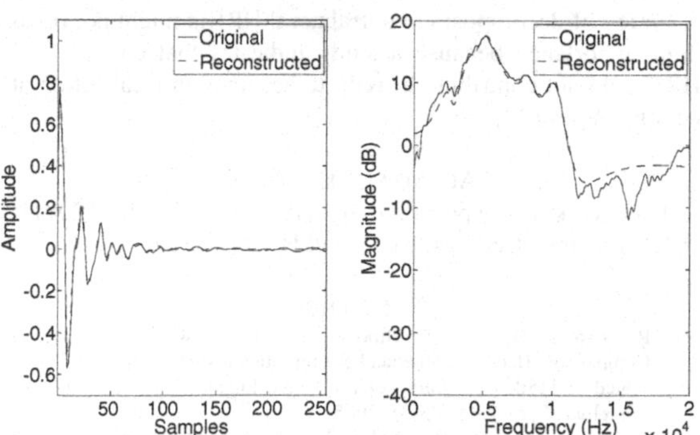

Fig. 9: Time domain (left) and magnitude response (right) plots of original (solid) and reconstructed (dashed) HRIRs for the left ear at azimuth = 90°, elevation = 10°.

V. Structural Pinna Model

Our previous model in [11] used a single resonance, which defined the frequency and damping factor of the DDSs for all the paths. However, recent findings have indicated that each DDS could have different characteristics. As a result, we extended the model to allow each DDS to have its own resonance (see Fig. 10). This does increase the complexity of the model but allows for more flexibility with the characteristics of the individual DDSs.

It was noticed that when HTLS is used to decompose HRIRs, the number of DDSs returned varied with elevation and azimuth. The number of DDSs estimated for the ipsilateral HRIRs used in this experiment was between 4 and 5 when K=20 and the number of DDSs for the contralateral HRIR stayed at a single DDS when K=10. The model shown in Fig. 10 can be used for HRIRs with less than 6 DDSs by setting some of the reflection coefficients (ρ_i) to zero.

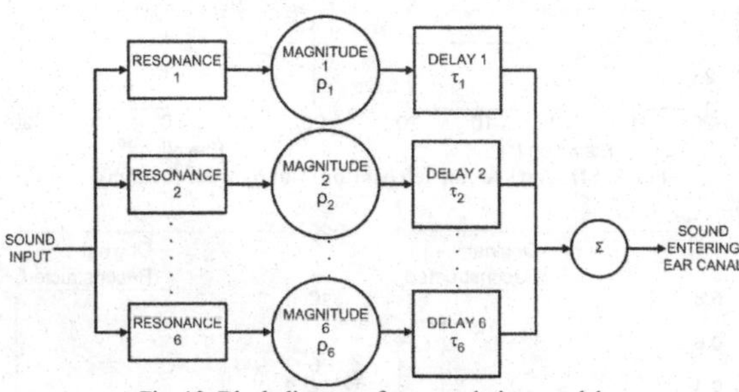

Fig. 10: Block diagram of structural pinna model.

VI. Conclusion

It has been shown that the HTLS method, augmented by the iterative zeroing algorithm, can adequately decompose HRIRs that are comprised of DDSs separated by short delays, which has proven to be a difficult task to achieve by previous methods. The method appears to be less successful at decomposing contralateral (the ear farthest from the sound source) HRIRs which could be due to the lower Signal-to-Noise (SNR) ratios and amplitudes when compared to ipsilateral (the ear closest to the sound source) HRIRs. However, the lower accuracy of decomposing contralateral HRIRs might not be of high importance because a study indicates that contralateral HRIRs do not contribute heavily to accuracy in localization at certain angles [20].

Acknowledgements

This work was sponsored by NSF grants CNS-0520811, HRD-0317692, CNS-0426125, and HRD-0833093.

References

[1] R. Kassier, H. Lee, T. Brookes, and F. Rumsey, "An Informal Comparison Between Surround-Sound Microphone Techniques," presented at 118th Convention of the Audio Engineering Society (AES), Barcelona, Spain, May 28-31, 2005.

[2] A. Berkhout, D. Devries, and P. Vogel, "Acoustic Control by Wave Field Synthesis," *Journal of the Acoustical Society of America*, vol. 93, pp. 2764-2778, 1993.

[3] L. Rayleigh, "On Our Perception of the Direction of a Source of Sound," *Proceedings of the Musical Association*, vol. 2, pp. 75-84, 1875.

[4] D. Batteau, "The Role of the Pinna in Human Localization," presented at Royal Society of London, Aug 15, 1967, pp. 158-180.

[5] J. Hebrank and D. Wright, "Spectral Cues Used in the Localization of Sound Sources on the Median Plane," *The Journal of the Acoustical Society of America (JASA)*, vol. 56, pp. 1829-1834, 1974.

[6] B. Gardner and K. Martin, "HRTF measurements of a KEMAR dummy-head microphone," Massachusetts Institute of Technology Media Laboratory Vision and Modeling Group, Last Accessed: Sept. 26, 2008; http://sound.media.mit.edu/KEMAR.html.

[7] V. Algazi, R. Duda, D. Thompson, and C. Avendano, "The Cipic HRTF database," 2001 IEEE Workshop on Applications of Signal Processing to Audio and Acoustics, Last Accessed: Sept. 26, 2008; http://interface.cipic.ucdavis.edu/.

[8] E. Wenzel, M. Arruda, D. Kistler, and F. Wightman, "Localization Using Nonindividualized Head-Related Transfer-Functions," *Journal of the Acoustical Society of America*, vol. 94, pp. 111-123, 1993.

[9] P. Satarzadeh, V. Algazi, and R. Duda, "Physical and Filter Pinna Models Based on Anthropometry," in 122nd Convention of the Audio Engineering Society, Vienna, Austria, 2007.

[10] J. Blauert, *Spatial hearing: The psychophysics of human sound localization*. Cambridge, MA: MIT Press, 1997.

[11] A. Barreto and N. Gupta, "Dynamic Modeling of the Pinna for Audio Spatialization," *WSEAS Transactions on Acoustics and Music*, vol. 1, pp. 77-82, January 2004.

[12] J. Proakis and S. Manolakis, *Digital Signal Processing: Principles, Algorithms and Applications*. Upper Saddle River, NJ: Prentice Hall, 1995.

[13] K. J. Faller II, A. Barreto, N. Gupta, and N. Rishe, "Decomposition and Modeling of Head-Related Impulse Responses for Customized Spatial Audio," *WSEAS Transactions on Signal Processing*, vol. 1, pp. 354-361, 2005.

[14] K. J. Faller II, A. Barreto, N. Gupta, and N. Rishe, "Decomposition of Head Related Impulse Responses by Selection of Conjugate Pole Pairs," in *Advances in Systems, Computing Sciences and Software Engineering*, K. Elleithy, Ed. Netherlands: Springer, 2007, pp. 259-264.

[15] S. Van Huffel, H. Chen, C. Decanniere, and P. Vanhecke, "Algorithm for Time-Domain NMR Data Fitting Based on Total Least-Squares," *Journal of Magnetic Resonance Series A*, vol. 110, pp. 228-237, 1994.

[16] A. Wevers, L. Rippert, J. M. Papy, and S. Van Huffel, "Processing of transient signals from damage in CFRP composite materials monitored with embedded intensity-modulated fiber optic sensors," *Ndt & E International*, vol. 39, pp. 229-235, 2006.

[17] R. Boyer, L. De Lathatrwer, and K. Abed-Meraim, "Higher order tensor-based method for delayed exponential fitting," *Ieee Transactions on Signal Processing*, vol. 55, pp. 2795-2809, 2007.

[18] R. Strum and D. Kirk, *First Principles of Discrete Systems and Digital Signal Processing*. Reading, MA: Addison-Wesley, 1988.

[19] H. L. Han, "Measuring a Dummy Head in Search of Pinna Cues," *Journal of the Audio Engineering Society*, vol. 42, pp. 15-37, 1994.

[20] T. Nishino, T. Oda, and K. Takeda, "Evaluation of head-related transfer function on the opposite side to a sound source," *The Journal of the Acoustical Society of America*, vol. 120, pp. 3094-3094, 2006.

Voiced/Unvoiced Decision for Speech Signals Based on Zero-Crossing Rate and Energy

Bachu R.G., Kopparthi S., Adapa B., Barkana B.D.
Department of Electrical Engineering
School of Engineering, University of Bridgeport
221 University Ave. Bridgeport, CT 06604, USA

Abstract--In speech analysis, the voiced-unvoiced decision is usually performed in extracting the information from the speech signals. In this paper, two methods are performed to separate the voiced and unvoiced parts of the speech signals. These are zero crossing rate (ZCR) and energy. In here, we evaluated the results by dividing the speech sample into some segments and used the zero crossing rate and energy calculations to separate the voiced and unvoiced parts of speech. The results suggest that zero crossing rates are low for voiced part and high for unvoiced part where as the energy is high for voiced part and low for unvoiced part. Therefore, these methods are proved effective in separation of voiced and unvoiced speech.

I. Introduction

Speech can be divided into numerous voiced and unvoiced regions. The classification of speech signal into voiced, unvoiced provides a preliminary acoustic segmentation for speech processing applications, such as speech synthesis, speech enhancement, and speech recognition.

"Voiced speech consists of more or less constant frequency tones of some duration, made when vowels are spoken. It is produced when periodic pulses of air generated by the vibrating glottis resonate through the vocal tract, at frequencies dependent on the vocal tract shape. About two-thirds of speech is voiced and this type of speech is also what is most important for intelligibility. Unvoiced speech is non-periodic, random-like sounds, caused by air passing through a narrow constriction of the vocal tract as when consonants are spoken. Voiced speech, because of its periodic nature, can be identified, and extracted [1]".

In recent years considerable efforts has been spent by researchers in solving the problem of classifying speech into voiced/unvoiced parts [2-8]. A pattern recognition approach and statistical and non statistical techniques has been applied for deciding whether the given segment of a speech signal should be classified as voiced speech or unvoiced speech [2,3,5, and 7]. Qi and Hunt classified voiced and unvoiced speech using non-parametric methods based on multi-layer feed forward network [4]. Acoustical features and pattern recognition techniques were used to separate the speech segments into voiced/unvoiced [8].

The method we used in this work is a simple and fast approach and may overcome the problem of classifying the speech into voiced/unvoiced using zero-crossing rate and energy of a speech signal. The methods that are used in this study are presented in the second part. The results are given in the third part.

II. Method

In our design, we combined zero crossings rate and energy calculation. Zero-crossing rate is an important parameter for voiced/unvoiced classification. It is also often used as a part of the front-end processing in automatic speech recognition system. The zero crossing count is an indicator of the frequency at which the energy is concentrated in the signal spectrum. Voiced speech is produced because of excitation of vocal tract by the periodic flow of air at the glottis and usually shows a low zero-crossing count [9], whereas the unvoiced speech is produced by the constriction of the vocal tract narrow enough to cause turbulent airflow which results in noise and shows high zero-crossing count.

Energy of a speech is another parameter for classifying the voiced/unvoiced parts. The voiced part of the speech has high energy because of its periodicity and the unvoiced part of speech has low energy. The analysis for classifying the voiced/unvoiced parts of speech has been illustrated in the block diagram in Fig.1.

At the first stage, speech signal is divided into intervals in frame by frame without overlapping. It is given with Fig.2.

A. End-Point Detection

One of the most basic but problematic aspects of speech processing is to detect when a speech utterance starts and ends. This is called end-point detection. In the case of unvoiced sounds occurring at the beginning or end of the utterance, it is difficult to detect accurately the speech signal from the background noise signal.

In this work, end-point detection is applied to the voiced/unvoiced algorithm at the beginning of the algorithm to separate silence and speech signal. A small sample of the background noise is taken during the silence interval just prior to the commencement of the speech signal. The short-time energy function of the entire utterance is then computed using Eq.4.

K. Elleithy (ed.), *Advanced Techniques in Computing Sciences and Software Engineering*,
DOI 10.1007/978-90-481-3660-5_47,

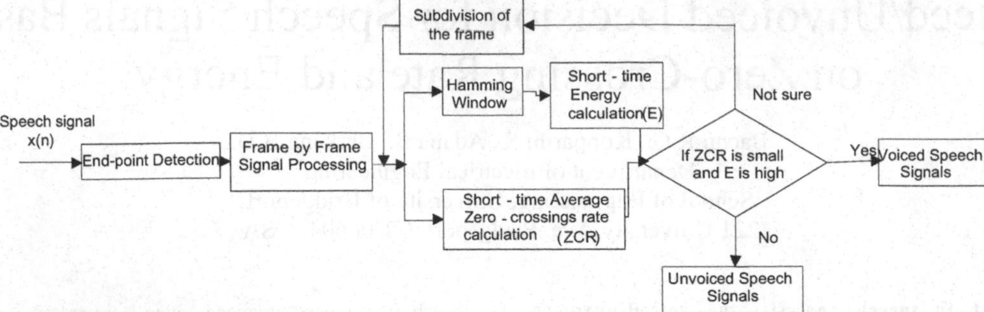

Fig.1: Block diagram of the voiced/unvoiced classification.

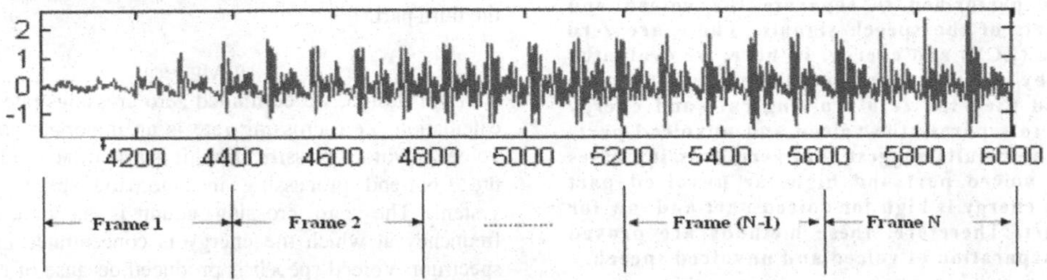

Fig. 2: Frame-by–frame processing of speech signal.

A speech threshold is determined which takes into account the silence energy and the peak energy. Initially, the endpoints are assumed to occur where the signal energy crosses this threshold. Corrections to these initial estimates are made by computing the zero-crossing rate in the vicinity of the endpoints and by comparing it with that of the silence. If detectable changes in zero-crossing rate occur outside the initial thresholds, the endpoints are re-designed to the points at which the changes take place [10-11]

B. Zero-Crossing Rate

In the context of discrete-time signals, a zero crossing is said to occur if successive samples have different algebraic signs. The rate at which zero crossings occur is a simple measure of the frequency content of a signal. Zero-crossing rate is a measure of number of times in a given time interval/frame that the amplitude of the speech signals passes through a value of zero, Fig3 and Fig.4. Speech signals are broadband signals and interpretation of average zero-crossing rate is therefore much less precise. However, rough estimates of spectral properties can be obtained using a representation based on the short-time average zero-crossing rate [12].

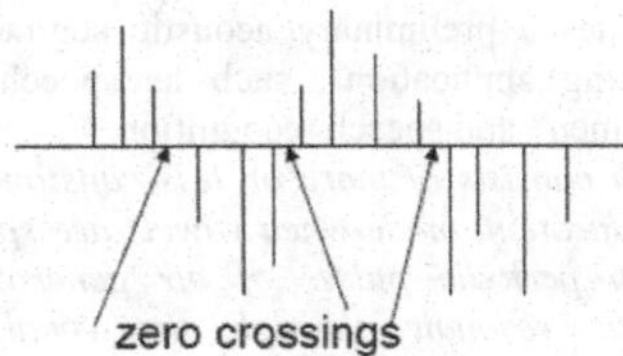

Fig. 3: Definition of zero-crossings rate

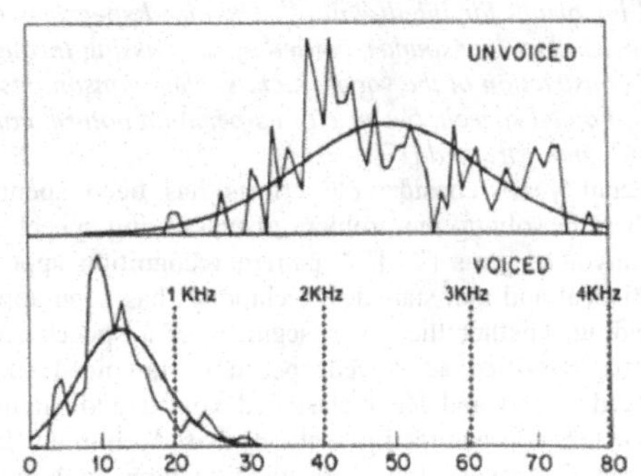

Fig. 4: Distribution of zero-crossings for unvoiced and voiced speech [12].

A definition for zero-crossings rate is:

$$Z_n = \sum_{m=-\infty}^{\infty} \left| \text{sgn}[x(m)] - \text{sgn}[x(m-1)] \right| w(n-m) \quad (1)$$

where

$$\text{sgn}[x(n)] = \begin{cases} 1, x(n) \geq 0 \\ -1, x(n) < 0 \end{cases} \quad (2)$$

and

$$w(n) = \begin{cases} \frac{1}{2N} \, for, 0 \leq n \leq N-1 \\ 0 \, for, otherwise \end{cases} \quad (3)$$

The model for speech production suggests that the energy of voiced speech is concentrated below about 3 kHz because of the spectrum fall of introduced by the glottal wave, whereas for unvoiced speech, most of the energy is found at higher frequencies. Since high frequencies imply high zero crossing rates, and low frequencies imply low zero-crossing rates, there is a strong correlation between zero-crossing rate and energy distribution with frequency. A reasonable generalization is that if the zero-crossing rate is high, the speech signal is unvoiced, while if the zero-crossing rate is low, the speech signal is voiced [12].

C. Short-Time Energy

The amplitude of the speech signal varies with time. Generally, the amplitude of unvoiced speech segments is much lower than the amplitude of voiced segments. The energy of the speech signal provides a representation that reflects these amplitude variations. Short-time energy can define as:

$$E_n = \sum_{m=-\infty}^{\infty} [x(m)w(n-m)]^2 \quad (4)$$

The choice of the window determines the nature of the short-time energy representation. In our model, we used Hamming window. The hamming window gives much greater attenuation outside the band pass than the comparable rectangular window.

$$h(n) = 0.54 - 0.46\cos(2\pi n/(N-1)), \qquad 0 \leq n \leq N-1 \quad (5)$$

$$h(n) = 0, \; otherwise$$

The attenuation of this window is independent of the window duration. Increasing the length, N, decreases the bandwidth, Fig 5. If N is too small, E_n will fluctuate very rapidly depending on the exact details of the waveform. If N is too large, E_n will change very slowly and thus will not adequately reflect the changing properties of the speech signal [12].

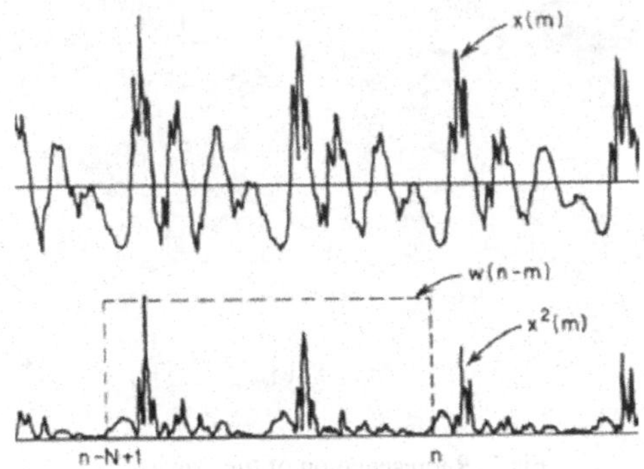

Fig.5. Computation of Short-Time Energy [12].

III. Results

MATLAB 7.0 is used for our calculations. We chose MATLAB as our programming environment as it offers many advantages. It contains a variety of signal processing and statistical tools, which help users in generating a variety of signals and plotting them. MATLAB excels at numerical computations, especially when dealing with vectors or matrices of data.

One of the speech signal used in this study is given with Fig.6. Proposed voiced/unvoiced classification algorithm uses short-time zero-crossings rate and energy of the speech signal. The signal is windowed with a rectangular window of 50ms duration at the beginning. The algorithm reduces the duration time of the window by half at each feedback if the decision is not clear. The results of voiced/unvoiced decision using our model are presented in Table 1.

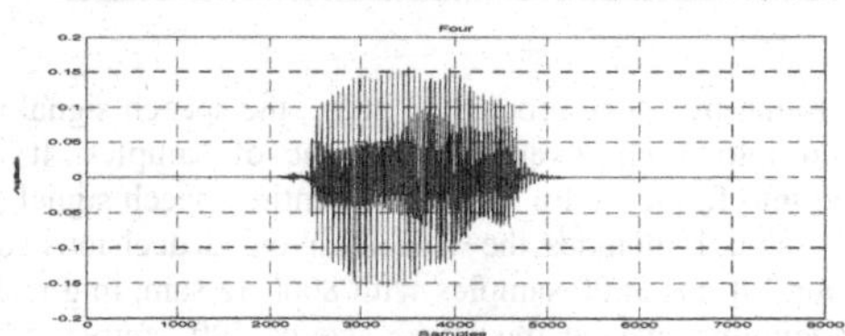

Fig.6: Original speech signal for the word "four."

The frame by frame representation of the algorithm is presented with Fig.7. At the beginning and the ending points of the speech signal, the algorithm decreases the window duration time. At the beginning, word starts with an "f" sound which is unvoiced. At the end, word ends with a "r" sound which is unvoiced.

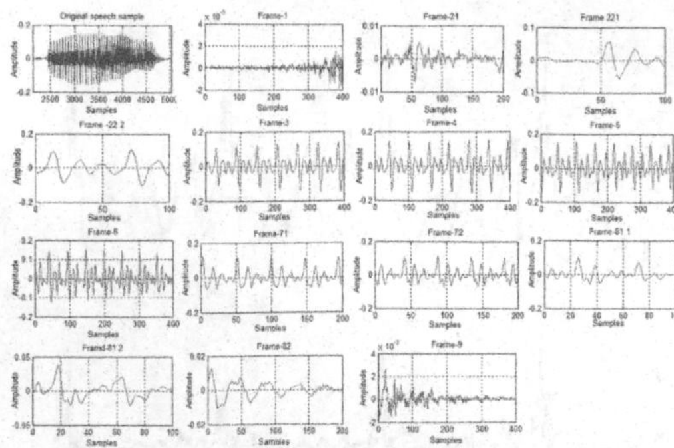

Fig.7. Representation of the frames.

TABLE I. VOICED/UNVOICED DECISIONS FOR THE WORD "FOUR" USING THE MODEL.

	ZCR	Energy (J)	Decision
Frame-1 (50 ms)	152	0.0018	u nvoiced
Frame-21(25 ms)	52	0.0543	unvoiced
Frame 22(25 ms)	19	21.1189	voiced
Frame-3 (50 ms)	41	186.6628	voiced
Frame-4 (50 ms)	41	230.5772	voiced
Frame-5 (50 ms)	43	252.98	voiced
Frame-6(50 ms)	56	193.70	voiced
Frame-71(25 ms)	31	27.2842	voiced
Frame-72(25 ms)	30	25.960	voiced
Frame-811(12.5 ms)	24	3.4214	voiced
Frame-812(12.5 ms)	11	0.4765	unvoiced
Frame-82(25 ms)	19	0.166	unvoiced
Frame-9 (50 ms)	89	0.0054	u nvoiced

In the frame-by-frame processing stage, the speech signal is segmented into a non-overlapping frame of samples. It is processed into frame by frame until the entire speech signal is covered. Table 1 includes the voiced/unvoiced decisions for word "four." It has 3600 samples with 8000Hz sampling rate. At the beginning, we set the frame size as 400 samples (50 ms). At the end of the algorithm if the decision is not clear, energy and zero-crossing rate is recalculated by dividing the related frame size into two frames. This phenomenon can be seen for Frame 2, 7, and 8 in the Table 1.

IV. CONCLUSION

We have presented an approach for separating the voiced /unvoiced part of speech in a simple and efficient way. The algorithm shows good results in classifying the speech as we segmented speech into many frames. In our future study, we plan to improve our results for voiced/unvoiced discrimination in noise.

REFERENCES

[1] J. K. Lee, C. D. Yoo, "Wavelet speech enhancement based on voiced/unvoiced decision", Korea Advanced Institute of Science and Technology The 32nd International Congress and Exposition on Noise Control Engineering, Jeju International Convention Center, Seogwipo, Korea, August 25-28, 2003.

[2] B. Atal, and L. Rabiner, "A Pattern Recognition Approach to Voiced-Unvoiced-Silence Classification with Applications to Speech Recognition," *IEEE Trans. On ASSP*, vol. ASSP-24, pp. 201-212, 1976.

[3] S. Ahmadi, and A.S. Spanias, "Cepstrum-Based Pitch Detection using a New Statistical V/UV Classification Algorithm," *IEEE Trans. Speech Audio Processing*, vol. 7 No. 3, pp. 333-338, 1999.

[4] Y. Qi, and B.R. Hunt, "Voiced-Unvoiced-Silence Classifications of Speech using Hybrid Features and a Network Classifier," *IEEE Trans. Speech Audio Processing*, vol. 1 No. 2, pp. 250-255, 1993.

[5] L. Siegel, "A Procedure for using Pattern Classification Techniques to obtain a Voiced/Unvoiced Classifier", *IEEE Trans. on ASSP*, vol. ASSP-27, pp. 83- 88, 1979.

[6] T.L. Burrows, "Speech Processing with Linear and Neural Network Models", Ph.D. thesis, Cambridge University Engineering Department, U.K., 1996.

[7] D.G. Childers, M. Hahn, and J.N. Larar, "Silent and Voiced/Unvoiced/Mixed Excitation (Four-Way) Classification of Speech," *IEEE Trans. on ASSP*, vol. 37 No. 11, pp. 1771-1774, 1989.

[8] J. K. Shah, A. N. Iyer, B. Y. Smolenski, and R. E. Yantorno "Robust voiced/unvoiced classification using novel features and Gaussian Mixture model", Speech Processing Lab., ECE Dept., Temple University, 1947 N 12th St., Philadelphia, PA 19122-6077, USA.

[9] J. Marvan, "Voice Activity detection Method and Apparatus for voiced/unvoiced decision and Pitch Estimation in a Noisy speech feature extraction", 08/23/2007, United States Patent 20070198251.

[10] T. F. Quatieri, Discrete-Time Speech Signal Processing: Principles and Practice, MIT Lincoln Laboratory, Lexington, Massachusetts, Prentice Hall, 2001, ISBN-13:9780132429429.

[11] F.J. Owens, Signal Processing of Speech, McGraw-Hill, Inc., 1993, ISBN-0-07-0479555-0.

[12] L. R. Rabiner, and R. W. Schafer, Digital Processing of Speech Signals, Englewood Cliffs, New Jersey, Prentice Hall, 512-ISBN-13:9780132136037, 1978.

A Survey of Using Model-Based Testing to Improve Quality Attributes in Distributed Systems

Ahmad Saifan, Juergen Dingel
School of Computing, Queen's University
Kingston, Ontario, Canada, K7L 3N6
Saifan@cs.queensu ca, Dingel@cs.queensu.ca

Abstract-This paper provides a detailed survey of how Model-Based Testing (MBT) has been used for testing different quality attributes of distributed systems such as security, performance, reliability, and correctness. For this purpose, three additional criteria are added to the classification. These criteria are: the purpose of testing, the test case paradigm, and the type of conformance checking. A comparison between different MBT tools based on the classification is also given.

I. Introduction

Every developer wants his software to have sufficient quality. However, the increasing size and complexity of software systems makes this difficult. Distributed systems (DSs) are no exception. Typically, DSs are heterogeneous in terms of communication networks, operating systems, hardware platforms, and also the programming language used to develop individual components. Consequently, testing of DSs is inherently difficult.

Models of software have been used during most stages of software development for a while. In general, by using model we facilitate the communication between the customers and the developers and between developers themselves. Moreover, models often enable abstraction and analysis. To address the challenges of testing software systems and to leverage models already produced and used in the context of existing model-oriented software development methods, model-based testing (MBT) has been suggested. In MBT there are four main phases: 1) build the model, 2) generate test cases, 3) execute test cases, and 4) check the conformance. We recently completed a thorough survey of the use of MBT to improve the quality of DSs with respect to performance, security, correctness, and reliability [1]. This paper contains an abbreviated summery of this survey. Due to space restrictions not all detail could be included. For more detailed account, the interested reader is referred to [1].

II. Using Mbt To Improve Different Quality Attributes In Distributed Systems

A. *Performance*

During performance testing, the tester should consider all performance characteristics to make sure that the DS meets the performance requirements of its users. These characteristics are: latency, throughput, and scalability. Scott Barber [2] identifies performance testing as "an empirical, technical investigation conducted to provide stakeholders with information about the quality of the product or service under test with regard to speed, scalability and/or stability characteristics". There might also be other information we need to measure when evaluating performance of a particular system, such as resource usage, and queue lengths representing the maximum number of tasks waiting to be serviced by selected resources. A lot of research in the literature has focused on building performance models. Various approaches have been proposed to derive different types of performance models from software architectures mostly presented using different types of UML diagrams. Some of these performance models are: Layered Queuing Networks (LQNs) which are an extension of the Queuing Network model (QN) presented in (e.g [4]), Stochastic Petri Nets (SPNs) [5], Stochastic Process Algebra (SPA) [6].

A.1 Sample Approach for Using MBT for Performance Testing:

As an example of using MBT to test the performance of distributed systems, in [8], the authors present an approach in which the architectural design of the distributed system is used to generate performance test cases. These test cases can be executed on the middleware that was used to build the distributed application. To check the feasibility of performance testing in the early stages of software development and the efficacy of their approach, the authors [8] performed an experiment based on Duke's Bank application (an application presented in the J2EE tutorial, consisting of 6,000 lines of Java code). In this experiment, the authors tried to compare the latency of the real implementation of the application with the latency of the test version of the same system (which is consists of the early available components) based on a specific use case, while varying the number of clients. In the first phase of the approach, a sample use case relevant to performance such as the transfer of funds between two bank accounts is selected. In order to map the use case to the middleware, in the second phase, the use case is augmented manually with some necessary information. In the third phase the test version of the Duke's Bank application is developed by implementing the needed stubs in order to realize the interactions for the use cases. After that, the real implementation of the system and the test version were executed to measure the latency of the test cases. To execute these systems, a workload with increasing number of clients starting from 1 to 100 (presenting the performance parameter of test cases) is generated. A workload generator is implemented and database is initialized with persistent data. The workload generator is able to activate a number of clients at the same time and takes care of measuring the average response time. Both the implementation and the test version are executed for the increasing number of clients and the average response time for

K. Elleithy (ed.), *Advanced Techniques in Computing Sciences and Software Engineering*,
DOI 10.1007/978-90-481-3660-5_48, © Springer Science+Business Media B.V. 2010

each of the test cases is measured. Each experiment is repeated 15 times.

Summary: As a result, the authors [8] found that latency times of the application and the test version are very similar. The result of this experiment suggests that this approach is suitable for performance testing of distributed applications in the early stages of software development. However, more experiments using other distributed applications need to be conducted before the general viability of this approach can be concluded. In particular, experiments using different use cases and different kinds of middleware and databases are necessary.

B. Security

A lot of research in the literature has focused on building security models. There are various approaches extending UML diagrams to specify the security requirements of the system. For example, in [9, 10], Jürjens presents *UMLsec* as an extension of UML diagrams to specify the security requirements. These security requirements are inserted into the UML diagrams as stereotypes with tags and constraints. *UMLsec* is also used to check whether or not the security requirements are met by an implementation. Moreover, it can be used to find violations of security requirements in the UML diagrams. A framework for specifying intrusions in UML called *UMLintr* is presented in [11]. In this framework UML diagrams are extended to specify security requirements (intrusion scenarios). Lodderstedt et al. [12] present a modeling language for the model-driven development of secure distributed systems as an extension of UML called *secureUML*. This language is based on Role Based Access Control (RBAC). RBAC is a model that contains five types of data: users, roles, objects, operations, and permissions. *SecureUML* can be used to automatically generate complete access control infrastructures.

There are two categories for testing security in general as presented in [13]:

- Security Functional Testing: used to test the conformance between the security function specification expressed in the security model and its implementation.
- Security Vulnerability Testing: identification of flaws in the design or implementation that can cause violations of security properties.

Example for both security functional testing and security vulnerability testing will be given below.

B.1 Sample Approach for Using MBT for Security Functional Testing:

Blackburn et al [15] developed a model-based approach to automate security functional testing called Test Automation Framework (TAF). The model is used for testing the functional requirements of centralized systems. However, according to the authors the model is extensible enough to be used for distributed systems. In private communication, one of the authors said "it would be fairly straightforward to model distributed system relationships in TTM (T-VEC Tabular Modeler) and generate vectors that could test those relationships, provided you had a testbed/test environment designed to set up, control, and record your distributed systems environment in the manner of the test vectors that would result from such an approach."

In this approach, a part of an Oracle8 security document [16] is used to build a functional specification model using the Software Cost Reduction (SCR) method [26]. More specifically, the security function of Granting Object Privilege Capability (GOP) is used in this approach. The SCR model is a table-based model, representing the system input variables, output variables, and intermediate values as term variables. Term variables are used to decompose the relationship between inputs and outputs of the system. Once these variables are identified, the behavior of the system can be modeled. In step 2 of this approach, the SCR tables are translated into a test case specification called T-VEC. For GOP requirements, about 40 test specification paths were generated. In step 3, the test cases (test vectors) are generated from the T-VEC test specification using particular coverage criterion similar to boundary testing. In order to generate test drivers that are executed against the system, the test driver generator needs the test driver schema, user-defined object mappings, and test vectors. So in step 4, test schemas are generated manually. These test schemas represent the algorithm for the test execution in the specific test environment. In the object mapping, the authors also manually map the object variables of the model to the interface of the implementation. After that in step 5, test drivers are automatically generated using the test driver generators by repeating the execution steps that are identified in the test schema for each test vector.

Summary: For the evaluation of their approach, two different test driver schemas are used (one for an Oracle test driver and another for an Interbase test driver) to test two different test environments. They found that the model executed without failure in the Oracle database driver schema, and results in test failures when using the Interbase database schema. But as described in the Interbase documentation, the failures are associated with restrictions on granting roles. These results demonstrate the feasibility of using MBT to improve security functional testing.

B.2 Sample Approach for Using MBT for Security Vulnerability Testing:

As an example of using MBT to test the system for security vulnerabilities of distributed systems, Wimmel et al. [17] presented a new approach to find test sequences for security-critical systems. These test sequences are used to detect possible vulnerabilities that violate system requirements. In this approach, the AUTOFOCUS tool is used to automatically generate test sequences. More specifically, mutation of the system specification and the attacker scenario is used to automatically generate the test sequences that are likely to lead to violations of the security requirements. In this approach, AUTOFOCUS is used to describe the structure and the interface of the system by using System Structure Diagrams (SSDs). An SSD presents the interaction between the system components (similar to UML component diagrams). Each component has two ports: source and destination for receiving and sending messages. These ports are connected via directed channels. Furthermore, AUTOFOCUS is used to express the behavior of the components by using State Transition Diagrams (STDs). Threat scenarios which represent the capability of the attacker are generated automatically by AUTOFOCUS based on the security attributes assigned to SSDs and STDs. There are five types of security attributes that are associated with components and channels in SSDs. These attributes are "critical", "public", "replace", and "node". For more information for these attributes see [17]. The security attribute "critical" is also associated to the transitions and states of the STDs as appropriate. After generating the security-critical model, the authors [17] use this model to generate test sequences in order to test the implementation of the system. These test sequences should cover all possible violations of the security requirements. In order to generate test sequences from the security-critical model, first the structural coverage criteria are needed. State or transition coverage is not suitable, because it does not take into account the security requirements. So a mutation testing approach is used. In mutation, we introduce an error into the specification of the behavior of the system, and then the quality of test suites is given by its ability to kill mutants. During mutation testing, one of the critical transitions (t) of the STD of the component to be tested is chosen and then a mutation function is applied to obtain a mutated transition t'. Mutation functions can be used to modify, for example, the precondition or post-condition of a transition. Next, threat scenarios are automatically generated from the mutated version of the component to be tested in order to obtain the mutated system model Ψ'. After that, each of the system requirements Φ_i is taken to compute a system run that satisfies $\Psi' \wedge \neg\Phi_i$ using test sequence generator. If it is successful, then mutating t into t' introduced a vulnerability with respect to Φ_i and the traces show how it can be exploited. In this technique, the original specification of the components is used as an oracle. Note that test sequences are used to test the security-critical model and not its implementation. To test the actual implementation, the test sequences should be translated to concrete test data. Concrete test data is generated from the abstract sequences by using an algorithm presented in [17]. This concrete test data is executed by the test driver that passes inputs to the component to be tested, and then checks whether or not the actual outputs satisfy the expected outputs.

Summary: This approach allows finding tests likely to detect possible vulnerabilities even in complex execution scenarios such as the Common Electronic Purse Specifications (CEPS) purchase transaction protocol.

C. Correctness

To check the correctness of the system under test we need to compare the actual outputs with expected outputs by using some kind of conformance relation such as input-output conformance (ioco)[1]. A more detailed description of conformance relation is omitted due to space limitations. The interested reader is referred to [1]. In general, conformance relations are based on comparing traces, i.e., sequences of states that the system as model exhibits during execution. Simply comparing initial and final execution states is insufficient for DSs, because systems may be designed never to terminate or correctness may depend on, e.g., the order in which messages are exchanged.

We could not find papers that use MBT activities as specified in the introduction for deadlock detection. Instead, we will describe the use of models to ensure the deadlock free execution of a DS.

C.1 Avoid Deadlock

In [18], a test control strategy is provided to control the execution order of synchronization events and remote calls. This test control strategy is used to ensure deadlock free execution of a distributed application that consists of a set of processes (which have one or more Java thread) communicating through CORBA. All synchronization events that could lead to deadlock are sent to a central controller first which permits or defers the event based on the control strategy. Synchronization events considered in [18] are: remote method invocation and its completion, and access to a shared object and its completion. The following steps are used in [18] to construct the test control strategy:

1) Build the test constraints using static analysis. These test constraints express the partial order of synchronization events.

2) Choose the thread model. In the CORBA middleware, there are three thread models: thread-per-request, thread pool, and thread-per-object. If thread-per-request is used, there is no deadlock, because for every request, a separate thread is created to handle the request. So we just choose thread pool or thread-per-object as a thread model for this technique.

3) Build the test model (M) using the test constraints and the thread model from the previous 2 steps. M is a finite state automaton describing the sequence of events.

4) Find and remove deadlock states from M to get M'.

5) Controller uses the deadlock free model M' to avoid leading the execution of the application into a deadlocked state. The test controller should decide which remote calls at each moment should be executed.

Summary: In this technique, 2 different thread models are allowed. These thread models may not be appropriate anymore when e.g., message passing is used instead of remote method invocation using CORBA. The approach relies on a partial order relation of events. If this relation is incomplete, then possible deadlocks may be overlooked.

---

[1] An implementation under test (IUT) conforms to its specification if the IUT never produces an output that cannot be produced by the specification.

D. Reliability

According to ANSI [19], Software Reliability is defined as: "the probability of failure-free software operation for a specified period of time in a specified environment".

D1. Sample Approach for Using MBT for Reliability Testing:

Guen et al. [20] present improved reliability estimation for statistical usage testing based on Markov chains. Statistical usage testing is used to test a software product from a user's point of view (usage model). The usage model represents how the user uses the system. Markov chains usage models consist of states representing states of use and transitions labeled with usage events (stimuli). Moreover, probabilities are also assigned to all the transitions that reflect how likely a transition is to occur. The new measure of estimating reliability presented in [20] improves on the two standard reliability measures proposed by Whittaker et al. [21] and Sayre et al. [22].

It has also been implemented together in a tool named MaTeLo (Markov Test Logic). The input of the MaTeLo tool is the usage model to the software. So before using this tool, the tester should develop the usage models. MaTeLo accepts models in three different notations. These notations are: MSC (Message Sequence Chart), UML sequence diagrams, or a statecharts. However, if the tester chooses one of the first two notations, the tool converts it to a statechart. After that, the probabilities of the transitions between states are assigned using a Markov usage editor component, in order to get the design of the usage models. These probabilities can be calculated automatically based on a pre-defined distribution or can be manually specified by the user. For more information about how to calculate the probabilities of the transitions, see [21, 22, 20]. Next, the tool checks the correctness of the usage models (e.g. the probabilities are arranged between 0 and 1, or terminal states have no outgoing transitions, etc) and then converts it into test cases. Test cases are then automatically generated using several test generation algorithms. These algorithms are based for example on the probability of the transitions, the test cases that have minimal test steps, or on randomization. The test cases that have been generated can be represented in the formats TTCN-3 or XML. After the generation of the test cases, they are executed against the system under test and then the results are recorded. The results of the test are analyzed to estimate the reliability probability of the usage model obtained from the test runs.

Summary: MaTeLo presents improved reliability estimation for statistical usage testing based on Markov chains. For example, when test cases are randomly generated from a profile, it is possible to estimate the reliability for all profiles associated to the model. As another improvement in this approach, if the test cases do not reveal a failure, the reliability is not equal to 1, as the Whittaker estimation.

III. MBT TOOLS FOR TESTING DISTRIBUTED SYSTEMS

We briefly describe two MBT tools. Other MBT can be found in our technical report [1].

A. SpecExplorer

Spec Explorer [23] is the second generation of MBT technology developed by Microsoft. Spec Explorer is used in Microsoft for testing e.g. operating system components, reactive systems, and other kinds of systems.

In the first step of Spec Explorer, the tester builds the model program of the system. The model program specifies the desired behavior of the system. It can be written in any high level programming language such as Spec#, C#, or VB. The model program is not the implementation of the system. The model program just focuses on the aspects of the system that are relevant for the testing effort. Furthermore, it represents the constraints that a correct implementation must follow. More precisely, the model program declares a set of action methods, a set of state variables, and the preconditions of these states. In the second step of this tool, Spec Explorer explores all the possible runs of the model program and represents them as a FSM, named model automaton. A state in this automaton corresponds to a state of the model program, and transitions represent the method invocations that satisfy their preconditions. This automaton can have a very large number of transitions. To solve this problem several techniques for selectively exploring the transitions of the model program are used (e.g. parameter selection, method restriction, state filtering, directed search, state grouping). The graph of the FSM can be viewed in different ways in step 3, e.g. based on how to group similar states. In step 4, test cases are generated from the model automaton. There are two different kinds of testing in this tool: offline testing and online testing. Test cases are generated using traversal algorithms based on different test criteria. The test criteria are: random walk by specifying bound of the walk, transition coverage, or using the shortest path between the initial and an accepting state. In step 5, test cases are executed against the system under test. To do that, an API driver can be used. The API driver (which is a program written in C# or VB) does whatever is necessary in order to invoke the individual actions of the test suites in the implementation under test (step 6). In other words, the corresponding methods and classes of the API driver should be mapped to the model. This process is called object binding. In step 7, the ioco conformance relation is used to compare the expected outputs of the model with the actual outputs of the implementation.

B. TorX:

The purpose of the TorX [24] tool is to implement the ioco theory. In TorX there are four main phases that are provided automatically in an on-the-fly manner. The phases are: test case generation from the specification of the system under test, translate the test cases into executable test scripts, test execution, and test analysis. In the first step of TorX, the behavior of the system under test is specified using labeled transition systems. TorX contains several explorers to explore the labeled transition system defined by specifications given in different formal notations such as LOTOS, PROMELA, or LTSA. TorX implements a test derivation algorithm that is used to generate test cases from the labeled transition systems.

Test cases are generated randomly based on a walk through the state space of the specification. To maximize the chance of finding errors in the system under test, two kinds of test coverage criteria are used. Heuristic criteria provide some assumptions about the behavior of the system under test such as the length of the test case. Additionally, TorX allows the test purpose criterion, that is, of a particular part of the behavior of the system. If the test purpose is given, the test case generation algorithm guarantees that generated test cases will exercise that part of the system. In TorX, the test generation and test execution phases are integrated into one phase since test cases are generated on-the-fly during the test execution. Also, offline test case generation is available in TorX. An adapter component is used to translate the input actions of the test case that have been generated from the specification to be readable by the real implementation of the system under test. It is also used to translate the outputs that are produced by the implementation back to the output action of the test case. Then the ioco relation is used to check the conformance between the actual outputs of the real implementation and the expected outputs of the test cases.

IV. A Comparison Between Different MBT Tools Based On The Classification

In [14], Utting et al. specify several criteria for the classification of MBT. Four of these criteria are related to phase 1 of the MBT process, two criteria to phase 2, and one criterion to phase 3. The classification is used to compare different MBT tools. In [14], the authors' classification does not include a criterion for phase 4 (type of conformance check). Moreover, the classification does not include the purpose of testing. In addition, the test case notation is not used as a criterion for phase 3. Therefore, three other criteria will be added to the classification. Moreover, other tools are added to the tools that are mentioned in [14]. To summarize, the new three criteria are:

- **Type of conformance check**: This criterion is used to classify how different MBT tools check the conformance between the actual and the expected outputs of the system under test. In other words, how the tools determine whether a test case is pass or fail. There are different ways to check conformance, for example, some tools use conformance relations. For example, TorX and Spec Explorer use ioco conformance relation. Other tools use test oracles to check conformance. Test oracle can be generated automatically by the tool or they can be provided manually by the user. Moreover, conformance can be decided by the human.

- **Purpose of testing**: As we have seen in the previous sections, systems can be tested for different purposes. For example:

  - Testing the functional properties of the system: The model in this kind of testing represents the behavior of the main functions of the system under test.

  - Testing non-functional properties of the system: we can test the performance, security, reliability, etc properties of the system. For example if we are going to test the security of the system, then the model will represent the security requirements of the system under test.

- **Test case notation**: This criterion is used to describe the notation in which test cases are given. There are different notations that can be used to describe the test cases e.g. FSM, sequence of operations, input output labeled transition systems, TTCN-3, etc.

Table I compares different MBT tools based on Utting et al's. modified classification. Also, three other tools are added to the tools presented in Utting et al. paper [13], these tools are: Spec Explorer [23], Conformiq Qtronic [3] and TAF [7].

TABLE I: Some of MBT Tools Based On The Classification
F: Functional, R: Reliability, S: Security, SUT: System Under Test

		Phase 1				Phase 2		Phase3		Phase 4
MBT tool	**Purpose of testing**	**Subject of the model**	**Model redundancy level**	**Model characteristics**	**Model paradigm**	**Test selection criteria**	**Test generation technology**	**Test case paradigm**	**Online/ Offline**	**Type to check conformance**
TorX [24]	F	SUT behavior model with some environmental aspects	Dedicated testing model	Non-deterministic, untimed, discrete	Input Output Labeled Transition System	A walk (done randomly or based on test purpose) through state space of the model	On-the-fly state space exploration	input output transition system	Both	ioco relation
LTG [27]	F	SUT behavior model	Dedicated testing model	Deterministic, untimed, discrete, and finite	UML state-machines, B notation	For Statemachines: state coverage, all transitions, all extended transitions and all transitions pairs. For B notation: all effects, and all pair effects	Symbolic execution, search algorithm	Sequence of operation	Offline	Expected outputs assigned with each test case
JUMBL [28]	R	Behavior environment model	Dedicated testing model	Untimed, and discrete	Markov chain usage models	Random and statistical criteria	statistical search algorithm	Test case markup language	Offline	Test oracle provided by user not the tool
AETG [29]	F	Data environment model	Dedicated model for test input generation only	Untimed, discrete	No modeling of the behavior just data	Data coverage criteria	N-way search algorithm	Table of inputs values	Offline	Test oracle provided by user not the tool
Spec Explorer [23]	F	SUT behavior model with environment model	Dedicated testing model	Non-deterministic, Untimed, discrete	FSM	Randomly, shortest path, transition coverage Traversal algorithms	Traversal algorithms	FSM with different views	Both	ioco relation
Conformiq Qtroinq [3]	F	SUT behavior model with environment model	Dedicated testing model	Non-deterministic, timed	UML state-machines with blocks of Java or C#	State coverage, arc coverage, branch coverage, condition coverage, requirement coverage	Symbolic execution algorithm	TTCN-3	Both	Expected outputs are generated in automatic way from the model
TAF [7]	S	SUT behavior model with environment model	Dedicated testing model	Deterministic, untimed, hybrid	SCR table converted to T-VEC	Domain testing theory	Test vector generation	Test vector	Offline	The test oracle is the model itself

V. Conclusion

Because of the heterogeneity of Distributed systems (DSs), testing DSs is more difficult than centralized systems. This paper summarizes how MBT can be used for testing different non-functional properties of DSs such as performance, security, correctness, and reliability. As we have seen in this paper, the classification of MBT proposed in [14] is suitable to compare between different MBT tools. However, three new different criteria are added to this classification in this paper. These criteria are: test purpose, test cases paradigm, and the way the tool checks the conformance between the expected outputs and actual outputs. We have gathered information about 7 different MBT tools and compared them (4 of them listed in [14]). Testing non-functional properties of DSs such as security, reliability, performance, etc. using MBT is an interesting field for research. For example, a lot of research has been done to specify the performance model of the system under test, but still we do not have a tool used to evaluate the performance of DSs. Moreover, Dias Neto et al. [25] point out that several types of non-functional attributes of DSs such as usability, maintainability, and portability are not supported by current MBT techniques.

References

[1] A. Saifan and J. Dingel: "Model-Based Testing of Distributed Systems". Technichal report 2008-548, school of computing, Queen's University (2008).

[2] S. Barber. What is performance testing?,2007.Available at: http://searchsoftwarequality.techtarget.com/tip/0,289483,sid92gci12475 94,00.html

[3] A. Huima. "Implementing Conformiq Qtronic". In TestCom/FATES, pages 1-12,Tallinn, Estonia, June 2007.

[4] E. Lazowska, J. Zahorjan, G. Graham, and K. Sevcik. "Quantitative System Performance: Computer System Analysis Using Queuing Network Models". Prentice-Hall, Inc., NJ, USA, 1984.

[5] S. Bernardi, S. Donatelli, and J. Merseguer. "From UML Sequence Diagrams and Statecharts To Analyzable Petri Net Models". In WOSP '02: Proceedings of the 3rd international workshop on Software and performance, pages 35-45, New York, NY, USA, 2002. ACM.

[6] R.Pooley. "Using UML to Derive Stochastic Process Algebra Models".In Proceedings of the 15th UK Performance Engineering Workshop, pages 23-33, 1999.

[7] R. Chandramouli and M. Blackburn. "Model-based Automated Security Functional Testing". In OMGs Seventh Annual Workshop on distributed objects and components security, Baltimore, Maryland, USA, April 2003.

[8] G. Denaro, A. Polini, and W. Emmerich. "Early Performance Testing of Distributed Software Applications". SIGSOFT Softw. Eng. Notes, vol. 29(1), pp. 94-103, 2004.

[9] J. Jürjens. "Towards Development oF Secure Systems USING UMLsec". In Fundamental Approaches to Software Engineering, 4th International Conference, FASE 2001, Genova, Italy, April, 2001, Proc., volume 2029 of Lecture Notes in Computer Science, pages 187-200. Springer, 2001.

[10] J. Jürjens. "Secure Systems Development With UML". SpringerVerlag, 2005.

[11] M. Hussein and M. Zulkernine. "UMLintr: A UML Profile for Specifying Intrusions". In 13th Annual IEEE International Conference and Workshop on Engineering of Computer Based Systems, 27-30, 2006, Potsdam, Germany, pages 279-288. IEEE Computer Society, 2006.

[12] T. Lodderstedt, D. Basin, and J. Doser. "Secureuml: A UML-Based Modeling Language for Model-Driven Security". In UML '02: Proc. of the 5th International Conference on The Unified Modeling Language, pages 426-441, London, UK, 2002. SpringerVerlag.

[13] R. Chandramouli and M. Blackburn. "Model-Based Automated Security Functional Testing". In OMGs Seventh Annual Workshop on distributed objects and components security, Baltimore, Maryland, USA, April, 2003.

[14] M. Utting, A. Pretschner, and B. Legeard. "A Taxonomy of Model-Based Testing". Technical Report 04/2006, Department of Computer Science, The Universiy of Waikato(New Zealand), 2006.

[15] M. Blackburn, R. Busser, A. Nauman, and R. Chandramouli. "Model-Based Approach to Security Test Automation", proceeding in quality week, June, 2001.

[16] Oracle Corporation. Oracle8 Security Target Release 8.0.5, April, 2000.

[17] G. Wimmel and J. Jürjens. "Specification-Based Test Generation for Security-Critical Systems Using Mutations". In ICFEM '02: Proceedings of the 4th International Conference on Formal Engineering Methods, pages 471-482,London, UK, 2002. Springer-Verlag.

[18] J. Chen. "On Using Static Analysis in Distributed System Testing". In EDO'00: Revised Papers from the 2nd International Workshop on Engineering Distributed Objects, pp. 145-162, London, UK, 2001. SpringerVerlag.

[19] IEEE Standard Glossary of Software Engineering Terminology. Technical report,1990.

[20] H. Guen, R. Marie, and T. Thelin. "Reliability Estimation for Statistical Usage Testing Using Markov Chains". In ISSRE '04: Proceedings of the 15th International Symposium on Software Reliability Engineering, pages 54-65, Washington, DC, USA, 2004. IEEE Computer Society.

[21] J. Whittaker and M. Thomason. "A Markov Chain Model for Statistical Software Testing". IEEE Trans. Softw. Eng., vol. 20(10), pp. 812-824, 1994.

[22] K. Sayre and J. Poore. "A Reliability Estimator for Model Based Software Testing". In Proc. of the 13th International Symposium on Software Reliability Engineering, page 53, Washington, DC, USA, 2002. IEEE Computer Society.

[23] C. Campbell, W. Grieskamp, L. Nachmanson, W. Schulte, N. Tillmann, and M. Veanes. "Model-Based Testing of Object-Oriented Reactive Systems With Spec Explorer". Technical report, Microsoft Research, Redmond, May 2005.

[24] G. Tretmans and H. Brinksma. "Côte de Resyste: "Automated Model Based Testing". In Proc. of the 3rd PROGRESS workshop on Embedded Systems, Veldhoven, The Netherlands, pages 246-255, Utrecht, 2002. STW Technology Foundation.

[25] A. Dias Neto, R. Subramanyan, M. Vieira, and G. Travassos. "Characterization of Model-Based Software Testing Approaches". Technical Report ES-713/07, PESC-COPPE/UFRJ,2007.

[26] C. Heitmeyer, R. Jeffords, and B. Labaw. "Automated Consistency Checking of Requirements Specifications". ACM Trans. Softw. Eng.Methodol.,vol. 5(3),pp.231-261, 1996.

[27] F. Bouquet, B. Legeard, F. Peureux, and E. Torreborre. "Mastering Test Generation from Smart Card Software Formal Models". In Procs. ofthe Int. Workshop on Construction and Analysis of Safe, Secure and Interoperable Smart devices (CASSIS'04), vol. 3362 of LNCS, pp. 70-85, Marseille, France, March 2004. Springer.

[28] S. Prowell. "JUMBL: A Tool for Model-Based Statistical Testing". In HICSS '03: Proceedings of the 36th Annual Hawaii International Conference on System Sciences (HICSS'03) - Track 9, page 337.3, Washington, DC, USA, 2003. IEEE Computer Society.

[29] D. Cohen, S. Dalal, M. Fredman, and G. Patton. "The AETG System: An Approach to Testing Based on Combinatiorial Design". Software Engineering, vol.23 (7), pp.437-444, 1997.

Contributions in Mineral Floatation Modeling and Simulation

B. L. Samoila, M. D. Marcu
University of Petrosani
20 University Str.
Petrosani, Hunedoara, Romania

Abstract: **The paper deals with the mineral floatation study by modeling and simulation. Some of the main functions characterising the floatation process were simulated using Matlab Simulink programs. By analysing the results of these simulations and comparing them, we reached a conclusion concerning the optimising factors of the floatation duration. We also elaborated a Visual Basic Application which allows the calculation of quantities and contents in every point of a simple floatation circuit, for any number of operations. It's an easy to use, conversational application that allows studying more configurations in order to find the optimum one, sparing the researchers' and designers' time and effort.**

1. INTRODUCTION

The term "Computer Aided Process Engineering" (CAPE) summarizes computer applications for process design, process synthesis and process control. Flow sheet simulation [1] systems are a major application in this area. They calculate the whole plant performance, enable an easy comparison of process alternatives, and allow for a purposeful process synthesis by formulation of design specifications. In mineral processing engineering such programs are widely used and at least application of steady-state simulators for complex fluid processes has become state of the art [2].

A few commercial available simulators [3], [4], [5] have extensions to handle solids, but these solids capabilities are by no means as sophisticated as the capabilities for pure fluid processes. As a consequence the use of flow sheet simulation is not yet familiar in mineral processing and in chemical engineering process alternatives with solids are often ignored during process design, because of the lack of a useful tool to handle solids processing steps within the usual CAPE-environment.

Flow sheet simulation of solids processes faces special problems. On the one hand models for processing in particle technology are not derived from first principles, i.e. for each apparatus several models of different degree of sophistication and different application domains exists. On the other hand the usual information structure representing a process stream consists of pressure, temperature, components, and their partial mass flows. This structure is not sufficient for solids processes, where a particle size distribution must be considered and also additionally solid properties should be handled simultaneously, because further solid properties may be required by the process under consideration. Examples are many processes in mineral processing, processes of the gravel and sand industry, and soil-washing, where at least liberation states, fractional densities, or fractional contamination, respectively, must be considered besides the particle size distribution.

The aim of this research project is the development of tools for the modeling and simulation of complex solids process. Currently a flow sheet simulation system for complex solids processes with following objectives is developed:

- a process stream structure which describes solids not only by particle size distribution.
- an extensible model library for solids processing steps with models for different apparatuses, different application areas and with short-cut and rigorous approaches.
- an object-oriented software design.

The information structure for process streams consists of the usual fluid stream representation and "particle types", which allows distinction of different solids phases, e.g. ore and gangue components. Each particle type is described by its components, its size distribution and other user-defined size independent attributes. For each attribute at this level further dependent attributes may be described. All attributes may be described as distributions or by mean values.

Models with different degree of sophistication and for different application domains are provided and model parameters are adjustable, to take into account empirical or semi-empirical model derivations [6]. A well-structured model library is therefore essential. Within the model library a general unit operation model with an appropriate interface has been defined. This leads to a flexible structure, supports further extensions and allows a simple intern program communication. A specific model is derived from the general model by adding functions and data through stepwise specialization. Alternative, the structuring of necessary functions and data may take place in two distinct class hierarchies. The first hierarchy is then used exclusively for specialization of apparatus features, while the second hierarchy organizes specialization of modeling approaches. A concrete process model is then obtained by combining these two branches in one object.

2. FLOATATION TWO PHASE MODEL

Analyzing, optimizing and designing flotation circuits using models and simulators have improved significantly over the last 15 years.

K. Elleithy (ed.), *Advanced Techniques in Computing Sciences and Software Engineering*,
DOI 10.1007/978-90-481-3660-5_49,

In order to calculate a floatation technological flow sheet, we should previously know the following elements [7]:

- the quantities of material in the feed, output products and all intermediate points;
- valuable mineral contents in all of the formerly mentioned products;
- water quantities, that is dilution, for each product.

It becomes necessary to calculate the flow sheet in the following situations:

- to design an installation or a new section in an installation;
- to know the products characteristics all along a technological flow sheet, for a certain functioning regime.

From time to time, we must know the characteristics of an installation, in order to be able to modify or to correct the existing technological scheme without supplementary investments as well as to improve the technical and economical indexes. That may be needed when the raw material quality has been changed or the customers requests are different.

Our model considers the froth and the pulp as two distinct phases dynamically equilibrated (fig. 1).

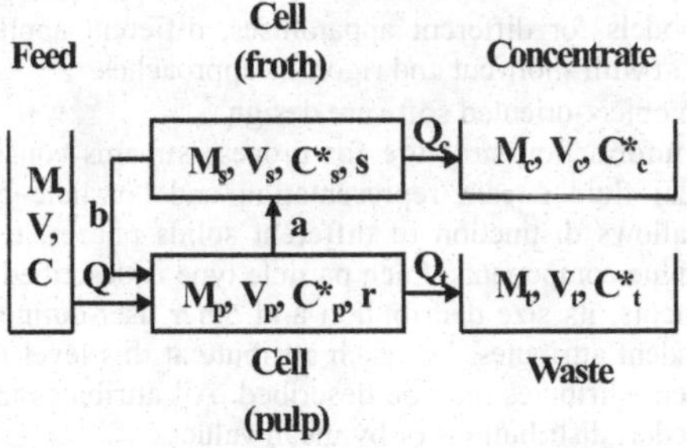

Fig. 1 Flotation cell block diagram

The significance of the notations in the figure 1 is:

M – floated material mass;

V – pulp volume, except air;

r, s – the ratio between the pulp volume (including air) and the airless one, respectively between the froth volume (including air) and the airless one;

Q – volumetric flow rate, except air;

C* - concentration (ratio between mass and volume unit, including the air in the pulp);

a - speed constant, regarding the concentrate transfer from the pulp to the froth;

b - speed constant, regarding the concentrate transfer from the froth to the pulp;

p, s, c, t – indexes, referring to the pulp, froth, concentrate and waste.

The differential equation reflecting the pulp composition change, for ideal mixing cells, is:

$$\frac{dM_p}{dt}=QC^*-rQ_tC_p-arV_pC_p+bsV_sC_s=\frac{dC_p}{dt}V_pr \quad (1)$$

The differential equation reflecting the froth composition change, for the same cells, is:

$$\frac{dM_s}{dt}=-sQ_cC_s-bsV_sC_s+arV_pC_p=\frac{dC_s}{dt}V_ss \quad (2)$$

We determined the transfer functions, which are:

$$H_1(s)=\frac{M_p(s)}{Q(s)}=\frac{k_1}{T_ps+1} \quad (3)$$

$$H_2(s)=\frac{M_p(s)}{C(s)}=\frac{k_2}{T_ps+1} \quad (4)$$

$$H_3(s)=\frac{M_s(s)}{Q(s)}=\frac{k_3}{T_ss+1} \quad (5)$$

$$H_4(s)=\frac{M_s(s)}{Q(s)}=\frac{k_4}{T_ss+1} \quad (6)$$

where we expressed the time constants T_p, T_s and the proportionality constants k_1, k_2, k_3, k_4, depending on V_p, V_s, Q_t, Q_c, Q, which have the significance according to fig. 1.

3. FLOATATION CELL SIMULATION USING MATLAB - SIMULINK SOFTWARE

We studied the mineral floatation of the raw material exploited in the Baia-Mare area, Romania, processed in the mineral processing plant from this town. We considered a primary floatation cell, for which we evaluated the characteristic constant values in the transfer functions (table 1).

Table 1. Values for time and proportionality constants in the transfer functions

Constant	Relation	Value
T_p	V_p/Q_t	0,004808
T_s	V_s/Q_c	0,025749
k_1	T_pC^*	0,003468
k_2	T_pQ	1,153920
k_3	T_sC^*	0,018572
k_4	T_sQ	6,179760

The response of the considered system was searched, for unit input signal of the feeding flow rate Q and of the concentration C*, as well as for "white noise" signal, in the case of C*.

The results of the simulation are presented in figure 2, figure 3 and figure 4.

Analysing the system response, we may conclude:

- the stabilising time for the pulp composition is 1.5-2 min., a sudden variation of the feeding flow rate or the feeding concentration being reflected in the system for about 2 min.; the concentration influence is more significant;
- the froth composition stabilisation time is about 6 min;
- the "white noise" stochastic variation of the flow rate influences more the froth composition.

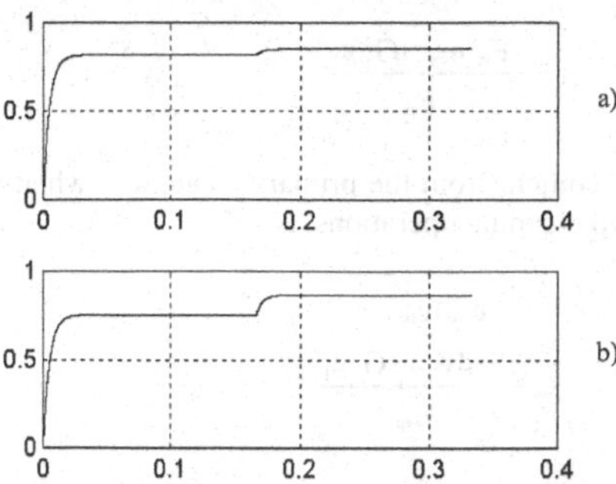

Fig. 2 System response concerning the pulp composition variation when the unit input is the feeding flow rate (a) and the feeding concentration (b)

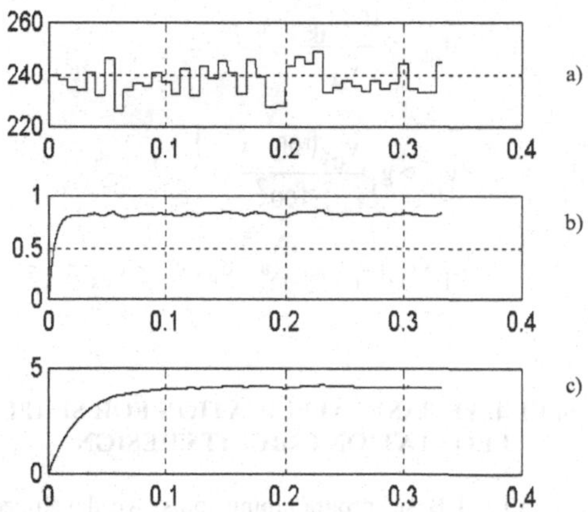

Fig. 3 System response to the stochastic variation of the feeding flow rate (a) concerning the pulp (b) and the froth (c) composition

The optimum duration of floatation in a cell can be determined using the transfer functions which allow studying the mechanism of the mineral particles transfer from the pulp to the froth and reverse, respectively the process stabilisation time.

Considering the estimating relations for the time and proportionality constants, it can be told that these ones depend on pulp flow rate, processed mineral composition, useful elements content in the raw material, weight recovery, aeration and dilution.

For a certain material and a given floatation technological line, our conclusion is that the optimisation parameters are the aeration and the feeding dilution.

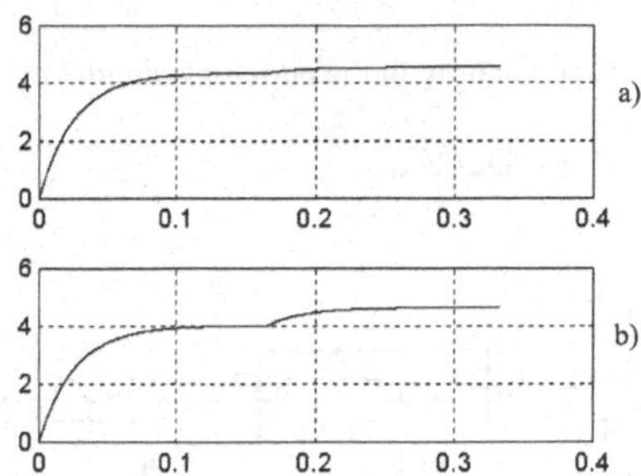

Fig. 4 System response concerning the froth composition variation when the unit input is the feeding flow rate (a) and the feeding concentration (b)

4. SIMPLE FLOATATION CIRCUITS MODEL

Using the balance equations [8], [9] written for different types of monometallic mineral simple floatation circuits, we calculated the quantities and valuable metal contents in each point of the circuit.

We expressed these parameters only depending on recoveries in weight for each operation (v_{ik}, v_{cj}), total weight recovery (v), feed quantity (A) and contents in: feed (a), concentrate (c), floated products of each enriching operation (d_k) and cleaning operation (g_j).

From our point of view, a floatation circuit can be represented as in figure 5, where the notation significance is:

k - enriching operation number;

j - cleaning operation number;

(x_k) - vector of enriching operation parameters;

(y_j) - vector of cleaning operation parameters.

Analyzing the calculation results, we observed the existence of some general relations. That was possible by noting the input and the output parameters in each block representing a floatation operation as in figure 6.

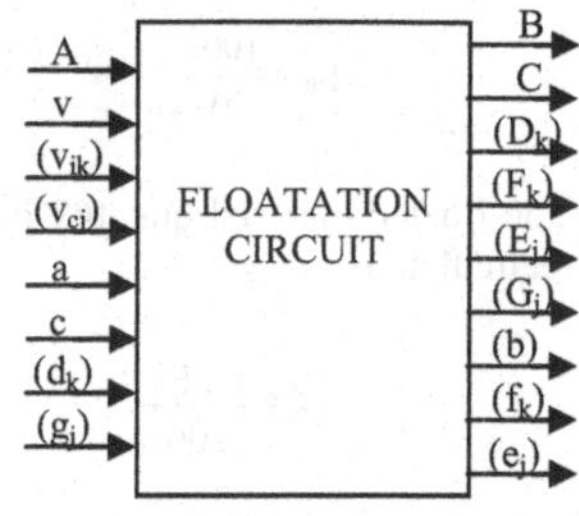

Fig. 5 Simple floatation flow sheet

The indexes significance is:

- "i" means "input";
- "e" means output.

It must be said that the operations were numbered from the

circuit end to the beginning, in order to express the relations in a more simple form [10].

The relation between the output material quantity and the input one, for an enriching block, is:

$$D_{ik}=\frac{100\,D_{ek}}{v_{ik}} \tag{7}$$

The recirculated quantity in the enriching circuit is:

$$F_k=D_{ik}-D_{ek} \tag{8}$$

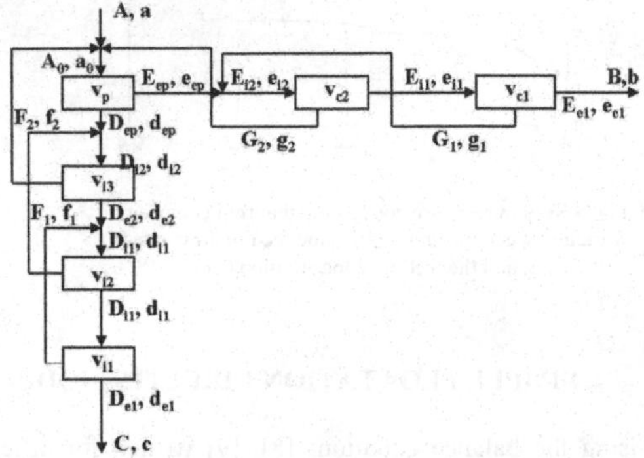

Fig. 6 Simple floatation flow sheet

The output quantity for the primary floatation is:

$$D_{ep}=D_{i,k-1}-F_{k-2} \tag{9}$$

The output material quantities from enriching blocks 2 and 3 (if they exist) are:

$$D_{e2}=D_{i1}\,;\quad D_{e3}=D_{i2}-F_1 \tag{10}$$

The relation between the input quantity for a cleaning block and the output one for the same block, k, is:

$$E_{ik}=\frac{100\,E_{ek}}{100-v_{ck}} \tag{11}$$

The floated material quantity in the cleaning circuit, which is recirculated, is:

$$G_k=\frac{v_{ck}\,E_{ik}}{100} \tag{12}$$

If there are two cleaning operations, the output 2 is equal to the input 1 and the waste quantity from the primary floatation is:

$$E_{e2}=E_{i1}\,;\quad E_{ep}=E_{i,k-1}-G_{k-2} \tag{13}$$

The waste content and the input content in an enriching operation are:

$$f_k=\frac{D_{ik}\,d_{ik}-D_{ek}\,d_{ek}}{F_k} \tag{14}$$

$$d_{ik}=\frac{D_{e,k+1}\,d_{e,k+1}+F_{k-1}\,f_{k-1}}{D_{ik}} \tag{15}$$

The input content for a cleaning operation is:

$$e_{ik}=\frac{E_{ek}\,e_{ek}+G_k\,g_k}{E_{ik}} \tag{16}$$

The waste content from the primary floatation, when there are one or two cleaning operations, is:

$$e_{ep}=e_{ik}$$

$$e_{ep}=\frac{E_{i2}\,e_{i2}-G_1\,g_1}{E_{ep}} \tag{17}$$

We identified also the restrictive conditions which have to be respected:

$$c<\frac{100a}{v} \tag{18}$$

$$d_{ek}<\frac{100d_{ik}}{v_{ik}} \tag{19}$$

$$e_{i2}>g_1\frac{v_{c1}(100-v_{c2})}{100^2} \tag{20}$$

$$g_j>e_{ej};\,g_j>g_{j-1};\,a<d_{ep}<\ldots<d_{ek}<\ldots<c \tag{21}$$

5. VISUAL BASIC APPLICATION FOR SIMPLE FLOATATION CIRCUITS DESIGN

Using Visual Basic programming tools, we developed an application that allows the calculation of any floatation circuit.

In the main window, the operator has to introduce the values of the input variables, including the number of both kinds of operations (figure 7).

The data are verified and, if they are not properly introduced, a message appears to worn the operator (figure 8).

After introducing the number of enriching operations, a window is opening to introduce the values of weight recoveries and contents in the floated outcomes (figure 9).

A library of flow sheet diagrams is included and, according

to the specified number of operations, the appropriate one can be visualized in order to see the specific notations (figure 10).

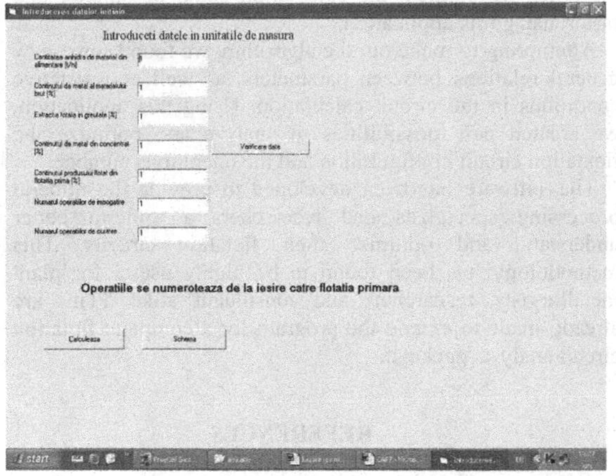

Fig. 7 Main window

After the operator introduces the number of cleaning operations, another window opens that allows him to introduce the weight recoveries and floated outcome contents in this kind of operations (figure 11).

The calculation results are displayed in a table (fig. 12).

Further development of this application let the data and the results be saved. Thus, the designer or the researcher can compare different sets of values in order to establish the optimum configuration of the technological line.

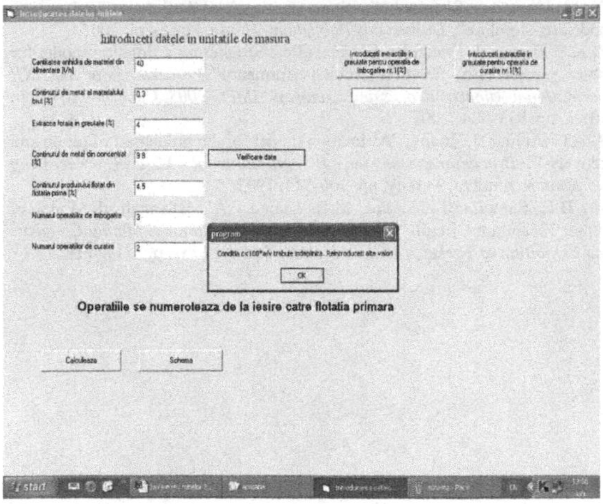

Fig. 8 Warning message

Using such an application, time can be spared in floatation technology design and the accuracy of the results is improved.

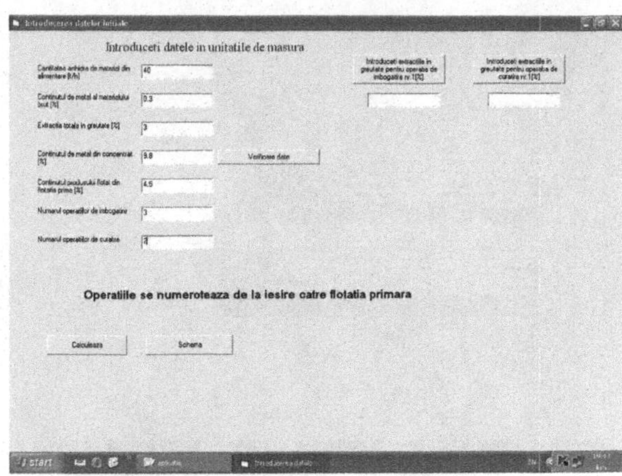

Fig. 9 Window for enriching operations

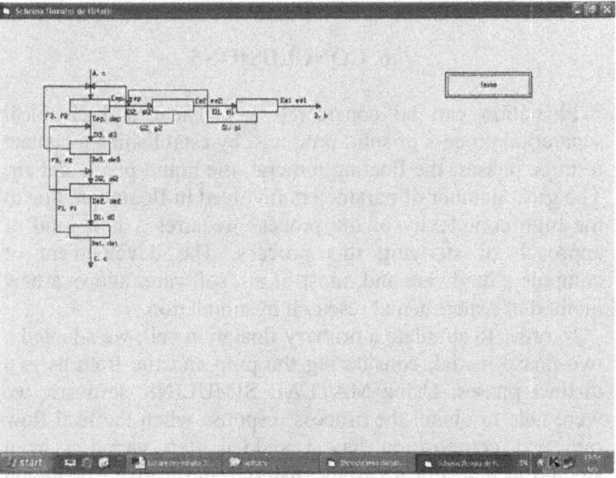

Fig. 10 Floatation flow sheet diagram

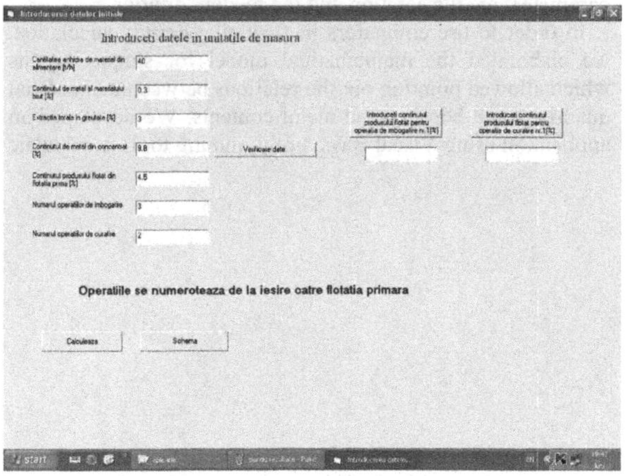

Fig. 11 Window for cleaning operations

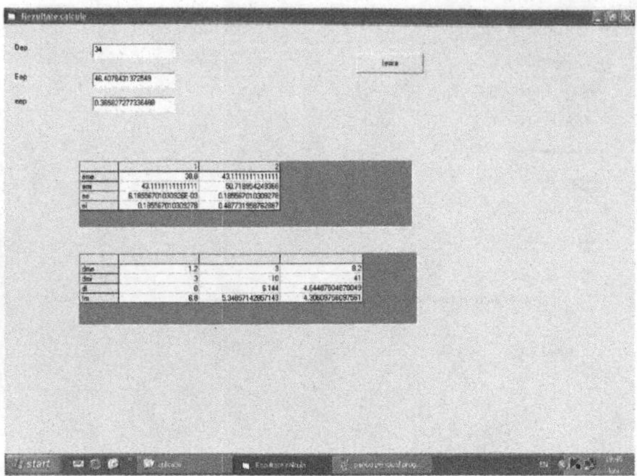

Fig. 12 Calculation results window

6. CONCLUSIONS

Floatation can be considered a physical and chemical separation process of solid products, by establishing a contact in three phases: the floating mineral, the liquid phase and air. The great number of parameters involved in floatation, due to the high complexity of the process, requires a new kind of approach of studying this process. The development of computers hardware and, most of all, software, allows a new method of experimental research by simulation.

In order to simulate a primary floatation cell, we adopted a two-phase model, considering the pulp and the froth as two distinct phases. Using MATLAB SIMULINK software, we were able to obtain the process response when the feed flow rate and composition has a sudden step variation or a stochastic one. For a certain material and a given floatation technological line, the conclusion was that the optimisation parameters are the aeration and the feeding dilution.

In order to use computers in floatation circuit calculation, we elaborated the mathematical model for simple circuits which allowed pointing out the relations between the material quantities and between the metal contents. We developed an application using Visual Basic programming tools to calculate any kind of circuit. As any real floatation circuit, no matter how complicated, may be decomposed in simple circuits, like the types we studied, the calculation could be simpler and faster, using this application.

Attempting to make out the algorithm, we found some new general relations between parameters, as well as restrictive conditions in the circuit calculation. Using this application, we created new possibilities to analyze and optimize the floatation circuit configuration and the operations number.

The software has been developed to provide the mineral processing specialists and researchers a tool to better understand and optimize their flotation circuits. This methodology has been found to be highly useful for plant metallurgists, researchers and consultants alike. Plans are already made to extend the program for a complete flotation circuit analysis package.

REFERENCES

[1] B.L. Samoila, "Introducerea tehnicii de calcul in conducerea procesului de flotatie a substantelor minerale utile", *Doctoral thesis*, University of Petrosani, 1999.

[2] M.C. Harris, K.C. Runge, W.J.,Whiten, R.D. Morrison, " JKSimFloat as a practical tool for flotation process design and optimization". *Proceedings of the SME Mineral Processing Plant Design, Practice and Control Conference*, SME, Vancouver, 2002, pp. 461-478.

[3] http://www.metsominerals.in.

[4] http://www.mineraltech.com/MODSIM.

[5] S.E.E. Schwarz, D. Alexander, "Optimisation of flotation circuits through simulation using JKSimFloat". *Proceedings of the 33rd International Symposium on the Application of Computers and Operations Research in the Minerals Industry. APCOM 2007*, Santiago, Chile, pp. 461-466, 2007.

[6] J. Villeneuve, J.C. Guillaneau, M. Durance, "Flotation modelling: a wide range of solutions for solving industrial problems", *Minerals engineering '94. International conference Nr.4*, Lake Tahoe NV, USA, vol. 8, pp. 409-420, 1995

[7] B.L. Samoila, "Sistemul tehnologic multivariabil de flotaţie. Traductoare, modelare, simulare", *Universitas Publishung House*, Petrosani, 2001.

[8] K.C. Runge, J. Franzidis, E. Manlapig, "Structuring a flotation model for robust prediction of flotation circuit performance" *Proceedings of the XXII International Mineral Processing Congress. IMPC 2003*, Cape Town, South Africa, pp. 973-984, 2003

[9] A.C. Apling, J. Zhang, "A flotation model for the prediction of grade and recovery." *Proceedings of the IV International Mineral Processing Symposium*, Antalya, Turkey, pp. 306-314, 1992

[10] B.L. Samoila, L.S. Arad, M.D. Marcu, „A Mathematical Model of Simple Floatation Circuits". *Proc. of the 17 th International Mining Congress and Exhibition of Turkey, IMCET*, Ankara, Turkey, 2001, pg. 711- 714.

Selecting the Optimal Recovery Path in Backup Systems

V.G. Kazakov, S.A. Fedosin
Mordovian State University
Bolshevistskaya street, 68
430005 Saransk, Russian Federation
E-mail: vitalykg@gmail.com

***Abstract*-This work examines the problem of creating a universal recovery path search algorithm in backup systems. The possibility of applying the graph theory is investigated. Various algorithms are examined and their selections are substantiated. Recommendations are presented for realizing the proposed method.**

I. Introduction

It is obvious that, for traditional backup schemes, the question of selecting an optimal path basically has no value. During the operation of, for example, an incremental backup algorithm, there exists a single path for recovery to any point, and its composition is elementary. The recovery path for a full backup is reduced into a singly constructed – completed copy of the data at the moment of recovery time.

Despite the fact that traditional approaches are used at the present moment universally, they represent the extremes in a selection of relationships between the basic characteristics of backup copying processes: the time of the copy's creation, the recovery time and the size of the repository for storing the copies. This appears to be an occasion making it necessary to doubt the effectiveness of the application of traditional backup schemes. Between the extremes of a full backup, requiring unacceptably large volumes of memory for storage, and incremental backup, with extremely low effectiveness of recovery, there are other backup algorithms which make it possible to find the necessary balance between the basic characteristics of the backup processes. See [1, 2, 3, 4].

During the operation of those more complex algorithms, there appears to be a need to find the optimal recovery path. For this, it is necessary to investigate the problem of the path's optimality and methods of their effective construction.

II. Problem Statement

Let us examine a certain backup system, which creates a copy of the data in successive moments of time *{t_k, k=0,1,2,…,T}*. Let us designate the data system for backup at a moment of time t_j as D_j. We will henceforth always assume that the initial state of the data D_0 – is empty, but D_l is not. Each created copy of data is stored in a certain data archive – repository. Each such copy will be designated as a repository's cell. A repository's cell, which contains the changes between the states of data from the moment t_l to the moment t_{l+n}, that is to say from D_l to D_{l+n} is designated as R_l^n or *R(l ,n)*, where *l* and *n* – are integers, while $n>0$, $l\geq 0$. Taking into account that the initial state of the data is empty, R_0^n actually denotes a full backup copy of the data's state at the moment of t_n, that is to say copy of D_n.

An operation to associate (in other words, to merge) the repository's cells makes sense. Not all cells can be associated with each other. It is not reasonable to associate cells containing data changes in nonsequential periods of time. Element *X* is introduced for the association of such unmergeable cells.

The set of all the repository's cells is designated *Rep*:

$$\boldsymbol{Rep} = \{X; R_i^j\}_{i=0,1,2\ldots;\, j=1,2,3\ldots} =$$
$$= \{X; R_0^1; R_0^2; R_0^3; \ldots; R_1^1; R_1^2; R_1^3; \ldots; R_k^1; R_k^2; R_k^3; \ldots\}.$$

The association operation of the repository's cells ⊕ in the set of all the repository's cells is introduced in the following manner:

$$1.\ X \oplus X = X;\ \ R_l^n \oplus X = X \oplus R_l^n = X;$$

$$2.\ R_i^{n_i} \oplus R_j^{n_j} = \begin{cases} R_i^{n_i+n_j}, \text{if } i<j \text{ and } j=i+n_i; \\ X, \text{otherwise.} \end{cases}$$

Notice that the introduced method to associate the repository's cells appears to be an algebraic associative operation.

The selection of the path for recovery to a certain moment of time t_j for a certain backup algorithm denotes the selection of the sequence of the cells from the repository, the

K. Elleithy (ed.), *Advanced Techniques in Computing Sciences and Software Engineering*,
DOI 10.1007/978-90-481-3660-5_50, © Springer Science+Business Media B.V. 2010

association of which in sequence, leads to the formation of a copy of data at the required moment of time t_j.

It is necessary to investigate the optimal recovery path selection problem for various backup schemes, to select a search algorithm which is universal as much as possible and at the same time sufficiently effective in the sense of computational complexity.

III. Presentation of the Backup Processing in Graph Form

The backup processing is conveniently represented in the form of oriented multigraphs. In this case, the state of the data is reflected in the form of vertices, and in the form of arcs – repository's cells, stored data about the change between the two states of data. In this case, it is convenient to assign numbers to vertices in a sequence corresponding with the moments of backup copy t_k. Thus, the beginning of an arc, representing repository cell $R(l, m)$, will have a vertex corresponding to D_l, and an end, corresponding to D_{l+m}.

The concept of a recovery path in such an interpretation is graphic and coincides with the concepts of path in oriented graphs. In this case, application of the graph theory becomes possible.

Let us illustrate the processing of an incremental backup, using a representation in the form of an oriented graph. See Fig. 1.

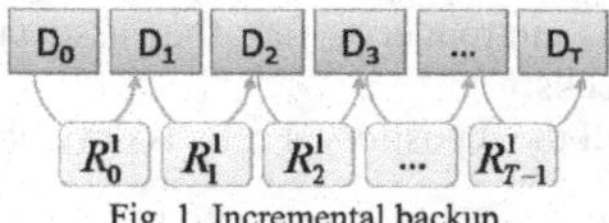

Fig. 1. Incremental backup.

And full backup will appear as shown in Fig. 2.

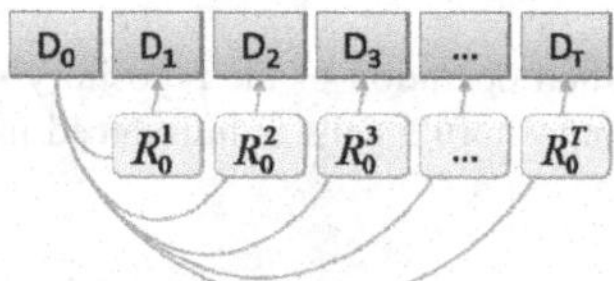

Fig. 2. Full backup.

Let us also illustrate the problem of associating repository's unmergeable cells by a simple example depicted in Fig. 3. Let us assume that the repository's cells were created, depicted in the figure, namely $R(0,1)$, $R(1,2)$ and $R(0,2)$. While, A^- in cell $R(1,2)$ indicates data about the deletion of A.

The association of unmergeable cells $R(0,2) \oplus R(1,2)$ is not equivalent to $R(0,3)= R(0,1) \oplus R(1,2)=\{C\}$. Although the necessary file C is contained in $R(0,2)$ and $R(1,2)$, but there is no necessary information about the deletion of B for accurate result.

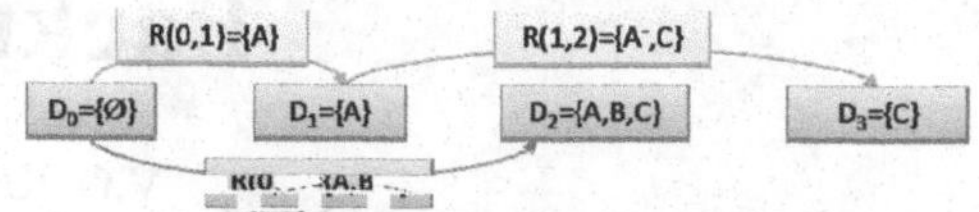

Fig. 3. Example of associating repository's unmergeable cells.

As an example of a nontrivial backup scheme, let us examine a multilevel backup (see Fig. 4). Copying is performed at several levels: full copies are created at level *0*, subsequently – files are copied, modified from the moment preceding a lower level backup copy.

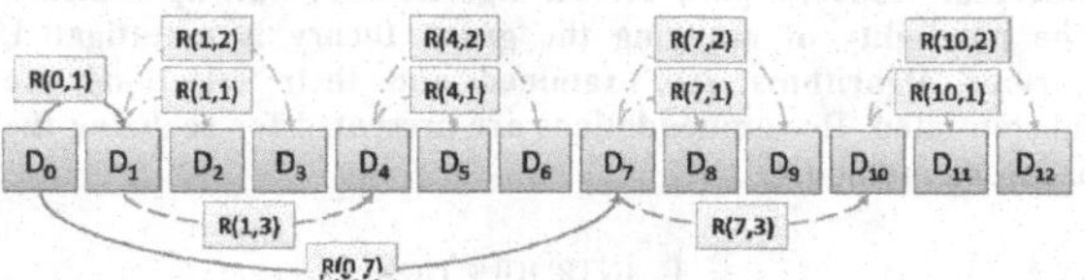

Fig. 4. Multilevel backup.

IV. Schemes with Explicit Duplication of Data

There is a possibility to reduce the number of cells in the recovery path, and thus also the recovery time, creating redundant repository cells. Let us examine the example represented in Fig. 5. This graph clearly shows that using redundant cells $R(0,2)$ and $R(2,2)$ it is possible to create a path twice as short from D_0 to D_4.

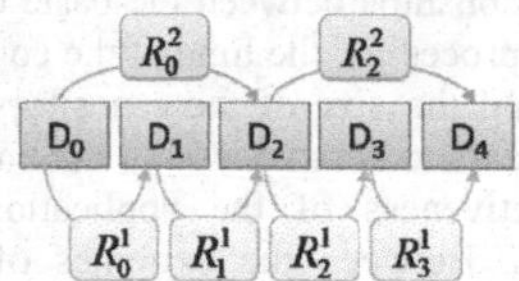

Fig. 5. Example of duplication.

A method for introducing redundant cells can be used for a substantial increase in the effectiveness of the recovery procedure [3].

The created duplicate cells can carry the index character. In this case, the creation of a redundant cell can occur without real duplication of the data.

The exclusion operation makes sense from a repository's created cell for changes attributed to a certain period, if it is known that, subsequently, this part of the data will not be required during recovery. Thus, it will be possible to reduce the volume of analysis during the recovery operation and thus increase its speed [4].

Let us examine an example which demonstrates this approach (see Fig. 6). At the moment of t_3, R_2^1 is created, and also a new cell is created from cell R_1^1 excluding the data about changes during the period from t_2 to t_3, included in R_2^1, a new repository cell is created during this designated $\dot{R}_1^1$. It possesses the special property: its association with R_2^1 gives a result as $R_1^1 \oplus R_2^1$, however $\dot{R}_1^1$ contains a smaller amount of data, and that means it is more advantageous for use than R_1^1. At the moment of t_4, data concerning changes during the period from t_3 to t_4 included in $\dot{R}_1^1$ is deleted, and a cell, designated $\ddot{R}_1^1$ is created.

The created duplicate cells can carry the index character. Several possible recovery paths appear.

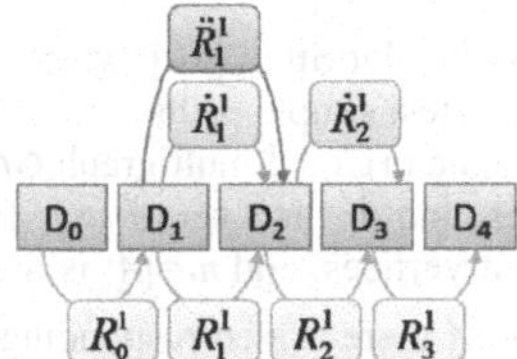

Fig. 6. Example of duplication.

It is evident that the most optimal path to D_4 appears to be the sequence $\{ R_0^1 ; \ddot{R}_1^1 ; \dot{R}_2^1 ; R_3^1 \}$.

It is necessary, however, to note that in a practical situation, when vertices in an orientated graph and connecting arcs turn into hundreds or even thousands, the location of an optimal path becomes not so simple of a task.

V. Repository Cells Association Operation

The speed of the repository cells association operation depends, first of all, on the number of repository cells under consideration in the recovery path. Less obvious appears to be the factor of number of contained files: the more files in the repository's cells, the more comparison operations needed to be carried out.

Significant processing time expenditures can be connected with the accessibility problem of different repository cells. Let us carry out this factor beyond the limits of examination, and we will assume that all repository cells are equally accessible. In this case, the difference in the recovery process work time will be determined by the computational speed of the recovery path's search algorithms and the operation's processing speed for determining the location of the necessary file versions.

VI. Size Factor of Repository Cells in the Recovery Path

An example illustrating the calculation's inefficiency for just one of the repository cell's minimum criterion in a recovery path is depicted in Fig. 7.

For recovery to the moment of time t_3 there are two paths: $P_1=\{R(0,1);\ R(1,2)\}$ and $P_2=\{R(0,2);\ R(2,1)\}$. Both paths consist of two repository cells each. However, for association of repository cells for P_1, it is necessary to analyze more on $\{A\}$ data, than for P_2. If we assume that $\{A\}$ is not a file, but a set of several thousand files, then the difference in the recovery procedure's processing time for the paths in question can prove to be significant.

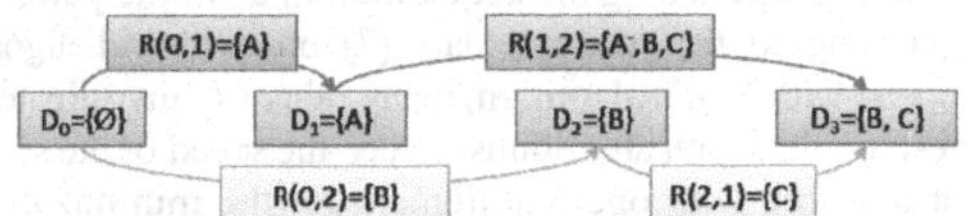

Fig. 7. Example of the importance of the repository cell size factor in a path.

Thus, the shortest path is necessary, which has the additional property of minimum total size of its repository's cells.

VII. Algorithm's Selection

A specialized algorithm for the location of the optimal path can be developed for each backup scheme taking into account its processing properties. It is practically necessary to know how to search for the optimal recovery path provided that some of the repository's cells can be proven to be inaccessible due to damage or other reason. This requirement further complicates their development.

It would be convenient to use a universal recovery path search procedure without relying on the selection of an algorithm for the creation of a backup copy.

It is possible to use the graph theory for solving the problem of locating the shortest path. A problem based on two named factors can be considered as a problem with some additional constraints or as a multipurpose problem. These complications will, generally speaking, strongly increase the computational work and, from a practical point of view, it will be simpler to find the shortest paths with the minimum length and to select among them that which possesses the necessary property [5].

The presence of cycles in the path indicates unsuitability for recovery, so therefore it is necessary to obtain simple paths.

There are many algorithms, in which there is a requirement for obtaining simple paths, however, as the research shows, this requirement considerably complicates the algorithm's work, and draws a notable increase in its processing time [6].

Usually, there is no possibility of getting a cycle in a recovery path for backup algorithms; however, in principal this is not excluded. Indeed there is a possibility to create the repository's cells which store the "reversed" changes relative to the previous state of data (e.g. reverse incremental scheme). In actuality, this needs to be considered in a universal recovery algorithm.

In that case, it makes sense to use hybrid algorithms. During the search for each subsequent optimal path, the initial working algorithm, which does not eliminate the appearance of cycles, checks the path found, and in case a cycle is detected, the optimal path is located with the use of another algorithm, guaranteeing the acquisition of a simple path.

According to the research data [7] on a hybrid algorithm composed with Yen's algorithm, many other (all investigated) K shortest loopless path algorithms exceed the speed of the search.

For a search with one requirement for the minimization of the path's length we will use Dijkstra's algorithm.

VIII. Dijkstra's Algorithm

In a general case, the method is based on adding to the vertices of temporary notations, so that the vertex's notation gives the path's upper boundary length from the beginning of path s to this vertex. These notations are gradually decreased with the aid of a certain iterative procedure, and at each step one of the temporary notations becomes permanent. The last one indicates that the notation is no longer the upper boundary, but gives the precise length of the shortest path to the vertex in question. The algorithm stops as soon as the notation, which belongs to the final vertex t of the desired path, becomes permanent. If we do not complete the algorithm's work on this, then the tree of shortest paths will be found from the beginning $s - T_s$ [5].

This algorithm can be applied for finding the tree of shortest paths T_t of all vertices to the final t. For this will sufficiently change the direction of all arcs to the direct opposite, and compute the final vertex – as the start [9].

Dijkstra's algorithm builds the shortest (s, t)-path for graph G in time $O(n^2)$, where $n=|G|$ – number of vertices [8].

IX. Observations on the Application of Dijkstra's Algorithm

During the algorithm's implementation in a recovery path search, the possibility of multiple arcs between two vertices needs to be considered. Therefore, when checking the vertices with permanent notations, all possible variants of the arcs need to be examined. But immediately during construction, paths consider not only the passage sequence of vertices, but also the arcs used in this case.

It is convenient to link with each vertex v one additional list of labels $\Theta(v)$, in which to carry the identifier indicated on v arc in (s, v)-path, having the minimal weight among all (s, v)-paths, passing through the vertices, the permanent labels acquired up to a given moment. There can be several such paths, therefore $\Theta(v)$ in general is a list.

After vertex t obtains its permanent label, with the aid of labels Θ, it is possible to easily indicate the sequence of vertices and arcs consisting of the shortest (s, t)-path.

Graph representation in a form of adjacency list is suited for the task.

Since it is required to find a path that is shortest in the sense of number of repository cells used, all weights should be considered equal to one.

X. Yen's Algorithm

Let us describe the algorithm [10], which makes it possible to find K of the shortest simple paths.

Let us examine the oriented multigraph $G(N, A)$, where N – is a set of vertices, and A – is a set of arcs. In this case $n=|N|$ – is the number of vertices, and $m=|A|$ is the number of arcs. For any arc $(x, y) \in A$ there is a corresponding weight $c_{xy} \in IR$.

Let $P^k = \{s, v_2^k, v_3^k, ..., v_{q_k}^k, t\}$ – be the k-th shortest (s, t)-path, where v_i^k is its corresponding vertex. Let P_i^k – be a deviation from path P^k at point i. The following is understood by this: P_i^k – is the shortest of the paths, coinciding with P^k from s to the i-th vertex, and then going to the next vertex along the arc, which is different from the arcs of those (previously already built) shortest paths P^j $(j=1,2,...,k)$, having the same initial sub-paths from s to the i-th vertex. P_i^k comes to vertex t along the shortest sub-path, while not passing through the vertices $\{s, v_2^{k-1}, v_3^{k-1}, ..., v_i^{k-1}\}$.

The algorithm begins to work from the location of shortest path P^1 with the aid of Dijkstra's algorithm. The next is to find the following optimal path P^2. There is a search for a collection of the assigned path's best deviations using Dijkstra's algorithm for this at each step. The shortest amongst found candidates P_i^1 will be P^2.

The next will pass to search P^3. And so on.

In order to find a more detailed description of Yen's algorithm, see for example [5, 10].

During the search for a recovery path, it is necessary to find all of the shortest paths, and the algorithm's work should be stopped, only when the following found path becomes longer than one already found. Then the optimum should be selected from the found paths for the second factor. For this, it is sufficient to find a path with the minimum total size of its repository's cells.

The algorithm's time complexity is evaluated to be $O(Kn^3)$.

XI. Hybrid Algorithm

The shortest (s, t)-path deviations P_i^k, which differs from P^k at vertex v_i^k, is searched in the form of $Root_{P^k}(s, v_i^k) \Diamond (v_i^k, j) \Diamond T_t(j)$, so that arc (v_i^k, j) belongs to none of the candidates P^j, found thus far. Where $Root_{P^k}(s, v_i^k)$ – is sub-path of P^k from s to v_i^k; the association operation of paths p and q, is designated as $p \Diamond q$, in this case the combined path is formed by path p and the following after it q.

The creation of each new candidate in this way is more effective, since the procedure for detecting the shortest path at each step is reduced to the problem of selecting the necessary arc (v_i^k, j), that can be made for time $O(1)$. However, this algorithm can return the path with cycles when $Root_{P^k}(s, v_i^k)$ and $T_t(j)$ contain the same nodes. In this case, the hybrid algorithm is switched over to use Yen's algorithm.

So that the selection of arc (v_i^k, j) would be fast, weights c_{xy} must be substituted by those reduced: $c^*_{xy} = c_{xy} - c(T_t(x)) + c(T_t(y))$ for all $(x, y) \in A$. So $c_{xy} \geq 0$ for any two vertices $(x, y) \in A$, and $c_{xy} = 0$ for $(x, y) \in T_t$, and the unknown arc, which begins with v_i^k, will be that which has the minimal reduced weight.

For the hybrid algorithm's work, it is necessary at the preliminary stage to determine T_t, and to replace the weights by those reduced.

In the worst case, the hybrid algorithm's work time will be identical to Yen's algorithm, i.e., $O(Kn^3)$, and in the best case, when there is no possibility of getting a cycle, you should expect $\Omega(Kn + n^2 + m\log n)$.

To find a more detailed description of the hybrid algorithm examined, see [7].

XII. Trial Application and Conclusions

The examined procedure for creating optimal paths was realized in the developed author's backup system, where different backup schemes were preliminarily introduced: full backup; incremental; differential; multilevel; schemes, realizing two methods of duplication, described above; and also other, experimental schemes. The trial application showed the usefulness of the described method. The universal search procedure's productiveness proved to be comparable with the work of specialized algorithms.

With the spread of the system to an idea of introducing new backup schemes, the task of developing specialized recovery algorithms does not stand now.

It is worthwhile to note that, from a practical point of view, it makes sense to disconnect the hybrid algorithm's procedure for trivial schemes and limit the use of Dijkstra's algorithm, in order to maximize productivity.

Acknowledgment

This work was supported in part by The Foundation for Assistance to Small Innovative Enterprises (FASIE).

References

[1] A. L. Chervenak, V. Vellanki, Z. Kurmas, and V. Gupta. "Protecting file systems: a survey of backup techniques", in proceedings, Joint NASA and IEEE Mass Storage Conference, 1998. www.isi.edu/~annc/papers/mss98final.ps

[2] Z. Kurmas, A. Chervenak. "Evaluating backup algorithms", proc. of the Eighth Goddard Conference on Mass Storage Systems and Technologies, 2000. http://www.cis.gvsu.edu/~kurmasz/papers/kurmas-MSS00.pdf

[3] A. Costello, C. Umans, and F. Wu. "Online backup and restore." Unpublished Class Project at UC Berkeley, 1998. http://www.nicemice.net/amc/ research/backup/

[4] Z. Kurmas. "Reasoning Behind the Z Scheme," 2002. http://www.cis.gvsu.edu/~kurmasz /Research/BackupResearch/rbzs.html

[5] Christofides, N., Graph Theory--An Algorithmic Approach, Academic Press, London, 1975.

[6] D. Eppstein. "Finding the k shortest paths", 35th IEEE Symp. Foundations of Comp. Sci., Santa Fe, 1994, pp. 154-165. http://www.ics.uci.edu/~eppstein/pubs/Epp-TR-94-26.pdf

[7] M. M. B. Pascoal, "Implementations and empirical comparison for K shortest loopless path algorithms," The Ninth DIMACS Implementation Challenge: The Shortest Path Problem, November 2006. http://www.dis.uniroma1.it/~challenge9/papers/pascoal.pdf

[8] Dijkstra E. W. (1959), "A note on two problems in connection with graphs," Numerische Mathematik, 1, p. 269.

[9] E. Q. V. Martins, M. M. B. Pascoal and J. L. E. Santos, "The K shortest paths problem," CISUC, 1998. http://www.mat.uc.pt/~marta/Publicacoes/k_paths.ps.gz

[10] J. Y. Yen. "Finding the K shortest loopless paths in a network," Management Science, 17:712–716, 1971.

Telemedicine Platform Enhanced visiophony solution to operate a Robot-Companion

Th. Simonnet, A. Couet, P. Ezvan, O. Givernaud, P. Hillereau
ESIEE-Paris,
Noisy le Grand – FRANCE
Email: {t.simonnet, coueta, ezvanp, givernao, hillerep}@esiee.fr

***Abstract* - Nowadays, one of the ways to reduce medical care costs is to reduce the length of patients hospitalization and reinforce home sanitary support by formal (professionals) and non formal (family) caregivers. The aim is to design and operate a scalable and secured collaborative platform to handle specific tools for patients, their families and doctors.**
Visiophony tools are one way to help elderly people to have relationships with their family or caregivers. But it is possible to use the same tool to remote control a Robot-Companion.

I. Introduction

In its 2002 European Health Report [1], the World Health Organization recommended hospital services reorganization, with an increased role for substitution between different levels of care, strengthening primary health care services, increasing patient choice and participation in health services and improving outcomes through technology assessment and quality development initiatives. According to these recommendations, the number of tele-surveillance implementations and pilot experimentations is growing in Europe, especially within the Northern and Scandinavian countries such as Norway, Finland, Sweden and Germany. A leading example in this field is the Philips Healthcare Centre in Düsseldorf, Germany.

II. Telecare platform

ESIEE-Paris (Ecole Supérieure d'Ingénieurs en Electronique et Electrotechnique) develops, operates and maintains a Telemedicine platform that offers medical and communication services to patients, especially for elderly people. This platform has two main components - central server and local equipment (PDI, Passerelle Domestique Internet - Domestic Internet Gateway) - both communicating using a secured IP Network over Internet (VPN).
Such a platform is easy to deploy, because all functions and related virtualized servers can be held on one physical server. It is actually operated at ESIEE-Paris but the aim is to install it in any hospital that needs it. Each patient, at home, will have a PDI that provides Internet services and records operational data such as agendas and monitoring data to prevent temporary Internet failure.

On telemedicine collaborative projects, it is useful to have at any time an operational platform to handle new developments and to evaluate the impact of each functionality with different patients. It is then possible to modify these functions according to medical protocols. Today one of the critical functions to deploy is visio-conferencing to keep the social link between patient, family and care holders.

The situation of health services, nowadays, makes it quite difficult to satisfy all needs and to maintain care quality without strongly increasing medical care's costs. One of the ways to reduce medical care costs is to reduce the length of patients' stay in hospitals, to reinforce home sanitary support by formal (professionals) and non formal (family) caregivers and to facilitate home hospitalization. Staying at home is a way to maintain the social link for people who need frequent medical care or constant surveillance in their everyday life, such as people suffering from cardiac or Alzheimer diseases, and/or frail elderly patients, as they could be separated from their entourage, due to their prolonged stay in a medical centre.

A telemedicine platform must integrate communication functions (like mail, instant messaging, visio-conferencing, agendas, and alarms) with specific medical tools (like cognitive exercises, patient monitoring and in a near future robot remote operation and increased reality, emotion recognition). Its scalability must allow further developments without building a new platform for each new project. This platform is the result of several telemedicine projects started in 2000 with the Mediville project [2], followed in November 2003 with the RNTS (Réseau National Technologies pour la Santé) project TELEPAT [3], [4] and by the TANDEM project (http://tandem.memosyn.org). The platform was designed in 2007 as a full collaborative platform for patients, family and medical people (geriatrists, paramedics) to integrate previous developments. This telemedicine platform is already available and operates main functions; but it doesn't operate yet all the

K. Elleithy (ed.), *Advanced Techniques in Computing Sciences and Software Engineering*,
DOI 10.1007/978-90-481-3660-5_51,

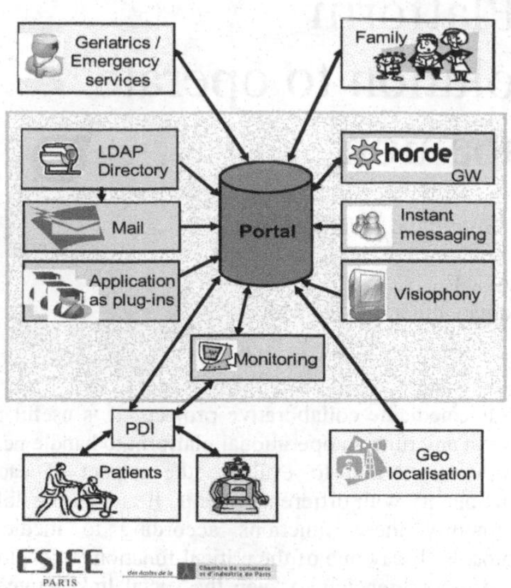

Fig. 1. Telemedicine platform

advanced and secured functionalities as simplified GUI for patients, anti-spam filter [5] and video bandwidth management [6]. It will integrate, before the end of the year, monitoring and geo-localization of elderly people.

A. Architecture

This platform (Fig. 1) is Open Source based and is available for two kinds of hardware platforms using both Debian Gnu/Linux: Xeon and Itanium 2 processors. This last one offers mathematical capabilities, useful for security and calculation purposes.

The platform architecture is Xen based. The use of Xen as para-virtualization tool is a means to keep a running platform over all the project developments. There is one virtualized server for each main function. It is an easy and costless way to manage and test versions; creating a new virtual server consists of copying some files without changing anything on other servers. It is also a way to manage platform scalability. When a more powerful server is needed, it is possible to add a new real server, migrate existing virtual servers or create new ones. Virtual servers can be 32 or 64 bits ones, running different version of Linux distribution, depending on developers needs.

B. Horde

Horde is a web based groupware, which was chosen in order to provide a set of basic applications: mail, calendar, address book and task manager accessible through a web browser. These applications are needed by both the patients and the medical staff. Task manager is useful for the monitoring of medical treatment. The calendar is helpful to plan exercise and appointments.

User management is a challenge since critical applications like geo-localization should only be accessed by medical staff and other like memory stimulation by patients. Horde permissions management system is used in this scope to manage application access based on the user's group.

Furthermore Horde also provides a framework in which to develop web applications which allows easy integration of specific applications developed to suit our needs in the groupware.

One of these needs is to provide a user-friendly and intuitive interface for elderly patients. Horde is not easy to use and our aim is to create a dynamic interface suitable for each user, patient or medical staff. An application has been developed permitting authorized users (doctors) to choose the interface for each group of users. Whilst a patient suffering pre-Alzheimer disease would have a simple custom interface, the medical staff would have the standard one. Each patient has a different level of comprehension of the interface, so the patient group can be divided into subgroups, each one having a different interface associated to it.

Horde provides the interface to the platform with the integration of all the applications, permitting access to most of the platform from a simple web browser.

III. Communication

It is necessary for pre-Alzheimer patients to keep a strong social link with people they know, such as family thus a great video quality visio-conferencing service is necessary. For a home based solution, ad-hoc wireless peripherals are convenient. The use of a motion compensated spatio-temporal wavelet transformation ([7]) can be a solution to keep the best video quality and to avoid latency and jitter. A set of optimized algorithms will be deployed on both the server side and the PDI side to improve quality and fluidity. In the near future, an emotion recognition tool will be also implemented.

There are two implemented solutions that work simultaneously: a lightweight, OpenMeetings based one and a more complex one using a central Asterisk PBX (Private Automatic Branch eXchange). Both solutions can be used for visio-conferencing, but they have different architectures and don't satisfy the same needs.

A. OpenMeetings

A Red5 Open Source flash® streaming server is hosted on a virtual server. Visio-conferencing is only a web solution and an adapted GUI (Graphical User Interface) can be easily implemented. The main advantage is that there is no need to install anything but a browser with flash® plug in on a client PC. But this solution can't allow specific QoS support. Therefore, it will be difficult to use the same

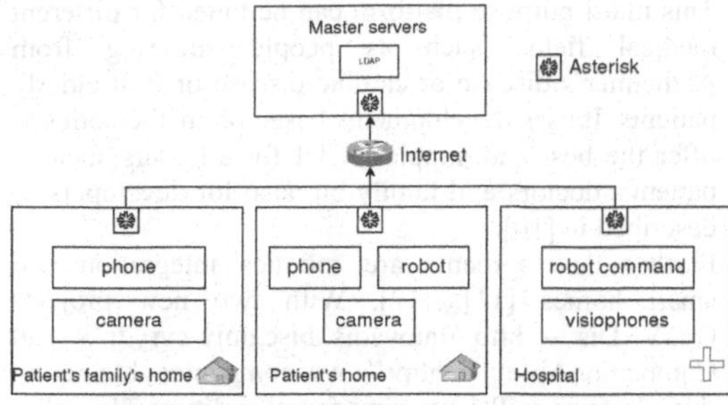

Fig. 2. Asterisk servers distribution schematic

communication channel with a future robot-companion. This solution is the good one to test different patient graphical interfaces for this service.

This solution was stopped mostly because Broca Hospital feedback [8] shows that elderly patients do not like OpenMeetings; only 42% of them said they would like to use Open Meetings as a visio-conference solution. OpenMeetings has also several limitations. It is difficult to provide an efficiency QoS (Quality of Service) support, flash does not allow a large bandwidth and this product can't be used for handling teleoperation functions .

B. *Asterisk*

i. SIP (Session Initiation Protocol) servers

Each patient will be equipped with a SIP phone (visiophone or softphone) and with one or more web cams. Robot will have its own equipment (softphone and cams) and will have its own SIP line. However SIP protocol identifies each line by its IP and then it is necessary to allow domestic equipment to manage more than a line behind Internet gateway. Two solutions are available to solve this. - Using a STUN [RFC 3489] server which implies restraining to UDP and non symmetric NAT. - Installing a local SIP server on each private network. This approach allows us to avoid all problems caused by the the different networks architectures (proxy, NAT, firewall). The only thing to do is to open the SIP port on the gateway allowing the local server to connect to the master one. In addition, having a local Asterisk allow to easily manage SIP phone lines which are often given with Internet services.

In order to keep the local configuration as simple as possible, the major part of it will be kept on the global server. Information from the authentication and dial-plan will be kept respectively on the LDAP and the master server (Fig. 2).

Asterisk was chosen as SIP server because it is OpenSource, well documented and has a lot of users. It also offers services like multilanguage voice-mail.

ii. SIP clients

SIP client has to provide support for visio-conferencing with a high quality: low latency and jitter. A complex system of QoS [9] has to be established. The SIP client will be able to dynamically balance the compression ratio and the quality of the different channels (video and audio) to adapt to the situation. Among other, load balancing of band with between sound and video stream. For example monitoring does not need a high quality of sound but a low latency, whereas the priority in a visio-conference is the voice.

Actually, Asterisk does translate codecs however it can use an external tool. This practice is inadvisable because of its high CPU cost. That is why the choice of the codecs and the soft-phone is critical. The SIP protocol permit to auto-negotiate the codec for encoding the video and audio. Ekiga, to choose the right codec, take in consideration the bandwidth and the latency to obtain the best quality.

Ekiga will be used as SIP client because :

- It is Open-Source and allows easy modification;
- It is portable (Linux, Windows);
- Of its comprehensiveness (support of h264, theora, PCMU, GSM,...);
- It offers an instant messaging channel.

Furthermore, the use of the libglade allows us to load a new GUI (Graphic User Interface) from XLM files. Those are created with glade (the Gnome Gui builder).

IV. ENHANCED COMMUNICATION PLATFORM

In the future smart homes and robot companion will be integrated to the patient home in the scope of two projects, QuoVadis and CompanionAble.
Handling robot companion and smart homes involves deploying new tools. Instead of creating a new platform it will be more efficient to use the existing one.

Due to this fact, Asterisk server has to be modified to allow opening a new channel. This channel will be first used to assign pre-programmed missions to robot by sending dial tone pulse to it. Since some SIP clients like Ekiga already send dial tone pulse when a button is pushed, it will be easier to do it.

The main challenge is to manage communication between a distant operator and the robot. Tele-operator will need to have the ability of taking control of the robot. One way of doing this is to integrate the robot as a virtual and automated user which can be called from a classical telephony client. Therefore, tele-operator will only have to call the robot to control it by sending pulses on line. Several IP cameras will be used to allow operator to switch between camera views.

The second step will be to create a new bidirectional communication channel to have a real robot remote control. Operator will send orders though keyboard or specific peripheral like joystick. Voice and video will be sent using classical VoIP mechanism. Orders will go through Asterisk using SIP protocol to robot softphone that will transmit these orders to the robot control unit. Then every sensor values will be sent back through Asterisk using the same protocol. Latency and jitter parameters must be real time managed have an acceptable QoS.

But since the elderly user also needs to communicate with his doctor and his family, a simple way of doing this is needed. Due to the fact that Ekiga SIP client has a complex user interface for elderly patients, its' interface has to be redesigned. This interface should be intuitive, pre-programmable and dynamically modifiable. Depending on which user uses it, a simpler one will be loaded for a pre-Alzheimer patient and the complete one will be loaded for an operator user.

The milestones of this project are :

- End of 2008: basic Asterisk configuration with dial plan, ekiga softphone (both linux and windows implementation), pulse tone transmission between 2 clients. Interface will be re-designed for elderly people.
- Mid 2009 : new softphone with 3 bidirectional channels (voice, video and data)
- End 2009: order transmission to the robot
- Mid 2010: Graphical GUI to pilot a robot through keyboard
- End 2010: specific interface for data channel to plug a specific peripheral like joystick or joypad and to connect robot control unit.

V. CONCLUSION

The setting up of a telemedicine platform will be an evolution in the treatment of disease where patients need, if they do not already have to stay in a hospital, to regularly see a doctor to control the progress of their cure. Firstly, by making the presence of a doctor optional during routine tests, it will release time for other more important consultations. It will also be beneficial for the patient: it is easier to find a few minutes to go on the web and do the control exercise than about an hour to go to the medical centre, wait for the appointment, do the exercise and return home. Since the PDI will provide as much security as we might find in a hospital, people will be able to stay at home without the need of the presence of medical care.

Due to the time the patient spends on his treatment, it's hard for him to maintain a social link with his friends or his family and this is where the second aspect of this project, the communication, becomes important. Moreover, this keeps the 'human contact' with doctors which is lost with the automation of routine tests.

This multi-purpose platform can be tuned for different medical fields such as people suffering from Alzheimer's disease or cardiac disease or frail elderly patients. It uses developments based on mathematics to offer the best and simplest GUI for all users such as patients, doctors and family but also for developers as described in [10].
Further developments are robotics integration and smart homes [11], [12]. With two new projects QuoVADis – http://quovadis.ibisc.univ-evry.fr - and CompanionAble – http://www.companionable.net -, this platform will have new functionalities: The robot (as a companion) helps patient in some activities like handling pill boxes, relaying alarms, relaying operator or doctor voices [13], [14], visualizing the home using augmented reality, emotion recognition, etc.

AKNOWLEDGMENT

ESIEE-Paris team works in close cooperation for platform validation, especially for Alzheimer disease people with:

- Broca Hospital (Assistance Publique Hôpitaux de Paris – leader of Tandem project)) for content generation for cognitive stimulation and day-time management as well as ethics,
- INT for monitoring and sensors,
- IBISC, Université d'Evry, leader of QuoVadis project,
- Private companies like Legrand and SmartHomes for smart homes,
- University of Reading, leader of CompanionAble project.

Some research leading to these results has received funding from the European Community's seventh Framework Programme (FP7/2007-2013) under grant agreement n° 216487.

REFERENCES

[1] World Health Organization, 2002, *The European Health Report*, European Series, N° 97.

[2] Baldinger J.L. et al., 2004, Tele-surveillance System for Patient at Home: the MEDIVILLE system, *9th International Conference, ICCHP* 2004, Paris , France, Series : Lecture Notes in Computer Science, Ed. Springer, July 2004.

[3] Lacombe A. et al., 2005, Open Technical Platform Prototype and Validation Process Model for Patient at Home Medical Monitoring System, *Conf. BioMedsim*, Linköping, Sweden.

[4] Boudy J. et al. 2006 Telemedecine for elderly patient at home : the TelePat project. *International Conference on Smart Homes and Health Telematics,*Belfast UK.

[5] Robinson G., 2003, A Statistical Approach to the Spam Problem, *Linux Journal*, 1st March 2003.

[6] Nelwamondo S. G., 2005, Adaptive video streaming over IEE 802.11, Master of Technology, F'SATIE, September 2005.

[7] Torrence C., Compo G.P., 1998, A practical guide to Wavelet Analysis, *Bulletin of the American Meteorological Society*, January 1998.

[8] Faucounau V., Ya-Huei Wu, Boulay M., Maestrutti M., Simonnet Th., Rigaud A.-S., 2008, Apport de l'informatique dans la prise en charge médico-psycho-sociale des personnes âgées, AMINA 2008, Monastir Tunisia.

[9] Serra A., Gaiti D., Barroso G., Boudy J., 2005, Assuring QoS differentiation and load balancing on Web servers clusters, *IEEE CCA conference on Command Control.*

[10] Scapin D.L., Bastien J.M.C., 1997, Ergonomic criteria for evaluating the ergonomic quality of interactive systems, *Behavior and Information Technology*, N°17 (4/5).

[11] Noury N., Virone G., Barralon P., Rialle V., Demongeot J., 2002, Du signal à l'information : le capteur intelligent – Exemples industriels et en médecine, Mémoire d'Habilitation à Diriger des Recherches, avril 2002.

[12] Noury N,., 2004, Maisons intelligentes pour personnes âgées : technologies de l'information intégrées au service des soins à domicile", *J3eA - Vol. 3, Hors-Série* pp1 – 20.

[13] Topp E., Huettenrauch H., Christensen H., Severinson Eklundh K., 2006, Bringing Together Human and Robotic Environment Representations – A Pilot Study, IROS, October 2006.

[14] B. Burger, I. Ferrané, F. Lerasle, 2008, Multimodal Interaction Abilities for a Robot Companion Int. Conf. on Computer Vision Systems (ICVS'08), Santorini, Greece.

Some Practical Payments Clearance Algorithms

Deniss Kumlander
Department of Informatics, TTU
Tallinn, Estonia
e-mail: kumlander@gmail.ee

***Abstract*—The globalisation of corporations' operations has produced a huge volume of inter-company invoices. Optimisation of those known as payment clearance can produce a significant saving in costs associated with those transfers and handling. The paper revises some common and so practical approaches to the payment clearance problem and proposes some novel algorithms based on graphs theory and heuristic totals' distribution.**

Keywords-payments clearance; netting; combinatorial optimisation; heuristic

I. Introduction

Globalisation of the world has opened new markets for many organisations and stimulated a lot the growth of multinational conglomerates and extremely large corporations owning branches and independent sub-organisations all over the world. A process of merging concerns into even larger corporations constantly increase the number of companies in groups and the depths of structure producing up to 4 000 companies and more inside one legal unit. Companies included into the same corporations share a lot of activities producing a huge amount of inter-company invoices between those. They often use others companies of the same group as subcontractors, which increases the volume of payments even more. There are different extra costs associated with each transaction multiplied by the high volume of those forcing corporations to deal with optimisation techniques in order to save quite a sufficient amount of money in total. Minimisation of transfers will sufficiently reduce the overall cost [1] by eliminate currency exchange costs, transfer costs, reduce time spent by the accounting department personnel to process those invoices and so forth.

A similar situation has been produced by the growing popularity of Internet and different web services correspondently. It has produced a high volume of sales over web generating a need to handle efficiently the billing process, which is challenging as extremely high volume of transactions is generated by quite low value payments and so the charge of each transaction, especially of cross-countries and banks transactions, add quite a sufficient extra cost to each payment.

The netting process (or payments clearance) is a process of optimising inter-group payments. Normally the number of transfers is reduced in the result of netting as well as the total amount paid between participants of this group, which are either companies belonging to the same legal group or having an agreement to participate in such netting scheme (directly or indirectly – for example, if banks have executed this scheme to reduce payments on the bank-end, i.e. invisible to actual parts involved into each payment transaction). The same process is also known as the bank clearing problem [2].

II. Netting

There are several optimisation goals. The first one asks to minimise the total transferred amounts. Let say we have a netting system $NS = (V,E,W)$, where V is a set of business units involved into the netting process, E is set of transfers between those and W is a set of amounts corresponding to each transfer. In this case the goal will be to minimise $\sum w_{ij}$, *where* $i, j \in V$.

Obviously the V set is a stable set that doesn't change during the netting process, while E and W are constantly updated.

The second optimisation goal asks to minimise the total number of payments that should be done to transfer all required amounts. Notice that this goal is always secondary, i.e. should be targeted only after the first minimum is acquired and should not corrupt it.

This goal can be described as: minimise $|E|$ constrained by $\min(\sum w_{ij})$.

There could be extra constraints defined by specific commercial environments complicating earlier stated goals. That is the main reason why the most often used netting method is netting center one: in order to avoid extra compilations, each company simply sum up inflow and outflow by currencies and do correspondent payments to the netting centre or receive payments from there. It is also possible to ignore currencies and do payments in the receiver company currency.

Finally it should be mentioned that the problem is NP-hard [3, 4] as this optimisation task can be easily transformed into the satisfiability problem or any other NP-complete class problem. Therefore in practise the payment clearance is solved by heuristic algorithms – algorithms that doesn't guarantee finding an optimal solution, but will generate a good enough time in a polynomial time.

III. Netting Based on Graph Theory Approaches

In case of the netting based on graph theory approaches the first step in the process is to transform the current payments' net into a graph. Companies are represented as vertices and payments as edges connecting those. The graph is weighted as each edge has a weight, which is the amount to be paid by the invoice corresponding to this edge. Notice that the graph is

K. Elleithy (ed.), *Advanced Techniques in Computing Sciences and Software Engineering*,
DOI 10.1007/978-90-481-3660-5_52, © Springer Science+Business Media B.V. 2010

oriented (directed) as the amount should be transferred not just between companies, but from one company to another.

Below a new approach called transactional closure rule based netting (where the second step is more or less novel – as is a well known graph theory technique) is be introduced. The algorithm is compiled using two major rules to be used: a summing up rule and a transactional closure rule, which are executed in cycle until no more transformations can be done.

A. *Summing up rule*

Consider for example the following case

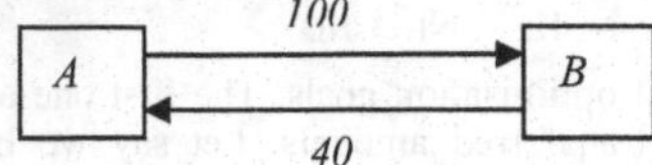

Figure 1. Summing up rule: state before

There are two companies (*A* and *B* correspondingly) and a set of transactions to be made between those in different directions. The summing up rule asks to sum up those transactions into one transaction with respect of the signs and produce only one in the result. Figure 1 shows an example. Here we have two transactions: 100 from *A* to *B* and 40 from *B* to *A*. It can be represented as 100 from *A* to *B* and -40 from *A* to *B* (notice the sign of the second transaction). Summing up those amounts we will have 60 from *A* to *B*.

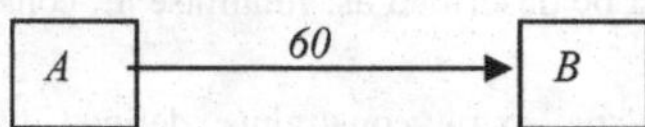

Figure 2. Summing up rule: state after

Generally saying the rule says that if there are more than one invoice between any two companies (*A* and *B*) that all those invoices should be summed up into one invoice considering a direction of each transaction (positive amount from *A* to *B* and negative from *B* to *A*). The total sum should be paid as the total's sign defines: if the total is negative then from *B* to *A*, otherwise from *A* to *B*.

Some organizations do use netting systems based just on this rule, i.e. they are not minimizing amounts to be transferred and transactions to be executed in global scope, but do it only between individual companies.

B. *Transactional closure rule*

This rule is applied after the summing up rule. The general rule says that if three companies (*A*, *B* and *C*) are organized in such chain by payments that company *A* has to pay to company *B* and company *B* has to pay to the company *C* then transactions can be transformed in order to minimise the transferred amount using the transactional closure rule (presented below).

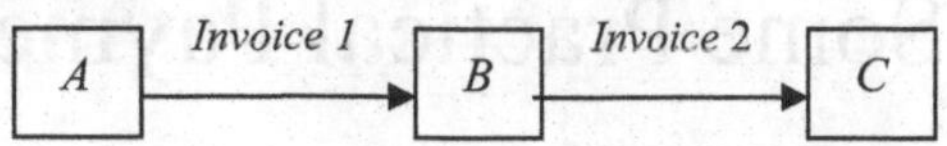

Figure 3. Trnsaction closure chain

If company *A* has to pay to company *B* and company *B* to company *C* then both existing invoices will be removed and two new will be produced. An amount of the first new invoice will equal to the smallest amount between *Invoice* 1 and *Invoice* 2: min(*Invoice1*, *Invoice2*) and will be paid by company *A* to company *C*. The second invoice will be produced instead of the invoice that had the maximum amount among those two: max(*Invoice1*, *Invoice2*). This second invoice amount will be equal to the difference between those invoices' amounts.

In other words invoices will be transformed by the following practical rules:

- If an amount of the *Invoice* 1 is larger than an amount of the *Invoice* 2 then the *Invoice* 2 is eliminated and the amount of the *Invoice* 1 will be decreased on the amount of the *Invoice* 2.
- If an amount of the *Invoice* 2 is larger than an amount of the *Invoice* 1 then the *Invoice* 1 is eliminated and the amount of the *Invoice* 2 is decreased on the amount of the *Invoice* 1.
- If invoices' amounts are equal then produce eliminate Invoice 2 and forward Invoice 1 to company C.

Consider the following two examples:

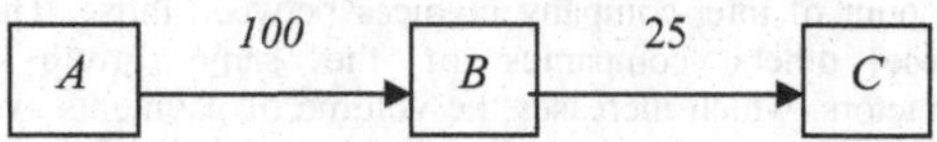

Figure 4. State before appliying the trnsaction closure rule

This chain will transform into the following:

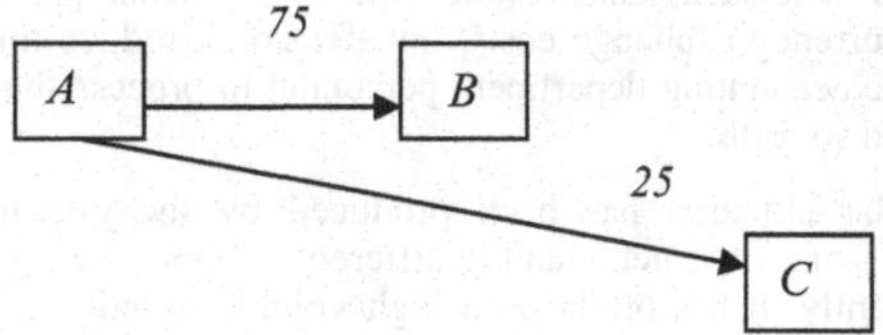

Figure 5. State after appliying the trnsaction closure rule

Explanation:

- The minimum between two invoices (100 and 25) is 25 and there will be a new transaction from *A* to *C* with this amount;
- The amount paid from *A* to *B* is larger than the one paid from *B* to *C*, so the *A* to *B* transaction will stay and the transaction from *B* to *C* will be gone. The amount of the invoice that was preserved will change to 100 – 25 = 75.

The next example is similar but the second transaction is larger:

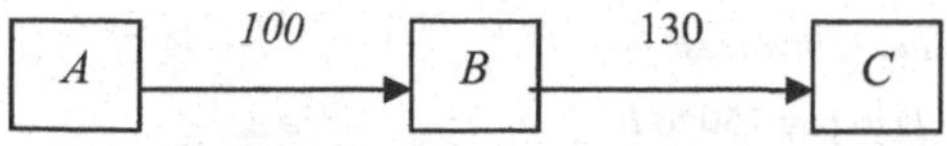

Figure 6. State before appliying the trnsaction closure rule

This chain will transform into the following:

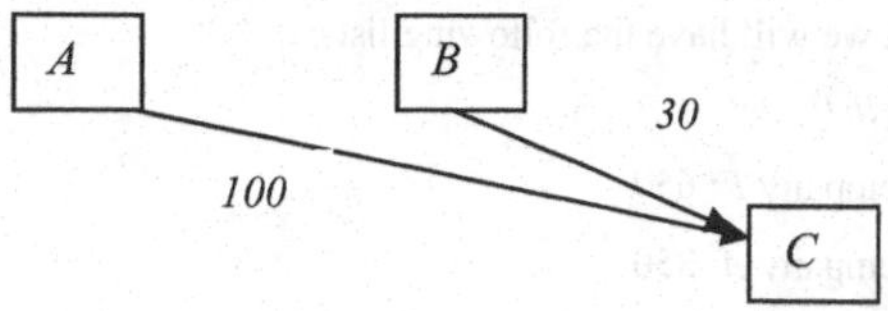

Figure 7. State after appliying the trnsaction closure rule

IV. NETTING BASED ON TOTALS

Graph theory based netting gives us the exact result, but doesn't guarantee the best solution as it uses the current configuration of invoices transforming those during the optimisation process. The process of finding a solution is very similar to the local optimum search and has a disadvantage of impossibility to find a global optimum locating beyond the local optimum as in this case the best solution can be found only if we temporary de-optimise the current solution.

Therefore the process of finding a better approach requires completely another system to be used. That is exactly the idea employed by the second (main) novel approach proposed by the paper, which is based on totals and described later.

Here the netting is started by constructing payments instead of optimizing existing. The in- and outflow is summed up per company calculating a total amount to be paid or received for each company. All companies are divided into two classes basing on the total – those that should receive in total (having positive flow) and those that should pay (having negative flow). All the system should do is to construct new transactions to match negative and positive flows and try to produce a minimum number of new transactions.

A. *Netting center based clearance*

Let's say that we have N companies and m of those have a positive flow and k of those have a negative flow. Formally saying $V=\{R, P\}$, where R is a set of receivers $\{r_1, r_2, \ldots, r_m\}$ and P is a set of payers $\{p_1, p_2, \ldots, p_k\}$. The problem of constructing new transactions is very similar to the problem of fitting things into available bags, with the only exceptions: in our case things can be broken into parts (i.e. if a company has to pay, for example 400 EUR, then the payment can be done in 2, 3 and so forth parts to different companies).

The maximum number of payments in this scheme equals to $(N\text{-}1)$ – this number of transactions can be achieved in the netting centre scheme. All companies have exactly one transaction – either to the netting centre or from the netting centre. As the netting centre doesn't have to pay to itself the total number of transactions is the number of companies minus 1. Unfortunately the amount transferred in total will be sufficient in this approach. If we assume, for simplicity, that companies could be optimally netted using such transfers (E) that all companies divide into pairs (p_i, r_j) such that both p_i and r_j will appear only once in E, then the netting centre based clearance will be nearly two times larger than the global optimum we are looking for.

B. *Heuristic, totals based distribution*

The minimum number of payments will equal to max(m,k) – to either m or k, depending on which one of those is larger. Obviously the number of payments cannot be less than the number of companies that either have to issue a transaction (pay) or to get it (receive).

Therefore the solution that we are looking for is located somewhere between (N-1) and max(m,k) or alternatively between (m+k-1) and max(m,k).

We propose the following algorithm to achieve a better transactions minimisation:

Execute the following steps until two lists are not empty:

1) Resort both categories of companies (to pay and to receive, i.e. in- and outflow companies) by amounts;

2) Compare amounts assigned to the top companies of each list (i.e. match the largest amount to be paid to the largest amount to be received) using the following sub-rules:

a) If amounts are equal then
- create a new invoice with this amount between those top companies;
- delete both companies from lists.

b) If the amount to be paid is larger then
- create a new invoice with the amount to be received between those top companies;
- delete the company to receive the payments from the receivers list;
- decrease the amount to be paid associated to the top payer company on the new invoice amount;

c) If amount to be received is larger then
- create a new invoice with the amount to be paid between those top companies;
- delete the company to pay from the payers list;
- decrease the amount to be received associated to the top receiver company on the new invoice amount.

C. Example of the heuristic, totals based distribution method

Let say that we have 6 companies: *A, B, C, D, E* and *F* and lets assume that the following list of companies / totals will appear after all transactions to be executed are summed up per company. Notice in the list that payers have positive amount (flow) and receivers' negative.

- Company *A*: total 350
- Company *B*: total -750
- Company *C*: total -600
- Company *D*: total 1000
- Company *E*: total -650
- Company *F*: total 650

1) First of all we need to divide companies into two sets

The payers list contains: company *A* (350), company *D* (1000) and company *F* (650). The receivers list contains company *B* (-750), company *C* (-600) and company *E* (-650).

2) The second step (and the first executed in the cycle in the first iteration) to sort companies inside each list by amounts.

The sorting result is the following:

Payers list:

- Company *D*: 1000
- Company *F*: 650
- Company *A*: 350

Receivers list:

- Company B: -750
- Company *E*: -650
- Company *C*: -600

3) The next step is to match largest amounts

On this step we will match an amount associated to company *D* and an amount associated to company *B*. As the amount to be paid (1000 [D]) is more than to receive (750 [B]) we

a) Produce a new invoice between those companies: D to pay B 750;

b) Exclude company *B* from the list of receivers;

c) Change *D* company amount from 1000 to 250 (=1000-750)

After this step we will have:

Payers list:

- Company *D*: 250
- Company *F*: 650
- Company *A*: 350

Receivers list:

- Company *E*: -650
- Company *C*: -600

Produced invoices:

- *D* to pay 750 to *B*

4) Now we start the next iteration and return to the resorting step.

After it we will have the following lists:

Payers list:

- Company *F*: 650
- Company *A*: 350
- Company *D*: 250

Receivers list:

- Company *E*: -650
- Company *C*: -600

Notice that the company *D* moved to the end after resorting.

5) Now we match again the largest amounts of both lists and produce a new invoice.

So, we match an amount of company *F* and company *E*. As the amounts are equal we

a) Produce a new invoice between those companies: *F* to pay E 650;

b) Exclude the company *F* from the list of payers;

c) Exclude the company *E* from the list of receivers.

After this step we will have:

Payers list:

- Company *A*: 350
- Company *D*: 250

Receivers list:

- Company *C*: -600

Produced invoices:

- *D* to pay 750 to *B*
- *F* to pay 650 to *E*

Continuing this way we will arrive to the following list of payments to be executed:

Produced invoices:

- *D* to pay 750 to *B*
- *F* to pay 650 to *E*

- *A* to pay 350 to *C*
- *D* to pay 250 to *C*

V. Tests and Analysis of Algorithms

A. *Efficiency tests*

The transaction closure based algorithm and the heuristic algorithm for netting based on totals distribution were developed by the author in 1998. Later the author started to work for Simple Concepts Eesti OÜ and those algorithms were implemented inside the cash management system targeted for large groups. Although the similar package for middle size and small organisation is available form the paper author, we will base our tests on the large groups' data acquired by the earlier mentioned company. The data represents one of the large groups (to be not named) made available to the company as a sample for testing and contains thousand of invoices per each financial period. Unfortunately the payments clearance problem doesn't offer fixed packages, which are for example offered by the second DIMACS challenge [5] for the maximum clique problem. Therefore tests will be rather local as results depend on the structure of the payments graph. The last depends on the business type, the group structure and standard business processes. Nevertheless the testing will be good enough to demonstrate base characteristics of the algorithms and therefore good enough to obtain understanding of how those perform.

The table 1 below shows an average result for 25 tests we have conducted, while an example of a test is outlined in the table 2. The table 1 shows the compression of amounts, i.e. the rate of decreasing the total sum to be transferred by dividing the total sum before netting on the total sum after. The compression of transactions shows the decrease rate of transactions.

TABLE I. TESTING RESULTS

Algorithms	Test results	
	Compression of transactions	*Compression of amounts*
Summing rule only	17.41	1.53
Netting center approach	124.22	1.50
Transaction closer based algorithm	61.18	2.93
Heuristic algorithm based on totals distribution	124.22	2.93

For example 17.41 in the compression of transactions column for "summing rules only" means that the algorithm based only on summing up existing transactions will generate (in average) 17.41 times less transactions than there were before the netting process. A number for the same row in the last column: 1.53 means that the total amount to be transferred after applying this rule will be 1.53 times less than originally.

The following table shows one case as an example to demonstrate the netting effect. Amounts in the last columns are presented in EUR.

TABLE II. AN EXAMPLE OF RESULTS BY PRESENTED ALGORITHMS

Name	Total number of transactions	Total value (amount)
Initial state	3 356	5 889 619 530
Summing rule only	227	3 820 303 522
Netting center approach	32	3 873 938 731
Transaction closer based algorithm	68	1 962 491 837
Heuristic algorithm based on totals distribution	32	1 962 491 837

B. *Analysis of algorithms and comparison to other approaches*

This subchapter is designed to analyse proposed approaches and find boundaries for both maximum number of transactions to be generated and complexity.

Theorem 1. The heuristic algorithm for netting based on totals distribution will require at most $(m+n-1)*(c+\log_2(\max(m,k)))$ steps to complete the netting process, where *c* is a constant.

Proof: The algorithm defines that on each high level step at least one company will be eliminated either from *P* or *R* set. The last step will eliminate two remaining companies at once and therefore the total number of steps required to complete the algorithm will be equal to the number of elements in the *P* plus number of elements in the *R* minus 1. As $|P|$ equals to *k* and $|R|$ equals to *m* then the maximum number of steps will equal to $m+k-1$. Each step contains a constant number of sub-steps to produce a new transaction and eliminate one company and extra steps to resort the list, where the company remained with a decreased amount. Notice that we can employ the binary search algorithm to re-sort lists having complexity $\log_2 T$, where *T* is the length of the list. In the worst case the maximum length of the list will equal to the largest number of companies in the beginning of the algorithm either in *P* or in *R* list and so will equal to $\max(m,k)$

Note: a smaller number of steps will be required if on any algorithm step both payer and receiver will be eliminated at once.

The theorem shows that the complexity of the netting heuristic algorithm based on totals distribution is $O(\max(m,k)*\log_2(\max(m,k)))$, i.e. $O(N\log_2 N)$

The maximum number of transactions generated by the algorithm can be calculated using the same logic.

Theorem 2. The heuristic algorithm for netting based on totals distribution will generate at most $(m+n-1)$ transaction.

Proof: The algorithm starts from no transactions. On each step the algorithm will generate exactly one transaction and eliminate at least one company. As the number of companies equal to $(m+k)$ and two companies are eliminated at the last step then the total number of high level steps is $(m+k-1)$ and the same is the total number of produced transactions.

Theorem 3. The transaction closure based netting generates at most $m*k$ transactions.

Proof: The algorithm work is based on eliminating cycles, i.e. on transforming the payments net into such a state, when each node will have either only incoming or only outgoing edges. The set of nodes that have only outflow will be represented by companies having to pay and therefore is *P*. The set of nodes that have only inflow edges will be represented by receivers and therefore equals to |*R*|. Each *P* node could have only one edge to each *R* node since the summing rule is constantly applied and therefore no more than *m* outgoing edges. As the total number of nodes in *P* set is *k* then the maximum number of existing transactions is $m*k$.

For the comparison we bring out a table [6] demonstrating some other approaches to solve the netting problem. Those algorithms mainly concentrate on minimising the number of invoices rather than minimising the transferred amount.

Some algorithms presented in the table use a concept of clusters – each cluster is a small group of companies that will be netted using the netting centre approach, so each cluster consumes a local netting centre. During algorithms work all companies are divided into sub-groups that can be netted locally to centres and invoices between those groups are netted via local netting centres. The total number of clusters is denoted as *p*.

TABLE III. AN EXAMPLE OF RESULTS BY PRESENTED ALGORITHMS

Name	Maximum number of transactions	Complexity
ICBR [7]	$m+k-1$	$O(m^2k^2)$
VCUBE-D	$m+k+3p-1$	$O(mk)$
DNA	$m+k+3p-1$	$O(m+k)$

VI. Conclusion

The paper reviewed some practical approaches to the netting problem, which is NP-hard and therefore is solved by heuristic algorithms. The choice of an algorithm to be applied is usually driven by extra constraints that corporation financial systems have. The paper proposed some new methods of payment clearance. The first one is based on graph theory and is quite similar to existing although is not quite the same. The second one, the heuristic approach based on totals distribution, is a novel approach allowing sufficiently decrease the total number of transactions.

References

[1] V. Srinivasan and Y. H Kim, "Payments netting in international cash management: a network optimization approach", *Journal of International Business Studies*, vol 17(2), 1986, pp. 1-20,

[2] M.M. Güntzer, D. Jungnickel, and M. Leclerc, "Efficient algorithms for the clearing of interbank payments," *European Journal of Operational Research*, vol. 106, pp. 212–219, 1998.

[3] M.R. Garey and D.S. Johnson, *Computers and Intractability: A Guide to the Theory of NP-completeness*, New-York: Freeman, 2003.

[4] Y.M. Shafransky, A.A. Doudkin, "An optimization algorithm for the clearing of interbank payments," *European Journal of Operational Research*, vol. 171(3), pp.743-749, 2006.

[5] D.S. Johnson and M.A. Trick, "Cliques, Colouring and Satisfiability: Second DIMACS Implementation Challenge," *DIMACS Series in Discrete Mathematics and Theoretical Computer Science*, American Mathematical Society, vol. 26, 1996.

[6] A. Datta, U. Derigs, and H. Thomas, "Scalable Payments Netting in Electronic Commerce", working paper, 2003.

[7] J. Chomicki, S. Naqvi, and M.F. Pucci, "Decentralized Micropayment Consolidation," *18th IEEE International Conference on Distributed Computing Systems (ICDCS'98)*, pp. 332-341, 1998.

Using LSI and its variants in Text Classification

(Shalini Batra, Seema Bawa)
Computer Science and Engg. Dept., Thapar University,

Patiala, Punjab.(India)

(sbatra@thapar.edu,seema@thapar.edu)

Abstract. Latent Semantic Indexing (LSI), a well known technique in Information Retrieval has been partially successful in text retrieval and no major breakthrough has been achieved in text classification as yet. A significant step forward in this regard was made by Hofmann[3], who presented the probabilistic LSI (PLSI) model, as an alternative to LSI. If we wish to consider exchangeable representations for documents and words, PLSI is not successful which further led to the Latent Dirichlet Allocation (LDA) model [4]. A new local Latent Semantic Indexing method has been proposed by some authors called "Local Relevancy Ladder-Weighted LSI" (LRLW-LSI) to improve text classification [5]. In this paper we study LSI and its variants in detail , analyze the role played by them in text classification and conclude with future directions in this area.

Key Words: Text Classification, LSI, PLSI, LDA, VLSI

1. Introduction

Traditional text classification is based on explicit character, and the common method is to represent textual materials with space vectors using Vector Space Model (VSM) [22, 23, 18, 16], which finally confirm the category of the test documents by comparing the degree of similarity. With more and more textual information available on the internet, conceptual retrieval has become more important than word matching retrieval [5].

Traditional information retrieval system such as VSM retrieves relevant documents by lexical matching with query. Support Vector Machines (SVD) have become a "workhorse in machine learning, data mining, signal processing, computer vision,..." [10]. Eckart and Young [8] proved that in a specific technical sense, the SVD yields the best possible approximation to any matrix A, given any target rank for the approximation A0. As a result of the SVD, each document can be viewed as a vector in a low dimensional space of a few hundred dimensions; the axes in the new space do not in general correspond to terms in the lexicon. The classic application of SVD's to text analysis stems from work of Dumais et al. [9, 7]. The authors adapted SVD to the term-document matrix, characterizing their method as latent semantic indexing (LSI). The principal application of LSI was to respond to text queries: following standard practice in text retrieval, each query (expressed as a set of terms) is viewed as a unit vector in the space in which each document is represented (whether the original set of terms as in A, or in the low-dimensional approximate space). Given a query, the system would identify the documents with the highest cosine similarity to the query (in the primary or approximate space) and return these as the best matches to the query. Experimentally, these and subsequent papers [11,12,14] showed that latent semantic indexing was effective not only in that it found a good approximation A0 of rank about 300, but further that for retrieval on queries the approximation A0 often yielded better (rather than almost as good) results than A [3]. It is widely reported that good approximations of rank 200-300 exist for typical document collections in text analysis. Computationally such approximations are typically found using the linear-algebraic technique of singular value decompositions (SVD's), a method rooted in statistical analysis [13]. The paper is divided into five sections with first section covering LSA, second concentrating on VLSA, third on PLSA, followed by LDA and LRLW-LSI. The paper concludes with analysis of all the variants being applied to text classification and future scope of these techniques in information retrieval.

2. Latent Semantic Analysis

Latent Semantic Analysis (LSA) is a theory and method for extracting and representing the contextual-usage meaning of words by statistical computations applied to a large corpus of text [7]. The similarity estimates derived by LSA are not simple contiguity frequencies, co-occurrence counts, or correlations in usage, but depend on a powerful mathematical analysis that is capable of correctly inferring much deeper relations (thus the phrase "Latent Semantic"). LSA, as currently practiced, induces its representations of the meaning of words and passages from analysis of text alone. It differs from some statistical approaches significantly. The input data "associations" from which LSA induces representations are between unitary expressions of meaning—words and complete meaningful utterances in which they occur—rather than between successive words. That is, LSA uses as its initial data not just the summed contiguous pairwise (or tuple-wise) co-occurrences of words but the detailed patterns of occurrences of very many words over very large numbers of local meaning-bearing contexts, such as sentences or paragraphs, treated as unitary wholes. It concentrates only on how differences in word choice and differences in passage meanings are related [1].

K. Elleithy (ed.), *Advanced Techniques in Computing Sciences and Software Engineering*,
DOI 10.1007/978-90-481-3660-5_53, © Springer Science+Business Media B.V. 2010

LSA is a fully automatic mathematical/statistical technique for extracting and inferring relations of expected contextual usage of words in passages of discourse. It does not uses any humanly constructed dictionaries, knowledge bases, semantic networks, grammars, syntactic parsers, or morphologies, or the like, and takes as its input only raw text parsed into words defined as unique character strings and separated into meaningful passages or samples such as sentences or paragraphs. The number of dimensions retained in LSA is an empirical issue. Although LSA has been applied with remarkable success in different domains including automatic indexing (Latent Semantic Indexing, LSI) [3], it has a number of deficits, mainly due to its unsatisfactory statistical foundation. The drawback of VSM is that it cannot retrieve the conceptually relevant documents with respect to query, and the semantic information may lose during the process of VSM [5].

3. Variable Latent Semantic Indexing

VLSI (Variable Latent Semantic Indexing) is a new query-dependent (or "variable") low-rank approximation that minimizes approximation error for any specified query distribution. With this it is possible to tailor the LSI technique to particular settings, often resulting in vastly improved approximations at much lower dimensionality. Probabilistic latent semantic indexing [2] and its cousins from statistics [19] use a generative probabilistic model for the entries of the matrix A, rather than for the query distribution. For any quantitative level of retrieval effectiveness, the number of dimensions in the low-rank approximation is dramatically lower for VLSI than for LSI. It has been experimentally proved in [2] that VLSI shows a 10% improvement at 10 dimensions, a 27% improvement at 50 dimensions, a 50% improvement at 125 dimensions, and an 80% improvement at 1000 dimensions. Stated alternatively, VLSI with just 40 dimensions is about equivalent in performance to LSI with 250 dimensions.

4. Probabilistic Latent Semantic Indexing

Probabilistic Latent Semantic Indexing is a novel approach to automated document indexing which is based on a statistical latent class model for factor analysis of count data. Fitted from a training corpus of text documents by a generalization of the Expectation Maximization algorithm, this model is able to deal with domain specific synonymy as well as with polysemous words. In contrast to standard Latent Semantic Indexing (LSI) by Singular Value Decomposition, the probabilistic variant has a solid statistical foundation and defines a proper generative data model. This implies in particular that standard techniques from statistics can be applied for questions like model fitting, model combination, and complexity control. In addition, the factor representation obtained by Probabilistic LSA (PLSA) allows to deal with polysemous words and to explicitly distinguish between different meanings and different types of word usage. The core of PLSA is a statistical model which has been called aspect model [17, 20]. The latter is a latent variable model for general co-occurrence data which associates an unobserved class variable. One advantage of using statistical models vs. SVD techniques is that it allows us to systematically combine different models [2].

Variants of Probabilistic Latent Semantic Indexing

Two different schemes to exploit PLSA for indexing have been investigated:

(i) as a context dependent unigram model to smoothen the empirical word distributions in documents (PLSI-U),

(ii) as a latent space model which provides a low dimensional document/query representation (PLSI-Q) [2].

While Hofmann's work is a useful step toward probabilistic modeling of text, it is incomplete in that it provides no probabilistic model at the level of documents. In PLSI, each document is represented as a list of numbers (the mixing proportions for topics), and there is no generative probabilistic model for these numbers. This leads to several problems:

(1) the number of parameters in the model grows linearly with the size of the corpus, which leads to serious problems with overfitting, and

(2) it is not clear how to assign probability to a document outside of the training set [4].

5. Latent Dirichlet Allocation

Latent Dirichlet allocation (LDA) is a generative probabilistic model for collections of discrete data such as text corpora. LDA is a three-level hierarchical Bayesian model, in which each item of a collection is modeled as a finite mixture over an underlying set of topics. Each topic is, in turn, modeled as an infinite mixture over an underlying set of topic probabilities. In the context of text modeling, the topic probabilities provide an explicit representation of a document [4].

LDA is based on a simple exchangeability assumption for the words and topics in a document; it is therefore realized by a straightforward application of de Finetti's representation theorem. It can be viewed as a dimensionality reduction technique, in the spirit of LSI, but with proper underlying generative probabilistic semantics that make sense for the type of data that it models. Exact inference is intractable for LDA, but any of a large suite of approximate inference algorithms can be used for inference and parameter estimation within the LDA framework.

Various methods used for text classification including LSI and PLSI are based on the "bag-of-words" assumption—that the order of words in a document can be neglected. In the language of probability theory, this is an assumption of *exchangeability* for the words in a document [11]. Moreover, although less often stated formally, these methods also assume that documents are exchangeable; the specific ordering of the documents in a corpus can also be neglected. A classic representation theorem due to de Finetti [18] establishes that any collection of exchangeable random variables has a representation as a mixture distribution—in general an infinite mixture. Thus, if we wish to consider exchangeable representations for documents and words, we need to consider mixture models that capture the exchangeability of both words and documents [4].

Document classification in LDA

In the text classification problem, aim is to classify a document into two or more mutually exclusive classes. In particular, by using one LDA module for each class, a generative model for classification is obtained. A challenging aspect of the document classification problem is the choice of features. Treating individual words as features yields a rich but very large feature set [15]. One way to reduce this feature set is to use an LDA model for dimensionality reduction. In particular, LDA reduces any document to a fixed set of real-valued features—the posterior Dirichlet parameters associated with the document [4].

6. LRLW-LSI: An Improved Latent Semantic Indexing (LSI) Text Classifier

When LSI is applied to text classification, there are two common methods. The first one is called "Global LSI", which performs SVD directly on the entire training document collection to generate the new feature space. This method is completely unsupervised, *i.e.*, it pays no attention to the class label of the existing training data. It has no help to improve the discrimination power of document classes, so it always yields no better, sometimes even worse performance than original term vector on classification [21]. The other one is called "Local LSI", which performs a separate SVD on the local region of each topic. Compared with global LSI, this method utilizes the class information effectively, so it improves the performance of global LSI greatly.

To overcome the drawbacks of Global LSI, authors in [5] proposed a local LSI method which has been named as "Local Relevancy Ladder-Weighted LSI (LRLW-LSI)". In local LSI, each document in the training set is first assigned with a relevancy score related to a topic, and then the documents whose scores are larger than a predefined threshold value are selected to generate the local region. Then SVD is performed on the local region to produce a local semantic space. In other words, LRLW-LSI gives same weight among a ladder-range and different weight located in different ladder-range to documents in the local region according to its relevance before performing SVD so that the local semantic space can be extracted more accurately considering both the local and global relevancy and more relevant documents can be introduced with higher weights, which make them do more contribution to SVD computation. Hence, the better local semantic space is available which results in better classification performance can be extracted to separate positive documents from negative documents.

7. Conclusion

Based on the study and analysis conducted it has been observed that Latent Semantic Analysis (LSA) is an approach to automatic indexing and information retrieval that attempts to overcome text classification problems by mapping documents as well as terms to a representation in the so called latent semantic space. LSI has often matched queries to (and only to) documents of similar topical meaning correctly when query and document use different words. When LSA was first applied in the text–processing problem in automatic matching of information requests to document abstracts there was a significant improvement over prior methods. LSA performs some sort of noise reduction and has the potential benefit to detect synonyms as well as words that refer to the same topic. [1]. Although LSA has been applied with remarkable success in different domains including automatic indexing [1, 3], it has a number of deficits, mainly due to its unsatisfactory statistical foundation.

Probabilistic Latent Semantic Analysis (PLSA) based on a statistical latent-class model utilizes the (annealed) likelihood function as an optimization criterion and Tempered Expectation Maximization is a powerful fitting procedure in PLSA. It is indeed a promising novel unsupervised learning method with a wide range of application in text learning, computational linguistics, information retrieval, and information filtering [6].

LDA illustrates the way in which probabilistic models can be scaled up to provide useful inferential machinery in domains involving multiple levels of structure and although it can be viewed as a competitor to methods such as LSI and PLSI in the setting of dimensionality reduction for document collections and other discrete corpora.. The major advantages of generative models such as LDA include their modularity and their extensibility. As a probabilistic module, LDA can be readily embedded in a more complex model— a property that is not possessed by LSI [4].The major thrust area visible in LDA is numerous possible extensions of LDA which includes expectation propagation Markov chain Monte Carlo algorithm, *etc*.

It has been observed that all documents in the local region are equally considered in the SVD computation in local LSI. But intuitively, first, more relevant documents to the topic should contributes more to the local semantic space than those less-relevant ones; second, tiny less local relevant documents may be a little more global relevant. So based on these ideas, local LSI method "Local Relevancy Ladder-Weighted LSI (LRLW-LSI)" selects documents to the local region in a ladder way so that the local semantic space can be extracted more accurately considering both the local and global relevancy which can be indeed beneficial in text classification.

Future Directions

All the LSA and its variants studied so far are relevant in text classification at up to certain extent and when applied on a large text corpora the result are not very promising.. Classification is done at syntactic levels only and in future we will try to explore their successful implementation when applied on semantically related data. VSM will be implemented not on terms but on concepts and text classification will concentrate on synonymy as well as on polysemous words clustering.

References

[1] Landauer, T. K., Foltz, P. W., & Laham, D. (1998). Introduction to Latent Semantic Analysis. *Discourse Processes*, **25**, 259-284.

[2] T. Hofmann. Probabilistic latent semantic indexing. In Proceedings of the 22nd ACM Conference on Research and Development in Information Retrieval (SIGIR), pages 50–57, 1999.

[3] Anirban Dasgupta, Ravi Kumar, Prabhakar Raghavan, Andrew Tomkins, Variable Latent Semantic Indexing , SIGKDD'05.

[4] David M. Blei, Andrew Y. Ng, Michael I. Jordan, Latent Dirichlet Allocation Journal of Machine Learning Research 3 (2003) 993-1022.

[5] Wang Ding, Songnian Yu, Shanqing Yu, Wei Wei, and Qianfeng Wang, LRLW-LSI: An Improved Latent Semantic Indexing (LSI) Text Classifier.Unpublished.

[6] Thomas Hofmann, Unsupervised Learning by Probabilistic Latent Semantic Analysis, Department of Computer Science, Brown University, Providence, RI 02912, USA Editor: Douglas Fisher.

[7] S. T. Dumais, G. Furnas, T. Landauer, and S. Deerwester. Using latent semantic analysis to improve information retrieval. In Proceedings of ACM Conference on Computer Human Interaction (CHI), pages 281–285, 1988.

[8] C. Eckart and G. Young. The approximation of a matrix by another of lower rank. Psychometrika, 1:211–218, 1936.

[9] S. Deerwester, S. T. Dumais, T. K. Landauer, G. W. Furnas, and R. A. Harshman. Indexing by latent semantic analysis. Journal of the Society for Information Science, 41(6):391–407, 1990.

[10] T. Hoffmann. Matrix decomposition techniques in machine learning and information retrieval. http://www.mpi-sb.mpg.de/_adfocs/adfocs04. slides-hofmann.pdf, 2004.

[11] Aldous. Exchangeability and related topics. In *E'cole d'e'te' de probabilite's de Saint-Flour, XIII—1983*, pages 1–198. Springer, Berlin, 1985

[12] T. Hofmann. Latent semantic models for collaborative filtering. ACM Transactions on Information Systems, 22(1):89–115, 2004.

[13] I. T. Jolliffe. Principal Component Analysis. Springer, 2002.

[14] Landauer, T. K. & Dumais, S. T. (1997). A solution to Plato's problem: The Latent Semantic Analysis theory of the acquisition, induction, and representation of knowledge. Psychological Review , 1 0 4 , 211-140.

[15] T. Joachims. Making large-scale SVM learning practical. In Advances in Kernel Methods – Support Vector Learning. M.I.T. Press, 1999.

[16]. Salton, Gerard.: Introduction to modern information retrieval, Auckland, McGraw-Hill (1983)

[17] Hofmann, T., Puzicha, J., and Jordan, M. I. Unsupervised learning from dyadic data. In Advances in Neural Information Processing Systems (1999), vol. 11.

[18] de Finetti. Theory of probability. *Vol. 1-2*. John Wiley & Sons Ltd., Chichester, 1990. Reprint of the 1975 translation.

[19] M. Tipping and C. Bishop. Probabilistic principal component analysis. Journal of the Royal Statistical Society, Series B, 61(3):611–622, 1999.

[20] Saul, L., and Pereira, F. Aggregate and mixed order Markov models for statistical language processing. In Proceedings of the 2nd International Conference on Empirical Methods in Natural Language Processing (1997).

[21] Torkkola, K.: Linear Discriminant Analysis in Document Classification. In: Proceedings of the 01' IEEE ICDM Workshop Text Mining (2001)

[22] Golub, G., Reinsch, C.: Handbook for automatic computation: linear algebra. Springer-Verlag, New York (1971)

[23]. Golub, G., Loan, C.V.: Matrix Computations. Johns-Hopkins, Baltimore, second ed. (1989)

On Optimization of Coefficient-Sensitivity and State-Structure for Two Dimensional (2-D) Digital Systems

Guoliang Zeng
Phone: (480) 727-1905; E-mail: gzeng@asu.edu
Motorola DSP Laboratory, Division of Computing Studies, Department of Engineering
Arizona State University Polytechnic Campus
7171 East Sonoran Arroyo Mall, Mesa, AZ 85212

***Abstract*-A new, tighter bound of coefficient-sensitivity is derived for 2-D digital systems. A new algorithm is also proposed which optimizes both the coefficient-sensitivity and the state-structure of 2-D digital systems.**

I. INTRODUCTION

After Roesser gave the state-space model of 2-D digital systems in 1975 [1], the majority researches has focused on the optimization of structure. But due to the inherent, finite-word-length effect, the sensitivity is, at least, as important as structure. In 1987, Kawamata suggested a definition and an optimization algorithm for the sensitivity to the coefficients in Roesser's model [3]. Recent years, more papers on the optimization of coefficient-sensitivity are published [6]. But in most cases, the sensitivity optimization and structure optimization are not compatible. In this paper, a new algorithm is proposed which optimizes both the sensitivity and the structure. A more accurate expression of sensitivity is also derived.

II. MODEL

A causal, linear, shift invariant, 2-D digital system with single-input and single-output can be modeled with a set of first-order, state-vector difference equations over R [1]:

(1) $$x^h(i+1,j) = A_1x^h(i,j) + A_2x^v(i,j) + B_1u(i,j)$$

(2) $$x^v(i,j+1) = A_3x^h(i,j) + A_4x^v(i,j) + B_2u(i,j)$$

(3) $$y(i,j) = C_1x^h(i,j) + C_2x^v(i,j) + du(i,j)$$

Where i and j are integer-valued horizontal and vertical coordinates; $x^h \in R^{m\times 1}$ conveys information horizontally; $x^v \in R^{n\times 1}$ conveys information vertically; $u \in R$ acts as input; $y \in R$ acts as output; $A_1 \in R^{m\times m}$, $A_2 \in R^{m\times n}$, $A_3 \in R^{n\times m}$, and $A_4 \in R^{n\times n}$ are matrices; $B_1 \in R^{m\times 1}$, $B_2 \in R^{n\times 1}$, $C_1 \in R^{1\times m}$, $C_2 \in R^{1\times n}$, are vectors; $d \in R$ is a scalar. In a more compact form,

(4) $$x'(i,j) = Ax(i,j) + Bu(i,j)$$

(5) $$y(i,j) = Cx(i,j) + du(i,j)$$

Here

(6) $$x'(i,j) = \begin{bmatrix} x^h(i+1,j) \\ x^v(i,j+1) \end{bmatrix},$$

$$x(i,j) = \begin{bmatrix} x^h(i,j) \\ x^v(i,j) \end{bmatrix},$$

$$A = \begin{bmatrix} A_1, A_2 \\ A_3, A_4 \end{bmatrix} = \{a_{ij}\}_{N\times N},$$

$$B = \begin{bmatrix} B_1 \\ B_2 \end{bmatrix} = \{b_i\}_{N\times 1}$$

$$C = [C_1, C_2] = \{c_j\}_{1\times N},$$

K. Elleithy (ed.), *Advanced Techniques in Computing Sciences and Software Engineering*,
DOI 10.1007/978-90-481-3660-5_54, © Springer Science+Business Media B.V. 2010

$N = m + n.$

The transfer function of the system is [2]

(7) $$H(z_1,z_2) = C[z_1 I_m \oplus z_2 I_n - A]^{-1} B + d$$

Here $\oplus$ is the symbol for direct sum. H is a scalar function of z_1 and z_2. And the $N\times N$ reachability and observability gramians matrices are [4]

(8) $$K = (2\pi j)^{-2} \oint_{|z_2|=1} \oint_{|z_1|=1} F(z_1,z_2) F^{*T}(z_1,z_2) z_1^{-1} z_2^{-1} dz_1 dz_2$$

(9) $$W = (2\pi j)^{-2} \oint_{|z_2|=1} \oint_{|z_1|=1} G^{*T}(z_1,z_2) G(z_1,z_2) z_1^{-1} z_2^{-1} dz_1 dz_2$$

Here $*T$ stands for complex conjugate and transpose, and the $N\times 1$ vector F and the $1\times N$ vector G are given by

(10) $F(z_1,z_2) = [z_1 I_m \oplus z_2 I_n - A]^{-1} B$,

(11) $G(z_1,z_2) = C[z_1 I_m \oplus z_2 I_n - A]^{-1}$

Clearly, $K = \{k_{ij}\}_{N\times N}$ and $W = \{w_{ij}\}_{N\times N}$ are two $N\times N$ matrices.

III. SENSITIVITY

The sensitivities of an $H(z_1,z_2)$ to the coefficients a_{ij}, b_i c_j and d are defined as the following derivatives [3]:

(12) $$S_{a_{ij}} = \frac{\partial H}{\partial a_{ij}},\ S_{b_i} = \frac{\partial H}{\partial b_i},$$

$$S_{c_j} = \frac{\partial H}{\partial c_j},\ S_d = \frac{\partial H}{\partial d}$$

Then the deviation of $H(z_1,z_2)$ due to the quantization (round-off or/and truncation) of its coefficients is as follows [8]:

(13) $$\Delta H = \sum_{i=1}^{N}\sum_{j=1}^{N} S_{a_{ij}} \Delta a_{ij} + \sum_{i=1}^{N} S_{b_i} \Delta b_i + \sum_{j=1}^{N} S_{c_j} \Delta c_j + S_d \Delta d$$

If any coefficient, say a_{ij} = 0 or -1, then Δa_{ij} = 0, i.e. no quantization error. Hence, for a_{ij} = 0 or -1, $S_{a_{ij}}$ can be set to zero. Any coefficient which is not 0 or -1 might cause quantization error. To take this observation into account, we define

(14) $$\delta(x) = \begin{cases} 1,,, x = 0,-1 \\ 0,,, x \neq 0,-1 \end{cases},$$

$$\bar{\delta}(x) = 1 - \delta(x) = \begin{cases} 0,,, x = 0,-1 \\ 1,,, x \neq 0,-1 \end{cases}$$

Obviously this $\delta(x)$ is a well defined function of x. With $\bar{\delta}(x)$, from the equation (7), we have

(15) $S_{a_{ij}} = \bar{\delta}(a_{ij}) G(z_1,z_2) e_i e_j^T F(z_1,z_2)$

(16) $S_{b_i} = \bar{\delta}(b_i) G(z_1,z_2) e_i$

(17) $S_{c_j} = \bar{\delta}(c_j) e_j^T F(z_1,z_2)$

(18) $S_d = \bar{\delta}(d)$

Here e_i is the unit column ($N\times 1$) vector with only nonzero element, 1 in the ith position. And the sensitivities of $H(z_1,z_2)$ to matrices A, B, C can be defined as the Frobenius norms of matrices $\partial H/\partial A$, $\partial H/\partial B$, $\partial H/\partial C$ [7]:

(19) $$S_A = \left\{ \sum_{i=1}^{N}\sum_{j=1}^{N} \left|S_{a_{ij}}\right|^2 \right\}^{1/2} = \left\{ \sum_{i=1}^{N}\sum_{j=1}^{N} \bar{\delta}(a_{ij}) \left|G(z_1,z_2)e_i\right|^2 \left|e_j^T F(z_1,z_2)\right|^2 \right\}^{1/2}$$

(20) $$S_B = \left\{ \sum_{i=1}^{N} \left|S_{b_i}\right|^2 \right\}^{1/2} = \left\{ \sum_{i=1}^{N} \bar{\delta}(b_i) \left|G(z_1,z_2)e_i\right|^2 \right\}^{1/2}$$

(21) $$S_C = \left\{ \sum_{j=1}^{N} \left| S_{c_j} \right|^2 \right\}^{1/2} = \left\{ \sum_{j=1}^{N} \bar{\delta}(c_j) \left| e_j^T F(z_1, z_2) \right|^2 \right\}^{1/2}$$

Clearly, all S's above are scalar functions of z_1 and z_2. To obtain a frequency-independent sensitivity for $H(z_1,z_2)$, an l_2 norm of $X(z_1,z_2)$ is defined by

(22) $$\|X(z_1,z_2)\|_2 = \left\{ (2\pi j)^{-2} \oint_{|z_2|=1} \oint_{|z_1|=1} |X(z_1,z_2)|^2 z_1^{-1} z_2^{-1} dz_1 dz_2 \right\}^{1/2}$$

The total sensitivity S of $H(z_1,z_2)$ can be well defined as $\|S_A\|_2^2 + \|S_B\|_2^2 + \|S_C\|_2^2 + \|S_d\|_2^2$. Since $\|XY\|_2^2 \le \|X\|_2^2 \|Y\|_2^2$, the total sensitivity S of $H(z_1,z_2)$ is suggested as the following upper bound:

(23) $$S = \sum_{i=1}^{N} \sum_{j=1}^{N} \bar{\delta}(a_{ij}) [(2\pi j)^{-2} \oint_{|z_2|=1} \oint_{|z_1|=1} |G(z_1,z_2) e_i|^2 z_1^{-1} z_2^{-1} dz_1 dz_2] \times$$
$$[(2\pi j)^{-2} \oint_{|z_2|=1} \oint_{|z_1|=1} |e_j^T F(z_1,z_2)|^2 z_1^{-1} z_2^{-1} dz_1 dz_2] +$$
$$\sum_{i=1}^{N} \bar{\delta}(b_i) (2\pi j)^{-2} \oint_{|z_2|=1} \oint_{|z_1|=1} |G(z_1,z_2) e_i|^2 z_1^{-1} z_2^{-1} dz_1 dz_2 +$$
$$\sum_{j=1}^{N} \bar{\delta}(c_j) (2\pi j)^{-2} \oint_{|z_2|=1} \oint_{|z_1|=1} |e_j^T F(z_1,z_2)|^2 z_1^{-1} z_2^{-1} dz_1 dz_2 + \bar{\delta}(d)$$

Using following two equations: $|G(z_1,z_2)e_i|^2 = e_i^T G^{*T}(z_1,z_2) G(z_1,z_2) e_i$ and $|e_j^T F(z_1,z_2)|^2 = e_j^T F(z_1,z_2) F^{*T}(z_1,z_2) e_j$, (23) becomes

(24) $$S = \sum_{i=1}^{N} \sum_{j=1}^{N} \bar{\delta}(a_{ij}) [(2\pi j)^{-2} \oint_{|z_2|=1} \oint_{|z_1|=1} e_i^T G^{*T}(z_1,z_2) G(z_1,z_2) e_i z_1^{-1} z_2^{-1} dz_1 dz_2] \times$$
$$[(2\pi j)^{-2} \oint_{|z_2|=1} \oint_{|z_1|=1} e_j^T F(z_1,z_2) F^{*T}(z_1,z_2) e_j z_1^{-1} z_2^{-1} dz_1 dz_2] +$$
$$\sum_{i=1}^{N} \bar{\delta}(b_i) (2\pi j)^{-2} \oint_{|z_2|=1} \oint_{|z_1|=1} e_i^T G^{*T}(z_1,z_2) G(z_1,z_2) e_i z_1^{-1} z_2^{-1} dz_1 dz_2 +$$
$$\sum_{j=1}^{N} \bar{\delta}(c_j) (2\pi j)^{-2} \oint_{|z_2|=1} \oint_{|z_1|=1} e_j^T F(z_1,z_2) F^{*T}(z_1,z_2) e_j z_1^{-1} z_2^{-1} dz_1 dz_2 + \bar{\delta}(d)$$

From (8) and (9),

(25) $$S = \sum_{i=1}^{N} \sum_{j=1}^{N} \bar{\delta}(a_{ij}) [e_i^T W e_i][e_j^T K e_j] + \sum_{i=1}^{N} \bar{\delta}(b_i) e_i^T W e_i + \sum_{j=1}^{N} \bar{\delta}(c_j) e_j^T K e_j + \bar{\delta}(d)$$
$$= \sum_{i=1}^{N} \sum_{j=1}^{N} \bar{\delta}(a_{ij}) w_{ii} k_{jj} + \sum_{i=1}^{N} \bar{\delta}(b_i) w_{ii} + \sum_{j=1}^{N} \bar{\delta}(c_j) k_{jj} + \bar{\delta}(d)$$

Replacing $\bar{\delta}(x)$ by $1-\delta(x)$ gives

(26) $$S = \{ \sum_{i=1}^{N} \sum_{j=1}^{N} w_{ii} k_{jj} + \sum_{i=1}^{N} w_{ii} + \sum_{j=1}^{N} k_{jj} + 1 \} -$$
$$\{ \sum_{i=1}^{N} \sum_{j=1}^{N} \delta(a_{ij}) w_{ii} k_{jj} + \sum_{i=1}^{N} \delta(b_i) w_{ii} + \sum_{j=1}^{N} \delta(c_j) k_{jj} + \delta(d) \} =: S_1 - S_2$$

Here

$$(27)\qquad S_1 = \sum_{i=1}^{N}\sum_{j=1}^{N} w_{ii}k_{jj} + \sum_{i=1}^{N} w_{ii} + \sum_{j=1}^{N} k_{jj} + 1 = tr(W)\,tr(K) + tr(W) + tr(K) + 1$$

$$(28)\qquad S_2 = \sum_{i=1}^{N}\sum_{j=1}^{N} \delta(a_{ij})w_{ii}k_{jj} + \sum_{i=1}^{N} \delta(b_i)w_{ii} + \sum_{j=1}^{N} \delta(c_j)k_{jj} + \delta(d)$$

Clearly, S_1 is the total sensitivity suggested in [2] and [3]. From the derivation above, both S and S_1 are upper bounds for the sensitivity, but S is tighter than S_1, because $S \le S_1$.
Also, from equation (26), S can be reduced by decreasing S_1 and increasing S_2. But S_2 is proportional to the number of coefficients which equal to 0 or -1. Naturally, our optimization method below will try to increase the number of zero coefficients.

IV. OPTIMIZATION

The state-space realization (*A*, *B*, *C*, *d*) of a given $H(z_1, z_2)$ is not unique. Different realizations have different sensitivities. The objective of sensitivity-minimization is to find a realization ($\hat{A}$, $\hat{B}$, $\hat{C}$, $\hat{d}$) with a minimal sensitivity. To do so, Let T be a nonsingular matrix and

(29) $T = T_1 \oplus T_4,\ T^{-1} = T_1^{-1} \oplus T_4^{-1}$.

Here, $T_1 \in R^{m\times m}$ and $T_4 \in R^{n\times n}$

We define the similar transformation of the system (*A*, *B*, *C*, *d*) as

(30) $\hat{A} = T^{-1}AT$, $\hat{B} = T^{-1}B$,
$\hat{C} = CT$, $\hat{d} = d$

Proposition 1: A similar transformation keeps the transfer function $H(z_1, z_2)$ unchanged, but the sensitivity (S_1) could be changed.

Proof: From (7) the new transfer function:

$$(31)\qquad \begin{aligned}\hat{H}(z_1,z_2) &= \hat{C}[z_1I_m \oplus z_2I_n - \hat{A}]^{-1}\hat{B} + \hat{d}\\ &= CT[z_1I_m \oplus z_2I_n - T^{-1}AT]^{-1}T^{-1}B + d\\ &= CT[T^{-1}(z_1I_m \oplus z_2I_n - A)T]^{-1}T^{-1}B + d\\ &= CTT^{-1}(z_1I_m \oplus z_2I_n - A)^{-1}TT^{-1}B + d = H(z_1,z_2)\end{aligned}$$

Therefore ($\hat{A}$, $\hat{B}$, $\hat{C}$, $\hat{d}$) has the same transfer function as (*A*, *B*, *C*, *d*).

To find the sensitivity for ($\hat{A}$, $\hat{B}$, $\hat{C}$, $\hat{d}$), we notice that

$$\hat{F}(z_1,z_2) = [z_1I_m \oplus z_2I_n - \hat{A}]^{-1}\hat{B} = T^{-1}F(z_1,z_2)$$
$$\hat{G}(z_1,z_2) = \hat{C}[z_1I_m \oplus z_2I_n - \hat{A}]^{-1} = G(z_1,z_2)T$$

From (8) and (9),

(32) $\hat{K} = T^{-1}KT^{-T}$, and $\hat{W} = T^TWT$

The sensitivity of ($\hat{A}$, $\hat{B}$, $\hat{C}$, $\hat{d}$) is

$$(33)\ \begin{aligned}\hat{S}_1 &= tr(\hat{W})tr(\hat{K}) + tr(\hat{W}) + tr(\hat{K}) + 1\\ &= tr(T^TWT)tr(T^{-1}KT^{-T}) + tr(T^TWT)\\ &\quad + tr(T^{-1}KT^{-T}) + 1\end{aligned}$$

Therefore ($\hat{A}$, $\hat{B}$, $\hat{C}$, $\hat{d}$) and (*A*, *B*, *C*, *d*) could have different sensitivities. #

Since the sensitivity S_1 is a function of T, we can minimize $\hat{S}_1$ by choosing the right T. A method to find the right T can be found in reference [5].

Now, let Q be an orthogonal matrix and

(34) $Q = Q_1 \oplus Q_4$ with

$Q_1 \in R^{m\times m}$ and $Q_4 \in R^{n\times n}$

(35) $Q_1^T = Q_1^{-1}$, $Q_4^T = Q_4^{-1}$, then $Q^T = Q^{-1}$

We define the orthogonal transformation of the system (*A*, *B*, *C*, *d*) as

(36) $\breve{A} = Q^T A Q$, $\breve{B} = Q^T B$,

$\breve{C} = CQ$, $\breve{d} = d$

Proposition 2: An orthogonal transformation keeps both, the transfer function $H(z_1, z_2)$ and the sensitivity (S_1) unchanged.

Proof: Obviously, an orthogonal transformation is a special similar transformation. Therefore, from (31) the new transfer function will not change. That is $\breve{H}(z_1, z_2) = H(z_1, z_2)$.

Since *Q* is orthogonal, from (32)

(37) $\breve{K} = Q^T K Q$, and $\breve{W} = Q^T W Q$

Because an orthogonal transformation of a matrix will not change the eigen values, and the trace of a matrix equals the sum of its eigen values, therefore from (27)

(38) $\breve{S}_1 = tr(\breve{W})tr(\breve{K}) + tr(\breve{W}) + tr(\breve{K}) + 1$

$= tr(W)tr(K) + tr(W) + tr(K) + 1 = S_1$ #

Now given a transfer function $H(z_1, z_2)$ with any realization (*A*, *B*, *C*, *d*), based on the above two propositions, we suggest the following algorithm to optimize both the sensitivity and structure.

Step 1: According to reference [5], carry out similar transformation on (*A*, *B*, *C*, *d*) to minimize the sensitivity (S_1). Assume that ($\hat{A}$, $\hat{B}$, $\hat{C}$, $\hat{d}$) is a resultant realization with minimal sensitivity (S_1).

Step 2: Carry out orthogonal transformation on ($\hat{A}$, $\hat{B}$, $\hat{C}$, $\hat{d}$) to get a realization ($\breve{A}$, $\breve{B}$, $\breve{C}$, $\breve{d}$) which reduces structure and total sensitivity (*S*).

Under very general conditions, it is proved in [7] that for a square matrix *M*, an orthogonal matrix *Q* can be formed such that

$$(39)\quad Q^T M Q = \begin{bmatrix} J_1, \times, \cdots\cdots, \times \\ 0, J_2, \times, \cdots, \times \\ \cdots\cdots\ddots\cdots\cdots \\ 0, \cdots\cdots, 0, J_s \end{bmatrix} =: \breve{M}$$

Here the J_i's are either 1×1 or 2×2 diagonal blocks. Each 2×2 block will correspond to a pair of complex conjugate eigen values of *M*. The eigen values of *M* will be eigen values of J_i's. In the case where *M* has a complete orthogonal set of eigenvectors, e. g. *M* is symmetric, $\breve{M}$ will be diagonal with all eigen values as the diagonal elements. The orthogonal matrix *Q* can be formed from *M* by the following iterative algorithm:

1. Let $M_1 = M$ and factor it into a product of an orthogonal matrix Q_1 and an upper triangular matrix R_1, say, by Givens transformations. That is $M_1 = M = Q_1 R_1$.
2. Let $M_2 = Q_1^T M_1 Q_1 = R_1 Q_1$ and factor it into a product of an orthogonal matrix Q_2 and an upper triangular matrix R_2. That is $M_2 = Q_1^T M_1 Q_1 = R_1 Q_1 = Q_2 R_2$.
3. Let $M_3 = Q_2^T M_2 Q_2 = R_2 Q_2$ and factor it into a product of an orthogonal matrix Q_3 and an upper triangular matrix R_3. Or,
$$M_3 = Q_2^T M_2 Q_2 = (Q_1 Q_2)^T M (Q_1 Q_2)$$
$$= R_2 Q_2 = Q_3 R_3.$$
4. In general, if $M_k = Q_k R_k$, then,
$$M_{k+1} = R_k Q_k = Q_k^T M_k Q_k.$$
$$= (Q_1 Q_2 \cdots Q_k)^T M (Q_1 Q_2 \cdots Q_k)$$
5. Continue in this manner until $M_{K+1} = \breve{M}$ as defined in (39). Set $Q = Q_1 Q_2 \cdots Q_K$.

Now using this algorithm to form Q_1 and Q_4 from $\widehat{A}_1$ and $\widehat{A}_4$, such that $Q_1^T \widehat{A}_1 Q_1 = \breve{A}_1$ and $Q_4^T \widehat{A}_4 Q_4 = \breve{A}_4$, where $\breve{A}_1$ and $\breve{A}_4$ are in the form defined in (39). Let $Q = Q_1 \oplus Q_4$ and $\breve{A} = Q^T \widehat{A} Q$. Then comparing with $\widehat{A}$, $\breve{A}$ will have more than $\frac{1}{2}(m^2 + n^2 - m - n)$ zero elements. In the case where $\widehat{A}_1$ and $\widehat{A}_4$ have a complete orthogonal set of eigenvectors, e. g. they are symmetric, $\widehat{A}_1$ and $\widehat{A}_4$ will be diagonal, and the number of zeros will be $m^2 + n^2 - m - n$. This is a big reduction in structure. It is also a big reduction in the total sensitivity S due to the increase of S_2.

V. Discussion

In this paper it is shown that both coefficient-sensitivity and state-structure of a 2-D system can be reduced simultaneously. The sensitivity bound derived in this paper is much tighter, but is not the lowest. The algorithm proposed in this paper is practical, but the two steps could be combined. More research is going on in this direction.

Notes

This paper is the further work of my other paper published in 1995 [9]. Both use orthogonal transformation. Unlike that paper trying to simplify $\widehat{A}_2$ or/and $\widehat{A}_3$, this paper simplifies $\widehat{A}_1$ and $\widehat{A}_4$. Unless $\widehat{A}_2 = \widehat{A}_3^T$, $\widehat{A}_2$ and $\widehat{A}_3$ can not be made diagonal simultaneously. Using the algorithm proposed in this paper, $\widehat{A}_1$ and $\widehat{A}_4$ will always converge to upper triangular, bi-diagonal or diagonal form simultaneously. Therefore, it always provides better result.

References

[1] R.P. Roesser, *A Discrete State Space model Jor Linear Image Processing*, IEEE Trans. Automat. Contr., Vol. AC-20, pp. 1-10, Feb. 1975.

[2] A. Zilouchian and R. L. Carroll, *A coefficient Sensitivity Bound in 2-D State-space Digital Filtering*, IEEE Trans. Circuits Systems, Vol. CAS-33, pp 665-667, July 1986.

[3] M. Kawamata, T. Lin, and T. Higuchi, *Minimization of sensitivity of 2-D state-space digital filters and its relation to 2-D balanced realizations*, Proc. IEEE ISCAS, pp 710-713, 1987.

[4] K. Premaratne, E. I. Jury, and M. Mansour, *An Algorithm for Model Reduction of 2-D Discrete Time Systems*, IEEE Trans. CAS-37, pp 1116-1132, Sept. 1990.

[5] B.C. Moore, *Principle component analysis in linear systems: controllability, observability, and model reductions*, IEEE Trans. AC-26, pp 17-32, 1981.

[6] T. Hinamoto, K. Iwata, and W. Lu, *L2-Sensitivity Minimization of One- and Two-Dimensional State-Space Digital Filters Subject to L2-Scaling Constraints*, IEEE Trans. On Signal Processing, Vol. 54, No. 5, pp 1804-1812, May 2006.

[7] Steven J. Leon, *Linear Algebra with Applications*. Prentice Hall. 2005

[8] S. K. Mitra, *Digital Signal Processing* 3rd edition. McGraw-Hill. 2006.

[9] Guoliang Zeng and N. T. Phung, *An Application of Matrix in the Minimization of 2-D Digital Filters*, Pro. of 5th SIAM Conference on Applied Linear Algebra, pp 222-226, 1995.

Encapsulating connections on SoC designs using ASM++ charts

Santiago de Pablo, Luis C. Herrero, Fernando Martínez
University of Valladolid
Valladolid, Spain
Email: sanpab@eis.uva.es, {lcherrer,fer_mart}@tele.uva.es

Alexis B. Rey
Technical University of Cartagena
Cartagena, Murcia, Spain
Email: alexis.rey@upct.es

***Abstract* – This article presents a methodology to encapsulate, not only the functionality of several SoC modules, but also the connections between those modules. To achieve these results, the possibilities of Algorithmic State Machines (ASM charts) have been extended to develop a compiler. Using this approach, a SoC design becomes a set of chart boxes and links: several boxes describe parameterized modules in a hierarchical fashion, other boxes encapsulate their connections, and all boxes are linked together using simple lines. At last, a compiler processes all required files and generates the corresponding VHDL or Verilog code, valid for simulation and synthesis. A small SoC design with two DSP processors is shown as an example.**

I. INTRODUCTION

System-on-a-Chip (SoC) designs integrate processor cores, memories and custom logic joined into complete systems. The increased complexity requires not only more effort, but also an accurate knowledge on how to connect new computational modules to new peripheral devices using even new communication protocols and standards.

A hierarchical approach may encapsulate the inner functionality of several modules on black boxes. This technique effectively reduces the number of components, but system integration becomes more and more difficult as new components are added every day. Thus, the key to a short design time, enabling "product on demand", is the use of a set of predesigned components which can be easily integrated through a set of also predesigned connections.

Because of this reason, Xilinx and Altera have proposed their high end tools named Embedded Development Kit [1] and SoPC Builder [2], respectively, that allow the automatic generation of systems. Using these tools, designers may build complete SoC designs based on their processors and peripheral modules in few hours.

On the language side a parallel effort has been observed. In particular, SystemVerilog [3] now includes an "interface" element that allow designers to join several inputs and outputs in one named description, so textual designs may become easier to read and understand. At a different scale, pursuing a higher level of abstraction, the promising SpecC top-down methodology [4] firstly describes computations and communications at an abstract and untimed level and then descend to an accurate and precise level where connections and delays are fully described.

The aim of this paper is to contribute to these efforts from a bottom-up point of view, mostly adequate for educational purposes. First of all, we present several extensions to the Algorithmic State Machine (ASM) methodology, what we have called "ASM++ charts", allowing the automatic generation of VHDL or Verilog code from this charts, using a recently developed ASM++ compiler. Furthermore, these diagrams may describe hierarchical designs and define, through special boxes named "pipes", how to connect different modules all together.

II. ASM++ CHARTS

The Algorithmic State Machine (ASM) method for specifying digital designs was originally documented at 1973 by Christopher R. Clare [5], who worked at the *Electronics Research Laboratory* of *Hewlett Packard Labs*, based on previous developments made by Tom Osborne at the University of California at Berkeley [5]. Since then it has been widely applied to assist designers in expressing algorithms and to support their conversion into hardware [6-9]. Many texts on digital logic design cover the ASM method jointly with other methods for specifying Finite State Machines (FSM) [10].

A FSM is a valid representation for the behavior of a digital circuit when the number of transitions and the complexity of operations are low. The example of fig. 1 shows a FSM for a 12x12 unsigned multiplier that computes *'outP = inA * inB'* through twelve conditional additions. It is fired by an active high signal named *'go'*, it signals the answer using a synchronous output named *'done'*, and it indicates through an asynchronous *'ready'* signal that new operands are welcome.

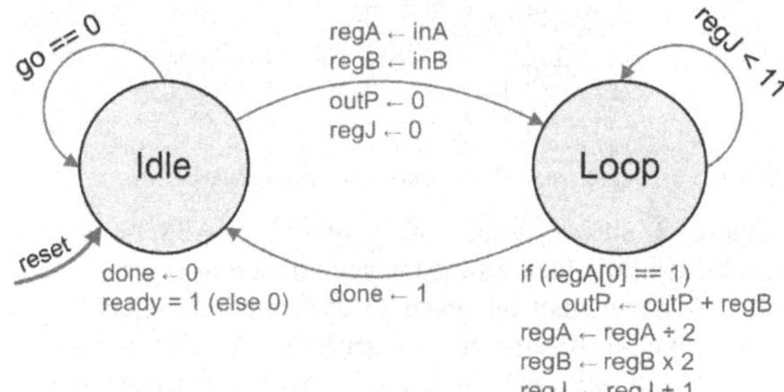

Fig. 1. An example of a Finite State Machine for a multiplier.

K. Elleithy (ed.), *Advanced Techniques in Computing Sciences and Software Engineering*,
DOI 10.1007/978-90-481-3660-5_55,

However, on these situations an ASM++ chart [11-12] may be more accurate, consistent and useful. These charts are an extension of traditional ASM charts [5] with different boxes to provide more clarity and flexibility, and additional boxes to allow automatic compilation. Moreover, they use standard HDL expressions –currently VHDL and Verilog are allowed– to increase designers' productivity.

The ASM++ chart equivalent to the previous FSM is shown at fig. 2. Its algorithmic section describes the circuit functionality, and starts with an oval box named *'Idle'* that represents the beginning (and the end) of each state (and clock cycle). Following, two boxes specify an asynchronous assertion to *'ready'* and a synchronous assignment to *'done'*. Thus, a DFF will be generated for the later signal and full combinational logic will be created for the former one. Later on, but on the same state, several synchronous operations are described after a diamond box that represents a decision. They are executed in parallel with previous operations, but only when previous conditions become valid. The following state would be *'Idle'* or *'Loop'*, depending on the value of *'go'* signal.

Comparing this representation with classical ASM charts [12], important differences arise: traditional ASM use rectangular boxes to describe the beginning of states and also any unconditional operation at that state, so a different shape must be used for conditional operations. Meanwhile, ASM++ makes no distinction among conditional and unconditional operations, and states are described using its own oval box.

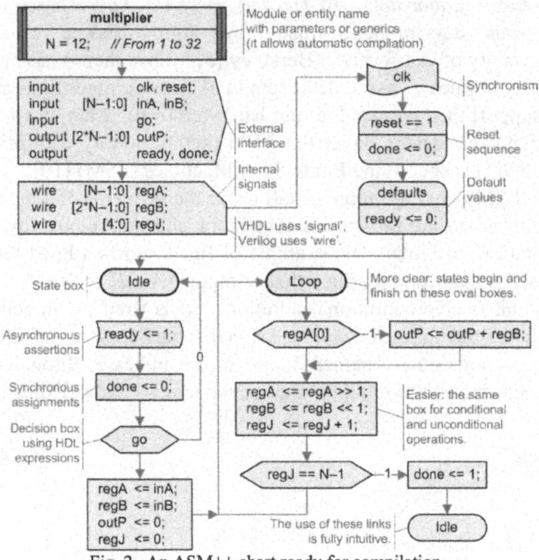

Fig. 2. An ASM++ chart ready for compilation.

Figure 2 shows additional features of ASM++ charts, included to allow their automatic compilation to generate HDL code. A name may be given to each module or entity; its implementation parameters or generics, the full peripheral interface, all internal signals, a reset sequence and a synchronization signal can also be fully specified in a very intuitive way. A box for *'defaults'* has been added to easily describe the circuit behavior when states leave any signal free. Furthermore, all boxes use well-known VHDL or Verilog expressions; the ASM++ compiler usually detects the HDL and then it generates valid HDL code using the same language.

III. HIERARCHICAL DESIGN USING ASM++ CHARTS

As soon as a compiler generates HDL code related to an ASM++ chart, several advanced features of modern VHDL and Verilog languages can be easily integrated on this methodology. The requirements for hierarchical design have been included through the following elements:

- Each design begins with a *'header'* box that specifies the design name and, optionally, its parameters or generics.
- Any design may use one or several pages on a MS Visio 2003/2007 document, saved using its VDX format. Each VDX document may include several designs identified through their header boxes.
- Any design may instantiate other designs, giving them an instance name. As soon as a lower level module is instantiated, the ASM++ compiler generates a full set of signals named "instance_name.port_name" (see fig. 4); these signals make connections with other elements simpler. Later on, any 'dot' will be replaced by an 'underline' because of HDL compatibility issues.
- When a description of an instantiated module is located on another file, a *'RequireFile'* box must be used before any header box to allow a joint compilation. However, the ASM++ compiler identifies any previously compiled design to avoid useless efforts and invalid duplications.
- VHDL users may include libraries or packages using their *'library'* and *'use'* sentences, but also before any header box.
- Nowadays, compiler does not support reading external HDL files, in order to instantiate hand written modules. A prototype of them, as shown at fig. 3, can be used instead.

Using these features, an example of a slightly improved multiplier can be easily designed. First of all, a prototype of a small FIFO memory is declared, as shown at fig. 3, so compilers may know how to instantiate and connect this module, described elsewhere on a Verilog file. Then three FIFO memories are instantiated to handle the input and output data flows, as shown at fig. 4, so several processors may feed and retrieve data from this processing element.

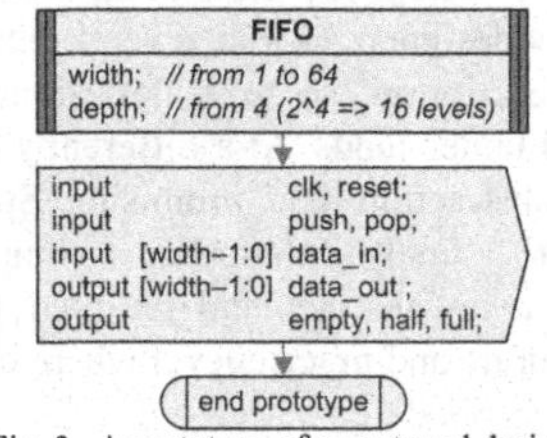

Fig. 3. A prototype of an external design.

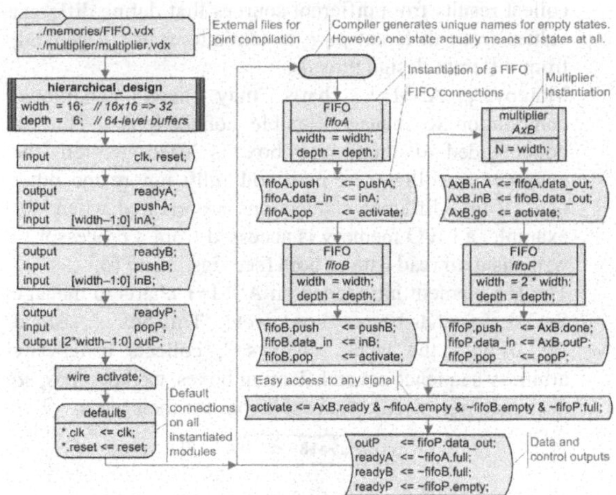

Fig. 4. An example of hierarchical design.

The ASM++ chart of fig. 4 can be compared with its compiled code, shown below. The advantages of this methodology on flexibility, clarity and time saving are evident. Not always a text based tool is faster and more productive than a graphical tool.

```
module hierarchical_design (clk, reset, inA, inB, outP,
                readyA, readyB, readyP, pushA, pushB, popP);
  parameter width = 16; // 16x16 => 32
  parameter depth =  6; // 64-level buffers
  input                     clk, reset;
  output                    readyA;
  input                     pushA;
  input       [width-1:0]   inA;
  output                    readyB;
  input                     pushB;
  input       [width-1:0]   inB;
  output                    readyP;
  input                     popP;
  output    [2*width-1:0]   outP;
  wire                      activate;
  wire                  fifoA_clk,        fifoA_reset;
  wire   [width-1:0]    fifoA_data_in,    fifoA_data_out;
  wire                  fifoA_push,       fifoA_pop;
  wire                  fifoA_empty,      fifoA_half,      fifoA_full;
  FIFO #(
    .width(width), .depth (depth)
  ) fifoA (
    .clk     (fifoA_clk),      .reset    (fifoA_reset),
    .data_in (fifoA_data_in),  .data_out (fifoA_data_out),
    .push    (fifoA_push),     .pop      (fifoA_pop),
    .empty   (fifoA_empty),    .half     (fifoA_half),     .full (fifoA_full)
  );
  wire                  fifoB_clk,        fifoB_reset;
  wire   [width-1:0]    fifoB_data_in,    fifoB_data_out;
  wire                  fifoB_push,       fifoB_pop;
  wire                  fifoB_empty,      fifoB_half,      fifoB_full;
  FIFO #(
    .width(width), .depth (depth)
  ) fifoB (
    .clk     (fifoB_clk),      .reset    (fifoB_reset),
    .data_in (fifoB_data_in),  .data_out (fifoB_data_out),
    .push    (fifoB_push),     .pop      (fifoB_pop),
    .empty   (fifoB_empty),    .half     (fifoB_half),     .full (fifoB_full)
  );
  wire                  AxB_clk,          AxB_reset;
  wire                  AxB_go,           AxB_ready,       AxB_done;
  wire   [width-1:0]    AxB_inA,          AxB_inB;
  wire [2*width-1:0]    AxB_outP;
  multiplier #(
    .N(width)
  ) AxB (
    .clk  (AxB_clk),   .reset (AxB_reset),
    .go   (AxB_go),    .ready (AxB_ready),    .done(AxB_done),
    .inA  (AxB_inA),   .inB   (AxB_inB),      .outP(AxB_outP)
  );
  wire                  fifoP_clk,        fifoP_reset;
  wire [2*width-1:0]    fifoP_data_in,    fifoP_data_out;
  wire                  fifoP_push,       fifoP_pop;
  wire                  fifoP_empty,      fifoP_half,      fifoP_full;
  FIFO #(
    .width(2 * width), .depth (depth)
  ) fifoP (
    .clk     (fifoP_clk),      .reset    (fifoP_reset),
    .data_in (fifoP_data_in),  .data_out (fifoP_data_out),
    .push    (fifoP_push),     .pop      (fifoP_pop),
    .empty   (fifoP_empty),    .half     (fifoP_half),     .full (fifoP_full)
  );
  assign fifoA_clk     = clk;      // Connections of 'defaults'
  assign fifoB_clk     = clk;
  assign AxB_clk       = clk;
  assign fifoP_clk     = clk;
  assign fifoA_reset   = reset;
  assign fifoB_reset   = reset;
  assign AxB_reset     = reset;
  assign fifoP_reset   = reset;
  assign fifoA_push    = pushA;    // User's connections
  assign fifoA_dataIn  = inA;
  assign fifoA_pop     = activate;
  assign fifoB_push    = pushB;
  assign fifoB_dataIn  = inB;
  assign fifoB_pop     = activate;
  assign AxB_inA       = fifoA_data_out;
  assign AxB_inB       = fifoB_data_out;
  assign AxB_go        = activate;
  assign fifoP_push    = AxB_done;
  assign fifoP_dataIn  = AxB_outP;
  assign fifoP_pop     = popP;
  assign activate      = AxB.ready & ~fifoA_empty
                         & ~fifoB_empty & ~fifoP_full;
  assign outP          = fifoP_data_out;
  assign readyA        = ~fifoA_full;
  assign readyB        = ~fifoB_full;
  assign readyP        = ~fifoP_empty;
endmodule   /// hierarchical_design
```

IV. Encapsulating Connections Using *Pipes*

Following this bottom-up methodology, the following step is using ASM++ charts to design full systems. As stated above, a chart can be used to instantiate several modules and connect them, with full, simple and easy access to all port signals.

However, system designers need to know how their available IP modules can or must be connected, in order to build a system. Probably, they need to read thoroughly several data sheets, try different combinations and fully verify the circuit behavior to finally match their requirements. Nonetheless, when they become experts on those modules, newer and better IP modules are developed, so system designers must start again and again.

This paper presents an alternative to this situation, called *"Easy-Reuse"*. During the following explanations, please, refer to figures 5 and 6.

- First of all, a new concept must be introduced: an ASM++ chart may describe an entity/module that will be *instantiated*, like 'multiplier' at fig. 2, but it may also be used for a description that will be *executed* (see fig. 5). The former will just instantiate a reference to an outer description, meanwhile the later will generate one or more sentences inside the module that calls them. To differentiate executable modules from those that will be instantiated, header boxes enclose one or more module names using '<' and '>' symbols. Later on, these descriptions will be processed each time a *'pipe'* or a *'container'* (described below) calls them.
- Additionally, the ASM++ compiler has been enhanced with PHP-like variables [13]. They are immediately evaluated during compilation, but they are available only at compilation time, so no circuit structures will be directly inferred from them. Their names are preceded by a dollar sign ('$'), they may be assigned with no previous declaration and they may store integer values, strings or lists of freely indexed variables.
- In order to differentiate several connections that may use the same descriptor, variables are used instead of parameters or generics (see fig. 5). The corresponding field at a header box, when using it to start a connection description, is used to define default values for several variables; these specifications would be changed by *pipes* on each instantiation (see fig. 6).
- Usual ASM boxes are connected in a sequence using arrows with sense; a new box called *"pipe"* can be placed out of the sequence and connect two instances through single lines.
- When compiler finishes processing the main sequence, it searches for *pipes*, looks for their linked instances, and executes the ASM charts related to those connections. Before each operation, it defines two automatic variables to identify the connecting instances. As said above, the *pipe* itself may define additional variables to personalize and differentiate each connection.
- As soon as several *pipes* may describe connections to the same signal, a resolution function must be defined to handle their conflicts. A tristate function would be used, but HDL compilers use to refuse such connections if they suspect contentions; furthermore, modern FPGAs do not implement such resources any more because of their high consumption, so these descriptions are actually replaced by gate-safe logic. Thus, a wired-OR has been chosen to collect results from different sources that define different values from different *pipe* instantiations or, in general, from different design threads.
- Moreover, ASM++ charts may need conditional compilation to manage flexible connections. Thus, a double-sided diamond-like box is used to tell the compiler to follow one path and fully ignore the other one. Thus, different connections are created when, for example, a FIFO memory is accessed from a processor to write data, to read data or both (see figs. 5 and 6).
- The last element introduced on ASM++ charts to manage SoC designs has been a *"container"*. This box, as seen at fig. 6 with the label "<Nexys>", collects using any arbitrary sequence several floating boxes, except *pipes*, so designers may decide the best layout for their ideas.

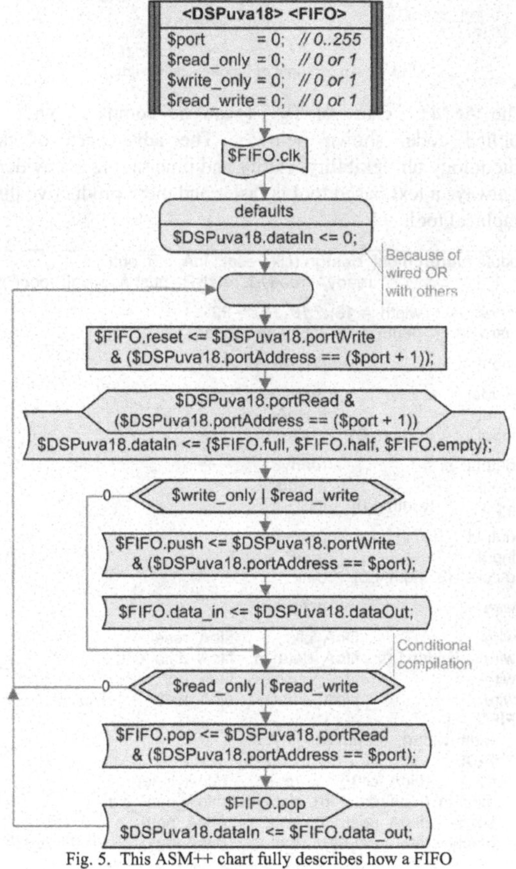

Fig. 5. This ASM++ chart fully describes how a FIFO must be connected to a DSP processor.

Using these ideas, a SoC design may now encapsulate not only the functionality of several components, but also their connections. The expertise acquired connecting own or third party modules can be encapsulated on easy to use elements, as shown on the following example at fig. 6.

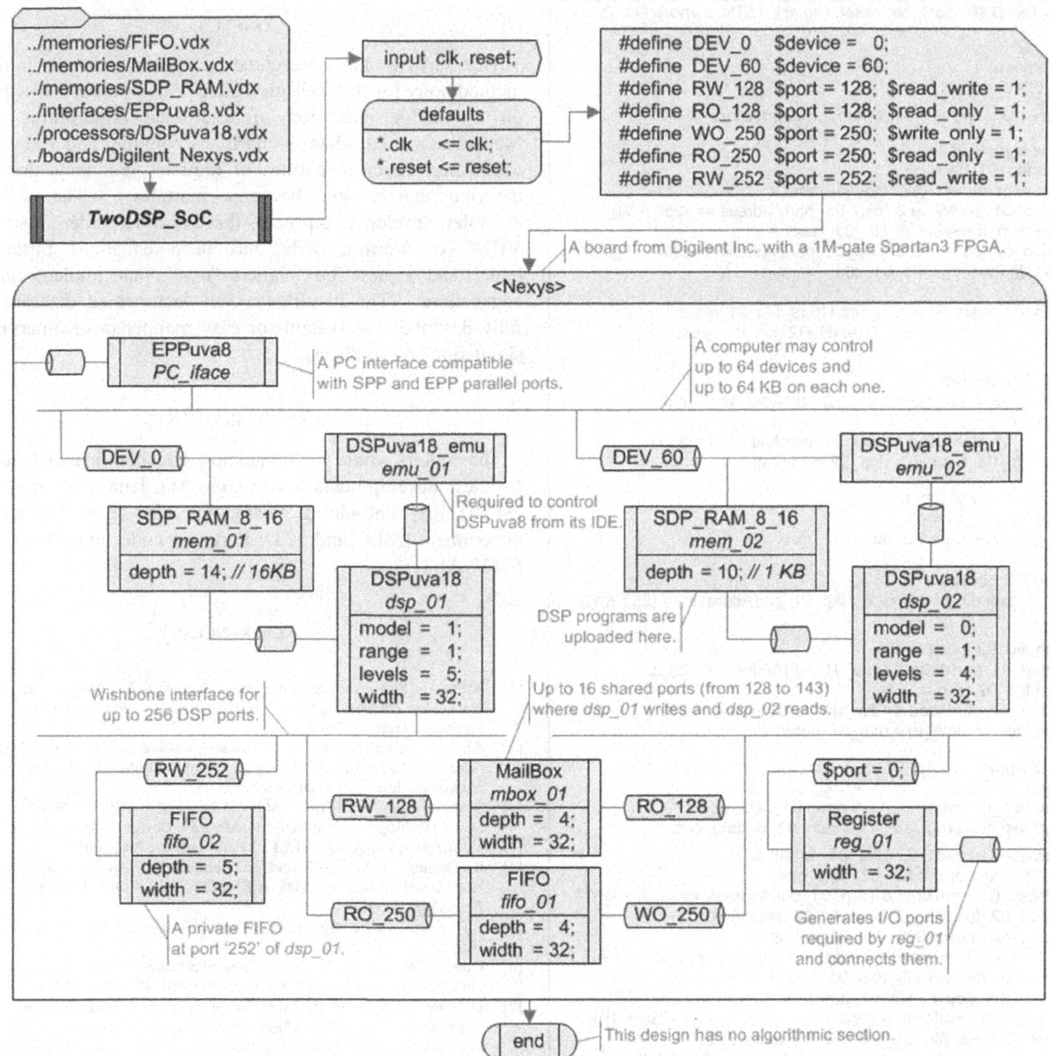

Fig. 6. A small SoC design example with two DSP processors using one 'container', eleven *'instances'* and fourteen *'pipes'*.

Figure 6 describes a small SoC designed to test all these features. It implements two Harvard-like DSP processors [11] named *"dsp_01"* and *"dsp_02"*, connected through 16 shared registers (*"mbox_01"*) and a FIFO (*"fifo_01"*). Additionally, these processors have a private synchronous dual port RAM (*"mem_01"* and *"mem_02"*) to store their programs and also an emulator to allow remote control from a PC, through parallel port. The first DSP has its own private FIFO to speedup the execution of filters [11], while the second one has a register connected to physical pads to display debug results.

On this SoC design, the chart from fig. 5 is executed three times: when *dsp_02* connects to *fifo_01* to write values (the *pipe* has a text "WO_250" that modifies variables $port and $write_only before chart execution); when *dsp_01* connects also to *fifo_01* to read from it (see the *pipe* with text "RO_250" at fig. 6); and finally when *dsp_01* connects to *fifo_02* to freely read and write values on it (see the *pipe* with text "RW_252").

Several sentences of the HDL code generated by the ASM++ compiler when processing these diagrams (and other ones located on six files, not included here for shortness) are displayed following.

This small example reveals that ASM++ charts are fully capable of describing the hardware of SoC designs using an intuitive, easy to use and consistent representation. Using this tool, designers may focus their attention on the software of the application, because adding more processors or peripherals is easy and does not require a large and difficult verification process.

```
module TwoDSP_SoC (clk, reset, reg_01_LEDs, /* more I/O */);

  input clk;
  input reset;
  output [7:0] reg_01_LEDs;

  /* more I/O ... wires ... instances ... connections */

  // Input signals of 'fifo_01':
  assign fifo_01_clk = clk;
  assign asm_thread1030_fifo_01_reset =
      dsp_01_portWrite & (dsp_01_portAddress == (250 + 1));
  assign asm_thread1032_fifo_01_reset =
      dsp_02_portWrite & (dsp_02_portAddress == (250 + 1));
  always @ (posedge fifo_01_clk)
  begin
      fifo_01_reset <= asm_thread1030_fifo_01_reset
                     | asm_thread1032_fifo_01_reset;
  end

  assign fifo_01_push =
      dsp_02_portWrite & (dsp_02_portAddress == 250);
  assign fifo_01_pop =
      dsp_01_portRead & (dsp_01_portAddress == 250);
  assign fifo_01_data_in = dsp_02_dataOut;

  // Input signals of 'fifo_02':
  assign fifo_02_clk = clk;
  always @ (posedge fifo_02_clk)
  begin
      fifo_02_reset <=
          dsp_01_portWrite & (dsp_01_portAddress == (252 + 1));
  end
  assign fifo_02_push =
      dsp_01_portWrite & (dsp_01_portAddress == 252);
  assign fifo_02_pop =
      dsp_01_portRead & (dsp_01_portAddress == 252);
  assign fifo_02_data_in = dsp_01_dataOut;

  // Input signals of 'dsp_01':
  assign dsp_01_clk        = clk;
  assign dsp_01_reset      = emu_01_dsp_reset;
  assign dsp_01_progData   = mem_01_B_data_out;

  assign asm_thread1020_dsp_01_dataIn =
      ((fifo_02_pop)) ? fifo_02_data_out :
      ((dsp_01_portRead & (dsp_01_portAddress == (252 + 1)))) ?
      {fifo_02_full, fifo_02_half, fifo_02_empty} : 0;
  assign asm_thread1029_dsp_01_dataIn =
      ((mbox_01_rw_read)) ? mbox_01_rw_data_out : 0;
  assign asm_thread1030_dsp_01_dataIn =
      ((fifo_01_pop)) ? fifo_01_data_out :
      ((dsp_01_portRead & (dsp_01_portAddress == (250 + 1)))) ?
      {fifo_01_full, fifo_01_half, fifo_01_empty} : 0;
  assign dsp_01_dataIn = asm_thread1020_dsp_01_dataIn
                       | asm_thread1029_dsp_01_dataIn
                       | asm_thread1030_dsp_01_dataIn;

  // Input signals of 'dsp_02':
  assign dsp_02_clk        = clk;
  assign dsp_02_reset      = emu_02_dsp_reset;
  assign dsp_02_progData   = mem_02_B_data_out;

  assign asm_thread1019_dsp_02_dataIn =
      ((dsp_02_portRead & (dsp_02_portAddress == 0))) ?
      reg_01_data_out : 0;
  assign asm_thread1031_dsp_02_dataIn =
      ((mbox_01_ro_read)) ? mbox_01_ro_data_out : 0;
  assign asm_thread1032_dsp_02_dataIn =
      ((dsp_02_portRead & (dsp_02_portAddress == (250 + 1)))) ?
      {fifo_01_full, fifo_01_half, fifo_01_empty} : 0;
  assign dsp_02_dataIn = asm_thread1019_dsp_02_dataIn
                       | asm_thread1031_dsp_02_dataIn
                       | asm_thread1032_dsp_02_dataIn;
endmodule   /// TwoDSP_SoC
```

V. CONCLUSIONS

This article has presented a powerful and intuitive methodology for SoC design named *"Easy-Reuse"*. It is based on a suitable extension of traditional Algorithmic State Machines, named ASM++ charts, its compiler and a key idea: charts may describe entities or modules, but they also may describe connections between modules. The ASM++ compiler, developed to process these charts in order to generate VHDL or Verilog code, has been enhanced further to understand a new box called *"pipe"* that implements the connections. The result is a self-documented diagram that fully describes the system for easy maintenance, supervision, simulation and synthesis.

ACKNOWLEDGMENTS

The authors would like to acknowledge the financial support for these developments from eZono AG, Jena, Germany, from ISEND SA, Valladolid, Spain, and also from the Spanish government (MEC and FEDER funds) under grant ENE2007-67417/ALT.

REFERENCES

[1] Xilinx, "Platform Studio and the EDK", on-line at *http://www.xilinx.com/ise/embedded_design_prod/platform_studio.htm*, last viewed on October 2008.

[2] Altera, "Introduction to SoPC Builder", on-line at *http://www.altera.com/literature/hb/qts/qts_qii54001.pdf*, May 2008, from *Quartus II Handbook*, last viewed on October 2008.

[3] SystemVerilog, "IEEE Std. 1800-2005: IEEE Standard for SystemVerilog - Unified Hardware Design, Specification, and Verification Language", IEEE, 3 Park Avenue, NY, 2005.

[4] R. Dömer, D.D. Gajski and A. Gerstlauer, "SpecC Methodology for High-Level Modeling", *9th IEEE/DATC Electronic Design Processes Workshop*, 2002.

[5] C.R. Clare, *Designing Logic Systems Using State Machines*, McGraw-Hill, New-York, 1973.

[6] Douglas W. Brown, "State-Machine Synthesizer - SMS", Proceedings of 18th Design Automation Conference, pp. 301-305, June 1981.

[7] D. Ponta and G. Donzellini, "A Simulator to Train for Finite State Machine Design", *Proceedings of 26th Annual Conference on Frontiers in Education Conference (FIE'96)*, vol. 2, pp. 725-729, Salt Lake City, Utah, USA, November 1996.

[8] J.P. David and E. Bergeron, "A Step towards Intelligent Translation from High-Level Design to RTL", *Proceedings of 4th IEEE International Workshop on System-on-Chip for Real-Time Applications*, pp. 183-188, Banff, Alberta, Canada, July 2004.

[9] E. Ogoubi and J.P. David, "Automatic synthesis from high level ASM to VHDL: a case study", *2nd Annual IEEE Northeast Workshop on Circuits and Systems (NEWCAS 2004)*, pp. 81-84, June 2004.

[10] D.D. Gajski, *Principles of Digital Design*, Prentice Hall, Upper Saddle River, NJ, 1997.

[11] S. de Pablo, S. Cáceres, J.A. Cebrián and M. Berrocal, "Application of ASM++ methodology on the design of a DSP processor", *Proc. of 4th FPGAworld Conference*, pp. 13-19, Stockholm, Sweden, September 2007.

[12] S. de Pablo, S. Cáceres, J.A. Cebrián, M. Berrocal and F. Sanz, "ASM++ diagrams used on teaching electronic design", *Innovative Techniques in Instruction Technology, E-learning, E-assessment and Education*, Ed. Springer, pp. 473-478, 2008.

[13] The PHP Group, on-line at *http://www.php.net*, last stable release has been PHP 5.2.6 at May 1st, 2008.

Performance Effects of Concurrent Virtual Machine Execution in VMware Workstation 6

Richard J Barnett[1] and Barry Irwin[2]
Security and Networks Research Group (SNRG)
Department of Computer Science
Rhodes University
Grahamstown, South Africa
[1]barnettrj@acm.org, [2]b.irwin@ru.ac.za

Abstract—**The recent trend toward virtualized computing both as a means of server consolidation and as a powerful desktop computing tool has lead into a wide variety of studies into the performance of hypervisor products.**

This study has investigated the scalability of VMware Workstation 6 on the desktop platform. We present comparative performance results for the concurrent execution of a number of virtual machines. A through statistical analysis of the performance results highlights the performance trends of different numbers of concurrent virtual machines and concludes that VMware workstation can scale in certain contexts.

We find that there are different performance benefits dependant on the application and that memory intensive applications perform less effectively than those applications which are IO intensive. We also find that running concurrent virtual machines offers a significant performance decrease, but that the drop thereafter is less significant.

I. Introduction

In the recent past, virtualization technologies have become increasingly popular. Full system virtualization and paravirtualization have become so commonplace that system builders are even offering solutions designed specifically for virtualization.

The sheer number of hypervisors on the market is a symptom of this popularity. VMware currently produces hypervisor products for a wide variety of platforms and its Workstation product is its flagship desktop solution.

In this paper we investigate the viability of the use of VMware Workstation 6 for the execution of multiple concurrent virtual machines. We explore the benefits of running a variety of concurrent virtual machine combinations and their respective performance attributes. We also investigate the popularity of virtualization technologies as reported by the wide variety of literature available on the subject.

The authors would like to acknowledge the support of the National Research Foundation and the financial support of Telkom SA, Business Connexion, Comverse SA, Stortech, Tellabs, Amatole, Mars Technologies, openVOICE and THRIP through the Telkom Centre of Excellence in the Department of Computer Science at Rhodes University.

A. Paper Organisation

The remainder of this paper is structured as follows: Section II discusses the background to virtualization and a selection of similar works. Section III discusses the process taken in evaluating Workstation 6, followed by the results obtained in Section IV. Finally conclusions which we have drawn from our research are presented in Section V.

II. Background and Related Work

Virtualization is a term which is used to describe a fairly broad suite of very different technologies. Whilst our research has only been concerned with one form of virtualization, we will briefly observe the presence of the others here.

There are three groups of virtualization technologies [1], *Hardware Level Virtualization*, *Operating System Level Virtualization* and *High Level Language Virtualization.*

The second, Operating System level virtualization, refers to technologies which offer process separation and isolation within a running system. *FreeBSD Jails* [1] are one example of this. The third refers to application level virtualization technologies such as the *Java Virtual Machine* and Microsoft's *.net CLR* [1].

Our interest, however, lies with hardware level virtualization which permits entire hardware systems to be virtualized. This is performed by making use of a small software layer, known as a hypervisor, which sits between the base hardware and the virtualized systems. Depending on the configuration, the hypervisor my be installed directly, or as an application program on top of an existing operating system [2].

This form of virtualization has existed since the 1960's [2] when it was commonplace on IBM systems. However, as the x86 architecture was not designed to support virtualization [3], the technology was underutilised for an extensive period. The recent adoption of extensions to the x86 architecture [3], [4] by both Intel [5] and AMD [6] and the wide variety of hypervisors on the market have led to massive growth and adoption of virtualization technologies in industry [7].

K. Elleithy (ed.), *Advanced Techniques in Computing Sciences and Software Engineering*,
DOI 10.1007/978-90-481-3660-5_56,

There is a wide variety of literature available on the performance of different hypervisors, but many of these focus on comparing two different technologies. Ahmad *et al.* [8] provide one study which is similar to our own, with the exception that it focuses exclusively on disk performance in VMware's enterprise ESX server.

VMware has conducted a number of its own studies which compare one of its products to an alternative. In the case of [9], VMware compares ESX Server to the open source Xen 3 server. The study concludes that customers should consider making use of ESX as Xen is unreliable on virtual SMP (symmetric multiprocessor) architecture. A variety of results are presented to confirm this.

III. Performance Testing

In our study, we constructed a test environment to asses the performance of multiple concurrent virtual machines so as to determine the effect this had on the performance of each virtual machine. A number of identical virtual machines were deployed to a university computer laboratory to obtain a significant number of results.

The results obtained from this initial process were statistically analysed to determine at what point the deployment of an additional virtual machine would negatively impact on the running of the remaining virtual machines.

This section describes the experiment we constructed and how we evaluated the results. It is subdivided into two subsections, the first (§ III-A) describes the configuration of our testing environment, the second (§ III-B) describes how we processed the data to obtain the relevant results.

A. Test Configuration and Deployment

In order to obtain an unbiased result, we deployed identically configured virtual machines to an entire laboratory of computers at Rhodes University. This permitted us to execute the test many times on identically configured hardware and to permit us to obtain many results quickly. (We made use of approximately 37 machines per test run.)

We achieved this by developing a small deployment and execution framework to allow the rapid transmission of virtual machine files and execution instructions to each computer. This framework was developed in Perl for two primary reasons. The first was to prevent having to compile code prior to (or after) deployment (so as to be able to use the framework in non-heterogeneous environments). The second as VMware's virtual machine control API VIX [10] is available with bindings for C++ and Perl.

The host machine was configured with Microsoft Windows as the base operating system. VMware Workstation 6 was installed and configured above this. This configuration was deployed to each workstation by making use of Norton Ghost.

TABLE I
Total Number of Test Results

Number of Concurrent VMs	Total Results Obtained
1	74
2	146
3	217
4	287
5	350

The virtual machine images were configured with Ubuntu Linux and had the *dbench* [11] and *Netperf* [12] benchmarks installed. dbench was configured to access the local (virtual) disk so as to determine the performance of the disk. Netperf was, however, used somewhat unconventionally as it was used to test the performance of the loopback interface, rather than using it to test an actual network interface. This was done to act as a CPU and memory test rather than as a network adaptor interface.

A standard Desktop install of Ubuntu (as we used it) installs a X11 server and whilst, due to the overheads of running X11, running this would have had a detrimental effect on the performance of our systems, we left it as we were only interested in the relationship between the results from differing numbers of concurrent machines, rather than the individual results themselves.

Deployment of the virtual machine images was done by making use of the Perl framework and the multicast file transfer application UFTP [13]. Hosts were configured to start one copy of the virtual machine, let its benchmark run, upload its results to the central server, and reboot. It then ran two concurrently and was repeated to five concurrent virtual machines executing.

VMware Workstation imposes an upper limit on the number of virtual machines that can be run concurrently based on the memory requirements of the virtual machine. Specifically we found that you could not run machines with more memory than was physically present in the machine. We also found that VMware Workstation warned about memory swapping at roughly two thirds of total memory utilisation.

As a result of this we limited our tests to five concurrent virtual machines each with 128MB of virtual memory, executing on a physical machine which had 1GB of memory. This was the limit before VMware Workstation warned about swapping. In theory we could have executed eight concurrent machines.

B. Data Processing

Results from the testing phase were uploaded automatically to a central server for processing. Both dbench and Netperf produced fixed format results files, from which it was possible to extract the relevant information.

Not all of the test cases succeeded. There were some cases where the transfer of the virtual machine image or results transfer failed, and as a result the number of results obtained

TABLE II
SHAPIRO-WILK TEST RESULTS FOR
DISK THROUGHPUT AND NETWORK LOOPBACK THROUGHPUT

Number of Concurrent VMs		W Statistic	p-value
1	Disk	0.9864	0.6161
	Net	0.4727	7.36×10^{-15}
2	Disk	0.8374	2.012×10^{-11}
	Net	0.8738	8.178×10^{-10}
3	Disk	0.7255	$< 2.2 \times 10^{-16}$
	Net	0.8738	8.178×10^{-10}
4	Disk	0.9405	2.375×10^{-9}
	Net	0.8507	5.64×10^{-16}
5	Disk	0.9604	4.004×10^{-8}
	Net	0.8693	$< 2.2 \times 10^{-16}$

is not perfectly divisible by the number of computers in the laboratory. Table I shows the number of results obtained. The data obtained was processed using the statistical programming language R [14]. The results are presented in the next section.

IV. TESTING RESULTS

In order to successfully ascertain the point at which adding an additional virtual machine had a significant effect on the performance of the other running virtual machines we performed statistical analysis on the benchmarking test results.

Our initial assessment was performed by analysing side-by-side boxplots of the test results. These boxplots allowed us to obtain an idea of the distribution and shape of the population. It can be seen from Figure 1 that our test results have both a skewed and longtailed distribution. From this we can estimate that the distribution is not normal.

Despite this, however, in order to determine if there was a significant difference in the performance as the number of concurrent virtual machines increased, we needed to know for sure if the data was normal. We made use of the Shapiro-Wilk test for data normality, the results of which are shown in Table II. The null hypothesis (H_0) for this test is that the population is normal. The alternate hypothesis (H_a) is that the data is not normal. The null hypothesis is rejected if the p-value is less than or equal to the significance level as in (1). For this experiment we held the significance level (α) at 5% or 0.05.

$$\text{p-value} \leq \alpha \tag{1}$$

It can be seen from Table II that the only case where the p-value is above α is for a single virtual machine's disk performance. Therefore the majority of the populations for this experiment are not normalised. As a result of this, we made use of the Kruskal-Wallis rank sum test to determine whether the effects were significant.

We performed the Kruskal-Wallis test with the null hypothesis (H_0) as shown in (2). The alternative hypothesis (H_a) is that at least two of the population means ($\tilde{\mu_i}$) show a significant difference ($\tilde{\mu_i}$ representing i concurrent virtual machines).

$$H_0\colon \tilde{\mu_1} = \tilde{\mu_2} = \tilde{\mu_3} = \tilde{\mu_4} = \tilde{\mu_5} \tag{2}$$

The results for this test showed that p-value $\leq \alpha$ is true for both tests (in both cases the p-value was $< 2.2 \times 10^{-16}$)[1] and as a result we find that there is a significant difference in performance when multiple concurrent virtual machines are executed. The Kruskal-Wallis test was performed with 4 degrees of freedom and the χ^2 value was 213.7616 and 260.7058 for disk throughput and loopback interface throughput respectively.

This result was, however, expected and of more interest to us was at what point the effect becomes significant. It therefore become appropriate for us to determine where the differences in population medians occur. We made use of the Wilcoxon rank sum test. As we were only comparing two of the five populations in each test we altered the significance in accordance with the Bonferoni procedure. The new α is shown in (3).

$$\alpha^* = \frac{\alpha}{\binom{n}{m}} = \frac{0.05}{\binom{5}{2}} = 0.005 \tag{3}$$

Table III illustrates the results of the Wilcoxon rank sum test for both Disk Throughput and Network Loopback Throughput. It can be seen that there is a significant effect in both tests when transitioning from one virtual machine to multiple concurrent virtual machines.

Of more interest, however, we found that the results for the network loopback throughput and disk throughput show different characteristics. We do, however, find that three virtual machines performed significantly worse than expected and indeed performed worse than four or five concurrent virtual machines.

The results of the Wilcoxon test suggest that if we are going to run multiple concurrent virtual machines, for disk throughput (or IO) we get a similar performance from five concurrent machines as two but for network loopback performance (or memory) there is a more significant effect.

V. CONCLUSIONS

In this paper we have analysed VMware Workstation 6 and have investigated the scalability of the product for use with a variety of concurrently executing virtual machines.

We have identified that it is increasingly popular to run several concurrent virtual machines and that there is some performance penalty to doing so. Despite this, should there be a need to deploy several virtual machines concurrently, we find that the ideal number of virtual machines which can be easily run concurrently differs dependant on the nature of the computation within the virtual machine.

[1]R's minimum resolution is 2.2×10^{-16} and results less than this are simply expressed as being $< 2.2 \times 10^{-16}$

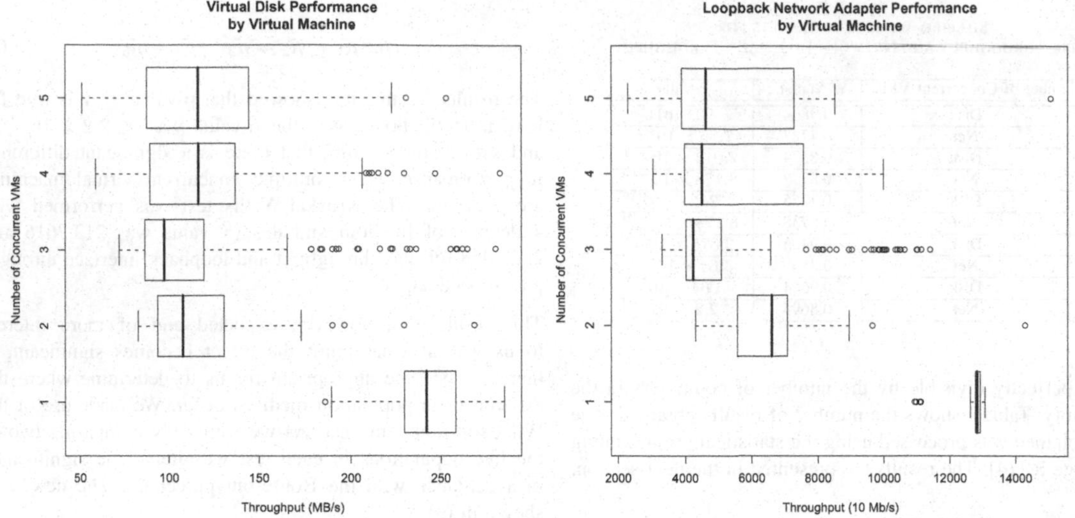

Fig. 1. Side-by-Side Boxplots of Disk Throughput and Loopback Interface Throughput

TABLE III
WILCOXON RANK SUM TEST RESULTS FOR DISK THROUGHPUT AND NETWORK LOOPBACK THROUGHPUT

W Statistic / p-value	Number of VM's ($\phi < 2.2 \times 10^{-16}$)									
Number of VMs	1		2		3		4		5	
1			10730	ϕ	16044	ϕ	21238	ϕ	25826	ϕ
2	10715	ϕ			26309	ϕ	25951	4.882×10^{-5}	33493	4.773×10^{-8}
3	15575	ϕ	19888	3.663×10^{-5}			28952	0.1767	36632	0.4789
4	21042	ϕ	20520	0.7266	25163	0.0002231			58327	0.0004555
5	25749	ϕ	24325	0.4	30920	0.0001987	49346	0.738		
Disk Throughput					Network Loopback Throughput					

We found that there is a significant performance penalty to running multiple concurrent virtual machines both for IO and for memory applications, however we found that once we increased from two concurrent virtual machines that memory and IO performed differently.

In IO intensive applications we found that the drop in performance from two to three, four and five concurrent virtual machines was statistically insignificant. For memory intensive applications, there was an effect moving from two to three machines, but no significant effect from three to four or five.

We also found that there is an upper limit on the number of concurrent virtual machines we can run. This limit is based on the memory requirements of the virtual machine as well as the capabilities of the physical hardware.

A. Limitations of the Results

Our results show a variety of beneficial effects to running concurrent virtual machines. They do however have a few limitations. VMware Workstation 6 is a desktop hypervisor product and is likely to be used simultaneously with the host system. Our results do not take this into consideration and as a result, the performance may be degraded when used in those circumstances.

Further, these tests pushed the virtual machines to their limit by performing benchmarks which stress test the respective component. Most running systems will not push the virtual machine to its limits.

Finally, it is also noted that virtual machines are unlikely to perform identical tasks, virtually simultaneously. In practice, this along with the previous point should cause performance to increase.

Future work may test this with benchmarks from SPEC [15] and VMware [7]. Both organisations are developing virtualization specific benchmarks.

ACKNOWLEDGMENT

The authors would like to thank Mr Jeremy Baxter of the Department of Statistics at Rhodes University for his advice when we were performing the analysis for this work.

REFERENCES

[1] M. Rosenblum, "The reincarnation of virtual machines," *Queue*, vol. 2, no. 5, pp. 34–40, 2004.

[2] "Ibm systems: Virtualization," IBM, Tech. Rep., 2005.

[3] K. Adams and O. Agesen, "A comparison of software and hardware techniques for x86 virtualization," in *ASPLOS-XII: Proceedings of the 12th international conference on Architectural support for programming languages and operating systems*. New York, NY, USA: ACM Press, 2006, pp. 2–13.

[4] J. Humphreys and T. Grieser, "Mainstreaming server virtualization: The intel approach," IDC, Tech. Rep., 2006.

[5] "Intel virtualization technology." [Online]. Available: http://www.intel.com/technology/platform-technology/virtualization/index.htm

[6] "Advanced micro devices." [Online]. Available: http://www.amd.com

[7] V. Makhija, B. Herndon, P. Smith, L. Roderick, E. Zamost, and J. Anderson, "Vmmark: A scalable benchmark for virtualized systems," VMware, Inc., Tech. Rep., 2006.

[8] I. Ahmad, J. M. Anderson, A. M. Holler, R. Kambo, and V. Makhija, "An analysis of disk performance in vmware esx server virtual machines," in *Proceedings of the Sixth Annual Workshop on Workload Characterization*, 2003.

[9] "A performance comparison of hypervisors," VMWare Inc., Tech. Rep., 2007.

[10] I. VMware, "Vix api reference documentation." [Online]. Available: http://pubs.vmware.com/vix-api/ReferenceGuide/

[11] A. Tridgell, "dbench readme." [Online]. Available: http://samba.org/ftp/tridge/dbench/README

[12] R. Jones, "Netperf manual." [Online]. Available: http://samba.org/ftp/tridge/dbench/README

[13] D. Bush, "Uftp homepage." [Online]. Available: http://www.tcnj.edu/~bush/uftp.html

[14] R Development Core Team, *R: A Language and Environment for Statistical Computing*, R Foundation for Statistical Computing, Vienna, Austria, 2007, ISBN 3-900051-07-0. [Online]. Available: http://www.R-project.org

[15] SPEC, "Spec to develop standard methods of comparing virtualization performance," 11 2006. [Online]. Available: http://www.spec.org/specvirtualization/

Towards Enterprise Integration Performance Assessment based on Category Theory

Victoria Mikhnovsky, Olga Ormandjieva
Concordia University, Montreal, Canada
E-mails: tori.mikhnovsky@hotmail.com, ormandj@cse.concordia.ca

Abstract - **A major difference between what we refer to as a "well-developed" science, such as civil engineering, and sciences which are less so, like enterprise engineering, is the degree to which nonfunctional requirements, such as performance, are integrated into the design and development process, and satisfaction of those requirements is controlled by theoretically valid measurement procedures. This paper introduces the preliminary results, which are aimed at developing a concise formal framework for enterprise performance modeling, measurement, and control during enterprise integration activities. The novelty of this research consists in employing the mathematical category theory for modeling purposes, an approach that is broad enough to formally capture heterogeneous (structural, functional and nonfunctional) requirements, by, for example, using the constructs from the graphical categorical formal language.**

Keywords - **Performance modeling, performance measurement, enterprise integration, category theory.**

I. INTRODUCTION

Enterprise integration occurs when there is a need to improve interaction among people, systems, departments, services, and companies (in terms of material flows, information flows, or control flows). The scenarios are immensely complex nowadays, as they frequently include mergers, fusions, acquisitions, or new partnerships. From an organizational standpoint, enterprise integration is concerned with facilitating information, control, and material flows across organization boundaries by connecting all the necessary functions and heterogeneous functional entities (e.g. information systems, devices, applications, and people) in order to improve communication (data and information exchanges at system level), cooperation (interoperation at application level), and coordination (timely orchestration of process steps at the business level) within the enterprise so that it behaves as an integrated unit [25].

In systems engineering and requirements engineering, nonfunctional requirements are requirements which specify criteria that can be used to judge the operation of a system. In the context of enterprise modeling and integration, non-functional requirements are typically referred to as performance requirements. We will be using both terms interchangeably.

The research presented in this paper is motivated by the need to build performance requirements into enterprise models, a need rooted in the current industrial trend towards developing complex integrated enterprises. The importance of enterprise integration compliance with the imposed performance requirements requires continuous control of the performance indicators in real time, which brings up another important research issue, namely, understanding complex and frequently ill-defined nonfunctional requirements, their roles, and their inter-relations in increasingly complex large-scale enterprise integration systems.

The goal of this research is to represent the enterprise integration structure and enterprise performance model in a single formal framework which is broad enough to encompass enterprise integration strategies, objectives, and communication structure. Our approach is to develop an enterprise modeling and measurement framework based on category theory. Through the application of this theory, we seek to avoid the undesirable results of decision-making in quality control based on the informal and sometimes arbitrary assignment of numbers, as proposed by many measures [1, 2, 3].

The rest of the paper is organized as follows. Section 2 introduces the related work in performance modeling and control in enterprise integration. Section 3 explains the use of category theory constructs for building performance into the overall enterprise architecture model and the theoretical basis for enforcing performance indicators in enterprise integration. The proposed enterprise integration performance model is outlined in section 4. The paper concludes with section 5, comprising a discussion of the approach and an outline of future work directions.

II. RELATED WORK

The background to this research extends back to the mid-1980s, when the need for better integrated performance measurement systems was identified [4, 5, 6, 7, 8]. Since then, there have been numerous publications emphasizing the need for relevant, integrated, balanced, strategic, and improvement-oriented performance measurement systems. None of the existing studies [9, 11, 12, 13, 14, 15, 16, 17] considers qualitative performance modeling and quantitative performance control from an enterprise integration perspective, however. In order to identify the most appropriate performance measurement system for enterprise integration, a review of performance measurement systems in supply chains, extended enterprises, and virtual enterprises was conducted [1, 18, 19, 20]. The majority of the enterprise integration measurements proposed so far have been adopted from the

K. Elleithy (ed.), *Advanced Techniques in Computing Sciences and Software Engineering*,
DOI 10.1007/978-90-481-3660-5_57,

traditional measurement procedures of single enterprises corresponding to supply chains (i.e. plan, source, make, deliver, etc.).

Despite the large expenditures on enterprise integration, few companies are able to assess the current level of performance of their business. Performance measurement at an individual enterprise level plays only a minor role, and at the integration level it is virtually nonexistent. In addition, those few enterprises that do have an enterprise measurement system in place usually take only financial and time-related aspects into consideration. The reason for this is the absence of suitable performance measurement procedures developed specifically for enterprise integration performance characteristics. In order to measure that performance, an enterprise integration model is needed, on the basis of which the measurement of performance can be defined.

The research presented here differs from the existing work in the area in at least two important ways: (i) the components, the measurement procedure, and the performance indicators are represented formally as categories within the same formal framework, which makes it possible to monitor the satisfaction of requirements automatically; (ii) the way in which the behavior of the overall system depends on the performance indicators from the perspective of the integrated enterprise system as a whole is established.

III. Generic Framework for Enterprise Integration and Performance Assessment

To develop a model of a complex system, the first problem is to derive a framework for discussion. Formal frameworks treat the system as a collection of mathematical objects where the behavior of each object could be described by various models like automata, Petri-nets, set theory, or process algebra, to name a few. That requires a sound scientific and theoretical basis, where system specifications, designs, and implementations are treated as mathematical objects (system properties can be investigated by formal analysis).

Because non-functional requirements like performance tend to have a wide-ranging impact on enterprise architecture, existing software modeling methods are incapable of integrating them into the software engineering process. In our work here, a number of important enterprise modeling concepts are given a theoretical foundation through the use of category theory as an advanced mathematical formalism which is independent of any modeling or programming paradigm.

A. Enterprise integration modeling

Enterprise integration is concerned with interoperability among business processes, either within a large (distributed) enterprise or within an enterprise network. In this paper, we present an enterprise integration model which describes the structure and behavior of a business process in terms of its properties and actions.

The state of the business process is the set of (usually static) properties plus the current (usually dynamic) values of each property. The behavior of the business process shows the way in which the business process acts and reacts, in terms of state changes and event-raising. The state space for a business process is a cross product of property domains, and consequently describes the current state of the process as a selection from this state space. It describes the events of a business process as transitions between states and allows events and actions to be defined. When two business processes are combined, the states for both processes are combined as if a natural join is preformed on the two state spaces where common attributes have the same values. The combined actions have to satisfy the preconditions and postconditions that have been joined. The process invariants are joined by logical conjunction, and it is possible for the two invariants to contain conflicting conditions such that the combination resolves to false (no states are possible). If this is the case, the business processes are deemed incompatible. The combined invariant is applied after the two unconstrained state spaces have been joined.

B. Categorical framework

The objective in this section is to illustrate performance assessment in enterprise architectures using the language of category theory. We use the formalism constructs of category theory to represent our conceptual framework, where a homomorphism of objects is called a *function* and a homomorphism of categories is called a *functor*. Informally, a category is a collection of heterogeneous objects and morphisms which model the social life of these objects, that is, their interactions. A category can be defined as zero or more objects bound together, where each object may be either a primitive or a category. A category may be produced by explicitly listing functions and objects or by invoking a function which returns a category. A category may also be augmented, diminished, or joined with other categories to produce a new category. Morphisms can be used to define interconnection within the enterprise business processes or between enterprises. The morphisms are represented graphically as arrows (see, for instance, Fig. 2). The categorical rules require the presence of: i) the identity morphism: ii) the operator composition of the morphisms; and iii) the associativity law.

Owing to lack of space, we only give a high-level sketch of the framework and omit the category theory basics that are assumed for an understanding of the model developed. Details on category theory can be found in [22, 23, 24]. It is to be noted that the concepts of soundness and completeness do not arise in categorical specification, because this specification uses algebraic theories rather than treating rules of deduction.

From the category theory point of view, Business Process and Performance Measurement are considered to be categories. Enterprise integration consists of the Business Process categories (see section 3.3 for relevant definitions and concepts), their integration (see section 3.4), and assessment of their performance (see section 3.5).

C. The Business Process

In this section, we model the empirical structure of a business process, which is the fundamental component of enterprise integration. A business process can be described in terms of its properties and its behavior. The *properties* of a business process are independent of their values at any given point in time. The *state* of a business process refers to the set of (usually static) properties plus the current (usually dynamic) values of each property. The *behavior* of a business process shows how the business process acts and reacts, in terms of state changes and the events being raised. A property is the name of a set of potential values, which is called a *property domain.* The business process policies may impose direct limitations on the possible values of a property. We refer to these limitations collectively as *process invariants*. The process invariant is a set of conditions taken in conjunction with one another which must be satisfied by every process in any *steady state* of the process. An event causes a process to transition from one state to another, or possibly to the same state, as determined by its behavior: if the source state satisfies the preconditions of the event *E*, then the business process will change state if and only if the postconditions of *E* are satisfied by the destination state.

The Business Process category consists of seven types of objects (Business Process Behavior, Business Process Property, Process Domain *PrDom,* Business Process Policy, Business Process State Space, Business Process State Transition, Business Process Event) and nine morphisms (arrows), defined as follows: *raise*: BP Behavior → BP Event; *value*: BP Property → PrDom; *constraint*: PrDom → BP Policy; *invariant*: BP State Space → BP Policy; *behavior*: BP Property → BP State Transition; *source, destination:* BP State Transition → BP State Space; *define*: PrDom: CI → BP State Space; *trigger:* BP Event →BP State Transition; and *precondition, postcondition*: BP Event → BP Policy. The diagram commutes as shown in Fig. 2, which ensures behavior compliance with the business process policies. The Identity morphisms are omitted from the Business Process category to simplify the presentation.

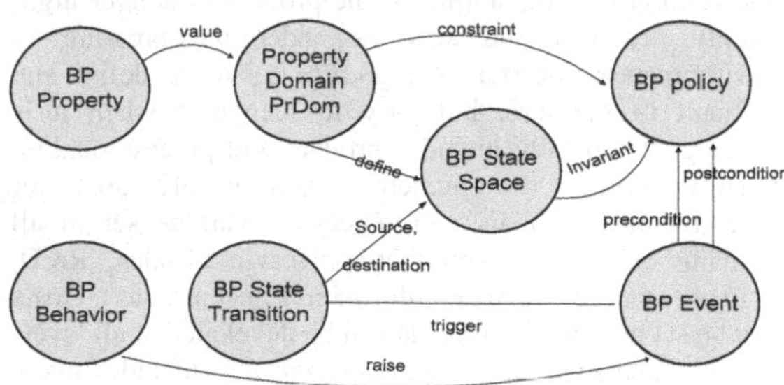

Fig. 1. Business Process Category Theory

D. Integrating Business Process categories

When two business processes are combined, the states for both processes are combined, as if a natural join is preformed on tables formed by the two state spaces. The rows in the new state space are created by combining rows in the two merged state spaces in which common attributes have the same values. The process invariants are joined by logical conjunction. If the two invariants contain conflicting conditions, such that the combination resolves to false (no states are possible), then the business processes are deemed *incompatible.* The combined invariant is applied after two unconstrained state spaces are joined. Event arrows are drawn between every pair of states for which the source state satisfies the preconditions and the target state satisfies the post conditions while duplicate arrows are dropped. The above explained integration of two business processes BP1 and BP2 can be implemented as a categorical product of the corresponding categories BP1×BP2 (all possible pairs <element from BP1, element from BP2>). Definitions for the resulting category are straightforward and have therefore been omitted.

E. Performance Measurement category

A major application of the representational theory of measurement is to facilitate decision-making. Making judgments based on performance indicators should satisfy certain conditions (axioms) about an individual's judgments, such as preferences, which make the measurement possible. To perform a measurement, we start with an observed or empirical relational structure E, representing entities to be judged (like, for instance, enterprise architecture), and we seek a mapping (here, performance measurement) to a numerical relational structure R, which "preserves" all the relations and operations in E. A mapping from one relational structure to another that preserves all relations is called a *homomorphism* (see Fig. 2).

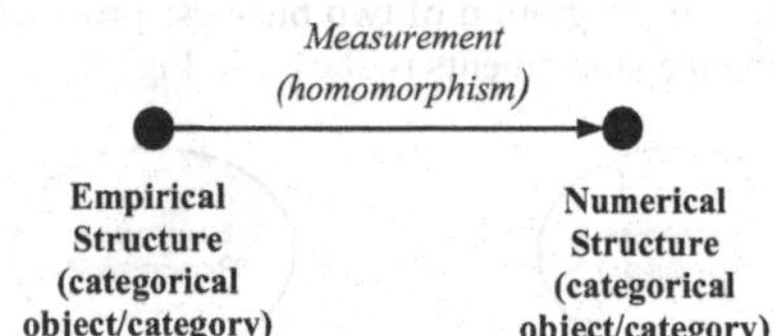

Fig. 2: Representing Measurement

Using the categorical theoretical framework, we will analyze the mathematical mappings from the enterprise integration structure to the performance assessment results (usually expressed as numbers) and the properties of those mappings. The mapping analysis is important for at least three reasons: (1) understanding the mathematical properties of mapping; (2) detecting the undesirable properties of mapping; and (3) correctly using the resulting numbers in the statistical and related performance analysis required for decision-making.

The Performance Measurement category represents high-level performance goals decomposed into more specific performance subcharacteristics called *indicators*, which lead to the definition of a hierarchical model for performance measurement (see section 4). The indicators are operationalized into measurement *procedures*. These procedures provide the operations, processes, data representations, structuring, and constraints in the target system required to quantify the current level of performance indicators. The measurement data are collected from the empirical structure while the measurement procedures are applied (see Fig. 3). The analysis model defines rules relating the quantitative information to the qualitative performance goals in the language of the measurement users who will make decisions based on the performance assessment feedback. The performance measurement procedure is expressed formally in terms of the composition operator •, assigning it to each pair of arrows: *Performance measurement* = *decompose* • *refine* • *apply* • *analyze*. The diagram commute, that is, a change in a performance goal will be traced to the indicators and to the corresponding operationalizations, as well as to the measurement data collection and analysis processes.

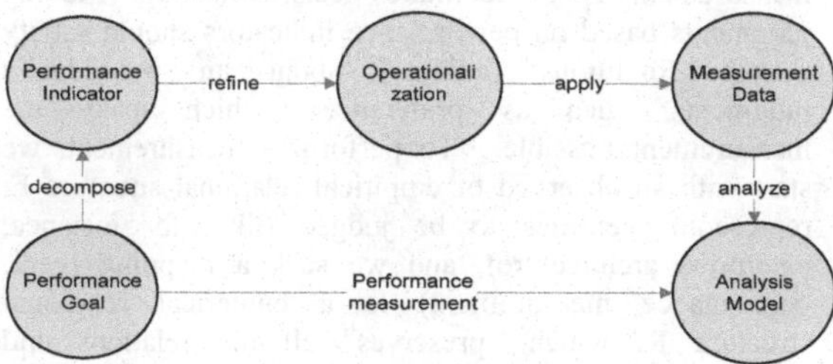

Fig. 3. Categorical representation of a performance model

F. Integrating Business Processes and Performance Measurement

A diagram of the integration of two business processes and their performance assessments is shown in Fig. 4.

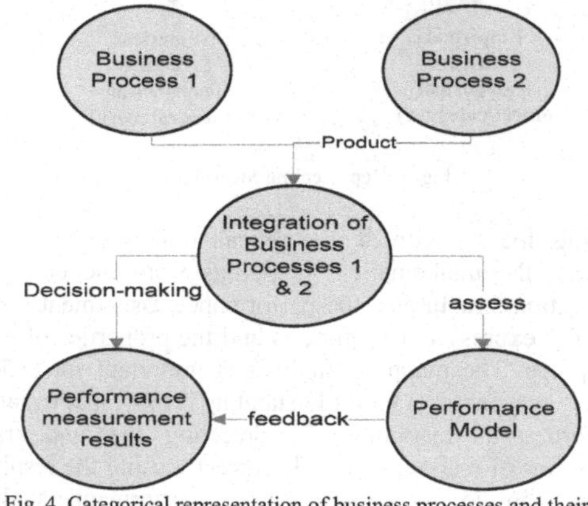

Fig. 4. Categorical representation of business processes and their performance assessments

The outcome of the measurement process are the Performance Measurement Results, which provide feedback on the performance level and help decide whether or not the business processes satisfice the performance requirements. The diagram commutes that ensure the preservation of the empirical relational structure (integrated business processes) in numbers (measurement results) through performance modeling and measurement will guarantee the theoretical validity of the performance assessment and thus feasible and reliable decision-making based on the performance results collected.

The goal in the next section is to build a systematic, quantitative, and formal approach to performance assessment in enterprise integration.

IV. Enterprise Performance Measurement Model (EPMM)

We present a model for enterprise performance decomposed into a hierarchy of measurable indicators in line with enterprise decision-making strategy and objectives.

The most difficult question in developing such a model is deciding what to measure and how to measure it. In our research, we used the goal-driven approach to decompose the performance of enterprise integration into eight qualitative performance factors. These factors are further refined into quantitative indicators derived from measurement procedures (the lowest hierarchical level), as illustrated in Fig. 5 with the example of dependability.

Process performance requirements can be modeled and measured along many dimensions. *Flexibility*, sometimes referred to as agility, is the ability of a process to change the way it operates or change its outputs, such as products or services, for example. In today's fast-changing environments, this is considered to be one of the most important qualities of any enterprise. The *dependability* of an enterprise reflects the degree of trust or confidence that its clientele, partners, or anyone else in the outside world places in that enterprise. When we are talking about dependability in the context of an enterprise, we are concerned not only with the reliability of a product, but also with reliability of the delivery of that product as an operational issue. Two important dimensions of *operational dependability* are: *availability* – the ability of a process to deliver products and services when requested, and *reliability* – the ability of the process to deliver high-quality products and services under the pressure of environmental uncertainties. *Quality* is hard to define and difficult to measure, but easy to recognize when it is missing. It typically includes product and process quality. Quality models are numerous and mostly goal- or objective-based. Goals and objectives can be set in all domains of activity (production, services, sales, R&D, human resources, finance, information systems, etc.). Goals can be set and quality models can be developed at all levels in an enterprise. *Timing Requirements* include many categories of requirements as a measure of enterprise performance, such as process lead time, process response time, and new product introduction lag, among others. The *lead time* of business process is the interval between the start and the end of the process. Reducing lead time will eliminate all non-value-adding activities and will free up resources, reduce cost, and possibly improve quality. *Cost* provides tremendous insights into enterprise problems and inefficiencies, and measuring the cost of the process provides avenues for process improvement strategies. Another important performance requirement is *asset utilization.* The term *asset* is very broad, meaning anything from human resources to manufacturing plants. In many companies, assets are worth billions of dollars, and optimizing their utilization by business and organizational processes is an important issue. The maximum output of the process is called its *capacity*. All the work processes and subprocesses at all levels must be balanced in terms of capacity; otherwise, there will be bottlenecks and delays. There must be strategic alliances in place among enterprises in integration in order for them to handle variable capacity. For example, a little overcapacity to meet rush demands can improve operational measurement procedures; overcapacity can be assumed as an operationalization of the Capacilty indicator.

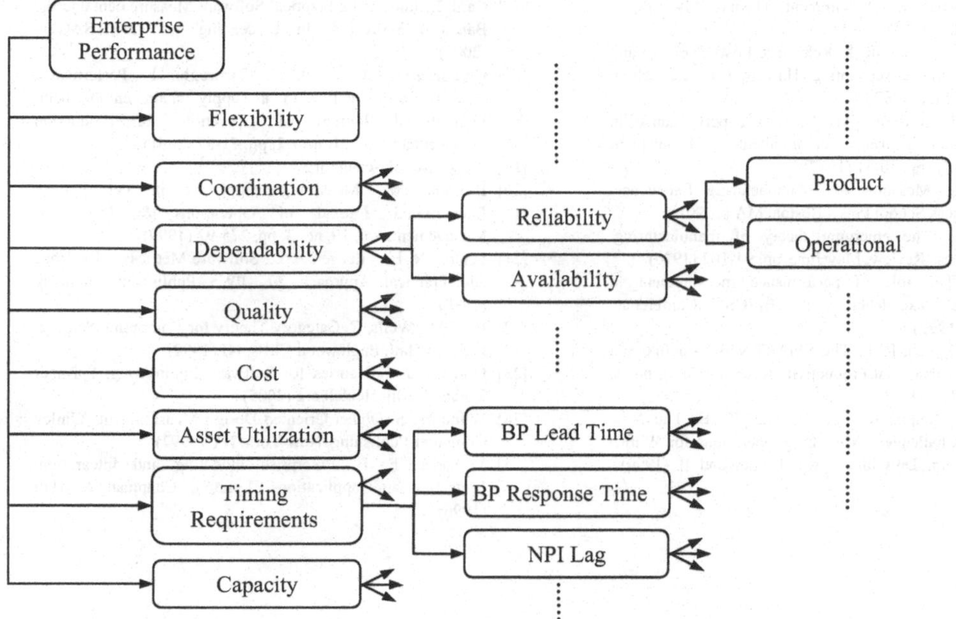

Fig. 5. Representing Measurement in Category Theory

V. Conclusions and Future Work

To conclude, this paper reports on the first results produced toward a formal and comprehensive performance measurement framework for enterprise integration founded on a solid mathematical theory, namely, category theory. Category theory is a branch of mathematics which is supremely capable of addressing the communications structure. This consideration was the main motivation behind our research: structure emerges from interactions between elements as captured by arrows, rather than extensionally as in set theory. The semantics of enterprise architecture can be captured in terms of a categorical framework, and consequently the formal functionality of the system can be specified using a language derived from the categorical framework. Thus, it is suitable for formalizing system-level integration through interactions between enterprise components, as well as for controlling system behavior conformance to nonfunctional requirements, such as performance, for enterprise integration quality control and tradeoff decision-making purposes. However, it is not immediately obvious which parts of category theory should be utilized for modeling large complex systems in general, or enterprise integration in particular. This work will continue exploring the semantic properties of a formal category theory framework to define categorical specification language constructs for enterprise integration modeling. The basic constructs will include the categorical rules, basic primitives and operators, and communication primitives for specifying processes. The appropriateness of the categorical objects and morphisms would become apparent when modeling the interactions and relations of an enterprise, such as recursive hierarchies and cooperative networks. The equivalence of the categorical framework and the categorical language will be established formally and illustrated through selected examples.

VI. REFERENCES

[1] Kochhar, A., Zhang, Y. A framework for performance measurement in virtual enterprises. In Proceedings of the 2nd International Workshop on Performance Measurement, 6-7 June, Hanover, pp. 2-11 (2002)

[2] Krantz, D. H., Luce, R. D., Suppes, P., Tversky, A.. Foundations of Measurement – Additive and Polynomial Representations, vol. 1. Academic Press, New York (1971)

[3] Narens, L. Abstract Measurement Theory. MIT Press, Cambridge, MA (1985)

[4] Johnson, H. T., Kaplan, R. S. Relevance Lost: The Rise and Fall of Management Accounting, Harvard Business School Press, Boston, MA (1987)

[5] McNair, C. J., Masconi, W. Measuring performance in advanced manufacturing environment. Management Accounting, July, pp. 28-31 (1987)

[6] Kaplan, R. S. Measures for Manufacturing Excellence. Harvard Business School Press, Boston, MA (1990)

[7] Druker, P. E. The emerging theory of manufacturing. Harvard Business Review, May/June, pp. 94-102 (1990)

[8] Russell, R. The role of performance measurement in manufacturing excellence. In BPICS Conference, Birmingham (1992)

[9] Cross, K. F., Lynch, R. L. The SMART way to define and sustain success. National Productivity Review, vol. 9, no. 1, pp. 23-33 (1988/89)

[10] Dixon, J. R., Nanni, A. J., Vollmann, T. E. The New Performance Challenge – Measuring Operations for World-class Competition, Dow Jones-Irwin, Homewood, IL (1990)

[11] Kaplan, R. S., Norton, D. P. The Balanced Scorecard: Translating Strategy into Action, Harvard Business School Press, Boston, MA (1996)

[12] EFQM. Self-assessment Guidelines for Companies, European Foundation for Quality Management, Brussels (1998)

[13] Beer, S. Diagnosing the System for Organizations, Wiley, Chichester (1985)

[14] Neely, A., Mills, J., Gregory, M., Richards, H., Platts, K., Bourne, M. Getting the Measure of your Business. University of Cambridge, Cambridge (1996)

[15] Neely, A., Adams, C. The performance prism perspective. Journal of Cost Management, vol. 15, no. 1, pp. 7-15 (2001)

[16] Capability Maturity Model Integration (CMMI[SM]), Version 1.1, Software Engineering Institute, Pittsburgh, March 2002, CMMI-SE/SW/IPPD/SS, V1.1

[17] Card, D. Integrating Practical Software Measurement and the Balanced Scorecard. In Proceedings of COMPSAC'03 (2003)

[18] Gunasekaran, A., Patel, C., Tirtiroglu, E. Performance measures and metrics in a supply chain environment. International Journal of Operations & Production Management, vol. 21, no.1/2, pp. 71-87 (2001)

[19] < www.supply-chain.org >

[20] Beamon, M. Measuring supply chain performance. International Journal of Operations & Production Management, vol. 19, no..3, pp. 275-92 (1999)

[21] Fenton, N. E., Pfleeger, S. L., Software Metrics: A Rigorous and Practical Approach 2/e, PWS Publishing Company (1998)

[22] Barr, M., Wells, C. Category Theory for Computing Science, Prentice-Hall, Englewood Cliffs, NJ (1990)

[23] Fiadeiro, J. Categories for Software Engineering, Springer Verlag, Berlin Heidelberg (2005)

[24] Whitmire, S. Object Oriented Design Measurement, Whiley Computer Publishing, New York, NY (1997)

[25] Vernadat, F. B. Enterprise modeling and integration: Principles and applications, London, Chapman & Hall (1996).

Some aspects of bucket wheel excavators driving using PWM converter – asynchronous motor

Marcu Marius Daniel[1], Orban Maria Daniela[2]
[1] Electric and Energetic Systems dept., University of Petrosani,, Romania
[2] Electric and Energetic Systems dept., University of Petrosani,, Romania
marcu@upet.ro

Abstract— **Static converters are very important element in controlled drive systems. The development of the power semiconductor elements has revolutionized the field of electric power control and has permitted the improvement and the diversification of the electric energy converter. Loads with highly distorted current waveforms also have a very poor power factor; because of this, they use excessive power system capacity and could be a cause of overloading. The power supply from the Oltenia open pit coal mine machinery consists of electrical cables from 20/6 kV transformer station rigged with 4 or 6 MVA transformers counting 188 pieces, providing 1263,3 MVA rated capacity.**

I. Introduction

Using the PWM inverters in motor drive applications is considered advantageous in many ways. For traction a.c. drives fed by a d.c. input power source, the PWM inverter is a practical solution which only involves a single power conversion. For industrial applications the PWM drive obtains its d.c. input through simple controlled rectification of the commercial a.c. line and is favored for its good power factor , good efficiency, its relative freedom from regulation sinusoidal current waveforms. The PWM drive is often preferred facility for its compatibility with the use of a single centralized d.c. rectifier facility providing a common d.c. bus serving a number of inverters, which in large installations is considered a favorable arrangement.

The open pit coal mines technological processes mechanization and automation trend has determined the rigging of above mentioned with bucket wheel excavators (BWE), large capacity conveyor belts (BC), stockpiling machines (SM) etc., so with continuous action machinery, with high productivity, with electric motor drives mainly asynchronous (wounded or squirrel cage motors).

The surface and depth expansion of the open pit coal mines mechanization and automation of the technological processes in continuous flow for extraction, transport and stockpiling, has determined the increase of rated capacity and productivity growth of the mining machinery because the lignite has become an intensive energetic product due to the over dimensioning of the technologies from the mining pits and use of the equipment under capacity and the preoccupation of SNLO Targu-Jiu in the domain of modernizing electric driving and electric installations for automation.

The power supply from the Oltenia open pit coal mine machinery consists of electrical cables from 20/6 kV transformer station rigged with 4 or 6 MVA transformers counting 188 pieces, providing 1263,3 MVA rated capacity. The bucket wheel excavator is continuous flow complex machinery that excavates the coal using the buckets fitted on the wheel and in the same time conveys the material to the transportation facility.

The supply of the equipments in the mining pit is made through 6kV electric cable (or 6kV E.A.L.) with outputs from the 6kV cells afferent to the 20/6kV transformation stations, this supply being made on the basis of a "Contract for providing electric energy to the big final, industrial, or similar consumers", which anticipates the limit regimes and parameters for supply with electric energy.

The working element – bucket wheel executes horizontal and vertical movements, the basic one being the rotation. The bucket wheel turns and in the same time swivels so a bucket moves on a helical trajectory (helix). The asynchronous motors in schortcut or induction motors are frequently used in electrical drives for open pits machines. The motor `s power are from 100 to 630 kW and the open pits costumers are supplied from transformer stations about 2x4MVA, 20/6 kV using different electrical cables. The lignite open pits from Oltenia are equipped with technologies in continuous flux characterized by: the lignite deposits are excavated by bucket wheel excavators (BWE) with the following capacities 470 l, 1300 l, 1400 l, 2000 l and productivity about 1680 m3/h - 6500 m3/h; the belt conveyer has the productivity about 1400 - 12500 m3/h; the laying down machines have capacity about 2500 -12500 m3/h.

Each belt conveyor is designed to have:

- driving with motors of 6kV/630kW (three-phase asynchronous motors with coiled
motor), start being made with liquid rheostats or with static rotor converters type CSR 630 P;
- three-phase transformer 6/0,4kV; S = 160kVA;
- force installations, command and automation for auxiliary and driving;
- illumination, heating and ventilation installations. [4]

II. About the Electric Drives Used in BWE'S

In the last years static converter asynchronous motor drives have become the preferred choice in the adjustable drives, displaying an annual increase rate of 13-14% compared with

K. Elleithy (ed.), *Advanced Techniques in Computing Sciences and Software Engineering*,
DOI 10.1007/978-90-481-3660-5_58, © Springer Science+Business Media B.V. 2010

only 3-4% for DC drives. This increase was generated by the new high productivity requirements that led to more complex operating cycles, higher maxim values of velocity and easier maintenance.

As execution element of adjustable drives, the advantages of the asynchronous motor over the DC motor are: higher overload capacity, higher feasibility and strength, lower specific rated power, higher values of velocity, easier maintenance, easier construction, etc.

The drive of bucket wheel at BWE is realized using a synchronous motor with wounded rotor, with rated power of 400 kW up to 4500 kW, using in general the asynchronous motor with rings of 630 kW, 6kV, 71A, 980 rot/min.

A. Simulation of the current and voltage harmonics for a PWM inverter

The three phase voltage, frequency and variable amplitude system from the outlet of the inverter is obtained by the Pulse Width Modulation method with the following advantages (regarding the classic methods): wider frequency variation range, reduction of the oscillation in the electromagnetic couple, lower content of superior harmonics in the diagram of the current, improved power factor irrespectively of the load, etc. [3]

The PWM strategies has constituted a complex research field, leading to the improvement of a large number of methods: sine-modulation with regular or natural sampling, symmetric or asymmetric, synchronic or a-synchronic, sub-harmonic sine modulation, sine modulation with the third harmonic insertion, biphasic sine modulation and more recently, Space Vector Modulation (SVM).

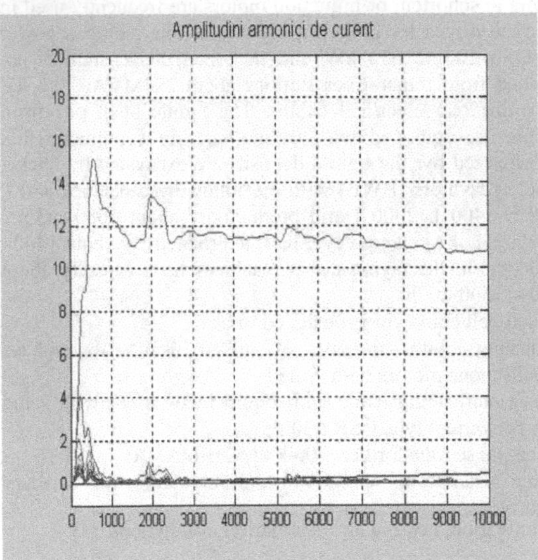

Fig. 1 The amplitudes adjustment of current harmonics

Static converters AC adjustable drives – AC asynchronous motors supplied with variable frequency represent nowadays the solution most widely accepted. The indirect voltage bridge inverter (with DC intermediary circuits) is the most advantageous diagram regarding to: reliability, speed adjustment range, dynamic behavior, and cost prize. [2] The simulations for the indirect voltage bridge PWM with DC intermediary circuits were made for a frequency of 1000 Hz and were shown in the below figures (fig.1-fig.10).

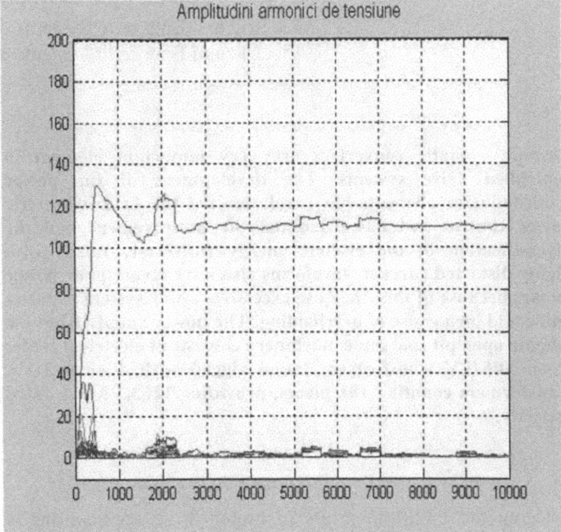

Fig. 2 The amplitudes adjustment of voltage harmonics

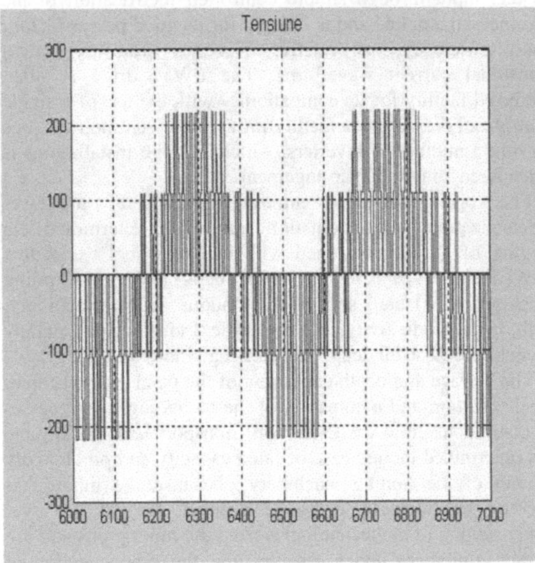

Fig. 3 The wave form of voltage from the output of inverter for f = 50 Hz

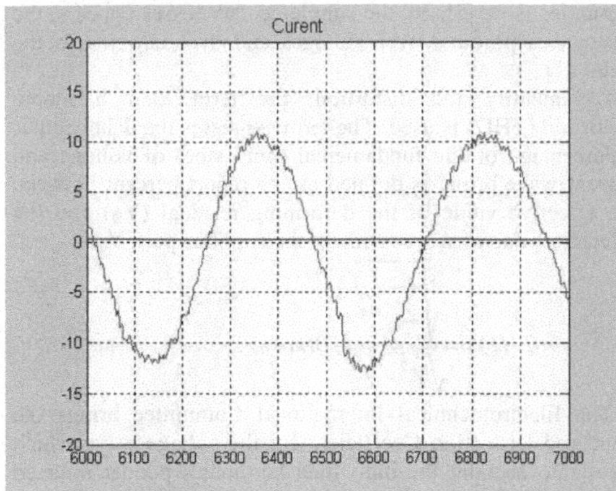

Fig. 4 The wave form of load current for f = 50 Hz

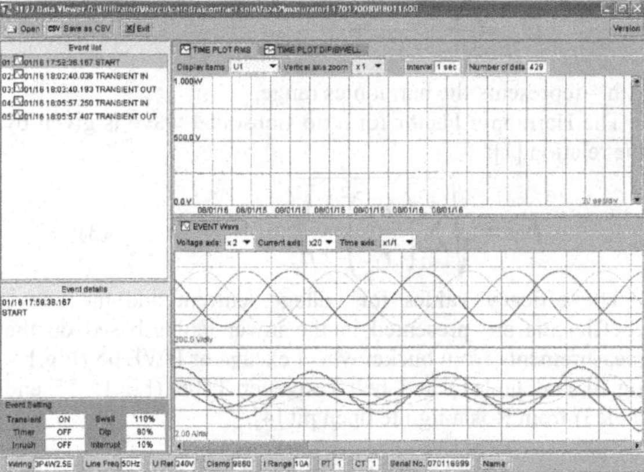

Fig. 5 The effective voltage adjustment

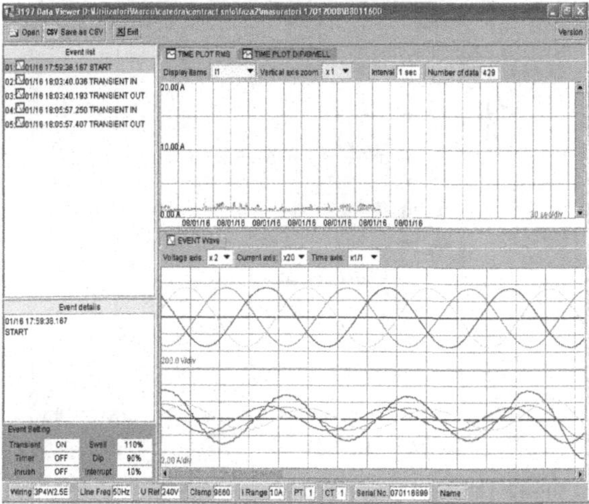

Fig. 6 The effective current adjustment

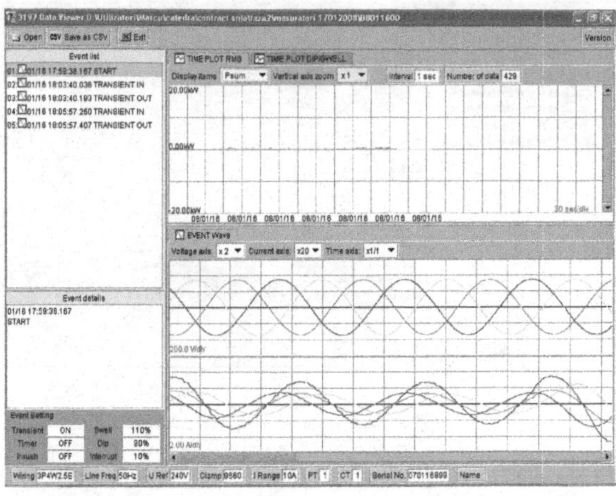

Fig. 7 The active power adjustment

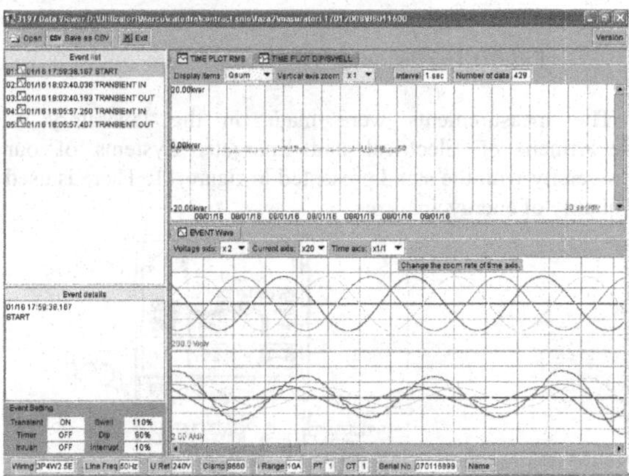

Fig. 8 The reactive power adjustment

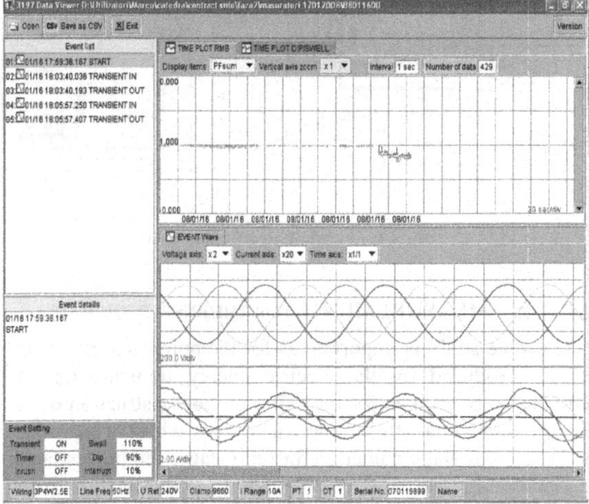

Fig. 9 The power factor adjustment

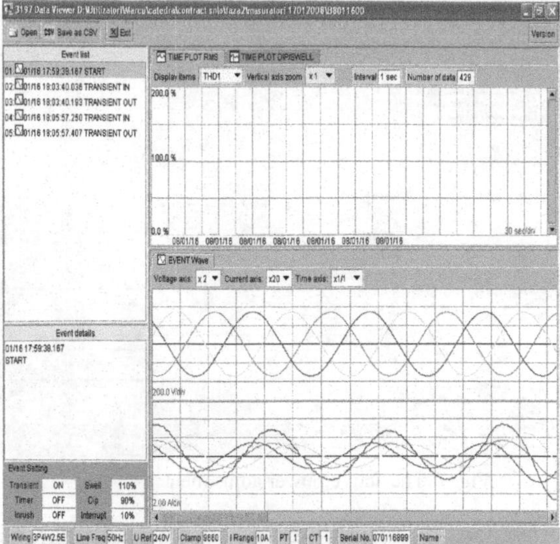

Fig. 10 The distortion factor adjustment

The measurements were made in the laboratory of Department of Electrical and Energetic Systems of our University with the stand presented in figure 11. There is used a quality of energy analyzer.

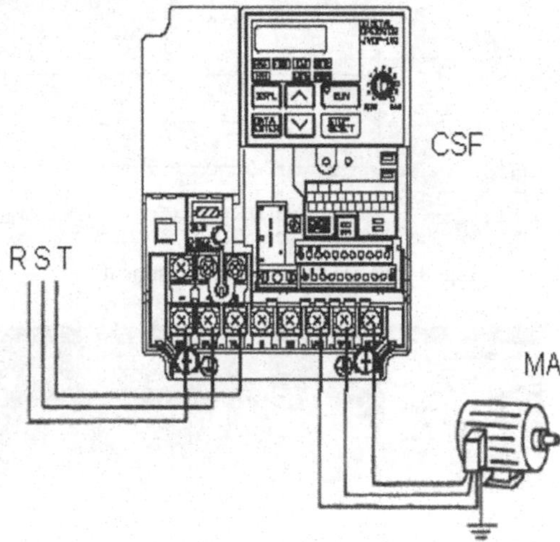

Fig.11 The measurements stand

III. MEASUREMENTS FOR THE ENERGY QUALITY

The state study is important as for the point of view of bad effects produced by the electric energy transmission and delivery webs well as for the using in the construction of the electric apparatus. In the electrical webs which work in the no sinusoidal stage named deforming stage, the power factor is decreasing and the reactive power compensation using capacities is very hard, the supplementary losses appears, the resonances produce over voltages and over currents in the webs. [1]

To quantify the distortion, the term total harmonic distortion (THD) is used. The term expresses the distortion as a percentage of the fundamental (pure sine) of voltage and current waveforms, is defined as the report percent between the effective value of the deforming residual (Y_D) and the effective value of wave (without the continue part, Y_0).

$$THD = \frac{\sqrt{\sum_{v=2}^{n} Y_v^2}}{\sqrt{Y^2 - Y_0^2}} 100\,(\%) \qquad (1)$$

The Electrotechnical International Committee brings the Pondered Distortion Coefficient for the voltage wave which takes into account the individual harmonics ponder induced by the specifically application:

$$D_w = \frac{\sqrt{\sum_{n=2}^{N} n^2 \cdot Y_n^2}}{Y_1} \qquad (2)$$

n – represents the harmonics range.

The Harmonic Factor for a no sinusoidal wave is given by the relation [1]:

$$FA = \sqrt{\sum_{n=2}^{\infty} \left(\frac{Y_n}{Y_1}\right)^2 \cdot \frac{1}{n}} \qquad (3)$$

The efficacy values for voltage, current and the above coefficients are presented in the lower tables based on the measurements from bucket wheel excavator BWE 05 (Fig.12-fig.14 and table1) and belt conveyer T500 (Fig.15-17 and table3) from Rosia lignite open pit [4].

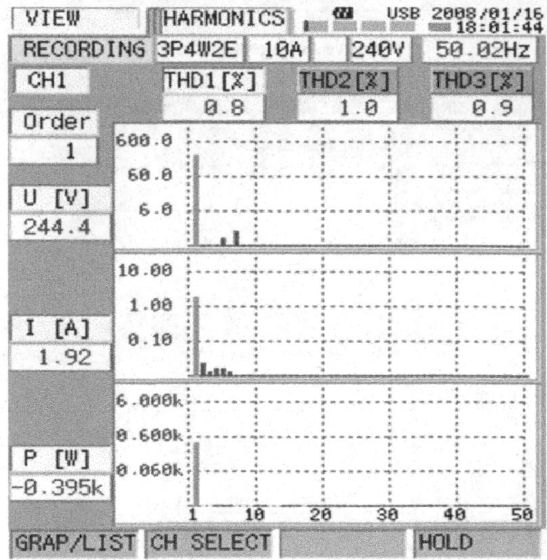

Fig.12 The harmonics range

VIEW | DMM | USB 2008/01/16 18:02:05

RECORDING | 3P4W2E | 10A | 240V | 50.02Hz

U	rms [V]	peak+[V]	peak-[V]	THD [%]
ch1	244.2	345.7	-344.9	0.8
ch2	246.6	347.4	-348.3	1.0
ch3	245.2	345.4	-344.8	0.9

I	rms [A]	peak+[A]	peak-[A]	KF
ch1	1.96	2.85	- 2.83	1.2
ch2	1.89	2.77	- 2.76	1.2
ch3	1.99	2.91	- 2.90	1.2
ch4	3.92	5.70	- 5.64	

	P [W]	S [VA]	Q [var]	PF
ch1	-0.447k	0.480k	0.173k	0.928
ch2	0.440k	0.467k	-0.156k	-0.939
ch3	0.463k	0.487k	-0.154k	-0.949
sum	0.45k	1.43k	- 1.36k	-0.959

Uave [V]	Iave [A]	Uunb [%]
245.4	1.95	0.6

HOLD

Fig. 13 The rms and peak values

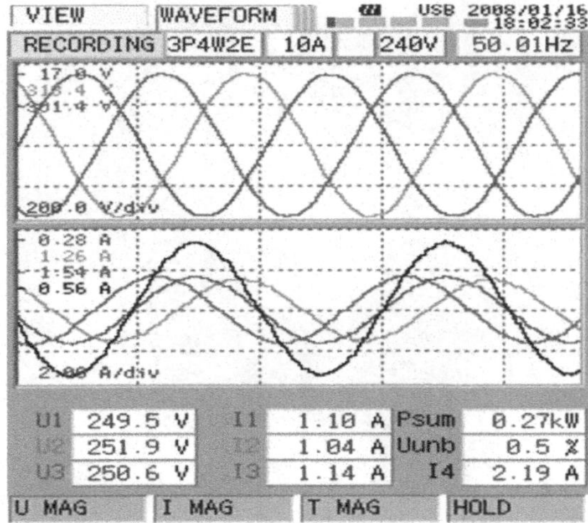

Fig. 14 The Waveform

TABLE.1A The Voltage Harmonics

Range	Value
1	6110
5	20,36
7	40,72

TABLE 1B The CurrentHharmonics

Range	Value
1	192
2	2,56
3	0,64
4	1,28
5	1,28
6	0,64

TABLE 1C Results

	RMS	TDH (%)	Dw	FA
U	6110,1696	0,74	0,049	0,003
I	192,0277	1,7	-	0,0106

The capture measurements for belt conveyers are presented below:

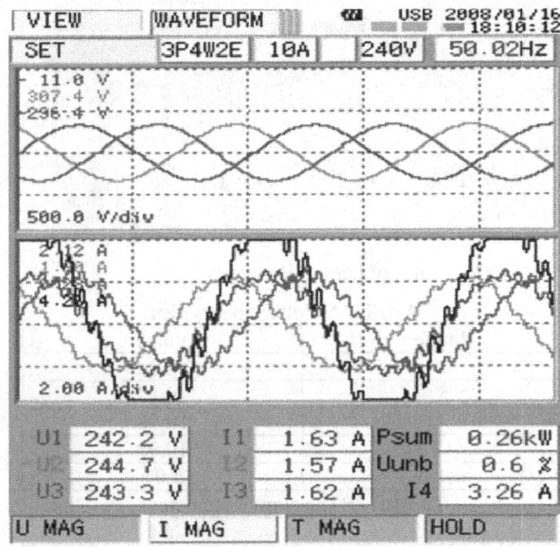

Fig. 15 The Waveform

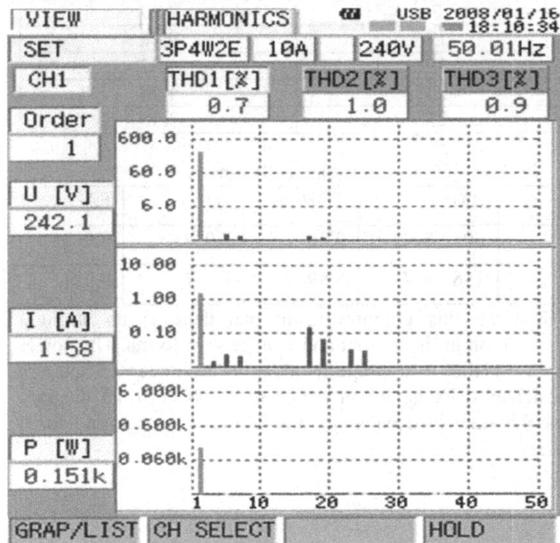

Fig. 16 The harmonics range

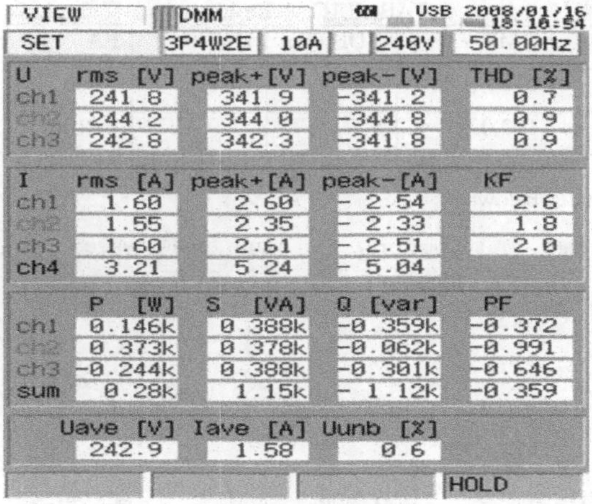

VIEW	DMM		USB 2008/01/16 18:10:54
SET	3P4W2E 10A	240V	50.00Hz

U	rms [V]	peak+[V]	peak-[V]	THD [%]
ch1	241.8	341.9	-341.2	0.7
ch2	244.2	344.0	-344.8	0.9
ch3	242.8	342.3	-341.8	0.9

I	rms [A]	peak+[A]	peak-[A]	KF
ch1	1.60	2.60	- 2.54	2.6
ch2	1.55	2.35	- 2.33	1.8
ch3	1.60	2.61	- 2.51	2.0
ch4	3.21	5.24	- 5.04	

	P [W]	S [VA]	Q [var]	PF
ch1	0.146k	0.388k	-0.359k	-0.372
ch2	0.373k	0.378k	-0.062k	-0.991
ch3	-0.244k	0.388k	-0.301k	-0.646
sum	0.28k	1.15k	- 1.12k	-0.359

Uave [V]	Iave [A]	Uunb [%]
242.9	1.58	0.6

HOLD

Fig. 17 The values of harmonics waves

TABLE 2A Voltage harmonics

Range	Value
1	6052,5
5	20,17
7	10,08
17	10,08
19	5,04

TABLE 2BThe current harmonics

Range	Value
1	158
3	1,58
5	3,16
7	2,37
17	9,48
19	6,32
23	4,74
25	3,95

Table 2C Results

	RMS	TDH (%)	Dw	FA
U	6052,5441	0,38	0,038	0,0016
I	158,5874	**8,59**	-	0,021

Studying this graphics result that there is no strongly perturbation in the current and voltage waveforms. The power factor variation is improperly taking values under 0,5 and less to increased loads. It can notice from the vectorial diagram that the state is unbalanced due to placing to ground for a phase.

REFERENCES

[1] M. Iordache, L. Dumitriu, "Electrical Circuits Theory," Editura Matrixrom, ISBN 978-973-755-174-0, Bucuresti 2007, pag.190-216

[2] M.D. Marcu, M.D.Orban, K.Yanakiev, E.Kartzelin, "Static converters and their applications"Univ.of Mining and Geology St.Ivan Rilski, ISBN 954-97748-35-9, Sofia 2001, pg.91-97

[3] R. Magureanu, C. Solacolu, A Maxim, D. Floricanu ``Inteligent PWM inverter for induction motor control`` Proc.IEE-Power Electronics and Variable Speed Drives, London 1988

[4] *** Research work no.20/2006 `` The energy quality study in the lignite open pits in conditions of introducing the power electronics for respecting the european standards`` University of Petrosani - S.N.L.Oltenia

Health Information Systems Implementation: Review of change management issues

Paulo Teixeira
Polytechnic Institute of Cávado and Ave
Barcelos, Portugal

Abstract **— It is common for information system projects' implementation to fail. One of the most mentioned reasons is implementation management, particularly the change management during the implementation cycle. This study provides a literature review that aims at supporting this assertion, and identifies some of the most evident issues on the subject. The study highlights the need for new approaches on the Change Management process in the implementation of health information systems.**

Index Terms**— Health, Implementation, Information Systems, Change management, Methodology.**

I. Introduction

Change is one of the characteristics that best describes organizations. Like in any living organism, it is part of their nature react and evolve. Market uncertainty and competitiveness, new technologies, the need to reduce costs and improve quality or access to new markets imply organizational changes, often translated in implementation or adaption of information systems.

In the context of a systems development project, Alan Dennis and Barbara Wixom [1] describe the Change Management as the process of helping the end users adopting the system under development, and to monitor their work processes without unnecessary pressure. Despite this, it is not a well-defined area, neither in theory nor in practice, and it needs further research and systematization [2].

The information systems introduction is usually presented as an organizational value. However, this introduction is not always translated into success. Like any process in change management derived from systems introduction, it must obey rules and follow the best practices to reduce the cases of failure. As such, this study follows an exploratory review of the literature concerning the change management in health information systems, in order to identify some of the raised issues on the subject.

Manuscript received October 15, 2008.

Paulo Adriano Marques Sousa Teixeira is with the Higher School of Technology of the Polytechnic Institute of Cávado and Ave, Urbanização Quinta da Formiga 4750 Barcelos, Portugal. (phone: 351-253802260; fax: 351-253812461; e-mail: pteixeira @ ipca.pt).

II. Systems Project Implementation

Systems project implementation, and for the purposes for which this study is intended, is understood as the set of stages that flow from the choice of an effective system and its entry into production. The systems introduction only benefits the organization if the implementation improves essential business processes. It is also clear that there is little benefit from systems with low use, hence to get investment return in information systems, its implementation should involve changes in people and processes: change for better, faster or cheaper.

When one wishes to make changes in an organization, the conduct of this change is as significant as the project itself. The expectations created by a process of change may fall substantially simply because the users do not understand or have resistance to the project [3]. Kolltveit [4], after a content analysis of the project management of information technologies literature, concludes that change management key contents are ignored. Moreover, Roger Pressman [5] suggests the change management is a key pillar of project management.

III. Change Management

Despite the imprecision of the area cited in the introduction, change management takes itself as a central element of an implementation and involves managing change and transition planning. For over 40 years the theory and practice of change management was dominated by the work of Kurt Lewin [6, 7]. This approach advises managing change as a series of planned phases for the transition, from the present until the desired state. More recently, during the last two decades, this approach to change has attracted much criticism, yet it continues being an important approach in the area. More recent approaches, often referred to as emerging, complement the model of Lewin and, together with this model, are combined according to each particular situation [8]. These approaches require the change management to be a continuous process.

K. Elleithy (ed.), *Advanced Techniques in Computing Sciences and Software Engineering*,
DOI 10.1007/978-90-481-3660-5_59, © Springer Science+Business Media B.V. 2010

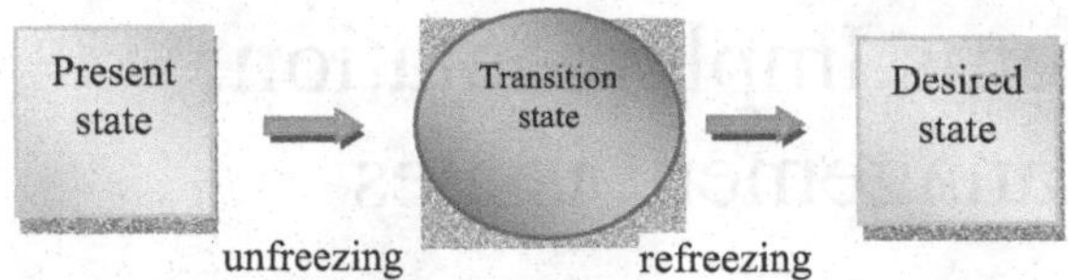

Figure 1: Change process based on the theory of Lewin [9]

Other studies show that the negative results of the change management initiatives tend to overcome the positive and indicate the need for modifications in the operationalization of theoretical models of change [10].

The need for software engineering methodologies regarding change management is mentioned in several components of this knowledge area. The SWEBOK (software engineering body of knowledge) [11] indicates this need in the requirements identification, as it will affect the whole life cycle of software. The documentation of these requirements will facilitate its entry into production, parameterization and further developments [12].

The change management is normally taken into consideration when implementing a new system but, as in SWEBOK, some authors stress the importance that this process should be taken at the beginning [13] and in the course [14] of new systems development. These concerns led to later facilitation of implementation projects.

The implementing process of a new or adapted information system is similar to the reengineering processes to fit the specific features of the system. This process requires a careful coordination between different organizational functions and the management of several issues concerning the organizational culture. The adoption of bit by bit approaches to the management of this process is highlighted by previous authors [15] as the reason for the failure of multiple deployments. Consequently, there is the need for an overview of the process of organizational change as a way to mitigate this failure.

The change management incorporates several skills associated with projects management and, according to PMBOOK [16], are as follows:

The project management body of knowledge;
Application area knowledge, regulation, and standards;
Understanding the project environment;
General management knowledge and skills; and
Interpersonal skills.

This finding reflects the need to combine different areas of expertise in implementing a strategy for managing change, or in the proposal of a new methodology. Figure 2 illustrates a range of skills associated with managing change in the implementation of a health information system.

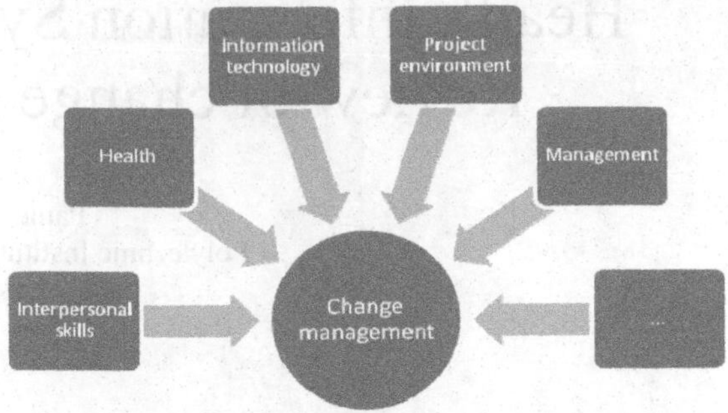

Figure 2: Required skills for the change management

These expertise areas need to be explored in specific issues of change management. Ziaul [15] presents a set of those particular issues:

Leadership issues;
Barriers to Change;
Communications;
Implementation of change and control;
People cultural factors; and
Change review.

In addition, there is a need to deepen the inter-relationship between the tools and techniques used in change management with the theories, models and concepts of change management [17].

IV. Reasons for Change

The implementation of information systems always involves complexity associated with risk, as this involves modifying procedures which may cause organizational instability. Managers must have a guide to effectively and efficiently lead the process of implementing systems, in order to minimize the risk associated with change and maximize its benefits.

Research on managing change focuses on the principles of continuous assessment and management of people, processes and systems, and organizational issues. The motivation for change can be caused by internal factors, such as strategy, structure, technology, or as result of the external environment, such as the economic situation, political climate and socio-cultural and technological advancement.

Albert Endres and Dieter Rombach [18] mention the Lehman's first law, according to which "a system that is used will be changed". This belief indicates the inevitability that the systems will change over time, by internal or external reasons to the organizations.

The available literature suggests a number of critical success factors associated with systems implementation, and in these critical success factors, change management and managing the process of change, are always present [19].

The primary objective of information systems implementation is to increase organizational efficiency, which means that achieving the desired situation is more important than the delivery of the system, unless it is properly implemented.

V. The Introduction of Information Systems in the Health Area

The success or not of systems implementation depends on the acceptance or rejection of such systems by health professionals [20]. The physician, for example, in general, is known to be resistant to initiatives based on computer systems [20-24]. The way the adverse reaction of health professionals is managed is of crucial importance to the success of a project's implementation. Moreover, Connell [25] believes that the requirements of health information systems are sufficiently specific to be considered a special case that requires different approaches from other systems.

The success of information systems in health is questioned by several authors and their failure rate is considered a major problem [12, 26, 27]. If, as appears to result from the consulted literature, there is a significant number of problems associated with the introduction of information systems in health care units, this implies the existence of a gap between the potential offered by these systems and existing reality. On the other hand, it also indicates that a substantial part of the directed investment to these systems ends up being wasted on inefficient solutions. This failure is justified by some authors as a result of cultural factors and resistance to change [28].

Like the above aspects for information systems in general, the majority of health information systems result in cases of failure [29]. The organizational issues are sometimes referred to as a barrier that leads to failure in the introduction of these systems. This and other factors should be systematized in the analysis of cases of failure to serve as references for new implementation strategies [30]. Regardless of the type of change, Prayer & Chahal [31] consider the alignment of organizational culture in the process of change management as vital to the success of change.

A. Critical success factors

The practices of change management are the most referenced critical success factors contributing to the success of the information systems implementation [32]. The combination of critical success factors can affect the success or failure of a proposed implementation of information systems in health [33]. The change management of information systems introduction may also have critical success and failure factors, which should be obtained from the existing models and theories, as well as by the analysis of systems introduction cases.

B. Failures in managing change

The information systems effective implementation can have a positive impact in economic terms, yet the literature shows a high rate of failure in their implementation [34, 35]. This is a phenomenon that worries both researchers and organizations since the information systems success is of vital importance for society. Heeks [12] stresses the importance to identify the factors that determine that success or failure in each particular situation.

Friedman [23] argues about the need, within the health sector, in adopting strategies for information systems integration that are based on cultural differences that characterize the industry as a whole, as well as the various professional groups involved.

VI. The Need for a Methodology

The managers of change processes need methodologies that provide measurable and structured implementation tools, techniques and approaches to manage and evaluate the change process [36, 37].

Recent years have been fruitful in the development of studies comparing the best practices. However, the change management is an area where there is still lack of good practices systematization [38]. Studies show research in this area [39], yet they are still too broad and do not regard specific sectors of activity. The references to cases of success and failure should be taken into consideration in developing new methodologies, attempting to optimize them based on the experience from other organizations.

Previous authors have proposed the adoption of evaluation system models [40, 41], yet few proposals are targeted at how to conduct such adoption, and when they occur in literature, such proposals are too simplified to be able to serve as a guide to in course implementation of change management [42, 43].

VII. Conclusions

It is necessary to optimize the change process to minimize the resulting dysfunctionalities: the purpose of implementation in the process of change agitation cannot be lost. Consulted references show that the bit by bit approaches to change management leverage the failure in information systems implementation. It is noticeable the wide variety of influences from different areas of knowledge that are incorporated in the change management This diversity enhances the bit by bit approaches and suggests the need for new methodologies that allow managers to lead more effectively and efficiently the process of implementing systems.

The development or application of managing change methodology should involve a combination of planned and emerging models. In the particular case of healthcare, these models should incorporate the sector's organizational culture.

References

[1] Dennis, A. and B.H. Wixom, Systems analysis and design. 2 ed. 2003, Toronto: John Wiley & Sons. 537.

[2] Young, C.G.E., Change Management 2003/2008: Significance, strategies, trends. 2003, Cap Gemini Ernst & Young: Berlin.

[3] LaClair, J. and R.P. Rao, Helping employees embrace change. The McKinsey Quarterly, 2002. 4: p. 17–20.

[4] Kolltveit, B.J., B. Hennestad, and K. Grønhaug, IS projects and implementation. Baltic Journal of Management, 2007. 2(3): p. 235 - 250.

[5] Pressman, R.S., Software Engineering: A Practitioner's Approach. 6 ed. 2005, Boston: McGraw - Hill International Edition.

[6] Burnes, B., Kurt Lewin and the Planned Approach to Change: A Re-appraisal. Journal of Management Studies, 2004. 41(6): p. 977-1002.

[7] Buchanan, D., et al., No going back: A review of the literature on sustaining organizational change. International Journal of Management Reviews, 2005. 7(3): p. 2005.

[8] Cameron, E. and M. Green, Making Sense of Change Management: A Complete Guide to the Models, Tools & Techniques of Organizational Change. 2004: Kogan Page.

[9] Lewin, K., Field theory in social science: selected theoretical papers. 1st ed. 1951, New York,: Harper & Brothers. 346 p.

[10] Clegg, C. and S. Walsh, Change management: Time for a change! European Journal of Work and Organizational Psychology, 2004. 13(2): p. 217 - 239.

[11] Swebok, Guide to the software engineering body of knowledge : 2004 version. 2004, Los Alamitos, CA: IEEE Computer Society Press.

[12] Heeks, R., Health information systems: Failure, success and improvisation. International Journal of Medical Informatics, 2006. 75(2): p. 125-137.

[13] Mumford, E., Designing Human Systems-The Ethics Method, M.B. School, Editor. 1983: Manchester, England.

[14] Sjøberg, D.I.K. Managing Change in Information Systems: technological Challenges. in IRIS 17 - Information Research Seminar. 1994. Olou, Finland.

[15] Ziaul, H., et al., BPR through ERP: Avoiding change management pitfalls. Journal of Change Management;, 2006. 6(1): p. 67.

[16] PMI, A guide to the project management body of knowledge (PMBOK guide). 3rd ed. 2004, Evanston, IL: Project Management Institute.

[17] Hughes, M., The Tools and Techniques of Change Management. Journal of change management, 2007. 7(1): p. 37 - 49.

[18] Endres, A. and H.D. Rombach, A Handbook of Software and Systems Engineering: Empirical Observations, Laws and Theories. The Fraunhofer IESE Series on Software Engineering. 2003, Kaiserslautern: Addison Wesley. 327.

[19] García-Sánchez, N. and L.E. Pérez-Bernal, Determination of critical success factors in implementing an ERP system: A field study in Mexican enterprises. Information Technology for Development, 2007. 13(3): p. 293-309.

[20] Lapointe, L. and S. Rivard, Getting physicians to accept new information technology: insights from case studies. Canadian Medical Association Journal, 2006. 174(11): p. 1573-1578.

[21] Sim, I., G.D. Sanders, and K.M. McDonald, Evidence-based practice for mere mortals: The role of informatics and health services research. Journal of General Internal Medicine, 2002. 17: p. 302-308.

[22] Thomas, M., Introducing Physician Order Entry at a Major Academic Medical Center: Impact on Organizational Culture and Behavior, in Evaluating the Organizational Impact of Healthcare Information Systems. 2005, Springer: New York. p. 253-263.

[23] Friedman, C.P., Information technology leadership in academic medical centers: a tale of four cultures. Academic Medicine - Journal of the Association of American Medical Colleges, 1999. 74(7): p. 795-800.

[24] James, G.A., Clearing the way for physicians' use of clinical information systems. Commun. ACM, 1997. 40(8): p. 83-90.

[25] Connell, N.A.D. and T.P. Young, Evaluating healthcare information systems through an "enterprise" perspective. Information & Management, 2007. 44(4): p. 433-440.

[26] Berg, M., Implementing information systems in health care organizations: myths and challenges. International Journal of Medical Informatics, 2001. 64(2-3): p. 143-156.

[27] Heeks, R., D. Mundy, and A. Salazar, Why Health Care Information Systems Succeed or Fail, in Information Systems for Public Sector Management Working Paper Series 9. 1999, Institute for Development Policy Management: Manchester

[28] Adel Ismail, A.-A., Investigating the Strategies for Successful Development of Health Information Systems: A Comparison Study. Information Technology Journal, 2006. 5(4).

[29] David, A. and Y. Terry, Time to rethink health care and ICT? Communications of the ACM, 2007. 50(6): p. 69-74.

[30] Gray, S., Lessons from a failed information systems initiative: issues for complex organisations. International Journal of Medical Informatics, 1999. 55(1): p. 33.

[31] Price, A.D.F. and K. Chahal, A strategic framework for change management. Construction Management and Economics, 2006. 24(3): p. 237 - 251.

[32] Michel, A., Reexamining Information Systems Success through the Information Technology Professionals Perspective. Sprouts: Working Papers on Information Environments, Systems and Organizations, 2003. 3(2).

[33] Day, K.J., Supporting the emergence of a shared services organisation: Managing change in complex health ICT projects. 2007, University of Auckland: Auckland. p. 254.

[34] Kelegai, L.K., Elements influencing IS success in developing countries: a case study of organisations in Papua New Guinea, in Faculty of Information Technology. 2005, Queensland University of Technology. Centre for Information Technology Innovation: Queensland. p. 290.

[35] Davenport, T.H., Putting the Enterprise into the Enterprise System. Harvard Business Review, 1998. 76(4): p. pg. 121, 12.

[36] Singh, M. and D. Waddell, E-Business Innovation and Change Management. 2004: Irm Press. 350.

[37] Kotter, John P., Why Transformation Efforts Fail Harvard Business Review, 1995. 2: p. 59-67.

[38] Teixeira, P. and Á. Rocha, Gestão da mudança de sistemas de informação em unidades de saúde: Proposta de investigação. RISTI - Revista Ibérica de Sistemas e Tecnologias de Informação, 2008. 1(1): p. 78-82.

[39] Clarke, A. and J. Garside, The development of a best practice model for change management. European Management Journal, 1997. 15(5): p. 537-545.

[40] Ammenwerth, E., C. Iller, and C. Mahler, IT-adoption and the interaction of task, technology and individuals: a fit framework and a case study. BMC Medical Informatics and Decision Making, 2006. 6(3).

[41] Delone, W.H. and E.R. McLean, The DeLone and McLean Model of Information Systems Success: A Ten-Year Update. Journal of Management Information Systems, 2003. 19(4): p. 9-30.

[42] Rita, K., Grounding a new information technology implementation framework in behavioral science: a systematic analysis of the literature on IT use. Journal of Biomedical Informatics, 2003. 36(3): p. 218.

[43] Joanne, L.C., Contextual Implementation Model: A Framework for Assisting Clinical Information System Implementations. Journal of the American Medical Informatics Association, 2008. 15(2): p. 255.

Cache Memory Energy Exploitation in VLIW Architectures

N. Mohamed, N. Botros and M. Alweh
Department of Electrical and Computer Engineering
Southern Illinois University, Carbondale, IL 62901-6603

ABSRACT: This is a comparative study of cache energy dissipations in Very Long Instruction Word (VLIW) and the classical superscalar microprocessors. While being architecturally different, the two types are analyzed in this work on the basis of similar underlying silicon fabrication platforms. The outcomes of the study reveal how energy is exploited in the cache system of the former which makes it more appealing to low-power applications with respect to the latter.

I. INTRODUCTION

In recent years, VLIW microprocessors have become more attractive to embedded systems due to their high performance and relatively low-power consumption which is attributed to relatively less complexities in their hardware design; a feature that is attributed to reliance on smarter compliers. VLIW architectures meet the need for high-performance of today's embedded applications by resorting to Instruction Level Parallelism (ILP) exploitation within system datapath more than increasing system clock rates which leads adversely to substantial increase in dynamic power consumptions in the hardware-centric superscalar counterparts. This leverage in low-energy budget is accredited to the simpler hardware implementations in VLIW which comes at the expense of condense code overhead. In fact, because of that, this type of architectures demand and exhaust system cache memory which could deleteriously impact energy consumption; a characteristic they are developed to optimize in the first place. This work focuses on the energy dissipation in cache memory hierarchies of VLIW and superscalar architectures and its impact in performance.

The rest of this work is organized as follows: In section II, we shed light on significant variation in instruction processing between superscalar and VLIW architectures. In section III, we review memory organization within the two types of architectures. In section IV, we developed experimentations in two representative prototypes and analyze the collected simulation results. Finally, the work concludes in section V.

II. SUPERSCALAR VERSUS VLIW ARCHITECTURE

Unlike the traditional superscalar processors that rely heavily on the hardware to dynamically schedule program sequential instructions in an out-of-order fashion (as long as program dependencies are not violated) to increase Instruction Level Parallelism (ILP) and therefore boosts system performance [1], VLIW-like microprocessors, on the other hand, functions differently in spite sharing quite similar silicon platforms. VLIW architectures alleviate the burden of achieving higher ILP from the hardware [2]. They depend on smarter compilers to expose and exploit parallelism within program instructions sequence early on at compilation time. In other words, their program is expected to provide explicit information about tasks that are to be accomplished and the means to do that: such as instructions dependency, the instructions dispatch logic, operands destinations, Register File (RF) renaming, buffer reordering, etc at time of compilation. A glimpse of this can be elucidated by figure 1 below.

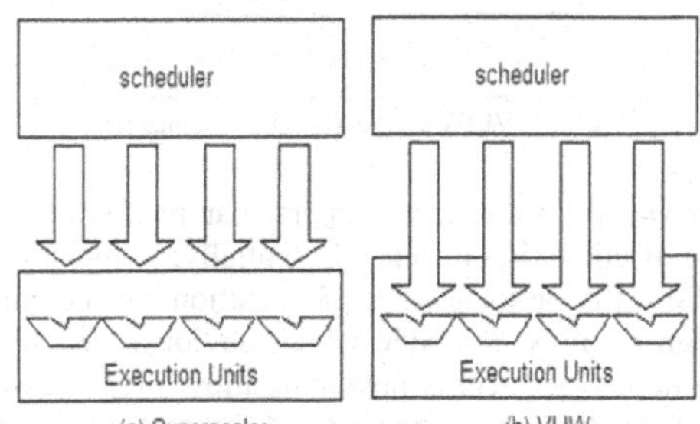

Figure 1: Superscalar and VLIW architectures

Dynamically scheduled superscalar instructions are dispatched in the pipeline without compiler pre-

K. Elleithy (ed.), *Advanced Techniques in Computing Sciences and Software Engineering*,
DOI 10.1007/978-90-481-3660-5_60, © Springer Science+Business Media B.V. 2010

knowledge of available hardware resources such as register files, reservation stations, functional units, etc. The task of assigning these instructions to specific resources is resolved later in the fly at the run time as shown in Figure 1 (a). With VLIW architectures on the other hand, the targeted resources are transparent to the compiler; a leverage that allows multiples instructions to be packaged simultaneously into a monolithic more efficient *Very Long Word*. Hence higher degree of ILP can be achieved as long as the hardware resources are available [3]. In essence, VLIW architectures lower the Dynamic-Static Interface (DSI): increasing the tasks that are done statically at compile time over those done dynamically by the hardware at the run time [4]. Recent Digital Signal Processors (DSP) can typically execute up to eight 32-bit parallel instructions per single cycle. Texas Instrument TMS 320C6410 [5] would provide a perfect representation to this category.

III. CACHE MEMORY ORGNANIZATION

This work uses the popular and widely used 2-level cache hierarchy with spilt equally sized instruction and data caches in the lower level and unified upper level in both types of architectures. The functional units are sandwiched between the instruction and data caches as shown architecturally in figure 2. While such types of organizations are

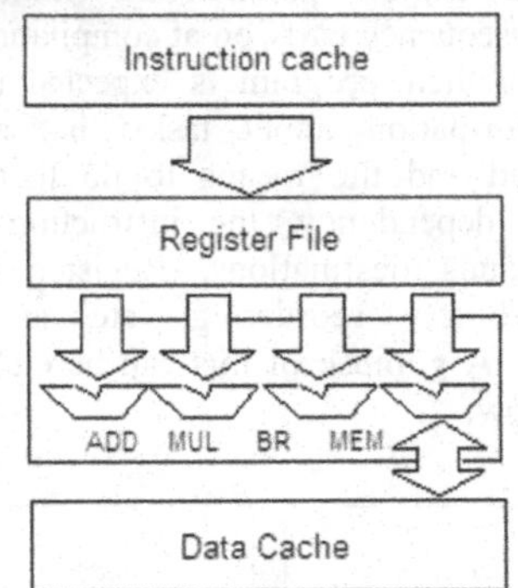

Figure 2: VLIW Cache Memory Organization

deja vue in precedent superscalar processors, the style of utilization differs. Essentially, in the case of VLIW, the associated sophistication in compiler design requires rich code which obviously translates into higher per access power to allow preprocessing of information about the targeted embedded hardware resources well in advance of instructions scheduling. Yet the effect of dense code utilization is greatly offset by a less complex targeted hardware. This can also lead to engaging multiple functional units in full capacity in each cycle and hence maximize ILP which translates into high instruction execution throughput within VLIW datapath. Conversely, having less dense and dynamically scheduled micro-operations traversing complex datapath, as the case in superscalar architectures, incurs relatively more energy in their ways in and out of the same types of embedded cache memories of superscalar processors (overlooking the fact that increasing superscalar machine width to enhance performance-which mandates high level of memory cells sharing among multiple units such as reorder and store buffers- does elevate interconnect power dissipation substantially).

For the superscalar and VLIW cache schemes, this work adopts the models provided by the academically widely used frameworks [6] and [7, 8] respectively. In essence, the setup consists of on-chip embedded cache system with 2-level hierarchy. The architectural simulators have been modified and configured to reflect the two targeted organizations. The details of this are the topic of the next section.

VI. EXPERIMENTATIONS AND RESULTS

In this section, we carried out experimentations in the two models of microprocessors with comparable hardware overhead and of yet different architectural organizations as provided by Table-1. The models qualify to represent the two classes under consideration. The first model mimic the AMD Athlon [9]; a commercially known RICS processor. The second model is similar to Texas Instrument

Table-I Simulated processor configurations

Processor core	
Technology	70nm
Functional Units	8 int ALUs,4 FP ALUs,1Multiplier/Divider
Cycle Time	100 cycles
Memory System	
L-1 Instruction Cache	64KB,32B block size, DM, WB
L-1 Data Cache	64KB, 32B block size, DM, WB
L-2 Unified Cache	Unified,512KB, 64B block size,4-way,WB
Memory	100 cycles latency

TMS 320 [5], a VLIW microprocessor that powers many of today's embedded systems. The cache systems energy consumption on both processors were assessed and quantified using M5 [8] and cacti [10] estimators respectively.

As for the workload on these systems, we developed a set of custom-made benchmarks whose overall characteristics are tailored to mimic diversified real-world applications. The list of these benchmarks with brief descriptions is given in Table-II. These applications were elected to represent different program behaviors.

Table- II: Benchmarks and Characterizations

Benchmark Name	*text*	*sort*	*bin_decode*	*nnet*	*encrypt*
Benchmark Type	integer	float	integer	float	integer
Characterization	Word Processing	Bubble Sorting	Binary Decoding	Neural Network	RSA encryption

Table-III: Performance measurement in terms of miss rates

Benchmarks	L1-cache				L-2 Cache	
	icache	icache	dcache	dcache		
	VLIW	Superscalar	VLIW	Superscalar	VLIW	Superscalar
text	0.0086	0.016556	0.0304	0.00001	0.5376	0.5101
bin_decode	0.0068	0.018078	0.0175	0.491228	0.5309	0.3333
sort	0.0098	0.014946	0.0123	0.053299	0.4252	0.318182
nnet	0.0286	0.007512	0.0439	0.077949	0.5535	0.323077
encrypt	0.0423	0.081967	0.0913	0.125000	0.5474	0.4375

The cache performances on the host superscalar and VLIW architectures have been quantified in terms of accurate cycle simulations using modified versions of simplescalar tool-set [6] and Trimaran [7] respectively. The simulation outcomes are shown in Table-III as well as in figures 3, 4 and 5. In almost all 5 benchmarks, VLIW processor demonstrates superiority in performance of both level-1 caches and inferiority with respect to the second level. The culprit for the performance decay in L-2 cache is

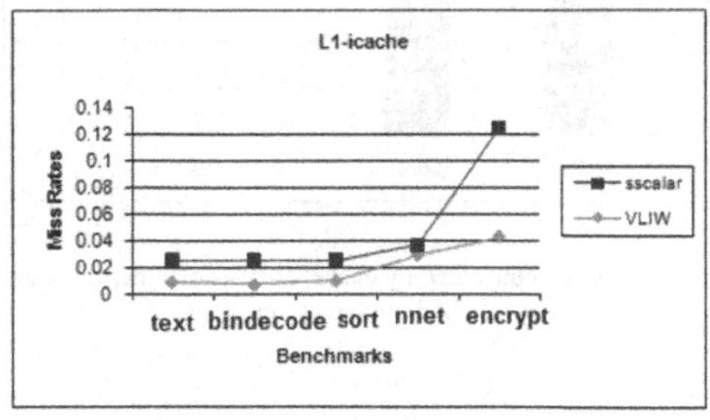

Figure 3: Level-1 instruction cache miss rates given as by benchmarks

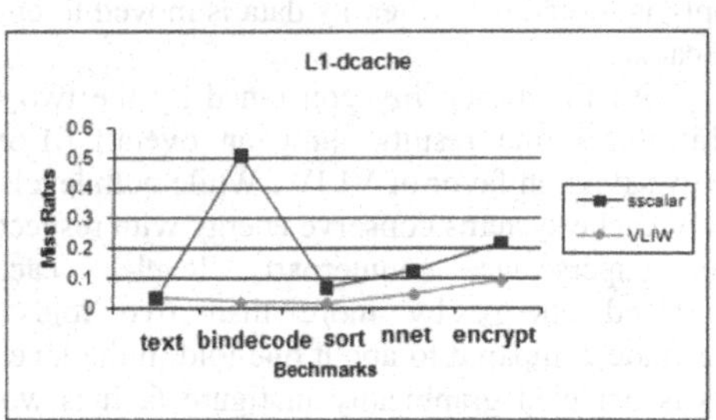

Figure 4: Level-1 data cache miss rates as given by benchmarks

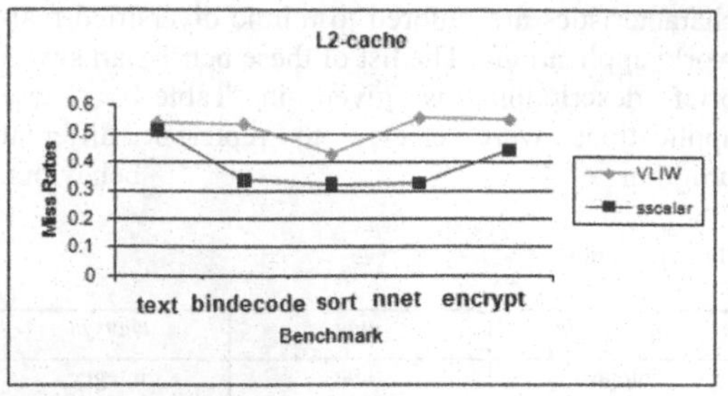

Figure 5: Level-2 data cache miss rates as given by benchmarks

attributed to adopting a state-destructive leakage control mechanism [11] where L-2 subblock power supply is turned off when its data is moved to either L-1 caches.

Table-IV: Cache Memory Energy Consumptions

Architecture	Cache-Level	Total Accesses	Leakage Energy (micro J)	Dynamic Energy (micro J)	Total Energy (micro J)
Superscalar	L-1 cache	44349	51.044	25.545	76.589
	L-2 cache	1431	153.132	0.85	153.982
VLIW	L-1 cache	7278	8.3768	4.192	12.5688
	L-2 cache	655	75.3912	0.389	75.7802

As for the energy consumed by the two are architectures, the results show an overall 61.68% energy saving in favor of VLIW. While both levels of VLIW cache systems conserve energy with respect to their superscalar counterpart, level-1 caches minimized energy by more than five folds of magnitude compared to about one fold in the level-2. This is depicted graphically in figure 6. It is worth mentioning here that employing the leakage control mechanism mentioned earlier has greatly slashed leakage energy in level-2 cache in both architectures since it drains supply voltage from selective cells-which results in elevated miss rates in both architectures as shown in figure 5. Moreover, the near perfect hit rates in level-1 caches further help minimizing leakage in level-2 cache memories.

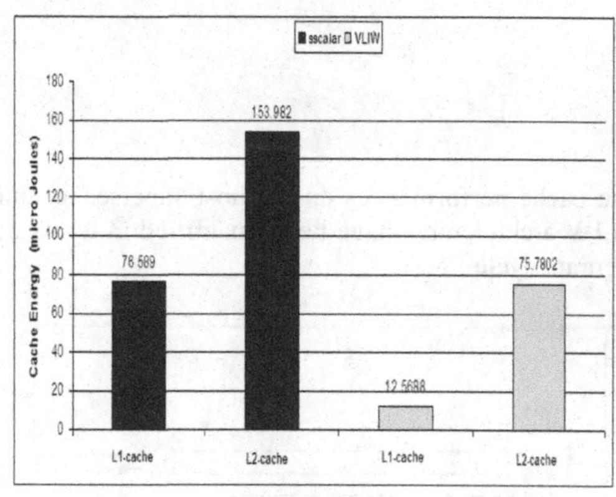

Figure 6: Energy consumed by cache systems of superscalar and VLIW architectures

V. CONCLUSION

In this work, we contrasted the cache energy consumption of two dominant classes of contemporary microprocessors. The work quantified and assessed how energy is exploited in VLIW organizations by lowering the DSI; a measure that translates into substantial energy savings in the datapath of that type of organizations relative to their classical superscalar counterparts. That makes the former to be more appealing to drive today's prevalent embedded systems. In addition, the work highlighted the ramifications of energy optimization within the cache hierarchy of the two classes, the tradeoff and impact in system performance.

REFERENCES

[1] D. Bourger and T. Austin, The simplescalar Tool Set Version 2.0, Computer Architecture News, pages 13-24, June 1997

[2] J. A. Fisher, "Very long instruction word architectures and the ELI-512," in *Twenty-Five Years of Proceedings of the International Symposia on Computer Architecture*, June 1998, pp. 263–273

[3] A. Hussain, "Power and Interconnect Analysis in Very Long Instruction Word Architecture," Thesis, University of Illinois at Urbana- Champagne, 2004

[4] Shen and Lipasti *Modern Processor Design, MiGrow-Hill, International Edition, 2005,pp -9*

[5]Texas Instruments Technical Staff, *TMS320C6000 CPU and Instruction Set Reference Guide*. Texas Instruments, Inc., Oct. 2000

[6] www.simplescalar.com

[7] www.trimaran.org

[8] www.m5sim.org

[9] AMD Athlon X2 white paper

[10] P. Shivakumar and N. Jouppi, "CACTI 3.0: An Integrated Cache Timing, Power and Area Model," WRL Research Report, 2003

[11] Kadayif et al. "Leakage Energy Management in Cache Hierarchies," PACT 2002

A Comparison of LBG and ADPCM Speech Compression Techniques

Rajesh G. Bachu, Jignasa Patel, Buket D. Barkana
Department of Electrical Engineering
University of Bridgeport
221 University Ave. Bridgeport, Connecticut, 06604 USA

***Abstract*— Speech compression is the technology of converting human speech into an efficiently encoded representation that can later be decoded to produce a close approximation of the original signal. In all speech there is a degree of predictability and speech coding techniques exploit this to reduce bit rates yet still maintain a suitable level of quality. This paper is a study and implementation of Linde-Buzo-Gray Algorithm (LBG) and Adaptive Differential Pulse Code Modulation (ADPCM) algorithms to compress speech signals. In here we implemented the methods using MATLAB 7.0. The methods we used in this study gave good results and performance in compressing the speech and listening tests showed that efficient and high quality coding is achieved.**

I. INTRODUCTION

Signals such as speech and image data are stored and processed in computers as files or collections of bits. Signal compression is concerned with reducing the number of bits required to describe a signal to a given accuracy. The compression of speech signals has many practical applications such as digital cellular technology. Compression allows more users to share the system than otherwise possible. Also, for a given memory size, compression allows longer messages to be stored than otherwise in digital voice storage such as answering machines. Speech compression is called speech coding. During the coding process, after high-rate data transforms low-rate data, loss or accuracy will be increased, relatively. The aim of the coding is to find the best possible accuracy for a given compression rate. Such as transform coding, predictive coding, or pulse code modulation techniques give good accuracy, relatively [1-2-3].

II. SPEECH CODERS

Speech coding techniques or speech coders is simply described by dividing them into three groups, namely: *waveform coders*, *source coders*, and *hybrid coders*. Wave-form coders lead to very good quality speech and are used at high bit rates. Source coders operate at very low bit rates and the reconstructed speech is often `robotic sounding. Hybrid coders use elements from both waveform coders and source coders. Hybrid coders lead to good reconstructed speech and average bit rates.

A. Waveform Coders

The simplest form of waveform coding is Pulse Code Modulation (PCM), which is simply sampling and quantizing. Narrow-band speech is normally band-limited to 4 kHz and sampled at 8 kHz. 12 (bits/sample) is needed for satisfactory linear quantization so the resulting bit rate is therefore 96kbits/s. The bit rate in PCM can be greatly reduced by using non-uniform quantization. A commonly used technique in waveform coding is to attempt to predict the value of the next sample based on the previous samples. This is possible due to correlations that occur in speech from vocal cord vibrations. If the error between the predicted samples and actual samples has less variance than the original speech itself, then the error can be quantized instead of the waveform. This is the basis of Adaptive Differential Pulse Code Modulation (ADPCM) where the difference between the original and the predicted samples are quantized [4].

Frequency domain based waveform coding can be used as well as time domain based coding. Sub-Band Coding (SBC) is where the input speech signal is split into a number of frequency bands, or sub-bands and each is coded separately using a DPCM based code. The receiver decodes each of these sub-bands and then combines them to create the reconstructed speech. Frequency domain based coding is effective because different sub-bands can be allocated more or less bits depending on their perceived importance. Filtering within SBC codes leads to a higher level of complexity when compared to time based codes.

B. Source Coders

Source coders use a model of how the source was created, and try to extract parameters from the original signal. These extracted parameters are transmitted. Source coders work by viewing the vocal tract as a time-varying filter, and they are excited by noise, unvoiced speech, or a series of pitch pulses spaced at pitch intervals. Along with the filter specifications a voiced/unvoiced flag, variance of excitation signal, and pitch period for voiced speech are sent in the transmission. The parameters are recalculated every 10-20ms to allow for the changes in speech. Source encoders can calculate the

K. Elleithy (ed.), *Advanced Techniques in Computing Sciences and Software Engineering*,
DOI 10.1007/978-90-481-3660-5_61,

parameters in different ways and employ both time domain and frequency domain methods. Source coders function at approximately 2.4kbits/s or less and produce satisfactory speech although it is not as natural sounding as most people would accept. Source coders have found a place in the military where natural sounding speech is not as important as bit rates [4].

C. Hybrid Coders

A good compromise between waveform coders and source coders is found within the group of hybrid coders. Hybrid coders produce good speech quality at relatively low bit rates. The most successful and the most common hybrid coders are time domain Analysis-by-Synthesis (AbS) coders. AbS coders use the same filter as the Linear Predictive Code (LPC), however they don't use a two state voiced/unvoiced model to find the filter inputs. The excitation comes from attempting to match the reconstructed to the original. Commonly used techniques in this grouping are Multi-pulse excited (MPE) codes, Regular-Pulse excited (RPE), and Code Excited Linear Predictive (CELP). CELP codes use a code book of waveforms as inputs into the filter. Originally the CELP code book contained white Gaussian sequences because it was found that this could produce high quality speech. However an analysis-by-synthesis procedure meant that every ex- citation sequence had to be passed through. This leads to a high level of complexity and processing [4]. Today CELP codes have reduced complexity and they are aided greatly by increases in processing, such as high speed DSP chips.

III. METHODS

The methods that we used to explain the speech coding techniques are LBG and ADPCM.

A. ADPCM

Adaptive Differential Pulse Code Modulation, or ADPCM, is a digital compression technique used mainly for speech compression in telecommunications. ADPCM is a waveform coder that can also be used to code other signals than speech, such as music or sound effects. ADPCM is simpler than advanced low bit-rate voice coding techniques and doesn't require as heavy calculations, which means encoding and decoding can be done in a relatively short time[5,6].

The principle of ADPCM is to predict the current signal value from previous values and to transmit only the difference between the real and the predicted value. In plain Pulse-Code Modulation (PCM) the real or actual signal value would be transmitted. In ADPCM the difference between the predicted signal value and the actual signal value is usually quite small, which means it can be represented using fewer bits than the corresponding PCM value [7].

Depending on the desired quality and compression ratio, a difference signal is quantized using 4, 8, 16 or 32 levels. ADPCM is usually used to compress an 8 kHz, 8-bit signal, with an inherent flow rate of 64 (Kbits/s). When encoded at the highest compression ratio, using only 2 bits to code the ADPCM signal, the flow rate is reduced to 16 (Kbits)/s, i.e. 25% of the original. Using 4-bit coding, the flow rate is 32 (Kbits/s), i.e. 50% of the original and the quality of the signal is fine for most applications [8].

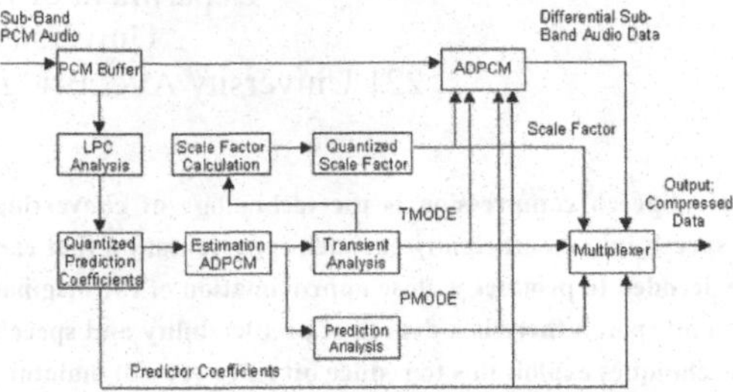

Fig.1. Block diagram of ADPCM [9].

The encoding process is based on a subtraction of a linear predictive value (LPC) from the input PCM signal. The resulting difference signal is then quantized and transmitted to the decoder [9]. At the decoder the predicted signal, which was removed at the encoder, can be exactly regenerated and added back to the residual signal, thereby recreating the original input signal.

B. LBG algorithm

An algorithm for a scalar quantizer was proposed by Lloyd (1957). Later, Linde et al. (1980) generalized it for vector quantization.

1) Bit Allocation and Codebook Design

The codebook design using LBG algorithm is a clustering algorithm method also known as the generalized Lloyd algorithm (GLA). The algorithm requires an initial codebook. This initial codebook is obtained by the splitting method. The technique used in the design of the codebook compares the input vectors to every candidate vectors of the codebook. Quantization distortion (D_m) is measured from the minimum mean-square error (MSE) between the centroid C_m and the input vector x_i (data at the ith vector).

$$D_m = \frac{1}{M}\sum_{i=0}^{M-1}\left(\frac{1}{N}\sum_{k=0}^{N-1} d[x_{ik}, c_{mk}]\right) \tag{1}$$

M is the number of the input vectors classified to the centroid and N is the number of points in a vector. For a B- bit codebook, it would have 2^B number of codebook vectors. Each codebook vector is assigned to a codebook cell C_i for $0 \le i \le (2^B - 1)$. C_0 (C_i at $i = 0$) is determined by averaging the entire input vectors. C_i is then split into two close vectors, C_i - ε and C_i - ε, where ε represents a small varying constant, Fig.2. This codebook design process is continued until the number of codebook vectors reach 2^B.

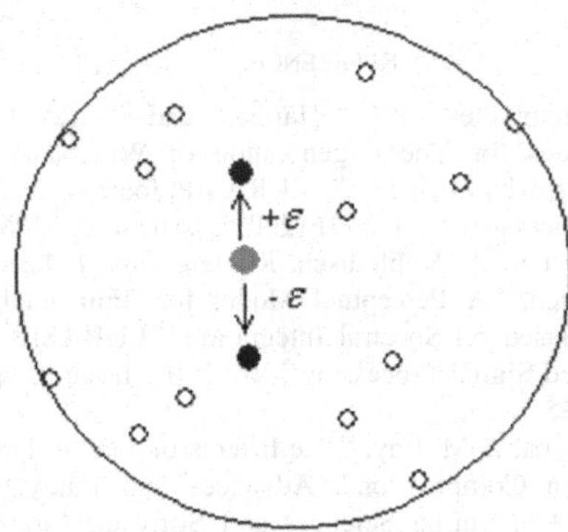

Fig.2. Preparation of the codebook using Pair-wise Nearest Neighbor.

We used the "vector bit-allocation algorithm [10]" to find the bits for the vectors. For following equations, mean square distortion (D_k) for k^{th} transform vector x_k (data at the kth vector).

$$D_k = Ae^{-\frac{4}{N}b_k} \quad k=0,1,2,\ldots,N-1 \tag{2}$$

A is a constant. b_k is a bit size that is allocated for k^{th} vector. Using the Eq.2, total distortion on the vectors,

$$D = \sum_{k=0}^{N-1} \frac{\sigma_k^2}{N} D_k = \frac{A}{N} \sum_{k=0}^{N-1} \sigma_k^2 e^{-\frac{4}{N}b_k} \tag{3}$$

σ_k^2 is variance for x_k. Allocated bit number for each kth vector is given with Eq.5. B is a total bit number for a speech block. R is an average bit number for each block.

$$R = \frac{B}{N} \text{ (Bit/dimension)} \tag{4}$$

$$b_k = \frac{N}{2}R + \frac{N}{4}\left(\log_2 \sigma_k^2 - \frac{1}{N}\sum_{k=0}^{N-1}\log_2 \sigma_k^2\right) \tag{5}$$

Calculated bits are not an integer. For this reason, they are rounded to the nearest integer. The first term in the bit-allocation equation, $\frac{N}{2}\times R$, is average bit number for each vector. The term in the parenthesis can be positive or negative depends on the variance value. There are three important things in the bit-allocation equation. These are;

1. b_k is generally not an integer.
2. If σ_k^2 is very small, b_k can be a negative.
3. If σ_k^2 is very big, b_k can be a positive integer [10].

The signal to noise ratio (SNR) of the reproduced speech signals is given with Eq.6.

$$SNR(dB) = 10\log_{10} \frac{\sum_{n=0}^{V-1} x[n]^2}{\sum_{n=0}^{V-1} (x[n]-\hat{x}[n])^2} \tag{6}$$

$x[n]$ is an original speech signal, $\hat{x}[n]$ is a reproduced speech signals.

IV. Results

In this section, results of the coding techniques in this study are presented and compared. Original speech signals that are studied are given with Fig. 3. It has 7372 samples at 8000 sampling frequency.

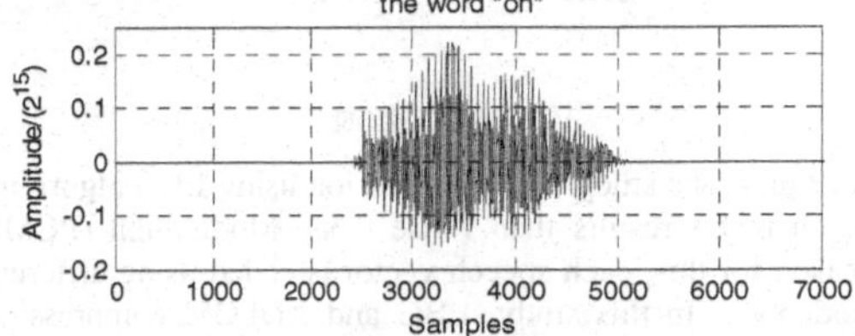

Fig.3. Original speech signals.

The results of the LBG design method using different number of bits are shown with Fig.4 and 5.

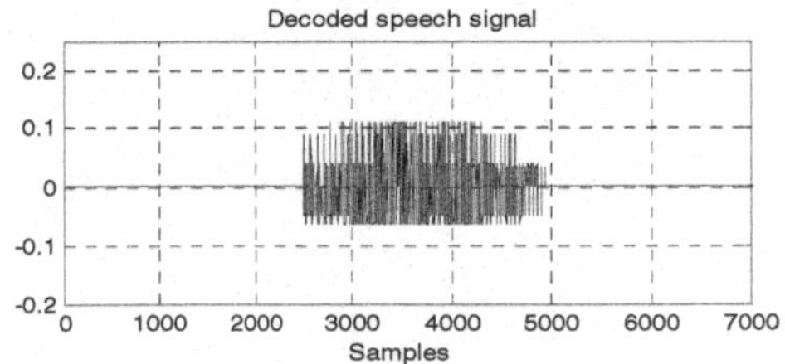

Fig.4. Decoded speech signal for 4-bits using LBG algorithm. SNR value is 8.44 dB.

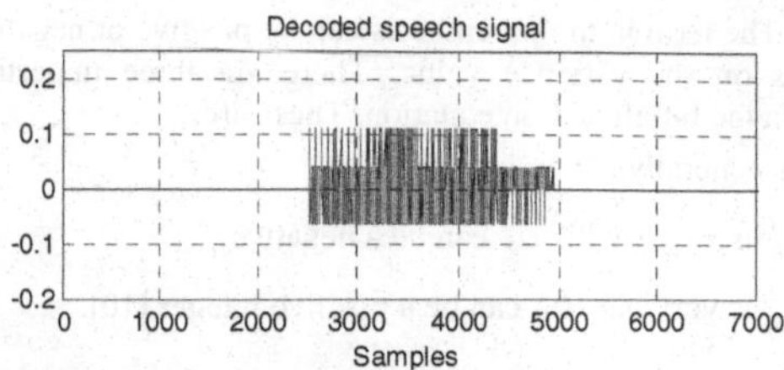

Fig.5. Decoded speech signal for 2-bits using LBG algorithm. SNR value is 7.6 dB.

The result of the ADPCM algorithm using 2-bits is shown with Fig.6.

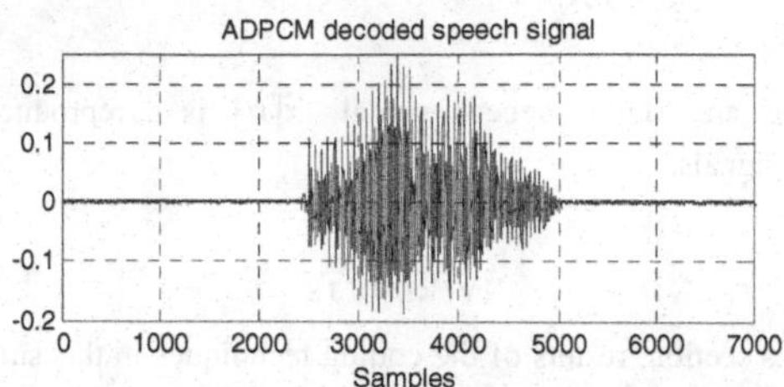

Fig.6. Decoded speech signal for 2-bits using ADPCM algorithm.
SNR value is 12.95 dB.

V. Conclusion

At our previous study [3], compression using LBG algorithm has much better results than Pulse Code Modulation (PCM). The reason for this, each speech vector is coded using different size codebook. In this study, LBG and ADPCM compression algorithm are compared. Listening tests showed that the speech quality was efficient to understand for both algorithms. From the results, the SNR of LBG algorithm is much lower than the SNR of the Adaptive Differential PCM for the same number of the bits. ADPCM method is more effective than LBG algorithm for coding speech signals.

References

[1] J.A. Fuemmeler, R.C. Hardie, and W.R. Gardner, "Techniques for The Regeneration of Wideband Speech from Narrowband Speech", EURASIP Journal on Applied Signal Processing, Vol. 2001 (2001), Issue 4, pp. 266-274.

[2] S. Van de Par, A. Kohlrausch, R. Heusdens, J. Jensen, and S.H. Jensen, "A Perceptual Model for Sinusoidal Audio Coding Based on Spectral Integration", EURASIP Journal on Applied Signal Processing , Vol.2005, Issue 9, pp.1292-1304, 2005.

[3] B.D. Barkana, A.M. Cay, "The Effects of Vector Transform on Speech Compression", Advances and Innovations in Systems, Computing Sciences and Software Engineering, Springer, pp: 67-70, 2006.

[4] A report on 'speech coding'-University of Southern Queensland Faculty of Engineering & Surveying.

[5] Atmel Application Note AVR335: Digital Sound Recorder with AVR ® and DataFlash ®. http://www.atmel.com.

[6] A.V. Oppenheim, R.W. Schafer, 'Discrete-Time Signal Processing', Prentice Hall, Englewood Cliffs, NJ, 1989.

[7] H.J. Choi, Y.H. Oh, "Speech Recognition Using an Enhanced FVQ Based on a Codeword Dependent Distribution Normalization and Codeword Weighting by Fuzzy Objective Function", Dept. of Computer Science Taejon, Korea.

[8]. ITU-T Recommendation G.726: 40, 32, 24, 16 kbit/s Adaptive Differential Pulse Code Modulation (ADPCM). http://www.itu.int

[9] ADPCM- http://www.telos.info/ADPCM.702.0.html

[10] W. Li, "Vector Transform and image coding" IEEE Transactions on Circuit and Systems for Video Tech., Vol.1, No.4, pp. 297-307, December 1991.

LCM: A new approach to parse XML documents through loose coupling between XML document design and the corresponding application code

Vaddadi P. Chandu
Citco Canada Inc.
Halifax, NS, Canada
chandu@terpalum.umd.edu

Abstract- **XML (eXtensible Markup Language) is a widely used language for communication between heterogeneous, homogeneous, and internet software applications. Currently available binding and parsing techniques for XML documents involve tight coupling of the document with the application code and require a redesign of the application code for any change in the document design. This paper proposes a 'Loosely Coupled Model (LCM)' for XML document parsing, which eliminates the need to redesign the application code for moderately large changes in the XML document design. LCM takes a token based approach, which is very different from the existing methodologies for XML parsing. LCM is implemented in Java using Sun's JDK 1.5. The dataset includes 24 XML documents categorized into two sets- (a) deeply nested and (b) non-deeply nested XMLs. Each set contains 12 documents ranging from 2KB up to 4MB. A performance comparison for LCM is obtained and is noticed to be comparable or better against this dataset with (a) Sun's Java XML Binding (JAXB), (b) Apache's XMLBeans, and (c) JDOM parser.**

I. INTRODUCTION

Owing to its characteristics of platform independence and extensibility through use of custom tags XML is widely used for information exchange between software applications. Industries such as healthcare, financial services, telecommunication, and several others use XML to implement proprietary protocols. In internet applications also, XML has become a de-facto standard for information interchange. Simple Object Access Protocol (SOAP) [1] a widely used protocol in internet applications, is built upon XML. Web services [2], which communicate through SOAP are again dependent on the XML indirectly. Due to its extensive usage in applications efficient and forgiving techniques to infer the data from an XML document are very important.

Simple API for XML (SAX) [3], Document Object Model (DOM) [4, 5], Streaming API (StAX) for XML [6], and JDOM [7] are most popular parsing techniques (that read an XML document) and Sun's Java Xml Binding (JAXB) [8] and Apache's XMLBeans [9] are examples of popular binding techniques (that bind an XML document to an Object).

Although no single technique is suitable to all scenarios, each one has its own advantages and suitability to a specific application depending on the response time, throughput, and available resources. For example, DOM allows both forward and backward traversal of an XML document but uses memory in abundance. SAX provides higher throughput but restricts the document traversal to be forward-only. StAX allows selective retrieval of the elements in the document. Binding techniques such as JAXB work very well when the XML structure remains constant because, there is a very tight binding between the XML document and the object it maps to. Apache XMLBeans is also similar to JAXB but provides the additional features of (a) generation of the binding-code on the fly, and (b) integration into Eclipse [10] based IDEs [11].

These techniques are proven to be efficient [12] but are not forgiving. In most business scenarios the applications are designed in a client-server mode where the server extends a service and provides an XML defining that service and clients consume the service by implementing the client application code around that XML document. To accommodate a new request from a specific client or to alter an existing business-functionality, the server needs to change the XML document structure by adding or removing one or more elements. Such a case which is very common in business scenarios, the techniques mentioned above for XML parsing/binding will require a redesign of the client application code. However, not all clients will be interested in the enhancement and will want to continue using the service without redesigning the application code.

LCM solves this problem by decoupling the client application code and the XML document design through a Layer of Abstraction (LoA). For changes in the XML document design other than a change in the cardinality of an element, the LCM client application code will not require redesign.

The remainder of this paper is organized as follows. Section II describes the existing methodologies for XML parsing and binding. Section III explains LCM in detail and presents LCM Application Programming Interface. Section IV presents the experimental results and finally, Section V presents conclusions and future work.

K. Elleithy (ed.), *Advanced Techniques in Computing Sciences and Software Engineering*,
DOI 10.1007/978-90-481-3660-5_62, © Springer Science+Business Media B.V. 2010

II. Existing Methodologies for Parsing and Binding of XML Documents

A. *SAX: Simple API for XML Parsing*

SAX API is based on event based parsing paradigm. In this paradigm the XML document is scanned from the beginning to the end pushing out values whenever a syntactical construction such as, an attribute, a text node, or an element is encountered. The application code that is using this parser needs to handle these callbacks from the parser. Listing 1 shows eleven callback methods exposed by the *ContentHandler* SAX API [13] which is widely used to parse an XML document.

```
void characters (char [], int, int)
void endDocument(void)
void endElement(String, String, String)
void endPrefixMapping(String)
void ignorableWhitespace(char [], int, int)
void processingInstruction(String, String)
void setDocumentLocator(Locator)
void skippedEntity(String)
void startDocument(String)
void startElement(String, String, String, Attributes)
void startPrefixMapping(String, String)
```

Listing. 1 Callback methods in SAX. These mehods are exposed by the ContentHandler API of the SAX paradigm.

SAX allows high throughput and low memory usage [12, 15] because the document is not stored in memory. However, this makes the document traversal forward-only. More importantly, the application code identifies an element by matching the incoming value from the parser to the name of a desired element in the XML document. This creates a tight coupling between the application code and the XML document design requiring a change in the application code with changes in XML document.

B. *Document Object Model for XML Parsing*

In this model, a given XML document is read in a single pass and is stored as an in-memory tree. See Listing 2 for a sample XML document and Fig. 1 for the corresponding in-memory tree. For a given XML document, a client application code has to first create this tree using the DOM APIs (See Listing 3) and then traverse this tree to retrieve the elements. Traversal, manipulation, insertion, and deletion of the nodes are possible with this technique through its tree based representation.

```
<ROOT>
        <ELEM_1>
                <ELEM_2>
                        VAL_ 2
                </ELEM_2>
                <ELEM_3>
                        VAL_3
                </ELEM_3>
        </ELEM_1>
        <ELEM_4>
                VAL_4
        </ELEM_4>
</ROOT>
```

Listing. 2 Sample XML Document

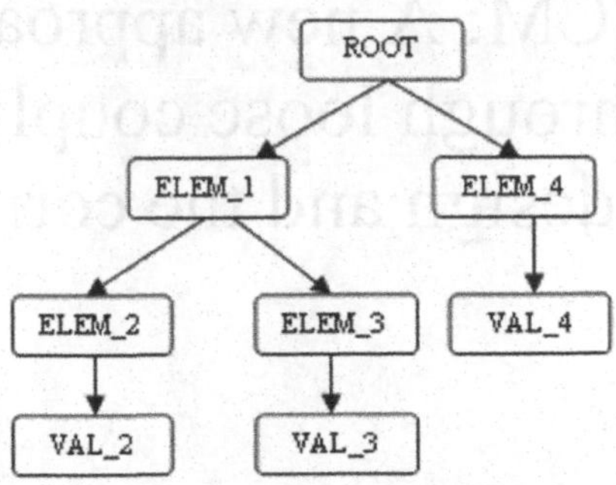

Fig. 1 DOM Tree for XML Document in Listing. 2

DOM is very easy to use due to its tree based traversal. However, in this case also there is a very tight coupling between the XML design and the corresponding client application code because the value of an element is extracted by comparing the incoming tag name to a known XML tag name (See recursive loop in Listing 2). Furthermore, it consumes 2 to 15 times memory and provides lesser throughput than the SAX parser [12, 15].

```
// Create a new document builder factory (DBF)
DocumentBuilderFactory dbf = new DocumentBuilderFactory();
// Set validation mode (if any)
...
// Create a document builder (DB) from DBF
DocumentBuilder db = dbf.newDocumentBuilder();

// Create a tree representation of the document
Document d = db.parse(<XML document>);

// Parse the XML document traversing each node in the tree
// starting from 0
visitRecursively(d, 0);
```

Listing. 3 XML parser pseudo application code using DOM technique

C. *Streaming API (StAX) for XML Parsing*

StAX is based on a streaming programming model and combines the advantages of both SAX (Section II.A) and DOM (Section II.B). In this model, the application code has the capability to explicitly ask for subsections in the XML document and to arbitrarily stop, suspend, or resume parsing.

StAX parser primarily offers two APIs, Cursor API (an easy-to-use and high efficiency) and Event Iterator API (easy pipelining of XML documents). Listings 4, 5 show code snippets of client application code implementing Event Iterator and Cursor APIs respectively.

StAX parser provides additional advantages over SAX and DOM parsers [12, 14, 15] such as, (a) application driven XML parsing, (b) simultaneous multiple-XML-documents read capability, (c) XML document hopping capability, which allows selective retrieval of elements, (d) less number of API resulting in easy coding interface, and (e) XML document read-write capability. However, StAX parser also involves tight coupling of the XML document with its application code because of the same reasons as explained for the SAX and the DOM parsers.

```
// Get an instance of the XML input factory
XMLInputFactory myf = XMLInputFactory.newInstance();
// Set the preferences- validating, error reporting etc
...
// Create new XML Stream Reader
XMLStreamReader er = myf.createXMLStreamReader(...);

// Traverse the document
while(er.hasNext()) {
    XMLEvent xe = er.nextEvent();
    if (xe.isStartElement()) // do this
    else // do this
}
```

Listing. 4 XML parser pseudo application code using the Event Iterator API of the StAX methodology

```
// Get an instance of the XML input factory
XMLInputFactory myf = XMLInputFactory.newInstance();

// Set the preferences- validating, error reporting etc
...
// Create new XML Stream Reader
XMLStreamReader str = myf.createXMLStreamReader(...);

// Traverse the document
while(streamedReader.hasNext()) {
    int xe = str.next ();
    if (xe == START_ELEMENT) {
        QName en = str.getQName();
    } else {
        // do the following
    }
}
```

Listing. 5 XML parser pseudo application code using the Cursor API of the StAX methodology

D. *JDOM Xml Parsing methodology*

JDOM [7] is a Java based document object model for XML parsing. It builds a JDOM XML tree from XML documents, SAX events, or DOM tree through use of existing XML parsers. Listing 6 shows a code snippet of client application code using JDOM technique for XML parsing. As it can be seen in the listing, the value of an element is retrieved by sending the name of the element in the XML document as an input argument to the JDOM API (elem.getAttribute(*<element_name>*)) making this model is also tightly coupled like the parsers described in Sections II.A to II.C.

```
SAXBuilder bld = new SAXBuilder();
Document doc = bld.build(...);
Iterator itr = doc.getRootElement().getChildren().iterator();
while (itr.hasNext()) {
    Element elem = (Element) itr.next();
    elem.getAttribute("<element_name>").getValue());
}
```

Listing. 6 XML parser client application code using JDOM

E. *Sun's Java XML Binding technology (JAXB)*

JAXB [8] is a binding technology that converts a given XML document into a corresponding Object Oriented representation in Java. JAXB mainly consists of two components (See Fig 2). The first component (XJC) builds the Java Objects for a given XML document and the second component (JAXB engine) uses the objects created by the XJC to read or write an XML document.

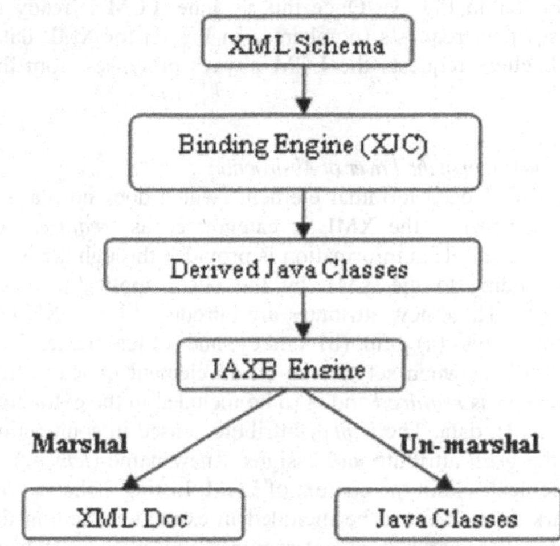

Fig. 2 JAXB control flow (XML marshal/un-marshal)

In order to develop a client application code for an XML document using this technique, first a corresponding XSD is to be passed through the Binding Engine (XJC) to generate Java objects for the XML. Next, using the API exposed by the JAXB (JAXBContext, ObjectFactory, Unmarshaller, Marshaller) code needs to be developed around the Java objects generated by XJC to read or write the XML document.

Traversing and manipulating the XML document is very easy because the XML is represented as Java Objects and the performance is very good because of very tight coupling through derived classes [15]. However, due to the very tight coupling, any change in the XML document requires regeneration of the derived classes and redesign of the client code.

F. *Apache's XMLBeans XML Binding technology*

XMLBeans [9] is also a binding technology and is similar to the JAXB technology (Section II.E). However, XMLBeans allows more convenience by integrating the engine into Eclipse based IDE [11] and eliminating the need to manually generate the Java classes outside the IDE. However, due to same reasons as explained for the JAXB technology, this methodology also suffers from the problem tight coupling between the client code and XML document design.

III. LCM- LOOSELY COUPLED MODEL FOR XML PARSING

In order to parse a given XML document LCM employs a totally different approach from the methodologies described in Section II. LCM introduces a Layer of Abstraction (LoA) between the client application code and the XML document that acts as a replacement to the XML document and absorbs the changes in XML document design. In the LCM approach, the LCM engine accepts an XML document and corresponding XSD and dynamically generates the LoA (numbers 1 in Fig. 3). Once this is done, LCM is ready to process client requests (numbers 2 in Fig. 3) for XML data. For all client requests the LCM always processes from the LoA.

A. *Generation of the Layer of Abstraction*

In LCM each terminal element (which does not have a child element) in the XML is categorized as '*required*' or '*not-required*'. This information is provided through the XSD corresponding to the XML by the client application code developer. Three new attributes are introduced in the XSD to implement this- (a) grab, (b) lcmkey, and (c) lcmmarker. The *grab* attribute when set to *true* on an element indicates that the element is *required* and is to be included in the extraction of the XML data. The *lcmkey* attribute is used in conjunction with the *grab* attribute and assigns a new name (*lcmkey*) to this element within the context of LCM. Listing 7 shows how to mark an element to be included in extraction. Setting the grab to *false* marks it as *not-required* and excludes it from extraction. The *lcmmarker* is used for looped elements. If an element is marked with a grab=true, and if it occurs inside a looped element, then the looped element is to be marked with *lcmmarker* as shown in Listing 8.

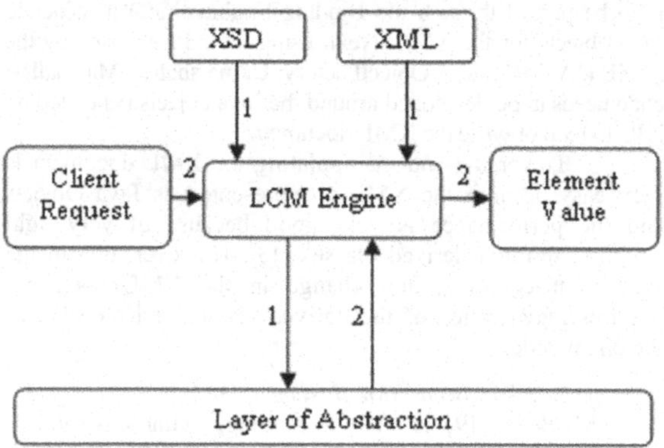

Fig. 3 LCM Architecture. Step 1 executes first and creates the Layer of Abstraction . In Step 2, the LCM engine is ready to process client requests.

This allows identifying blocks/loops in the XML.

```
<xsd:element name="elem_name1" type="xsd:string" grab="true"
lcmkey="book_author"/>
```

Listing. 7 Using lcmkey and grab attributes to mark an element to be included or excluded

```
<xsd:element name="elem_name2" minOccurs="0" maxOccurs =
"unbounded" lcmmarker=" b1">
```

Listing. 8 Using lcmmarker to mark loops. This is used in conjunction with the lcmkey. All elements in the loop will contain the lcmmarker as part of its token.

For a given XML the LCM parser first reads the corresponding XSD (fully marked, with each element categorized as explained above) and creates a map (*xsd-token-key-map*) of *lcmtoken* to the corresponding *lcmkey* of all *required* elements. An *lcmtoken* for any element is a unique string that is generated by appending the names of all elements starting from root down to that element.

The LCM parser then reads the XML document element-by-element, constructing tokens for each element and attribute. Since the token generation methodology is the same in XSD and XML, a token generated for an element in the XML will automatically match the token generated for the same element in the XSD. Hence, if an element is marked as *required*, the token generated for the element from the XML will have a corresponding *lcmkey/lcmmarker* value in the *xsd-token-key-map* created above. LCM performs this check and if the token maps to a value in the *xsd-token-key-map*, the *lcmkey*-element_value or *lcmmarker*-element_name pair is stored in the LoA. Listing 9 presents the algorithm for LoA generation.

1. Start with an empty Layer of Abstraction
2. Accept an input XML and corresponding XSD
3. For each element in the XSD that is marked with a grab as 'true' create a token and add it to the xsd-token-key-map
4. For each element and attribute in the XML create the corresponding token
5. Hash the xsd-token-key-map with the token generated in Step 4 to see if there is a key value for the token
6. If Step 5 results in a 'YES', then add the lcmmarker-element_name or lcmkey_element_value pair to the LoA
7. Repeat from Step 4 until no more elements are present in the XML

Listing. 9 Algorithm for generation of Layer of Abstraction

Since the LoA is generated dynamically, renaming, adding, and removing elements from the XML document will not need a code redesign as long as the *lcmkey* stays unaltered. However, in the only case where the cardinality of a *required* element changes the client application code will need to be changed.

B. *LCM Application Pramming Interface*

LCM API is very simple and exposes only one object *LCMEngine* with its five methods. Each one of those five methods is described below and Listing 10 demonstrates an XML, corresponding XSD and the LCM client application code.

*1) boolean exists ():*This method indicates if there is more data available in the XML document from the current location. Returns *true* if yes, *false* otherwise.

2) boolean moreMarkersInRange (String, int): This method indicates if there are more markers of the given type from the current location to given range. The name of the maker is the first argument to the method and the range is specified in the second argument. Returns *true* yes, false otherwise

3) int fetchNextLocOfMarker (String, int, int): This method returns the next location of a marker within a given range. First argument specifies the marker and the second and third arguments provide the range. If there is no marker in the given range -1 is returned.

4) String fetchTokenValue (String): This method returns the value of the first element from the current location for corresponding *lcmkey,* which is the argument to the method. Returns *null* if no such element exists.

5) String fetchTokenValue (String, int): This method returns the value of the first element from the current location within a given range for a given *lcmkey*. The *lcmkey* is the first argument to the method and the range is specified in the second argument. Returns *null* if no such element exists.

```
<catalog>
  <book id="Id1">
    <author>Author1</author>
  </book>
</catalog>
```

Listing 10a. Sample XML document with an element and attribute

```
<xsd:element name="catalog">
  <xsd:complexType>
    <xsd:sequence>
     <xsd:element name="book" minOccurs="0"
           maxOccurs="unbounded"
               dx2jmarker="book_block">
      <xsd:complexType>
        <xsd:sequence>
     <xsd:element name="author" type="xsd:string"
              grab="true" dx2jkey="book_author"/>
         <xsd:attribute name="id" type="xsd:string"
              grab="true" dx2jkey="book_id"/>
    </xsd:sequence>
  </xsd:complexType>
</xsd:element>
```

Listing 10b. Corresponding XSD to the XML document in Listing 10a

```
// Instantiate an LCM engine by giving XML and its XSD
// This creates the xsd-token-key-map and the lcmstring
LCMEngine lcm = new LCMEngine(xsd, xml);

// Elements that are not in the loop can be extracted using // the lcmkey,
else if loop, then first get the loop limits
int bookBlock = xxe.fetchNextLocOfMarker("book_block", bookBlock,
xxe.MAX_LENGTH);
while (bookBlock!=-1) {
    String author = xxe.fetchTokenValue("book_author",bookBlock);
    String title = xxe.fetchTokenValue("book_title",bookBlock);
}
```

Listing 10c. LCM client application code to the XML and XSD in Listing 10a, 10b

IV. EXPERIMENTAL RESULTS

The experimental environment consisted of Sun's Java, JDK 1.5, Oralcle BEA's Weblogic 9.2 Server, 1.6GHz Intel Centrino dual core processor, and 2 GB RAM. The test bed consisted of two sets of XML documents- (a) deeply nested and (b) non-deeply nested. Each set consisting of 12 XML documents of sizes starting from 2 KB up to 4 MB. The deeply nested XML document set consisted of hypothetical airline data and the other set consisted of hypothetical data on books (author, name, title, publisher etc). Comparison is made against this test bed with Sun's JAXB [8], Apache's XMLBeans [9], and JDOM [7] parser.

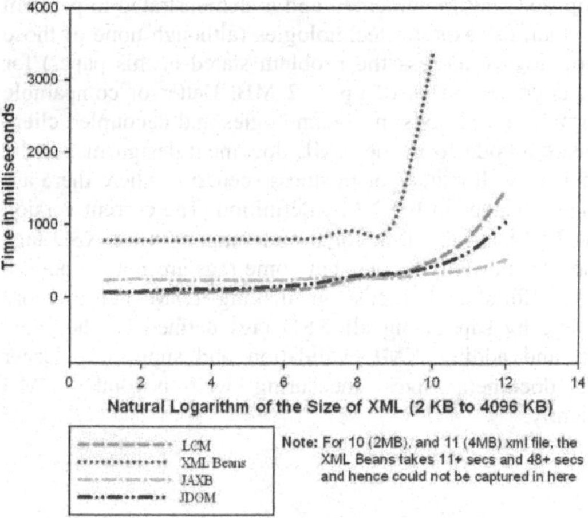

Fig. 4 Performance Comparison of LCM against non-deeply nested XML documents in the dataset

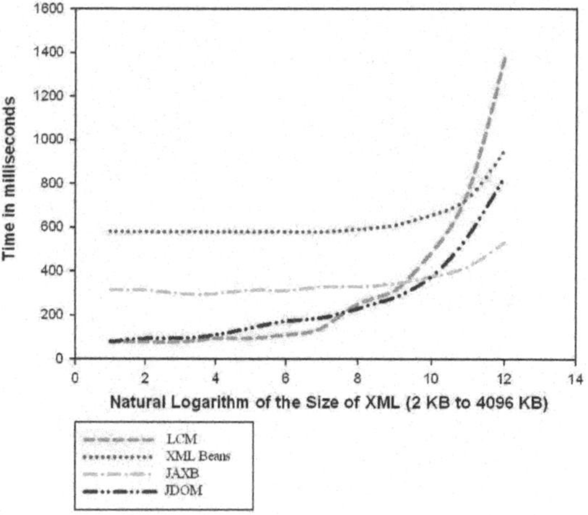

Fig. 5 Performance Comparison of LCM against deeply nested XML documents in the dataset

The results for the non-deeply nested XML are shown in Fig. 4 and for the deeply nested XML are shown in Fig. 5. It can be seen that for LCM performs better than the other methodologies in spite of more computation (due to the LoA generation and token comparison) up to 2 MB. After 2 MB the performance becomes comparable and slightly degrades for 4 MB in the case of deeply-nested XML documents. XMLBeans is seen to be slowest (by the metric of execution time) and JAXB comes next. JDOM and LCM compare very close to each other in both cases for up to sizes of 2 MB.

V. CONCLUSION AND FUTURE RESEARCH

This paper presented and addressed a problem that has never been presented so far. A solution (LCM) to the problem is proposed and implemented and is demonstrated to perform better than the existing technologies (although none of those methodologies address the problem stated in this paper) for XML document sizes of up to 2 MB. Better or comparable performance with existing technologies and decoupled client application code form the XML document design makes the LCM very well suited for business scenarios where there are frequent changes in the XML definition. The current version of the LCM is fully functional and supports most XSD tags defined by the W3C group but some tags are not supported. Future efforts will focus on making LCM parser more complete by supporting all XSD tags defined by the W3C group and adding XML validation and supporting larger XML document sizes measuring well beyond 4 MB efficiently.

REFERENCES

[1] SOAP Version 1.2, http://www.w3.org/TR/soap12-part1/.
[2] Web Service @ W3C, http://www.w3.org/2002/ws/.
[3] SAX Project, http://www.saxproject.org/.
[4] Document Object Model, http://www.w3.org/DOM/.
[5] DOM API, http://www.w3.org/TR/REC-DOM-Level-1/.
[6] JSR 173 StAX, http://www.jcp.org/en/jsr/detail?id=173.
[8] JSR 31, XML Data Binding Specification, http://www.jcp.org/en/jsr/detail?id=31.
[9] Apache XMLBeans, http://xmlbeans.apache.org/.
[10] Eclipse.org, http://www.eclipse.org/.
[11] Using XML Beans in the IDE, http://edocs.bea.com/wlw/docs101/guide/ideuserguide/conUsingXMLBeans.html.
[12] Determine the correct XML parser for a Java Application, Padma Apparao, Intel Software Network, http://softwarecommunity.intel.com/articles/eng/3151.htm.
[13] Interface ContentHandler, http://www.saxproject.org/apidoc/org/xml/sax/ContentHandler.html.
[14] Streaming API for XML Parsers, Java Web Services Performance Team, Aug 2005, Sun Microsystems, http://java.sun.com/performance/reference/whitepapers/StAX-1_0.pdf.
[15] XML and Java Technologies: Data binding, Part 2: Performance http://www.ibm.com/developerworks/xml/library/x-databdopt2/.

A Framework to Analyze Software Analysis Techniques

Joseph T. Catanio
La Salle University
Mathematics and Computer Science Department
Philadelphia, PA 19141
1.215.951.1142
catanio@lasalle.edu

***Abstract* – The software community uses a multitude of varying analysis techniques to define the "what" of software artifacts. How do different analysis techniques compare and contrast with each other? This paper presents a new analysis framework to describe and characterize software analysis techniques employed during the specification process.**

I. INTRODUCTION

The Institute of Electrical and Electronics Engineers, Inc. (IEEE) defined the term software engineering in 1993 as the application of a systematic, disciplined, quantifiable approach to the development, operation, and maintenance of software; that is, the application of engineering to software. Software engineers have developed methodologies that assist them to develop software products. These methodologies encompass the entire systems development life-cycle process. Which begins with a statement of requirements and ends with the product being retired (IEEE, 1998) (Blum, 1994). The software development process is the progression from the identification of some application specific domain need to the creation and delivery of a software product to fulfill that need. To understand the need, one must first understand the application domain. Analysis exists at the application domain level and conceptual models are used to explain the application need and describe domain concepts. Models are prescriptive in nature and are intended to provide clear and concise software requirements, which are then utilized to construct the software system. The most fundamental software development process activities are specification, development, validation, and evolution (Sommerville, 2001). These activities can be realized using a varying number of techniques and methods. This paper presents a new analysis framework to help describe various techniques used during the specification process.

II. ANALYSIS FRAMEWORK

The goal of developing the analysis framework is to make it general enough to accommodate different analysis techniques and provide the ability to describe each technique in a systematic, repeatable method. Developing the framework in this manner makes it possible to describe issues, limitations, problems, and opportunities with various analysis techniques.

The analysis framework is based upon the component-oriented approach to the software system development process. Component-based development or component-based software engineering is a re-use based approach to software systems development (Sommerville, 2001). A component is an independent entity that provides services and may be described at different levels of abstraction. The analysis phase views the system at a high-level of abstraction and identifies major system components. These components interact with each other to create system functionality. The goal of the analysis is to describe the software system in its entirety by decomposing it into its relevant components. It is possible to describe a software system at different levels of abstraction and detail. The aggregate of these components comprise the software system. A more refined description yields a more detail-oriented software system description.

A component interaction is described by function, communication, and behavior characteristics (Wieringa, 1998) and can best be viewed graphically as depicted in Figure 1.

Figure 1 Component Interaction Characteristics

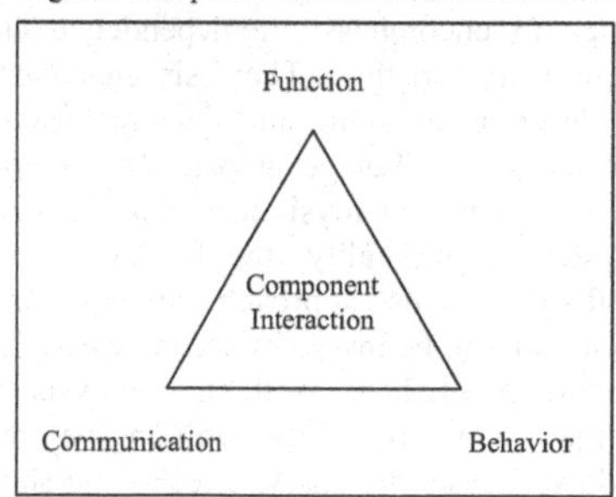

K. Elleithy (ed.), *Advanced Techniques in Computing Sciences and Software Engineering*,
DOI 10.1007/978-90-481-3660-5_63, © Springer Science+Business Media B.V. 2010

The function characteristic describes the actions or functionality of a component interaction. Communication characteristics describe how information is exchanged among components. Behavior describes how the component responds to an event. Function, communication, and behavior characteristics can act either independently or dependently with each other to describe a component interaction.

III. ANALYSIS FRAMEWORK EXTENDED

Extending the framework to incorporate the dependent nature of the aforementioned characteristics yields three additional characteristics depicted in Figure 2. These additional characteristics result from the intersection of the three primary characteristics.

Figure 2 Component Interaction Analysis Framework

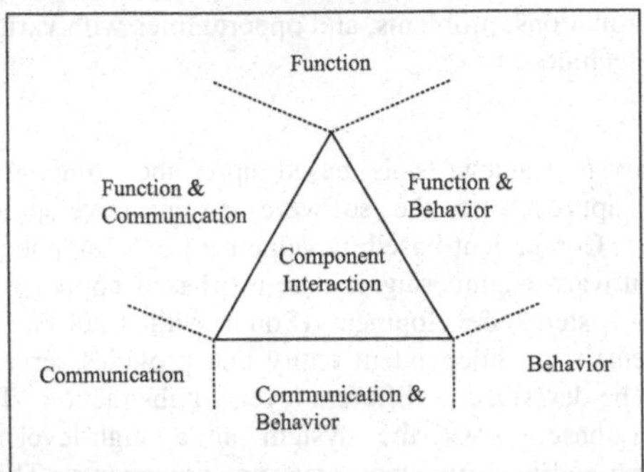

A component interaction may consist of one or more functions that can communicate and share information with each other. The way functions communicate with each other helps to capture relationships among the functions. Function & behavior, represents the time-ordered behavior of a function. Communication & behavior, describes the sharing of information with respect to time. These six characteristics comprise a component interaction and form the analysis framework that can be used to compare analysis techniques. The literature does not explicitly categorize systems using all these characteristics in a framework. Wieringa's paper is an original approach that provides a framework to compare analysis techniques. This paper extends the framework and adds three additional characteristics to encompass interdependence among the original three characteristics. These six characteristics can be used to describe the static and dynamic features of an analysis technique. For example, to determine the effectiveness of how an analysis technique identifies major software system functionality, the function characteristic can be utilized. This approach permits comparisons between techniques to be made. A second example involves the determination of how well the analysis technique identified the scope of work utilizing a re-usability approach. In this case, the analysis technique should focus on the function and communication characteristics of the framework since re-usability is most successful when major system functionality can be encapsulated into a component that communicates through an interface.

Both of these examples show how the framework could be used to compare and contrast concepts utilizing various analysis techniques. This framework allows analysis techniques to be compared and contrasted in a variety of ways utilizing the function, communication, and behavior system properties in either an independent or dependent manner.

IV. STRUCTURED ANALYSIS & DESIGN (SAD)

Structured analysis is a methodology that aids the practitioner during the analysis phase of the software system development life-cycle. Structured analysis is a process-oriented system definition approach to the description of the software system in a top-down fashion. The top-down approach decomposes the software system in a leveled manner whereby each level provides more details until a primitive, atomic level is reached. This is a conceptual decomposition as opposed to a physical decomposition. A conceptual decomposition partitions the software system in terms of components that correspond to domain entities. In contrast, a physical decomposition is defined in terms of the actual software system components and is realized during the implementation phase. The conceptual decomposition is a way to make the demands of external functionality explicit without yet worrying about implementation decisions (Wieringa, 1998). Therefore, the conceptual decomposition is a top-down leveled approach that organizes views of the software system into a hierarchical structure. This structure defines the software system based on the system functionality and behavior, emphasizing both data and control flow. The structured analysis perspective is to generate a detailed, logical description of tasks and operations by focusing on the control flow and data processing of information.

Within the context of structured analysis, there are many prominent variants to analyze information flows. These variants suggest different ways to approach analysis in a structured manner, sharing the common goal of improving the understanding of the software system. Ross developed the "Structured Analysis and Design Technique" (SADT), which begins the process of analysis by determining the why and what of the software system components before progressing to the implementation or how phase (Ross, 1986). The SADT graphically depicts interactions of data and activities by utilizing activity diagrams.

There were other developments of structured analysis techniques around the same time period. For example, DeMarco's technique for analyzing information flow is based upon the process flow chart developed by Taylor and Gilbreth (Couger, 1973). These process flow charts graphically depict the movement of materials in a manufacturing or service-oriented capacity. The process flow chart is an abstraction and defines the key points and activities of the system processes. DeMarco and Yourdon extended the process flow chart concept to include the analysis component of the software engineering process (DeMarco, 1978). Gane and Sarson developed a method similar to DeMarco's process and data flow oriented-technique but it focuses more on the data view by emphasizing the identification of the data components of the software system (Gane & Sarson, 1979). To this end, their technique utilizes data access diagrams to describe the contents of the software system data stores. These data access diagrams depict the entities and links of a data store. This data centric technique builds from Chen's unified view of data concept (Chen, 1976). Chen identifies major data components of the system by utilizing the characteristics of entities, attributes, and relationships. The characteristics of these components are captured in an entity-relationship-attribute (ERA) model. These entity relationship (E/R) diagrams have become the basic building blocks to database design techniques and model the characteristics of the database system to be designed.

Models and structural analysis techniques can be used to produce software system specifications that describe the software system. All versions of structured analysis utilize a top-down decomposition approach to develop conceptual abstractions that lead to concrete software components. These structural analysis techniques can produce software system specifications by utilizing the constructs previously described. The specifications consist of various diagrams depicting the systems processes and data flow in both a static and dynamic nature. Traditional structured analysis of business-oriented software systems utilizes data flow diagrams, data dictionary, mini-specifications, and structured walkthrough components to identify the requirements (Svoboda, 1990).

V. APPLYING THE ANALYSIS FRAMEWORK TO SAD

The following list summarizes the structured analysis technique utilizing the characteristics of the analysis framework.

- Function Description: Explicitly realized through the use of functional decomposition.
- Function & Communication Description: Depicted through data flow diagrams that show function and data flow.
- Communication Description: Depicted through data flow diagrams that show data input and output.
- Communication & Behavior Description: Depicted through state diagrams.
- Behavior Description: Depicted through data and control flow.
- Function & Behavior Description: Depicted through state transition diagrams.

Structured analysis does address the six characteristics of the analysis framework presented. In particular, the system entities and relationships among them are identified by data flow diagrams. As the name implies the nature of the relationships among entities are data flow centric. These relationships are tightly coupled with corresponding functions thus making relationship identification dependent on function identification. This relationship identification is solely based upon the ability of the development team to identify system functions. However, many functions are not identified until the implementation phase causing many relationships to be missed during the analysis phase (Catanio & Bieber, 2006) (Catanio et al., 2004). This could lead to inadequate problem domain understanding and an incomplete analysis process.

VI. OBJECT-ORIENTED ANALYSIS (OOA)

Object-oriented analysis (OOA) is a method of analysis that examines requirements from the perspective of the classes and objects found in the vocabulary of the problem domain (Booch, 1994). With respect to software systems, OOA is a method that develops software engineering requirements and specifications utilizing an object model approach. An object model represents the software system by providing a description of the major software components or objects comprising the system. An object is a real world concept or abstraction that represents a portion of the problem that is to be solved. An object is an entity that has a state and a set of operations that access the state.

Objects are comprised of two sets of components: state information and operations. An object's state is defined by a set of attributes and the operations performed on that state are called methods. Consequently, the object model is a collection of interacting objects that maintain their own state and provide operations that permit access to this state information. These objects help to encapsulate an abstract concept into a self-contained unit. This unit or component-based approach provides object-oriented analysis powerful modeling techniques. Therefore, the principle behind object modeling is encapsulation and abstraction (Booch, 1996) (Rumbaugh, 1991). Booch defines a spectrum of abstraction for objects that closely model problem domain entities:

- Entity abstraction is an object that represents a useful model of a problem domain or solution-domain entity.
- Action abstraction provides a generalized set of operations, all of which perform the same kind of function.

- Virtual-Machine abstraction is an object that groups together control operations.
- Coincidental abstraction is an object that packages a set of operations that have no relation to each other.

These types of objects are the building blocks of the object-oriented paradigm, which incorporate the object-oriented strategy throughout the software development process. Sommerville breaks the object-oriented development process into three main components: analysis, design, and implementation (Sommerville, 2001).

- Object-oriented analysis develops an object-oriented model of the application domain.
- Object-oriented design develops an object-oriented model of a software system to implement the identified requirements.
- Object-oriented programming realizes a software design using an object-oriented programming language.

Each stage of the object-oriented development process uses the same notation, thereby eliminating transition gaps. These uniform principles apply throughout the software development process. Objects identified during the analysis phase map directly into the design and implementation phases. This similar notation dependency has both positive and negative aspects. Since objects encapsulate a portion of the problem to be solved, tracing requirements become easier since manipulation of object entities is a more natural approach to problem solving (Nerson, 1992). Conversely, if during the analysis phase the objects are incorrectly created, it negatively impacts the design and implementation phases and could result in a final architecture that reflects the poor decisions made during the analysis phase.

The object-oriented analysis problem solving method differs from the structured analysis process-oriented method in two major respects (Bailin, 2000):

- The method in which a software system is portioned into subsystems and components.
- The way in which the interactions between these subsystems or components are described.

The object-oriented (OO) paradigm takes the data and procedure components, discussed in structured analysis, but de-emphasizes the procedures, stressing instead the encapsulation of data and procedural features together. A fundamental goal in defining objects is to group data items together with methods that read and write to these data items. This kind of grouping makes each object a cohesive set of methods and data thereby helping to encapsulate problem domain concepts into a collection of self-contained units.

Encapsulation and abstraction are the principles behind object-oriented data modeling encompassing the fundamental abstraction concepts of :

- Classification: Grouping entities that share common characteristics
- Generalization: Extracting from one or more objects the description of a more general object that captures the commonalities but suppresses the differences
- Aggregation: Treating a collection of objects as a single object
- Association: Considering set of member objects as an object
- Attribution: Identification of properties or attributes of an object

Therefore, encapsulation helps to decentralize object-oriented architectures resulting in software systems to be more understandable, reliable, and easier to maintain (Anderson, 1989) (Sun, 2002) (Booch, 1994) (Rumbaugh, 1991).

Object-oriented analysis views the software system as a collection of interacting objects. These objects are part of the object-oriented model that features an approach based on abstraction, encapsulation, classification, and inheritance. This approach identifies objects and encapsulates data and operations together. In addition, similar objects are grouped together to form classes. This type of decomposition of a problem into objects and classes depends on judgment and the nature of the problem. There is no one correct representation. All variants of object-oriented analysis methods were developed to represent or view the software system in terms of object identification, communication, behavior, and operations. These views help to outline the static architectural structure of the software system as well as the system's dynamic behavior.

Booch, Rumbaugh, and Jacobson have unified their object-oriented analysis and design techniques and provide a method to describe the development of a software system's static and dynamic architecture in detail. The systems development life-cycle is titled the Rational Unified Process (RUP) and uses an iterative method development process as opposed to the traditional sequential development process offered by the Waterfall model (Boehm, 1988) (Sommerville, 2001) (Booch et al., 1998). The iterative development process treats the project as a series of small Waterfalls. Each one is designed to encompass a subset of the entire project. Each subset or project piece is large enough to mark the completion of an integral component of the project, but small enough to minimize the need for backtracking. The RUP process provides specific process steps, guidelines, and workflows that can be used during the development process. These steps have helped support the iterative approach to the development of a system using object-oriented techniques. The RUP life-cycle approach helps to break a problem into smaller more manageable pieces, which in turn makes these components more re-usable, maintainable, and extensible.

VII. APPLYING THE ANALYSIS FRAMEWORK TO OOA

The following list summarizes the object-oriented analysis technique utilizing the characteristics of the analysis framework.

- Function Description: Explicitly realized through the use of object identification.
- Function & Communication Description: Depicted through class diagrams that depict attributes and properties of objects.
- Communication Description: Depicted through sequence, collaboration and interaction diagrams.
- Communication & Behavior Description: Depicted through state-transition diagrams.
- Behavior Description: Depicted through state-transition diagrams.
- Function & Behavior Description: Depicted through state-transition diagrams.

Object-oriented analysis does address the six characteristics of the analysis framework discussed. In particular, the system objects or entities are identified through the use of object identification techniques. The concepts of classification, generalization, aggregation, association, and attribution are key principles behind object-oriented modeling. Object identification is an intuitive process in which entities and relationships are determined by examining the noun phrases contained within the problem domain narrative description. Once objects are identified, they are grouped together to form a class, which is a collection of interacting objects. The interaction among the objects represents the relationship structure. As with object identification, relationships among the objects comprising a class as well as the relationships among different classes are also determined by examining the narrative description of the problem domain. Determining the entity and relationship structure of a problem domain using object-oriented analysis is an implicit process. As with structured analysis, an implicit process can cause many relationships to be missed during the analysis phase (Catanio & Bieber, 2006) (Catanio et al., 2004). This could lead to an inadequate problem domain understanding and an incomplete analysis process.

VIII. SUMMARY

Software engineers have a wide assortment of analysis techniques that can be used to help analyze and design software systems. The underlying commonality or objective of different analysis techniques is to gain a better understanding of the problem domain by identifying entities and relationships.

The types of techniques employed are largely based on personal choice as well as corporate guidelines. If a software development organization finds their established analysis techniques insufficient, then more appropriate tools should be utilized. To that end, this paper presents a framework to analyze software development analysis techniques. By using the framework, software engineers should be able to objectively characterize other analysis techniques and select the best approach based on their need and method of analyzing and specifying software design artifacts.

ACKNOWLEDGMENTS

Special thanks to John and Shadow for their continual support and encouragement.

REFERENCES

[1] Anderson, J. "Automated Object-Oriented Requirements Analysis and Design," The Association for Computing Machinery, (1989), 265-271.

[2] Bailin, S. "Object-Oriented Requirements Analysis," Software Requirements Engineering, Second Edition, IEEE, Los Alamitos, California, (2000), 334-355.

[3] Blum, B. "A Taxonomy of Software Development Methods," Communications of the ACM, Vol. 37, No. 11, (1994), 82-94.

[4] Boehm, B. "A Spiral Model of Software Development and Enhancement," IEEE Computer, Vol. 21, No. 5, (1988), 61-72.

[5] Booch, G. Object-Oriented Analysis and Design, Second Edition, Benjamin/Cummings Publishing Company, California, (1994).

[6] Booch, G. "Object-Oriented Development," IEEE Transactions on Software Engineering, SE-12, 2, (1996), 211-221.

[7] Booch, G., Jacobson, I., & Rumbaugh, J. The Unified Modeling Language Users Guide, Addison Wesley, Massachusetts, (1998).

[8] Catanio, J., & Bieber, M., "Improving the Software Development Process by Improving the Process of Relationship Discovery," Information Resources Management Association International Journal (IRMA), May, (2006).

[9] Catanio, J., Nnadi, N., Zhang, L., Bieber, M., & Galnares, R., (2004) "Ubiquitous Metainformation and the WYWWYWI* Principle," Journal of Digital Information, Volume 5, Issue 1, April, (2004). (*What you want, when you want it)

[10] Chen, P. "The Entity-Relationship Model – Toward a Unified View of Data," ACM Transactions on Database Systems, Vol. 1, No. 1., (1976).

[11] Couger, J.D. "Evolution of Business System Analysis Techniques," Comput. Surv., Vol. 5, No. 3, (1973), 167-198.

[12] DeMarco, T. Structured Analysis and System Specification, Yourdon Press, New York, (1978).

[13] Gane, C., & Sarson, T. Structured Systems Analysis: Tools and Techniques, Prentice-Hall, Englewood Cliffs, New Jersey, (1979).

[14] IEEE. "IEEE Guide for Information Technology-System Definition-Concept of Operations (ConOps) Document," IEEE-SA Standards Board, (1998).

[15] Nerson, J. "Applying Object-Oriented Analysis and Design," Communications of the ACM, Vol. 35, No. 9, (1992) 63-74.

[16] Ross, D. "Classifying Ada Packages," Ada Letters, Vol. 6, No. 4, (1986).

[17] Rumbaugh, J. Object-Oriented Modeling and Design, Prentice-Hall, New Jersey, (1991).

[18] Sommerville, I. Software Engineering, Sixth Edition, Addison-Wesley Publishers, Massachusetts (2001).

[19] Sun, L. "An Experimental Comparison of the Maintainability of Structured Analysis and Object-Oriented Analysis," Department of Information and Software Engineering Archive, George Mason University, (2002) 1-11.

[20] Svoboda, C. "Structured Analysis," System and Software Requirements Engineering, IEEE, Los Alamitos, California, (1990), 218-237.

[21] Wieringa, R. "A Survey of Structured and Object-Oriented Software Specification Methods and Techniques," ACM Computing Surveys, Vol.30, No. 4, (1998), 459-527.

Economic Path Scheduling for Mobile Agent System on Computer Network

E. A. Olajubu
emmolajubu@oauife.edu.ng
Department of Computer Science & Engineering
Obafemi Awolowo University,
Ile-Ife, Nigeria.

Abstract

Mobile agent technology has a lot of gains to offer network-centric applications. The technology promises to be very suitable for narrow-bandwidth networks by reducing network latency and allowing transparent per-to-per computing. Multi-agent technology had been proposed for many network-centric applications with little or no path scheduling algorithms. This paper describes the need for path scheduling algorithms for agents in multi-agent systems. Traveling salesman problem (TSP) scheme is used to model ordered agents and the unordered agents schedule their path based on random distribution. The two types of agents were modeled and simulated based on bandwidth usage and response time as performance metrics. Our simulation results shows that ordered agents have superior performance against unordered agents. The ordered agents exhibit lower bandwidth usage and higher response time.

Keyword: TSP, Ordered Agent, Bandwidth Usage, Response time, Routing model

I INTRODUCTION

Mobile agent technology promises to offer omnipotent solutions to network-centric application problems. With the mobility infrastructure, mobile agents can start execution on a computer system, and autonomously transfer itself and the know-how to a remote system on computer network resumes execution from where it stopped. The mobility advantage reduces mobile agent network latency and traffic. This factor is fueling researchers attention (especially from developing nations where bandwidth is very expensive) on mobile agent technology against other communication technology such as remote procedure call. Highly scalable and flexible distributed systems are easily developed and deployed using mobile agent technology. Thus, mobile agent is an easy and comfortable paradigm for developing complex distributed system, in such a system, each task is viewed as an agent, thc whole system now form a multi-agent system.

The intelligent agent and multi-agent systems had been proposed for many network-centric applications such as electronic commerce [5], network management [6] and system administration, information retrieval from data-intensive remote application[9] etc. with little or no consideration for the route scheduling of the agent on computer network. While there are research efforts that have compared the performance of mobile agent and RPC based applications using network resources usage as performance metrics [11][12], not much work have been done on path planning models for mobile agent systems. Many other researches on mobile agent routing [13] focused on routing model that emphasis completion time, [14] is based on comparing two agents from different platforms (MAR and RIP). Though, mobile agent promised to offer low-bandwidth utilization, when there is no appropriate route scheduling for the agent(s), more bandwidth could be used than necessary due to unordered path planning. The motivating factor for this work is to have an economic path planning for platform independent agent oriented systems that optimize bandwidth usage and at the same time offers real time response. It is a known fact that as the intelligence of a mobile agent increases, the more complicated is migration pattern. Therefore, this paper presents a TSP based path planning for platform independent (Java) mobile agent system by comparing ordered agent routing against unordered agent routing on computer network using bandwidth usage and response time as performance metrics. The rest of the article are arrange in this order. Section two briefly discusses existing work on path planning models for mobile agent. Section three presents the two routing models while section four gives simulation results based on the performance metrics used for comparison. Section five concludes the discussion.

II PATH PLANNING FOR MOBILE AGENT

The deployment of multi-agent systems has encouraged a few researchers to study the path planning problem for intelligent mobile agent and agents in multi-agents systems. Research outputs are in favor of multi-agents usage for data storage and retrieval in communication network [3][4][10]. These proposed multi-agents architectures paid no attention or due considerations for the implications of many unordered agents routing the

K. Elleithy (ed.), *Advanced Techniques in Computing Sciences and Software Engineering*,
DOI 10.1007/978-90-481-3660-5_64, © Springer Science+Business Media B.V. 2010

network to accomplish computational task. Though mobile agent is a widely acclaimed technology that reduces network loads (minimizes bandwidth usage) through remote computation, yet when the agents path are not planned, the code size of these agents may constitute unnecessary bottleneck for the network during their itinerancy [7][8]. Therefore within the context of mobile agent or multi-agents system s there is the need to order the path scheduling for agents so that there will be economic bandwidth usage during their itinerancy on communication network. When the path scheduling of an agent is not ordered, its itinerancy on network will lead to bandwidth wastages i.e. the agent will tend to use more bandwidth than necessary. Mobile agent unlike RPC that holds on to network resources during its communication has been viewed as an omnipotence tool for implementing code mobility for network management systems [2]. The concept of mobile agent path planning has been investigated in [10] which finds the shortest route between adjacent nodes on the network but allows multiple visits to a node in the same itinerancy is different from TSP concept. Also, the argument of "succeed-and-stop" which was also investigated in [9] does not hold for a TSP model. This model stops an agent when it has got result and therefore does not allow the agent to complete its journey to the remaining nodes. [1] has argued that it is necessary to optimize the path planning of mobile agent in order to improve its performance on communication network. We therefore modeled two types of agents (ordered and unordered agents) under our consideration which must visit every node on the network once for computational task to comply with the classical TSP model. The ordered agent is model based on TSP (each node on the network is visited once in an itinerancy) while unordered agent's path planning is based on randomization (the path of the agent does not follow any pattern and a node may be visited than once in one itinerancy).

III ROUTING MODEL FOR THE SYSTEM

Path Planning Model for TSP Ordered agent

In this work, ordered agent is defined as an agent whose routing path on computer network strictly follow TSP scheme. The solution to ordered mobile agent routing problem consists of specifying the order in which the nodes on the network are visited, which can be referred to as permutation $<i_1, i_{2...}i_m>$ from the first node through m. The path planning of a mobile agent is very synonymous with the TSP which could be stated as follows: A mobile agent wishes to visit an m distinct node on the network and return to the home node. The latency between node k and node k+1 is given as $_{k,k+1}$ and if $_{k,k+1}$ = $_{k+1,k}$ then the problem is symmetric. The agent assignment is to find the sequence of the tour so that the overall distance traveled is minimized which is an optimization problem. To formulate the symmetric TSP for a network of m nodes, there is the need to introduce zero-one variables which describe the sub tour elimination. In this paper, this permutation is referred to as tour. There are m nodes ($i_1, i_2 \ldots i_m$) where S is the home node. Each node has a computational time required for the agent to perform the necessary task at node m_i. Latencies for the agent to move between nodes m_i and m_j are assumed to be known. The execution time from the first node is zero. The problem is to minimize the length off distance covered by the agent. The distance travel to complete the task or inspect all the nodes for the tour T= $<i_1, i_2, i_m>$ should not be less than the longest tour in the network. As the agent moves, it appends a digital

```
Initialize Agent
∀ x=1 to N do
Agent move from S (home node) to the first node
while !eof() do
AgentMinPath <- active.MinPath()
AgentPath<- Paths(agentMinPath)
If (agent MinPath) {
        return AgentPath
}
For each node n on AgentMinPath{
    Agentpath<- Agentpath[N]
    if !eof(N)
    n    N∀x N
            currentpath <-Npath
        append digital sigure on the node
        add node to agent list
select n from N with priority equals Npath's cost
        }
    }
agentMinPath end?
}
Agent return to S the home node
Agent deactivated
```

signature on the node visited and the node is added to the agent's list. The list is updated on each node during agent's itinerancy.

The sequential movement of the ordered mobile agent is defined by

$$Mini: \int_1^m TX_{k,k+1} \quad (1)$$

where T = *f*(,) is the latency between adjacent nodes on the network while is the required computational time on each node on the network.

Subject to:

$$X_{k,k+1} \in \{1,0\} \text{ for all K, K+1= 1...m} \quad (2)$$

eliminate all sub tours that exist in the agent path.

Path planning pseudocode for Unordered agent

```
Initialize agent
∀x=1 to N do
Agent move from S(home node) to the first node
while !eof() do
From current node repeat {
    If node is not yet visited then

        Append digital signature
    If (ni is in N), Then
        Xk+1 = ((α *mod)/ N ) /* Select 'Xk+1'
        node randomly from the N */
        Else, /* Xk+1is not in N */
                if Xk+1 ∉ N /* Xk+1 is not in N */
        Repeat random selection of Xk+1
        End
    End
        if Xk+1 ∈ N /* Xk+1 is a member of N */
        append digital signature

        Advance to the next node
}
Agent return to S the home node
Agent deactivated
```

$_{k,k+1}$= Latency between node k and node k+1
$_k$= Computational time on the node k
m_i= Node identifier
N= Total number of nodes on the network

X = Binary variable that is either 1or 0

Path Planning Model for Unordered agent

The path of unordered agent is determined by pseudo random number generator. At the home node, the pseudo random generates an integer number which correspond to the Internet protocol (ip) address of a node in the hash table. The agent finds its way to the node. The hash table transform the key (pseudo number generated) using a hash function into the *ip* address of the systems on the network indexed in an array. The table is constantly been refresh in the memory of the agent. In its itinerancy, a node can only be visited once. It is assumed, that the latency between two adjacent nodes on the network are equal such that $_{k,\ k+1}$ = $_{k+1,\ k}$ irrespective of the direction of agent movement. When the agent gets to a node, the next node to visit is determined by the number generator. Lehmer multiplicative congruential algorithm [15] is used to model unordered agent itinerant on the network. The agent sequentially follow the random number as they are generated. Lehmer random number generator which is presented as follows:

$$x_{k+1} = \alpha x_k + \beta \bmod N \quad (3)$$

The generator involves three integers α β and N,
Where
α =13
β =0 is the x_0 is the seed.
N= n+1; n is the last node on the network
This generates integer value between node 1 (the first node) and the last node N. Where repetition of integer value occurs they are removed. This implies a node will be visited once in an agent itinerant. When duplicate of any integer value occurs, the agent knows that all nodes on the network have been visited. Now, a simple case scenario for the random number generator is presented for a computer network of 30 nodes. Using the Lehmer multiplicative congruential algorithm, with α = 13, β = 0, N = 31 and x_0 = 1, the sequence of random number generated is presented as:[1; 13; 14; 27; 10; 6; 16; 22; 7;29;5;3;8;11;19;30;18;17;4;21;25;15;9;24;2;26;28;23; 20;12] this implies every node on the network will be visited sequentially in this order in an itinerancy.

VI ANALYSIS OF SIMULATION RESULT

Bandwidth Usage

The simulation result shows the difference between ordered mobile agent and an agent that determine its path by randomly selecting the nodes to visit on the network. The bandwidth usage for the two types of agent is quiet striking. The ordered agent conducts its itinerancy on the network using TSP algorithms while unordered agent randomly selects the nodes to visit until all nodes are visited. The bandwidth usage is optimized while the unordered agent's itinerancy which is based randomly selecting nodes to visit on the network. The bandwidth used the two types of agent increases with increase on the nodes on the network, but the bandwidth used by unordered agent is significantly higher than the bandwidth used by the ordered agent. It shows that the itinerant of unordered agent results in bandwidth wastage. In a large network, the difference is going to be astronomical. Developing Internet or network-centric applications using unordered agent will increase the overhead of such application while ordered agent minimizes bandwidth usage.

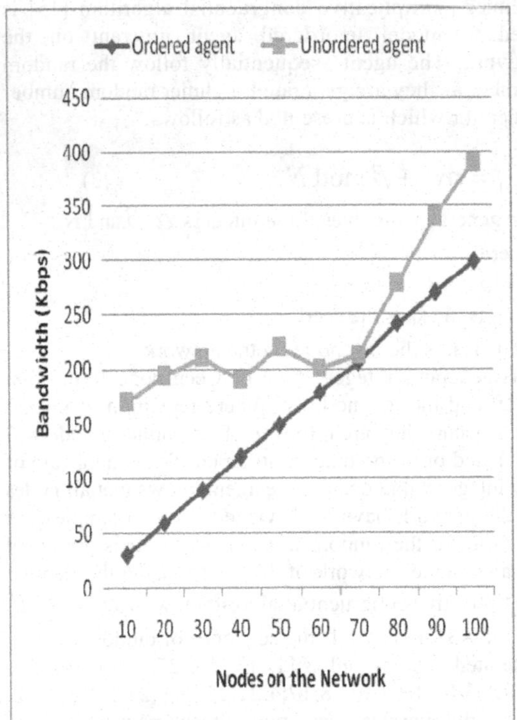

One of the advantages promised by mobile agent based application is reduced traffic and bandwidth usage, but this advantage may not be truly realized especially in a multi agents system where there are many agents routing the network if the agent(s) are not ordered.

Response Time

Response time is defined as the time between the beginning of a computational task and the time the user get the result of the request. In this simulation, the response time includes the length of time for agent migration and the time required to perform computational task (information retrieval or any other computational task). The time required for unordered agent to send the result of its computation back to its user is significantly higher than the ordered agent this is due to the random movement of the agent. This gives an idea that, there more delay will be experienced in any application that is based on unordered agent; because more time will be required before the end user can receive the result of processed information, therefore real time application will suffer some time

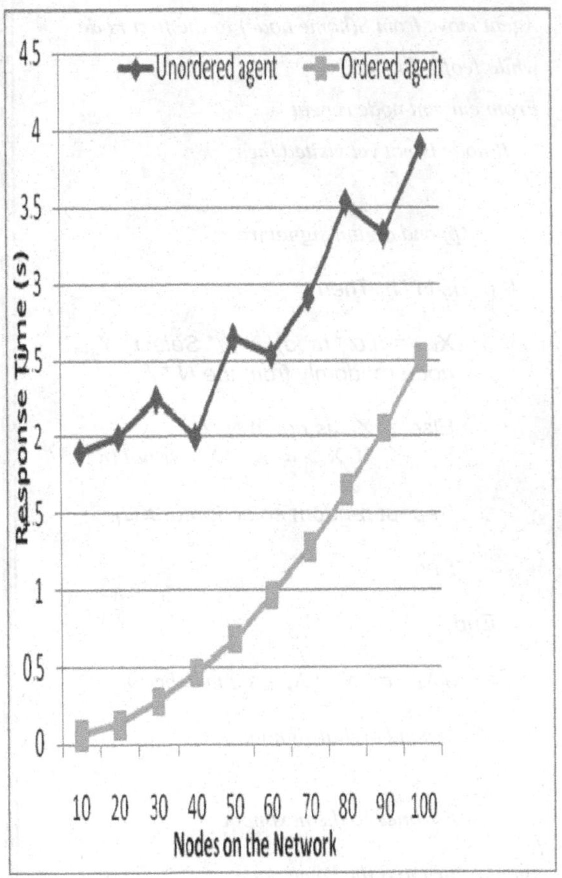

lag. The normal network delay is ignored in this simulation since it affected the two types of agent equally. The simulation result shows what happens when mobile agent are left on their own to seek their itinerant path without any intelligence that dictate how to find the shortest route on the network.

V Conclusion

In this paper, the consequent of unordered agent on computer network and the gains of ordered agent is presented. To take adequate advantage(s) which mobile agent technology offers over RPC based application, it is necessary that agent's intelligence includes ability to find the shortest route on the network. It also shows that mobile agent applications use for real time applications should contain intelligence that will specify their path during their itinerancy on communication network so that bandwidth usage can be minimized and the response time can be minimal as much as possible. It should be noted, the observation in this paper is based on simulation and not real implementation of the software application systems. Based on this simulation result, it is important that agent's path scheduling are properly ordered so as to enjoy the maximum advantages promised by mobile agent based systems.

Reference

[1] Nehra N, Patel R. B. and Bhat V. K (2007) Distributed Parallel Resource Co-Allocation with Load Balancing in Grid Computing International Journal of Computer Science and Network Security 7(1): 282-291

[2]A. Bieszczad, B. Pagurek, and T. White Mobile Agents for Network Management IEEE Communication Surveys. Available at www.comsoc.org/pubs/surveys [Aug. 10, 2006]

[3]A.S. Torrellas Gustavo and A.V. Vargas Luis Modeling a flexible Network Security Systems Using Multi-Agent Systems: Security Assessment Considerations. Proceedings of the 1st International Symposium on Information and communication technologies pp. 365-371, 2003.

[4]M. F. De Castro, H. Lecarpenttie, L. Merghem and D. Gaiti An Intelligent Network Simulation PlatformEmbedded with Multi-Agents Systems for Next Generation *Internet. Telecommunications and Networking-ICT* pp. 1317-1326, 2004.

[5]Vogler H., Moschgatt ML, and Kunkelmann T. Enhancing Mobile Agents with Electronic Commerce Capabilities. Proc. Of 2nd International workshop on cooperative Information Agents (CIA-98) pp. 148-159, 1998.

[6]Gabri G. Leonardi L and Zambonelli F. Mobile agent Coordination for Distributed Network Management. Journal of Network Management 9(4): 435-456, 2001.

[7]Boutaba R., Iraqi Y., and Mehaoua A. A Multi-Agent Architecture for QoS Management in Multimedia Networks. *Journal of Network and System Management Vol.11* (1): 83-107, 2003.

[8]Baek, J., Kim, J., and Yeom, H.. Timed Mobile Agent Planning for Distributed Information Retrieval Proceedings of the fifth international conference on Autonomous agents pp 120 - 121 , 2001.

[9]K. Moizumi and G. Cybenko The Traveling Agent Problem *Mathematics of Control, Signals and System*, 1998.

[10]Das S., Shuster K., Wu C. and Levit I. Mobile agents for Distributed and Heterogeneous Information Retrieval *Information Retrieval* 8(3): 383-416, 2005.

[11]Aderounmu G.A. Performance Comparison of remote procedure call and mobile agent approach to control and data transfer in distributed computing environment. Journal of Network and Computer Applications 27, 2004,:pp113-129.

[12]Olajubu E. A, G. A. Aderounmu and E.R. Adagunodo (2008): Optimizing Bandwidth usage and Response Time using Lightweight Agents on Data Communication Network. T. Sobh et al. (eds), *Novel Algorithms and Techniques in Telecommunications, Automation and Industrial Electronics*, pp. 335-340.

[13]Manvi S. S. and Venkataram P. An Agent-Based Best Effort Routing Technique for Load BalancingInformatica, 2006, Vol. 17, No. 3, 407–426

[14]Szczypiorski K, Margasinski I and Mazurczyk W (2007) Steganographic Routing in Multi Agent System Environment . Journal of Information Assurance and Security 2 pp. 235-243.

[15]http://www.csse.monash.edu.au/courseware/cse3142/2006/Lnts/Crng.pdf

A Database-Based and Web-Based Meta-CASE System

Erki Eessaar, Rünno Sgirka
Department of Informatics, Tallinn University of Technology,
Raja 15, 12618 Tallinn, Estonia
eessaar@staff.ttu.ee, runno.sgirka@gmail.com

***Abstract*-Each Computer Aided Software Engineering (CASE) system provides support to a software process or specific tasks or activities that are part of a software process. Each meta-CASE system allows us to create new CASE systems. The creators of a new CASE system have to specify abstract syntax of the language that is used in the system and functionality as well as non-functional properties of the new system. Many meta-CASE systems record their data directly in files. In this paper, we introduce a meta-CASE system, the enabling technology of which is an object-relational database system (ORDBMS). The system allows users to manage specifications of languages and create models by using these languages. The system has web-based and form-based user interface. We have created a proof-of-concept prototype of the system by using PostgreSQL ORDBMS and PHP scripting language.**

I. INTRODUCTION

Nowadays researchers and developers pay a lot of attention to *model-driven development,* according to which models can be used as the basis in order to fully or partially generate program code. There exists *general purpose* modeling languages like Unified Modeling Language (UML). However, the use of domain-specific modeling languages that have higher expressive power allows developers to create more precise specifications. The use of this kind of languages improves the understandability of specifications because the languages allow developers to use concepts from the problem domain [1]. We need Computer Aided Software Engineering (CASE) systems in order to create models in these new languages. In addition, it should be as easy as possible to adapt a CASE system if it underlying language changes. Isazadeh and Lamb [2] write: "CASE tools are large, complex, and very labour-intensive to produce". How could we simplify the creation of CASE tools?

In a CASE system the specification of a language is hard-coded to the system and it is not possible to change it. It means that in a CASE system we have to use one or more fixed languages in order to create models. It is only possible to *extend* the languages if these languages contain built-in extension mechanisms. For instance, it is possible to extend UML in a limited manner by using profiles.

A meta-CASE system is a system that allows us to develop new CASE systems. It is possible to specify the abstract syntax of a language by using a metamodel [3]. An important part of a meta-CASE system is *persistence layer* where data about models and metamodels must be stored. Karagiannis and Kühn [4] write that a meta-CASE system must contain the metamodel base, the model base, and the mechanism base. These bases contain information about metamodels, models, and functionalities that could be applied to models and metamodels, respectively. Many meta-CASE systems are file-based systems that do not use the services of a database system (DBMS). The direct use of files means that developers of meta-CASE systems have to work out solutions to the problems that have already been solved in DBMSs.

In this paper, we are interested in meta-CASE systems that use object-relational DBMSs as their enabling technology. In these DBMSs it is possible to use SQL database language that conforms more or less to SQL:2003 standard [5]. We denote this kind of DBMS as $ORDBMS_{SQL}$ and a database that is created by using an $ORDBMS_{SQL}$ as an OR_{SQL} database. Some reports of existing CASE systems refer to DBMSs, which use a database language that conforms to SQL:1992 or earlier version of the SQL standard. We denote this kind of DBMS as $RDBMS_{SQL}$ and a database that is created by using a $RDBMS_{SQL}$ as a R_{SQL} database.

The *goal* of the paper is to introduce a meta-CASE system, the data of which is stored in an OR_{SQL} database. The system allows users to manage metamodels and models by using web-based and form-based user interface. According to the classification of CASE products [6], the CASE systems that are created by using the proposed system are analysis or design workbenches that support analysis or design activities.

The rest of the paper is organized as follows. *Firstly*, we introduce some of the existing CASE and meta-CASE systems. It allows us to illustrate the need of the current research. *Secondly*, we propose a new system and demonstrate its use with a small case study. *Finally*, we draw conclusions and point to the future work with the current topic.

II. RELATED WORK

Some CASE systems use a DBMS as their *enabling technology* that "provides services and functionalities for their operation in an integrated and homogeneous environment"[6].

Miguel et al. [7] see many advantages of this kind of data/knowledge centric architecture.

Each tool that belongs to the system uses the set of views that present the required data in the required format.

"The data base becomes the medium of communication and coordination between tools." [7]

K. Elleithy (ed.), *Advanced Techniques in Computing Sciences and Software Engineering*,
DOI 10.1007/978-90-481-3660-5_65,

System developers don't have to implement the features that are available in the DBMS. It increases their productivity.

Gray and Ryan [8] present a set of CASE tools, the database of which is a R_{SQL} database. They also introduce a meta-CASE environment ToolBuilder, which is not based on a R_{SQL} database. They note that the meta-CASE environment is complex and lacks many of the features of a $RDBMS_{SQL}$. For instance, its query mechanism is "crude compared to that of SQL". It means that access to the ToolBuilder repository by using non-ToolBuilder tools is possible but more complex than just sending SQL statements to a database server. NutCASE is [9] another example of a CASE system, which stores models in a R_{SQL} database. In addition, it provides lightweight web-based user interface for creating and managing UML class diagrams. "It can be controlled by the user without using plug-ins or applets."[9] Mackay et al. [9] note that the advantages of a web-based system are a possibility to distribute it widely, its support to group collaboration, and portability. For instance, the use of a system does not depend on the operating system that is in the computer of a user. Eessaar [10] proposes a web-based and form-based CASE system, the database of which must be implemented by using a $RDBMS_{SQL}$ or an $ORDBMS_{SQL}$.

Our goal is to develop a meta-CASE system. It should take advantage of an $ORDBMS_{SQL}$ on the server and should provide a web-based user interface that does not require installation of any plug-in in a computer of a user.

Are there any meta-CASE systems that record their data in a R_{SQL} or an OR_{SQL} database and provide web-based user interface? Lutteroth [11] proposes AP1 platform for the development of model-based CASE tools. These systems store models in a R_{SQL} database. The CASE tools will interact with a $RDBMS_{SQL}$ either directly or through an object-oriented interface. However, the proposed version of the system doesn't allow us to manage metamodels and models by using a web-based user interface. General Modeling Environment (GME) [12] stores data in the Microsoft (MS) repository or in the files, which are created based on a proprietary binary file format. MS repository is an object-oriented layer that encapsulates MS SQL Server or MS Access $RDBMS_{SQL}$s. However, GME doesn't provide web-based user interface. Isazadeh and Lamb [2] evaluate meta-CASE tools. They note that systems Metaview and 4thought use a Prolog-based deductive DBMS and system Toolbuilder uses an object-oriented DBMS. MetaBuilder[13] is another example of a meta-CASE system, which use the help of an object-oriented DBMS. Some meta-CASE systems like development environments of visual modeling environments GenGED[14] and Pounamu [15] store their data in XML files. A system based on MetaL metamodeling language [16] supports the use of XML, RDF, or JDO files. openArchitectureWare is a system that can be used in order to create model driven development tools [17]. The system uses XML files in order to store metamodels. Systems that are created by using [17] store models in XML files. ConceptBase [18] is a deductive database system that can be used as a repository system in meta-CASE systems. $ATomM^3$ is a meta-CASE tool that stores models "as Python functions that contain the executable statements" [19]. The system Pounamu/Thin provides web-based user interfaces.

Developers of meta-CASE systems do not prefer to use $ORDBMS_{SQL}$s as an enabling technology. What could be the reason of such state of affairs? Eessaar [20] presents an analysis of SQL standard and different $RDBMS_{SQL}$s and $ORDBMS_{SQL}$s. He concludes that these systems have limitations that restrict their use in software engineering systems. For instance, these database systems impose restrictions to declarative integrity constraints and updatable views. Therefore, one of our motivations is to use the creation of a prototype as an experiment that helps us to evaluate the suitability of $ORDBMS_{SQL}$s as an enabling technology of meta-CASE systems.

III. A Web-Based and Database-Based Meta-CASE System

How should we design the schema of the database of a meta-CASE system? A possibility is to use the *universal database design,* based on which models and metamodels must be recorded in a small set of base tables (tables in short) that represent generic concepts – *entity type*, *entity*, *attribute*, *attribute value*, and *relationship*. If an administrator wants to create a new CASE system, then he/she should record the corresponding metamodel in the tables at the knowledge level – *entity type* and *attribute*. This data determines the structure of data about models that we can record in the tables at the operational level – *entity*, *attribute value*, and *relationship*. However, the design has at least twelve drawbacks like difficulties in enforcing constraints, complexity of queries, increasing size of data, and reduced query speed [21].

Therefore, we decided not to use the universal design in order to develop the entire database. Instead, the proposed meta-CASE system will create tables based on specifications of metamodels. Metamodels are recorded in the tables that resemble the universal database design. The user interface of each CASE system is generated dynamically based on the specification of its underlying metamodel.

Fig. 1 presents the general architecture of the meta-CASE system. We have implemented a proof-of-concept prototype of the system by using PHP 5.0.5 scripting language. Its database is implemented by using $ORDBMS_{SQL}$ PostgreSQL 8.0.4. The system is intended to be a research vehicle and not an industrial strength system.

A. The Structure of the System

The database of the proposed system consists of exactly one metamodel base and zero or more model bases (see Fig. 1). Each metamodel that is specified in the system has exactly one corresponding model base. It is possible to implement subsets of a SQL-database as *SQL-schemas*. SQL-schema is a persistent, named collection of descriptors of SQL schema objects [5]. Schemas and schema objects are database objects.

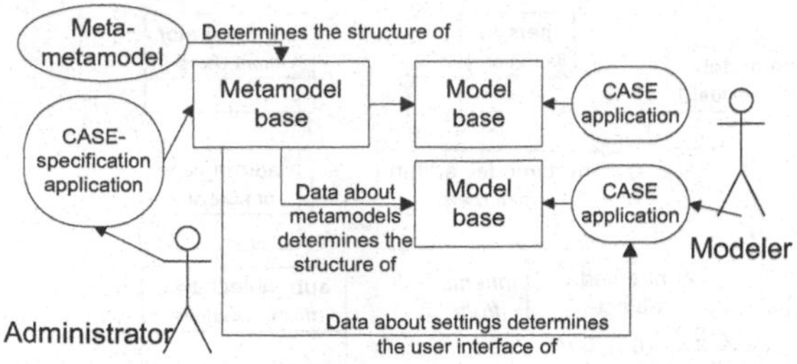

Fig. 1. An architectural view of the proposed system.

Tables that belong to the metamodel base are in exactly one schema. In addition, each model base has exactly one corresponding schema in the database. Each schema that corresponds to a model base contains tables that are necessary for the recording of models. All the models that are recorded in a model base *b* have the common metamodel. For instance, if an administrator defines the metamodel *Use_cases*, then the system will create schema *Use_cases*.

Each DBMS has its own rules for identifiers of database objects. For example, SQL regular identifiers cannot contain spaces. For each metamodel or its element, the meta-CASE system has to create the correct identifier of the corresponding database object. The system creates the identifiers based on the names (identifiers) of the metamodels and their elements. In case of our simplified prototype system, the names of metamodels and metamodel elements must also be correct identifiers in a PostgreSQL database. The system doesn't allow us to use names of metamodels and their elements that don't correspond to these rules.

Schema is a namespace. Different metamodels could contain elements that have the same name. On the other hand, their corresponding tables are in different schemas and therefore there will be no name conflicts.

The metamodel base can be divided into the following parts: *metamodels*, *settings of metamodels*, *users*, and *classifiers*. Fig. 2 presents a conceptual data model of the metamodel base. It illustrates the meta-meta modeling language, based on which it is possible to create new metamodels. Next, we will discuss each of these parts. We also illustrate the use of the system with the help of a case study about creating a CASE system for managing use cases.

B. Metamodels

Each metamodel contains zero or more objects. Each object has zero or more associated sub-objects. Each object has a type. Possible types of objects are *main*, *inherited main*, *relationship*, and *classifier*. The classification scheme is similar to the scheme that Codd [22] has proposed in order to classify entity types. The objects with the type *main* or *inherited main* are kernels in the sense of Codd classification. Classifiers are used in order to characterize other data in a database. Management of classifiers is a task of administrators in our proposed system. Modelers can use the values of classifiers that administrators have specified. However, classifiers are special case of kernels. Relationship objects correspond to associatives in the classification of Codd. The entities, which are associatives, interrelate entities of other types.

We propose not to store models as values with type CLOB or XML. Instead, different model elements will be recorded in different tables in order to simplify queries based on a model. Therefore, each object *o* has exactly one corresponding base table *t* in exactly one model base *b*. Each sub-object of *o* has exactly one corresponding column of *t*. A sub-object of *o* can be used in order to represent a relationship between *o* and some other object *o'* in the same metamodel. In this context *o'* is called foreign object.

Why we need information about types of objects and sub-objects? The type of an object determines how the primary key of its corresponding base table will be found. Each table that is created based on a main or a classifier object has the surrogate key. It means that each such table has a column where the values are system-generated unique identifiers. Each table that is created based on a relationship object must contain two or more foreign keys. The primary key of this table involves the foregoing foreign key columns. Therefore, the proposed system hides the complexity of specifying constraints from an administrator who defines metamodels. The type of a sub-object determines the type of the corresponding column that is created in the database.

Let us assume that an administrator defines that metamodel *Use_cases* contains object *use_case*, a sub-object of which is *name*. Object *use case* has type *main*. Sub-object *name* has type *text*. Objects and sub-objects at the metamodel level have corresponding instances (object values, model elements) at the model level. We can specify that if a user of a CASE system has to select an instance of an object *o* (for example *use_case*) at the model level, then the system must provide for this purpose a combo box that presents instances of a particular sub-object of *o* (for example *name*).

In this case the system creates table *use_case* in schema *Use_cases* and determines that the table has column *name*, the type of which is *text*. In addition, the table has column *use_case_id*, the type of which is *integer*. The latter column has associated sequence generator that generates unique identifiers. The primary key of the table is (use_case_id).

Let us assume that an administrator defines that metamodel *Use_cases* contains object *importance*, a sub object of which is *name*. Object *importance* has type *classifier*. Sub-object *name* has type *text*. In addition, object *use_case* has sub-object *importance* that refers to object *importance*. In this context object *importance* is a foreign object.

In this case the system creates table *importance* in schema *Use_cases*. This table has columns *name* (with type *text*) and *importance_id* (primary key column with type *integer* and associated sequence generator). In addition, the system adds column *importance* to table *use_case*. This column is a foreign key column. Possible values in this column are identifiers of different levels of importance.

Let us assume that an administrator defines that metamodel *Use_cases* contains object *actor*, a sub-object of which is *name* (with type *text*). Object *actor* has type *main*.

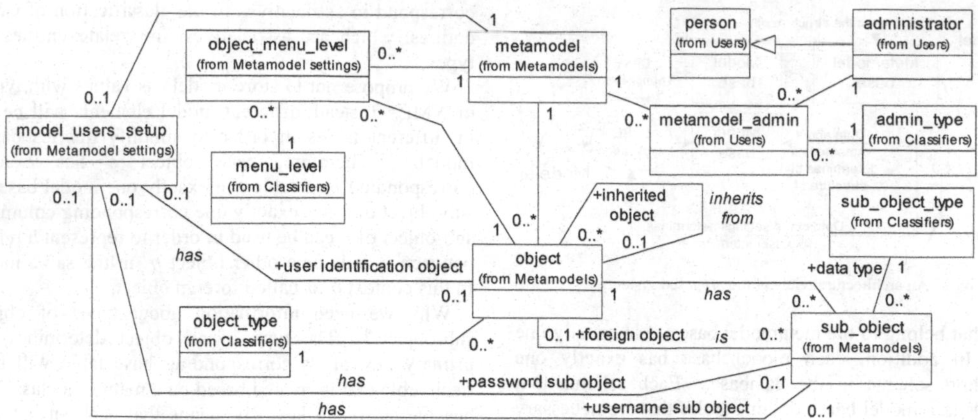

Fig. 2. A conceptual data model of the metamodel base.

In addition, the metamodel contains object *interest* (with type *relationship*), the sub-objects of which are *description* (with type *text*), *actor*, and *use_case*. *actor* and *use_case* have associated foreign objects *actor* and *use_case*, respectively.

In this case the system creates table *actor* in schema *Use_cases*. This table has columns *name* (with type *text*) and *actor_id* (primary key column with type *integer* and associated sequence generator). In addition, the system creates table *interest* that has columns *description*, *actor*, and *use_case*. Columns *actor* and *use_case* are foreign key columns that refer to tables *actor* and *use_case*, respectively. The primary key of table *interest* must involve columns *actor* and *use_case*.

Modelers have to distinguish different use case models. Therefore, an administrator should define main object *model*, a sub-object of which is *name*. In addition, he/she has to determine that main objects *use_case* and *actor* have both a sub-object that refers to object *model*. This kind of sub-object is not needed in case of object *importance* because classifier values that represent different levels of importance can be used in different use case models.

In this case the system creates table *model* in schema *Use_cases*. It adds foreign key columns and constraints to tables *use_case* and *actor*.

Each main object can inherit its sub-objects from zero or one main object. The type of this kind of object is *inherited main*. Each inherited main object can have additional sub-objects compared to the object from which it inherits.

Let us assume that an administrator defines that metamodel *Use_cases* contains object *fully_dressed_use_case,* which inherits from object *use_case*. Object *fully_dressed_ use_case* has sub-object *extensions* in addition to inherited sub-objects (*name* and *importance*).

In this case the system creates in schema *Use_cases* table *fully_dressed_use_case* by using INHERITS clause in the table definition statement. The unstandardized INHERITS clause specifies "a list of tables from which the new table automatically inherits all columns" [23]. The table has column *extensions* that is not inherited.

In conclusion – modification of a specification of a metamodel in the metamodel base causes changes in the structure and constraints of the schema of one model base.

C. Settings of Metamodels

The system is able to construct the user interface of a CASE system based on the information about settings of a metamodel. Firstly, for each main and inherited main object an administrator has to determine the menu level (first, second, or third). This level determines the navigation path that a modeler has to take in order to reach to the form for managing instances of the object (model elements). Corresponding class in Fig. 2 is *object_menu_level*.

For example, an administrator can determine that object *model* will be presented at the first level. After a user of the CASE system selects a model he/she can manage data about use cases, actors, and interests. We claim that specification of the user interface of a CASE system is declarative in nature because an administrator doesn't have to write any procedural code.

Secondly, an administrator has to register the values of classifiers. For example, in case of our use case metamodel an administrator has to register different levels of importance of use cases. Remember, object *importance* has the type *classifier* in our example.

By default, access to a CASE system is not restricted and anyone can manage models by using this system. An administrator can specify that an object in a metamodel *m* is so-called user-object (see class *model_users_setup* in Fig. 2). The corresponding table of this object must contain usernames and passwords. If a user wants to manage models that conform to *m*, then he/she must firstly register in the system and identify himself/herself.

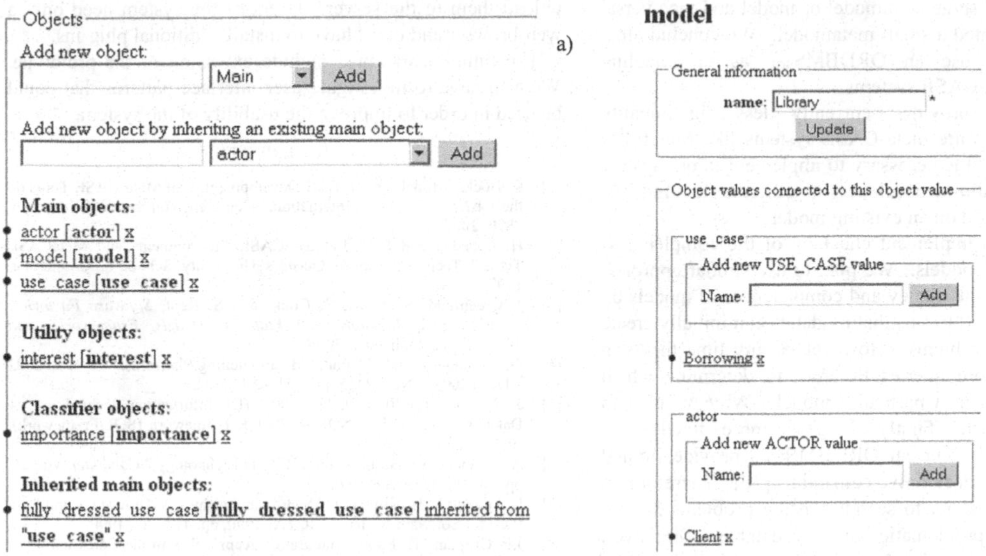

Fig. 3. Screenshots of the user interface of the prototype.

For instance, an administrator can determine that metamodel *Use_cases* contains object *system_user*, the sub-objects of which are *username* and *password*. The system creates table *system_user* in schema *Use_cases*.

D. Users

Administrators manage metamodels. If an administrator creates a new metamodel *m*, then he/she will become the owner of this metamodel. The owner can determine that zero or more other administrators are moderators of *m*. Corresponding class in Fig. 2 is *metamodel_admin*. The moderators have the same rights in *m* as the owner except that they cannot delete *m*. Deletion of *m* is the exclusive right of the owner of *m*.

E. Classifiers

menu_level, *admin_type*, *object_type*, and *sub_object_type* (see Fig. 2) are classifiers. They are used in order to characterize other data about metamodels. These classifiers are systemic – it means that the proper work of the system depends on the values of these classifiers. Therefore, administrators can only view the values of these classifiers and change their textual explanations. Administrators cannot add or remove the values of these classifiers. The values of these classifiers must be recorded during the setup of the meta-CASE system.

F. Discussion

Why is our proposed meta-CASE system important? Gray et al. [24] write that tool architecture is an issue that still requires investigation in case of meta-CASE systems. Current paper proposes a possible architecture of meta-CASE systems.

Gray et al. [24] note that there are two different approaches of creating new CASE systems by using a meta-CASE – *assemble from components* and *specify and generate*. The proposed system uses a variant of *specify and generate*. An administrator must specify a metamodel and determine settings of the CASE system. Based on this information the system is able to create data structures and dynamically generate web pages, through which it is possible to manage models. Dynamic generation means that if a modeler requests a web page, then the system determines the structure of the page based on data that is stored in the metamodel base. The system is in this sense similar to Oracle Application Express development environment [25], which stores specifications of entire applications in an Oracle database.

Users of the system need only a web-browser and don't have to install additional plug-ins. Fig. 3 presents two screenshots of the user interface of the prototype. Part a) is a fragment of the user interface through which administrators can specify metamodels. We see the objects that belong to metamodel *Use_cases*. Part b) is from the user interface of a CASE system through which modelers can manage use case models. We can see that use case model named *Library* has associated use case *Borrowing* and actor *Client*.

An administrator doesn't have to recompile files or upload them to server in order to modify a CASE system. He/she only has to change specification of the structure of a metamodel or settings of a metamodel in the metamodel base.

We have created a prototype of the system. An advantage of using PostgreSQL ORDBMS$_{SQL}$ is that the system uses *multiversion concurrency control* and therefore allows users to "read consistent data during writer activity" [23]. Hence, viewing of a metamodel or a model doesn't block the modification of the same metamodel or model and vice versa. We have implemented a small metamodel. We conclude that it is possible to use an $ORDBMS_{SQL}$ as an enabling technology of meta-CASE systems.

The prototype provides currently less functionality compared to the mature meta-CASE systems like MetaEdit+ [26]. For instance, it is necessary to implement in our system functionality that allows developers to generate new models, code, or reports based on an existing model.

We also have to implement checking of the completeness and consistency of models. We plan to use similar approach as [10] and check consistency and completeness of models by using SQL queries. It means that modelers can initially create models that have problems. However, at any time they can execute one or more queries in order to determine which problems still exist in a particular model. Why we plan to select this approach? Firstly, it gives more freedom to modelers. Secondly, current $ORDBMS_{SQL}$s provide limited support for creating declarative constraints [20]. Therefore, in these systems it is easier to search existing problems than to prevent storing of problematic data in a database. Moreover, queries can be used in order to perform data analysis based on models and metamodels.

The proposed system doesn't currently allow us to specify models and metamodels by using a visual language. Instead, we have to use form-based user interface, the usability of which needs improvement. The user interface is more similar to the generic editor of AP1 platform [11] than to the visual modeling tools.

A problem of the current system is its dependence on DBMS-specific features of SQL (like INHERITS clause). It is possible to overcome this problem by instead using updatable views in order to implement generalization relationships. The problem is that SQL-standard and DBMSs impose restrictions to updatable views [20]. As you can expect, the problems of an $ORDBMS_{SQL}$ influence the design decisions of the meta-CASE system and make it more difficult.

IV. CONCLUSIONS

We have presented a meta-CASE system, the database of which is implemented with the help of an object-relational database system. We have created a proof-of-concept prototype of the system. The system allows users to manage metamodels and models through a web-based and form-based user interface. Each metamodel has a corresponding database schema. The system creates schema objects (like base tables) based on the specifications of elements of a metamodel. Models, which are created based on this metamodel, are recorded in these tables. The web pages, which are used for the management of models, are generated dynamically based on the specification of a metamodel. However, before it is possible to create models it is also necessary to determine settings of their metamodel. All the changes that are made in the specification of a metamodel are after completion instantly visible to the modelers. It is not necessary to compile files or upload them to the server. Users of the system need only a web-browser and don't have to install additional plug-ins.

The future work must include extension of the prototype. We also have to investigate user interface patterns that could be used in order to improve the usability of the system.

REFERENCES

[1] C. Bock, "Model-Driven HMI Development: Can Meta-CASE Tools do the Job?," 40th Hawaii International Conference on System Sciences, p. 287b, 2007.

[2] H. Isazadeh and D.A. Lamb, "CASE Environments and MetaCASE Tools," Technical report, Queen's University School of Computing, 1997.

[3] J. Greenfield, K. Short, S. Cook, and S. Kent, *Software Factories: Assembling Applications with Patterns, Models, Frameworks, and Tools*. Wiley Publishing, 2004.

[4] D. Karagiannis and H Kühn, "Metamodelling Platforms," EC-Web 2002 – Dexa 2002, LNCS 2455, pp. 451–464, 2002.

[5] J. Melton, ISO/IEC 9075-1:2003 (E) Information technology — Database languages — SQL — Part 1: Framework (SQL/Framework). August, 2003.

[6] A. Fugetta, "A Classification of CASE Technology," *Computer*, vol. 26, pp. 25–38, December 1993.

[7] L. Miguel, M.H. Kim, and C.V. Ramamoorthy, "A Knowledge and Data Base for Software Systems," ICTAI 1990, pp. 417-423, 1990.

[8] J.P. Gray and B. Ryan, "Integrating Approaches to the Construction of Software Engineering Environments," SEE 1997, pp. 53–65, 1997.

[9] D. Mackay, J. Noble, and R. Biddle, "A Lightweight Web-Based Case Tool for UML Class Diagrams," Fourth Australasian User interface Conference on User Interfaces, ACM International Conference Proceeding Series, vol. 36, pp. 95–98, 2003.

[10] E. Eessaar, "Integrated System Analysis Environment for the Continuous Consistency and Completeness Checking," JCKBSE 2006, pp. 96–105, 2006.

[11] C. Lutteroth, "AP1: A Platform for Model-Based Software Engineering," TEAA 2006, LNCS 4473, pp. 270–284, 2007.

[12] A. Lédeczi, M. Maroti, A. Bakay, and G. Karsai, "The Generic Modeling Environment," WISP'2001, 2001.

[13] M. Gong, L. Scott, Y. Xiao, and R. Offen, "A Rapid Development Model for Meta-CASE Tool Design," ER'97, LNCS 1331, pp. 464–477, 1997.

[14] R. Bardohl, C. Ermel, and I. Weinhold, "GenGED – A Visual Definition Tool for Visual Modeling Environments," AGTIVE 2003, LNCS 3062, pp. 413–419, 2004.

[15] N. Zhu, J. Grundy, J. Hosking, N. Liu, S. Cao, and A. Mehra, "Pounamou: A meta-tool for exploratory domain-specific visual language tool development," *The Journal of Systems and Software*, vol. 80, pp. 1390–1407, August 2007.

[16] V. Englebert and P. Heymans, "Towards More Extensible MetaCASE Tools," CAISE 2007, LNCS 4495, pp. 454–468, 2007.

[17] S. Effinge et al., openArchitectureWare User Guide, Version 4.3, 2008.

[18] M.A. Jeusfeld and C. Quix, "Meta Modeling with ConceptBase," Workshop on Meta-Modelling and Corresponding Tools, 2005.

[19] J. d. Lara and H. Vangheluwe, "AToM3: A Tool for Multi-Formalism and Meta-Modelling," FASE 2002, LNCS 2306, pp.174–188, 2002.

[20] E. Eessaar, "Using Relational Databases in the Engineering Repository Systems," ICEIS 2006, vol. Databases and Information Systems Integration, pp. 30–37, 2006.

[21] E. Eessaar and M. Soobik, "On Universal Database Design," Baltic DB & IS 2008, pp. 349–360, 2008.

[22] E. F. Codd, "Extending the Database Relational Model to Capture More Meaning," *ACM Transactions on Database Systems*, vol. 4, pp. 397–434, 1979.

[23] PostgreSQL 8.3.4 Documentation.

[24] J.P. Gray, A. Liu, and L. Scott, "Issues in software engineering tool construction," *Information and Software Technology*, vol. 42, pp. 73–77, 2007.

[25] Oracle Application Express 3.1 Documentation.

[26] J. Tolvanen and M. Rossi, "MetaEdit+: defining and using domain-specific modeling languages and code generators," OOPSLA 2003, pp. 92–93, 2003.

On Dijkstra's Algorithm for Deadlock Detection

Youming Li, Ardian Greca, and James Harris
Department of Computer Sciences
Georgia Southern University
Statesboro, GA 30460
Email: yming@georgiasouthern.edu, agreca@georgiasouthern.edu, jkharris@georgiasouthern.edu

***Abstract*-We study a classical problem in operating systems concerning deadlock detection for systems with reusable resources. The elegant Dijkstra's algorithm utilizes simple data structures, but it has the cost of quadratic dependence on the number of the processes. Our goal is to reduce the cost in an optimal way without losing the simplicity of the data structures. More specifically, we present a graph-free and almost optimal algorithm with the cost of linear dependence on the number of the processes, when the number of resources is fixed and when the units of requests for resources are bounded by constants. The algorithm is readily used to improve the running time of Banker's algorithm for deadlock avoidance.**

Categories and Subject Descriptors:
D.4.1 Operating Systems: Process Management;
General Terms: Deadlock, algorithms, performance, optimal algorithm, complexity

I. INTRODUCTION

We first introduce the context of deadlock detection and review the Dijkstra's algorithm dealing with the problem.
In a computer system with a shared address space, let $S = \{P_i : i = 1, \ldots, n\}$ be a set of n processes, and $\{Q_j : j = 1, \ldots, d\}$ be a set of d resources shared by the processes. All the resources are reusable. Each P_i is holding the resources, grouped as a vector $H_i = (h_{i1}, \ldots, h_{id})$. At the same time, it is requesting additional resources, grouped as a vector $R_i = (r_{i1}, \ldots, r_{id})$.

The set S is said to be in a deadlocked state if there is no schedule of execution by which every process in S can satisfy its resource requests. In other words, for every $\sigma \in S_n$, the permutation group on the set $\{1, \ldots, n\}$, there exists i: $1 \le i \le n$ such that

$$R_{\sigma(i)} \geq V + \sum_{1 \leq k \leq i-1} H_{\sigma(k)}.$$

Here $V = (v_1, \ldots, v_d)$ is the free resource vector, and the addition and comparison for vectors is done component-wise.

The problem of deadlock detection is tractable. In fact, a polynomial time algorithm for deadlock detection is given in [1]. The algorithm, which we will call Dijkstra's algorithm in this paper, has the running time of $\Theta(dn^2)$. The algorithm is the same as safe state detection in Banker's algorithm proposed by Dijsktra [3]; also see [5], and Operating Systems textbooks such as [11] and [12].

The data structures used by Dijkstra's algorithm are arrays of integers. No graph model of any kind is involved. However, the quadratic dependence on n of the running time of the algorithm makes it unsuitable for systems with large numbers of processes.

To deal with deadlock detection for systems with large numbers of processes, an ideal solution would be to have an algorithm with similar simplicity of data structures and with running time of linear dependence on the number of processes. In this paper, the ideal solution is conditionally obtained. More specifically, we present a new algorithm with the cost of

$$O[(n + \textstyle\sum_{1 \leq j \leq d} M_j)d].$$

Here M_j are the total number of units of the resources Q_j. Therefore, when d and M_j are constants, the running time is of order $O(n)$. This is particularly true when $M_j = 1$, which corresponds to the critical section problems or lock applications. Since linear running time is also the lower bound for the complexity of the problem, (see the next subsection for terminologies,) we see that the algorithm is asymptotically optimal.

With the same assumptions, the new algorithm is readily applicable to Banker's algorithm for deadlock avoidance. The running time for deadlock avoidance can then be controlled in $O(n^2)$, and thus the Banker's algorithm is asymptotically optimal for the problem of deadlock avoidance.

In Subsection A, the concepts of complexity and almost optimality are introduced. In Section 2, the new algorithm for solving the deadlock problem and its cost analysis are given. In Section 3, we discuss some related perspectives.

A Assumptions and Terminologies

For the deadlock detection problem, we assume that the resource types Q_j ($1 \le j \le d$) are fixed, reusable and independent, and that h_{ij}, r_{ij} and v_i are nonnegative integers for all i and j. Furthermore, we assume that all the request vectors are bounded, namely there exist integers M_j for j: $1 \le j \le d$ such that

$$R_i \leq (M_1; : : : ;M_d) \text{ for all } i: 1 \leq i \leq n.$$

Definition. *Let A be an algorithm that solves the deadlock detection problem, for any n.*

1. We define cost(A) *to be the number of arithmetic and logical operations, and assignments used by A.*

2. The complexity comp(n) *of the problem of deadlock detection is the minimal cost of all algorithms that solve the problem.*

3. A is almost optimal if $\text{cost}(A(n)) = \Theta(\text{comp}(n))$ *as n goes to infinity.*

K. Elleithy (ed.), *Advanced Techniques in Computing Sciences and Software Engineering*,
DOI 10.1007/978-90-481-3660-5_66,

II. An Algorithm for Deadlock Detection

Now we present the algorithm. To simplify the presentation of the algorithm, we will mainly use *C*-like style statements, and some code such as composite statements will be described using statements enclosed in two '/'s.

A. *Algorithm B*

Input: A system of *n* processes with the following state information: the resource request matrix, the resource allocation matrix, and the resource available vector. All entries are general nonnegative integers.

Output: TRUE if the system is deadlocked, FALSE otherwise

Data Structures: Resource allocation matrix $[h_{ij}]$ $(1 \le i \le n,\ 1 \le j \le d) = (H_1, ..., H_n)^t$, resource request matrix $[r_{ij}]$ $(1 \le i \le n,\ 1 \le j \le d) = (R_1, ..., R_n)^t$, and available resource vector $V = (v_1, ..., v_d)$. Permutations σ_j $(1 \le j \le d)$ will be used for tracking relations of processes. Permutations are computed using arrays, though function-like notations will be used for accessing. An array *Mark*[*n*] of the Boolean type is used for marking processes that satisfied their requests, they are initialized to be FALSE. Finally two vector variables *U* and *W* are used.

In the algorithm, we use counting sort to sort the *d* columns of the request matrix. To facilitate the process, we compute an array $E_j = (e[j, 1], ..., e[j, M_j])$ of integers for each resource type Q_j. Here $e[j, s]$ is the number of r_{ij} for $i = 1, ..., n$ that are less than or equal to *s*. As a matter of fact, such data structures can be incorporated and are actually used in the counting sort algorithm [2].

```
Begin Algorithm B
for each j from 1 to d
        w_j := M_j ;
end for
for each i from 1 to n
        Mark[i] := FALSE;
end for
for each j from 1 to d
        / Use counting sort to find permutation σ_j, such that /
        / r_σj(1),j ≤ ... ≤ r_σj(n),j /
        /record the arrays E_j ; j = 1; : : : ; d from the sorting /
        /Compute σ_j^-1 as array/
end for
while (/there exist q for which e [q, v_q ]< n /)
        if(/ there exist j for which  e[j, v_j ] == 0/)
                return TRUE;
        end if
        if(/ for all j; e[j, v_j]== e[j, w_j] /)
                return TRUE;
        end if
        U := V ;
        for each j from 1 to d
                if (v_j > M_j)
                        v_j := M_j ;
                end if
                for each p from e[j, w_j] + 1 to e[j, v_j]
                        if(Mark[σ_j(p)] ==FALSE &&
                           /all k, e[k, σ_kσ_j^-1(p)] > 0/)
                                V := V + H_σj (p);
                                Mark[σ_j(p)] :=TRUE;
                        end if
                end for
        end for
        W := U;
end while
return FALSE;

End Algorithm B
```

The algorithm first, for each resource, sorts the request array by the processes, in nondecreasing order. The sorted request arrays provide a locally optimal allocation of resources to processes. Namely, for each resource, to search for which processes are scheduled for resource allocation, we need only to check the last position in the request array to verify that the request of the resource type by the process is less than or equal to the current free resource units of the type.

The permutations $\sigma_k\sigma_j^{-1}$ establishes a map between the processes of two sorted requests for the resources Q_k and Q_j. Of course, a process can be scheduled only if its requests for all the resources can be satisfied. In order to identify such a process P_j, the algorithm needs to locally check the feasibility of allocation for every resource type, as mentioned above. The algorithm accomplishes this job using the permutations.

If such a process is found, the algorithm deallocates all resources allocated to the process, and finds the next and all such processes, and continues with the next iteration. The algorithm detects the deadlocked state when no progress can be made on $e[j, v_j]$ in this next iteration (and under the while condition that not all processes having satisfied their requests.) The algorithm detects the deadlock-free state when the condition for the while failed, namely, there exists a sequence of schedule by which all process satisfy their requests.

B. Cost Properties of B

Now we estimate the cost for the algorithm ***B***. The initialization is of cost $O(d + n)$. The computation for permutations takes $O(dn)$. In fact, we have

Lemma. *Let s and t be any two permutations on set* $\{1, ..., n\}$. *Then the number of operations to perform* t^1s *can be achieved in* $O(n)$.

Proof. We implement a permutation *s* as an array of length *n* by letting *s*(*i*) be the *i*-th element in the array.

The following code computes t^1:

```
for each i from 1 to n
        t^1 s (i)) := i;
end for
```

The following code computes t^1s :

for each i from 1 to n

$f^{1}s(i) := f^{1}(s(i))$;

end for

The total number of operations for each computation is obviously $O(n)+O(n) = O(n)$.

The counting sort part has the cost of $\sum_{1 \le j \le d} O(M_j + n) = O(dn + \sum_{1 \le j \le d} M_j)$. For the while loop, notice that for each process P_j, the checking for the satisfaction of its requests for all the resources is performed in disjoint position intervals with the previous iteration. This is due to the facts that request arrays are sorted and that available resource vector is increasing. Base on this observation, it is easy to show that the loop has cost $O(dn) + O(d) \sum_{1 \le j \le d} M_j$. We have

Theorem. *The cost for the algorithm B satisfies*

$$\text{cost}(\boldsymbol{B}) = O(dn + d\textstyle\sum_{1 \le j \le d} M_j).$$

Since d and Mj are constants, we have

$$\text{cost}(\boldsymbol{B}) = O(n):$$

Observe that the cost of any algorithm is at least $\Theta(n)$. Thus we have $\text{cost}(\boldsymbol{B}) = \Theta(n)$, which in turn implies that $\boldsymbol{B}$ is almost optimal.

III CONCLUSIONS

We have demonstrated an improved algorithm for Dijkstra'a algorithm for deadlock detection for systems with reusable resource types. This new algorithm enjoys two major properties. First it is $\Theta(n)$ faster than Dijkstra's algorithm. As a matter of fact, it is almost optimal. This is particularly desirable when the number of processes is large. Secondly, the new algorithm only uses straightforward arrays as its data structures. In particular, no graph model of any kind is used. The purely algebraic nature of the algorithm makes its implementation flexible and simple.

For Banker's algorithm for deadlock avoidance, if $\boldsymbol{B}$ is used in the part of safe state detection, then we have an algorithm of cost $\Theta(n^2)$, which is also the complexity of the problem of deadlock avoidance. Thus the algorithm is almost optimal.

For general systems in which resources maybe consumable, deadlock detection has been studied by researchers for many years [3, 5, 6, 7, 8, 10, 13], to name a few. In general, deadlock detection in the optimal way has been proven to be an NP-complete problem [4]. Similar results hold for deadlock avoidance [13]. When the number of resources and the number of units of resources vary, the following upper bound is obtained in [9], for the complexity of the deadlock detection for systems with reusable resources

$$\text{comp}(d, n) = O(dn(\log(n) + d)).$$

We conjecture that

$$\text{comp}(d, n) = \Theta(d\, n\, (\log(n) + \log(d))).$$

REFERENCES

[1] E.G. Co®man, M.J.Elphick, and A. Shoshani (1971): "System Deadlocks", *Computing Surveys*, 3.2, 67-78.

[2] T.H. Cormen, C.E. Leiserson, and R.L. Rivest (1998): *Introduction to Algorithms*, The MIT Press, Cambridge, MA.

[3] E.W. Dijkstra (1965): "Cooperating Sequential Processes", *Technical Report, Technological University*, Eindhoven, the Netherlands, 43-112.

[4] E.M. Gold (1978): "Deadlock Prediction: Easy and Dificult Cases", *SIAM Journal of Computing*, Vol. 7, 320-336.

[5] A.N. Habermann (1969): "Prevention of System Deadlocks", *Communications of the ACM*, 12.7, 373-377, 385.

[6] R.C. Holt (1971): "Comments on Prevention of System Deadlocks", *Communications of the ACM*, 14.1, 179-196.

[7] R.C. Holt (1972): "Some Deadlock Properties of Computer Systems", *Computing Surveys*, 4.3, 179-196.

[8] A.J. Jammel, and H.G. Stiegler (1980): "On Expected Costs of Deadlock Detection", *Information Processing Letters* 11, 229-231.

[9] Y. Li, and R. Cook (2007): "A New Algorithm and Asymptotical Properties for Deadlock Detection Problem for Computer Systems with Reusable Resource Types", *Advancesand Innovations in Systems, Computing Sciences and Software Engineering*, Springer 509-512.

[10] T. Minura (1980): "Testing Deadlock-Freedom of Computer Systems", *Journal of the ACM*, 27.2, 270-280.

[11] A. Silberschatz, P.B. Galvin, and G. Gagne (2002): *Operating System Concepts*, John Wiley & Sons, Inc., NY.

[12] W. Stallings (1997): *Operating Systems, Internals and Design Principles*, Prentice-Hall, NJ.

[13] Y. Suguyama, T. Araki, J. Okui, and T. Kasami (1977): "Complexity of the Deadlock avoidance Problem", *Trans. Inst. of Electron. Comm. Eng. Japan* J60-D, 4, 251-258.

Analysis of Ten Reverse Engineering Tools

Jussi Koskinen[1,*], Tero Lehmonen
[1] Department of Computer Science and Information Systems, University of Jyväskylä,
P.O. Box 35, 40014 Jyväskylä, Finland, koskinen@jyu.fi
[*] Corresponding author

***Abstract*-Reverse engineering tools can be used in satisfying the information needs of software maintainers. Especially in case of maintaining large-scale legacy systems tool support is essential. Reverse engineering tools provide various kinds of capabilities to provide the needed information to the tool user. In this paper we analyze the provided capabilities in terms of four aspects: provided data structures, visualization mechanisms, information request specification mechanisms, and navigation features. We provide a compact analysis of ten representative reverse engineering tools for supporting C, C++ or Java: Eclipse Java Development Tools, Wind River Workbench (for C and C++), Understand (for C++), Imagix 4D, Creole, Javadoc, Javasrc, Source Navigator, Doxygen, and HyperSoft. The results of the study supplement the earlier findings in this important area.**

I. INTRODUCTION

Software maintenance is a very important part of the software life-cycle because maintenance is often elaborate, costly, and inevitable. The relative amount of maintenance costs has traditionally been estimated to be around 50-75% of the total software life-cycle costs in case of successful software systems and more recently sometimes even higher [1].

Maintenance is inevitable and challenging especially in case of large *legacy systems*, which typically are successful software systems with a long lifetime. Unfortunately, they often also have poor documentation, maintainability, and technical quality. Because user requirements and technical environments tend to change during their lifetimes source code changes are constantly needed [2]. Legacy systems are often large investments and contain valuable business logic and information. Due to their business value they are difficult to be replaced cost-effectively. Due to their technical weaknesses they are also difficult to maintain and their size makes it difficult to comprehend their structure and operation.

Program comprehension is a necessary precondition for making changes safely to computer programs. Ideally changes should be made to the source code only in such ways that the quality of the system can be preserved and validated. For example, the system should not be adversely affected by the made changes regarding its reliability. However, these goals are often difficult to achieve in practice in case of large systems due to *software complexity* (regarding, for example, control dependencies, data dependencies, procedure calls and cross-references). Therefore, in non-trivial cases, human understanding focusing on critical parts of the system is very important for the preservation of the system quality.

Program comprehension in turn is often difficult in case of large systems due to the cognitive limitations of humans. Safe changes require sufficient knowledge and understanding of the components and dependencies which are relevant to the specific maintenance situation. Legacy systems are famous of having poor or non-existent documentation. Therefore, there is a great need for tools which provide various kinds of support mechanisms for the program comprehension process.

Reverse engineering [3] means the process of identifying system's components and their interrelations and of creating representations of the system in another form or at a higher level of abstraction. Nowadays, there exists numerous so-called *reverse engineering tools*. They provide an automated analysis of the software system, regarding e.g. its structure, function, and operation. Typically these tools use source code as an input and generate and represent various kinds of abstracted and visual views and graphs for the tool user.

Ideally, these representations enable getting information about the software components and their dependencies which are relevant in particular situations during the program comprehension process [4]. Ideally, these tools provide effective support for satisfying the typical *information needs* [5] which software maintainers have while trying to understand legacy systems. The potential effectiveness of the support is naturally affected by the provided tool features.

One of the authors of this paper has earlier studied the empirically validated information needs of maintainers as such, developed a framework for classifying information retrieval capabilities of reverse engineering tools, and surveyed in general level tool support for information needs in [5]. He has also studied reverse engineering support for some of those needs [6], as well as effectiveness [7] and evaluation aspects [8] of HyperSoft, which is one of the well-founded reverse engineering approaches [9]. Among other things, these studies; especially [5,8], have revealed a need to analyze systematically and in detail the information retrieval capabilities provided by reverse engineering tools. This paper provides such an analysis in case of selected representative tools based on the framework presented earlier in [5].

Section II shortly describes the framework and characterizes the aspects which are relevant in this context. Section III describes the tool selection criteria and introduces the selected tools. Section IV analyzes the capabilities of those tools. The analysis is followed by Section V which discusses

K. Elleithy (ed.), *Advanced Techniques in Computing Sciences and Software Engineering*,
DOI 10.1007/978-90-481-3660-5_67, © Springer Science+Business Media B.V. 2010

other related works and briefly relates the main characteristics and focus areas of the performed analysis to the other most similar kinds of scientific surveys and analyses. Finally, the paper is concluded in Section VI.

II. Framework of Tool Capabilities

Software maintenance and reverse engineering tools support information retrieval, which is a wide concept. It deals with information representation, storage, organization, and access. The main aspects of important features have been classified in [5]. Basically the tools may provide:

A) Data structures (including internal data structures, and more abstract structures investigated by the tool users).

B) Visualization mechanisms for the source code and the formed data structures.

C) Information request specification mechanisms (such as selection mechanisms and query languages).

D) Navigation features (such as additional hypertext capabilities for the already formed data structures).

A. Data structures

Reverse engineering tools typically first create internal data structures (such as parse trees, abstract syntax trees, and program dependence graphs) which are primarily not meant to be investigated by humans. They typically contain such data which can be determined during the compile time of the maintained program and which is most necessary for later formation of more abstract and sophisticated structures targeted directly to the tool users. These internal data structures are typically very detailed, large, stored statically into a program database or repository, and used especially to enhance the speed of the tool by eliminating the otherwise constant need for recreating parts of the data during the creation of the various kinds of needed abstracted structures.

In this paper we will focus on the abstracted data structures and support mechanisms which are directly visible to the tool users, instead of the more technical back-end infrastructure. The abstracted data structures provide, in an ideal case, well-focused and relevant information to the tool users [8]. They are tailored to meet the general information needs of their users. In most cases they contain the same information as the well-known, standardized types of abstracted graphs which are in common use in the general software engineering context.

These structures include the conventional popular graphical representations of system documentation (most notably UML-diagrams) as well as some of the other graphs which can be generated by automated program analysis. These include call graphs, control-flow graphs, data-flow graphs, and program slices. For example, program slicing [10] produces a focused data structure, which includes relevant statements in a given situation, based on their control-flow, and data-flow dependencies to support debugging and impact analysis.

B. Visualization mechanisms

Software visualization typically relies on visualizing the above-mentioned graph contents to the tool user, and on code visualization. Code visualization may include high-lighting of selected code contents, color-coding, elision, zooming, execution profiles etc. Even though that the typically formed abstracted data structures are smaller than the internal data structures, they are often nevertheless very large in case of very large systems. Therefore there is a strong need to visualize especially those structures.

C. Information request specification mechanisms

Specification of the selections as a basis for creating the abstracted data structures in the first place has to be organized somehow in the tools. The simplest (and in some cases sufficient) way is to enable the user to make selections from a generated fixed list of options. Other possible mechanisms include QBE-style semi-graphical selections, regular expressions, and dedicated query languages, such as [11], which have the highest level of expressional power to retrieve complexly specified data from program databases.

D. Navigation features

Finally, navigation features enable non-linear browsing of the information contents of the already formed access structures. Hypertext features are useful since program comprehension typically requires extensive browsing of the source code and the related relevant documentation.

III. Reverse Engineering Tools

In this section we will describe the applied tool selection criteria, the reasons why we have selected particular tools into the study, and their basic characteristics.

A. Tool selection criteria

Because there are numerous tools for reverse engineering purposes it is not possible to analyze all of them in a single study. We have decided to focus on some of the well-known freely available tools which support C, C++ or Java languages. The languages have been selected since they are among both the most commonly used and supported ones. The selected tools should also be either under active current development or be related to scientific publications of software maintenance.

The C programming language is still very important in this context since it is used in numerous important legacy systems which are under maintenance. It is also the only language for which there exists multiple empirical studies on information needs [4]. Object-orientation (OO) is important in the development of new systems which will be legacy systems in the future. The most commonly used OO-languages include C++ and Java. Most of the reverse engineering tools support C language. Some of them support also at least some of the OO-languages, most notably C++ or Java.

B. The selected tools

This subsection describes the basic properties of the tools which were selected based on the above-described criteria.

Table I lists the selected tools, the programming languages that they support, and scientific citations. The first four tools are commercial. Others, except HyperSoft, are based on open source software (OSS).

More tools could not be reported within the conference paper size-limitations. Although this sample is only a relatively small subset of the totality of the tools developed in the field it nevertheless includes many of the most popular and versatile ones. The study is comparable to other similar studies in terms of the size of the sample, as to be described in Section V.

TABLE I. THE SELECTED TOOLS

Tools	Programming languages	Citations
Eclipse JDT	Java	-
WRW	C, C++, Java, FORTRAN, Ada, Assembly	[5,12,13,14]
Understand	C, C++, Java, FORTRAN, Ada, Delphi, JOVIAL	-
Imagix 4D	C, C++	[5,12,16,17]
Creole	Java	[13,15]
Javadoc	Java	[5]
Javasrc	Java	[5]
Source Navigator	C, C++, Java, FORTRAN, COBOL, Tcl	-
Doxygen	C, Objective-C, Python, IDL, C#, D	
HyperSoft	C, ESQL	[5,6,7,8,9]

B.1. Eclipse JDT (v. 3.2.1) (http://www.eclipse.org/jdt/). Eclipse JDT (Java Development Tools) has been selected since Eclipse is one of the currently most popular software development environments and many reverse engineering tools are implemented as its extensions. Eclipse JDT is based on OSS. There are extensions to many programming languages of which the JDT is the best known one. Many reverse engineering tools (such as: Wind River Workbench, Creole and Javadoc) also function as extensions of Eclipse.

B.2. Wind River Workbench (for C and C++) (v. 1.1.0) (http://www.windriver.com/products/workbench/) is a commercial Eclipse-based program development environment. WRW has been selected since it includes very versatile reverse engineering capabilities, and the WRW tool family supports many programming languages. WRW provides many reverse engineering capabilities which are not provided by the free Eclipse-environment. It has mainly been targeted at supporting engineering embedded software. WRW has in essence included into itself most of the versatile capabilities provided by the SNiFF+ tool, which is cited e.g. in [5,12,13,14].

B.3. Understand (for C++) (v. 1.4) (http://www.scitools.com/products/understand/cpp/product.php) is a commercial reverse engineering tool. The Understand tool family provides different variants to support reverse engineering of many programming languages. UfC++ has been selected since it includes versatile features for the important and complex C++ language. It is also possible to integrate the tool with many other tools including a compiler.

B.4. Imagix 4D (v. 5.0) (http://www.imagix.com/products/products.html) is a commercial reverse engineering tool. It has been selected since it includes versatile reverse engineering capabilities and is actively developed further. It includes also versatile visualization and selection mechanisms.

B.5. Creole (v. 1.3.1) (http://www.thechiselgroup.org/creole) is a versatile reverse engineering tool for Java. It has been selected since it is based on the SHriMP representation which has been cited in many scientific studies including [13,15]. In essence Creole implements the visualization capabilities of SHriMP as an extension of Eclipse.

B.6. Javadoc (v. 1.5) (http://java.sun.com/j2se/javadoc/) is an OSS-based redocumentation tool for Java programs. It has been selected since it is used very widely and developed actively. E.g., Java's class library has a Javadoc-based documentation. Javadoc creates documentation based on comments in the source code which follow special mark-up.

B.7. Javasrc (v. 12.17.00) (http://javasrc.sourceforge.net/) is a redocumentation tool for Java programs. It has been selected since it provides very specialized support for viewing source code. It creates documentation based on source code.

B.8. Source Navigator (v. 5.1.4) (http://sourcenav.sourceforge.net/) is an OSS-based reverse engineering tool. It has been selected since it includes versatile integrated reverse engineering and programming features and supports many programming languages. It includes an integrated debugger, a compiler and a text editor.

B.9. Doxygen (v. 1.4.7) (http://www.stack.nl/~dimitri/doxygen/) is an OSS-based redocumentation tool. It has been selected since it supports many programming languages, is a popular redocumentation tool in OSS-projects, and is under active further development. Doxygen utilizes specially marked-up comments to create documentation.

B.10. HyperSoft (v. 1.0) (http://www.cs.jyu.fi/~koskinen/hypersys.htm) is an experimental reverse engineering tool supporting transient source code level hypertext linking [8,9]. It has been selected since there are many scientific articles which discuss the information needs of maintainers and are related to the tool. HyperSoft provides strong and versatile capabilities especially to support source code browsing.

IV. ANALYSIS OF THE TOOL CAPABILITIES

In this section we will analyze the tool capabilities as grouped into the described four categories. Table II (Appendix) shows a summary of the capabilities of the tools.

A. *Data structures*

Call graph is the most important abstract data structure provided by these tools (80% of the tools generate them). Class diagrams are provided by most of these tools (70%). 30% of the tools focused on automated redocumentation. Additionally, there is a large set of different kinds of structures (such as deltas, element information, and metrics) each provided typically by only 10-20% of the tools.

A.1. Graphs. Call graphs are generated by all of these tools except Javadoc and Javasrc. Class diagrams are generated by all of these tools except Javadoc, Javasrc, and HyperSoft. Intra-procedural control flow diagrams are provided by UfC++. Intra-procedural data flow diagrams are provided by

Imagix 4D. Collaboration diagrams are provided by Doxygen. HyperSoft generates also program slices.

A.2. Redocumentation structures. Documents which describe classes, inner classes, interfaces, constructors, methods, and fields can be generated by Javadoc. Documents which show classes, objects, class variables, and methods are generated based on source code by Javasrc. Doxygen, UfC++, and Imagix 4D are able to produce system documentation.

A.3. Deltas. The changes (effectively the corresponding deltas) which are made to the source code via a code editor during a tool usage session are stored by Eclipse JDT. Eclipse JDT and Source Navigator are integrated, e.g., with CVS for comparing differences between different versions of a system.

A.4. Special data structures. The so-called element tree of Eclipse JDT shows elements like package fragments, compilation units, binary classes, types, methods, and fields. Detailed additional information related to identifiers is provided by UfC++. Imagix 4D calculates various metrics. Similarly, UfC++ calculates metrics whose values are appended into the automatically produced documentation. All these tools, except WRW and Creole, store browsing history.

B. *Visualization mechanisms*

Almost all of the tools (90%) provided an editor or a viewer for the source code. Most of the tools (80%) focus heavily on call graph (or tree) based visualizations. Most of the visualizations are conventional, but there are also notable exceptions (including Imagix 4D's 3d-view). UfC++ and Imagix 4D provide also other kinds of graphs.

B.1. Source code viewing. All the tools in the sample, except Javadoc, provide either a text editor or some other means to view source code, whereas Javadoc focuses solely on supporting software documentation. Eclipse JDT and Javadoc provide an editor which is specially tailored to creating documentation. Eclipse JDT, WRW, and HyperSoft are also able to show selected identifiers as color-coded text fragments.

B.2. Graph visualizations. Most of the tools focus on visualizing the program information based on call graphs. The data contents of the formed call graphs are visualized in more or less traditional ways by most of these tools. Most of the created call graph types resemble each other consisting of visual objects and arrows connecting them.

Eclipse JDT and WRW are exceptions because their call graphs are actually trees. UfC++ provides both the traditional call graph and a tree-structure for the same information. Creole shows in the graph also the implementation code of the relevant methods by user request. It also provides both the traditional call graph and a combined class diagram and a call graph. Imagix 4D provides both a two-dimensional and a three-dimensional variant of the call graph. The 3d-variant helps in the common problem of having a very dense and thereby confusing call graph to be viewed. The 3d-variant can be rotated and zoomed to further help its viewing.

Most of the tools apply their own specific notations to show the class diagrams (instead of UML). Imagix 4D and Doxygen use the UML-notation. Eclipse JDT, WRW, and Javadoc show the inheritance hierarchies as trees. Others use notations close to UML, except some of them use different conventions for inheritance (reversed arrows).

HyperSoft provides both a graphical representation of the call graph and the call graph as embedded hypertext, which always shows also the context of the hypertext nodes supporting program comprehension (procedure names serve as hypertext nodes). Additionally, all the generated hypertext structures of HyperSoft can be viewed by a call graph variant which shows the abstract external interface of the procedure-level element which is the current focus of browsing the source code. The graph shows those functions which are directly related to the selected one via calling dependencies.

Some of the tools, including UfC++ and Imagix 4D, provide also other kinds of graphs. They are represented in the conventional 2d-format, mainly following the standard notations of UML and the traditional ways of representing information in system documentation. All the hypertext structures which can be generated by HyperSoft can be viewed via module dependence graphs which show on abstract-level the modules which contain the generated hypertext nodes.

C. *Information request specification mechanisms*

Most of the tools (80%) support simple filtering and expansion of the call graph contents. HyperSoft's speciality is program slicing, which supports retrieval of information based on very complex analyses. Most of the tools provide only conventional text search mechanisms. Regular expressions are supported only by 40% of the tools.

C.1. Graph filtering. Eclipse JDT, WRW, UfC++, Imagix 4D, Javasrc, Source Navigator, Doxygen, and HyperSoft are able to provide a list of procedures which call some specified procedure. Eclipse JDT, WRW, and Source Navigator enable smooth expansion of the shown call graph nodes by enabling the user to simply retrieve more node contents from the call tree by selecting them. Imagix 4D also provides versatile options to filter the call graph based on the level of shown details, such as procedures, statements and variables. The data-flow diagrams of Imagix 4D can be specified to be shown as containing conditional statements, function and method calls or all statements based on the tool user's choice. The tree structure of UfC++ enables focusing on specific functions and their interfaces. The control-flow diagrams of UfC++ can be filtered similarly as the data-flow diagrams of Imagix 4D.

C.2. Program slicing. Formation of a slice in HyperSoft is initiated by the tool user by specifying a slicing criterion, which is a seed for the slice. The slicing criterion is a variable occurrence, which is selected by directly pointing at it and selecting the desired slicing variant.

C.3. Search mechanisms. Most of the tools contain only the conventional text search. In its basic form it covers only one file at a time and includes also comments matching the search criteria. More advanced searches mechanisms enable checking multiple files during a single search (like e.g. in grep) and to filter out comments when necessary based on the stored syntactic information. For example, HyperSoft forms a list of the occurrences of a selected identifier based on global search and the visibility scope of the identifier. Source Navigator also enables selection of the procedures to be included into the call graph via its text-based search-function.

C.4. Regular expressions. Eclipse JDT, WRW, UfC++, and Source Navigator include a query mechanism based on regular expressions. Their use enables creating very refined queries but the formation may be elaborate. These tools also enable making queries to the already parsed and indexed source code which enables filtering out the comments, taking into account the visibility scope of the variables and to enhance the speed of retrieving the results.

D. Navigation features

Navigation among methods and procedures, other kinds of data elements, and browsing history are supported by almost all of these tools (80-90%), but the precise nature and quality of the support mechanisms varies among the tools. HyperSoft focuses on navigation features, including e.g. hypertext links between source code elements and the visual elements in the provided views and graphs supporting source code browsing.

D.1. Methods and procedures. The call tree of Eclipse JDT enables direct navigation from method implementations to the places where they are called from (i.e. reversed hypertext links). WRW provides similar features. Imagix 4D provides links from all the call graph elements to the corresponding elements in the source code. Creole provides a similar feature for methods. In Hypersoft all the module, procedure, and statement-level objects within the provided graphical views contain a hypertext link(s) to the corresponding place(s) in the source code. In case of multiple links the user is supported by a list of possible selections for each such hypertext node.

Javasrc and Doxygen are able to create a cross-referenced hypertext representation based on source code. Cross-references are provided for classes, class variables, and methods. Uses are linked to the corresponding definitions, and definitions are linked to a list which includes all the uses. Doxygen is also able to link together an UML class diagram and the corresponding documentation. Eclipse JDT provides links from the uses of methods to their declarations. UfC++ provides very versatile features for navigating among methods and functions appearing in the source code. UfC++ opens on request, related to a selected identifier, a menu from which the user can select hypertext links to the occurrences. Source Navigator provides similar kinds of navigation capabilities.

D.2. Other data elements. UfC++, Javasrc, Source Navigator, and Doxygen provide versatile features also for browsing variables in general. Eclipse JDT, and Javadoc enable direct moving to the documentation associated to a selected variable. Imagix 4D enables browsing between variable declarations, definitions, and uses. HyperSoft lists the occurrences of the variables and enables navigating among them. HyperSoft presents slices as high-lighted embedded hypertext (identifiers serve as primary hypertext nodes).

UfC++ provides hypertext links for navigating among the occurrences of header files, macros, and enumerations. HyperSoft provides cross-references from the occurrences of user-defined types, and macros to their declarations. Additionally, HyperSoft provides so-called structured map view for showing all the generated data structures and to select modules, procedures, and statements which are all hyperlinked to the corresponding elements in the source code.

D.3. Browsing history. UfC++, Javadoc, Javasrc, and Doxygen are redocumentation tools. The produced HTML-documentation is usually viewed by some Internet browser. The accociated page history and the linked list of the browsed pages can be used while browsing the documentation. Eclipse JDT, UfC++, Imagix 4D, Source Navigator, and HyperSoft also enable similar kind of browsing.

V. Related Works and Discussion

In this paper we have aimed at performing an analysis which focuses on the information retrieval capabilities of selected, actively developed or studied, well-known, and well-available tools. The study complements the earlier studies in this area. SNiFF+ (which is used in WRW) is evaluated also in [5,12,13,14]. SHriMP (which is used in Creole) is evaluated also in [13,15]. Imagix (4D) is evaluated also in [5,12,16,17].

The journal article [5] cited 24 tools (HyperSoft, HyperPro, Whorf, SHriMP, Javasrc, SNiFF+, DIF, SLEUTH, PROTEUS, PAS, Javadoc, EDATS, Ciao, Acacia, SCA, GENOA, SPYDER, SLICE, an unnamed slicer, Imagix 4D, Refine/C, Rigi, Small Worlds, and Shimba). It also listed their main features, and data structures. The main conclusions were that effective maintenance support benefits from visualization and navigation features, that scope of the provided information could be extended, and that versatile visualization and navigation features could be integrated into a single tool. Our current study focuses on the selected tools, includes six of these tools, and provides a more detailed analysis.

The paper [16] compared 16 tools. Seven of them were commercial (Software Refinery, McCabe VRT, Imagix 4D, Xinotech Research, Logiscope, Ensemble Software, and Design Maintenance System), and nine were research tools (PAT, COBOL/SRE, DECODE, LANTRN, Maintainer's Assistant, REDO Toolset, Rigi, AutoSpec, and RMTool). The survey focused on comparing the tools in terms of their by-products (constructed artifacts), revealed the variety of the tools and that different approaches can provide complementary information. We studied mostly other tools (except Imagix 4D), more modern tools and from a different perspective.

The journal article [17] studied empirically nine call graph extractors (cflow, cawk, CIA, Field, GCT, Imagix, LSME, Mawk, and Rigiparse) in detail. The main result of the study was the revelation that the call graphs extracted by the different tools have much variance. Our study has a wider scope (although Imagix was investigated by both studies).

The paper [14] evaluated five tools (Rigi, Dali, Software Bookshelf, CIA, and SNiFF+) based on their data extraction, classification, and visualization capabilities. That paper focused solely on architectural recovery. Instead, we studied more tools from a wider perspective. The features of SNiFF+ were analyzed by both studies. SNiFF+ was stated to be the best one of the evaluated tools in [14].

The journal article [12] analyzed four tools (Refine/C, Imagix 4D, SNiFF+, and Rigi). That article provided a detailed evaluation of the benefits and shortcomings of the studied tools. The features of Imagix 4D and SNiFF+ were studied also by us. The benefits of Imagix 4D stated in [12] were the variety of its views, and support for creating documentation. The stated benefits of SNiFF+ were its cross-references, and its fast and flexible parser.

Finally, the paper [13] compared empirically three tools (SHriMP, SNiFF+, and Rigi). The paper reported how these tools supported program comprehension. The features of SHriMP and SNiFF+ were studied also by us. The strong side of SHriMP stated in [13] was especially the capability to switch seamlessly between high-level views. The strong side of SNiFF+ was its capability to filter the provided information.

VI. CONCLUSION

This paper has provided an analysis of the information retrieval capabilities of ten representative reverse engineering tools regarding four aspects (provided data structures, visualization mechanisms, information request specification mechanisms, and navigation features). This study complements the earlier similar analytical studies on the field [5,12-14,16,17] and earlier studies on creating new ways to support information retrieval for software maintenance [5-9]. Possible future research options include an even more detailed analysis of some of the most versatile tools (such as: Eclipse, WRW, Understand, Imagix 4D, and Creole) focusing, e.g., on the provided data structures based on the framework [8] or to the 24 information needs presented in [5] and whether they can or can not be effectively satisfiable by those tools.

REFERENCES

[1] R. Seacord, D. Plakosh, and G. Lewis, *Modernizing Legacy Systems: Software Technologies, Engineering Processes, and Business Practices*. Addison-Wesley, 2003.

[2] M.M. Lehman, D.E. Perry, and J.F. Ramil, "Implications of evolution metrics on software maintenance", *Proc. Int. Conf. Softw. Maint.* (ICSM 1998). IEEE Computer Society, 1998, pp. 208-217.

[3] E.J. Chikofsky, and J.H. Cross, II, "Reverse engineering and design recovery: a taxonomy", *IEEE Sw.*, vol. 7, number 1, pp. 13-17, 1990.

[4] A. von Mayrhauser, and A.M. Vans, "Industrial experience with an integrated code comprehension model", *Software Engineering Journal*, vol. 10, number 5, pp. 171-182, 1995.

[5] J. Koskinen, A. Salminen, and J. Paakki, "Hypertext support for the information needs of software maintainers", *Journal of Softw. Maint. Evol.*, vol. 16, number 3, pp. 187-215, 2004.

[6] J. Koskinen, and A. Salminen, "Supporting impact analysis in HyperSoft and other maintenance tools", *Proc. 2nd IASTED Int. Conf. Softw. Eng.* (SE 2005). Anaheim, CA: Acta Press, 2005, pp. 187-192.

[7] J. Koskinen, "Experimental evaluation of hypertext access structures", *Journal of Softw. Maint. Evol.*, vol. 14, number 2, pp. 83-108, 2002.

[8] J. Koskinen, "Evaluation framework of hypertext access for program comprehension support", *Innovative Techniq. Instruction Techn., E-learning, E-assessment, and Education*. Springer, 2008, pp. 235-240.

[9] J. Paakki, A. Salminen, and J. Koskinen, "Automated hypertext support for software maintenance", *The Computer Journal*, vol. 39, number 7, pp. 577-597, 1996.

[10] M. Weiser, "Program slicing", *IEEE Transactions on Software Engineering*, vol 10, number 4, pp. 352-357, 1982.

[11] S. Paul, and A. Prakash, "A query algebra for program databases", *IEEE Transactions on Software Engineering*, vol. 22, number 3, pp. 202-217, 1996.

[12] B. Bellay, and H. Gall, "An evaluation of reverse engineering tool capabilities", *Journal of Softw. Maint.*, vol. 10, pp. 305-331, 1998.

[13] M.A. Storey, K. Wong, and H. Muller, "How do program understanding tools affect how programmers understand programs?", *Proc. 4th Working Conf. Reverse Engineering* (WCRE 1997). IEEE Computer Society, 1997, pp. 12-21.

[14] M. Armstrong, and C. Trudeau, "Evaluating architectural extractors", *Proc. Fifth Working Conf. Reverse Engineering* (WCRE 1998). IEEE Computer Society, 1998, pp. 30-39.

[15] M. Storey, and H. Müller, "Manipulating and documenting software structures using SHriMP views", *Proc. 11th Int. Conf. Softw. Maint.* (ICSM 1995). IEEE Computer Society, 1995, pp. 275-284.

[16] G.C. Gannod, and B.H.C. Cheng, "A framework for classifying and comparing software reverse engineering and design recovery techniques", *Proc. Sixth Working Conf. Reverse Engineering* (WCRE 1999). IEEE Computer Society, 1999, pp. 77-88.

[17] G. Murphy, D. Notkin, W. Griswold, and E. Lan, "An empirical study of static call graph extractors", *ACM Transact. Softw. Eng. Meth.*, vol. 7, number 2, pp. 158-191, 1998.

APPENDIX: SUMMARY

TABLE II. SUMMARY OF THE CAPABILITIES OF THE TOOLS

Tools	Data structures	Visualization	Information requests	Main navigation features
Eclipse JDT	call graph, deltas, element tree, browsing history	source code viewer, document editor, color-coding, call tree, inheritance tree	syntactic search, regular expressions, call graph filtering and expansion	method implementation => method calls, method use => method declaration, variable => document
WRW (including SNiFF+)	call graph	source code viewer, color-coding, call tree, inheritance tree	syntactic search, regular expressions, call graph filtering,expansion	method implementation => method calls
Understand	call graph, control-flow diagram, identifier-centered documents, metrics, browsing history	source code viewer, traditional call graph, call tree for interfaces, control-flow	syntactic search, regular expressions, call graph filtering, filtering of control-flow	occurrence (method, other variable) => list of occurrences
Imagix 4D	call graph, data-flow diagram, metrics, browsing history	source code viewer, traditional and 3d call graph, data flow	text search, call graph filtering, filtering of the details of calls and data-flow	call graph element (function) => source code element, declarations => definitions => uses
Creole (including SHriMP)	call graph	source code viewer, code-based call graph, combined call graph and class diagram	text search	call graph element (method) => source code element, variable => document
Javadoc	class-centered documents, browsing history	document editor, traditional call graph, inheritance tree	text search	variable => document
Javasrc	class-centered documents, browsing history	source code viewer, traditional call graph	text search, call graph filtering	use (class, method, other variable) => definition => list of uses
Source Navigator	call graph, deltas, browsing history	source code viewer, traditional call graph	syntactic search, regular expressions, search-based selections, call graph filtering and expansion	occurrence (method, other variable) => list of occurrences
Doxygen	call graph, collaboration diagram, system documentation, browsing history	source code viewer, traditional call graph, UML-notation	text search, call graph filtering	use (class, method, other variable) => definitions => list of uses
HyperSoft	call graph, program slices (backward and forward), list of identifiers, browsing history	source code viewer, color-coding, traditional and abstracted call graph, module dependence graph	syntactic identifier search, call graph filtering, program slicing (backward and forward analysis)	all elements in the graphs (module, procedure, statement) => source code element, list of hypertext links

The GeneSEZ approach to model-driven software development

Tobias Haubold, Georg Beier, Wolfgang Golubski, Nico Herbig
Zwickau University of Applied Sciences, Informatics
Zwickau, Germany
toh@fh-zwickau.de, geobe@fh-zwickau.de, golubski@fh-zwickau.de, nihe@fh-zwickau.de

***Abstract*-This paper presents an approach to model driven software development. It first covers some basics about model driven approaches and then the main concepts of the GeneSEZ[1] approach are presented and discussed to regard the needs of software architects and developers. A meta model is introduced to provide reusable transformation chains across different projects and its model driven development process is covered. The GeneSEZ approach is applied and discussed on an example application.**

***Index Terms*-MDSD, MDA, UML, DSL**

I. INTRODUCTION

Nowadays model driven (MD-) approaches are highly regarded. This can be seen on the increasing number of acronyms: MDA (Model Driven Architecture), MDSD (Model Driven Software Engineering), MDE (Model Driven Engineering), MDTD (Model Driven Test Development), etc. Model driven approaches are applied with success in software projects [1, 2].

There are two kinds of model driven approaches: the interpretative and the generative approach. With an interpretative approach, the model is used at runtime and is executed or at least interpreted. Generative approaches are using code generators to generate a part of the source code of an application from one or more models during the software development process. In this paper the generative approach is covered in more detail. For more information about interpretative and generative approaches see [3].

The code generators used in generative approaches can be grouped in two categories: active and passive code generators [4]. Passive code generators are only used once to generate an initial amount of source code for the software which later can be modified by developers. If the generator is used again, all changes made are lost. Passive code generators are popular in the area of application frameworks. They are usually embedded within a framework to generate further source code from a definition made using the framework. Good examples are Hibernate[2] and Grails[3]. Hibernate is a persistence framework for Java. It has a meta model on its core and can use various input formats, e.g. it can use hibernate mapping files or EJB (Enterprise Java Beans) to generate a database schema. Or it can use an existing database schema to generate e.g. EJB. Grails is a web framework which can use a domain model definition to generate a complete web application with CRUD (create, retrieve, update, delete) operations. One pitfall with passive generators is that they only provide an initial value by creating new source code artefacts. There is no support for an incremental software development process to reflect changes or new features for the software. The presented approach is an active code generator.

Further differentiation of an MDSD process can be done by the used meta models, the target platforms, supported transformations and the use of open-source, freeware or commercial tools. The meta model is the foundation of a model driven transformation chain. A transformation chain takes a model as input and either modifies the model, creates a new model or creates some textual output. Within an MDSD process there can be several meta models and any number of transformation chains. To be successful, the transformation chains of an MDSD process have to be reusable across projects. The time needed to build a transformation chain to generate a special kind of artefact is usually higher than writing the artefact manually. Some transformation chains are already profitable when applied on one software project while other need to be applied on more than one software project. So, the choice of a meta model is an important and serious decision.

Basically there are efforts to use UML as well as DSLs (Domain Specific Languages) within model driven approaches and there are pros and cons for both. The UML is a standardized graphical notation for object oriented software. Unfortunately the UML has grown quite complex through its standardization process [5]. It also does not provide a mapping to programming languages [6]. DSLs on the other hand are usually more handy and precise [7]. But they can only be used for a certain field of application. At the moment, DSLs applied in model driven approaches are not standardized. The combination of both, UML and DSLs seams to be a key role in model driven approaches.

This paper provides an overview about our MDSD approach and covers the meta model as the core component of our approach in more detail. The frameworks and tools our approach is based on are covered in section 4. Then the model driven development process of the meta model as well as one possible model driven development process of an example project is covered. The paper ends with a discussion of related work, a summary and future prospects.

---

1 Generative Software Engineering Zwickau, http://genesez.de/.
2 Hibernate, object-relational persistence (Java), http://www.hibernate.org/.
3 Grails is a web framework based on the dynamic language Groovy which is itself based on Java. See http://grails.org/, http://groovy.codehaus.org/.

K. Elleithy (ed.), *Advanced Techniques in Computing Sciences and Software Engineering*,
DOI 10.1007/978-90-481-3660-5_68, © Springer Science+Business Media B.V. 2010

II. GeneSEZ MDSD Approach

The GeneSEZ approach consists of the following four main concepts illustrated in Fig. 1:

- *Model adapters*: are used to transform certain models into GeneSEZ models for further processing and eventually code generation
- *GeneSEZ meta model*: is the used meta model within the GeneSEZ MDSD process
- *Components*: several components supporting the MDSD process, e.g. for executing model transformations, to ease the development of platform projects, etc
- *Platform projects*: support the code generation for a particular programming language

The GeneSEZ approach decouples the meta models, used in modeling and code generation from each other. Our approach does not prescribe a particular meta model to model the application. Instead, the authors believe that the meta model used to model the software should be chosen by software architects not the MDSD process. For that reason model adapters are used to populate a particular GeneSEZ model. So UML models as well as models built with a particular DSL

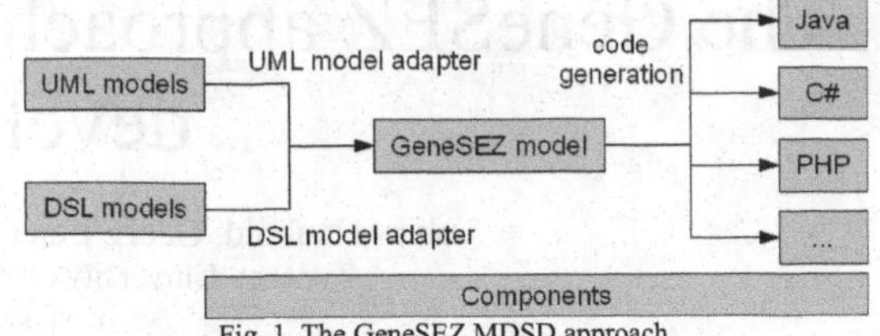

Fig. 1. The GeneSEZ MDSD approach.

can be used. The GeneSEZ meta model is the fixed meta model within our MDSD process and the foundation for reusable transformation chains. The components consist of several artefacts which are used to ease the creation of transformation chains. Every transformation chain is part of a platform project to support the code generation for a particular programming language. There are also transformation chains which perform code generation for particular frameworks of the target programming language.

III. GeneSEZ MEta Model

The GeneSEZ meta model is a general purpose meta model for object oriented code generation. Its main purpose is to

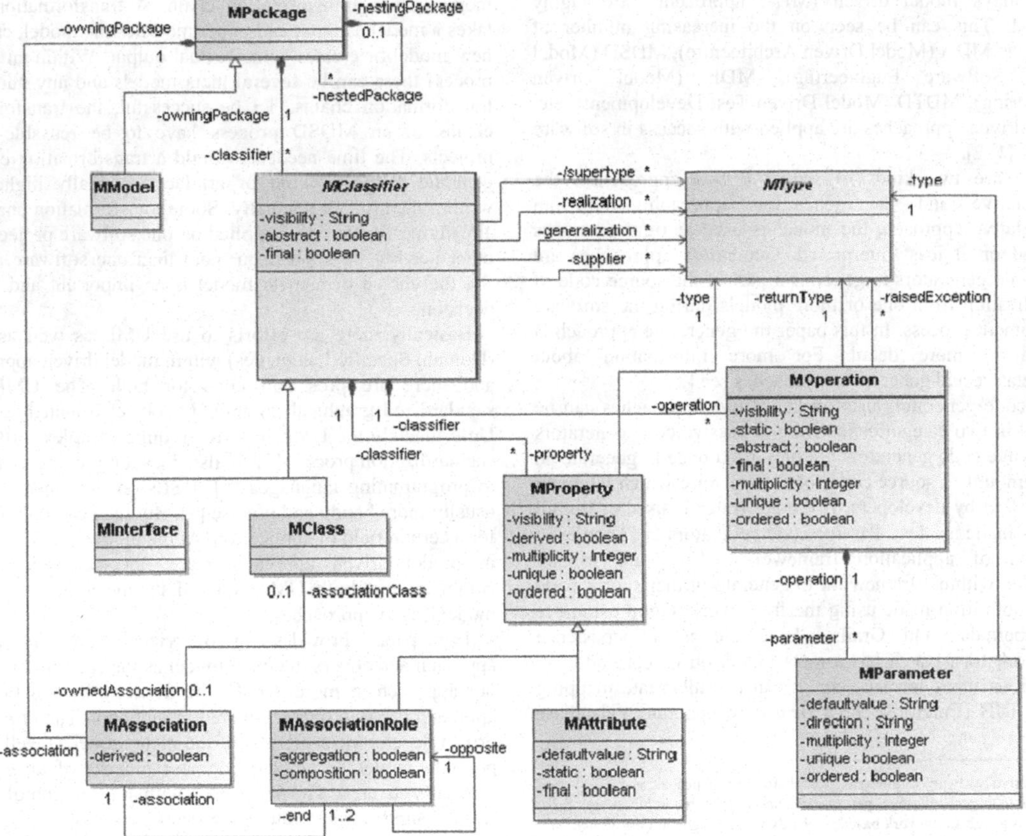

Fig. 2. The structure of the meta model.

provide a view to the needed information from a code generation perspective. That means it contains only the information which matters to generated code. Developers of transformation chains have direct access to the aspects they care about and don't have to deal with to much unnecessary information in the model. Due to its reduced complexity compared to the UML it is easier to write model modifications and model to model transformations. It is inspired by the "classic" meta model of openArchitectureWare [8], but has a revised structure, an implementation driven feature set and a model driven development process.

The core of the GeneSEZ meta model is shown in Fig. 2. The structure is similar to the UML meta model but with a remarkable reduced amount of meta classes. The constructs taken from UML have the same semantics as in the UML meta model. Three constructs are renamed UML constructs to have a common name for their meaning. The *final* attribute of the classifier specifies if the classifier can be inherit or not. The *final* attribute of an operation means if it can be overridden or not. The *final* attribute of a property means if its value is changeable or read-only. One distinction to the UML meta model is a separate meta model element for association roles (association ends). In the UML association roles and attributes are both properties [9] and distinct if there is a linked association or not. The GeneSEZ meta model introduces a separate model element for association roles to address the different semantics [9] in code generation.

In Fig. 3 an overview of types in the GeneSEZ meta model is shown. There are two differences to the UML meta model. First, the UML template concept to specify type parameters is replaced by the concept of generics. Second a new type *MExternal* is introduced to specify already existing types, e.g. types of a library or a software development kit (SDK). Only types which inherit from classifier are defined in the target programming language.

IV. BASE TECHNOLOGY

Our approach is based on the following open-source frameworks and technologies: openArchitectureWare, the Eclipse UML2 project [10] and the Eclipse Modeling Framework (EMF) [11].

EMF is a modeling framework and code generation facility for structured data models based on a meta model named ecore [12]. Ecore is a meta model which is similar to the OMG EMOF (Essential Meta Object Facility), which is itself part of the OMG MOF (Meta Object Facility) [13] specification. Both, the OMG MOF as well as Ecore can be used to describe meta models. The benefits of Ecore are the reference implementation and the available tooling.

The Eclipse UML2 project is an implementation of the OMG UML2 [9] meta model based on the Ecore meta model. The lack of the missing implementation of OMGs UML meta model leads to non-interoperability of UML models from different UML tools[4]. The Eclipse UML2 addresses this issue by providing a reference implementation which gets more and more support by UML tool vendors.

openArchitectureWare is a model driven generator framework. It supports a couple of meta models out of the box and comes with three statically typed languages: a declarative language for model constraints, a functional language for model-to-model transformations or model modifications and a

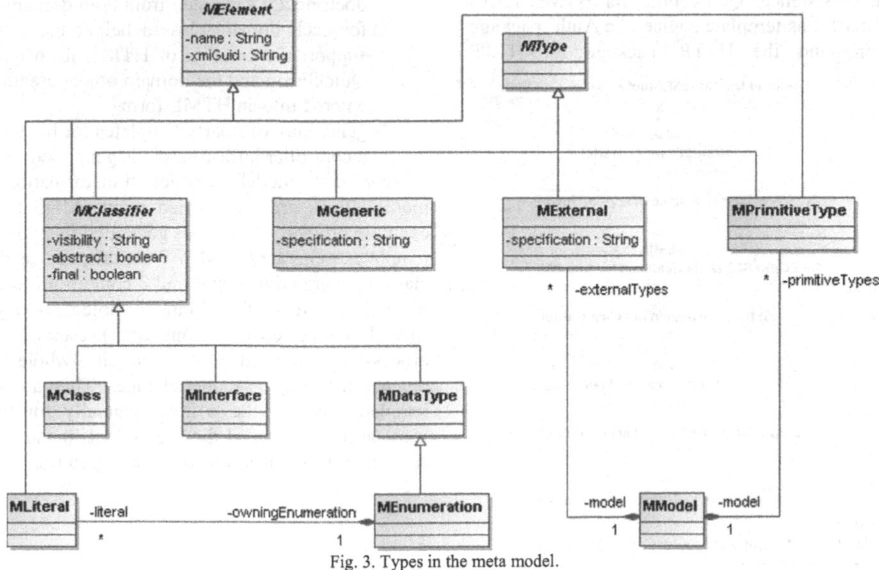

Fig. 3. Types in the meta model.

[4] One handicap for model driven approaches based on the UML meta model is not standardized serialization format. So they have to provide an adapter for each particular UML tool and version.

template language for model-to-text transformations. The latter two have support for polymorphism and aspect oriented programming (AOP). Finally there is support for non-invasive meta model extensions, it is well integrated within the Eclipse Modeling Project [14] and completely open source. Both EMF and openArchitectureWare are used in industry [1, 15] and provide a good foundation for an MDSD process.

V. GENESEZ META MODEL DEVELOPMENT PROCESS

The GeneSEZ meta model is developed in a kind of bootstrapping process as shown in Fig. 4: 1) A UML tool is used to model the meta model in an intuitive and common manner. 2) Then the UML tool provides an export wizard to the Eclipse UML2 XMI format. 3) This representation of an UML model is transformed into an Ecore based model using a slightly modified version of the uml2ecore transformation shipped with openArchitectureWare. 4) In the last step EMF is used to automatically generate a Java implementation from the Ecore model. This implementation can be directly used within openArchitectureWare to work with GeneSEZ models. There is no need for further manual implementation. If development of the meta model continues, the implementation can be regenerated.

VI. EXAMPLE PROJECT DEVELOPMENT PROCESS

Our MDSD approach has been applied in a couple of software projects (see section 9). For a clear and better understanding a simple example PHP web application with CRUD functionality is covered here. It consist of two domain classes as shown in Fig. 5 and uses the following frameworks: doctrine[5] for the persistence, QuickForm[6] for HTML forms and validation, Smarty[7] as template engine, the Auth[8] package for authentication, and the HTTP[9] package for HTTP

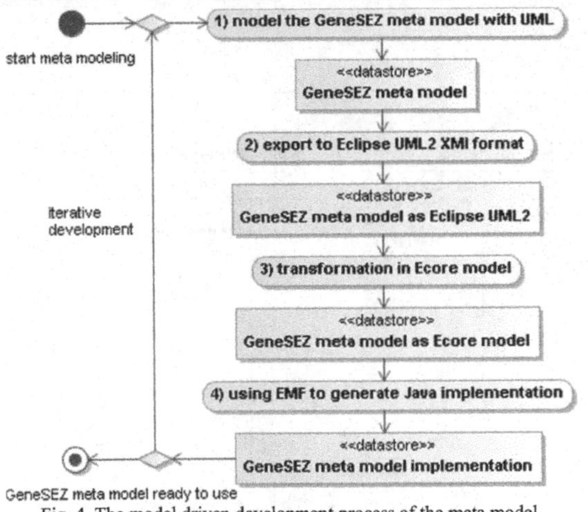

Fig. 4. The model driven development process of the meta model.

Fig. 5. The domain model of the example application.

redirects.

The Auth package uses an authentication provider to check the provided user authentication. A couple of authentication providers for different data sources are shipped with the package, but none for doctrine. For that reason there is a base class to build custom authentication providers. Fig. 6 shows the UML model of the authentication provider. The class *Auth_Container* has the stereotype *external* to mark it as an already existing class of a framework. This class is transformed into an instance of *MExternal* in the GeneSEZ model. The class *DoctrineAuthProvider* has to override the *fetchData* method. It returns a boolean value which indicates if the access for a user is allowed (true) or denied (false).

The basic structure of the GeneSEZ MDSD process is shown in Fig. 7. The concrete steps for the example application are as follows and shown with notes in Fig. 7:

1) the UML model is validated to ensure the transformation into a GeneSEZ model
2) the UML model is transformed into a GeneSEZ model using the UML model adapter
3) the GeneSEZ model is validated to ensure if PHP source code can be generated from it
4) the class definition of the authentication provider for the doctrine framework is generated
5) the class definitions of the persistent entities using doctrine are generated from both domain classes
6) for each domain class a helper class is generated to support the creation of HTML form definitions using QuickForm and the domain object creation from values entered into an HTML form
7) generation of smarty templates for list views
8) a controller infrastructure is generated

Note that model-to-model transformations and model modifications are not applied in the MDSD process of the example project. The classes generated in step 5 and 6 and the templates generated in step 7 are already ready to use. The classes generated in step 4 and 8 contain protected regions for manual implementation. Manual implementations within such special source code regions are preserved if the MDSD process is executed again and the whole application is regenerated to apply model changes. Then a couple of smarty templates have to be written manually for the whole web application layout and the views to edit the domain objects. After that the web application is ready to use.

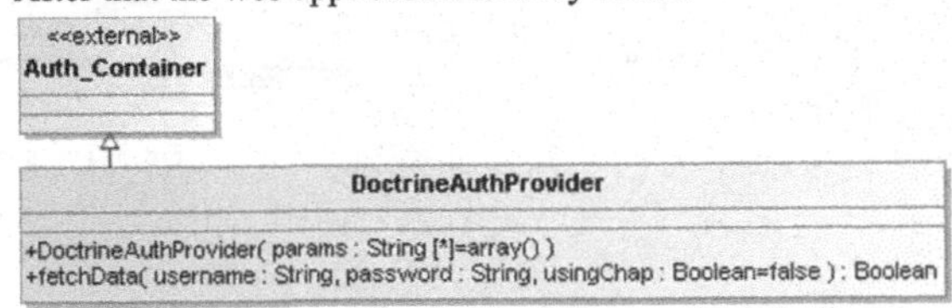

Fig. 6. The authentication provider for the example application.

[5] Doctrine PHP persistence framework, http://www.doctrine-project.org/
[6] PHP QuickForm, http://pear.php.net/package/HTML_QuickForm/
[7] Smarty Template Engine, http://www.smarty.net/
[8] PEAR Auth package, http://pear.php.net/package/Auth
[9] PEAR HTTP package, http://pear.php.net/package/HTTP

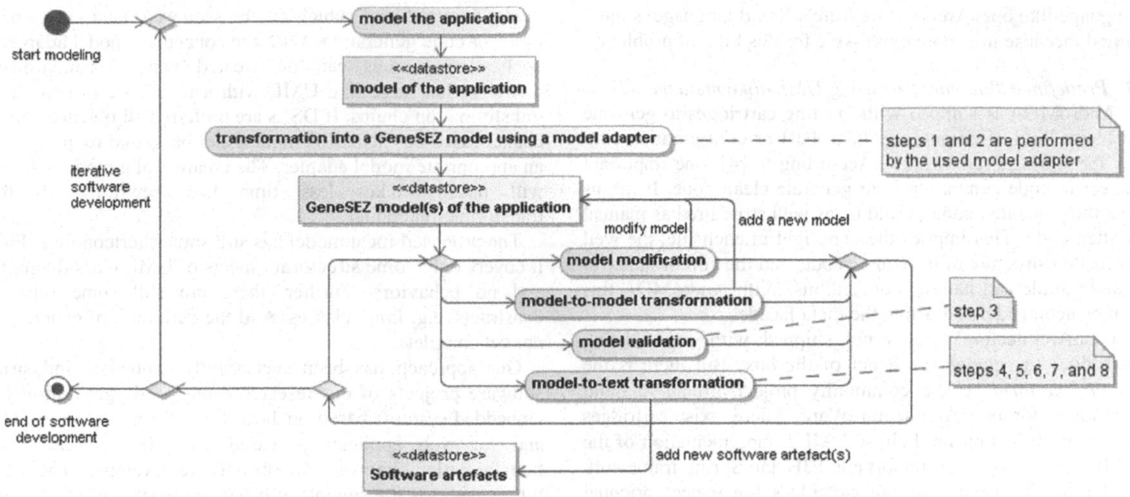

Fig. 7. The GeneSEZ MDSD process including the process of the example application.

VII. Evaluation of the GeneSEZ Approach

With the PHP web application presented in the last section the following comparison can be done: Compared on the basis of the whole web application 49% of the lines of code were generated. For a more realistic comparison the user interface (UI) part (the smarty templates) is left out and only the PHP source code is evaluated. In this case 60% of the lines of code were generated. The UI part can be left out because the appropriate way to generate the source code of an UI is to use a WYSIWYG[10] editor instead of a UML model. The PHP platform project is currently in an early stage of development. In other software projects (see section 9) a ratio of 70% was reached. These values show the relation between generated and manual written code if only the structure of an application is targeted to a concrete platform, i. e. an architecture with a set of frameworks. The generative part can be still increased by using an appropriate MVC[11] framework for the controller infrastructure. Furthermore it can be increased by evaluating dynamic aspects modeled with activity or state charts.

By now, our approach offers the following advantages: 1) A particular meta model is not prescribed to model the application. It can be chosen by the software architects. 2) The transformation chains can be reused across software projects despite that other meta models are used to model the application. 3) A couple of transformation chains already exist. 4) The transformation chains are manageable because of the simplified meta model. 5) The whole GeneSEZ meta model can be printed on one usual paper in letter format to provide a handy overview (that is not possible with UML), e.g. useful for developers of transformation chains.

VIII. Related Work

Two popular open source frameworks for generative MDSD approaches are openArchitectureWare [8] and AndroMDA [16]. They both have an own meta model abstraction for the UML meta model. AndroMDA has an out of the box platform support. For openArchitectureWare the community provides platform support.

A. UML alike Meta Models

openArchitectureWare supports a couple of meta models and is shipped with a so called "classic" meta model which is a simplified version of the UML meta model implemented in Java. The developed templates for code generation contain less filter expressions and navigation logic using the simplified meta model instead of the normalized UML meta model. To use the UML profile mechanism the meta model must be extended with implementations of the stereotypes and tagged values. Other meta model extensions can be done in a non intrusive way using the functional Xtend language [17] of openArchitectureWare. The "classic" meta model is still supported but the future is unclear.

The latter tool AndroMDA also contains a special implementation of the UML meta model written in Java. AndroMDA uses the Velocity template language[12] to generate code from a Java model. A template set which generates code is called *cartridge*. For meta model adjustments, AndroMDA has the concept of so called *Metafacades*. These are Java interfaces and classes which provide the special view to the UML model which is needed by the templates. They provide only the information of the UML model needed by a special cartridge. Usually every cartridge has its own meta facades. The meta facades of a cartridge use the UML model directly or meta facades of other cartridges. So developers who create or maintain a cartridge usually have to work with the UML meta model. Model transformations can also be implemented by these meta facades. Currently everything has to be implemented using Java, which is sometimes a bit intricately, e.g. iterating over the model or calculating the full qualified name of a model element. For such calculations, a functional

[10] WYSIWYG = What You See Is What You Get
[11] MVC = Model View Controller, Design Pattern

[12] The Velocity Template Language, http://velocity.apache.org/

language like openArchitectureWare's Xtend language is more suited, because it is more expressive for this kind of problems.

B. Predefined Template Sets using UML alike meta models

AndroMDA is shipped with existing cartridges to generate different kind of output, e.g. Java, EJB or configuration files for the Spring Framework[13]. According to [4], one important aspect in code generation is to generate clean code. It means that the generated code should be as well structured as manual written code. This implies the copyright in each file, the well formatted structure of the source code, and the conformance to a style guide and naming conventions. With AndroMDA this can be achieved by adjusting the meta facades.

openArchitectureWare is not shipped with any existing cartridges to generate code out of the box. But there is the *fornax platform* [18], a community project aiming to build cartridges for openArchitectureWare. There exist cartridges which are based on the Eclipse UML2 implementation of the UML meta model and support e.g. EJB, the Spring framework or Grails. To customize the cartridges the aspect oriented programming (AOP) feature of the openArchitectureWare languages XPand [19] and Xtend [17] can be used.

C. DSL based approaches

An alternative to UML are DSLs. An open source approach using a DSL is the *Sculptor* [20] project, which can be found on the fornax platform. It is based on openArchitectureWare and can be used to generate web applications based on the Spring Framework, Spring Web Flow[14], Hibernate, Java Server Faces (JSF)[15] and JavaEE (Java Enterprise Edition). Sculptor uses openArchitectureWare's xtext language [21] to describe the DSL used to model the application. A commercial product for DSL based approaches is e.g. MetaCase [22].

Compared to the GeneSEZ approach AndroMDA and the fornax cartridges for openArchitectureWare uses the UML meta model directly in their transformation chains. This turns the creation and maintenance of transformation chains into a complex and tedious task. With sculptor this task is easier but the transformation chains are tied to the domain specific meta model and so the reusability is restricted to the domain. The GeneSEZ meta model has both advantages: easy creation and maintenance of transformation chains which are only tied to the domain of object oriented software. The customization of the transformation chains in AndroMDA is done in a tightly coupled way compared to the non-intrusive way the GeneSEZ approach used, offered by openArchitectureWare with AOP.

IX. CONCLUSION

There was a big effort to establish the standardized graphical notation UML for describing object oriented software and the use of standards is always preferable. Because its complexity the UML isn't an appropriate meta model for transformation chains. Therefore a simpler meta model is introduced, which can be seen as a meta model of a DSL for code generation. With the concept of model adapters, application models can be created with a standardized modeling language like UML with a reduced effort to write transformation chains. If DSLs are preferred all transformation chains based on GeneSEZ models can be reused by providing an appropriate model adapter. The creation of a model adapter will usually takes less time than rewriting all the transformation chains.

The proposed meta model has still some shortcomings. First it covers only some structural aspects of UML class diagrams and no behaviors. Further, there are still some missing constructs, e.g. inner classes. And the definition of generics is not yet complete.

Our approach has been successfully applied in industrial software projects of our research group: code generation for embedded systems based on Java, for software written in C#, and for web applications based on PHP. Currently the transformation chains to support EJB are developed. The main future aspect is the support of behaviors to support UML state machines.

[13] A popular Java Application Framework, http://springframework.org/

[14] A Spring web framework, http://www.springframework.org/webflow

[15] A web UI framework, http://java.sun.com/javaee/javaserverfaces/.

REFERENCES

[1] Usage of openArchitectureWare in the software industry, http://www.openarchitectureware.org/index.php?topic=success

[2] Usage of AndroMDA in the software projects, http://galaxy.andromda.org/index.php?option=com_content&task=blogcategory&id=26&Itemid=40

[3] T. Stahl, M. Völter, Model-Driven Software Development - Technology, Engineering, Management. Wiley, 2006.

[4] Sven Efftinge, Peter Friese, Jan Köhnlein, Best Practices for Model-Driven Software Development, Article on InfoQ, June 25, 2008, http://www.infoq.com/articles/model-driven-dev-best-practices

[5] Dave Thomas: "Unified or Universal Modeling Language?", in Journal of Object Technology, vol. 2, no. 1, January-February 2003, pp. 7-12, http://www.jot.fm/issues/issue_2003_01/column1/column1

[6] Martin Fowler, Language Workbenches and Model Driven Architecture, 06/2005, http://martinfowler.com/articles/mdaLanguageWorkbench.html

[7] Krzysztof Czarnecki, Overview of Generative Software Development, in UPP, pp. 326-341, 2004

[8] openArchitectureWare, http://openarchitectureware.org/

[9] Unified Modeling Language, http://www.omg.org/spec/UML/2.1.2/

[10] Eclipse UML2 project, http://www.eclipse.org/modeling/mdt/?project=uml2

[11] Eclipse Modeling Framework, http://www.eclipse.org/modeling/emf/

[12] Ecore, meta model of EMF, http://download.eclipse.org/modeling/emf/emf/javadoc?org/eclipse/emf/ecore/package-summary.html

[13] OMG, Meta Object Facility (MOF), http://www.omg.org/spec/MOF/2.0/

[14] The Eclipse Modeling Project, http://www.eclipse.org/modeling/

[15] F. Budinsky, D. Steinberg, E. Merks, R. Ellersick, T. J. Grose, Eclipse Modeling Framework - A Developer's Guide, Pearson Education, 2004

[16] AndroMDA, http://andromda.org/

[17] openArchitectureWare's functional expression language Xtend, http://www.eclipse.org/gmt/oaw/doc/4.2/html/contents/core_reference.html#Xtend

[18] The Fornax Platform, http://fornax-platform.org/

[19] openArchitectureWare's template language XPand, http://www.eclipse.org/gmt/oaw/doc/4.2/html/contents/core_reference.html#xpand_reference_introduction

[20] Sculptor, a DSL based approach to MDSD, http://www.fornax-platform.org/cp/display/fornax/Sculptor+(CSC)

[21] openArchitectureWare's xtext language for creating textual DSLs, http://www.eclipse.org/gmt/oaw/doc/4.3/html/contents/xtext_reference.html

[22] MetaCase, a company selling tools to build DSL based solutions, http://www.metacase.com/

JADE: A Graphical Tool for Fast Development of Imaging Applications

J. A. Chávez-Aragón[1], L. Flores-Pulido[1,2], E. A. Portilla-Flores[3], O. Starostenko[2], G. Rodríguez-Gómez[4]

[1]Universidad Autónoma de Tlaxcala, Facultad de Ciencias Básicas, Ingeniería y Tecnología
Laboratorio de Tecnologías Visuales Calzada Apizaquito s/n, Km. 1.5,
Apizaco, Tlaxcala, México, C.P. 90300.
e-mail: albertochz@gmail.com[1], aicitel_tryn@yahoo.com.mx[1,2]

[2]Universidad de las Américas Puebla, Escuela de Ingeniería y Ciencias,
Sta. Catarina Mártir Cholula, Puebla, México C.P. 72820.
e-mail: oleg.starostenko@udlap.mx

[3]Instituto CIDETEC - Instituto Politécnico Nacional, Departamento de Posgrado,
Area de Mecatrónica, Av. Juan de Dios Batiz s/n, Esq. Miguel Othon de Mendizabal,
Unidad Profesional Adolfo López Mateos, México D.F. C.P. 07700.
e-mail: eportilla@ipn.mx

[4]Instituto Nacional de Astrofísica, Óptica y Electrónica, Departamento de Ciencias Computacionales
Luis Enrique Erro 1. Santa Maria Tonatzintla, Apdo. Postal 51 y 216, Puebla. México.
e-mail: grodrig@ccc.inaoep.mx

***Abstract*—This paper presents a novel graphic tool to develop imaging applications. Users interact with this tool by means of constructing a DIP graph, which is a series of nodes and edges indicating the processing flow of the images to be analyzed. Solutions created using our tool can run inside the developing environment and also we can get the equivalent Java source code; so that, we can reused the code in other platforms. Another advantage of our software tool is the fact that users can easily propose and construct new algorithms following the Java beans rules. Our proposal can be seen as a DIP compiler because our tool produces fullfunctional Java programs that can solve an specific problem. The program specification is not a text based one, but a graphic specification and that is one of the main contributions of this work.**

Index Terms— **Graphical user interfaces, Image analysis, Image processing, JADE.**

I. Introduction

IMAGE processing technology is present more and more in industrial, educational and entertainment products. Thus, there is an increasing necessity of new software tools for the development of imaging applications. The creation of these solutions has not been changed in the last years. That means that, the process continuing being the same. Programmers built a series of libraries and use them to construct their solutions. Different implementations of the same visual operator can produce complete different results due to technology used and programming technique. In order to reduce the former problem we proposed an integral imaging application development environment. Among the outline advantages of our proposal we can mention: fast prototyping based on a graph solution, environment independency, automatic code generation and reuse of software in stand alone applications. This paper is organized as follows: section II presents related works analyzing advantages and disadvantages of similar proposals. Section III details the graph construction process. The transformation stage from PDI graph to JAVA source code is explained in section IV, as well as the results and evaluation of system which is called JADE. JADE stands for JAva imaging software Development Environment. Finally, in the last section we present concluded remarks and future work.

II. Related Work

In this section two well-known related works are analyzed. The first tool is called Neatvision© [1]. Neatvision is a free java development environment for designing PDI graphs. NeatVision [1] provides an intuitive interface that make use of block diagrams. Construction of that diagrams follows the drag & drop interaction style. Each block which represents: images, operations and flow control structures, could be connected to others in order to form DIP graphs. Neatvision graphs represent prototyping imaging applications. User can observe the results of his/her product in the Neatvision environment. Nevertheless, it is not possible to obtain this code to be used in other applications. Perhaps this is the main

Manuscript received October 15, 2008. This work was supported in part by the Universidad Autónoma de Tlaxcala, Universidad de las Américas Puebla, Instituto Politécnico Nacional, Instituto Nacional de Astrofísica, Óptica y Electrónica.

K. Elleithy (ed.), *Advanced Techniques in Computing Sciences and Software Engineering*,
DOI 10.1007/978-90-481-3660-5_69, © Springer Science+Business Media B.V. 2010

disadvantage of this work. Additionally, Neatvision provides a way for building user's operators, but the form of producing new components it is not intuitive, as a result, the complexity for new component implementation increases drastically.

Another work in this area is WiT©. WiT [5] is a platform of visual design for prototyping programs on DIP. This tool is handled in visual environment easy to use, each algorithm is defined by a block. Image algorithms are placed in an accessible library in a drag & drop form [6], [7], [8], [9], [10]. These process is made trough a graph. WiT environment allows incorporating new algorithms by means of an integration tool. This is made with tool "WiTEngine". We can conclude that the version WiT Demo is a tool that fulfills basic requirements of DIP. However, WiT is not independent of its environment and code generation is not provided.

III. OUR PROPOSAL: JADE

The tools analyzed previously offer different options to create solutions for the DIP problem. However, none of them allow us to obtain the source code to be used in other applications. In JADE we combine characteristic of different tools to create a visual environment for Digital Processing of Images (DIP). One of the main contributions of our application is the Java code generation. Former function was developed taking into account the lack of tools that contain this functionality. Following sub-section explains the concepts of DIP graph that is the a main concept in our proposal since a DIP graph is the input specification for our tool.

A. DIP Graph

Definition. **DIP** Graph is a directed graph, each node in the graph represents an operator of DIP or a supported variable. Each directed arch represents transference of data among nodes. The supported variables represent primitive types that can be transferred. So we can says that JADE is a visual environment based on components that offer support for the creation of DIP Graphs. It is necessary to mention that if each node is consider as a variable or as a DIP operator, the rules for the creation of a DIP graph and the DIP graph by itself can be seen as a visual programming language. Based on the input DIP graph the application can carry out two possible tasks:

1. To show the behavior of the DIP graph by means of an execution process inside the development environment.
2. To generate a file of JAVA source code equivalent to the input DIP graph.

B. DIP Component

Definition. **A** DIP Component ([2], [3], [4]) is a graphic representation of a Digital Image Processing program that follows certain convention of notation. A Component DIP can represent a variable or an operator and it can have *n* ports of entrances and *n* output ports. There is at least one entrance port or one output port. The entrance ports are distributed in the left side of the component while the exit ports are distributed in the right side.

C. Generation of Imaging Application Source Code

JADE has the ability to transform a DIP graph into a Java class that observes the same behavior of the DIP graph when this graph is executed in the environment. The class generated by JADE is based on the following structure:

```
<imports>
public class <nombreClase>{
<declaracionVariables>
public<nombreClase>{
<CreacionInstancias>

  nombreClase>{<CreacionInstancias>
   }
   public void ProcesarDatos(){
   <ModificacionDePropiedades>
   <TransferenciasDeDatos>
       }
}
```

Table 1. Structure of a DIP Graph java class.

Each node (DIP component) is already implemented. So, it is necessary to indicate the path of that pieces of software, *<imports>*ψsentences do this work. As a result, we will be able to instance any component inside the class. The method $ProcesarDatos\leftarrow\leftarrow$ in the structure of the class model is a representation in java code of the DIP graph behavior.

D. System Architecture

The system architecture is divided into different modules. We use the View-Controlled Model (VCM) [13], [14] that allows us to isolate the logical applicative from the GUI. Figure 1 shows a block diagram of our system.

The architecture shows the separation among the pattern, the view and the controller. Using VCM is easier to substitute new modules and also to do some maintenance work. Principal modules that make up the systems are listed above.

1. Module for creation and supervision of DIP graphs
2. Module of JAVA code generation
3. Module for supervision of components
4. Module for instance administration
5. Module for execution of DIP graphs

Creation and Supervision Modules for DIP Graphs

This module is in charge of the process of creating a DIP graph which involves dragging components from the component tree to the workspace by means of the Drag & Drop paradigm.

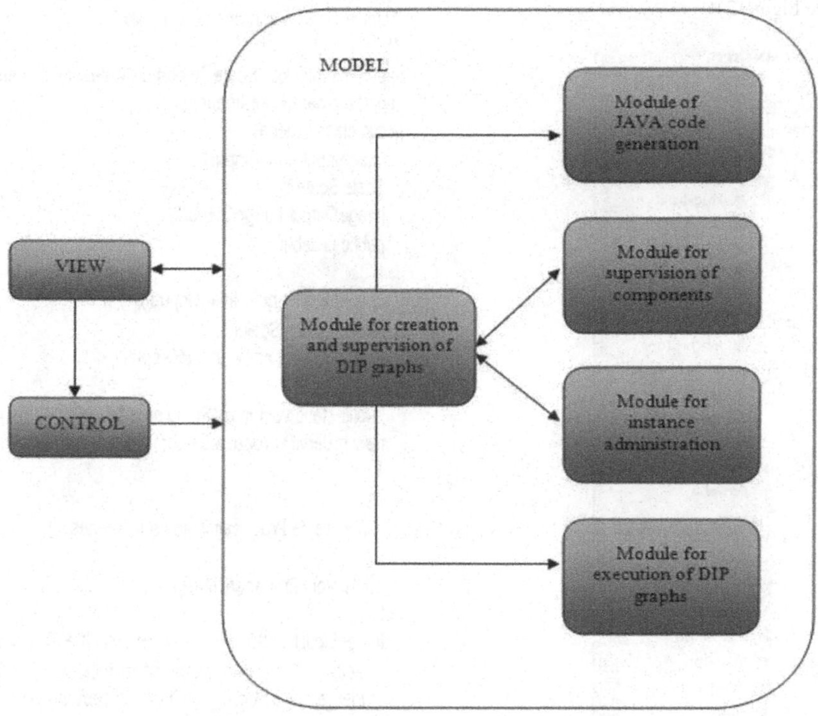

Fig. 1. JADE Block Diagram.

Each component has a set of features that are visible in their graphical representation.

Component Inspection Module

The Component Inspection Module makes different functions, loads the components available in the environment and obtains features of them by means of a process of introspection of JavaBeans. This module has a high interaction with the Management Instance Module and also with the Creation and Supervision of DIP graphs Module.

Management Instance Module

The management instance module is in charge of the creation of new instances of the DIP graphs components.

Java Source Code Generation Module

The Java Source Code Generation Module works once a DIP graph has been constructed, first creates an XML file intermediate representation of DIP graph and then transforms the specification into a Java file that has the same behavior of the DIP Graph.

Graph Execution Module

It runs each algorithm that is represented as a vertex in the DIP graph.

IV. RESULTS

We present the results that we had in the creation of components and later we show the source code generation:

A. Visual Environment

JADE visual environment has an intuitive way to work. It is based on the Drag & Drop paradigm. User can build his/her application using those components contained in the component area generating the DIP graph which is the visual specification of a program that solve an specific problem.

B. Component Creation Support

This tool permits the creation of new components. It is adaptable to users with experience in the field of DIP programming, since it allows us to implement our own algorithms as visual components. Then, we will be able to test their operation by means of the construction of DIP graphs in combination with other components that already exist. Also this tool provides to novice users the possibility of creating graphs with the components available and verifying the own processes of DIP (see Figure 2).

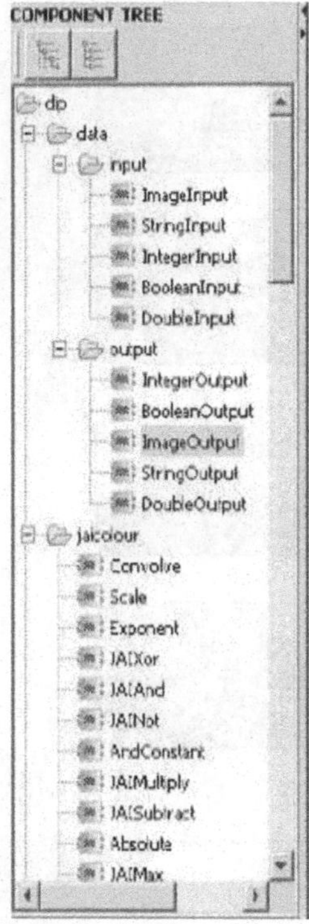

Fig. 2. Visual components of DIP graphs.

C. Code Generation

This tool is able to generate a Java class on the basis of a DIP graph. The generated classes have the same behavior that DIP graph when it is executed within JADE. The classes generated by the system are based on the internal bookstore of components. The generated code is a solution of high level DIP problem.

D. Graph Development

Next we show an example of the implementation of a DIP graph in our system. We have divided this section into steps that orient the reader in the process of builting a DIP graphs in JADE.

Fig. 3. A DIP Graph Example.

```
/**
 * Código generado por JADE
 *(Java Imaging Software Development Environment)
 * Universidad Autónoma de Tlaxcala
 **/
import dip.jaicolour.Scale; import dip.data.output.ImageOutput;
import dip.data.input.ImageInput;
public class prueba{
  ImageInput ImageInput1;
  Scale Scale2;
  ImageOutput ImageOutput3;
  public prueba()
  {
  ImageInput1 =new ImageInput();
  Scale2 =new Scale();
  ImageOutput3 =new ImageOutput();
  }
  public static void main(String args[]){
  new prueba().ProcesarDatos();
  }
  /**
   * Tareas de Procesamiento de Imágenes.
   */
  public void ProcesarDatos(){
  try{
  ImageInput1.setUrl(new java.net.URL("file:/C:/Documents and
     Settings/Oliver/Mis documentos/Buck/CORE 2007_2_archivos
     /der_archivos/core_logo2007%20copy.jpg"));
  Scale2.setScale(2.0F);
  ImageInput1.Ejecutar();
  Scale2.Ejecutar(ImageInput1.getImage());
  ImageOutput3.Ejecutar(Scale2.getImagenResultante());
  /*Desplegando resultados en Frames*/
  javax.swing.JFrame frameImageOutput3 = new javax.swing.JFrame();
  javax.media.jai.PlanarImage imageImageOutput3=
                ImageOutput3.ObtenerImagenMostrable();
  com.sun.media.jai.widget.DisplayJAI djImageOutput3=
    new com.sun.media.jai.widget.DisplayJAI(imageImageOutput3);
  frameImageOutput3.getContentPane().add(djImageOutput3);
  frameImageOutput3.pack();
  frameImageOutput3.show();
  frameImageOutput3.setTitle("ImageOutput3");
  frameImageOutput3.setDefaultCloseOperation(javax.swing.JFrame.
                     EXIT_ON_CLOSE);
  }catch(Exception ex){
    System.out.println("Error: "+ex.toString());
    }
  }
}
```

Fig. 4. Java Source Code generated by JADE

E. Graph Construction

After initiating the application, in the properties (Figure 2) is shown the available components and they are contained in folders. We can create some new components with the intention to solve a particular problem. After creating a new workspace, we dragged the components into the workspace from the tree of components. Then the components are connected using the input and output ports. Connections are indicated by arrows which show the data flow into the DIP graph as is shown in Figure 3.

F. Graph Execution

Once DIP Graph is constructed it is possible to use the tool to execute the program and the system automatically generates the output the user expected.

G. Code Generation from a DIP Graph

We concluded showing the results that the module of generation of code produced, Figure 4. As you can see, the structure is based on the set of rules mentioned previously.

V. CONCLUSIONS

Visual programming tools for Digital Image Processing are very helpful for programmers. These kind of tools provide an environment to develop visual programs that includes operators of DIP, flow control structures and utilities tailored to the imaging applications area. Therefore, it is possible to design complex applications very quickly connecting components. There already exists systems like the ones described above; however, most of time these systems do not allow us to get the source code from the area of development.

They just permit run the applications inside the workspace area. That is why we propose a new development tool that overcome that problems. Among the outline advantages of our proposal we can mention: fast prototyping based on a graph solution, environment independency, automatic code generation and reuse of software in stand alone applications. In order to test our proposal we introduced in this paper a system called JADE. JADE stands for JAva imaging software Development Environment.

JADE GUI (see Figure 5) offers a programming tool for the DIP problem in a visual environment as well as the advantages that represents the source code generation for external use.

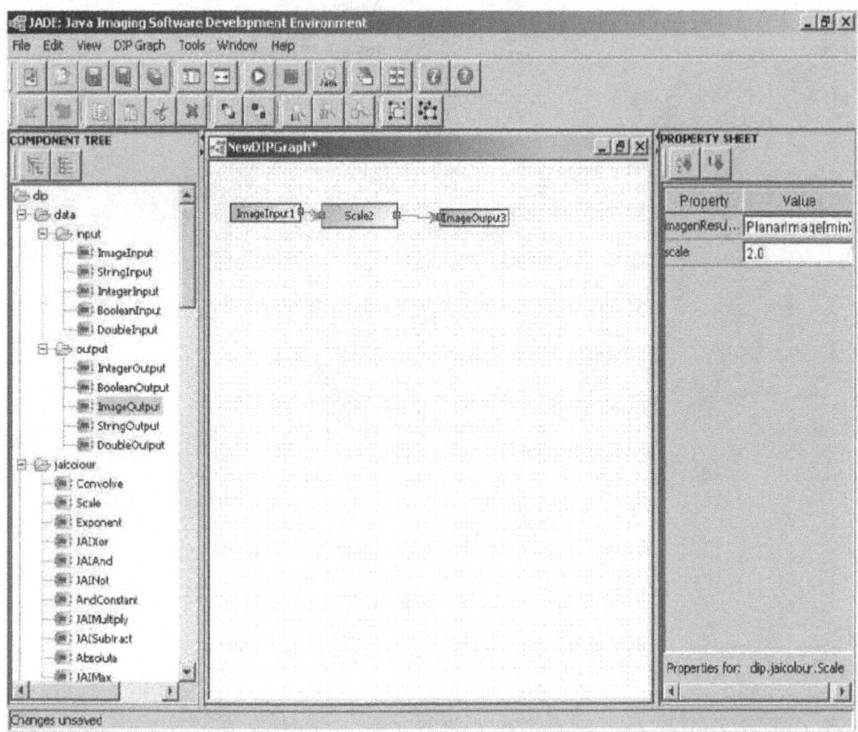

Fig. 5. JADE GUI showing a DIP graph.

A user can work in JADE at different levels:

1. using the visual paradigm for creating graphs by means of drag and drop components available in the environment and test their behavior.
2. adding new components to JADE and test their operation in combination with those already available.
3. users can also generate their solutions in JAVA code avoiding tedious process o writting a program in a traditional compiler.

Additionally, JADE permits the creation of new components expanding this way the capacities of the system. Moreover, JADE has a potential application in the academic environment, because when a student begins an image processing courses the knowledge about the API Java and Advanced Imaging (JAI) is limited. JADE overcome this problem. However, JADE has also some limitations. At this time, there is not support for complex data flow and cycles. The programming concept is limited due to the construction of acyclic graphs. Consequently, code generated by JADE is limited the characteristics mentioned above. Generated classes describe a sequential execution of DIP operators. All this problems can be solved in the next version of JADE.

REFERENCES

[1] An Image Analysis & Software Development Environment. NeatVision. Available, http://www.neatvision.com.2005.

[2] Espeso M., P. J. Diseño de componentes Software de tiempo real (2002) España, 2002, pp 1-3.

[3] Bramley A., Fox G., Programming the Grid: Distributed Software Components, P2P and Grid Web Services for Scientific Applications, Journal of Cluster Computing, 2002, Vo. 5, pp. 325-336.

[4] Mohammad Akif. "Java y XML: Referencia para el programador". Anaya Multi- media.2001. CORECO ImagingInc.

[5] WiTDemo. Available: http://www.wit-igraph.com , 2005.

[6] JavaBeans API Specification. Available en http://java.sun.com/beans. 1998.

[7] The Java Tutorial, a practical guide for programmers.Available http://www.sun.com.2005.

[8] P.G.CableLaurence. "Drag and Drop subsystem for the Java Foundation Class- es". Available http://java.sun.com/2000.

[9] P.F. Whelan and R.J.T. Sadleir (2004), "A Visual Programming Environment for Machine Vision Engineers", Sensor Review, 24(3), pp 265-270.

[10] P.F. Whelan and R.J.T. Sadleir, (2004) "NeatVision - Visual Programming for Computer Aided Diagnostic Applications" Radiographics, Nov 2004

[11] Robert J. T. Sadleir, Paul F. Whelan, Padraic MacMathuna, and Helen M. Fenlon (2004) "Informatics in Radiology (infoRAD): Portable Toolkit for Providing Straightforward Access to Medical Image Data" Radiographics. 2004 Jul-Aug:24(4):1193-1202.

[12] P.F. Whelan and D. Molloy (2000), Machine Vision Algorithms in Java: Techniques and Implementation, Springer (London), 298 Pages. ISBN 1-85233-218-2

[13] Ingeniería de Software (sexta edición), Ian Sommerville. Addison Wesley.

[14] Reenskaug, The model-view-controller(mvc) its past and present and mvc pattern language, 2003.

A View on Power Efficiency of Multimedia Mobile Applications

Marius Marcu, Dacian Tudor, Sebastian Fuicu
"Politehnica" University of Timisoara, Computer Science and Engineering Department
2 V. Parvan, Timisoara, Romania

Abstract - *Multimedia applications running on mobile devices and using wireless networks become more and more demanding with respect to processing speed and power consumption. Power saving is one of the most important feature that network interface and mobile CPU must provide in order to prolong battery lifetime. The multitude and complexity of devices that implement a large spectrum of multimedia and wireless protocols requires closer evaluation and understanding in respect to power efficiency. As there is very little analysis on the relationship between multimedia applications and their power profile in the context of mobile wireless devices, we investigate the landscape of multimedia communication efficiency methods. The proposed evaluation scenarios address CPU power consumption, wireless communication power consumption and video and sound interface power consumption..*

Keywords - ***power consumption, energy efficiency, multimedia applications, wireless communication***

I. Introduction

Recent development in areas like multimedia mobile systems and wireless communications as well as the trend to incorporate more functionalities (such as WLAN, Bluetooth, GPS, multimedia, VoIP) have led to the their acceptance in almost every domain of activity. The evolution of portable and mobile computation systems towards an increased feature set and increasing hardware and software requirements demand, is raising complex problems from a reasonable energy consumption level point of view under different usage scenarios [1]. The element that has underlined the development of this domain was the reduction of the commercialization price of intelligent mobile systems (PocketPC, SmartPhone, PDA and even notebooks) together with their increase in performance and the pronounced development of wireless communication technologies (WLAN, WiMax, VoIP) [2]. In the near future, portable embedded devices have to run multimedia applications with large computational requirements at low energy consumption [3]. These applications demand extensive CPU power, memory capacity and wireless communication bandwidth. All these demands bring an increasing power consumption.

The power consumption problem of computing systems is in general a very complex one [4] because each physical component from the system has its own consumption values depending especially on the execution operation type, so that together with the physical components, the software applications has a big influence on the energy consumption [5]. The proportional ratio of a system consumption between the physical components vary a lot, so according to the results presented in Fig. 1 for a PocketPC [6], we can observe the consumption distribution between different types of circuits: processor, memory, communication, display, etc. There are some substantial consumption differences at GSM communication, TCP/IP communication over WiFI or Bluetooth communication, which show that certain types of wireless communication still do not fit well with current mobile systems or they are not optimized for thsese systems [6, 7].

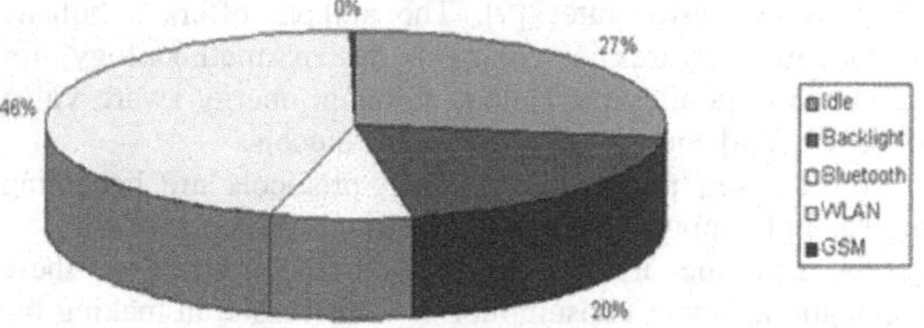

Fig. 1 Power consumption distribution for a mobile device [6]

One important fact is that the energy demands of battery powered devices and their applications substantially exceed the capacity of the actual batteries that power them [8]. To respond to this negative characteristic, development of new architectures and design methodologies for battery-efficient, thermal-aware and power-aware systems have been pursued. In order to implement power-aware applications, their power profiles have to be studied.

The main goal of our work is to design, implement and validate a software framework for power-aware mobile applications in order to reduce overall power consumption and increase the efficiency of the energy usage. In order to achieve this goal we have to study and profile each component of a mobile device in order to introduce these profiles in the final framework. In this paper we present our results of the multimedia power efficiency study on different battery powered devices, the metrics used for multimedia and wireless energy-efficient communication and the relations between different types of communication patterns and the proposed metrics. The remainder of this paper contains in the next section a summary on current advances in multimedia and wireless power efficiency analysis and measurements. Power consumption measurement, energy efficiency metrics and the

K. Elleithy (ed.), *Advanced Techniques in Computing Sciences and Software Engineering*,
DOI 10.1007/978-90-481-3660-5_70, © Springer Science+Business Media B.V. 2010

monitoring framework is presented in section III. Section IV contains power profiles for different multimedia applications and their energy-efficiency metrics obtained in our experiments. The conclusions and future work are presented in the last section.

II. Multimedia and Wireless Communication Evaluation Methods

Multimedia systems represent a very special class of complex computing systems. Multimedia mobile devices design process should start by taking into consideration their unique characteristics which are dominated by the huge amount of data that needs to be processed and transmitted in a continuous manner, and the timing constraints that need to be satisfied in order to have an informational message meaningful to the end-user [9]. For example, due to the large amount of data that needs to be processed, the video streams require consistently high throughput, but can tolerate reasonable levels of jitter and packet errors. In contrast, the audio applications manipulate a much smaller volume of data (therefore do not require such a high bandwidth), but place tighter constraints on jitter and error rates [9]. The authors offers a holistic perspective towards a coherent design methodology for multimedia mobile applications through: energy-aware video streaming and energy-aware routing protocols.

WiFi based phones using VoIP protocols are becoming increasingly popular due to the ubiquitous presence of wireless LANs and the use of unlicensed spectrum. For these applications power consumption is a vital issue in making the usage of these phones widespread for a long period of time. Efforts are underway in improving power conservation in these phones and thus increasing the duration between recharging the battery [10]. The authors in [10] provide a detailed anatomy of the power consumption by various components of WiFi-based phones, they have analyzed the power consumption for various VoIP workloads at various components.

Many wireless network interfaces, especially wireless LAN cards, consume a significant amount of energy not only while sending and receiving data, but also when they are idle with their radios powered up and able to communicate. Consequently, there has been much research in the last years into how to lower the energy consumption of mobile devices. Researchers have focused on developing low-power techniques at all levels, from designing energy-efficient circuits to managing power dynamically to adapting CPU frequencies to enhancing and modifying network protocols [11].

In [12], the authors investigate per packet energy consumption of an IEEE 802.11b card in various modes: transmit, receive, broadcast, and idle modes. Their data reveals how complexities introduced by the IEEE 802.11 protocol can impact the overall energy consumed by the card during its operation. The IEEE 802.11 WLAN specification describes a power-saving mode (PSM) that periodically turns the network interface off to save energy, and on to communicate [13]. In the infrastructure mode a mobile device using PSM communicates with a wired access point (AP), the AP buffers data destined for the device. Once every beacon period, the AP sends a beacon containing a traffic indication map that indicates whether or not the mobile device has any data waiting for it. The mobile device wakes up to listen to beacons at a fixed frequency and polls the AP to receive any buffered data. Significant reduction in consumption is obtained using this static PSM method [13].

One very popular application on a battery powered mobile devices is multimedia content visualization over a wireless link. The authors in [14] evaluate the impact on three encoding parameters on the CPU cycles during the decoding process. A total cost in terms of CPU cycles is introduced and a relationship to power consumption is suggested, but power consumption measurements are not performed, thus the cost model cannot be verified in practice.

A more generic approach towards application-level prediction of power consumption and battery dissipation has been taken in [15], where an estimation methodology is introduced. Based on a set of drain-rate curves from a set of benchmarks, estimation for an arbitrary program is composed. The estimation is limited to CPU and memory dimensions and it does not take into account the display, wireless communication or I/O components, but could be extended to include wireless basic benchmarks to assess wireless based communication applications.

One of the implications of the application move towards unified communication systems is that most of them require secure communications that are provided through cryptographic components realized in both hardware and software. As one might suspect, there should be a link between security protocols and energy consumption on battery powered embedded systems. The authors of [16] have conducted thorough measurements and proved that security processing has a significant impact on battery life.

Some proof on the increased usage of wireless mobile devices has been reported in [17], where wireless data transfer and battery lifetime for Wi-Fi connected mobile terminals has been logged in several networks in the US. It is striking that 99% of the survey participants were under cellular networks while 49% were under Wi-Fi networks on everyday life. Based on application usage type (email, web feeds, web browsing and multimedia) and usage interval time, a relationship between data-size transfer, transfer interval and battery life is presented from the obtained logged data.

VoIP usage is evolving for mobile devices such as smartphones with multiple wireless interfaces. However, the high energy consumption, presents a significant barrier to the widespread adoption of VoIP over Wi-Fi. To address this issue in [18] is presented a practical and deployable energy management architecture that leverages the cellular radio on a smartphone to implement wakeup for the high-energy consumption Wi-Fi radio.

III. POWER CONSUMPTION MONITORING FRAMEWORK

A. *Power consumption metrics*

The monitoring application we used in our tests to evaluate application level power efficiency is based on the power benchmark concept we defined in [19]. The power benchmark is an extension of the benchmarking concept by applying power consumption metrics. Our proposed power benchmark software is based on the information measured or estimated by the battery driver or the operating system. Battery status is based on several parameters: battery capacity [mWh], maximum battery capacity [mWh], charge/discharge rate [mW], current drawing [mA], battery remaining life time [s], battery temperature [oC], etc. All these parameters provide an image over the application level power consumption.

As a battery discharges, its terminal voltage decreases from an initial voltage V_{max}, given by the open circuit voltage of the fully charged battery, to a cut-off voltage V_{cut}, the voltage at which the battery is considered discharged. Because of the chemical reactions within the cells, the capacity of a battery depends on the discharge conditions such as the magnitude of the current, the duration of the current, the allowable terminal voltage of the battery, temperature, and other factors. The efficiency of a battery is different at different discharge rates. When discharging at low rate, the battery's energy is delivered more efficiently than at higher discharge rates [20]. Battery *lifetime* under a certain load is given by the time taken for the battery terminal voltage to reach the cut-off voltage. The *standard capacity* of a battery is the charge that can be extracted from a battery when discharged under standard load conditions. The *available capacity* of a battery is the amount of charge that the battery delivers under an arbitrary load, and is usually used (along with battery life) as a metric for measuring battery efficiency [21].

Different power consumption metrics presented before can be used in benchmark, but these can be grouped in two classes:

- Power consumption metrics – battery discharge rate, current consumption;
- Remaining battery metrics – remaining life time, remaining battery capacity, etc.

For power efficiency we used energy consumed to send/ receive a bit of information or the number of information bits transferred between two battery capacity values.

B. *Wireless implications for TCP/IP*

Wireless links are an important part of the Internet today. In the future the number of wireless or mobile hosts is likely to exceed the number of fixed hosts in the Internet. Despite its increasing usage, TCP/IP stack implementations are not optimized for wireless particularities and power management aspects. For example, the Collision Detection mechanism is suitable on wired LAN, but it cannot be used on a WLAN for two reasons: a) implementing a collision detection mechanism would require the implementation of a full duplex radio; b) in wireless environments we cannot assume that all stations hear each other. Therefore 802.11 WLAN uses a Collision Avoidance (CA) mechanism together with a positive acknowledgement mechanism.

Errors rates on wireless links are a few orders of magnitudes higher then on fixed links. These errors cannot always be compensated by the Data Link Layer. The fundamental design problem for TCP is that a packet loss is always classified as a congestion problem. TCP cannot distinguish between a loss caused due to a transmission error and a loss caused by a network overload.

Therefore we can say that TCP/IP protocols stack implementations are not well optimized for wireless communication medium and management of power consumption.

C. *Monitoring application*

The general architecture of the application presented in figure 2 has a modular structure divided in several abstracting levels. On the low level of the framework application will use the operating system's drivers of different physical components took into account in the optimizing process of energy consumption: the processor, the battery, wireless chipset, main-board chipset, the memory etc. The kernel of the execution framework takes the available measures through the monitoring drivers, calculates the energy consumption of the used applications and through the application interface, it communicates using specific messages for consumption control.

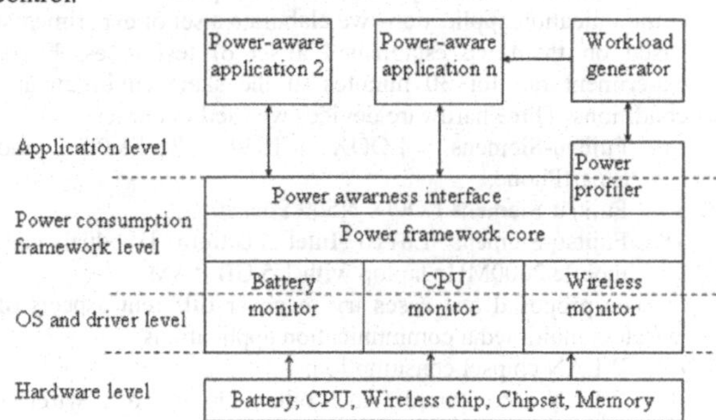

Fig. 2 Power framework architecture

In order to show how different types of application level patterns influence the power consumption of a mobile device at the application level, we implemented a prototype of the framework. The prototype was written in C++ using MS Visual Studio 2005. The execution framework prototype source code is portable so it was built and tested on different Microsoft Windows platforms: Win32, Window Mobile 5.0 PocketPC and Windows Mobile 5.0 Smartphone.

The framework application is composed from a number of specialized modules (Fig. 2):

- Battery monitor - is a software module running at OS and drivers level, used to achieve real-time on-line power consumption measures from battery device;
- CPU monitor - is a software module used to monitor CPU parameters such as load, temperature, etc.;

- Wireless monitor - is a software module implemented to monitor different parameters of wireless communication: signal power strength (RSSI), bandwidth, data transferred, etc.;
- other types of monitoring modules could also be implemented.
- Workload generator - Generates different types of workload patterns. For example it contains a traffic generator - client/ server classes used to generate TCP/UDP traffic on wireless network card. The generator could be configured to use different values for communication parameters: data block size, data transfer direction (download/ upload), transport protocol (TCP/ UDP), data patterns, etc.; CPU workload generator - run different types of CPU benchmarks.
- Power profiler - logging and profiling module to save all monitoring values from all modules for future analysis. This module is used in profiling power consumption of applications with respect to the used workload pattern.
- Power framework core and power framework API extract relevant monitoring data and provide these data to application level. These modules are used to implemented auto-adaptable mobile applications aware of their power consumption.

D. *Experimental testcases*

In order to evaluate power efficiency profiles for wireless communication applications we elaborate a set of experiments, based on them we established a set of test cases. Every experiment ran for 30 minutes in the same environmental conditions. Three hardware devices we used in our tests:

- Fujitsu-Siemens LOOX T830 PocketPC and SmartPhone;
- Fujitsu-Siemens LOOX N560 PocketPC;
- Fujitsu-Siemens E8110 Intel Pentium IV dual core mobile 2000MHz laptop with 1.5 GB RAM.

The proposed test cases try to cover different aspects of wireless multimedia communication applications:

- WLAN chipset consumption;
- WLAN compared to other kind of wireless communication standards;
- ad-hoc and infrastructure wireless communication;
- reliable (TCP) and non-reliable (UDP) wireless communication;
- wireless communication data patterns;
- wireless communication data block sizes;
- wireless security usage.

Other multimedia applications' power profiles were evaluated: video and sound players, multimedia instruction sets: FPU, SSE.

IV. ENERGY-EFFICIENCY CHARACTERIZATION OF MULTIMEDIA WIRELESS COMMUNICATION

We used the monitoring framework application to emphasis power consumption of wireless communication on mobile system related to different communication patterns.

A. *Wireless communication devices power consumption*

In order to measure the power consumption of different wireless communication devices existing in a mobile system we used the same LOOX T830 PocketPC. The power signatures obtained for this system when each of the following subsystems like GSM, Bluetooth and WLAN are respectively switched on are presented in Fig. 3. For each power signature in Fig. 3 the communication device was turned on but no communication traffic was initiated. It can be observed that the GSM communication hardware is very well designed for battery powered devices because its power consumption is quite the same with the GSM hardware deactivated. Bluetooth and WLAN hardware switched on imply more power consumption than with no communication hardware activated. Bluetooth device activation add around 13% to the total idle power consumption, and WLAN card activation consumes around 50% more than the idle state.

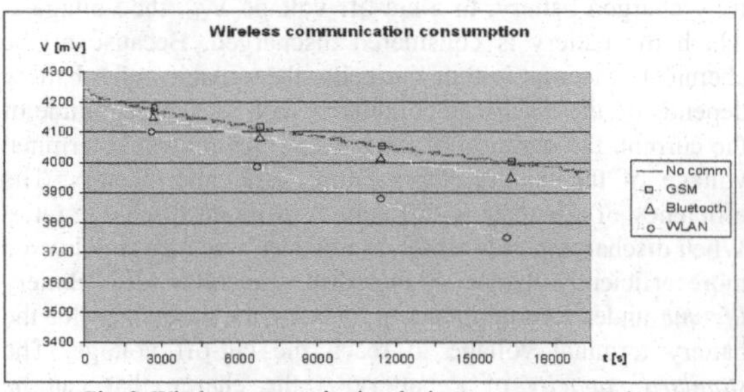

Fig. 3. Wireless communication devices power consumption

B. *WLAN chipset power consumption*

From Fig. 1 it can be easily observed that WLAN chipset and communication represents an important part from total power consumption of the device. In order to show the different aspect of WLAN chipset power consumption the following experiments were executed on a Loox N560 PocketPC:

- WLAN chipset off;
- WLAN chipset on and not connected to the network;
- WLAN chipset on and trying to connect to the network;
- WLAN chipset on and connected to the network;
- WLAN chipset on and transferring data on the network.

Related to the idle state power consumption with WLAN chipset switched off, WLAN chipset set on and WLAN communication show an increase in power consumption as presented in Fig. 4. WLAN chipset switched on has the same power consumption independent of the WLAN network status: not connected, trying to acquire network connection and connected.

Current consumption of the PocketPC when WLAN chipset is switched on is three times larger than the case WLAN is off. When communication is started over the wireless, the increase in current consumption is 6 times greater than the case WLAN is switched off. We can state that wireless communication has three states of power consumption observed at application level: WLAN-OFF power state; WLAN-IDLE power state; WLAN-COMM power state.

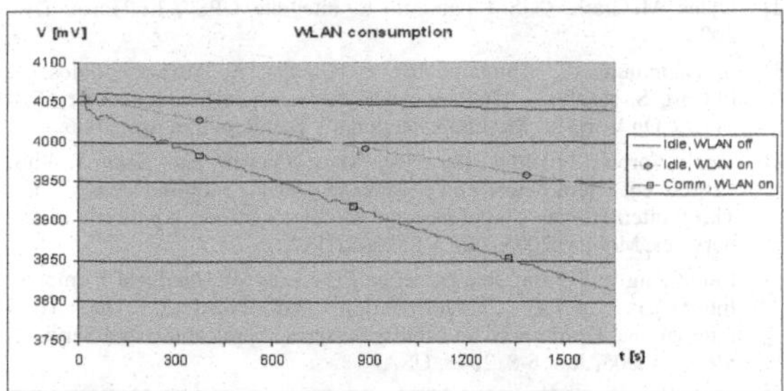

Fig. 4. WLAN chipset power consumption

C. WLAN communication power consumption

In this test case we tried to find out how different communication parameters influence the power consumption and energy efficiency. The first test was selected to emphasis the relationship between communication block sizes and power consumption and efficiency. This test involved two battery powered devices: a laptop and a PocketPC. The monitoring application was deployed and configured as server on the laptop and as client on the PocketPC. The monitoring application ran for 30 minutes for each test and in each test different block size transfers were effectuated: 1KB, 4 KB, 10 KB and 100KB. From our tests it can be observed that power consumption is the same no matter the size of blocks transferred (all plots are overlapped). Communication bandwidth depends on the blocks' size transferred therefore it can be observed that power efficiency of wireless communication is directly proportional with communication bandwidth (Fig. 5).

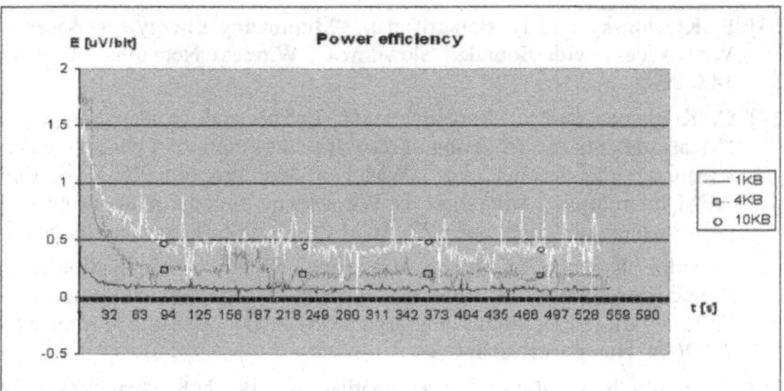

Fig. 5. Power efficiency for different transferred block sizes

D. WLAN security power consumption

Security is an important aspect of WLAN communication. Different encryption methods can be used but each of them has impact on overall power consumption. We show energy efficiency for no encryption case and WPA security usage (Fig. 6). An increase in power consumption due to CPU load increase is observed.

E. Multimedia CPU instruction set

Another aspect of multimedia application is related to power consumption of specific instruction CPU instruction set. We run the same workload implemented with ordinary FPU instructions and then with SSE instructions on the laptop device. (Fig. 7).

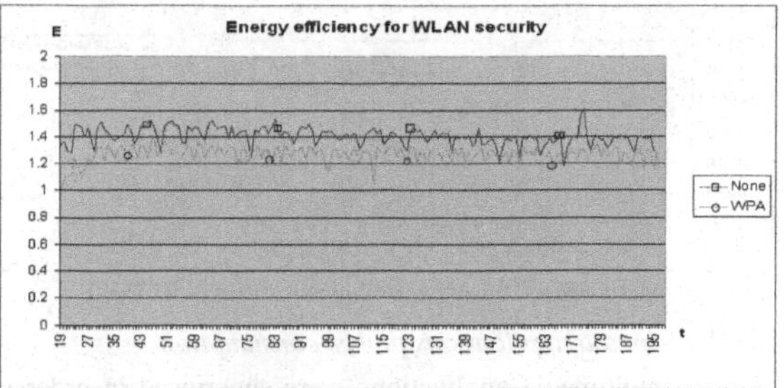

Fig. 6. Energy efficiency decrease when encryption is used

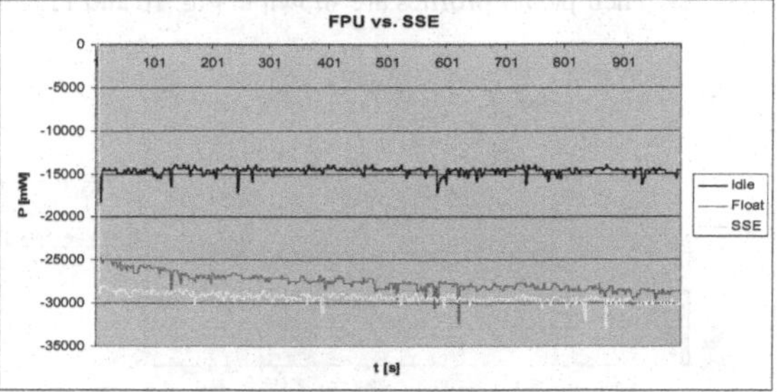

Fig. 7. Power consumption of FPU and SSE

F. Wireless multimedia communication power consumption

When wireless communication is performed more power is consumed than the case when only the communication hardware is turned on. Two types of wireless voice communication are presented in Fig. 8 and Fig. 9: GSM and WLAN VoIP using Skype application for laptop and pocketpc in tests. It can be observed that GSM communication is very well suited for mobile devices because when GSM chipset is switched on the total system power consumption does not increase and when a voice call over GSM consumes less energy than WLAN VoIP communication.

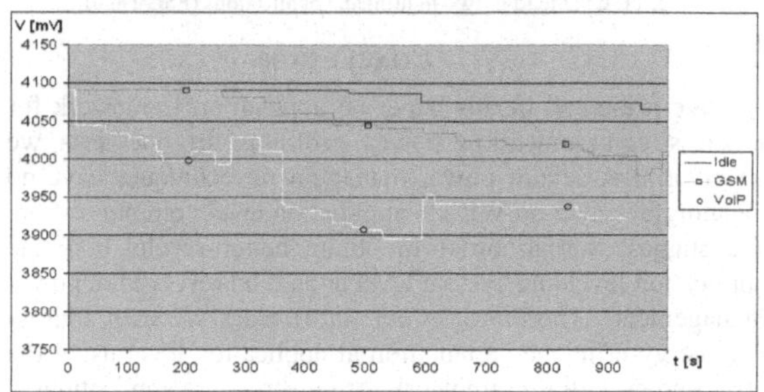

Fig. 8. GSM vs. WLAN voice communication (FSC E8110)

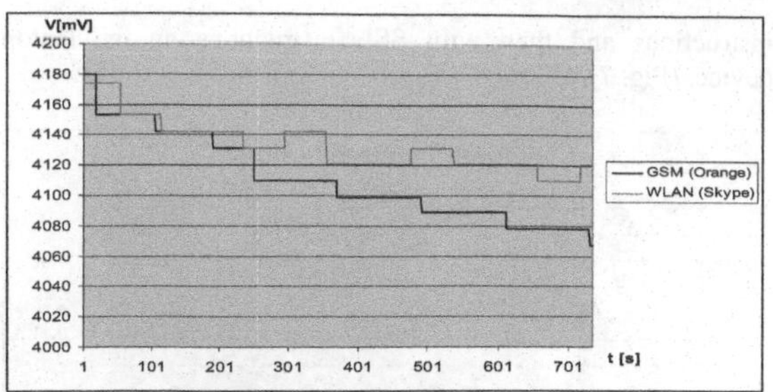

Fig. 9. GSM vs. WLAN voice communication (FSC T830)

G. *Multimedia applications power consumption*

Two multimedia applications were supervised in order to emphasis their power profiles: video (wmv) display and sound mp3 play. Their power profiles are shown in Fig. 10 and 11.

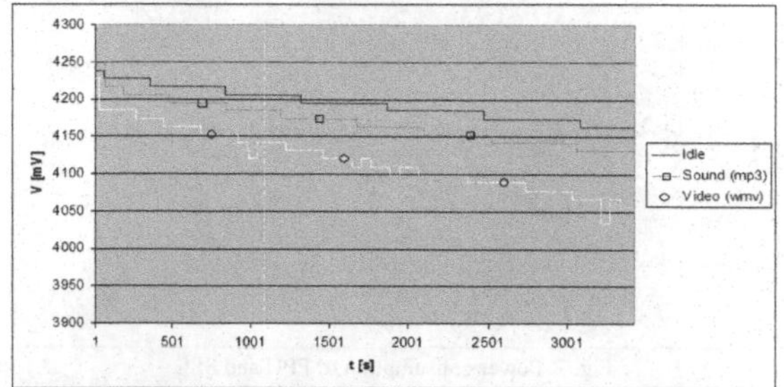

Fig. 10. Multimedia applications power consumption (FSC T830)

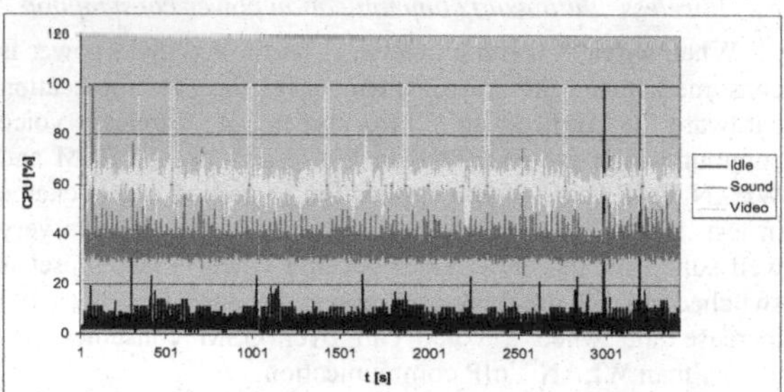

Fig. 11. CPU load of two multimedia applications (FSC T830)

V. Conclusions

We presented in this paper an application framework for wireless communication power profiling. In our tests we observed that default power management techniques have no meaningful effect on WLAN application level communication. We suggest that in order to obtain better results both the application level and system level should be involved in power management. Therefore, in our future work we shall address the energy-efficiency adaptation at application level using the proposed execution framework for power-aware applications.

References

[1] David Haskin, "Top Mobile and Wireless Trends for 2007", PCWorld, Dec. 2006.

[2] Dylan McGrath, "GSM handsets to integrate GPS", EETimes, Dec. 2007.

[3] S. Giannoulis, C. Antonopoulos, E. Topalis, A. Athanasopoulos, A. Prayati, S. Koubias, "TCP vs. UDP Performance Evaluation for CBR Traffic On Wireless Multihop Networks", TechRepublic Apr. 2006.

[4] Jacob Sorber, Nilanjan Banerjee, Mark Corner, and Sami Rollins, "Turduken: Hierarchical Power Management for Mobile Devices", The Third International Conference on Mobile Systems, Applications and Services, Mobisys2005, Jun. 6-8, 2005, USA.

[5] Lin Zhong and Niraj Jha, "Energy Efficiency of Handheld Computer Interfaces: Limits, Characterization and Practice", The Third International Conference on Mobile Systems, Applications and Services, Mobisys2005, Jun. 6-8, 2005, USA.

[6] M. Marcu, D. Tudor, H. Moldovan, M. Micea, Power Profile Evaluation of Battery-Powered Mobile Applications, 14th IEEE International Conference on Electronics, Circuits and Systems, ICECS 2007, Dec. 11-14, 2007 Marrakech, Morocco, pp. 1015-1018, ISBN 1-4244-1378-8.

[7] Jeongjoon Lee, Catherine Rosenberg, Edwin Chong, "Energy efficient schedulers in wireless networks: design and optimization", Mobile Networks and Applications, Vol. 11, No. 3, Jun. 2006, pp. 377-389.

[8] K. Lahiri and S. Dey, "Efficient Power Profiling for Battery-Driven Embedded System Design", IEEE Transactions on Computer-Aided Design of Integrated Circuits and Systems, Vol. 23, No. 6, Jun. 2004, pp. 919-932.

[9] R. Marculescu, M. Pedram, J. Henkel, "Distributed Multimedia System Design: A Holistic Perspective", Proceedings of the Conference on Design, Automation and Test in Europe, 2004.

[10] Ashima Gupta and Prasant Mohapatra, "Power Consumption and Conservation in WiFi Based Phones: A Measurement-Based Study", Fourth Annual IEEE Communications Society Conference on Sensor, Mesh and Ad Hoc Communications and Networks SECON 2007, Jun. 2007.

[11] E. Shih, P. Bahl and M.J. Sinclair, "Wake on Wireless: An Event Driven Energy Saving Strategy for Battery Operated Devices", MOBICOM '02, September 23–28, 2002, Atlanta, Georgia, USA, pp. 160-171.

[12] L. M. Feeney and M. Nilsson. "Investigating the Energy Consumption of a Wireless Network Interface in an Ad Hoc Networking Environment". In IEEE INFOCOM 2001, 2001.

[13] R. Krashinsky and H. Balakrishnan, "Minimizing Energy for Wireless Web Access with Bounded Slowdown", Wireless Networks 11, 135–148, 2005.

[14] C. Koulamas and A. Prayati and G. Lafruit and G. Papadopoulos., "Measurements and modeling of resource consumption in wireless video streaming: the decoder case", WMuNeP '06: Proceedings of the 2nd ACM international workshop on Wireless multimedia networking and performance modeling, pp67-72, ACM Press, 2006, Terromolinos, Spain

[15] Chandra Krintz, Ye Wen, and Rich Wolski, "Application-level Prediction of Battery Dissipation", ACM/IEEE International Symposium on Low Power Electronics and Design (ISLPED), pp224-229, August 9-11, 2004, Newport Beach, CA

[16] Potlapally, N.R. Ravi, S. Raghunathan, A. Jha, N.K., "Analyzing the energy consumption of security protocols", Proceedings of the 2003 International Symposium on Low Power Electronics and Design, 2003, pp30-35, August 25-27, 2003, Seoul, Korea.

[17] Ahmad Rahmati and Lin Zhong., "Context-for-wireless: context-sensitive energy-efficient wireless data transfer", Proceedings of the 5th international conference on Mobile systems, applications and services, MobiSys '07, pp165-178, June 11-14, 2007, Juan, Puerto Rico.

[18] Y. Agarwal, R. Chandra, A. Wolman, P. Bahl, K. Chin and R. Gupta, "Wireless Wakeups Revisited: Energy Management for VoIP over Wi-Fi Smartphones", MobiSys'07, June 11-14, 2007, San Juan, USA.

[19] D. Tudor and M. Marcu, "A Power Benchmark Experiment for Battery-Powered Devices", IEEE Int. Workshop on Intelligent Data Acquisition and Advanced Computing Systems, 6-8 Sep. 2007, Germany.

[20] D. Linden and T. Reddy, "Handbook of Batteries", McGraw-Hill, 2001.

[21] M. Hachman, "SPEC Developing Benchmark To Measure Power", ExtremeTech, May 2006.

Trinary Encoder, Decoder, Multiplexer and Demultiplexer Using Savart Plate and Spatial Light Modulator

Amal K Ghosh
Department of Applied Electronics & Instrumentation Engineering, Netaji Subhas Engineering College, Techno City, Garia, Kolkata-700 152, India.
Email: amal_k_ghosh@rediffmail.com

Souradip Singha Roy & Sudipta Mandal
Department of Electronics and Communication Engineering, Netaji Subhas Engineering College, Techno City, Garia, Kolkata-700 152, India.
Email: soura_87@yahoo.com

Amitabha Basuray
Department of Applied Optics & Photonics, University of Calcutta, 92, A.P.C.Road, Kolkata-700 009, India
Email: abasuray@rediffmail.com

Abstract:

Optoelectronic processors have already been developed with the strong potentiality of optics in information and data processing. Encoder, Decoder, Multiplexers and Demultiplexers are the most important components in modern system designs and in communications. We have implemented the same using trinary logic gates with signed magnitude defined as Modified Trinary Number (MTN). The Spatial Light Modulator (SLM) based optoelectronic circuit is suitable for high speed data processing and communications using photon as carrier. We also presented here a possible method of implementing the same using light with photon as carrier of information. The importance of the method is that all the basic gates needed may be fabricated based on basic building block.

Index Terms: Trinary, dibit, trit, polarization, savart plate, SLM and MTN.

1. INTRODUCTION

Trinary system finds its importance in the modern optical computation. A carry-free mathematical operation is possible with such MTN system[1-4]. However, the practical implementation of such systems calls for development of basic processors based on such logic[5-9]. In this communication we present a method for implementation of such processors as encoder, decoder[10-11], multiplexer and demultiplexer [12-17].

2. TRINARY LOGIC AND THE BASIC BUILDING BLOCK FOR IMPLEMENTATION

Before going into details of the implementation it is felt a short introduction of the basic logical system based on trinary along with the process of implementation. The three states of the trinary representation are classified as the true, false and contradiction as shown in table-1.

Table-1: Trinary Logic System

Logical state	Represented by	Dibit representation	State of polarization
True/ Complete information	1	01	Vertical polarization
False/ Wrong information	$\bar{1}$	10	Horizontal polarization
Contradiction/ Partial information	0	11	Presence of both the horizontal & vertical polarization
Don't care state		00	-

The truth table for the basic trinary gates required to implement the encoder, decoder, multiplexer and demultiplexer are mentioned in table-2. The basic logic gates and their output functions have been defined earlier[18]. These basic gates have been used to implement the circuits mentioned here.

Table-2: Truth Table for OR, AND, XOR, NOR, NAND, XNOR, Complement, True Selector, False Selector, Exclusive True Selector and Exclusive False Selector Gates.

A	: 1	1	1	0	0	0	$\bar{1}$	$\bar{1}$	$\bar{1}$	Operation
B	: 1	0	$\bar{1}$	1	0	$\bar{1}$	1	0	$\bar{1}$	
$A \vee B$	: 1	1	0	1	0	$\bar{1}$	0	$\bar{1}$	$\bar{1}$	A OR B
$A \wedge B$	: 1	0	0	0	0	0	0	0	$\bar{1}$	A AND B
$A \oplus B$	: 0	1	0	1	0	$\bar{1}$	0	$\bar{1}$	0	A XOR B
$\overline{A \vee B}$	: $\bar{1}$	$\bar{1}$	0	$\bar{1}$	0	1	0	1	1	A NOR B
$\overline{A \wedge B}$	: $\bar{1}$	0	0	0	0	0	0	0	1	A NAND B
$\overline{A \oplus B}$	: 0	$\bar{1}$	0	$\bar{1}$	0	1	0	1	0	A XNOR B
$\bar{A}$	: $\bar{1}$	$\bar{1}$	$\bar{1}$	0	0	0	1	1	1	Complement
$A\uparrow$	: 1	1	1	1	1	1	0	0	0	True Selector
$A\downarrow$	: 0	0	0	$\bar{1}$	$\bar{1}$	$\bar{1}$	$\bar{1}$	$\bar{1}$	$\bar{1}$	False Selector
A♠	: 1	1	1	0	0	0	0	0	0	Exclusive True Selector
A♣	: 0	0	0	0	0	0	$\bar{1}$	$\bar{1}$	$\bar{1}$	Exclusive False Selector

The gates required as mentioned may be implemented[18] by using the basic building block discussed hereunder.

2.1 THE BASIC BUILDING BLOCK AND OTHER REQUIRED GATES

For implementation of these logical operations we have used a basic building block given in fig.1. A light from a laser source L through a polarizer P is incident on the first savart plate S_1.

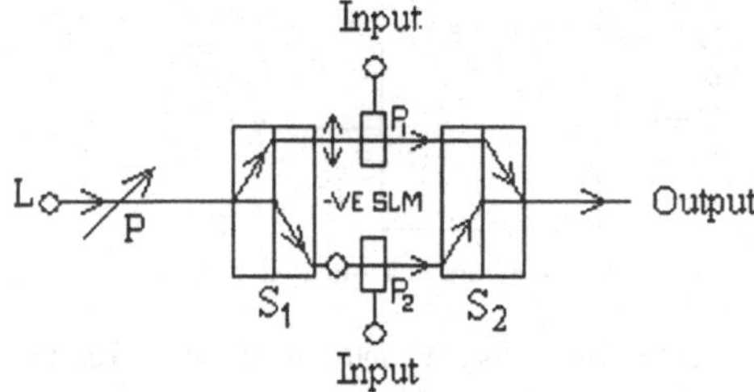

Fig. 1. The basic building block.

The basic property of a savart plate is that if a light polarized at a direction making 45^0 with the vertical axis as shown in the figure, is incident on it the output will be two parallel beams shifted between themselves. The state of polarization of the output beams are orthogonal to each other as shown. The output beams may be controlled (the presence or absence) by using two input signals through spatial light modulators (SLMs) P_1 and P_2. The SLMs are of

K. Elleithy (ed.), *Advanced Techniques in Computing Sciences and Software Engineering*,
DOI 10.1007/978-90-481-3660-5_71, © Springer Science+Business Media B.V. 2010

two kinds – positive and negative. The nature of the negative SLM is such that it is transparent when there is no electric voltage applied on it and it becomes opaque when an electric voltage is applied on it. The property of positive SLM is just reverse. In the output two beams are combined by using a second savart plate S_2 for which the crystal axes are opposite that of first savart plate S_1. Different logic gates mentioned in Table-2 may be implemented by combining this very basic module.

3. TRINARY ENCODER

An Encoder is a combinational logic circuit used to convert an active input signal into a coded output signal. Out of n input lines only one remains active at any time and it has m output lines. It encodes one of the active inputs to a coded trinary output with m trits. As in the case of a binary system, the number of outputs of an encoder is always less than the number of inputs. The block diagram of an encoder is shown in fig.2.

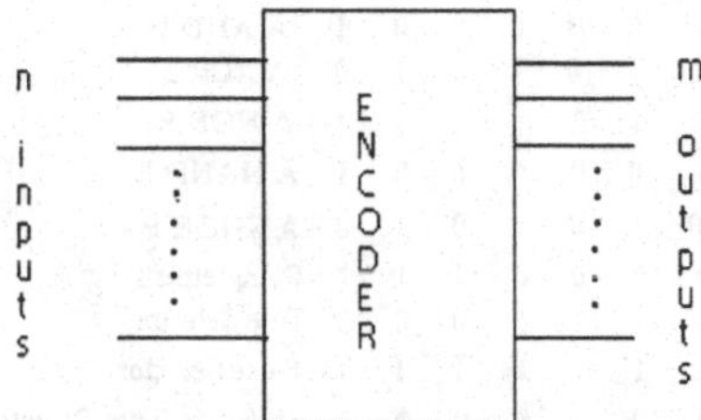

Fig.2: Block diagram of an Encoder

3.1 BASIC 9-TO-2 TRINARY ENCODER

It is well known that a 9-to-2 encoder accepts 9 inputs and produces a 2-trit output code corresponding to the activated input. The basic block diagram of a 9-to-2 encoder is shown in fig.3 and the truth table is given in Table-3.

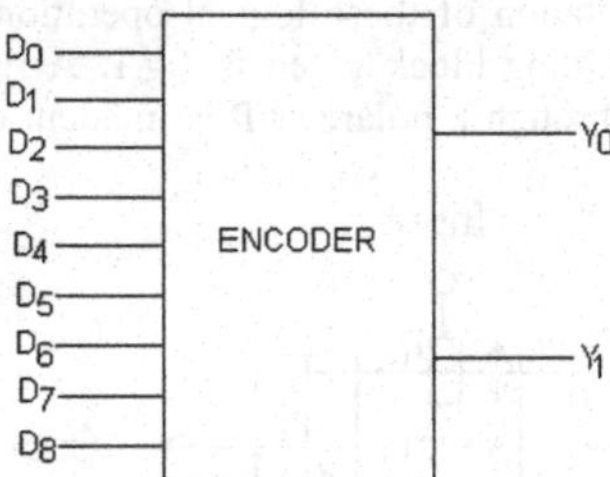

Fig. 3: Basic block diagram of a 9-to-2 encoder.

The truth table shows that Y_0 must be 1 whenever the input D_2 OR D_5 OR D_8 is HIGH & Y_0 must be $\bar{1}$ whenever the input D_0 OR D_3 OR D_6 is HIGH. Thus,

$$Y_0 = \overline{D}_0 + D_2 + \overline{D}_3 + D_5 + \overline{D}_6 + D_8$$

Similarly,

$$Y_1 = \overline{D}_0 + \overline{D}_1 + \overline{D}_2 + D_6 + D_7 + D_8$$

Using the above expressions, the 9-to-2 trinary encoder can be implemented by using the basic trinary gates of opto-electronics system is shown in Fig.4.

Table-3: Truth Table of 9 - to - 2 Encoder.

Inputs									Outputs	
D_0	D_1	D_2	D_3	D_4	D_5	D_6	D_7	D_8	Y_1	Y_0
1	0	0	0	0	0	0	0	0	$\bar{1}$	$\bar{1}$
0	1	0	0	0	0	0	0	0	$\bar{1}$	0
0	0	1	0	0	0	0	0	0	$\bar{1}$	1
0	0	0	1	0	0	0	0	0	0	$\bar{1}$
0	0	0	0	1	0	0	0	0	0	0
0	0	0	0	0	1	0	0	0	0	1
0	0	0	0	0	0	1	0	0	1	$\bar{1}$
0	0	0	0	0	0	0	1	0	1	0
0	0	0	0	0	0	0	0	1	1	1

The circuit is designed in such a way that, when D_0 is HIGH, the trinary code $\bar{1}\,\bar{1}$ is generated; when D_1 is HIGH, the trinary code $\bar{1}\,0$ is generated, and so on. It is implemented by inverter or NOT gates only as shown in fig.4.

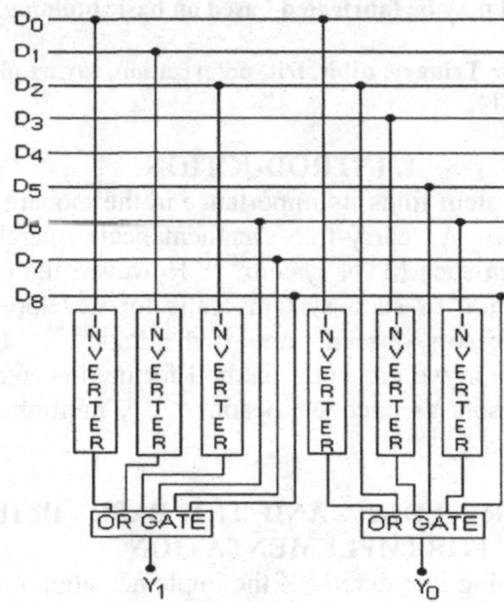

Fig.4: Encoder Circuit.

4. TRINARY DECODER

A decoder is a logic circuit that converts an n-trinary input code (data) into 3^n output lines such that each output line will be activated for only one of the possible combinations of inputs. As in binary decoder system, the number of output is greater than the number of inputs. A block diagram of a decoder circuit is shown in fig.5.

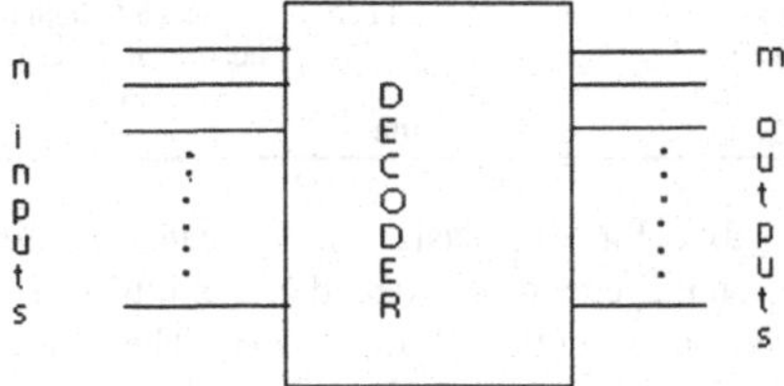

Fig.5: Block diagram of a Decoder.

4.1 BASIC 2-TO-9 TRINARY DECODER

The block diagram of a 2-to-9 decoder is shown in fig.6. A 2-to-9 decoder has two inputs (S_1, S_0) and nine outputs (D_0 to D_8). Based on the 2 inputs, one of the nine outputs is selected.

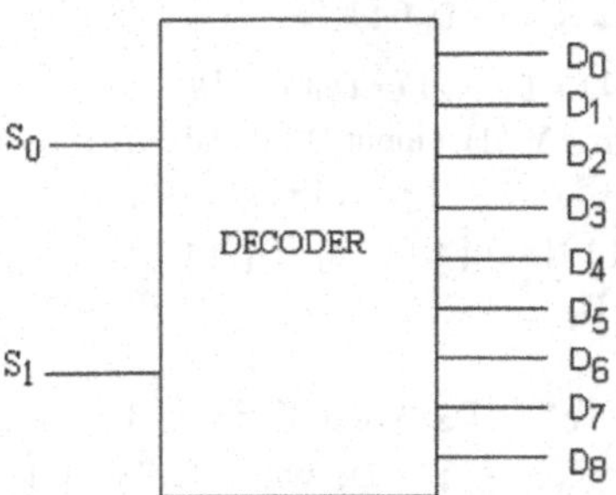

Fig. 6: Basic block diagram of a 2-to-9 decoder.

The truth table for the 2-to-9 decoder is shown in Table-4.

Table- 4: Truth Table of Decoder.

Inputs		Outputs								
S_1	S_0	D_0	D_1	D_2	D_3	D_4	D_5	D_6	D_7	D_8
$\bar{1}$	$\bar{1}$	1	0	0	0	0	0	0	0	0
$\bar{1}$	0	0	1	0	0	0	0	0	0	0
$\bar{1}$	1	0	0	1	0	0	0	0	0	0
0	$\bar{1}$	0	0	0	1	0	0	0	0	0
0	0	0	0	0	0	1	0	0	0	0
0	1	0	0	0	0	0	1	0	0	0
1	$\bar{1}$	0	0	0	0	0	0	1	0	0
1	0	0	0	0	0	0	0	0	1	0
1	1	0	0	0	0	0	0	0	0	1

From the above truth table, it is clear that only one of the possible nine outputs (D_0 to D_8) is selected based on the two select inputs. Also, from the truth table, the logic expressions for the output are obtained as follows:

$$D_0 = (\overline{S_1\downarrow}.\overline{S_0\downarrow})$$
$$D_1 = [\overline{S_1\downarrow}.(S_0\uparrow.\overline{S_0\downarrow})]$$
$$D_2 = (\overline{S_1\downarrow}.S_0\uparrow)$$
$$D_3 = [(S_1\uparrow.\overline{S_1\downarrow}).\overline{S_0\downarrow}]$$
$$D_4 = [(S_1\uparrow.\overline{S_1\downarrow}).(S_0\uparrow.\overline{S_0\downarrow})]$$
$$D_5 = [(S_1\uparrow.\overline{S_1\downarrow}).S_0\uparrow]$$
$$D_6 = (S_1\uparrow.\overline{S_0\downarrow})$$
$$D_7 = [(S_1\uparrow.(S_0\uparrow.\overline{S_0\downarrow})]$$
$$D_8 = [(S_1\uparrow.S_0\uparrow]$$

Using the above expressions, the circuit of a 2-to-9 decoder can be implemented using AND, NOT, True Selector, Exclusive True Selector, False Selector and Exclusive False Selector gates as shown in Fig.7.

The two inputs S_1, S_0 are decoded into nine outputs and each

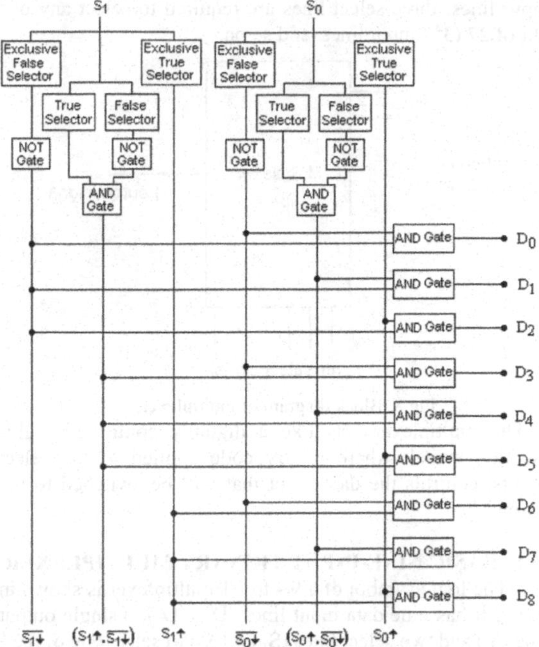

Fig.7: Decoder Circuit.

output representing one of the minterms of the 2-input variables. The Exclusive False Selectors provide the false value if and only if the input information contain any false value and it is complemented by NOT gate to get the true value, the Exclusive True Selectors provide the true value if and only if the input information contains any true value, the True Selector provide only true value of the input, the False Selector provide only false value of the input and it is complemented by the NOT gate to get the true value and then it is ANDed with the output of the True Selector so that they provide true value only and only when the input is in contradiction i.e. 0. This decoder circuit can be used for decoding any 2-trit code to provide nine outputs, corresponding to nine different combinations of the input codes.

This is also called a 2-to-9 decoder, since only one of nine output lines is HIGH for a particular input combination. For example, when $S_1S_0 = \bar{1}\ 0$, then only the D_1 is HIGH. Similarly, when $S_1S_0 = 1\ \bar{1}$ then also only the D_6 is HIGH.

5. TRINARY MULTIPLEXER

Multiplexer is the most common term in electronics. It is the process of transmitting a large number of information over a single line. A multiplexer is also called data selector since it can select any one out of many inputs and steers the information to the output.

The basic block diagram of a multiplexer with n input lines, m select lines and one output line is shown in Fig.8. The select lines decide the number of possible input lines of a particular multiplexer. In the trinary system if the number of select lines are m, then they may be used to select any of input lines 3^m which is equal to n. For example, in trinary two select lines are required to select any one out of $9(3^2)$

input lines; three select lines are required to select any one out of 27 (3^3) input lines, and so on.

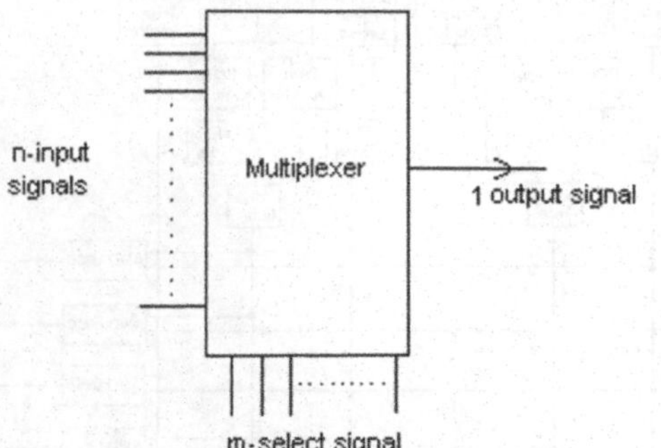

Fig.8: Block diagram of multiplexer.

The multiplexer acts like a digitally controlled multi-position switch where trinary code applied to the select inputs, controls the data input that will be switched to the output.

5.1 BASIC NINE-INPUT TRINARY MULTIPLEXER

The logic symbol of a 9 - to - 1 multiplexer is shown in Fig.9. It has nine data input lines ($D_0 - D_8$), a single output line (Y) and two select lines (S_1 and S_0) to select one of the nine input lines. The truth table for a 9 - to -1 multiplexer is shown in table- 5.

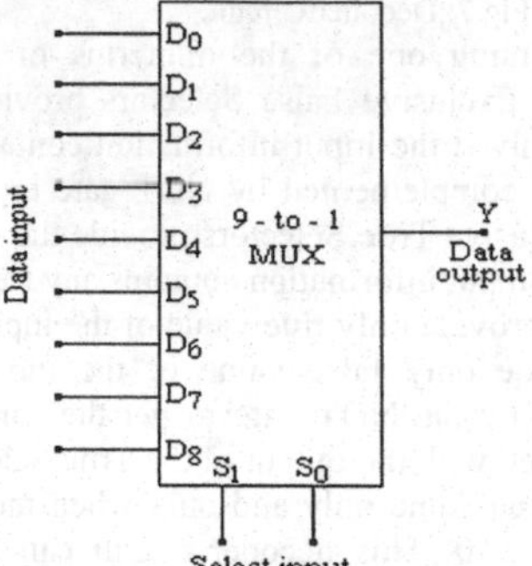

Fig.9: Multiplexer Logic Symbol

From the truth table-5 a logical expression for the output in terms of the data input and select input can be derived as follows:

Table-5: Truth table of 9-to-1 multiplexer

Data select inputs		**Outputs**
S_1	S_0	Y
$\bar{1}$	$\bar{1}$	D_0
$\bar{1}$	0	D_1
$\bar{1}$	1	D_2
0	$\bar{1}$	D_3
0	0	D_4
0	1	D_5
1	$\bar{1}$	D_6
1	0	D_7
1	1	D_8

The data output Y= data input D_0, if and only if $S_1=\bar{1}$ and $S_0=\bar{1}$.

Therefore,

$$Y = D_0(\overline{S_1\downarrow}.\overline{S_0\downarrow}) = D_0(\overline{\bar{1}\downarrow}.\overline{\bar{1}\downarrow})$$

$$= D_0(\bar{\bar{1}}.\bar{\bar{1}}) = D_0(1.1) = D_0(1) = D_0$$

The data output Y=data input D_1, if and only if

$$Y = D_1[\overline{S_1\downarrow}.(S_0\uparrow.\overline{S_0\downarrow})] = D_1[\overline{\bar{1}\downarrow}.(0\uparrow.\overline{0\downarrow})]$$

$$= D_1[\bar{\bar{1}}.(1.\bar{\bar{1}})] = D_1[1.(1.1)] = D_1(1.1)$$

$$= D_1(1) = D_1$$

Similarly,

$$Y = D_2(\overline{S_1\downarrow}.S_0\uparrow) = D_2 \text{ when } S_1S_0 = \bar{1}\,1$$

$$Y = D_3[(S_1\uparrow.\overline{S_1\downarrow}).\overline{S_0\downarrow}] = D_3 \text{ when } S_1S_0 = 0\,\bar{1}$$

$$Y = D_4[(S_1\uparrow.\overline{S_1\downarrow}).(S_0\uparrow.\overline{S_0\downarrow})] = D_4 \text{ when } S_1S_0 = 0\,0$$

$$Y = D_5[(S_1\uparrow.\overline{S_1\downarrow}).S_0\uparrow] = D_5 \text{ when } S_1S_0 = 0\,1$$

$$Y = D_6(S_1\uparrow.\overline{S_0\downarrow}) = D_6 \text{ when } S_1S_0 = 1\,\bar{1}$$

$$Y = D_7[(S_1\uparrow.(S_0\uparrow.\overline{S_0\downarrow})] = D_7 \text{ when } S_1S_0 = 1\,0$$

$$Y = D_8[(S_1\uparrow.S_0\uparrow] = D_8 \text{ when } S_1S_0 = 1\,1$$

If the above terms are ORed, then the final expression for the data output is given by:

$$Y = D_0(\overline{S_1\downarrow}.\overline{S_0\downarrow}) + D_1[\overline{S_1\downarrow}.(S_0\uparrow.\overline{S_0\downarrow})] + D_2(\overline{S_1\downarrow}.S_0\uparrow) + D_3[(S_1\uparrow.\overline{S_1\downarrow}).\overline{S_0\downarrow}] + D_4[(S_1\uparrow.\overline{S_1\downarrow}).(S_0\uparrow.\overline{S_0\downarrow})] + D_5[(S_1\uparrow.\overline{S_1\downarrow}).S_0\uparrow] + D_6(S_1\uparrow.\overline{S_0\downarrow}) + D_7[(S_1\uparrow.(S_0\uparrow.\overline{S_0\downarrow})] + D_8[(S_1\uparrow.S_0\uparrow]$$

Using the above expression the 9-to-1 multiplexer can be implemented using NOT gates, 2-input AND gates, true

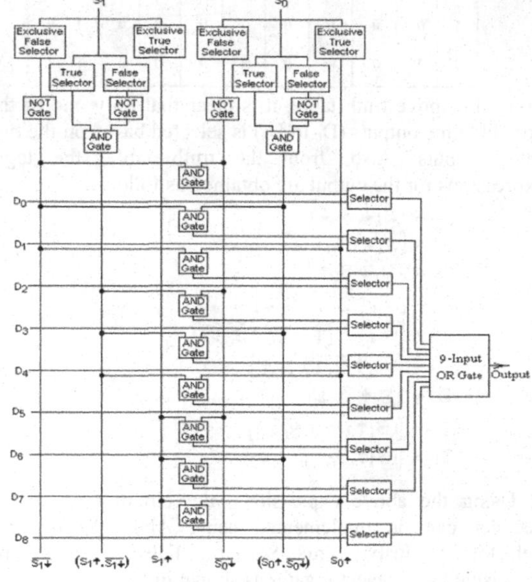

Fig.10 Logic design of 9-to-1 multiplexer.

selector, false selector, exclusive true selector and exclusive false selector gates, selector networks and one 9-input OR gate as shown in Fig 10. The selector outputs are connected with the inputs of the OR gate to generate the output Y.

Here we have to notice the case when S_1 and S_0 both are equal to 1 and then $S_1 . S_0 = 1$. Now depending on the data input D = 1 or 0 the output Y = 1 or 0 respectively. But in the case when $D = \bar{1}$ then the output Y = 0, as a result we do not get the desired output. So, to overcome this problem we have designed a 2-input selector network as shown in fig.11. It consists of two 2-input AND gates, two NOT gates and one 2-input OR gate. The data select inputs S_1 and S_0 are ANDed and feed to the input S of this selector network and the data input is directly feed to the another input D of this network.

To demonstrate the operation of this multiplexer circuit, let us consider the case when $S_0\ S_1 = \bar{1}\ \bar{1}$. If this value is applied to the select lines, then the output of the AND gate = 1. Now it is feed to S and the data input D_0 to D of the selector network.

Here S= 1 but D may take any values among $\bar{1}$, 0, 1. Consider the case when $D=\bar{1}$, then the two inputs of AND Gate1 are 1, $\bar{1}$. So the output of AND Gate1=0. Again, the two inputs of AND Gate2 are 1,1. So the output is 1, which

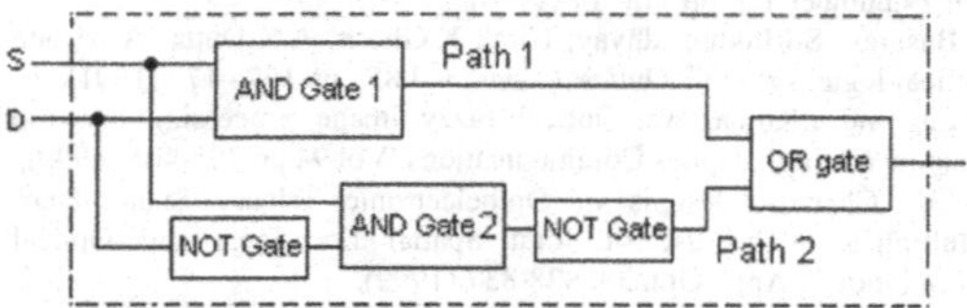

Fig..11: Logic diagram of selector network

after passing through the NOT Gate will give the output $\bar{1}$. The two outputs paths - Path 1 and Path 2 are then ORed and generate the network selector output $\bar{1}$.

When D= 0 the two inputs of AND Gate1 are 1,0. So the output of AND Gate1=0. Now, the two inputs of AND Gate2 are 1, 0 and so the output is 0 which after passing through the NOT gate will give the output 0. The two outputs are then ORed and generate the network selector output 0.

When D= 1 the two inputs of AND Gate1 are 1,1. So the output of AND Gate1=1. Now, the two input of AND Gate2 are $\bar{1}$, 1 and So the output is 0 which after passing through the NOT gate will give the output 0. The two outputs is then ORed and generate the network output 1.

6. TRINARY DEMULTIPLEXER

Demultiplexing is the process of taking information from one input and transmitting the same over one of several outputs. A demultiplexer is a logic circuit that receives information on a single input and transmits the same over one of several (3^n) output lines.

The block diagram of a demultiplexer which is opposite to a multiplexer in its operation is shown in fig 12. The circuit has one input signal, m select signals and n output signals. The select input determines to which output the data input will be connected. As the serial data is changed to parallel data, i.e. the input causes to appear on one of the n output lines, the demultiplexer is also called a distributor or a serial-to- parallel converter.

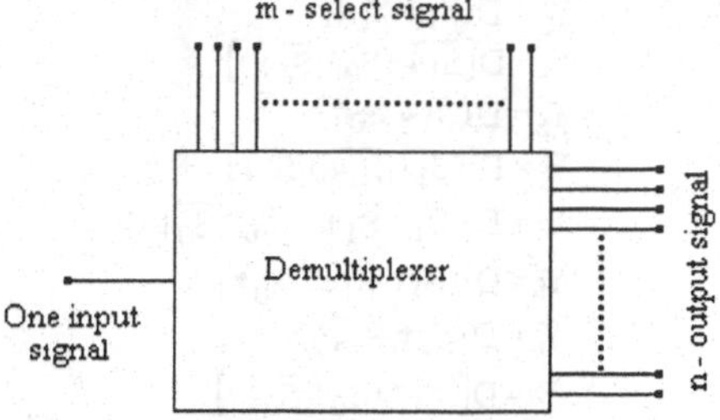

Fig. 12: Block diagram of demultiplexer.

6.1 1-TO- 9 TRINARY DEMULTIPLEXER

The block diagram of a 1-to-9 demultiplexer is shown in fig.13. A 1-to-9 demultiplexer has a single input (D), nine outputs (Y_0 to Y_8) and two select inputs (S_1 and S_0).

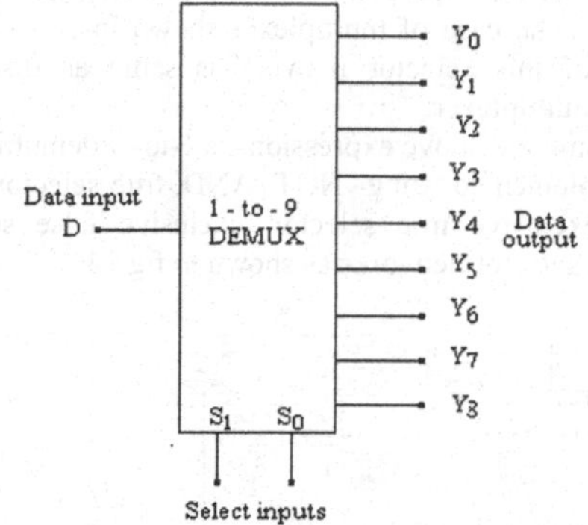

Fig.13: Block diagram of a 1-to-9 demultiplexer.

The truth table of the 1-to - 9 demultiplexer is shown in Table 6. From the truth table it is clear that the data input is connected to output Y_0 when $S_1 = \bar{1}$ and $S_0 = \bar{1}$. And the data input is connected to output Y_1 when $S_1 = \bar{1}$ and $S_0 = 0$ and

Table-6: Demultiplexer 1- to - 9

Data input	*Select inputs*		*Outputs*								
D	S_1	S_0	Y_0	Y_1	Y_2	Y_3	Y_4	Y_5	Y_6	Y_7	Y_8
D	$\bar{1}$	$\bar{1}$	D	0	0	0	0	0	0	0	0
D	$\bar{1}$	0	0	D	0	0	0	0	0	0	0
D	$\bar{1}$	1	0	0	D	0	0	0	0	0	0
D	0	$\bar{1}$	0	0	0	D	0	0	0	0	0
D	0	0	0	0	0	0	D	0	0	0	0
D	0	1	0	0	0	0	0	D	0	0	0
D	1	$\bar{1}$	0	0	0	0	0	0	D	0	0
D	1	0	0	0	0	0	0	0	0	D	0
D	1	1	0	0	0	0	0	0	0	0	D

so on. From the truth table, the expressions for the outputs can be written as follows.

$$Y_0 = D(\overline{S_1\downarrow}.\overline{S_0\downarrow})$$
$$Y_1 = D[\overline{S_1\downarrow}.(S_0\uparrow.\overline{S_0\downarrow})]$$
$$Y_2 = D(\overline{S_1\downarrow}.S_0\uparrow)$$
$$Y_3 = D[(S_1\uparrow.\overline{S_1\downarrow}).\overline{S_0\downarrow}]$$
$$Y_4 = D[(S_1\uparrow.\overline{S_1\downarrow}).(S_0\uparrow.\overline{S_0\downarrow})]$$
$$Y_5 = D[(S_1\uparrow.\overline{S_1\downarrow}).S_0\uparrow]$$
$$Y_6 = D(S_1\uparrow.\overline{S_0\downarrow})$$
$$Y_7 = D[(S_1\uparrow.(S_0\uparrow.\overline{S_0\downarrow})]$$
$$Y_8 = D[(S_1\uparrow.S_0\uparrow]$$

We find the equation as similar to the binary logic. But it is to be noted here when we consider the case for S_1 and S_0 both are equal to 1 then S_1. $S_0 = 1$ and then for $D = 1$ or 0 the output $Y_8 = 1$ or 0 respectively. But the case when $D = \bar{1}$ then $Y_8 = 0$ and we do not get the desired result. So to overcome this we have designed a 2-input selector network same as in the case of multiplexer shown in fig. 11. The activities of this selector network is same as discussed earlier for multiplexer.

Now using the above expressions, a 1-to- 9 demultiplexer can be implemented using NOT, AND, true selector, false selector, exclusive true selector, exclusive false selector gates and selector networks as shown in fig.14.

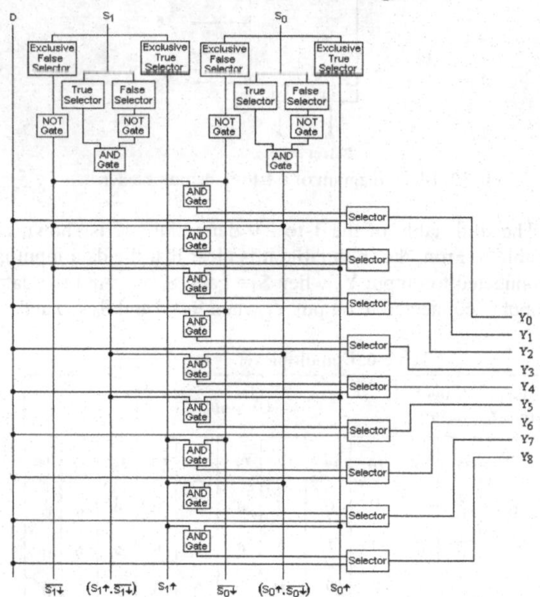

Fig.14 Logic design of 1-to- 9 demultiplexer.

The two select lines S_1 and S_0 enable only one gate at a time and the data that appears on the input line passes through the selected gate to the appropriate output lines.

7. CONCLUSION

We have discussed the implementation of encoder, decoder, multiplexer and demultiplexer circuit by using trinary optical logic gates comprises of savart plates and spatial light modulators. It is also suitable for VLSI implementation due to the repetitions of basic gate. In this paper the very basic trinary encoder, decoder, multiplexer and demultiplexer circuits and their practical implementations have been discussed by using the trinary logic gates of opto-electronic devices for the fast operation. Due to signed bit implementation the mathematical operations are also very simple. The trinary logic also find its applications in gray image processing, cellular automata, fuzzy logic systems, fractals and other emerging areas where the fast operations are needed.

8. REFERENCES

[1] A.Avizienis "Signed-digit number representation for fast parallel arithmetic", IRE Trans. Electron. Comp. EC-10 pp 389-400 (1961).

[2] A.K.Datta, A.Basuray and S.Mukhopadhyay,"Arithmetic operations in optical computations using a modified trinary number of system", Optics Letters 14 pp 426-428 (1989).

[3] Fyath R.S.; Alsaffar A.A.W.; Alam M.S., "Optical two-step modified signed-digit addition based on binary logic gates", Optics Communications, Volume 208, Number 4, pp. 263-273(2002).

[4] R. S. Fyath, A. A. W. Alsaffar and M. S. Alam,"Nonrecoded trinary signed-digit multiplication based on digit grouping and pixel assignment", Optics Communications,Vol.230, Issue1-3,pp.35-44 (2004).

[5] Amal K. Ghosh and A.Basuray, "Trinary optical logic processors using shadow casting with polarized light",*Optics Communications*, vol.79,number 1,2 pp 11 – 14 (1990).

[6] A.Basuray, S.Mukhopadhyay, Hirak K.Ghosh, A.K.Dutta,"A tristate optical logic system", *Optics Comm.*,vol.85, pp 167 –170 (1991).

[7] S.Lin and I.Kumazawa,"Optical fuzzy image processing based on shadow casting", Optics Communications, Vol 94 pp.397-405 (1992).

[8] A. K. Cherri, "Designs of Optoelectronic Trinary Signed-Digit Multiplication by use of Joint Spatial Encodings and Optical Correlation ", Appl. Opt. 38, 828-837 (1999).

[9] A.K.Cherri and M.S. Alam,"Algorithms for Optoelectronic Implementation of Modified Signed-Digit Division, Square-Root, Log-arithmic, and Exponential Functions",Appl. Opt. 40,1236-1243 (2001).

[10] Thoidis, I.M.;Soudris, D.; Karafyllidis, I.; Thanailakis, A., "The design of low power multiple-valued logic encoder and decoder circuits," Proceedings of ICECS apos;99, The 6th IEEE International Conference on Electronics,Circuits and Systems,1999, vol.3, pp.1623-1626 (1999).

[11] Mozammel H.A. Khan, and Marek Perkowski, "Quantum Realization of Ternary Encoder and Decoder," Proc. International Symposium on Representations and Methodologies for Emergent Computing Technologies, Tokyo, Japan, (September 2005).

[12] Brackenbury,L.E.M.,"Multiplexer as a universal computing element for electro-optic logic systems," IEE Proceedings-J(Optoelectronics), 137(5),pp.305-310 (1990).

[13] S. Yasuda, Y. Ohtomo, M.Ino, Y.Kado and T.Tsuchiya, "3-Gb/s CMOS 1:4 MUX and DEMUX Ics", IEICE Trans. Electron, vol.E78-C, no.12, pp.1746-1753, (December 1995).

[14] T.Yamamoto, E.Yoshida and M.Nakazawa,"Ultra fast non-linear optical loop mirror for demultiplexing 640Gbits/s TDM signals", Electron Lett. 34, pp. 1013 – 1014 (1998).

[15] I .Glesk, R.J.Runser and P.R.Prucnal "New generation of devices for all-optical communication," Acta Phys Slov, 5l, pp.151-162(2001).

[16] A. Tanabe, Y. Nakahara, A. Furukawa and T. Mogami, "A Redundant Multivalued Logic for a 10-Gb/s CMOS Demultiplexer IC", IEEE J. Solid-State Circuits, vol.38, pp.107-113(January 2003).

[17] Kim, Jeong Beom and Ahn, Sun Hong, "An 11Gb/s CMOS Demultiplexer Using Redundant Multi-valued Logic", 13th IEEE International Conference on Electronics, Circuits and Systems, (ICECS'06), pp.838-841(2006).

[18] Amal K. Ghosh, P.Pal Choudhury and A.Basuray,"Modified Trinary Optical Logic Gates and their Applications in Optical Computation", CISSE 2007, IEEE, University of Bridgeport and published in the proceeding on Innovations and Advanced Techniques in Systems, Computing Sciences and Software (Springer, 2008) pp. 87-92 (2008).

Regional Competitiveness Information System as a Result of Information Generation and Knowledge Engineering

Aleksandras Vytautas Rutkauskas
Department of Finance Engineering
Vilnius Gediminas Technical University
Sauletekio ave. 11. CR-618, Vilnius
LT-10223 Lithuania

Viktorija Stasytyte
Department of Finance Engineering
Vilnius Gediminas Technical University
Sauletekio ave. 11. CR-605, Vilnius
LT-10223 Lithuania

***Abstract.* The main idea of the paper is the formation of regional competitiveness information system conception, structurization of this system, determining the sources of its information supply and its application to decisions management. Proper knowledge on competitiveness contents and various attributes of competitiveness ensure the effective functioning of the system. Fully developed regional competitiveness information system can be applied for formation and application of regional development strategies.**

***Keywords and expressions:* competitiveness, regional competitiveness information system, regional development strategy, risk management competitiveness measures and factors, competitiveness management.**

I. Introduction

Competitiveness, as an attribute, is linked to the following levels of socio-economic aggregation: micro (firms), mezzo (industries), macro (countries), mega (number of countries).

However, any distinguished aggregation level does not possess an unambiguous definition (perception) of competitiveness. In fact, the content of competitiveness conception can be and should be different as the subject itself changes, as well as the activity environment of this subject [9]. Thus, for the analysts, as well as for the holders (subjects) of competitiveness attribute, the content of competitiveness category is different. The chosen perception of competitiveness on the same aggregation level, as the highest generated value for residential customers and shareholders, allows unambiguously understanding not only the goals of competitiveness development, but also the complex of means for reaching these goals.

The polysemy of previously mentioned competitiveness content has also determined the abundance of its measurement indicators. This allows perceiving adequately the content and development complexity of competitiveness, however, the ambiguity of competitiveness measurement can provide presumptions for false choice of competitiveness development means. Also, there is no doubt that selection of inadequate definition of competitiveness can mostly harm while generating competitiveness development strategies (Fig. 1).

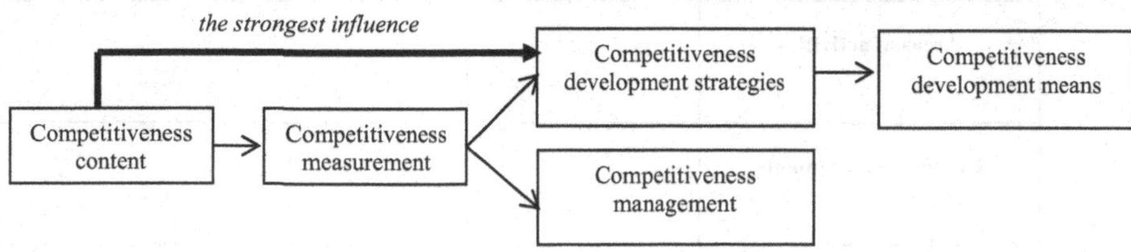

Fig. 1. Conceptual scheme of competitiveness management

K. Elleithy (ed.), *Advanced Techniques in Computing Sciences and Software Engineering*,
DOI 10.1007/978-90-481-3660-5_72,

Thus, inadequate perception of competitiveness content often provokes selection of inadequate system of means for competitiveness management.

In essence, the preparation of the competitiveness information system as instrumentary for regional competitiveness management will be discussed in the paper, as well as possibilities of system's application for the generation of regional development strategies.

The main outputs of the paper should be information about:

1. Conceptualization of regional (state) competitiveness information system and its structure;
2. Elaboration of the main principles of system functioning;
3. Determining the necessity of REGCIS for regional development strategies.

II. Conceptualization of the Problem

Region is a geographic term that is used in reference to various coexistence ways among the different activities. In general, a region is a medium-scale area of land, Earth or water, smaller than the whole areas of interest (which could be, for example, the world, a nation, a river basin, mountain range, and so on), and larger than a specific site or location. A region can be seen as a collection of smaller units (as in "the New England states") or as one part of a larger whole (as in "the New England region" of the United States).

In the paper "region" will be perceived as the area of people community social dislocation and economic activity. Example of the typical region will be territory of small countries, or the areas of bigger ones and even areas of the big countries.

Competitiveness will be perceived as the coexistence form of separate forms of life, business units, branches of activity, countries or regions, pursuing certain advantages and ensuring possibilities of survival. Competitiveness is the degree of ability of the certain subject or of the whole system to compete in the mentioned coexistence.

If understanding of increase of regional welfare is based on amount of value used, the concept of residential value seems to be applicable. Residential value could be understood as residual value for resident customers and shareholders in the region. Maximization of residential value is the main motive and mean for regional competitiveness development if that is regarded in structurized manner (Table I).

Regional competitiveness refers to a dynamic process of usage (exploitation) of own or acquired from aside resources, transforming them into residential value (utility) [14]. So, competitiveness of region (country) is associated with their ability to generate more residential value from own and attracted resources, including financial.

Development of competitiveness – it is the identification of competitiveness factors and their appearance circumstances, - and creation and realization of their fostering and development mechanism. Sustainable competitiveness development or competitiveness sustainability insurance – it is the realization of earlier mentioned circumstances and mechanisms, after choosing a certain competitiveness development guarantee and risk management tools.

TABLE I
Structuring of Regional Competitiveness

Types of competitiveness / Subjects asserting competitiveness	**Financial competitiveness**	**Social-economical competitiveness**	**Survival competitiveness**
Business units			
Areas of activities			
Country governments			

The paper is intended for region as an area of social dislocation. In this area the conditions for economic activity are provided, i.e. self-sufficient business units exist, corporations and separate areas of activity are formed, federal and local governments are established. Thus regional competitiveness is perceived as interaction of business units, branches of activities and state or local government competitiveness. The dimension of competitiveness measurement or nature of competitiveness will be analysed as financial competitiveness, social-economical competitiveness and survival competitiveness.

Thus for perception of competitiveness research and management structure Table I could be useful.

The paper will be based on region perception, concretizing holistic conception of competitiveness and paying attention to region multiaspectiveness, stochastic nature of competitiveness powers formation, cyclical nature of interdependency competitiveness determinants (indicators) and factors, as well as competition and competitiveness, interdependency character. This allows to develop quantitative models of regional competitiveness management with the help of expert decisions information supply and management systems. These models would be applied to effective regional strategies development and implementation.

Regional development strategies would be formulated to determine the way in which the region could manage its competitiveness abilities and change current position to a new stronger competitive position. But such goals would be reached only possessing highly developed understanding of competitiveness sense, measurement indices (determinants) and factors, as well as adequately assessed interdependencies between competence level and amount of factors or being in line resources.

Competition is a component of dynamics and attribute of coexistence. Competitiveness as level of ability to compete also serves as indicator for arrangement of competing units or systems according this ability. Competitiveness in the project would be treated as ability to compete but in the sense to overcome when own and attracted resources are transformed into residential utility. While regional development strategies help as instrumentary for direction of movement to future competitiveness management would be used as component of these strategies.

For countries, where resident business do not dispose higher technologies, or countries which do not have strategically important vast natural resources and which have just soaked up fundamentals of market economics (e.g. Lithuania, Latvia), it is necessary to insure, that every business unit, every decision of government, every feet of its territory would become competitive and completely responsible for their survival and value, if these countries want to become equal EU members [6], [13]. Along with that, the sustainability of the competitive powers growth is the cornerstone of success.

There are many different competing abilities, and they play different role when resources are transformed into residential utility. In the paper main attention would be paid for financial and especially investment resources. So, utility measured as function from efficiency, risk and reliability would become the target for management.

Regional competitiveness information system (REGCIS) ought to be accepted as a platform for generation of the adequate concepts, information, decision making and management ways and methods when resources are transformed into residential utility.

III. Regional or Country Competitiveness and its Sustainability

The competitiveness of a specific country depends upon its ability to make market participants efficiently use available resources, as well as upon its ability to introduce innovations and positively change environment in order to guarantee the development sustainability. The results of incurring pressure and demands of a market help companies to increase their advantages in the competition struggle with the strongest competitors of the global market. These companies have advantage of existing strong local competitors, aggressive suppliers and demanding local consumers [7].

In modern economics, as the competition in the world is increasing, the role of country (government) becomes more and more important. Competitive advantage is being created and maintained through processes, which are strongly localized. The difference of national values, structure of evolution – everything brings a certain contribution on the way to successful competitiveness. Substantial differences exist in competitiveness structure of each country, because it is impossible for a country to be competitive in all or most of the fields. Specific countries reach better results in certain fields, because their inner conditions are more favourable, more dynamic and prospective.

There are three attributes of a country, which constitute the base of country's competitive advantages:

1. Environment adequacy for development of competitiveness.
2. Technological and organizational perfection of the fields of activity.
3. Utility and efficiency of the international relations [9].

These factors along with the regional features most often guarantee regional resources, labour qualifications for the selected field of activity. They also in general determine the formation of national environment, when company is set up and learns to compete (Fig. 2).

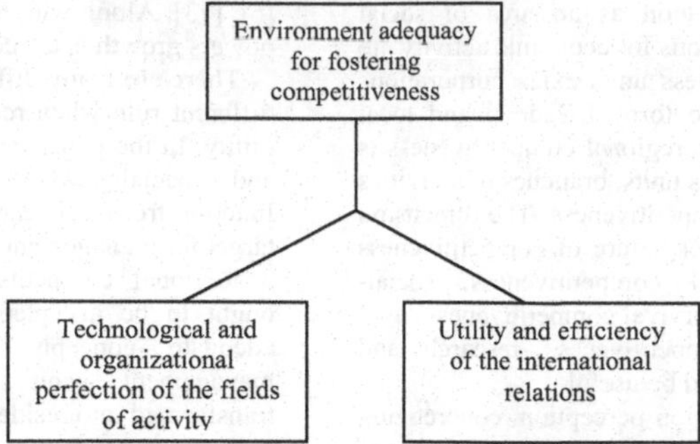

Fig. 2. Determinants of competitive advantages of a country

The apex of each triangle, shown on Fig. 2, and all rhomb illustrates the key points of competitive success achievement in the international scale: accessibility of resources and qualified labour force, needed for achievement of competitive advantages; information that forms good opportunities, which are felt by companies and directions, where they use the resources and qualification of associates, goals of directors, managers and separate company's employees; and, the most important aspect - experienced pressure upon the company, - making it introduce and realize innovations.

In general, the presumption of competitiveness of a separate country and the whole of competitiveness factors can be shown in Fig. 3.

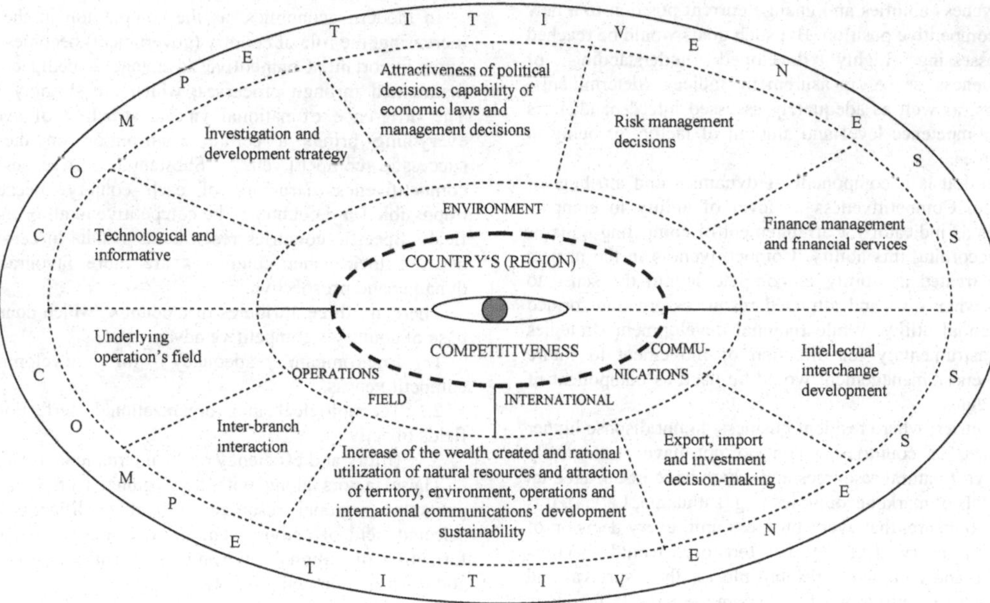

Fig. 3. Presumptions of sustainable competitiveness of a country (region)

IV. The Influence of Information Supply on REGCIS

Regional Competitiveness Information System (REGCIS) is a certain development of the regional business risk informative system (REGBRIS), which was described in some former papers of the author [8], [10]. Regional business risk monitoring (REGBRMON) should prove to be a significant element of the REGBRIS. The monitoring of region business risk is understood as important means, which creates the presumptions of risk management. There are several basic functions of the region business risk monitoring: verification and improvement of settled conceptions, criteria and risk management models; specification of the influence of risk factors on particular objects hierarchy; information accumulation and arrangement in order to meet the needs of decision-making.

The creation of regional competitiveness information system is perceived as a process of conceptualization, mining and structurization of information on competitiveness while the competitiveness portfolio, comprised of different investments, intended for the subject's competitive power improvement, is understood as instrument for competitiveness assessment and management [11].

Capability to provide and use adequate information is the most important component of separate individual's intellectual capital [2]. Also, information supply is a key problem in competitiveness management and development. The concept of information is not perceived homogeneously and this should be constantly improved by analyzing further its content and seeking that this category would become constructive instrument for systems' interaction research. At the moment information category is objectified increasingly separating its content from all data content. When separating information and data concepts' content, it is very important to consider qualities – the knowledge (intelligence) of the subject, which is using it. Perhaps, the best way of depicting this process would be a simple scheme reflecting data collection, information development and knowledge deepening necessity concept interaction: herein information is perceived as a tool, which converts data into knowledge and deepens the level of individual's perception [12].

Adequate requirements of information expansion and perception of regularity are necessary for correct information system development. The capability to convert internal and external data into knowledge assure information accumulation, create conditions for optimal decision-making, and is considered as exceptional quality of information system [1], [5], [15]. Adequate knowledge, together with information system, functioning properly, are the competent conditions for regional development strategies formation.

Talking about business information systems, three systems' functions are usually emphasized: improvement of business operations' interaction; improvement of knowledge for decision-making process; amelioration of business competitiveness.

Any business is inevitably related with risk. Risk could be stipulated by uncertainty of present activities' future as well as by possibly wrong business managers or decisions. Management of risk requires exceptionally good special knowledge about risk factors and ways of influence [3], [4]. Successful risk management leads to effective use of regions' resources as well as to connections with external partners. Hence, in order to create REGCIS, we need heuristic approach, first of all emphasizing correct formulations of system objects and operation principles, creation presumptions and practical ways of implementation prevision. Only after solving all these problems properly, we can talk about successful execution of developed competitiveness information system's functions, as well as about formation and application of regional development strategies.

Conclusions

The content of competitiveness conception can be and should be different depending on the subject. Inadequate to particular situation selection of competitiveness content and definition mostly negatively influences the competitiveness development strategies.

Maximization of residential value out of resources available in the region, i.e. the sum of consumers' value and shareholders' value, is an adequate indicator for regional competitiveness definition.

Business information or management systems, including REGCIS, must be risk-oriented, because any business is inevitably related with risk.

Regional competitiveness information system (REGCIS) ought to be accepted as a platform for generation of the adequate concepts, information, decision making and management ways and methods when resources are transformed into residential utility. The creation of regional competitiveness information system is perceived as a process of conceptualization, mining and structurization of information on competitiveness and its risk.

Adequate knowledge, together with information system, considering risk, constitutes the competent conditions for regional development strategies formation.

Further research of the problem analysed in the paper should concern the following provisions:

- Buildng up the set of methods and models for competitive ability drivers detection.
- Exploration and systematization of the principles and methods for exploitation of REGCIS for competitiveness management when preparing development strategies.
- Implementation of worked out REGCIS for experimental strategies planning.

- Practical exploitation of regional development strategies.
- Building up information communication technologies (ICT) for "Interneting" REGCIS as instrument to be experimentally used and permanently developed.

References

1. D. Acemoglu and F. Zilibotti, "Information accumulation in development", *Journal of Economic Growth*, Vol. 4, No. 1, 1999, pp. 5-38.
2. J. L. Aguirre, R. Brena and F. J. Cantu, "Multiagent-based knowledge networks", *Expert Systems with Applications,* Vol. 20, No. 1, January 2001, pp. 65-75.
3. M. H. Bouchet, E. Clark and B. Groslambert, *Country Risk Assessment*, Willey, 2003.
4. P. C. Kinsinger, "The „business intelligence" challenge in the context of regional risk", *Thunderbird International Business Review*, Vol. 49, No. 4, 2007, pp. 535-541.
5. R. Paci and S. Usai, "Knowledge flows across European regions", *The Annals of Regional Science*, 2008.
6. M. A. Weresa. *Poland. Competetiveness Report 2007*, Warshaw School of Economics, 2007.
7. M. Porter, *On Competition*. Harvard Business Press 1998.
8. A. V. Rutkauskas, V. Rutkauskas and A. Miečinskienė, "Regional business risk informative system", *Zeszyty naukowe, Kolegium gospodarki swiatowej, Szkola glowna handlowa, Warszawa* 2002, p. 148-166.
9. A. V. Rutkauskas, „On the sustainability of regional competitiveness development considering risk", *Technological and Economic Development of Economy*, Vol. 14, No. 1, 2008, pp. 89-99.
10. A. V. Rutkauskas, "Regional business risk E-informative system", *IV International Scientific Conference "Management and Engineering '06"*, June 19-23, 2006, Sozopol, Bulgaria 2006, p. 268-271.
11. F. M. Santoro, M. R. S. Borges and E. A. Rezende, "Collaboration and knowledge sharing in network organizations", *Expert Systems with Applications,* Vol. 31, No. 4, November 2006, pp. 715-727.
12. Shu-Hsien Liao, "Knowledge management technologies and applications – literature review from 1995 to 2002", *Expert Systems with Applications,* Vol. 25, No. 2, August 2003, pp. 155-164.
13. G. Tumpel-Gugerell, "The competitiveness of European financial markets", *Business Economics*, Vol. 42, No. 3, 2007.
14. J. Wilkins, B. van Wegen and R. de Hoog, "Understanding and valuing knowledge assets: Overview and method", *Expert Systems with Applications,* Vol. 13, No. 1, July 1997, pp. 55-72.
15. Wu Lin-Jung, Hsiao Hsien-Sheng, "Using a knowledge-based management to design a web-based creative problem solving system", *Lecture Notes in Computer Science, Advances in Web-Based Learning – ICWL 2004*, Vol. 3143, 2004, pp. 225-232.

A Case Study on Using A Knowledge Management Portal For Academic Departments Chairmen

Mohammad A. ALHarthi [1] and Mohammad A. ALGhalayini [2]

[1]School Of Education
King Saud University, Riyadh, Saudi Arabia

[2]Director, Computer and Information Unit, Vice Rectorate for Knowledge Exchange and Technology Transfer
King Saud University, Riyadh, Saudi Arabia

Abstract

Today, many institutions and organizations are facing serious problem due to the tremendously increasing size of documents. This problem is further triggering the storage and retrieval problems due to the continuously growing space and efficiency requirements. This problem is becoming more complex with time and the increase in the size and number of documents in an organization. Therefore, there is a growing demand to address this problem. This demand and challenge can be met by developing a web-based database to enable specialized document imaging people to upload the frequently used forms and related information to use when there is a need. This automation process, if applied, attempts to solve the problem of allocating the information and accessing the needed forms to some extent. In this paper, we present an automation experience which is applied in King Saud University[1] to assist Academic Departments Chairmen finding all needed information and periodically used forms on an intranet site which proved to be very practical and efficient as far as optimizing the effort and time consumed for information and documents retrieval.

Introduction

Problem Definition

KSU has about 28 colleges. Each college consists, on the average, of 4 to 6 academic departments. Each of these departments is managed by a department chairman and an administrative team. The newly assigned department chairman is usually a faculty member who sometimes lacks many administrative skills. In addition, the academic department work process includes following certain regulations as well as completing certain preprinted forms for both employees/faculty and students. Many preprinted forms and regulations have to be present at all times to take some action when facing the various daily cases in the department. In the past, many cases took longer time to be decided on because the chairman has to know the exact process and has to retrieve the suitable form(s) used for such cases. This situation was affecting the effectiveness of the department chairman main duty and hence, the whole department administrative status.

A local web site which contains the processes and their corresponding regulations and related forms to be used was a must to make it easier and faster to refer and review any situation(s) regulations and related information, in addition to the related forms and documents[2] which are necessary to be used or/and filled for that specific situation.

Solution Steps

1 - A work force was formed to gather the cases the academic departments usually encounter.

2 - All the forms related to those cases where gathered and reprinted.

3 - The process to be followed by the department chairman is clarified and printed in detailed steps.

4 - A web-based site is prepared (Knowledge Management KSU Web Site) to refer to whenever needed (usually by the academic department administration).

5 - On the Knowledge Management web site, all gathered processes and information were

---

[1] More information about King Saud University is available at www.ksu.edu.sa

[2] Documents may include preprinted forms, reports forms, students transfer application forms, hardware/software request forms, etc.

K. Elleithy (ed.), *Advanced Techniques in Computing Sciences and Software Engineering*, DOI 10.1007/978-90-481-3660-5_73, © Springer Science+Business Media B.V. 2010

entered in addition to uploading all the related forms used for that case[3].

6 - Each process has a group of related processes which are links to it and specified by the web site administrator.

Special Notes About the KSU Knowledge Management Web Based Database

1 - This web site was designed using JAVA, PHP, and FrontPage.
2 - It was taken into consideration making this web site a very user friendly and totally in Arabic language and easy to use as much as possible.
3 - All KSU chairmen could access the web site on the local intranet from their office pc's with no difficulties.
4 - The web site server is central and is located in KSU administration building (The Internet Unit) for administration purposes.
5 - It is possible to print any page or pages of the forms or the documents stored on the site on the local printer.

Contents And Data Security

Since the web site contents do not contain private information which could be confidential, it was not necessary to the website designers to include a multi level security access layers; however, it was enough to have a dedicated administrator user name and password for data and contents manipulation purposes.

A monthly backup of the whole web site contents including the stored forms and documents is performed on a removable writable CD.

Hardware Used

1 - PC with the following specifications :
3.2 Mhz Intel processor
2048 MB DDR RAM
80 GB WD Hard Disk
256 MB ATI display adaptor
10/100 3Com network adaptor
52 X CD writer
17 inch display monitor

2 - Image scanner with the following specifications
40-60 pages per minute
Simplex / duplex
Black and white and color scanning

3 - Central file server with the following specifications :
2.8 Mhz Intel processor
1024 MB DDR RAM
80 GB WD Hard Disk
128 MB ATI display adaptor
10/100 3com network adaptor
40 X CD writer
17 inch display monitor

Software Used

1 - Linux operating system on the web site server
2 - PaperPort 11.0 professional[4] version for scanning images on the data entry PC.
3 - Acrobat writer (for creating the PDF files from word).

Time Consumed

No.	Details	Time consumed
1	Designing and implementing the web site	6 week
2	Buying and setting up the hardware.	2 weeks
3	Gathering the processes related to the department duties.	6 months
4	Gathering the forms related to the processes.	1 month
5	Data entry and forms upload	2 months

Screen Photos Of The KSU Knowledge Management Intranet Web Site

Figure (1) Shows the **"KSU Knowledge Management"** Intranet Web Site main page.

[3] Different form types where uploaded for each case (Word, PDF, on line-filled Flash forms).

[4] PaperPort is a trade name for ScanSoft Inc.

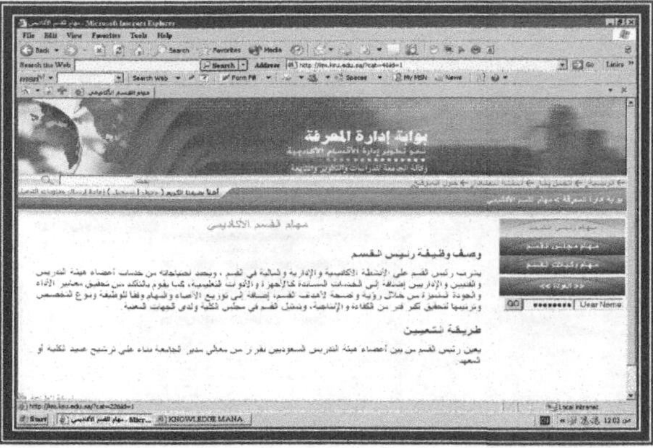

Figure (2) Shows the academic department description screen.

Figure (3) Shows the **"Department Chairman"** job description screen.

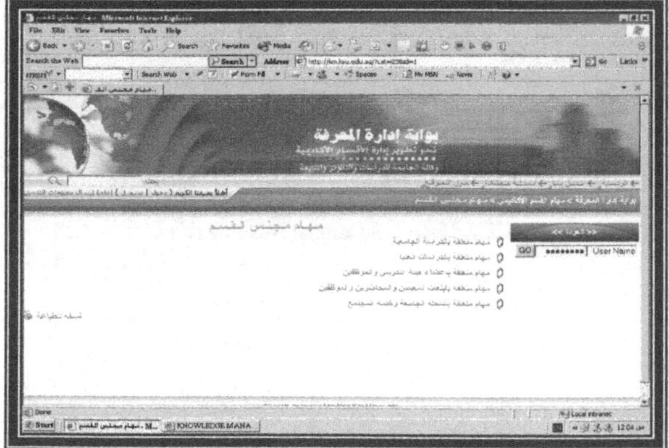

Figure (4) Shows the **"Department Council"** different tasks description screen.

Figure (5) Shows the detailed description for department council (university related issues).

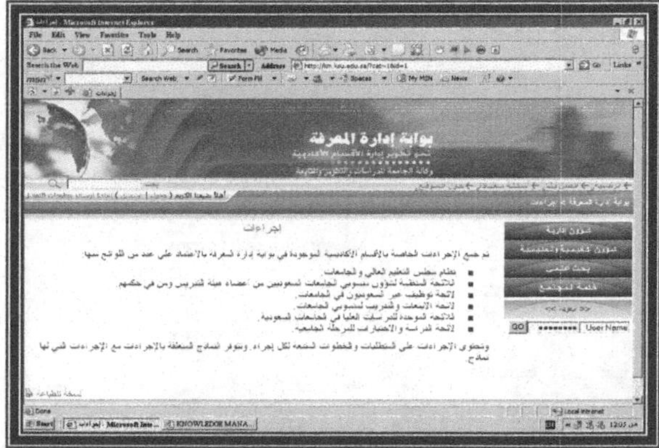

Figure (6) Shows the **"Processes"** main selection screen.

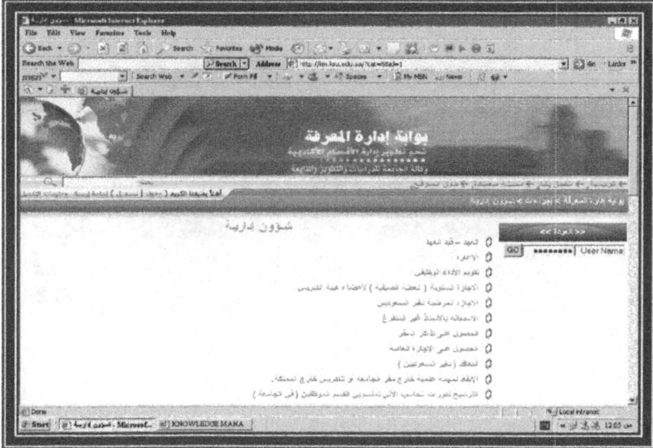

Figure (7) Shows the **"Administrative"** type processes selection list.

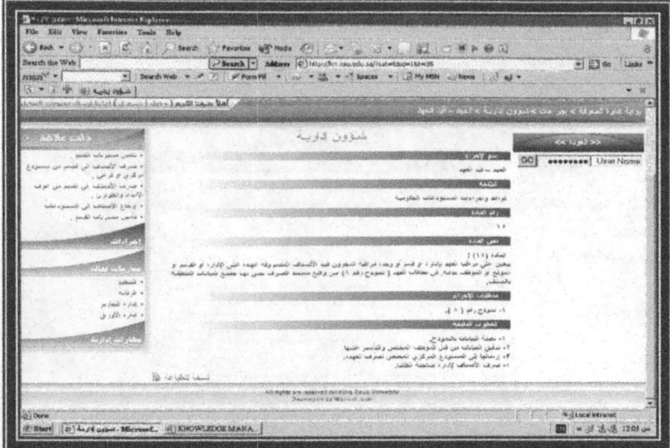

Figure (8) Shows one of the administrative processes detailed steps and related regulations.

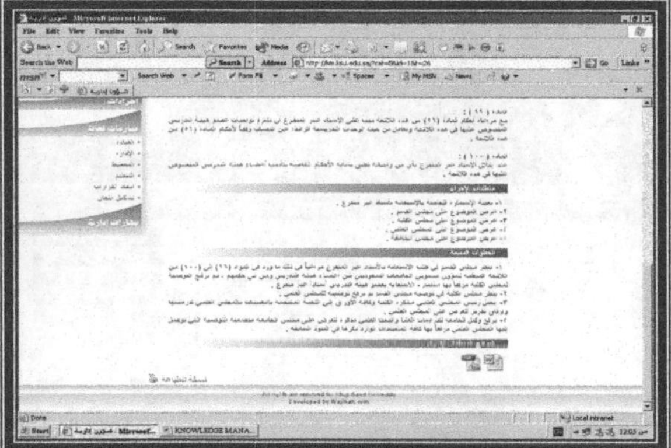

Figure (9) Shows the uploaded documents related to the active process being reviewed. In addition, it also shows the related process(s) and effective practices.

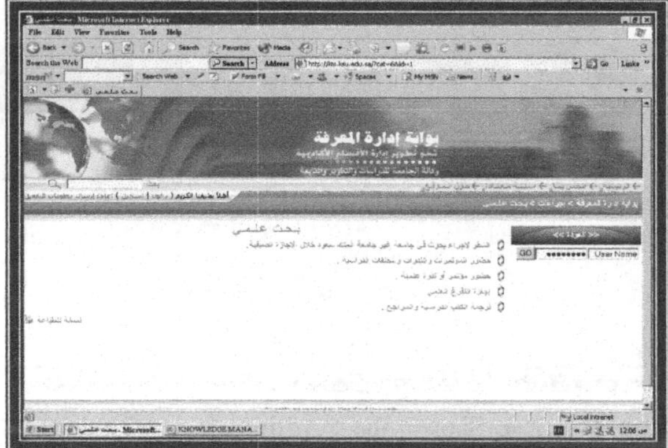

Figure (10) Shows the "**Scientific Research**" related processes.

Figure (11) Shows the "**Attending Conferences**" process steps and regulations.

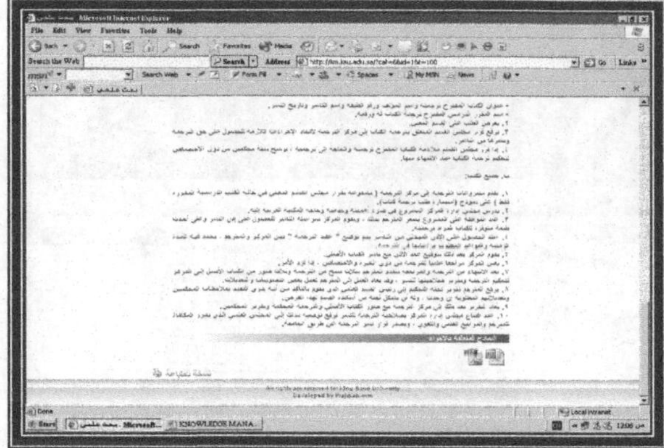

Figure (12) Shows the uploaded documents related to the "Attending Conferences" process being reviewed.

Figure (13) Shows the "**Academic and Educational**" type processes selections list.

Figure (14) Shows the "**Administrative Skills**" list to assist the chairman in his/her administrative operation.

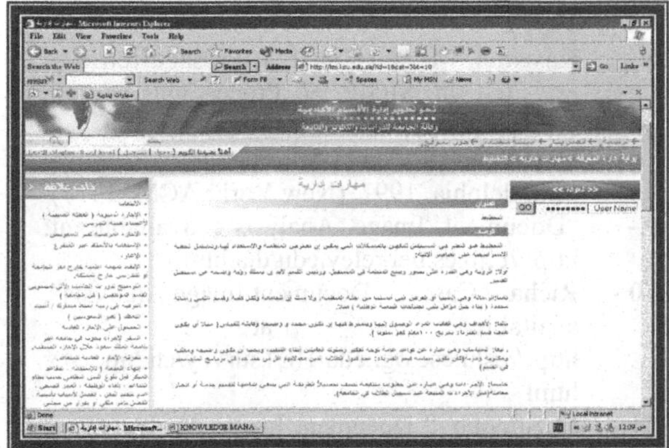

Figure (15) Shows the planning as an administrative skill in details.

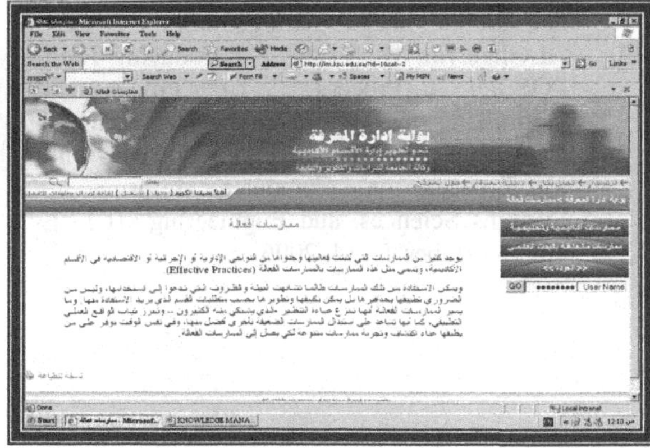

Figure (16) Shows the "**Effective Practices**" main screen.

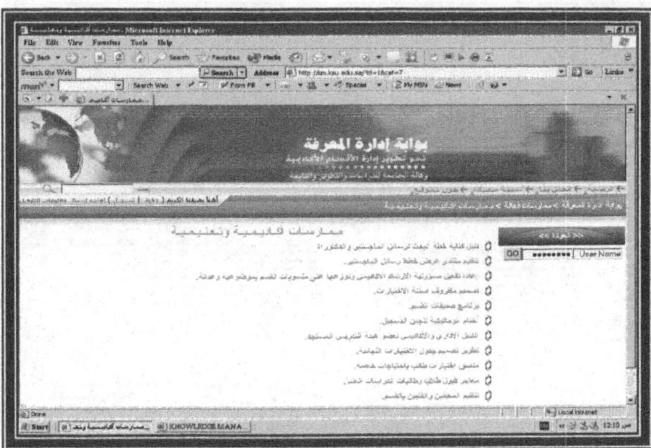

Figure (17) Shows the academic and educational effective practices selection list.

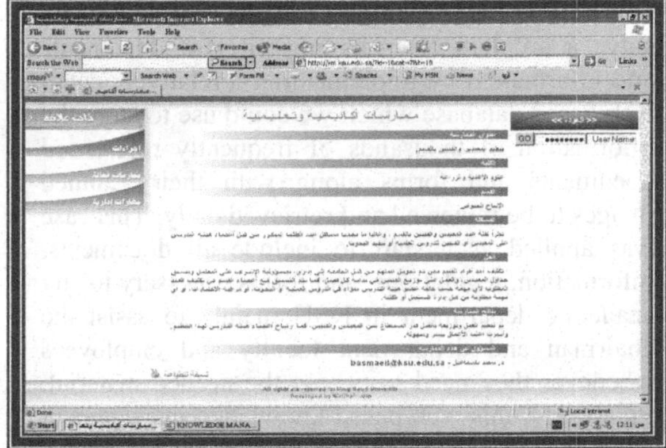

Figure (18) Shows the details of one of the academic and educational effective practices.

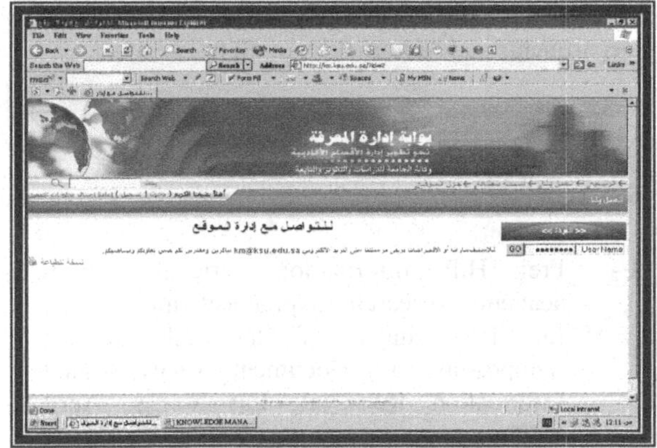

Figure (19) Shows the "**Contact Information**" screen.

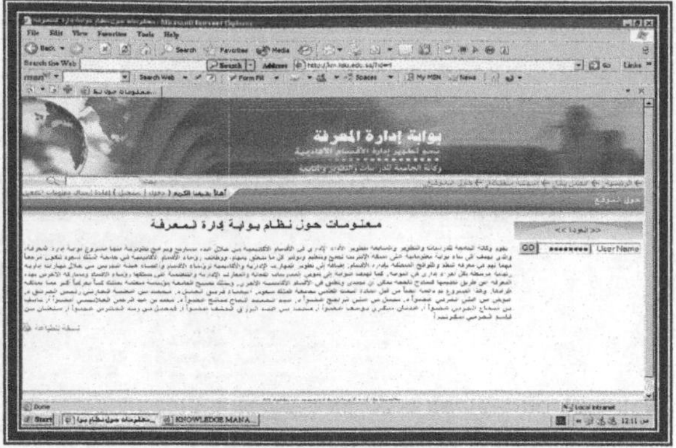

Figure (20) Shows the web site "**About**" screen.

Conclusion

Our goal in this paper was to show how we could solve a problem of managing piles of documents and information by implementing a relatively small web-based database which we could use to store the information of thousands of frequently referenced documents and forms along with their scanned images to be reviewed and retrieved daily. This case was applied as a start to include all documents, information, regulations, and forms necessary to any academic department in KSU mainly to assist the chairman and department faculty and employees whenever they need to access the needed material. Practically, this web-based knowledge management site was able to save the KSU academic departments members a lot of effort and time in referring to any specific document, information, or form and information retrieval which added a major efficiency credit to the role of KSU Academic Departments and "automation of all tasks" trend.

References

1 - Frei, H.P. Information retrieval - from academic research to practical applications. In: Proceedings of the 5th Annual Symposium on Document Analysis and Information Retrieval, Las Vegas, April 1996.

2 - Clifton, H., Garcia-Molina, H., Hagmann, R.: ''The Design of a Document Database'', Proc. of the ACM

3 - ISO/IEC, Information Technology - Text and office systems - Document Filing and Retrieval (DFR) -

4 - Fuhr, N. Toward data abstraction in networked information systems. Information Processing and Management 5(2) (1999) 101-119.

5 - Jacobson, I., Griss, M. and Jonsson, P. Software reuse: Architecture, process and organization for business success. (New York: ACM Press, 1997).

6 - Davenport, T.H. Information ecology: Mastering the information and knowledge environment. (Oxford: Oxford University Press, 1997).

7 - Bocij, P. et al., Business information systems: Technology, development and management. (London: Financial Times Management, 1999).

8 - Kaszkiel, M. and Zobel, J. Passage retrieval revisited. In: SIGIR'97: Proceedings of 20th ACM-SIGIR Conference on Research and Development in Information Retrieval Philadelphia, 1997. (New York: ACM, 1997).

9 - "Document Image Analysis", available at: http://elib.cs.berkeley.edu/dia.html

10 - Richard Casey, " Document Image Analysis", available at : http://cslu.cse.ogi.edu/HLTsurvey/ch2node4.html

11 - M Simone, " Document Image Analysis And Recognition" available at : http://www.dsi.unifi.it/~simone/DIAR/

12 - "Read and Display an Image" available at : http://www.mathworks.com/access/helpdesk/help/toolbox/images/getting8.html

13 - ALGhalayini M., Shah A., "Introducing The (POSSDI) Process : The Process of Optimizing the Selection of The Scanned Document Images". International Joint Conferences on Computer, Information, Systems Sciences, and Engineering (CIS^{2}E 06) December 4 - 14, 2006

Configurable Signal Generator Implemented on Tricore Microcontrollers

M. Popa, I. Silea, M. Banea
"Politehnica" University from Timisoara, Romania

Abstract—**Most of the electronic measurements laboratories need signal generators capable to generate several types of signals with different shapes, frequencies and amplitudes. Unfortunately the more versatile such a signal generator will be the higher its cost will be. The alternative is the PC based instrumentation. There are classical tools, as Lab VIEW, but sometimes a better approach is to develop a PC based instrumentation system targeted to one or few of the electronic measurement operations.**

This paper describes a configurable signal generator implemented with the PWM method on the advanced 32 bit Tricore microcontroller family. Several types of signals can be generated: sinus, pulse, saw tooth and custom. The proposed configurable signal generator is targeted to low frequency applications (1 Hz – 600 Hz) useful in motor control, laboratory experiments etc. The system offers the flexibility and versatility of the PC based electronic instrumentation, the performance of a 32 bit microcontroller and low cost.

I. Introduction

Electronic measurements are necessary in various fields from industrial, research, education to domestic ones. Most of the electronic measurements laboratories need signal generators capable to generate several types of signals with different shapes, frequencies and amplitudes. Unfortunately the more versatile such a signal generator will be the higher its cost will be.

The alternative is the PC based instrumentation. There are important achievements in this area, for instance Lab VIEW. The name comes from **Lab**oratory **V**irtual **I**nstrumentation **E**ngineering **W**orkbench and is a platform and development environment for a visual programming language from National Instruments. Lab VIEW includes a virtual signal generator and is commonly used for data acquisition, instrument control, and industrial automation on a variety of platforms including Microsoft Windows, UNIX, Linux and Mac OS.

Lab VIEW and other classical PC based instrumentation achievements cover multiple sides of the electronic measurement process and consequently have high prices. In many situations a better approach is to develop a PC based instrumentation system targeted to one or few of the operations involved in the electronic measurement. Such a system will be less performing than a classical one but its cost will be significantly reduced.

This paper describes a configurable signal generator implemented with the PWM method on the advanced 32 bit Tricore microcontroller family. Several types of signals can be generated: sinus, pulse, saw tooth and custom. The next section presents related works, the third section details the proposed signal generator, the fourth section presents experimental results and the last section outlines the conclusions.

II. Related Work

Electronic measurements achievements and applications based on them are described in many papers.

Reference [1] approaches the problem of digital control of DC-to-DC power converters. It is shown that flexible control algorithms were realized using microcontrollers or DSPs. The paper evaluates the closed loop performance benefits that the nonlinear algorithms can bring to DC-to-DC power converters and to evaluate digital control opportunities. Variable PWM frequency will be used as a method of improving low power converter efficiency.

Reference [2] presents the use of FPGAs, DSPs and microcontrollers in the implementation of discrete time control of motor driven actuator systems. Very few of such devices are produced to support any level of radiation or harsh environment encountered on spacecraft. The paper is focused on creating a fully digital, flight ready controller design that utilizes an FPGA for implementation of signal conditioning for control feedback signals, generation of commands to the controlled system, and hardware insertion of adaptive control algorithm approaches.

The paper from reference [3] describes another application of PWM, a low-cost single-chip PI-type fuzzy logic controller design and an application on a permanent magnet dc motor drive. The presented controller application calculates the duty cycle of the PWM chopper drive and can be used to dc–dc converters as well. The contribution of this paper is to present the feasibility of a high-performance non-linear fuzzy logic controller which can be implemented by using a general purpose microcontroller without modified fuzzy methods.

A virtual signal generator with a specific destination is described in reference [4]. A hard disk read back signal generator designed to provide noise-corrupted signals to a channel simulator has been implemented on a Xilinx VirtexTME FPGA device. The generator simulates pulses sensed by read heads in hard drives. All major distortion and noise processes, such as intersymbol interference, transition noise, electronics noise, head and media nonlinearity, intertrack interference, and write timing error, can be generated by the user. Another electronic measurement application is described in reference [5], an integrated test core for mixed-signal circuits. It consists of a completely digital implementation, except for a simple reconstruction filter and a comparator. It is capable of both generating arbitrary band-limited waveforms (for excitation purposes) and coherently digitizing arbitrary periodic analog waveforms (for DSP-based test and measurement). A prototype IC was fabricated in a 3.3 V 0.35 μm CMOS process.

K. Elleithy (ed.), *Advanced Techniques in Computing Sciences and Software Engineering*,
DOI 10.1007/978-90-481-3660-5_74, © Springer Science+Business Media B.V. 2010

Reference [6] presents experiments that are part of a virtual laboratory. The activity is organized around a spectrum analyzer and a function generator, as real devices mirrored into Web (Internet) nodes. A logically structured replica of the Advantest spectrum analyzer is realized on a Lab VIEW platform, in a client-server approach. Students from different locations and enrolled in different can access the virtual laboratory.

The present paper describes a PWM based configurable signal generator implemented on a 32 bit Tricore microcontroller connected to a PC. The system offers the flexibility and versatility of the PC based electronic instrumentation, the performance of a 32 bit microcontroller and low cost.

III. The Configurable Signal Generator

The proposed configurable signal generator is targeted to low frequency applications (1 Hz – 600 Hz) useful in motor control, laboratory experiments etc.

The system can generate different types of signals: sinus, pulse, saw tooth, and custom with the maximum amplitude of 3.3 V. This maximum value comes from the "high" logic value of the microcontroller and can be increased with extra hardware. The signals can be generated on maximum 8 PWM outputs of an advanced 32 bit Tricore microcontroller. The first four outputs generate normal signals and the other four outputs generate inversed signals.

Fig. 1 presents the block diagram of the system. Through the PC a user sends the parameters of the signals to be generated to the Development Kit based on the TC1766 Tricore microcontroller. The Development Kit generates the requested signals which can be visualized on an oscilloscope.

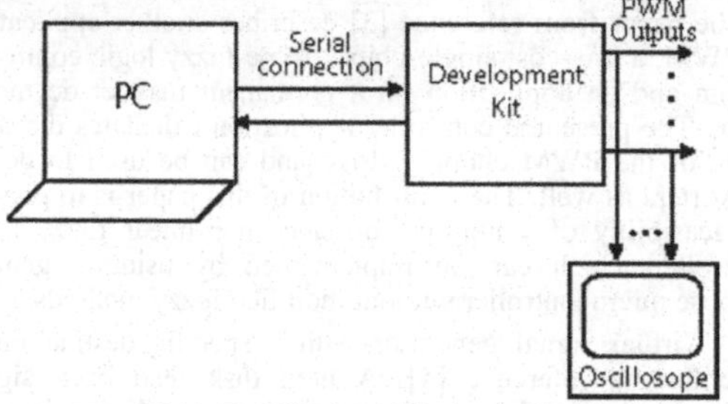

Fig. 1. Block diagram of the signal generator

A. *The hardware implementation*

The hardware implementation is based on the TC7166 microcontroller, [7]. It is an advanced 32 bit microcontroller from the Tricore family with RISC architecture, operations and addressing modes specific to DSPs and on-chip memories and peripherals. Its main features are:

- high performing 32 bit CPU;
- a 4 GB unified address space;
- fast task switching and flexible power management;
- on-chip memories: program memory (1504 KB Flash, 16 KB Scratch-Pad RAM, 8 KB Instruction Cache and 16 KB Boot ROM), data memory (56 KB Local data RAM, 32 KB data Flash), SRAMs with parity error detection;
- interrupts system: 103 sources, 256 priority levels, fast or normal interrupt service;
- 8 DMA channels, 81 general purpose inputs/outputs;
- powerful on-chip peripherals: two USART serial interfaces, CAN interface, SSC interface, two Micro Link interfaces, timers, 32 inputs ADC;
- emulation, debug and test modes.

The communication with the PC is done on its RS 232 interface connected to the CAN interface of the microcontroller through two converters: a RS 232 – k line converter followed by a k line – CAN converter.

The PWM signals are obtained by program. Internal counters are used and their content (C) is continuously compared with programmed values (PW). If $C < PW$, 0 logic will be sent on the PWM output and if $C \geq PW$, 1 logic will be sent on the PWM output. This signal is passed through a filter for obtaining the desired sinus, saw tooth or custom signal.

B. *The software implementation*

The user interface permits to the user to choose among:

- SIN Generator: Generate, Fill parameters, Exit;
- PULSE Generator: Generate, Fill parameters, Exit;
- SAW TOOTH Generator: Generate, Fill Parameters, Exit and
- CUSTOM Generator: Generate, Fill parameters, Save/Load parameters, Draw wave, Exit.

The application was conceived in Visual C++ based on MFC (Microsoft Foundation Class) library which permits easy implementation of the Windows based programs. The application has a graphical interface with dialog windows specific to the Microsoft Visual C++ environment.

Fig. 2 presents the architecture of the application running on the PC.

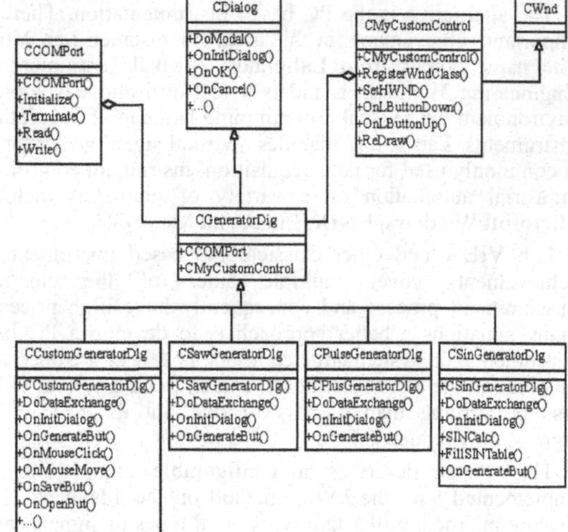

Fig. 2. Architecture of the application running on PC

The four dialog types interacting with the user inherit the *CGenerator.Dlg* class. This, at its turn, inherits the *CDialog* class included in MFC. The *CGeneratorDlg* class contains

CCOMPort objects for communicating through the serial (COM1) port with the microcontroller and a *CMyCustomControl* object used as output for illustrating the signals generated by the microcontroller or as input for receiving a customized signal.

Sinus signal generator

An instance of the *CSinGeneratorDlg* class is called in order to fulfill this functionality, fig. 3.

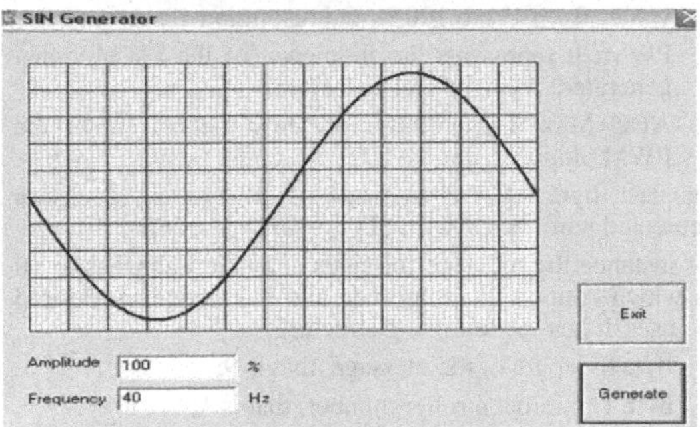

Fig. 3. CSinGeneratorDlg

The required parameters are given and the Generate operation is chosen. The application sends the two parameters to the microcontroller. The waveform is determined with the method *FillSINTable()*:

```
for (i=0;i<SCREEN_SIZE_H;i++)
    {
    value = (float)(2*i*PI)/(SCREEN_SIZE_H));
    result = (float)SINCalc(value,11);
    result1 = (int)((result)*115);
    point.x = i+1;
    point.y = (int)result1;
    m_trace.AddTail(point);
    }
```

The function which calculates the value for the sin function in a certain point is:

```
while (n<iteration){
    count = 0;
    neg1 = 1;
    while (count<n){
            neg1 = -1 * neg1;
            count = count + 1;
    }
    count = 0;
    x2nplus1 = 1;
    while (count<(2*n)+1){
            x2nplus1 = x2nplus1 * Radians;
            count = count + 1;
    }
    count = 2 * n +1;
    factorial = 1;
    while (count>0){
            factorial = factorial * count;
            count = count – 1;
    }
    sine = sine + neg1*(x2nplus1/factorial);
    n = n + 1;
}
```

Saw tooth signal generator

An instance of the *CSaw toothGeneratorDlg* class is called for this functionality, fig. 4.

The application sends the three parameters to the microcontroller.

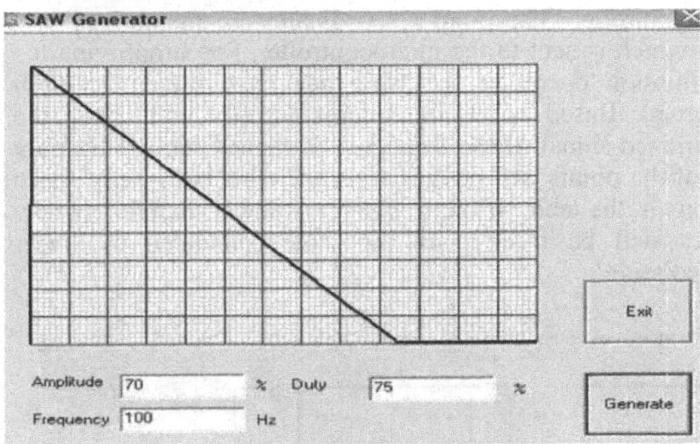

Fig. 4. CSaw toothGeneratorDlg

The slope of the wave is calculated with the following formula:

$m = (y - y_1)/(x - x_1)$

and the points from the line are given with the formula:

$y_1 = m(x - x_1)/y$.

The waveform is generated with the method *OnGenerateBut()*:

```
j = (int)((SCREEN_SIZE_H * (floati_d)/(float)100);
slope = (float)((-1) * (1./j));
point.x = 0;
point.y = SCREEN_SIZE_V;
m_trace.AddTail(point);
for (i=1;i<j;i++);
{
    value = (float)(slope*(i-0));//y1 = [m*(x1 – x)]/y;
    value = (float)(value/1);
    result1 = (int)((value)*SCREEN_SIZE_V);
    result1 = (int)((result1+SCREEN_SIZE_V);
    point.x = i+1;
    point.y = (int)(SCREEN_SIZE_V – result1);
    m_trace.AddTail(point);
}
for (i=j;i<SCREEN_SIZE_H;i++)
{
    point.x = i+1;
```

point.y = SCREEN_SIZE_V;
m_trace.ADDTail(point);

}

Pulse signal generator

For this functionality an instance of the *CPulseGeneratorDlg* class is called, fig. 5. The application sends the three parameters to the microcontroller. The waveform is determined with the method *OnGenerateBut()* which will hold the signal on high level for a Duty time duration and on low level for the time duration Period (1/Frequency) – Duty.

Custom signal generator

An instance of the *CCustomGeneratorDlg* class is called for this functionality, fig. 6. The parameters are sent to the microcontroller. The point's coordinates are memorized in a table which is sent to the microcontroller. For simplifying the transmission operation the table has fixed dimension (200 elements). Based on it, the microcontroller will create the customized signal. If the frequency is higher than 40 Hz only part of the points will be written in the table because of space reasons in the table. If the frequency is lower than 40 Hz new values will be inserted in the table calculated by linear interpolation.

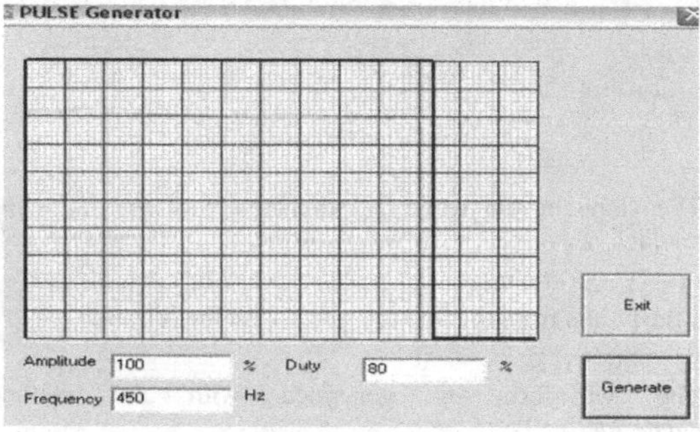

Fig. 5. CPulseGeneratorDlg

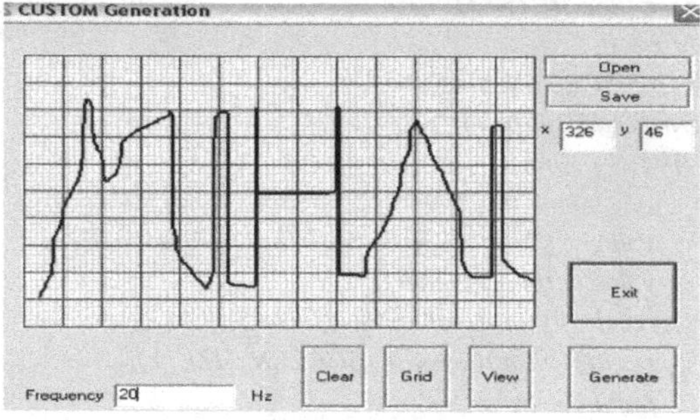

Fig. 6. CSaw toothGeneratorDlg

The custom wave is assumed from the user through a panel which can be also output, for displaying the PC generated signal.

Next, several details will be given concerning the PC – microcontroller communication. For accessing the physical resources of the serial port, in order to send or receive bytes, the Windows API interface was used. The serial port is opened with the CreateFile function, data are transferred with WriteFile and ReadFile functions and the port is closed with the function CloseHandle.

The PC sends the following parameters:

- the frequency: it can be from 10 Hz to 600 Hz with a resolution of 1 Hz;
- amplitude: it represents the PWM duty-cycle with values from 1% to 100 % and resolution 1 %;
- PWM: it represents the frequency of the PWM signal generated; it can be 2, 4 or 8 KHz;
- ADC Max. Current: it represents the current at the PWM output.

The last byte of the message is always a checksum implemented with the EXCLUSIVE – OR operation.

For instance the message for sinus activation consists in 10 bytes which contain the amplitude and the frequency desired by the user. It has the following structure:

- byte 0: length of the message, that is 9;
- byte 1: microcontroller number, that is 1;
- byte 2: command, that is 38 for sinus generation;
- byte 3: MSB of the frequency;
- byte 4: LSB of the frequency;
- byte 5: amplitude;
- byte 6: PWM;
- byte 7: MSB of ADC max. current;
- byte 8: LSB of ADC max. current and
- byte 9: checksum.

The message for saw tooth signal activation has 10 bytes, the same as for pulse signal activation, while the message for custom signal activation has 209 bytes, 200 of them representing the waveform draw by the user with the mouse.

The microcontroller verifies the request type, the parameters and the checksum and if all the bytes are according to the protocol, it executes the request and sends back an acknowledge indicating "No error".

IV. Experimental Results

Fig. 7 presents sinus signals.

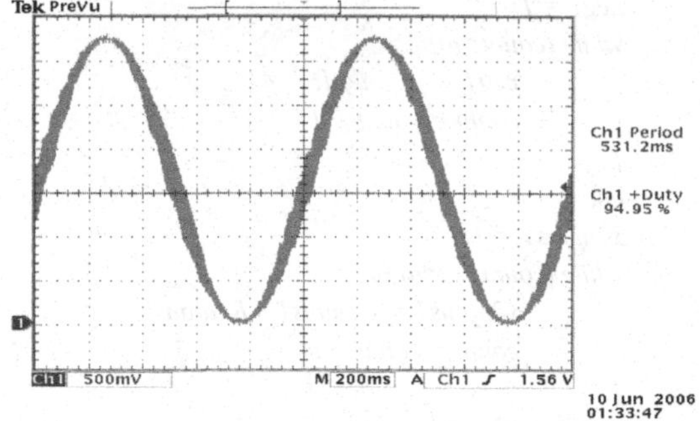

A = 100%, F = 1 Hz, 8000 points/period

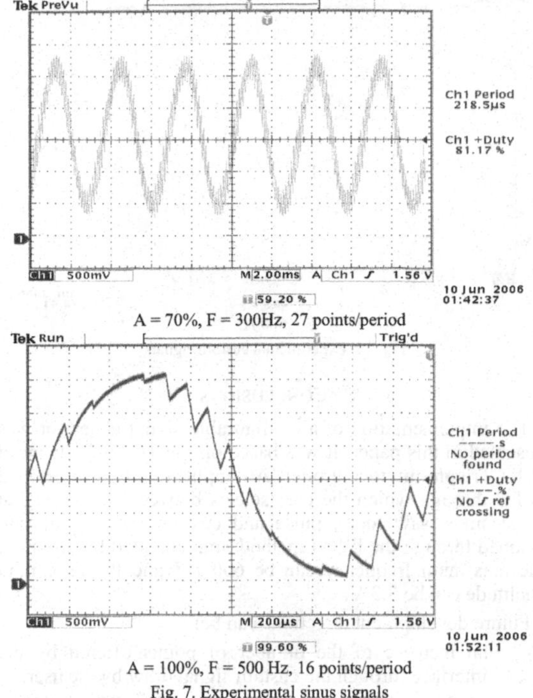

A = 70%, F = 300Hz, 27 points/period

A = 100%, F = 500 Hz, 16 points/period

Fig. 7. Experimental sinus signals

The last picture shows the signal obtained by PWM and passed through the filter.

Fig. 8 presents experimental saw tooth signals.

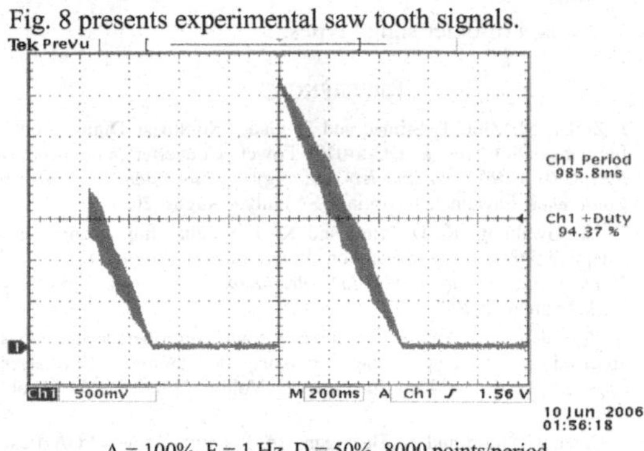

A = 100%, F = 1 Hz, D = 50%, 8000 points/period

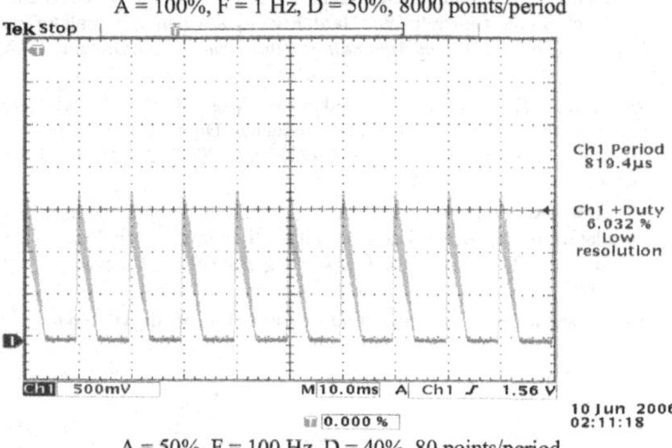

A = 50%, F = 100 Hz, D = 40%, 80 points/period

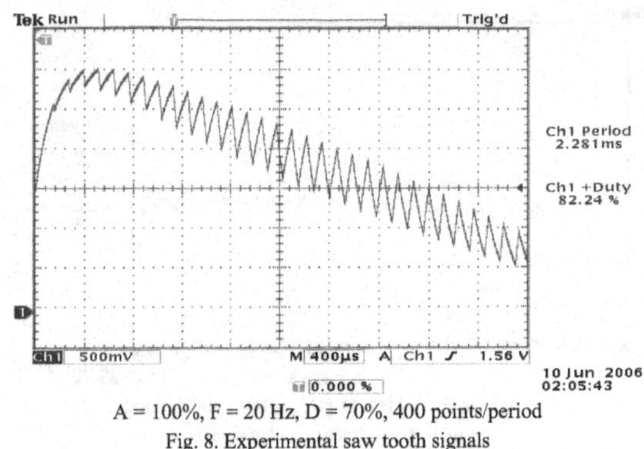

A = 100%, F = 20 Hz, D = 70%, 400 points/period

Fig. 8. Experimental saw tooth signals

The last picture shows the signal obtained by PWM and passed through the filter.

Fig. 9 presents experimental pulse signals.

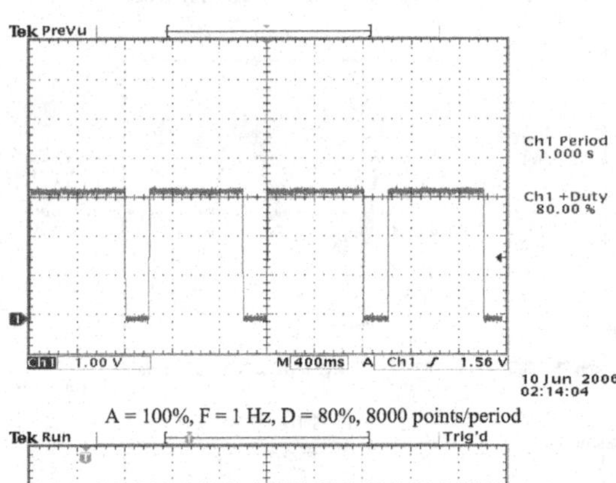

A = 100%, F = 1 Hz, D = 80%, 8000 points/period

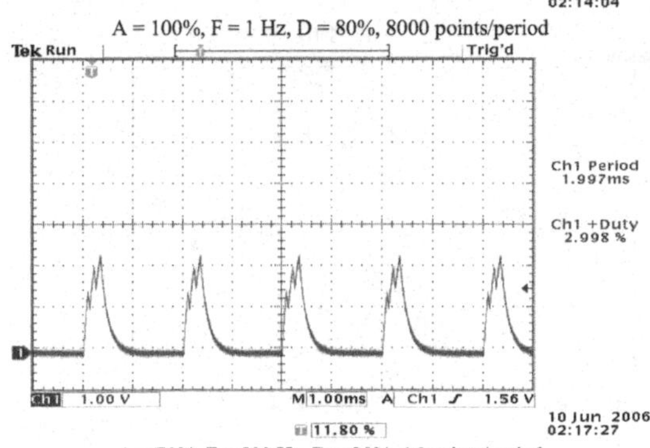

A = 70%, F = 500 Hz, D = 20%, 16 points/period

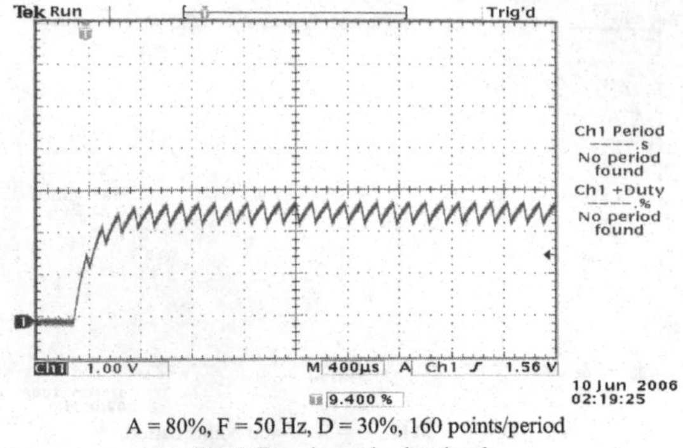

A = 80%, F = 50 Hz, D = 30%, 160 points/period

Fig. 9. Experimental pulse signals

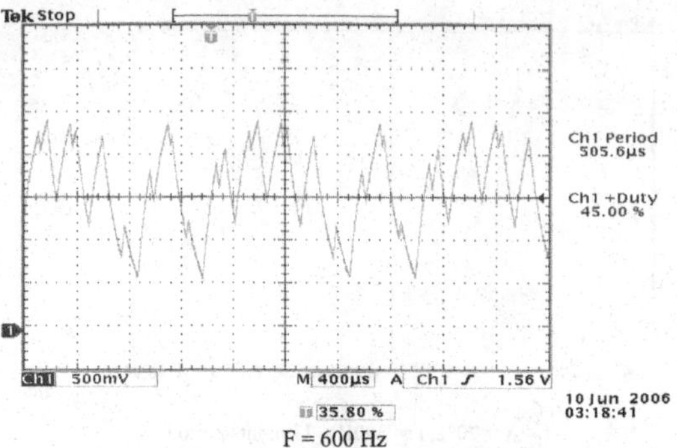

F = 600 Hz

Fig. 10. Experimental custom signals

The last picture shows the signal obtained by PWM and passed through the filter.

Fig. 10 presents experimental custom signals. The customized signal corresponds to the MIHAI word.

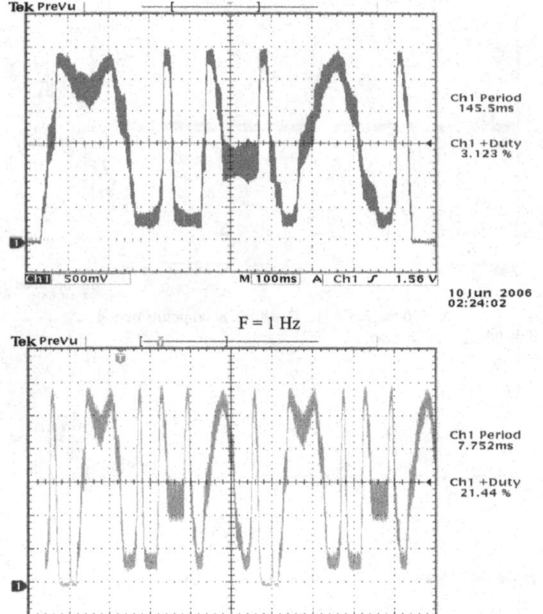

F = 1 Hz

F = 20 Hz

V. Conclusions

The implementation of a configurable signal generator was presented in this paper. It was based on the TC1766 advanced 32 bit Tricore microcontroller. An application was developed on PC through which the user selects between several signal types: sinus, saw tooth, pulse and custom. The signals are obtained through the PWM method, implemented by program. The maximum frequency can be 600 HZ and the maximum amplitude can be 3.3 V.

Future development directions can be:

- an increase of the number of points offered by the interface through the custom signal draw by the user;
- a faster transmission between the PC and the microcontroller by eliminating the RS 232 - k – line converter;
- the add of other signal types.

References

[1] J. Zeller, M. Zhu, T. Stimac and Z. Gao, "Nonlinear Digital Control Implementation for a DC-to-DC Power Converter", in *Proc. of IECEC'01 36th Intersociety Energy Conversion Engineering Conference,* Savannah, Georgia, USA, July – August 2001

[2] D. A. Gwaltney, K. D. King and K. J. Smith, "Implementation of Adaptive Digital Controllers on Programmable Logic Devices", in *Proc. of the 5th Annual MAPLD International Conference,* Maryland, USA, September 2002

[3] S. Pravadalioglu, "Single – chip fuzzy logic controller design and an application on a permanent magnet dc motor", *Engineering Applications of Artificial Intelligence*, Volume 18, Issue 7, October 2005

[4] J. Chen, J. Moon and K. Bazargan, "A Reconfigurable FPGA Based Readback Signal Generator For HardDrive Read Channel Simulator", in *Proc. of The 39th Design Automation Conference,* New Orleans, USA, June 2002

[5] M. Hafed, N. Abaskharoun and G.W. Roberts, "A Stand-Alone Integrated Test Core for Time and Frequency Domain Measurements", in *Proc. of International Test Conference 2001,* Baltimore, USA, October – November 2001

[6] M. Albu, K. Holbert and F. Mihai, "Online Experimentation and Simulation in a Signal Processing Virtual Laboratory", in *Proc. of the International Conference on Engineering Education,* Valencia, Spain, July 2003

[7] http://www.infineon.com/upload/Document/TC1766_ds_v0_6.pdf

Mathematic Model of Digital Control System with PID Regulator and Regular Step of Quantization with Information Transfer via the Channel of Plural Access

Abramov G.V., Emeljanov A.E., Ivashin A.L.
Voronezh State Technological Academy
agw@vgta.vrn.ru

Abstract: **Theoretical bases for modeling a digital control system with information transfer via the channel of plural access and a regular quantization cycle are submitted. The theory of dynamic systems with random changes of the structure including elements of the Markov random processes theory is used for a mathematical description of a network control system. The characteristics of similar control systems are received. Experimental research of the given control systems is carried out.**

I. Introduction

The development of automated production processes and production is aimed at the use of open Internet/Intranet system solutions of Ethernet technologies[1, 2]. The application of this solution on the low level of automation will bring some specifity in the process of digital control systems functioning. Nowadays there is no adequate mathematical description of digital control systems (DCS) which use as information media multiple access channel (MAC) in particular the Ethernet realizing plural access method with the carrier-sense control and collisions detection (CSMA/CD).

The article deals with structural mathematical modeling of digital control system with information transfer via multiple access channel. The network Ethernet is considered as the MAC.

II. The Mathematical Model

Fig. 1 provides a functional diagram of the digital control system.

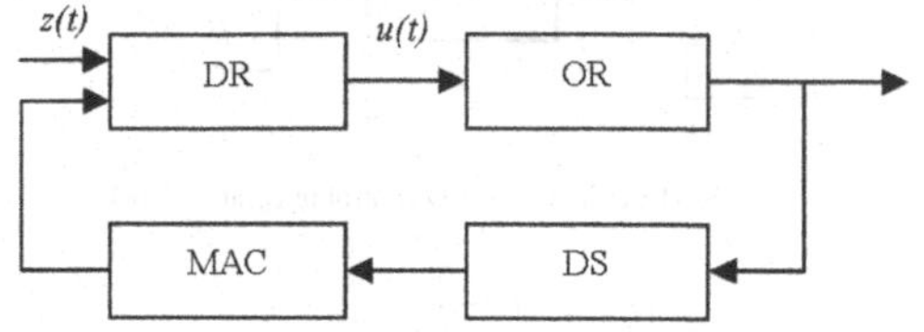

Fig. 1 Functional diagram of the digital control system.

The system works in the following way: a digital sensor (DS) with a given quantization cycle T_0 quantizises an output signal of an object of regulation (OR). After reading out the data DS transmits them to digital regulator (DR) via MAC. If during T_0, i.e. before a new quantization of the OR output signal no data have been transmitted to DR, then DS carries out a new quantization and replaces the old data with the new ones. In doing so, new data should be transferred via MAC. DR at each quantization cycle makes a control action on OR. In doing so, work DS and DR should be synchronized.

The peculiarity of the given NCS operation is that if the data received from the DS by DR during the quantization cycle T_0, then they will be taken into account when generating control action in the next cycle. If the data are not received then for calculating a control action the previously received data are used. Data transfer via the MAC is accidental as transfer time depends on the channel traffic and the number of conflicts is characterized by the distribution law $f_r(t)$. Data transfer via MAC during T_0 with a probability p is:

$$p = \int_0^{T_0} f_r(t)dt.$$

Thus, this NCS under consideration is stochastic. For a mathematical description of stochastic systems the Markov random processes theory has been widely spread and in particular, the theory systems with random changes in the structure [3]. To use the approaches and methods of these theories for the NCS being considered we assume that simultaneously with the control action the white noise $\zeta(t)$ is fed in the output OR.

Let us build a structural model of the given NCS assuming that a mathematical apparatus of the mentioned theories will be used for its description. Taking into account that the regulating action from the DR is issued only in moments of quantization we shall assume that the data from DS read out from the output OR at the moment $t=(k-1)T_0$ pass to DR at the moment $t=kT_0$ with the probability p where they are processed instantly and are taken into account when generating control action issued to

K. Elleithy (ed.), *Advanced Techniques in Computing Sciences and Software Engineering*,
DOI 10.1007/978-90-481-3660-5_75,

the OR at the time point $t=kT_0$. Assume that DR realizes PID law regulation. We shall submit this law as follows:

$$u[kT_0]=u_P[kT_0]+u_I[kT_0]+u_D[kT_0]. \quad (1)$$

Where $u[kT_0]$ - output DR,
$u_P[kT_0]$ - proportional component of the control action,
$u_I[kT_0]$ - integral component,
$u_D[kT_0]$ - differential component.

For individual components of regulating action $u[kT_0]$ we have:

$$u_P[kT_0]=k_P e[kT_0],$$
$$u_I[kT_0]=u_I[(k-1)T_0]+k_I T_0 e[kT_0],$$
$$u_D[kT_0]=k_D\frac{e[kT_0]-e[(k-1)T_0]}{T_0}.$$

Where k_P, k_I, k_D - configuration options DR, $e[kT_0]$ - mismatch for the time $t=kT_0$, $k=0, 1, 2, \ldots$.

Let us draw up a block diagram for implementation of the law regulation components. To do this in the beginning we shall study a structural pattern of error formation block (EFB). Fig. 2 presents a block diagram of the error formation block where $z(t)$ — representation of DR, $y_1(t)$ — output of DS, $y_2(t)$ - MAC output (cell memory).

$$e_1(t)=z(t)-y_1(t);$$
$$e_2(t)=z(t)-y_2(t).$$

Discrete part of the error block consists of the error quantizer k_1 and hold element which includes an integrating element.

The error $e_1(t)$ takes into account the data from the DS which with the probability p will be transferred via MAC in the next quantization cycle. The error $e_2(t)$ takes into account the data from the DS which are already received via MAC at the previous quantization cycle.

Thus, $e_1(t)$ and $e_2(t)$ - possible errors which with some probability can be realized in the next quantization cycle, $e_3(t)$ – is the error put into the previous quantization cycle.

In time $t=kT_0$ quantizer k_1 with probability p quantizises output *1* and $e_3(t)=e_1(t)$. This corresponds to the occasion of error $e_1(t)$ realization taking into account the transfer of data DS to DR via MAC. With the probability $(1-p)$ quantizer k_1 quantizeses output *2* and $e_3(t)=e_2(t)$. This corresponds to the occasion of error $e_2(t)$ realization without data transfer via MAC.

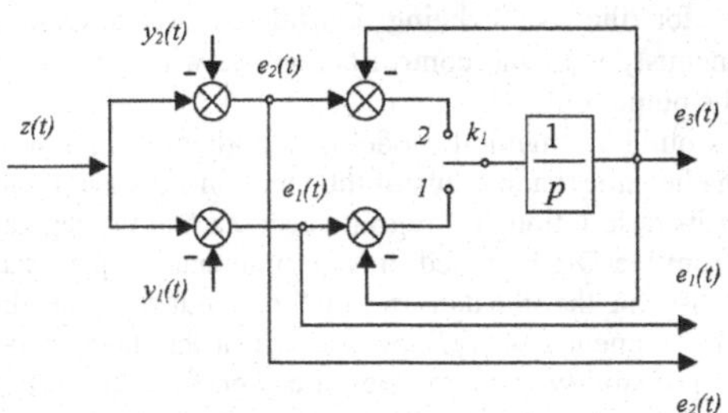

Fig. 2 The block diagram of the error formation block.

For the error $e_3(t)$ we have:

$$\frac{de_3(t)}{dt}=(e_1(t)^- - e_3(t)^-)g_1(t)- -(e_2(t)^- - e_3(t)^-)g_2(t). \quad (2)$$

Here a sign "-" above a symbol of the error $e(t)$ indicates that at the quantization time their values are taken in tend to this point of time on the left. This designation will be further used in case of phase coordinate system. It reflects the peculiarity of the quantizer and hold element operation [4].

$g_1(t)=\sum_{i=-\infty}^{\infty}\delta(t-t_i)$ and $g_2(t)=\sum_{j=-\infty}^{\infty}\delta(t-t_j)$ are random sequences of delta-pulses. Moments of pulses appearance t_i and t_j match respectively time points of outputs *1* and *2* quantization (fig. 2). In doing so, as it have been mentioned these points do not coincide that is $t_i \neq t_j$. In addition, the data points multiple of quantization cycle T_0. It should be noted that:

$$\sum_{i=-\infty}^{\infty}\delta(t-t_i)+\sum_{j=-\infty}^{\infty}\delta(t-t_j)=\sum_{k=-\infty}^{\infty}\delta(t-kT_0).$$

Where $\sum_{k=-\infty}^{\infty}\delta(t-kT_0)=g_3(t)$ is regular sequence of delta-impulses. Thus, a regular sequence of issuing control action on the object of regulation is modeled by a sum of two random sequences.

For the proportional part of PID control we have: $u_P(t)=k_P e_3(t)$. Here $u_P(t)$ modifies its value in discrete points of time $t=kT_0$ according to the (2). Fig. 3 displays the diagram integrating block of PID control.

Quantizer k_2 working synchronously with the quantizer k_1 and exits quanta of outputs respectively. For an integral component we have:

$$\frac{du_I(t)}{dt}=k_I T_0 e_1(t)^- g_1(t)+k_I T_0 e_2(t)^- g_2(t).$$

Fig. 4 represents a block diagram of differentiating PID control block. Quantizer k_3 works synchronously with the quantizer k_1 and k_2.

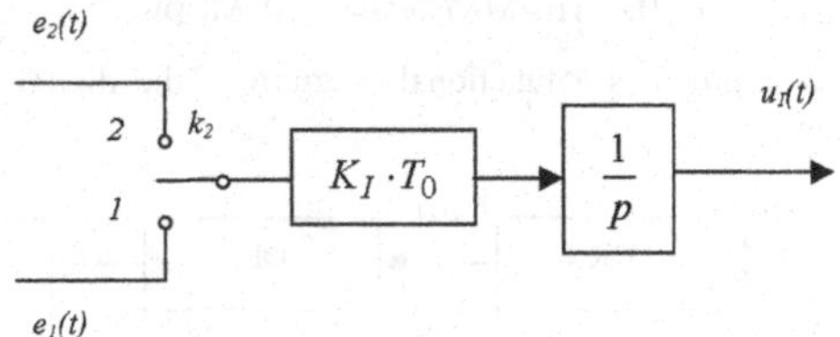

Fig. 3. The diagram of PID control integrating block.

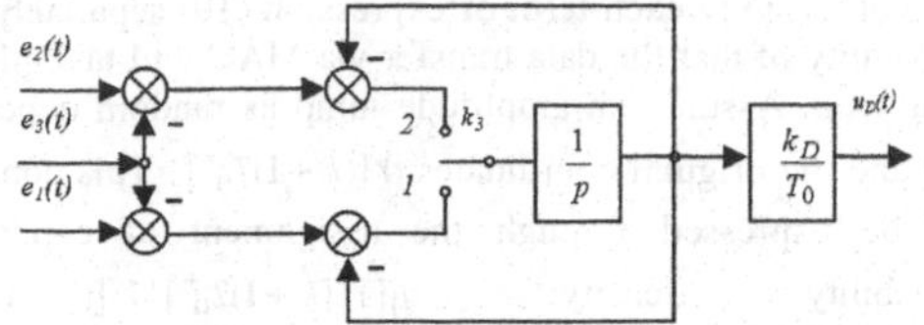

Fig. 4. Block diagram of differentiating PID control block.

For differentiating part we have:

$$\frac{du_D(t)}{dt} = \frac{k_D}{T_0}(e_1(t)^- - e_3(t)^- - \frac{T_0}{k_D}u_D(t)^-)g_1(t) +$$

$$+\frac{k_D}{T_0}(e_2(t)^- - e_3(t)^- - \frac{T_0}{k_D}u_D(t)^-)g_2(t).$$

The obtained equations for PID control parts are in conformity with the relevant parts of (1). In doing so, $u_P(t)$, $u_I(t)$, $u_D(t)$ retain their value over the quantization cycle T_0.

A represented generalized block diagram of the NCS under consideration is shown in fig. 5.

Structural scheme EFB and DR considered above and shown in fig. 5 are represented by rectangles to simplify the images. In future we shall assume that the object of regulation being considered is described by a linear differential equation of an appropriate order. At the entrance of the object with control action $u(t)$ the white noise $\zeta(t)$ is fed.

DS is modeled by the combination of continuous and discrete parts. Here $W^s(p)$ - continuous function of DS (sensitive element). Quantizer k_4 regularly quantizes output signal $x(t)$ with the cycle T_0.

Thus

$$\frac{dy_1(t)}{dt} = \left[x(t)^- - y_1(t)^-\right]g_3(t).$$

Quantizer k_5 quantizes the DS output if the data have been transferred via MAC to DR during T_0. Together with the hold element it forms a memory cell for saving data and their further use in EFB.

Thus

$$\frac{dy_2(t)}{dt} = \left[y_1(t)^- - y_2(t)^-\right]g_1(t).$$

On the structural pattern MAC is not represented in explicit form. Its functions are performed partially by EFB, DR and quantizer k_5 with hold element. In general the developed block diagram adequately describes the process of functioning of the NCS under consideration. Let us build a mathematical model of the NCS with the information transfer via MAC which correspondes to fig. 5. We shall designate the phase system of variables including the impact posed by y_i.

Then the NCS under consideration will be described by continuous-discrete stochastic equation which in vector-matrix form is:

$$\dot{Y}(t) = Z(t) - A\cdot Y(t) - \sum_{j=1}^{3} L_j\cdot Y^-(t)\cdot g_j(t) + C\cdot V(t). \quad (3)$$

Where $Y(t)$ - n-dimensional phase vector,

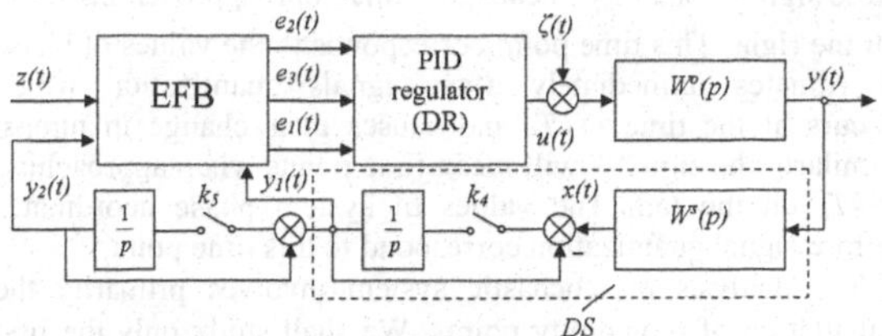

Fig. 5. Generalized block diagram of the NCS.

$Z(t)$ - n-dimensional vector of regular impacts,
A - $n\times n$-matrix of the system continuous part,
L_j- $n\times n$-matrix of the system discrete part,
C - $n\times n$-matrix,
$V(t)$ - n-dimensional vector of white noises with symmetric $n\times n$-matrix spectral densities $S(t)$.

Taking into account that the regular sequence of delta-pulses $g_3(t)$ can be represented as the sum of $g_1(t)$ and $g_2(t)$ of random sequences and imposing symbols:

$$B_1 = L_1 + L_3,$$

$$B_2 = L_2 + L_3,$$

we can represent (3) as follows:

$$\dot{Y}(t) = Z(t) - A\cdot Y(t) - \sum_{j=1}^{2} B_j\cdot Y^-(t)\cdot g_j(t) + C\cdot V(t). \quad (4)$$

The equation (4) describes a system in which phase coordinates have jumps in the moments of quantization signals kT_0.

Fig. 6 represents possible realization of the vector random process $Y(t)$. At the interval $]kT_0, (k+1)T_0[$ vector random process $Y(t)$ is continuous. The system at this point of time is described by the following stochastic equation:

$$\dot{Y}(t) = Z(t) - A\cdot Y(t) + C\cdot V(t).$$

This multidimensional Markov process has a continuous function of the probability density $f(Y, t)$ satisfying the equation of Fokker-Planck-Kolmogorov (FPK):

$$\dot{f}(Y,t) = -\left(\frac{\partial}{\partial Y}\right)^T \cdot \Pi(Y,t), \quad (5)$$

where $\Pi(Y, t)$ – vector of flux probability density in the initial conditions: $f(Y,t)|_{t=kT_0^+} = f(Y,kT_0^+)$.

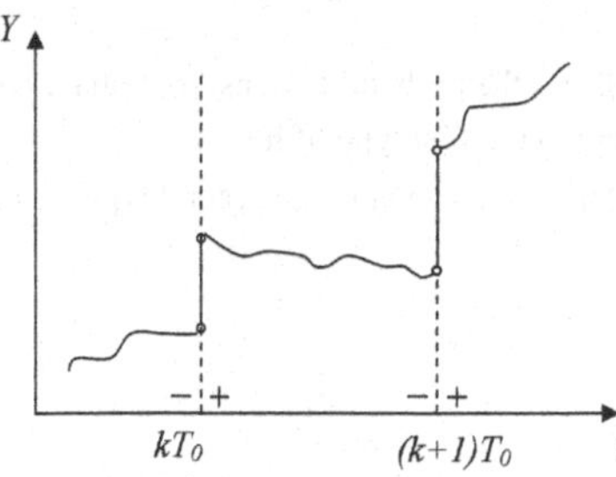

Fig. 6 Realization of the vector random process.

Here sign "+" in kT_0^+ means the time point approaching $t=kT_0$ on the right. This time point corresponds to the values of phase coordinates immediately after signals quantization which occurs at the time $t=kT_0$ and causes their change in jumps. Similarly the sign "-" will mean time points when approaching $t=kT_0$ on the left. The values of system phase coordinates before signal quantization correspond to this time point.

The analysis of stochastic system involves primarily the calculation of probability points. We shall study only the first two points of probability: mathematical expectations and correlation moments as they have found widespread use in engineering practice of automatic control systems calculation.

Using the equation FPK (5) we shall find the equation of probability moments for the time period $]kT_0, (k+1)T_0[$ under consideration:

$$\dot{M}(t) = Z(t) - A \cdot M(t), \tag{6}$$

$$\dot{\Theta}(t) = -A \cdot \Theta(t) - \Theta(t) \cdot A^T + C \cdot S(t) \cdot C^T. \tag{7}$$

Here:

$M(t)$ - n-dimensional vector of mathematical expectations,

$\Theta(t)$ - $n\times n$-matrix of correlation moments.

Assuming the proposed quantization cycle T_0 as small we shall represent (6) and (7) in the finite-difference form and proceed on to discrete equation.

$$M\left[(k+1)T_0^-\right] = T_0 \cdot Z[kT_0^+] + (I - T_0 \cdot A) \cdot M[kT_0^+], \tag{8}$$

$$\begin{aligned}\Theta\left[(k+1)T_0^-\right] &= (T - T_0 \cdot A) \cdot \Theta[kT_0^+] - \\ &- \Theta[kT_0^+] \cdot A^T \cdot T_0 + T_0 \cdot C \cdot S[kT_0^+] \cdot C^T\end{aligned} \tag{9}$$

Here I – identity $n\times n$-matrix; $k = 0,1....$

We shall study the time point $t=(k+1)T_0$. At this point in time the phase coordinates jump takes place. Thus we get the equation of the probability density $f(Y,(k+1)T_0^+)$ for the moment $t=(k+1)T_0^+$, that is immediately after the phase coordinates change.

At time point $t=(k+1)T_0$ there are two types of jumps: the first - in the case when the transfer via the MAC has taken place. In this case the time point $t=(k+1)T_0$ refers to the sequence $g_1(t)$, second - when there was no transfer via MAC and time point $t=(k+1)T_0$ refers to the sequence $g_2(t)$.

Since these options are not joint, then the density can be found as:

$$f(Y,(k+1)T_0^+) = f_1(Y,(k+1)T_0^+) + f_2(Y,(k+1)T_0^+). \tag{10}$$

Where:

$f_1(Y,(k+1)T_0^+)$ - the probability density of phase coordinates corresponding to the first type of jumps,

$f_2(Y,(k+1)T_0^+)$ - referring to the second type of jumps.

Let us study each term of expression (10) separately. The probability of that the data transfer via MAC will take place is equal to p. Assume an amplitude jump as random depending on t and the original coordinates $Y[(k+1)T_0^-]$. This condition can be expressed through the assignment of conditional probability density: $q_1[Y[(k+1)T_0^+] \mid Y'[(k+1)T_0^-]]$ determining the law of process amplitude distribution $Y[(k+1)T_0^+]$ after a jump on the condition that the jumps has occurred from the level $Y'[(k+1)T_0^-]$. The set function is standardized by a variable $Y[(k+1)T_0^+]$:

$$\int_{-\infty}^{+\infty} q_1[Y[(k+1)T_0^+] \mid Y'[(k+1)T_0^-]]dY[(k+1)T_0^+] = 1.$$

Then the density $f_1(Y,(k+1)T_0^+)$ can be presented:

$$\begin{aligned}f_1(Y,(k+1)T_0^+) &= p\int_{-\infty}^{+\infty} q_1[Y[(k+1)T_0^+] \mid Y'[(k+1)T_0^-]] \cdot \\ &\cdot f(Y',(k+1)T_0^-)dY'[(k+1)T_0^-]\end{aligned}$$

Under the assumed scheme at the quantization time point $t=(k+1)T_0$ the output process alters their values juddering from the level $Y'[(k+1)T_0^-]$ to the level:

$$Y[(k+1)T_0^+] = Y'[(k+1)T_0^-] - B_1 \cdot Y'[(k+1)T_0^-]. \tag{11}$$

Assuming symbol $F_1=I-B_1$ the equation (11) can be represented as:

$$Y[(k+1)T_0^+] = F_1 \cdot Y'[(k+1)T_0^-].$$

Conditional density of probability we can represent in the form:

$$q_1[Y[(k+1)T_0^+] \mid Y'[(k+1)T_0^-]] = \delta(Y[(k+1)T_0^+] - F_1 \cdot Y'[(k+1)T_0^-]).$$

Here

$$\delta(Y[(k+1)T_0^+] - F_1 \cdot Y'[(k+1)T_0^-]) = \prod_{i=1}^{n} \delta(y_i[(k+1)T_0^+] - \sum_{j=1}^{n} f_{ij} y_j[(k+1)T_0^-]).$$

Where y_i - the coordinates of the phase vector Y, f_{ij}- elements of the matrix F_1.

Taking into account the structure of this function we have the equation for the first term of a desired density:

$$\begin{aligned}f_1(Y,(k+1)T_0^+) &= p \cdot \int_{-\infty}^{+\infty} \delta(Y[(k+1)T_0^+] - F_1 \cdot Y'[(k+1)T_0^-)] \cdot \\ &\cdot f(Y',(k+1)T_0^-)dY'[(k+1)T_0^-].\end{aligned} \tag{12}$$

After integrating the right side of (12) we receive:

$$f_1(Y,(k+1)T_0^+) = p \cdot \left|F_1^{-1}\right| \cdot f(F_1^{-1} \cdot Y[(k+1)T_0^+],(k+1)T_0^-).$$

Similarly for the second term of (10) we have:

$$f_2(Y,(k+1)T_0^+) = (1-p) \cdot \left|F_2^{-1}\right| \cdot f(F_2^{-1} \cdot Y[(k+1)T_0^+],(k+1)T_0^-) \cdot$$

Here: $F_2=I-B_2$.

Then equation (13) will look like this:

$$f(Y,(k+1)T_0^+) = p\cdot\left|F_1^{-1}\right|\cdot f(F_1^{-1}\cdot Y[(k+1)T_0^+],(k+1)T_0^-) + \\ +(1-p)\cdot\left|F_2^{-1}\right|\cdot f(F_2^{-1}\cdot Y[(k+1)T_0^+],(k+1)T_0^-). \quad (13)$$

Then we calculate the probability moments of phase coordinate of the system for $t=(k+1)T_0$ using (13).

For mathematical expectations:

$$M[(k+1)T_0^+] = \int_{-\infty}^{+\infty} Y[(k+1)T_0^+]\cdot f[Y,(k+1)T_0^+]dY[(k+1)T_0^+] =$$

$$=[p\cdot F_1+(1-p)\cdot F_2]\cdot M[(k+1)T_0^-].$$

For the correlation moment:

$$\Theta[(k+1)T_0^+] = \int_{-\infty}^{+\infty}(Y[(k+1)T_0^+]-M[(k+1)T_0^+])\cdot(Y[(k+1)T_0^+]-M[(k+1)T_0^+])^T\cdot f(Y,(k+1)T_0^+)dY[(k+1)T_0^+] =$$

$$= p\cdot\{F_1\cdot\Theta[(k+1)T_0^-]\cdot F_1^T+(F_1\cdot M[(k+1)T_0^-]-M[(k+1)T_0^+])(F_1\cdot M[(k+1)T_0^-]-M[(k+1)T_0^+])^T\}+$$

$$+(1-p)\cdot\{F_2\cdot\Theta[(k+1)T_0^-]\cdot F_2^T+(F_2\cdot M[(k+1)T_0^-]-M[(k+1)T_0^+])\cdot(F_2\cdot M[(k+1)T_0^-]-M[(k+1)T_0^+])^T\}.$$

Using the equations (8) and (9), we obtain the recurrence equations for probability moments. In doing so, the left and right part of equation will match time point with the sign "+", so in the future the sign will be omitted.

$$M[(k+1)T_0] = T_0\cdot[pF_1+(1-p)F_2]\cdot Z[kT_0]- \\ +[I-T_0\cdot A]\cdot[pF_1+(1-p)F_2]\cdot M[kT_0], \quad (14)$$

$$\Theta[(k+1)T_0] = p\cdot F_1\cdot\{(I-T_0\cdot A)\cdot\Theta[kT_0]- \\ -\Theta[kT_0]\cdot A^T\cdot T_0+T_0\cdot C\cdot S[kT_0]\cdot C^T\}\cdot F_1^T+ \\ + M[(k+1)T_0]=T_0\cdot[pF_1+(1-p)F_2]\cdot Z[kT_0]+ \\ +[I-T_0\cdot A]\cdot[pF_1+(1-p)F_2]\cdot M[kT_0]+ \\ +p\cdot(1-p)\cdot(F_1-F_2)\cdot[T_0\cdot Z[kT_0]+(I-T_0\cdot A)\cdot M[kT_0]]\cdot \\ \cdot[T_0\cdot Z[kT_0]+(I-T_0\cdot A)\cdot M[kT_0]]^T\cdot(F_1-F_2)^T. \quad (15)$$

This system of (14) and (15) is a mathematical model of the considered NCS which should be solved under appropriate initial conditions: *M [0]*, *Θ [0]* and *S [0]*, *Z [0]*.

III. EXPERIMENTAL DATA

The system of the illumination intensity maintenance can serve as an example of network control system realization [5].

The control system functions as follows: the photo cell sensor will transform illumination created by a bulb, in voltage. Further on this signal gets to the microcontroller, transforming an analog signal in a digital type and passing it as a data package via network Ethernet through hub to a regulator. It's functions consist in the controlling influence on an electric bulb for stabilization of illumination under the influence of interferences in the system. There are not less than 10 stations connected by the help of hub with each other for generation of the traffic in a network.

Then we shall study the NCS with PID-control, object and sensor which have the following functions

$$W_o(p)=\frac{k}{T\cdot p+1};W_s(p)=k_s.$$

Fig. 7 shows graphics of probability moments received from the calculations by the model, where *y(t)* represents OR output.

The presented mathematical model is able to describe either stable or unstable digital control system with information transfer via multiple access channel and regular quantization cycle. By varying the parameters of channel access, control object or regulator you can receive different control systems. At this stage more experimental confirmation of the obtained results is required.

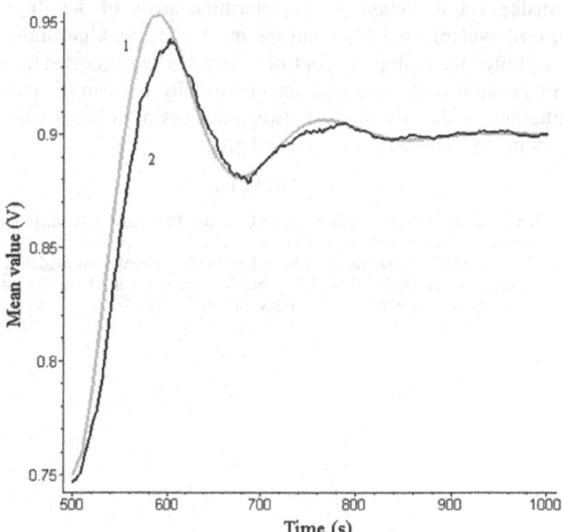

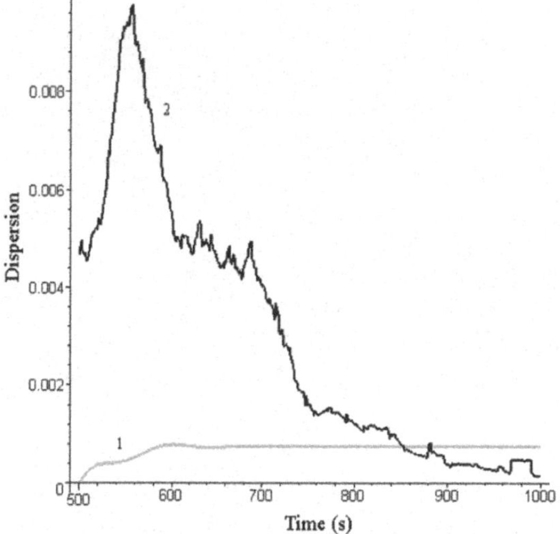

Fig. 7. The first and second moments of $y(t)$ coordinate where $k=0.6$, $k_p=0.1$, $k_d=0.0$, $k_s=1.0$, $T=0.05$, $T_0=0.005$, $S_0=0.0001$ (1 - model data, 2 – experimental data).

IV. RESULTS AND CONCLUSION

As a result of structural and mathematical modeling the NCS model with information transfer via MAC as continuous - discrete with random structure has been developed which allows to take into account stochastic character of time of data transfer via the channel, to determine areas of stability of control systems and to obtain the most effective algorithms of distributed technological control systems. The suggested model can be applied to evaluate an opportunity of using standard Ethernet for data exchange between devices of network control systems by concrete technological process.

REFERENCES

[1] Kruglyak K. Local Ethernet network in the PCS: faster, farther, more reliable // STA, 2003 - pp 1 6-13.

[2] Abramov G.V., Yemelyanov A.E., Ivliev M.N. Mathematical modeling of digital control systems with information transfer via MAC // control systems and information technology, in 2007 - pp 3 27-32.

[3] Dvoryannikov Y.V., Tumanov M.P. Various lag in network component and its influence on the stability control systems // control systems and information technology, in 2007 - pp 4 32-35.

[4] Artemev V.M., Ivanovsky A.V. The digital control system with a random period of quantization. - Moscow: Energoatomizdat, 1986 - 96.

[5] Abramov G.V., Emelyanov A.E., Ivliev M.N. Research of Network Control Systems with Competing Access to the Transfer Channel [Текст] // Advances in Computer and Information Sciences and Engineering (CISSE' 2007) / University of Bridgeport, CT, USA - pp. 178-183.

Autonomic Printing Infrastructure for the Enterprise

Riddhiman Ghosh[1], Andre Rodrigues[2], Keith Moore[1], Ricardo Pianta[2], Mohamed Dekhil[1]
[1]Hewlett-Packard Laboratories, Palo Alto, CA, USA
[2]Hewlett-Packard Co., Porto Alegre, RS, Brazil
riddhiman.ghosh@hp.com

Abstract— **In this paper we describe a solution to introduce autonomic behavior to the enterprise printing infrastructure. The techniques proposed do not need an overhaul to replace existing print devices—our solution introduces fail-over print capability to the millions of printers that comprise the installed base. We describe techniques for "printer neighborhood-awareness" and for enabling an easy path to the fail-over solution deployment without requiring custom PC software or new driver installs. We also discuss our experiences deploying this solution in a live enterprise IT environment.**

I. Introduction

Printing is one of the last unregulated IT costs in an enterprise and all predictions over the last few decades of a paperless office notwithstanding, it constitutes a significant part of an organization's IT spend. Managed Print Service (MPS) infrastructures, such as those provided by the authors' employer, and other organizations, centrally manage and optimize customers' imaging and printing environments to reduce cost and increase operational efficiencies. Most contracts with large organizations have Same-Day (4 hour) or Next-Day Service Level Agreements (SLA). Aggressive SLAs, combined with the fact that up to 20 percent of all help-desk calls are printer related [8] has required MPS to maintain dedicated service representatives on-call for each customer, sometimes located on-site, to perform hardware and software servicing—an increasingly expensive proposition given rising IT labor costs.

Consider a typical printing scenario at present: a user sends a print job to be printed on a printer. He walks up to the printer at a later instance to collect his job, and realizes that his job was not fulfilled as the printer is out of toner. He now has to walk back to his computer and resend the print job to a different printer. He might have to spend time locating the correct file on his computer to re-print. He might also have to add new printers on his machine to be able to print to them, and in this process may have to walk back and forth between the printers and the PC. This experience can be especially frustrating because the user usually sees another printer situated close to the malfunctioning printer—ideally he should be able to press a "detour" button on the printer and have the job automatically re-routed to the printer close-by, but that is not the case. It should be noted that this "user inconvenience" in the printing experience is not an intangible, but rather a concrete cost that could be quantified as a contributing component of the Total Cost of Ownership for MPS device fleets [7].

Honoring typical printing Service Level Agreements (Same-Day / 4 Hour repair) at reduced cost, lowering printer related IT help-desk call volume, and significantly improving the end-user printing experience can all be simultaneously achieved by applying a realization of the vision of autonomic computing as described in [1] to enterprise printing—a self-managing and healing print infrastructure that guarantees job fulfillment even when printers are down, and whose behavior can be configured based on end-user or administrator specified policies for print fulfillment (print cost, speed, location, job capabilities etc.). This would introduce efficiencies and allow for the elimination or reduction of the need for dedicated service representatives on-call at customer sites to perform hardware and software servicing of enterprise print infrastructures.

A. Problem Statement

The Detour Printing Solution on the OpenPrint Platform (DETOUR), which we describe in this paper, would allow jobs intended for a printer that is unavailable or malfunctioning (is out of toner, has a paper jam or other service error) to be transparently re-routed to another printer "close-by" (with or without user intervention). We can thus guarantee print output even though printers are down. DETOUR gives MPS providers significant cost-savings and a competitive edge while bidding for contracts, by having

K. Elleithy (ed.), *Advanced Techniques in Computing Sciences and Software Engineering*,
DOI 10.1007/978-90-481-3660-5_76, © Springer Science+Business Media B.V. 2010

SLAs that guarantee print output, and not necessarily same-day on-site servicing, which can be deferred.

However there are several technical challenges that DETOUR needs to successfully address:

- First, from an IT management perspective, one must be able to deploy this solution in customer environments without requiring custom software to be installed on every client PC (zero client software foot-print) or requiring the installation of new print drivers on every PC (IT departments typically balk at updating, testing, certifying and deploying new sets of drivers).
- Second, the solution relies on us being able to reliably monitor and interpret printing state across various printer models, and yet the ability to store and redirect print jobs must be independent of, and unaffected by, printer failure.
- Finally, when a job needs to be detoured, a decision has to be made regarding which printer the job should be detoured to. This decision must be based on suitably matching the print job characteristics to printer capabilities.
- Apart from matching job characteristics to device capabilities, it is desirable that the job is detoured to a printer located in the vicinity of the malfunctioning printer. This requires DETOUR to be "location-aware".

II. DETOUR

The DETOUR solution is deployed on the *OpenPrint Platform* (OPP), which is a service enablement and accessorization platform designed for HP printers. The OPP device is an independent computing unit—a Windows CE based card with its own CPU, memory and I/O that plugs into the printer either through its Enhanced IO slot, or is accessible to it via the network interface, cost for which in service contracts would be marginal. While we focus on DETOUR in this paper, it should be noted that it is but one of the possible solutions and services that can be deployed. Deploying DETOUR on this auxiliary platform gives us two advantages:

- first, detour intelligence can be added to any of the millions of existing printers that comprise the installed base;
- second, detour capability is independent of the failure modes of the printer.

DETOUR acts as a printer "proxy" standing in the middle of the communication path between the print driver and the device. It impersonates the printer by "passing through" messages between the printer and the user's PC and vice versa, and can intercept and modify those messages if needed. DETOUR caches print jobs as they are being sent to the device, and continuously monitors printer state. If printing fails for some reason, it is able to reroute the job to any one of the detour printers it has been configured to work with. By default, this configuration is done automatically, where the OPP device discovers and adds suitable printers to its detour list, but can be overridden by manually specifying which printers to use for detour. Before performing a detour, a match is done between job characteristics and capabilities of printers in the detour list, to route jobs to the correct printer. We currently limit ourselves to the following job characteristics: color/monochrome, duplex/single-sided, and printer language family. The re-route decision matrix however can be extended to include other parameters such as current printing load, consumable availability, policies for printing cost or speed. Since DETOUR runs on a processor and memory constrained device, a network folder is used to cache print jobs in case they need to be detoured. Figure 1 gives an overview of the DETOUR system.

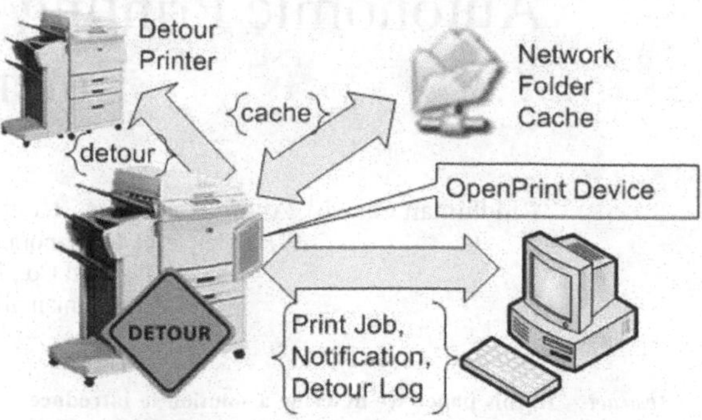

Figure 1: Overview and major components of DETOUR

DETOUR is a human-in-the-loop autonomic system that needs to address users' expectation of control over their print jobs (see Section IV). Currently notification that a detour has occurred is displayed on the printer device panel, on the print driver UI on the user's PC and is also recorded on a web-accessible detour log, from where a user can determine if his

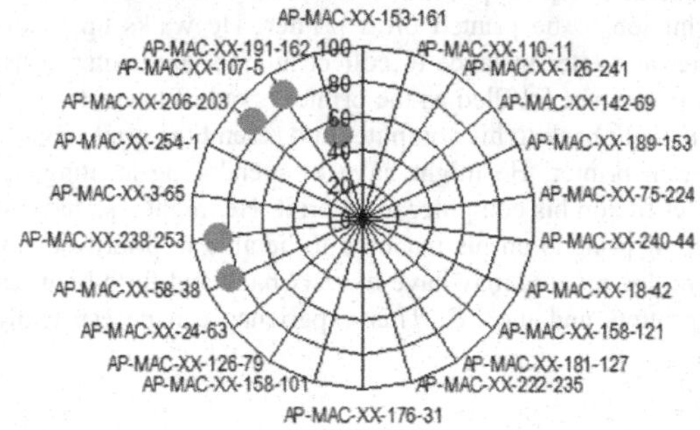

Figure 2: A Sample Wi-Fi fingerprint

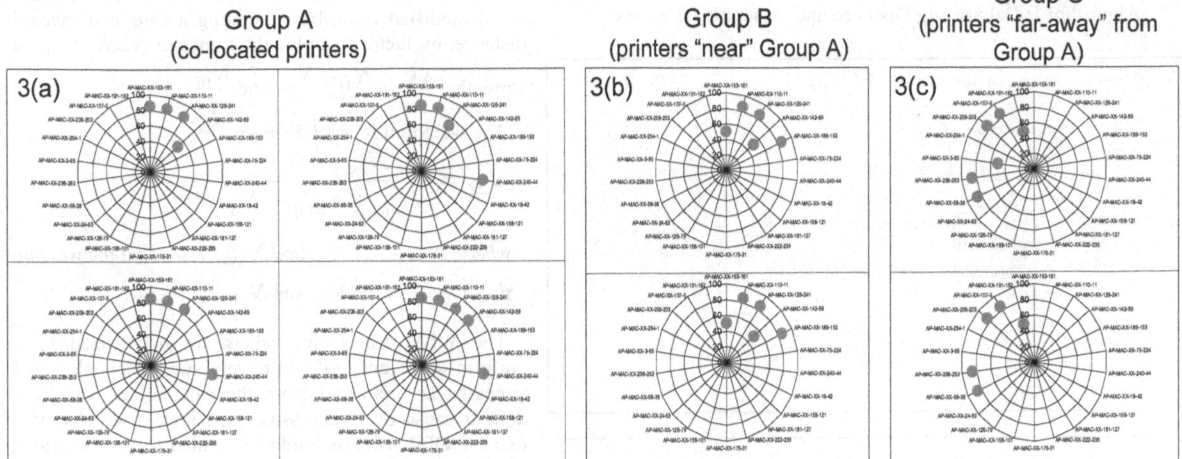

Figure 3: Fingerprints of some OPP printers at HP Labs Palo Alto

job was detoured. Alternate client-side notification schemes are also being considered, such as via email.

III. Neighborhood Awareness

The Printers with OPP devices are *neighborhood-aware*—they are aware of which printers they are co-located with and which are near-by. In most enterprise environments, workspaces are mounted with dozens of wireless access points (AP) to allow employees Wi-Fi access. We leverage this fact, along with the integrated Wi-Fi capability on OPP devices to achieve neighborhood awareness. The OPP device generates a snapshot of all the Wi-Fi access points it can "see" by recording a unique identifier (MAC address), and signal strength received (over several measurements). This is called the printer's "Wi-Fi fingerprint".

Figure 2 shows a sample Wi-Fi fingerprint, with the access points plotted on the circular axis, and an indicator of signal strength percentage marked by the orange dots on the radii. The Wi-Fi fingerprints are sent to a service, which uses a clustering algorithm to identify which fingerprints are very similar, thus allowing us to infer which printers are co-located such as in the same cluster. Fig. 3(a) shows the fingerprints of an actual group of co-located printers. Dissimilar fingerprints, but with strong signals from common APs indicate that the printers are reasonably close, even though not co-located, such as fingerprints of printers in Fig. 3(a) and Fig. 3(b). Fingerprints in Fig 3(c) are different from the other 2 groups, and don't have APs in common, and thus one can deduce they are situated far-away. This scheme of "neighborhood-awareness" is novel, and different from other Wi-Fi location schemes that use path-loss based triangulation [5], [6]. Unlike their focus on *location* determination, we focus on *neighborhood* determination.

B. *Algorithm for Neighborhood Awareness*

We now describe our method of analyzing the printer neighborhood. The fingerprints sent by the OPP printers are sent to a database, where the records are analyzed by a clustering algorithm. The objective of the algorithm, given a printer fingerprint, is to generate clusters of other printers that can be inferred to be co-located with, reasonably close, or faraway from the printer in question. We use a modified form of the Jaccard similarity coefficient [2] as the basis of clustering. Given two sets X and Y, the Jaccard similarity coefficient $J(X,Y)$, and hence the corresponding dissimilarity measure, the Jaccard distance $J_\delta(X,Y)$, is given by:

$$J_\delta(X,Y) = 1 - J(X,Y) = \frac{|X \cup Y| - |X \cap Y|}{|X \cup Y|}$$

In our case X and Y represent the set of APs given by any two printer fingerprints FP-I and FP-II. We treat the presence or absence of particular APs in the fingerprints, as an attribute, with a value of 1 or 0 respectively. Thus in our case the Jaccard distance between 2 printers, J' can be calculated as follows:

$$J' = \frac{N_{01} + N_{10}}{N_{01} + N_{10} + N_{11}},$$

Algorithm 1: Calculating Distance and Generating Clusters

```
procedure generate_clusters(PrintersTable):
begin
    while running is true
    begin
        for each P in PrintersTable
            for each Q in PrintersTable
                Dist ⟵ jaccard_dist(P,Q)
                record Dist for P w.r.t. Q
                assign Q to cluster based on thresholds
            end for
        end for
        sleep for cluster_update_interval
    end while
end procedure
```

```
procedure jaccard_dist(fingerprint A, fingerprint B):
begin
    normalize(A)
    normalize(B)
    Shared ⟵ 0
    NonShared ⟵ 0
    for each attribute in A
    begin
        if attribute of A and attribute of B not 0 then
            W ⟵ weight based on attribute values
            Shared ⟵ Shared + 1*W
        else
            if attribute of A or attribute of B not 0 then
                W' ⟵ weight based on non-shared attribute values
                NonShared ⟵ NonShared + 1*W'
            end if
        end if
    end for
    Dist ⟵ NonShared / (NonShared + Shared)
    return Dist
end procedure
```

where,

N_{11} *represents total number of attributes where X and Y both have value 1,*

N_{01} *represents total number of attributes where the attribute of X is 0 and Y is 1,*

N_{10} *represents total number of attributes where the attribute of X is 1 and Y is 0,*

N_{00} *represents total number of attributes where X and Y both have value 0.*

While this algorithm based on the Jaccard distance would suitably "reward" presence of shared APs between fingerprints, it would not take into account the similarity of the received signal strengths from the APs in question. Values of received signal strength are strongly correlated with location/proximity with respect to an AP. We therefore use a modified form, by computing a weighted Jaccard distance, by factoring in the difference of received signal strengths ΔS in N_{11}, N_{01} and N_{10}.

The weighted Jaccard distance $J_{\Delta S}'$,

$$J_{\Delta S}' = \frac{N_{\Delta S01} + N_{\Delta S10}}{N_{\Delta S01} + N_{\Delta S10} + N_{\Delta S11}},$$

where $N_{\Delta S01}$, $N_{\Delta S10}$ *and* $N_{\Delta S11}$ *represent the weighted versions of* N_{11}, N_{01} and N_{10}

The Wi-Fi fingerprint values are normalized to a uniform scale and APs with very low received signal strengths are removed from the calculation. This was done because over successive experiments we observed that the OPP wireless hardware tended to give spurious values for APs that it received very low signal from.

The description of the algorithm is given below in Algorithm 1. A current shortcoming of this algorithm is that a printer that is in a Wi-Fi "shadow", i.e. experiences significantly attenuated Wi-Fi reception across-the-board for some reason, will appear to this algorithm to be farther from its neighbors than it actually is. This can largely be overcome by choosing suitable thresholds for clustering.

This algorithm and approach would not be able to account for printers situated in non-Wi-Fi environments. However it is increasingly rare to find non-wireless environments in today's enterprise. Given that our solution is targeted towards the enterprise, we believe assuming the presence of a wireless environment is a reasonable assumption to make.

IV. Pilot Deployment Experience

A prototype of the DETOUR solution has been deployed on printers in the authors' workgroups in Palo Alto and Brazil for the past year, and in that time it has been successfully used internally. In addition, DETOUR has been deployed in the production IT environment of a large enterprise HP customer in Sao Paulo, Brazil. Here DETOUR is managing printers with monthly usage cycles in excess of 5000 pages, and approximately 10% of these printed pages in a month benefit from the autonomic behavior of the print infrastructure. The pages benefiting from the autonomic behavior of DETOUR also resulted in a marked reduction in the printer-related IT trouble tickets. A detailed benefit-analysis of DETOUR using IT asset monitoring tools and print log analysis is underway. A key feature of the deployment and management of the infrastructure—which was greatly appreciated by the customer's IT department—was that the deployment was entirely printer-side and no custom software or new print drivers were needed to be installed by users on their client machines. Our deployment experiences raised, among others, interesting user-interface

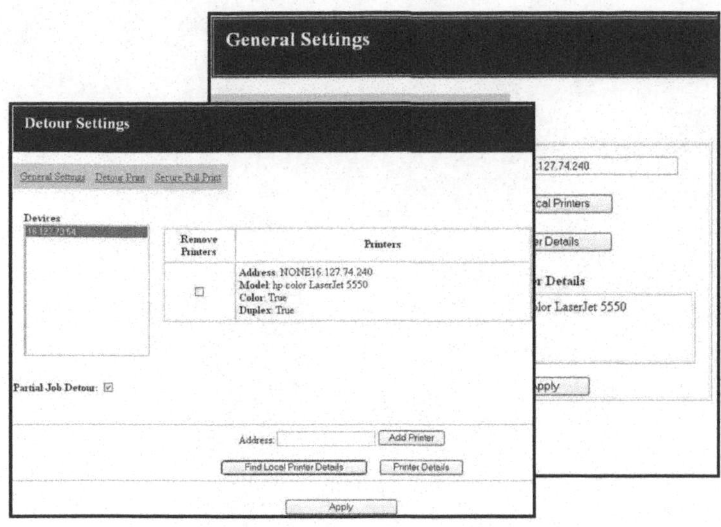

Figure 4: Screenshot of Detour configuration web page of a printer

issues: the trade-offs between the advantages of automated fail-over and re-routing versus the users' expectation of control over their printed pages; how to treat sensitive print jobs differently; what to have as the user touch-point to interact with the autonomic infrastructure—the PC/driver interface, or the printer device interface or a combination of both. Subsequent iterations of the system had to address many of these issues. For brevity, we do not discuss the human factors considerations of the solution in this paper, apart from noting that careful thought is required in the usability and design of a human-in-the-loop autonomic computing system.

V. Related Work

Printers from SHARP provide a "bypass printing" feature [3] that allows the printer to proceed with remaining jobs in its print queue, if the paper for a particular job (e.g. A3) is not available. This not only represents extremely limited fail-over, but also only works with *one* device—the printer is not able to re-route jobs to other printers for fulfillment.

INTELLIscribe [4] is a third-party Windows-based printing software application that once installed on users' machines can be configured with a set of back-up printers. If the first printer is offline INTELLIScribe tries one of the other printers in round-robin fashion. It is not able to monitor printer state or errors such as paper jams, toner depletion, etc. It requires every user to install custom software on his machine, which is not desirable from an IT deployment perspective. Unlike DETOUR, it is unable to intelligently match jobs to printers based on device characteristics and capabilities, nor are the printers sentient to their neighborhood to better route jobs. INTELLIscribe and its ilk are more suited for batch-printing with homogeneous device clusters, rather than for office printing environments.

VI. Conclusion

In this paper we describe a solution to introduce autonomic behavior to enterprise printing infrastructures. Fail-over capability is added to the millions of *existing* printers that constitute the installed base (rather than requiring new devices). Increasing efficiencies in managed print service environments is a very relevant problem space, especially since approximately 50% of that total cost of printing in an enterprise is due to services and support, and approximately 20% of all calls to IT help-desks are printer related [8]. Our solution also does not require changes to the IT client infrastructure—our solution has a zero client software foot-print since we do not require installation of custom software on users' PCs, nor do we require installation of new print drivers; the entire deployment is printer side and is designed to work with the installed base. We presented a novel neighborhood-awareness scheme for printers using the enterprise wireless networking infrastructure that brings significant additional value to the autonomic printing behavior.

We are presently continuing our research efforts around leveraging capabilities from the latest print drivers to enhance detour user notification on client machines, incorporating multicast-DNS based network printer discovery for ease in configuration, and augmenting matching of jobs to printers by extending the decision matrix to include other factors such as status of consumable availability on a printer, its printing load and policies related to job fulfillment cost and speed.

References

[1] Kephart, J., Chess, D., The vision of autonomic computing, in *IEEE Computer Journal*, 36(1), January 2003.

[2] Han, J., Kamber, M., *Data mining: concepts and techniques*, Morgan Kaufmann, San Francisco, 2000.

[3] SHARP Corporation, Bypass printing, *Convenient usages for Sharp MFPs* (Product feature documentation).

[4] INTELLIscribe LPR print client, www.intelliscribe.net.

[5] Savarese, C., Rabaey, J., Beutel, J., Location in distributed ad-hoc wireless sensor networks, in *IEEE International Conference on Acoustics, Speech and Signal Processing,* Salt Lake City, May 2001.

[6] Wang, Y., Jia, X., Lee, H., Li, G., An indoors wireless positioning system based on wireless local area network infrastructure, in *6th International Symposium on Satellite Navigation Technology including Mobile Positioning and Location Services,* Melbourne, July 2003.

[7] Morciniec, M., Rahmouni, M., Yearworth, M., Operating cost calculation for printing solution cost-benefit analysis tool, in *HPL Technical Report HPL-2004-194(R.1),* January 2005.

[8] Fernandes, L., Print Management: Taking Control of Hidden Costs, in *Association for Information and Image Management (AIIM) E-DOC Magazine,* April 2007.

Investigating the Effects of Trees and Butterfly Barriers on the Performance of Optimistic GVT Algorithm

Abdelrahman Elleithy, Syed S. Rizvi, and Khaled M. Elleithy
Computer Science and Engineering Department, University of Bridgeport, Bridgeport, CT USA
{aelleithy, srizvi, elleithy}@bridgeport.edu

Abstract- ***There is two approaches for handling timing constraints in a heterogeneous network; conservatives and optimistic algorithms. In optimistic algorithms, time constraints are allowed to be violated with the help of a time wrap algorithm. Global Virtue Time (GVT) is a necessary mechanism for implementing time wrap algorithm. Mattern [2] has introduced an algorithm for GVT based computation using a ring structure. which showed high latency. The performance of this optimistic algorithm is optimal since it gives accurate GVT approximation. However, this accurate GVT approximation comes at the expense of high GVT latency. Since this resultant GVT latency is not only high but may vary, the multiple processors involve in communication remain idle during that period of time. Consequently, the overall throughput of a parallel and distributed simulation system degrades significantly In this paper, we discuss the potential use of trees and (or) butterflies structures instead of the ring structure. We present our analysis to show the effect of these new mechanisms on the latency of the system.***

I. Introduction

Many GVT algorithms were introduced in the literature. In [1] Chen *at. al.*, provided a comparison between 15 GVT algorithms. Table 1 [1] shows a detailed comparison between the different algorithms.

Mattern's GVT algorithm [2] proposed a 2-cut algorithm to avoid synchronizing all processors at the same wall clock. The two cuts define a past and a future point. In a consistent cut, no transient jobs can travel from the future to the past. Messages crossing the second cut from the future to the past do not need to be taken into account because these messages are guaranteed to have a timestamp larger than the GVT value.

Mattern's GVT algorithm uses a token passing to construct the two cuts. It uses two cuts C1 and C2. C1 is intended to inform each processor to

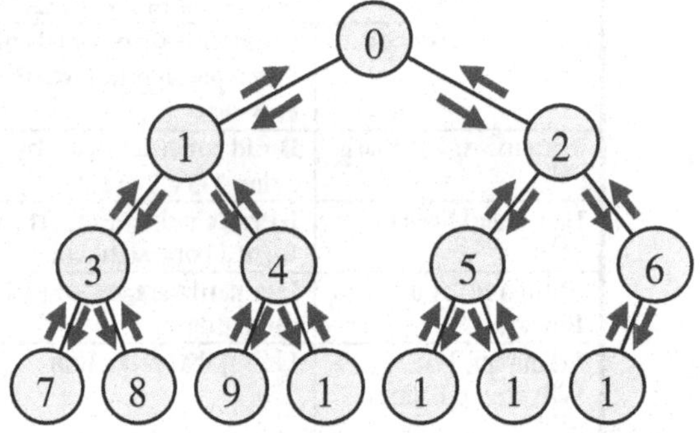

Fig.1. Tree barrier mechanism for synchronization among the logical processes, Green font arrow lines represent the LBTS computation and the new GVT announcement

begin recording the smallest time stamp where as C2 guarantees that no message generated prior to the first cut is in transient. A vector clock passed between processors monitors the number of transient messages sent to every processor. The token can leave the current processor only after all

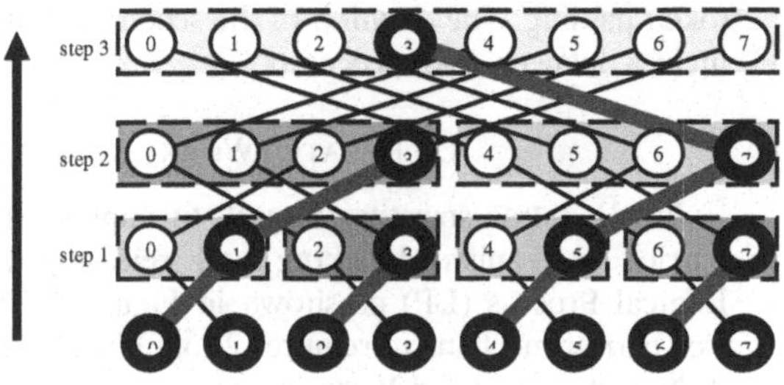

Fig. 2. Butterfly barrier mechanism between 8 LPS. Three steps are needed to complete the synchronization. The red font represents the synchronization for LP3

K. Elleithy (ed.), *Advanced Techniques in Computing Sciences and Software Engineering*,
DOI 10.1007/978-90-481-3660-5_77, © Springer Science+Business Media B.V. 2010

Table 1: Comparison between Different GVT Algorithms [1]

Authors	Idea	Ack	Vector	Channel	Scalability
Samadi [19]	Broadcast START and STOP messages to form overlapping intervals	Yes	No	Any	N/A
Bellenot [12]	Use message routing graph instead of broadcast	Yes	No	Any	104 [26]
Das and Sarkar [14]	Optimize the computation for hypercube topology	No	No	Maximum Delay	N/A
Baldwin, Chung and Chung [10]	Pass a token to form overlapping intervals	Yes	No	FCFS	N/A
Lin and Lazowska [16]	Send valley messages to reduce acknowledgement traffic	Implicit	No	Any	N/A
Mattern [17]	Construct two cuts such that no transient messages sent before the first cut exist	No	Yes	Any	12 [13]
Choe and Tropper [13]	Create multiple rounds of token passing to form the two cuts	No	No	Any	12 [13]
Tomlinson and Garg [22]	Build consistent cuts by using TGVT events	No	Yes	Any	N/A
Bauer and Sporrer [11]	Identify pairs of reports that form a consistent cut	No	No	FCFS	N/A
Srinivasan and Reynolds [20]	Use hardware-based global reduction	No	No	Any	N/A
Steinman, Lee, Wilson, and Nicol [21]	Use global reduction	No	No	Any	64 [21]
Perumalla and Fujimoto [18]	Use global reduction	No	No	Any	16 [18]
D'Souza, Fan, and Wilsey [15]	Report stragglers to a GVT manager	Yes	No	Any	2 [15]
Bauer, Yuan, Carothers, Yuksel, and Kalyanaraman [23]	Extend Fujimoto's shared-memory GVT algorithm with the notion of network atomic operations	No	No	Maximum Delay	16 [23]
Deelman and Szymanski [24]	Use vector and matrix clocks to keep track of messages in transit	No	Yes	Any	16 [24]

messages destined to it have been received. The second cut can be built with only one round of token passing. The creation of the second cut may incur a delay on each processor.

II. Related Work

In [3], a tree structure is used to implement a barrier mechanism blocking and releasing for Logical Process (LP) as shown in Figure 1. The tree barrier mechanism requires 2 $\log_2 N$ steps and 2 (N-1) messages for *N* processors.

A butterfly mechanism is discussed in [3] to eliminate the need for broadcasting as shown in Figure 2. The butterfly mechanism requires log N steps to complete and N * Log N messages for N processors.

III. Analytical Model

In comparing centralized barriers, it is noticed that the butterfly mechanism has a better performance when comparing the required time as it needs half the number of steps (Figure 3). Butterfly barrier has the butterfly mechanism in terms of the required exchanged messages (Figure 4). It should be clearly noted in Fig. 3 that the performance of the tree barrier is much better than the butterfly barrier for all values of N. This is due to the fact that the time complexity of the tree barrier is much lower than the butterfly barrier.

IV. USING TREE AND BUTTERFLIES

By analyzing the ring structure used in Mattern's we notice that the ring works as follows:

1. C1 is constructed by sending a control message around the ring. Once the control message is received, the color of the processor changes from white to red then passes the message. This step of the algorithm will take (N-1) steps.
2. C2 is constructed by sending the control message around the ring. This step of the algorithm will take (N-1) steps.

We assume that initially all processors (nodes) and their neighbors that are organized in a minimal tree (i.e.., no cycles) based structure are colored white. In addition, we also assume that there should be one initiator of GVT computation that may also be considered as a root of the tree (i.e.., the node where message transmission starts). The moment initiator processor initiates GVT computation, it becomes red from white. At the same time, it starts a broadcast scheme to indirectly (i.e.., from node to edges) send control messages to all connected processors. Thus, this first transmission (the process of making red) of broadcast from root (i.e.., the initiator processor) to all its connected nodes is intended for the first cut C1.

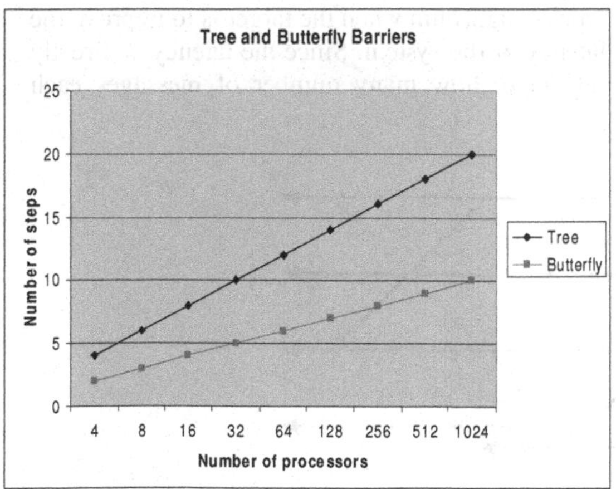

Fig. 3: Time Comparison between the Tree and the Butterfly Barriers with a random number of message transmissions with a large number of processors

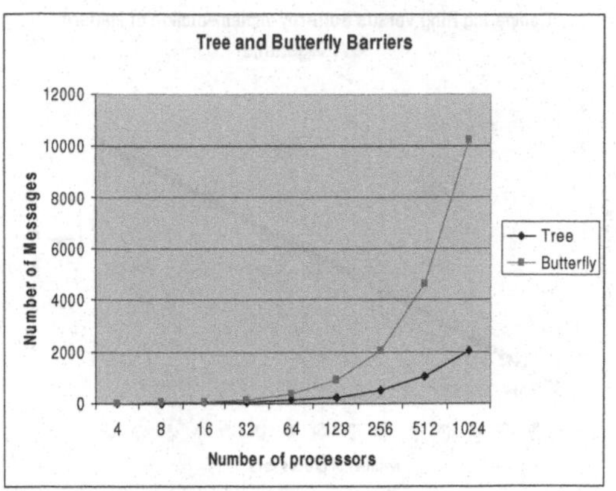

Figure 4: Communication Messages Comparison between the Tree and the Butterfly barriers for a large number of processors

According to our initial assumptions, Mattern's algorithm does not require acknowledgement messages but it does require the construction of the second cut C_2. We assume that, in order to construct the second cut C_2, we need the same number of messages that will propagate from processors (i.e.., the edges of the tree) to the initiator (i.e.., the root of the tree). Therefore, this implies that any processor in the given design which is the part of a balanced minimal tree must process two messages; one for constructing the first cut C_1 and the other for constructing the second cut C_2. The total number of steps in implementing the ring is 2 * (N-1).

Instead of using a ring structure, we can use a tree structure. The number of steps using the tree structure to implement Mattern algorithm is 2 * $\log_2$ (N) as per our discussion in section III. One can clearly observe in Fig. 4 that the number of messages transmitted with the tree barrier is much lower than the number of messages required for the butterfly barrier. This is especially true for a large number of processors. In other words, as we start increasing the number of processors in the system, the performance differences between the tree and the butterfly barrier is obvious. For instance, the number of messages transmitted for the tree barrier do not exceed to 2000 messages for even a large value of processors (typically 1000 processors) as shown in Fig. 4. Furthermore, if we use butterflies, the numbers of steps is $\log_2$ N. Figure 5 shows a comparison between using a ring and a butterfly in implementing the Mattern GVT algorithm. Fig. 6 represents the implementation of the butterfly barrier where four processors are organized and sending/receiving messages to each other. When compare the

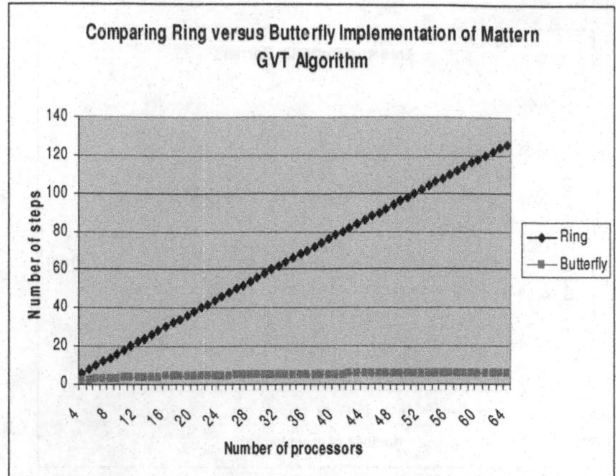

Figure 5: Comparison of using a ring and a butterfly in implementing Mattern GVT algorithm

performance of butterfly barrier with the tree barrier, one can clearly observe that the performance of butterfly is overlapping the tree barrier for a small number of processors (typically for 150 processors) as shown in Fig. 4. However, as we start increasing the number of processors (LPs > 150) in the system, the performance of butterfly degrades significantly than the tree barrier. This is due to the fact that the time complexity of the butterfly barrier is slightly higher than the tree barrier. On the other hand, when the performance of butterfly barrier is compared with the ring structure, the simulation results of Fig 5 suggest that the butterfly is clearly a better choice for using as a synchronization mechanism with the Mattern's GVT algorithm.

Although the actual number of steps has decreased significantly by using a butterfly compared to a ring in implementing Mattern's GVT algorithm, a large number of messages have been created. In the original ring implantation, there are only two control messages that are circulating in the ring. For a butterfly implementation, the number of messages is N * Log N.

To make it more clear, this barrier requires steps with the transmission of messages, since each processor must send and receive one message in each step of the algorithm. Thus, the asymptotic complexity of this barrier is clearly higher than the tree or ring structures which in turn give a higher value of latency.

It is worth mentioning that the asymptotic latency of butterfly is exactly the same as the merge algorithm where the total of N number of comparisons are analogous to the total number of N messages transmitted in one direction. From message complexity point of view, it is obvious that the latency of butterfly barrier for Mattern's GVT algorithm exists in a logarithmic region with a constant N.

In addition, our analysis demonstrates that one can achieve the same latency for Mattern's algorithm if we assume that two rounds of messages propagate from initiator to all processors (i.e.., intended for C1) and from all processors to the root (i.e.., intended for C2) in a tree barrier. However, the latency can be improved if parallel traversal of connected processors is allowed. The above discussion can be extended for a tree structure where the left and the right sub trees have different length.

V. CONCLUSION

In this paper, we have investigated on the possibility of using Trees and Butterfly barriers with the Mattern GVT algorithm. The simulation results have verified that the use of butterfly barriers is inappropriate with an asynchronous type of algorithm when the target is to improve the latency of the system. Since the latency is directly related to how many number of messages each

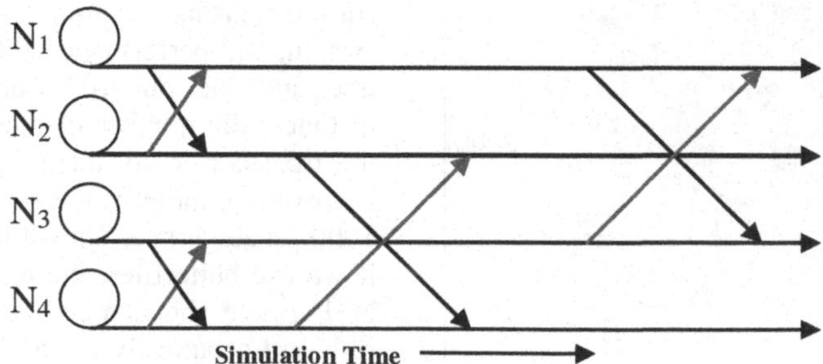

Fig. 6: Butterfly Barrier Organization: Arrows in the figure show that the node is arriving/reaching barrier to other processors. Once the LBTS computation is initiated by all processors, the execution of the messages will be halt unless all the processors achieve synchronization by knowing a Global minimum value

processor is sending, butterfly barrier may not be a good candidate to improve the latency of the GVT computation. However, we have shown that the latency of the GVT computation can be improved if the tree based structure is organized in a way that allows parallel traversing of each left and the right sub trees. The improvement in the latency has a higher cost of communications.

REFERENCES

1. Gilbert G. Chen and Boleslaw K. Szymanski, Time Quantum GVT: A Scalable Computation of the Global Virtual Time in Parallel Discrete Event Simulations, Scientific International Journal for Parallel and Distributed Computing, pages 423–435, Volume 8, no. 4, December 2007.
2. F. Mattern, Efficient algorithms for distributed snapshots and global virtual time approximation, J. Parallel and Distributed, Computing 18(4) (1993) pp. 423–434.
3. Fujimoto, R., Parallel and Distributed Simulation Systems, Willey Series on Parallel Distributed Computing, 2000.

Authors Biographies

Abdelrahman Elleithy has received his BS in Computer Science in 2007 from the Department of Computer Science and Engineering at the University of Bridgeport, Connecticut, USA . Abdelrahman is currently a MS student and expected to receive his MS in Computer Science in December 2008. Abdelrahman has research interests in wireless communications and parallel processing where he published his research results papers in national and international conferences.

SYED S. RIZVI is a Ph.D. student of Computer Engineering at University of Bridgeport. He received a B.S. in Computer Engineering from Sir Syed University of Engineering and Technology and an M.S. in Computer Engineering from Old Dominion University in 2001 and 2005 respectively. In the past, he has done research on bioinformatics projects where he investigated the use of Linux based cluster search engines for finding the desired proteins in input and outputs sequences from multiple databases. For last one year, his research focused primarily on the modeling and simulation of wide range parallel/distributed systems and the web based training applications. Syed Rizvi is the author of 45 scholarly publications in various areas. His current research focuses on the design, implementation and comparisons of algorithms in the areas of multiuser communications, multipath signals detection, multi-access interference estimation, computational complexity and combinatorial optimization of multiuser receivers, peer-to-peer networking, and reconfigurable coprocessor and FPGA based architectures.

DR. KHALED ELLEITHY received the B.Sc. degree in computer science and automatic control from Alexandria University in 1983, the MS Degree in computer networks from the same university in 1986, and the MS and Ph.D. degrees in computer science from The Center for Advanced Computer Studies at the University of Louisiana at Lafayette in 1988 and 1990, respectively. From 1983 to 1986, he was with the Computer Science Department, Alexandria University, Egypt, as a lecturer. From September 1990 to May 1995 he worked as an assistant professor at the Department of Computer Engineering, King Fahd University of Petroleum and Minerals, Dhahran, Saudi Arabia. From May 1995 to December 2000, he has worked as an Associate Professor in the same department. In January 2000, Dr. Elleithy has joined the Department of Computer Science and Engineering in University of Bridgeport as an associate professor. Dr. Elleithy published more than seventy research papers in international journals and conferences. He has research interests are in the areas of computer networks, network security, mobile communications, and formal approaches for design and verification.

Implementation of Tree and Butterfly Barriers with Optimistic Time Management Algorithms for Discrete Event Simulation

Syed S. Rizvi[1] and Dipali Shah
Computer Science and Engineering Department
University of Bridgeport
Bridgeport, CT 06601
{srizvi, dipalis}@bridgeport.edu

Aasia Riasat
Department of Computer Science
Institute of Business Management
Karachi, Pakistan 78100
Aasia.riasat@iobm.edu.pk

***Abstract*: The Time Wrap algorithm [3] offers a run time recovery mechanism that deals with the causality errors. These run time recovery mechanisms consists of rollback, anti-message, and Global Virtual Time (GVT) techniques. For rollback, there is a need to compute GVT which is used in discrete-event simulation to reclaim the memory, commit the output, detect the termination, and handle the errors. However, the computation of GVT requires dealing with transient message problem and the simultaneous reporting problem. These problems can be dealt in an efficient manner by the Samadi's algorithm [8] which works fine in the presence of causality errors. However, the performance of both Time Wrap and Samadi's algorithms depends on the latency involve in GVT computation. Both algorithms give poor latency for large simulation systems especially in the presence of causality errors. To improve the latency and reduce the processor ideal time, we implement tree and butterflies barriers with the optimistic algorithm. Our analysis shows that the use of synchronous barriers such as tree and butterfly with the optimistic algorithm not only minimizes the GVT latency but also minimizes the processor idle time.**

I. Introduction

The main problem associated with the distributed system is the synchronization among the discrete events that run simultaneously on multiple machines [4]. If the synchronization problem is not properly handled, it can degrade the performance of parallel and distributed systems [7]. There are two types of synchronization algorithms that could be used with the parallel and discrete-event simulation (PDES): conservative and the optimistic synchronization algorithms. The conservative synchronization ensures that the local causality constrain requirement must not be violated by the logical processes (LPs) within the simulation system [5]. On the other hand, optimistic synchronization allows the violation of the local causality constraint requirement. However, such violation can not only be detected at run time but can also be dealt by using the rollback mechanism provided by optimistic algorithms [1, 2, 6]

The Time Wrap [2, 3] is one of the mechanisms of optimistic time management algorithm (TMA) which includes rollback, anti-message, and GVT computation techniques [1, 4]. The rollback mechanism is used to remove causality errors by dealing with straggling events. The straggling events are referred to those events whose time-stamp is less than the current simulation time of an LP. In addition, the occurrence of a straggling event may cause the propagation of incorrect events messages to the other neighboring LPs. Optimistic TMA demands the cancellation of all such event messages that might have been processed by the other LPs. The anti-message is one of the techniques of the Time Wrap algorithm that deals with the incorrect event messages by cancelling them out.

Formally, the GVT can be defined as a minimum time-stamp among all the unprocessed and partially processed event messages and anti-messages present in the simulation time at current clock time *Ts*. This implies that those events whose time-stamp is strictly less that the value of GVT can be considered as safe event message whose memory can be reclaimed. In addition to normal event messages, the anti-messages and the state information associated with the event messages whose time-stamp is less than the GVT value would also be considered as safe event which in turns allows the global mechanism to reclaim the memory. This GVT is a global function which might be computed several times during the execution of distributed simulation. In other words, the success of global control mechanism is dependent on the factor that how fast the GVT is computed level of the front of you who the Therefore, the time required to compute the GVT value is critical for the optimal performance of the optimistic TMA. If the latency for computing GVT is high, which is true in the case of the optimistic algorithms, the performance of optimistic TMA degrades significantly due to a lower execution speed. The increase in the latency also increases the processor idle time since the processor has to wait unless the global control mechanism knows the current value of GVT. Once it comes to know the value of GVT, the

[1]Contact author: srizvi@bridgeport.edu,

K. Elleithy (ed.), *Advanced Techniques in Computing Sciences and Software Engineering*,
DOI 10.1007/978-90-481-3660-5_78, © Springer Science+Business Media B.V. 2010

event processing will then be resumed by the individual processors of the distributed machines.

Once the GVT computation is initiated, the controller freezes all the LPs in the system. The freezing of LP indicates that all LPs have entered into the *find mode* and they can not send and receive messages to the other neighboring LPs unless the new GVT value is announced by the controller. For this particular case, there is a possibility that there might be one or more messages that are delayed and stuck somewhere in the network. In other words, these messages were sent by the source LP but did not yet receive by the destination LP. These event messages are referred as transient messages which should be considered in the GVT computation by each LP. Since the LPs are freeze during the GVT computation, they might not be able to receive any transient message that arrives after they enter into the find mode. This problem is generally referred as transient message problem as shown in Fig. 1. In order to overcome this problem, message acknowledgement technique is used in which the sending LP needs to keep track that how many event messages has been sent and how many acknowledgements has received. The LP remains block as long as it receives all the acknowledgements for the transmitted packets.

Other problem faced at the time of computing GVT is the simultaneous reporting problem as shown in Fig. 2. For this particular scenario, once the controller sends a message to the LP asking to start computing the local minimum, there is always a possibility that the LP may receive one or more event messages from the other neighboring LPs. If this happens, there is no way that the LP can compute the correct value of the local minimum since the LBTS computation did

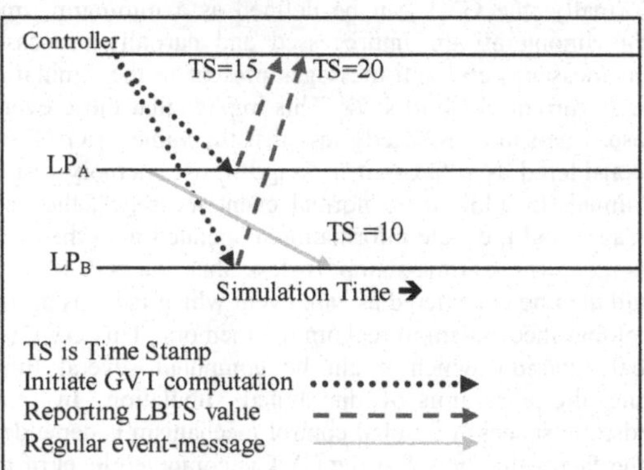

Fig.1. Representation of transient message problem between two LPs and the controller. X-axis represents the global current simulation time of the system. The first two messages are transmitted from controller (indicated by full dotted lines) to initiate GVT computation. LP_A and LP_B compute its LBTS values as 15 and 20, respectively and reported to the controller (indicated by partially cut lines). A transient message arrived with the time stamp 10 at LP_A. If this time stamp is not considered, the value of GVT will be incorrectly computed as 15 rather than 10.

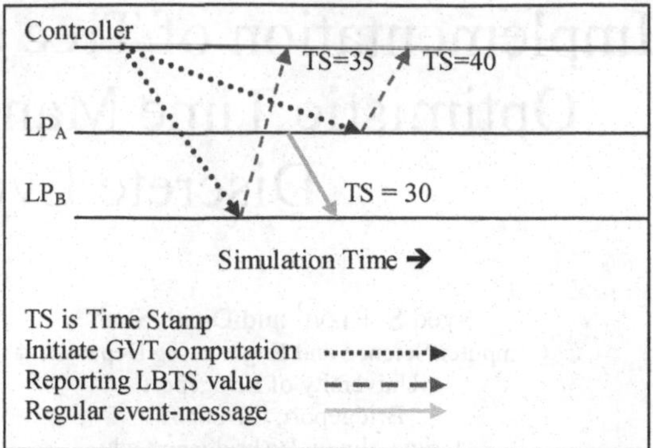

Fig.2. Representation of simultaneous reporting problem between 2 LPs and controller. X-axis represents the global current simulation time of the system.

not account the time-stamp of those event messages that arrived at LP during the LBTS computation. The result would be a wrong value of the LBTS which in turns may cause an error in the GVT value. For instance, as shown in Fig. 2, the controller sends the message to both LP_A and LP_B asking to compute their local minimum. Once LP_B receives this message, it immediately reports its local minimum to controller as 35. However, the second message from controller to LP_A was delayed. LP_A sends one event message to LP_B with the time-stamp 30. Later, it reports its local minimum (40) to the controller after processing an event message with the time-stamp 40. The controller computes the global minimum (GVT) as 35 rather than 30. This problem was raised due to the fact that LP_B did not account an event message with the time-stamp 30.

II. THE ORIGINAL SAMADI'S GVT ALGORITHM

In the Samadi's algorithm [8], sender has to keep track that how many messages have been sent so far, how many acknowledgments have been received for the transmitted messages, and how many sent message are still unacknowledged. The transient message problem as cited above will be solved by keeping track of all such messages.

However, in order to solve the issue of simultaneous reporting problem, the Samadi's algorithm requires that an LP should place a flag on each acknowledgement message that it transmits to the other LPs after start computing its local minimum. This acknowledgement message with the flag indicates that the time-stamp of this message was not considered by the sending LP in its LBTS computation. If the receiving LP receives this message before it starts computing its local minimum, the receiving LP must consider the time-stamp of the received message. In this way, we ensure that not only the unprocessed messages but also the transient and anti-messages are considered by all LPs while computing their local minimum values. This in turns leads us to the correct GVT computation. For instance, as shown Fig. 3, if

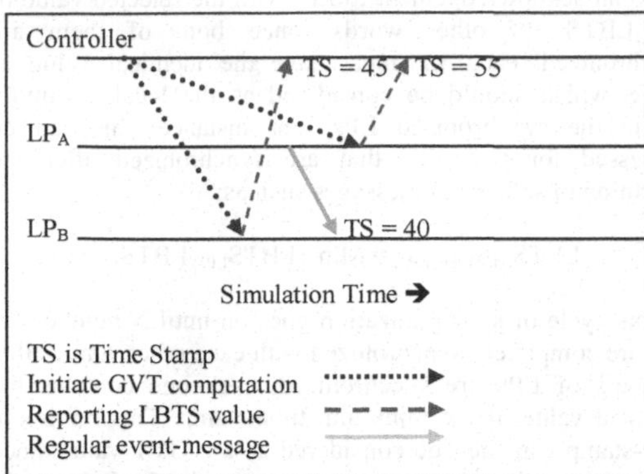

Fig.3. Representation of Samadi's GVT algorithm for dealing with transient message problem between two LPs and the controller, X-axis represents the global current simulation time of the system

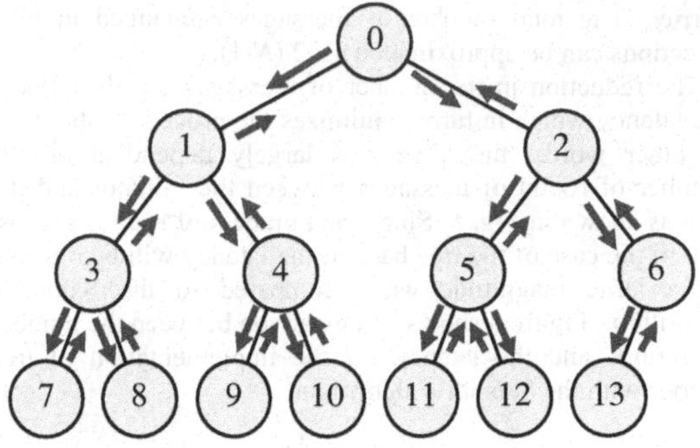

Fig.4. The organization of 14 LPs in a tree barriers structure. LP_0 is a controller which computes global minimum. LP_1 to LP_6 are non-leaf nodes. LP_7 to LP_{13} represents leaf nodes. Green lines represent LBTS computation and GVT announcement whereas red lines represent synchronization messages

an event-message with the time-stamp 40 reaches to P_B from P_A, the receiving LP has to send an acknowledgement back to the sending LP. This acknowledgement must be flagged, so that the receiving LP must include that value if it does not start computing the LBTS value. If this does not properly manage, the GVT computation will not only be delayed due to the transient messages but also produced the incorrect GVT values.

This implies that the role of GVT computation is critical in the performance of optimistic TMA. Samadi's algorithm provides a foolproof solution to all problems cited above as long as the algorithm is implemented as described. However, the algorithm itself does not guarantee that the GVT computation is fast enough that it can minimize the processors idle time. Therefore, in order to minimize the latency involves in GVT computation, we deploy the synchronous barriers.

III. Comparative Analysis Of Tree Barrier

In the tree barrier (see Fig. 4), the LPs are categorized into three different levels. The first level is the root LP which is responsible to initiate and finish the process of GVT computation. In the second level, we might have several non-leaf LPs that could be a parent or child. Finally, at the lowest level of tree, we may have several leaf LPs which do not have any child LP. The GVT computation will be carried out in both directions going from root to leaf via non-leaf nodes. The root LP initiates the GVT computation by broadcasting a single message that propagates all the way from the root node to the leaf LPs via the non-leaf nodes.

When this initial broadcast message is received by the LPs, each LP starts computing its local minimum value and sends a message to the respective parent node. Similarly, a parent node does not send a message to its parent unless it receives a message from each child LP. The leaf nodes are exempted from this condition. Once the non-leaf node receives a message from all child LPs with the LBTS values, it can then determine the local minimum and send the final value to its parent. This cycle of LBTS computation goes on unless the root LP receives the message from its child LPs. The final computation will be done by the root LP in which it determines the minimum of all LBTS values that it receives from its child LP. Finally, a message will be broadcasted to all LPs with the new GVT value.

This analysis shows that the number of messages transmitted in both directions (i.e., from root to leaf LPs and vice versa) will be the same. In other words, based on our hypothesis, each LP in the tree barrier processes two messages, one for receiving message whereas the other for acknowledging the received message. After receiving the acknowledgement from all the child nodes, the parent node starts computing the local minimum which automatically solves the problem of simultaneous reporting. Since there is no freezing LP exists in the system, there is no way that a message might left in the network until the new GVT value is computed. Therefore, this solves the transient message problem. This discussion implies that the tree barrier offers a complete structure that one can use to solves the problems of optimistic TMA while at the same time it offers a low latency to compute the GVT value.

The total number of messages transmitted in the tree barrier is clearly less that the Samadi's algorithm. This is simply because in the proposed tree barrier, there is no message containing events as well as all the messages are processed with respect to the number of edges via they are connected with each other. If there are N LPs exist in the system, they need typically N-1 number of edges. This implies that the total number of messages need to be transmitted in order to compute the GVT value in both directions (i.e., from root LP to leaf and vice versa) can not exceed to twice of the number of edges exist in the tree barrier. The total number of messages exchanged in both directions can be approximated as: 2 (*N*-1).

The reduction in the number of messages clearly reduces the latency which in turns minimizes the processor idle time. In other words, the latency is largely dependent on the number of round of messages between the initiator and the LPs as shown in Fig. 5. Since the numbers of rounds are just two in the case of the tree barrier, the latency will be reduced by a large magnitude when compared to the Samadi's algorithm. Figure 5 shows a comparison between the number of rounds and the latency for the implementation of tree barrier with the Samadi's algorithm.

IV. Comparative Analysis Of Butterfly Barrier

Butterfly barrier is a technique where we eliminate the broadcast messages. Assume that there are *N* numbers of LPs present in the LP. For instance, if LP_2 wants to synchronize with the other neighboring LPs once it is done with the execution of safe event messages, it can send a synchronization message to one of the neighboring LPs whose binary address differ in one bit. The LP_2 can only initiate this synchronization once it is done with the processing of safe event messages so that it can compute its own local minimum. Therefore, the first message send from LP_2 to the other neighboring LPs will have its LBTS value. When one of the neighboring LPs receives such synchronization message, it first determines its own status to see weather it has unprocessed safe event messages. If the receiving LP has unprocessed safe event messages present, it firsts executes them before it goes for the synchronization with LP_2. However, if the receiving LP has already finished executing the safe event messages, it computes its own LBTS value and compares it with the one it receives from LP_2. The minimum of the two LBTS will be selected as the local minimum for the two synchronized LPs. The receiving LP sends an acknowledgement to LP_2 with the selected value of the LBTS. In other words, once both of them are synchronized, each LP must have the identical value of LBTS which should be considered as the local minimum within the synchronized LPs. For instance, this can be expressed for two LPs that are synchronized after the execution of safe event messages such as:

$$LBTS_{(LP1 \text{ and } LP2)} = \text{Min}\ \{LBTS_{LP1}, LBTS_{LP2}\}$$

This cycle of synchronization goes on until *N* numbers of LPs are completely synchronized with each other. Once all *N* numbers of LPs are synchronized, each one contains the identical value of the minimum time-stamp. This minimum time-stamp can then be considered as the GVT value since this is the smallest local minimum among all LPs. In butterfly barrier, each LP typically processes two messages. One is for the transmission of the synchronization message and one for its acknowledgment. In other words, each LP must send and receive one message in each step of the algorithm. Since the synchronization is achieved in a pair wise fashion, a total of $\log_2 (N)$ numbers of steps are required to complete the algorithm. In other words, each LP is released from the barrier once it has completed $\log_2 N$ pair wise barriers. Taking these factors into account, one can approximate the total numbers of messages that we must send and receive before the GVT value is computed, such as: $N \log_2 (N)$. Figure 6 shows a comparison between the number of rounds and the GVT latency for the implementation of butterfly barrier with the Samadi's algorithm. It should be noted in Fig. 6 that the performance of butterfly barrier is not as impressive as the tree barrier due to the high latency for a large number of rounds.

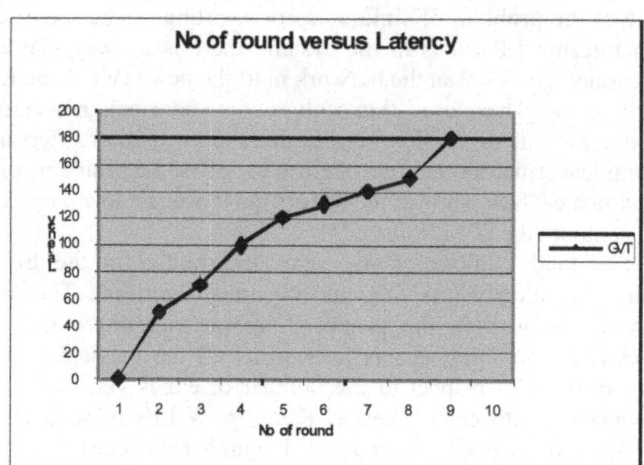

Fig.5. Tree implementation with Samadi's algorithm, Number of rounds versus latency for *N*=10.

V. Performance Evaluation Of Tree And Butterfly Barriers

Figure 7 shows a comparison between the Samadi's algorithm and the tress and butterfly barriers for *N* = 9. Based on the simulation results of Fig. 7, one can clearly infer that the tree barrier outperforms the Samadi's algorithm in terms of the numbers of message that one may need to transmit in order to compute the GVT value. This reduction in the number of messages not only minimizes the GVT latency but also improves the CPU utilization by reducing the processor idle time. However, the implementation of butterfly barrier for computing the GVT values results in high latency than the Samadi's algorithm. This is mainly due to the fact that the butterfly barrier may not perform well in the presence of network error or delayed acknowledgement. In other words, the simultaneous reporting problem causes the performance degradation for the butterfly barrier.

Next, we discuss that how the butterfly barrier deals with the common problems of the optimistic TMA such as the transient messages, and simultaneous reporting. Like the tree barrier, butterfly barrier also avoid the possibility of any

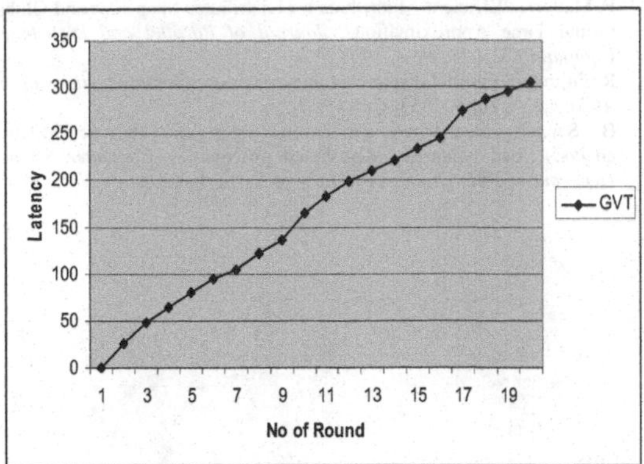

Fig.6. Butterfly barrier implementation with the Samadi's algorithm, GVT computation with number of rounds versus latency for N=20

freezing LP. Since the LP may not get involve in the deadlock situation, there is no way that a system may have a transient message somewhere in the network. Therefore, the transient message should not be a problem in the case of butterfly barrier. In addition to transient message problem, the butterfly barrier needs to address the problem of simultaneous reporting. The simultaneous reporting might cause a serious problem in the case of butterfly barrier especially when communication network is not reliable. For instance, let LP_2 sends a synchronization message to the other LP which differs in one bit position with the LBTS value. Also, assume that the receiving LP has finished executing the safe event messages and has successfully computed its own LBTS value. After comparing the two LBTS values, it sends an acknowledgement with the local minimum to LP_2. If that acknowledgement message is not

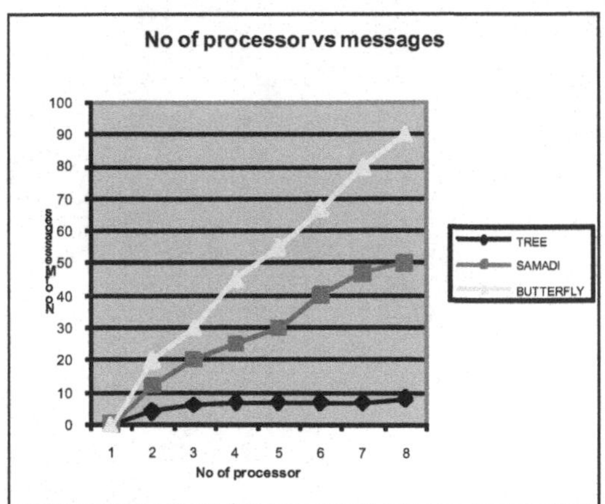

Fig.7. Comparison of Samadi's algorithm with the Tree and Butterfly barriers for $N = 9$

arrived to the destination LP (i.e., LP_2) due to the network error or it is delayed, the pair wise synchronization will be failed. As a result, when the sending LP did not hear anything from the receiving LP, it will eventually look for the other LPs (only LPs whose binary addresses differ in one bit) and send a new synchronization message.

If we assume that the LP_2 has successfully synchronized with one of the other neighboring LPs, then what would happen if the delayed acknowledgement message arrived at LP_2. Obviously, LP_2 had an impression that the previous LP was not interested in the synchronization process where as the previous LP had an impression that I was synchronized with LP_2. For this particular scenario, the butterfly barrier will fail to successfully compute the GVT values. However, if we assume that the network is completely reliable and all the links between the LPs are working without malfunctioning, the simultaneous reporting problem can not be raised within the pair wise synchronization of the butterfly barrier structure.

VI. CONCLUSION

In this paper, we presented the implementation of synchronous barriers such as tree and butterfly with the optimistic TMA. This approach is quite new since it combines two different families of algorithms (conservative and optimistic) to go for the same task of synchronization. We started our discussion from the optimistic algorithm in general and Time Wrap and Samadi's algorithm in particular. We also presented an analysis that shows why an optimistic algorithm must have the capability to deal with some common problems like rollback, reclaiming memory, transient messages, and simultaneous reporting. Finally, we showed that how the tree and butterfly barrier can be implemented with the optimistic algorithm to compute the GVT value. Both our theoretical analysis and the simulation results clearly suggest that the tree barrier performs well than the pure optimistic algorithm in terms of the number of messages that one may need to transmit to compute the GVT value. In addition, we also discussed that how these two barriers can deal with the common problems of optimistic algorithms. For the future work, it will be interesting to design the simulation where we can compare the performance of these barriers with the Time Wrap algorithm.

REFERENCES

[1] D. Bauer, G. Yaun, C. Carothers, S. Kalyanaraman, "Seven-O' Clock: A new Distributed GVT Algorithm using Network Atomic Operations," *19th Workshop on Principles of Advanced and Distributed Simulation (PADS'05)*, PP 39-48, 2005.

[2] F. Mattern, H. Mehl, A. Schoone, Tel, G. Global Virtual Time Approximation with Distributed Termination Detection Algorithms. *Tech. Rep. RUU-CS-91-32, Department of Computer Science, University of Utrecht*, The Netherlands, 1991.

[3] F. Mattern. Efficient algorithms for distributed snapshots and global virtual time approximations. *Journal of Parallel and Distributed Computing*, 18:423--434, 1993

[4] R. Fujimoto, "Distributed Simulation system," *proceeding of the 2003 winter simulation conference*. College of Computing, Georgia Institute of Technology, Atlanta.

[5] S. Rizvi, K. Elleithy, and A. Riasat, "Trees and Butterflies Barriers in Mattern's GVT: A Better Approach to Improve the Latency and the Processor Idle Time for Wide Range Parallel and Distributed Systems", *IEEE International Conference on Information and Emerging Technologies (ICIET-2007),* July 06-07, 2007, Karachi, Pakistan.

[6] F. Mattern, "Efficient Algorithms for Distributed Snapshots and Global virtual Time Approximation," *Journal of Parallel and Distributed Computing,* Vol.18, No.4, 1993.

[7] R. Fujimoto, Parallel discrete event simulation, *Communications of the ACM*, v.33 n.10, p.30-53, Oct. 1990.

[8] B. Samadi, Distributed simulation, algorithms and performance analysis (load balancing, distributed processing), *Computer Science Department, PhD Thesis, University of California, Los Angeles*, 1985.

Improving Performance in Constructing specific Web Directory using Focused Crawler: An Experiment on Botany Domain

Madjid Khalilian, Farsad Zamani Boroujeni, Norwati Mustapha
Universiti Putra Malaysia (UPM),Faculty of Computer Science and Information Technology(FSKTM)

Abstract-Nowadays the growth of the web causes some difficulties to search and browse useful information especially in specific domains. However, some portion of the web remains largely underdeveloped, as shown in lack of high quality contents. An example is the botany specific web directory, in which lack of well-structured web directories have limited user's ability to browse required information. In this research we propose an improved framework for constructing a specific web directory. In this framework we use an anchor directory as a foundation for primary web directory. This web directory is completed by information which is gathered with automatic component and filtered by experts. We conduct an experiment for evaluating effectiveness, efficiency and satisfaction.

Keywords: Botany, Ontology, Focused Crawler, Web Directory, Document Similarity

I. Introduction

One of the main challenges for a researcher is how to organize vast amount of data and facilities for accessing them through the web. With the explosive growth of the web, the need for effective search and browse is apparent. User overwhelmed with a lot of unstructured or semi-structured information. One solution for this problem is creating and organizing data in a specific domain directory. There are two main approaches for creating a web directory:

1-Creating web directory manually, 2-creating and organizing automatically. Each of them has its own disadvantages. For manual web directory we need lots of editors whose knowledge is different from one another. On the other hand, it is not scalable because web is being growth but our resources are limited. Using automatic approach has own difficulties. Most experiments show that information which gathered by this approach does not have enough quality. We proposed a semi-automatic framework for constructing high quality web directory for experiment.

We select plants classification and identification domain because of the importance of this domain for agricultural science. Besides having own special properties, this domain uses a literature includes specific terms in Latin and it is standard for all languages. In addition, this domain has a perfect hierarchy. These two properties of plant classification affect methods and frame work.

Two main components have been used in our framework; one of them refers to human knowledge while the second one is about automatic component. We choose to use vertical search engine instead of multipurpose search engine. Vertical search engine uses a specific domain for doing search. There is a main component in vertical search engine that is called focused crawler. In contrast with regular crawler, in this kind of crawler specific domain web pages are gathered. It helps us to avoid less quality web pages which are not related to specific domain. Some techniques have been used for increasing precision and recall. These two parameters are described by Fig. 1.

Our interest is to increase the intersection of A and B; this can be done by increasing precision and recall. In the next section a survey on related works is given. Then in section 3 our research model is explained. Section 4 describes the methodology. Experiment, results and discussion are explained

K. Elleithy (ed.), *Advanced Techniques in Computing Sciences and Software Engineering*,
DOI 10.1007/978-90-481-3660-5_79,

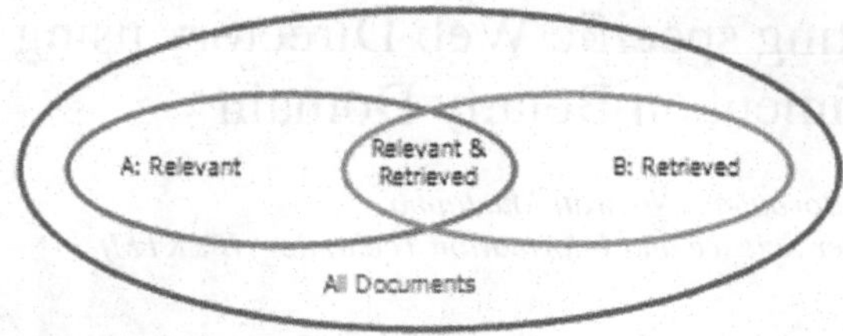

Fig.1. **A:** specific domain web pages, **B:** gathered web pages by crawler, **B-A:** gathered web pages which are not relevant, **A-B:** not gathered web pages which are relevant web pages.

in sections 5 and 6 and finally we present our conclusion and future works.

II. Related Works

A very large number of studies over the years have been done to improve searching and browsing web pages. There is a general categorization for search systems: traditional text-based search environments [1] and Web directory. Development organizing knowledge for domain specific applications has challenged academics and practitioners due to the abundance of information and the difficulty of the categorizing the information. In this review of related research, we maintain our focus on research concerning web directory search environments. In this review we study three main topics: Ontology, automatic and semiautomatic search environments, plants web directory.

1-Ontology and web directory

In the first section we review the ontology concept. In fact knowledge is classification and without classification there could be no thought, action or organization. Reference [2] mentioned that one of the most important problems for web is huge amount of information without any structure among them. Ontology is an agreement between agents that exchange information. The agreement is a model for structuring and interpreting exchanged data and a vocabulary that constraints these exchanges. Using ontology, agents can exchanges vast quantities of data and consistently interpret it.

Many applications contain leverage ontology as an effective approach for constructing web directory. Reference [3] states that ontology is used for web mining of home pages. In this work the employed ontology is hand made. Some of projects use ontology to do online classification of documents [4]. To do so for different meaning of query terms different classes are created and documents in the result list are put in appropriate classes. The used ontology in this case is WordNet which is a general ontology. Reference [5] defined ontology for email and dose classification. It also uses protégé for ontology definition and CLIPS is employed reasoning on it. In most of this works a ready made ontology is used and prevents extendibility of the proposed methods for new domains. Reference [6] proposed an approach that uses an automatic method for creation of ontology. Although it might not be suitable for reasoning proposes, it is a good facility for extending the search for the whole web space.

2-Automatic and semiautomatic search environments

One Approach for creating effective ontology is using vertical search engine as a tool. Vertical search engines use focused crawler for specific domain. The motivation for focused crawling comes from the poor performance of general-purpose search engines, which depended on the result of generic web crawlers. The focus crawler is a system that learns the specialization from examples, and then explores the web, guided by a relevance and popularity rating mechanism.

Reference [7] suggested a framework using combination of link structure and content similarity. Experiments in this paper show that considering link structure and content together can improve quality of retrieved information. Reference [8] proposed a new method based on content block Algorithm to enhance focused crawler's ability of traversing tunnel. The novel algorithm not only avoid granularity becoming too coarse when evaluation on the whole page but also avoid granularity becoming too fine based on link context. An intelligent crawling technique has been used which can dynamically adapt ontology to the particular structure of the relevant predicate. In reference [9] adaptive focused crawler based on ontology has used which is included two parts: topic filter and link forecast. The topic filter can filter the web pages having already been fetched and not related to topic ; and the link forecaster can predict topic link for the next crawl , which can guide our crawler to fetch topic page as many as possible in a topic group by traversing the unrelated link . Proposed method use ontology for crawling and during this process it is able to concrete ontology. It also use ontology learning concept that it can be useful for using for our project as a future work. Reference [10] proposed a frame work with semi-automatic approach. In fact it combines the ability of human and machine together and avoids disadvantages of each of these two methods separately. It seems that the automatic part of this framework needs to be improved. A meta-search section has been employed. In order to gather information and web pages for this purpose, a multipurpose search engine is used. [11]. Using multipurpose search engine has own disadvantages: 1) Enormous amount of unusable web pages can reduce the speed of the process. 2) Precision is not acceptable because a huge amount of web pages should be searched. 3) Multipurpose repository should be up to dated and it consumes time. On the other hand only a limited section of web has been covered by even the biggest search engine. Reference [12] mentioned that using traditional methods of information retrieval can help us to increase efficiency based on these methods. Using vector space model, web pages which are related to specific domain had been selected and all links inside them had been crawled. Reference [13] proposed a novel approach based on manifold ranking of document blocks to re-rank a small set of documents initially retrieved by some existing retrieval function. The proposed approach can make full use of the intrinsic global manifold structure of the document blocks by propagating the ranking scores between the blocks on a weighted graph. The manifold ranking process can make use of the relationships among document blocks to improve retrieval performance. However, link structure among web pages has not been considered. Reference [14] has used domain graph concept for web page ranking. For every page in kernel a graph is constructed from pages that point to selected page. Multipurpose search engine is leveraged in this method because they can determine list of web pages which point to specific page. After extraction of this graph web pages are ranked and base on this ranking web pages are arranged.

3-Plants web resources

There is some web directories such as USDA (Protection American agriculture) is the basic charge of the U.S. Department agriculture (www.usda.gov). APHIS (Animal and Plant Health Inspection Science) provides leadership in ensuring the health and care of animal and plants. None of them have been concentrated on botany topic. The Internet Directory of Botany (IDB) is an index to botanical information available on the internet (www.botany.net). It is created by some botany specialists manually but it is a well organized alphabetic web directory which we choose it as a benchmark.

III. Research Model

In this study we improve the framework proposed by [10] which includes three main steps. Instead of using meta- search we use a vertical search engine with focused crawler. For this reason we need to have a kernel for our crawler and a vector that shows domain literature. So we change step1 and step2 of this framework and step3 remains unchanged, As shown in Fig.2. In our framework, after constructing anchor directory we initiate a new set of URLs which are related to plant domain. In third step we design a focused crawler that works based on cosine similarity. For this purpose, we developed a software tool that crawls links from the kernel using cosine similarity method. Gathered web pages are put in a repository and finally a heuristic filtering is done. In this section we describe the proposed framework.

1. Anchor directory

Based on our review on web search engines and directories, we choose DMOZ directory as the anchor directory (www.Dmoz.com). In most search engine this directory is employed. In addition, expert people refer to this web directory for special search in Botany. For this reason some refinement should be done to construct anchor web directory.

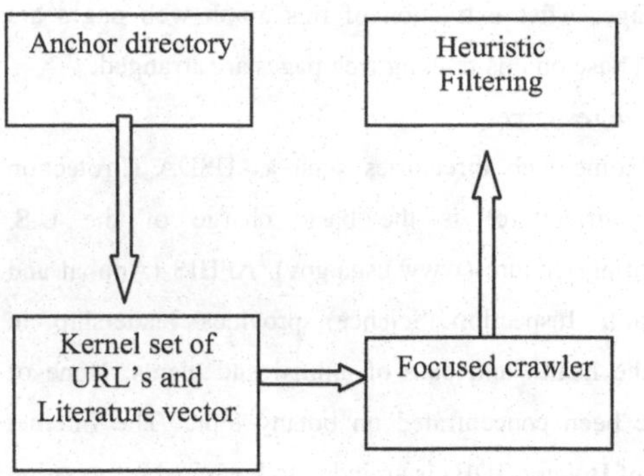

Fig.2.An improved semi-automatic framework for constructing web directories.

2. Kernel & literature vector creation for the next section

We need to have a kernel for gathering information through the web. For this purpose, a set of URLs should be created. It also requires a vector with specific terms in Botany domain. Creation of the set and vector has been done by experts. For this purpose 110 web pages have been selected and a vector with 750 terms in botany domain has been created.

3. Vertical search engine

To fill in the plant web directory we use a focused crawler that works based on similarity search among web pages and vector that describes plant domain. For example, in focused crawler based on content of current web page, a vector of existing terms is created after that value of similarity between current page and vector domain is calculated. If this similarity was less than a threshold value, this page would be saved in a repository and crawling would be continued, otherwise it would be deleted and would be crawling stopped.

4. Heuristic filtering

Finally we evaluate output of step 2 and step 3 by a group of experts. Expert people can score the output of crawling.

IV. Methodology

Using mentioned framework, we developed a web directory called PCI web directory (Plant Classification and Identification) which includes plant web sites. Based on our view on web search engines, we have chosen the DMOZ directory as the anchor directory because of its comprehensive plant directory and its wide acceptance, as seen in the fact that is used by such major search engines as Google, AOL and Netscape search [10]. We use English language in our research. At the first trying we aggregated 1180 nodes from DMOZ, and after filtering the nodes we got 331 nodes corresponding to plant classification and identification. Table 1 Summarize statistics of PCI web directory.

In order to compare our proposed approach, we selected a benchmark web directory called IDB (internet directory for botany) www.botany.net/IDB/botany.html, it is an index to botanical information available on directory.

V. Experimental Procedure

A. Design and Participants

We have measured effectiveness, efficiency and satisfaction of our PCI web directory with compare IDBAL as a benchmark. It is one of the best well designed web site for botany but it is not scalable, user friendly and efficient. Ten experts in botany have been invited and by them we measured the criteria that we mentioned. Two groups of experts, graduated and undergraduate with different age and experience have been selected, for testing and comparing PCI and IDBAL,

We interested in observing effectiveness, efficiency and satisfaction between two groups of PCI and IDBAL via experts' testing. Basically we measure effectiveness by three values of precision, recall and f-value. Efficiency is time taken by the users to get some questions. Based on some questions we are able to measure satisfaction. For calculating efficiency, we organize an experiment on the search of a specific topic by experts in both web directories (benchmark and proposed web directory). At the end, using a questionnaire about different

TABLE 1
SUMMERY STATISTICS OF PCI WEB DIRECTORY

STATISTICS	PCI DIRECTORY
Total number of categories	331
Total number of unique web sit URLs	2736
Average number of pages per category	8.2
Maximum depth	7

aspects of web directory, i.e. being helpful and positive and negative points, we are able to measure the satisfaction parameter (table2). We ranked each question between 1 and 5 where 1 is for completely satisfied and 5 for not satisfied.

B. Hypothesis testing

Our objective focused on using a frame work that combines experts' knowledge and automatic approaches to help for searching and browsing information in specific domain like botany and improves efficiency and speed up in this situation. Constructing a web directory for botany can help specialists to search and browse more efficient, effective and it is able to satisfy users than benchmark.

C. Performance measure

We defined follow terms for evaluating proposed framework:

- Precision: the percentage of retrieved documents that are in fact relevant to the query (i.e., "correct" responses)

$$precision = \frac{|\{Relevant\} \cap \{Retrieved\}|}{|\{Retrieved\}|}$$

- Recall: the percentage of documents that are relevant to the query and were, in fact, retrieved

$$recall = \frac{|\{Relevant\} \cap \{Retrieved\}|}{|\{Relevant\}|}$$

- F-value = 2 * recall * precision / (recall + precision).

This parameter is used to balance between recall and precision. With these three parameters we can define the quality.

- Efficiency (Speed up): comparing speed of users when they use web directory for searching and browsing a specific topic with multipurpose search engines.
- Satisfaction: how much web directory is useful for users?

TABLE 2

COMMENTS BASED ON EXPERIMENTS

Comments on PCI Web directory	Number of subjects expressing the meanings
Easy and nice to use	1
Useful and with relevant results	4
Good way to organization information	1
Not enough information for browsing	4
No comments/unclear comments	0

Comments on IDBAL Web directory	Number of subjects expressing the meanings
Not useful/not good	2
Not clear for browsing/gives irrelevant results	2
Easy to navigate or browse	3
No comments/unclear comments	3

VI. Results and discussion

In this section, we report and discuss the results of our experiment. For this experiment (Table3) a web directory has been organized (PCI) and compared with a benchmark (IDBAL). As it is seen hypothesis supported in most criteria. Efficiency is not better in PCI than IDBAL. It refers to sorted and well organized web pages in IDBAL. It seems that is difficulty in structure and organization of web directory (PCI).

VII.Conclusions and Future Works

Constructing a specific domain directory which is included high quality web pages has challenged developers of web portals. In this paper we proposed a semiautomatic framework that combines knowledge of human with automatic techniques. Experts evaluate outline web directory with compare a benchmark. Most expected results has been satisfied.

For future work we will investigate new intelligent techniques for improving precision and recall in focused crawler. It seems that using dynamic vector instead of vector with constant terms can improve effectiveness. It is feasible with adding and deleting terms in reference vector. Another direction includes refining PCI web directory and makes it more efficient. Other specific domain can be applied in this framework.

TABLE 3

STATISTICAL RESULTS OF HYPOTHESIS TESTING (S=supported)

	PCI	IDBAL	
MEASURE	Mean	Mean	Results
Precision	0.87	0.75	S
Recall	0.29	0.17	S
f-value	0.435	0.255	S
Efficiency	173	160	NOT S
Satisfaction	3.1	3.3	S

References

[1] H. Topi, W. Locase, Mix and Match: combining terms and opratores for successful web searches, information processing and management 41(2005) 801-817

[2] H.P.ALSSO and F.SMILL, Thinking on the Web, John WILLEY New Jersey 2006.

[3] Ee -Peng Lim and Aixin Sun: Web Mining- The Ontology Approach, The International Advanced Digital Library Conference in Nagoya Noyori Conference Hall Nagoya University, Japan August 25-26, 2005

[4] EW De Luca, A Nürnberger: Improving Ontology-Based Sense Folder Classification of Document Collections with Clustering Methods Proc. of the 2nd Int. Workshop on Adaptive Multimedia. 2004.

[5] Taghva, K. Borsack, J. Coombs, J. Condit, A. Lumos, S. Nartker, T: Ontology-based classification of email, ITCC 2003. International Conference on Information Technology: Coding and Computing 2003.

[6] M. Khalilian, K. Sheikh, H. abolhassani (2008), Classification of web pages by automatically generated categories, Innovations and Advanced Techniques in Systems, Computing Sciences and Software Engineering, springer,ISBN: 978-1-4020-8734-9

[7] M. Jamali et .al . A Frame Work using Combination of link structure and Content similarity.

[8] N.LUO, W-ZUO. F.YUON, A New Method for Focused Crawler Cross Tunnel, RSKT2006. pp 632-637

[9] Chage Su. J.yang,An efficient adaptive focused crawler based on ontology learning . 5th ICHIS IEEE 2005

[10] Wingyan Chung, G. Lai, A. Bonillas, W. Xi, H. Chen, organizing domain-specific information on the web: An experiment on the Spanish business web directory, int. j. human computer studies 66 (2008) 51-66

[11] M. Khalilian, K. Sheikh, H. abolhassani (2008), Controlling Threshold Limitation in Focused crawler with Decay Concept, 13th National CSI Conference Kish Island Iran

[12] F. Menczer and G. Pant and P. Srinivasan. Topic-driven crawlers: Machine learning issues, ACMTOIT, Submitted 2002.

[13] X. Wan, J. Yang, J. Xiao, Towards a unified approach to document similarity search using manifold ranking of blocks, Information processing and Management 44 (2008) 1032-1048

[14] M. Diligenti, F. Coetzee, S. Lawrence, C. Giles and M. Gori, Focused Crawling Using Context Graphs, In Proceedings of the 26th International Conference on VLDB Egypt (2000)

[15] Cai, D., Yu, S., Wen,J., & Ma., W, -Y. (2003) ;VIPS ; A vision based page segmentation algorithm. Microsoft Technical Report, MSRTR- 2003-79.

"Security Theater" in the Pediatric Wing: The Case for RFID Protection for Infants in Hospitals

David C. Wyld
Southeastern Louisiana University
Department of Management – Box 10350
Hammond, LA 70402-0350

***Abstract*- In a typical year, five infants are abducted by strangers from hospitals. These are devastating events for the families involved and for the health care facilities' staff and executives. This article looks at the nature of newborn and infant abductions, analyzing data on these kidnappings from the FBI and from the National Center for Missing & Exploited Children. Then, the article examines the potential for RFID (radio frequency identification) based systems to improve security in the pediatric area of hospitals, providing an overview of the technology on the market today for infant protection. The article concludes with an analysis of the ROI (return on investment) equation for health care administrators to consider in weighing their options on how to prevent a statistically unlikely, but potentially cataclysmic occurrence, at their facility. In the end, RFID-based infant protection systems can be seen as a form of "security theater," serving as a "palliative countermeasure" that will indeed work – both substantively and psychologically – to promote a more secure hospital environment for moms and their newborns.**

I. INTRODUCTION

A. *The "Worst Case Scenario"*

The new mom is elated, holding her new daughter in her arms in the hospital's birthing center. Her husband is down in the hospital gift shop buying batteries for his digital camera, having shot dozens of pictures to document their little girl's first 24 hours of life. After a shift change, a new nurse comes in and chats with the mom, offering to take the newborn back to the nursery to allow her to take her first shower since the delivery yesterday morning. The new mom gives her baby girl a kiss on her rosy cheek and hands the baby over to the nurse. She then proceeds to take a long, luxurious (as can be in a hospital) shower, dressing as a refreshed, new woman. When hubby finally gets back from the gift shop, with flowers, a stuffed animal, and, oh yes, batteries, the couple decide to walk down to the nursery to see their daughter. When they pass the nurses station, the wife does not spot the nurse who carried her daughter out of her room fifteen minutes ago. When they arrive in the nursery, they spot three babies – but not their precious child. The couple both blurt out almost simultaneously to the nurse standing there - herself a bit perplexed - "Where's our daughter?" They see the nurse instantly become very pale, racing to grab the phone. In an instant, alarms are whirring – but it's all too late. That "nurse" had a fifteen minute head start, heading somewhere with their most precious treasure. This has got to be every new mother's nightmare – and that of every hospital administrator as well. And it does happen….all too much.

B. *Baby Snatching*

According to statistical data from the National Center for Missing & Exploited Children (NCMEC) [1] show that there have been 252 infants abducted in the United States over the past 25 years, with just under half of these kidnappings – 123 in all - taking place in the hospital environment. As can be seen in Figure 1, the scenario above is the most common type of in-hospital infant abduction. The U.S. Department of Justice reports that on average each year, 115 children become the victims of "stereotypical" kidnapping – where crimes involve someone the child does not know or someone of slight acquaintance, who holds the child overnight, transports the child 50 miles or more, kills the child, demands ransom, or intends to keep the child permanently [2]. Thus, on average, the "nightmare scenario" of in-hospital abductions of newborns has been occurring at the rate of 5 each year.

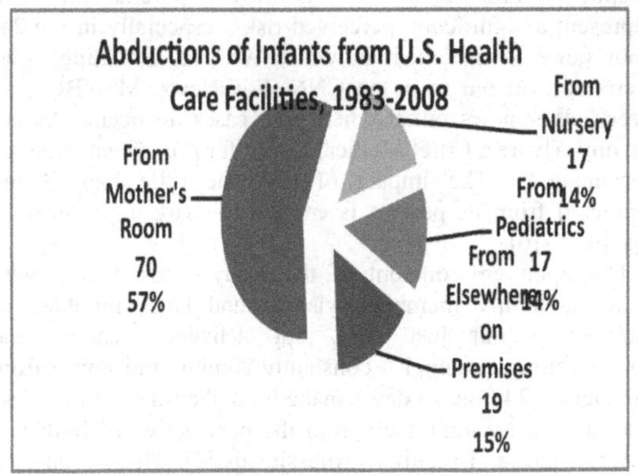

Fig. 1. U.S. Infant Abduction Data.
Source Data: National Center for Missing & Exploited Children, 2008.

Two recent cases illustrate the gravity of the problem when it does occur, both in the U.S. and abroad. In March, at Covenant Lakeside Hospital in Lubbock, Texas, newborn Mychael Darthard-Dawodu was taken from his mother's room by a woman posing as a hospital employee on the

K. Elleithy (ed.), *Advanced Techniques in Computing Sciences and Software Engineering*,
DOI 10.1007/978-90-481-3660-5_80, © Springer Science+Business Media B.V. 2010

pretense of taking the baby for some tests. The child was located a day later by police in Clovis, New Mexico – over 300 miles away. The accused kidnapper was Rayshaun Parson, a 21 year-old woman, who was described by relatives as being deeply depressed following a miscarriage [3]. Likewise, in November 2007, in Sudbury, Ontario Canada, Brenda Batisse stands accused of kidnapping an unidentified newborn by posing as a nurse and slipping out of the hospital room when the new mother was distracted in conversation. In this case, the credit for alerting the staff to the incident went to a member of the hospital's cleaning staff, who took note of the fact that the kidnapper was carrying the baby in her arms down a hallway, rather than transporting it in a bassinet, which would have been the protocol for any nurse or staff member at the facility. After a massive search, the child was found unharmed in Elliot Lake, Ontario a town almost 200 miles away. Again, the young woman accused of committing the crime was believed to have recently suffered a miscarriage. When police arrived at her home, she attempted to convince the police that the child was her own, even showing the officers what she claimed were suture marks from a recent cesarean section [4].

Some have criticized the need for RFID-based in-hospital protection systems due to the relatively low instance of such crimes. Katherine Albrecht, the founder and director of CASPIAN - Consumers Against Supermarket Privacy Invasion and Numbering – recently observed that "Baby snatching from hospital facilities is a diaper full of nonsense" [5]. However, hospital executives – and their patient/ customers – are finding unique value from this unique RFID application. While statistics showing the rarity of infant abductions from health care facilities should be comforting to hospital administrators and reassuring to parents, they still represent a significant, perceived risk – especially in the 24-hour news environment in which we live. Breaking news alerts scream out to us on CNN, Fox News, MSNBC, and every other news outlet when such cases do occur. As Dr. Manny Alvarez, Chief Medical Editor for Fox News, recently commented: "The impact of just one little baby being abducted from its parents is enough to spark a nation-wide manhunt" [6].

The open environment of the today's hospitals – with patients, family members, visitors and large numbers of workers – not just staff, but delivery, vendor, and construction personnel – constantly coming and going from the facility 24 hours a day – make these facilities nothing less than a security nightmare from the perspective of both law enforcement and security professionals [7]. Thus, even as a great deal of effort is being made to emphasize security in the design and layout of new facilities, most hospitals have to deal with the campuses that are not intended to provide a modern airport-level equivalent level of security [8]. However, security can be vastly improved – both in reality and in the realm of what might best be called "security theater."

II. RFID Security Solutions for Pediatrics

What to do? Certainly, today's health care executives are exploring and implementing a wide panoply of security-oriented options. RFID is presently being used all around today's hospitals, with increasing use for protecting and locating both high-dollar medical and electronic equipment [9] and highly valued patients (including those suffering from Alzheimer's and dementia) [10]. In point of fact however, one of the most long-standing RFID applications is in the pediatric area of hospitals [11]. Yet, the overall penetration of such RFID systems is still low, and there is a great deal of potential for using these systems to not just provide security for newborns, but to provide value-adds to both the hospital and – most importantly – to new Moms and Dads.

The most established brand in the market today is the "Hugs" system. It is marketed by Ottawa, Ontario-based Xmark, which today is a wholly-owned subsidiary of VeriChip. The Hugs system works by having an active, tamper-proof RFID tag attached as an ankle bracelet on the baby's leg. The tag constantly checks in with readers, reporting every ten seconds that it is present and functioning. The Hugs system can be tied into the hospital's security system, set to activate security cameras, trip electronic door locks, and shutdown elevators for a "lockdown" of the facility in the event of an alarm. The Hugs bracelet is also designed to set-off an alarm if it is loosened or cut-off from the newborn's ankle [5]. The Hugs system garnered a great deal of media attention in 2005, when it was credited with helping hospital security personnel prevent a couple attempting to pull-off an infant abduction from the nursery at Presbyterian Hospital in Charlotte, North Carolina. When the baby was removed from the nursery in an unauthorized manner, the Hugs system set-off a "Code Pink" alert in the hospital that prevented the kidnappers from succeeding [12].

The complimentary, optional "Kisses" component for the Hugs system adds another level of security to prevent the occasional "mismatch" between mother and child. With this add-on, a "Kisses" bracelet is attached to the mother's wrist. Then, when the baby is given to its Mom and the two bracelets come within range, the signal that all is "OK" and that the right baby had been given to her is the sound of a lullaby that automatically plays. On the other hand, in the event of a mismatch, where the wrong baby is brought to the mother, an audible alarm sounds [11]. In Detroit, Michigan, the Hugs and Kisses system is being used at St. Joseph Mercy Hospital. New mother Michelle McKinney, who had delivered a child two years earlier at the facility, prior to the system being put in place, recently commented to the Detroit News on the security and psychological benefits of the system, saying: "You always feel safer knowing they are bringing you the right kid especially when they're gone for an hour or so. Who wants to chance it?" Plus, Mrs. McKinney said the whole lullaby thing was comforting, reporting that: "It was kind of cute. I looked at him (her newborn son, Colin) and said, 'It's nice to know you belong to me'" [13].

Last year, the Hugs system protected over 1 million newborns in American hospitals, and it is used by over 5,000 facilities globally [14]. Still, with many facilities not having adopted RFID-based security measures, the market potential for growth is quite significant. Industry estimates show that market penetration may be as low as half of all pediatric and newborn wings of hospitals have such RFID-based systems in place. This is drawing competitors into the hospital infant protection market. These include the BlueTag system, created by the French firm, BlueLinea, and marketed in North America by RFID ProSolutions, which is installing its first infant protection systems in the U.S. this spring. The start-up's BlueTag system, which can also be used to track Alzheimer's and dementia patients in health care facilities, holds many of the same operational characteristics as the Xmark offering. Based on BlueLinea's installations of the system in Paris hospitals, the system also provides another significant benefit for hospitals, combating problems with what is termed "premature leaving." According to Jebb Nucci, RFID ProSolutions' Vice President of RFID: "Both [of the French] hospitals had experienced a high level of mothers and babies who would leave the ward before being properly discharged," Nucci explains. "This was a major problem for nurses, because they would spend so much time looking for mothers and babies that were already gone. With the system in place, a mother and her baby must go see the nurses before leaving so they can deactivate and remove the baby's tag to avoid sounding the alert on their way out" [15].

Another significant competitor is the Safe Place Infant Security System, marketed by RF Technologies. The system is in place in numerous facilities, including Shawnee Mission Medical Center in Kansas City, Missouri. Recently, this hospital installed the RF Technologies system to actually upgrade its prior infant security system. Because it makes use of high frequency RFID signals (operating at 262 kHz and 318 MHz), the system experiences less interference with the vast array of other electronic medical devices and personal electronics (including cell phones and even electric toothbrushes). This has led to far less frequent false alarms from the new system. In fact, according to administrators at Shawnee Mission, the false alarm rate fell from up to a hundred a day to approximately five. This significant decline means that when a baby is brought too near a monitored doorway or another cause to trigger an alarm, staff responds much more earnestly to all alarms generated by the system, rather than seeing them as likely false and a "nuisance" alarm [16].

III. Analysis: The Real ROI for Infant Security

What is the ROI for such infant security measures? This is one instance where it is very difficult to speak just in terms of "hard" numbers, due to the nature of the threat of infant abduction. With the incidence being rare (computed by experts as being a 1 in 375,000 chance of abduction), the likelihood of any individual facility and their moms and babies falling victim to an infant abduction case is exceedingly small [17]. And, the price of such security systems can range into the hundreds of thousands of dollars, with costs varying substantially depending on a variety of factors. These include the layout, size, and birth volume of the facility, as well as the level desired for integration with other security and asset/patient tracking RFID systems at the health care facility.

However, as we have seen here, it only takes one case. One baby kidnapped from a facility can not just devastate the family involved and terrify the staff, but as more than one health care executive pointed out, it can cause long-lasting damage to the reputation and regard for the hospital itself. It can devastate the hospital's "brand" – causing spillover effects far beyond its maternity and pediatric wings, discouraging parents who have a choice from "choosing" that facility for their births, even perhaps raising not-so-subtle questions about the security of the entire facility due to a single case. Thus, when health care executives come to make decisions on whether or not RFID-based security for infants – and as increasingly common now, for their entire pediatric patient population, they must ask themselves whether they and their facility could stand – both monetarily and psychologically – the "worst case scenario" for their youngest patients and their families. And today, they must also look at the intrinsic legal issues, for without the implementation of proven technology to safeguard their infant population, legal counsel would surely advise that a facility and its executives could face substantial liability concerns for not being vigilant in safeguarding the newborns in their care.

Security expert Robert Schneier recently categorized the RFID-enabled bracelets worn by newborns and their moms in more and more hospitals today as the ultimate example of what might he dubbed "security theater" - security taken against an unlikely threat that is primarily designed to make you feel more secure. He compared the security theater of the pediatric ward to that of tamper-resistant pharmaceutical packaging and airport security after September 11th [18]. The dollars and cents ROI behind security theater is that it provides a very real way to simultaneously heighten security and curb the legal threat (assuming staff remain trained and vigilant and not over reliant on the technology of protection). However, the true, even more tangible benefit of such infant protection systems is that they are a "palliative countermeasure," visibly making the new mom feel more secure in her bed, knowing that her new baby is better protected when she hands-off her newborn baby to the nurse for the night.

References

[1] National Center for Missing & Exploited Children, *Newborn/Infant Abductions - June 2008*. Retrieved June 30, 2008, from http://www.missingkids.com/en_US/documents/InfantAbductionStats.pdf.

[2] A.J. Sedlak, D. Finkelhor, H. Hammer, and D.J. Schultz, "National estimates of missing children: An overview," in *National Incidence Studies of Missing, Abducted, Runaway, and Thrownaway Children*,

October 2002. Washington, DC: U.S. Department of Justice, Office of Juvenile Justice and Delinquency Prevention.

[3] H. Brickley, "Family: Parson had miscarriage," *Lubbock Avalanche Journal*, March 14, 2007. Retrieved June 19, 2008, from http://www.lubbockonline.com/stories/031407/loc_031407074.shtml.

[4] J. Cowan, "Man freed, woman still in custody in baby abduction," *National Post*, November 5, 2007. Retrieved June 20, 2008, from http://www.canada.com/topics/news/story.html?id=6b93216d-3f86-4dd9-bd8e-5b2bb851b8c3&k=38071.

[5] J.R. Corsi, "Life with Big Brother: Hospitals tagging babies with electronic chips - Privacy advocates protest as half of Ohio birthing centers turn to tracking technology," *World Net Daily*, January 15, 2008. Retrieved June 27, 2008, from http://www.worldnetdaily.com/news/article.asp?ARTICLE_ID=59690.

[6] M. Alvarez, "Hospitals and your baby: How safe are you?" *FoxNews.com*, March 14, 2007. Retrieved June 4, 2008, from http://www.foxnews.com/story/0,2933,258309,00.html.

[7] J. Aldridge, "Hospital security: The past, the present, and the future," *Security Info Watch*, July 8, 2008. Retrieved July 28, 2008, from http://www.securityinfowatch.com/article/printer.jsp?id=5425.

[8] S. Slahor, "Designing hospital solutions," *Security Solutions*, February 1, 2001. Retrieved July 26, 2008, from http://securitysolutions.com/mag/security_designing_hospital_security/

[9] B. Bacheldor, "Medical distributor puts RFID tags on equipment," *RFID Journal*, July 25, 2006. Retrieved October 2, 2008, from http://www.rfidjournal.com/article/articleview/2513/1/1/.

[10] C. Swedberg, "Alzheimer's care center to carry out VeriChip pilot: The Florida facility will implant RFID chips in 200 volunteers this summer to test the VeriMed system's ability to identity patients and their medical histories," *RFID Journal*, May 25, 2007. Retrieved June 20, 2008, from http://www.rfidjournal.com/article/articleprint/3340/-1/1/.

[11] G. Baldwin, "Emerging technology: Hospitals turn to RFID. *SnapsFlow Healthcare News*, August 26, 2005. Retrieved May 26, 2008, from http://snapsflowhealthcarenews.blogspot.com/2005/08/emerging-technology-hospitals-turn-to.html.

[12] L. Sullivan, "RFID system prevented a possible infant abduction: VeriChip's 'Hugs' infant protection system sounded an alarm when the parents of an infant attempted to remove their baby without authorization from a hospital's nursery," *Information Week*, July 19, 2005. Retrieved June 2, 2008, from http://www.informationweek.com/news/mobility/RFID/showArticle.jhtml?articleID=166400496.

[13] C. Stolarz, "High-tech bracelets ease moms' fears," *The Detroit News*, June 9, 2007. Retrieved June 29, 2008, from http://www.detnews.com/apps/pbcs.dll/article?AID=2007706090307&template=printart.

[14] VeriChip Corporation, *Press Release: VeriChip Corporation's infant protection systems now installed in more than half of all birthing facilities in Ohio*, December 12. Retrieved June 28, 2008, from http://www.verichipcorp.com/news/1197464691.

[15] B. Bacheldor, "BlueTag patient-tracking comes to North America: The active UHF RFID system, marketed by RFID ProSolutions, is designed to protect newborn babies, Alzheimer's patients and other individuals staying in health-care facilities," *RFID Journal*, February 13, 2008. Retrieved May 30, 2008, from http://www.rfidjournal.com/article/articleview/3906/1/1/.

[16] C. Swedberg, "Shawnee Mission Medical Center expands pediatric tracking: The hospital has upgraded its 10-year-old system for tracking young patients, and also plans to begin tagging assets," *RFID Journal*, May 15, 2008. Retrieved May 28, 2008, from http://www.rfidjournal.com/article/articleview/4082/.

[17] S. Goodman, "Guarding against infant abduction with new technology," *Associated Content*, July 7, 2005. Retrieved July 13, 2008, from http://www.associatedcontent.com/article/4977/guarding_against_infant_abduction_with.html.

[18] R. Schneier, "In praise of security theater," *Wired*, January 25, 2007. Retrieved May 26, 2008, from http://www.wired.com/politics/security/commentary/securitymatters/2007/01/72561.

Creating a Bootable CD with Custom Boot Options that contains Multiple Distributions

James D. Feher, Kumud Bhandari, Benjamin York
Computer Science Department
McKendree University

ABSTRACT

It is often common for people to have more than one operating system (OS) that they use on a regular basis. This can necessitate carrying multiple live or installation CD or DVDs. When multiple OSs are installed on a given machine, boot loaders on the system hard drive allow the users to access the various platforms. However, when the system is located in a public environment such as a computer lab, it is often desirable to hide the existence of any platform other than the default standard platform to avoid confusing the general user. This paper details how to create a CD that contains not only a boot menu to reach all of the OSs installed on your system, but also installation, live and network distributions that can be accessed directly from the single CD media.

Keywords: boot loader, Linux, computer lab, open source

I. INTRODUCTION

It is often advantageous to run multiple platforms on a single computer, especially in an environment where resources such as space and funding are limited. Boot loaders can be installed on the hard drive of a system allowing it to boot into a multitude of platforms. In an open lab environment, it is often desirable to insure that only the default system used by the general population be accessible. This helps prevent those without knowledge of the other OS platforms from becoming confused when presented with something else. While GRUB can boot with a default OS without displaying any boot options for the user, the possibility still exists that the other resident platforms could be accessed by the general population. This could result in confusion for the general user or the corruption of other platforms on the system. With the boot partition of the hard drive pointing directly to the default OS, the means of accessing other system platforms without extra boot media is limited. Additionally, as memory is available on the CD media, the boot media can also contain helpful tools for system maintenance, extra live distributions, and installable distributions. The result is a veritable Swiss army knife for maintaining the computer lab.

II. CONFIGURATION FOR THE COMPUTER SCIENCE LABORATORY

To fully utilize the computer laboratory resources, the systems were configured to run many different platforms. The example provided here was for the computer science lab at McKendree University. However, the same principles can be used to create boot media for other purposes. In this case, the default system for the general purpose lab was WindowsXP. However, other systems were also used. To illustrate the functionality of a thin client, the systems were configured to use the *Linux Terminal Server Project (LTSP)* [1]. During other times, the lab was configured to operate in a clustered environment using the default Rocks Cluster platform [2]. While LTSP does not require any space on the hard drive, Rocks required several partitions on the hard drive. If Rocks is installed first on the system, with WindowsXP installed second, Windows will overwrite the master boot record (MBR) to point to Microsoft's boot loader for the system. Even though Rocks will still exist on the hard drive, after this point the means of booting the system into Rocks will be with the use of the external media. Configuring Rocks and Windows in this manner carries with it several challenges [3].

A boot loader could have been installed directly onto the hard drive of these systems. However, in order to insure that the general users only had access the default Windows system, the MBR of the system hard drive was left to point to the Windows installation only. Other platforms were loaded via boot CD media. This could have been accomplished with the use of floppy disk media, but the extra space provide by using a CD allowed for additional platforms to be provided. With the extra space, the DSL live distribution was included along with SystemRescueCD which contain utilities available for maintaining the lab such as the Partimage program for creating system images.

Other boot loaders such as ISOLINUX, LILO or the Microsoft NTLDR exist. In order to use an open source solution that is easily configurable, well documented and used by many current distributions, the GNU *GRUB* boot loader was used [4]. As these platforms were configured to boot WindowsXP by default, the Microsoft boot loader is actually used on the hard drive to boot WindowsXP. Lastly, keep in mind, in order to utilize the boot CD media, the systems must have their BIOS settings set to boot from the CD/DVD drives first. With this configuration, unless the CD is provided, the systems will use the secondary boot option, in this case WindowsXP on the hard drive.

K. Elleithy (ed.), *Advanced Techniques in Computing Sciences and Software Engineering*,
DOI 10.1007/978-90-481-3660-5_81, © Springer Science+Business Media B.V. 2010

III. GRUB

The GRUB boot loader is well documented and contains many different options for configuration of the boot media. Basically, GRUB provides a bare bones platform that contains pointers to the kernels of the various operating systems that the system might run. *GRUB* even has a basic command line structure that can be used to load parameters at boot from the keyboard. The systems can be loaded from a system hard drive, CD/DVD drive, USB thumb drive or on another networked server. Specifically in the case of Rocks and WindowsXP, the operating system resides on the system hard drive, for LTSP, the networked Linux Terminal Server, and lastly DSL and SystemRescueCD are contained on the CD media itself.

The main item to configure for GRUB is the text file called menu.lst which is used to provide the *GRUB* boot menu. This file contains configuration options for the boot menu, boot parameters for each platform as well as descriptions that are displayed for of each boot option. As the configuration options for GRUB are already well documented in GRUB manual [5], only the mechanism for determining the options required to boot each system will be discussed here.

3.1 Rocks: Rocks, like many other Linux distributions already uses GRUB as the default boot loader. For these systems, the files for booting are located in the /boot directory. The items of interest are located within the */boot/grub* directory. In that directory are located various files that *GRUB* needs to execute with names like: *e2fs_stage1_5, fat_stage1_5, minix_stage1_5, reiserfs_stage1_5, menu.lst, stage1, stage2, stage2_eltorito*. Of particular importance at this point is the *menu.lst* file which will contain the boot options. As the *menu.lst* file is a text file, it can be examined with the text editor of one's choice. In this case, the lines of interest found in lines 5 through 8 on the sample *menu.list* file provided in Figure 1. The title line is what is displayed for the user in the boot menu. The second line contains the location of the kernel code. In this case, it contains the first partition of the local hard drive. The next line specifies the name of the kernel code along with the appropriate path on the hard drive and specifications for loading the kernel. The last line starting with *initrd* specifies the name and path to the image of a *Linux RAM* disk used by kernel during boot. These specifications will vary for each system, so after they are located on a given installation, copy them and paste them into the new menu.lst file that will be used for the creation of the boot CD.

```
1  default=0
2  timeout=30
3  splashimage=(cd)/boot/grub/splash.xpm
4
5  title   Rocks(2.6.18-53.1.14.el5)
6          root    (hd0,0)
7          kernel  /boot/vmlinuz-2.6.18-53.1.14.el5 ro root=/dev/sda1 dom0_mem=1024M
8          initrd  /boot/initrd-2.6.18-53.1.14.el5.img
9
10 title   Windows XP
11         rootnoverify (hd0,3)
12         chainloader +1
13
14 title   LTSP Realtek Etherboot
15         kernel  /Etherboot/rtl8139.zli
16
17 title   System Rescue CD (x86-1.0.4)
18         root (cd)
19         kernel  /isolinux/rescuecd
20         initrd  /isolinux/initram.igz
21
22 title   DSL (4.4.10)
23         root      (cd)
24         kernel  /boot/isolinux/linux24 ramdisk_size=100000 init=/etc/init lang=us
              apm=power-off vga=791 nomce noapic quiet BOOT_IMAGE=knoppix
25         initrd /boot/isolinux/minirt24.gz
```

Figure 1: The snapshot of menu.lst

3.2 WindowsXP: As already mentioned, Windows does not use GRUB. GRUB however can pass control along to the Windows boot loader. In this case, as control is passed along to another boot loader and not to the *Windows* kernel directly, the specifications within the menu.lst file will change. The lines required for this particular installation are 10, 11 and 12 in Figure 1. The title line serves the same purpose as it did for Rocks, just to provide message for the boot menu. The next line indicates that the target system is located on the fourth partition of the hard drive (0 is the first partition, 1 the second, etc.) and the last line indicates that control will be passed to a different bootloader, in this case the Microsoft bootloader. These lines may need to be changed if Windows is installed on a partition other than the fourth.

3.3 LTSP: LTSP was configured to load the system through the server over the network, with entries in the menu.lst file found in lines 14 and 15 of Figure 1. Notice that the code is located on the CD itself in the */Etherboot* directory. The Rom-o-matic website provides a script that will accept the configuration options for your thin client and LTSP server and generate the code required for your system [6]. Once this code is obtained, it must be copied to the appropriate place for creation of the CD and the title needs to be placed within the menu.lst file. Notice that if the CD is to be used on more than one type of physical system, where portions of the hardware such as the network card may differ, an entry for each system along with the required code for LTSP must be provided on the CD.

3.4 Live Distributions: In this case, DSL and SystemRescueCD were included on the media. The setup for this is slightly different as the code required for the system is located on the media itself. First, portions of the code will need to be copied to the appropriate place on the CD, this will be described in a later section. Secondly, they will need entries much like the ones provided in the sample menu.lst file on lines 22 through 26 for DSL and 17 through 20 for SystemRescueCD. The first two lines of the menu.lst entry are still fairly obvious. Notice that this is like the LTSP configuration in that it does not specify a partition on the hard drive, but rather the location of the files within the CD. The next two lines specifying the location of the kernel along with the boot options takes a bit more work to locate. The configuration options provided above are quite detailed, but a complete understanding of GRUB parameters is not necessary to develop the boot media.

Many live Linux distributions, like those provided here, use the ISOLINUX boot loader. Configuration options for the menu.lst file can be obtained by closely inspecting those in the *isolinux.cfg* file. This file provides the boot menu options for ISOLINUX, much like *menu.lst* does for GRUB. The *isolinux.cfg* file is found in the */isolinux* directory that is usually located near the root of the directory tree on the file system for the installation media. From the multiple boot options that live distribution may contain within its menu, select the one(s) that you wish to use for your media. For the DSL version included, the lines directly from the isolinux.cfg file are shown in figure 1.

```
LABEL dsl
KERNEL linux24
APPEND ramdisk_size=100000 \
init=/etc/init lang=us \
apm=power-off vga=791
initrd=minirt24.gz \
nomce noapic \
quiet BOOT_IMAGE=knoppix
```

Figure 2: Lines from isolinux.cfg file

The lines of interest will begin with *KERNEL* and *APPEND*. The *KERNEL* line indicates the code for the kernel. The *APPEND* line will contain various parameters for the kernel, that will be added to the kernel line for the menu.lst file. One exception is the initrd specification. The RAM disk location should not be included with the kernel line, but rather be placed upon its own line. When these lines are translated in the proper format for GRUB, the location of the files must be changed to specify where the code exists within the media that is being created, rather than the location from which it is being copied. An excellent example of translating the various lines from the ISOLINUX format to GRUB exists online [7].

IV. PREPARING BOOT MEDIA

The *mkisofs* command will create an iso image of the boot media which can then be used to create as many boot CDs as necessary [8]. On some systems the command *genisoimage*, a fork of the *mkisofs* command, will be found and should be used instead. Items that are necessary to produce the boot media include the following.

- *menu.lst* file
- GRUB code
- Other code required, including those for any live or network loaded distributions
- Proper *mkisofs* or *genisoimage* command to create the bootable iso

4.1 menu.lst: This is the file created in the previous section.

4.2 GRUB Code: The code for GRUB can be obtained in more than one way. It can be downloaded and compiled. However, if the boot media is being prepared using a Linux machine, chances are that this machine already has grub installed and this code can be copied directly. The necessary files are found in the

/boot/grub directory and can be used directly. If these files are copied, care should be taken not to use the same *menu.lst* file and the *stage2_eltorito* file is included. The eltorito specification is used when booting CD media.

4.3 Other Code Required: Other items that will be required include any code for live or install distributions as well as bootstrapping code for any network boot systems. Depending upon what is included in the boot CD, a directory structure that contains the appropriate files required by the various components will need to be created and copied. A word of caution, the location of these files is not arbitrary, be careful when copying files over. First, care must be taken to insure that this code is placed in the same location that was previously specified for it in the *menu.lst* file.

In addition, some files must retain the same directory structure, location and name as they do on the boot media from which they are obtained. Nearly any location can be specified on the new menu.lst file. However, if a distribution contains a RAM disk, the RAM disk image specified by the *initrd* line may have the locations of certain portions of code preloaded in that image. While it is feasible that a new ram disk image could be created to point to any location, this is beyond the scope of this paper. For that reason, it is advised that the locations of the various portions of code required for any live distributions be copied with the same file structure, location and name. Note that this may prevent the inclusion of distributions that require components that have the same name and location or for the inclusion of multiple versions of the same distributions where this is the case as well.

Keeping all that in mind, what follows is the procedure for creating the proper directory structure as well as copying the files necessary. The following instructions assume that the commands specified below are run on a Linux platform.

4.3.1: Create space for the iso: In a working directory that contains enough space for all of the live distribution code, create a directory that will contain the contents of the iso. For example, if the contents are going to be placed into the *myiso* directory, issue the following command.

mkdir myiso

4.3.2: Copy over GRUB code: In this directory, create a boot folder and within the boot directory, and create grub folder in a similar manner. Place all of the code for GRUB in this *myiso/boot/grub* folder, including the menu.lst file that contains the specifications for the boot media being created.

4.3.3: Copy over additional code: Any additional code required by the live distributions included in the boot CD as well as the code for any network booting systems must be added to the root directory of the *myiso* directory and care must be taken to insure that the menu.lst file correctly specifies the location of this code. The root directory of the DSL CD displayed by the *ls* command is shown in figure 2.

```
$ ls -l /media/cdrom/
total 7
dr-xr-xr-x 3 root root 2048 2004-07-25 03:13 boot
-r--r--r-- 1 root root  391 2004-05-03 00:00 index.html
dr-xr-xr-x 2 root root 2048 2008-08-10 08:11 KNOPPIX
dr-xr-xr-x 2 root root 2048 2004-08-06 19:16 lost+found
```

Figure 3: The root directory of the DSL CD

The entire KNOPPIX directory along with the *isolinux* directory within the boot directory was copied over to the *myiso* directory. Moreover, the exact location of the kernel and RAM disk were then specified in the menu.lst file. A similar procedure was followed for the SystemRescueCD. At the conclusion of copying all the relevant files, the contents of the *myiso* directory are shown in figure 4.

```
$ ls -l myiso
total 167152
drwxr-xr-x 4 user group      4096 2008-08-22 16:36 boot
drwxr-xr-x 2 user group      4096 2008-06-25 03:51 bootdisk
-r--r--r-  1 user group     32606 2008-08-22 14:41 cdrom.ico
drwxr-xr-x 2 user group      4096 2008-08-22 15:16 etherboot
drwxr-xr-x 3 user group      4096 2008-08-22 15:04 isolinux
dr-xr-xr-x 2 root  root      4096 2008-08-10 08:11 KNOPPIX
-rw-r--r-- 1 user group       601 2008-08-22 16:33 README
-rw-rw-rw- 1 user group 170930176 2008-06-25 03:58 sysrcd.dat
-rw-r--r-- 1 user group        45 2008-06-25 03:59 sysrcd.md5
```

Figure 4: The contents of myiso directory

4.4 mkisofs: Once the menu.lst file has been created and all of the files and directories have been copied over, the *mkisofs* command can be used to create the iso image of the boot media. The command that we used is shown in figure 5. Note that the *stage2_eltorito* is specified for boot. Consult the *mkisofs* manual page for a detailed description of the *mkisofs* parameters. The iso image can now be burned to a CD. If your system uses the *genisoimage* command, just substitute it for *mkisofs* listed above.

```
$mkisofs -R -b \
boot/grub/stage2_eltorito \
-no-emul-boot -boot-load-size 4 \
-boot-info-table \
-o YOURBOOTMEDIA.iso myiso
```

Figure 5: The mkisofs command

V. CONCLUSION

This paper demonstrated how to produce a custom boot CD that will boot a variety of live, install, network or system contained operating systems. Should you feel adventurous, you can even add your own custom splash image to your new boot CD [9]. This CD is a handy administration tool that allows us to hide platforms from the general population, while providing access to a multitude of different configurations.

VI.REFERENCES

[1] Linux Terminal Server Project [Online] Available: http://www.ltsp.org/ (accessed August 21, 2008).

[2] Rocks Group, Rocks User Guide, CA: UC Regents, 2007

[3] James Feher and Kumud Bhandari, "An Inexpensive Approach for a Scalable, Multi-Boot Computer Lab", presented at the International Conference on Engineering Education, Instructional Technology, Assessment, and E-learning (EIAE 07), December 10, 2007.

[4] GNU GRUB, [Online]. Available: http://www.gnu.org/software/grub/ (accessed August 21, 2008).

[5] GNU GRUB Manual. Available: http://www.gnu.org/software/grub/manual/grub.html (accessed August 21, 2008).

[6] ROM-o-matic [Online] Available: http://rom-o-matic.net/ (accessed August 21, 2008).

[7] How to Boot several CD iso files in an DVD [Online] Available: http://justlinux.com/forum/showthread.php?p=872569 (accessed June 2, 2008).

[8] mkisofs [Online] Available: http://www.linuxcommand.org/man_pages/mkisofs8.html (accessed August 21, 2008).

[9] Bacon, Jono and Petreley, Nicholas, Linux Desktop Hacks, page 18, O'Rielly Media Inc., 2005.

Architecture of COOPTO Remote Voting Solution

Radek Silhavy, Petr Silhavy, Zdenka Prokopova
Faculty of Applied Informatics, Tomas Bata University in Zlin
Nad Stranemi 4511, 76005 Zlin, Czech Republic
rsilhavy@fai.utb.cz; psilhavy@fai.utb.cz; prokopova@fai.utb.cz

Abstract- **This contribution focuses on investigation of remote electronic voting system, named COOPTO. Researching of suitability of electronic voting solution is forced by necessity of the improvement election process. The COOPTO is based on topical investigation of voting process and their implementation of using modern information and communication technology. The COOPTO allows voters, who are not in their election district, to participate in the democracy process. The aim of this contribution is to describe results of the development of the COOPTO solutions.**

I. Introduction

The electronic government, the actual point of the computer-social investigation, uses several methods for improving governmental processes in Europe. These electronic methods allow the improvement in direct democracy. Probably, the most relevant solution is remote internet voting. COOPTO research project has been investigating the benefits of remote electronic voting. The remote voting solution allows participation in election process with respect to personal conditions and to the physical accessibility of polling stations, which may possibly prevent citizens from casting their votes. Therefore, remote internet voting should effectively support voters, who are resident abroad. These voters use at present the embassy election rooms only.

The aim of this contribution is to introduce architecture of COOPTO voting system, which is based on web-based application. Probably, the most significant results is remote internet voting in appropriate form, which understandable and clearly usable.

The organization of this contribution is as follows. Chapter 2 describes the electronic voting approaches and conditions for electronic voting systems. Chapter 3 describes the COOPTO architecture. Finally chapter 4 is the discussion.

II. Electronic Voting Approaches

Internet voting solutions are usually divided the tree basic categories – poll site, kiosk and remote voting.

In the poll site voting, election technology is located in the election rooms. Comparing poll site voting to traditional paper-form voting, poll site brings more flexibility and usability, because voters are allow to vote from elections room up to their choice. There is no restriction to geographical locations. Poll site voting represents concept of the electronic voting. Poll site voting is effective in votes casting and tallying, because it allows certain and quicker processes.

Internet concepts allow expanding poll site voting to self-service kiosks. These kiosks should be placed in various locations. Elections room is usually monitored by authorities – local election committee, kiosk should be monitored by physical attendance or by using security cameras.

Remote internet voting is probably the most attracting methods of using internet voting process. Remote voting expands remote voting schemas which are used in some countries. These schemas compared to postal voting, offers improved casting ballots from remote locations. Voter are allows to vote from home, office or other places, which are equipped by computers and internet connections.

Conditions for voting systems

There are several conditions for electronic voting systems. The law in the country has to support the electronic voting systems. The internet voting solution has to follow the technical and process conditions listed below:

1. Participation in the voting process is granted only for registered voters.
2. Each voter has to vote only once.
3. Each voter has to vote personally.
4. Security and anonymity of voters and voting.
5. Security for the electronic ballot box.

The first condition for electronic voting means, the voter should be registered by voting committee in the list of voters. This list is used as the basis for distribution of log-in information. If the voter is registered, they will be able to display the relevant list of parties and candidates.

Voters could also vote more than once, but only the last attempt will be included in the final results of the election. This possibility varies in different e-voting systems. If it is not possible to vote more than once, there should be more complicated protection for the election against manipulation and assisted voting.

The third condition – Right to vote personally – is closely connected to the previous. On the other hand this is the basic responsibility of each voter to protect his private zone for voting – in the case of the internet-based remote voting. In the "in-site" voting the system of privacy protection will be similar to the current situation.

Security and anonymity of voters and voting is probably the most important issue in the electronic voting process. The appropriate voting system should be realized in two separate parts. The first part should be responsible for authorization of the voter and the second for storing votes. Therefore the system will support anonymity. The voter should check his

K. Elleithy (ed.), *Advanced Techniques in Computing Sciences and Software Engineering*,
DOI 10.1007/978-90-481-3660-5_82, © Springer Science+Business Media B.V. 2010

vote by the list of collected votes. The unique identification of vote will be known by the voter only. The voting process will be protected by using a cryptographic principle. One of the many applicable solutions is Private Key Infrastructure. This approach deals with two pairs of keys in the first part of voting system – for authorization. In the second part of voting system – storing votes – it should deals with a public key for protection of the vote in the transport canal.

The electronic ballot box should form as a database. Votes in the database will be ciphered by the public key of the election committee. The private key, which is necessary for decrypting votes, will be held by members of the committee. Each member will hold only part of the key.

By investigation of these conditions and by the determination of the initial technological principles, authorities will be able to establish law to support the electronic voting system. The voting public's consensus to the electronic voting is quite important for the parliament process too.

III. COOPTO Architecture Description

The web-based approach is useful for electronic voting

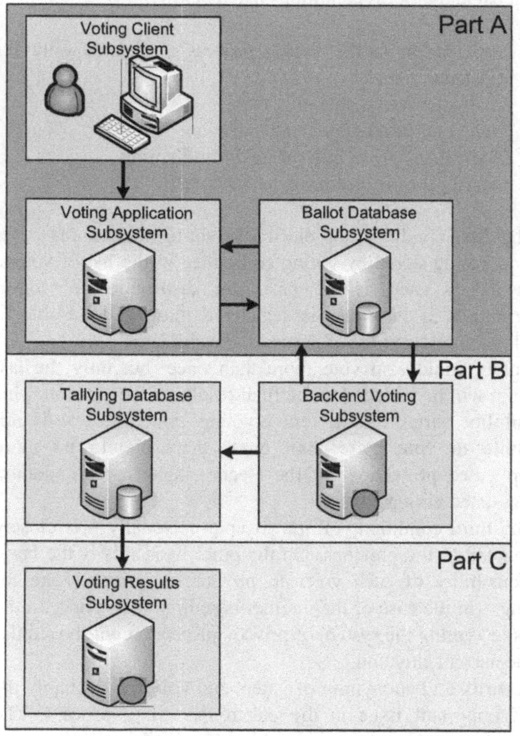

Fig 1: COOPTO Voting System Architecture

systems. This technology is based on a client-server. The client-server technology has advantages in the field of support and installation.

COOPTO system consists four main parts:

1. Voting Client Subsystem (VCS)
2. Voting Application Subsystem (VAS) and Ballot Database Subsystem (BDS).
3. Voting Backend Subsystem (VBS) and Tallying Database Subsystem (TDS).
4. Voting Results Subsystem (VRS)

In the figure 1 can be seen COOPTO Voting System Architecture. COOPTO Voting Architecture three separate parts are recognized. Part A is used for casting votes and contains Voting Client Subsystem, Voting Application Subsystem and Voting Database Subsystem. Voting clients represent voting terminal in elections rooms, kiosk voting or voters own computers.

Voting Application Subsystem is represented by web-based application, which contains user interface for voters, voter validation services and communication interface for Ballot Database Subsystem.

There are two most significant tasks for BDS. Votes are cast there and default ballots are generated for individual voter. Votes are cast in encrypted form, which depends on cryptographic methodology adopted for the election. For the protection against manipulation with votes in BDS HASH algorithm is implemented. HASH value is calculated irregularly based on votes, which are cast. Default ballots are generated for individual voter with respect to the election district he belongs to.

Part B represents Backend Voting Subsystem and Tallying Database Subsystem. The part B is securely connect to BDS from part A. The BVS is used my electoral committee. The BDS is responsible for auditing elections by comparing HASH based on votes and stored HASH value. The BVS deals with decryption of votes, validating of them and storing in TDS. The TDS is used for storing votes in open form. Part B is realized as web-based application and relational database server. Final part – part C – is responsible for counting final Results of the election. Part C is realized as web-based application.

In the figure 2 is shown workflow description of individual vote through the COOPTO Voting System.

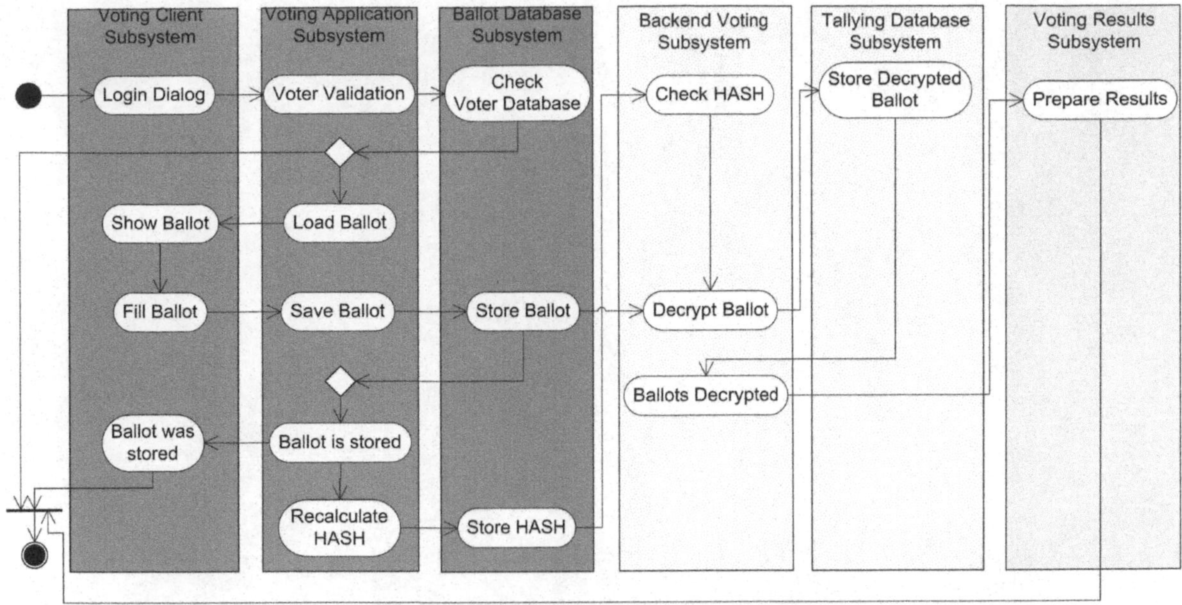

Fig 2: COOPTO Voting Workflow

IV. Discussion

The main task of this contribution was to introduce the idea of COOPTO Voting System. COOPTO is under the investigation as the basic implementation of remote voting conditions in the Czech Republic.

Contribution focused on architecture concept of the web-based system because this system is more flexible and comfortable for voters.

The next obvious advantage of electronic voting systems and especially of web-based systems is related to the cost of elections.

Many countries have researched the benefits of e-voting solutions. Every country uses individual ways to solve e-voting problems. Only a few are seriously studying internet voting. Internet voting solutions represent the future of electronic voting in Europe.

Increasing internet access supports the e-Government and e-Democracy. People are used to communicating through the Internet. In the future people will be used to electronic elections too.

References

1. Alexander Prosser, Robert Krimmer (Eds.): Electronic Voting in Europe - Technology, Law, Politics and Society, Workshop of the ESF TED Programme together with GI and OCG, July, 7th-9th, 2004, in Schloß Hofen / Bregenz, Lake of Constance, Austria, Proceedings. GI 2004, ISBN 3-88579-376-8
2. Leenes, R., Svensson, K.: Adapting E-voting in Europe: Context matters.Proceedings of EGPA, 2002.
3. Commission on Electronic Voting: Secrecy, Accuracy and Testing of the Chosen Electronic Voting System. Dublin, 2006, available at http://www.cev.ie/htm/report/download_second.htm accessed on 2007-06-06.
4. Chevallier, M.: Internet voting: Status; perspectives and Issues, ITU E-Government Workshop, Geneva, 6 June 2003, available at: http://www.geneve.ch/chancellerie/EGovernment/doc/UIT_6_6_03_we b.ppt accessed on 2007-06-06.
5. VVK [online]. 2007 [cit. 2007-11-02]. Accessible on WWW: <http://www.vvk.ee/english/results.pdf>.

A Framework for Developing Applications Based on SOA in Mobile Environment with Security Services

Johnneth de Sene Fonseca & Zair Abdelouahab
Federal University of Maranhão, CCET/DEEE
Av. Dos portugueses, Campus do Bacanga, São Luis – MA 65080-040
johnneth.sfonseca@gmail.com, zair@dee.ufma.br

***Abstract-* Constant evolution of technologies used in mobile devices allows an increase of capabilities related to storage, processing and transmission of data, including more than one type of technology of transmission in a same device. These factors enable a greater number of applications but also it gives rise to a necessity to find a model of service development. One of the best options which currently exist is SOA (Service Oriented Architecture). This article aims to present a framework that allows the development of SOA in mobile environment. The objective of the framework is to give developers with all necessary tools for provision of services in this environment.**

I. Introduction

Over the last years there is a considerable increase of capabilities of mobile device, both in its storage capacity and in processing. This development has enabled a greater popularization of these devices which now offer a large number of applications to end users. In addition, new communication technologies allow these devices to access the Internet more efficiently and to communicate with each other.

There is also a possibility of installing in these equipments other applications and/or services beyond those already coming from factory; they can act not only as service consumers but also as service providers. Thus, there is a need to use a pattern of development that allows developers to create and provide its services more quickly and efficiently. Service-Oriented Architecture (SOA) emerge as a solution to this problem

The aim of this paper is to describe a framework to the development of SOA in mobile environment drawing the complexity of their development, with mechanisms to perform all necessary functions for provision of services, such as describing services, carry messages from the parser with specific format, creating a channel of communication to receive and send messages. Another mechanism allows adding security properties developed with the service.

This paper is structured of the following form. The first section presents the motivations of the work. The second section describes the SOA architecture and its main components. The third section describes the provision of services in the mobile environment. The fourth section shows the mains problems related to security in the mobile environment. The fifth section describes the proposed architecture and the sixth section shows some work related to the proposed work. Finally the last section presents the conclusions and suggestions for future work.

II. SOA

SOA describes the keys concepts of software architecture and their relations, where a service and its use are the key concepts that are involved, following a model of publishing services and applications and their universal access [5]. SOA has an interface that describes a collection of operations accessible over the network via a standardized format (e.g. XML). These requirements are activated anywhere in a dynamic computing environment and/or pervasive computing where service providers offer a range of services.

SOA creates an environment in which distributed applications and components may create independently of language and platform and focuses on the use of a relatively widespread pattern of communication between operations, enabling thus a model for homogeneous distribution and composition of components.

SOA is a model of components, providing an environment for building distributed systems [6]. SOA applications communicate functionally as a service to the end user's applications and other services, bringing the benefits of low coupling and encapsulation for the integration of enterprises applications. SOA defines the rules of the participants as provider of services, customer of services and registry of services. SOA is not a rating and many new technologies such as CORBA and DCOM at least already had this idea. Web services are new to developers and are the best way to achieve and develop an SOA.

A. *Architecture*

The architecture of SOA is built with three components [3]:

- Service Requestor (Client) – this entity requires certain functions to perform some task, application or service that relies on interaction with a boot or some service;
- Service Provider – this entity creates and provides the service, it also makes a description of the service and publish it in a central registry;

K. Elleithy (ed.), *Advanced Techniques in Computing Sciences and Software Engineering*,
DOI 10.1007/978-90-481-3660-5_83, © Springer Science+Business Media B.V. 2010

- Service Registry (Broker) - location of service description, that is where the Service Provider published a description of the service;

The basic architecture of SOA is shown in Figure 1.

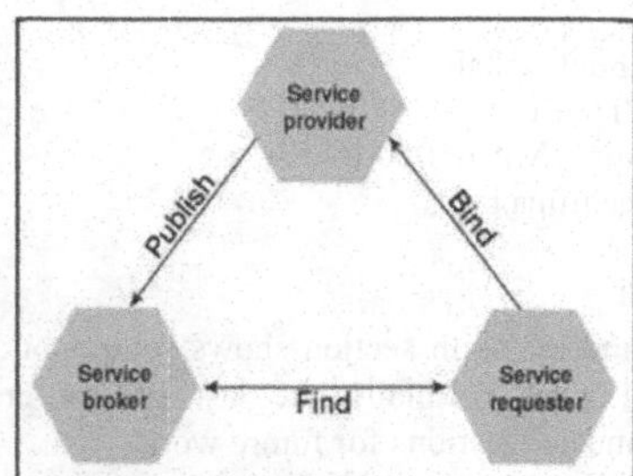

Figure 1 – Basic architecture of a SOA

Although a service provider is implemented in a mobile device, the standard WSDL can be used to describe the service, and the standard UDDI registry may be used to publish and make the service available. A challenge is in developing mobile terminal architectures such the one in standard desktop system, taking into account low resources of mobile device [7].

B. *Operations*

The components of SOA interact with each other through a series of operations (Figure 1) which are described below:

- Publish - record a description of the service in directory services, covering the registration of its capabilities, interface, performance and quality that it offers offers;
- Find – search for services registered in directory services, provided they meet the desired criteria and it can be used in a process of business, taking into account the description of the published service;
- Bind - this operation relies on the service requested or boots an interaction with the service at runtime using the information obtained in discovery of the service;

III. Mobile Host (Mobile Web Service Provision)

Mobile Host is a provider of services (***Light Weight***) built to run on mobile devices such as smart-phones and PDAs [1], developed as a *Web Service Handler* built on top of a normal Web server. Mobile Host opens a new set of applications yet little explored [2]. They may be used in areas such as location-based services, community support for mobile and games. It also allows smaller mobile operators increase their business without resorting to a stationary infrastructure. However, these additional flexibilities generate a large number of interesting questions for surveys which require further investigation.

The design of a "Mobile Host" is going through many things, some issues where there is very little research; so far set up service provisioning is very limited to devices. The work in [4] describes a model for the development of a Mobile Host system in general.

Traditionally, mobile systems have been designed as client-server systems where **thin clients** such as PDAs or phones are able to use wireless connections to gain access to resources (data and services) provided by central servers [2]. With the emergence of wireless networks, Ad-Hoc and powerful mobile devices it becomes possible to design mobile system using an architecture peer-to-peer.

According with [2] the following characteristics must be guaranteed so that SOA can be built in the mobile environment:

1) The interface must be compatible with the interface of SOA used in the desktop environment for customers not note the difference;
2) The space used by the service should be small in relation to available on mobile device;
3) The service should not affect the normal operation of the device;
4) A standard Web server that handle requests of network;
5) A provider of basic services for treatment of requests for SOA;
6) Ability to deal with competing requests;
7) Support the deployment of services at runtime;
8) Support for the analysis of performance;
9) Access the local file system, or any external device like a GPS receiver, using infrared, Bluetooth etc.

IV. Security

Security in wireless networks always is evolving. With adequate time a persistent cracker is capable of invading a wireless system. Moreover, some attitudes need to be taken to hinder as much as possible the work of an intruder, allowing basic services of security are met.

Risks already common in wired networks are incorporated into the wireless networks, new arise due to differences in physical structure of these and how they operate. Thus, any solution targeted for wireless networks are to be built in compliance with these new risks because that they are unique to wireless networks.

The greatest threat to a mobile network is the possibility of installing wires through doors in phone calls and data traffic. This threat can be remedied in part with the use of encryption. Consequently, the probability of threat depends on the strength of the encryption algorithm. This resistance is an exit that becomes questionable in the GSM system. Another critical threat, although more hypothetical, is amending the original mobile traffic. In this case the attacker overwrites the data with their own information.

The monitoring of traffic between the device and base station can get the position, speed, duration of traffic, duration, identification of a mobile device. However, the scenarios of exploitation by intruders are the greatest benefit from limited information can be possibly details of location and profile of the user.

Since a SOA is implemented as a Mobile Host, the services are prone to different types of security breaches: such as denial of service attacks, man-in-the-middle, and spoofing of intrusion, and so on. SOA in mobile environment using technologies based on message (such as SOAP over HTTP) for complex operations in several areas. Also, there may be many legitimate services intermediaries in the communication between doing composes a particular service, which makes the context of a security requirement end-to-end.

The need for sophisticated message-level security *end-to-end* becomes a priority for a mobile web service. Figure 2 illustrates some of the typical violations of security in SOA environments in wireless [12].

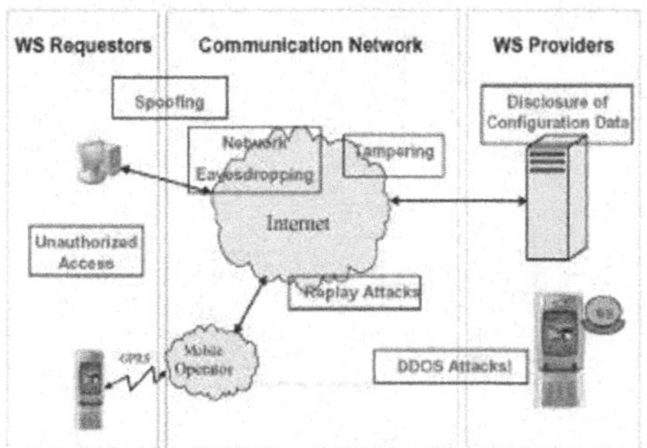

Figure 2 – Typical breaches of security in SOA Mobile

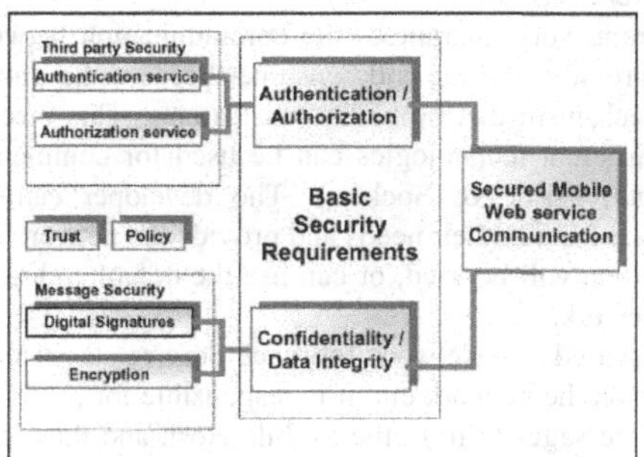

Figure 3 – Basic requirements for the safety of Mobile SOA

Considering the breaches of security, the SOA mobile communication must contain at least the basic requirements of security, as shown in Figure 3. Secure transmission of messages is achieved by ensuring the confidentiality and integrity of the data, while the authentication and authorization will ensure that the service is accessed only by trusted requesting. After the success of the implementation of such basic requirements of security, confidence and politicies may be considered as services for mobile field. Political trust can ensure a correct choreography of services. It sets any general policy statements on security policy of insurance SOA, while building relationships of trust in SOA allows safe for the exchange of security keys, providing an appropriate standard of safety.

V. Overview of Architecture

The proposed solution is the development of a framework based on a Mobile Host. Figure 4 shows the main components of the proposed architecture. A developer can implement the proposed framework and thus make their services (with multiple services in one device) available to the general public without the need of additional deployments to the activities necessary for the provisioning of service, such as receive and send messages, conduct parser of messages to/from the format SOAP/XML, publication of services, generation of WSDL containing information of public services, creation and tracking of an interface for communication to transmit messages on different types of existing technologies (Bluetooth, wireless, HTTP), to identify and implement a requested service. Furthermore, the architecture provides a model of communication secure end-to-end based on public keys.

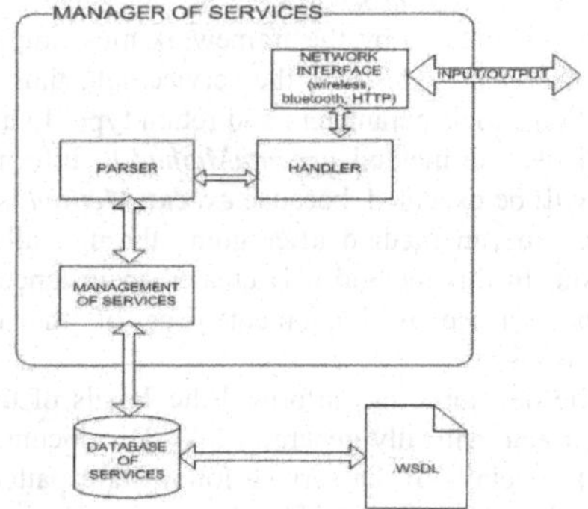

Figure 4 – Core Mobile Host Proposed

The proposed architecture can be structured into a model composed of the following layers (see in Figure 5):

1) Network Layer Interface - responsible for creating a channel of communication between the provider and consumer service, receiving and transmitting messages and can use different types of available communication technologies;
2) Events Layer - verify what kind of message is being transmitted and therefore can perform activities related to it, is also responsible for conducting the parser of messages to/from the format SOAP/XML;
3) Service Layer - through this layer a service can be described, made available for publication, obtain and enforce a method when it is invoked;
4) storage Layer – place where all created services and the latest user requests are stored; this layer stores as well WSDL descriptions and the document object containing the information of each created department;
5) Security Layer - ensures that there is security end-to-end in every step since the transmission of a request and the response of the request is made by the end user;

6) Management Layer - allows the management of all services offered by the developed framework;

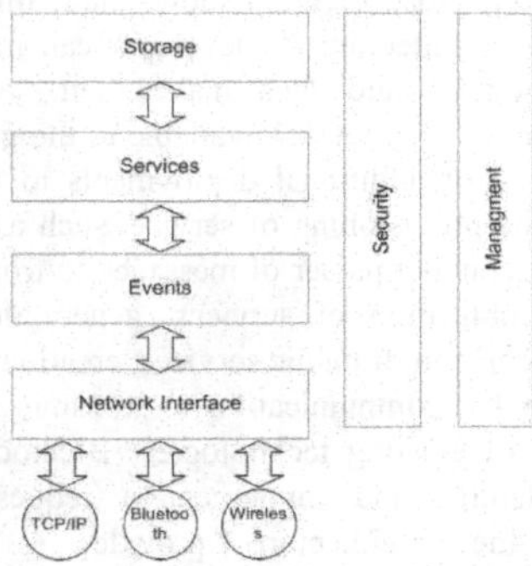

Figure 5 – Model in Layers

A. *Process Development Services*

A service is created by the framework must implement the interface ***iService*** to describe the service, and thus informing what methods, their parameters and return type. In addition, A user must use the method ***executeMethod*** to inform how the methods will be executed, because ***executeMethod*** is called by the system to run method after going through all stages of verification. In this method it is created an instance of object where the service is implemented or of the class that implements *iService*.

After the developer has informed the details of the service, the system automatically generates a WSDL document, which contains the details of the service following a pattern already known, and stores in a specific folder on the device itself for further verification, or can be sent to a consumer. Figure 6 shows the mapping of a Java class to a WSDL document that contains a description.

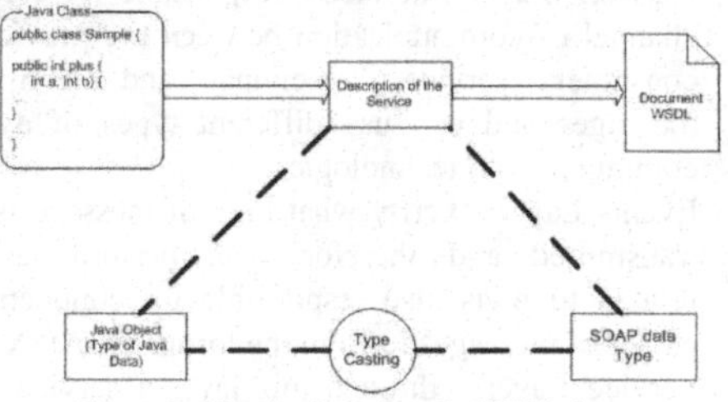

Figure 6 – Mapping Java class for WSDL document

In case the developer wants to add security services the developer must instantiate to class "***SymmetricKey***" responsible for the generation of a pair of symmetric keys (public and private) used to create a digital certificate (which can be sent to Final consumers) for the service and create a digital signature for each message sent, to guarantees its integrity.

Following the steps outlined above the service will be stored in a repository specifically for its further use. Several services can be created and stored, allowing the developer holds a range of services to make them available to their consumers, but the restrictions are limited memory capacity of the device and where it will be allocated. Figure 6 shows a diagram of activities with for the startup process of a Mobile Host and creating new services.

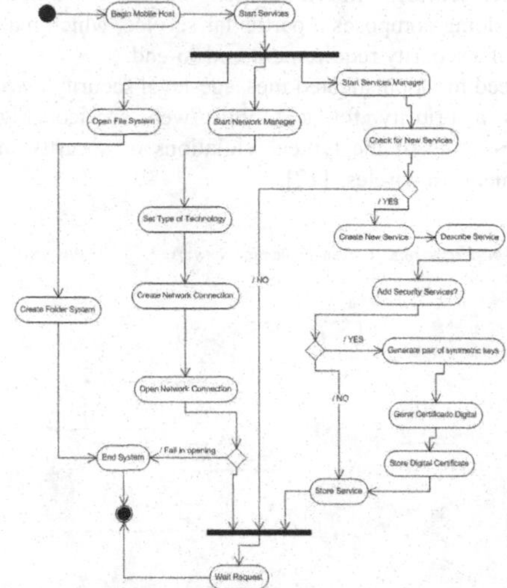

Figure 7 – Activity Diagram – Mobile Host -- creation of services

B. *Network Interface*

The framework guarantees the communication between the service provider and the end consumer because the framework has a mechanism that provides this features. This mechanism allows different technologies can be used for communication (Bluetooth, HTTP or Sockets). The developer can choose which best fits that their needs and provide the network address and port that will be used, or can use the default values set by the framework.

Once started, the Network Interface Service stay listening to demands on the network and it is responsible for receiving and sending messages to/from the Mobile Host, and thus behaving a standard Web server. Each received request a new line of enforcement is designed to meet the request to identify if it is a SOAP request or a normal web request.

C. *Invocation Service*

When the message received is a normal web request the web server treats normally,; however if the request is a SOAP request its treatment is passed to a handler for the SOAP requests, which performs the message parsing, extracting the information necessary for the invocation of a service. During the process of parsing, it is checked if the invoked service is compound all the requirements of the method invoked (quantity and types of variables, besides the type of return, according to the SOAP specification) according to the data reported by the developer in the process of creating the service.

If the message is not compatible with the requirements, a error message is generated by the system and send it to the customer, otherwise the method is executed through the invocation of the ***executeMethod*** to which are passed the method name and parameters from the message. Once in possession of the response of the implementation of the method invoked a reply message is generated and sends it to the customer. If the developer has added security services, a digital signature is generated for the response of the Mobile Host and sent along the message. Figure 8 shows the diagram of activities with the process of invocation of a service.

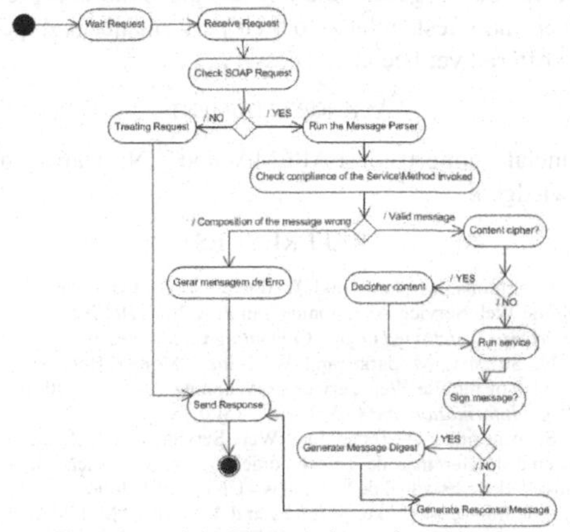

Figure 8 – Activity Diagram – invocation of service

D. *Parsing of messages*

An important feature of the Mobile Host is the ability to perform the parsing of messages received as a SOAP envelope to an object that contains the details of the request and that is used to obtain the information necessary to implement the request. Also, the reverse process is possible, that is the response is converted into a document SOAP before being sent to the requester. The framework also generates error messages (SOAP FAULT) if any failure is identified in the request.

E. *Security Service*

The model described in this paper proposes a security mechanism based on the content of messages, which is engaged in protecting the communications end-to-end between the consumer and service provider, ensuring that messages delivered are not corrupted by third party. This mechanism provides tools for generation of digital signatures and encryption of transmitted messages. The framework can create pair of keys (public and private) for each service provided by the Mobile Host during the process of creating the service. Though, this is optional the final developer decides whether or not this service can be added to its system. The encryption based on public key can also be used in the process of digital signature, which ensures the authenticity of who sends the message, associated with the integrity of its contents. Through these keys it can be created a digital signature to be sent along with a message, and also for allowing the transmitted messages to be encrypted and decrypted as observed in Figure 9.

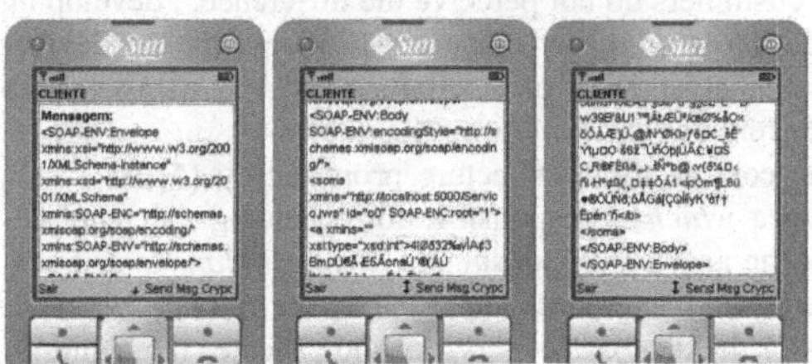

Figure 9 – Sample of Encrypted Message in Mobile Device

To make the public key available for a possible consumer, the system adds the generation of digital certificate that can be sent to mobile consumers, through which they get the public key of the services which he wants to communicate, and thus can encrypt the messages and verify the authenticity of digital signatures received by them. An example of a Digital Certificate generated by the system can be seen below:

---------- Begin Certificate ----------
Type: X.509v1
Serial number: 35:32:35:38:30:39
SubjectDN: MobileHost/
IssuerDN: MobileHost/
Start Date: Wed Aug 13 17:37:58 UTC 2008
Final Date: Sat Aug 23 17:37:58 UTC 2008
Public Key: RSA
modulus:
9337497433152274214014446837013394548439412
0408072691949673228126709006135379056199872
85614350948448889003887517758 ...
public exponent:65537
Signature Algorithm: RSA
Signature:
2673219510732109272348323117338219317732109
2162271232116242582531891517420121224291208
4101821296251673769631166216239 ...
---------- End Certificate ----------

VI. RELATED WORKS

The idea of provisioning SOA for mobile devices is explored in [4]. According with [4], the basic architecture of a mobile terminal as a service provider can be established through a: Service Request, Service Provider (Mobile Host) and service registry. The service providers implemented in the mobile device are used to describe the standard service (WSDL), and a standard (UDDI) is used to publish and remove publication services. Once you can see that this architecture follows the pattern of desktop systems, but taking into account the low resources of the device.

According with [4] with the advent of wireless networks and ad-hoc increase in capabilities of equipment that allows to be created entirely pure P2P networks, where the key piece of this architecture is to maintain the Mobile Host completely compatible with the interfaces used by (WS) SOA in such a way customers do not perceive the difference; , developing the Mobile Host with few resources available on the device and limit the performance of functions of SOA in such a way as not to interfere in the functions of the device.

The core of the architecture proposed by [4] is: a *network interface which* is responsible for receiving the requests and sends the answers to consumers, *Request Handler* extracts the contents of the request and sent to the *Service handler* that accesses the database of services and executes the requests.

In [9], it is proposed a mechanism for hierarchical service overlay, based on the capabilities of the hosting devices. The Middleware of [9] is called PEAK (Pervasive Information Communities Organization) which enables the transformation of resources into services. Based on the widespread hierarchical "overlay service" by developing a mechanism for composition of services which is able to make dynamically complex services using the availability of basic services.

In [10] it is developed a mobile middleware technology, motivated primarily by projects related to mobile applications. Using this Middleware, mobile devices become a part of SOA (Web Services) located on the Internet. The Middleware enables not only access to SOAs on the Internet, but also the provisioning of services. These services can be used by other applications of the Internet (mobile call back services) or any other fixed or mobile device (P2P services).

In [11], the author investigates mechanisms to support dynamic m-services oriented architecture context with the objective of service publication and dynamic discovery of mobile users anywhere and anytime discussing the characteristics of SOAs for wireless and wired networks. Moreover, [11] investigates the availability of technologies for mobile device and SOA paradigm; proposing a m-services architecture to support dynamic and scalable mobile services. [11] proposes an entity management services with a service record as an intermediate layer between service providers and mobile users. The management service is responsible for coordination of interactions between service providers and mobile users. This interaction produces information services, providing and delivering services for mobile users at any time. [11] is also investigating the use of dynamic invocation interfaces as a mechanism of communication between departments of the description and invocation at runtime.

VII. Ends Considerations

This paper proposes the use of devices as service providers, however its development requires an implementation work that can become extremely complex. Therefore, we see the need for a tool for development and provision of services on mobile devices in an automated way.

This work introduces the Mobile Host as part of a Middleware for construction of SOA in mobile environment. This paper presents an overview of the concepts, and proposes an architecture. The architecture proposal aims to provide a developer with a tool for rapid development of services in mobile environment and aggregates all services necessary for the provision of the service for invocations in an infra-structured network such as in P2P networks.

A security service is included in the framework, where the developer has the option of whether or not to use it. While it is a very important requirement it use requires a greater use in processing power of the device and the response time of a request, then you should consider whether its use is feasible for each service being developed. The framework presented proved to be a good solution to the problem presented. However, more tests related to their performance and quality of service offered yet become necessary.

Acknowledgment

Financial supports of FAPEMA and CNPq are gratefully acknowledged.

REFERENCES

[1] S. N. Srirama, M. Jarke and W. Prinz. "A Mediation Framework for Mobile Web Service Provisioning", in Proc 2006 *10th IEEE International Enterprise Distributed Object Computing Conference Workshops.*

[2] S. N. Srirama, M. Jarke and W. Prinz. "Mobile Host: A feasibility analysis of mobile Web Service provisioning", in proc 2006 *Ubiquitous Mobile Information and Collaboration Systems.*

[3] S. S. Amorim. "A Tecnologia Web Service e sua Aplicaçãoo num Sistema de Gerência de Telecomunicaçõeses. Dissertação de Mestrado, Universidade Estadual de Campinas - UNICAMP, 2004.

[4] S. N. Srirama, M. Jarke, Matthias and W. Prinz. "Mobile Web Service Provisioning". In proc 2006 IEEE *Proceedings of the Advanced International Conference on Telecommunications and International Conference on Internet and Web Applications and Services.*

[5] N. Dokovski, I. Widya and A. van Halteren. "Paradigm: Service Oriented Computing". University of Twente, 2004.

[6] S. N. Srirama, M. Jarke and W. Prinz. "Security analysis of mobile web service provisioning". In proc 2007 *Int. J. Internet Technology and Secured Transactions*. Inderscience Enterprises.

[7] P. Pawar, S. Srirama, B. J. van Beijnum and A. van Halteren. "A Comparative Study of Nomadic Mobile Service Provisioning Approaches". In proc 2007 *IEEE The 2007 International Conference on Next Generation Mobile Applications, Services and Technologies*

[8] C. Candolin. "A Security Framework for Service Oriented Architectures". IEEE, 2007.

[9] S. Kalasapur, M. Kumar and B. Shirazi. "Seamless Service Composition (SeSCO) in Pervasive Environments". In proc 2005 ACM *MSC 2005.*

[10] L. Pham and G. Gehlen. "Realization and Performace Analysis of a SOAP Server for Mobile Devices". RWTH Aachen University, Chair of Communication Networks.

[11] E. S. Nilsen, S. M. Ruiz and J. R. Pedrianes. "An Open and Dynamical Service Oriented Architecture for Supporting Mobile Service". In Proc 2006 ACM *ICWE'06.*

[12] J. D. Meier, A. Mackman, M. Dunner, S. Vasireddy, R. Escamilla, and A. Murukan. "Improving web application security: threats and countermeasures", Microsoft Corporation, MSDN. Jun, 2003.

Towards a New Paradigm of Software Development: an Ambassador Driven Process in Distributed Software Companies

Deniss Kumlander
Department of Informatics, TTU
Tallinn, Estonia
e-mail: kumlander@gmail.ee

***Abstract*—The globalization of companies operations and competitor between software vendors demand improving quality of delivered software and decreasing the overall cost. The same in fact introduce a lot of problem into software development process as produce distributed organization breaking the co-location rule of modern software development methodologies. Here we propose a reformulation of the ambassador position increasing its productivity in order to bridge communication and workflow gap by managing the entire communication process rather than concentrating purely on the communication result.**

Keywords-software development; distributed; outsourcing; ambassador

I. Introduction

The increased competition among software vendors and globalization of operations demand both improving quality and decreasing the cost of software projects on the constant base. Unfortunately customers still have to pay a very high price for the delivered software packages mainly because of high rate of failed projects. The overall percentage of functionality considered by customers as failed to be implemented is more than one third in average [1, 2]. The software quality depends a lot on teams' ability to communicate, avoid misinterpretation of requirements and messages posted during their co-work on the project from one work-cycle step to another. Knowing that communication and feedback are key parts of any team performance [3] we need to carefully consider all possible restrictions for those flows. Those will negatively affect two important factors of the team performance: communication and feedback. Dividing the process of software development into steps greatly simplified the software development process and the resultant software generally better aligning the process to the market expectations [4]. At the same time such simplification by dividing into steps produces certain barriers between those affecting earlier mentioned key factors.

The distribution of team members all over the world produces ever wider troubles [5]. It is crucial for the organization to understand in advance problems that could arise in such environment and deal with those on proactive rather than reactive base in order to minimize the project cost or even avoid a failure of the project.

II. Distributed Organisations

In this paper under distributed organization we mean a company that satisfies to the following criteria:

- The company has more than one office;
- Offices are located on a sufficient distance from each other (for example a company is not distributed if different departments are located on different floors on in different rooms);
- All offices' teams participate in the same core business activity and none of those can be omitted without affecting the work-cycle.

As you can see it is a mixture of traditional distributed organization with a team centric point of view. We will not consider a company to be distributed if it is a group of companies were branches' operations never cross.

Although each company is likely to be started as a co-located one, the larger it grows the higher is probability that it eventually will be split into parts located in different towns, countries or continents. Below we compile a short list of major reasons why distributed organizations occur. Reasons are grouped by direct or indirect factors, where the factor is called "direct" if the decision was to become distributed to derive some advantages of such organization or "indirect" if distributing is a side effect or some kind other decision or influence on the organization..

1. Direct factors
   - The software development process will be cheaper;
   - A misfit of a skilled workforce and product markets (management core of the company). Notice that here the distribution doesn't necessary produce an advantage to be cheaper, but certainly will increase the income as the new location will unlock the product development;
2. Indirect factors
   - Merging with other organizations;
   - Acquisition of other companies;

K. Elleithy (ed.), *Advanced Techniques in Computing Sciences and Software Engineering*,
DOI 10.1007/978-90-481-3660-5_84, © Springer Science+Business Media B.V. 2010

- Expanding to new markets – establishing an office in new location and leading the product evolution from this new branch (not just selling an universal product via this office)
- Different branches have to cooperate to increase productivity, integrate their products etc;
- Identified requirement to cooperate with partners been in different groups from the legal perspective.

In most approaches so far the case of none co-located development was either omitted or greatly simplified in order not to destroy the internal logic of approaches. The typical example is to cut distributed edges out assuming that the team is either co-located or the distance edge team (development, consultants etc) will constantly generate a correct communication flow. Because of reasons described earlier we can state that the distributed organizations' pattern is a common form of many serious organizations nowadays and therefore should be carefully explored. It is important to examine either how distributed organizations affect common approaches or do a step further and propose some methods balancing software development process in such specific environment.

III. Communication and Workflow Gaps

A narrow definition of the communication gaps comes from the traditional point of view on them and states that it is a problem occurring in communication between people (group of people) where the information is either lost or sufficiently corrupted.

This situation is well known in the agile software development methodologies [6, 7, 8] since those base a lot on internal communication and therefore any problem occurring in it will be very critical. Unfortunately classical agile methods of co-located teams are not always possible due earlier defined reasons. Therefore there was a suggestion to send a person oversea who will improve the communication process and this person is nominated to the ambassador position.

The difference between the traditional definition of the ambassador and the one proposed by this paper lies in tasks officially assigned to this person. Let's examine the current approach.

Agile methods rely on the face to face communication. If everyone cannot be co-located, then moving key persons helps a lot [9]. The problem arises mostly in offshore types of developing trying to implement agile approaches. The initial approach that most companies do implement is to move to the oversea location just one person handling the project management process. Mostly, it is the person responsible for the project and s/he is moved from the management location to the offshore location. This person will mostly add to the distance edge team own contacts and experience as either knows requirements' reasons or knows somebody in the main office who can be questioned on those. In other words s/he starts to cut communication process corners. At the same time the conventional agile methods define that there should be a way to communicate results, targets and requirements freely and massively. So such person activities rarely fit into this definition. Having realized that, the company will start to migrate to the next stage of bridging communication gaps: the bridging process is expanded to several levels by sending overseas ambassadors of different sub-teams like analysts, developers, testers and so forth. Finally the approach is sometimes strengthened by sending ambassadors in backward direction as well. Here ambassadors' targets are transfer communication flow messages from one site to another and backward increasing the level of understanding and decreasing misinterpretations of what was said by participating in both location teams' meetings or connecting both teams providing sometimes also a possibility to communicate in both teams' native languages.

A similar definition is given for DSDM approach [10]. Here a special type of contributor is recognized: an "Ambassador User". This person responsibility is to bring the knowledge of end-users community into the project, ensures that the developers get enough information on logic and reasons of requested features. Besides s/he ensures a feedback during the development process. For example, in Scrum projects, this role is mostly assigned to the "product owner" role [8]. This person is in charge of managing the feedback process although s/he could be not involved directly in formulating the feedback.

Unfortunately this way doesn't guarantee the perfect result. Simplifying a lot arising troubles, we can state that companies (and teams) will still face the following problems [9, 11]. There is a lot of informal communication, decisions on higher level etc that is still missed. Although it is not possible to say exactly what will be missed and prevent other people to understand correctly the communicated message, it is still true that adding the entire context improve result sufficiently. Besides companies have to rotate ambassadors every few months as many people don't want to spend several months away from their homes.

IV. Applying Ambassadors: When Do We Need to Do That?

A. Cases when there are no reasons to applied ambassadors driven development

Here we are going to do a short review of cases that we are not covered by our proposal in this paper. This chapter is designed to highlight major differences between covered and uncovered cases giving to readers a solid understanding where the bound is and why. It will also highlight reasons why ambassador positions are established

First of all, there is no need to apply ambassadors if the team is co-located. As the information is flowing freely during meetings and ideally each member of the team have either equal knowledge of the subject or knows who s/he can provide required information and this person is somewhere close to his/her desk. Notice that under co-location of the team we mean a co-location the entire team including consultants, analysts, testers, management and developers. In case of any exceptions ambassadors could become important to be used.

Secondly there is no need to establish such special position if the team has good internal relations and can easily self-organize over a distant edge. This traditionally means constant face-to-face meetings, reluctant schedules in order to leave enough time for informal communication and an average pressure of tasks to keep the distance edge tasks both visible and prioritized.

B. *Complex reality*

Unfortunately there are a lot of cases when reality differs sufficiently from what were described above. It is possible to state that those cases in nowadays world are rather exceptional than traditional. There are several reasons and the globalization of operations leading to clustering teams into sub-teams located in different places as it was described earlier, is just one of them. Let's review others.

The more project and pressure the team gets the less important looks tasks that come from a distance edge. Unfortunately in nowadays world the high pressure became rather the rule than an occasion. There wee a lot of attempts to virtualize the distance by using web cams and improving communication by periodic face-to-face meetings, but those failed with grows of pressure on each team in modern environment. In this condition, the physical distance transforms into communicational and work-flow tasks distance affecting the organization performance.

The communication process complicates a lot the more is the time difference between communicating locations. If the difference is huge then each person could prioritize his/her current tasks over other team communication tasks as nobody waits for an immediate reaction (as it is, for example, a night on the second team location). In the result of this communication tasks can be pushed out from the daily schedules and become permanently pending.

There are organizations that do not practice agile methods of software development and probably will never do that. In the absence of massive communication workflow process tasks become loosely coupled and a lot of communication gaps arise between different steps of the cycle [12, 13].

Unfortunately it is nearly impossible to have a person travelling constantly from one to another office to push the communication and his/her work becomes impossible if s/he has to do it every single moment of work-time.

V. Ambassador Driven Software Development

Starting explanation of the new ambassador set of responsibilities we would like to reformulate the communication gap term meaning expanding it. Instead of concentrating only on the result of communication we should concentrate also on the process of communication. Moreover the process should be considered in its context – mostly it occurs as a part of some kind process like for example software development work-cycle and the communication process occurs between different work-flow stages. As you can see we also try to move this definition out of agile type teams where the communication happens from each one to each one (mostly at once) to the well-defined, traditional etc. work-processes as well.

Moreover the communication process can be examined not only from the result perspective, but also from the timing perspective. The communication should be considered as failed if the expected feedback or information came later that it was expected producing inefficient consumption of available resources up to now and extra costs on reworking the current work-cycle step deliveries.

In the new approach to the ambassador position, the person in that position should be responsible for the following:

1. Speed up the communication process by managing the communication flow

   a. Select a right person the communication flow should be directed to for any particular messages, either directly or advising if none meaningful response is got from the default person;

   b. Chase people [14] – as we have described earlier it is hard to prioritize work that come from a distance edge even if it is a stopper for another team. As the distance edge team have no ways to promote any activities pushing in the face-to-face communication, then there is a need to have somebody actively dealing with pending issues in that particular location;

   c. Ensure none-zero communication: asking for a feedback, i.e. expanding the communication (as you can see we cannot say that the transferred information is broken as none information is moving and some persons could say that it is not expected to be here while others would disagree with it);

2. Serve as a communication flow supervisor ensuring that it flows to the right person, is complete and comprehensive. An incorrect (but complete) communication can lead to an infinite loop when involved partners either struggle to understand each other or have opposite opinions and need supervising to arrive to some common understanding of the problem and alternative solutions;

3. Supervise context of the flow and guarantee that the context is also transferred ensuring complete understanding of the transferred message. Traditionally the conversional communication gaps' reasons are either technically restricted communication (described earlier) or differences in persons frames (like background, personal skills etc). By sometimes there is something else beyond such static factors. Context is a sufficient part of the information, which is rarely communicated assuming that everybody acquired it already and the communication goes in its scope. In fact it is missed sometimes and it makes it impossible to understand some portion of the communicated statement, allow to misinterpret it or completely miss.

4. Synchronize different teams' efforts. Consider for example self-organizing agile Scrum teams working on the same project. Having nothing special defined in the methodology, teams in practice start to search for a solution for the synchronization problem and normally end

up visiting other meetings been "chickens" [8] in those. At the same time having no right to talk they have a very limited ability to affect the other team activities. If a member of the other team will be nominated to another team ambassador then this will improve synchronization and fit into the agility practice.

Previously describing set of tasks is assigned to a *communication improving ambassador position*. Notice that this role does not always mean 100% workload and can be seen sometimes as an honorary ambassador title. This role can be extended if required into a *managing ambassador position* by adding the following extra responsibilities:

5. Monitoring and reporting project statuses over the edge. Clearly managers have no or limited visibility what is happening in the team physically located far away and therefore need somebody to increase transparency of processes on another end of the edge. Notice that the biggest danger here is to transform this position into a "spy" kind of it. The person should be trusted by the team and his/her feedback should be visible to the team in order to promote cooperation and visibility instead of motivating conflicts.
6. Expose informational about the team, like for example each team member abilities, experience and skills, i.e. information that management sometimes need planning tasks and projects.

VI. Conclusion

Communication and workflow gaps existing in none co-located teams generate a lot of mistakes and can easily lead to the project failure. Researches have shown that the number of failed project is still high despite all modern methodologies. This clearly indicates persistence of certain problems. A "none co-located teams" case arising in distributed organizations is one potential reason of earlier mentioned problems and requires active dealing with it in order to improve the situation. The paper proposes nominating certain members of distance teams to either communication improving ambassador or managing ambassador positions in order to improve the software development process, balance it and rapidly progress. From our point of view such ambassadors' position is the first step toward a new paradigm of software development locating beyond the modern agile approaches.

References

[1] E. N. Bennatan, K.E. Emam, "Software project success and failure," *Cutter Consortium*, 2005, http://www.cutter.com/press/050824.html.

[2] A.A. Khan, "Tale of two methodologies for web development: heavyweight vs agile," *Postgraduate Minor Research Project*, 2004, pp. 619-690.

[3] R.A. Guzzo, M.W. Dickson, "Teams in organizations: Recent research on performance and effectiveness," *Annual Review of Psychology*, vol. 47, pp. 307-338, 1996.

[4] A. Hoffer, J.F. George, and J.S. Valacich, *Modern system analyses and design*, Addison Wesley, 1999.

[5] E. Carmel, *Global Software Teams*, Prentice Hall, 1999.

[6] A. Cockburn, *Agile Software Development*, Reading, MA: Addison Wesley, 2002.

[7] K. Beck, *Extreme Programming Explained: Embrace Change*, Reading, MA: Addison Wesley, 2000.

[8] M. Beedle, and K. Schwaber, *Agile Software Development with SCRUM*, Englewood Cliffs, NJ: Prentice Hall, 2001.

[9] K. Braithwaite, and T. Joyce, "XP Expanded: Distributed Extreme Programming," *6th International Conference on eXtreme Programming and Agile Processes in Software Engineering*, Springer, 2005, pp. 180-188.

[10] J. Stapleton, *DSDM Dynamics System Development Method*, Reading, MA: Addison Wesley, 1997.

[11] T. Chau, F. Maurer, "Knowledge Sharing in Agile Software Teams," in *Proc. Logic versus Approximation*, 2004, pp.173-183.

[12] D. Kumlander, "Towards dependable software: researching work-flow gaps in the distributed organisations," *Proceedings of the 7th Conference on 7th WSEAS International Conference on Applied Informatics and Communications*, pp. 206-210, 2007.

[13] D. Kumlander, "Supporting Software Engineering," *WSEAS Transactions on Business and Economics*, vol. 3(4), 2006, pp. 296-30.

[14] D. Kumlander, "Collaborative software engineering of the life-critical systems," *Proceedings of the 6th WSEAS International Conference on Software Engineering, Parallel and Distributed Systems*, pp. 35-38, 2007.

An Equivalence Theorem for the Specification of Asynchronous Communication Systems (SACS) and Asynchronous Message Passing System (AMPS)

A.V.S. Rajan, A.S.Bavan, and G. Abeysinghe
School of Engineering and Information Sciences,
Middlesex University,
The Burroughs, Hendon,
London NW4 4BT, United Kingdom
{a.rajan, s.bavan, g.abeysinghe}@mdx.ac.uk

Abstract

Formal semantics have been employed in the specification and verification of programming languages. Language for Implementing Parallel/distributed Systems (LIPS) is an asynchronous message passing parallel programming language which handles communication and computation parts independently. The communication part of LIPS can be specified using a process algebraic tool, Specification of Asynchronous Communication Systems (SACS), and is implemented using Asynchronous Message Passing System (AMPS). An implementation is said to be complete only when we prove that it meets its specifications. To achieve that we need to prove an equivalence relation between a program's specification and its implementation. This means that it is necessary to study the proof of equivalence of SACS and AMPS to prove the completeness of AMPS. The semantics of both SACS and AMPS have been defined using Structural Operational Semantics (SOS) in terms of Labelled Transition Systems (LTS). So we have two labelled transition system semantics : one for SACS and one for AMPS. In this paper we are proving the bisimilarity of these two labelled transition systems to prove that SACS and AMPS are equivalent.

Keywords *: Asynchronous message passing, Structural Operation Semantics, Labelled Transition Systems, equivalence, bisimilarity.*

1. Introduction

Language for Implementing Parallel/distributed Systems (LIPS) [2] is an asynchronous message passing parallel programming language which handles communication and computation independently. The operational semantics for LIPS has been defined using a two step strategy to adequately provide implementation information for both computation and communication parts.

1. Firstly, big-step semantics or evaluation semantics has been used to define the computation part of a LIPS program.

2. Secondly, the defined semantics has been extended with Structured Operational Semantics (SOS) to describe the communication part.

The asynchronous communication has been implemented using Asynchronous Message Passing System (AMPS) conceptualised by Bavan [4]. AMPS is based on a simple architecture comprising of a Data Structure (DS), a Driver Matrix (DM), and interface codes. In a LIPS program, a message is sent and received using simple assignment statements and the program is not concerned with how the data is sent or received. With the network topology and the guarded process definitions, it is easy to identify the variables participating in the message passing. The DS and the DM for the AMPS are defined using these variables. A detailed explanation of AMPS can be found in [4] and its Structured Operational Semantics can be found in [10].

Process algebra is considered as a formal framework to model concurrent systems of interacting processes and their behaviour [12]. Few of the well known process algebraic tools include Communicating Sequential Processes (CSP) [7], Calculus of Communicating Systems (CCS) [9], Synchronous Calculus of Communicating Systems (SCCS) [6], and Language Of Temporal Ordered Systems (LOTOS) [8]. Since the development of CCS many extensions have been proposed to model different aspects of concurrent processing [5]. Specification of Asynchronous Communicating Systems (SACS) [1], [3] is one of them. SACS is a point-to-point message passing system which is an asynchronous variant of SCCS developed to specify the communicating

K. Elleithy (ed.), *Advanced Techniques in Computing Sciences and Software Engineering*,
DOI 10.1007/978-90-481-3660-5_85, © Springer Science+Business Media B.V. 2010

part of LIPS. The main objective of SACS is to separate the specification of communication from the computation part of LIPS programs so that they can proceed independently. The behaviour of process algebra can be described using Structural Operational Semantics (SOS) which is defined using Labelled Transition Systems (LTS). Rajan et al. [11] describes the Structured Operational Semantics (SOS) for SACS. They use Labelled Transition Systems (LTS) to describe the operational semantics of SACS.

An imlementation is said to be complete only when we prove that it meets its specifications and to prove that we need to prove an equivalence relation between the specification and its implementation. SACS has been used for the high level specification of the communication part of LIPS programs and is implemented using the Asynchronous Message Passing Systems (AMPS). It is necessary to study the proof of equivalence of SACS and AMPS to prove the completeness of AMPS. The operational semantics of both SACS and AMPS are based on Structural Operational Semantics (SOS) using Labelled Transition Systems. We then have two labelled transition system semantics: one for SACS and one for AMPS. To prove that they are equivalent, it is enough if we can prove the bisimilarity of these two labelled transition systems. This paper presents the proof of equivalence between SACS and AMPS.

2. Summary of the Proof of Equivalence Between SACS and AMPS

We consider the weak bisimulation equivalence between SACS and AMPS. The reason is that when we compile a language or specification to another, it is very unlikely that we can faithfully preserve the operational semantics. This means that a transition from $P' \xrightarrow{\alpha} Q'$ in SACS may become a sequence of transitions in AMPS, namely $P' \xrightarrow{\alpha 1} ... \xrightarrow{\alpha n} Q'$ where most of $\alpha 1$, ..., αn are silent transitions. It might also be that we may not reach Q but a process equivalent to Q. In such situations, the weak bisimulation which compares only the external behaviours of the processes is preferred.

Definition 1. (Weak Bisimulation Equivalence)

A binary relation over a set of states of the LTS system is a weak bisimulation, $P \approx Q$:

- if $P \xrightarrow{\alpha_1} P'$, then either
  - $\alpha_1 = 1$ and $P' \approx Q$ (or)
  - for some Q', $Q \stackrel{\alpha_1}{\Rightarrow} Q'$ and $P' \approx Q'$

and conversely,

- if $Q \xrightarrow{\alpha_1} Q'$ then either
  - $\alpha_1 = 1$ and $P \approx Q'$ (or)
  - for some P', $P \stackrel{\alpha_1}{\Rightarrow} P'$ and $P' \approx Q'$

where $P, Q \in \Re$ and $\alpha_1 \in \alpha$.

The proof of equivalence between SACS and AMPS depends on the Labelled Transition System (LTS) and the definition for LTS can be found in Definition 2.

Definition 2. Labelled Transition System

A Labelled Transition System (LTS) is a triplet {S, K, T} where:

- S is a set of states,
- K is a set of labels where $\overline{K}$= {$\overline{k}$— k ∈ K},
- T = {$\xrightarrow{k}$, k ∈ K } is a transition relation where $\xrightarrow{k}$ is a binary transition relation on S.

The main results for the proof of equivalence between SACS and AMPS are summarised in Figure 1.

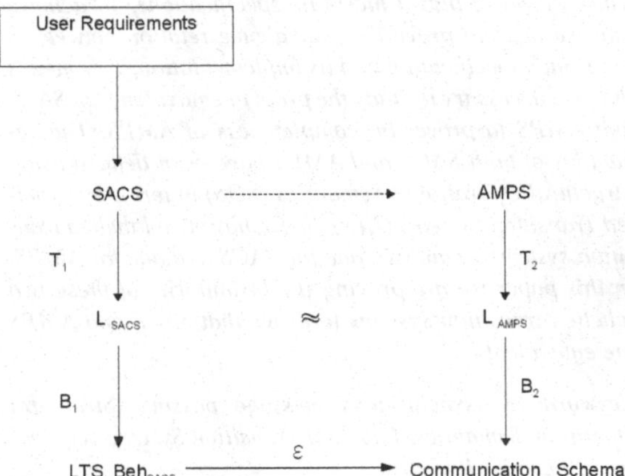

Figure 1. Summary of the proof of equivalence between SACS and AMPS

Given the user requirements, the SOS for SACS known as L_{SACS} can be defined using the LTS. The function T_1 mapping the SACS to its LTS is shown below:

$$T_1 : SACS \rightarrow L_{SACS}$$

The behaviour of SACS is defined using the sequences of LTS configurations used to define the SOS for SACS. It can be represented using a mapping B from LTS, L_{SACS}, to its set of behaviours, LTS_Beh_{SACS}, which is shown below:

$$B_1 : L_{SACS} \rightarrow LTS\_Beh_{SACS}$$

The communication part of LIPS program written using its SACS specification is implemented using the AMPS. The relationship between the SACS and AMPS is shown using dotted lines in Figure 1. The SOS for AMPS is the set of LTS configurations denoted by L_{AMPS}. The function T_2 mapping the AMPS to its LTS is shown below:

$$T_2 : AMPS \rightarrow L_{AMPS}$$

The behaviour of AMPS is defined using its communication schema implemented using a set of functions. The mapping, B_2, from AMPS to the communication schema is shown below:

$$B_2 : L_{AMPS} \rightarrow Communication\_Schema$$

The behaviour of AMPS depends on the communication schema described in [4], [10]. When the specifications of SACS have to be implemented, the SACS transitions have to be extended to include the communication schema of the AMPS. The extension function, ε, to be included to the SACS behaviour for its implementation purposes is shown below:

$$\varepsilon : LTS\_Beh_{SACS} \rightarrow Communication\_Schema$$

We can say that the set of configurations of L_{AMPS} is the set of configurations of L_{SACS} and a set of functions used to implement asynchronous message passing.

Set of configurations of SACS is given as:

$$L_{SACS} = \{Guard,\ Guarded\,Process,\ Node,\ Concurrency\,Composition,\ System\}$$

Set of configurations of AMPS is given as:

$L_{AMPS} = L_{SACS} \cup$

$\{Is\_input\_available,\ Is\_ok\_to\_send,\ Send\}$

An equivalence relation has to be constructed between the configurations, L_{SACS} and L_{AMPS}. This can be expressed using the behaviour of their configurations which can be shown as below:

$$Communication\_Schema \approx \varepsilon \cup LTS\_Beh_{SACS}$$

where ε is the set of functions, $Is\_input\_available$, $Is\_ok\_to\_send$, $Send$.

Definition 3. (Bisimilarity between two Labelled Transition Systems)

Two equally labelled transition systems L_1 and L_2 are bisimilar (written as $L_1 \approx L_2$) if and only if

$$L_i \stackrel{def}{=\!=} (S_i,\ K,\ T_i) \text{ for } i = 1, 2$$

and there exists a relation $R \subseteq S_1 \times S_2$ such that $k \in K$:

1. $p \in S_1 \Rightarrow \exists q . q \in S_2 \wedge p, q \in R$ and $q \in S_2 \Rightarrow \exists p . p \in S_1 \wedge p, q \in R$
2. $\forall p\, q\, p' . p, q \in R \wedge p \rightarrow p'$ $\Rightarrow \exists q' . q \rightarrow q' \wedge p', q' \in R$
3. $\forall p\, q\, q' . p, q \in R \wedge q \rightarrow q'$ $\Rightarrow \exists p' . p \rightarrow p' \wedge p', q' \in R$

We show the equivalence relation between SACS and AMPS using Theorem 1.

Theorem 1. Equivalence Relation for the SACS and AMPS

For every correctly labelled AMPS specification,

$$L_{AMPS} \approx L_{SACS}$$

Proof: The first step in the proof is to identify related configurations for

$(S_1, k, t), (S_2, k, t) \in R$ and

$(S_2, k, t) = (S_1, k, t) \cup set\ of\ functions.$

Then bisimilarity is proved using the three conditions stated in Definition 3.

Condition 1: Every pair $(S_1, k, t), (S_2, k, t) \in S_1 \times S_2$ must be in R. That is we must show that

$$(S_2, k, t) = (S_1, k, t) \cup set\ of\ functions$$

Let 'G' be a guard, 'GP' be a guarded process, 'N' be a node, 'S' be a set of parallel nodes, 'f_1, f_2, f_3' are the functions $Is\_input\_available$, $Is\_ok\_to\_send$, and $Send$ respectively.

$$\begin{aligned}
(S_1, k, t) &= (\{G, GP, N, S\}, k, t) \\
(S_2, k, t) &= (\{G, GP, N, S, f_1, f_2, f_3\}, k, t) \\
&= (S_1 \cup set\ of\ functions, k, t) \\
&\quad \text{where set of functions} = \{f_1, f_2, f_3\} \\
&= (S_1, k, t) \cup set\ of\ functions
\end{aligned}$$

Condition 2 and **Condition 3** state that from any pair $(S_1, k, t), (S_2, k, t) \in R$ every L_{SACS} transition to (S_1', k', t') must have a corresponding L_{AMPS} transition and every L_{AMPS} transition to (S_2', k', t') must have a corresponding L_{SACS} transition. Also, the resulting pair should satisfy the condition,

$$((S_1, k, t), (S_2, k, t)) \in R.$$

For each configuration of AMPS and SACS,

1. Find the (S_1', k', t') reached by α transition.
2. Assign

$$(S_2, k, t) = (S_1 \cup set\ of\ functions, k, t)$$

so that the configurations are in R.

3. Find the (S_2', k', t') reached by the α transition.
4. Check

$$((S_1', k', t'), (S_2', k', t')) \in R.$$

This is done by checking that

$$Communication\_Schema \approx \varepsilon \cup LTS\_Beh_{SACS}$$

for each of the configurations of SACS and AMPS and they are listed below:

1. **Guard:**

The L_{SACS} and L_{AMPS} configurations for a guard are shown below:

$$L_{SACS} : (\alpha_1 \circ \alpha_2 \circ \cdots \circ \alpha_k, s_1)$$

$$\xrightarrow{T} (\underline{T}\{\alpha_1, \alpha_2, \cdots, \alpha_k\}, s_2)$$

$$L_{AMPS} : ((fch_{i1} \wedge fch_{i2} \wedge \cdots \wedge fch_{im}), s_1)$$

$$\xrightarrow{T} (\underline{T}\{ch_{i1}, ch_{i2}, \cdots, ch_{im}\}, s_2)$$

The function $Is\_input\_available$ will be called when an input channel in a guard checks for the availability of a new value. If a new value is available, the function will return a 1. The configuration for the function is given as:

$$(Is\_input\_available(m, n), s_1) \xrightarrow{m, n} (1, s_1')$$

where m, n specify the node number and variable number respectively.

When this function returns a 1, the $Send$ function will be called to send the data from the DS of the AMPS to the respective channel. The configuration for $Send$ function is given as:

$$(Send(m, n, t, d), s_1) \xrightarrow{m, n, t, d} (1, s_1')$$

These two functions will be called consecutively for all the input channels and they are inter transitions. As a result the system will move from state s_1 to s_2.

The equivalence relation checks only the initial and final states and they are one and the same as that of the L_{SACS}.

$$\Rightarrow Communication\_Schema \approx \varepsilon \cup LTS\_Beh_{SACS}$$

Therefore,

$$(\alpha_1 \circ \alpha_2 \circ \cdots \circ \alpha_k, s_1) \xrightarrow{T} (\underline{T}\{\alpha_1, \alpha_2, \cdots \alpha_k\}, s_2)$$

$$\approx ((fch_{i1} \wedge fch_{i2} \wedge \cdots \wedge fch_{im}), s_1)$$

$$\xrightarrow{T} (\underline{T}\{ch_{i1}, ch_{i2}, \cdots, ch_{im}\}, s_2)$$

2. **Guarded Process:**

The L_{SACS} and L_{AMPS} configurations for a guarded process are shown below:

$$L_{SACS} : (\eta\ \mathtt{Proc}, s_1) \xrightarrow{\eta} (\mathtt{Proc}, s_2\{\sigma\})$$

where η is a guard which becomes true to execute the associated process $proc$ and generate values of the set of output characters σ.

$$L_{AMPS} : (if\ G_i\ then\ P_i, s_1) \xrightarrow{CH} (OCH_i, s_2)$$

When G_i becomes true, the associated process body is executed to generate 0 or more values for the output channels OCH_i. For every value available in the output channel, $Is\_ok\_to\_send$ function is called to find the value that can be sent to the data structure of AMPS.

$$(Is\_ok\_to\_send(m, n), s_1) \xrightarrow{m, n} (1, s_1')$$

where m, n specify the node number and variable number respectively.

If it is ok to send, the $Send$ function is called to send the value to the DS.

$$\Rightarrow Communication\_Schema \approx \varepsilon \cup LTS\_Beh_{SACS}$$

Therefore,

$$(\eta\ \mathtt{Proc}, s_1) \xrightarrow{\eta} (\mathtt{Proc}, s_2\{\sigma\})$$

$$\approx (if\ G_i\ then\ P_i, s_1) \xrightarrow{CH} (OCH_i, s_2)$$

3. **Node:**

The L_{SACS} and L_{AMPS} configurations for a node are shown below:

$$L_{SACS} : (\sum_{i=1}^{k} \eta_i\ \mathtt{Proc}_i, s_1) \xrightarrow{for\ any\ i=1\ to\ k, \eta_i}$$

$$(\mathtt{Proc}_i, s_2\{\sigma_i\})$$

where $\eta_i\texttt{Proc}_i$ is a guarded process GP_i. In other words,

$$\Re = GP_1 + GP_2 + ... + GP_k$$

where $\Re$ is a node and $GP_1, GP_2, ..., GP_k$ are guarded processes.

L_{AMPS} :

$$(while\,(\underline{T})\,do\,(GP_1\,else\,GP_2\,else\,...\,else\,GP_k),\ s_1)$$
$$\xrightarrow{T} ((if\ (\underline{T})\,then\,GP_i;\ while\ (\underline{T})\ do$$
$$(GP_1\,else\,GP_2\,else\,...\,else\,GP_k)),\ s_2\{OCH_i\})$$

The AMPS configuration shows the actual implementations of the SACS and expresses the guarded processes involved in making a node. In actual implementation, this is an infinite *while loop*. To execute a guarded process, its associated guard should become true. When a guard is true, its associated process body is executed to generate values for 0 or more output channels.

$\Rightarrow Communication\_Schema \approx \varepsilon \cup LTS\_Beh_{SACS}$

Therefore,

$$\left(\sum_{l=1}^{k} \eta_i\texttt{Proc}_i,\ s_1\right) \xrightarrow{\text{for any } i=1\,to\,k,\,\eta_i} (\texttt{Proc}_i,\ s_2\{\sigma_i\})$$

$$\approx$$

$$while\,(\underline{T})\,do\,(GP_1\,else\,GP_2\,else\,...\,else\,GP_k),\ s_1)$$
$$\xrightarrow{T} ((if\ (\underline{T})\ then\ GP_i; while\ (\underline{T})\ do$$
$$(GP_1\,else\,GP_2\,else\,...\,else\,GP_k)),\ s_2\{OCH_i\})$$

4. **System/Network:**

The L_{SACS} and L_{AMPS} configurations for a node are shown below:

Let there be n nodes in a network,

L_{SACS}:

$$(\Re_1 x \Re_2 x ... x \Re_m,\ s_1) \xrightarrow{\mu} (\Re'_1 x \Re'_2 x ... x \Re'_m,\ s_2)$$

where $\Re_1, \Re_2, ..., \Re_n$ denote the nodes and μ is a prefix.

If $\Re_1$ and $\Re_2$ are two nodes to be executed concurrently, their concurrency composition is denoted by

$$\Re_1\ x\ \Re_2.$$

Upon concurrently executing the nodes, the system changes its state from s_1 to s_2.

L_{AMPS}:

The network definition part of a LIPS program is defined using a set of connect statements. The configuration for a set of connect statements is given below:

$$\forall i : 1 \le i \le n((R_i(ich_{i1} \wedge ich_{i2} \wedge ... \wedge ich_{im}),\ s_1)$$
$$\longrightarrow ((och_{i1},\ och_{i2},\ ...,\ och_{is}),\ s_2))$$

Based on these connect statements, the DS and DM of AMPS is created and initialised. The connect statements are implemented using the nodes definition. When these nodes execute, they make use of the set of input channels and produce values for the set of output channels using guarded processes. Once all the nodes finish executing to produce the intended output, the system will change its state from s_1 to s_2.

This implies that,

$$Communication\_Schema \approx \varepsilon \cup LTS\_Beh_{SACS}$$

Therefore,

$$(\Re_1 x \Re_2 x ... x \Re_m\,,\ s_1) \xrightarrow{\mu} (\Re'_1 x \Re'_2 x ... x \Re'_m\,,\ s_2)$$

$$\approx$$

$$\forall i : 1 \le i \le n((R_i(ich_{i1} \wedge ich_{i2} \wedge ... \wedge ich_{im}),\ s_1)$$
$$\longrightarrow ((och_{i1},\ och_{i2},\ ...,\ och_{is}),\ s_2))$$

Thus for every $(S_1,\ k,\ t)$ in R there exists a $(S_2,\ k,\ t)$ in R as required.

□

3. Conclusion

SACS is a formal specification tool to specify the asynchronous communication in a LIPS program. This is implemented using the Asynchronous Message Passing System (AMPS) which is created during the execution of a LIPS program. We have derived an equivalence theorem to show that the implementation of asynchronous communication using AMPS satisfies its formal specification defined using SACS. The proof is derived by creating a weak bisimilarity relation between the LTS of SACS and AMPS. Having verified their equivalence, SACS and AMPS can be implemented in any asynchronous communicating systems with minimal modifications.

References

[1] A. S. Bavan and E. Illingworth. Design and implementation of reliable point-to-point asynchronous message passing system. In *Proceedings of The 10th International Conference on Computing and Information ICCI '2000, Kuwait, (18th–21st November 2000)*. ICCI, 2000.

[2] A. S. Bavan and E. Illingworth. *A Language for Implementing Parallel and distributed Systems using asynchronous point-to-point communication*. Nova Science Publishers, Inc., Commack, NY, USA, 2001.

[3] A. S. Bavan, E. Illingworth, A. V. S. Rajan, and G. Abeysinghe. Specification of Asynchronous Communicating Systems (SACS). In *Proceedings of IADIS International Conference Applied Computing 2007, Salamanca, Spain (18th–20th February 2007)*. IADIS, 2007.

[4] A. S. Bavan, A. V. S. Rajan, and G. Abeysinghe. Asynchronous message passing architecture for a distributed programming language. In *Proceedings of IADIS International Conference Applied Computing 2007, Salamanca, Spain (18th–20th February 2007)*. IADIS, 2007.

[5] V. Galpin. *Equivalence semantics for concurrency: comparison and application*. PhD thesis, ECS-LFCS-98-397, Department of Computer Science, University of Edinburgh, 1998.

[6] D. Gray. *Introduction to the Formal Design of Real-Time Systems*. Springer, Paperback, 2000.

[7] C. A. R. Hoare. Communicating sequential processes. *Communications of the Association of Computing Machinery*, 21(8):666–677, 1978.

[8] L. Logrippo, T. Melanchuck, and R. J. D. Wors. An algebraic specification language lotos: An industrial experience. In *Proc. ACM SIGSOFT Int'l. Workshop on Formal Methods in Software Development*, pages 59–66, 1990.

[9] R. Milner. *A Calculus of Communicating Systems*. Springer Verlag, New York Inc.,, 1982.

[10] A. V. S. Rajan, A. S. Bavan, and G. Abeysinghe. *Semantics for an Asynchronous Message Passing System*, volume XVIII of *Advances and Innovations in Systems, Computing Sciences and Software Engineering*. Springer, 2007.

[11] A. V. S. Rajan, A. S. Bavan, and G. Abeysinghe. *Semantics for the Specification of Asynchronous Communicating Systems (SACS)*, pages 33–38. Advances in Computer and Information Sciences and Engineering. Springer Netherlands, 2008.

[12] E. Tuosto. *Non Functional Aspects of Wide area Network Programming*. PhD thesis, Dipartimento di Informatica, Univ. Pisa, 2003.

A Framework for Decision Support Systems Based on Zachman Framework

S. Shervin Ostadzadeh
Computer Engineering Department,
Faculty of Engineering,
Science & Research Branch of
Islamic Azad University, Tehran, Iran
ostadzadeh@sr.iau.ac.ir

Jafar Habibi
IT Department,
Faculty of Computer Engineering,
Sharif University of Technology,
Tehran, Iran
jhabibi@sharif.edu

S. Arash Ostadzadeh
Computer Engineering Laboratory,
Microelectronics and CE Department,
Delft University of Technology,
Delft, The Netherlands
arash@ce.et.tudelft.nl

Abstract - **Recent challenges have brought about an inevitable tendency for enterprises to lunge towards organizing their information activities in a comprehensive way. In this respect, Enterprise Architecture (EA) has proven to be the leading option for development and maintenance of information systems. EA clearly provides a thorough outline of the whole information system comprising an enterprise. To establish such an outline, a logical framework needs to be laid upon the entire information system. Zachman framework (ZF) has been widely accepted as a standard scheme for identifying and organizing descriptive representations that have critical roles in enterprise management and system development. In this paper, we propose a framework based on ZF for Decision Support Systems (DSS). Furthermore, a modeling approach based on Model-Driven Architecture (MDA) is utilized to obtain compatible models for all cells in the framework. The efficiency of the proposed framework is examined through a case study.**

Keywords: Decision Support System, Zachman Framework, Enterprise Architecture Framework, Model Driven Architecture.

I. Introduction

Enterprises face a lot of challenges nowadays, however, fully utilizing and integrating rapid improvements of information technology remains one important challenge. This is due to the fact that proper development of information systems would have substantial impact on tuning activities in enterprises. The dynamic nature of activities in an enterprise involves changes in corresponding information systems and this imposes considerable maintenance costs to enterprises. These costs usually make managers reluctant to commit to regular changes, degrading gradual developments in enterprises. Enterprise Architecture (EA) is introduced to specifically address this problem by organizing business processes and information technology infrastructure reflecting the integration and standardization requirements of an enterprise operating model [13].

To have a thorough understanding of activities going on in an enterprise, one should perceive different views of components contained in that enterprise. Each view represents an architecture and refers to one entity. For example, data architecture represents a single view of the data structures used by an enterprise and its applications. They contain descriptions of data in storage and data in motion; descriptions of data stores, data groups and data items; and mappings of those data artifacts to data qualities, applications, locations, etc. Enterprise Architecture represents a collection of views from different perspectives, which is expected to create a comprehensive overview of the entire enterprise when put together. It should be noted that handling this large amount of information is quite challenging and needs a well-developed framework. To address this problem, various enterprise architecture frameworks have emerged, such as, FEAF [10], TEAF [11], and C4ISR [12]. Zachman Framework (ZF) [7] originally proposed by John Zachman, is often referenced as a standard approach for expressing the basic elements of enterprise architecture.

In this paper we propose a framework for Decision Support Systems (DSS) inspired by ZF. Previously, Sprague and Watson [5] proposed a framework for DSS. Their framework has three main levels: (1) Technology levels, (2) People involved, and (3) The developmental approach. However, we opted to apply an enterprise architecture framework for DSS. This resulted in a comprehensive structure for all aspects of DSS. ZF is widely accepted as the main framework in EA. Compared to other proposed frameworks, it has evident advantages to list [1]: (1) using well-defined perspectives, (2) using comprehensive abstracts, (3) normality, and (4) extensive usage in practice. They were the motivations for ZF adoption in our work, nevertheless; there are challenges to overcome, among them is the absence of an integrated language to model cells in the framework [1]. MDA is mature enough to address various aspects of EA [2,15,16]. So, we aspire to resolve the problem by utilizing MDA concepts to model all cells in our proposed framework.

The rest of this paper is organized as follows. In Section 2, we introduce some basic concepts and principles. We present the proposed framework in section 3. The framework's cell modeling concepts are discussed in section 4. Section 5 presents a case study. Finally, we make conclusions and suggest some comments for future works.

II. Basic Concepts

In this section we briefly introduce some basic concepts and principles. We believe these remarks can help readers to clearly understand what we mean by the concepts that are used throughout the paper.

K. Elleithy (ed.), *Advanced Techniques in Computing Sciences and Software Engineering*,
DOI 10.1007/978-90-481-3660-5_86,

A. Decision Support Systems (DSS)

Due to existence of many approaches in decision-making and the wide range of domains in which decisions are made, the concept of decision support system is very broad. The term decision support system has been used in many different ways and has been defined in various ways depending on the author's point of view [6]. Finlay [4] defines a DSS as a computer-based system that aids the process of decision making. Turban [3] defines it more specifically as an interactive, flexible, and adaptable computer-based information system, especially developed for supporting the solution of a non-structured management problem for improved decision making that utilizes data, provides an easy-to-use interface, and allows for the decision maker's own insights.

B. Zachman Framework (ZF)

The Zachman Framework is a framework for enterprise architecture, which provides a formal and highly structured way of defining an enterprise. In 1987, the initial framework, titled "*A Framework for Information Systems Architecture*", was proposed by John A. Zachman [8]. An update of the 1987 original work was extended and renamed to the Zachman Framework in the 1990s. In essence, the Framework is a two dimensional matrix consisting of 6 rows and 6 columns.

Each row represents a total view of the enterprise from a particular perspective. These rows starting from the top include: Planner's View (Scope), Owner's View (Enterprise or Business Model), Designer's View (Information Systems Model), Builder's View (Technology Model), Subcontractor's View (Detail Representation), and Actual System View (The Functioning Enterprise).

The columns describe various abstractions that define each perspective. These abstractions are based on six questions that one usually asks when s/he wants to understand an entity. The columns include: The Data Description (What?), The Function Description (How?), The Network Description (Where?), The People Description (Who?), The Time Description (When?), The Motivation Description (Why?). Further information and cell definitions of ZF can be found in [9].

C. Model-Driven Architecture (MDA)

The Object Management Group (OMG) founded in 1989 by eleven companies, originally aimed at setting standards for distributed object-oriented systems with currently more that 800 companies as members, is focusing on modeling and model-based standards. OMG evolved towards modeling standards by creating the standard for Unified Modeling Language (UML) followed by related standards for Meta-Object Facility (MOF), XML Metadata Interchange (XMI), and MOF Query/Views/Transformation (QVT). These together provide the foundation for MDA. More information can be found in [17-19].

Model-driven architecture is a software design approach for the development of software systems. It provides a set of guidelines for the structuring of specifications, which are expressed as models. These specifications will lead the industry towards interoperable, reusable, and portable software components.

MDA separates certain key models of a system, and brings a consistent structure to these models [17]. A Computation Independent Model (CIM) is a view of a system from the computation independent viewpoint. A Platform Independent Model (PIM) is a view of a system from the platform independent viewpoint. A PIM exhibits a specified degree of platform independence so as to be suitable for use with a number of different platforms of similar type. A Platform Specific Model (PSM) is a view of a system from the platform specific viewpoint. Figure 1 depicts how these models will be created in an MDA development process.

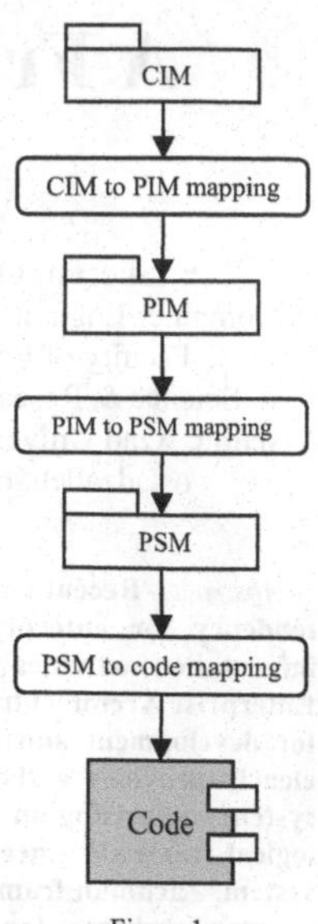

Figure 1.
MDA Process [1]

III. A Framework For DSS

As mentioned earlier, currently there are some frameworks for DSS. However, none of them is based on EA. In this section, we introduce our framework proposal based on ZF intended to support all aspects of a decision support system as a whole.

Figure 2 depicts the proposed framework schema. Similar to ZF, it is a two dimensions matrix. The columns are the same as ZF. They are based on six basic interrogatives that are asked to understand an entity. The columns include:

- ***Data (What is it made of?):*** This focuses on the material composition of the information. In the case of DSS, it focuses on data that can be used for decision-making.
- ***Function (How does it work?):*** This focuses on the functions or transformations of the information used for decision-making.
- ***Network (Where are the elements?):*** This focuses on the geometry or connectivity of the data used for decision-making.
- ***People (Who does what work?):*** This focuses on the people and the manuals and the operating instructions or models one uses to perform their decision-making.
- ***Time (When do things happen?):*** This focuses on the life cycles, timings and schedules used for decision-making.
- ***Motivation (Why do things happen?):*** This focuses on goals, plans and rules that prescribe policies and ends that guide an organization for decision-making.

The rows represent various perspectives of those who are involved in decision-making. These perspectives, starting from the bottom, are:

DSS Framework	*Data*	*Function*	*Network*	*People*	*Time*	*Motivation*
Tactic						
Decision & Management						
Knowledge						
Operation						

Figure 2. A framework for DSS based on Zachman Framework

- ***Operation:*** This is the lowest level of abstraction in the decision framework. It refers to Transaction Processing Systems (TPS) which are used for decision-making in the operational level. This is the perspective of executive staff in an enterprise.
- ***Knowledge:*** This row stands above the Operation layer. In this row, the knowledge that is contained in transactions will be extracted. This is the perspective of an enterprise's expertise.
- **Decision & Management:** Decision & Management row is the first row of decision-making. It stands above the Knowledge layer, and focuses on knowledge controlling, monitoring, and management. It maps knowledge to decision. Decision & Management row is the perspective of an enterprise managers.
- ***Tactic:*** This is the highest level of abstraction in the framework. It is the second row of decision-making that focuses on strategies for decision-making. This row is the perspective of chief enterprise's managers.

IV. MDA-BASED MODELS FOR THE FRAMEWORK

One of the difficulties we are facing to use ZF, and consequently a framework inspired by it, is the problem of incoherent models [1]. ZF expresses what information must be created for each cell of the framework; however, it does not indicate how this information must be created and which model(s) must be applied to present that information. In addition, a set of coherent and consistent models across the entire cells does not exist. In fact, these can not be considered as ZF weak points, since ZF is just a framework, not a methodology. Anyhow, an architect who uses ZF has to overcome these problems. To address the mentioned challenges, which our framework will undoubtedly face, we have investigated the use of MDA in a similar way that we previously suggested in [1]. This can improve the usability of the proposed framework for the developers who intend to put it into practice.

Now, let's examine the MDA-based approach to model the framework. Figure 3 depicts the initial scheme with respect to the mapping between the rows in DSS framework and different model types in MDA. In the primary plan, CIM is employed for the first row. As a result, all the models related to the cells of the first row (tactic), to be proposed later; have to be described in the CIM level. In the second row, we use PIM, therefore all the models for the decision row cells have to be in PIM level. PSM is recommended for the third and forth rows and all the cells constituting the knowledge and operation rows will be designed in the PSM level.

CWM is the best option for Data column modeling; however it should be noted that using simpler models is recommended for the higher rows in the framework. We use Business profile for the tactic/data cell. Relations between data are also demonstrated using Specialization /Generalization, Aggregation, and Inclusion relationships. In order to model the remaining rows we adopt CWM. Since the stakeholders of these rows are familiar with formal methods of software modeling, it is expected that no problem will occur using CWM. However, it should be noted that the CWM models for the second row have to be in the PIM level and for the 3rd and 4th rows they should be in the PSM level.

We prefer employing BPDM to model the Function column. As with the Data column, we try to use simpler models for the higher rows in the framework. We utilize Business Use Case diagram for the tactic/function cell. Each process in the enterprise is modeled with a Use Case and organized with Aggregation and Inclusion relationships. Activity diagram is also used to present the workflow of processes. We utilize BPDM for the decision/function cell. Since knowledge/function cell is considered as design of applications, we employ UML 2.x [20] diagrams to model this cell. Finally, for the operation/function cell we refer to PSM-specific profiles. Considering chosen technologies, it's possible to use WSM and CORBA, EJB, and .Net profiles.

The tactic/network cell is modeled with the Organization Unit stereotype contained in UML Packages. Relations between organization sections are depicted using the Dependency relationships and their essences are clarified with extra comments. To model the decision/network cell, we employ the EDOC profile. It enables us to model different

DSS Framework	*Data*	*Function*	*Network*	*People*	*Time*	*Motivation*
Tactic	CIM					
Decision	PIM					
Knowledge	PSM					
Operation						

Figure 3. Applying MDA models to the DSS Framework rows

DSS Framework	*Data*	*Function*	*Network*	*People*	*Time*	*Motivation*
Tactic	Business profile	Business Use Case	Organization Unit stereotype	Business profile/use case	Timing diagram	BSBR BMM
Decision & Management	CWM	BPDM	EDOC	BPDM	Timing & State diagrams	PRR
Knowledge	CWM	UML 2.x	Deployment diagram	Interaction Overview diag.	Timing diagram	OCL
Operation	CWM	CORBA, EJB, .Net, WSM	CORBA, EJB, .Net, WSM	Interaction Overview diag.	Timing diagram	OCL

Figure 4. An MDA-based modeling approach for the DSS Framework

parts of organizational computations which are distributed within an enterprise. Deployment diagram is used for the knowledge/network cell. This gives us the opportunity to model hardware resources and their relations. We also utilize WSM and PSM-specific profiles (CORBA, EJB, and .Net) for the operation/network cell.

In order to model the People column in the first row, we employ the Use Case diagrams along with the Business profile. Organization structure is depicted with nested Packages and Organization Unit stereotype. Worker, Case Worker, Internal Worker, External Actor or Work Unit symbols are used for organizational roles. Communication diagram is also employed to show the workflow. We use BPDM for the decision/people cell. Interaction Overview diagram included in UML 2.x is utilized to model the knowledge/people cell. Unfortunately, modeling the operation/people cell causes a problem since no particular diagram in MDA is suitable for defining the operational characteristic of the workflow. However, utilizing the method proposed in the knowledge/people cell, and with the help of extra comments, it's possible to model this cell to some extent.

Time column can be modeled with Timing diagram and Scheduling profile which are presented in UML 2.x. In the past, proposed solutions often utilized Sequence diagram to model this column, however it is not possible to correctly demonstrate the duration of an event with this diagram. Using the Timing diagram allows us to specify the exact start and end times and the duration for each decision. In the second row, it is possible to employ the State diagram along with the Timing diagram to show different decision states.

OCL is often recommended for the Motivation column modeling. However, it should be noted that using this language causes trouble in higher rows. The OCL language is a formal language to specify rules and constraints. Its thorough understanding requires familiarity with some concepts of formal specifications in mathematics. This requirement leaves it vague for the managements who are probably unfamiliar with formal concepts. Besides, OCL is designed for stating rules and constraints upon objects, nevertheless the abstraction level related to the higher rows of the framework is beyond the concept of an object and using objects to model these levels is inappropriate [21]. We use Business Motivation Model (BMM) diagrams related to BSBR, to model the tactic/motivation cell. By defining goals, viewpoints, missions, objects, facts, strategies, and so on along with their relations, BMM diagrams try to model the concepts governing the processes within an enterprise. To model the decision/motivation cell, Production Representation Rule (PRR) is employed. This meta-model, let us specify the rules which are defined in the first row in the form of CIM as a presentation of production rules in PIM level. OCL is used to state the rules supervising current objects for the third and fourth rows.

Figure 4 summarizes our proposed method. All models can be integrated, since they are MOF-compliant. This characteristic guarantees that all models used in an MDA system can communicate with every other MOF-compliant model.

V. Case Study

Traditionally, the evaluation of a new solution regarding software engineering problems, is conducted via case studies as it has been proven that these solutions do not fit well into formal verifications. With this respect, a thorough evaluation of a solution can only be verified after extensive usage in real world applications, through which the performance gain of the solution and the pros and cons are cleared.

A *functional architectures* of FAVA project [22] in the I.R.I presidential office was the target of our proposed framework. FAVA architectures consist of two main parts: *Fundamental Architectures* and *Functional Architectures.*

One of the functional architectures is the Chief Management Functional Architecture,on which this case study is based. FAVA project is designed using CHAM [23]. We do not intend to specify the details of CHAM, however it is worthwhile to note that in CHAM, the architecture design is done in two phases: Macro Architecture design and Detailed Architecture design.

In the macro architecture design phase we specify the architecture essence. This phase for the *Chief Management Functional Architecture* is presented in [24].

In the detailed architecture design phase we clearly describe the architecture details and the implementation methods. In this phase, we were required to use one of the enterprise architecture frameworks and we opted to utilize the DSS framework based on our proposed method.

Figure 5 depicts a part of Tactic models. We use BMM to model means-end. A part of Decision model is shown in Figure 6. It depicts the "Request" entity. Figure 7 and 8 present parts of Knowledge and Operation models, respectively. Complete detailed architecture design of the Chief Management Functional Architecture is presented in [25].

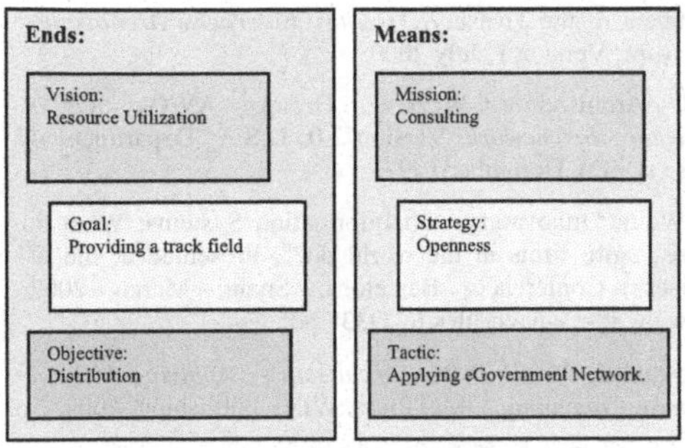

Ends:
Vision:
Resource Utilization
Goal:
Providing a track field
Objective:
Distribution
Means:
Mission:
Consulting
Strategy:
Openness
Tactic:
Applying eGovernment Network.

Figure 5: Part of Tactic models

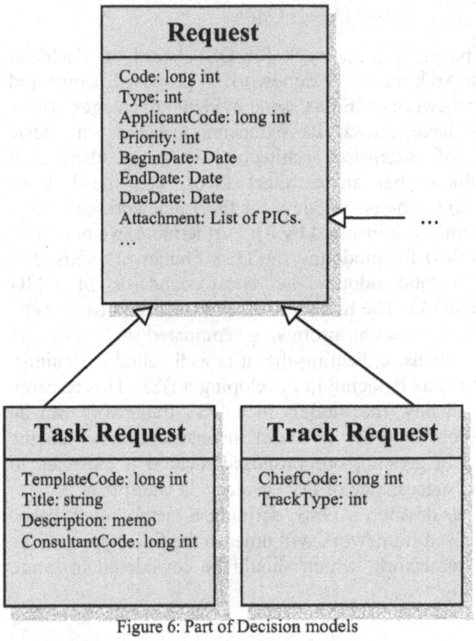

Request
Code: long int
Type: int
ApplicantCode: long int
Priority: int
BeginDate: Date
EndDate: Date
DueDate: Date
Attachment: List of PICs.
…
…
Task Request
TemplateCode: long int
Title: string
Description: memo
ConsultantCode: long int
Track Request
ChiefCode: long int
TrackType: int

Figure 6: Part of Decision models

Expert
Study
Send to
Consultant
Report
Send to
Applicant
Send to
Chief

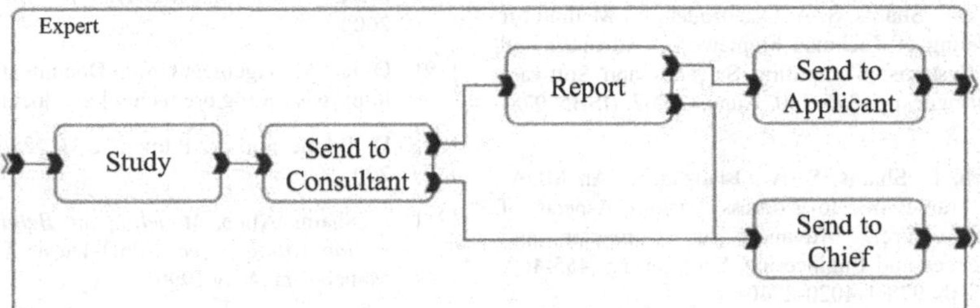

Figure 7: Part of Knowledge models

execution environment
Web Application Server
UI Layer
Business Logic Layer
Database Layer
execution environment
Portal Server
Portal MVC
Portal WSRP
SOAP

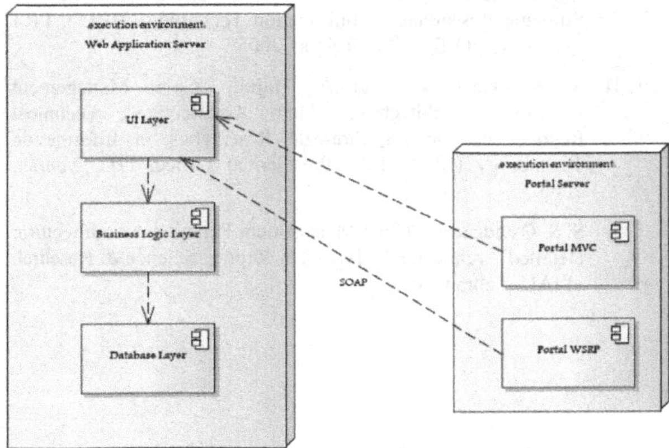

Figure 8: Part of Operation models

VI. Conclusions

In this paper, a framework for DSS based on Zachman (Enterprise Architecture) Framework is proposed. Compared to other frameworks, ZF has some evident advantages. These advantages have caused its extensive usage as the basic framework of enterprise architecture However, there is a major problem that an architect faces, i.e., the lack of consistent and coherent models for the framework cells (and also any framework inspired by it). Furthermore, we presented a novel method for modeling the DSS Framework cells. The presented method adopts the latest standards of OMG modeling (MDA). The method was evaluated in a case study. We found the results promising compared to previously proposed methods, indicating that it is well suited to diminish the problems one is facing in developing a DSS. This research demonstrated how the models in a DSS framework can be made coherent in order to avoid inconsistencies within the framework. In general, our proposed method is expected to increase the success rate of DSS projects in enterprises.

In order to develop a DSS, defining a methodology based on the proposed framework will unleash the full support of the presented framework, which should be considered in future work.

References

[1] S. S. Ostadzadeh, F. Shams, S. A. Ostadzadeh, "A Method for Consistent Modeling of Zachman Framework," Advances and Innovations in Systems, Computing Sciences and Software Engineering, Springer, pp. 375-380, August 2007. (ISBN 978-1-4020-6263-6)

[2] S. S. Ostadzadeh, F. Shams, S. A. Ostadzadeh, "An MDA-Based Generic Framework to Address Various Aspects of Enterprise Architecture," Advances in Computer and Information Sciences and Engineering, Springer, pp. 455-460, August 2008. (ISBN 978-1-4020-8740-0)

[3] E. Turban, *Decision support and expert systems: management support systems*, Englewood Cliffs, N.J., Prentice Hall, 1995.

[4] P. N. Finlay, *Introducing decision support systems*, Oxford, UK Cambridge, Mass., NCC Blackwell, Blackwell Publishers, 1994.

[5] R. H. Sprague and H. J. Watson, *Decision support systems: putting theory into practice*, Englewood Clifts, N.J., Prentice Hall, 1993.

[6] M. J. Druzdzel, and R. R. Flynn, *Decision Support Systems*, Encyclopedia of Library and Information Science, A. Kent, Marcel Dekker, Inc., 1999.

[7] J. A. Zachman, *The Zachman Framework: A Primer for Enterprise Engineering and Manufacturing*, 2003.

[8] J. A. Zachman, "A Framework for Information Systems Architecture", IBM Systems Journal, Vol. 26, No. 3, 1987.

[9] J. A. Zachman, "The Framework for Enterprise Architecture – Cell Definitions", ZIFA, 2003.

[10] Chief Information Officers (CIO) Council, Federal Enterprise Architecture Framework, Version 1.1, September 1999.

[11] Department of the Treasury, *Treasury Enterprise Architecture Framework*, Version 1, July 2000.

[12] C4ISR Architecture Working Group (AWG), *C4ISR Architecture Framework*, Version 2.0, U.S.A. Department of Defense (DoD), December 1997.

[13] Peter Weill, "Innovating with Information Systems: What do the most agile firms in the world do?", Presented at the 6th e-Business Conference, Barcelona, Spain, March 2007. http://www.iese.edu/en/files/6_29338.pdf

[14] D. S. Frankel, *Model Driven Architecture: Applying MDA to Enterprise Computing*, OMG Press, Wiley Publishing, 2003.

[15] G. Booch, B. Brown, S. Iyengar, J. Rumbaugh, and B. Selic, "An MDA Manifesto", MDA Journal, May 2004.

[16] D. D'Souza, "Model-Driven Architecture and Integration", Kinetium, March 2002.

[17] MDA Guide, OMG document, 2003. http://www.omg.org/

[18] S. J. Mellor, K. Scott, A. Uhl, and D. Weise, *MDA Distilled: Principles of Model-Driven Architecture*, Addison Wesley, 2004.

[19] Object Management Group Document, http://www.omg.org/technology/documents/

[20] D. Pilone and N. Pitman, *UML 2.0 in a Nutshell*, O'Reilly, 2005.

[21] F. Shams Aliee, *Modeling the Behavior of Processes Using Collaborating Objects*, PhD Thesis, University of Manchester, Manchester, May 1996.

[22] A. Majidi, "FAVA: Strategic Plan & Architecture", Technical Report, I.R.I Presidential Office, ITC, Tehran, 2004.

[23] A. Majidi, "CHAM: National Framework & Methodology for Macro/Micro Systems", Technical Report, Institute for Strategic Researches in Information Technology (IRIT), I.R.I Presidential Office, ITC, Tehran, 2005.

[24] S. S. Ostadzadeh and A. Majidi, "Chief Management Functional Architecture: Macro Architecture", Technical Report, Institute for Strategic Researches in Information Technology (IRIT), I.R.I Presidential Office, ITC, Tehran, 2006.

[25] S. S. Ostadzadeh, "Chief Management Functional Architecture: Detailed Architecture", Technical Report, Science & Research of IAU, Tehran, 2006.

In – line determination of heat transfer coefficients in a plate heat exchanger

S. Silva Sotelo [a*], R. J. Romero Domínguez [b]

[a] Posgrado en Ingeniería y Ciencias Aplicadas, Centro de Investigación en Ingeniería y Ciencias Aplicadas
[b] Centro de Investigación en Ingeniería y Ciencias Aplicadas, Universidad Autónoma del Estado de Morelos, Av. Universidad 1001, Col. Chamilpa, Cuernavaca Morelos, México. C.P. 62209
Tel.Fax +52(777)3297084.
Corresponding author: *sotsil_silva@uaem.mx*

***Abstract*—this paper shows an in – line determination of heat transfer coefficients in a plate heat exchanger. Water and aqueous working solution of lithium bromide + ethylene glycol are considered. Heat transfer coefficients are calculated for both fluids. "Type T" thermocouples were used for monitoring the wall temperature in a plate heat exchanger, which is one of the main components in an absorption system. Commercial software Agilent HP Vee Pro 7.5 was used for monitoring the temperatures and for the determination of the heat transfer coefficients. There are not previous works for heat transfer coefficients for the working solution used in this work.**

Index terms: absorption system, heat transformer, heat transfer coefficients, in line determination, plate heat exchanger.

I. INTRODUCTION

For improving the performance of absorption systems is necessary a better knowledge of the heat transfer mechanisms. Absorption systems could be used in air conditioned or upgrade energy systems (absorption heat transformers). These systems as the advantage of diminish the electric energy consumption and CO_2 emissions; compared with traditional compression systems. An Absorption Heat Transformers (AHT) allows to use alternative energy (like solar or geothermal energy), increasing the temperature from this source to a higher useful level, which can be applied in different processes. One of the main components in absorption systems is the steam generator. For this work heat transfer coefficients in a steam generator are calculated.

II. STEAM GENERATOR

An AHT consists of an evaporator, a condenser, a generator, and an absorber. Fig. 1 shows a schematic diagram of an AHT. A working solution (water – lithium bromide + ethylene glycol) flows between absorber and generator. This solution is called water – Carrol™, it was created by Carrier Co. for air conditioner systems.

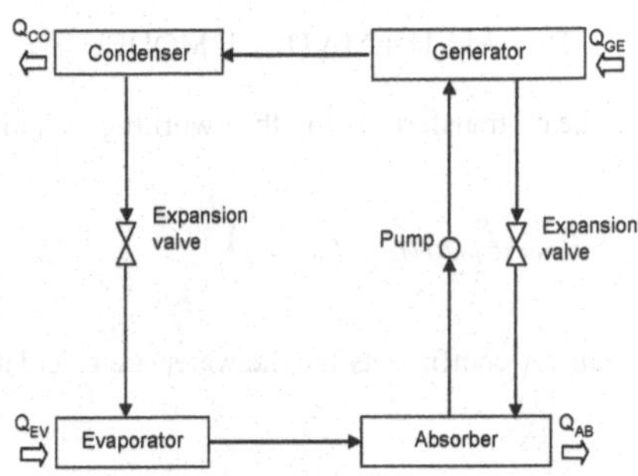

Fig. 1. Schematic diagram of an AHT.

Diluted solution leaves the absorber and goes to the generator inlet. A part of the water is evaporating and a concentrated solution flows to return at the absorber. External hot water is used to heat the working solution (Fig. 2).

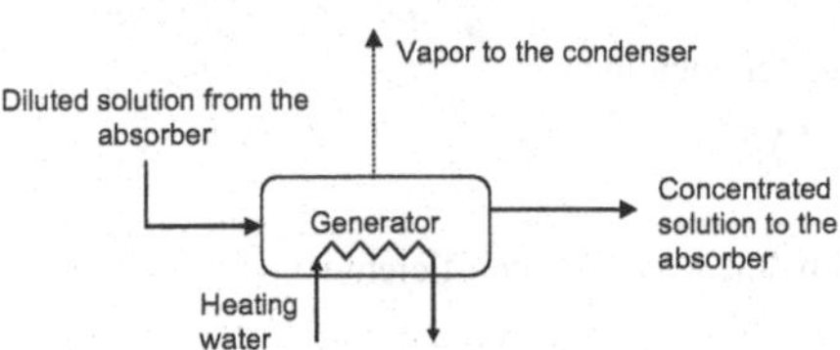

Fig. 2. Schematic diagram of the steam generator.

III. EXPERIMENTAL SET – UP

The generator is a plate heat exchanger made of stainless steel 316, of seven plates, thermal capacity by design of 1000W. Type T thermocouples were used, in one plate are placed 8 thermocouples. The thermocouples are connected to a data logger for monitoring the temperature every ten seconds.

K. Elleithy (ed.), *Advanced Techniques in Computing Sciences and Software Engineering*,
DOI 10.1007/978-90-481-3660-5_87, © Springer Science+Business Media B.V. 2010

The temperature measures are sending in line to an Excel® file. Fig. 3 shows the location of the thermocouples.

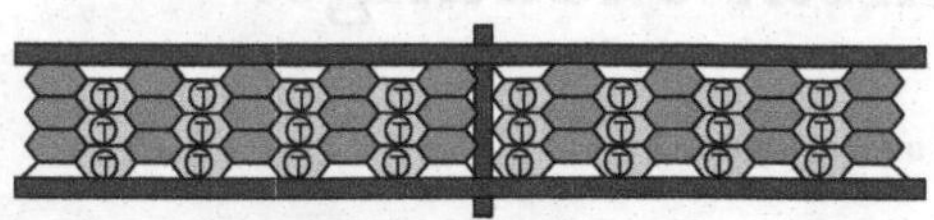

Fig. 3. Location of the thermocouples.

A channel is the space between two plates. Green color represents the three channels were working solution flows, blue color are the other three channels were heating water flows, in counter current with the working solution.

IV. MATHEMATICAL MODEL

The total heat transferred to the working solution is calculated by:

$$Q = m_{wat} Cp_{wat} (T_{wat(out)} - T_{wat(in)}) \tag{1}$$

The heat transfer coefficients for the water are calculated by [1]:

$$h_{wat} = 0.295 \left(\frac{k_{wat}}{D_h} \right) \mathrm{Re}^{0.64} \mathrm{Pr}^{0.32} \left(\frac{\pi}{2} - \beta \right)^{0.09} \tag{2}$$

Where k, is thermal conductivity, D_h is hydraulic diameter and β is the angle of the plate of 30°.

The transfer equation is:

$$q = UA\Delta T_{lm} \tag{3}$$

The global heat transfer coefficient is:

$$U = \frac{1}{\frac{1}{h_{wat}} + \frac{\Delta Z}{k_{SS}} + \frac{1}{h_{sol}}} \tag{4}$$

Thermal resistance of the wall is negligible due to its small thickness; the equation (4) reduces to:

$$U = \frac{1}{\frac{1}{h_{wat}} + \frac{1}{h_{sol}}} \tag{5}$$

The log mean temperature is:

$$\Delta T_{lm} = \frac{(T_{wat(in)} - T_{sol(out)}) - (T_{wat(out)} - T_{sol(in)})}{\ln \frac{T_{wat(in)} - T_{liq(out)}}{T_{wat(out)} - T_{liq(in)}}} \tag{6}$$

The heat transfer coefficient for the working solution is obtained for the Eq. (5). The simulation and the data acquisition are programmed in the commercial software HP Vee Pro, by Agilent Technologies.

V. RESULTS AND DISCUSSION

Experimental data were obtained in line, for steady state. Fig. 4 shows a plot of the thermocouples measures along the plate. Its length is 0.40m. Fig. 5 shows the heat transfer coefficient for the working solution against the tube length. It can be observed that the values are between 1.46 and 1.53. Those coefficients increase according to the heating of the solution.

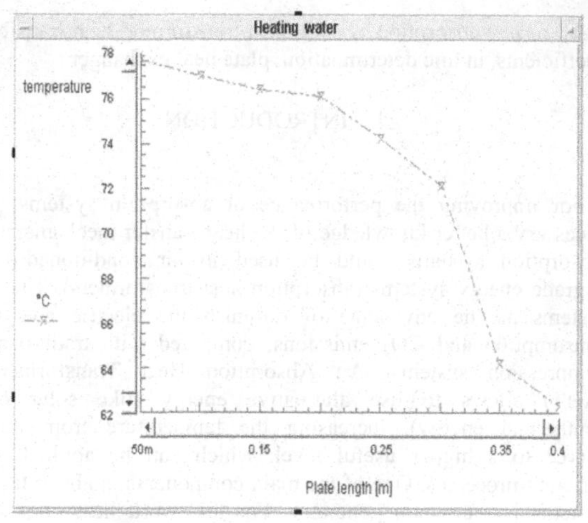

Fig. 4. Temperature in the plate.

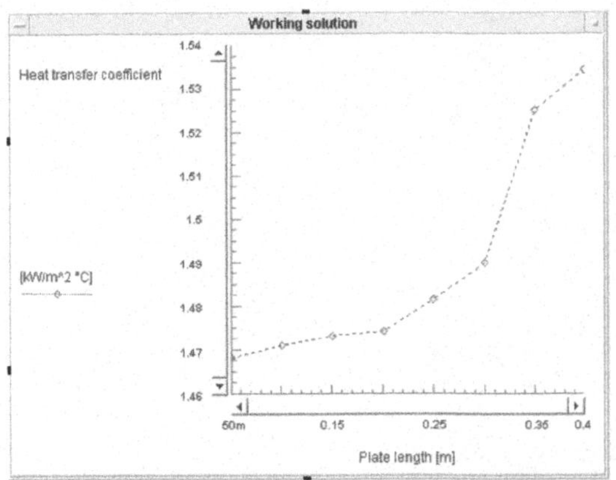

Fig. 5. Heat transfer coefficients for the working solution.

For this paper, 58% weight concentration of Carrol™ was tested. The heat transfer coefficient will be modified by a change in the concentration. There are not previous works about heat transfer coefficients for this working solution. In previous works experimental data were obtained for the solution water – lithium bromide [2], which is one of the more common solutions used in AHT. For Carrol™ 52% concentration was obtained heat transfer coefficients between 1.1 and 1.4 kW/m^2 °C. One of the advantages of the solution water - Carrol™ is its higher solubility, compared to water lithium – bromide [3]. This last has the disadvantage of crystallization risk at high concentrations, water – Carrol™ allows a bigger range of concentration for operation, without the risk of crystallization, above 75% weight concentration.

Fig. 6 shows the heat transfer coefficients obtained for heating water, it can be observed that these obtained values are higher that values for the working solution.

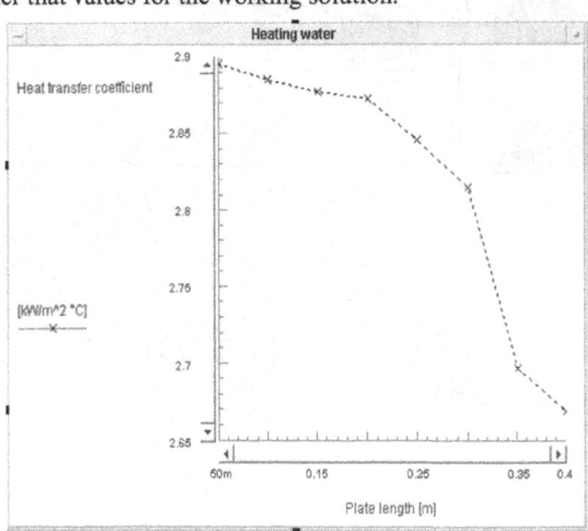

Fig. 6. Heat transfer coefficients for the heating water.

Fig. 7 shows the detail program presented in this work, however is possible to create a friendly panel for final user, this is presented in Fig. 8.

VI. CONCLUSIONS

A mathematical model for in – line determination of heat transfer coefficients was presented. The experimental set – up allows the measure of temperatures along one plate of a heat exchanger, 8 thermocouples were placed. The experimental data was obtained for one water - Carrol™ concentration of 58% weight. The heat transfer coefficients for the heating water are higher than the coefficients for the working solution. In future works it will be obtained experimental data for several concentrations.

VII. NOMENCLATURE

A	*Area [m^2]*
Cp	*specific heat [kJ/Kg °C]*
D_h	*Hydraulic diameter [m]*
h	*Heat transfer coefficient [kW/m^2 °C]*
k	*Thermal conductivity [kW/m^2 °C]*
m	*Mass flow rate [kg/s]*
Nu	*Nusselt number [dimensionless]*
Pr	*Prandtl number [dimensionless]*
Q	*Heat flux [kW]*
q	*Heat flux [kW/m^2]*
Re	*Reynolds number [dimensionless]*
T	*Temperature [°C]*
U	*Global heat transfer coefficient [kW/m^2 °C]*

Greek symbols

β	*chevron angle [radian]*
ΔZ	*Plate thickness [m]*
ΔT	*Log mean temperature difference [°C]*

Subscripts

in	*inlet*
out	*outlet*
sol	*Solution (working solution)*
ss	*Stainless Steel*
wat	*water of the heating system*

VIII. REFERENCES

[1] Experiments on the characteristics of evaporation of R-410A in brazed plate heat exchanger with different geometric configurations, Dong-Hyouck Han, Kyu-Jung Lee, Yoon-Ho Kim, Applied Thermal Engineering 23, 2003, pp. 1209–1225.

[2] W. Rivera, A. Xicale, Heat transfer coefficients in two phase flow for water/lithium bromide solution used in solar absorption refrigeration systems, Solar Energy Materials & Solar cells, 70, 2001, 309 – 320.

[3] R. J. Romero, Sotsil Silva – Sotelo, Comparison of instrumental methods for in-line determination of LiBr concentration in solar absorption thermal systems, Solar Energy Materials And Solar Cells, Vol. 90 ,Issue 15, 2006, pp. 2549 – 2555.

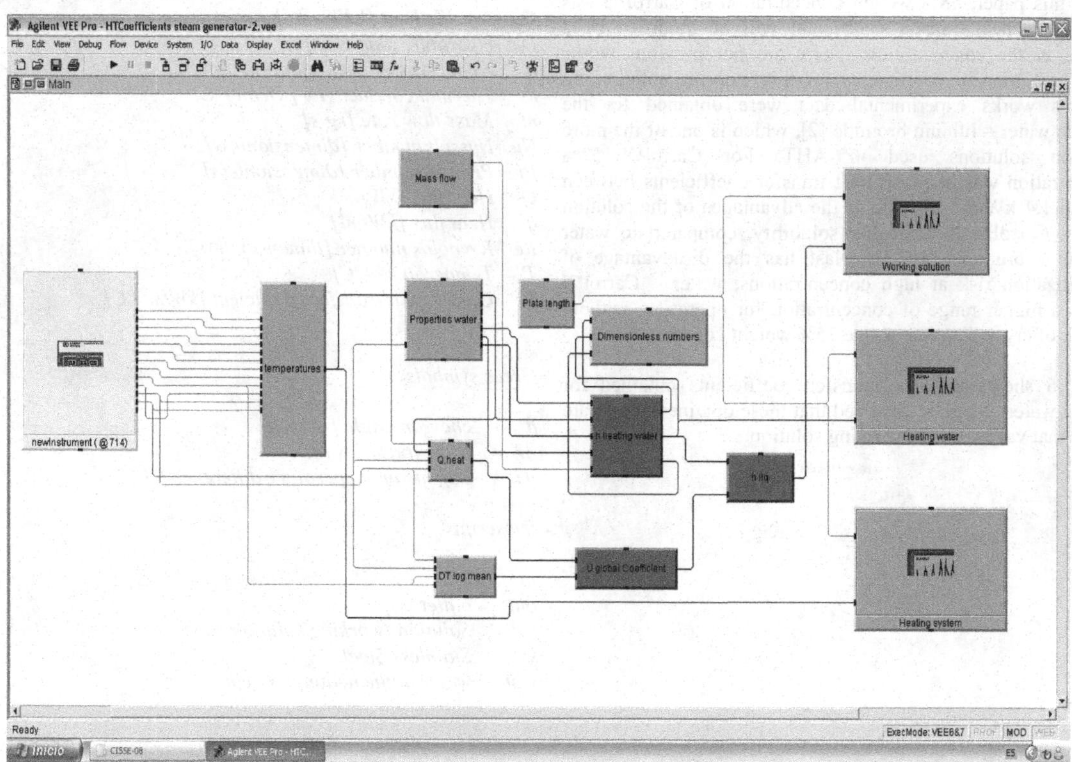

Fig. 7. Program for obtaining heat transfer coefficients

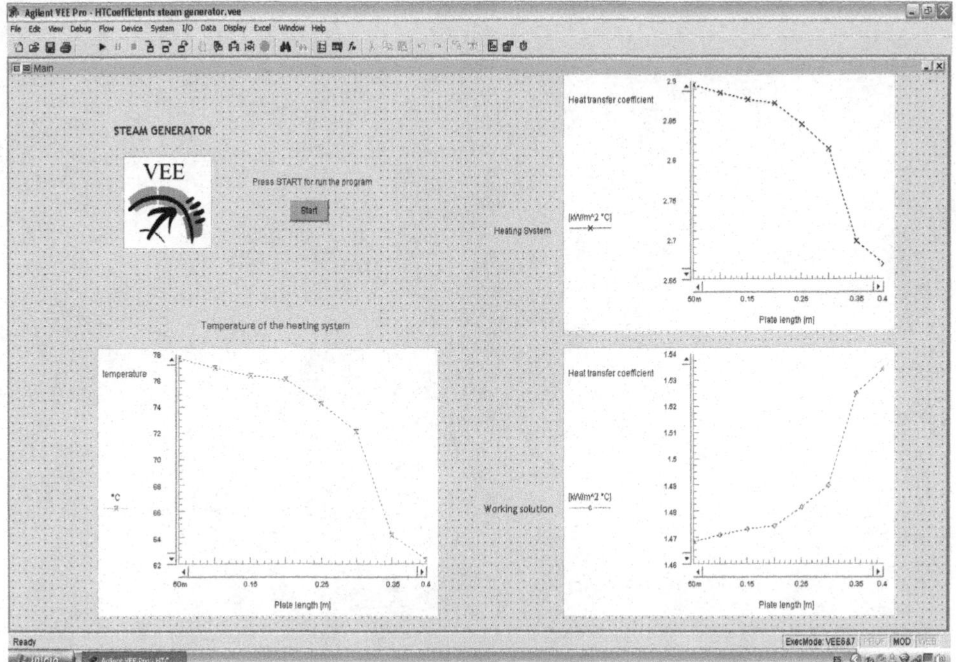

Fig. 8. Panel for the program

Model Based Control Design Using SLPS "Simulink PSpice Interface"

Saeid Moslehpour[#1] ***Member, IEEE,*** **Ercan K. Kulcu**[#2] **and Hisham Alnajjar**[#3]

[#1,#2,#3]Department of Electrical and Computer Engineering
University of Hartford, West Hartford, CT, USA
[#1]moslehpou@hartford.edu
[#2]kulcu@hartford.edu
[#3]alnajjar@hartford.edu

***Abstract*—This paper elaborates on the new integration offered with the PSpice SLPS interface and the MATLAB simulink products. SLPS links the two widely used design products, PSpice and Mathwork's Simulink simulator. The SLPS simulation environment supports the substitution of an actual electronic block with an "ideal model", better known as the mathematical simulink model. Thus enabling the designer to identify and correct integration issues of electronics within a system. Moreover, stress audit can be performed by using the PSpice smoke analysis which helps to verify whether the components are working within the manufacturer's safe operating limits. It is invaluable since many companies design and test the electronics separately from the system level. Therefore, integrations usually are not discovered until the prototype level, causing critical time delays in getting a product to the market.**

The system chosen in this case is a model based design of ambulatory rehabilitation suspension system. Ambulatory suspension system [1] is an apparatus that is utilized for the purpose of helping patients during gait rehabilitation and provides safety support during exercise. The device can lift a patient from a sitting position in a wheel chair to a standing position, has the provisions to remove a percentage of the patients' body weight and recognize subtle changes in the elevation. This model was used for mathematical and electrical comparison using the SLPS interface.

***Index Terms*— Design and test in Simulink, Mathematical PSpice Model, Analog PSpice model, SLPS modeling, Comparison of SLPS and Simulink Output.**

I. INTRODUCTION

The PSpice SLPS simulation environment supports the substitution of an actual simulink block with an equivalent analog PSpice electrical circuit. It allows co-simulation with Matlab and Simulink and thereby enables the designer to identify and correct integration issues of electronics within a system. This helps identify errors earlier in the design process, which can save time and money that may otherwise be spent in debugging trial boards within system designs. With the PSpice Smoke option, designers can perform a stress audit to verify that electrical components are operating within the manufactures' safe operating limits or de-rated limits.

Model Based Control Design of Ambulatory Rehabilitation Suspension System and Observer Design for Control System applications are used for testing SLPS software. Model Based Control Design of Ambulatory Rehabilitation System is ongoing research and development in the area of assisted ambulatory rehabilitation at the University of Hartford. Different types of ambulatory system design are studied at the University of Hartford. By each new design, system became better and more efficient. These researches are improved now by using a model-based control design for Ambulatory Suspension System. This prototype provides electrical motion to the X, Y and Z axis. The system allows for large vertical displacement and effortless walking for the patient. The system also provides fall prevention and ensures safety for the patients. Competition between companies and engineers motivate them to produce faster machines with smaller components. Moreover, increasing fuel costs and government regulations require higher efficiency for businesses. This means that companies need higher performing control systems. Thus, Model-based control design provides a solution.

II. DESCRIPTION

Model based control design consists of the following four phases;

1) Developing and analyzing a model to describe a plant.
2) Designing and analyzing a controller for the dynamic systems
3) Simulating the dynamic system, and
4) Developing the controller

K. Elleithy (ed.), *Advanced Techniques in Computing Sciences and Software Engineering*,
DOI 10.1007/978-90-481-3660-5_88, © Springer Science+Business Media B.V. 2010

Control design is a process which involves developing mathematical models that describe a physical system. It analyzes the models to learn about their dynamic characteristics, and creates a controller to achieve certain dynamic characteristics. Control systems contain components that direct, command, and regulate the physical system, also known as the plant. In this setting, the control system refers to the sensors, the controller, and the actuators.

Two types of controllers are used for such systems. These are respectively called, open-loop controller and the closed-loop controller. A closed loop controller is shown in Figure 1.

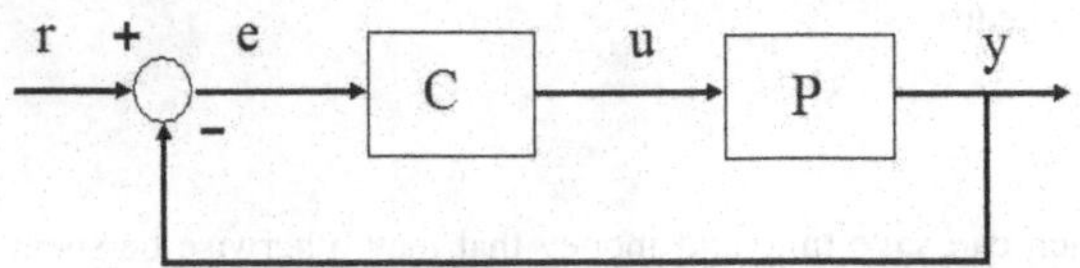

Fig. 1 Closed-loop Control System

The controller operates to make sure that there is no direct connection between the output of the system and input. The open-loop controller is often used in simple processes because of its lack of complexity, especially in systems where feedback is not critical. Nevertheless, to obtain a more accurate or more adaptive control, it is necessary to feed the output of the system back to the inputs of the controller. This type of system is called a closed-loop controller.

Ambulatory control system's purpose is to control the motion of the carriage. This means to observe the patient's horizontal movements in maximum accuracy. Independent analysis for each axis is made by using the same tools as well as the same modeling logic. The Ambulatory system is shown in Fig. 2. When the carrier moves in a way that keeps the angle Ø at a minimum value the control system activates the motor. Tilt sensor measures the control variable (Ø), the control algorithm C(s) processes the difference or error. [2]

Control system parameters are given as: F ; Lifting Force, L ; Length of Cable, Y ; Carriage Position, ΔY ;Walking Distance
Ø ; Tilted Angle, H(s) ; Tilt Sensor, C(s) ; Control Algorithm G(s) ; System of Motor Belt Carrier, PWM ; Output Amplifier

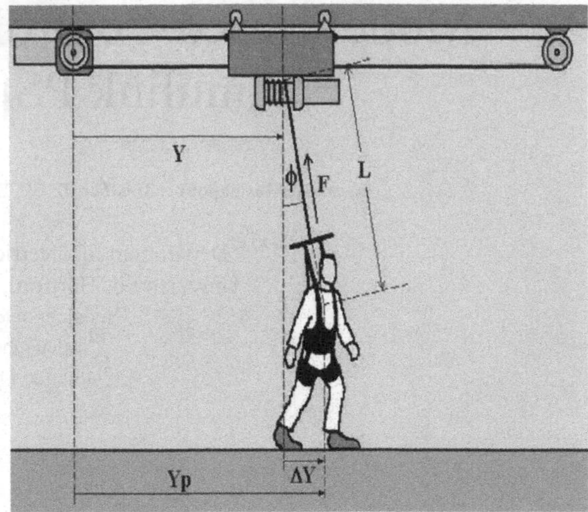

Fig.2- Y Axis Variables of the System

The Control system block diagram is given in Fig.3 above.

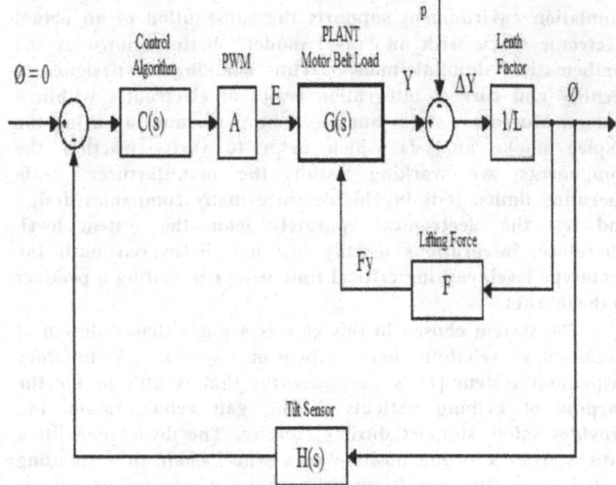

Figure -3 Control System Block Diagram

The input of the closed-loop system is zero and main feedback of the system is provided by the tilt sensor. Fy provides mechanical feedback for inner loop. The specific values of the parameters of the plant model were gathered by experimental estimation, and the data is used to design Simulink and PSpice model. Also, the transfer function of the plant is designed by using the data. . Transfer function of the plant is given as follows.

$$G(s) = \frac{Y(s)}{E(s)} = \frac{17807}{s^4 + 130.3s^3 + 9098s^2 + 681504s}$$

The block diagram of the plant is given in Fig. 4 below.

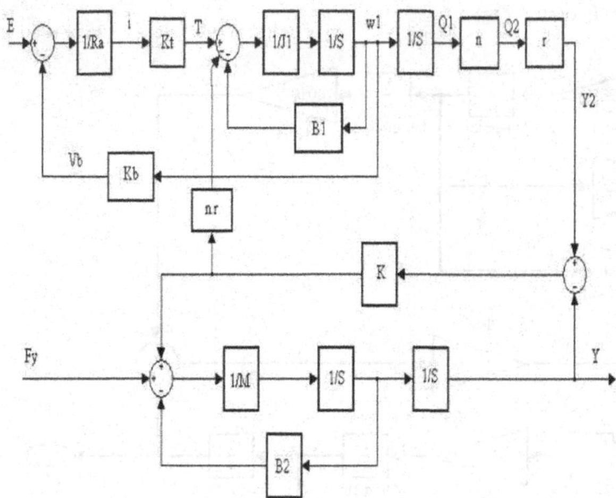

Fig. 4- Block Diagram of the Plant

III. PROCEDURE

A. *DESIGN AND TEST IN SIMULINK*

Fig. 5 is a screenshot of the simulink model of the system

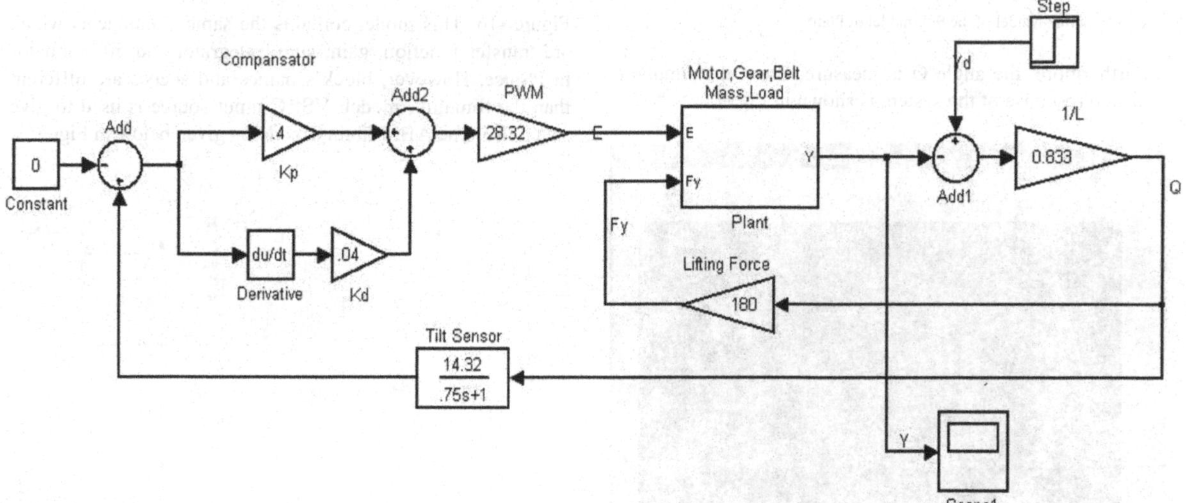

Fig. 5-Simulink Model of the System

Simulink simulates the system plant, sensor, and the other components by using mathematical blocks. Motor-Belt-Load system is created as a subsystem. As a Tilt Sensor Block, H(s) function is used and the PWM is replaced by its gain A = 28.32.

Values of F and L are considered with constant values L = 1.2 m (3.9 ft), and F = 180.5 N. These values are considered average values of the patients. In the simulation program, all values are designed for average patients. The proportional gain Kp is set to 0.4, and the derivative gain Kd was set to 0.04. The simulink model for the plant or Sub Model is shown below in Fig. 6[2] [3]

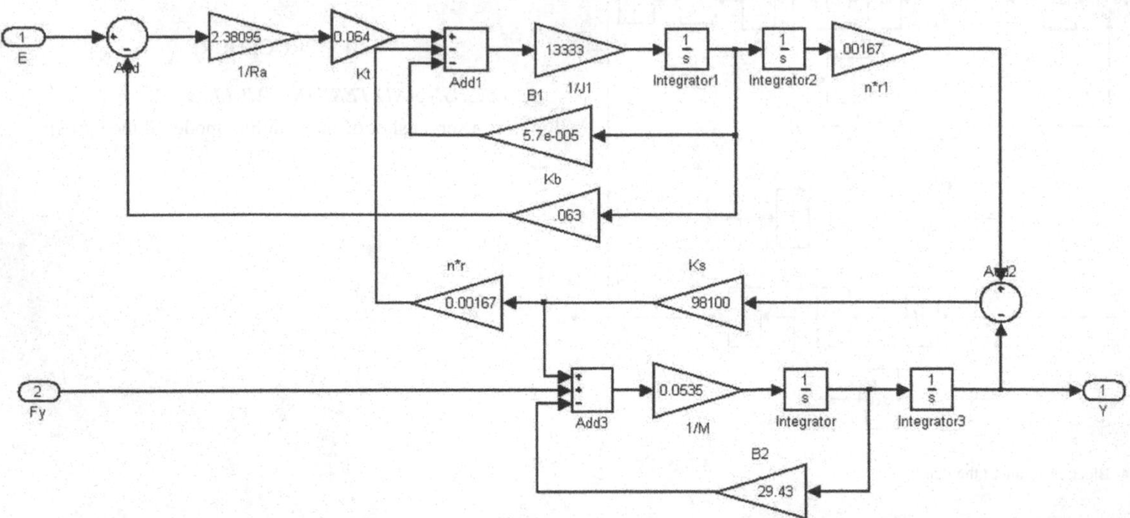

Fig. 6- Simulink Model of the Sub-model or Plant

Furthermore, the angle Ø is measured from the simulation and step response of the system is shown in Fig. 7

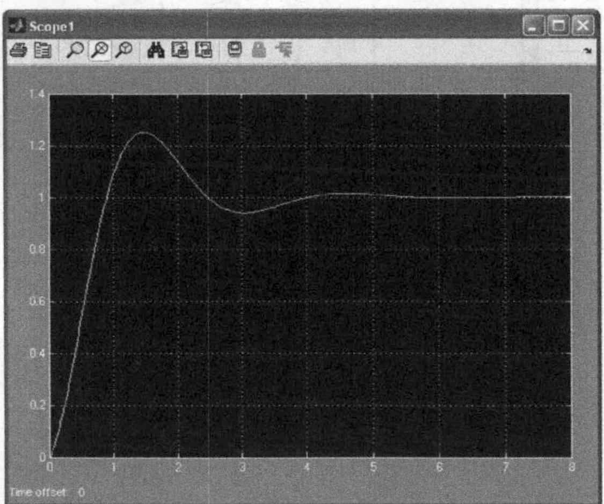

Fig. 6-Simulink Output of the System

Maximum value of step response is around 1.25 and step response becomes stable after 6 seconds.

B. MATHEMATICAL PSPICE MODEL

The Mathematical PSpice model is constructed using the ABM library. Analog behavioral modeling feature of Pspice are used to make flexible descriptions of electronic components in term of transfer function or lookup table. In other words, a mathematical relationship is used to model a circuit segment [4]. PSpice ABM library model is designed by using the same structure used in Simulink as it is shown in Figure -16. This model contains the same components which are transfer function, gain, sum, integrator and differentiator in PSpice. However, block's names and shapes are different than the Simulink model. VSRC input source is used to give step pulse. The ABM library model is given below in Fig. 7.

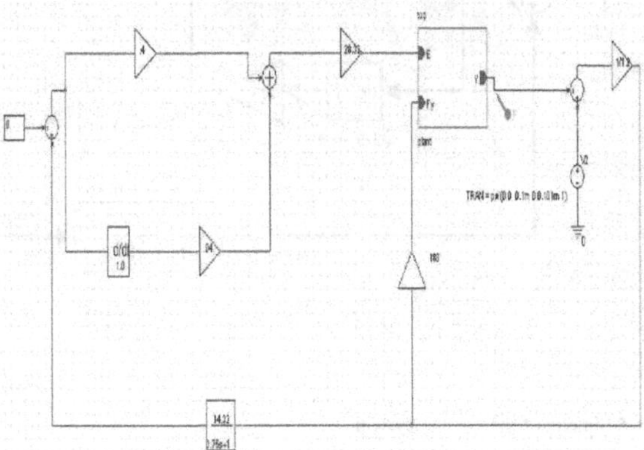

Fig. 7- PSpice ABM Library Model of the System

Output of the PSpice is the same with Simulink output. PSpice ABM contains mathematical model of gain, sum, integrator, transfer functions, and differentiator; therefore, Simulink and PSpice models give the same result and graph. Fig. 8 shows the sub Model.

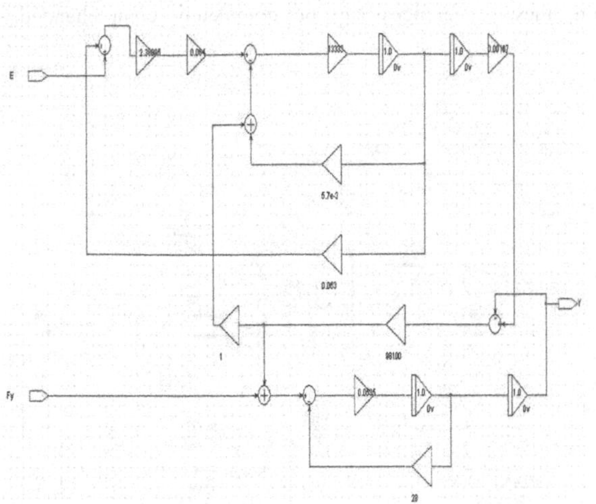

Fig. 8- PSpice ABM Library Model of the Sub-model

Maximum value of the step response is a little bit different than the Simulink output. The output is shown in Fig. 9 below

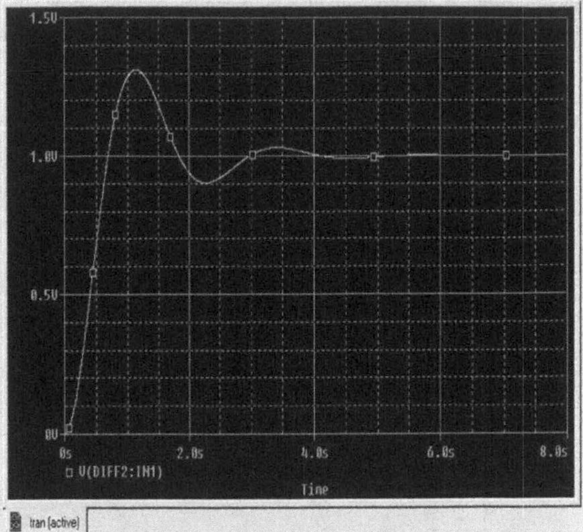

Fig.9- PSpice ABM Library Output

C. ANALOG PSPICE MODEL

Ideal Op-Amp models are used to design analog circuits. Gain, integrator, differentiator, sum, and the transfer function models are replaced by using ideal Op-Amp circuit [5]. PSpice analog circuit is shown in Fig.10.

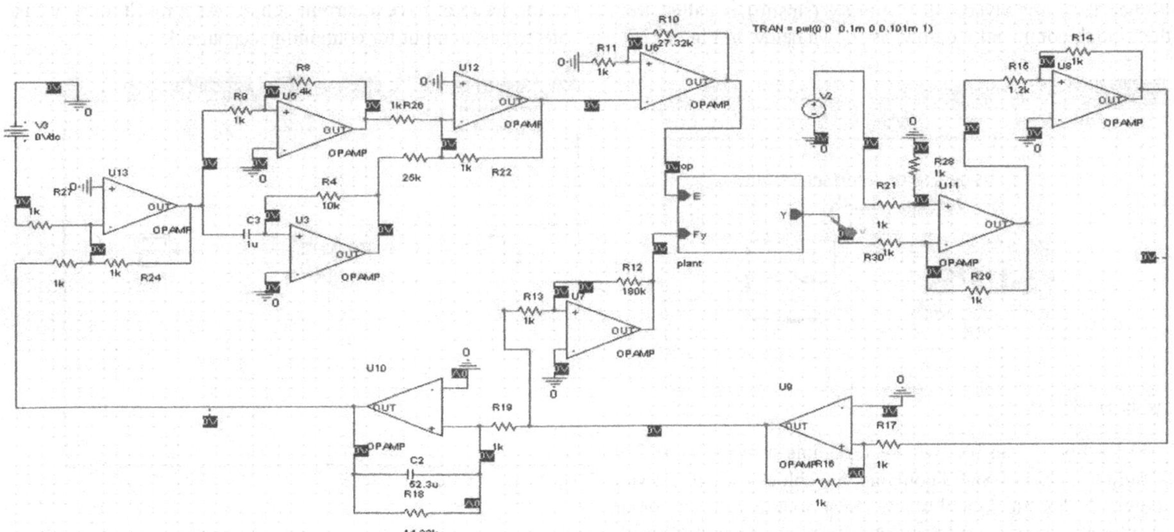

Fig. 10-PSpice Analog Model of the System.

The peak value of the step response is close to 1.4 V. This peak value is different than Simulink and PSpice ABM model. Moreover, step response of the analog output is more oscillating than the other outputs. Furthermore, step response of analog model becomes stable after 7 seconds. However, step responses of other models become stable after 6 seconds. Testing analog circuit is similar to real environmental parameters such as real parts, delay time, heat, smoke options, and metal losses. Therefore, output of the electronic model has a dissimilar response. However, the ideal model does not have these kinds of losses or effects because it just uses the mathematical model of the parts. When SLPS is tested and run, these differences can be completely comprehended.

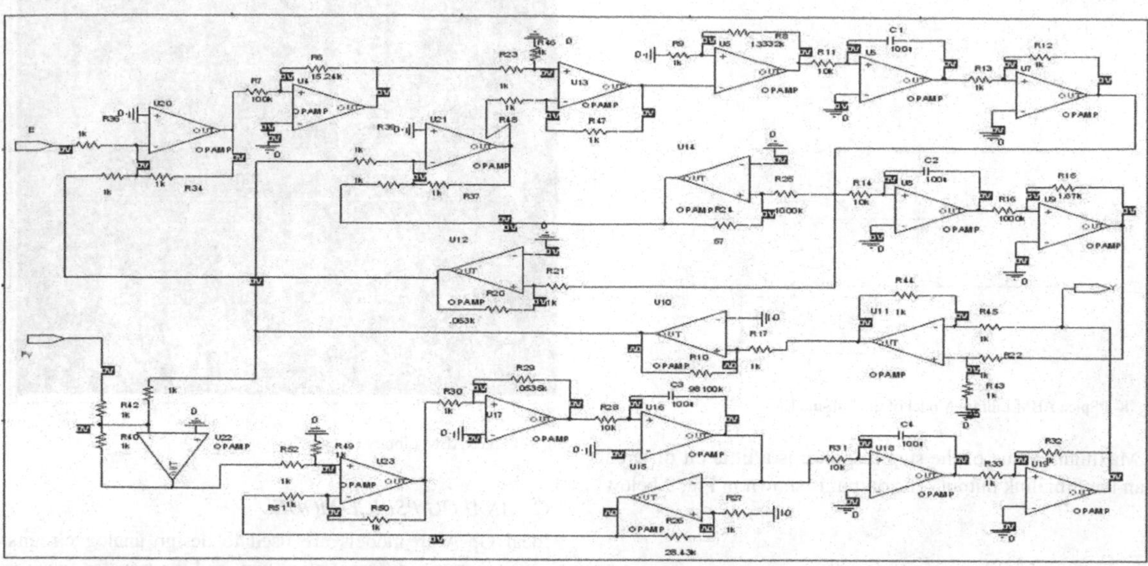

Fig. 11- PSpice Analog Circuit of the Plant

D. SLPS MODELLING.

SLPS is a Simulink library; however, this library is unlike than real Simulink library files. Therefore, the designer has to set the current directory of Matlab to the work directory where PSpice files are saved. SLPS library (slpslib) is called from the Matlab command window, as shown below in Fig.-12 [6].

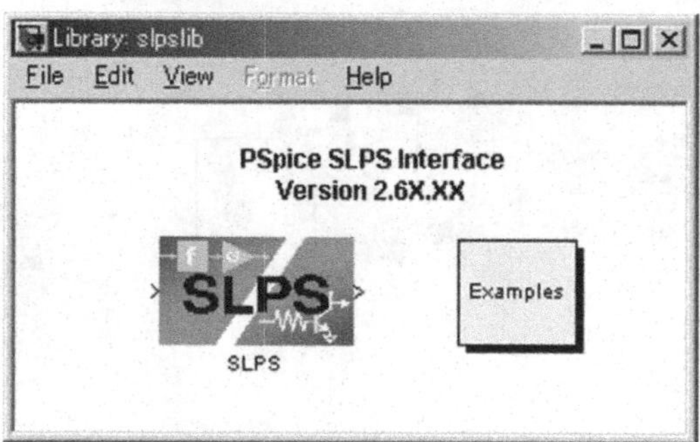

Fig. 12-SLPS Library

SLPS library can be used for designing a model now. The Simulink model of SLPS contains three blocks which are step source, SLPS block and scope. Step source is used for an input and it is set at a 0.1 second delay to get the same input in the PSpice model. Scope is used for displaying output of the model. SLPS model is shown in Fig. 13.

The setting window is opened by double-clicking on the SLPS block as shown in Fig. 14. When a project is designed, ambulatory control files included in the project are listed as "PSpice Circuit File"; thus, designate the file to be used [7].

Here, only the file "trana.cir" is listed and, therefore, it is automatically selected. Input and output source are selected for desired inputs and outputs. If voltage source is selected, the input data will be set to circuit as a voltage value, and if a current source is selected, it will be supplied as a current value. The input sources are listed here as V1 and V2. This circuit does not have a current source. As a result, only V2 is selected as an input for containing step signal.

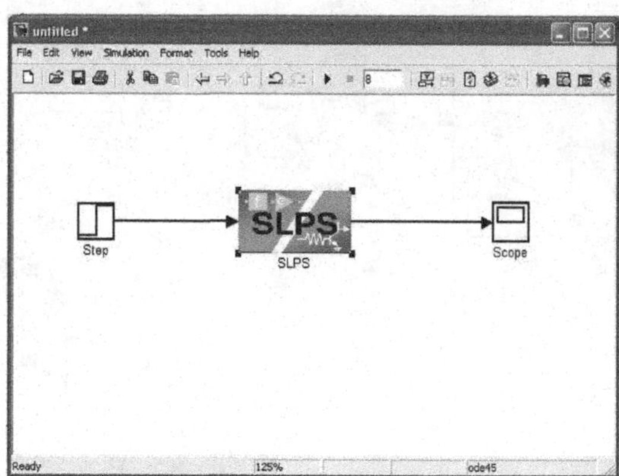

Fig. 13-Simulink Model of SLPS

Fig. 14-SLPS Setting

When the output select button is pressed, all output variables in the file are listed. N557082 node is selected as a desired output for this application because this node is used for output in the PSpice circuit. After completing of SLPS setting, simulation time is entered at 8 seconds.

When this model is run, Figure -28 occurs. This output is the same as the output of the analog model. Nonetheless, despite the similarity, we need to look at the dissimilarities rather than the ideal (mathematical) model.

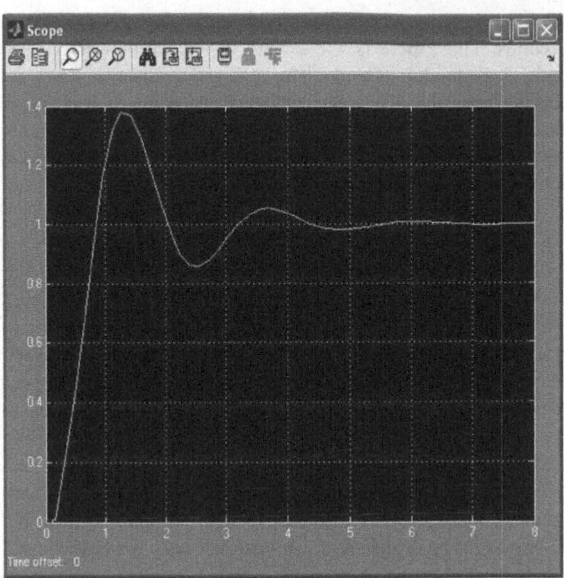

Fig.15-Output of the SLPS

E. COMPARISON OF SLPS AND SIMULINK OUTPUT

This comparison is shown in Fig. 16.

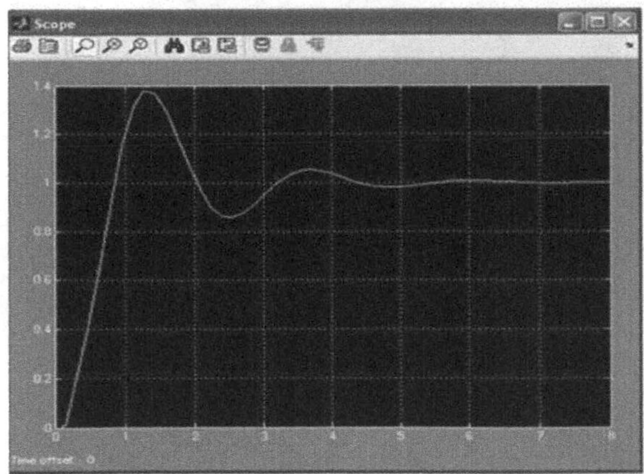

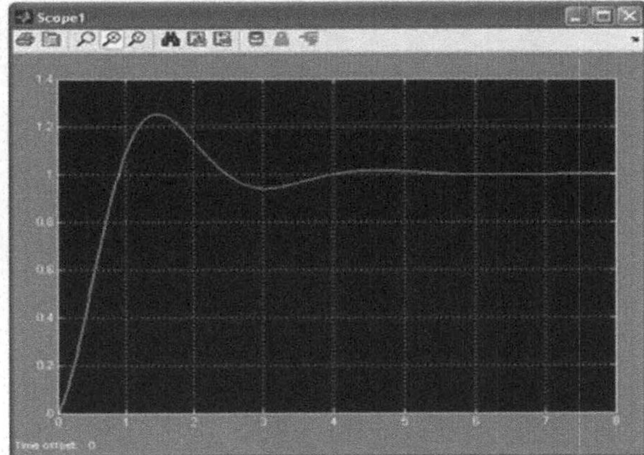

Figure -1 Comparisons of the SLPS and Simulink Outputs (The left scope is SLPS output, whereas the right one is Simulink)

When the two step responses are compared, the differences between Simulink and SLPS can be easily captured. These are the peak value of the signal, oscillating of the signal shape and the stability time.

As it can be easily seen in this Fig, the ideal and electronic model behavior is different. One of them has the peak value of the signals in that while SLPS peak value is close to 1.4 Volts, the peak value of the ideal model is close to 1.3 Volts. There is 0.1 Volts difference between these models. Secondly, SLPS signal shape is bigger and sharper than the ideal model. Furthermore, stability time of SLPS becomes after 7 seconds, whereas stability time of the ideal model is reached after 6 seconds. Thus, there is a 1 second difference between SLPS and ideal model. These differences demonstrate that the analog model have one second delay and some losses. These are due to the fact that analog model uses real parts and parameters. Hence, it is possible to grasp the ideal and real model behaviors to improve the circuit efficiency.

IV. CONCLUSION

SLPS is efficient and a useful software for testing and comparing the ideal and the real output of a circuit. It offers a significant opportunity to use people the most common product in the industry. The user can identify and correct integration issues of electronics within a system. This helps to identify errors earlier in the design process, which can save time and money often spent in debugging trial boards within system designs.

Like any other new software, SLPS has some problems and incomplete aspects. Therefore, SLPS software should be tested by using a different application such as Model-based Control Design of Ambulatory Rehabilitation Suspension System.

This application is more powerful example for testing SLPS and it applications presents an opportunity to better understand the SLPS software due to the fact that each application has different outputs and problems. Consequently, this makes it easier to have a deeper insight into the similarities and dissimilarities along with the causes that lie behind them.

What seems to be a serious problem is that designers have to use one SLPS block for each model.

REFERENCES

[1] D. Shetty, C. Campana, "Ambulatory Rehabilitation System", Presented at the International ASMI Congress, November 2003, Washington DC.

[2] O. Aksoy, "Model-based Control Design and Optimization for the Ambulatory Suspension System ", University of Hartford, 2007

[3] U. Arifoglu, 2005, "Simulink and Engineering Applications", Alfa Co.

[4] Cadence Design Systems, "PSpice User Guide", June 2003, pp. 257-308

[5] A. M. Davis, 1998, "Linear Circuit Analysis", PWS Publishing Company , pp. 189-205

[6] EMA Design and Automation. PSpice MATLAB/Simulink Interface (Accessed May 10, 2007) http://www.ema-eda.com/products/orcad/tech.matlabsimulink.aspx.

[7] Cybernet Systems Co.,Ltd. "PSpice SLPS Interface Version 2.60 User's" November 2004, pp. 9-34

Design of RISC Processor Using VHDL and Cadence

Saeid Moslehpour[#1], Chandrasekhar Puliroju[#2], Akram Abu-aisheh[#3]
[#1,#2#3]Department of Electrical and Computer Engineering, University of Hartford, West Hartford, CT

[#1]moslehpou@hartford.edu
[#2]puliroju@hartford.edu
#3abuaisheh@hartford.edu

Abstract— **The project deals about development of a basic RISC processor. The processor is designed with basic architecture consisting of internal modules like clock generator, memory, program counter, instruction register, accumulator, arithmetic and logic unit and decoder. This processor is mainly used for simple general purpose like arithmetic operations and which can be further developed for general purpose processor by increasing the size of the instruction register. The processor is designed in VHDL by using Xilinx 8.1i version. The present project also serves as an application of the knowledge gained from past studies of the PSPICE program. The study will show how PSPICE can be used to simplify massive complex circuits designed in VHDL Synthesis. The purpose of the project is to explore the designed RISC model piece by piece, examine and understand the Input/ Output pins, and to show how the VHDL synthesis code can be converted to a simplified PSPICE model. The project will also serve as a collection of various research materials about the pieces of the circuit.**

I. INTRODUCTION

The present project serves as an application of the knowledge gained from past studies of the PSPICE program. The study will show how PSPICE can be used to simplify massive complex circuits. In other words, the study will involve the breakdown of a complex circuit designed by a graduate student at the University of Hartford. The circuit is a VHDL Synthesis Model of an 8-bit processor. The purpose of the project is to explore the designed processor model piece by piece, examine and understand the Input/ Output pins, and to show how the VHDL synthesis code can be converted to a simplified PSPICE model. The project will also serve as a collection of various research materials about the pieces of the circuit. Background information will be given to give the reader a basic understanding of the parts to support the further exploration in the project methods. This report is useful for students who would like to know more about the PSPICE program and how it is useful for many educational purposes.

II. SIGNIFICANCE OF THE PROJECT

The study of VHDL and PSPICE is important because technology students are using computer software to design and analyze circuits. A better understanding of computers and computer languages can be gained when exploring these programs. It is also important to show that a VHDL model can be represented in a working PSPICE schematic. Each part of a processor, whether it be an Arithmetic Logic Unit or a simple Program Counter- has a specific set of input and output logic that can be emulated from PSPICE. If the logic is already known to the user, then the user does not have to build the part using each individual logic gate. Instead, a circuit designer can create a PSPICE part that gives the desired logical outputs. Most importantly, this study can help bridge a gap between those who simulate with VHDL and those who use PSPICE.

III. METHODOLOGY

The following VHDL synthesis model provided is an attempt toward a RISC processor. First, a data path module of the processor will be shown which clearly shows how the individual modules are going to send the data from one module to the other and then a screenshot will be shown of the top-module view to show each piece of the processor from an outside perspective (with view of all modules)

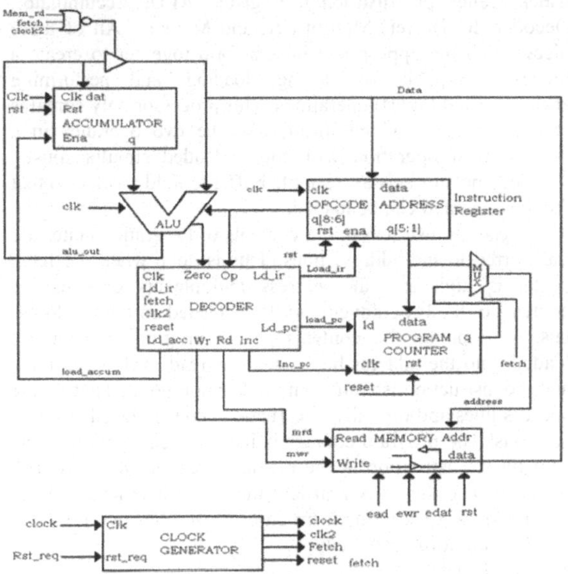

Figure 1 Data path module of the processor

K. Elleithy (ed.), *Advanced Techniques in Computing Sciences and Software Engineering*,
DOI 10.1007/978-90-481-3660-5_89, © Springer Science+Business Media B.V. 2010

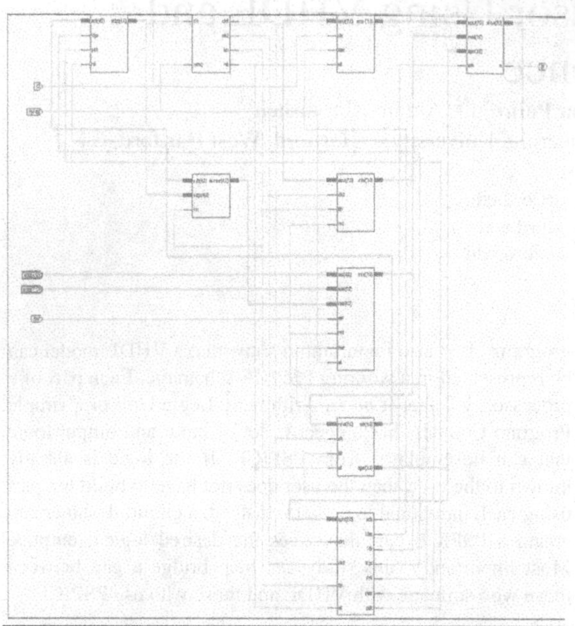

Figure 2 Screenshot of VHDL synthesis top module

If one is familiar with PSPICE, it would not be hard to recognize the style of the schematic. However this is not a PSPICE model, yet only a graphical representation of the VHDL synthesis. It may not be clear in the above screen shot, however the complete processor contains nine separate modules. These modules include: Program Counter (PC), Clock Generator, Instruction Register, ALU, Accumulator, Decoder, IO Buffer, Multiplexer, and Memory. All of these pieces with the appropriate busses form together to create a processor capable of storing, loading, and performing arithmetic and logical operations. This processor only contains one accumulator which means that the two operands in a mathematical operation will not be loaded simultaneously. The accumulator acts as a simple buffer to hold a value so that the next operand can be loaded.

To read an instruction, the contents of Program counter are transferred to the address lines. This is done when the fetch signal is high and the address multiplexers chooses the contents of the Program counter to be loaded on to the address bus. As soon as the contents of the program counter are loaded onto the address bus a memory read cycle is initiated and the instruction is read from the location pointed out by the address lines and the micro instruction code is placed onto the data bus. The program counter is incremented to point to the next micro instruction in the memory location of the control memory. The data bus transfers the micro instruction to the Instruction Register. The instruction register has two fields, in the different formats namely,

1. Opcode, data operand.
2. Opcode, Address of data operand.

During the first case the opcode is given to the ALU and decoder for decoding and a series of micro operation are generated. The data operand is loaded on to the data bus and transferred to the ALU for its respective micro operations as specified by its opcode. In the second case the address of the data operand is loaded onto the address bus (As the fetch signal is low and the multiplexer loads the IR's address contents onto the address lines) and a memory read cycle is initiated. Here the memory location in the main memory specified by the address lines is read and the data is transferred onto the data bus and thus given the ALU to undergo the operations specified by its opcode.

The results of the ALU are stored in the accumulator. Data operations may be combined with the memory contents and the accumulator and the result is transferred back to the accumulator. The function of the NOR gate is that when ever all inputs are low the output is high and at all other times remains low. It is attached to tri-state buffer. When the tri-state buffer is enabled the data from ALU is fed to the memory thus allowing the data to be stored into the memory. When disabled the data is given to all and cut off from being written onto the data bus. Whenever there results a zero in the ALU a zero flag is set.

A. CLOCK GENERATOR

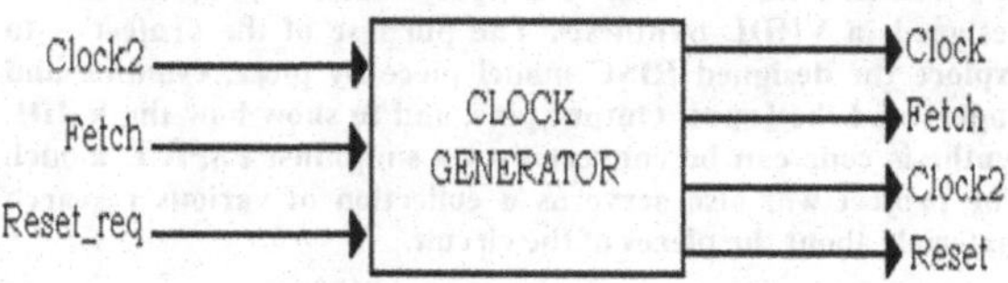

Figure 3 Clock generator

Clock generator generates clock, clock2, fetch signals. For every negative edge of clock, clock2 is generated and for every positive edge of clock2, fetch signal is generated. Clock2 is generated from clock and fetch is generated from clock2. They are used as the inputs to the decoder which controls the operation of the processor. It generated reset pulse which must be active low. The reset should allow the rest of the signals to go high on falling edge of clock2 when fetch is low.

B. INSTRUCTION REGISTER

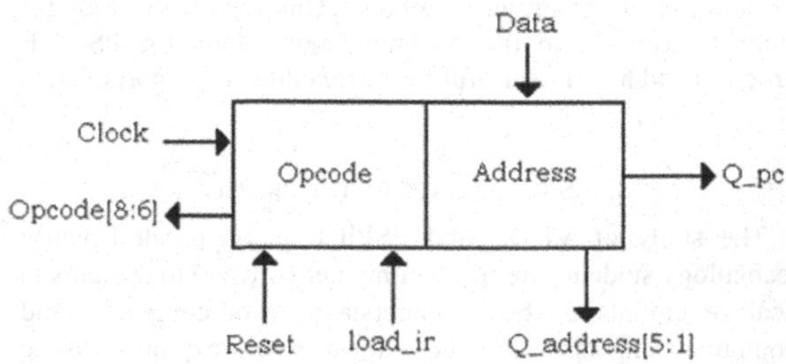

Figure 4 Instruction Register

In instruction register instructions are fetched from and stored. It performs the action always at positive edge of clock. It has 3 instructions clock, reset, load_ir, and data as inputs and opsode and address as outputs. If load_ir and reset both are high, data in instruction register splits into upper 3 bits as opcode and lower 5 bits as address.

C. ACCUMULATOR

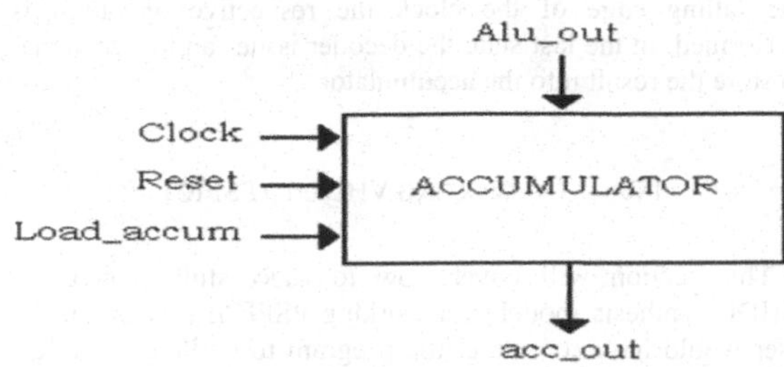

Figure 5 Accumulator

Accumulator is a register, the result from Arithmetic and logic unit is stored back in the accumulator. It has clock, reset, load_acc and alu_out as inputs and accumulator as output. It is activated only at the positive edge of clock. If load_acc and reset both are high, data in the accumulator is loaded into alu_out.

D. MEMORY

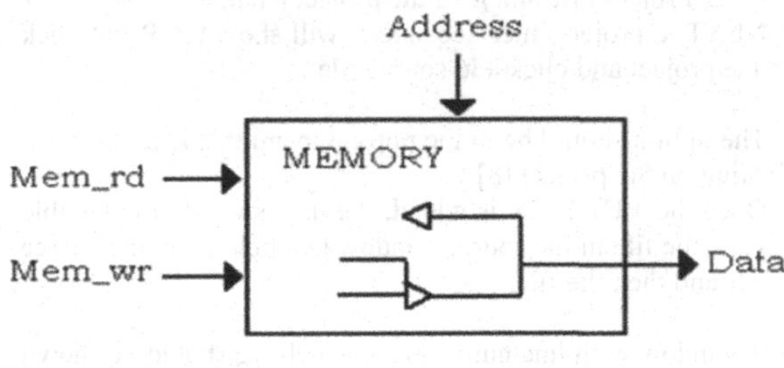

Figure 6 Memory

The memory should be 8-bit wide, 32-bit location deep. Each instruction retrieved from the memory will have its upper 3 bits as the opcode and lower 5-bits as the address. Memory has mem_rd, mem_wr and address as inputs and data as output. If mem_rd is high, it reads the data of memory to the data register, if mem_wr is high, data is written to the memory.

E. ARITHMETIC AND LOGIC UNIT

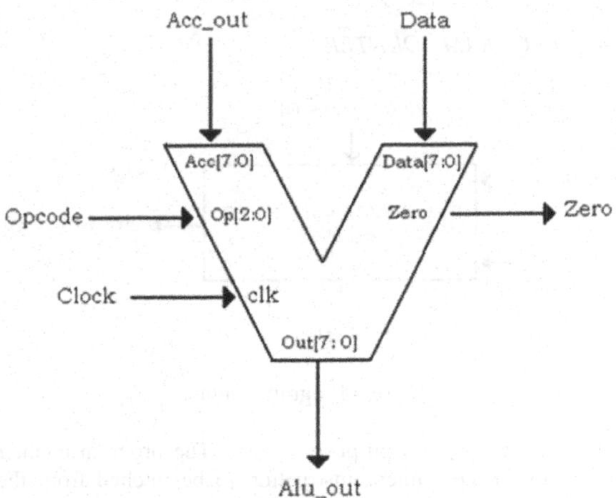

Figure 7 Arithmetic and logic unit

Arithmetic and logic unit (ALU) is a multiplexer, performs standard arithmetic and logic operations. The operations performed are listed below. ALU operations should be synchronized to the negative edge of the clock. At each negative edge, the ALU should perform the appropriate operation on the incoming data and accumulator, placing the result in alu_out. The 3-bit opcode decodes as follows:

000: Pass Accumulator.
001: Pass Accumulator.
010: Add (data + Accumulator).
011: And (data & Accumulator).
100: Xor (data ^ Accumulator).
101: Pass data.
110: Pass Accumulator.
111: Pass Accumulator.

F. MULTIPLEXER

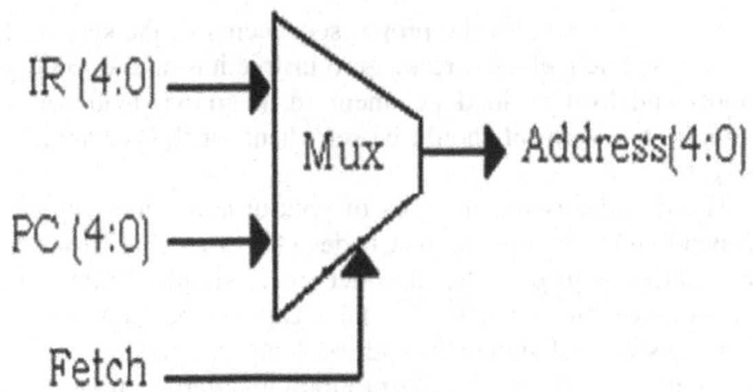

Figure 8 Multiplexer

The address multiplexer decides one output out of the two given inputs. When the fetch signal is high, the address of the program counter is transferred on to the address buses and hence instruction is fetched. But if low, the operand address specified in the address field of the instruction register is transferred onto the address bus and consequently fetched.

G. PROGRAM COUNTER

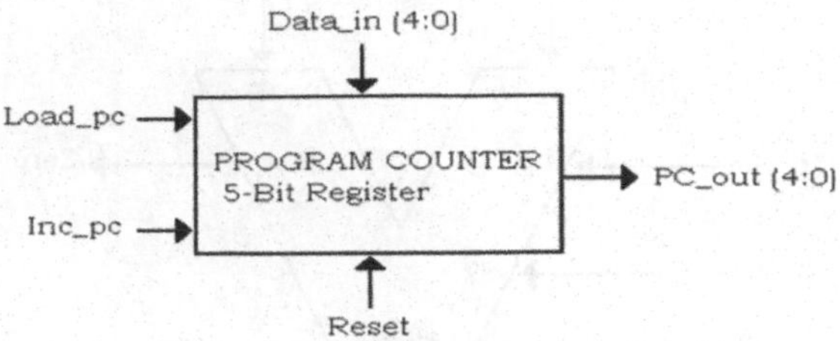

Figure 9 Program counter

It is a 5-bit general purpose register. The program counter points to the next micro instruction to be fetched from the memory. In case of an unconditional branch the said address is loaded into the program counter for fetching of that instruction. Normally, after the fetch cycle is completed the program counter is incremented and now thus points to the next instruction.

H. DECODER

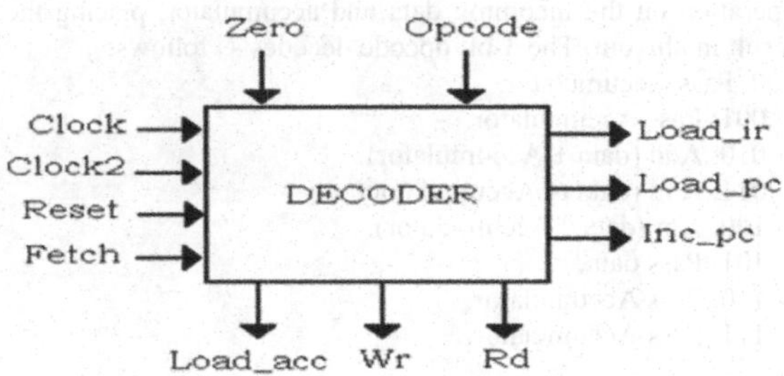

Figure 10 Decoder

A decoder provides the proper sequencing of the system. It has clock, fetch, clock2, reset, zero instructions and opcode as inputs and load_ir, load_pc, mem_rd, mem_wr, load_acc as outputs. The zero bit should be set whenever the accumulator is zero.

The decoder issues a series of control and timing signals. Depending on the opcode it decodes after it receives from the instruction register. The decoder is a simple finite state machine which consists of states. During the first state it generates control signal for address setup. The address bus is setup and the contents of the program counter are transferred on to the address bus. The instruction fetch is generated in the second state and the instruction is read from the memory with the memory read signal and transferred on to the data bus. When the third state starts the instruction is loaded with the ld_ir being high into the instruction register. The opcode is sent to the decoder and the appropriate control and timing signals are initiated for the execution cycle. This is done in the next state in which it remains idle during the decoding of the opcode.

The fetch cycle ends and the program counter is incremented with the inc_pc signal. The execution cycle starts and the address is again setup in the fifth state, but this time instruction registers address field is loaded onto the address bus. The operand is fetched in the sixth state with the mem_rd signal being high and the data id is transferred onto the data bus and given to the ALU for processing. In the seventh state the ALU is given its alu_clock amd in synchronization with the falling edge of the clock the respective operation is performed. In the last state the decoder issues an ld_acc signal to store the result into the accumulator.

IV. CONVERTING VHDL TO PSPICE

This section will cover how to successfully convert a VHDL synthesis model to a working PSPICE part. First, the user would have to extract the program to C:\Program Files. This part is important because the program will not work unless it is extracted to the specified directory. It will create a directory in C:\Program Files called VHDL2PSPICE. The contents of the folder will show all the necessary files.

The folders created are Capture Symbols, DelphiExamples, Library, and Source. These folders are necessary for the program to save the appropriate Capture Model symbols and library files.

- Now, the user should open the Lattice Synplify program. Once in the program, click FILE -> NEW -> and start a new project.

The window shown above should pop-up onto the screen.

- Click Project File and give the project a name.
- NEXT, a project interface screen will show up. Right click on the project and click add source file.

The option should be in the pop-up menu that appears when clicking on the project.[8]

Once the VHDL file is added, the user should now be able to view the file in the source window by clicking on the source folder and then the file.

A window with line numbers shown above should be shown on the screen. However before synthesis, certain conditions need to be set up by the user in order for the synthesis to properly work.

- Right click on the icon in the project window that appears as a 'blank screen'. It should be the last item in the project window. Click on Implementation options. A window should appear with several tabs and options.

Make sure the Technology option is set to LATTICE MACH, and the part option is set to MACH111. Click the box in the Device Mapping Options window that says "Disable I/O insertion."

NEXT, go to the implementation tab and click the Write Mapped VHDL Net list check box.

After this is done, the program is ready to synthesize.

Click the RUN button back on the project menu. On the right window, several files will be generated.

We are interested in the gate level VHDL file. This is the file with the extension VHM on the right list.

Double click on the VHM file to view its contents. Once there, the file should be saved. Click File-> SAVE AS and save the file in C:\ProgramFiles\Vhdl2Pspice\Source.

- Once the file is saved, the user can now exit the Synplify program as it is no longer needed for the conversion.

NOW, in C:\ProgramFiles\Vhdl2Pspice directory, run the Vhdl2Pspice.exe utility.

A window will pop-up on screen with a button labeled "Convert Vhdl 2 Pspice"

Enter the name of the VHDL file that was saved into the SOURCE directory. There is no need to type the extension, just the name of the file.

After OK is clicked, the user should enter the name of the top level entity (in this example the device is called COUNTER). Click OK again and almost immediately, the conversion will take place. The user should see a success window appear.

NOW, the program has created output files that can be used in PSPICE. Go to the LIBRARY folder in VHDL2PSPICE to view the converted files. The file that the user should look for will be the same name, however with the word "copy" starting off the filename. The extension of the file is .LIB.

There is also a file called LIBPOINTER.LIB, which is a library file that tells PSPICE where to look for the created parts.

NEXT-> The user should open the PSPICE Model Editor program and open the created library file that was converted. It should still be found in the VHDL2PSPICE library directory.

Click on the part name to see the structural representation of the VHDL code that was converted.

The next part is very important. Click FILE and click Export to Capture Part Library. The library file is already created, however the program will not know how to display the part unless it is properly exported.

Export it to the VHDL2PSPICE library called "Capture Symbols". It is ok to copy the files here because the Libpointer.lib file will tell PSPICE where these files are saved.

When the Capture program is opened, the user should be able to place the symbol onto the schematic. The part should work with PSPICE simulation as well.

The picture above shows the VHDL device successfully working in PSPICE.

V. Applying the Conversion Steps to the Processor Model

The next objective will be to successfully synthesize and convert the code of each VHDL model to a PSPICE library and object file.

The following files will be converted in this project:

These files are named pul (which I used the first three letters of the creators name) followed by a word that represents the part. This makes it easier to follow.

Example

The first file we will synthesize will be the ALU file. The file is called pulalu.vhd. This file can be saved anywhere on the computer, as the VHDL2PSPICE program will not need this. The file that the VHDL2PSPICE will use will be the synthesized version.

By following the steps, the first file is synthesized. As can be seen on the right window, the file PULALU.VHM has been created. This is the file that will be saved to C:\program files\vhdl2pspice. Now the synthesis program is finished and it is time to run the VHDL2PSPICE program to create a library file.

This window shows that we are ready to enter our filename. After entering the name of the part, a message box will appear and tell the user that the conversion was successful. Note: The Entity name must match the entity title located in the VHDL file, it cannot be a made up label. It should be remembered that the library file generated will say copy_filename_lib. In this case, the filename in the LIBRARY folder is named "copypulalu.lib". Now, the file is ready to be opened with PSPICE Model Editor. This program will allow the library file to be exported to the Capture Library. Once in the model editor, the copypulalu.lib file should contain one part named "alu". When exporting to Capture Part Library, the .olb file should be exported to the Capture Symbols folder that is a part of the VHDL2PSPICE folder. When placing a part, the library will be located in the same capturesymbols directory. This is the .olb file and not the .lib file.

A. *Converted ALU*

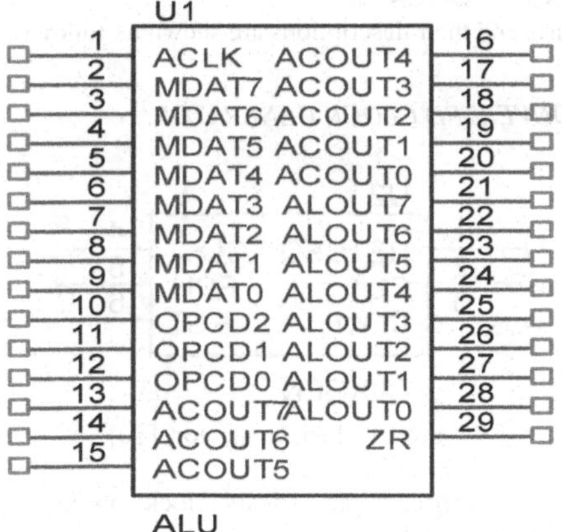

Figure 11 Converted ALU, [10]

This is the converted ALU. The device, just like the VHDL model has all of the input/output pins. The only difference is the orientation of the 8-bit input/outputs. This circuit can be simplified further by placing it into a hierarchal block and sending a bus line that covers all 8 bits. This device has an ACLK pin, which means that the opcode is read and the ALU performs at the RISING edge of the clock.

PINS	WHEN ASSERTED	WHEN DE-ASSERTED
ACLK	Clock trigger	Clock trigger
MDAT 7-0	8 bit data code (1 or 0 depends on data word)	8 bit data code (1 or 0 depends on data word)
OPCD 2-0	3-bit opcode that determines the operation the ALU must perform	(same)
ACOUT 7-0	8-bit data coming from accumulator to serve as an operand.	
ALOUT	The 8-bit result output from the ALU	
ZR (zero bit)	ALU result is 00000000	ALU result is anything other than ZERO

Figure 12 ALU assertion chart, [10]

The remaining parts of the processor will be implemented using the same steps as the ALU since it is known that the program works successfully.
These parts and their descriptions are shown as follows:

B. CONVERTED CLOCK GENERATOR:

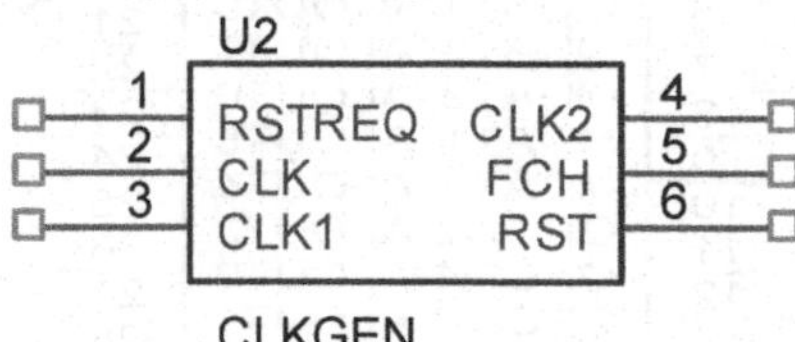

Figure 13 Clock Generator, [10]

RSTREQ: (input)Reset pin. Clears clock on high signal. Inactive on low signal
CLK, CLK1, CLK2: Clock generators necessary for processing cycles.
RST: (output) Sends high signal for other components to be reset
FCH: (output) used as interface with program counter.

C. CONVERTED PROGRAM COUNTER:

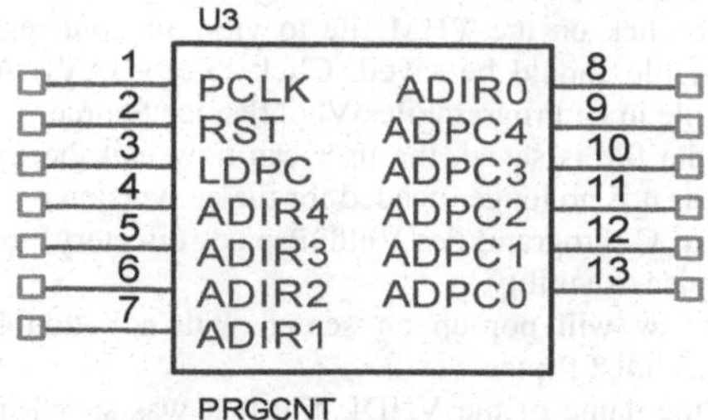

Figure 14 Program Counter, [10]

PCLK: Clock signal is sent to this input pin to allow the PC to count upward.
RST: This line resets the program counter with a high signal
LDPC: When set to high signal, the program counter sets the PC with the value at ADPC 4:0
ADIR 4:0: The value of PC that is passed to the instruction register.
ADPC4:0: The value that is returned to the PC.

D. CONVERTED INSTRUCTION REGISTER

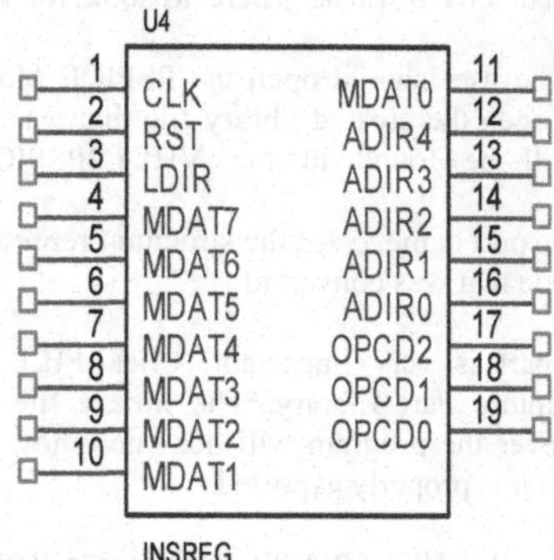

Figure 15 Instruction Register, [10]

CLK: The clock pin is necessary for the device works with a rising clock edge.
RST: On a rising clock edge, the Instruction register becomes reset
LDIR: Load Instruction Register. High signal allows the signal from the PC to be input into ADIR4:0
OPCD: This is the three bit operation code used for operations of the ALU.

E. CONVERTED ACCUMULATOR

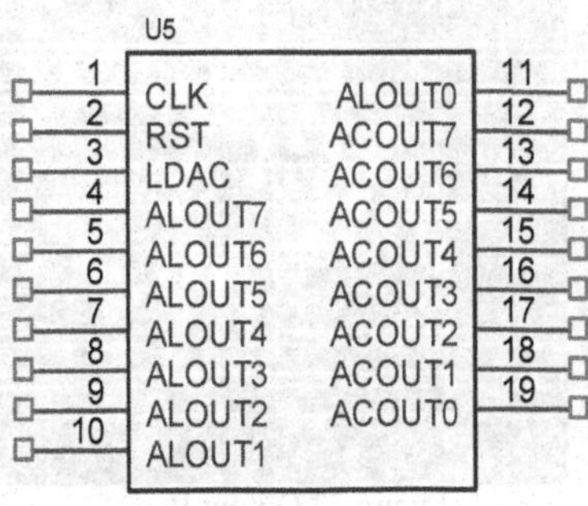

Figure 16 Accumulator, [10]

CLK: clock pin runs the accumulator, since in theory the accumulator is made up of latches.
RST: Resets the accumulator to an 'empty state'
LDAC: When this pin is asserted the accumulator is loaded with the value that flows through ALOUT (input pins)
ALOUT: This 8-bit set of pins is the data passed from the ALU
ACOUT: This 8-bit set of output pins sends a value to the ALU for operation.

F. CONVERTED DECODER

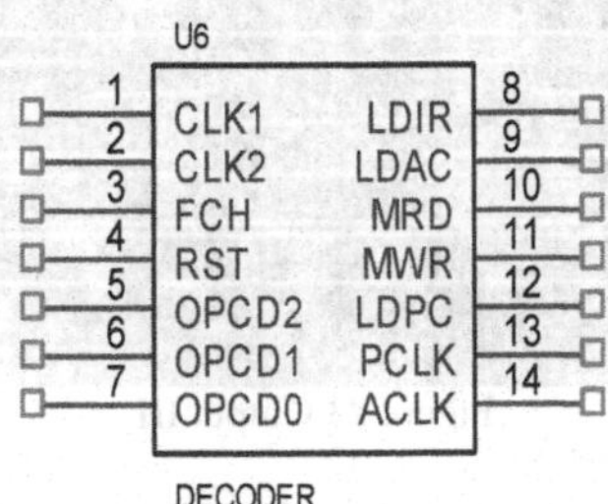

Figure 17 Instruction Decoder, [10]

The decoder contains pins that enable the respective control lines.

G. CONVERTED IO BUFFER

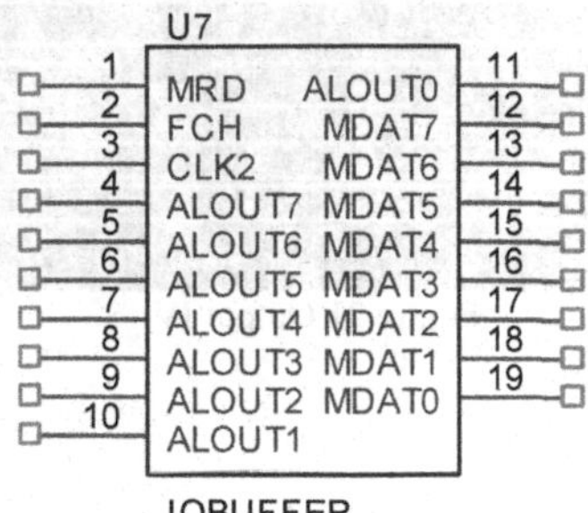

Figure 18 Input/ Output Buffer, [10]

MRD: Memory Read, allows the data held in the buffer to be passed out of MDAT 7:0.
FCH: Allows the device to be written to with the 8-bit data word of ALOUT 7:0
CLK2: Clock signal that runs the part (rising clock edge)

H. CONVERTED MULTIPLEXER

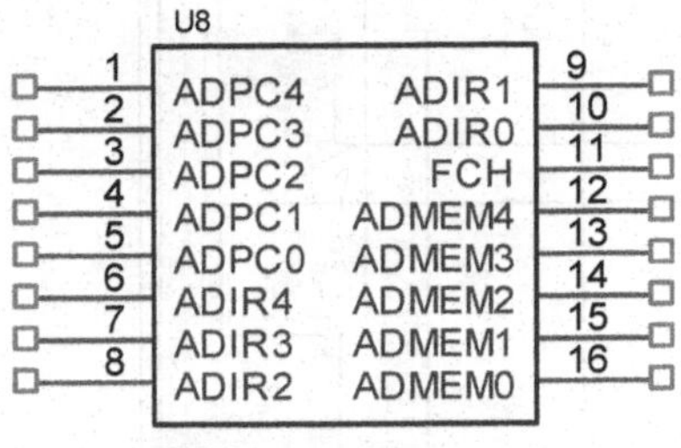

Figure 19 Multiplexer, [10]

ADPC 4:0: PC address to be sent to memory and Instruction register.

I. CONVERTED MEMORY

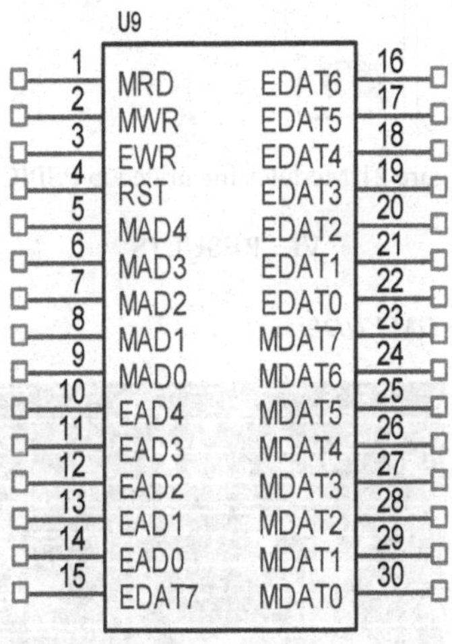

Figure 20 Memory, [10]

J. MODULES INTEGRATED TOGETHER IN PSPICE

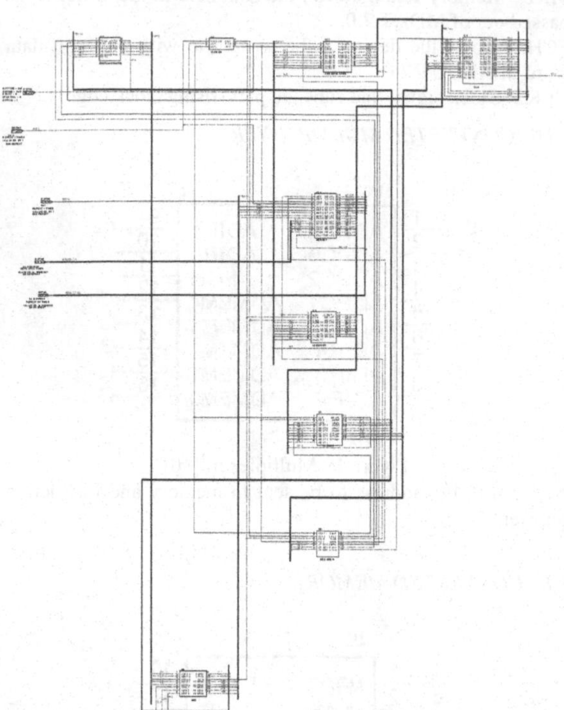

Figure 21 Modules integrated in PSPICE

VI. RESULTS

RESULTS FROM VHDL:

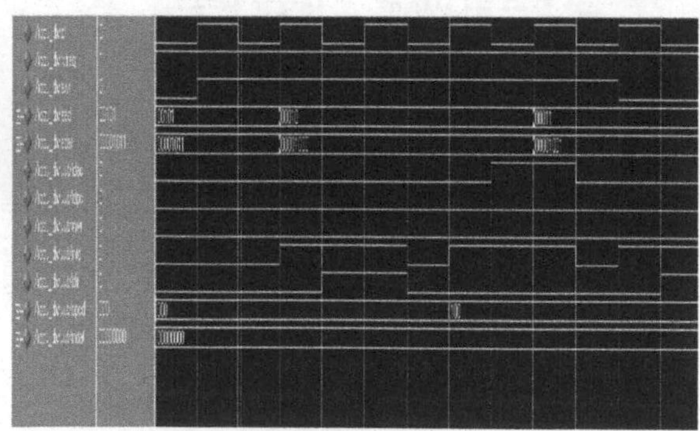

Figure 22 Output I

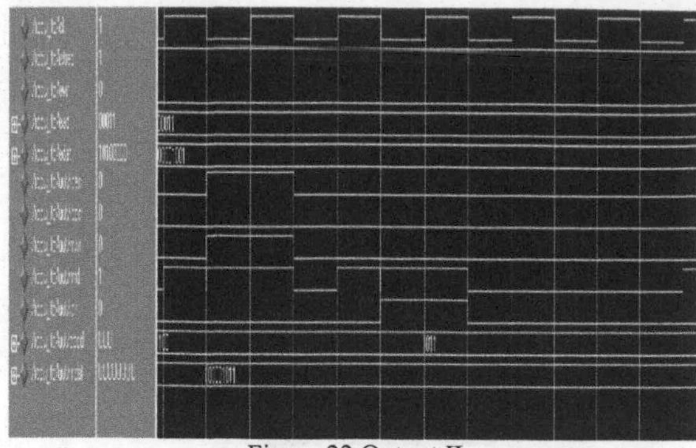

Figure 22 Output II

RESULTS FROM PSPICE

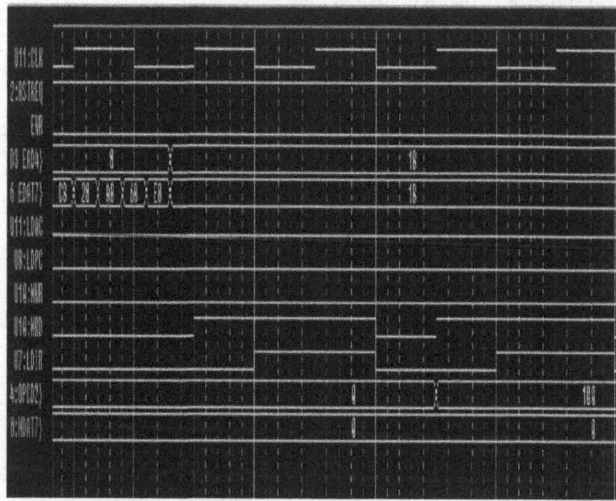

Figure 23 Output III

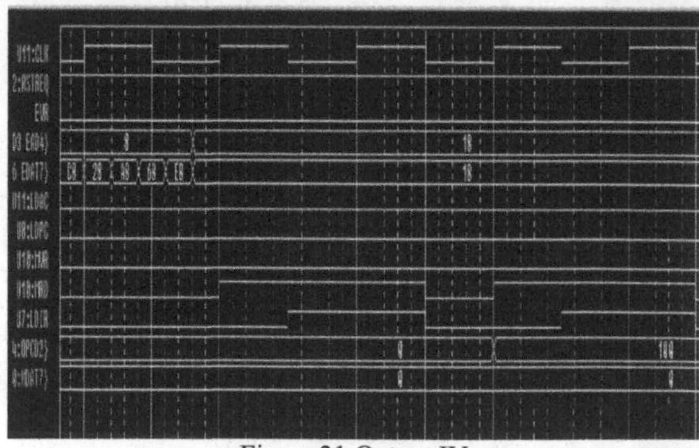

Figure 21 Output IV

VII. Conclusion

The first problem to note was that halfway through the components; the synthesis program gave "warnings" yet still synthesized the files. These warnings only existed with the IO BUFFER, MULTIPLEXER, and MEMORY. The program stated that there were illegal statements/declarations. However, the parts of the project that were successfully synthesized proved to work with the CAPTURE program just fine. Another problem lies in the program's disregard to the synthesis errors. When a part is created from the synthesis file, the running part will crash the software. This error was not fixed, as it wasn't a PSPICE software error, or an error with the VHDL2PSPICE program. The problem existed in the VHDL errors. It is not known whether the errors are small or large, or whether or not the parts are complete. However, the project goal has still been met. VHDL parts can successfully be converted to PSPICE and can be displayed and simulated in a schematic. Given fully functional code, a complete processor would have been assembled and simulated. It is possible to connect all of the modules just as the VHDL top module, however without working code; there is no reason to attempt to simulate the modules. Also, the RISC design provided contained no set of pre-programmed code or language. The model seemed to only be an empty shell of what a RISC processor looks like from an outside perspective. The project can be deemed a failure in terms of building a fully functional processor, however the bridge between PSPICE and VHDL has been met. The knowledge has been gained on how to convert the VHDL modules to PSPICE. However the VHDL programmer must be sure that the code is fully functional. There is no software known as of yet that will convert PSPICE to VHDL.

The first problem to note was that halfway through the components; the synthesis program gave "warnings" yet still synthesized the files. These warnings only existed with the IO BUFFER, MULTIPLEXER, and MEMORY. The program stated that there were illegal statements/declarations. However, the parts of the project that were successfully synthesized proved to work with the CAPTURE program just fine. Another problem lies in the program's disregard to the synthesis errors. When a part is created from the synthesis file, the running part will crash the software. This error was fixed with the VHDL program. VHDL parts can successfully be converted to PSPICE and can be displayed and simulated in a schematic. The fully functional code helped to complete the processor and it has been assembled and simulated. It is possible to connect all of the modules just as the VHDL top module. The knowledge has been gained on how to convert the VHDL modules to PSPICE. However the VHDL programmer must be sure that the code is fully functional. There is no software known as of yet that will convert PSPICE to VHDL.

References

[1] mackido.com – What is Risc. Retrieved November 2006 http://www.mackido.com/Hardware/WhatIsRISC.html

[2] Risc Architecture. Retrieved: November 2006 http://www.geocities.com/SiliconValley/Chip/5014/arch.html

[3]aallison.com – Brief History of RISC. Retrieved: November 2006 http://www.aallison.com/history.htm

[4] VHDL and Verilog. Retrieved November 2006 http://course.wilkes.edu/Engineer1/

[5] IEEE.ORG – Vhdl Synthesis model. Retrieved November 2006 http://www.ewh.ieee.org/soc/es/Nov1997/01/INDEX.HTM

[6] RCore54 Processor. Retrieved: November 2006 http://www.ht-lab.com/freecores/risc/risc.html

[7] PSPICE (Copyright © 1985 - 2004 Cadence Design Systems, Inc)
-PSPICE Online Manual Retrieved: September 2006 (As shown in Appendices)
(Copyright © 1985 - 2004 Cadence Design Systems, Inc)

[8] Sreeram Rajagopalan – Mixed Level and Mixed Signal Simulation using PSpice A/D and VHDL
VHDL2PSPICE Utility.
VHDLPSpice_CDNLive_Sreeram.pdf

[9] Chandrasekhar Puliroju –VHDL synthesis models used in this study.

[10] Christopher L Spivey , "Creating PSPICE Parts with VHDL Models." EL 482: Senior Project Report, Fall 2006.

The Impact of Building Information Modeling on the Architectural Design Process

Thomaz P. F. Moreira, Neander F. Silva, Ecilamar M. Lima
Universidade de Brasília
Laboratório de Estudos Computacionais em Projeto – LECOMP
ICC Norte, Ala A, Subsolo, CEP 70910-900, Brasília, DF, BRASIL
http://lecomp.fau.unb.br/, thomaz.moreira@camara.gov.br, neander@unb.br, ecilamar@unb.br

Abstract – Many benefits of Building Information Modeling, BIM, have been suggested by several authors and by software vendors. In this paper we describe an experiment in which two groups of designers were observed developing an assigned design task. One of the groups used a BIM system, while the other used a standard computer-aided drafting system. The results show that some of the promises of BIM hold true, such as consistency maintenance and error avoidance in the design documentation process. Other promises such as changing the design process itself seemed also promising but they need more research to determine to greater extent the depth of such changes.

I. Research Problem

Building Information Modeling, BIM, is not a new concept. Some authors have argued it is not even a technological advancement, but rather an "attitude" [1], a chosen method of working. Although there may be claims that the term BIM is relatively new, let us say 10 years old, the concepts it conveys can be traced back to at least 33 years ago [2]. Besides, applications which meet the present BIM criteria have been marketed for more than 20 years [3] even the term itself have been coined much later.

However, it was only in recent years that implementations of such concepts became more significantly popular due to the advent of more powerful hardware and their more affordable prices.

The concept of BIM, as defined by Eastman et al in [4] refers to a modeling technology and processes that present some essential features: the building components are digitally represented as objects that include not only geometry but also data describing what they are made of and how they behave. Their data is represented coordinately in a three-dimensional (3D) space so that changes to one view are automatically updated in all the other views. The relationships between components are parametrically represented [4], so that changes to one of them result automatically in changes to all the related ones.

Many benefits from using BIM tools have been suggested by several authors [5][6]. However, in this paper we will focus in three of the design benefits: earlier and more accurate visualizations of a design; the automatic low-level corrections when changes are made to a design; the generation of accurate and consistent two-dimensional (2D) drawings at any stage of the design from a single 3D representation.

In this context, our main research questions are: can the use of BIM tools be associated with changes of human strategy in the design process? Do the users of BIM tools adopt different design methods if compared with those who use non BIM tools? Can the benefits of BIM, such as consistent representation and error avoidance, be observed in the practice of those who use it?

II. Hypothesis

We believe that the traditional design method tends strongly to be replaced by a new way of designing when architects adopt BIM tools. Designers will predominantly work from 3D holistic representations, even if such tools allow for partial 2D views to be used, and they will be less prone to representational errors and to produce inconsistent design documentation.

III. Research Method

In this paper we describe an experiment of observing and comparing the design methods of two groups of 15 architects each: the first group was composed of users of a traditional computer-aided drafting tool. The second group was composed of users of a BIM tool.

The first group of architects was composed of regular users of the software AutoCAD, while the second one was composed of users of the software Revit Architecture (both from Autodesk, www.autodesk.com).

A standard design task was defined and assigned to each of the designers in the two groups of users. The experiment was designed in order to be simple enough for completion in a maximum of one hour and a half and comprehensive enough to include a number of day to day architectural design situations.

This task was presented to each user as a set of written instructions, rather than through graphic representations. The

K. Elleithy (ed.), *Advanced Techniques in Computing Sciences and Software Engineering*,
DOI 10.1007/978-90-481-3660-5_90, © Springer Science+Business Media B.V. 2010

main reason for that was to preclude the induction of a particular type of design strategy by a particular type of graphic representation.

The adopted research method has drawn inspiration from several sources, but protocol analysis, as described in [7], was the major one. Data about the design method used by each user, particularly the representation processes and associated thought, for accomplishing with the standard task was registered through direct observation, note taking and by enquiring the designers during the development of the task. This process was aided by the adoption of a standard form which contained reminders of the important points to be observed.

The first part of the standard task consisted in designing a small tourist information bureau with an area of 24 square meters and with a total height of 4.5 meters. It was also specified the construction and finishing materials as well as the type of structural system to be used. The objective was to detect the predominant ways of representing during the design process.

The second part of the standard task consisted of making some changes to the initial design, such as changing the height and width of the building and changing a door. The objective was to track the level of consistency in each architects work and verify how prone they would be to unnoticed errors.

The third part of the task consisted of inserting a block of furniture, object or person into the building design. The objective was to assess the knowledge of such resources and the ability of using parametric blocks if they were available in the software at hand.

The fourth and last part of the standard task consisted of generating spreadsheets with non graphic information about the produced design. The objective was to assess the ability to quickly extract important design and construction information such as specification lists, bills of quantities and simulations.

IV. Results

The answers given by each architect and the observations were consolidated into the Table I shown below:

TABLE I

Questions And Consolidated Answers

Questions and answers	AutoCAD		Revit Arc.	
	#	%	#	%
1. The knowledge of BIM				
1.1 What is your level of knowledge about BIM – Building Information Modeling tools?				
None	15	100.0	5	33.3
Basic	0	0.0	5	33.3
Intermediate	0	0.0	0	0.0
Advanced	0	0.0	5	33.3
Questions and answers	AutoCAD		Revit Arc.	
	#	%	#	%
1.2 What is your level of knowledge about parametric functions and commands?				
None	11	73.3	2	13.3
Basic	3	20.0	8	53.3
Intermediate	1	6.7	0	0.0
Advanced	0	0.0	5	33.3
2. First part of standard task				
2.1 What graphic products will be developed first to accomplish with the task? 2D or 3D? What will be the order of their production?				
Floor plan, cross sections, facades, 3D views.	15	100.0	0	0.0
3D model and simultaneously the floor plan, cross sections, perspective views.	0	0.0	6	40.0
3D model, 3D views, floor plan, cross sections and elevations.	0	0.0	9	60.0
2.3 Will the first part of this task (designing a small tourist information bureau) be developed only with the software under assessment?				
Yes.	14	93.3	14	93.3
No.	1	6.7	0	0.0
Yes, but in the day to day life I would take the orthogonal views to AutoCAD and the 3D model to 3D Studio Max.	0	0.0	1	6.7
2.4 What software will be used to help in the development of this first part of this task?				
AutoCAD only.	13	86.7	0	0.0
3D Studio Max	1	6.7	0	0.0
Sketchup	1	6.7	0	0.0
Revit Arc. only.	0	0.0	15	100.0
2.5 Does this software allow for graphic information of the element wall, for example, to be associated with non graphic information (physical features and behavior)?				
No.	15	100.0	0	0.0
Yes.	0	0.0	15	100.0
2.6 What non graphic information could be extracted from the representation of the element wall, for example?				
None.	15	100.0	0	0.0
Materials properties	0	0.0	9	60.0
The real features of a wall and others that might be necessary.	0	0.0	4	26.7
Several (physic features, costs, manufacturer, identification of construction phase, etc.).	0	0.0	2	13.3
2.7 Does this software also allow the association of the other elements (doors, windows, slabs, etc.) with non graphic information?				
No.	15	100.0	0	0.0
Yes.	0	0.0	15	100.0
2.8 Does this software require that non graphic information of building components must be defined at the outset of the design process?				
No.	15	100.0	0	0.0
Yes.	0	0.0	4	26.7
Yes, but you can choose a default or basic option and change it later.	0	0.0	5	33.3
No, the program allows studies	0	0.0	6	40.0

based on generic entities without the definition of constructive information.				
Questions and answers	AutoCAD		Revit Arc.	
	#	%	#	%
2.9 Did the default library of components have the non graphic information needed for the task or they had to be added or configured?				
Not applicable.	15	100.0	-	-
No.	-	-	2	13.3
No, they had to be adapted, with difficulties.	-	-	2	13.3
No, they had to be adapted, but without difficulties.	-	-	8	53.3
No, it's possible to begin with generic entities and to specify them later.	-	-	1	46.7
Yes.	-	-	2	13.3
2.10 Does this software offer construction components (walls, doors, windows, roof tiles, etc.) that fit the specifications of traditional construction systems and materials of Brazil as part of its original options?				
Not applicable.	15			
No.	-	-	8	53.3
No, but you gradually build your own components.	-	-	5	33.3
No. This is a great problem, particularly with simple structural components such as concrete beans.	-	-	1	6.7
Yes, with some initial adaptations.	-	-	1	6.7
2.11 The inclusion of new non graphic information in this software is a friendly procedure?				
Not applicable.	15	-	-	-
Yes, but the procedure is not graphic. It is based on filling forms.	-	-	1	6.7
Yes for the simple information, but difficult for complex information.	-	-	1	6.7
Yes.	-	-	13	86.7
2.12 Does this software allow the first part of the task to be executed in different ways?				
Yes.	2	13.3	10	66.7
Perhaps.	1	6.7	1	6.7
Yes, but it is necessary to have advanced knowledge and this is rarely used.	1	6.7	0	0.0
Yes, it is possible to import the floor plan from AutoCAD and extrude the walls.	0	0.0	2	13.3
Yes. In this experiment it was used a technique based on generic entities that it would not be usual.	0	0.0	2	13.3
No.	11	73.3	0	0.0
2.13 Was the set of graphical and non graphic information provided for the task enough to complete the first part of the task?				
Yes.	15	100.0	15	100.0
No.	0	0.0	0	0.0
3. Second part of standard task				
3.1 What graphic products from the first part of this task will have to be individually altered to make the changes required for the task?				
All of them.	15	100.0	0	0.0
Only the 3D model.	0	0.0	14	93.3

Questions and answers	AutoCAD		Revit Arc.	
	#	%	#	%
The simple update of the product will result in the automatic updating of all the others.	0	0.0	1	6.7
3.2 Briefly, how will the proposed alterations be carried out with this software?				
All products from the first part will have to be individually changed.	15	100,0	0	0.0
The 3D model will be altered and automatically floor plan, cross section, elevations and perspectives will be changed.	0	0.0	9	60.0
Since everything is linked, one needs only to alter it once, be it through 2D or 3D view.	0	0.0	5	33.3
The internal dimensions will be altered through a 2D view of the floor plan. The room height will be altered in a side view and the door will be altered in the information box of the element.	0	0.0	1	6.7
4. Third part of standard task				
4.1 Does the software offer a good original library of elements in 2D and 3D?				
No. The software does not have its own library.	15	100.0	0	0.0
No. The library is too poor.	0	0.0	6	40.0
It is extensive, but it was not built for Brazilian standards and the furniture library is not nice.	0	0.0	1	6.7
It is simple, but it can be edited.	0	0.0	1	6.7
No, but it is not necessary more than what it offers.	0	0.0	1	6.7
No.	0	0.0	5	33.3
Yes.	0	0.0	1	6.7
4.2 Have new blocks been made available by the market (Internet, industry, etc.)?				
No.	0	0.0	3	20.0
I don't know.	0	0.0	4	26.7
Some are available, but they are not of good quality.	1	6.7	0	0.0
Yes, through user exchange in the Internet.	3	20.0	3	20.0
Not in Brazil, but in foreign sites.	0	0.0	2	13.3
Yes.	11	73.3	3	20.0
4.3 Is the procedure for inserting blocks simple and friendly?				
Yes.	15	100.0	14	93.3
Yes. However the basic components can be also considered blocks.	0	0.0	1	6.7
4.4 Can the inserted blocks be associated with non graphic information?				
No.	15	100.0	0	0.0
I don't know.	0	0.0	1	6.7
A few, with restritions.	0	0.0	4	26.7
yes	0	0.0	10	66.7

Questions and answers	AutoCAD		Revit Arc.	
	#	%	#	%
4.5 How much of non graphic information could be associated with the inserted blocks?				
Not applicable.	15	-	-	-
I don't know.	-	-	5	33.3
It was inserted a wooden bench and, besides the dimensions, information such as manufacturer and cost could be included.	-	-	1	6.7
Did not answer.	-	-	9	60.0
5. Fourth part of standard task				
5.1 Does the software have resources to generate automatically additional non graphic information, such as spreadsheets, specification lists and bills of quantities?				
No.	15	100.0	0	0.0
Yes, they were easily generated.	0	0.0	2	13.3
Yes.	0	0.0	13	86.7
5. 2 What non graphic information was generated in this software?				
Not applicable.	15	-	-	-
Several: room areas list, bills of quantities, specification lists, etc.	-	-	3	20.0
Several, but I could have generated more, if I had more time.	-	-	3	20.0
The project was small, therefore, only one spreadsheet was generated	-	-	2	13.3
Did not generate.	-	-	3	20.0
Did not answer.	-	-	4	26.7

We have stated as our first hypothesis the belief that the traditional design method tends strongly to be replaced by a new way of designing when architects adopt BIM tools. This seems to be confirmed by the data collected in our experiment.

The answers to question 1.1 indicate a lack of knowledge about BIM at considerable levels: 66% of the subjects using Revit had either no knowledge of BIM or had a basic knowledge about it. However, this may indicate a lack of theoretical knowledge rather than of an intuitive understanding of what a BIM tool allows the designer to do.

The answers to question 2.1 given by Revit users seem to corroborate this view: 60% of the subjects manifest the intention of designing from the 3D model and only later taking care of 2D representations. This also comes in support to our hypothesis that the traditional design method tends strongly to be replaced by a new way of designing when architects adopt BIM tools.

One could even argue that the amount of subjects answering to this question by saying they are going to use the 3D model simultaneously with 2D views, 40%, weakens our hypothesis. But this is not true for a number of reasons: firstly, they did not say that they were going to work primarily from 2D views, but simply that they would use them along with the 3D model. Secondly, even when they are seeing their project in a 2D view, the data representation remains three-dimensional. A 2D view in a BIM tools is not the same of a 2D drawing in a standard computer-aided drafting tool.

The answers to question 2.1 given by the AutoCAD users also seem to support our view. This system is obviously a non BIM tool, but it provides the means for geometric 3D modeling if the user wants to use them. Of course there are significant interface problems which discourage AutoCAD users from doing 3D modeling in it. The answers given by the AutoCAD users to question number 2.4 may be symptomatic of this discouragement: even though the sample is relatively small two users chose other 3D modelers to overcome AutoCAD limitations. However, this again just seems to support our view: the use of BIM tools seems indeed related to a tendency to design differently from those currently using standard computer-aided drafting tools.

The answers given by the Revit users to questions 2.5 through 2.7 indicates that, although most of them claim either to have never heard of BIM or having only a vague notion of what it means (question 1.1), they seem to have a good understanding of some of most important features of Building Information Modeling, such as the existence of data and behavior attached to the geometric objects.

It seems also from the answer given by Revit users to questions 3.1 and 3.2 that the greater control over the design process and consistency management is one of the greatest attractions of BIM systems. More than 90% of its users understood very well this BIM benefit when they answered that only the 3D model needed to be redone.

One of the biggest problems with BIM seems to be the slow pace of its library extendibility, particularly for those living in emerging economies. Most of BIM tools have native libraries based on North American and European standard construction components and systems. The answer given by Revit users to question 2.10 shows this problem, with 53.3% of subjects saying that the native library does not satisfy the needs of the Brazilian construction industry. The same problem surfaces again in the answers given to question 4.1 where 40% of users said the native library was too poor.

The answers to questions 5.1 and 5.2 indicates that the users of BIM tools are greatly aware of some of its essential features, particularly the ability to generate valuable non graphic information such as bills of quantities, cost estimates and specifications lists.

The design process and consistency management, as mentioned before, is central to the BIM approach with important implications in the design process. Table II shows the implications and results of this in process of making changes in an ongoing design process.

In the experiment with the user's samples, three proposed changes were required: building dimensions, height and door dimensions. Therefore, each modification in a constructive element (wall, slab, door and window) was considered um step of local change independently from the number of commands necessary to execute it. The resulting simplified comparison is shown in Table II.

TABLE II
REQUIRED NUMBER OF STEPS IN IMPLEMENTING CHANGES

	Non BIM Software				BIM Software
	Proposed Changes			Resulting Changes	Proposed Changes
Design Representation	Building Dimensions	Height	Door Dimension	Window Dimensions	
Floor Plan	2	0	1	1	0
Cross Section 1	2	1	1	1	0
Cross Section 2	2	1	0	1	0
Front Elevation	2	1	1	0	0
Back Elevation	2	1	0	1	0
Left Elevation	2	1	0	0	0
Right Elevation	2	1	0	0	0
3D Model	2	1	1	1	4
Subtotal	16	7	5	5	4
Total	**33 steps**				**4 steps**

It is possible to observe that an alteration in the dimensions of the building in the BIM system was executed by users changing only two walls of the 3D model. The non-BIM system required the users to change all the same elements in all fragmented representations.

If all necessary changes required for the task are carried out, the numbers of steps in the non-BIM system amounts to 33, whilst in the BIM tool to just 4 steps.

V. CONCLUSIONS

This brings us back to our research question: can the use of BIM tools be associated with changes of human strategy in the design process? Do the users of BIM tools adopt different design methods if compared with those who use non BIM tools? We believe that the data collected through our experiment and the arguments presented in the previous session have shown that our hypothesis is a promising one. The traditional design method tends to be replaced by a new way of designing when architects adopt BIM tools. The designers who use such tools are increasingly working from 3D holistic models and this makes their work less prone to representational inconsistencies or errors.

There still are lots to be researched in the field of Building Information Modeling and a more extensive survey could be carried out to investigate the questions raised above in a broader way. We reached the conclusion that there is a need for a more profound analysis of BIM regarding its influence on design creativity, one that could find ways of making it more compatible with discovery and innovation.

REFERENCES

[1] Chris Yessios, *"Are We Forgetting Design?"*, in AECBytes Viewpoint, number 10, http://www.aecbytes.com/viewpoint/2004/issue_10.html, November, 2004.

[2] C. Eastman, *"The use of computers instead of drawings in building design"*, in Journal of the American Institute of Architects, pp. 46-50, March 1975.

[3] Chuck Eastman, Paul Teicholz, Raphael Sacks and Kathleen Liston, *"BIM Handbook – A Guide to Building Information Modeling"*, John Wiley & Sons, New Jersey, p. 59, 2008.

[4] Chuck Eastman, Paul Teicholz, Raphael Sacks and Kathleen Liston, *"BIM Handbook – A Guide to Building Information Modeling"*, John Wiley & Sons, New Jersey, pp. 13-14, 2008.

[5] Eddy Krygiel and Bradley Nies, Green BIM – Successful Sustainable Design With Building Information Modeling, Sybex Wiley Publishing, Indianapolis, pp. 34-35, 2008.

[6] Chuck Eastman, Paul Teicholz, Raphael Sacks and Kathleen Liston, *"BIM Handbook – A Guide to Building Information Modeling"*, John Wiley & Sons, New Jersey, pp. 16-21, 93-206, 2008.

[7] Nigel Cross, Henri Christiaans and Kees Dorst, Analysing Design Activity, John Wiley & Sons, Chichester, UK, 1996.

An OpenGL-based Interface to 3D PowerPoint-like Presentations of OpenGL Projects

Serguei A. Mokhov
SGW, EV7.139-2
Department of Computer Science and Software Engineering
Concordia University, Montreal, Quebec, Canada
Email: mokhov@cse.concordia.ca

Miao Song
SGW, EV8.245
Department of Computer Science and Software Engineering
Concordia University, Montreal, Quebec, Canada
Email: m_song@cse.concordia.ca

***Abstract*—We present a multimedia 3D interface to power-point-like presentations in OpenGL. The presentations of such kind are useful to demonstrate projects or conference talks with the demonstration results of a 3D animation, effects, and others alongside the presentation 'in situ' instead of switching between a regular presentation software to the demo and back – the demo and the presentation can be one and the same, embedded together.**

***Index Terms*—slide presentations, 3D, OpenGL, real-time, frameworks, tidgets**

I. Introduction

We present a re-usable open-source[1] framework's design and implementation in C++ to prepare slides for OpenGL-based presentations. We briefly go over the visual and software designs and show some examples and argue why one may want such a framework for their OpenGL [1], [2], [3] project as opposed to the defacto standard widespread generic presentation software [4], [5], [6].

II. Design

The design of this OpenGL slide presentation framework can be split into two parts – the visual and the software. We briefly discuss both of them in the sections that follow.

A. Visual

The actual visual design aspect of the OpenGL-based slides is up to a particular "designer", i.e. the project developer or artist that prepares the demo together with the main project concept. The said designer can set up the main colors of the fonts, the text of the items and the title bar, as well as the display of frame rate, elapsed time, etc. as needed. The main body of each slide is the scene being displayed, potentially with some animation – the design of that scene is completely determined by the nature of the project's scene itself and what is being displayed. The colors of the power points and other textual widgets, let's call them *tidgets*, can be adjusted to match the main background theme. All textual components, the tidgets, can be disabled from being displayed such that they don't clutter the slide screen if there is something happening at the main scene of that slide.

[1]to be officially released, unofficially available

One of such designs in illustrated through some select slides of an actual presentation. In Figure 1 is an example of the title slide; in Figure 2 is an example of a table of contents (TOC). One can see the basic structure of the slides – the slide title is on the top left corner in bold, followed by navigational help line on how to move between the slides, and some help on the default slide control keys at the bottom left, then some state information about the status of the vertex and fragment shaders, and the frames per second (FPS) and the elapsed time metrics tidgets that are updated at run-time, and often are asked about to be shown while doing demo to show how efficient (or inefficient) some real-time graphics algorithms are. Thus far, we have seen "text-only" slides. In Figure 3 is an example of a simple scene with a 3D mesh (that can be interacted with as instructed on the slide and to turn on the animation). The slides in Figure 4, Figure 5, Figure 6, and Figure 7 show progressively the increase of the level-of-detail (LOD) of the graphical information shown on the slide scene from that particular project.

B. Software Architecture

We are not going to cover the complete software design and architecture here as it deserves a separate work to describe as well as this information is detailed within the project documentation. We, therefore, concisely go over the essential core concepts.

Every slide is a scene that can feature its own geometry hierarchy according to the procedural modeling techniques [7], lighting setup (if necessary), pixel and vertex shaders [8], [9], [10], or even custom navigation keys alongside the general callback controls. We use object-oriented design and implementation paradigm in C++ with some software engineering design patterns in place, such as a Composition and Observer [11], [12]. Every slide is a C++ class derived from `Slide`. All slides are grouped under a `Presentation` class instance, which is a collection of slides. The callback controls of the OpenGL state machine, such as `idle()`, `functionKeys()`, `keyboard()`, and `mouse()` typically found in OpenGL programs, are also handled by the `Presentation`-, and `Slide`-derived classes. Typically, the callback handling is done top-down, i.e. the callbacks are activated in the `main()` program, which may do any handling

K. Elleithy (ed.), *Advanced Techniques in Computing Sciences and Software Engineering*,
DOI 10.1007/978-90-481-3660-5_91, © Springer Science+Business Media B.V. 2010

of its own, then pass the handling to the `Presentation` instance, which passes it down to the current slide. This order, however, is not hardwired, and can be changed depending on the requirements of a particular project demo. This layout makes the framework really minimal and easy to integrate with existing or new OpenGL projects written in C++ to enhance project demo interaction and illustration right within the project itself.

The framework as a side bonus has also support for the vertex and fragment shaders [13], [14], [15], [16] that are now popular among graphics community of GPU programmers and beyond that can be enabled and disabled during the demo.

The framework is designed to be source-code portable for the largest part, i.e. it can compile and produce a runnable executable in Windows, Linux, and MacOS X platforms, wherever the OpenGL library and a C++ compiler are available. A corresponding building automation provided in the form of project, workspace, and Makefiles [17].

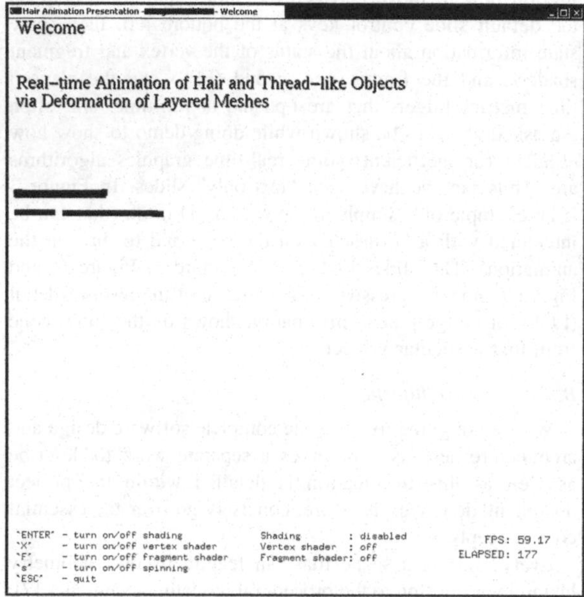

Fig. 1. Welcome Slide Example

III. EXAMPLE

One of the examples of the projects that used the framework is the preliminary real-time hair animation modeling done by students [18] at the early stages of the development as a progress report. The figures in this work illustrate the examples of the framework's usage from that project (some select slides). There were overall 11 slides with a variety of keys and animation options in that presentation. The figures in this work were processed to be optimized for printing on the black-and-white printers. The original color scheme had a black background, teal title and power points, green shader

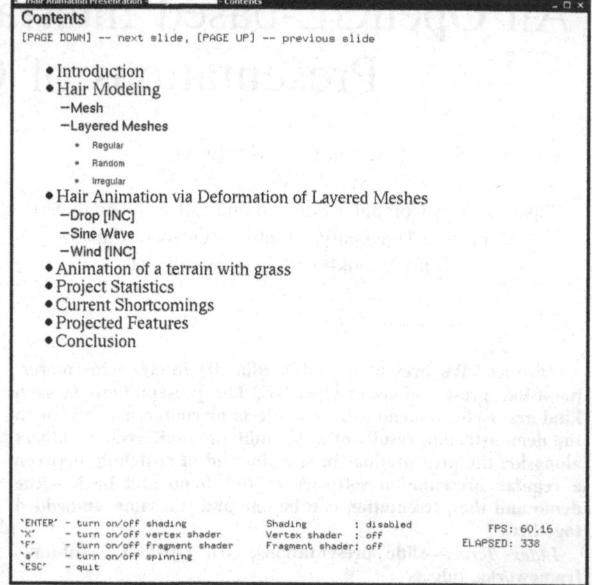

Fig. 2. TOC Slide Example

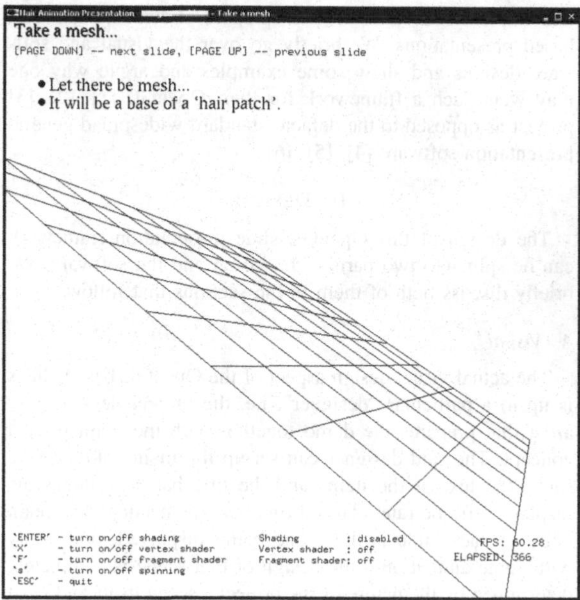

Fig. 3. Content Slide Example 1

status at the bottom, general controls in white a bottom-left, FPS and elapsed time in red at bottom right. The remaining color scheme depended on a particular scene. The particular presented slides are described earlier in Section II-A.

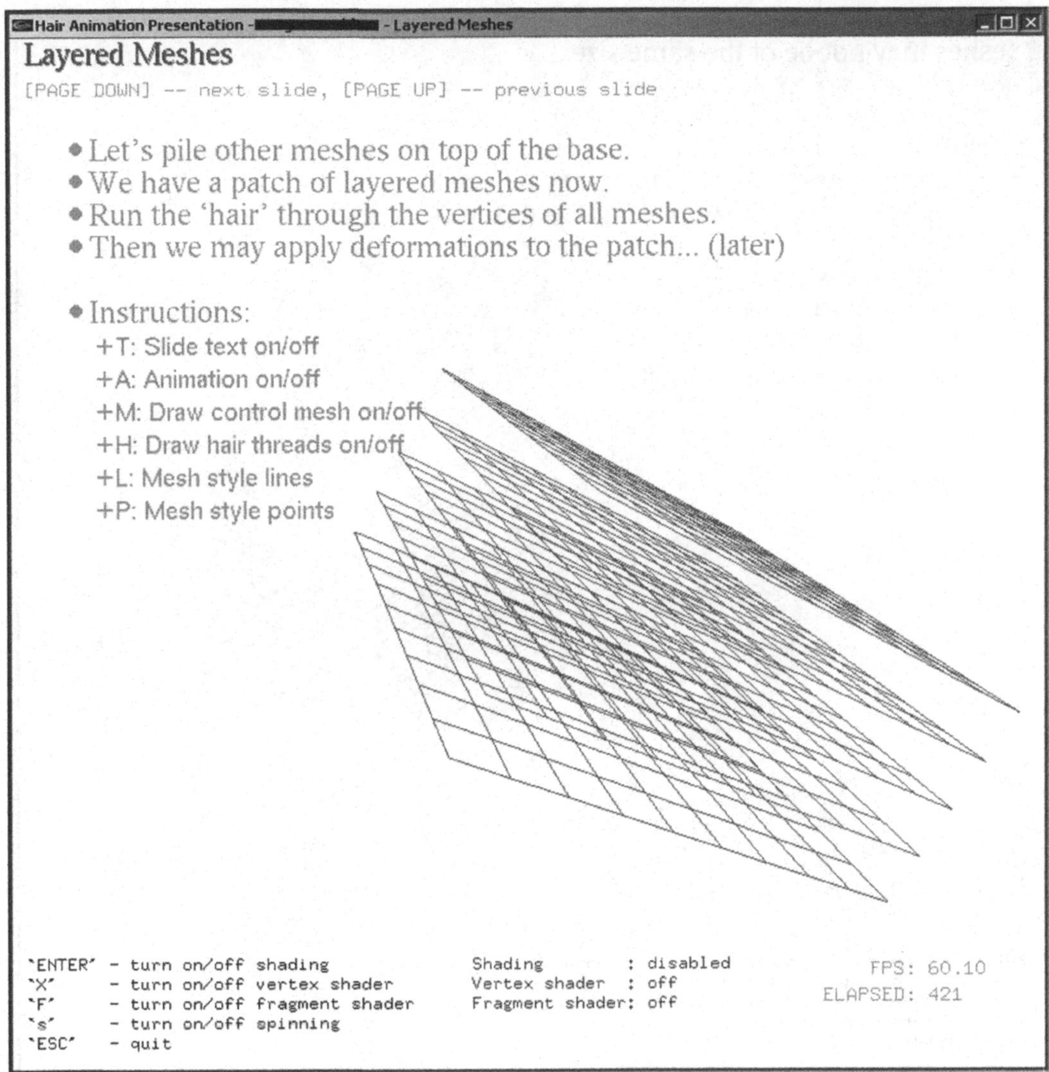

Fig. 4. Content Slide Example 2

IV. Conclusion

Through a collection the design concepts, we have presented a C++-based development framework for OpenGL-based programs to make power-point-like presentation interfaces to computer graphics projects (e.g. for students in computer graphics and beyond), where each slide has a set of navigation capabilities, power points, and tidgets alike. Additionally, each slide includes an OpenGL-based scene that can be interacted with within that slide, with shader support, and other features. It allows the presenters to integrate their computer graphics project with their presentation saving demo and presentation time to illustrate various stages of the project evolution as well as to demo them. A possible downside of this approach to presenting OpenGL project is the additional amount programming effort required to make and troubleshoot the slides. While the slides development may require more effort to produce as they are required to be programmed, they integrate much better and can express much better the animation and modeling aspects, even with movies and sounds that OpenGL and its extensions allow with the demo that other presentation software tools, such as OpenOffice Impress [4] and Microsoft Power Point [6] (their power lies in other aspects that we are not competing with) cannot allow seamless integration with one's own projects while giving the presenter all of the control and interactivity they need.

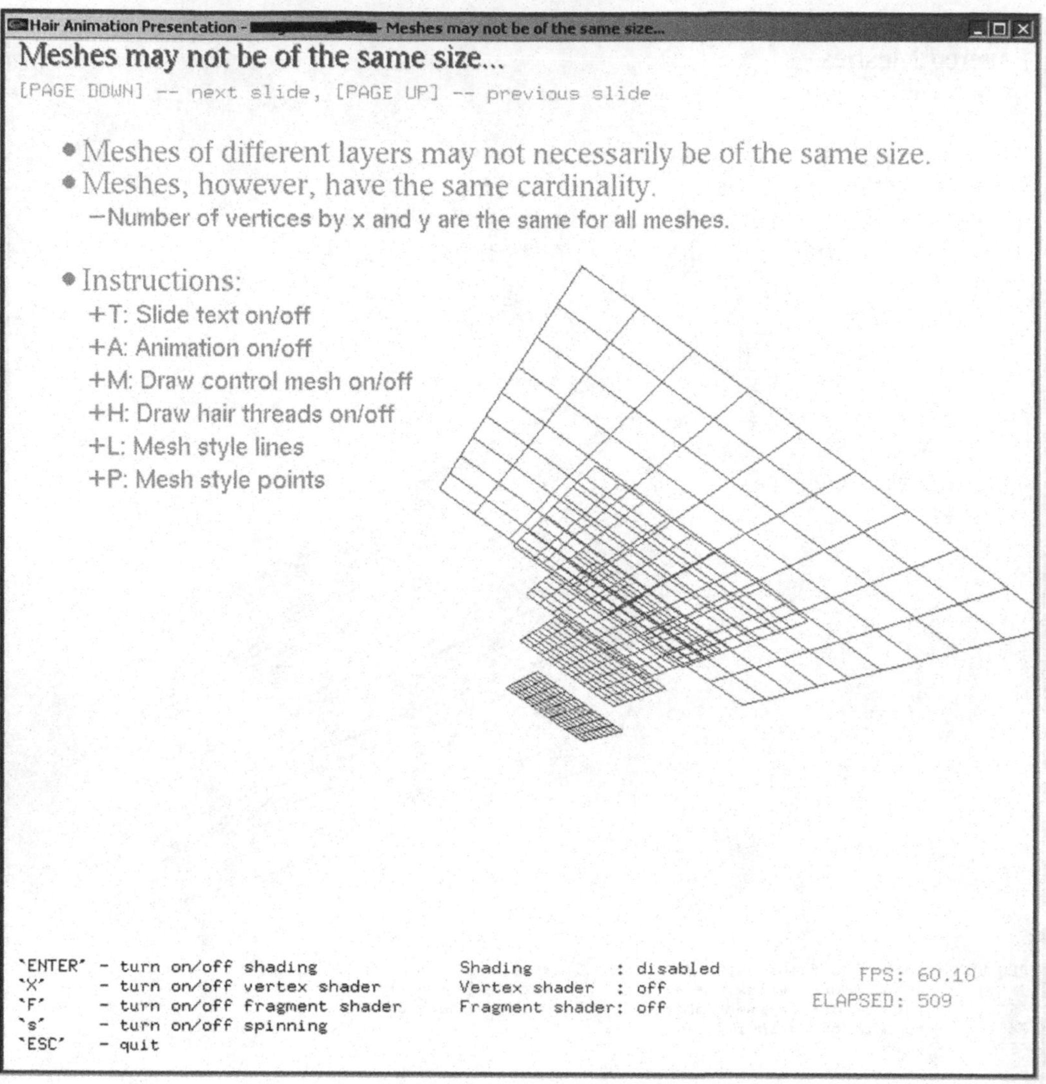

Fig. 5. Content Slide Example 3

V. Future Work

There are a number of items to improve on in the framework to make it more usable and to validate it further with a variety of projects the authors are involved in:

- Allow for XML-based or .properties based configuration files to be imported and exported to generate the slide content and instantiate the appropriate scene, i.e. to reduce the programming effort required thereby making the OpenGL Slide Framework more accessible to non-programmers (e.g. majority users of major 3D modeling software packages such as Maya, Blender, 3DS Max, etc.).
- Add an optional GLUI interface [19] for navigation and control when not in the full-screen mode.
- Further streamline code and make an official open-source release on SourceForge.net.
- Extend the framework to other languages that support OpenGL, e.g. Java, Python, and others by providing the corresponding interfacing code.
- Provide export functions to allow exporting presentations as a sequence of images automatically.
- Integrate the framework with CUGL [20].
- Make the Softbody Simulation Framework [21], [22], [23] demonstration as a set of OpenGL slides and provide the means of integration of the two projects.

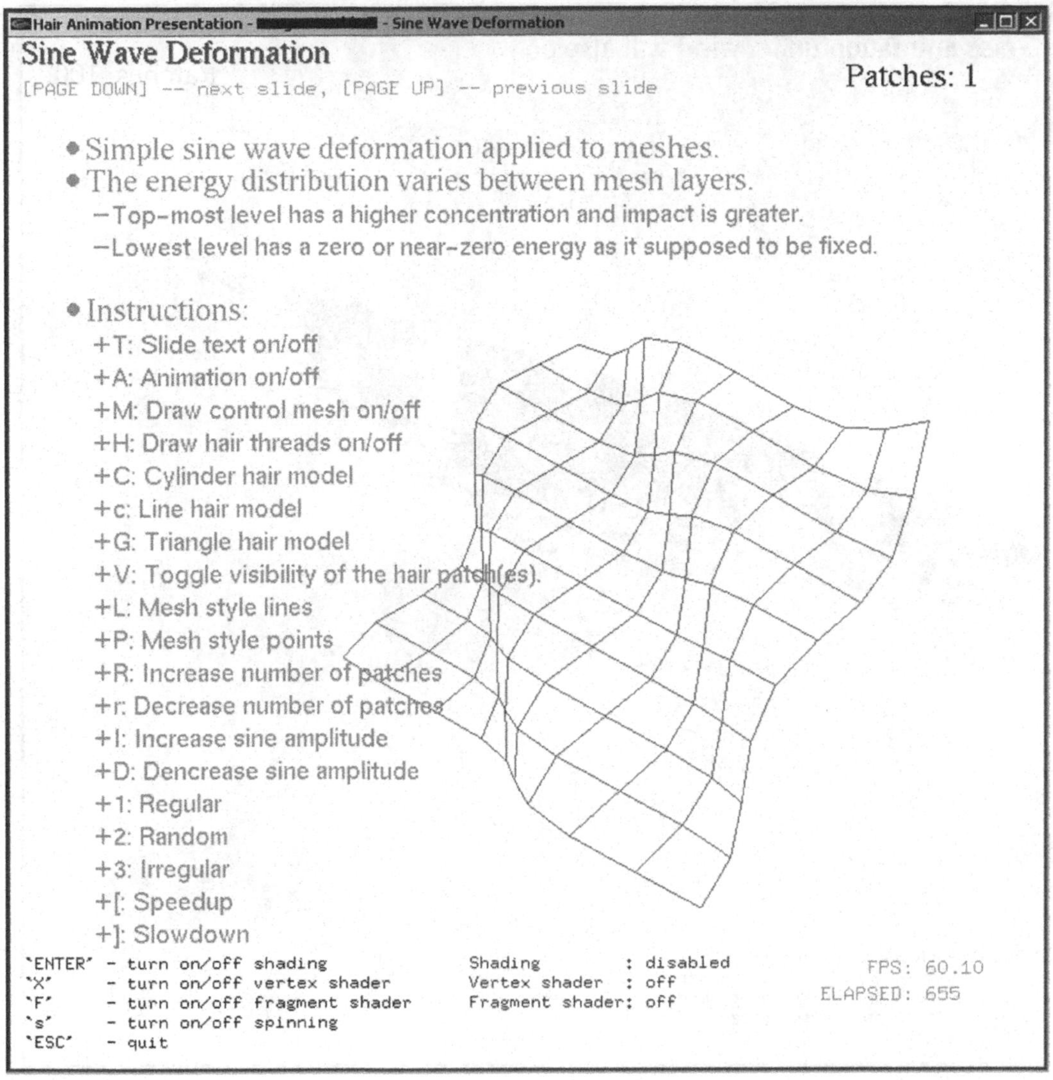

Fig. 6. Content Slide Example 4

Acknowledgment

We acknowledge the reviewers of this work and their constructive feedback. We also acknowledge Dr. Sudhir Mudur for his advice and support, Dr. Peter Grogono, and Ramgopal Rajagopalan. This work was sponsored in part by the Faculty of Engineering and Computer Science, Concordia University, Montreal, Canada.

References

[1] OpenGL Architecture Review Board, "OpenGL," [online], 1998–2008, http://www.opengl.org.

[2] E. Angel, *Interactive Computer Graphics: A Top-Down Approach Using OpenGL.* Addison-Wesley, 2003.

[3] M. Woo, J. Neider, T. Davis, D. Shreiner, and OpenGL Architecture Review Board, *OpenGL Programming Guide: The Official Guide to Learning OpenGL, Version 1.2*, 3rd ed. Addison-Wesley, Oct. 1999, ISBN 0201604582.

[4] SUN Microsystems, Inc., "OpenOffice Impress," [online], 2008, openoffice.org.

[5] The KPresenter Team, "KOffice KPresenter," [online], 1998–2008, www.koffice.org/kpresenter/.

[6] Microsoft, Inc., "Microsoft Office Power Point," [digital], 2008, microsoft.com.

[7] Wikipedia, *Procedural Modeling*. http://en.wikipedia.org/wiki/, 2007.

[8] Various contributors, NVIDIA Corporation, "ARB_vertex_program, Revision 46," NVIDIA Corporation, 2002–2003.

[9] Various contributors, Microsoft Corporation, "ARB_fragment_program, Revision 26," Microsoft Corporation, 2002–2003.

[10] M. J. Kilgard, "All about OpenGL extensions," OpenGL.org, 1998–1999, http://www.opengl.org/resources/features/OGLextensions/.

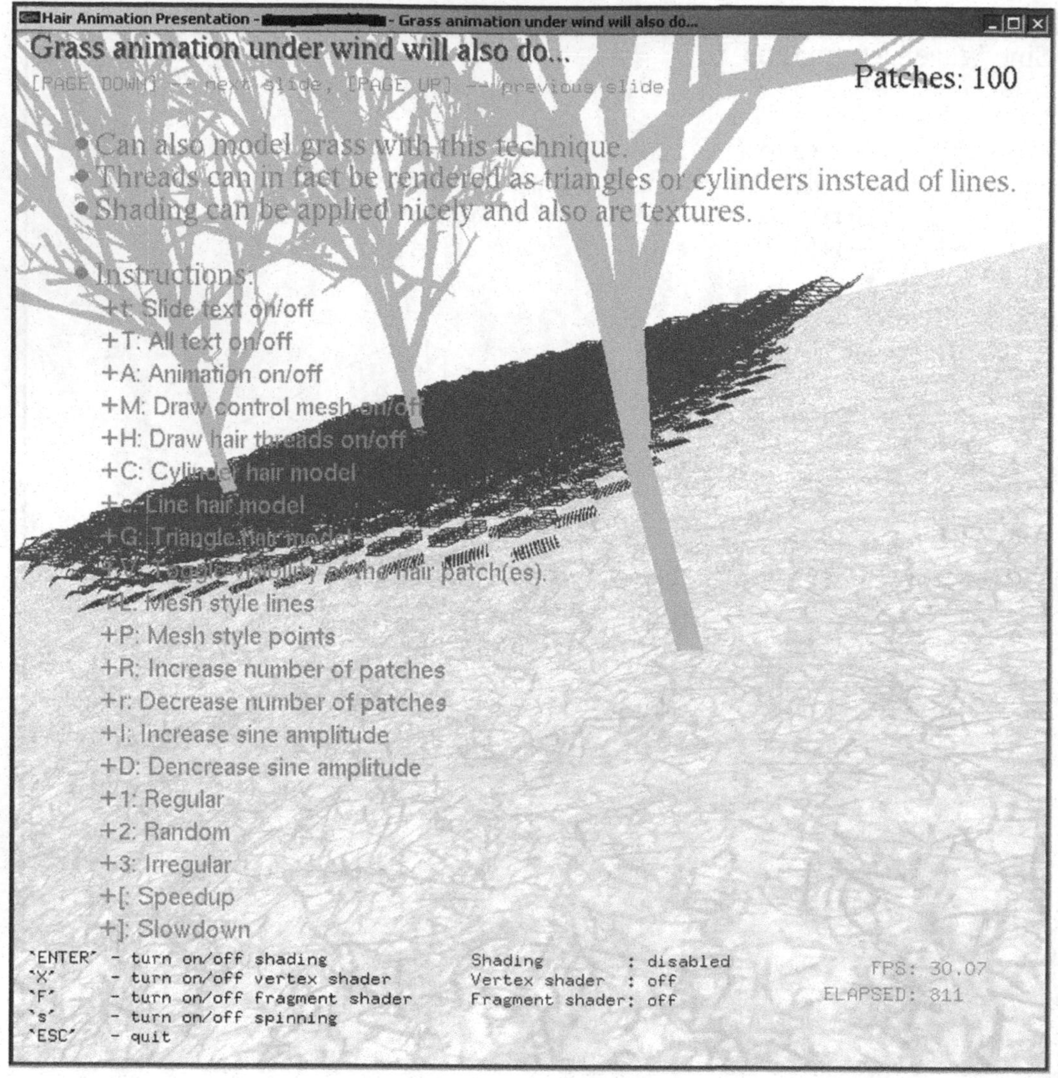

Fig. 7. Content Slide Example 5

[11] E. Gamma, R. Helm, R. Johnson, and J. Vlissides, *Design Patterns: Elements of Reusable Object-Oriented Software*. Addison-Wesley, 1995, ISBN: 0201633612.

[12] S. R. Schach, *Object-Oriented and Classical Software Engineering*, 6th ed. McGraw-Hill, 2005.

[13] R. J. Rost, *OpenGL Shading Language*. Pearson Education, Inc., Feb. 2004, ISBN: 0-321-19789-5.

[14] C. Everitt, "OpenGL ARB vertex program," NVIDIA Corporation, 2003.

[15] 3D Labs, "OpenGL Shading Language demo and documentation," 3D Labs, Inc., 2004, http://developer.3dlabs.com/openGL2/downloads/index.htm.

[16] ——, "OpenGL Shading Language shader examples and source code," 3D Labs, Inc., 2004, http://3dshaders.com/shaderSource.html.

[17] R. Stallman, R. McGrath, P. Smith, and the GNU Project, "GNU Make," Free Software Foundation, Inc., [online], 1997–2006, http://www.gnu.org/software/make/.

[18] S. A. Mokhov, "Real-time animation of hair and thread-like objects via deformation of layered meshes," Department of Computer Science and Software Engineering, Concordia University, Montreal, Canada, 2004, project and report.

[19] P. Rademacher, "GLUI - A GLUT-based user interface library," SourceForge, Jun. 1999, http://glui.sourceforge.net/.

[20] P. Grogono, "Concordia University Graphics Library (CUGL)," [online], Dec. 2005, http://users.encs.concordia.ca/~grogono/Graphics/cugl.html.

[21] M. Song, "Dynamic deformation of uniform elastic two-layer objects," Master's thesis, Department of Computer Science and Software Engineering, Concordia University, Montreal, Canada, Aug. 2007.

[22] M. Song and P. Grogono, "A framework for dynamic deformation of uniform elastic two-layer 2D and 3D objects in OpenGL," in *Proceedings of C3S2E'08*. Montreal, Quebec, Canada: ACM, May 2008, pp. 145–158, ISBN 978-1-60558-101-9.

[23] ——, "An LOD control interface for an OpenGL-based softbody simulation framework," in *Proceedings of CISSE'08*. University of Bridgeport, CT, USA: Springer, Dec. 2008, to appear.

ASPECTS REGARDING THE IMPLEMENTATION OF HSIAO CODE TO THE CACHE LEVEL OF A MEMORY HIERARCHY WITH FPGA XILINX CIRCUITS

*O. Novac, *St. Vari-Kakas, *F.I. Hathazi, **M. Curila and *S. Curila

*University of Oradea, Faculty of Electrical Engineering and Information Technology,
1, Universităţii Str., Oradea, Romania, email: ovnovac@uoradea.ro
** University of Oradea, Faculty of Environmental Protection,
26, Gen. Magheru Street, Oradea, România

***Abstract*-In this paper we will apply a SEC-DED code to the cache level of a memory hierarchy. From the category of SEC-DED (Single Error Correction Double Error Detection) codes we select the Hsiao code. The Hsiao code is a odd-weight-column SEC-DED code. For correction of single-bit error we use a syndrome decoder, a syndrome generator and the check bits generator circuit.**

1. Introduction.

There are many possible fault types in cache memories (that reduce the reliability): Permanent faults (Stuck-at, Short, Open-line, Bridging, Stuck-open, Indetermination), Transient faults, and Intermittent faults, [1]. Anothert metod to solve the problem of faults is the method of testing/correcting errors inside the chip, [2]. If we will use a error correcting code inside the chip, we will obtain memory chips with high reliability, low cost and greater memory capacity, [3]. We must give a special attention, if we use this technique, at the size of the memory chip when we test and correct the memory cells, and we must take count of additional cells for testing bits, [4], [5]. Correction of single error in the cache level of a memory hierarchy, is a correction of single error at parallel transfer and data memorizing, this correction can be done using the scheme presented in figure 3.

In figure 1, we present a typical memory hierarchy.

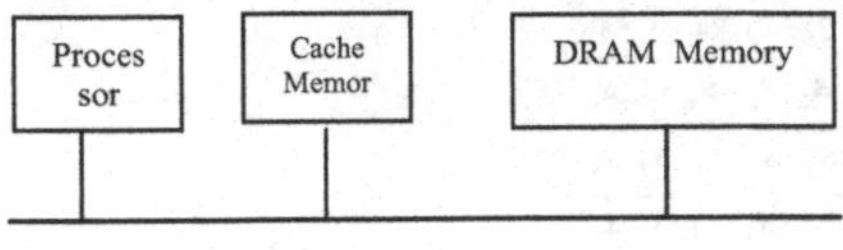

Figure 1 Memory hierarchy

In modern computer systems, at the cache level of the memory hierarchy, we can succesfuly apply multiple error correction codes. This codes for detection and correction of errors are added to memories to obtain a better dependability. In high speed memories the most used codes are Single bit Error Correcting and Double bit Error Detection codes (SEC-DED), [6]. This codes can be implemented in parallel as linear codes for this type of memories. We have choose this Hsiao code, because his properties. An important application of this code is in recovery from multiple errors. This is done with an automatic reconfiguration technique that uses the concept of address skewing to disperse such multiple errors into correctable errors. The Hsiao code is a modified Hamming code, with an odd-weight-column, because every H matrix column vector is odd weight. This code has the minimum number of 1's in the H matrix, which makes the hardware and the speed of the encoding/decoding circuit optimal. Matrix H is presented above in figure 2.

2. Block diagram of a cache memory

În figure 2.a) is presented the block diagram of cache memory. This block diagram is presented because we want to present how is splited the cache memory in two parts. First part presented in figure 2.a) is the cache Tag memory (also named Tag RAM), and the second

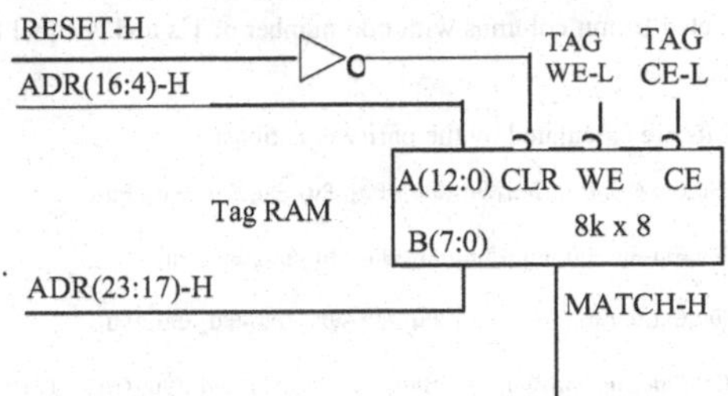

Figure 2.a)

part, presented in figure 2. b) is the cache RAM memory.

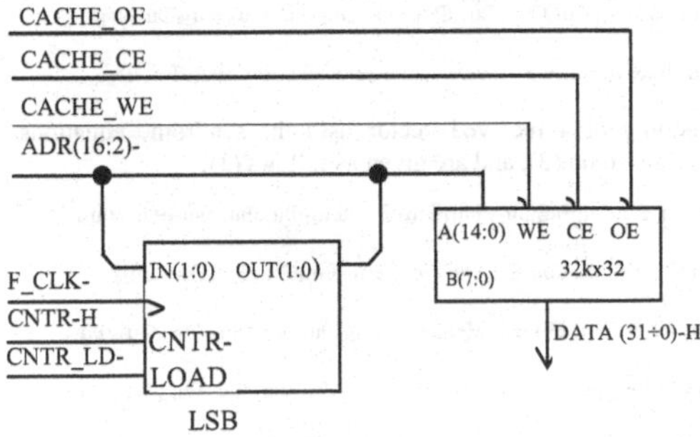

Figure 2. b)

K. Elleithy (ed.), *Advanced Techniques in Computing Sciences and Software Engineering*,
DOI 10.1007/978-90-481-3660-5_92, © Springer Science+Business Media B.V. 2010

For the Cache RAM part of the cache memory we will use a (39,32,7), Hsiao code, this code has 7 control bits and 32 data useful bits. For this code (39,32,7), we have k = 7 control bits, u = 32 useful (data) bits and the total number of code is t = 39. In this case for correcting single bit error, between this two values, u and k, we have satisfied the condition , $2^k > u+k+1$. Usualy it is enough a number of k= 6 control bits, but we will use k = 7 control bits, the last bit is used for double bit error detection.
The Hsiao code used for the part of cache RAM, is defined by matrix H, given by (1):

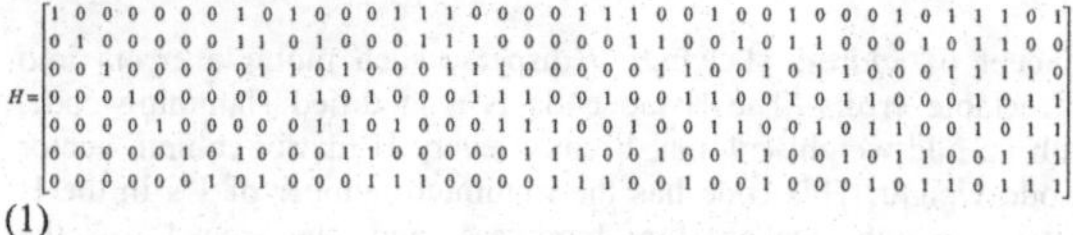
(1)

Figure 3. Hsiao Matrix

3. Applying SEC-DED code to cache RAM

A typical codeword, of this matrix has the folowing form: $u=(c_0c_1c_2c_3c_4c_5c_6u_0u_1u_2u_3u_4u_5u_6u_7u_8u_9u_{10}u_{11}u_{12}u_{13}u_{14}u_{15}u_{16}u_{17}u_{18}u_{19}u_{20}u_{21}u_{22}u_{23}u_{24}u_{25}u_{26}u_{27}u_{28}u_{29}u_{30}u_{31})$ and has parities in position 1,2,3,4,5,6 and 7 and data bits elsewhere (from position 8 to 39). We have select for matrix H, this placement for 1's and 0's to obtain well-balanced equations for control bits. H matrix has on first six rows and six columns a unit matrix, and in continuare this matrix is constructed fromn columns with odd number of 1's and is equal to 3 or 5.

Control bits are calculated by the parity equations:

$$c_0=u_0\oplus u_2\oplus u_6\oplus u_7\oplus u_8\oplus u_{13}\oplus u_{14}\oplus u_{15}\oplus u_{18}\oplus u_{21}\oplus u_{25}\oplus u_{27}\oplus u_{28}\oplus u_{29}\oplus u_{31}$$

$$c_1=u_0\oplus u_1\oplus u_3\oplus u_7\oplus u_8\oplus u_9\oplus u_{15}\oplus u_{16}\oplus u_{19}\oplus u_{21}\oplus u_{22}\oplus u_{26}\oplus u_{28}\oplus u_{29}$$

$$c_2=u_1\oplus u_2\oplus u_4\oplus u_8\oplus u_9\oplus u_{10}\oplus u_{16}\oplus u_{17}\oplus u_{20}\oplus u_{22}\oplus u_{23}\oplus u_{27}\oplus u_{28}\oplus u_{29}\oplus u_{30}$$

$$c_3=u_2\oplus u_3\oplus u_5\oplus u_9\oplus u_{10}\oplus u_{11}\oplus u_{14}\oplus u_{17}\oplus u_{18}\oplus u_{21}\oplus u_{23}\oplus u_{24}\oplus u_{28}\oplus u_{30}\oplus u_{31} \quad (2)$$

$$c_4=u_3\oplus u_4\oplus u_6\oplus u_{10}\oplus u_{11}\oplus u_{12}\oplus u_{15}\oplus u_{18}\oplus u_{19}\oplus u_{22}\oplus u_{24}\oplus u_{25}\oplus u_{28}\oplus u_{30}\oplus u_{31}$$

$$c_5=u_0\oplus u_4\oplus u_5\oplus u_{11}\oplus u_{12}\oplus u_{13}\oplus u_{15}\oplus u_{19}\oplus u_{20}\oplus u_{23}\oplus u_{25}\oplus u_{26}\oplus u_{29}\oplus u_{30}\oplus u_{31}$$

$$c_6=u_1\oplus u_5\oplus u_6\oplus u_7\oplus u_{12}\oplus u_{13}\oplus u_{14}\oplus u_{17}\oplus u_{20}\oplus u_{24}\oplus u_{26}\oplus u_{27}\oplus u_{29}\oplus u_{30}\oplus u_{31}$$

The decoding of a received vector uses the syndrome equations, which are based on (2), and are given as follows (3):

$$s_0=c_0\oplus u_0\oplus u_2\oplus u_6\oplus u_7\oplus u_8\oplus u_{13}\oplus u_{14}\oplus u_{15}\oplus u_{18}\oplus u_{21}\oplus u_{25}\oplus u_{27}\oplus u_{28}\oplus u_{29}\oplus u_{31}$$

$$s_1=c_1\oplus u_0\oplus u_1\oplus u_3\oplus u_7\oplus u_8\oplus u_9\oplus u_{15}\oplus u_{16}\oplus u_{19}\oplus u_{21}\oplus u_{22}\oplus u_{26}\oplus u_{28}\oplus u_{29}$$

$$s_2=c_2\oplus u_1\oplus u_2\oplus u_4\oplus u_8\oplus u_9\oplus u_{10}\oplus u_{16}\oplus u_{17}\oplus u_{20}\oplus u_{22}\oplus u_{23}\oplus u_{27}\oplus u_{28}\oplus u_{29}\oplus u_{30}$$

$$s_3=c_3\oplus u_2\oplus u_3\oplus u_5\oplus u_9\oplus u_{10}\oplus u_{11}\oplus u_{14}\oplus u_{17}\oplus u_{18}\oplus u_{21}\oplus u_{23}\oplus u_{24}\oplus u_{28}\oplus u_{30}\oplus u_{31} \quad (3)$$

$$s_4=c_4\oplus u_3\oplus u_4\oplus u_6\oplus u_{10}\oplus u_{11}\oplus u_{12}\oplus u_{15}\oplus u_{18}\oplus u_{19}\oplus u_{22}\oplus u_{24}\oplus u_{25}\oplus u_{28}\oplus u_{30}\oplus u_{31}$$

$$s_5=c_5\oplus u_0\oplus u_4\oplus u_5\oplus u_{11}\oplus u_{12}\oplus u_{13}\oplus u_{15}\oplus u_{19}\oplus u_{20}\oplus u_{23}\oplus u_{25}\oplus u_{26}\oplus u_{29}\oplus u_{30}\oplus u_{31}$$

$$s_6=c_6\oplus u_1\oplus u_5\oplus u_6\oplus u_7\oplus u_{12}\oplus u_{13}\oplus u_{14}\oplus u_{17}\oplus u_{20}\oplus u_{24}\oplus u_{26}\oplus u_{27}\oplus u_{29}\oplus u_{30}\oplus u_{31}$$

We will apply this SEC-DED code to cache RAM memory of the system with the capacity of 128K x 8 bits. When we will read the information from the cache RAM, we will read the useful data bits (u_0 u_1 u_2 u_3 u_4 u_5 u_6 u_7 u_8 u_9 u_{10} u_{11} u_{12} u_{13} u_{14} u_{15} u_{16} u_{17} u_{18} u_{19} u_{420} u_{21} u_{22} u_{23} u_{24} u_{25} u_{26} u_{27} u_{28} u_{29} u_{30} u_{31}) and also the control bits (c_0 c_1 c_2 c_3 c_4 c_5). We will implement with XOR gates, the equations (2). We will generate the control bits c_0' c_1' c_2' c_3' c_4' c_5' and c_6', from data bits that we have read from the cache Tag. For example, to generate the control bit c_0', we will use the equation:

$$c_0'=u_0\oplus u_2\oplus u_6\oplus u_7\oplus u_8\oplus u_{13}\oplus u_{14}\oplus u_{15}\oplus u_{18}\oplus u_{21}\oplus u_{25}\oplus u_{27}\oplus u_{28}\oplus u_{29}\oplus u_{31},$$

and to implement this equation we will use 14 XOR gates with 2 inputs, situated on four levels, aspect presented in figure 4. We will do in the same mode to generate all control bits, c_1', c_2', c_3', c_4', c_5', c_6'. The generated control bits (c_0' c_1' c_2' c_3' c_4' c_5' c_6') are compared with control bits that we have read from the cache RAM (c_0 c_1 c_2 c_3 c_4 c_5 c_6), also with 2 input XOR gates, and we get as result syndrome equations: $s_0= c_0 \oplus c_0'$, $s_1= c_1 \oplus c_1'$, $s_2= c_2 \oplus c_2'$, $s_3= c_3 \oplus c_3'$, $s_4= c_4 \oplus c_4'$, $s_5= c_5 \oplus c_5'$, $s_6=c_6\oplus c_6'$. We will connect 7 NOT gates, on each syndrome line, and we will construct with 32 AND gates with 7 inputs, the syndrome decoder. The equations to built the syndrome decoder are:

$$u'_0 = s_0 \cdot s_1 \cdot \overline{s_2} \cdot \overline{s_3} \cdot \overline{s_4} \cdot s_5 \cdot \overline{s_6}$$
$$u'_1 = \overline{s_0} \cdot s_1 \cdot s_2 \cdot \overline{s_3} \cdot \overline{s_4} \cdot \overline{s_5} \cdot s_6$$
$$u'_2 = s_0 \cdot \overline{s_1} \cdot s_2 \cdot s_3 \cdot \overline{s_4} \cdot \overline{s_5} \cdot \overline{s_6}$$
$$u'_3 = \overline{s_0} \cdot s_1 \cdot \overline{s_2} \cdot s_3 \cdot s_4 \cdot \overline{s_5} \cdot \overline{s_6}$$
$$u'_3 = \overline{s_0} \cdot \overline{s_1} \cdot s_2 \cdot \overline{s_3} \cdot s_4 \cdot s_5 \cdot \overline{s_6}$$
$$u'_5 = \overline{s_0} \cdot \overline{s_1} \cdot \overline{s_2} \cdot s_3 \cdot \overline{s_4} \cdot s_5 \cdot s_6$$
$$u'_6 = s_0 \cdot \overline{s_1} \cdot \overline{s_2} \cdot \overline{s_3} \cdot s_4 \cdot \overline{s_5} \cdot s_6$$
$$u'_7 = s_0 \cdot s_1 \cdot \overline{s_2} \cdot \overline{s_3} \cdot \overline{s_4} \cdot \overline{s_5} \cdot s_6$$
$$u'_8 = s_0 \cdot s_1 \cdot s_2 \cdot \overline{s_3} \cdot \overline{s_4} \cdot \overline{s_5} \cdot \overline{s_6}$$
$$u'_9 = \overline{s_0} \cdot s_1 \cdot s_2 \cdot s_3 \cdot \overline{s_4} \cdot \overline{s_5} \cdot \overline{s_6}$$
$$u'_{10} = \overline{s_0} \cdot \overline{s_1} \cdot s_2 \cdot s_3 \cdot s_4 \cdot \overline{s_5} \cdot \overline{s_6}$$
$$u'_{11} = \overline{s_0} \cdot \overline{s_1} \cdot \overline{s_2} \cdot s_3 \cdot s_4 \cdot s_5 \cdot \overline{s_6}$$
$$u'_{12} = \overline{s_0} \cdot \overline{s_1} \cdot \overline{s_2} \cdot \overline{s_3} \cdot s_4 \cdot s_5 \cdot s_6$$
$$u'_{13} = s_0 \cdot \overline{s_1} \cdot \overline{s_2} \cdot \overline{s_3} \cdot \overline{s_4} \cdot s_5 \cdot s_6$$
$$u'_{14} = s_0 \cdot \overline{s_1} \cdot \overline{s_2} \cdot s_3 \cdot \overline{s_4} \cdot \overline{s_5} \cdot s_6$$
$$u'_{15} = s_0 \cdot s_1 \cdot \overline{s_2} \cdot \overline{s_3} \cdot s_4 \cdot \overline{s_5} \cdot \overline{s_6}$$
$$u'_{16} = \overline{s_0} \cdot s_1 \cdot s_2 \cdot \overline{s_3} \cdot \overline{s_4} \cdot s_5 \cdot \overline{s_6} \quad (4)$$
$$u_{17} = \overline{s_0} \cdot \overline{s_1} \cdot s_2 \cdot s_3 \cdot \overline{s_4} \cdot \overline{s_5} \cdot s_6$$
$$u_{18} = s_0 \cdot \overline{s_1} \cdot \overline{s_2} \cdot s_3 \cdot s_4 \cdot \overline{s_5} \cdot \overline{s_6}$$
$$u'_{19} = \overline{s_0} \cdot s_1 \cdot \overline{s_2} \cdot \overline{s_3} \cdot s_4 \cdot s_5 \cdot \overline{s_6}$$
$$u'_{20} = \overline{s_0} \cdot \overline{s_1} \cdot s_2 \cdot \overline{s_3} \cdot \overline{s_4} \cdot s_5 \cdot s_6$$
$$u'_{21} = s_0 \cdot s_1 \cdot \overline{s_2} \cdot s_3 \cdot \overline{s_4} \cdot \overline{s_5} \cdot \overline{s_6}$$

$u'_{22} = \overline{s_0} \cdot s_1 \cdot s_2 \cdot \overline{s_3} \cdot s_4 \cdot \overline{s_5} \cdot \overline{s_6}$

$u'_{23} = \overline{s_0} \cdot \overline{s_1} \cdot s_2 \cdot s_3 \cdot \overline{s_4} \cdot s_5 \cdot \overline{s_6}$

$u'_{24} = \overline{s_0} \cdot \overline{s_1} \cdot \overline{s_2} \cdot s_3 \cdot s_4 \cdot \overline{s_5} \cdot s_6$

$u'_{25} = s_0 \cdot \overline{s_1} \cdot \overline{s_2} \cdot \overline{s_3} \cdot s_4 \cdot s_5 \cdot \overline{s_6}$

$u'_{26} = \overline{s_0} \cdot s_1 \cdot \overline{s_2} \cdot \overline{s_3} \cdot \overline{s_4} \cdot s_5 \cdot s_6$

$u'_{27} = s_0 \cdot \overline{s_1} \cdot s_2 \cdot \overline{s_3} \cdot \overline{s_4 \cdot s_5} \cdot s_6$

$u'_{28} = s_0 \cdot s_1 \cdot s_2 \cdot s_3 \cdot s_4 \cdot \overline{s_5} \cdot \overline{s_6}$

$u'_{29} = s_0 \cdot s_1 \cdot s_2 \cdot \overline{s_3} \cdot \overline{s_4} \cdot s_5 \cdot s_6$

$u'_{30} = \overline{s_0} \cdot \overline{s_1} \cdot s_2 \cdot s_3 \cdot s_4 \cdot s_5 \cdot s_6$

$u'_{31} = s_0 \cdot \overline{s_1} \cdot \overline{s_1} \cdot s_3 \cdot s_4 \cdot s_5 \cdot s_6$

We will connect the output of AND gates to an input of an XOR gate, the second input of the gate has the data bit read from the cache Tag. To correct the data bits we will use 32 XOR gates with 2 inputs, this correction is realised with the following equations:

$$\begin{aligned}
u_{0cor} &= u_0 \oplus u_0', \\
u_{1cor} &= u_1 \oplus u_1', \\
u_{2cor} &= u_2 \oplus u_2', \\
u_{3cor} &= u_3 \oplus u_3', \\
u_{4cor} &= u_4 \oplus u_4', \\
u_{5cor} &= u_5 \oplus u_5', \\
u_{6cor} &= u_6 \oplus u_6', \\
u_{7cor} &= u_7 \oplus u_7' \\
u_{8cor} &= u_8 \oplus u_8', \\
u_{9cor} &= u_9 \oplus u_9', \\
u_{10cor} &= u_{10} \oplus u_{10}', \\
u_{11cor} &= u_{11} \oplus u_{11}', \\
u_{12cor} &= u_{12} \oplus u_{12}', \\
u_{13cor} &= u_{13} \oplus u_{13}', \\
u_{14cor} &= u_{14} \oplus u_{14}' \\
u_{15cor} &= u_{15} \oplus u_{15}' \qquad (5) \\
u_{16cor} &= u_{16} \oplus u_{16}', \\
u_{17cor} &= u_{17} \oplus u_{17}', \\
u_{18cor} &= u_{18} \oplus u_{18}', \\
u_{19cor} &= u_{19} \oplus u_{19}', \\
u_{20cor} &= u_{20} \oplus u_{20}', \\
u_{21cor} &= u_{21} \oplus u_{21}', \\
u_{22cor} &= u_{22} \oplus u_{22}' \\
u_{23cor} &= u_{23} \oplus u_{23}', \\
u_{24cor} &= u_{24} \oplus u_{24}', \\
u_{25cor} &= u_{25} \oplus u_{25}', \\
u_{26cor} &= u_{26} \oplus u_{26}', \\
u_{27cor} &= u_{27} \oplus u_{27}', \\
u_{28cor} &= u_{28} \oplus u_{28}', \\
u_{29cor} &= u_{29} \oplus u_{29}' \\
u_{30cor} &= u_{30} \oplus u_{30}' \\
u_{31cor} &= u_{31} \oplus u_{31}'
\end{aligned}$$

We generate the Hsiao matrix, so that the column vectors corresponding to useful information bits to be different one from other. In figure 4 we will use three shift registers and one exclusive-or gate to implement the Hamming matrix. We present in figure 4 a scheme used for single error correction.

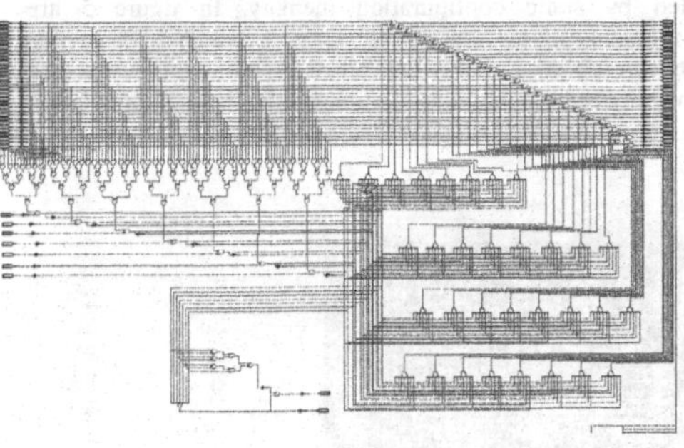

Figure 4. Single error correction scheme

If we write in cache RAM memory the control bits are generated from useful data bits, and they are obtained from data bus using (2), in cache RAM memory we will write codewords with followimg :

$u=(c_0c_1c_2c_3c_4c_5c_6u_0u_1u_2u_3u_4u_5u_6u_7u_8u_9u_{10}u_{11}u_{12}u_{13}u_{14}u_{15}u_{16}u_{17}u_{18}u_{19}$ $u_{20}u_{21}u_{22}u_{23}u_{24}u_{25}u_{26}u_{27}u_{28}u_{29}u_{30}u_{31})$.

3. Implementation with FPGA Xilinx circuits of Hsiao code to the cache RAM memory.

In this part we will implement the error correction scheme from figure 4, with FPGA Xilinx circuits. The designing process with FPGA Xilinx circuits is fast and efficient. Internl structure of a FPGA circuit is has a matrix composed from Configurable Logic Blocks (CLB) and Programable Switch Matrices (PSM), sourended by I/O pins.

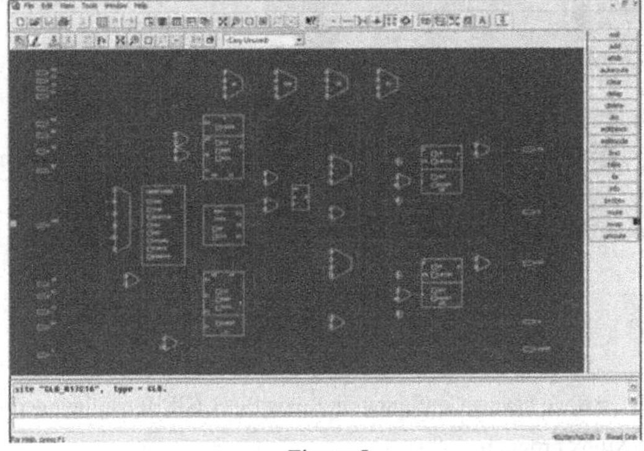

Figure 5.

Programable internal structure include two configurable elements: Configurable Logic Blocks, with functional elements and implement the designed logical structure and Input Output Blocks (IOB), witch realise the interface between internal signals and the outside of circuit, using pins. The logical function realised by CLB is implemented by static configuration memory. In figure 5 are presented elements of a Configurable Logic Block (CLB). In this figure, we can observe that CLB block is not configurate, the connections between block will be figurated after the implementation of scheme from figure 4. .

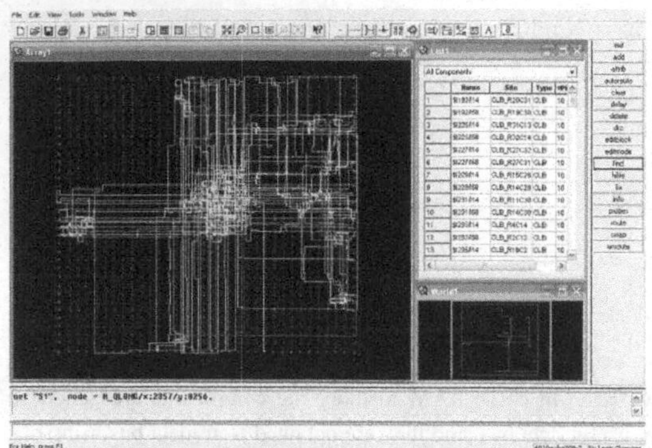

Figure 6.

We have implemented the Hsiao code (39,32,7), for cache RAM memory, with FPGA Xilinx circuits from XC4000 family. This implementation is presented în figures 6, 7 şi 8. In figure 6 we can observe board covering with CLB's and also the connections between CLB's.

In figure 7 we can observe a part of CLB's and the connections between CLB's.

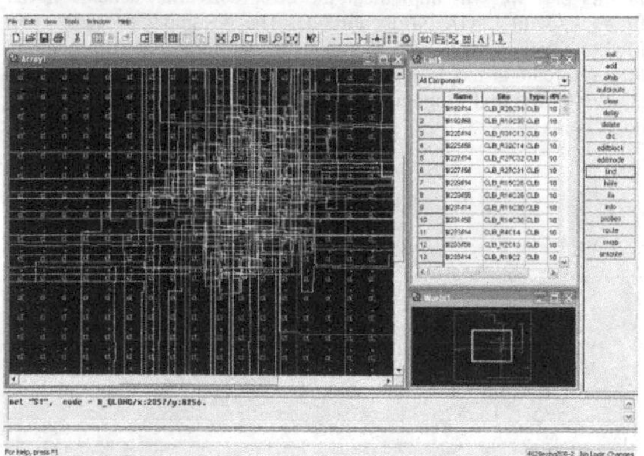

Figure 7.

In figure 8 we have a Configuration Logic Block (CLB) configurated with Xilinx program.

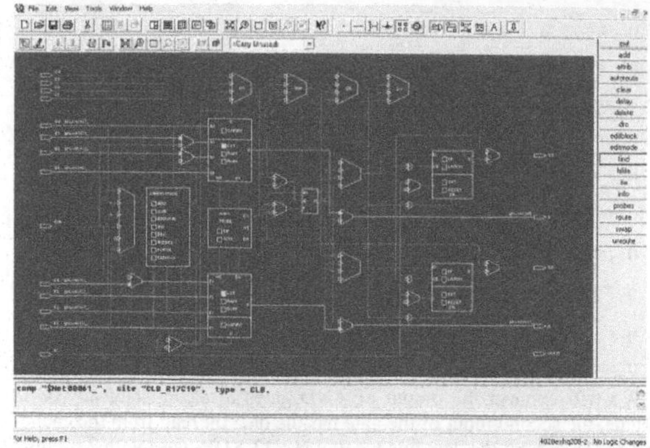

Figure 8.

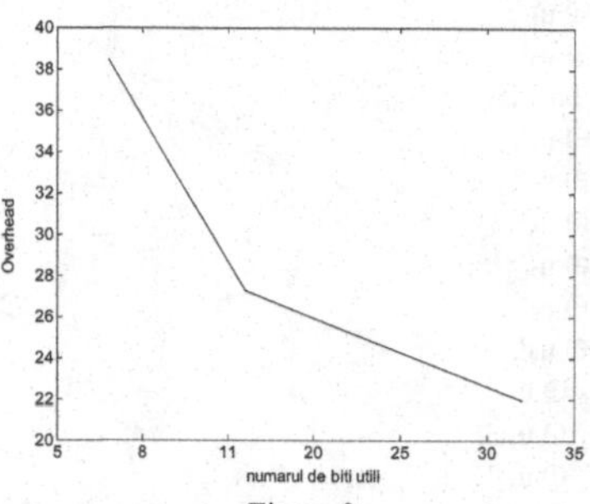

Figure 9

We can observe in figure 9, that if we increase the number of data bits of Hsiao code, the overhead is decreasing.

CONCLUSIONS

In the first implemetation of cache RAM memory with capacity 8K x 8 bits, without Hsiao code, we have 2 groups of 2 memory circuits and each circuit store 4 bits, so we will store 4 bits x 2 circuits = 8 bits. In second implementation of cache RAM with Hsiao code, we have 2 circuits that store 5 bits (4 data bits and one parity bit), and 2 circuits that stores 6 bits (4 data bits and 2 parity bits), so we will store 10 bits + 12 bits = 22 bits. In this case we have an overhead of 27,27 % is obtained from the next calculus (22-16)x100/22= 27,27 %, [7]. Overhead induced by the supplementary circuits for the error correction (AND gates, NOT gates, XOR gates, OR gates, is 6,25 %, this result is obtained with from Xilinx using a report named map.mrp, this report can give us the number of CLB used for implementation, from total number of CLB. In our case the implementation of Hsiao code use 64 CLB's from a total number of 1024 CLB's (Number of CLBs 64 out of 1024).

ACKNOWLEDGMENT

*"University of Oradea", Department of Computers, Faculty of Electrical Engineering and Information Technology, Oradea, Romania,

**"Politehnica" University of Timişoara, Department of Computers, Timişoara, Romania.

***"University of Oradea", Department of Electrotechnics, Measurements and using of Electrical Energy, Faculty of Electrical Engineering and Information Technology, Oradea, Romania.

REFERENCES

[1] David A. Paterson, John L. Henessy – "Computer architecture. a quantitative approach", *Morgan Kaufmann Publishers*, Inc. 1990-1996.

[2] Ovidiu Novac, Gordan M, Novac M., Data loss rate versus mean time to failure in memory hierarchies. Advances in Systems, Computing Sciences and Software Engineering, *Proceedings of the CISSE'05,* Springer, pp. 305-307, University of Bridgeport, USA, 2005.

[3] Ovidiu Novac, Vladutiu M, St. Vari Kakas, Novac M., Gordan M., A Comparative Study Regarding a Memory Hierarchy with the CDLR SPEC 2000 Simulator, Innovations and Information Sciences and Engineering, *Proceedings of the CISSE'06*, University of Bridgeport, Springer, pp. 369-372, USA, 2006.

[4] Ooi, Y., M. Kashimura, H. Takeuchi, and E. Kawamura, Fault-tolerant architecture in a cache memory control LSI, *IEEE J. of Solid-State Circuits*, Vol. 27, No. 4, pp. 507-514, April 1992.

[5] Philip P. Shirvani and Edward J. McCluskey, "PADded cache: a new fault-tolerance technique for cache memories", *Computer Systems Laboratory, Technical Report* No. 00-802, Stanford University, Stanford, California, December 2000.

[6] T.R.N. Rao, E. Fujiwara,"Error-Control Coding for Computer Systems ", Prentice Hall International Inc., Englewood Cliffs, New Jersey, USA, 1989

[7] Ovidiu Novac, Şt. Vari-Kakas, O. Poszet „Aspects Regarding the use of Error Detecting and Error Corecting Codes in Cache Memories", EMES'07, University of Oradea, 2007

Distributed Environment Integrating Tools for Software Testing

Anna Derezińska, Krzysztof Sarba
Institute of Computer Science, Warsaw University of Technology
A.Derezinska@ii.pw.edu.pl

Abstract – **This work is devoted to problems of testing many programs using various testing tools of different origin. We present a distributed system, called Tester, which manages and automates testing process. The system is based on client-server architecture and integrates testing tools, including commercial tools, open-source and own applications. It uses repository for storing projects to be tested and database with test results. Different performance issues concerning test automation are discussed and experimentally evaluated. The system was used in mutation testing of C# programs using object-oriented mutation operators, in experiments investigating the relation between line and assembly instruction coverage on the set of C++ programs, and statistic analysis of different program characteristics.**

I. Introduction

Software testing is counted as a very labor-consuming activity in the software development. Much effort is devoted into developing various testing techniques and tools [1]. However many of them are still not much widely adopted throughout industry [2]. According to surveys [3] knowledge on the importance of testing in general and unit testing in particular seem low. Developers prefer using simple methodologies and tools, also new approaches should be kept simple and easy to learn – (KISS) Keep It Simple Stupid.

Full automatic testing is recognized as a one of the main target ideas in the testing research roadmap [4]. Automation in software testing can be applied in three main areas: administration of testing process, tests execution and test cases generation [5]. In this paper, we consider only automation of two first areas. It is especially important for dealing with huge number of test runs and test data. Effective management of test resources, like projects under tests, testing scripts and test results, has a big influence on requirements on the system performance.

Besides the test automation and test execution performance, this paper addresses the problem of integration of testing tools [5]. There are many comprehensive testing environments providing different testing tools, but usually originating from the same vendor. Such commercial systems are often not flexible enough to integrate easily with the tools of different origin. The most problems arise, when different kinds of tools, i.e. commercial or open-source ones, should effectively cooperate.

Most testing activities carried on in industry involve retesting already tested code to ascertain that changes did not adversely affect system correctness [2]. Therefore, using regression testing, often the same test suites are run many times, and the code of consecutive projects under tests differ only a little. Similar needs encounter when mutation testing is performed [6]. Mutation testing is a very labor-intensive activity. It requires an automated tool support for generation of mutated programs (so-called mutants) and for tests evaluation. The number of created mutants depends on the selected mutant operators that specify the kind of inserted mistakes of programmers. In typical experiments there can be many (hundreds, thousands) of mutants.

Facing the above mentioned problems, a system, called Tester, has been developed [7], which supports execution of tests. It was designed as a distributed environment integrating different testing tools. The integrated tools can be commercial, open-source or our developed tools. The general feature of the Tester is its flexibility and extensibility. The system was intended as a labor-saving support in different testing activities.

The implemented version of Tester environment integrates tools for code coverage, performance and static analysis, unit testing. It supports automatic testing process, in particular regression testing and mutation testing. It can be used for programs written in different languages, like C++, C#, Java. Test results of different programs obtained from different tools can be analyzed with statistic methods.

The system includes also specialized applications, like HNunit and C++ Coverage Analyzer that combine test results of different tools. HNUnit (Hybrid NUnit) perform unit and coverage tests of a program. C++ Coverage Analyzer compares code coverage in assembler instructions level and source line level taking into account C++ programs [8].

Two kinds of experiments will be discussed in the paper. The first group of experiments concerned performance analysis of Tester environment components and cooperating tools. Taking into account the Tester environment, the efficiency analysis of distributed system configurations, repository communication protocol, and usage of HNUnit was performed. The results can be used for the effective application of the system.

The second group of experiments illustrated the application of the system in different tasks of software testing. Mutation testing of several C# programs was performed using object-oriented mutation operators. On that basis the qualities of test suites were evaluated. Experiments of other programs concerned statistical analysis of C++ programs code coverage. Assembly instruction coverage and line coverage were

K. Elleithy (ed.), *Advanced Techniques in Computing Sciences and Software Engineering*
DOI 10.1007/978-90-481-3660-5_93, © Springer Science+Business Media B.V. 2010

compared showing a significant discrepancy of results. Finally, performance analysis in regression testing and statistic analysis of program profiles and performance measures were performed.

The Tester environment, its architecture, processes and realization are presented in the next section. In Section III we discuss the performance evaluation of the system, comparing different configurations, communication protocols and different testing tools. Next, the application of the system is illustrated by various testing experiments. Section V concludes the paper.

II. TESTER ENVIRONMENT

This section describes architecture and processes of Tester environment and its exemplary realization.

A. *Tester architecture*

General requirements of the Tester system were dealing with the integration of different test tools, high system flexibility, and efficient support for the test automation [7].

The Tester system was designed using modular client-server architecture. Several communicating modules of the system realize different tasks. This solution allowed distributing system modules among different computers, in order to improve efficiency of task realization.

The system consists of the following basic modules: system server, project repository, data base server and system clients. System server module is responsible for fetching a project under test from the project repository, performing tests of the project using selected testing tools, receiving test results and storing them to the data base of test results.

The Tester system can include various client modules, a general purpose client, as well as clients dedicated for the specialized kinds of tests. A client module prepares projects for testing, delivers them to the repository that stores projects and presents test results to a user.

The general structure of the system server is shown in Fig. 1. A structure of a system client is very similar to the server's one. The server and any client have many common sub-modules, but use different parts of the functionality of these sub-modules.

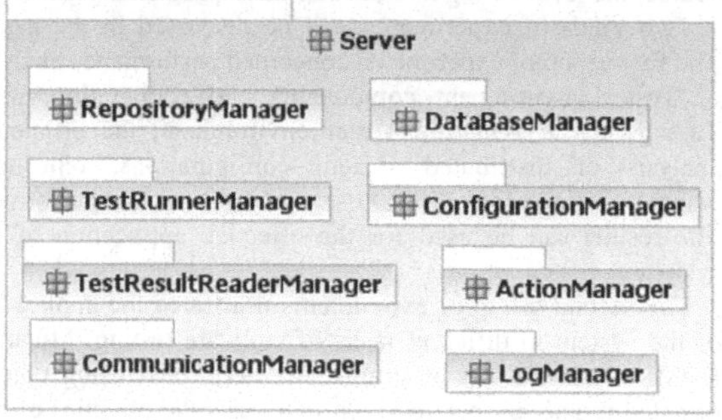

Fig. 1. Architecture of the system server.

Sets of projects under test are stored in the project repository. For any project, only its first version is stored in the repository as entire code. Further versions of the same project are identified by the changes. Using such repository we can spare memory needed for storing of programs. It is especially beneficial when many versions of the same project have to be stored, like for example in regression testing or mutation testing.

Test results are stored in a separate data base. The data base can be asynchronously accessed (locally or remotely) by many system clients. This solution supports parallel execution of different test tasks and therefore assures good system performance.

The Tester system cooperates with different external tools, (Sec. II.C). The architecture is open for the usage of different kinds of tools, taking advantage of existing solutions, as much as possible. In order to automate the testing processes of many programs, we selected such variants of tools that are run in the console (batch) mode. It allowed limiting the necessary interaction with a user to the preparation phases of a testing task, without time-consuming assistance during its realization.

B. *Tester processes*

In the Tester environment three main processes can be distinguished:

1) preparation of test projects and scripts,
2) test realization,
3) presentation and statistical evaluation of results.

During the first process the projects to be tested are prepared and stored in the project repository. A given directory, placed on a disc, in web resources, or on any other storage, can be automatically searched and projects to be tested are selected. A user can accept all projects found during the search procedure or a subset of them. A selected project can be supplemented with additional information about the project type, its arguments, directories with test files, etc. All these data are recorded in a script that can be used for running and testing the given project. Projects and their scripts are stored in the project repository.

The second, main system process consists of the following steps:

a) fetching of a set of projects from the repository,
b) selecting of testing tools,
c) testing the projects using given tools, test scripts, test data, etc.
d) storing of test results in the data base.

Once the Tester system is informed about selected projects and testing tools, other activities are preformed in an automatic way. The testing process is supported by the test scripts. Various testing tools are run in the console mode (batch mode). Therefore a big number of tested programs can be executed without user assistance.

The third process of the Tester system is responsible for processing of test results. A user can select sets of test results stored in the data base. Statistic calculations pointed out by the user are performed on given data sets. The results can be interpreted and presented to a user in a textual or graphical form.

C. *System realization*

The Tester system integrates different commercial, open-source and our own tools for testing programs. They support code coverage and performance analysis, unit testing, and automation of testing process, especially for regression and mutation testing. The projects of different platforms, like C++, .NET and Java were foreseen for testing.

The current version of Tester system cooperates with the following tools:

- IBM Rational PureCoverage [9] - measuring function and line coverage of programs,
- IBM Rational Purify [9] - searching for program errors during static analysis,
- IBM Rational Quantify [9] - measuring program execution times and program profiles of one- and multithreaded programs,
- Golden Runner (a part of FITS tool [10]) - providing assembly instruction coverage of C++ programs,
- CoverageAnalyzer [8] - integrating different analyzers for C++ programs and combining their outcomes (line coverage, assembly instruction coverage and fault injection results).
- NUnit [11] - executing unit tests for programs of .NET platform (e.g. C# programs),
- HNUnit - a hybrid tool combining functionality of NUnit and PureCoverage.

Apart from testing tools, the Tester system uses other general purposes tools. A system module for storing projects to be tested is based on SVN repository [12]. A data base for keeping test results was built using MS SQL Server. Statistic calculations are performed with help of GNU Octave program [13]. We can calculate for example mean values, standard deviation, variance, covariance, Pearson, Spearman and Kendal correlation coefficients, as well as different statistic tests, like Kolomogorov-Smirnow, Chi-quadrat, Kruskall-Wallis, Wilcoxon, t-Student, Mann Whitney [14]. Results of these calculations can be graphically presented using GNU Plot tool.

In order to automate the testing process, for all programs their variants with console interfaces or API were used.

One of a Tester client was dedicated for the automation of mutation testing. Mutation technique can be used for selecting effective test cases, for investigating the quality of software with available test suites, and for comparing the effectiveness of testing methods [6, 15-17]. Mutations are simple changes inserted into a source code. A modified program, so-called "mutant", can be further tested with given test cases. Well designed test suite should detect an error and "kill" the mutant. If the mutant was not killed, either error detecting capabilities of the test suite are not good enough, or the mutant is an equivalent one. An equivalent mutant has the same behavior and generates the same output as the original program, and cannot be killed by any test case.

The specialized Tester client cooperates with the Cream tool [18], a system for creation of mutated programs in C# language. The Cream system is the first tool that can generate mutated C# programs using object-oriented mutation operators. The generated mutants can be stored in the project repository. The Tester runs the mutants for the given unit test suites. Test results are delivered to the data base. Next, the test results are compared with the oracle data generated by the original program. Comparison of results shows whether the test run had detected a mutation introduced in the program (i.e. killed the mutant) or not.

III. PERFORMANCE EVALUATION OF TESTER

This section discusses different experiments devoted to performance evaluation of the Tester environment.

A. *Environment distribution*

The advantage of the distributed architecture is possibility of placing system modules on different computers. System modules, like the Tester system server, the data base server, the repository, and any system client can be run in many possible configurations. For example, each module runs on a different computer, a pairs or triples of modules on the same computer, etc. The distribution of tasks should allow realizing the testing processes in more effective way than running on a single computer. However, using different computers the bigger communication overhead should be taken into account.

The influence of system configuration on the system performance was examined. We made some experiments in which the basic configurations were compared. In order to limit the number of possible configurations, the considered systems included only one client module. Three examined configurations are shown in Fig. 2. Each configuration consists of two computers, A and B, on which the system is allocated. In the first configuration, all but one module are placed on computer A, only the client module runs on computer B. In the second configuration the tasks are distributed more equally, two tasks (system server and project repository) run on computer A and the remaining two tasks run on computer B. Finally, in the last case the system server runs separately on one computer, whereas the client, test data base and project repository are situated on computer B.

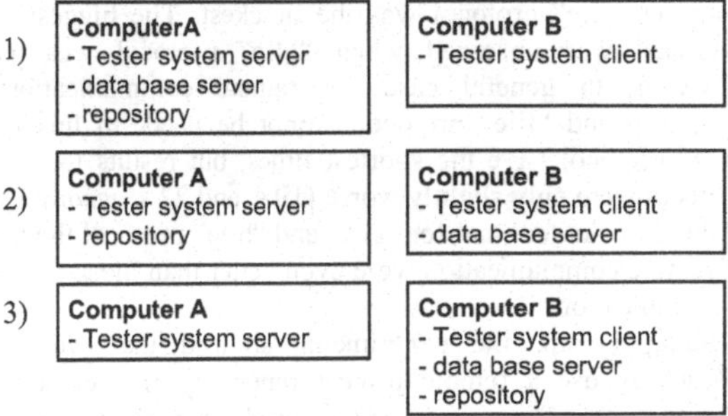

Fig. 2. Comparison of three configurations.

The impact of system configuration on the execution time was experimentally evaluated. Therefore the same bundle of tasks was realized three times using above configurations. Mutated programs (all projects and a series of variants reflecting program changes) were stored in the repository. Each mutant was taken from the repository and tested with a given set of unit tests. Results were compared against an oracle and stored in the data base. The total execution times were measured and compared. Both computers in all experiments were dedicated only for the Tester purposes and did not realized any others applications.

Results of experiments showed that the system distribution has a significant influence on the performance. The execution time was the longest for the first configuration (average task execution time about 62 sec). The slightly better result was obtained for the second configuration (61 sec on average). The execution time was evidently shorter for the third configuration (about 47 sec, improvement of 24%). In this case only Tester system server run on computer A, all other tasks were situated on the second computer.

B. Communication with project repository

The important performance factor is communication time between the repository that stores projects to be tested and other system modules. There are two possible situations, depending on the system configuration. Either the SVN repository is placed on the same computer as such a Tester module that interacts with the repository (so-called local communication), or the repository and the module are on different computers (remote communication). Local communication can be realized using one of three communications protocols: "file", "svn", or "http". "File" protocol can be used only if the system components are placed on the same computer. Two remaining protocols need additional software, but they can be used also in case of remote communication. In order to communicate using "svn" protocol, the SVN server should be running on the given computer. Usage of "http" protocol requires installation of Apache server and special facilities of SVN program.

Influence of the communication protocol on the system performance was compared. Average times of fetching projects from the repository were measured (Fig. 3, 4). If both modules are on the same computer and communication can be local, usage of "file" protocol was the quickest. The biggest time overhead was observed when "http" protocol was used. However, in general case the remote communication is necessary and "file" protocol cannot be used. In this case, "svn" protocol gave the shortest times, but results for "http" protocol were only slightly worse (19 s, and 22 s, accordingly). It should be noted, that for "svn" and "http" protocol the times of remote communication were even better than those for local communication.

Summing up, the experiments showed that we could effectively use a remote project repository if necessary. If possible, "svn" protocol was used in subsequent test experiments. Although differences between the protocols are not big, they are of some importance especially when many (e.g. thousands) of projects are processed during test experiments.

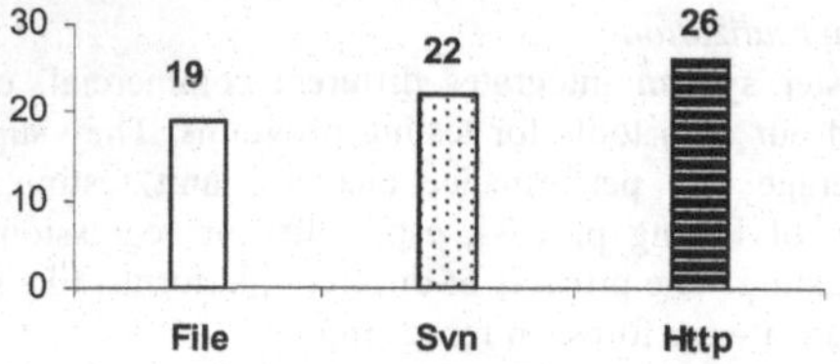

Fig. 3. Average execution times for local communication.

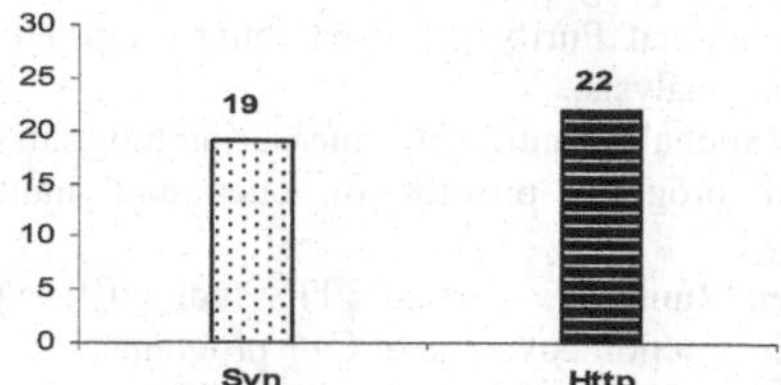

Fig. 4. Average execution times for remote communication.

C. Performance evaluation of HNunit

NUnit tool [11] can be used for running unit tests of programs for .NET platform. Tested programs with the sets of their unit tests are executed under control of NUnit. During testing process, information about test library (stored in *.dll file) and a reference to the module under test are delivered. Test execution is performed using the reflection mechanism. However, if the code coverage should be also measured the PureCoverage tool takes an executable task as an input argument. Taking NUnit as an argument, the obtained result refers to the NUnit program and test library, but not to the program under test.

In order to solve this problem, HNUnit tool (a hybrid NUnit) was created. HNUnit uses the modules of NUnit, but it allows performing unit tests and measuring of the code coverage of a tested program. The performance of this solution was evaluated in experiments. We compared times of unit tests execution performed by original NUnit and the times of the same functionality realized by HNUnit. In experiments about 100 of mutated programs of Castle.Dynamic.Proxy were tested with unit tests [7]. The average times were shorter for HNunit (about 1.4 s) in comparison to NUnit (2.3 s). The experiment results confirmed that the usage of HNUnit that combined two tools was justified; we achieved the desired functionality and the performance was improved.

IV. Software Testing Experiments

Different testing experiments were performed with support of the Tester environment [7,19]. In this section, we present several kinds of them and the results obtained during these experiments.

A. *Mutation testing of C# programs*

One of Tester clients was specialized for cooperation with the Cream testing tool [18]. Using the Cream system mutated C# programs with the selected object-oriented faults can be generated.

Integration of Cream system and Tester environment helped in dealing with several problems. The first one is storing of many mutated programs. Mutants are similar versions of the same project. Therefore they can be effectively stored in the project repository, one of Tester components. Secondly, Tester environment can be helpful in managing test experiments on many programs and their test suits in an automated way. Using appropriate test scripts, the projects are fetched from the repository, tests are run, test results compared against the given oracle, and the mutation results evaluated.

Integration of many tools in one environment enables to use results of one testing procedure in other tests of the same program, in order to obtain better fault detection or better test performance. For example, at first a program coverage for given test suite could be measured. The coverage results are stored in the data base. The Tester client specialized for Cream can deliver this data to the Cream system together with the appropriate project to be tested. The Cream system can use coverage information during mutant generation process. The mutation operators can be applied only for code areas covered by the tests. Therefore a smaller number of equivalent mutants can be generated.

The whole testing process consists of the following steps:

1) An original project is added to the project repository with its test suite. The testing script is created.

2) (optional) Code coverage of the program and the given test suite is measured. Test and coverage results are stored in the data base.

3) Using the specialized client, the Cream system takes the appropriate project from the repository, and coverage results from the data base (if required).

4) The Cream system generates set of mutants (may be using coverage data) and prepares the mutated projects. Information about valid mutants is stored in the project repository.

5) Mutants are tested with their test suites in the Tester environment. Generated results are compared against the given oracle. Test results are stored in the data base.

6) Mutation results are evaluated using stored data. Mutation score is calculated as a ratio of the number of mutants killed by the test suite over the total number of the non-equivalent mutants.

According to the above scenario five C# projects with selected object-oriented mutation operators were tested (Tab. I). The projects are commonly used and available in the Internet. For all programs more than 8.5 thousand of mutants were generated, stored in the repository and tested with appropriate test suites. In experiments, quality of given test suites were measured. The best mutation score was obtained for Dynamic Proxy (53%), and the worse for the Core project (about 16.7% of killed mutants). The last column summarizes the times of performing the whole experiments (together with mutants generation). Automated tool support made possible to perform such time-lasting experiments. It should be noted that experiments were run only on two computers. Bigger number of computers used in the distributed environment would be decrease the execution times.

TABLE I
MUTATION RESULTS

Programs	Number of tests	Number of mutants	Killed mutants	Mutation score [%]	Experiment time [h]
Core	165	222	37	16.7	0.9
Cruise	1279	2488	532	21.4	103.6
Dynamic Proxy	78	166	88	53.0	1.4
Iesi	153	83	31	37.8	0.6
AdapDev	14	5599	1714	30.6	253.7

A. *Line and assembly instruction coverage*

Assembly instruction coverage should correspond to coverage at the higher level, i.e. program lines. To support the coverage analysis a tool, Coverage Analyzer, was created [8]. It integrates measures collected by the IBM Rational Pure Coverage tool [9] as well as static and dynamic coverage calculated by the FITS fault injection systems [10]. Coverage Analyzer was incorporated into the Tester environment. It was used in experiments on the set of 31 C++ programs. In experiments, relation between line coverage and assembly instruction coverage was investigated. We compared numbers of executions (in short multiplicity) of code coverage for line code and assembly instructions.

The experiments showed a discrepancy between the results of two kinds of coverage in more than 50% of code. Three situations were observed: the number of execution of an assembly instruction is bigger than the number of execution of the enclosing code line, the number is less than the corresponding one, or the numbers are equal. Only in less than 50% of code lines the multiplicity of line coverage was the same as for the assembly instruction coverage. In 40% of cases assembly instruction coverage had higher multiplicity than line coverage. The relation was opposite one in the remaining 15% of cases.

B. *Regression testing*

Another type of experiments concerned regression testing. During program evolution different program versions are created. They can be effectively stored in the project repository and taste using various tools.

For example, two versions of a Game project were developed in C#. Both versions had similar functionality, but different data structures were used. Object DataTable was substituted by ArrayList in the second version. The second program had also optimized number of function calls.

Using the Tester environment, different performance and test measures were compared and the correlation between programs calculated. Execution times were highly correlated (according to Spearman and Pearson correlation coefficients), but the second version of the program was on average more effective than the first one. Code coverage showed high linear correlation between programs. Calculation of statistical hypothesis confirmed, that the second program had better code coverage than the first one.

C. Dependency analysis of program performance

Different experiments were devoted to performance evaluation of programs.

For example, two Java programs using the same enterprise data base were compared. Both programs performed the some bookkeeping operations in a small company. In both cases the same data base storing data of the company was used. Each of programs was run 15-times, performing all calculation procedures and data base interactions. We compared numbers of calls to the data base, times of function execution, minimal, maximal and average times, etc. Calculating different statistical tests we showed, that the programs although different, had correlated numbers of function calls concerning data base access. However execution times were independent.

Other experiments examined the relation between performance measures given by IBM Rational Quantify [9] and execution profile measured by Golden Runner (a part of FITS tool [10]). The profile included data about numbers of assembly instruction calls (statically in the code and dynamically during code execution), CPU resources statistics (usage of registers, numbers of read and write operations), etc.

A set of 36 C++ programs was tested. Different measures were compared, looking for existing correlations. For example, the most correlated usage of assembly instruction was instruction JMP with the times of function calls. It was confirmed by statistical tests, like Manna Whitney, Wilcoxon. Evaluation of register activity showed, that activity of EAX register is more correlated with a reading operation than with a writing one. The observed features confirmed the principle of locality of calls in a program.

V. CONCLUSIONS

The paper presented the environment for automation of testing processes. While integrating test tools of different origin, various test experiments could be effectively performed. The distributed and flexible architecture allowed easily extension of the system with variety of test tools. We investigated system performance, comparing different system configurations, communication protocols of transmitted projects, and variants of tools. Results of these experiments benefited in adjustment of the system for efficient running of huge number of test cases. Test performed on various sets of programs showed usefulness of the system for different kinds of testing, e.g. mutation, unit, coverage, regression testing, as well as storing and processing of test data. Further development of the Tester system should takes into account more test experiments in the industrial context and extension with other testing tools.

ACKNOWLEDGMENT

The research within Sections II and III was supported by the Polish Ministry of Science and Higher Education under grant 4297/B/T02/2007/33.

REFERENCES

[1] B. Beizer, *Software testing techniques,* second ed. Van Nostrand Reinhold, 1990.

[2] S. Allott, "Testing techniques - Are they of any practical use?", Testing: Academic and Industrial Conference - Practice and Research Techniques TAIC-PART, IEEE Comp. Soc., 2007.

[3] R. Torkar, S. Mankefors, "A survey on testing and reuse", *Proc. of the IEEE Intern. Conf. on Software - Science, Technology & Engineering,* SwSTE'03, IEEE Comp. Soc., 2003.

[4] A. Bertolino, "Software testing research: achievements, challenges, dreams", in *Proc. of Future of Software Engineering* FOSE'07, IEEE Comp. Soc. 2007.

[5] M. Polo, S. Tendero, M. Piattini, "Integrating techniques and tools for testing automation", *Software Testing Verification and Reliability*, vol. 17, pp. 3-39, J.Wiley & Sons, 2007.

[6] W.E Wong (Ed.) *Mutation testing for the new century*, Kluwer Acad. Publ., 2001.

[7] K. Sarba, *Distributed environment integrating tools supporting software testing*, Master Thesis, Inst. of Computer Science, Warsaw University of Technology, 2007 (in polish).

[8] A. Derezińska, K. Sarba, P. Gawkowski, "On the analysis of relation between line coverage and assembly instruction coverage", *Annals Gdańsk University of Technology Faculty of ETI*, No 6, Information Technologies, vol. 16, pp. 269-274, 2008.

[9] IBM Rational tools: http://www-306.ibm.com/software/rational/

[10] P. Gawkowski, J. Sosnowski, "Experiences with software implemented fault injection", *Proc. of the 20th Int'l Conf. on Architecture of Computing Systems*, March 2007, pp. 73-80.

[11] NUnit, http://www.nunit.org

[12] Subversion svn, http://subversion.tigris.org

[13] Octave, http://www.octave.org

[14] T. Hill, P. Lewicki, *Statistical methods and applications*, StatSoft Inc., 2006.

[15] S. Kim, J. Clark, J. A. McDermid, "Investigating the effectiveness of object-oriented testing strategies with the mutation method", *Journal of Software Testing, Verification, and Rel.*, vol 11(4), 2001, pp. 207-225.

[16] Andrews, L. C. Briand, Y. Labiche, A. S. Namin, "Using mutation analysis for assessing and comparing testing coverage criteria", *IEEE Trans on Software Engineer.*, vol. 32, no. 8, Aug. 2006, pp. 608-624.

[17] S. Elbaum, D. Gable, G. Rothermel, "The impact of software evolution on code coverage information", *Proc. of IEEE Intern. Conf. on Software Maintenance*, 7-9 Nov. 2001, pp. 170-179.

[18] A. Derezińska, A. Szustek, "Tool-supported mutation approach for verification of C# programs", *Proc. of Inter. Conf. on Dependability of Computer Systems*, DepCoS-RELCOMEX, IEEE Comp. Soc. USA, 2008, pp. 261-268.

[19] A. Derezińska, "Experiences from an Empirical study of Programs Code Coverage", in *Advances in computer and information sciences and engineering*, T. Sobh, Ed. Springer Science+Business Media B.V. 2008, pp. 57-62,

Epistemic Analysis of Interrogative Domains using Cuboids

Cameron Hughes
Ctest Laboratories
ctestlabs@ctestlabs.org

Tracey Hughes
Ctest Laboratories
ctestlabs@ctestlabs.org

Abstract-We are interested in analyzing the propositional knowledge extracted by an epistemic agent from interrogative domains. The interrogative domains that have our current focus are taken from transcripts of legal trials, congressional hearings, or law enforcement interrogations. These transcripts have be encoded in XML or HTML formats. The agent uses these transcripts as a primary knowledge source. The complexity, size, scope and potentially conflicting nature of transcripts from interrogative domains bring into question the quality of propositional knowledge that can be garnered by the agent. Epistemic Cuboids or Cubes are used as a knowledge analysis technique that helps determine the quality and quantity of the propositional knowledge extracted by an epistemic agent from an interrogative domain. In this paper we explore how 'Epistemic Cubes' can be used to evaluate the nature of the agent's propositional knowledge.

I. Introduction

We are interested in analyzing the propositional knowledge extracted by an epistemic agent from interrogative domains. The interrogative domains that have our current focus are taken from transcripts of legal trials, congressional hearings, or law enforcement interrogations. These transcripts have been encoded in XML or HTML formats. The agent uses these transcripts as a primary knowledge source. The complexity, size, scope and potentially conflicting nature of transcripts from interrogative domains bring into question the quality of propositional knowledge that can be garnered by the agent. Epistemic Cuboids or Cubes are used as a knowledge analysis technique that helps determine the quality and quantity of the propositional knowledge extracted by an epistemic agent from an interrogative domain. The quality of the agent's knowledge can be examined by the classification of individual propositions. We utilize the properties of the cubes by mapping its attributes to the classifications of the propositions of our Epistemic Structure [6], E_s. In this paper we explore how *Epistemic Cubes* can be used to evaluate the nature of the agent's propositional knowledge.

There are many important questions that must be answered when evaluating a rational or epistemic agent's knowledge. For example:

- What kinds of things does the agent know?
- How much does the agent know?
- Where did the agent obtain its knowledge?
- Is the agent's knowledge justified? If so how?
- What is the agent's commitment level to (belief in) its knowledge?

At the core of these questions is a more central question. How useful is an agent's knowledge? Put another way, what is the quality of an agent's knowledge? How does *agent epistemology* proceed when the primary source of knowledge is derived from complex and potentially voluminous transcripts? Transcripts may have many incomplete, inconsistent or incoherent statements. This makes the extraction of propositional knowledge particularly treacherous. Before we can consider any conclusions or answers that an agent offers we need to know what epistemic landscape the agent is working from.

At Ctest Laboratories, we are investigating the use of *epistemic cuboids* or *cubes* to evaluate and understand the propositional quality of a transcript and the agent's knowledge that is based on that transcript. The agents that we used for this paper are part of the NOFAQS system. NOFAQS is an experimental system currently under development at Ctest Laboratories. NOFAQS is designed to answer questions and draw inferences from interrogative domains (e.g. hearing, interviews, trials, surveys/polls, and interrogations). NOFAQS uses rational agents and Natural Language Processing (NLP) to aid users during the process of deep analysis of any corpus derived from any context that is question and answer intensive. The rationalization for the epistemic cube relies on the application of the Tripartite Analysis of knowledge [15] and the notion of an E*pistemic Structure* [6].

II. Background

The United States Government and many European countries are in the process of standardizing the digital representation of Legislative and legal documents [21]. Thousands of documents and transcripts are being encoded using XML standards such as LegalXML and GOVXML. Among these documents are hearings, trials, and interrogations. The United States Government Printing Office (GPO) is now a source for freely available transcripts that have been encoded in either an XML or HTML standard. The availability and easy accessibility of digital transcripts

K. Elleithy (ed.), *Advanced Techniques in Computing Sciences and Software Engineering*,
DOI 10.1007/978-90-481-3660-5_94,

from the GPO and many other public web sources provide substantial resources to researchers with a focus on computational linguistics, Semantic Web, and information retrieval agents. The fact that many of these documents are now available in a standardized digital format reduces some of the primary obstacles to the research. We are interested in the idea of viewing the digital transcript of interrogative domains or processes as a knowledge space [1] and a primary knowledge source for rational agents. In particular, questions and answers that occur during the process of interrogations, examinations, cross-examinations, interviews or hearings are used as micro-sources for contingent truth. Here we use the notion of contingent truth as it is used in modal logic introduced by Lewis [2] and some of its variations formalized by Kripke [3] and Hintikka [4]. The questions and answers are taken together as pairs and combined to construct contingently true statements [5], [20]. For example:

Q. Do you recall particular items that you removed from the cab area of the truck?

A. We removed a dash board cover, carpet-type dash board cover, and we also removed the floor mats.

Statement: A carpet-type dash board cover and floor mats were removed from the cab area of the truck.

This statement becomes a candidate proposition to be added to the agent's knowledge. During the process of knowledge extraction from the transcript [7] all question and answer pairs are combined to make propositions that will be considered to be contingently true by the agent and accepted as knowledge. Many statements made during the course of a hearing, interrogation, etc., are only possibly true and therefore their basis as propositional knowledge has to be qualified. Reference [3] and [4] give formalizations for this kind of qualification.

The prism of the propositional account of knowledge allows us to consider the combination of question and answer pairs in the context of justified true belief [8]. Once the agent has mined the propositions from the transcript(s) we can then begin to classify and analyze the agent's knowledge. In this case we are interested in which statements are justification for other statements. We are interested in whether the justification can be understood as internal or external justification [9]. We look at the nature of the propositions. Are they analytic or synthetic in nature? We further divide the propositions into a priori and posteriori categories. We've introduced the notion of epistemic cubiods or cubes to help us perform this analysis. Before we can use epistemic cubes to analyze agent's knowledge we must convert the XML or HTML semi-structured transcript to its epistemic structure [6].

A. The Epistemic Structure of the transcript and the agent's knowledge

The epistemic structure [6] of the transcript and thus the agent's knowledge is formally denoted by E_s. Let E_s be the structure:

$$E_s = <G_1, G_2, J, C, F>$$

Where:

G_1 is a Graph of *a priori* propositions
G_2 is a Graph of *posteriori* propositions
J is a Set of justification propositions
C is a Vector of Commitment
F is a non-monotonic truth maintenance function on E_s.

For each logical domain in the transcript, we have a distinct E_s:

Let $d = \{E_{s1}, E_{s2}, E_{s3} \ldots E_{sn}\}$

Where d is a set of epistemic structures representing a particular domain, or a collection of domains. Then we have:

$$K_s = \bigcup_{i=1}^{N} d$$

where K_s is the total knowledge space of the agent..

Using the epistemic structure as a knowledge representation scheme and the Tripartite Analysis as the definitional basis of the agent's knowledge we are able to use cuboids to analyze the quality, quantity and nature of propositional knowledge contained in digital transcripts.

III. Method

To generate an epistemic cuboid or cube for a transcript, we build a First Order Logic (FOL) model of the content of the transcript [10], [14]. $\mu\psi$ will form the initial vocabulary of the agent's ontology and frame system [11], [12]. $\mu\psi$ gives us the vocabulary of the transcript decomposed into a set of relations persons, and objects [13]. These three components serve as the basic dimensions of our cuboid. The length, height, and depth of the cuboid is determined by the number of Relations (R), Persons (P), and Objects (O) mined from the transcript. If R_n, is a relation in $\mu\psi$ and P_n is a person in $\mu\psi$ and O_n is an object in $\mu\psi$ then $\phi_1(R_n, P_n, O_n)$ is a boolean function that maps $\mu\psi$ into the three dimensional space of our cuboid. Since we have associated the dimensions of our cuboid with R, P, and O, then the logical volume of our cuboid is R*P*O and is proportionately related to K_s.

In addition to the dimension and volume of our epistemic cuboid, it has four other important attributes:

- 6 Surfaces
- 8 Vertices,
- 3 Pairs of Rectangles
- 12 Edges

We map the propositions from E_s to each of these attributes. If we let ρ = {propositions formed from question & answer pairs} then ϕ_2: $\rho\psi \rightarrow \phi_1$. ϕ_3 is a boolean function that maps each member of $\rho\psi$ to a proposition surface on our epistemic cuboid. For example, posteriori propositions are mapped to Surface1, a priori propositions are mapped to Surface 2, synthetic propositions are mapped to Surface 3 and so on. Figure 1 shows how the $\mu\psi$ constrains the dimensions of our cuboid and how the proposition classifications are mapped to the surfaces. If it turns out that the number of relations, persons, and objects are equal, then we have an *epistemic cube* rather than a *cuboid.*

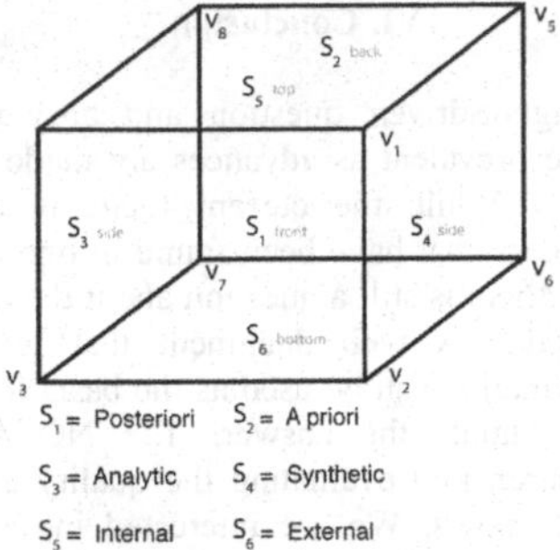

Figure 1. Shows how the $\mu\psi$ constrains the dimensions of our cuboid and how the proposition classifications are mapped to the surfaces.

There is one epistemic cuboid or cube for each E_s in K_s
Each of the eight vertices is incident with 3 edges and each edge with an associated surface. Let $S_1 \ldots S_6$ denote the surfaces of our cuboid and Table 1 give the mappings of $S_1 \ldots S_6$.

Table 1 maps a surface to a classification that gives us information about the nature of the propositions that are true at that surface. For example, propositions that are true at S_5 are internally justified [16]. Propositions that are true at S_3 have analytic truth [8] and so on. ϕ_3: $\rho\psi \rightarrow S_1 \ldots S_6 \triangleright \psi$ The nature of the cuboid suggests the surfaces are comprised of the points taken from the three dimensions of the transcript's μ.

Table 1. Proposition Surfaces

Surface	Proposition Classification
S_1	Posteriori
S_2	A priori
S_3	Analytic
S_4	Synthetic
S_5	Internal
S_6	External

To illuminate the semantics of the interaction of $S_1 \ldots S_6$ and the three dimensions of the transcript's FOL, we show the mappings of the eight vertices of our cuboid to $S_1 \ldots S_6$. Each of the eight vertices represents an intersection of three proposition surfaces. Let $V_1 \ldots V_8$ denote the eight vertices and $P_1 \ldots P_8$ denote our eight propositional subtypes

Table 2. Proposition Types

Vertex	Surface Intersections (3 Dimensions)	Proposition Sub-types
V_1	$\{S_5,S_4,S_1\}$	P_1
V_2	$\{S_6,S_4,S_1\}$	P_2
V_3	$\{S_6,S_3,S_1\}$	P_3
V_4	$\{S_5,S_3,S_1\}$	P_4
V_5	$\{S_5,S_4,S_2\}$	P_5
V_6	$\{S_6,S_4,S_2\}$	P_6
V_7	$\{S_6,S_3,S_2\}$	P_7
V_8	$\{S_5,S_3,S_2\}$	P_8

Table 2. The mappings of $V_1 \ldots V_8$ to the proposition sub-types.

then ϕ_4 : $\rho\psi \rightarrow P_1 \ldots P_8$. Where each proposition subtype has three dimensions. Once we discover that a proposition is true at some surface, we can also look at two other dimensions of that proposition. For example, we may discover that a proposition that is true at S_4 may also be analytic and externally justified, that is it may have a S_5 and S_6 dimension.

Table 2 shows a second level decomposition of proposition types $S_1 \ldots S_6$. If the boolean function ϕ_3 maps the relations from the $\mu\psi$ decomposition of the transcript to one of the surfaces in $S_1 \ldots S_6$ and ϕ_2 is a function that maps every proposition that was generated from a question and answer pair to a relation in μ then wherever $\phi_{2\psi} \leftarrow \phi_1) =$ True, we can use ϕ_4 to further classify the proposition as one of $P_1 \ldots P_8$. For example, the proposition:

A carpet-type dashboard cover and floor mats were removed from the cab area of the truck.

was classified as P_6. This means that it was true at S_4, S_2, and S_6.

We take the 12 edges from the three rectangles. Let R_1 = (A priori, posteriori), R_2 = (analytic, synthetic), and R_3 = (internal, external). R_1 identifies the origin of the proposition. R_2 identifies a truth attribute of the proposition. R_3 identifies a justification attribute of the proposition. Although we formally have three pairs of rectangles a proposition can only be classified in 3 of them at a one time. We can classify the edges as universal affirmative, universal negative, particular, affirmative, and particular negative. This classification gives us a logical view of our propositions in $P_1 \ldots P_8$.

Now using the epistemic cube E_c, we can determine where an agent's knowledge is internally or externally justified. We can identify which propositions are a priori true for the agent or which are posteriori true. We can examine which propositions the agent holds as synthetically true, or analytically true.

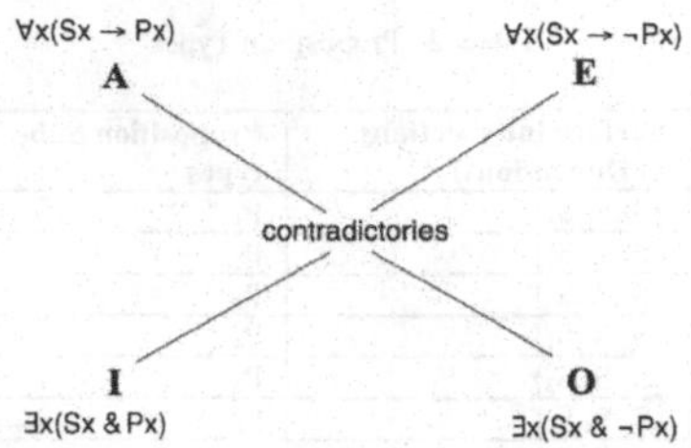

Figure 2. A modified square of opposition is mapped to R_1, R_2, and R_3.

IV. Transcript Corpus

The corpus was obtained from the Internet as a collection of Hypertext Mark up Language (HTML) files. The corpus is for a famous court case USA v. Usama Bin Laden. The trial lasted for approximately 76 days and there was one HTML file generated for each day. So for the purpose of this paper our digital transcript consisted of:

- 76 Files formatted in standard HTML (one for each day of the trial)
- 25,979 Questions
- 25,930 Answers

From the 25,979 questions and their corresponding answers, we were able to generate 12,200 statements. The difference between the original number of questions and the statements is due to redundancy of answers. That is many question and answer pairs are combined to make the same statement. In addition to redundancy there were many instances where the combined question and answer did not form a legitimate language construction. So the transcript data available for the context of this paper consisted of 12,200 statements that were distributed among 76 HTML files.

V. Summary

In this paper we restricted our epistemic analysis to day 5,11, and 12 from the 76 days. These three days yielded the highest quality of data for our purposes. Prior to constructing an epistemic cuboid for analysis, the data has to survive the usual obstacles found in natural language processing and computational linguistics [10] [17] [18] [19]. We loose data because of words that are out of lexicon bounds. For instance in the corpus we used for this paper several of the witnesses were Arabic speakers who used terminology that was not present in our lexicons. Further we lost data during syntax analysis and semantic analysis. Any data that survived these stages in our pipeline was then converted to its epistemic structure [6]. Once we have an epistemic structure we are able to look at those propositions that the agent is committed to. It is these propositions that we build epistemic cuboids for. In this case over 90% of the propositions that the agent was committed to were classified as P_7 a priori analytic, and externally justified. These classifications were done with a 19% error rate.

VI. Conclusion

The agent-driven question and answer system is becoming more prevalent as advances are made in Semantic Web research. While the current focus is on returning answers to a query that have been found in one or more web documents [5], there is still a question about the quality of the answer returned. A web document that gives false or erroneous information can be used as the basis for an answer. Obviously this taints the answer. The NOFAQS system represents research into evaluating the quality and nature of what an agent knows. We are interested in far more than certainty factors for conclusions or responses that a question & answering system agent may offer. We want to know what epistemic and logical landscape an agent is operating from.

The Epistemic Cuboid in conjunction with the Tripartite Analysis gives us a small step in the direction of evaluating the quality of an agent's knowledge. If we can understand the quality of an agent's knowledge then we can make informed judgments about the quality of answers or conclusions that an agent may offer. The epistemic cube relies heavily on logical analysis of propositional knowledge. This means that it cannot be applied to other paradigms of knowledge or in situations where logic breaks down or is not applicable. This is partially what invigorates our choice of interrogative domains. One of the ultimate artifacts of transcripts taken from interrogative domains is a collection of statements that can be understood in the context of the Tripartite Analysis account of propositional knowledge. If these statements are used as primary knowledge sources for rational agents then we can use epistemic cuboids as an analysis technique.

VII. References

[1] J.P. Doignon and J.C. Falmagne, "Knowledge Spaces," Hiedelberg; Springer ISBN-3-540-64501-2, 1999.

[2] Lewis, C.I., "A Survey of Symbolic Logic," Berkeley: University of California Press, 1918.

[3] Kripke, S., "Semantical Considerations on Modal Logic," Acta Philosophica Fennica, pp. 16, 83-94, 1963.

[4] Hintikka, J., "Individuals, Possible Worlds and Epistemic Logic," Nous, pp. 1, 33-62, 1967.

[5] Galitsky, B., "Natural Language Question Answering System Technique of Semantic Headers," Advanced Knowledge International Pty Ltd, 2003

[6] T. Hughes, C. Hughes, and A. Lazar, " Epistemic Structured Representation for Legal Transcript Analysis," Advances in Computer and Information Sciences and Engineering Springer, pp. 101-107, 2008

[8] P.K. Moser, D.H. Mulder, and J.D. Trout, "The Theory of Knowledge: A Thematic Introduction," Oxford University Press, 1998.

[9] Casullo A. "A Priori Justification," Oxford University Press, 2003.

[10] P. Blackburn and J. Bos, "Representation and Inference for Natural Language," CSLI Publications, 2005.

[11] S. Nirenburg, V. Raskin, "Ontological Semantics," MIT Press, 2004.

[12] M. Minsky, "A framework for representing knowledge," *The psychology of computer vision*. New York: McGraw-Hill, pp. 211-277, 1975.

[13] R. Brachman, H Levesque "Knowledge Representation and Reasoning", Elsevier, 2004.

[14] R. Fagin, J. Halbern and M. Vard, " Model Theoretic Analysis of Knowledge," *Journal of Association for Computing Machinery Vol 38. No 2*, 1991.

[15] E. Gettier, "Is Justified True Belief Knowledge?," Analysis, Vol.23, pp. 121-23. 1963.

[16] Swinburne R, "Epistemic Justification",Clarendon Press Oxford, 2001.

[17] J. Cussens and S. Pulman, "Incorporating Linguistics Constraints into Inductive Logic Programming," *Proceedings of CoNLL-2000 and LLL-2000*, pp. 184-193. Lisbon, Portugal, 2000.

[18] J. F. Sowa, "Semantics of Conceptual Graphs," *Proceeding of the 17th Annual Meeting of the Association for Computation Linguistics*, pp. 39-44, 1979.

[19] M.Covington, "Natural Language Processing For Prolog Programmers," Prentice Hall, 1994.

[20] W.G. Lehnert, "The Process of Question Answering," Lawrence Erlbaum Associates, Inc., Publishers, 1978.

[21] M. Biasiotti, E. Francesconi, M. Palmirani, G. Sartor, F. Vitali, "Legal Informatics and Management of Legislative Documents," Global Centre for ICT in Parliament Working Paper No.2, January 2008.

Dynamical Adaptation in Terrorist Cells/Networks

D. M. Akbar Hussain & *Zaki Ahmed

Department of Electronic Systems

Niels Bohrs Vej 8, 6700, Aalborg University, Esbjerg, Denmark

Tel: (45) 99 40 77 29

Email: akbar@aaue.dk

** Institute of Laser and Optics (PILO)*

Islamabad, Pakistan

Email: zaki424@hotmail.com

Abstract — **Typical terrorist cells/networks have dynamical structure as they evolve or adapt to changes which may occur due to capturing or killing of a member of the cell/network. Analytical measures in graph theory like degree centrality, betweenness and closeness centralities are very common and have long history of their successful use in revealing the importance of various members of the network. However, modeling of covert, terrorist or criminal networks through social graph dose not really provide the hierarchical structure which exist in these networks as these networks are composed of leaders and followers etc. In this research we analyze and predict the most likely role a particular node can adapt once a member of the network is either killed or caught. The adaptation is based on computing Bayes posteriori probability of each node and the level of the said node in the network structure.**

Key terms — **Social Networks Analysis, Bayes Theorem, Hierarchical Structure, Dynamical.**

I. INTRODUCTION

Drug dealers, terrorist and covert networks are typically, represented through social graphs. Since 9-11 terrorist attacks, a great deal of research is taking place firstly to understand the dynamics of these terrorist networks (analysis) and secondly, developing methods to either destabilize or disintegrate these networks. Insight visualization of any social network typically focuses on the characteristics of the network structure.

Social Network Analysis (SNA) is a mathematical method for 'connecting the dots', it allows us to map and measure complex relationships/connections between human groups, animals, computers or other information/knowledge processing entities and organizations [1]. These relationships can reveal unknown information about these dots and the network itself. Jacob Moreno invented "Sociometry" which is the basis of SNA, he utilized "sociograms" to discover leaders and map indirect connections in 1934 [2]. The two basic elements of SNA are connections and nodes. Connections are ties between individuals or groups and nodes are the individuals or groups involved in the network. There are different dynamics of social networking for example Kin-based (father, husband), Role-based (office), Interactions (chatting) and Affiliations (clubs etc). Analysts have applied SNA in many fields to reveal hidden informal links between nodes [3]. For example in businesses SNA have been used to analyze email patterns to determine which employees are overloaded, similarly, law enforcement and national security organizations are using various method of SNA to identify important nodes and connections of terrorist organizations [4].

Many traditional social network measures and the information processing network measures can help in revealing importance and vulnerabilities of the nodes/agents in the network [5], [6], [7], [8]. Application of existing tools on these complex socio-technical networks/systems is very demanding to winkle out the required information. Most of the measures and tools work best when the data is complete; i.e., when the information is inclusive about the interaction among the nodes. However, the difficulty is that covert and terrorist networks are typically distributed across many boundaries for example from cities or countries and data about them is never complete or correct at a certain instant of time. Normally, a sampled snapshot data is available some of the links may be intentionally hidden. Also data is collected from multiple sources for example news (print/tv), open source internet data, security agencies, etc., and at different time instants. In addition inclusive and correct information may be prohibitive because of secrecy. Obviously, there could be other difficulties but even these provide little guidance for what to expect when analyzing these complex socio-technical systems with the developed tools.

K. Elleithy (ed.), *Advanced Techniques in Computing Sciences and Software Engineering*,
DOI 10.1007/978-90-481-3660-5_95,

II. LITERATURE SURVEY

We start with the strength and limitations of SNA and subsequently discuss many important research topics in this area.

- Strengths

The most fundamental strength of SNA is that it provides a visual representation of the network structure. It allows the analysts to compare and identify previously unknown links. The knowledge gained through this process can be used to forecast not only the individual activities of the actors but also of network/organization.

- Limitations

SNA is data dependent like most analytical software, therefore, correct and up to date data is essential for true analysis of a network/organization. This means if the data is incomplete or incorrect final product will be inaccurate. Generally it is believed that SNA is used as a tool only and one should not be relied upon to provide an absolute depiction of a network. Another important point of its limitation is that it is time consuming it takes a great deal of time to research a topic in order to find the appropriate information.

SNA has been used to analyze for example in the network qualitative studies, the facilitators of link establishment and in quantitative studies, the use of statistical methods to measure existing network. Most studies in link establishment have been carried out in sociology and criminology [9]. Statistical analysis mostly dealt with exploring the key actors using standard centrality measures. In contrast to this, the dynamic social network analysis methods have been dealing with network recovery and network measurement. In network recovery multiple instantaneous network representation are recovered from longitudinal data to model the evolving network. In dynamical network measurement three types of techniques are used, deterministic measure, probabilistic measures and temporal measures. In deterministic measures network size, degree, betweenness and closeness measures are computed whereas in probabilistic measures degree distribution and clustering coefficient are measured. As the network development is a continuous process so temporal measure deals with this continuous process by considering a time variable. Statistical analysis typically studies and explains the topologies of networks.

Paramjit and Swartz [10] have used random-effects models to incorporate dependence between the dyads, originally this idea was proposed by Wong [11] in which the likelihood of ties in terms of the nodal attributes rather than in terms of network structural properties for example transitivity and cyclicity are expressed. Bayesian approach has been used in network modeling. Markov chain Monte Carlo (MCMC) simulation technique has also been used to determine the characteristic marginal posterior distribution which allows for complicated modeling and inference independent of sample size. This is in contrast with analyses which focus only on the estimation of primary parameters and their asymptotic standard errors. MCMC has been used by Gill and Swartz for Bayesian analysis of round robin interaction data where the response variable was continuous [12], [13]. Nowicki and Snijders [14] used MCMC Bayesian analysis for block model structures where the relationship between two nodes depends only on block membership. Statistical analysis and Network measurement based on Bayesian approach have also been used in dynamical SNA issues [15], [16], [17], [18]. How the basic Bayesian model can be modified to cater to special settings are presented by Holland and Leinhardt [19]. Paramjit [10] demonstrated and introduced covariates and the stochastic block models to the basic Bayesian model [11] and how MCMC simulation output can be used in model selection for Bayesian analysis of directed graphs data. Our method of analysis using Bayes posterior probability theory is straight forward as we compute the posterior probability for each node of the network. The following subsection provides the detailed description of our approach for the network structure analysis.

A. Our Approach for the Analysis

As pointed out earlier terrorist cells/networks have dynamical structure which evolves according to the situations and they are quite robust. They adapt to situations for example when a member of the cell/network is eliminated due to any reason (killed/captured). They may also adjust their role if required to manipulate the security agencies surveillance. Our study here therefore, analyzes these cells/networks and predicts the most likely network structure which can evolve after elimination of a particular member. We are using the probabilistic measure based on Bayes probability theory in analyzing the structure of the network however, the technique is much simpler compared with probabilistic measures where degree distribution and clustering coefficient is measured. Once a member of the network disappeared due to any reason the system first analyze its behavior by computing the posterior probabilities of each node. As it is quite obvious that in most cases there would be more than one candidate for assuming the role so that node is selected which has posterior probability nearer to the disappeared node. It is also important to note as we pointed out earlier that terrorist networks have hierarchical structure so the node assuming the role of the disappeared node should be of similar level (ranking) in the network structure otherwise it will not be an efficient transition.

III. BAYES THEOREM

Bayes' theorem is a simple mathematical formula used for calculating conditional probabilities. Bayes' theorem originally stated by Thomas Bayes and it has been used in a wide variety of contexts, ranging from marine biology to the development of "Bayesian" Spam blockers for email systems. Through the use of Bayes' theorem precise measures can be obtained by showing how the probability that a theory is correct is affected by new evidence [15], [20]. In a Bayesian framework the conditional and marginal probabilities for stochastic events for example A and B are computed through this relationship:

$$P(A \mid B) = \frac{P(B \mid A)P(A)}{P(B)} \tag{1}$$

where P(A) is the prior probability or marginal probability of A, P(A|B) is the conditional probability given B also called posterior probability. P(B|A) is conditional probability given A, P(B) is prior probability and considered as normalizing constant. Basically, it is derived from,

$$P(A \mid B) \propto P(B \mid A)P(A) \tag{2}$$

P(B|A) is normally equal to L(A|B) which is the likelihood of A given fixed B. Most times likelihood L is multiplied by a factor so that it is proportional to, but not equal probability P. It should be noted that probability of an event A conditional on another event B is generally different from the probability of B conditional on event A, however, there is a unique relationship between the two which is provided by Bayes theorem. We can formulate the above relationship as:

$$posterior = \frac{likelihood \times prior}{normalizing\ cons\tan t} \tag{3}$$

We can re-write equation 1 as the ratio P(B|A)/P(B) which is typically called as standardized likelihood or normalized likelihood so it can be written as:

$$posterior \;=\; normalized \;\; likelihood \;\; \times \; prior \tag{4}$$

Suppose we have a network of nodes (graph) and we are interested in calculating the posterior probability P(A|B) of any node A given some information B, so that we can determine its position (level) relative to other nodes in the network. Bayes probability theory provides such possibility through its conditional probability theorem, for this reason we have expanded the above expression for convenience to interpret various terms according to our implementation model. Therefore,

$$P(A \mid B) = \frac{P(B \mid A)P(A)}{P(B \mid A)P(A) + P(B \mid N)P(N)} \tag{5}$$

P (A) is the prior probability or marginal probability of node A regardless of any information, which is computed by considering the total number of nodes present in the network. For example if there are ten nodes in the network and we assume that the node under consideration is connected with 4 nodes then it has prior probability value of 0.4 having a certain level (rank) in the network structure. P(N) is the probability that this node does not have this level in the network structure given by (1 - P(A)). P(B|A) is conditional probability given A, it is a function of network size and the highest level (rank), for example if a node has level 3 and the size of the network is 10 and maximum rank in the network is 4 then P(B|A) will be 0.5. P(B|N) is conditional probability given N and is obtained by computing (1 - P(B|A)) which means that it does not believe the hypothesis given by P(B|A). Basically, we are interested in evaluating the theory or hypothesis (equation 1) for A based on B which is the new information (evidence) that can verify the hypothesis and P(A) is our best estimate of the probability (known as the prior probability of A) prior to considering the new information. We are interested to discover the probability that A is correct (true) with the assumption that the new information (evidence) is correct.

IV. ANALYSIS & RESULTS

To describe our analysis and the strategy of our computation we have selected a small network comprising of 10 nodes as shown in figure 1(a), although it looks simple however, it has all the in gradients for our investigation. Figure 1(b) is re-structured to show the 5 levels based on the network connectivity structure. Now if we look at the network structure it is not possible at least for node number 4, 6, 7, 8, 9 and 10 to predict if any of these are leader/head or follower/leaf nodes of the network as there are many possibilities some of which are shown in figure 2. For example in the first case (top figure) node 10 could be the leader node and the rest as followers and so on.

Now our analysis systematically compute Bayes posterior probability starting from node number 1 and predicts which node can assume the role if a node is removed due to any reason. This sequence of starting from node number 1 is rudimentary it can be started from any node. Figure 3 shows the result for first 4 nodes, it can be seen that for node number 1 if it is being killed, which belongs to level 2 in the network so obvious choice should be any member from the same level for replacement. This is based on the assumption that terrorist network members have specific and very well defined responsibilities based on their trade/calibre. Therefore, if a member is killed/captured then the replacement must be from similar category. As for this case there is no other node in this level so the most probable choice should be a member from other levels having its probability value near to its own value, in this case our system has predicted node 2, which is the most appropriate replacement.

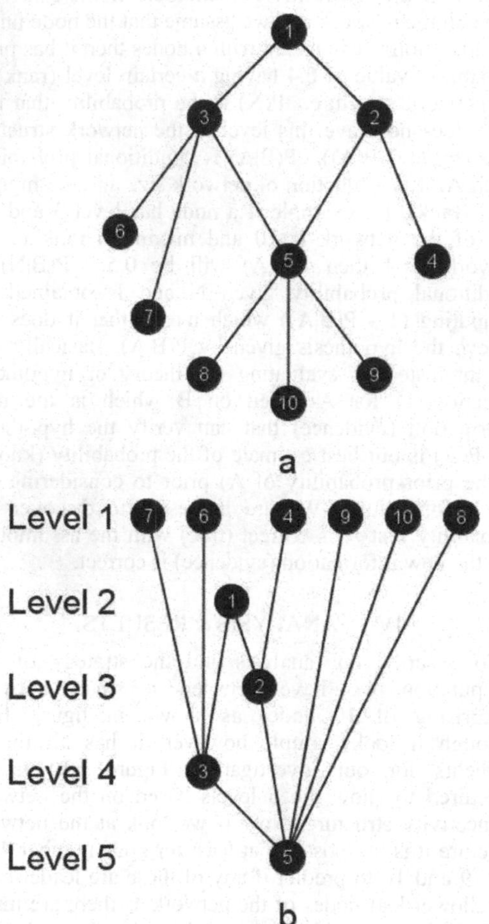

Figure 1: Sample Network

Similarly, for node 2 the predicted node for replacement is node 3 due to similar reasons but for node 3 there are two possibilities node number 2 and 5 and the most suitable choice is node 2 as the network structure evolved with this replacement is more balanced and obviously efficient. Now node 4 is interesting as it is at level 1 and our system correctly predicted that node 6, 7, 8, 9 and 10 are most likely candidates but node 6 or 7 are the best choices in terms of compact/robust network structure.

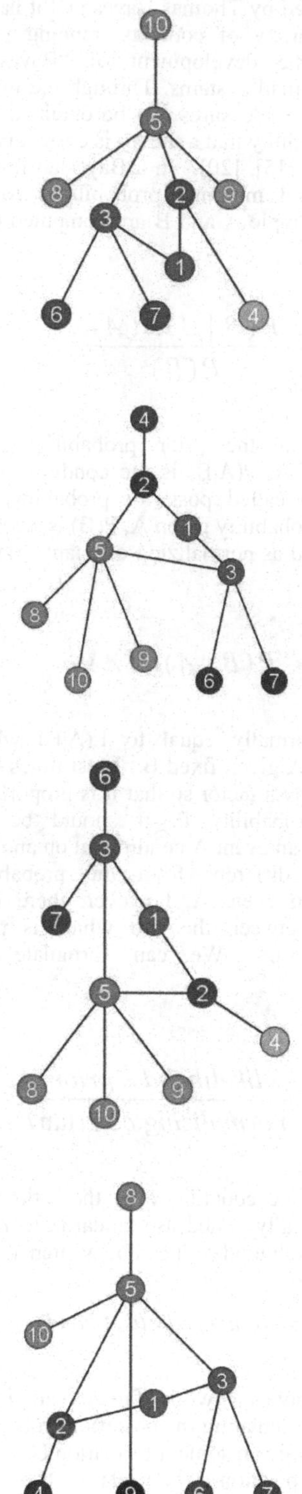

Figure 2: Head and Leaf Nodes Combination

Figure 4 shows the results for node 5 to 8, it can be seen that node 5 is at the highest level in the network and system has predicted that if this node is removed the most suitable replacement should be node 3 as that is the only node having level suitable to work and keep the functioning of the network in similar fashion before the elimination of node 5. The replacement for nodes 6 to 10 are similar as each of these nodes are at same level so any of these nodes can assume the role once a node at this level is eliminated and it can be seen that similar probability graphs are obtained for nodes 6 to 10 however, only 6, 7 and 8 are shown here.

V. CONCLUSION

The standard statistical solution for SNA has been matured for long time and being used in the studying social behaviour. Networks visualization is semantically presented in the form of a graph in which the nodes represent entities and the arcs represent relationship among nodes. Classification of nodes and its distinctiveness is a challenging task. However, illuminating the pattern of connections in social structure is very challenging and the conventional social network analysis may not be enough to reveal the structural pattern of nodes. Most terrorist cells and networks are small world networks however they have varying complexity posing difficulty in analysis. The purpose of this paper is to investigate how a network would shape up after a certain member of the network is killed or captured. The computation is based on the underlying assumption philosophy of Bayesian Posterior Probability that uncertainty and degree of belief can be measured as probability. The initial investigation shows that it is possible to predict such structure and these are preliminary results and one may not draw a final verdict from the investigation. However, our investigations here have shown that it is possible to predict how nodes can assume various roles once a member is eliminated. These results are based on a random network but interestingly it has shown consistency in its computation. In future we would like to extend this framework for more complex and well known network structures available as open source to test the consistency of our proposed computational model.

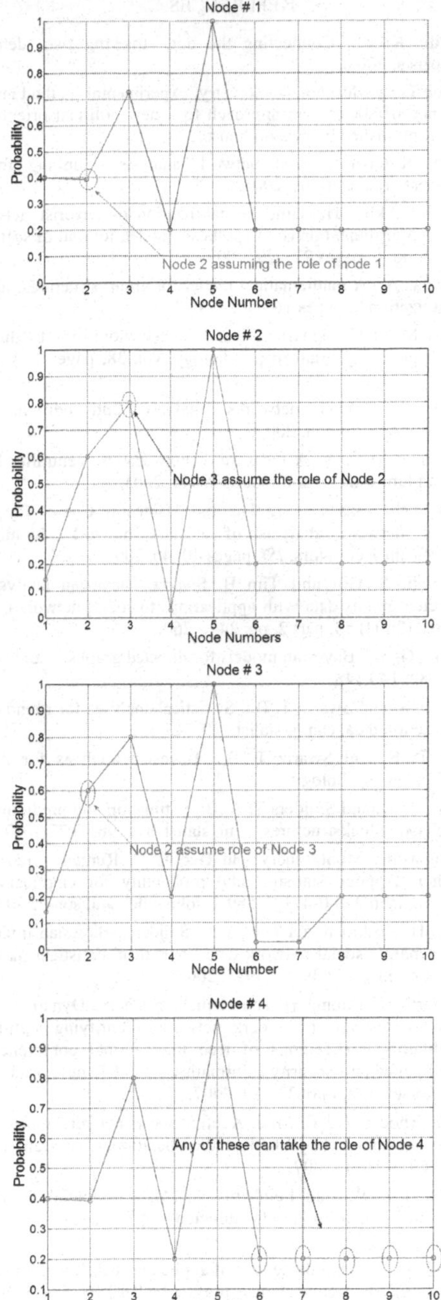

Figure 3: Probability Graph for Nodes 1 – 4

REFERENCES

[1] Valdis Krebs:. Connecting the dots, tracking two identified terrorists, 2002.

[2] Jacob Levy Moreno:. Sociometry, experimental method and the science of society, an approach to a new political orientation, 1951, published by beacon house.

[3] Cory Kutcher:. Social network analysis - linking foreign terrorist organizations, 2008.

[4] Dil M. Akbar Hussain:. Destabilization of terrorist networks through argument driven hypothesis model, journal of software, vol. 2, no. 6, pages 22 - 29, 2007.

[5] Bavelas A:. A mathematical model for group structures, human organization 7, pages 16 - 30, 1948.

[6] Shaw M. E:. Group structure and the behaviour of individuals in small groups, journal of psychology, vol. 38, pages 139 - 149, 1954.

[7] Scott J:. Social networks analysis. 2nd edition, sage publications, london, 2003.

[8] Newman M. E. J:. A measure of betweenness centrality based on random walks, cond-mat/0309045, 2003.

[9] Siddharth Kaza Daning Hu and Hsinchun Chen:. Dynamic social network analysis of a dark network: Identifying significant facilitators, *ISI*, pages 40–46, 2007.

[10] Paramjit S. Gill and Tim B. Swartz:. Bayesian analysis of directed graphs data with applications to social networks, appl. statist. (2004) 53, part 2, pp. 249 - 260.

[11] Wong G. Y.:. Bayesian models for directed graphs. j. am. statist. ass., 82, 140 148.

[12] Gill P. S. and Swartz T. B.:. Statistical analyses for round robin interaction data. can. j. statist., 29, 321 331.

[13] Gill P. S. and Swartz T. B.:. Bayesian analysis for dyadic designs in psychology.

[14] Nowicki K. and Snijders T. A. B:. Estimation and prediction for stochastic blockstructures. j.am. statist. ass., 96, 1077 1087.

[15] Douglas C. Montgomery and George C. Runger:. George c. runger: Applied statistics abd probability for engineers, 4th edition, isbn 978-0-471-74589-1, john wiley and sons, 2006.

[16] John H. Koskinen and Tom A. B. Snijders:. Bayesian inference for dynamic social network data, journal of statistical planning and inference 137 3930 - 3938, 2007.

[17] Siddarth K. Daning H. and Hsinchum Chen:. Dynamic social network analysis of a dark network: Identifying significant facilitators, proceedings of ieee international conference on intelligence and security informatics, isi 2007, new brunswick, new jersey, usa, may 23 - 24, 2007.

[18] C. J. Rhodes and E. M. J. Keefe:. Social network topology: a bayesian approach, journal of the operational research society 58, 1605 - 1611, 2007.

[19] Holland P. W. and Leinhardt S.:. An exponential family of probability distributions for directed graphs. j. am. statist. ass., 76, 33 65.

[20] Oliver C. Ibe:. Fundamentals of applied probability and random processes, elsevier academics press, isbn 0-12-088508-5, 2005.

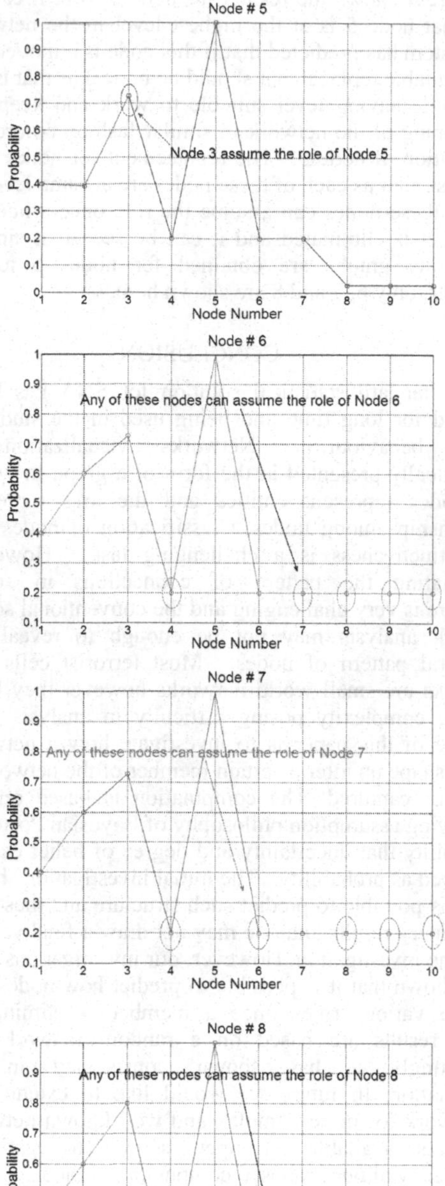

Figure 4: Probability Graph for Nodes 5 - 8

Three-dimensional Computer Modeling and Architectural Design Process – A Comparison Study

Miguel C. Ramirez, Neander F. Silva, Ecilamar M. Lima
Universidade de Brasília
Laboratório de Estudos Computacionais em Projeto – LECOMP
ICC Norte, Ala A, Subsolo, CEP 70910-900, Brasília, DF, BRASIL
http://lecomp.fau.unb.br/, miguelramirez99@hotmail.com, neander@unb.br, ecilamar@unb.br

Abstract – In this paper we describe the assessment of three categories of CAD systems with focus on three dimensional computer modeling and its suitability for the architectural design process. The results show that there are significant differences in their performance and that some are more adequate to be incorporated into the design process than others.

I. Research Problem

The development and availability of new computer programs offer new possibilities in the field of architectural design. These tools have brought a significant change in the role of computers in architecture over the last two decades.

The use of interactive three-dimensional representations for designing in architecture has been central in this transformation. Even before the rise of modern computing, some authors [1] have stressed the need for a holistic representation of architecture, one that goes beyond the fragmented and misleading two-dimensional (2D) drawings created from orthogonal projections.

A virtual and interactive three-dimensional model is not just an equivalent to a scaled model. Besides representing the four dimensions of architecture, described by Zevi [1], as width, height, depth and time (motion in time and space), it allows for many possibilities which cannot be provided by a scaled model or any other traditional media. Among these possibilities are interactive walk-through, object animation, solar animation, global illumination calculation, designing complex non Euclidian geometries, digital fabrication and mass customization [2][3][4][5], just to name a few.

Different implementations of interactive three-dimensionality in computer graphics have been developed with varying degrees of success regarding their suitability for the architectural design process. In this research we have selected three of those software categories on the basis of representing the most popular types of three dimensional (3D) computer-aided design (CAD) systems in contemporary architectural practices. The first of them are the early two-dimensional drafting systems which slowly evolved to incorporate some 3D capabilities. The limitations of such systems to produce 3D models outside the scope of vertical extrusions of 2D entities, led them to be nicknamed "2 ½ CAD systems".

The second category is that of the simpler surface modelers, i.e., systems which model 3D entities without representing them as solids, but just as empty spaces bounded by surfaces. Most of the early 3D CAD systems would fit in this category and they did so because of the hardware limitations of their time. Such equipment restrictions are by now overcome. However, such paradigm is still adopted by some very popular and modern CAD implementations. This is due mainly to the benefit of producing relatively very small files which are suitable for Internet use. These systems are efficient in allowing for real time interaction. However, they achieve that efficiency based on the use of simplified sketchy rendering algorithms. They also have significant limitations regarding the representation of complex, non Euclidian geometries.

The third category is that of the true solid modelers, those that model 3D entities as real solids, with volume and mass. A true solid modeler must be able also, by definition, to model any thinkable geometry that may be found in the real world. They resort to both faceted as well as to real curve surfaces to model such complex geometries. These systems usually provide powerful rendering capabilities, including accurate lighting calculations, which result in much more realistic representations than the other two categories mentioned above. The disadvantages of such systems are their higher hardware demand compared with the other types of CAD systems mentioned before.

Other important categories of 3D implementations could be mentioned such as Building Information Modeling systems (BIM) [6], but they are, at the moment, beyond the scope of the research presented in this paper.

Although many architects still use computers only for automating the production of construction drawings, a few of them have found, in some software available since early 90's, new possibilities that have changed the role of computing in the design process.

K. Elleithy (ed.), *Advanced Techniques in Computing Sciences and Software Engineering*,
DOI 10.1007/978-90-481-3660-5_96, © Springer Science+Business Media B.V. 2010

They have also found that these tools, with friendlier interfaces and more powerful resources, have brought about new architectural solutions. They allow for greater creativity, experimentation, simulation and facilitate the decision making process from the conceptual design to the communication to construction site. Authors in [4], [5], [6] and [7] have demonstrated these possibilities in great detail.

As such computer systems continue to be developed rapidly, significant variations in the possibilities and performances may arise and become clearer among them. However, for the newcomers it is not ease to perceive which system may offer more advantages for the design process.

Therefore, our initial research question was: what types or categories of CAD systems are more appropriate for the architectural design process taking into consideration the different interfaces and resources offered by them?

II. Hypothesis

A comparative study of the most used software could be beneficial to architects in taking decisions about which type or category of CAD systems to choose. We believe that this research will show that some types of CAD systems are more suitable than others to the exploration, experimentation and decision making characteristic of the architectural design process.

III. Method of Investigation

Three CAD systems were chosen for comparison purposes, each of them representing one of the three categories described earlier: a 2 ½ CAD system, AutoCAD (www.autodesk.com), a 3D surface modeler, Sketchup (sketchup.google.com) and a 3D solid modeler, formZ (www.formz.com). The choice was guided by two principles: each system should be a well known software in the international architectural design arena and should offer the main features of the category it belongs to as described earlier in this paper.

Two design/modeling tasks were defined in order to test our hypothesis through the five groups of parameters described later in this paper.

Task 1 had the objective of providing a means of assessing the interactivity of each system through the production of a 3D model of a terrain. The information needed to construct such model was provided through a bitmapped image containing the limits of the land plot and the topographic contours. The product should be a terrain 3D model with the most realistic finishing that could be provided by each assessed system. The assessment parameters were applied to the process of constructing the model as well as to its final product.

Task 2 had the objective of assessing the 3D modeling capabilities of the chosen systems. Information about a small building was provided through plans, sections and elevations. A 3D model of this building was created in each chosen system, materials were assigned to its components and natural and artificial lights were created and configured. At the end of the modeling and configuration process, files in the format .dwg were exported from each software and imported into each of them to verify their compatibility and interoperability. Another objective of this task was to simulate situations that are routine in a cutting edge architectural office such as the interactive visualization of the building design and the production of construction drawings from a single source 3D model.

With the aforementioned hypothesis in mind, five parameters were built, each one involving a set of questions for comparative purposes. They include interface, precision in the resolution of a specific task, final product, relationship with the computer platform and time spent on the chosen task. Each one of these groups is briefly described below. These criteria were further detailed into the questions which can be seen in Table I later in this paper.

A. Interface

The interface of each CAD system was assessed taking into consideration its intuitiveness, the degree of simplicity or complexity, the excess or lack of information available to user and the level of organization through categories of operations.

B. Precision in the resolution of a specific task

This group of parameters has the purpose to assess each chosen system in the execution of an architectural task. They identify what tools are used during the assessment and the level of precision they offer to the user. These parameters were organized into three subsets. The first subset is comprised of tools used for the production of two-dimensional architectural drawings. The second subset contains the precision tools, including Cartesian system, numerical input and snaps. The third subset comprises the tools for three-dimensional modeling, including geometry construction and manipulation, rendering and interactive capabilities.

C. Final product

This group comprises the possible final products that each system allows the user to produce. This includes the production of bitmapped synthesized images containing static views of the building design, walkthrough, solar and object animations, VRML files exportation, and the production of two-dimensional drawings.

D. Relationship with the computer platform

This group of parameters allows for the assessment of each system performance in the machine used for the development of the tasks. The objective of this assessment is to verify if the system`s minimum configuration recommended by the software maker is adequate for running it alone or simultaneously with other applications.

E. Time spent and number of steps needed for a chosen task

This group of parameters comprises measuring the total time in minutes needed for the execution of each task and its subtasks. Since all the tasks were performed in the same machine and in similar conditions, we believe this provides a reasonably fair comparison between the chosen systems. Additionally, "time spent" was also measured in relative terms by counting the number of steps needed in each chosen systems to complete a task or subtask. This allows an assessment which is less dependent on the conditions of the machine used. It also allows an assessment less dependent on the personal speed of the user in executing operations and personal knowledge of each system.

IV. RESULTS

The results of our experiment are shown in Table I below.

TABLE I
ASSESSMENT CRITERIA AND EXPERIMENTATION RESULTS

Assessment Criteria ● Yes ○No n/a not applicable	AutoCAD	Sketchup	FormZ
1. INTERFACE			
1.1 Does the program have a friendly user interface?	○	●	●
1.2 Is it easy to find the commands?	○	●	●
1.3 Is there an excess of information in the initial interface?	●	○	○
1.4 Are the commands grouped by clear categories?	●	●	●
1.5 Is it easy to find one tool already used in the program?	●	●	●
1.6 Is it easy to find one tool never user (by the user) in the program?	○	●	○
1.7 Is it easy to configure the working area?	●	●	●
1.8 Are the buttons self explained?	○	●	●
2. ABOUT PRECISION IN THE EXECUTION OF THE TASK			
2.1. TWO-DIMENSIONAL DRAWING			
2.1.1 Does it allow importing image files into the drawing?	●	●	●
2.1.2 Does it allow importing image DWG into the drawing?	●	●	●
2.1.3 Does it have tools for drawing?	●	●	●
2.2. PRECISION TOOLS			
2.2.1. Does it have precision tools such as grid, object or directional snap?	●	●	●
2.1.2 Does it allow configuring the precision tools?	●	○	●
2.2.3 Does it allow inserting numerical parameters with precision?	●	●	●
2.2.4 Does it work with Cartesian coordinates?	●	○	●
2.2.5 Does it work with Polar coordinates?	●	○	●
2.2.6 Can it edit the DWG imported file?	●	●	●
2.3. THREE-DIMENSIONAL MODELING			
2.3.1 Does it work with extrusion tools?	●	●	●
2.3.2 Does it allow the construction of non Euclidian curved surfaces?	○	○	●
2.3.3 Is it a surface modeler only?	○	●	○
2.3.4 Is it a true solid modeler?	●	○	●
2.3.5 Can it create solids with precision?	●	●	●
2.3.6 Can it create complex forms with precision?	○	○	●
2.3.7 Does it allow simultaneous visualization of the tree-dimensional and two-dimensional views?	●	○	●
2.3.8 Does it have its own library of blocks and symbols?	●	●	●
2.3.9 Was its default library enough to complete the task one?	○	●	●
2.3.10 Was its default library enough to complete the task two?	○	●	○
2.3.11 Does it allow to create new objects and add them to the library?	●	●	●
2.3.12 Does it allow to change the view point interactively?	●	●	●
2.3.13 Does it allow to assign materials to the objects?	●	●	●
2.3.14 Does it allow to assign internal lights to the object?	○	○	●
2.3.15 Does it allow to assign external lights to the object?	●	●	●
2.3.16 Does it allow to change the view point without having to reprocess the textures?	○	●	●
2.3.17 Does it allow to change the view point and to see the lights simultaneously?	○	●	●
2.3.18 Does it allow to change the view point and to see the lights and the texture in real time?	○	●	●
2.3.19 Does it convert schematic drawings and sketches in walls?	○	●	●
2.3.20 Does it allow to draw sketches of conceptual ideas?	○	●	●
2.3.21 Can it create a representation of the sky through parameters?	○	●	●
3. FINAL PRODUCT			
3.1. STATIC IMAGES AND ANIMATION			
3.1.1 Is the quality of the image of the object satisfactory?	○	○	●
3.1.2 Can it export the image in BITMAP file?	●	●	●
3.1.3 Can it export the file to another 3D modeler?	●	●	●
3.1.4 Was there any loss of information when it exported the file?	●	●	○
3.1.5 Is it necessary to use another program for image post production?	●	●	○

Assessment Criteria ● Yes ○No n/a not applicable	AutoCAD	Sketchup	FormZ
3.1.6 Does the program produce animation?	○	●	●
3.1.7 Does the program do object animation?	○	○	●
3.1.8 Does the application allow modifying the modeled solid?	●	●	●
3.1.9 Can the program do sun animation?	○	●	●
3.1.10 Can the application configure the geographic coordinates?	○	●	●
3.1.11 Can the program produce VRML file?	○	○	●
3.2. THE PRODUCTION OF WORKING DRAWINGS			
3.2.1 Can the application produce 2D file for floor plans, section and elevations?	●	●	●
3.2.2 Is it necessary to use another program for the production of 2D drawing with hyper realistic renderings?	○	●	●
3.2.3 Can it produce fast and easily the floor plans?	○	●	●
3.2.4 Can it produce fast and easily the elevations?	○	●	●
3.2.5 Can it produce fast and easily the sections?	○	●	●
3.2.6 Was there any difficulty to produce working drawings according to Brazilian legislation?	○	●	●
3.2.7 When modifying one part of the project the other ones are automatically updated?	○	●	●
3.2.8 When exporting from AutoCAD to the program under assessment was there any loss of information?	n/a	○	○
3.2.9 When exporting from FormZ to the program under assessment was there any loss of information?	○	○	n/a
3.2.10 When exporting from Sketchup to the program under assessment was there any loss of information?	○	n/a	●
4. PLATAFORM RELATED PERFORMANCE			
4.1 Did the program crash performing the task 1?	○	●	●
4.2 Did the program crash performing the task 2?	○	○	●
4.3 Was there any problem with the computer while working the task one simultaneously with other applications?	○	●	●
4.4 Was there any problem with the computer while working the task two simultaneously with other applications?	○	●	●
4.5 Is it possible to print the file produced in a 3D printer?	○	○	●
4.6 Is it the application compatible with different platforms?	●	●	●
4.7 Was the minimal recommended platform enough for the task one?	●	●	●

Assessment Criteria ● Yes ○No n/a not applicable	AutoCAD	Sketchup	FormZ
4.8 Did the minimal platform recommended by the manufacturer was enough for performing the task two?	●	●	●
5. TIME SPENT			
5.1. TASK ONE			
5.1.1 How much time needed to model the land?	9m30	6m25	5m08
5.1.2 How much time needed to apply and configure texture?	2m53	1m54	2m04
5.1.3 How much time needed to apply and configure the sun light?	N/A	1m03	1m13
5.1.4 How much time needed to insert the vegetation (five trees)?	5m23	2m35	4m49
5.1.5 How much time needed to modify the sky?	3m10	1m32	1m55
5.1.6 How much time needed to configure views, render and save images?	12m17	6m43	7m29
5.1.7 How much time needed to configure and generate the sun animation?	N/A	9m53	5m12
5.1.8 How much time needed to configure and generate the walkthrough animation?	N/A	12m22	1h12m 05s
5.1.9 How much time needed to configure and generate the object animation?	N/A	N/A	1h23m 15s
5.1.10 How many steps needed to model the land?	88	93	65
5.1.11 How many steps were needed to apply and configure texture?	21	9	13
5.1.12 How many steps were needed to apply and configure light?	N/A	7	14
5.1.13 How many steps were needed to insert the vegetation (five trees)?	34	15	23
5.1.14 How many steps were to needed to modify the sky?	54	21	36
5.1.15 How many steps were needed to configure views, render and save images?	112	65	93
5.1.16 How many steps were needed to configure and generate sun animation?	N/A	14	6
5.1.17 How many steps were needed to configure and generate walkthrough animation?	N/A	18	21
5.1.18 How many steps were needed to configure and generate object animation?	N/A	N/A	16
5.2. TASK TWO			
5.2.1 How much time was needed to model the land and the building?	29m15	13m32	26m64
5.2.2 How much time was necessary to apply and configure the texture?	18m02	3m25	14m52
5.2.3 How much time was needed to apply and configure sun light?	N/A	2m22	2m26

Assessment Criteria ● Yes ○ No n/a not applicable	AutoCAD	Sketchup	FormZ
5.2.4 How much time was needed to apply the blocks according to the project?	26m12	7m22	17m33
5.2.5 How much time was needed to model the sofa?	N/A	N/A	7m23
5.2.6 How much time was needed to make changes in the model?	4m02	2m39	3m42
5.2.7 How much time was needed to make changes in the sky?	3m32	2m20	2m08
5.2.8 How much time was needed to configure the views, render and save the images?	16m32	4m08	5m46
5.2.9 How much time was needed to configure and generate the sun animation?	N/A	2m54	9m33
5.2.10 How much time was needed to configure and generate the walkthrough animation?	N/A	3m45	1h07m 20s
5.2.11 How much time was needed to configure and generate the object animation?	N/A	N/A	2h15m 56s
5.2.12 How much time was needed to generate the floor plans, sections and elevations and to save them?	N/A	4m19	8m43
5.2.13 How many steps were necessary to model the land and the building?	85	48	73
5.2.14 How many steps were necessary to apply and configure the texture?	45	15	112
5.2.15 How many steps were necessary to apply and configure the sun light?	N/A	9	23
5.2.16 How many steps were necessary to apply the blocks according to the project?	195	85	184
5.2.17 How many steps were necessary to model the sofa?	N/A	N/A	77
5.2.18 How many steps were necessary to make the changes in the model?	57	63	71
5.2.19 How many steps were necessary to make the changes in the sky?	55	24	31
5.2.20 How many steps were necessary to configure the views, render and save the images?	123	62	78
5.2.21 How many steps were needed to configure and generate sun animation?	N/A	32	6
5.2.22 How many steps were needed to configure and generate walkthrough animation?	N/A	15	25
5.2.23 How many steps were needed to configure and generate object animation?	N/A	N/A	28
5.2.24 How many steps were needed to generate floor plans, sections and façades and to save them?	N/A	12	34

Space allows us to comment on the above results only on a selective basis.

One could argue that we are comparing things which are so different that this render them incomparable. However, it is necessary to stress that the three CAD systems assessed in this paper are very often object of comparison among architects that are less familiar with design computing. They do that because when they think of a tool they use as first reference the usual design tasks performed by an architect on a day to basis, rather than the specific features of the tool itself. Therefore, it was necessary to compare these categories of CAD systems to make the distinctions clearer to the newcomer.

On the other hand, the history of CAD systems shows that the frontiers among the categories tend to become blurred over time.

The focus of our experiment was the ability to model artifacts three-dimensionally. This has put the category of the 2 ½ CAD system into disadvantage according to the data above. AutoCAD, by being the product of a slow metamorphosis, does not provide significant facilities the other two programs do, as the 'not applicable' in the above table indicates (for example items 5.1.3, 5.2.9, 5.2.10, 5.2.11, 5.2.17, etc.). It cannot model complex solids, it cannot animate and it does not render realistic views. It is very good in what was originally thought for, i.e., the fast production of 2D drafting.

On the other hand, Sketchup is a much newer program, with a very interactive and fast interface with many resources AutoCAD does not have. The times and number of steps necessary to perform specific task has put Sketchup in a leading position in several items of the above table (for example items 5.1.11, 5.1.12, 5.1.13, 5.1.17, etc.). However, it hides a major problem behind these advantages: it is a reduction of reality. It does things faster and requiring less steps because it also does less things than the other two CAD systems categories. It cannot model complex geometries (for example item 5.2.17 which concerns modeling complex curved surfaces). Therefore, it poses significant restrictions on the designer's creativity. It does not support high quality rendering and it cannot calculate lighting accurately (3.1.1). Therefore, it is not a good tool for more powerful design analysis.

The third category of CAD systems was represented here by formZ. This a program which demands high performance hardware, and this has put it in disadvantage in relation to the other software assessed here, particularly regarding time spent on performing some tasks and some eventual crashes (items 4.1, 4.2, 4.3). However, with the availability of affordable powerful hardware, formZ is a good design tool. It models complex geometries (items 2.3.2, 2.3.6), it produces high quality renderings and it can calculate lights accurately (item 3.1.1). It can produce animated walkthroughs and it can animate lights, objects and textures (items 3.1.6, 3.1.7). It also handles significant interaction through the use of interactive shading which allows for real time walkthroughs (items 2.3.12, 2.3.16, 2.3.17, 2.3.18).

V. CONCLUSIONS

We have made a contribution by comparing three major categories of CAD systems. We believe that this research showed that some types of CAD systems are more suitable than others to the exploration, experimentation and decision making that is characteristic of the architectural design process. The results show that each of these categories has its place into the day to day activities of an architect. However, significant differences exist between such categories. If the objective is to look for a type of CAD system which would allow incorporating the computer into the design decision making process, then we think the best option is a true solid modeler.

We acknowledge that more needs to be done. A broader analysis including more software in each category would allow assessing their suitability for the design decision making process. Also it would be important to incorporate other categories of CAD systems such as BIM tools in order to acquire a better understanding of the implications of using these resources into the design process.

REFERENCES

[1] Bruno Zevi, *"Architecture as space – How to look at architecture"*, 1st American edition, Horizon Press, New York, pp. 22, 23, 26, 1957.

[2] Loukas Kalisperis, *"CAD in Education: Penn State University"*, in ACADIA Quarterly, vol. 15, number 3, 1996.

[3] William Mitchell & Malcolm McCullough, *"Digital Design Media"*, John Wiley & Sons, New York, 1994.

[4] Branko Kolarevic, *"Architecture in the Digital Age – Design and Manufacturing"*, Taylor & Francis, New York, 2003.

[5] D. Schodek, M. Bechthold, K. Griggs, K. M. Kao & M. Steinberg, *"Digital Design and Manufacturing – CAD/CAM Applications in Architecture and Design"*, John Wiley & Sons, New York, 2005.

[6] Chuck Eastman, Paul Teicholz, Raphael Sacks and Kathleen Liston, *"BIM Handbook – A Guide to Building Information Modeling"*, John Wiley & Sons, New Jersey, 2008.

[7] Michael Stacey, Philip Beesley & Vincent Hui, "Digital Fabricators", University of Waterloo School of Architecture Press, 2004.

Using Decision Structures for Policy Analysis in Software Product-line Evolution – A Case Study

Nita Sarang
CMC Limited, Mumbai, India
nita.sarang@cmcltd.com

Mukund A Sanglikar
University Department of Computer Science, University of Mumbai, Mumbai, India
masanglikar@rediffmail.com

Abstract- **Project management decisions are the primary basis for project success (or failure). Mostly, such decisions are based on an intuitive understanding of the underlying software engineering and management process and have a likelihood of being misjudged. Our problem domain is product-line evolution. We model the dynamics of the process by incorporating feedback loops appropriate to two decision structures: staffing policy, and the forces of growth associated with long-term software evolution. The model is executable and supports project managers to assess the long-term effects of possible actions. Our work also corroborates results from earlier studies of E-type systems, in particular the FEAST project and the rules for software evolution, planning and management.**

I. Introduction

The role of software in complex systems has increased phenomenally over the last couple of decades. Software is now a critical factor of most businesses. As the global software footprint grows, the need to continually support software installations for the ever changing needs of the businesses becomes inevitable. Fielded software used for real-world solutions must be continuously upgraded to ensure fitness of use to the application domain. Long-term software (product) evolution is a challenge especially when maintaining large complex systems.

The software evolution context is unique. Schedule is a constant factor in software maintenance: business drives the software release plans and calls for extensive planning on the part of the software maintenance teams to ensure that the committed new functionality gets out to the end-users in a timely manner. The software is progressively fixed, adapted and enhanced. System performance is not an afterthought and must be ensured with every release of software. In software evolution, the defined application architecture and design also define the overall boundary for software changes. We target zero post release defects; under-performance causes a disruption in operations and loss to the business. As the software evolves over time, the structural and functional complexity of the software increases; this affects overall stability of the product and may cascade to an eventual decline in the functional growth rate.

Software evolution is affected not only by the local forces of growth, technical and management controls; but by a larger system of forces outside the process linked to organizational, competitive or marketplace pressures, business and usage. We therefore consider the individual, local, and collective global behavior of process constituents, impacted by positive and negative feedback loops and control mechanisms. In studying the behavior of such systems we observe the systems from outside the process (generally referred to as global behavior).

We attempt to understand the dynamic consequences of process and policy decisions, and the challenges that face project managers in the software maintenance phase. We use a blend of hybrid and combined approaches to model post-implementation maintenance processes [11], Our model takes roots in the global process feedback [1] phenomena associated with the evolution of fielded software product-lines as these are maintained over time. The Feedback, Evolution And Software Technology hypothesis / project (FEAST/1) [6], was the first focused efforts towards software evolution, and generated wide interest among researchers.

We establish a model for software evolution of product-lines. The unified, integrated model is original but uses model elements and builds on previous models of Abdel-Hamid [2], Kellener [3], Lehman [4], Kahen [5], and Ramil [10]. We overlay our policy decision structures on the base model.

The authors thank CMC Limited higher management, HR division, and project managers from the Securities practice for providing the data set that served as the basis for this study.

K. Elleithy (ed.), *Advanced Techniques in Computing Sciences and Software Engineering*, DOI 10.1007/978-90-481-3660-5_97, © Springer Science+Business Media B.V. 2010

II. The System Modeling Context

It is necessary to recognize the importance and influence of context variables, the environment and the domain when interpreting the results of empirical studies. This research and associated analysis is based on E-type systems; that is, systems implemented to support applications operating in the real world.

The product-lines under consideration are from the Securities domain. We consider three distinct product-lines, all of which have been in the field for over a decade and continue to undergo evolution. New allied business applications continue to be added to the base product, and these follow a complete software development life-cycle up to implementation. Post warranty support (three months, in our case), these applications are merged with others under maintenance. Along with functional enhancements, these applications have also undergone migrations for new technologies and platforms; including architecture makeover.

During the software maintenance phase, change requests can be generalized to broadly fall within the below categories:

- Wrongly understood and / or implemented functionality.
- Functional enhancements and the ripple effect for further adaptations and customizations; there is a need to continuously get new business content (change requests) into the products.
- Technology migration of one or more application components happens over a period of time. This also includes new versions of system / layered software and databases.
- The sunset of older platforms and technologies demands re-design, and
- Defect correction, including latent defect in fielded software.

On the defect front, we aim for zero post-release defects – there are severe penalties associated with business opportunity losses due to software defects and / or down times. The projects follow reasonably strong review and testing practices. All software work products are reviewed / tested. There are multiple test environments. Release to production is tightly controlled and based on a successful real-world mock test by a controlled group of users.

From a future model replication and adaptability perspective, it would be contextual to record that all three product lines (associated projects) had fully embraced the organizations process improvement initiatives around the CMMI®, the organization was assessed in 2006 against the continuous representation with the verification, validation, project planning and project monitoring and control process areas at capability level 4. The organization implemented a coherent measurement program to drive its quantitative management activities. We record that all data points showed greater than 78% process compliance.

III. Policy Decision Structures

Process focus in long-term product evolution is around the key product attributes, namely, reliability, performance, scalability, portability [14]; and the overall total cost of maintenance and ownership of the software systems deployed. We examine two specific organizational policies: staffing policy and work policy related to progressive and anti-regressive work allocation. Both policies have a bearing on growth productivity.

Organizational and project knowledge building and retention are a very significant facet of growth productivity and must be successfully managed in the backdrop of the constraints of project management. The organization's attrition rate, staffing policy and hiring delays have bearing on growth productivity. We also assess team member transfers, and the need to provide job rotation as we deal with hiring delays.

Another very important and non-compromising consideration with respect to continually evolving software is to ensure that the software continues to be evolvable [10]. We consider two elements of change – progressive work (functional and technology change requests to ensure continued fitness of purpose) and anti-regressive work (to control and manage the structural complexity as more and more progressive work is incorporated). Undertaking anti-regressive work is an implicit expectation of the customer and more often than not, must be managed within the same pool of resources. Achieving controlled and effective management of such complex multi-loop feedback systems and improving its reliability and effectiveness is a major challenge.

We study the feedback phenomena associated with the evolution of software as it is maintained over time. We target to define an optimal and required ratio for distributing progressive and anti-regressive work (and associated staffing), and implement the same through a formal contract with our clients.

We selectively apply model elements to our specific context. For areas not directly related to policy investigation (like testing and rework), we use a macro perspective of the system and associated processes. For other areas directly linked to the policy investigation (like growth productivity), we delve into the low level details of how the system (for maintenance activity) is implemented along with the associated feedback loops. We note that the effort perspective and defect reliability and defect localization perspectives within our problem domain have already been studied [11], [12], [13].

Figure 1 is our integrated model for product-line evolution using iThink® modeling package. The model developed is an executable simulation of the software maintenance phenomena and can facilitate managers to assess the effects of changed processes and management policy decisions before they can commit to software engineering plans.

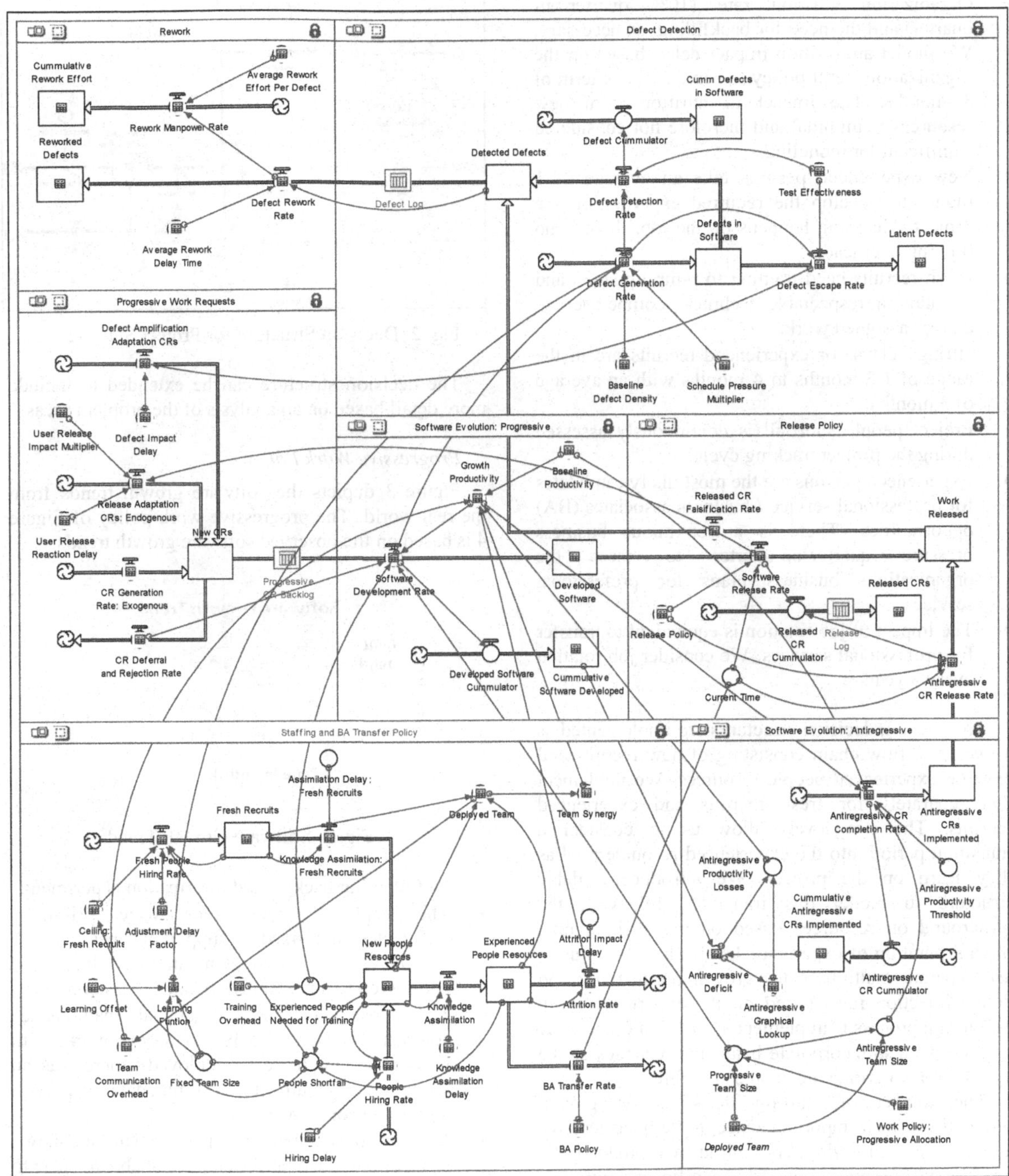

Fig. 1: A System Dynamics Model for Policy Analysis of Software Product-line Evolution

We describe the model decision structures below.

Staffing Policy:

The staffing policy of Figure 2 is based on the following considerations:

- Organization attrition rate (10% quarter-on-quarter) and the need for backfilling, as necessary. We model an attrition impact delay based on the organization's exit policy to serve a notice term of 3 months. The impact of attrition from new resources is minimal and therefore not considered significant for modeling.
- New experienced persons take an average of 1 month to develop the required expertise on the project – learning happens on the job; there is no training overhead.
- Fresh recruits need training to ramp up skills and maintain a respectable technical competence to deliver assigned work.
- Hiring delays for experienced recruits are in the range of 1.5 months to 4 months with an average of 3 months.
- Project people shortfall is continuously assessed during the project tracking cycle.
- Experienced persons are the most likely candidates for professional service / business associates (BA) opportunities. There is a continuous business pressure quarter-on-quarter to meet the organizations business plans for professional services.
- The impact of job rotation is equivalent to transfer for professional services. We consider job rotation every 3 years.

The staffing decision structures are implemented as a two-level flow chain consisting of new recruits and existing experienced people resources. We implement this separately for fresh recruits and experienced recruits. The two levels allow us to consider a transition period into the experienced resource pool as they learn on the project. We introduce a delay structure to model this transition. Likewise, the turnaround on resources based on the attrition rate. Such separation of experience levels also allows us to use separate multipliers for growth productivity and defect injection rates based on the experience level. The hiring process is implemented as a first order delay system; we also incorporate balancing feedback as the system tries to bridge the people shortfall.

The impact of staffing decisions on growth productivity is an important aspect to be managed. We associate several multipliers with growth productivity. From a staffing perspective, the significant multipliers are communication overhead and the efforts to train new recruits, attrition rate and team synergy. We associate Communication Overhead directly with the total team size. We do not consider the nullification effects of establishing sub-teams.

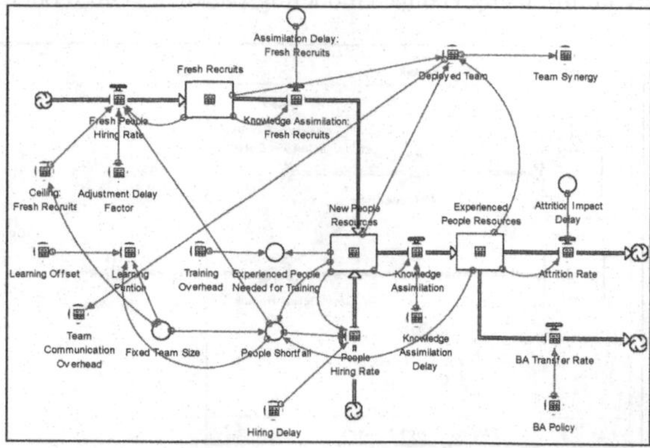

Fig. 2: Decision Structure for Project Staffing

The decision structure can be extended to include more detail based on an analysis of the problem areas.

Progressive Work Policy:

Figure 3 depicts the software growth trends from the real-world. The progressive work policy of Figure 4 is based on the observed software growth trends.

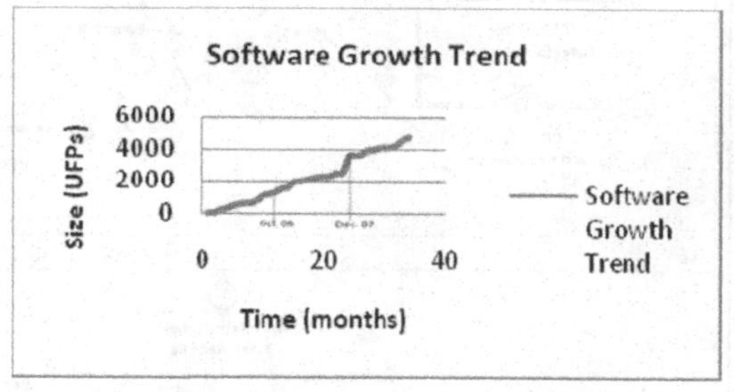

Fig. 3: Software Growth Trends

The following background information is pertinent:

- The deployed applications were originally architected and sized to support business growth for 7 years post implementation; client had committed to an extended maintenance contract.
- The subsequent renewed annual maintenance contracts focused on only progressive work (with fixed man months to be deployed); there was no scope contractually to accommodate any anti-regressive component.
- 10 years post implementation, system has shown signs of degraded performance – unable to scale to meet the demands of the business, peak transaction

handling capability, performance issues, penalties attributed to latent defects, reduced flexibility of the system to accommodate change requests. It was almost as if the system had served its life in the field and needed to be phased out. Client was considering a complete overhaul and migration – COTS components was an available option.

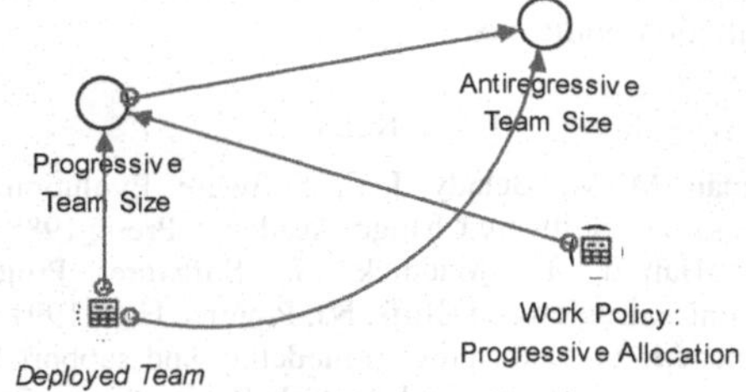

Fig. 4: Decision Structure for Progressive Work Policy

The progressive work policy decision structure is based on specifying the ratio of progressive to anti-regressive work; the team sizes are also deployed in the same ratio. We model fixed people allocation strategies across the progressive and anti-regressive work components. The structure allows us to assess the impact of a fixed anti-regressive work component on growth productivity.

The anti-regressive work is modeled using a productivity threshold. When the productivity falls below the threshold, we pump in anti-regressive CRs.

IV. Policy Analysis

The model is used by the project team proactively to simulate various policy scenarios and assess the long-term effects of such decisions on the project.

Figure 5 is an output of such simulation runs for staffing policy evaluation. We ascertain sensitivity of growth productivity to BA transfer policy for 0%, 20%, 50% and 70% protection. To balance the business pressure from Higher Management to push BA request through, we implement a protection for 20% of the experienced people on the project through the BA transfer policy. These named resources are held back within the project to ensure a sustained level of service and quality of deliveries.

The reality of the staffing position indicates a need to implement resource buffering or find innovative ways for quick staffing. We draw upon an internal channel to provide these resources. In early 2007, the project teams teamed up with the Education and Training line of business to staff its projects with near zero hiring delays.

While the overall educational profile of the people is vastly different from the normal requirements of open recruitment, the training courses were upgraded to bridge this gap with adequate project experience and a finishing module. In most cases, the people are available to the project prior to their course completion as project trainees (with no direct cost overheads). The impact of training on growth productivity as a result of this was incorporated into the model and studied. We fix this ratio to not more than 10% of the total team size – the balancing feedback structure provides an adequate implementation towards equilibrium.

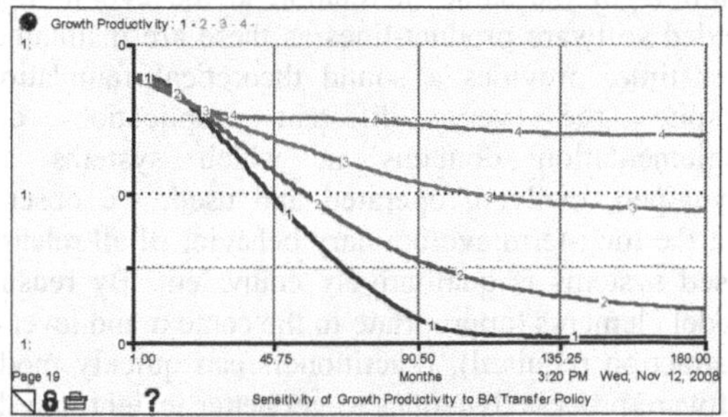

Fig. 5: Project Staffing Policy Response

Figure 6 is an output simulation run for the effects of anti-regressive work policy on cumulative CRs released over-time. We implement exogenous CRs as a random Poisson process [5]. The release policy drives the new change request activity and particularly controls the quantum of anti-regressive work required. We execute the model for 0%, 20%, 40% and 60% anti-regressive work. An anti-regressive component of 40% provides maximum long-term growth; anything more does not add value and is wasted.

We started with a 40% anti-regressive component in year 2006, taking a major share of that on our account. A year hence, system has improved significantly; customer satisfaction levels have risen.

We were able to take our results to the customer and request modifications to contract terms. The model indicates the need for a continued 40% anti-regressive componnet; customer has agreed to take on 25% anti-regressive work into the contracts.

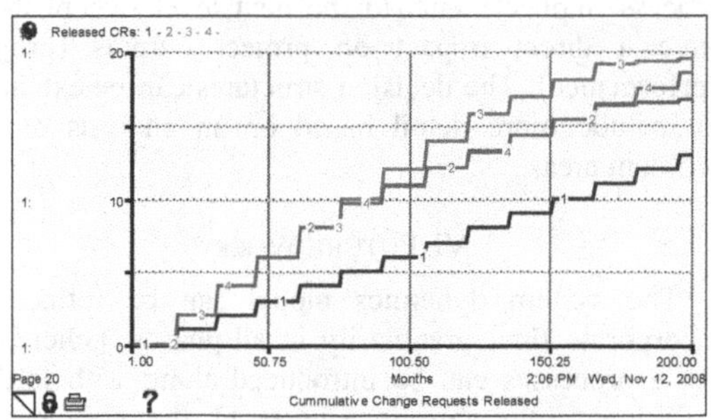

Fig. 6: Project Anti-regressive Policy Response

Our annual maintenance contracts for software maintenance now include two distinct components: functional enhancements to meet the demands for functional product evolution, and the associated (equivalent components) of refactoring to ensure continued system maintainability.

V. Conclusion

The global process feedback model and subsequent detailing of the same in relation to the evolution of fielded software product-lines as these are maintained over time, provides a sound theoretical foundation. Despite the very different application and implementation domains in which systems are developed, evolved, operated and used, we observe that the long-term evolutionary behavior of all release-based systems is qualitatively equivalent. By reusing model elements (appropriate to the context and level of abstraction required), practitioners can quickly model software process dynamics to get better insight into the many influencing factors that contribute to a project progressing at a rate different from the plan.

Dynamic modeling supports project risk assessment and management. Dynamic models facilitate and enable project managers to understand the long-term implications of policy and take proactive steps to maintain and / or restore productivity growth trends. We successfully modeled and implemented decision structures for policy analysis for the management of product-line evolution.

When considering the real-world, it is imperative to consider the role of factors such as, organizational and managerial policies, strategies, and other forces and constraints placed on the process by the environment. Appropriate policies are easily integrated into the model. The impact of different policies can be studied through simulation – the results of the simulation support project management decision: in particular, the long-term effects of those decisions.

Models incorporating all of the above can support detailed and realistic decision making that takes process improvements to the next level: where these have a direct impact on project outputs (project performance). The decision structures can be extended to include more detail based on an analysis of the problem areas.

VI. Future Work

The system dynamics model can be refined to incorporate finer granularity in all phases. Other life-cycle processes can be introduced along with further drill-down into various aspects of the system that warrant investigation.

We recommend additional sensitivity analysis of growth productivity to various model parameters like attrition impact delay, hiring delays, knowledge assimilation delays, and others. Such experiments can provide leads to organizational level processization initiatives. By adding the project monitoring and control aspects and synchronizing the policy assessment with the project tracking cycles, we model more realistic conditions.

References

[1] Lehman M M, Belady J F, Software Evolution – Processes of Software Change, Academic Press, 1985.

[2] Abdel-Hamid T, Madnick S, Software Project Dynamics, Englewood Cliffs, NJ, Prentice Hall, 1991.

[3] Kellner M, Software process modeling and support for management planning and control, Proceedings of the first international conference on the software process, pp.8028, IEEE Computer Society, 1991.

[4] Lehman M M, Ramil J F, The impact of feedback in the global software process, 1999.

[5] Kahen G, Lehman M M, Ramil J F, Wernick P, An approach to system dynamics modeling of aspects of the global software process, 2001.

[6] Wernick P, Lehman M M, software process dynamic modeling for FEAST/1, Software Process Improvement and Pracyice, 7(3-4), 2002.

[7] Lakey P, A hybrid software process simulation model for project management, ProSim Workshop 2003.

[8] Lehman M M, Ramil J F, Software evolution: background, theory, practice, Information Processing Letters archive, special issue contributions to computing science, 88(1-2), 2003.

[9] Boehm B, Brown A W, Madachy R, Yang Y, A software product line life cycle cost estimation model, ISESE '04:The 2004 International Symposium on Empirical Software Engineering, pp 156-164, IEEE Computer Society, 2004.

[10] Ramil J F, Lehman M M, Cohen G, Simulation process modeling for managing software evolution, in Acuna S T and Juristo N (Eds), Software process modeling, New York: Springer Science + Business Media Inc., 2005.

[11] Sarang N, Benchmarking product and service quality – a unified approach, Proceedings of the European Software Engineering Process Group Conference, 2004.

[12] Sarang N., Sanglikar M, An analysis of effort variance in software maintenance projects, Proceedings of the International Conference on Systems, Computing Sciences and Software Engineering, Springer, 2007

[13] Sarang N., Sanglikar M., Defect based reliability analysis model for business systems deployed on a large scale, Proceedings Software Process Improvement and Capability Determination Conference, 2008.

[14] Syanhberg M., Wohlin C., Lundberg L., Mattsson M., ACM International Conference Proceeding Series; Vol. 27, Processdings of the 14th International Conference on Software Engineering and Knowledge Engineering, Workshop on software engineering decision support methodology, 2002.

Fusion of Multimedia Document Intra-Modality Relevancies using Linear Combination Model

Umer Rashid
Faculty of Computing
Riphah International University,
Islamabad, Pakistan
UmerR@riphah.edu.pk

Iftikhar Azim Niaz
Faculty of Computing
Riphah International University,
Islamabad, Pakistan
ianiaz@riphah.edu.pk

Muhammad Afzal Bhatti
Department of Computer Science
Quaid-i-Azam University,
Islamabad, Pakistan
mabhatti@.qau.edu.pk

Abstract: **Information retrieval systems that search within multimedia artifacts face inter-modality fusion problem. Fuzzy logic, sequential and linear combinational techniques are used for inter-modality fusion. We explore asymptotically that Fuzzy logic and sequential techniques have limitations. One major limitation is that they only address fusion of documents coming from different modalities, not document relevancies distributed in different modality relevancy spaces. Linear combinational techniques fuse document relevancies distributed within different relevancy spaces using inter-modality weights and calculate composite relevancies. Inter-modality weights can be calculated using several offline and online techniques. Offline techniques mostly use machine learning techniques and adjust weights before search process. Online techniques calculate inter-modality weights within search process and are satisfactory for general purpose information retrieval systems. We investigate asymptotically that linear combination technique for inter-modality fusion outperforms fuzzy logic techniques and workflows. We explore a variation of linear combination technique based on ratio of average arithmetic means of document relevancies. Our proposed technique smoothes the effect of inter-modality weights and provides a moderate mechanism of inter-modality fusion.**

Keywords: inter-modality, intra modality, multimedia artifact, retrieval model, relevancy, digital libraries and multimodal fusion.

I. INTRODUCTION

Information retrieval systems like digital libraries and museum collections not only consist of text documents, their information resources may also hold any of multimedia data types such as image, audio and video objects [1,2]. Multimodal information is associated with multimedia artifacts. Information retrieval systems that perform search with in multimedia artifacts actually search with in multiple sources of information or modalities [3]. Multimodal search is distributed among different modalities. Final results of retrieval process are strongly based on fusion of results that are from different modalities associated with each object [4]. Constituents of multimedia objects are searched using diverse search mechanism e.g. video objects can be searched by using image content based and text based search mechanisms [5]. Multimedia documents search spaces or modalities are scattered in divergent spaces where varying search mechanisms are required for relevancy measures. Fusion techniques are used for the combination of results from different modalities. Fusion of relevancy weights come from different modalities of a document is called inter-modality fusion. Inter-modality fusion techniques are used for the calculation of final relevancies of multimedia documents. There are different techniques for multimodal fusion like fuzzy logic, sequential or workflows and linear combinational techniques [6]. Fuzzy logic and sequential fusion techniques are not satisfactory for documents relevancy calculations when multiple modalities are involved in the retrieval process. Sequential techniques also suffer from performance and sequence problems that affect precision and recall of overall retrieval process [7]. Fuzzy logic and sequential fusion techniques discuss fusion of documents not their relevancies.

Linear combinational techniques [5] for multimodal retrieval are simple, easy to implement in general purpose retrieval systems. They do not have limitations like sequential and fuzzy logic techniques. This paper discuses different multimodal fusion techniques, their limitations and propose a variation of linear combinational technique that calculates dynamically inter-modality weights for multimodal fusion.

II. EVALUATION OF INTER-MODALITY FUSION TECHNIQUES

In this section we discuss different fusion techniques and evaluate their limitations in the context of multimodal retrieval.

A. *Fuzzy Logic Techniques*

Fuzzy logic techniques are based on Boolean operations [8]. Conjunctions or disjunctions of results that are from different modalities are examples of fuzzy logic techniques [9, 10]. Formally disjunction and conjunction of results can be calculated by using following equations.

$$Rs = \bigcup_{i=1}^{n} Rs(i) \tag{1}$$

$$Rs = \bigcap_{i=1}^{n} Rs(i) \tag{2}$$

Where $Rs(i)$ consist of documents returned when search is performed on i[th] modality of documents.

K. Elleithy (ed.), *Advanced Techniques in Computing Sciences and Software Engineering*,
DOI 10.1007/978-90-481-3660-5_98, © Springer Science+Business Media B.V. 2010

Fuzzy logic techniques are simple and easy to implement. Retrieval function just requires searching of relevant documents from different modalities involved in the search process. Finally a merge function is used to merge documents from different modalities. Retrieval functions that perform search with in different modalities are independent and their distributed processing is possible. So time complexity of a fuzzy logic technique is sum of retrieval functions and a merge function. We explain time complexity by the following equation.

$$Tc = \sum_{i=1}^{n} Tc(Rf(i)) + Tc(Mf) \quad (3)$$

Where Tc is time complexity, $R_{f(i)}$ is i[th] retrieval function performed on i[th] modality involved in search process. M_f is merge function. We further simplify the above equation as:

$$Tc \geq nTc((Max(Rf))) + Tc(Mf) \quad (4)$$

Where n is the number of functions involved in the search process. Number of modalities and retrieval functions are always equal. So from above equation we conclude that time complexity of fuzzy logic techniques is always in the order of:

$$O(\max(Tc(Rf(i)))) + Tc(Mf) \quad (5)$$

According to equation 5 performance of fuzzy logic technique is satisfactory when they are used in multimodal search process. Fuzzy logic techniques have limitations that they cannot address fusion of weights that comes from different modalities. They are not appropriate for calculation of documents combined probabilistic relevancies. So system cannot give composite probabilistic relevancy of multimodal query with a multimedia object. By using these techniques probabilistic inter-modality fusion is not possible.

B. *Multimodal Fusion Using Sequential Techniques*

Sequential fusion techniques are basically workflows [6, 11] that exactly follow pipes and filter architectural approach [15]. Retrieval functions are applied on set of documents one by one such that output of one retrieval function will become input of next retrieval function. We better explain this approach with the help of an equation.

$$Rs = f1(f2(f3...fn(Ds))) \quad 6$$

Where *Rs* is result set, n is number of modalities, $f1$, $f2$, $f3$, $f4$… fn are retrieval function applied on modalities and *Ds* is document set. From above equation we analyze that by using sequential techniques retrieval function on document modalities are applied one by one in such a way that search process exactly works like pipes and filters architecture as shown in “fig. 1.”.

This technique is simple and easy to implement but distributed query processing is not possible. This technique is not suitable for general purpose information retrieval systems when large collections of multimedia digital repositories are available for search. By using sequential technique nested retrieval functions are executed. Time complexity can be calculated as

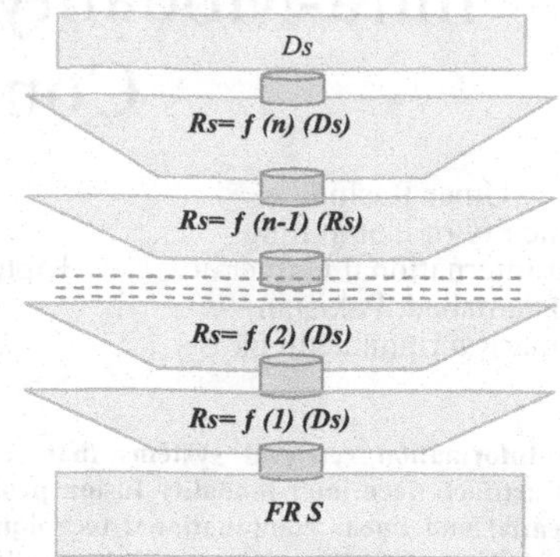

Fig. 1. Multimodal fusion using sequential/workflow model.

f(1), f(2),f(3)...f(n) are retrieval functions, Ds is document set on which retrieval functions are applied and FRS is final result set.

$$Tc = \sum_{i=1}^{n} Tc(Rf(i)) \quad (7)$$

Equation can also be written as

$$Tc \geq Tc(Rf(1)) + Tc(Rf(2) + Tc(Rf(3)) + ... + Tc(Rf(n) \quad (8)$$

Time complexity of retrieval process is equal to the sum of time complexities of all retrieval functions involved in the search process. So time complexity for sequential techniques is always order of:

$$O((Tc(Rf(1)) + Tc(Rf(2) + Tc(Rf(3)) + ... + Tc(Rf(n)) \quad (9)$$

According to equation 9 performance of sequential multimodal retrieval model is influenced by execution of retrieval functions. Retrieval functions are strongly interdependent so their distributed processing is not possible. Workflows or sequential models are feasible for simple query processing but they are not satisfactory for complex query processing [13] like multimodal query processing. In sequential techniques retrieval is also strongly influenced by order of functions performed on multiple modalities. If this order of function execution is not appropriate and function executed first on weaker modalities then retrieval process is strongly influenced [7]. This technique faces sequence problems. Machine learning techniques are used to learn sequences [16] for different query type sets but this approach is not applicable on vast amount of data and diverse query types. Offline learning of sequences of modalities for query sets has applications when domains are limited and query types are not so divergent [6]. When we know stronger and weaker modalities in advance then we can easily define cluster of documents and their corresponding sequences using machine learning techniques. But we want to explore some generic solution that has applications in different domains. Sequential techniques cannot give mechanism for probabilistic fusion [5] of document relevancies from different modalities.

C. *Multimodal Fusion Using Linear Combination Model*

Linear combinational techniques are mostly used for multimodal fusion. They cannot face problems that we discuss in other multimodal fusion techniques. Probabilistic inter-modality fusion is possible by using inter-modality weights [6]. Inter-modality weights are basically balancing factors or probabilistic factors that can be used to fuse probabilistic relevancies of a document from more than one modalities of information. Inter-modality weight shows probabilistic strength of each modality involved in the retrieval process.

Following equation explains inter-modality fusion using linear combinational model.

$$Sim\ (Q, D) = \sum_{i=1}^{n} \alpha\, iSim\ \ i(Qi, Di) \qquad 10$$

Where Q is multimodal query, D is multimodal document, αi is i^{th} inter-modality fusion factor and sum of all factors must be equal to one, Qi is i^{th} component of multimodal query, Qi is i^{th} modality of a document and $Sim\ i$ is probabilistic similarity of ith modality of a document with respect to i^{th} modality component of a query by using i^{th} similarity function.

From equation 10 we analyze that linear combinational model fuse probabilistic weights of a document from different modalities. These weights from different modalities are called intra-modality weights [6]. This model probabilistically fuses intra-modality weights for relevancy calculations while other two techniques combine relevant documents from different modalities. So this technique is considered as an ideal approach for inter-modality fusion.

By using linear combinational model distributed query processing is possible [11]. Time complexity depends upon processing time of retrieval function, inter-modality fusion function that we describe in equation 10 and function that calculates inter-modality weights, so time complexity will be

$$Tc = \sum Tc(Rf(i)) + Tc(Ff) + Tc(Wf) \qquad 11$$

Where fusion-Fun fuse relevancies come from different modalities and Weight-Fun calculates inter-modality weights but it depends upon on-line and off-line weight calculation techniques [6]. Time complexity will be mostly order of:

$$O(\max(Tc(Rf) + Tc(Wf) \qquad 12$$

Where Tc (Wf) = 0 when off-line weight calculation techniques are used.

As we noticed from equations 5, 9 and 12 that time complexity of fuzzy logic technique and linear combinational model is approximately equal except the overhead of merge and weight functions respectively. Time complexity of both techniques outperform sequential model.

Inter-Modality Weights: Inter-modality weights can be calculated using off-line and on-line techniques. Off-line weight calculation techniques are based on heuristics like speech modality associated with video artifact is stronger than visual modality [12] or all modalities have same strength. Machine learning techniques [7] are also used for the offline calculation of inter-modality weights. But using off-line techniques inter-modality weights are considered as constant [5] or predefined set of weights are calculated for predefined set of possible query types [14]. These weight calculation techniques are satisfactory for small specified domains but not a good approach when domains are vast and user queries are unpredictable.

On-line weight calculation techniques calculate inter-modality weights for each new search [6]. These techniques follow simple probabilistic approaches or simple online learning techniques for the calculation of inter-modality weights. Currently researchers try to calculate inter-modality weights using relevance feedback [17]. It is a multi-step process, first system shows relevant results then user gives relevance feedback and in last step system shows document relevancies by fusing newly calculated inter-modality weights with documents intra-modality weights.

III. PROPOSED INTER-MODALITY FUSION USING INTRA-MODALITY WEIGHTS

We investigate a new single step online approach for the calculation of inter-modality weights by using summarized information of documents intra-modality weights. Inter-modality fusion [6] requires inter-modality weights. We purpose a new technique for the calculation of inter-modality weights by exploiting intra-modalities weights. When search process starts query components are distributed in search space and relevancies are calculated for these distributed query components with in specified modalities of multimedia documents. So for a composite query different relevancy spaces are created. Probabilistic fusion process probabilistically fuses document relevancies from these relevancy spaces one by one and calculates composite relevancy for each document and creates composite relevancy space which is final result set. Main philosophy of our approach is that stronger modalities contribute more than weaker modalities in inter-modality fusion process.

We calculate modality strength by using relevancy spaces calculated for each modality in the search process. Our assumption is simple if relevancy space consists of high relevancies this makes corresponding modality strong and if relevancy space consists of low relevancies this makes corresponding modality weak.

Now we introduce two new possible methods for the calculation of probabilistic inter-modality weights by using above mentioned philosophy. We investigate arithmetic mean and ratio of average arithmetic mean method for the fusion of inter-modality weights.

Arithmetic Mean (AM): In this method we calculate inter-modality weights by taking ratio of arithmetic mean of all modalities relevancy spaces. Arithmetic mean takes average of each intra-modality weight set calculated during the information search process by retrieval function. Ratio of proportion of average arithmetic mean for each modality is used as inter-modality weight for that particular modality. Suppose we have n number of modalities $m_1, m_2, m_3 ... m_n$ and n number of relevancy spaces $s_1, s_2, s_3 ... s_n$ have been calculated when search is performed on modalities. Each relevancy space contains intra-modality weights. Inter-modality weights $w_1, w_2, w_3 ... w_n$ for inter-modality fusion using arithmetic mean can be calculated as:

$$w_1=AM(S_1)/(AM(S_1)+AM(S_2)+AM(S_3)+...+AM(S_n))$$
$$w_2=AM(S_2)/(AM(S_1)+AM(S_2)+AM(S_3)+...+AM(S_n))$$
$$W_3=AM(S_2)/(AM(S_1)+AM(S_2)+AM(S_3)+...+AM(S_n))$$
$$\ldots$$
$$w_n=AM(S_n)/(AM(S_1)+AM(S_2)+AM(S_3)+...+AM(S_n))$$

Average Arithmetic Mean (AVG (AM)): In this method we divide each relevancy space into equal regions and take ratios of average arithmetic means of these regions as inter-modality weights. Suppose we have n number of modalities $m_1, m_2, m_3 ... m_n$. Relevancy spaces $s_1, s_2, s_3 ... s_n$ have been calculated when search is performed on modalities. Each relevancy space contains intra-modality weights. By using average arithmetic mean method each relevancy space is divided into M periodic regions. Each periodic region contains relevancies in specified periodic limits. Average arithmetic mean of relevancy spaces can be calculated as

$$\text{Average } AM_1=(AM(S_{11})+AM(S_{12})+AM(S_{13})+...+AM(S_{1M}))/M$$
$$\text{Average } AM_2=(AM(S_{21})+AM(S_{22})+AM(S_{23})+...+AM(S_{2M}))/M$$
$$\text{Average } AM_3=(AM(S_{31})+AM(S_{32})+AM(S_{33})+...+AM(S_{3M}))/M$$
$$\ldots$$
$$\ldots$$
$$\ldots$$
$$\text{Average } AM_n=(AM(S_{n1})+AM(S_{n2})+AM(S_{n3})+...+AM(S_{nM}))/M$$

Average AM_1, Average AM_2, Average AM_3,..., Average AM_n are average arithmetic of relevancy spaces $s_1, s_2, s_3, ..., s_n$ respectively. Each relevancy space is divided into M regions. Inter-modality weights $w_1, w_2, w_3 ... w_n$ for inter-modality fusion using average arithmetic mean can be calculated as:

$$w_1= \text{Average } AM_1 / (\text{Average } AM_1 + \text{Average } AM_2 + \text{Average } AM_3 +... +\text{Average } AM_n)$$
$$w_2= \text{Average } AM_2 / (\text{Average } AM_1 + \text{Average } AM_2 + \text{Average } AM_3 +... +\text{Average } AM_n)$$
$$w_3= \text{Average } AM_3 / (\text{Average } AM_1 + \text{Average } AM_2 + \text{Average } AM_3 +... +\text{Average } AM_n)$$
$$\ldots$$
$$\ldots$$
$$\ldots$$
$$w_n= \text{Average } AM_3 / (\text{Average } AM_1 + \text{Average } AM_2 + \text{Average } AM_3 +... +\text{Average } AM_n)$$

We explain these methods with the help of an example represented in "fig. 2.".

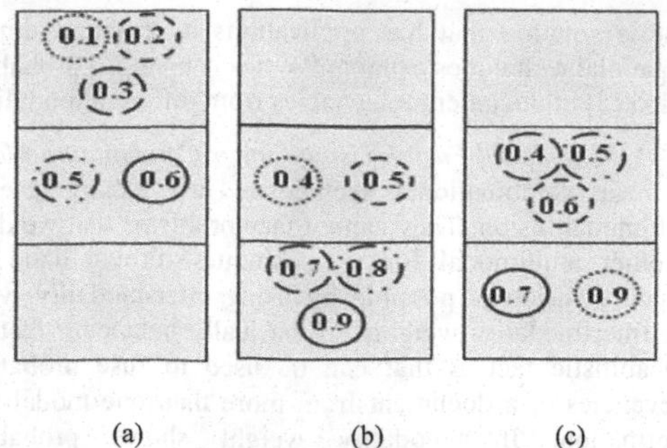

Fig. 2. Three relevancy spaces (a), (b) and (c), each having 5 relevancies, each relevancy space divided into three regions. Circles with same boundary styles represent relevancies of same document with in different relevancy spaces.

TABLE 1
INTER-MODALITY WEIGHTS USING AM AND AVERAGE (AM)

	Ratio(AM)	Ratio (Avg(AM))	constant
α1	0.21	0.23	0.33
α2	0.41	0.38	0.33
α3	0.38	0.39	0.33

In table 1 we calculate inter-modality weights of given relevance spaces using arithmetic mean and average arithmetic mean.

We notice from table 1 that by using arithmetic mean method more high relevancies in a relevancy space of a modality try to strongly influence strength of other modalities or decreases inter-modality weights of other modalities. Relevancy space (b) in "fig. 2." influences other relevancy spaces when we calculate inter-modality weights.

Average arithmetic mean method moderates effect of high and low relevancy regions in relevancy spaces for the calculation inter-modality weights. Both approaches act same when distribution of relevancies with in regions of relevancy spaces is uniform. When distribution of relevancies is not uniform average arithmetic mean method try to smooth effects of non-uniformities in relevancy spaces. This approach acts an intermediate approach between arithmetic mean and equal constant inter-modality weights.

TABLE 2
DOCUMENTS COMPOSITE RELEVANCIES BY USING ARITHMETIC MEAN, AVERAGE ARITHMETIC MEAN AND CONSTANT METHODS

	Relevancies	Ratio (AM)	Ratio (Avg(AM))	Const
Doc1	0.1, 0.4, 0.9	0.53	0.53	0.46
Doc2	0.2, 0.5, 0.6	0.47	0.47	0.43
Doc3	0.3, 0.7, 0.4	0.50	0.49	0.46
Doc4	0.5, 0.8, 0.5	0.62	0.61	0.59
Doc5	0.6, 0.9, 0.7	0.76	0.75	0.72

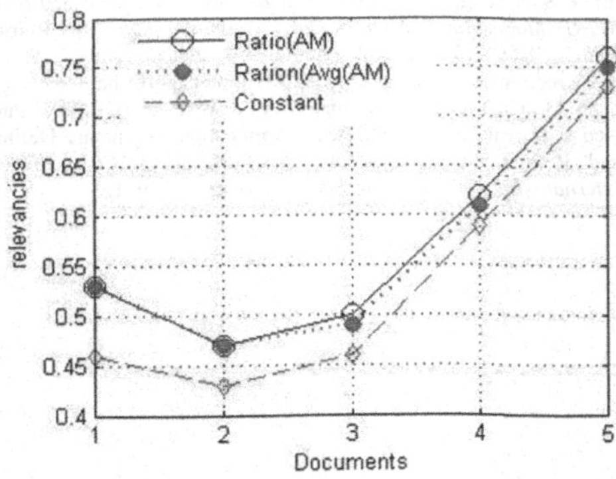

Fig. 3. document relevancies using three methods

Table 2 and "fig. 3." show that average arithmetic mean is a moderate approach between other two approaches. It smoothes the effects of non-uniform relevancies with in relevancy spaces for the calculation of inter-modality weights.

IV. EVALUATION

Average arithmetic mean is simple, easy to implement approach for online calculation of inter-modality weights. Average arithmetic mean can be calculated for relevancy spaces in non-polynomial time. Time complexity will be order of O (n) so it is performance wise feasible for online processing. This approach does not require relevance feedback so document relevancies can be calculated in a single step. We perform different tests for the evaluation of our proposed approach. We use randomly generated datasets using Mat Lab by assuming that original retrieval models also throw similar random frequencies.

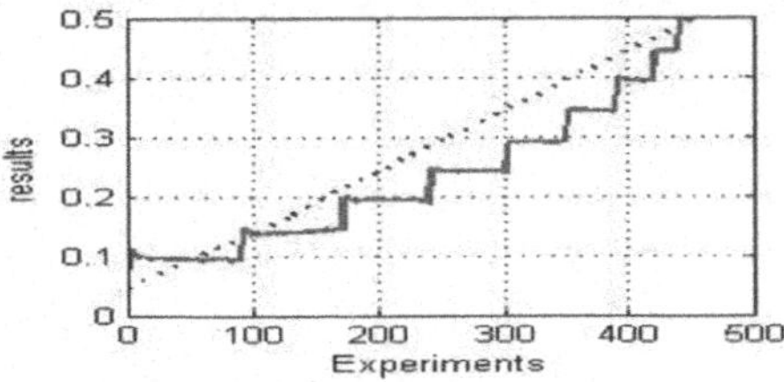

Fig. 4. (a) Solid line shows inter-modality weights using ratio (Avg (AM)) method (b) dotted line shows inter-modality weights using ratio (AM) method

Test1: In this test we take a relevancy space of hundred relevancies and divide space into ten regions. We initially place all hundred relevancies in first region of space and then move relevancies in next regions one by one and plot their corresponding ratio (Avg (AM)) and ratio (AM) such that in last step each region consists exactly ten relevancies. This test shows behaviour of relevancy spaces for inter-modality weights at each step when relevancies move among relevancy regions.

From "fig. 4." we conclude that both methods are equivalent when relevancy distribution is uniform among relevancy regions in relevancy space. Both lines intersect each other in case of uniform distribution. Dotted line is consistently influenced by the movement of relevancies in high relevancy regions while solid line is not consistently influenced. Solid line show smooth behaviour when relevancies move across regions.

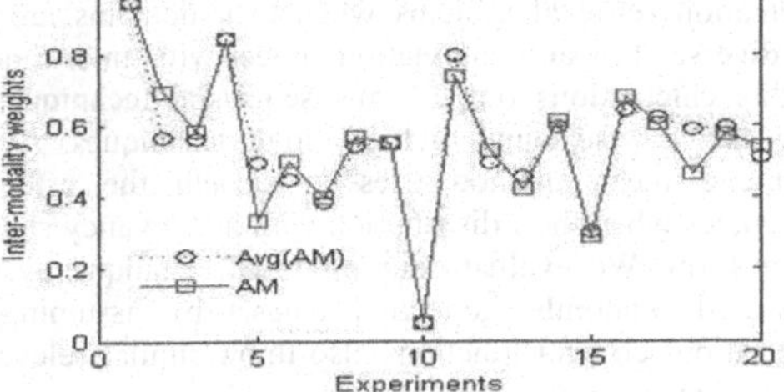

Fig. 5. inter-modality weights using ratio (avg(AM)) method

Test 2: In this test we take three relevancy spaces each having ten regions. Each region consists of less than or equal to ten randomly generated relevancies. We plot inter-modality weights of relevancy spaces. Experiment is repeated twenty-five times by refreshing relevancies in relevancy spaces as shown in "fig. 5".

From "fig. 5." we analyze that inter-modality weights mostly intersect each other. Intersection points are basically weights that are balanced. So from this graph we conclude that our proposed approach for inter-modality weight calculation mostly tries to smooth the effects of non-informalities in relevancy spaces.

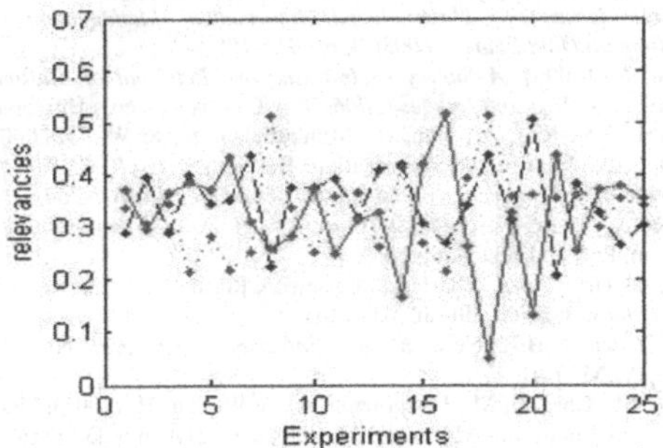

Fig. 6. Inter-modality weights by (a) Avg (AM) (b) AM and (c) constant method

Test 3: in this test we take relevancy space having ten regions. Each region consists of one hundred relevancies. We plot inter-modality weights by repeating experiment 20 times using both methods and also plot constant line as shown in figure 6.

From "fig. 6." we analyze that inter-modality weights calculated by using Avg (AM) methods mostly lies between weights calculated by other two methods. This validates our assumption that average arithmetic mean is a moderate approach between other two approaches.

V. CONCLUSION

For inter-modality fusion online inter-modality weight calculation techniques are satisfactory. A performance of our proposed technique is appropriate for general purpose information retrieval systems when their domains are large and diverse. Linear combination model with inter-modality weights calculations outperforms sequential techniques and performance wise equal to fuzzy logic techniques. Average arithmetic mean method tries to smooth the effects of relevancies when their distribution with in relevancy spaces is not uniform. We evaluate our proposed technique by using automated randomly generated dataset by assuming that information retrieval functions also throw similar relevancies in relevancy spaces.

REFERENCES

[1] Marchionini, G. & Geisler, G. 2002. "The Open Video Digital Library". *D-Lib Magazine, Volume 8 Number 12.*

[2] Wilfred Owen digital archive to expand. (1993, Feb 22). "BLUEPRINT the newsletter of the University of Oxford*", Oxford Press, pp. 4.*

[3] Mahmood, T.S., Srinivasan, S., Ami, A., Ponceleon, D., Blanchard, B., and Petkovicstf. D. (2000). "CueVideo: A System for Cross-Modal Search and Browse of Video". In CVPR'00, *IEEE Computer Society Conference on Computer Vision and Pattern Recognition, pp 2786.*

[4] Bruno, E., Loccoz, N.M., & Maillet, S.M. (2005). "Interactive video retrieval based on multimodal dissimilarity Interactive video retrieval based on multimodal dissimilarity". In *MLMM'05, workshop Machine Learning Techniques for Processing Multimedia Content.*

[5] Westerveld, T., Vries, A.P.D., A.R.V., Ballegooij, A.R.V., Jong, F.M.G.D., & Hiemstra, D.2003. "A Probabilistic Multimedia Retrieval Model and its Evaluation". *Journal on Applied Signal Processing, special issue on Unstructured Information Management from Multimedia Data Source, 2003(3), pp. 186-198.*

[6] Chen, N. (2006). *A Survey of Indexing and Retrieval of Multimodal Documents: Text and Images (2006-505).* Canada, Queen's University.

[7] Srihari, R.K, Rao, A., Han, B., Munirathnam, S., & Wu, X.2000. "A Model for Multimodal Information Rettrieval". In *ICME'00, IEEE International Conference on Multimedia and Expo*, pp. *701-704.*

[8] James L. Hein. *Discrete Structures, Logic, and Computability. 2nd* Edition, Portland State University.

[9] Thomas H. Cormen, Charles E. Leiserson, Ronald L. R. *Introduction to Algorithms.* Second Edition, MIT press.

[10] R.B. Yates & B.R. Neto. *Modern Information Retrieval.* New York, 1999, ACM Press.

[11] Chen, M., Christel, M., Hauptmann, A., & Wactlar. H. (2005). "Putting Active Learning into Multimedia Applications: Dynamic Definition and Refinement of Concept Classifiers". In *MM'05, 5th ACM International Conference on Multimedia pp. 902-911*

[12] Christel, M.G. & and Conescu, R.M.(2005). "Addressing the Challenge of Visual Information Access from Digital Image and Video Libraries". In *JCDL'05, ACM/IEEE-CS Joint Conference on Digital Libraries. pp. 69-78.*

[13] Chen, J.Y, Carlis, J.V, Gao,N. (2005). "A Complex Biological DataBase Quering Method". In *SAC'05, Symposium on Applied Computing, pp. 110-114.*

[14] Bruno_b, E., Loccoz, N.M., & Maillet, S.M. (2005). Learning User Queries in Multimodal Dissimilarity Spaces. In *AMR'05, 3rd International Workshop on Adaptive Multimedia Retrieval.*

[15] Bushman, F., Menuier, R., Rohnert, H., Sommerland, P., & Stal, M. (1996). *PATTERN-ORIENTED SOFTWARE ARCHITECTURE A System of Patterns.* Singapore, Wiley, volume 1.

[16] Li, C.S., Chang, Y.C, Smith, J.R. & Hill, M. (2001). *SPIRE/EPI-SPIRE Model-Based Multi-modal Information Retrieval.* Retrieved form IBM T. J. Watson Research Centre Website: www.rocq.inria.fr/imedia/mmcbir2001/FinalpaperLi.pdf

[17] Chen, M., Christel, M., Hauptmann, A., & Wactlar. H. (2005). Putting Active Learning into Multimedia Applications: Dynamic Definition and Refinement of Concept Classifiers. In *MM'05, 5th ACM International Conference on Multimedia*, pp. 902-911.

An Interval-based Method for Text Clustering

Hanh Pham

Department of Computer Science
State University of New York at New Paltz
pham@cs.newpaltz.edu

Abstract

This paper describes a clustering approach for classifying texts where dynamic text are evaluated with intervals (interval-valued numbers or IVN) rather than numbers. This new interval measurement allows us to capture the text relevance by multiple keywords at the same time and thus can provide multi-dimension text ranking. We build several comparison mechanisms to rank the intervals based on interval multiple values, their distributions and relationships. Then, the dynamic texts in information systems can be pro-actively clustered into collections usingthis ranking results.

1. Introduction

Texts are the most popular form to represent and store information and knowledge. With billions of emails alone being created everyday worldwide [14,16], the existing trillions of emails and other documents cannot be read by humans only anymore we must now try to analyze and process them automatically, at least partially. Therefore, evaluating the relevance of text documents to different topics is one of the most important and critical steps in managing modern information systems [1][2].

Various methods have been used to represent, evaluate, and compare texts. However, while these methods may produce some results with fixed texts they do not work with texts which are changed dynamically.

In this paper we propose a soft-computing method where interval-valued numbers (IVN) are used to represent ranges of contents or characteristics of dynamic texts. Then, a fuzzy comparison method is proposed to rank these interval-valued characteristics of various documents in regards of different topics.

The rest of this paper is organized as follows. Section 2 describes the problem of text evaluation and the special features of the dynamic text case. In section 3, areview of text representation and evaluation methods is provided. Our measurement and representation method using IVN is depicted in section 4. The fuzzy ranking for IVN characteristics of dynamic texts is explained in section 5. Finally, a brief assessment and possible directions of future work are discussed in section 6.

2. Text Evaluation Problem

The problem of evaluating texts in regards to given topics can be stated as the followings.

K. Elleithy (ed.), *Advanced Techniques in Computing Sciences and Software Engineering*,
DOI 10.1007/978-90-481-3660-5_99, © Springer Science+Business Media B.V. 2010

Given:a set of texts $T=\{t_1,t_2, .., t_n\}$ where $t_i \in \check{T}$, i=1..n, and a list of topics $K=\{k_1,k_2, .., k_m\}$ where $k_j \in \acute{K}$, j=1..m.

Question:how to evaluate each text t_i, i=1..n, in T in regards to the topics given in K ? That is to find the values of elements in the relevance matrix RM:

Relevance Matrix

T/K	k_1	k_2	…	k_m
t_1	r_{11}	r_{12}	…	r_{1m}
t_2	r_{21}	r_{22}	…	r_{2m}
…	…	…	…	…
t_n	r_{31}	r_{32}	…	r_{3m}

Where, r_{ij}shows how text $t_i$relates to topic k_j, i=1..n, j=1..m. This general problem may have different derivations for requirements:

- *Simple Evaluation:*$r_{ij} \in \{1,0\}$ that is "yes" or "no" for the answer of whether a text $t_i$relates to topic $k_j$or not.
- *Relative Evaluation:*$r_{ij} \in \check{R}$ a given common domain$\forall$i=1..n, j=1..m and therefore can also be used to rank or compare the relevance of different texts to one topic.
- *Composition Evaluation:*the relevance matrix RM can be extended by measuring the relevance of the given texts to different combinations of topics, for example: how a given text $t_i$may relate to several combined topics $\{k_1,k_2,k_5\}$?

When the texts in T are dynamic, that is any $t_i$in $T=\{t_1,t_2, .., t_n\}$ may be changed. This leads to dynamic values of r_{ij}, i=1..n, j=1..m. Thus, it would be very difficult to evaluate and rank their relevance since the relevance computation, even by simplest method, would create delays. Traditional evaluation methods for static or fixed texts would not work. There are two approaches to solve this problem more effectively: (i) use the available data and the texts at time moment$\tau$to predict the relevance ranks of texts for momentτ+1, or (ii) evaluate and capture possible relevancy in a range foreach text and then rank the texts based on these ranges.

3. Related Works

In reviewing related works for solving the text evaluation problem we divide them into two fundamental and related sub-areas: (i) text representation and (ii) measurement and evaluation mechanisms.

Text Representation

A text or document can be represented in different ways:

- a number [6]
- a vector or a set of numbers [7]
- a title [6]
- keywords or terms [8,15]
- indexes (words, phrases) [5,16]
- an abstract or summary[4]
- document "fingerprints" [2]
- contextual maps [2]

Measurement and Evaluation Mechanisms

In order to measure and evaluate how text documents may relate to a given topic the following approaches have been developed:

- Search for keywords or terms [8]
- Search for word aliases [9]
- Define vectors [7]
- Use metadata such as "tags" [2]
- Build classification trees [1,12]
- Indexing [5]
- Compare based on similarity [2]
- Fuzzy logics [3]
- Probabilistic estimation [5]
- Logic-based model [11]
- By clusters [1,13]

However, these methods would not work efficiently for dynamic texts when the content of the texts may be modified or changed constantly such as weather or stock market webpages, etc. In this case if we use any of the mentioned above ones the system would need to repeat all the computations for measuring and evaluating all the time. Here there could be two problems:either thecomputing overhead may exceed the deadline then the obtained evaluation result would not be valid anymore; or even if the deadline is met the system would be overloaded with the required computing power.

4. Measuring Relevance by Interval-Valued Numbers

In order to represent or describe how a dynamic text relates to a topic we propose to use the Interval-Valued Numbers (IVN) which were investigated in [9,10]. An IVN is a value X which could fall to an interval [a,b] and is represented by the two bounders [a,b] and any information about past values, their possible distribution and behaviors. When more information about the past or possible values of X in [a,b] is available this kind of information may be used to define the approximate value of X. In that case an IVN is more flexible than a regular number but more precise than a regular interval.

Assume that there is a list of texts $T=\{t_1,t_2, .., t_n\}$ where $t_i \in \check{T}$, i=1..n, and a list of topics $K=\{k_1,k_2, .., k_m\}$ where $k_j \in \acute{K}$, j=1..m. A vocabulary V is also available which contains links of synonyms and antonyms with fuzzy estimations.This vocabularycan be employed by several functions:

- $\vartheta 1(w)$ returns two sets of strings {{sw1,sw2, ...}, {aw1,aw2, ...}} where the first set is the set of synonyms and the second one is the set of antonyms to the given word*w*.
- $\vartheta 2(w1,w2)$ returns an interval-valued number (IVN)$\bar{u}$which shows the similarity degree of*w2*to*w1*and is denoted as:

$\vartheta 2(w1,w2) \cong \bar{u}$: {[a,b],P,Vp={x1,x2, ...xP}}

The IVN$\bar{u}$is defined in the vocabulary V basedon more than one estimation, for exampleexperts may define the similarity of w2 to w1 differently in different contexts. Thus, for each IVN$\bar{u}$we have the bounds [a,b] and a distribution of estimated similarity values in different cases {x1,x2, ...xS},$\forall x_i \in [a,b]$, i =1..S. These values are attached to$\bar{u}$and can be used later for ranking the relevance of texts by the fuzzy mechanism which is proposed in section 5.

We also assume that a text*t*consists of a string of words*t*=(s1, s2, .., sL). Suppose that$\Phi$is a function that measures the relevance of a text*t*to a single topic*k*and$\Psi$is a function that measures the relevance of a text to a set of topics K*. We have:

- $\Phi(t,k)=\phi$
- $\Psi(t,K^*)=\psi$

where,
$K^*=\{(k_1,z_1),(k_2,z_2),..(k_g,z_g)\}$, and $k_1,k_2, \ldots k_g$are the topics and $z_1,z_2,.., z_g$are their weights.

These functions can be defined as the followings:

$$\Phi(t,k)=\sum_{i=1}^{L}\vartheta 2(k,s_i)$$

and

$$\Psi(t,K^*)=\sum_{i=1}^{L}\frac{\sum_{j=1}^{G}(z_j \times \vartheta 2(k_j,s_i))}{\sum_{j=1}^{G} z_j}$$

sincethe function$\vartheta 2(w1,w2)$returns interval-valued numbers (IVN), the estimations of how the given text*t*relates to a topic*k*or a set of topics*K**are also in IVN form and can be defined using the interval operators.

Assume that the relevance of each string$s_j$in the text*t*in regards to a topic*k*is:

$\vartheta 2(k,s_i) \cong \bar{u}_j$: $\{[a_j,b_j],P_j,Vp_j=\{x^j1,x^j2, \ldots, xP_j\}\}$

where, j = 1..L. Then, the combined relevance of the whole text*t*to the topic*k*can be defined as a sum of those*L*IVNs:

$\Phi(t,k)\cong\{[af,bf],Pf, Vpf=\{xf1,xf2, \ldots xfP\}\}$

where,

$$a=\sum_{j=1}^{L}a_j, b=\sum_{j=1}^{L}b_j,$$

$$Vp=\bigcup\oplus_L^L(\chi 1,\chi 2,...,\chi L), x^j\in Vp_j, j=1..L$$

where, $U\oplus^L{}_L$ produces the set of any possible sums of L elements each from one Vp_j, j=1..L.

For example, given the following text*t* = *"The New York Times reported that up to today all 50 top banks of the world have their offices in New York"*.This text could be transformed into the following list of terms or main strings: t = {"New York", "New York Times","reported", "world top banks", "banks", "world", "top", "office", "today"} = {s1, s2, s3, s4, s5, s6, s7, s8, s9}. Suppose that the interested topics are k = "New York" and K*={("world",3), ("bank",7)}. Then, the vocabulary V may return the following IVNs:

ϑ2(k,s1)≅{[1,1],1,1}
ϑ2(k,s2)≅{[0.8,1],3,0.9,0.8,1}
ϑ2(k,s3)≅{[0,0.1],2,0.1,0}
ϑ2(k,s4)≅{[0.2, 0.5],5,0.2,0.35,0.3,0.4,0.5}
ϑ2(k,s5)≅{[0.1,0.3],2,0.3,0.1}
ϑ2(k,s6)≅{[0.2,0.4],3,0.2,0.35,0.4}
ϑ2(k,s7)≅{[0.1,0.2],2,0.1,0.2}
ϑ2(k,s8)≅{[0.1,0.25],3,0.1,0.2,0.25}
ϑ2(k,s9)≅{[0,0.1],4,0.6,0.5,0.9,0.8}

Then,

Φ(t,k)≅{[2.5, 4.55], P, Vp={2.5, …,4.55}}

5. Fuzzy Ranking of Relevance

Using IVN to represent the relevance of texts in T allows us to keep the past values and other information about the possible relevance which can be used at any stage of computation. However, comparison and ranking of IVN are not easy since instead of comparing two numbers now we need to compare two pairs of bounds, for example, [a1,b1] and [a2,b2], aswell as comparing the past estimated values or distribution of possible values of each IVN. In order to rank these IVN we combine the three approaches proposed in [9].

Assume that for each text in T={t_1,t_2, .., t_n} where t_i∈Ť, i=1..n, we obtained the relevance degrees by the method proposed in section 4 and have the following IVNs:

$\Phi(t_1,k)\cong\{[a_1,b_1],P_1, Vp_1=\{x^1 1,x^1 2, .. ,x^1 P_1\}\}$
$\Phi(t_2,k)\cong\{[a_2,b_2],P_2, Vp_2=\{x^2 1,x^2 2, .. ,x^2 P_2\}\}$
…
$\Phi(t_N,k)\cong\{[a_N,b_N],P_N, Vp_N=\{x^n 1,x^n 2, \ldots x^n P_N\}\}$

Now, we need to compare and rank these IVNs so that we can:

- define the most relevance text to the topic k from T, or
- defineall the related texts in T which relate to a given topic k (i.e. their relevance degree exceed some given threshold).

In order to do so we first develop a comparison mechanism to compare a pair of texts$t_1$and$t_2$and measure how much more relevant one text is compared with the other text in term of agiven topic*k*.

We need to develop a new comparison method because the traditional mathematical comparison, which uses only three estimation degrees (>,<,=), can not work efficiently for IVNs. When the two intervals have overlapped bounds such as$\bar{u}_1$:[1,5] and$\bar{u}_2$:[3,8], any traditional statement, for instance$\bar{u}_1 < \bar{u}_2$or$\bar{u}_1 > \bar{u}_2$could never be a guaranteed true. To deal with this problem we use afuzzy measurementof ">" more or "<" less with a degree D to describe the possible relationship between the two intervals. For instance, we have$\bar{u}_1$>(D)$\bar{u}_2$instead of just$\bar{u}_1 > \bar{u}_2$.

The degree D is defined based on the bounds of the IVNs and on the estimated similarity values as the follows.

First, we define a so-called pseudo-distribution Pdi of the IVN, which may show the possible distributions of $x_j$in $[a_j,b_j]$, j =1..Pi. For each data setVp$_i$= $\{x^j1, x^j2, .. ,x^jP_i\}$of the IVN$\bar{u}_i$:$[a_i,b_i]$, i=1..N, the pseudo-distribution is defined the by the following steps:

- Split each IVN$\bar{u}_i$:$[a_i,b_i]$ into Q subintervals $\{sX^1_i, sX^2_i,.., sX^Q_i\}$.
- Define the frequency f_i^t of the estimated similarity values which appear in each subinterval sX^t_i, t=1..Q.
- Represent the pseudo-distribution of the given IVN by a frequency set: Fi $=\{f_i^1, f_i^2,.., f_i^Q\}$, i=1..N.

The next step is to measure the pseudo-distribution of each IVN and calculate the degree D of relevance comparison.

The investigation in [9,10] showed that for different distribution types of the estimated values one comparison scheme may be more efficient than the others. In order to increase the efficiency and reduce the computation time, we compare the IVNs by three alternative schemes depending on the features and availability of data: one-peak, many-peak, and none-peak distributions as the followings.

One-Peak Frequency

When the pseudo-distribution has one peak we find the mid point Mi of the frequencies Fi $=\{f_i^1, f_i^2,.., f_i^Q\}$, i =1..Nas the following:

$$\text{Mi}=x_i^{\text{mod}}: F_i(x_i^{\text{mod}}) = \max_{i=1}^{Q} \{f_i^t\}$$

Where, function*Fi (x)*returns the frequency of estimated similarity values in an IVN [ai,bi]. For a pair of IVN:$\bar{u}_1$and$\bar{u}_2$we find the balanced point Xo with a value of$\bar{x}^p$ so that:

- $\bar{x}^p \in$ [Mmin,Mmax] , where Mmin = min{M1,M2} and Mmax = max{M1,M2}, and
- $\Delta_0 + |Ft_1(\bar{x}^p) - Ft_2(\bar{x}^p)|$ is minimum,

where,

+$Ft_1$and$Ft_2$are the fittest function for*F1*and*F2*in [Mmin,Mmax]

$$+\begin{cases} \Delta_0 = 1 & if \quad \bar{x}^p \in \bar{u}_1 \cup \bar{u}_2 \\ \Delta_0 = 0 & if \quad \bar{x}^p \notin \bar{u}_1 \cup \bar{u}_2 \end{cases}$$

Then, wehave $D* = \frac{1}{2}\left(\frac{\alpha_1}{Vp_1} + \frac{\alpha_2}{Vp_2}\right)$,

where$\alpha_1$and$\alpha_2$are defined by the functionΦ(a,b) which returns the sum of frequencies in [a,b] as the following :

- if M1>M2:α_1=Φ(Xo,b1);α_2=Φ(a2,Xo);Ω=">" ;
- if M1<M2:α_1=Φ(a1,Xo);α_2=Φ(Xo,b2);Ω="<";
- if M1 = M2:

+)α_1=Φ(Xo,b2);α_2=Φ(a2,Xo);Ω= ">" :
+)α_1=Φ(a1,Xo);α_2=Φ(Xo,b2);Ω= "<" ;

Suppose thatΔ1=b1-a1;Δ2=b2-a2. We define the degree D as the following:

- if $0 < \frac{b2 - a1}{\Delta 1 + \Delta 2} < 1$ then D($\bar{u}_1\Omega\bar{u}_2$) = D*;
- if $\frac{b2 - a1}{\Delta 1 + \Delta 2} \geq 1$ or $\frac{b2 - a1}{\Delta 1 + \Delta 2} \leq 0$ thenD($\bar{u}_1\Omega\bar{u}_2$) = |M1-M2|;

Many-Peak Frequency

When the pseudo-distribution has many peaks or "chaotic" we use the following procedure. For each frequencyset :Fi$=\{f_i^1, f_i^2,.., f_i^Q\}$, i=1..N we calculate the comparison values as the following:

$$\Phi_i = \frac{a1 \times (f_i^1)^h + ... + (a1 + (Vp_1 - 1)d_i) \times (f_i^Q)^h}{(f_i^1)^h + ... + (f_i^Q)^h}$$

$$= \frac{\sum_{t=1}^{Q} (a1 + (t-1)d_i) \times (f_i^t)^h}{\sum_{t=1}^{=Q} (f_i^t)^k}$$

where,*h*=1,2,3, ..and*h*will be chosen based on the data features of each particular problem, di=(bi-ai)/Q.

Suppose thatΔ1=b1-a1;Δ2=b2-a2. We define the degree D as the following:

- if $0 < \frac{b2 - a1}{\Delta 1 + \Delta 2} < 1$

then $D(\bar{u}_1\Omega\bar{u}_2) = \Psi1-\Psi2$;

- if $\frac{b2-a1}{\Delta1+\Delta2} \geq 1$ or $\frac{b2-a1}{\Delta1+\Delta2} \leq 0$

then $D(\bar{u}_1\Omega\bar{u}_2) = |\Psi1-\Psi2|$;
where,
Ω= ">" if $\Psi1>\Psi2$ and Ω= "<" otherwise.

No-Peak Frequency

When the estimated similarity data is not repeated, that is when their frequency is 1, the pseudo-distribution is flat and has no peak. We use the following procedure.

Define the means Me_i, i = 1..R, of the sets of frequencies by the following:

$$Me_i = \sum_{i=1}^{R} \frac{F_i(\overline{x}'_i)}{r_i}$$

For a pair $\bar{u}_1$ and $\bar{u}_2$ define the middle-point Xm of a value $\overline{x}^{me}$ so that :

- $\overline{x}^{me} \in$ [Memin,Memax] ,

where, Memin = min{Me_1,Me_2} and Memax =max{Me_1,Me_2}, and

- Δ_p +(Γ1+Γ2) is maximum,

where: $\Gamma1=\Phi$(Mmin,O) and $\Gamma2=\Phi$(Mmax,O) ; and $\begin{cases} \Delta_p = 0 & if \ \overline{x}^{me} \in \bar{u}_1 \cup \bar{u}_2 \\ \Delta_p = 1 & if \ \overline{x}^{me} \notin \bar{u}_1 \cup \bar{u}_2 \end{cases}$

We have $D = \frac{1}{2}\left(\frac{\beta1}{Vp_i} + \frac{\beta2}{Vp_2}\right)$, where $\beta1$ and $\beta2$ are defined similarly as α_1 and α_2.

Suppose that $\Delta1$=b1-a1 ;$\Delta2$=b2-a2; We build the degree D as the following:

- if $0 < \frac{b2-a1}{\Delta1+\Delta2} < 1$ then $D(\bar{u}_1\Omega\bar{u}_2) = D$;
- if $\frac{b2-a1}{\Delta1+\Delta2} \geq 1$ or $\frac{b2-a1}{\Delta1+\Delta2} \leq 0$

then $D(\bar{u}_1\Omega\bar{u}_2)$ = | Me1-Me2|;

Ranking Relevance

Thus, the texts in T={t_1,t_2, .., t_n} have relevance degrees to a topic k represented by the following IVNs:

$t_1 \rightarrow k : \bar{u}_1\{[a_1,b_1],P_1, Vp_1=\{x^1 1,x^1 2, .. ,x^1 P_1\}\}$
$t_2 \rightarrow k : \bar{u}_2\{[a_2,b_2],P_2, Vp_2=\{x^2 1,x^2 2, .. ,x^2 P_2\}\}$
...
$t_n \rightarrow k : \bar{u}_n\{[a_n,b_n],P_n, Vp_n=\{x^n 1,x^n 2, \ldots x^n P_n\}\}$

Since theses IVNs may have different frequency distributions we can apply the proposed fuzzy ranking mechanism for different cases to define the followings:

$\bar{u}_1\Omega(D_{1,2})\bar{u}_2$, where $D_{1,2}$= $D(\bar{u}_1\Omega\bar{u}_2)$
$\bar{u}_1\Omega(D_{1,3})\bar{u}_2$, where $D_{1,3}$= $D(\bar{u}_1\Omega\bar{u}_3)$
...
$\bar{u}_1\Omega(D_{1,N})\bar{u}_2$, where $D_{1,N}$= $D(\bar{u}_1\Omega\bar{u}_N)$

As these are regular numbers we can now rank them by traditional comparison as usual.

Then, the most relevance text to the topic k from T is t*where :

$$D(\bar{u}_1\Omega\bar{u}_*) = MAX\{ D(\bar{u}_1\Omega\bar{u}_2),\ldots, D(\bar{u}_1\Omega\bar{u}_N) \}$$

If the threshold for relevance to a topic k is λ_k then any text t_r whose $D(\bar{u}_1\Omega\bar{u}_r) > \lambda_k$ can be selected as "related" texts.

6. Conclusions

In this paper we have developed and described a soft-computation approach to dynamic text evaluation where interval-valued numbers (IVN) are used to represent the relevance of texts to given topics and fuzzy logics is employed to rank their relevance. The use of IVNs allows us to store different values of the estimation of how a string and a text may relate to a topic. These multiple values are used later in the computation which is carried through a fuzzy ranking of these IVNs. The proposed approach works when the changes of each dynamic text fall into one interval but would not work if otherwise or if the intervals all are too wide. Our possible future work includes the development of similar mechanism for theComposition Evaluation of dynamic texts as well as research on the effects of the frequency distributions of the estimated values on the results and on the computation time.

References

[1]. Moens M.,*"Automatic Indexing and Abstracting of Document Texts"*,Kluwer Academic Publishers, 265p, 2000.

[2]. Howlett R.J., Ichalkaranje N.S., L.C., Jain, Tonfoni G.,*"Internet-Based Intelligent Information Processing Systems*", World Scentific, 2003.

[3]. Bodenhofer U.,*"Fuzzy "Between" Operators in the Framework of Fuzzy Orderings"*,in B. Bouchon-meunier, L. Foulloy, and RR. Yager (Editors), "Intelligent Systems for Information Processing: from Representation to Applications", Elsevier Science, 2003.

[4]. Lancaster F.W.,*Indexing and Abstracting in Theory and Practice*, The Library Association, London, 1991.

[5]. Harter S.P.,*"A probabilistic approach to automated keyword indexing"*,JASIS, 26(4), 197-206, 1975.

[6]. Spark Jones K.,*"What might be a summary"*, Information Retrieval 1993.

[7]. Spark Jones K.,*"Evaluating Natural Language Processing: An Analysis and Review"*, Springer, 1995.

[8]. Suzuki G., Iizuka, and Kasuga S.,*"Integration of Keyword-Based Source Search and Structure-Based Information Retrieval"*,pp.149-158, in Heterogeneous Information Exchange and Organization Hubs, (Bestougeff et al edited), Kluwer Academic Publishers, 2002.

[9]. Hanh Pham,*Fuzzy Estimation for Interval-Valued Numbers with Statistical Information*, Proceedings of the 3rd International Symposium on Fuzzy Logic and Applications (ISFL'99), pp62-68, RIT New York, USA, June 1998.

[10]. Hanh Pham,*Fuzzy Comparisons for Interval-Valued Numbers,*in proceedings of the 6th International Conference on Fuzzy Theory and Technology, North Caroline, USA, 24-28 October 1998.

[11]. Radecki T.,*"Fuzzy set theoretical approach to document retrieval"*,Information Processing & Management Volume 15, Issue 5 , pages 247-259, Elsevier Science, 1979.

[12]. Giuseppe Manco , Elio Masciari , Andrea Tagarelli, Mining categories for emails via clustering and pattern discovery, Journal of Intelligent Information Systems, v.30 n.2, p.153-181, April 2008

[13]. R. Bekkerman, H. Raghavan, J. Allan, and K. Eguchi. Interactive Clustering of Text Collections According to a User-Specified Criterion. In Proceedings of IJCAI 2007

[14]. Danyel Fisher , A. J. Brush , Eric Gleave , Marc A. Smith, Revisiting Whittaker & Sidner's "email overload" ten years later, Proceedings of the 2006 20th anniversary conference on Computer supported cooperative work, November 04-08, 2006, Banff, Alberta, Canada

[15]. Giuseppe Carenini , Raymond T. Ng , Xiaodong Zhou, Summarizing email conversations with clue words, Proceedings of the 16th international conference on World Wide Web, May 08-12, 2007, Banff, Alberta, Canada

[16]. Gabor Cselle , Keno Albrecht , Roger Wattenhofer, BuzzTrack: topic detection and tracking in email, Proceedings of the 12th international conference on Intelligent user interfaces, January 28-31, 2007, Honolulu, Hawaii, USA

A GVT Based Algorithm for Butterfly Barrier in Parallel and Distributed Systems

Syed S. Rizvi, Shalini Potham, and Khaled M. Elleithy
Computer Science Department, University of Bridgeport, Bridgeport, CT 06601 USA
{srizvi, spotham, elleithy}@bridgeport.edu

Abstract-Mattern's GVT algorithm is a time management algorithm that helps achieve the synchronization in parallel and distributed systems. This algorithm uses ring structure to establish cuts C1 and C2 to calculate the GVT. The latency of calculating the GVT is vital in parallel/distributed systems which is extremely high if calculated using this algorithm. However, using synchronous barriers with the Matterns algorithm can help improving the GVT computation process by minimizing the GVT latency. In this paper, we incorporate the butterfly barrier to employ two cuts C1 and C2 and obtain the resultant GVT at an affordable latency. Our analysis shows that the proposed GVT computation algorithm significantly improves the overall performance in terms of memory saving and latency.

Keywords-Time management algorithm, latency, butterfly barrier

I. INTRODUCTION

A parallel and distributed system is an environment where a huge single task is being divided into several sub-tasks and each terminal getting a sub-task to execute. The main problem that is being faced here is the synchronization. All the processes need to be synchronized as the main aim of distributed system is that the final output after execution of entire task should be exactly the same as that of the output attained when the same task is executed sequentially on a single machine.

Mattern's GVT algorithm helps keep all the processes in synchronization by finding the minimum of time stamps of all the messages at a point. It also makes sure that there are no transient messages in the process of execution as it waits for the processes to receive all the messages that are destined for it. The backlog of this algorithm is that the latency is high. This keeps the algorithm away from its widespread usage. The performance of a parallel/distributed system can be degraded if the latency for computing the GVT is high. The Mattern's GVT algorithm uses several variables which in turn increase the number of memory fetches.

In the proposed algorithm, we implement the similar mechanism structure suggested by the Mattern's [1] with the use of a matrix. This utilization of the matrix eliminates two of the variables and an array as used by the original Mattern's GVT algorithm. Consequently, the use of matrix with the Mattern's algorithm provides several advantages such as it reduces the number of memory fetches, saves memory, increases the processor speed, and improves the latency. We incorporated the butterfly barrier as it has great performance when compared to the other barriers such as broadcast and the centralized barriers [7]. When we finish implementing the barrier with the proposed algorithm, the current simulation time is updated. This implies that there is no need to communicate the minimum time or the simulation time reducing the message exchanges. This, therefore, improves the latency at affordable rate.

II. RELATED WORK

The term distributed refers to distributing the execution of a single run of a simulation program across multiple processors [2]. One of the main problems associated with the distributed simulation is the synchronization of a distributed execution. If not properly handled, synchronization problems may degrade the performance of a distributed simulation environment [5]. This situation gets more severe when the synchronization algorithm needs to run to perform a detailed logistics simulation in a distributed environment to simulate a huge amount of data [6].

Event synchronization is an essential part of parallel simulation [2]. In general, synchronization protocols can be categorized into two different families: conservative and optimistic. Time Warp is an optimistic protocol for synchronizing parallel discrete event simulations [3]. Global virtual time (GVT) is used in the Time Warp synchronization mechanism to reclaim memory, commit output, detect termination, and handle errors. GVT can be considered as a global function which is computed many times during the course of a simulation. The time required to compute the value of GVT may result in performance degradation due to a slower execution rate [4].

K. Elleithy (ed.), *Advanced Techniques in Computing Sciences and Software Engineering*,
DOI 10.1007/978-90-481-3660-5_100,

On the other hand, a small GVT latency (delay between its occurrence and detection) reduces the processor's idle time and thus improves the overall throughput of distributed simulation system. However, this reduction in the latency is not consistent and linear if it is used in its original form with the existing distributed termination detection algorithm [7].

Mattern's [1] has proposed GVT approximation with distributed termination detection algorithm. This algorithm works fine and gives optimal performance in terms of accurate GVT computation at the expense of slower execution rate. This slower execution rate results a high GVT latency. Due to the high GVT latency, the processors involve in communication remain idle during that period of time. As a result, the overall throughput of a discrete event parallel simulation system degrades significantly. Thus, the high GVT latency prevents the widespread use of this algorithm in discrete event parallel simulation system.

However, if we could improve the latency of the GVT computation, most of the discrete event parallel simulation system would likely to get advantage of this technique in terms of accurate GVT computation. In this paper, we examine the potential use of butterfly barriers with the Mattern's GVT structure using a ring. Simulation results demonstrate that the use of the tree barriers with the Mattern's GVT structure can significantly improve the latency time and thus increase the overall throughput of the parallel simulation system. The performance measure adopted in this paper is the achievable latency for a fixed number of processors and the number of message transmission during the GVT computation.

Thus, the focus of this paper is on the implementation of butterfly barrier structures. In other words, we do not focus on how the GVT is actually computed. Instead, our focus of study is on the parameters (if any) or factors that may improve the latency involved in GVT computation. In addition, we briefly describe that what changes (if any) may introduce due to the implementation of this new barrier structure that may have an impact on the overall latency.

III. Proposed Algorithm

In this section, we present the proposed algorithm. For the sake of simplicity, we divide the algorithms for both cuts C1 and C2.

A. *The Proposed Algorithm*

```
ALGO GVT_FLY(V[N][N],Tmin,Tnow,Tred,Ts,n)
 Begin
   n = log_2 N
   Loop n times
    Begin
      //Green message sent by LP_i to LP_j
       V[i][j] = V[i][j]+1
      //Green message received by LP_i
       V[i][i] =V[i][i]+1
      //Calculate minimum time stamp
       Tmin = min (Tmin,Ts)

  CUT C1:
   //Messages exchanged at this point are the red messages
   // Red message sent by LP_i to LP_j
    V[i][j] = V[i][j]+1
   // Red message received by LP_i
    V[i][i] = V[i][i]+1
   //Calculate minimum time of red messages
    Tred = min (Tred,Ts)
   Forward token to appropriate LP

  CUT C2:
   //Wait until all messages are received
   Wait until(V[i][i] = Σ_{j=1}^{N} V[j][i] -V[i][i])
   Forward token to appropriate LP
  Tnow= min(Tred,Tmin)
END LOOP
END ALGO
```

B. *A Detailed Overview of the Proposed Algorithm*

The Mattern's algorithm uses *N* vectors of size *N* to maintain a track of the messages being exchanged among the LPs. It also uses an array of size *N* to maintain a log of number of messages a particular LP needs to receive. On the whole, it uses (*N*+1) vectors of size *N*. This increases the number of fetches to memory resulting in more processor idle time.

In our proposed algorithm, we implement an *N* x *N* matrix to calculate the GVT whose flow can be explained as shown in Fig.1. Firstly, the LPs exchange green messages (i.e., green messages represent those messages that are safe to process by LP). Whenever an LP_i sends a green message to LP_j, the cell V[i][j] of the matrix gets updated as shown in Fig.2. On the other hand, if LP_i receives a message, the cell V[i][i] of the matrix is updated. At this point, we also calculate the minimum of all the time stamps of the event messages. After a certain period of time, when the first cut point C1 is reached, the LPs start exchanging the red messages (i.e., the red messages represent those messages that are referred as the straggler or the transient messages). These messages are handled as shown in the Fig. 3.

When an LP_i sends a red message to LP_j, the cell V[i][j] of the matrix is updated. On the other hand, if LP_i receives a message, the cell v[i][i] of the matrix is updated. At this cut point, we also calculate the minimum timestamp of all the red messages and then the control is passed to the appropriate pair-wise LP. Next, at second cut point C2, the LPs have to wait until all the messages destined to them are being received and then calculate the current simulation time as the minimum of the minimum time stamps calculated for red and green messages.

The control token is then forwarded to the appropriate pair-wise LP. Since we are using the butterfly barrier, the entire process is repeated $\log_2 N$ times. In other words, the condition for this algorithm is that the number of processes involved in the system should be a multiple of 2 (i.e., $N=2^x$).

For the sake of a comprehensive explanation of the proposed algorithm, let us take an example of four LPs communicating with each other as shown in the Fig. 5. It can be seen in Fig. 5 that the four LPs are exchanging messages with respect to the simulation time. Let us see how it modifies the cells of a matrix which are initialized to zero. From the Fig.5, let us understand how the cells are modified with respect to time. The first message is sent by LP_1 to LP_3. As a result, the cell V[1][3] of the matrix is incremented and the message is immediately received by the LP_3 that will increment the cell V[3][3] of the matrix. In

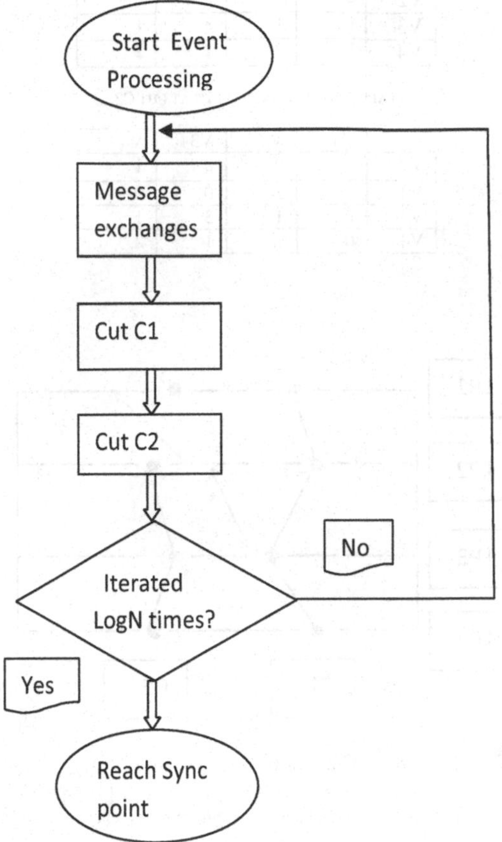

Fig. 1. A high level architecture of the proposed algorithm that shows the flow of data with the matrix and butterfly barrier

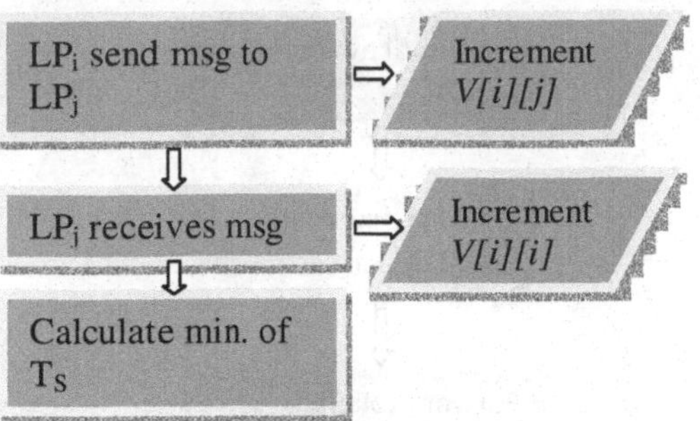

Fig. 2. Handling green messages

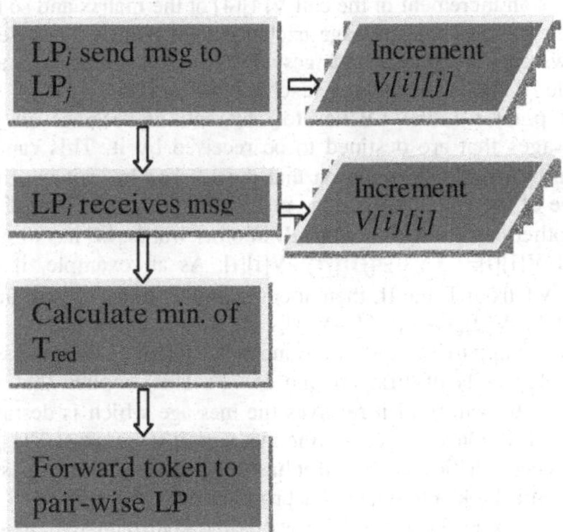

Fig.3 Cut *C1* handling Red messages that represent the transient or struggler messages

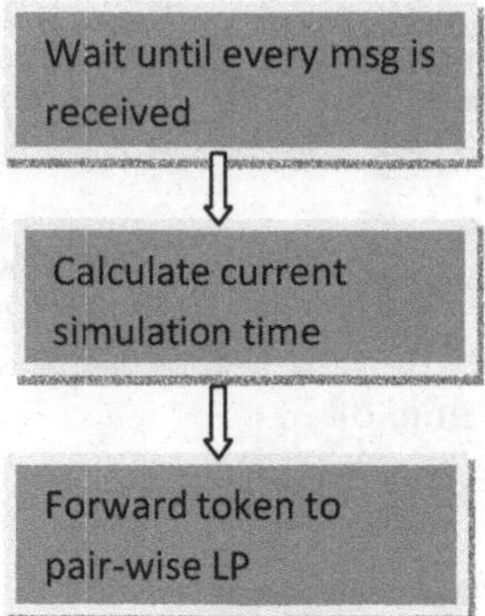

Fig 4: Cut C2 handling green messages for synchronization

the second round, the next message is sent by LP_1 to LP_4. Consequently, the cell V[1][4] of the matrix is incremented and since the message is immediately received by the LP_4, the cell V[4][4] of the matrix is incremented. The next message is sent by LP_3 to LP_4 that will increment the cell V[3][4] of the matrix. However, before this message could be received by LP_4, the next message is sent by LP_1 to LP_3. The result of this transmission would be an increment in the cell V[1][3] of the matrix. As time progresses, the LP_4 receives the message that results an increment in the cell V[4][4] of the matrix and so on. Table I shows the message exchanges till point C1. Table II shows the message exchanges after C1 and before C2 and Table III shows the message exchanges after C2.

At point C2, the LP has to wait until it receives all the messages that are destined to be received by it. This can be done by using the condition that the LP_i has to wait until the value of the cell V[i][i] of the matrix is equal to the sum of all the other cells of the column 'i'. In other words, LP_i has to wait until V[i][i]=$(\sum_{j=1}^{N} V[j][i])$ -V[i][i]. As an example, if we take V1 from Table II, then at cut point C2, it has to wait until V[1][1]=V[2][1]+V[3][1]+V[4][1].

According to Table II, the value of V[1][1] is '1' and the sum of other cells of first column is '2'. This implies that the LP_1 has to wait until it receives the message which is destined to reach it. Once it receives the message, it increments V[1][1] and again verifies weather if it has to wait. If not, it then passes the control token to the next appropriate pair-wise LP.

Every time the process forwards the control token, it also updates the current simulation time and as a result, we do not require additional instructions as well as time to calculate the GVT. This eliminates the need of communicating the GVT time among the different LPs exchanging messages. This saves

TABLE I: MATRIX OF 4 LPS EXCHANGING GREEN MESSAGES

	V1	V2	V3	V4
V1	1	0	2	1
V2	1	0	0	0
V3	0	0	1	1
V4	0	0	1	2

TABLE II: MATRIX OF 4 LPS AT CUT C1

	V1	V2	V3	V4
V1	1	0	2	1
V2	2	1	0	0
V3	0	1	3	1
V4	0	0	1	2

TABLE III: MATRIX OF 4 LPS AT CUT C2

	V1	V2	V3	V4
V1	2	0	2	1
V2	2	2	0	0
V3	0	1	3	1
V4	0	1	1	2

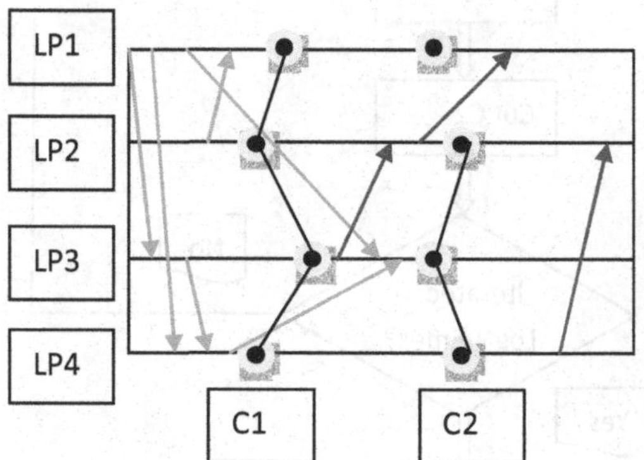

→ Represents Cut points
Green-font → green messages (safe events)
Red-font → red messages (transient/struggler messages)

Fig.5. Example of message exchanges between the four LPs. The C1 and C2 represent two cuts for green and red messages.

time which in turns improves the GVT latency. This algorithm proves helpful in upgrading the system performance of the parallel and distributed systems.

IV. Conclusion

In this paper, we present an algorithm that helps us to optimize the memory and processor utilization by using matrices instead of using *N* different vectors of size *N* in order to reduce the overall GVT latency. The improved GVT latency can play a vital role in upgrading the parallel/distributed system's performance. In the future, it will be interesting to develop an algorithm to calculate GVT using the tree barriers.

References

[1] Mattern, F., Mehl, H., Schoone, A., Tel, G. *Global Virtual Time Approximation with Distributed Termination Detection Algorithms*. Tech. Rep. RUU-CS-91-32, Department of Computer Science, University of Utrecht, The Netherlands, 1991.

[2] Friedemann Mattern, "Efficient Algorithms for Distributed Snapshots and Global virtual Time Approximation," *Journal of Parallel and Distributed Computing*, Vol.18, No.4, 1993.

[3] Ranjit Noronha and Abu-Ghazaleh, "Using Programmable NICs for Time-Warp Optimization," *Parallel and Distributed Processing Symposium., Proceedings International, IPDPS 2002, Abstracts and CD-ROM*, PP 6-13, 2002.

[4] D. Bauer, G. Yaun, C. Carothers, S. Kalyanaraman, "Seven-O' Clock: A new Distributed GVT Algorithm using Network Atomic Operations," *19th Workshop on Principles of Advanced and Distributed Simulation (PADS'05)*, PP 39-48.

[5] Syed S. Rizvi, Khaked. M. Elleithy, Aasia Riasat, "Minimizing the Null Message Exchange in Conservative Distributed Simulation," *International Joint Conferences on Computer, Information, and Systems Sciences, and Engineering, CISSE 2006*, Bridgeport CT, pp. 443-448 ,December 4-14 2006,

[6] Lee A. Belfore, Saurav Mazumdar, and Syed S. Rizvi et al., "Integrating the joint operation feasibility tool with JFAST," *Proceedings of the Fall 2006 Simulation Interoperability Workshop*, Orlando Fl, September 10-15 2006.

[7] Syed S. Rizvi, Khaled M. Elleithy, and Aasia Riasat, "Trees and Butterflies Barriers in Mattern's GVT: A Better Approach to Improve the Latency and the Processor Idle Time for Wide Range Parallel and Distributed Systems", *IEEE International Conference on Information and Emerging Technologies (ICIET-2007)*, July 06-07, 2007, Karachi, Pakistan.

Index

K. Elleithy (ed.), *Advanced Techniques in Computing Sciences and Software Engineering*,
DOI 10.1007/978-90-481-3660-5, © Springer Science+Business Media B.V. 2010

P

Q

R

S

T